SI PREFIXES

Submultiple	Prefix	Symbol	Multiple	Prefix	Symbol
10^{-1}	deci	d	10	deca	da
10^{-2}	centi	c	10^2	hecto	h
10^{-3}	milli	m	10^3	kilo	k
10^{-6}	micro	μ	10^6	mega	M
10^{-9}	nano	n	10^9	giga	G
10^{-12}	pico	p	10^{12}	tera	T
10^{-15}	femto	f			
10^{-18}	atto	a			

CONVERSION FACTORS

1 in.	0.0254 m	$\pi = 3.1416$
1 lb	0.45359 kg	$e = 2.7183$
1 torr (mm Hg)	133.32 N m^{-2}	$\ln x / \log x = 2.3026$
1 atm	101 325 N m^{-2}	
1 erg	10^{-7} J	
1 cal (thermochem.)	4.184 J	
1 BTU	1055.1 J	
1 kWh	$3.6 \cdot 10^6$ J	
1 eV	$1.6022 \cdot 10^{-19}$ J	
1 Hz	$6.6262 \cdot 10^{-34}$ J	
1 cm^{-1}	$1.9865 \cdot 10^{-23}$ J	
1 K	$1.3807 \cdot 10^{-27}$ J	
1 kg	$8.9876 \cdot 10^{16}$ J	
1 u	$1.4924 \cdot 10^{-10}$ J	
1 D (debye)	$3.3356 \cdot 10^{-30}$ Ams	

Physical Chemistry

Physical Chemistry

Vojtech Fried

Hendrik F. Hameka

Uldis Blukis

MACMILLAN PUBLISHING CO., INC.
New York
COLLIER MACMILLAN PUBLISHERS
London

Copyright © 1977, Macmillan Publishing Co., Inc.
Printed in the United States of America

Chapters 10 through 17 have been reprinted from *Quantum Theory of the Chemical Bond* by H. F. Hameka, copyright © 1975 Hafner Press, a division of Macmillan Publishing Co., Inc.

Macmillan Publishing Co., Inc.
866 Third Avenue, New York, New York 10022

Collier Macmillan Canada, Ltd.

Library of Congress Cataloging in Publication Data

Fried, Vojtech.
    Physical chemistry.

    Includes bibliographies and index.
    1. Chemistry, Physical and theoretical.
I. Hameka, Hendrik F., joint author.   II. Blukis,
Uldis, joint author.   III. Title.
QD453.2.F74   541'.3   75-26773
ISBN 0-02-339760-8

Printing: 1 2 3 4 5 6 7 8   Year: 7 8 9 0 1 2 3

To our wives, without whose
encouragement and moral support
the task never would have been completed.

This book focuses on the fundamental principles of physical chemistry, most of which are presented in Parts I to IV—"States of Matter" (including kinetic molecular theory), "Classical Thermodynamics," "Quantum Mechanics," and "Statistical Thermodynamics." The last two parts—"Chemical Kinetics" and "Electrochemistry"—apply the principles introduced earlier (supplemented with additional ones) to two very important and currently active areas of chemistry not covered in the first four parts. These two parts approach the frontiers of physical chemistry more closely than the first four parts. The macroscopic and especially the microscopic descriptions that are necessary to apply the fundamental principles to particular systems, for example, solids or liquids, are introduced where necessary. However, in order to keep the text to a reasonable size we have generally deemphasized broad coverage of particular systems.

Worked-out sample problems are provided in the body of the text. About 80 percent of the end-of-chapter exercises which illustrate the fundamental principles are provided with answers.

The following advice is intended for the student in difficulty or in need of information not found in this text. Besides consulting other texts in physical chemistry, the student should keep texts in general physics and in differential plus integral calculus at his fingertips while using this book. In addition it may be necessary to consult the specialized topical references given at the end of each chapter, as well as the following:

Collections of Problems
A. W. Adamson: *Understanding Physical Chemistry.* W. A. Benjamin, Inc., New York, 1964, 2 vols.
J. Bareš, Č. Černý, V. Fried, and J. Pick: *Collection of Problems in Physical Chemistry.* Pergamon Press, Oxford, 1962.
L. C. Labowitz and J. S. Arents: *Physical Chemistry.* Academic Press, New York, 1969.

Experimental Physical Chemistry
F. Daniels, J. W. Williams, P. Bender, C. D. Cornwell, J. E. Harriman, and R. A. Alberty: *Experimental Physical Chemistry*, 7th ed. McGraw-Hill, New York, 1969.
D. P. Shoemaker, C. W. Garland, and J. Steinfeld: *Experiments in Physical Chemistry*, 3rd ed. McGraw-Hill, New York 1974.
J. M. White: *Physical Chemistry Laboratory Experiments.* Prentice-Hall, Englewood Cliffs, N.J., 1975.

Advanced and Comprehensive Text
H. Eyring, D. Henderson, and W. Jost (eds.): *Physical Chemistry.* Academic Press, New York, 1967–?, 11 volumes.

Reviews
Accounts of Chemical Research, Advances in Chemical Physics, Annual Review of Physical Chemistry, MTP International Review—Physical Chemistry (A. D. Buckingham, consultant ed., Butterworths, London, Series One in 1971, 13 vols. (From 1974 on to be published biannually.)

Articles
Journal of Physical Chemistry, Journal of Chemical Physics;
Data: See D. R. Lide, Jr., and S. A. Rossmassler: _Ann. Rev. Phys. Chem.,_ **24**: 135 (1973).

At present the physical sciences are in transition toward a single system of units—the International System (SI, see Appendix IX). This text is in "transition" as well: in each area of physical chemistry we have tried to use the units most prevalent at the time of writing. In fast-developing areas where clear preferences for units are not apparent, we have used SI units.

We are very grateful to everyone who has helped with this book. However, we would like to single out the following colleagues who have been especially instrumental in the preparation of this text. Sincere thanks go to Professor Ira Levine for constructive suggestions made about the manuscript, and to Professor Orest Popovych (both of Brooklyn College) who was very helpful in writing the part on electrochemistry. We thank Professors Richard Pizer, Avigdor Ronn, Grace Wieder, and Mrs. Arlene Pollin (all of Brooklyn College) and Dr. Paul Solomon (Hunter College) who helped with chemical kinetics. One of us (U. B.) is grateful to Professor N. W. Gregory, Chairman of the Chemistry Department, for permission to use the chemistry library at the University of Washington.

The excellent support from the publisher, especially our editors Mr. James Smith and Mr. Leo Malek, is acknowledged.

Vojtech Fried
Hendrik F. Hameka
Uldis Blukis

contents

Part I.   STATES OF MATTER   1

Chapter 1.   EQUATIONS OF STATE   3

    1-1. Introduction   3
    1-2. The Ideal Gas Laws   3
    1-3. Intermolecular Forces   9
    1-4. The Nonideal Behavior of Gases   13
    1-5. Deviations from Ideal Behavior   13
    1-6. Some Equations of State of Real Gases   15
    1-7. Liquefaction of Gases   21
    1-8. The Principle of Corresponding States   26
    1-9. Mixtures of Gases   30
    1-10. Equations of State of Liquids   33
    Problems   34
    References and Recommended Reading   38

Chapter 2.   THE MOLECULAR BASIS OF THE EQUATIONS OF
            STATE   40

    2-1. Introduction and the Ideal Gas Laws   40
    2-2. The Kinetic Molecular Theory of Gases and the
        van der Waals Equation   44
    2-3. The Law of Atmospheres   46
    2-4. The Distribution of Molecular Velocities   48
    2-5. Experimental Verification of Maxwell's Distribu-
        tion Law   55
    2-6. The Most Probable and Mean Speed of Molecules 56
    2-7. Collision of Molecules with a Solid Surface:
        Effusion   57
    2-8. Collisions Between Molecules and the Maxwell
        Mean Free Path   59
    Problems   61
    References and Recommended Reading   62

Chapter 3.   TRANSPORT PROPERTIES OF GASES   64

    3-1. Viscosity   64
    3-2. Thermal Conductivity   71
    3-3. Diffusion   72
    Problems   75

Chapter 4.   MISCELLANEOUS TOPICS   77

                     4-1. The Maxwell Distribution Law in Terms of
                            Energies   77
                     4-2. The Equipartition Principle   78
                     Problems   83

Part II.   CLASSICAL THERMODYNAMICS   85

Chapter 5.   THE FIRST LAW OF THERMODYNAMICS   87

                     5-1. Introduction   87
                     5-2. Basic Definitions   88
                     5-3. Exact Differentials   91
                     5-4. Mathematical Formulation of the First Law of
                            Thermodynamics   95
                     5-5. Reversible and Irreversible Processes   96
                     5-6. Calorimetry   99
                     5-7. Enthalpy   100
                     5-8. Heat Capacity   101
                     5-9. The Application of the First Law of Thermo-
                            dynamics to Chemical Reactions: Thermo-
                            chemistry   104
                     5-10. The Standard Enthalpy of Formulation   106
                     5-11. Bond Energies   107
                     5-12. Variation of Enthalpy Change of Reaction with
                            Temperature   111
                     5-13. Application of the First Law of Thermodynamics
                            to an Ideal Gas   116
                     Problems   122
                     References and Recommended Reading   127

Chapter 6.   THE SECOND LAW OF THERMODYNAMICS   129

                     6-1. Introduction   129
                     6-2. Carnot's Cycle   131
                     6-3. The Thermodynamic Temperature Scale   134
                     6-4. The Mathematical Formulation of the Second
                            Law of Thermodynamics: Entropy   135
                     6-5. Entropy as a Measure of Spontaneity of Processes 136
                     6-6. The Combined First and Second Laws of Ther-
                            modynamics   138
                     6-7. The Helmholtz and Gibbs Free Energies   139
                     6-8. The Variation of the Helmholtz Free Energy with
                            Volume and Temperature and the Gibbs Free Energy
                            with Temperature and Pressure   140
                     6-9. Evaluation of Entropy Changes   142
                     6-10. The Thermodynamic Criterion of Equilibrium   146

6-11. Some Applications of the First and Second Laws
of Thermodynamics to Gases — 146
Problems — 153
References and Recommended Reading — 156

Chapter 7.   APPLICATION OF THERMODYNAMICS TO
PHASE EQUILIBRIA OF SINGLE COMPONENT
SYSTEMS — 157

7-1. Introduction — 157
7-2. The Condition of Equilibrium in Single-Component Multiphase Systems — 158
7-3. Phase Diagram of a Single-Component System — 159
7-4. The Clausius–Clapeyron Equation — 163
7-5. Vapor Pressure as a Function of the External
Pressure — 168
7-6. The Effect of the Size and Shape of the Surface
Area on Vapor Pressure — 169
7-7. Experimental Determination of the Vapor
Pressure — 172
7-8. Second-Order Phase Transitions — 173
Problems — 176
References and Recommended Reading — 178

Chapter 8.   APPLICATION OF THERMODYNAMICS TO
PHASE EQUILIBRIA OF MULTICOMPONENT
SYSTEMS — 179

8-1. Introduction — 179
8-2. Partial Molal Properties — 180
8-3. The Gibbs–Duhem Equation — 182
8-4. Determination of Partial Molal Properties — 183
8-5. Partial Molal Enthalpies and the Differential
Molal Enthalpies of Solution — 185
8-6. The Chemical Potential — 186
8-7. The Derivation of the Gibbs Phase Rule — 188
8-8. Auxiliary Thermodynamic Functions — 189
    8-8.1 Fugacity — 189
    8-8.2 Activity — 194
8-9. Ideal Mixtures — 200
8-10. Vapor–Liquid Equilibrium in an Ideal Solution — 202
8-11. Behavior of Dilute Solutions — 207
    8-11.1 The solubility of gases in liquids — 207
    8-11.2 Colligative properties — 211
8-12. Nonideal Solutions — 219
8-13. Vapor–Liquid Equilibrium in Nonideal Solutions — 220
8-14. Regular Solutions — 228
8-15. Solutions of Partially Miscible Liquids — 229
8-16. Vapor Pressure of Immiscible Liquids — 233
8-17. Fractional Distillation — 234

8-18. Phase Diagrams of Condensed Systems   235
    8-18.1   Solid–liquid diagrams   236
    8-18.2   Phase diagrams of reacting components   239
    8-18.3   Phase equilibrium in three-component systems   242
    8-18.4   The distribution of a solute between two mutually insoluble solvents   244
    8-18.5   Equilibrium between solid and liquid phases in three-component systems   247
8-19. Thermodynamics of Surfaces   247
Problems   250
References and Recommended Reading   253

Chapter 9.   APPLICATION OF THE FIRST AND SECOND LAW OF THERMODYNAMICS TO CHEMICAL REACTIONS   255

9-1. Introduction   255
9-2. Equilibrium Constants of Chemical Reactions   255
    9-2.1   Pressure dependence of the equilibrium constant   256
    9-2.2   Temperature dependence of the equilibrium constant   257
9-3. Standard Gibbs Free Energies of Formation   259
9-4. Relationships Among Equilibrium Constants   261
9-5. The Effect of Pressure, Inert Gas, and Temperature upon the Yield of a Gaseous Reaction   262
9-6. Chemical Equilibrium in Heterogeneous Reactions   265
9-7. The Third Law of Thermodynamics   266
9-8. Some Consequences of the Third Law of Thermodynamics   269
9-9. The Principle of Unattainability of Absolute Zero   271
9-10. The Experimental Verification of the Third Law of Thermodynamics   272
9-11. Enthalpy Functions $(\mathscr{H}_T^\circ - \mathscr{H}_0^\circ)/T$ and Gibbs Free Energy Functions $(\mathscr{G}_T^\circ - \mathscr{H}_0^\circ)/T$ 273
Problems   277
References and Recommended Reading   283

Part III.   QUANTUM MECHANICS   285

Chapter 10.   THE PRINCIPLES OF QUANTUM MECHANICS   287

10-1. Introduction   287
10-2. The Wave Function   290
10-3. The Bohr Atom   295
10-4. Stationary States and the Particle in the Box   297
10-5. The Color of Conjugated Organic Molecules   307
Problems   310
References and Recommended Reading   311

Chapter 11.   SOLUTIONS OF THE SCHRÖDINGER EQUATION   312

11-1. Introduction   312
11-2. The Harmonic Oscillator   313
11-3. The Rigid Rotor   320
11-4. Angular Momentum   326
11-5. The Hydrogen Atom   332
11-6. Approximate Methods   346
Appendixes   351
11-A. Eigenfunctions of the Harmonic Oscillator   351
11-B. The Ratio of Two Integrals   354
11-C. Eigenfunctions of the Rigid Rotor   355
11-D. Angular Momentum   357
11-E. Eigenfunctions of the Hydrogen Atom   359
Problems   361
References and Recommended Reading   363

Chapter 12.   ATOMIC STRUCTURE   364

12-1. Introduction   364
12-2. The Electron Spin   372
12-3. Exclusion Principle and Spin   378
12-4. Atomic Orbitals   382
Problems   387
References and Recommended Reading   388

Chapter 13.   LIGHT AND SPECTROSCOPY   389

13-1. The Nature of Light   389
13-2. Transition Probabilities   396
13-3. Transition Probabilities and Quantum Theory   398
13-4. The Maser and the Laser   402
Problems   405
References and Recommended Reading   405

Chapter 14.   THE SPECTRA OF DIATOMIC MOLECULES   407

14-1. Experimental Information   407
14-2. The Molecular Wave Function   411
14-3. Molecular Symmetry   420
14-4. Selection Rules and Spectral Intensities in Elec-
      tronic Bands   422
14-5. Intensities in the Infrared and Microwave
      Regions   432
Appendix 14-A   434
Problems   435
References and Recommended Reading   436

Chapter 15.   THE CHEMICAL BOND   437

15-1. Introduction   437
15-2. The Hydrogen Molecular Ion   438
15-3. The Hydrogen Molecule   444
15-4. Diatomic Molecules   449
15-5. Electronegativity   454
15-6. Hybridization   458
15-7. Unsaturated Molecules   465
15-8. Conjugated and Aromatic Molecules   469
      Appendixes   477
15-A. Evaluation of the Integrals I, J, and S, by Use of
      Elliptical Coordinates   477
15-B. Analytic Solutions of the Hückel Equation   478
Problems   480
References and Recommended Reading   482

Chapter 16.   MAGNETIC RESONANCE   484

16-1. Introduction   484
16-2. Relaxation Phenomena   489
16-3. Chemical Shifts   496
16-4. Spin–Spin Coupling   500
16-5. Electron-Spin Resonance   506
      Appendix 16-A. Two-Proton Spin System   511
Problems   513
References and Recommended Reading   515

Chapter 17.   THE SOLID STATE   516

17-1. Crystal Structures   516
17-2. Ionic Crystals   518
17-3. Metals and Semiconductors   521
17-4. Molecular Crystals   530
Problems   534
References and Recommended Reading   535

Chapter 18.   SCATTERED RADIATION AND MOLECULAR
      GEOMETRY   536

18-1. Introduction   536
18-2. Interactions of Radiation and Matter   536
18-3. Intensity and Amplitude   539
    18-3.1   Addition of amplitudes   540
18-4. Scattering by Atoms   544
    18-4.1   X-ray scattering by electrons   546
    18-4.2   Coherent scattering of X rays by atoms   547
    18-4.3   Incoherent scattering of X rays by atoms   548
    18-4.4   Electron and neutron scattering   550

18-5. Scattering Cross Sections 552
18-6. Diffraction by Molecules 553
    18-6.1 Diffraction by gases 553
    18-6.2 Diffraction by solids 560
18-7. Molecular Geometry 561
    18-7.1 Experimental geometries and molecular vibrations 561
    18-7.2 Theoretical geometries 565
Problems 566
References and Recommended Reading 567

Part IV. STATISTICAL THERMODYNAMICS 569

Chapter 19. STATISTICAL THERMODYNAMICS 571

19-1. Introduction 571
19-2. The Distribution Laws 574
19-3. Negative Absolute Temperatures 578
19-4. Some More Discussion of the Partition Function 579
19-5. Thermodynamic Functions in Terms of the Partition Function 583
19-6. The Molecular Interpretation of the Basic Laws of Thermodynamics 586
19-7. Evaluation of the Partition Function 589
    19-7.1 Translational partition function 589
    19-7.2 Rotational partition function 591
    19-7.3 Vibrational partition function 595
    19-7.4 Electronic partition function 600
    19-7.5 Nuclear partition function 602
19-8. Statistical Thermodynamics of Gaseous Mixtures 604
19-9. Statistical Interpretation of the Equilibrium Constant 606
19-10. Statistical Thermodynamics of an Ideal Crystal 610
19-11. Ideal Lattice Gas 614
19-12. Statistical Derivation of the Equation of State for Nonideal Fluids 615
19-13. Electron Gas 619
Problems 621
References and Recommended Reading 624

Part V. CHEMICAL KINETICS 627

Chapter 20. RATES AND MECHANISMS OF REACTIONS 629

20-1. Introduction 629
20-2. Rates of Homogeneous Reactions 631
    20-2.1 Reaction rate 633
    20-2.2 Determination of rate law 634
    20-2.3 Temperature-dependence of rate constant 640

20-3. Mechanisms and Elementary Steps   643
   20-3.1   Complex reactions   645
   20-3.2   Rice–Herzfeld chain reactions   649
   20-3.3   Explosions: $H_2 + O_2$ reaction   651
20-4. Detailed   Balancing: Kinetics   and   Ther-
      modynamics   655
20-5. Activator-initiated Reactions   661
   20-5.1   The yield   662
   20-5.2   Deposition of energy   665
   20-5.3   Mechanisms of photochemical reactions   669
   20-5.4   Mechanisms of radiochemical reactions   672
20-6. The Measurement of Fast Reaction Rates   676
   20-6.1   Classical methods   680
   20-6.2   Fast mixing methods   680
   20-6.3   Field-jump methods   680
   20-6.4   Pulse methods   685
   20-6.5   Lifetime methods   687
20-7. Intermediates   697
   20-7.1   Identification   697
   20-7.2   Detection in reacting systems   699
   20-7.3   Measurement of concentrations   701
   20-7.4   Inferences from indirect methods   702
20-8. Heterogeneous Reactions   708
   20-8.1   Types of reactions   708
   20-8.2   Mass transfer   709
   20-8.3   Adsorption and desorption   712
   20-8.4   Composition and structure of surfaces   723
   20-8.5   Mechanisms of catalytic surface reactions   737
Problems   741
References and Recommended Reading   754

Chapter 21.   STATISTICAL THEORIES OF KINETICS   757

21-1. Introduction   757
21-2. Collision Theory   758
   21-2.1   Relative motion of two molecules   759
   21-2.2   Reaction cross section   762
   21-2.3   The rate constant   764
   21-2.4   Reactions without activation energy   766
   21-2.5   Reactions in liquids   768
21-3. Molecular Dynamics: the $H + H_2$ Reaction   771
21-4. Energy Transfer   780
   21-4.1   Unimolecular reactions   780
   21-4.2   Combination reactions   784
21-5. Activated Complex Theory   785
   21-5.1   Thermodynamic   expression   for   rate
        constant   787
   21-5.2   Rate constants from molecular properties   788
   21-5.3   Refinements: the transmission coefficient
        and the activity   792

21-6. Unimolecular Reaction Theory 794
   21-6.1 The RRKM model 795
   21-6.2 Evaluation of vibrational partition functions 796
   21-6.3 The RRKM rate constants 798
   21-6.4 Activation energy, fall-off region, and RRKM theory 804
Problems 806
References and Recommended Reading 810

Chapter 22.   RESOLUTION OF REACTANT AND PRODUCT STATES 812

22-1. Introduction 812
22-2. Translational States 816
   22-2.1 Directions of reactant and product momenta 816
   22-2.2 Translational energy and reaction cross section 816
22-3. Internal States of Reactants 819
   22-3.1 Reactions of vibrationally excited species 821
   22-3.2 Intermolecular energy transfer 823
22-4. Internal States of Products 826
Problems 828
References and Recommended Reading 829

Part VI.   ELECTROCHEMISTRY 831

Chapter 23.   INTRODUCTION TO SOLUTIONS OF ELECTROLYTES 833

23-1. The Colligative Behavior of Electrolytic Solutions 833
23-2. Ion–Solvent Interactions 834
23-3. Ion–Ion Interactions 836
References and Recommended Reading 840

Chapter 24.   TRANSPORT AND THERMODYNAMIC PROPERTIES IN SOLUTIONS OF ELECTROLYTES 841

24-1. Electrolysis and Faraday's Law 841
24-2. Transference Numbers 844
24-3. Electrolytic Conductance 849
24-4. Arrhenius Theory and the Equivalent Conductance 854
24-5. The Theory of Ionic Interactions 857
24-6. The Ionic Atmosphere and the Theory of Conductance 863
24-7. Wien and Debye–Falkenhagen Effects 865

24-8. Temperature Dependence of the Equivalent
      Conductance   867
24-9. Some Applications of Conductance Measure-
      ments   867
24-10. Diffusion in Solutions of Electrolytes   869
24-11. Chemical Potential and the Standard State in
       Solutions of Electrolytes   870
24-12. Partial Molal Enthalpies of Ions in Solution   874
24-13. Experimental Determination of Activity
       Coefficients   876
24-14. Theoretical Calculation of Activity Coeffi-
       cients: The Debye–Hückel Law   879
24–15. Hydrogen Ion Concentration: The pH, pa_H,
       and ptH Scales   883
24–16. The Effect of Interionic Forces on Reaction
       Rates   884
Problems   888
References and Recommended Reading   891

Chapter 25.   GALVANIC CELLS   893

25-1. Introduction   893
25-2. Half-Cell Potentials   893
25-3. Determination of the Standard Electromotive
      Force of Cells   901
25-4. Evaluation of Some Thermodynamic Properties
      from Electromotive Force Measurements   902
25-5. Liquid Junction Potential   906
25-6. Classification of Half-cells   907
    25-6.1   Metal–metal ion half-cells   908
    25-6.2   Amalgam half-cells   908
    25-6.3   Gas-ion half-cells   908
    25-6.4   Metal–insoluble salt–anion half-cells   910
    25-6.5   Oxidation-reduction half-cells   911
25-7. Types of Galvanic Cells   912
25-8. Fuel Cells   917
25-9. Determination of pa_H 917
25-10. Glass Electrode   920
25-11. Membrane Equilibrium   921
25-12. Additional Comments on Electrode Processes   924
Problems   929
References and Recommended Reading   932

Appendixes

I. Partial Differentiation and Exact Differentials   933
II. Derivatives and Integrals   936
III. Frequently Used Expansions   938
IV. Complex Numbers   940
V. The Error Function   942

VI. The Method of Lagrange Undetermined Multipliers 943
VII. Stirling's Approximation 945
VIII. Method of Least Squares 946
IX. SI Units and Fundamental Constants 947
Answers to Problems 955
Index 963

Part I

States of Matter

chapter 1

Equations of State

1-1 INTRODUCTION

It is generally recognized that matter can exist in three different states of aggregation: solid, liquid, and gas. The plasma state—a highly ionized gaseous state—and the molten salt state are sometimes considered two additional states of matter.

The gaseous state is the simplest to analyze in terms of an atomic model and is essentially completely understood today. The solid state is more complicated, but great progress has been made toward understanding it fully. The structure of the liquid state is the most complicated and rather meager results have been obtained up to the present in attempts to analyze it. Liquids and gases, as well as plasma and molten salts, flow in an external force field and are therefore called **fluids**.[1] We shall refer to liquids and solids only peripherally here, but a detailed theory of gases will be outlined.

1-2 THE IDEAL GAS LAWS

A gas is a form of matter that possesses the property of filling a container completely and to a uniform density. Hence, a gas has neither a definite shape nor volume.

It is a fact of experience, that the three basic properties—pressure, P; volume, V; and temperature,[2] T—that describe the state of a gas (if the effects of the gravitational, electric, and magnetic fields are neglected) cannot be chosen independently. For a given amount of a substance, the three basic properties are related by an equation of state of the form

$$f(P, V, T) = 0 \qquad \textbf{(1-1)}$$

Similar equations are also applicable to liquids and solids.

Many experiments and theoretical attempts have been carried out to determine the form of this function. The measurements of the volume of a gas at different pressures and constant temperature [$V = F'(P)$] go back to the classical work of R. Boyle (1662). Boyle discovered that, at a constant temperature, the volume of a given quantity of gas is inversely proportional to the applied pressure [3]. In mathematical terms, **Boyle's law** is expressed by the following equation

$$V = \frac{A}{P} \qquad \textbf{(1-2a)}$$

---

[1] In this text the term *fluid* will be used for liquids and gases only.

[2] For the definition of these properties, see pages 91 and 188.

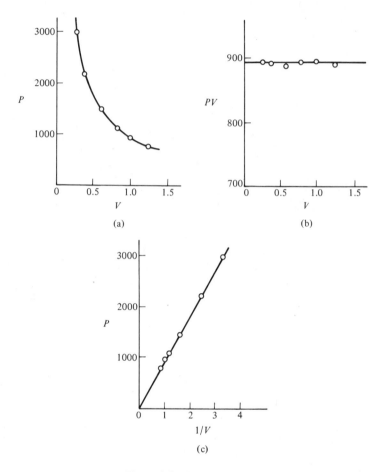

Figure 1-1   Graphical representation of Boyle's law.

where A is a constant depending on the amount of gas and on the particular temperature at which corresponding values of P and V are determined. This follows from the fact that equal volumes of different gases at the same temperature and pressure contain equal numbers of molecules (Avogadro, 1811) [4].[3]

The familiar plot of P against V (hyperbola) as shown in Figure 1-1a is not a very good way to verify whether Eq. (1-2a) accurately represents the experimental P-V data (Table 1-1). A better way of plotting the data for testing Boyle's law is suggested by the simple rearrangement of Eq. (1-2a) to read

$$PV = A \tag{1-2b}$$

or

$$P = A\frac{1}{V} \tag{1-2c}$$

---

[3] At a pressure of 1 atm and temperature of 0°C, the volume occupied by 1 mole of a gas is 22,414 cm³. One mole or 1 gram-molecular weight contains N molecules, where N is Avogadro's number, equal to 6.02×10^{23} molecules; 1 cm³ = 1 ml.

Eq. (1-2b) represents PV as a function of P (or V) at a constant temperature (Figure 1-1b), and Eq. (1-2c) represents P as a function of $1/V$ at constant temperature (Figure 1-1c). Both these forms of Eq. (1-2a) are straight lines.

Table 1-1 Constant temperature pressure-volume data collected by Boyle (1662)

Volume (expressed in arbitrary units)	Pressure (mm Hg)
1.2	740
1.1	811
1.0	898
0.9	998
0.8	1122
0.7	1278
0.6	1476
0.5	1796
0.4	2232
0.3	2986

Accurate experimental data on P-V indicate that linear relationships are attained only at very low pressures. The lower the pressure, the better Boyle's law describes the P-V behavior of a gas; and, in the limit of the pressure approaching zero, all gases obey Boyle's law practically with the same accuracy. It is more correct, therefore, to write Boyle's law in the form

$$\lim_{P \to 0} (PV) = A \tag{1-3}$$

Gay-Lussac (1802) studied experimentally the variation of the volume of a given quantity of gas with temperature at a constant pressure. His observations may be expressed quantitatively as

$$V = V_0(1 + \alpha t) \tag{1-4}$$

where V_0 is the volume at 0°C, t is the temperature expressed on the centigrade (Celsius) scale, and α is the cubic expansion coefficient defined by the equation

$$\alpha = \frac{1}{V_0}\left(\frac{\partial V}{\partial T}\right)_P \tag{1-5}$$

According to Gay-Lussac's observations α has the same value for all gases ($\alpha = 1/273.15 = 3.66099 \times 10^{-3} \text{ K}^{-1}$). Note that for a gas obeying Gay-Lussac's law, $\alpha = \gamma$, where $\gamma = (1/P)(\partial P/\partial T)_V$.

Introducing this value for α, Eq. (1-4) takes the form

$$V = V_0\frac{273.15 + t}{273.15} = V_0\frac{T}{T_0} \tag{1-6a}$$

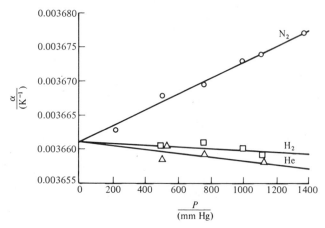

Figure 1-2 The cubic expansion coefficient of nitrogen, hydrogen, and helium as a function of pressure at constant temperature.

where T is the temperature expressed in the absolute temperature scale (Kelvin); $T = t + 273.15$.[4] Eq. (1-6a) may also be rewritten as

$$\frac{V_0}{T_0} = \frac{V_1}{T_1} = \frac{V_2}{T_2} = k \tag{1-6b}$$

or

$$V = kT \tag{1-6c}$$

At a constant pressure, the volume of a certain quantity of gas is directly proportional to the absolute temperature.

According to Gay-Lussac, this law was observed by Charles in 1787, although Charles never published his observations. The law is therefore often called Charles' law. The value of the constant k depends on the quantity of gas involved in the measurement and on the pressure. It is obvious that k and α are related to each other by the equation

$$\alpha = \frac{1}{V_0}\left(\frac{\partial V}{\partial T}\right)_P = \frac{k}{V_0} \tag{1-7}$$

and at low pressures k has the same value for all gases.

As experimental facts prove, α becomes a nonvariable quantity only at very low pressures (Figure 1-2); thus Gay-Lussac's law applies only under this restriction. We therefore write

$$\lim_{P \to 0} \frac{V}{V_0} = 1 + \alpha t \tag{1-8}$$

---

[4] Later we will prove that the ideal gas temperature scale expressed in T is equivalent to the thermodynamic temperature scale.

On combining Boyle's law (Eq. 1-3) and Gay-Lussac's law (Eq. 1-6), we obtain for 1 mole of a gas[5]

$$\lim_{P \to 0} (P\mathcal{V}) = RT \tag{1-9}$$

where $\mathcal{V}$ is the molar volume of the gas ($\equiv V/n$) and R is the universal gas constant for 1 mole of gas. For n moles, Eq. (1-9) takes the form

$$\lim_{P \to 0} (PV) = nRT \tag{1-10a}$$

Since in some aspects gases behave similarly at low pressure and high temperature (if ionization does not occur), Eq. (1-10a) may also be written as

$$\lim_{T \to \infty} (PV) = nRT \tag{1-10b}$$

The numerical value of R will of course depend on the choice of units in which the various quantities are expressed. Since, by definition, pressure is force per unit area, the product PV has the dimension of force times length, (force/length^2)(length^3), characteristic of energy in mechanics. Thus, dimensionally R must be of the form: energy per mol per degree Kelvin. Using $P_0 = 1$ atm, $\mathcal{V}_0 = 22,414$ cm^3 mol^{-1}, and $T_0 = 273.15$ K, we obtain

$$R = \frac{P_0 \mathcal{V}_0}{T_0} = \frac{1 \text{ atm} \times 22,414 \text{ cm}^3 \text{ mol}^{-1}}{273.15 \text{ K}}$$

$$= 82.057 \text{ cm}^3 \text{ atm K}^{-1} \text{ mol}^{-1}$$

The values of R for systems of units most often used are as follows:

R	Units[a]
82.057	cm$^3 \cdot$ atm K^{-1} mol^{-1}
0.082057	liter $\cdot$ atm K^{-1} mol^{-1}
8.31441	J K^{-1} mol^{-1}
8.31441×10^7	ergs K^{-1} mol^{-1}
1.98719	cal K^{-1} mol^{-1}

[a] In the SI system of units, cubic decimeters (dm^3) are used instead of liters and joules (J) are used instead of calories; 1 J = Newton (N) × meter (m).

As seen from Figure 1-3 R becomes a real constant only at low pressures.

---

[5] For the derivation see page 93.

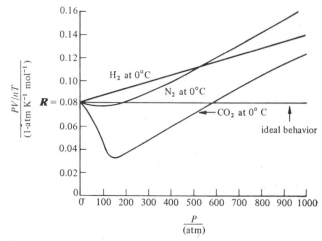

Figure 1-3   The ratio PV/nT as a function of pressure for a few gases; $\lim_{P\to 0}PV/nT = 0.08206$ liter·atm mol^{-1} K^{-1} for all the gases.

EXAMPLE 1-1

For the density of trimethyl amine as a function of pressure at a constant temperature, $t = 0°C$, C. J. Arthur and U. A. Felsing [17] give the following data:

P	ρ
(atm)	(g liter^{-1})
0.2	0.5336
0.4	1.0790
0.6	1.6363
0.8	2.2054

Determine the molecular weight of trimethyl amine.
Solution: Eq. (1-10a) is applied to the calculation

$$\lim_{P\to 0}(PV) = n\boldsymbol{R}T = \frac{w}{M}\boldsymbol{R}T$$

or

$$M = \boldsymbol{R}T\lim_{P\to 0}\left(\frac{w}{VP}\right) = \boldsymbol{R}T\lim_{P\to 0}\left(\frac{\rho}{P}\right)$$

where w is the mass and w/V is the density, ρ.

The value of $\lim_{P\to 0}(\rho/P)$ is best determined from a graph of ρ/P against P by extrapolation to zero pressure. The extrapolated value from Figure 1-4 is 2.6382.

$$M = 2.6382\times 0.08206\times 273.15 = 59.137 \text{ g mol}^{-1}$$

The theoretical value is 59.112 g mol^{-1}.   ●

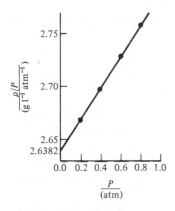

Figure 1-4 The dependence of ρ/P on pressure for trimethyl amine at 0°C.

EXAMPLE 1-2

The following equation defines the residual volume of a gas [10].

$$\Delta \mathcal{V} = \mathcal{V}_{ideal} - \mathcal{V}_{real} = \frac{RT}{P} - \mathcal{V}_{real}$$

At low pressures all the gases are supposed to behave ideally; consequently as $P \to 0$, $\Delta \mathcal{V} = 0$. Show how such a conclusion depends on the nature of the gas under consideration.

Solution: At pressures approaching the limit of zero, the preceding equation yields

$$\lim_{P \to 0} \Delta \mathcal{V} = \lim_{P \to 0} \frac{RT}{P} - \lim_{P \to 0} \mathcal{V}_{real} = \infty - \infty$$

which is an indeterminate form. We therefore rewrite the equation into the form

$$\mathcal{V}_{real} + \Delta \mathcal{V} = \frac{RT}{P}$$

As $P \to 0$, $\mathcal{V}_{real} \to \infty$ and $\Delta \mathcal{V}$ may become negligibly small in comparison with $\mathcal{V}_{real}$. The value of $\Delta \mathcal{V}$ as $P \to 0$ varies however, for the different gases, and only experiments can give quantitative information about its magnitude. Nonetheless, for all practical purposes, the ideal gas law may be applied as P approaches zero. ●

1-3 INTERMOLECULAR FORCES

A gas that exactly obeys the equations discussed in Section 1-2 in a wide pressure range is called an ideal (perfect) gas. The behavior of a real gas is different from that of an ideal one. Nevertheless, at pressures approaching zero, or at very high temperatures, all gases approach ideal behavior (see Example 1-2).

The explanation of the nearly ideal behavior of all gases at low pressures is based on two well known experimental facts. First, at low pressures, the volume occupied by a gas is largely empty space. The molecules are far apart and, consequently, the attractive forces acting between molecules are negligibly small on the average. Second, at low pressures, the repulsive forces are negligible, that is, the effective volume of the molecules themselves is negligibly small when compared with the volume of the vessel containing the gas.

Since an ideal gas behaves in the same manner as a real gas at low pressures, we may conclude that there is no attraction between the molecules of an ideal gas and also that the molecules of an ideal gas may be considered as volumeless, point-particles.

In real gases at pressures different from zero, however, intermolecular forces are present, and molecules do have a volume. If there were no forces of attraction, molecules would never cohere, as they manifestly do in the condensed states of matter—solids and liquids. (Note that an ideal gas cannot be liquefied.) On the other hand, if there were no forces of repulsion, molecular annihilation would result. Molecular behavior is largely determined by the balance struck between the forces that tend to pull molecules together (attractive) and those that tend to push them apart (repulsive).

Forces of different kinds may exist between atoms and molecules. Common to all these forces is their electrostatic and electrodynamic origin. We generally distinguish between long-range and short-range forces. A force that depends on r^{-2} (coulombic forces between ions) will be effective over a longer range than one dependent on r^{-5} (interaction forces between an ion and a dipole) or on r^{-7} (van der Waals forces). All these forces are related to the interaction potential energy, ε, through the equation

$$f = -\frac{\partial \varepsilon(r)}{\partial r} \qquad (1\text{-}11)$$

where f denotes the force, r is the interparticle distance, and $\varepsilon(r)$ is the potential energy as a function of distance.

The intermolecular potential energy, present only between molecules with permanent electric dipoles, arises from the attractions and orientations such dipoles exercise upon each other. For molecules with a dipole moment μ, this energy of interaction (averaged over all possible orientations) between a pair of molecules has been shown to be

$$\varepsilon_{\text{or}}(r) = -\left(\frac{2}{3}\right)\frac{\mu^4}{kT}\frac{1}{r^6} \qquad (1\text{-}12)$$

where $k(\equiv R/N)$ is Boltzmann's constant, T is the absolute temperature, and N is Avogadro's number.

The presence of a permanent dipole in one molecule may cause polarization in neighboring molecules and thereby induced dipoles—the inductive effect. Interaction of the induced and permanent dipoles leads to an attractive energy

$$\varepsilon_{\text{in}}(r) = -2\alpha\mu^2\frac{1}{r^6} \qquad (1\text{-}13)$$

In this equation r and μ again represent the interparticle distance and dipole moment respectively; α is the polarizability of the molecule and can be calculated from the molar refraction of the substance. The value of α is proportional to the volume of the molecule; the inductive effect is therefore greater for large than for small molecules.

The orientation and induction effects cannot be responsible for the interaction energy between molecules that do not have dipole moments. F. London (1930) solved the problem of the interaction energy between two such particles quantum mechanically. He showed that, due to the vibration of the electron clouds with respect to the nuclei of atoms in a molecule, tiny instantaneous dipoles of a specific orientation are formed. These dipoles induce dipole moments in neighboring molecules, and the latter interact with the original dipoles to produce an attraction between the molecules—the dispersion effect. For a pair of molecules the dispersion energy is given by

$$\varepsilon_d(r) = -\left(\frac{3}{4}\right) h\nu\alpha^2 \frac{1}{r^6} \tag{1-14}$$

where h is Planck's constant[6] and ν is the characteristic frequency of oscillation of the charge distribution. For many simple molecules, $h\nu$ is equal to the ionization energy of the molecule. The dispersion effect is also greater for large than for small molecules. As seen from Table 1-2, the dispersion interaction contributes the greatest part to the interaction (except with NH_3 and H_2O) even if the molecules have a dipole moment.

The forces responsible for the orientation, inductive, and dispersion energies are denoted as van der Waals forces.

Table 1-2 Relative magnitudes of the three constituents of the van der Waals forces between two identical molecules at 0°C

Substance	Dipole Moment (Debyes)	Orientation	Induction	Dispersion
		(erg cm$^6 \times 10^{60}$)[a]		
CCl_4	0	0	0	1460
CO	0.10	0.0018	0.0390	64.3
HBr	0.80	7.24	4.62	188
HCl	1.08	24.1	6.14	107
NH_3	1.47	82.6	9.77	70.5
H_2O	1.84	203	10.8	38.1
cyclohexane	0	0	0	1560

[a] $1\ J = 10^7\ erg$.

A very important contribution to the total energy of a pair of molecules arises from the interaction due to a close approach of the electron clouds and nuclei of the atoms in one molecule with those in another. As a result of this, a repulsive energy, ε_r, is developed. The magnitude of ε_r can be approximated within a narrow temperature range:

$$\varepsilon_r(r) = \frac{B}{r^n} \tag{1-15}$$

where B is a constant for a given substance and n may range from 9 to 15. No exact theoretical equation has been derived so far for the repulsive contributions. Lennard-Jones (1926) suggested the value of 12 for n. Provided the repulsive energy

[6] For its value see the inside back cover.

is allowed to vary with a large inverse power of the distance, the exact value of n does not seem to be very critical, especially for thermodynamic calculations.

On adding all the contributions, the total energy acting between a pair of molecules is obtained as

$$\varepsilon(r) = \varepsilon_{or}(r) + \varepsilon_{in}(r) + \varepsilon_d(r) + \varepsilon_r(r)$$

$$= -\left(\frac{2}{3}\frac{\mu^4}{kT} + 2\alpha\mu^2 + \frac{3}{4}h\nu\alpha^2\right)\frac{1}{r^6} + \frac{B}{r^{12}}$$

$$= -\frac{A}{r^6} + \frac{B}{r^{12}} \tag{1-16}$$

Figure 1-5 shows the interaction potential energy as a function of the separation for three different cases. If the particles were hard spheres (rigid sphere model), the energy would become infinite when the spheres came into contact, as shown in curves (a) and (b). Curve (a) represents hard spheres without attraction. Curves (a) and (b) will be recognized as less realistic than curve (c) for "soft spheres," although these sometimes represent a very useful approximation.

The potential energy equation (1-16) is more instructive when put in terms of the natural parameters ε^*, the depth at the minimum, and r^*, the distance at this point (see Figure 1-5c); ε^* equals $-\varepsilon$ when r equals r^*.

The fact that, at the bottom of the potential well, $\partial\varepsilon(r)/\partial r = 0$ results in the relation

$$B = \frac{A}{2}(r^*)^6 \tag{1-17}$$

and from the condition at the bottom of the curve, $\varepsilon = -\varepsilon^*$, we have

$$A = 2\varepsilon^*(r^*)^6 \tag{1-18}$$

On combining Eqs. (1-16), (1-17), and (1-18), we obtain

$$\varepsilon = -\varepsilon^*\left[2\left(\frac{r^*}{r}\right)^6 - \left(\frac{r^*}{r}\right)^{12}\right] \tag{1-19}$$

Since at a distance equal to the collision diameter, σ, the interaction energy is equal

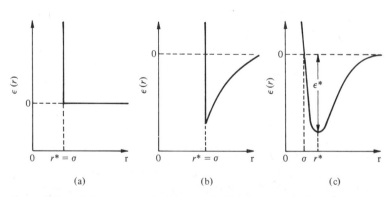

Figure 1-5 Different potential functions. (a) Hard sphere molecules. (b) Hard sphere molecules with attraction. (c) Lennard-Jones molecules.

to zero, we may also derive

$$r^* = 2^{1/6}\sigma \tag{1-20}$$

and consequently,

$$\varepsilon = -4\varepsilon^*\left[\left(\frac{\sigma}{r}\right)^6 - \left(\frac{\sigma}{r}\right)^{12}\right] \tag{1-21}$$

ε^* can be evaluated theoretically or experimentally. σ is estimated from measurements, such as density of the solidified gases, viscosity of gases, or from the second virial coefficients of gases. As is obvious the equilibrium distance, r^*, and the collision diameter, σ, are not equal, except for hard sphere molecules.

1-4   THE NONIDEAL BEHAVIOR OF GASES

We have seen in the previous sections that the ideal gas laws precisely represent the behavior of nonideal gases only at low pressures. At higher pressures, real gases exhibit deviations from ideal behavior. It is unrealistic to specify exactly the pressure below which a gas obeys the ideal gas laws. This depends, as we shall see next, on the physical and chemical properties of the individual gases and also to a large extent on the temperature.

The compressibility factor, z, defined by the equation

$$z = \frac{PV}{nRT} \tag{1-22}$$

represents a simple and convenient method for expressing the P-V-T behavior of nonideal gases. If the gas were ideal, z would be unity; deviations from $z = 1$ represent departures from ideality in gas behavior. In the limit of $P \to 0$ (vanishing densities), z approaches unity for all gases.

A value of z greater than 1 means that the gas is less compressible than predicted by the ideal gas law ($PV > nRT$); z less than 1 indicates that the gas is more compressible than predicted by the ideal gas law ($PV < nRT$). Commonly, at low and moderate pressures and low temperatures, $z < 1$; and, at high pressures, regardless of the temperature, $z > 1$. Realize also that the variation of z with pressure is not the same for different gases (Figure 1-6); it also differs for the same gas at different temperatures (Figure 1-7). The shape of the curves in Figures 1-6 and 1-7 is determined by the competition between the intermolecular potential energy and the thermal energy of the molecules.

1-5   DEVIATIONS FROM IDEAL BEHAVIOR

Some gases exhibit deviations from ideal behavior even at very low pressures. Such irregularities are observed especially in systems containing unstable molecules. At the conditions at which the P-V-T experiment is performed, association or decomposition of the molecules may occur in the system and this may result in a change in the number of moles in the system.

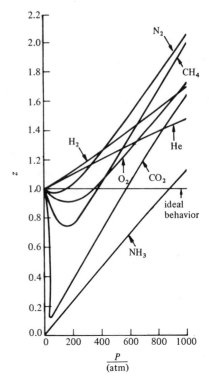

Figure 1-6 The compressibility factor of various gases as a function of pressure at 0°C.

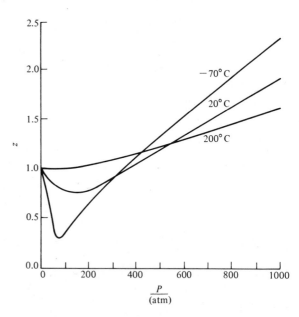

Figure 1-7 The compressibility factor of methane as a function of pressure at different temperatures.

To illustrate this, let us consider a vessel containing n moles of gaseous phosphorus pentachloride, PCl_5, at a certain P, V, and T. This compound may undergo decomposition according to the equation

$$PCl_5(g) \rightleftharpoons PCl_3(g) + Cl_2(g)$$

into phosphorus trichloride, PCl_3, and molecular chlorine. If α is the degree of decomposition (α is temperature and pressure dependent) the gaseous mixture contains $n(1-\alpha)$ moles of PCl_5, $n\alpha$ moles of PCl_3, and $n\alpha$ moles of Cl_2. The total number of moles present in the system at equilibrium is therefore different from the original n;

$$n_{tot} = n(1-\alpha) + n\alpha + n\alpha = n(1+\alpha)$$

Then, for the mixture of the three gases, we write

$$\frac{PV}{RT} = n_{tot} = n(1+\alpha) \tag{1-23}$$

Experimental data on n, P, V, and T enable one to evaluate α. From the dependence of α on T and P, the equilibrium constant of the chemical reaction and its pressure and temperature dependence may be evaluated.

1-6 SOME EQUATIONS OF STATE OF REAL GASES

As is evident from Figures 1-6 and 1-7 the compressibility factor, z, expressing the deviation of a real gas from ideal behavior, is strongly temperature and pressure dependent. For this reason and also because the simplest potential function (Eqs. 1-19 and 1-21) contains two adjustable parameters, all the equations of state of real gases must contain more than one constant.

Kamerlingh-Onnes (1901) recommended the following expansions for expressing z as a function of P and T and V and T, respectively.

$$z(P, T) = \frac{PV}{RT} = 1 + \mathcal{B}P + \mathcal{C}P^2 + \mathcal{D}P^3 + \cdots \tag{1-24}$$

and

$$z(V, T) = \frac{PV}{RT} = 1 + \frac{\mathcal{B}'}{V} + \frac{\mathcal{C}'}{V^2} + \frac{\mathcal{D}'}{V^3} + \cdots \tag{1-25}$$

where the constants $\mathcal{B}$, $\mathcal{C}$, $\mathcal{D}$, $\mathcal{B}'$, $\mathcal{C}'$, $\mathcal{D}'$,...., are called virial coefficients. For pure gases they are temperature dependent only (see Figure 1-8); for mixtures they are functions both of temperature and composition.

The number of constants used in the series depends on the applied pressure range and on the required accuracy. It is important to stress here, however, that the virial equations of state are not useful at high pressures, especially at pressures near and above the critical pressure, even when a great number of constants are used. The virial equations converge slowly and the higher terms are not accurately known. In the critical region (see page 21), the whole series fails to converge.

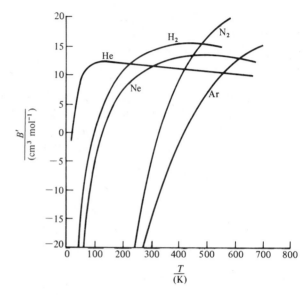

Figure 1-8 The second virial coefficient of various gases as a function of temperature.

The treatment of real gases by the methods of statistical mechanics leads to an equation of state in the form of a virial expansion (see page 616). Statistical mechanics relates the virial coefficients to the intermolecular potential energy; when the potential energy function is known, the virial coefficient may be evaluated. It is recognized that the second virial coefficients, $\mathscr{B}$ and $\mathscr{B}'$, are responsible for the deviation due to molecular pair interactions; the third virial coefficients, $\mathscr{C}$ and $\mathscr{C}'$, for the deviation due to trimolecular interactions; and so on. Since the probability of simultaneous multiparticle interactions is very small, the higher order virial coefficients have, in most cases, negligibly small values.

We can prove, from Eqs. (1-24) and (1-25), that the second and third virial coefficients of the pressure and volume series are related to each other through the equations

$$\mathscr{B} = \frac{\mathscr{B}'}{RT} \tag{1-26}$$

and

$$\mathscr{C} = \frac{\mathscr{C}'}{R^2 T^2} - \frac{\mathscr{B}\mathscr{B}'}{RT} = \frac{\mathscr{C}' - \mathscr{B}'}{R^2 T^2} \tag{1-27}$$

EXAMPLE 1-3
Derive Eqs. (1-26) and (1-27).
Solution: Eq. (1-24) can be rearranged into the form

$$\frac{1}{\mathscr{V}} = \frac{P}{RT}(1 + \mathscr{B}P + \mathscr{C}P^2 + \cdots)^{-1} \tag{a}$$

Since $\mathscr{B}P$ and $\mathscr{C}P^2$ are only correction terms, it may be assumed that their sum is less than 1 and therefore[7]

$$(1 + \mathscr{B}P + \mathscr{C}P^2)^{-1} \cong 1 - \mathscr{B}P - \mathscr{C}P^2 + \mathscr{B}^2P^2 + 2\mathscr{B}\mathscr{C}P^3 + \mathscr{C}^2P^4$$

Substituting this into Eq. (a), we have[8]

$$\frac{1}{\mathscr{V}} = \frac{P}{RT}(1 - \mathscr{B}P - \mathscr{C}P^2 + \mathscr{B}^2P^2)$$

$$= \frac{P}{RT} - \frac{\mathscr{B}P^2}{RT} - \frac{(\mathscr{C} - \mathscr{B}^2)P^3}{RT} \tag{b}$$

Substitution of Eq. (b) into Eq. (a) results in[9]

$$\frac{P\mathscr{V}}{RT} = 1 + \frac{\mathscr{B}'P}{RT} - \frac{\mathscr{B}'\mathscr{B}}{RT}P^2 + \frac{\mathscr{C}'}{R^2T^2}P^2$$

$$= 1 + \frac{\mathscr{B}'}{RT}P + \left(\frac{\mathscr{C}'}{R^2T^2} - \frac{\mathscr{B}'\mathscr{B}}{RT}\right)P^2$$

Equating the coefficients of P in this equation to the coefficients of P in Eq. (1-24), we obtain

$$\mathscr{B} = \frac{\mathscr{B}'}{RT}$$

and

$$\mathscr{C} = \frac{\mathscr{C}'}{R^2T^2} - \frac{\mathscr{B}'\mathscr{B}}{RT} = \frac{\mathscr{C}' - \mathscr{B}'^2}{R^2T^2} \qquad \bullet$$

If the assumption of pairwise additivity in the potential energies is considered, statistical mechanics gives, for a spherical monatomic gas, the following relationship between the second virial coefficient and the intermolecular potential energy function (see page 616):

$$\mathscr{B}' = 2\pi N \int_0^\infty (1 - e^{-\varepsilon(r)/kT})r^2 \, dr \tag{1-28}$$

where N is Avogadro's number, k is Boltzmann's constant, and $\varepsilon(r)$ is the potential energy as a function of the distance between the two molecules.

Unfortunately, virial coefficients are hard to compute because the potential functions are not yet well enough known. The calculation of $\mathscr{B}'$ for a simple case is shown in Example 1-4.

EXAMPLE 1-4

Calculate the second virial coefficient of a gas obeying the hard-sphere potential function with attraction. Assume that the attractive part of the potential function is given as $-A/r^n$.

---

[7] $(1-x)^{-1} \approx 1 - x + x^2$, where $x = \mathscr{B}P + \mathscr{C}P^2$.
[8] Disregarding terms involving P raised to any power greater than two.
[9] Disregarding again terms involving P raised to any power greater than two.

Solution: For a monatomic gas obeying the hard-sphere potential with attraction, Eq. (1-28) may be rewritten into

$$\mathscr{B}' = 2\pi N \left[\int_0^{\sigma} (1 - e^{-\infty/kT}) \right] r^2 \, dr$$

$$+ 2\pi N \left[\int_{\sigma}^{\infty} (1 - e^{A/kTr^n}) \right] r^2 \, dr$$

$$= \tfrac{2}{3} \pi N \sigma^3 + 2\pi N \int_{\sigma}^{\infty} (1 - e^{A/kTr^n}) r^2 \, dr$$

since $e^{-\infty/kT} = 0$. The value of the second integral cannot be evaluated in a simple analytical form.

At high temperatures however,[10] $e^{A/kTr^n} \approx 1 + A/kTr^n$, and the integral can be changed into

$$-2\pi N \frac{A}{kT} \int_{\sigma}^{\infty} r^{2-n} \, dr = \frac{2\pi NA}{kT(n-3)} \left[\frac{1}{r^{n-3}} \right]_{\sigma}^{\infty}$$

The integral converges only if $n > 3$, giving the result

$$\mathscr{B}'' = -\frac{2}{3} \pi N \sigma^3 \left[\frac{3A}{(n-3)\sigma^3 kT} \right]$$

Upon addition of both integrals, we have

$$\mathscr{B}' = \mathscr{B}'_0 + \mathscr{B}'' = \frac{2}{3} \pi N \sigma^3 - \frac{2}{3} \pi N \sigma^3 \left[\frac{3A}{(n-3)\sigma^3 kT} \right]$$

$$= \frac{2}{3} \pi N \sigma^3 \left[1 - \frac{3A}{(n-3)\sigma^3 kT} \right]$$

where $\mathscr{B}'_0 = \frac{2}{3} \pi N \sigma^3$ is the second virial coefficient of hard spheres without attraction and may have only positive values. The attractive part of the potential is responsible for the negative values of $\mathscr{B}'$ and for its temperature dependence.   ●

Many other equations of state have been proposed for real gases [1, 2, 5, 6] and we shall mention just a few. Historically, the most important is the van der Waals equation of state [8].

$$\left(P + \frac{a}{\mathscr{V}^2} \right)(\mathscr{V} - b) = RT \tag{1-29}$$

where the term $a/\mathscr{V}^2$, called the internal or cohesive pressure (see page 45), stands for the correction in pressure due to the attractive forces, and the term b, called the covolume, is related to the volume of the molecules themselves. The volume represented by b is actually somewhat greater than the molar volume of the liquefied gas.

The van der Waals equation contains, in addition to the universal gas constant R, two empirical parameters, a and b, whose values depend on the particular gas [9]. If

---

[10] For $x \ll 1$, $e^x = 1 + x + \frac{1}{2}x^2 + \cdots$.

pressure is expressed in atmospheres and volume in liters, the units of a are liter^2 atm mol^{-2} and those of b are liter mol^{-1}.

An alternative analytical representation for real gases is provided by the Berthelot equation [7]. This relation takes into consideration the temperature dependence of the attractive correction term. Thus

$$\left(P+\frac{a}{\mathcal{V}^2 T}\right)(\mathcal{V}-b)=\boldsymbol{R}T \tag{1-30}$$

For its high accuracy, especially at low pressures, the Berthelot equation is applied in gas thermometry.

An equation of state of a more complex form has been recommended by Dieterici:

$$(Pe^{a/\mathcal{V}\boldsymbol{R}T})(\mathcal{V}-b)=\boldsymbol{R}T \tag{1-31}$$

At higher temperatures and low pressures, when the intermolecular potential energy is much less than the kinetic energy, $a/\boldsymbol{R}T\mathcal{V}\ll 1$, the exponential term may be written as

$$e^{a/\boldsymbol{R}T\mathcal{V}}=1+\frac{a}{\boldsymbol{R}T\mathcal{V}}\approx 1+\frac{a}{P\mathcal{V}^2} \tag{1-32}$$

because, at such conditions, the error introduced by setting $\boldsymbol{R}T=P\mathcal{V}$ is negligibly small and the van der Waals and the Dieterici equations become identical.

The equation proposed by Redlich and Kwong

$$\left(P+\frac{a}{T^{1/2}\mathcal{V}(\mathcal{V}+b)}\right)(\mathcal{V}-b)=\boldsymbol{R}T \tag{1-33}$$

has proved to be surprisingly accurate.

The parameters a and b that appear in Eqs. (1-29), (1-30), (1-31), and (1-33) are not necessarily related to each other; in fact, some of them do not even have the same dimensions. Recognize also that a and b are not strictly constant for a given substance; for, if we assume that the equation of state is exact and put into them observed pressures, molar volumes, and temperatures, we find that the calculated a and b for the substance vary somewhat with temperature and pressure.

The van der Waals constants a and b for some gases are listed in Table 1-3.

It is difficult to find quantitative information, especially in a textbook, that shows how well an equation of state will describe the P-V-T relations of any given gas. The degree of success of an equation in describing P-V-T relationships depends primarily on the complexity of the equation—the number of constants in the equation—and also on the temperature, pressure, and chemical and physical behavior of the gas. It is reasonable to expect, that the larger the number of constants in the equation, the better the results will be. A large number of constants, however, makes an equation less practical, especially when applied to thermodynamics.

A common feature of all the equations of state, regardless of whether they contain two or more constants, is that at low pressures they must reduce to the ideal gas equation:

$$\lim_{P\to 0}\frac{P\mathcal{V}}{\boldsymbol{R}T}=1$$

Table 1-3 Van der Waals constants of some gases

Substance	a (liter^2 atm mol^{-2})	b (liter mol^{-1})
He	0.0341	0.0237
Ar	2.32	0.0398
H$_2$	0.244	0.0266
N$_2$	1.39	0.0391
O$_2$	1.36	0.0318
Cl$_2$	6.49	0.0562
CO	1.49	0.0399
CO$_2$	3.59	0.0427
N$_2$O	3.78	0.0441
CH$_4$	2.25	0.0428
SO$_2$	6.71	0.0564
H$_2$O	5.46	0.0305
NH$_3$	4.17	0.0371

To prove this, we rearrange the van der Waals equation into the form

$$z = \frac{P\mathcal{V}}{RT} = 1 + \frac{Pb}{RT} - \frac{a}{RT\mathcal{V}} + \frac{ab}{RT\mathcal{V}^2} \tag{1-34}$$

At pressures that are not very high, the last term approaches zero ($\lim_{P \to 0} ab/RT\mathcal{V}^2 = \lim_{\mathcal{V} \to \infty} ab/RT\mathcal{V}^2 = 0$) and $\mathcal{V}$ in the next to last term may be replaced by RT/P. At such conditions, we may write

$$z = \frac{P\mathcal{V}}{RT} = 1 + \frac{bP}{RT} - \frac{aP}{R^2T^2} = 1 + \frac{1}{RT}\left(b - \frac{a}{RT}\right)P \tag{1-35}$$

and in the limit of P approaching zero, $z = 1$.

From comparison of Eq. (1-35) with Eqs. (1-24) and (1-26) it is obvious that

$$\mathcal{B} = \frac{1}{RT}\left(b - \frac{a}{RT}\right) \tag{1-36}$$

and

$$\mathcal{B}' = b - \frac{a}{RT} \tag{1-37}$$

The result is the same as for a hard-sphere model with attraction.

At high temperatures, when $b > a/RT$, the virial coefficients are positive and become increasingly negative at lower temperatures. This fact also explains the different compressibilities of gases at low and high pressures and temperatures.

Using the same procedure as before, the Berthelot equation gives the following results for the second virial coefficients

$$\mathcal{B} = \frac{b}{RT} - \frac{a}{R^2T^3} \tag{1-38}$$

and

$$\mathscr{B}' = b - \frac{a}{RT^2} \tag{1-39}$$

The Dieterici equation gives the same expression as the van der Waals equation for the second virial coefficients but much better results for the third virial coefficients.

An equation that represents the P-V-T data of gases with high accuracy has been proposed by Beattie and Bridgeman. This equation, when simplified, may be written in the form

$$z = \frac{P\mathscr{V}}{RT} = 1 + \frac{\mathscr{B}'}{\mathscr{V}} + \frac{\mathscr{C}'}{\mathscr{V}^2} + \frac{\mathscr{D}'}{\mathscr{V}^3} \tag{1-40}$$

where

$$\mathscr{B}' = B_0 - \frac{A_0}{RT} - \frac{c}{T^3}$$

$$\mathscr{C}' = -B_0 b + \frac{A_0 a}{RT} - \frac{B_0 c}{T^3}$$

$$\mathscr{D}' = B_0 b c$$

A_0, B_0, a, b, and c are constants characteristic of the individual gases.

The relations between the volume series and pressure series virial coefficients are very complicated for the Beattie–Bridgeman equation.

1-7 LIQUEFACTION OF GASES

The liquid state is usually defined as that state which has a fixed volume for a given amount of substance but not a fixed shape. It is a state in which the molecules are very close to each other, a high density state; consequently liquids are hard to compress.

Due to the fact that the distance between molecules in the liquid state is much smaller than in the gaseous state, gas molecules must be squeezed together before liquefaction can occur. Although an increase in pressure can reduce the distance between the gas molecules, pressure alone would not be sufficient to keep the molecules together. The thermal kinetic energy of the molecules, which is directly related to the temperature, keeps them apart. In order to liquefy a gas the thermal energy (temperature) must be reduced to such a level[11] that the energy resulting from the attractive intermolecular forces will at least be equal to the thermal energy. High pressure and low temperature conditions are therefore required for liquefaction.

Because different gases follow different potential functions, the conditions at which a particular gas can be liquefied will vary with the behavior of the individual gas. As Andrew determined in 1869, for each gas there is a temperature above which it is impossible to induce liquefaction, no matter how great the pressure. This temperature is called the **critical temperature** [11, 14, 16].[12] The pressure necessary

---

[11] Techniques for achieving the requisite low temperatures utilize thermodynamic principles, for example the Joule–Thomson phenomenon, see page 150.

[12] If the system is a mixture of gases, its behavior may be more complex than that of a single gas.

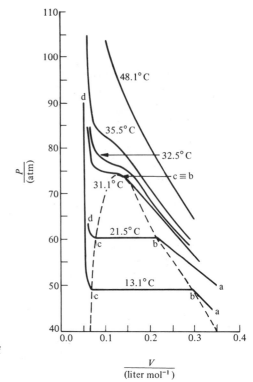

Figure 1-9 The liquefaction isotherms of carbon dioxide.

to liquefy a gas at its critical temperature is its **critical pressure**. The volume occupied by a gas at its critical temperature and critical pressure is the **critical volume**. The farther (lower) the temperature is from its critical value, the lower the pressure required for liquefaction (see Figures 1-9 and 1-10). At the critical point, there are no differences in the behavior of the gas and the liquid. The indexes of refraction, densities, and molar volumes of the two phases become identical.

Figure 1-9 shows the P-V relationship for carbon dioxide at, below and above the critical temperature. Each curve (isotherm) shows the dependence of P on V of a fixed amount of carbon dioxide at a constant temperature. At very high temperatures and low pressures, carbon dioxide exists only in the gaseous phase and the P-V behavior is well represented by Boyle's law. At temperatures closer to the critical temperature (but still above it), the P-V curves no longer obey Boyle's law, due to the deviations from ideal behavior. Still, the pressure increases continuously at constant temperature as the volume decreases. Below the critical temperature (isotherm 31.1°C), the curves are made up of three parts: the high-volume part a-b, where only gas exists; the horizontal part b-c, where liquid and vapor[13] exist in mutual equilibrium; and finally, the low-volume part, c-d, where only liquid exists. The liquefaction begins at point b, and ends at point c. The pressure corresponding to the horizontal part of the curve—the volume range in which the two phases coexist in mutual equilibrium—is the saturated vapor pressure of the liquefied gas. As long as the two phases are in mutual equilibrium, the pressure remains unchanged. The slope

---

[13] The word vapor is often used for a gas at any temperature below the critical temperature, though the distinction between gas and vapor is often ignored.

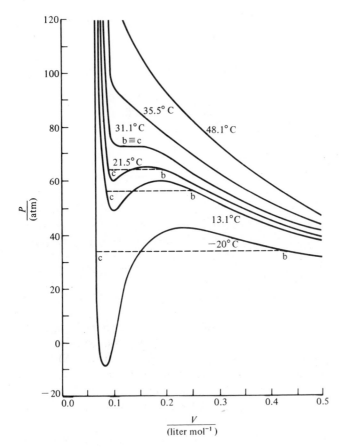

Figure 1-10   Van der Waals isotherms of carbon dioxide.

of the c-d curve is steeper than the slope of the a-b curve, since liquids are less compressible than gases.

In the region in which the two phases coexist, the volume is continuously changing while both the pressure and temperature are kept constant. No equation of state can account for this fact. The van der Waals equation, cubic with the molar volume, may account for three volumes, the three roots of the equation. Only one of these roots is real above the critical temperature; all three roots are real below the critical temperature. The van der Waals isotherm must therefore have a maximum and a minimum in the liquefaction region. The curves in Figure 1-10 represent the typical van der Waals isotherms of carbon dioxide.

The closer the temperature is to critical, the shorter the horizontal part of the curve. At the critical temperature, 31.1°C for CO_2, the liquefaction begins and ends at the same point (b≡c) (Figures 1-9 and 1-10). The dashed curve in Figure 1-9, joining all beginning and end points of the liquefaction, represents the two-phase region where liquid and vapor coexist in mutual equilibrium. To the right of the dashed curve only gas may exist, whereas to the left of it only liquid may exist. The two-phase solid and liquid region as well as the one-phase solid region are not shown in the diagram.

The horizontal part of a curve in Figure 1-10 can be mathematically expressed as

$$\left(\frac{\partial P}{\partial V}\right)_T = 0 \tag{1-41}$$

The critical temperature represents the highest temperature (maximum) at which such a horizontal line may still exist on the curve, and therefore,

$$\left(\frac{\partial^2 P}{\partial V^2}\right)_{T_c} = 0 \tag{1-42}$$

Thus Eqs. (1-41) and (1-42) characterize the critical isotherm completely.

Although the critical temperature and pressure may easily be measured, there are some difficulties in the accurate measurement of the critical volume. Hakala discovered an accurate empirical rule linking the densities of the mutually coexisting liquid and vapor phases at temperatures less than critical to the critical density. It has the form

$$\frac{\rho_l + \rho_v}{2} = \rho_c + K(\rho_l - \rho_v)^{10/3} \tag{1-43}$$

where ρ_l and ρ_v are the densities of the liquid and vapor phase, respectively, ρ_c is the critical density, and K is a constant whose value changes from substance to substance. If $(\rho_l + \rho_v)/2$ is plotted on a graph against $(\rho_l - \rho_v)^{10/3}$, a straight line is observed. The intercept of this line is equal to the critical density. Figure 1-11 illustrates such a graph for ethyl alcohol. The critical density for ethyl alcohol obtained by extrapolation is 0.276 g cm^{-3} and the critical molar volume $\mathcal{V}_c = M/\rho_c = 166.74 \text{ cm}^3$.

As we have shown before, a substance at its critical temperature exhibits a continuity of state, and thus any equation of state applicable to a nonideal gas may also be used for the interpretation of the critical state.

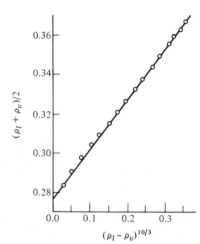

Figure 1-11 The evaluation of the critical density of ethanol by Hakala's method.

For example, the van der Waals equation, written for the critical state, has the form

$$P_c = \frac{RT_c}{V_c - b} - \frac{a}{V_c^2} \tag{1-44}$$

Applying the mathematical conditions Eqs. (1-41) and (1-42) to the van der Waals equation, we obtain

$$\left(\frac{\partial P}{\partial V}\right)_{T_c} = -\frac{RT_c}{(V_c - b)^2} + \frac{2a}{V_c^3} = 0 \tag{1-45}$$

and

$$\left(\frac{\partial^2 P}{\partial V^2}\right)_{T_c} = \frac{2RT_c}{(V_c - b)^3} - \frac{6a}{V_c^4} = 0 \tag{1-46}$$

Solution of Eqs. (1-45) and (1-46) results in the expressions

$$V_c = 3b \tag{1-47}$$

and

$$T_c = \frac{8a}{27Rb} \tag{1-48}$$

If we substitute the expressions from Eqs. (1-47) and (1-48) into Eq. (1-44) we obtain

$$P_c = \frac{a}{27b^2} \tag{1-49}$$

As is evident from the last three equations, the critical parameters may be evaluated from the known values of constants a and b. This is not, however, very common; as a matter of fact, more frequently the constants a and b are evaluated from the experimentally found values of the critical parameters. (The best method however, is to determine a and b from experimental P-V-T data.) The proper relations

$$a = \tfrac{9}{8}RT_c V_c = 3P_c V_c^2 \tag{1-50}$$

$$b = \tfrac{1}{3}V_c \tag{1-51}$$

are obtained from Eqs. (1-47), (1-48), and (1-49).

From Eqs. (1-47), (1-48), and (1-49) the critical value of the compressibility factor, z_c, can be evaluated:

$$z_c = \frac{P_c V_c}{RT_c} = \frac{(a/27b^2)3b}{R(8a/27Rb)} = \frac{3}{8} = 0.375 \tag{1-52}$$

The value of the critical compressibility factor predicted for any gas by the van der Waals equation is 0.375, which is a considerable improvement over the ideal gas value of unity but still somewhat above the average experimental value of 0.267. (The average is taken over a large number of gases.) Values of the critical compressibility factor of some substances are listed in Table 1-4.

Table 1-4 Values of critical constants and critical compressibility factors

Substance	T_c (K)	P_c (atm)	$\mathcal{V}_c$ (liter mol^{-1})	$\dfrac{P_c \mathcal{V}_c}{RT_c} = z_c$
helium	5.3	2.26	0.0578	0.300
hydrogen	33.3	12.80	0.0650	0.305
oxygen	154.8	50.10	0.0780	0.307
nitrogen	126.2	33.5	0.0901	0.291
methane	190.7	45.8	0.990	0.290
ethylene	282.4	50.0	0.124	0.268
benzene	562.0	48.6	0.260	0.274
carbon dioxide	304.2	72.9	0.094	0.274
ammonia	405.6	111.5	0.072	0.242
water	647.4	218.3	0.056	0.231

1-8 THE PRINCIPLE OF CORRESPONDING STATES

On introducing the expressions for R from Eq. (1-52) and for a and b from Eqs. (1-50) and (1-51) into the van der Waals equation, the following expression is obtained

$$\left(P + \frac{P_c \mathcal{V}_c^2}{\mathcal{V}^2}\right)\left(\mathcal{V} - \frac{\mathcal{V}_c}{3}\right) = \frac{8}{3}\frac{P_c \mathcal{V}_c T}{T_c} \tag{1-53}$$

Multiplication by $3/P_c \mathcal{V}_c$ yields

$$\left(P_r + \frac{3}{V_r^2}\right)(3V_r - 1) = 8T_r \tag{1-54}$$

where the dimensionless variables, P_r, V_r, and T_r, are the reduced pressure, $P_r = P/P_c$; reduced volume, $V_r = \mathcal{V}/\mathcal{V}_c$; and reduced temperature, $T_r = T/T_c$, respectively.

The important thing about Eq. (1-54) is that it does not contain any constants peculiar to the individual gases, although values of the critical constants are implied. According to this equation, there exists a universal function

$$f(P_r, T_r, V_r) = 0 \tag{1-55}$$

that is valid for all substances. To express this concept in other words, we may say that, at the same T_r, P_r, and V_r, all gases exhibit the same deviation from ideal behavior. Thus, at the same T_r, P_r, and V_r, the gases are in their corresponding states. Eq. (1-54) represents only one of the many forms of the principle of corresponding states.

We may apply the same mathematical procedure to the Berthelot equation[14] to obtain the following dimensionless equation

$$\frac{P\mathcal{V}}{RT} = z = 1 + \frac{9PT_c}{128P_cT}\left(1 - 6\frac{T_c^2}{T^2}\right)$$

$$= 1 + \frac{9}{128}\frac{P_r}{T_r}\left(1 - 6\frac{1}{T_r^2}\right) \tag{1-56}$$

The principle of the corresponding states may also be stated as

$$z = \frac{P\mathcal{V}}{RT} = f(T_r, P_r) \tag{1-57}$$

or

$$P_r = f'(T_r, V_r) \tag{1-58}$$

Of the three parameters that might vary from one gas to another (P_c, T_c, $\mathcal{V}_c$), one is fixed by knowing that, as $V_r \to \infty$, Eq. (1-58) must approach $P = RT/\mathcal{V}$ (which fixes the gas constant R). Thus all fluids that "correspond" obey the same equation of state, an equation containing only two parameters which can be adjusted to fit the particular gas.

The molecular basis of the law of corresponding states as expressed by Eq. (1-57) lies in the fact that the intermolecular potential curves of a great many kinds of molecules have the same general shape, and their analytic formula contains only two adjustable parameters, ε^* and r^* (see Eq. 1-19). Such gases would be expected to obey the simple law of corresponding states (Eqs. 1-57 and 1-58).

Relationship (1-57) is graphically presented in Figure 1-12. The accuracy obtained from this graph lies within 2–5%, except for the very polar gases like $NH_3(g)$ and $H_2O(g)$, where the accuracy is not as good.

The intermolecular potential function of nonspherical gaseous molecules requires three or more adjustable parameters. It is clear that in such cases three or more parameters would be needed to express the equation of state. For this reason, for a gas obeying a three-parameter potential function, the law of corresponding states must be modified into

$$z = \frac{P\mathcal{V}}{RT} = F(P_r', T_r, \Theta) \tag{1-59}$$

where Θ is a property depending on the nature of the individual gases. Various authors define Θ differently. Lydersen, Greenhorn, and Hougen identify Θ with the critical compressibility factor, $z_c = P_c\mathcal{V}_c/RT_c$, whereas Pitzer and others assume Θ to be equal to the so called acentric factor, ω, defined by the equation

$$\omega = -\log\frac{P}{P_c} - 1.000 \tag{1-60}$$

---

[14] In order to get a better agreement between the calculated and experimentally found $P\text{-}V\text{-}T$ data, semiempirical relations have been used between a and b and the critical constants

$$a = \frac{27}{64}\left(\frac{RT_c}{P_c}\right)P_cT_c = \frac{16}{3}P_c\mathcal{V}_c^2T_c \qquad \text{and} \qquad b = \frac{\mathcal{V}_c}{4} = \frac{9}{128}\frac{RT_c}{P_c}$$

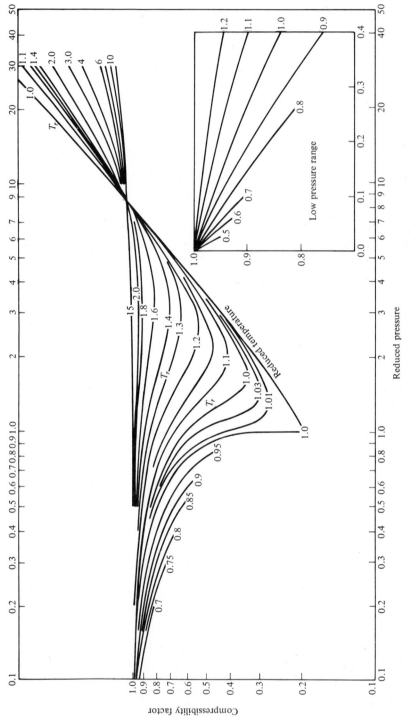

Figure 1-12 The compressibility factor as a function of the reduced pressure and reduced temperature. The theorem of corresponding states.

where P is the saturated vapor pressure of the substance at $T_r = 0.7$ and P_c is the critical pressure.

EXAMPLE 1-5

Derive the relations between the constants a and b of the Dieterici equation and the critical parameters and calculate the compressibility factor in the critical state.

Solution: The Dieterici equation may be written as

$$P = \frac{RT}{V - b} e^{-a/RTV}$$

Taking the first derivative with respect to V at constant temperature, we obtain

$$\left(\frac{\partial P}{\partial V}\right)_T = RT e^{-a/RTV}\left[\frac{a}{V^2 RT(V - b)} - \frac{1}{(V - b)^2}\right] = P\left[\frac{a}{V^2 RT} - \frac{1}{(V - b)}\right]$$

The second derivative is

$$\left(\frac{\partial^2 P}{\partial V^2}\right)_T = P\left[-\frac{2a}{V^3 RT} + \frac{1}{(V - b)^2} + \left(\frac{a}{V^2 RT} - \frac{1}{(V - b)}\right)^2\right]$$

Applying the critical isotherm conditions, we may write

$$P_c\left[\frac{a}{V_c^2 RT_c} - \frac{1}{(V_c - b)}\right] = 0$$

and

$$P_c\left[-\frac{2a}{V_c^3 RT_c} + \frac{1}{(V_c - b)^2} + \left(\frac{a}{V_c^2 RT_c^2} - \frac{1}{(V_c - b)}\right)^2\right] = 0$$

Simultaneous solution of these two equations results in

$$b = V_c/2 \qquad a = 2V_c RT_c$$

At the critical point, the Dieterici equation is then

$$P_c = \frac{RT_c}{V_c - b} e^{-a/RT_c V_c} = \frac{RT_c}{V_c - V_c/2} e^{-2V_c RT_c/V_c RT_c} = \frac{2RT_c}{V_c} e^{-2}$$

and the value of the compressibility factor in the critical state is

$$z_c = \frac{P_c V_c}{RT_c} = \frac{2}{e^2} = 0.271$$

This value is very close to the experimentally found average value of 0.267.   ●

EXAMPLE 1-6

At a pressure of 100 atm and a temperature of 198°C, the value of the ratio $PV/P_0 V_0$ for carbon dioxide is 1.5820 [18]. The density of carbon dioxide under normal conditions ($P_0 = 1$ atm, $T_0 = 273.15$ K) is 1.9768 g liter^{-1} [19].

Calculate the density of carbon dioxide at a pressure of 100 atm and a temperature of 198°C using the generalized compressibility factor diagram (Figure 1-12) and compare the result with the experimentally found value.

Given: $t_c = 31.1$°C; $P_c = 73.0$ atm.

Solution: Making use of the compressibility factor the density may be expressed as

$$\rho = \frac{w}{V} = \frac{PM}{z\textbf{R}T} \tag{a}$$

From the compressibility factor diagram (Figure 1-12), we obtain $z = 0.90$ when $T_r = T/T_c = 471.2/304.3 = 1.55$ and when $P_r = P/P_c = 100/73.0 = 1.37$. Substituting this value for z into Eq. (a), we obtain

$$\rho = \frac{100 \times 44}{0.90 \times 0.082 \times 471.2} = 126.54 \text{ g liter}^{-1}$$

For comparison, we calculate the density from the experimental data

$$\frac{PV}{P_0 V_0} = \frac{P\rho_0}{P_0\rho} = 1.5820$$

since the density is inversely proportional to the volume. The quantities with index zero relate to normal conditions. Upon substituting the given data, we get

$$\rho = \frac{100 \times 1.9768}{1 \times 1.5820} = 124.97 \text{ g liter}^{-1}$$

The value determined from Figure 1-12 differs from that found experimentally by 1.2%, whereas the ideal gas equation gives $113.70 \text{ g liter}^{-1}$, which differs by more than 9% from the value obtained experimentally.   ●

1-9   MIXTURES OF GASES

When we deal with mixtures of gases, an additional variable, concentration, must be considered. Consequently, the equation of state for a mixture has the form

$$f(P, V, T, \text{concentration}) = 0 \tag{1-61}$$

The simplest to treat analytically is the ideal mixture of ideal gases (to be distinguished from an ideal mixture of gases in which the constituents of the mixture do not behave ideally), for which each component contributes independently to the pressure.

$$PV = n_{\text{tot}}(\textbf{R}T) = (n_1 + n_2 + n_3 + \cdots)\textbf{R}T \tag{1-62}$$

or

$$P = \frac{n_1\textbf{R}T}{V} + \frac{n_2\textbf{R}T}{V} + \frac{n_3\textbf{R}T}{V} + \cdots \tag{1-63}$$

where n_{tot} denotes the total number of moles of the gaseous mixture, and $n_1, n_2, n_3, \ldots$, denote the number of moles of each of the constituents of the gaseous mixture. Writing $p_i = n_i\textbf{R}T/V$, we obtain

$$P = p_1 + p_2 + p_3 + \cdots = \sum_i p_i \tag{1-64}$$

According to Eq. (1-64) the total pressure of a mixture of gases is equal to the sum of the pressures—partial pressures—that each of its constituents would exert if alone in the volume of the mixture at the same temperature. Eq. (1-64) represents the mathematical formulation of the Dalton law (1801). Although Eq. (1-63) is valid for an ideal mixture of ideal gases, Eq. (1-64) has a much broader application. It requires only that the total pressure be a sum of the partial pressures; the partial pressures, however, do not necessarily have to obey the equation $p_i = n_i RT/V$. A system obeying Dalton's law exactly is known as an ideal mixture irrespective of whether its components individually behave as ideal gases.

Another expression of Dalton's law may be obtained by division of $p_i = n_i RT/V$ by $P = n_{tot} RT/V$. Thus

$$\frac{p_i}{P} = \frac{n_i RT/V}{n_{tot} RT/V} = \frac{n_i}{n_{tot}} = y_i$$

or

$$p_i = P y_i \tag{1-65}$$

Following from Eq. (1-65), the partial pressure of component i is equal to the product of the total pressure and the mole fraction of component i, y_i.

An alternative approach to the properties of gaseous mixtures leads to Amagat's law (1880): The volume of a mixture of gases is the sum of the volumes of its constituents, each at the pressure and temperature of the mixture. Mathematically,

$$V = V_1 + V_2 + V_3 + \cdots = \sum_i V_i \tag{1-66}$$

where the quantities V_i are called partial volumes. Similar to Eq. (1-65), Amagat's law can also be expressed as

$$V_i = V y_i \tag{1-67}$$

We may easily prove that for mixtures of ideal gases where $V_i = n_i RT/P$ and $p_i = n_i RT/V$, Dalton's and Amagat's laws supply identical results.

For a gaseous mixture of nonideal constituents, the partial pressures may be expressed as follows:

$$p_i = \frac{n_i z_i(p_i, T) RT}{V} \tag{1-68}$$

where $z_i(p_i, T)$ is the compressibility factor of component i evaluated at the temperature of the system and partial pressure of component i. Summing over all the constituents of the mixture, we obtain Dalton's law

$$\sum_i p_i = P = \sum_i \frac{n_i z_i(p_i, T) RT}{V} = \frac{RT}{V} \sum_i n_i z_i(p_i, T) \tag{1-69}$$

Comparison with $PV = n_{tot} z_m(P, T) RT$ results in

$$n_{tot} z_m(P, T) = \sum_i n_i z_i(p_i, T) \tag{1-70}$$

Consequently

$$z_m(P, T) = \sum_i y_i z_i(p_i, T) \tag{1-71}$$

$z_m(P, T)$ is the compressibility factor of the mixture. As is obvious from Eq. (1-71) Dalton's law implies that the compressibility factor of the mixture is approximated by the weighted average of the compressibility factors of the individual components, each evaluated at the approximate partial pressure. Since the partial pressures are not known, an iterative method (trial and error method) must be applied.

By a similar procedure, using Amagat's law, we derive

$$z_m(P, T) = \sum_i y_i z_i(P, T) \tag{1-72}$$

When comparing Eq. (1-71) with Eq. (1-72) we find that, according to Dalton's law, the component compressibility factors are evaluated at the respective partial pressures, whereas according to Amagat's law the component compressibility factors are evaluated at the total pressure of the mixture. Amagat's law, therefore, represents a considerably simpler way of expressing the P-V-T relation of gaseous mixtures.

Remember also that any equation of state applicable to single gases may also be used for gaseous mixtures. Different combination rules are available in the literature that enable the evaluation of the constants of these equations for the mixture from the similar constants of the pure gases.

The virial equation is the only equation of state for which rigorous "mixing rules" are available. The mixing rules for an m-component mixture are

$$\mathscr{B} = \sum_{i=1}^{m} \sum_{j=1}^{m} y_i y_j \mathscr{B}_{ij} \tag{1-73}$$

$$\mathscr{C} = \sum_{i=1}^{m} \sum_{j=1}^{m} \sum_{k=1}^{m} y_i y_j y_k \mathscr{C}_{ijk} \tag{1-74}$$

where the y's are mole fractions. The coefficients $\mathscr{B}_{ij}$, $\mathscr{C}_{ijk}$, are called the cross coefficients and have the properties:

$$\mathscr{B}_{ij} = \mathscr{B}_{ji} \tag{1-75}$$

$$\mathscr{C}_{ijk} = \mathscr{C}_{ikj} = \mathscr{C}_{jik} = \mathscr{C}_{jki} = \mathscr{C}_{kij} = \mathscr{C}_{kji} \tag{1-76}$$

Two types of coefficients appear in the series (1-73) and (1-74); those in which the successive subscripts are identical, $\mathscr{B}_{ii}$, $\mathscr{B}_{jj}$, $\mathscr{C}_{iii}$, $\mathscr{C}_{jjj}$, $\mathscr{C}_{kkk}$, and those for which at least one of the subscripts differs from the other, $\mathscr{B}_{ij}$, $\mathscr{C}_{iij}$, $\mathscr{C}_{iik}$, and so on. The first kind of coefficient refers to pure components; the second type is a mixture property.

For a binary mixture, Eq. (1-73) may be expanded into the form

$$\mathscr{B} = y_1^2 \mathscr{B}_{11} + y_1 y_2 \mathscr{B}_{12} + y_2 y_1 \mathscr{B}_{21} + y_2^2 \mathscr{B}_{22}$$
$$= y_1^2 \mathscr{B}_{11} + 2 y_1 y_2 \mathscr{B}_{12} + y_2^2 \mathscr{B}_{22} \tag{1-77}$$

It seems to be convenient to introduce a new factor, defined by the equation

$$\delta_{12} = 2\mathscr{B}_{12} - \mathscr{B}_{11} - \mathscr{B}_{22} \tag{1-78}$$

Making use of this factor, Eq. (1-77) assumes the form

$$\mathscr{B} = y_1 \mathscr{B}_{11} + y_2 \mathscr{B}_{22} + y_1 y_2 \delta_{12} \tag{1-79}$$

For an ideal gaseous mixture $\delta_{12} = 0$ and therefore[15]

$$\mathscr{B} = y_1\mathscr{B}_{11} + y_2\mathscr{B}_{22} \tag{1-80}$$

and

$$\mathscr{B}_{12} = \frac{\mathscr{B}_{11} + \mathscr{B}_{22}}{2} \tag{1-81}$$

For better understanding see Example 1-7.

EXAMPLE 1-7

The second virial coefficient of a mixture of methane (1) and *n*-hexane (2) in the molar ratio 1 : 1 is -517 cm^3 mol^{-1} at 50°C. Find the value of $\mathscr{B}$ of a mixture in which the molar ratio of methane to *n*-hexane is 1 : 3.

Given: $\mathscr{B}_{11} = -33$ cm^3 mol^{-1}, $\mathscr{B}_{22} = -1512$ cm^3 mol^{-1}.

Solution: Eq. (1-77) and the data given in the example enable one to evaluate the cross coefficient, $\mathscr{B}_{12}$:

$$\mathscr{B}_{12} = \frac{\mathscr{B} - y_1^2\mathscr{B}_{11} - y_2^2\mathscr{B}_{22}}{2y_1y_2}$$

$$= \frac{-517 - 0.5^2(-33) - 0.5^2(-1512)}{2 \times 0.5 \times 0.5}$$

$$= -262 \text{ cm}^3 \text{ mol}^{-1}$$

Upon substituting this value into Eq. (1-77) we obtain, for a 1 : 3 molar ratio mixture

$$\mathscr{B} = 0.25^2(-33) + 2 \times 0.25 \times 0.75(-262) + 0.75^2(-1512)$$

$$= -951 \text{ cm}^3 \text{ mol}^{-1}$$

If the mixture were ideal then

$$\mathscr{B} = y_1\mathscr{B}_{11} + y_2\mathscr{B}_{22}$$

$$= 0.25(-33) + 0.75(-1512) = -1142 \text{ cm}^3 \text{ mol}^{-1} \qquad \bullet$$

1-10   EQUATIONS OF STATE OF LIQUIDS

As we have mentioned before, there is a continuity between the gaseous and liquid states. The equations of state of gases, however, supply very poor results when applied to liquids. This is explained by the fact that the high density liquid state requires a different kind of—a more complex—potential function. The molecules are so closely packed that in the great majority of liquids a two-parameter potential function is insufficient to describe the molecular behaviour of liquids. Due to the close packing of molecules, liquids exhibit extremely high internal pressures.[16] For

---

[15] A mixture is ideal only when $(\varepsilon_{11} + \varepsilon_{22})/2 = \varepsilon_{12}$. This means that in an ideal mixture the energy between two unlike molecules ε_{12}, is given as the arithmetic mean of the energies of the pure constituents, ε_{11} and ε_{22}. .

[16] The internal pressure results from the large intermolecular forces. For a van der Waals gas, the internal pressure is simply given by the correction term $a/\mathscr{V}^2$. For more details see page 45.

this reason, for precise calculations of P-V-T of liquids, specially developed equations of state must be applied.

One of the best and most commonly used equations of state for liquids is the Tait equation.

$$-\left(\frac{\partial V}{\partial P}\right)_T = \frac{D}{E(T)+P} \tag{1-82}$$

in which D is a constant, $E(T)$ is a function only of temperature and P is the pressure. Integration of Eq. (1-82) at constant temperature yields

$$-\int_{V_1}^{V_2} dV = \int_{P_1}^{P_2} \left(\frac{D}{E(T)+P}\right) dP$$

$$V_1 - V_2 = D \ln\left(\frac{P_2+E(T)}{P_1+E(T)}\right) \tag{1-83}$$

This equation gives remarkable agreement with experimental values for a variety of liquids, polar as well as nonpolar, over a wide range of temperatures and pressures up to 3000 atm.

P-V-T data of liquids are commonly expressed in terms of the coefficient of thermal expansion, α:

$$\alpha \equiv \frac{1}{V}\left(\frac{\partial V}{\partial T}\right)_P \tag{1-84}$$

and coefficient of compressibility, β:

$$\beta \equiv -\frac{1}{V}\left(\frac{\partial V}{\partial P}\right)_T \tag{1-85}$$

The minus sign is introduced in order to make β a positive quantity, since $(\partial V/\partial P)_T$ is always negative.

_____ *Problems*

1. G. P. Baxter and H. W. Starkweather [20] studied the compressibility of oxygen at a low pressure at 0°C. The results of the measurement are as follows:

P (atm)	V (liter g^{-1})
1.00	0.69981
0.750	0.93328
0.500	1.40027
0.250	2.80120

Determine whether, under the given conditions, oxygen behaves according to Boyle's law.

2. F. Henning and W. Heuse [21] measured the coefficient of cubic expansion, α, of several gases at different pressures. The results of their measurements follow:

Helium		Nitrogen	
P (mm Hg)	$\alpha \cdot 10^6$ (K^{-1})	P (mm Hg)	$\alpha \cdot 10^6$ (K^{-1})
504.8	3658.9	511.4	3667.9
520.5	3660.3	1105.3	3674.2
760.1	3659.1		
1102.9	3658.2		
1116.5	3658.1		

Prove the validity of Gay-Lussac's law at low pressures and calculate the value of absolute zero in degrees centigrade from the data given.

3. The following data were obtained during a Victor Meyer experiment[17] for diethyl ether: $w = 0.1023$ g, the volume of the displaced air 35.33 cm³, temperature 32.5°C, and pressure 743.95 mm Hg. Calculate the molecular weight of the ether.

4. A badly functioning barometer (imperfect vacuum) shows a pressure of 755 mm Hg when the actual atmospheric pressure is 765.0 mm Hg; the evacuated part of the barometric tube is 25 mm long. Calculate the actual atmospheric pressure if the barometer shows a pressure of 745.0 mm Hg.

5. By heating from a temperature of 12°C to t°C, 169.1 cm³ of air, measured at 10°C, is driven from a closed 197.8 cm³ vessel (filled with air) that has permeable walls. The pressure during the experiment is 747 mm Hg. Assuming that the volume of the vessel does not change with temperature, calculate the temperature.

6. The highest temperature in a gas-holder in the summer is 42°C, the lowest winter temperature is −38°C. How many more kilograms of hydrogen can a gas-holder of 2000 m³ capacity take at the lowest winter temperature than at the highest summer temperature, if the pressure in the gas-holder is always 780 mm Hg?

7. At 220°C, and a pressure of 747 mm Hg, 1.3882 g of a certain organic substance in vapor form occupies a volume of 420 cm³. The elementary chemical analysis of this substance is as follows: C, 70.60%; H, 5.88%; and O, 23.52%. What is the molecular weight of the substance and what is its chemical formula?

8. If 5 g of ammonium carbamate ($NH_4CO_2NH_2$) dissociates at 200°C, the substance occupies a volume of 7.66 liters at a pressure of 740 mm Hg. Calculate the degree of dissociation, if dissociation takes place as follows:

$$NH_4CO_2NH_2(s) \rightleftharpoons 2\,NH_3(g) + CO_2(g)$$

---

[17] A very commonly used method for molecular weight determination of volatile liquids.

9. The dependence of the density of oxygen on the pressure was studied at a temperature of 0°C [22]. Results of measurements:

P (atm)	ρ (g liter^{-1})
0.25	0.356985
0.50	0.714154
0.75	1.071485
1.00	1.428962

Calculate the second virial coefficient, $\mathcal{B}'$, of oxygen and discuss whether at this temperature oxygen molecules may be treated as hard-sphere particles without attraction (*Hint*: evaluate P/ρ to five decimal places).

10. For a hard-sphere gas without attraction derive the expression for the second virial coefficient.

11. The value of the second virial coefficient for helium at 175 K is $12.24 \text{ cm}^3 \text{ mol}^{-1}$. Making use of the hard-sphere potential function, calculate the diameter of a helium molecule.

12. To what temperature can a 20-liter oxygen pressure flask containing 1.6 kg of oxygen be heated if the highest allowed pressure is 150 atm? The constants of the van der Waals equation for oxygen are $a = 1.360 \text{ liter}^2 \text{ atm mol}^{-2}$; $b = 31.83 \text{ cm}^3 \text{ mol}^{-1}$.

13. If 10 moles of ethane are confined to a volume of 4860 cm^3 at 300 K, make use of the Dieterici equation to calculate the pressure: $a = 6.97 \times 10^6 \text{ (cm}^3)^6$ atm mol^{-2}; $b = 69.4 \text{ cm}^3 \text{ mol}^{-1}$.

14. Calculate the pressure of 453.6 g of chlorine in a 10-liter vessel at 100°C using (a) the van der Waals equation and (b) the generalized Berthelot equation. *Given*:

$a = 6.493 \text{ liter}^2 \text{ atm mol}^{-2}$ $P_c = 76.1 \text{ atm}$
$b = 0.05622 \text{ liter mol}^{-1}$ $T_c = 417.3 \text{ K}$

15. The density of nitrogen oxide under normal conditions is $1.3402 \text{ g liter}^{-1}$. Assuming that under these conditions the nitrogen oxide follows the generalized Berthelot equation, calculate its molecular weight. *Given*: $t_c = -96.1°C$; $P_c = 64 \text{ atm}$.

16. Derive an expression for the residual volume for a gas obeying the equation

$$z = 1 + \frac{\mathcal{B}'}{\mathcal{V}} + \frac{\mathcal{C}'}{\mathcal{V}^2} + \frac{\mathcal{D}'}{\mathcal{V}^3} + \cdots$$

17. What volume is occupied by 5 kg of chlorine compressed to 10 atm at 30°C? Use the generalized compressibility factor diagram. *Given*: $t_c = 143.8°C$; $P_c = 76.1 \text{ atm}$.

18. Derive an expression that relates the van der Waals constants a and b, the cubic expansion coefficient $\alpha[\equiv (1/V_0)(\partial V/\partial T)_P]$ and the isothermal compressibility coefficient, $\beta[\equiv -(1/V_0)(\partial V/\partial P)_P]$.

19. The Boyle temperature of a gas is the temperature at which

$$\lim_{P \to 0} \left[\frac{\partial (PV)}{\partial P} \right]_T = 0$$

For each of the following equations of state evaluate the Boyle temperature in terms of the constants.

(a) $\quad z = \dfrac{P\mathcal{V}}{RT} = 1 + \dfrac{1}{RT}\left(b - \dfrac{A}{RT^{3/2}} \right) P$

(b) $\quad z = \dfrac{P\mathcal{V}}{RT} = 1 + \dfrac{9 T_c P}{128 T P_c}\left(1 - \dfrac{6 T_c^2}{T^2} \right)$

20. A hypothetical gas has the equation of state

$$P = \frac{RT}{\mathcal{V} - b} - \frac{a}{\mathcal{V}}$$

where a and b are constants distinct from zero. Ascertain whether this gas has a critical point. If it has, express the critical constant in terms of a and b.

21. Prove that a and b of the Dieterici equation may be expressed as

$$b = \frac{RT_c}{P_c e^2} \qquad a = \frac{4 R^2 T_c^2}{P_c e^2}$$

22. A gas following the hard-sphere potential without attraction obeys the equation

$$P(\mathcal{V} - b) = RT$$

Derive $(\partial V/\partial T)_P$ and $(\partial V/\partial P)_T$.

23. Derive the Redlich Kwong equation in the form

$$z = \frac{1}{1 - b/\mathcal{V}} - \frac{a}{RT^{3/2}(\mathcal{V} + b)}$$

24. The orthobaric densities of CCl_4 liquid and vapor at a series of temperatures have the following values:

$\dfrac{t}{(°C)}$	150	250	270	280
$\dfrac{\rho(l)}{(\text{g cm}^{-3})}$	1.3215	0.9980	0.8666	0.7634
$\dfrac{\rho(g)}{(\text{g cm}^{-3})}$	0.0304	0.1754	0.2710	0.3597

Using Hakala's procedure find the value of the critical volume of CCl_4, if its critical temperature is 283.1°C.

25. Water gas has the following composition in weight per cent: H_2, 6.43; CO, 67.82; N_2, 10.71; CO_2, 14.02; and CH_4, 1.02. Express the composition of the gas in volume and mole per cent and in mole fractions and calculate the density of the gas mixture at a temperature of 400°C and 1.5 atm pressure, assuming that under these conditions the mixture behaves like an ideal gas.

26. 100 g of a gas mixture of nitrogen and methane, containing 31.014 wt% of nitrogen, occupies a volume of 0.99456 liters at a definite pressure and a temperature of 150°C. Assuming that the mixture follows Dalton's law, calculate the total pressure of the mixture of gases and the partial pressures of the different constituents.

27. A mixture of 3 kg of methane and 2 kg of ethylene is compressed to 50 atm at a temperature of 38°C. What volume does the gas mixture occupy? Make use of the Amagat approximation: $z_m = \sum_i z_i(P, T)y_i$.
    Given:

$$CH_4, t_c = -82.5°C; \qquad P_c = 45.8 \text{ atm}$$
$$C_2H_4, t_c = 9.7°C; \qquad P_c = 50.9 \text{ atm}$$

References and Recommended Reading

[1] J. Joffe: *Chem. Eng. Progr.*, **45**:160 (1949). Critical review of equations of state.

[2] K. K. Shah and G. Thodos: *Ind. Eng. Chem.*, **57**:30 (1965). Critical review of equations of state.

[3] R. G. Neville: *J. Chem. Educ.*, **39**:356 (1962). The discovery of Boyle's law.

[4] N. Feifer: *J. Chem. Educ.*, **43**:411 (1966). The relationship between Avogadro's principle and the law of Gay-Lussac.

[5] J. B. Ott, J. R. Goates, and H. T. Hall: *J. Chem. Educ.*, **48**:515 (1971). Different equations of state compared with respect to goodness of fit to experimental *P-V-T* relations.

[6] W. Dannhauser: *J. Chem. Educ.*, **47**:126 (1970). *P-V-T* behavior of real gases. Computer assisted exercises for physical chemistry.

[7] A. F. Saturno: *J. Chem. Educ.*, **39**:464 (1962). Daniel Berthelot's equation of state.

[8] E. S. Swinbourne and P. D. Lark: *J. Chem. Educ.*, **32**:366 (1955). The van der Waals gas equation.

[9] S. S. Winter: *J. Chem. Educ.*, **33**:459 (1956). A simple model for van der Waals *a*.

[10] Y. Rasiel and W. Freeman: *J. Chem. Educ.*, **46**:310 (1969). Properties of real gases in the limit of zero pressure.

[11] J. G. Roof: *J. Chem. Educ.*, **34**:492 (1957). How should we define the critical state.

[12] R. W. Hakala: *J. Chem. Educ.*, **41**:380 (1964). Dimensional analysis and the law of corresponding states.

[13] A. G. DeRocco: *J. Chem. Educ.*, **46**:365 (1969). Intermolecular potential and virial of fluids.

[14] F. L. Pilar: *J. Chem. Educ.*, **44**:284 (1967). The critical temperature: a necessary consequence of gas nonideality.

[15] R. W. Hakala: *J. Chem. Educ.*, **45**:16 (1968). A derivation of the virial equation based on hard-sphere model of gases.

[16] F. Vaslow: *J. Chem. Educ.*, **40**:485 (1963). A simple condensation isotherm for a hard-sphere gas.

[17] C. J. Arthur and U. A. Felsing: *J. Am. Chem. Soc.*, **68**:1883 (1946).

[18] J. R. Partington: *An Advanced Treatise on Physical Chemistry.* Longmans, Green, London, 1949, Vol. I, p. 571.

[19] J. H. Perry: *Chemical Engineer's Handbook.* McGraw-Hill, New York, 1950.

[20] G. P. Baxter and H. W. Starkweather: *Proc. Nat. Acad. Sci.,* **14**:50 (1928).

[21] F. Henning and W. Heuse: *Z. Physik,* **5**:285 (1921).

[22] G. P. Baxter and H. W. Starkweather: *Proc. Nat. Acad. Sci.,* **12**:699 (1926).

chapter 2

The Molecular Basis of the Equations of State

2-1 INTRODUCTION AND THE IDEAL GAS LAWS

Thus far, our discussion of the properties of gases as well as of the equations of state, with a few exceptions, has been primarily descriptive. Such an empirical approach, however accurate, is obviously not satisfactory. It cannot explain, for example, why different gases behave differently under the same conditions.

A better understanding of the *P-V-T* relations in gases has been the goal of investigators for many years (D. Bernoulli, 1738; Joule, 1848; Clausius, 1857; Maxwell, 1860; Boltzmann, 1859; and others). As a result of these investigations a new theory called the kinetic molecular theory of gases has been developed.[1]

According to this theory, the molecules of any particular gas move about in a completely random manner at velocities that are constantly changing due to molecule-molecule and molecule-wall collisions. The assumption often made about the perfect collision (the kinetic energy remains unchanged during the collision) is for convenience rather than necessity; it has no effect on the final result [1, 2, 3, 4, 5].

The size of the molecule is assumed to be small in comparison with the average distance between molecules and with the size of the vessel containing the gas. Consequently, molecules do not exert appreciable forces upon each other except at the instant of collision.

To derive the simplest gas law—Boyle's relationship—from the kinetic molecular theory of gases let us consider a container of length l and cross-sectional area A, containing N' molecules, each of mass m (Figure 2-1). For the sake of simplicity, we will first observe only one molecule of the many; we will assume that this molecule is moving along the positive direction of the x axis with a velocity component, u_1. After a certain period of time, depending on u_1, the molecule strikes the right-hand wall of the container and rebounds. Following from Newton's law, the net change in the momentum due to the elastic collision is $mu_1 - (-mu_1) = 2mu_1$. After the rebound, the molecule moves along the negative direction of the x axis. It collides now with the left-hand wall of the container and after a period of time it returns to the right-hand wall, the collisions being repeated again and again.

It is assumed that the molecule on its way from one wall to the other does not collide with other molecules; this assumption, however, has no effect on the final result. Between two successive collisions of the molecule with the same wall, a distance equivalent to two lengths of the container, $2l$, is traveled; the time interval between the two collisions is therefore $\Delta \tau = 2l/u_1$. The change in momentum per unit time, which represents the force acting on the wall by the molecule,

---

[1] The most intensive study of the kinetic molecular theory of gases was performed between 1840 and 1890.

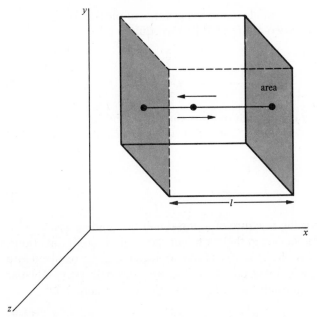

Figure 2-1 A molecule moving along the x axis with a velocity component u_1.

is equal to[2]

$$f_1 = \frac{2mu_1}{\Delta\tau} = \frac{2mu_1}{2l/u_1} = \frac{mu_1^2}{l} \tag{2-1}$$

All the molecules in the container moving along the positive x direction exert a similar force on the wall so that the total force acting on the wall is

$$f = \sum_{i=1}^{N'} f_i = \frac{m}{l} \sum_{i=1}^{N'} u_i^2 \tag{2-2}$$

Since the pressure is defined as force per unit area, we write

$$P = \frac{f}{A} = \frac{m}{lA} \sum_{i=1}^{N'} u_i^2 = \frac{m}{V} \sum_{i=1}^{N'} u_i^2 \tag{2-3}$$

where A is the area of the wall and $V(\equiv lA)$ is the volume of the container. Eq. (2-3) may be rewritten as

$$PV = m \sum_{i=1}^{N'} u_i^2 = \text{constant} \tag{2-4}$$

because, at constant temperature, the sum over u_i^2 is constant. Eq. (2-4) represents Boyle's law. Thus, the kinetic theory of gases predicts the $P\text{-}V$ behavior of an ideal gas exactly.

We define u_i^2 as the square of the velocity component in the x direction of molecule i. This term may vary from molecule to molecule and, for the sake of mathematical

---

[2] If a single molecule were in the container one would expect to observe a succession of impacts on the wall, in time intervals $\Delta\tau = 2l/u_1$, rather than a continuous force.

simplicity it seems to be more reasonable to use the average of the square value,[3] $\overline{u^2}$, instead of the sum over u_i^2. The two are related by the equation

$$\overline{u^2} = \frac{u_1^2 + u_2^2 + u_3^2 + \cdots u_{N'}^2}{N'} = \frac{\sum\limits_{i=1}^{N'} u_i^2}{N'} \tag{2-5}$$

$\overline{u^2}$ is the mean square velocity component in the x direction, and $(\overline{u^2})^{1/2}$ is the root mean square velocity component in the x direction, u_{rms}. On combining Eq. (2-4) with Eq. (2-5), we obtain

$$P = \frac{N' m \overline{u^2}}{V} \tag{2-6}$$

This is the final expression for the pressure of a one-dimensional ideal gas.[4]

The important fact, that molecules in the container move in different directions, has to be included in Eq. (2-6). The velocity[5] of any molecule may be resolved into components with magnitude u, v, and w, parallel to the three mutually perpendicular axes x, y, and z, so that its magnitude—speed—is given by (see Figure 2-2)

$$c^2 = u^2 + v^2 + w^2 \tag{2-7}$$

If Eq. (2-7) is averaged over all the molecules, it becomes

$$\overline{c^2} = \overline{u^2} + \overline{v^2} + \overline{w^2} \tag{2-8}$$

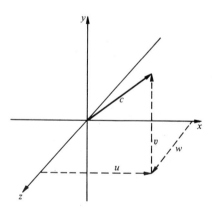

Figure 2-2 Components of the velocity vector.

[3] Due to the random distribution, one half of the molecules moves in the positive and one half in the negative direction of the x axis. Thus, the average velocity, defined as

$$\bar{u} = \sum\limits_{i=1}^{N'} u_1/N'$$

is equal to zero. Consequently $(\bar{u})^2$ is also zero. Since negative values squared are positive, the value of the mean square velocity, $\overline{u^2}$, is different from zero. Remember $(\bar{u})^2 = 0 \neq \overline{u^2}$.

[4] Such a gas moves in one dimension only.

[5] Velocity is a vector quantity, it has magnitude and direction. Speed is a scalar quantity; it is defined by the magnitude only.

Because of the random motion, $\overline{u^2} = \overline{v^2} = \overline{w^2}$, and $\overline{u^2} = \overline{c^2}/3$. Consequently, Eq. (2-6) takes the form

$$PV = \frac{N'm\overline{c^2}}{3} \tag{2-9}$$

where $\overline{c^2}$ is the mean square speed of the molecules.

On introducing the average kinetic energy per molecule, $\bar{\varepsilon}_k = \frac{1}{2}m\overline{c^2}$, into Eq. (2-9), we obtain

$$PV = \frac{2}{3}N'\bar{\varepsilon}_k \tag{2-10}$$

For 1 mole of a gas, N' is equal to Avogadro's number, N, and $P\mathcal{V} = RT$. Thus

$$\mathscr{E}_k = N\bar{\varepsilon}_k = \frac{3}{2}P\mathcal{V} = \frac{3}{2}RT \tag{2-11}$$

or

$$\bar{\varepsilon}_k = \frac{3}{2}RT/N = \frac{3}{2}kT \tag{2-12}$$

where $\mathscr{E}_k$ is the molar kinetic energy associated with the motion of N molecules and k is Boltzmann's constant[6] [6].

Eqs. (2-11) and (2-12) relate the kinetic translational energy of the random motion to the absolute temperature.[7] The higher the temperature, the higher is the kinetic energy of the gas. As is obvious, temperature is not associated with the kinetic energy of one molecule, but with the kinetic energy of an enormous number of molecules; that is, temperature is a statistical concept. Temperature may therefore be defined as a measure of the average translational kinetic energy of molecules[8] [7].

If we use $\bar{\varepsilon}_k = \frac{1}{2}m\overline{c^2}$ in Eq. (2-11), we find

$$\overline{c^2} = \frac{3RT}{Nm} = \frac{3RT}{M} \tag{2-13}$$

and for the root-mean-square speed

$$(\overline{c^2})^{1/2} = c_{rms} = (3RT/M)^{1/2} \tag{2-14}$$

where M is the molecular weight of the gas ($\equiv Nm$).

It is obvious from Eq. (2-14) that, at the same temperature, gases of lower molecular masses exhibit higher speeds. The ratio of the root mean square speeds of two molecules of different masses is equal to the square root of the inverse ratio of the masses and also (given equal temperatures and pressures) to the square root of the inverse ratio of the densities, ρ, as is evident from

$$\frac{(c_{rms})_1}{(c_{rms})_2} = \frac{(3RT/M_1)^{1/2}}{(3RT/M_2)^{1/2}} = \frac{(3P/\rho_1)^{1/2}}{(3P/\rho_2)^{1/2}}$$

$$= \left(\frac{M_2}{M_1}\right)^{1/2} = \left(\frac{m_2}{m_1}\right)^{1/2} = \left(\frac{\rho_2}{\rho_1}\right)^{1/2} \tag{2-15}$$

---

[6] $k = R/N = 1.3806 \times 10^{-23}$ J mol^{-1} K^{-1}.

[7] Since no gas can occupy a volume V less than zero at any pressure P, equation $P\mathcal{V} = RT$ implies the inequality $T \geqq 0$. See also page 576.

[8] This is true for nonideal as well as for ideal systems.

Thus, gaseous constituents of different molecular weights can be separated from each other on the basis of their different speeds[9] [8, 9, 10].

In our gas model we have assumed that the gas molecules do not interfere with each other. If this is true, the kinetic energy of a gaseous mixture must be equal to the sum of the kinetic energies of its constituents. Thus

$$\bar{\varepsilon}_k = \frac{N_1' \bar{\varepsilon}_{k_1} + N_2' \bar{\varepsilon}_{k_2} + N_3' \bar{\varepsilon}_{k_3} + \cdots}{N'} \qquad (2\text{-}16)$$

where $\bar{\varepsilon}_k$ is the average kinetic energy per molecule in the mixture and $N' = N_1' + N_2' + N_3' + \cdots$ represents the sum of the number of molecules of the individual constituents of the mixture.

According to Eq. (2-10), $N' \bar{\varepsilon}_k = \frac{3}{2} PV$ and $N_i' \bar{\varepsilon}_{k_i} = \frac{3}{2} p_i V$, so that Eq. (2-16) assumes the form

$$E_k = N' \bar{\varepsilon}_k = \tfrac{3}{2} PV = \tfrac{3}{2} V(p_1 + p_2 + p_3 + \cdots) \qquad (2\text{-}17)$$

Consequently

$$P = p_1 + p_2 + p_3 + \cdots = \sum_i p_i \qquad (2\text{-}18)$$

where p_i denotes the partial pressure of component i. The kinetic molecular theory predicts Dalton's law of partial pressures exactly.

For different gases at the same pressure and volume, Eq. (2-9) may be written as

$$\frac{N_1' m_1 \overline{c_1^2}}{3} = \frac{N_2' m_2 \overline{c_2^2}}{3} = \cdots = \frac{N_i' m_i \overline{c_i^2}}{3} \qquad (2\text{-}19)$$

Since at the same temperature the average kinetic energy of the different gases are equal

$$\frac{m_1 \overline{c_1^2}}{2} = \frac{m_2 \overline{c_2^2}}{2} = \cdots = \frac{m_i \overline{c_i^2}}{2} \qquad (2\text{-}20)$$

Eq. (2-19) reduces to the form

$$N_1' = N_2' = \cdots = N_i' \qquad (2\text{-}21)$$

Thus, equal volumes of gases at the same temperature and pressure contain an equal number of molecules; this is Avogadro's principle (Avogadro, 1811).

The fact that such well known relationships may be derived from the fundamental kinetic molecular theory implies that the postulates upon which this theory is based must be substantially correct.

2-2 THE KINETIC MOLECULAR THEORY OF GASES AND THE VAN DER WAALS EQUATION

Two of the assumptions made in the discussion of the previous results of the kinetic theory of gases—the absence of forces between molecules and the availability of the

[9] See also page 58.

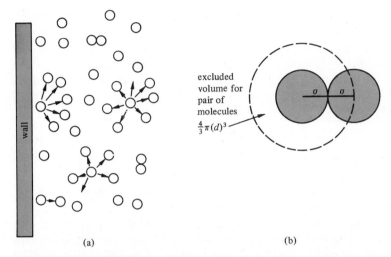

Figure 2-3 (a) Collision of the molecules with the walls of the container. The effect of intermolecular forces. (b) Volume swept out by moving molecules.

total volume of the container for the free motion of molecules—are unjustified in the case of more dense gases. Because of the attractive forces, a molecule that approaches a wall is decelerated; consequently, fewer molecules will hit the walls of the container in unit time. For the same reason a molecule just about to make an impact on the wall will exert a smaller force on the wall. According to the van der Waals theory, the retarding force acting on the impacting molecule is proportional to the number of moles of gas per unit volume,[10] n/V (see Figure 2-3). The number of molecules almost hitting the walls is also proportional[10] to n/V. Since the pressure is the consequence of impacts of molecules on the walls, the retarding force results in a reduction of the pressure. (The ignorance of the wall molecules—gas molecules interactions—is only partially justified.)

The decrease in the pressure from the value it would have if there were no intermolecular forces—the cohesive pressure—is proportional to the product of $n/V \times n/V$, that is $a(n/V)^2$. The process of interaction of molecules with the walls of the container is, however, not as simple as described here qualitatively. Therefore, the so-called proportionality constant, a, is not a real constant; it varies with many factors, but mainly with temperature. To correct the ideal gas law for the effect of the intermolecular forces, the cohesive pressure, $a(n/V)^2$, must be added to the observed pressure. Thus,

$$P_{ideal} = P_{observed} + a\left(\frac{n}{V}\right)^2$$

The cohesive pressure, resulting from the attractive force, is responsible for the negative deviation of gases from the ideal behavior. The larger a is, the larger the cohesive pressure. A gas following the hard-sphere potential without attraction has a value of a equal to zero and, therefore, no cohesive pressure. The value of a for hydrogen and helium is relatively small; consequently, at ordinary temperatures no negative deviations from the ideal behavior are observed with these gases.

[10] The proportionality is not exactly true.

To correct for the positive deviations from the ideal behavior, van der Waals' theory introduces another parameter, b, called the excluded volume per mole of gas.

The meaning of the excluded volume may be clarified by Figure 2-3. For a pair of molecules, this volume is $\frac{4}{3}\pi d^3$, where d is the distance between the centers of the molecules at the instant of collision. The excluded volume for one molecule is therefore $\frac{2}{3}\pi d^3$ and, for Avogadro's number of molecules, $\frac{2}{3}\pi N d^3 (\equiv b)$. For a hard-sphere model, $d = 2\sigma$, and consequently b is given by

$$b = \frac{2}{3}\pi N (2\sigma)^3 = \frac{16}{3}\pi N \sigma^3 = 4 \times \frac{4}{3}\pi N \sigma^3 \tag{2-22}$$

σ denotes the radius of the molecule. As seen, b is not actually equal to the volume occupied by N hard-sphere particles $(\frac{4}{3}\pi N\sigma^3)$ but to four times that volume. The free volume available for the molecules (n moles) to move around is therefore $V - nb$, and not V as required by the ideal gas equation.

At higher temperatures, when the average translational energy becomes very large, the violent collisions between molecules may result in shortening the distance between the centers of two molecules at the moment of collision; $d < 2\sigma$. The particles become deformable—soft spheres. This makes b temperature dependent. Furthermore, molecules of gases having a large value of a have a tendency to form clusters. The complexity of the clusters—the number of molecules participating in the cluster formation—is temperature dependent. This may have a further effect on the temperature dependence of b.

2-3 THE LAW OF ATMOSPHERES

The assumption of the uniform pressure of a gas within a container, upon which our previous discussions have been based, is correct only in the absence of force fields, such as gravity. For laboratory conditions, the effect of the gravitational forces on low density fluids may be fully ignored. Nevertheless, when large-scale systems are considered, the earth's atmosphere for example, the effect of the gravitational forces must be taken into account. Due to these forces, the density and consequently the pressure of the earth's atmosphere is not uniform; the density (pressure) decreases with increasing altitude.

The effect of gravity on fluids, both gases and liquids, can be represented by a model shown in Figure 2-4. A column of fluid at a constant temperature is subjected to a gravitational field of acceleration g. Due to this field, the forces acting upon the lower and upper boundaries of a thin layer (at z and $z + dz$) are not the same. The difference is equal to the weight of the fluid within the layer. Thus,

$$-df = \pi r^2 \rho g\, dz \tag{2-23}$$

The difference in the force per unit area is equal to the pressure difference between the lower and upper boundaries of the layer:

$$-\frac{df}{\pi r^2} = -dP = g\rho\, dz \tag{2-24}$$

where g is the gravitational acceleration and ρ is the density. The minus sign indicates

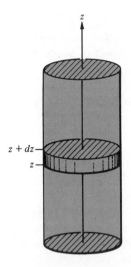

Figure 2-4 Column of fluid in gravitational field.

that the pressure decreases with increasing altitude. At constant g,[11] the pressure difference is proportional to the length of the column. Gases have low densities and therefore under laboratory conditions, where the length of the column is limited by the size of the laboratory, the effect of gravity upon the pressure (density) of gases is negligibly small.

The densities of liquids are not very sensitive to pressure changes and Eq. (2-24) may be integrated as

$$P - P_0 = -\rho z g \tag{2-25}$$

where P is the pressure at the height z above the bottom of the column and P_0 is the pressure at the bottom of the column ($z = 0$). Eq. (2-25) defines the hydrostatic pressure in liquids.

When dealing with gases, the effect of pressure on densities must be considered. For an ideal gas $\rho = PM/RT$ and integration of Eq. (2-24) yields

$$\ln \frac{P}{P_0} = -\frac{Mgz}{RT} = -\frac{\mathscr{E}_p}{RT} \tag{2-26}$$

Mgz is simply the gravitational potential energy, $\mathscr{E}_p$, of 1 mole of a gas at height z. Eq. (2-26) may also be written as

$$P = P_0\, e^{-\mathscr{E}_p/RT} \tag{2-27}$$

The pressure is directly proportional to the density and to the number of molecules per cubic centimeter, and therefore we may also write

$$\rho = \rho_0\, e^{-\mathscr{E}_p/RT} \qquad N' = N_0'\, e^{-\mathscr{E}_p/RT} \tag{2-28}$$

ρ and N' denote the density and the number of particles per cubic centimeter at height z, whereas ρ_0 and N_0' denote the same quantities at the bottom of the column ($z = 0$).

---

[11] g is not generally constant; it varies with the altitude and, to a small extent, with the latitude.

Eqs. (2-26), (2-27), and (2-28) represent the barometric formula. They describe the distribution of gas molecules in the atmosphere as a function of their potential energy. The pressure and consequently the density and the number of particles per cubic centimeter decreases exponentially with the altitude.

The barometric formulas (Eqs. 2-26, 2-27, 2-28) are only approximate, since the assumption of uniform temperature of the atmosphere is incorrect (see Problem 4).

2-4   THE DISTRIBUTION OF MOLECULAR VELOCITIES

As we have stated, gaseous molecules, even of the same kind, exhibit different velocities at constant temperatures. It is reasonable to expect that the distribution of velocities among the different molecules is such that the overwhelming majority of molecules exhibits velocities near the average velocity, whereas molecules having very great or very small velocities appear rarely in the gaseous system.

Instead of attempting to identify molecules with particular velocities, it seems more reasonable to choose molecules exhibiting velocities in a certain velocity range. To simplify our problem, we first consider a gas in which all the molecules move only in one direction, let us say the x direction. What is the probability of finding a molecule with an x component of velocity between u and $u + du$? It is understandable that the probability will depend on the velocity component, u, itself and on the velocity interval, du, and may be written as

$$P(u) = \frac{dN'(u)}{N'} = f(u^2) \, du \qquad (2\text{-}29)$$

where $dN'(u)$ denotes the number of molecules having velocities within the range of u and $u + du$ and N' is the total number of molecules in the container. Then $f(u^2)$, called the distribution function, is responsible for the distribution of the molecules among the different velocities.

Similar equations may also be written for the probability of finding molecules with y and z components of velocities within ranges v and $v + dv$ and w and $w + dw$, respectively

$$P(v) = \frac{dN'(v)}{N'} = f(v^2) \, dv \qquad (2\text{-}30)$$

$$P(w) = \frac{dN'(w)}{N'} = f(w^2) \, dw \qquad (2\text{-}31)$$

The form of the distribution functions $f(u^2)$, $f(v^2)$, and $f(w^2)$ is still unknown. The random distribution of molecules, however, implies that the three distribution functions have identical forms.[12] For reasons that follow and are proved in Example 2-1, the probability and velocity must be related exponentially. Therefore we write

$$P(u) = \frac{dN'(u)}{N'} = A \, e^{-\beta u^2} \, du \qquad (2\text{-}32)$$

---

[12] This is not true in a gravitational field.

$$P(v) = \frac{dN'(v)}{N'} = A\, e^{-\beta v^2}\, dv \tag{2-33}$$

$$P(w) = \frac{dN'(w)}{N'} = A\, e^{-\beta w^2}\, dw \tag{2-34}$$

where A and β are constants whose significance will be explained later. The exponential term, called the Boltzmann factor, is very important in statistical treatment of systems.

The square of the velocity component in the exponent takes care of the fact that, for a random distribution, the probabilities of finding a molecule with a velocity component within u and $u + du$ and within $-u$ and $-(u + du)$ must be identical. The negative sign in the exponent assures finite probabilities even for very high velocities.

Since the probability of finding a molecule with a velocity component within a certain range, u and $u + du$, for example, is independent of the other two velocity components, the probability of finding a molecule having velocity components simultaneously within u and $u + du$ and v and $v + dv$ is given as the product of the two individual probabilities

$$\frac{dN'(uv)}{N'} = \frac{dN'(u)}{N'} \times \frac{dN'(v)}{N'} = f(u^2)\, f(v^2)\, du\, dv$$

$$= A^2\, e^{-\beta(u^2 + v^2)}\, du\, dv \tag{2-35}$$

where $dN'(u, v)$ denotes the number of molecules having velocity components simultaneously within u and $u + du$ and v and $v + dv$.

Figure 2-5 represents a two-dimensional velocity space diagram. The location of a point representing a certain molecule in the two-dimensional velocity space diagram is given by the coordinates of u and v. The number of molecules having velocity components within u and $u + du$ and v and $v + dv$ is given by the number of representative points in the rectangle of area $du\, dv$. The number of points per unit area, the representative point density, is given as

$$\frac{dN'(uv)}{du\, dv} = N'f(u^2)\, f(v^2) = N'A^2\, e^{-\beta(u^2 + v^2)} \tag{2-36}$$

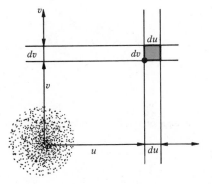

Figure 2-5 Two-dimensional velocity space.

Analogously to the two-dimensional velocity distribution, for the three-dimensional velocity distribution we write

$$\frac{dN'(uvw)}{N'} = \frac{dN'(u)}{N'} \times \frac{dN'(v)}{N'} \times \frac{dN'(w)}{N'}$$

$$= A^3 e^{-\beta(u^2+v^2+w^2)} \, du \, dv \, dw \tag{2-37}$$

and for the representative point density

$$\frac{dN'(uvw)}{du \, dv \, dw} = \frac{dN'(uvw)}{dV} = N'A^3 e^{-\beta(u^2+v^2+w^2)}$$

$$= N'A^3 e^{-\beta c^2} \tag{2-38}$$

since $c^2 = u^2 + v^2 + w^2$, and dV is the volume of the small parallelepiped ($du \, dv \, dw$), as is seen from Figure 2-6. N', A, and β are constants for a certain system.

As seen from Eq. (2-38), the representative point density depends only on c^2, the square of the speed, and not in any way on the particular direction. Thus, the point density has the same value everywhere on the sphere of radius c in the velocity space. By multiplying the point density on a sphere of radius c by the volume of the shell between spheres of radii c and $c + dc$, the number of molecules exhibiting speeds between c and $c + dc$, regardless of the direction, is obtained. Thus,

$$dN'(c) = N'A^3 e^{-\beta c^2} \, dV_{\text{shell}} \tag{2-39}$$

The volume of a shell between spheres of radii c and $c + dc$ is[13]

$$dV_{\text{shell}} = \tfrac{4}{3}\pi[(c+dc)^3 - c^3] = 4\pi c^2 \, dc \tag{2-40}$$

and Eq. (2-39) takes the form

$$dN'(c) = 4\pi N'A^3 e^{-\beta c^2} c^2 \, dc \tag{2-41}$$

Eq. (2-41), which contains two constants, A and β, represents the famous **Maxwell distribution law** for molecular speeds (Maxwell, 1860).

The values of the constants may be evaluated in the following way:

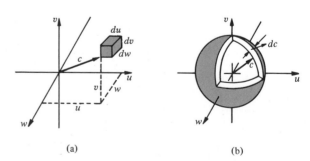

(a)   (b)

Figure 2-6   (a) Three-dimensional velocity space. (b) Spherical shell of volume $dV = 4\pi c^2 \, dc$.

---

[13] $3c \, dc^2 + dc^3 \approx 0$.

The fact that the integral of Eq. (2-41), taken over all the speeds from zero to infinity, must yield the total number of particles in the container, enables the evaluation of the constant A in terms of β. Hence

$$\int_{c=0}^{c=\infty} dN'(c) = N' = 4\pi N' A^3 \int_{c=0}^{c=\infty} c^2 e^{-\beta c^2} dc \qquad (2\text{-}42)$$

Consequently

$$4\pi A^3 \int_{c=0}^{c=\infty} e^{-\beta c^2} c^2 dc = 1 \qquad (2\text{-}43)$$

The value of the integral is equal to $\pi^{1/2}/4\beta^{3/2}$ (see Appendix II), and therefore

$$A^3 = \frac{1}{4\pi} \frac{4\beta^{3/2}}{\pi^{1/2}} = \left(\frac{\beta}{\pi}\right)^{3/2} \qquad (2\text{-}44)$$

According to the average property law

$$\bar{g} = \frac{\int_0^\infty g\, dN'(g)}{\int_0^\infty dN'(g)} = \frac{\int_0^\infty gf(g)\, dg}{\int_0^\infty f(g)\, dg} = \frac{1}{N'} \int_0^\infty g\, dN'(g) \qquad (2\text{-}45)$$

the average kinetic energy per molecule can be written as $[g = \tfrac{1}{2}mc^2,\ dN'(g) = dN'(c)]$

$$\bar{\varepsilon}_k = \frac{\int_{c=0}^{c=\infty} \tfrac{1}{2}mc^2\, dN'(c)}{N'} \qquad (2\text{-}46)$$

On introducing for $dN'(c)$ from Eq. (2-41), we obtain

$$\bar{\varepsilon}_k = \frac{\int_{c=0}^{c=\infty} \tfrac{1}{2}mc^2 4\pi N'(\beta/\pi)^{3/2} c^2 e^{-\beta c^2}\, dc}{N'}$$

$$= 2\pi m (\beta/\pi)^{3/2} \int_{c=0}^{c=\infty} c^4 e^{-\beta c^2}\, dc \qquad (2\text{-}47)$$

The value of this integral is $3\pi^{1/2}/8\beta^{5/2}$ (see Appendix II) and the average kinetic energy becomes

$$\bar{\varepsilon}_k = 2\pi m \left(\frac{\beta}{\pi}\right)^{3/2} \left(\frac{3\pi^{1/2}}{8\beta^{5/2}}\right) = \frac{3m}{4\beta}$$

Therefore

$$\beta = \frac{3m}{4\bar{\varepsilon}_k} = \frac{3m}{4 \times \tfrac{3}{2}kT} = \frac{m}{2kT} \qquad (2\text{-}48)$$

since according to Eq. (2-12), $\bar{\varepsilon}_k = \tfrac{3}{2}kT$.

After introducing expressions for A and β from Eqs. (2-44) and (2-48), respectively, into Eq. (2-41), the final form of the Maxwell distribution law is obtained.

$$dN'(c) = 4\pi N' \left(\frac{m}{2\pi kT}\right)^{3/2} c^2 e^{-mc^2/2kT}\, dc \qquad (2\text{-}49)$$

This can also be written as

$$\frac{dN'(c)}{N'\, dc} = \frac{1}{N'} \lim_{\Delta c \to 0} \left(\frac{\Delta N'(c)}{\Delta c}\right) = 4\pi \left(\frac{m}{2\pi kT}\right)^{3/2} c^2 e^{-mc^2/2kT} \qquad (2\text{-}50)$$

where $(1/N')[dN'(c)/dc]$ is the probability of finding a molecule with a speed between c and $(c+dc)$.

If one plots $(1/N')[dN'(c)/dc]$ against c, the curves exhibit a characteristic shape (Figure 2-7). The curves become broader and less peaked at higher temperatures, as the most probable speed (see page 56) increases and the distribution about the most probable speed becomes wider. The area under the curve remains unaffected by the change in temperature because it always represents all the molecules in the container. Thus, temperature is a measure of the broadening of the Maxwell distribution curve. As the temperature is lowered, the maximum on the distribution curve is shifted toward the y axis.

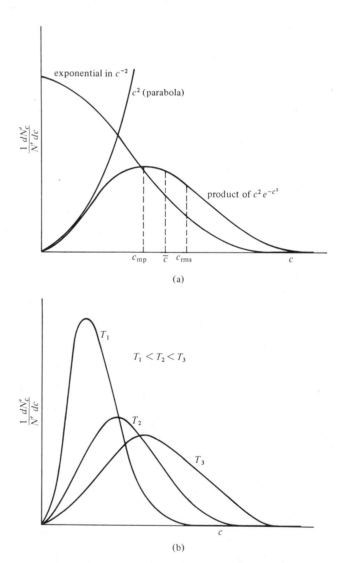

(a)

(b)

Figure 2-7 (a) The contribution of c^2, e^{-c^2} and $c^2 e^{-c^2}$ to the Maxwell speed distribution curve. (b) Distribution curves at three different temperatures.

EXAMPLE 2-1
Derive the form of the distribution function.
Solution: Following from Eqs. (2-29), (2-30), (2-31), and (2-37), we have

$$\frac{dN'(uvw)}{N'} = f(u^2) f(v^2) f(w^2) \, du \, dv \, dw \qquad \text{(a)}$$

If the molecular motion is isotropic, the distribution of the molecular velocities can be conveniently formulated in a spherical coordinate system.

$$\frac{dN'(c)}{N'} = f'(c^2) \, dc \qquad \text{(b)}$$

Since Eqs. (a) and (b) deal with the same distribution expressed in different ways, the two distribution functions must be identical. Consequently

$$f'(c^2) = f'(u^2 + v^2 + w^2) = f(u^2) f(v^2) f(w^2) \qquad \text{(c)}$$

because $c^2 = u^2 + v^2 + w^2$.

This is the basic equation from which the form of the distribution function can now be derived. For the sake of simplicity we will consider a case of only two independent variables: $f'(k) = f(m)f(n)$, where $k = m + n$ and $f(m)$ and $f(n)$ each is a function of only one variable.

Partial differentiation of $f'(k)$ with respect to m at constant n and with respect to n at constant m results in

$$\left(\frac{\partial f'(k)}{\partial m} \right)_n = f(n) \frac{df(m)}{dm} = \frac{df'(k)}{dk} \left(\frac{\partial k}{\partial m} \right)_n \qquad \text{(d)}$$

and

$$\left(\frac{\partial f'(k)}{\partial n} \right)_m = f(m) \frac{df(n)}{dn} = \frac{df'(k)}{dk} \left(\frac{\partial k}{\partial n} \right)_m \qquad \text{(e)}$$

But $(\partial k/\partial m)_n = (\partial k/\partial n)_m = 1$ and therefore

$$f(n) \frac{df(m)}{dm} = f(m) \frac{df(n)}{dn} \qquad \text{(f)}$$

Separating the unknowns in this partial differential equation yields

$$\frac{1}{f(m)} \frac{df(m)}{dm} = \frac{1}{f(n)} \frac{df(n)}{dn} \qquad \text{(g)}$$

The condition set by Eq. (g) is that both sides must be equal to the same constant. Denoting the constant as $-\beta$,[14] Eq. (g) may be transformed into

$$d \ln f(m) = -\beta \, dm \qquad \text{(h)}$$

because $df(m)/f(m) = d \ln f(m)$. Integration of Eq. (h) yields

$$\ln f(m) = -\beta m + \ln A \qquad \text{(i)}$$

---

[14] The necessity of the minus sign in the exponent was discussed on page 49.

or

$$f(m) = A e^{-\beta m} \tag{j}$$

where $\ln A$ is an integration constant.
  Similarly

$$f(n) = A e^{-\beta n} \tag{k}$$

Following from this

$$f'(m+n) = f(m) f(n) = A e^{-\beta m} A e^{-\beta n} = A^2 e^{-\beta(m+n)} \tag{l}$$

Thus, we have proved that only the exponential function satisfies the requirement

$$f'(m+n) = f(m) f(n)$$

for all values of m and n.
  For a problem with three variables Eq. (l) can be expanded into

$$f'(m+n+o) = f(m) f(n) f(o) = A^3 e^{-\beta(m+n+o)} \tag{m}$$

  If we substitute the square of the three different velocity components for m, n, and o, we have

$$f'(u^2 + v^2 + w^2) = f(c^2) = A^3 e^{-\beta(u^2+v^2+w^2)} = A^3 e^{-\beta c^2} \quad \bullet \tag{n}$$

EXAMPLE 2-2
  Calculate the fraction of molecules having speeds in excess of some $c = c_0$. Use the Maxwell distribution law.
Solution: From Eq. (2-50) this is given as

$$P(c \geq c_0) = \frac{dN'(c \geq c_0)}{N'} = 4\pi \left(\frac{m}{2\pi kT} \right)^{3/2} \int_{c_0}^{\infty} c^2 e^{-mc^2/2kT} \, dc$$

Changing variables to $x^2 = mc^2/2kT$, we have

$$P(c \geq c_0) = \frac{4}{\pi^{1/2}} \int_{x_0}^{\infty} x^2 e^{-x^2} \, dx$$

  Integrating by parts and recognizing the fact that

$$\frac{d}{dx}(e^{-x^2}) = -2x e^{-x^2} \qquad \text{and} \qquad \lim_{x \to \infty} (x e^{-x^2}) = 0$$

we have

$$P(c \geq c_0) = \frac{4}{\pi^{1/2}} \int_{x_0}^{\infty} x d\left[-\tfrac{1}{2} e^{-x^2} \right]$$

$$= \frac{4}{\pi^{1/2}} \left\{ \left[-\frac{x}{2} e^{-x^2} \right]_{x_0}^{\infty} + \frac{1}{2} \int_{x_0}^{\infty} e^{-x^2} \, dx \right\}$$

$$= \frac{2}{\pi^{1/2}} x_0 e^{-x_0^2} + \frac{2}{\pi^{1/2}} \int_{x_0}^{\infty} e^{-x^2} \, dx$$

The last integral is found from the relation

$$\frac{2}{\pi^{1/2}} \int_{x_0}^{\infty} e^{-x^2} \, dx = \frac{2}{\pi^{1/2}} \int_{0}^{\infty} e^{-x^2} \, dx - \frac{2}{\pi^{1/2}} \int_{0}^{x_0} e^{-x^2} \, dx$$

$$= 1 - \operatorname{erf} x_0$$

where the error function [11] is

$$\left[\operatorname{erf} x_0 = \frac{1}{\pi^{1/2}} \int_{-x_0}^{x_0} e^{-x^2} \, dx = \frac{2}{\pi^{1/2}} \int_{0}^{x_0} e^{-x^2} \, dx \right]$$

On substituting back the original variable, we obtain finally

$$P(c \geq c_0) = \frac{2}{\pi^{1/2}} \left(\frac{mc_0^2}{2kT} \right)^{1/2} e^{-mc_0^2/2kT} + 1 - \operatorname{erf} \left(\frac{mc_0^2}{2kT} \right) \qquad \bullet$$

2-5 EXPERIMENTAL VERIFICATION OF MAXWELL'S DISTRIBUTION LAW

Many experiments have been carried out in order to prove the validity of Maxwell's distribution law. The one we shall describe briefly here was reported by Estermann, Simpson, and Stern in 1947. A schematic diagram of their apparatus is given in Figure 2-8.

An oven O, with a horizontal narrow slit, heated to 450 K, served as a generator of molecular beams of cesium. The channel A and the foreslit F produced a well-defined beam. Slit C, which was exactly equidistant from the foreslit and detector D, acted as a collimator. In order to avoid collisions between molecules the apparatus between the slit of the oven and the detector was evacuated; the portion between the foreslit and detector had a vacuum of $\approx 10^{-8}$ mm Hg.

The working principle of the apparatus is as follows: Each molecule of an initially horizontal beam, passing the distance between the foreslit and detector, will be deflected downward by the gravitational field. The slower molecules, which have a longer time of passage, will be deflected much more by the gravitational field than the faster molecules. Consequently, the particles may be separated according to their horizontal speeds, and the number having speeds within any given range may then be compared with the value predicted from Maxwell's law. The apparatus is designed in

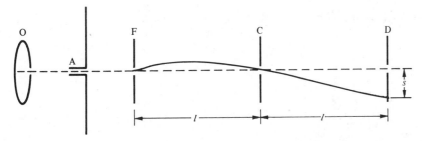

Figure 2-8   Schematic diagram for the deflection of a molecular beam by gravity.

such a way that the vertical displacement, s, of a molecule at the detector may be readily expressed in terms of the horizontal component of the velocity.

Good agreement was observed between the experimentally found values and values predicted by the Maxwell distribution law.

2-6 THE MOST PROBABLE AND MEAN SPEED OF MOLECULES

The distribution curves shown in Figure 2-7 have some common features. At low speeds the c^2 term dominates the distribution, and the curves are parabolic; at high speeds the exponential term dominates the distribution. Due to these two competing terms, the distribution curve exhibits a maximum at a speed called the most probable speed, c_{mp}. This speed corresponds to the maximum in the distribution curve and can therefore be calculated by differentiating the Maxwell distribution law (Eq. 2-50), and setting the derivative equal to zero. Thus,

$$4\pi\left(\frac{m}{2\pi kT}\right)^{3/2}\left[c\,e^{-mc^2/2kT}\left(2-\frac{mc^2}{kT}\right)\right]=0 \qquad (2\text{-}51)$$

The condition set by Eq. (2-51) requires that the curve has three horizontal tangents: at $c=0$, at $c=\infty$, and at $2-mc^2/kT=0$. The condition that corresponds to the maximum on the curve is therefore

$$2-\frac{mc_{mp}^2}{kT}=0$$

Consequently,

$$c_{mp}=\left(\frac{2kT}{m}\right)^{1/2}=\left(\frac{2RT}{M}\right)^{1/2} \qquad (2\text{-}52)$$

According to the law of averages, Eq. (2-45), the mean speed is given as

$$\bar{c}=\frac{\int_{c=0}^{c=\infty}c\,dN'(c)}{N'} \qquad (2\text{-}53)$$

On substituting the expression for $dN'(c)$ from the distribution law, we obtain

$$\bar{c}=4\pi\left(\frac{m}{2\pi kT}\right)^{3/2}\int_{c=0}^{c=\infty}c^3\,e^{-mc^2/2kT}\,dc \qquad (2\text{-}54)$$

which can be further simplified by introducing a new variable,

$$x=\frac{1}{2}\frac{mc^2}{kT}$$

to the form

$$\bar{c}=\left(\frac{8kT}{\pi m}\right)^{1/2}\int_{c=0}^{c=\infty}x\,e^{-x}\,dx=\left(\frac{8kT}{\pi m}\right)^{1/2}=\left(\frac{8RT}{\pi M}\right)^{1/2} \qquad (2\text{-}55)$$

since the value of the integral is 1.

The average speed is somewhat smaller than the root mean square speed (Eq. 2-14) and the most probable speed is even smaller (see Figure 2-7). The three different speeds are related to each other by the formula

$$c_{rms} : \bar{c} : \bar{c}_{mp} = \left(\frac{3RT}{M}\right)^{1/2} : \left(\frac{8RT}{\pi M}\right)^{1/2} : \left(\frac{2RT}{M}\right)^{1/2} \tag{2-56}$$

2-7 COLLISION OF MOLECULES WITH A SOLID SURFACE: EFFUSION

The rate at which molecules collide with a solid surface is frequently required in chemistry, especially in the field of adsorption and catalysis. A brief discussion of this problem follows.

Let us assume that a certain volume, V, is occupied by N' particles [2]. The molecular concentration is thus

$$n = \frac{N'}{V}$$

Further, we consider an element of a solid surface of unit area in the y–z plane. The shape of this area is unimportant, but for definiteness we assume that it is a circle (see Figure 2-9). The number of molecules that strike the element during 1 s with velocity components in the intervals u to $u + du$, v to $v + dv$ and w to $w + dw$ is equal to the number of such molecules initially contained in a cylinder with unit base and with a height of u[15] perpendicular to the surface.

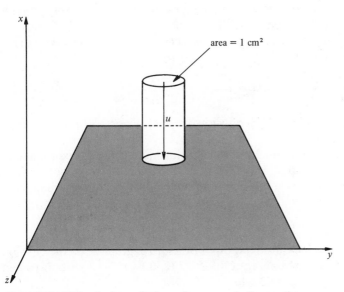

Figure 2-9 Molecules in a cylinder strike an area in the y-z plane.

[15] All the molecules with velocity components u, which are initially at a distance u or less from the unit area, can reach this area in 1 s if their v and w components are sufficiently small.

As is obvious from Eq. (2-37) this number is

$$dn(x) = nf(u^2)f(v^2)f(w^2)\, du\, dv\, dw \cdot u \qquad (2\text{-}57)$$

where n denotes the number of molecules in 1 cm^3 and the volume of the cylinder is u cubic centimeters.

The total number of molecules striking the unit area per second is obtained by integrating over all the possible values of $v(-\infty$ to $+\infty)$ and $w(-\infty$ to $+\infty)$ and over all positive values of $u.$[16] Thus

$$n(x) = n\int_0^\infty u\, f(u^2)\, du \int_{-\infty}^{+\infty} f(v^2)\, dv \int_{-\infty}^{+\infty} f(w^2)\, dw \qquad (2\text{-}58)$$

Since the probability of finding molecules within the complete velocity range ($-\infty$ to $+\infty$) is equal to 1, the last two integrals equal unity and, therefore

$$n(x) = n\int_0^\infty u\, f(u^2)\, du = n\int_0^\infty uA\, e^{-\beta u^2}\, du \qquad (2\text{-}59)$$

On introducing values for A and β from Eqs. (2-44) and (2-48), respectively, we have

$$n(x) = n\left(\frac{m}{2\pi kT}\right)^{1/2}\int_0^\infty u\, e^{-mu^2/2kT}\, du$$

$$= n\left(\frac{m}{2\pi kT}\right)^{1/2}\frac{kT}{m} = n\left(\frac{kT}{2\pi m}\right)^{1/2} = n\left(\frac{RT}{2\pi M}\right)^{1/2}$$

$$= \tfrac{1}{4}n\bar{c} \qquad (2\text{-}60)$$

since the value of the integral is kT/m and $\bar{c} = (8RT/\pi M)^{1/2}$.

If there is a small[17] hole of area S in the container, we expect that molecules that hit the hole will escape from the container. This type of flow is called effusion or molecular flow. The rate of molecular flow in molecules per second $(dn(x)/d\tau)$ is given by $n(x)S$. Following from this

$$\frac{dn(x)}{d\tau} = \tfrac{1}{4}n\bar{c}S = nS\left(\frac{kT}{2\pi m}\right)^{1/2} = \frac{PS}{kT}\left(\frac{kT}{2\pi m}\right)^{1/2} \qquad (2\text{-}61)$$

According to this equation the rate of effusion of a gas is inversely proportional to $m^{1/2}$.

Eq. (2-61) provides the basis of the Knudsen effusion method for the experimental determination of very low vapor pressures—the vapor pressures of solids. The pressure P may be calculated at a given temperature provided that the molecular weight is known.

EXAMPLE 2-3

L. Brewer and coworkers measured the vapor pressure of graphite by the effusion method and found that, at a temperature of 2603 K, 0.648 mg of carbon passed through an opening of 3.25 mm^2 in 3.5 hr.

---

[16] Only molecules moving toward the unit area (positive direction) can collide with the element.

[17] The hole must be very thin and very small in comparison with the average distance that a molecule travels between successive collisions.

Assuming that carbon is monatomic in the vapor phase, calculate the vapor pressure of graphite at 2603 K.

Solution: An integrated form of Eq. (2-61) is applied to the calculation

$$n(x) = z = \frac{P}{\sqrt{2kT\pi m}} S_T \tag{a}$$

Changing z from molecular units to molar units, $z'(\equiv z/N)$, and rearranging Eq. (a) results in

$$P = \frac{z'N\sqrt{2kT\pi m}}{S_T} = \frac{z'\sqrt{2RT\pi M}}{S_T} \tag{b}$$

Upon substituting the numerical data into Eq. (b) we have (taking 1 atm $= 1.013 \times 10^6$ dynes cm^{-2})

$$P = \frac{\dfrac{6.48 \times 10^{-4}}{12}\sqrt{2 \times 3.14 \times 12 \times 2603 \times 8.314 \times 10^7}}{(3.25 \times 10^{-2})(1.26 \times 10^4)(1.013 \times 10^6)}$$

$$= 5.24 \times 10^{-7} \text{ atm} \quad \bullet$$

2-8 COLLISIONS BETWEEN MOLECULES AND THE MAXWELL MEAN FREE PATH

The assumption made earlier in the chapter about the molecular size—volumeless particles—is improper when collision between molecules is discussed. Molecules do have size; therefore, they occupy volume. As a result of this fact, molecules collide with one another whenever one molecule touches the sphere of another. The number of collisions of molecules per unit time and unit volume is a very important factor in chemistry and especially in reaction kinetics. The mathematical expression for the number of collisions is derived as follows.

To make our derivation as simple as possible, let us assume that the molecules are hard spheres with a diameter[18] σ, and that they do not attract each other. The hard-sphere model is useful for a gas at high pressure if the temperature is high enough so that the effective attractive force is negligible. As is evident from Figure 2-10, a collision of molecule A with any other molecule will occur only when the distance between the centers of molecule A and the other molecule is as small as σ. As molecule A moves with an average speed $\bar{c}$ it sweeps out a certain cylinder. The volume of the cylinder swept out in unit time is $\pi\sigma^2\bar{c}$, and molecule A will collide with all the molecules present in the cylinder. If the number of molecules per unit volume is n, the number of molecules in the cylinder is $\pi\sigma^2\bar{c}n$ and the total number of collisions suffered by molecule A in unit time is also $\pi\sigma^2 n\bar{c}$.

The actual number of collisions is, however, different from that given by $\pi\sigma^2 n\bar{c}$. In fact, molecule A does not collide with all the molecules present in the cylinder. If

---

[18] The diameter of the hard-sphere model of a molecule is often called the collision diameter.

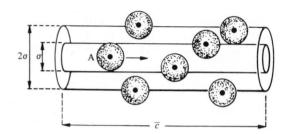

Figure 2-10 Exclusive volume, $\pi\sigma^2\bar{c}$ cm^3, swept out by molecule A traveling a distance $\bar{c}$ in unit time.

molecule A and any other molecule move with the same speed in the same direction, they will never collide. Apparently $\bar{c}$ is not the right speed to use in the evaluation of the number of collisions. Relative speed is more appropriate.

As can be seen from Figure 2-11, if two molecules travel with the same speed, parallel, but in reverse directions, their relative speed c_r is $2\bar{c}$. When the two molecules move with the same speed, parallel, and in the same direction, their relative speed is zero. A compromise between the two extreme cases—the two molecules move perpendicular to one another—leads to the relation: $c_r = 2^{1/2}\bar{c}$.[19] The number of collisions suffered by one molecule in unit time is therefore

$$z_A = \pi\sigma^2 n c_r = 2^{1/2}\pi\sigma^2 n\bar{c} \tag{2-62}$$

Consequently, the total number of collisions suffered by all the molecules in 1 cm^3 per unit time is

$$z_{AA} = \frac{z_A n}{2} = \frac{1}{2} 2^{1/2}\pi\sigma^2 n^2\bar{c} = 2n^2\sigma^2\left(\frac{\pi kT}{m}\right)^{1/2} \tag{2-63}$$

The factor $\frac{1}{2}$ is introduced so that each collision is not counted twice, once as A collides with B and once as B collides with A.

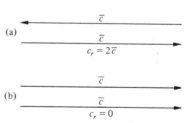

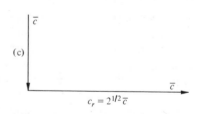

Figure 2-11 The relative speed of two molecules. (a) Same speed but reverse direction. (b) Same speed but parallel direction. (c) Same speed but perpendicular direction.

[19] The same result is obtained if the average over the relative speeds of identical molecules is taken exactly [2].

Putting $\sigma \approx 10^{-8}$ cm, $n \approx 10^{19}$ cm^{-3}, $\bar{c} \approx 10^4$ cm s^{-1} at ordinary pressure and temperature, z_A and z_{AA} are found to be 10^8 s^{-1} and 7×10^{28} s^{-1} cm^{-3}, respectively.

In the case of unlike molecules, the number of collisions is given as

$$z_{AB} = n_A n_B \pi \sigma_{AB}^2 \left(\frac{8kT}{\pi\mu}\right)^{1/2} \tag{2-64}$$

σ_{AB} is the average of the diameters of the two kinds of molecules, $(\sigma_A + \sigma_B)/2$, and $\mu = m_A m_B/(m_A + m_B)$ is the reduced mass.

Thus, the collision number is proportional to the square of the molecular density, n^2, to the cross-sectional area of the molecules $\pi\sigma^2$, and to a relative speed, c_r.

Another quantity of great importance is the Maxwell mean free path. This is a measure of the average distance that a molecule travels between successive collisions. Maxwell defined this quantity as

$$\bar{\lambda} = \frac{\bar{c}}{z_A} = \frac{\bar{c}}{2^{1/2}\pi\sigma^2 n\bar{c}} = \frac{1}{2^{1/2}\pi\sigma^2 n} \tag{2-65}$$

Since $n = NP/RT$, where N is Avogadro's number, the last equation may also be written in the form

$$\bar{\lambda} = \frac{RT}{2^{1/2}\pi\sigma^2 NP} = \frac{kT}{2^{1/2}\pi\sigma^2 P} \tag{2-66}$$

As is evident, the lower the pressure, the smaller is the collision number per unit time and the longer is the mean free path.

_____ *Problems*

1. Derive the Maxwell distribution law for molecular speeds in a two-dimensional gas.
2. Prove that the average values of the individual velocity components are

$$\bar{u} = \bar{v} = \bar{w} = 0$$

3. Assuming that the atmosphere is isothermal at 0°C and that the average molecular weight of the air is 29, calculate the atmospheric pressure at (a) 20,000 ft and (b) 500 miles above sea level. Make use of the barometric formula.
4. The temperature of the lower atmosphere decreases linearly with altitude according to the equation, $T = T_0 - bz$, where b is a constant, z is the altitude, T_0 is the temperature at ground level, and T is the temperature at altitude z. Derive the nonisothermal barometric formula.
5. The ratio of the number of molecules with a speed three times the average speed, $\bar{c}$, to the number with the average speed is a guide to the number of fast molecules that are present. Calculate this ratio for a gas at 25°C.

6. Calculate the root mean square speed, mean speed, and most probable speed (in $m\,s^{-1}$) of hydrogen molecules at a temperature of (a) 0°C, (b) −200°C, (c) 1000°C.

7. (a) What is the root mean square speed of methane molecules at 27°C? (b) At what temperature do ethane molecules have the same speed as methane molecules at 27°C?

8. Calculate the average kinetic translational energy of a molecule of ideal gas in joules and calories at a temperature of (a) 100°C, (b) 0°C, (c) −200°C.

9. Calculate c_{rms} for mercury atoms whose kinetic energy is 1.00 kcal mol^{-1}.

10. During molecular distillation, a substance evaporates on a warm metal plate and directly condenses on a cold plate, 10^{-3} m away from the first. Calculate to what pressure the distillation apparatus must be evacuated if the substance is to be separated by distillation when the molecules of the substance have a collision diameter σ approximately 10^{-9} m and the temperature of the heated plate is 50°C.

11. For good insulation, the pressure in the evacuated space of a thermos flask should be reduced to the point where the mean free path is greater than the distance between the walls. What should the pressure of air be at 25°C if the mean free path equals 5×10^{-3} m? (σ_{av} is 3.70×10^{-10} m).

12. One of the lowest pressures ever measured is $\approx 10^{-13}$ mm Hg; this corresponds to approximately 10^3 molecules per cm^3 at 1027°C. How long would it take at this pressure for aluminum atoms to cover a mirror with a close-packed film one atom thick if the area of an atom is effectively $\approx 5\,\text{Å}^2$, and the atoms stick on every collision and migrate to an empty site?

13. The molecular diameter of CO is 3.19×10^{-10} m. At 300 K and a pressure of 100 mm Hg what will be (a) the number of molecules colliding per cubic centimeter per second and (b) the mean free path of the gas? (c) Repeat the calculation for 300 K and a pressure of 1 atm. (d) How pronounced is the effect of pressure on the quantities sought?

14. Derive an expression for the mean free path for any gas in terms of the collision diameter, the temperature, and the pressure. Prepare a convenient graph showing the variation of the mean free path with pressure for 0°C and the pressure range 10^{-6} mm Hg to 1 atm if the gas is oxygen.

15. T. H. Swan and E. Mack measured the vapor pressure of crystalline naphthalene by the effusion method. They found that, at a temperature of 30°C, 15.47 mg of naphthalene passed through a small orifice of 6.39×10^{-4} cm^2 in 1 hr. Calculate the vapor pressure of naphthalene at this temperature.

_____ *References and Recommended Reading*

[1] L. F. Koons: *J. Chem. Educ.*, **44**:288 (1967). Teaching kinetic molecular theory by the factor change method.

[2] W. Kauzmann: *Kinetic Theory of Gasses.* Benjamin, New York, 1966.

[3] R. D. Present: *Kinetic Theory of Gases.* McGraw-Hill Book Co., New York, 1958.

[4] N. G. Parsonage: *The Gaseous State.* Pergamon Press, London, 1966.

[5] J. O. Hirschfelder, C. F. Curtiss, and R. B. Bird: *Molecular Theory of Gases and Liquids.* John Wiley, New York, 1954.

[6] J. C. Aherne: *J. Chem. Educ.*, **42**:655 (1965). Kinetic energies of gas molecules.

[7] D. K. Carpenter: *J. Chem. Educ.*, **43**:332 (1966). Kinetic theory, temperature, and equilibrium.

[8] E. A. Mason and B. Kronstadt: *J. Chem. Educ.*, **44**:740 (1967). Graham's law of diffusion and effusion.

[9] E. A. Mason and R. B. Evans, III: *J. Chem. Educ.*, **46**:358 (1969). Graham's laws: simple demonstration of gases in motion.

[10] A. D. Kirk: *J. Chem. Educ.*, **44**:745 (1967). The range of validity of Thomas Graham's law.

[11] E. Jahnke and F. Emde: *Tables of Functions with Formulae and Curves.* Dover Publications, New York, 1945.

[12] R. Hulme: *J. Chem. Educ.*, **34**:459 (1957). Kinetic derivation of the gas equation and collision frequency.

chapter 3

Transport Properties of Gases

Up to this point our discussion has been limited to gases in equilibrium. However, in this chapter we shall assume that the gas under consideration is not in equilibrium; the intensive properties, such as pressure, temperature, and concentration (density) are not uniform through the entire system. As a result of this nonuniformity, a physical quantity, such as momentum, energy, or mass, is transferred from one region of the system to another.

The following transport properties are discussed in this chapter: viscosity as transport of momentum, thermal conductivity as transport of energy, and diffusion as transport of mass. The transport of charge is discussed in Part VI, Electrochemistry (see page 839).

In all these cases the **flow**, that is, the amount of the physical quantity transported in unit time through a unit area perpendicular to the direction of flow, is directly proportional to the difference in the particular physical quantity per unit distance (called the **gradient**). Thus, according to Fick's first law

$$I(x) = -C\frac{dY}{dx} \tag{3-1}$$

where $I(x)$ is the flow in the x direction, C is the proportionality constant and dY/dx is the gradient of Y in the direction of flow. When Y is temperature, dY/dx is the temperature gradient and $I(x)$ represents the heat flow along the x axis. When Y is velocity, $I(x)$ represents the flow of momentum and dY/dx is the velocity gradient; and when Y is concentration (pressure, density), $I(x)$ represents the flow of mass (diffusion), and dY/dx is the concentration gradient.

3-1 VISCOSITY

One of the transport properties is viscosity, which is a characteristic property of all fluids, liquids as well as gases. **Viscosity** is defined as the resistance that one part of a fluid offers to the flow of another part of the fluid; it is a measure of resistance to flow. A force must be applied to a fluid to cause it to flow—pressure difference—and in the absence of such force, flow stops. Viscosity is not in evidence at all in a fluid at rest but produces important effects when the fluid flows.

To make viscosity more understandable, let us consider a fluid flowing in the x direction with a nonuniform linear velocity u; the velocity increases in the y direction, as is illustrated in Figure 3-1a. As a result, the fluid as a whole is broken up into thin layers. Each layer moves with a different velocity. Consider now two layers separated by a distance dy, as in Figure 3-1a. According to our velocity condition, the upper layer (the one closer to the moving plane) moves faster than the lower one. Consequently, the upper layer is slowed down by the adjacent lower layer, and the lower layer is speeded up by the faster moving upper layer. In order that the two layers keep their velocities unchanged, a force must be applied. The force necessary

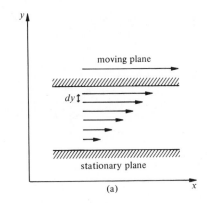

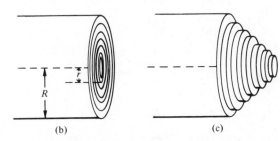

Figure 3-1 (a) Velocity gradient in a fluid due to a shearing action. (b) Fluid divided into imaginary layers at start of flow (time = 0). (c) Position of the layers at time τ.

per unit area, called the shear stress, S, must be proportional to the velocity gradient. Thus,

$$S = \frac{f}{A} = \eta \lim_{\Delta y \to 0} \left(\frac{\Delta u}{\Delta y}\right) = \eta \frac{du}{dy} \tag{3-2}$$

where du/dy is the velocity gradient and η is the viscosity coefficient. As is obvious, S has the dimensions of pressure. The difference between pressure and stress is that the former is force per unit area acting perpendicular to the area, whereas stress acts parallel to the area. The viscosity coefficient may be thought of as the stress—force per unit area—required to move a layer of fluid with a velocity difference of $1\ \text{cm s}^{-1}$ past another parallel layer 1 cm away. The dimensions of η in the cgs system[1] are grams per centimeter per second ($\text{g cm}^{-1}\ \text{s}^{-1}$). Such a unit is called a **poise**. Since this unit is rather large, viscosities are usually given in **centipoises** (10^{-2} poise).

The viscosities of some gaseous and liquid substances are given in Table 3-1.

Eq. (3-2) is applicable to a variety of experimental situations. Examples are coplanar circular plates rotating at different speeds, similarly rotating concentric cylinders, the velocity of a sphere falling through a fluid, or the rate of flow through a capillary tube. Any one of these may serve for the determination of η.

We shall deal in detail with the rate of flow of a fluid through a capillary tube. Consider an incompressible fluid flowing laminarily[2] through a tube of radius[3] R and

---

[1] In the SI system the dimension of η is kilograms per meter per second ($\text{kg m}^{-1}\ \text{s}^{-1}$).

[2] The linear velocity of flow, u, decreases from a maximum value at the center of the tube to a value of zero in a layer adjacent to the wall of the tube.

[3] The diameter of the tube is large compared to the mean free path in the fluid.

Table 3-1 The viscosities of some gases and liquids

Substance	$\dfrac{\eta}{(\text{g cm}^{-1}\,\text{s}^{-1})}$
Ar(g)[a]	2.099×10^{-4}
Ne(g)[a]	2.967×10^{-4}
N$_2$(g)[a]	1.663×10^{-4}
CO$_2$(g)[a]	1.365×10^{-4}
CH$_4$(g)[a]	1.030×10^{-4}
Benzene(l)[b]	0.652
Ethyl Alcohol(l)[b]	1.200
Water(l)[b]	1.002
Mercury(l)[b]	1.554
Glycerine(l)[c]	9.541

[a] At 1 atm and 0°C.
[b] At 20°C.
[c] At 25°C.

length *l*, Figures 3-1b and 1c. As the fluid exhibits a resistance to flow, a layer of fluid at a distance *r* from the center of the tube shows a viscous drag of the magnitude[4]

$$f = -\eta A \frac{du}{dr} = -\eta 2\pi r l \frac{du}{dr} \tag{3-3}$$

where $A \, (\equiv 2\pi r l)$ is the area of the cylindrical layer. Since the fluid in the layer is flowing steadily, the viscous force must be numerically equal to the driving force, which is given as the product of the pressure difference at the beginning and the end of the tube and the area of the cross section of the cylinder. Thus,

$$\pi r^2 (P_1 - P_2) = -\eta 2\pi r l \frac{du}{dr} \tag{3-4}$$

and

$$du = -\frac{(P_1 - P_2)r}{2l\eta} \, dr \tag{3-5}$$

The velocity of flow as a function of *r* is obtained by integration of Eq. (3-5)

$$u = -\frac{(P_1 - P_2)r^2}{4\eta l} + B \tag{3-6}$$

Since at $r = R$, $u = 0$, the integration constant is

$$B = \frac{(P_1 - P_2)R^2}{4\eta l}$$

---

[4] The velocity gradient is negative.

and

$$u = \frac{(P_1 - P_2)}{4\eta l}(R^2 - r^2) \qquad (3\text{-}7)$$

Thus, for laminar flow the graph of velocity as a function of r is a parabola.

The volume of fluid passing through the area between two concentric circles of radius r and $r + dr$, respectively, in time τ, is given by[5]

$$dV = 2\pi u\tau r\, dr \qquad (3\text{-}8)$$

Integration of Eq. (3-8) within the limits of $r = 0$ and $r = R$ yields the total volume of fluid flowing through a cross section of the tube in time τ

$$V = \int_0^R 2\pi\tau u r\, dr \qquad (3\text{-}9)$$

Making use of the value of u from Eq. (3-7) in the integral of Eq. (3-9), we obtain

$$V = \frac{\pi\tau(P_1 - P_2)}{2\eta l}\int_0^R (R^2 - r^2)r\, dr$$

$$= \frac{\pi R^4\tau(P_1 - P_2)}{8\eta l} \qquad (3\text{-}10)$$

which is Poiseuille's equation. Eq. (3-10) was derived for an incompressible fluid [V independent of $(P_1 - P_2)$] and therefore may be satisfactorily applied to liquids but not to gases. For gases, the strong volume dependence on pressure must be considered (see Example 3-1).

The commonly used type of a viscosimeter is shown in Figure 3-2. With this viscosimeter the time τ is measured for a fixed volume of liquid to flow through a vertical capillary tube. As seen from Eq. (3-10), to get η, the radius and length of the capillary tube as well as the pressure difference along the tube must be known. To avoid this, the rate of flow of a liquid of known viscosity—the reference liquid—is

Figure 3-2 Ostwald viscosimeter.

[5] $dV = \pi[(r + dr)^2 - r^2]\tau u = \pi\tau u(2r\, dr + dr^2) = 2\pi r u\tau\, dr$, since $dr^2 \to 0$.

measured with the same viscosimeter. Then the unknown viscosity coefficient is obtained from the following equation:

$$\eta = \eta_r \frac{\Delta P \tau}{(\Delta P)_r \tau_r} = \eta_r \frac{\tau \rho}{\tau_r \rho_r} \tag{3-11}$$

where $\Delta P = \rho g l$, ρ is the density of the measured liquid, g is the acceleration of gravity, and l is the length of the capillary tube; ρ_r, τ_r and η_r are the density, flow time and viscosity of the reference substance.

It has been experimentally demonstrated that the temperature dependence of the viscosity is quite different for liquids and gases. The viscosity of liquids decreases with increasing temperature approximately according to the equation

$$\eta = A \, e^{\mathscr{E}^*/RT} \tag{3-12}$$

where $\mathscr{E}^*$ is the activation energy of flow. The rate of flow depends on the net rate at which the molecules pass over an energy barrier, characterized by the activation energy. However, as we shall show next the viscosity of gases increases with increasing temperature.

On the basis of the different temperature dependence of the viscosities of liquids and gases, one must conclude that the mechanism of flow is different for gases than for liquids. The concept of momentum transfer by molecular collisions, which may be successfully applied to gases, is less justified and may not even be applicable to liquids.

The kinetic theory of gases ascribes viscosity to a transfer of momentum from one moving layer to another. As a result of the thermal motion, molecules may cross from one layer to another. The friction force between two layers of different velocities results from the fact that molecules crossing from the faster to the slower layer transport more momentum, on the average, than the molecules passing in the reverse direction. Momentum is flowing, therefore, from the faster to the slower layer. This transport of momentum tends to counteract the velocity gradient set up by the shear forces acting in the gas.

Now consider two layers separated by a distance equal to the mean free path, $\bar{\lambda}$. Since the velocity difference between the two layers is $\{[u + \bar{\lambda}(du/dy)] - u\} = \bar{\lambda}(du/dy)$, the change in momentum due to a molecule crossing from one layer into the other is $\bar{\lambda} m \, (du/dy)$. Because the number of molecules crossing a unit area in unit time is $\frac{1}{2}n\bar{c}$, the change in momentum per unit time and unit area is $\frac{1}{2}nm\bar{c}\bar{\lambda}(du/dy)$. This is, however, equal to the friction force per unit area:

$$\eta \frac{du}{dy} = \frac{1}{2}nm\bar{c}\bar{\lambda} \frac{-du}{dy} \tag{3-13}$$

Consequently,

$$\eta = \frac{1}{2}nm\bar{c}\bar{\lambda} = \frac{1}{2}\rho\bar{c}\bar{\lambda} = \frac{m\bar{c}}{2^{3/2} \pi \sigma^2} \tag{3-14}$$

because the product nm is equal to the density ρ, and $\bar{\lambda} = 1/2^{1/2} \pi n \sigma^2$.

It is evident from Eq. (3-14) that the viscosity coefficient of a hard-sphere gas is independent of the pressure (density). Maxwell came to this conclusion before quantitative measurements of gas viscosity had been made. The experimental verification of this fact at low pressures[6] was one of the greatest triumphs of the kinetic theory of gases.

On introducing for $\bar{c}$ from Eq. (2-55) into Eq. (3-14), the temperature dependence of the viscosity coefficient is obtained

$$\eta = \frac{m}{2^{3/2}\pi\sigma^2}\left(\frac{8kT}{\pi m}\right)^{1/2} = A'T^{1/2} \tag{3-15}$$

Thus the viscosity coefficient of a gas is directly proportional to $T^{1/2}$. The actual rise, at low temperatures especially, is more rapid than $T^{1/2}$ indicates. This is explained by the fact that molecules do not behave as hard spheres, and their collision diameter σ is temperature dependent. Making use of the empirical formula

$$\sigma^2 = \sigma_0^2\left(1+\frac{C'}{T}\right) \tag{3-16}$$

we obtain

$$\eta = AT^{1/2}\left(1+\frac{C'}{T}\right)^{-1} \tag{3-17}$$

where A and C' are constants; C' is called the Sutherland constant and is in some way related to the intermolecular potential.

In any case, the study of the temperature dependence of η and also of $\bar{\lambda}$, leads to information on the form of the molecular force field and, consequently, to evaluation of the gas behavior, such as the second virial coefficient.

EXAMPLE 3-1

The following table shows the results of a viscosity measurement of gaseous n-octane at 150.6°C.

Temperature of thermostat with capillary tube	150.6°C
Length of capillary tube	52.2 cm
Radius of capillary tube	0.01449 cm
Pressure before capillary tube	3.04 cm Hg
Pressure beyond capillary tube	0.80 cm Hg
Rate of flow	0.03996 g hr^{-1}
Mercury density at 0°C	13.596 g cm^{-3}
Gravity	981.0 cm s^{-2}
Molecular weight of n-octane	114.23 g mol^{-1}

Using these data, calculate the collision diameter of n-octane molecules for the temperature at which the viscosity was measured.

Solution: The volume of fluid flowing through a capillary tube per unit time can be obtained from Eq. (3-10)

$$\frac{dV}{d\tau} = \frac{\pi(P_1-P_2)R^4}{8l\eta} \tag{a}$$

---

[6] It fails completely at higher pressures.

For gases the volume is strongly pressure dependent. For this reason the difference in the pressure $(P_1 - P_2)$ is multiplied by the average pressure along the capillary tube $(P_1 + P_2)/2$. If P^0 is the pressure at which the volume of the gas is measured, Eq. (a) can be written as

$$\frac{dV}{d\tau} = \frac{\pi(P_1 - P_2)R^4}{8l\eta P^0}\frac{P_1 + P_2}{2} = \frac{\pi(P_1^2 - P_2^2)R^4}{16l\eta P^0} \tag{b}$$

Integration yields

$$V = \frac{\pi(P_1^2 - P_2^2)R^4}{16l\eta P^0}\tau \tag{c}$$

In our case the volume of the gas passing through the capillary is not given and must be evaluated from P^0, temperature of the thermostat in which the capillary tube is immersed, and the density of n-octane at STP, ρ_0

$$V = \frac{w}{M}\frac{RT}{P^0} \tag{d}$$

$$\rho_0 = \frac{M}{R}\frac{P_0}{T_0} \tag{e}$$

Substitution from (e) and (d) into (c) yields

$$\eta = \frac{\pi(P_1^2 - P_2^2)R^4\tau}{16lwP^0}\frac{T_0}{T}\rho_0 \tag{f}$$

where w is the amount of n-octane that has passed through the capillary tube in time τ.

Upon substituting the numerical data and the value of ρ_0

$$\rho_0 = \frac{MP_0}{RT_0} = \frac{114.23 \times 1}{82.06 \times 273.2} = 5.0953 \times 10^{-3}\ \text{g cm}^{-3}$$

we get

$$\eta = \frac{3.14(981 \times 13.596)^2(3.04^2 - 0.8^2)0.01449^4}{16 \times 52.2 \times 1.013 \times 10^6 \times 0.03996}$$

$$\times \frac{3600 \times 273.2 \times 5.0953 \times 10^{-3}}{423.8} = 7.45 \times 10^{-5}\ \text{g cm}^{-1}\ \text{s}^{-1}$$

For the collision diameter of molecules, we find from Eq. (3-14)

$$\sigma = \sqrt{\frac{m\bar{c}}{2^{3/2}\pi\eta}} = \sqrt{\frac{114.23}{6.02 \times 10^{23}}\sqrt{\frac{8 \times 8.314 \times 10^7 \times 423.8}{3.14 \times 114.23}}}{2^{3/2} \times 3.14 \times 7.45 \times 10^{-5}}$$

$$= 9.00 \times 10^{-8}\ \text{cm} \qquad \bullet$$

3-2 THERMAL CONDUCTIVITY

As we have seen previously, the viscosity of a gas can be interpreted in terms of momentum transfer by molecular collisions. In an analogous way, the thermal conductivity can be understood as transport of kinetic energy through a gas by molecular collisions as a result of a temperature gradient.

The rate at which heat is transported by gas molecules from a warmer to a cooler wall is proportional to the temperature gradient dT/dy. Thus

$$\frac{dq}{d\tau} = \dot{q} = \kappa \frac{dT}{dy} \tag{3-18}$$

where $\dot{q}$ is the heat transfer in unit time per unit area and κ is the thermal conductivity coefficient. This is defined as the heat flow in unit time, per unit temperature gradient across a unit cross-sectional area, with dimensions of joules per meter per second per degree Kelvin ($J\, m^{-1}\, s^{-1}\, K^{-1}$).

The rate of heat flow is directly proportional to the energy gradient, and therefore Eq. (3-18) may also be written as

$$\dot{q} \propto \lambda \frac{-d\varepsilon}{dy} \propto \lambda \frac{-d\varepsilon}{dT} \frac{dT}{dy} \tag{3-19}$$

because $d\varepsilon/dy = (d\varepsilon/dT)(dT/dy)$. Then $\bar{\lambda}(d\varepsilon/dy)$ is the difference in the energy of gas molecules separated by a distance equal to the mean free path, $\bar{\lambda}$. Since the number of molecules crossing a unit area in unit time is $\frac{1}{2}n\bar{c}$, the heat transfer per unit time and unit area is

$$\dot{q} = \frac{1}{2}n\bar{c}\bar{\lambda}\frac{-d\varepsilon}{dT}\frac{dT}{dy} \tag{3-20}$$

On comparing with Eq. (3-18), we find

$$\kappa = \frac{1}{2}n\bar{c}\bar{\lambda}\frac{-d\varepsilon}{dT} = \frac{1}{2}nm\bar{c}\bar{\lambda}c_V = \frac{1}{2}\rho\bar{c}\bar{\lambda}c_V \tag{3-21}$$

because $d\varepsilon/dT = mc_V$, where c_V is the specific heat capacity, m is the molecular mass, and the product nm is the density, ρ. From Eqs. (3-13) and (3-21), it follows that

$$\kappa = \eta c_V \tag{3-22}$$

This equation relates the heat transfer coefficient, a thermal property, to the viscosity coefficient, a mechanical property. It should be emphasized that, even for monatomic gases with only translational degrees of freedom, Eq. (3-22) is approximate because it assumes that all the molecules are moving with the same speed, $\bar{c}$, and that energy is exchanged completely at each collision. A more accurate equation has the form

$$\kappa = K\eta c_V \tag{3-23}$$

where K is a constant dependent on the particular gas. The value of this constant has been evaluated theoretically as well as experimentally for many gases. The

Chapman–Enskog approach for the thermal conductivity of monatomic molecules—rigid spheres—yields, in its first approximation

$$\kappa = 0.499 \times \tfrac{5}{2} n m \bar{\lambda} \bar{c} c_V \tag{3-24}$$

and hence it gives $K = \kappa / \eta c_V = 0.499 \times 5.$[7] This is within 4% of the experimentally found value for monatomic gases. This method has not, however, been successfully extended to deal with polyatomic molecules.

3-3 DIFFUSION

The diffusion of one kind of a gas into another is a phenomenon that depends on the transport of molecules themselves. It leads to the slow mixing of two gases, of the same kind or of different kinds, that are originally separated but in contact. Such a process of mixing is slow because of the innumerable collisions of the molecules.

The rate of flow of molecules across a unit area of a certain plane is directly proportional to the concentration (pressure) gradient, dx/dy. Thus,

$$\dot{n} = \frac{dn}{d\tau} = -D \frac{dx}{dy} \tag{3-25}$$

where D is the diffusion coefficient. Eq. (3-25) represents Fick's first law of diffusion. The dimensions of D in SI units are square meters per second ($\text{m}^2\,\text{s}^{-1}$).

Analogously to the viscosity coefficient and to the heat transfer coefficient, we can prove that

$$D \propto \bar{c}\bar{\lambda} \tag{3-26}$$

Comparison of Eq. (3-26) with Eqs. (3-14) and (3-21) shows that

$$\eta \propto \rho D \tag{3-27}$$

and

$$\kappa \propto \rho D c_V \tag{3-28}$$

The proportionality constants differ from gas to gas, and their values can only be obtained from experimental data, except for hard-sphere molecules, where the proportionality constants can be obtained from theoretical considerations. The values are 0.599 for Eq. (3-26), 0.833 for Eq. (3-27), and 2.11 for Eq. (3-28).

At low pressures gaseous mixtures can be assumed to behave ideally. If this is the case, then

$$D_{\text{mix}} = y_A D_A + y_B D_B \tag{3-29}$$

where y_A and y_B are the mole fractions of component A and B, respectively.

For a hard-sphere model, $D = 0.599\bar{c}\bar{\lambda}$ and Eq. (3-29) can be written as

$$D_{\text{mix}} = 0.599(y_A \bar{c}_A \bar{\lambda}_A + y_B \bar{c}_B \bar{\lambda}_B) \tag{3-30}$$

[7] The average experimental value for 20 gases, monatomic as well as polyatomic, is 1.89.

To find the distance y a molecule may diffuse in a certain period of time, Fick's second law of diffusion must be applied

$$\frac{dx}{d\tau} = D\frac{d^2x}{dy^2} \tag{3-31}$$

This is a second-order, linear, and homogeneous differential equation. Its solution yields concentration, x, as a function of time and distance

$$x = \beta\tau^{-1/2}\, e^{-y^2/4D\tau} \tag{3-32}$$

where β is a constant. It may be shown that at $\tau = 0$, $x = 0$ everywhere except at the origin, $y = 0$, where $x \to \infty$. The constant β can be evaluated from the condition that at any time all the molecules must be somewhere between $y = -\infty$ and $y = +\infty$; consequently

$$N = \int_{-\infty}^{+\infty} x\, dy = \beta\int_{-\infty}^{+\infty} \tau^{-1/2}\, e^{-y^2/4D\tau}\, dy$$

$$= 2\beta(\pi D)^{1/2} \tag{3-33}$$

or

$$\beta = \frac{N}{2(\pi D)^{1/2}} \tag{3-34}$$

Substitution for β from Eq. (3-34) into Eq. (3-32) yields

$$x = \frac{N}{2(\pi D\tau)^{1/2}}\, e^{-y^2/4D\tau} \tag{3-35}$$

N is the number of molecules initially present at $y = 0$, sometimes called the strength of the diffusion source, and x is the concentration expressed as the number of molecules per unit distance.

Diffusion occurs because molecules drift around under random kinetic motion and more move from a high concentration (pressure, density) region than from a low concentration region. It is purely as a statistical effect that net flow occurs down a concentration gradient.

The probability that a molecule will be found (as a result of diffusion) between a distance y and $y + dy$ is simply given as the number of molecules in this region divided by the total number in the original source, N. Thus, the probability, P, is

$$P = f(y)\, dy = \frac{x(y)\, dy}{N}$$

$$= \frac{1}{2(\pi D\tau)^{1/2}}\, e^{-y^2/4D\tau}\, dy \tag{3-36}$$

The mean distance which a molecule diffuses, $\bar{y}$, is zero, because diffusion in $+$ and $-$ directions is equally probable. The mean square distance, $\overline{y^2}$ is, however, different from zero and can be obtained from the formula

$$\overline{y^2} = \int_{-\infty}^{+\infty} y^2 f(y)\, dy$$

$$= \int_{-\infty}^{+\infty} y^2 \frac{1}{2(\pi D\tau)^{1/2}}\, e^{-y^2/4D\tau}\, dy \tag{3-37}$$

$$= 2D\tau$$

This approximate relation, frequently used in the field of heterogeneous catalysis, enables one to evaluate the mean diffusion distance in gases as well as liquids and solids (see Problem 7).

Collision diameters calculated from viscosity as well as from diffusion data are given for a few gases in Table 3-2, which shows that the values of σ calculated from the two different properties are fairly close to each other. The slight differences result from the inadequacy of the derived relationships rather than from the inaccuracy of the experimental data.

Table 3-2 Molecular diameters calculated from viscosity and diffusion data

Gas	$\dfrac{\eta}{(\mu\text{poise})}$	$\dfrac{D}{(\text{cm}^2\,\text{s}^{-1})}$	$\dfrac{\sigma \times 10^8}{(\text{cm})}$	
			from η	from D
Ne	296.7	0.452	2.58	2.42
Ar	209.9	0.156	3.64	3.47
N_2	166.3	0.185	3.75	3.48
O_2	197.8	0.187	3.61	3.35
CH_4	103.0	0.206	4.14	3.79
CO_2	136.6	0.0974	4.63	4.28
NH_3	98.2		3.64	

SOURCE: J. O. Hirschfelder, C. F. Curtiss, and R. B. Bird: *Molecular Theory of Gases and Liquids.* John Wiley & Sons, New York, 1954.

EXAMPLE 3-2

The viscosity coefficient of N_2 is 178.0 μpoise at 25°C and 1 atm.

Using the hard sphere model theory, calculate the average speed, mean free path, collision diameter, and collision numbers of nitrogen at the given condition.

Solution: The average speed is calculated from Eq. (2-55)

$$\bar{c} = \left(\frac{8RT}{\pi M}\right)^{1/2} = \left(\frac{8 \times 8.314 \times 10^7 \times 298.15}{3.14 \times 28}\right)^{1/2}$$

$$= 4.74 \times 10^4 \text{ cm s}^{-1}$$

The mean free path is defined by Eq. (3-14) as

$$\bar{\lambda} = \frac{2\eta}{\rho \bar{c}} = \frac{2\eta}{\bar{c}PM/RT} = \frac{2 \times 1.78 \times 10^{-4}}{\dfrac{1 \times 28}{82.06 \times 298.15} \times 4.74 \times 10^4} = 6.55 \times 10^{-6} \text{ cm}$$

Using this value for the mean free path, the collision diameter is obtained from Eq. (2-65) as

$$\sigma = \left(\frac{1}{2^{1/2}\pi\bar{\lambda}n}\right)^{1/2} = \left(\frac{1}{2^{1/2}\pi\bar{\lambda}NP/RT}\right)^{1/2}$$

$$= \left(\frac{1}{\frac{(2^{1/2}\times3.14)(6.55\times10^{-6})(6\times10^{23})1}{82.06\times298.15}}\right)^{1/2}$$

$$= 3.74\times10^{-8}\ \text{cm}$$

The collision numbers are given by Eqs. (2-62) and (2-63), respectively, as

$$z_A = 2^{1/2}\pi\sigma^2n\bar{c} = 2^{1/2}\times3.14\times3.74^2\times(10^{-8})^2\times\frac{6\times10^{23}\times1}{82.06\times298.15}\times4.74\times10^4$$

$$= 7.22\times10^9\ \text{collisions s}^{-1}$$

and

$$z_{AA} = \frac{z_An}{2} = \frac{7.22\times10^9\times\dfrac{6\times10^{23}\times1}{82.06\times298.15}}{2}$$

$$= 8.85\times10^{28}\ \text{collisions s}^{-1}\ \text{cm}^{-3} \qquad \bullet$$

_____ *Problems*

1. Solution viscosities are sometimes measured by determining the rate at which a metal sphere falls through the liquid, a kind of sedimentation. If the ball has a density ρ, radius r, and falls under the influence of gravity, derive an equation for the liquid viscosity in terms of the velocity of fall.

2. The second virial coefficient of helium is 12.24 cm^3 mol^{-1} at 175 K. Calculate (a) the mean free path, (b) the viscosity coefficient, (c) collision numbers, and (d) heat conductivity coefficient of helium at 175 K and 10^{-2} mm Hg. Use the hard-sphere model theory.

3. The viscosity coefficient of helium is 2.79×10^{-4} g cm^{-1} s^{-1} at 500 K. (a) Calculate the collision diameter of helium in Angstrom units. Estimate also (b) the second virial coefficient of helium.

4. The time required for a given volume of N_2 to effuse through an orifice is 35 s. Calculate the molecular weight of a gas that requires 50 s to effuse through the same orifice under identical conditions.

5. The viscosity coefficient of argon has the values 1.878×10^{-4} and 2.270×10^{-4} dyne $\cdot$ s cm^{-2} at 240 K and 300 K, respectively. Using the Chapman–Enskog approach, calculate the thermal conductivity coefficients of argon at both temperatures and compare them with the experimentally found values: $\kappa_{240} = 3.595\times10^{-5}$ cal cm^{-1} s^{-1} K^{-1} and $\kappa_{300} = 3.97\times10^{-5}$ cal cm^{-1} s^{-1} K^{-1}.

6. A certain volume of n-heptane flows through a viscosimeter capillary tube in 83.8 s. The same volume of water flows through the same viscosimeter in 142.3 s. The temperature during the experiment is 20°C. What is the viscosity of n-heptane at 20°C? The following data are given: density of n-heptane $\rho_4^{20} = 0.6890$ g cm^{-3}, and the viscosity of water $\eta^{20} = 1.005$ centipoises.

7. The diffusion coefficient of a certain gas in a certain solid is 10^{-8} cm^2 s^{-1}. How far would a gas molecule be expected to diffuse in the solid in a million years?

8. The viscosity coefficient of gaseous Cl_2 at 1 atm pressure and 20°C is 147.0 μpoise. Find the molecular diameter of the chlorine molecule.

9. Consider two parallel layers of NH_3 gas, one of large area and stationary, and the other 1×10^{-3} m^2 in area and moving at a fixed distance of 1×10^{-8} m above the first. What force will be required to keep the upper layer moving with a velocity of 5×10^{-3} m s^{-1} when the pressure of the gas is 10 mm Hg and the temperature is 300 K. The molecular diameter of the NH_3 molecule is 3.0×10^{-10} m.

10. An astronaut's spacesuit has a pinhole leak. If the suit contains air of normal composition (79% N_2 and 21% O_2), what should be the composition of the gas diffusing into space from the pinhole.

11. W. A. Hare and E. Mack measured the viscosity of gaseous diphenyl ether with a capillary viscosimeter, using the method described in Example 3-1. The conditions under which one of the measurements was carried out are given in the following table.

Temperature of thermostat with capillary tube	252.2°C
Pressure before capillary tube	2.475 cm Hg
Pressure beyond capillary tube	0.17 cm Hg
Rate of flow	0.4550 cm^3 hr^{-1}
Length of capillary tube	27.53 cm
Average diameter of capillary tube	0.02510 cm
Mercury density at 0°C	13.596 g cm^{-3}
Gravity	981.0 cm s^{-2}
Density of diphenyl ether at 250°C	0.9650 g cm^{-3}

Using these data, calculate the collision diameter of the molecules for the temperature at which the viscosity was measured.

_____ *References and Recommended Reading*

See the references and recommended reading at the end of Chapter 2.

Miscellaneous Topics

4-1 THE MAXWELL DISTRIBUTION LAW IN TERMS OF ENERGIES

The Maxwell speed distribution law may be easily converted into an energy distribution law. Using the well-known relation, $\varepsilon = \frac{1}{2}mc^2$, we may write

$$c = \left(\frac{2}{m}\right)^{1/2} \varepsilon^{1/2} \tag{4-1}$$

and

$$dc = \left(\frac{1}{2m}\right)^{1/2} \varepsilon^{-1/2}\, d\varepsilon \tag{4-2}$$

Substitution from Eqs. (4-1) and (4-2) into the speed distribution law, Eq. (2-49), yields

$$dN'(\varepsilon) = 2\pi N' \left(\frac{1}{\pi kT}\right)^{3/2} \varepsilon^{1/2}\, e^{-\varepsilon/kT}\, d\varepsilon \tag{4-3}$$

where $dN'(\varepsilon)$ is the number of molecules having kinetic energies between ε and $\varepsilon + d\varepsilon$.

Eq. (4-3), expresses the kinetic energy distribution (translational energy) between the many molecules of the system. It applies to the translational energy regardless of whether the gas is monatomic or polyatomic. In the case of polyatomic molecules, however, other forms of energy must also be considered (see page 79). A total energy distribution law will be derived later from the principles of statistical mechanics (see page 569).

The kinetic energy distribution curve, plotted in Figure 4-1, is evidently different in shape from the speed distribution curve (Figure 2-7b). The energy distribution

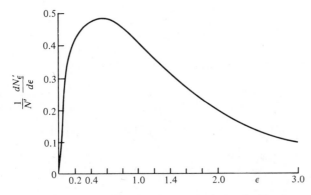

Figure 4-1 The distribution curve for the molecular kinetic energy (units of **k**T).

curve has a vertical tangent, whereas the speed distribution curve exhibits a horizontal tangent at the origin. The energy distribution curve rises much faster and falls off more gently than the speed distribution curve.

4-2 THE EQUIPARTITION PRINCIPLE

As has been stated before and can easily be proved from the distribution law, the average value of the velocity component in any one direction is zero. This is obvious because, for a random distribution, equal numbers of molecules move along both the positive and negative directions of any particular direction.

The average kinetic energy associated with a particular velocity is, however, different from zero, since the square of the velocity component appearing in the kinetic energy is always positive, regardless of whether the molecule exhibits a positive or a negative velocity component.

The average value of the kinetic energy associated with a particular velocity component may be obtained from the Maxwell distribution law. On substituting from Eq. (2-37) into the law of averages (Eq. 2-45), we obtain for the x component contribution

$$\bar{\varepsilon}(x) = \frac{1}{N'} \int_{-\infty}^{\infty} \int_{-\infty}^{\infty} \int_{-\infty}^{\infty} \tfrac{1}{2}mu^2 \, dN'(uvw)$$

$$= \tfrac{1}{2}mA^3 \int_{-\infty}^{\infty} \int_{-\infty}^{\infty} \int_{-\infty}^{\infty} u^2 \, e^{-\beta(u^2+v^2+w^2)} \, du \, dv \, dw$$

$$= \tfrac{1}{2}mA^3 \int_{-\infty}^{\infty} u^2 \, e^{-\beta u^2} \, du \int_{-\infty}^{\infty} e^{-\beta v^2} \, dv \int_{-\infty}^{\infty} e^{-\beta w^2} \, dw \qquad \text{(4-4)}$$

The value of the first integral[1] is

$$\int_{-\infty}^{\infty} u^2 \, e^{-\beta u^2} \, du = \frac{\pi^{1/2}}{2\beta^{3/2}} \qquad \text{(4-5)}$$

The values of the last two integrals are independent of the particular velocity components and are equal

$$\int_{-\infty}^{\infty} e^{-\beta v^2} \, dv = \int_{-\infty}^{\infty} e^{-\beta w^2} \, dw = \left(\frac{\pi}{\beta}\right)^{1/2} \qquad \text{(4-6)}$$

The average kinetic energy in the x direction may therefore be written as

$$\bar{\varepsilon}(x) = \frac{mA^3 \pi^{3/2}}{4\beta^{5/2}} \qquad \text{(4-7)}$$

On introducing the values for A^3 and β from Eqs. (2-44) and (2-48), we obtain

$$\bar{\varepsilon}(x) = \tfrac{1}{2}kT \qquad \text{(4-8)}$$

---

[1] See Appendix II.

Similar equations may also be written for the y and z components of the energy

$$\bar\varepsilon(x)=\bar\varepsilon(y)=\bar\varepsilon(z)=\tfrac{1}{2}kT \tag{4-9}$$

The kinetic energy associated with the translational motion of a particle is thus $\tfrac{3}{2}kT$. Since a point-particle cannot exhibit any other kind of motion, the translational kinetic energy is equal to the total energy of the particle.

Eq. (4-9) is known as the equipartition principle for a point particle. It states that the total energy of a point-particle is equally distributed among the three independent components of motion (Clausius, 1857). A point-particle has three degrees of freedom.

The number of independent coordinates required to locate all the atoms in a molecule represents the number of degrees of freedom of the molecule. As a result of this, a diatomic molecule has 6 (3×2) and an n-atomic molecule $3n$ degrees of freedom. Since only three degrees of freedom are translational, an n-atomic molecule must have, in addition to the translation, $3n-3$ more degrees of freedom available. These are associated with the internal motions of the molecule, such as rotation and vibration. The assumption generally made in molecular mechanics about the separability of the internal degrees of freedom into rotation and vibration is in most cases unjustified, since very often, rotation and vibration interact (see page 407).

How these additional $3n-3$ degrees of freedom are subdivided between rotation and vibration is now the question to be answered.

The maximum number of rotational degrees of freedom a molecule may possess is 3; this is the case of a nonlinear molecule. Such a molecule can rotate around the three coordinate axes. From elementary mechanics, the kinetic energy of rotation[2] is $\tfrac{1}{2}I\omega^2$, and therefore the rotational energy around the three axes is given by

$$\varepsilon_{rot}=(\varepsilon_{rot})_x+(\varepsilon_{rot})_y+(\varepsilon_{rot})_z$$
$$=\tfrac{1}{2}(I_x\omega_x^2+I_y\omega_y^2+I_z\omega_z^2) \tag{4-10}$$

where ω_x, ω_y, ω_z, are the angular velocities and I_x, I_y, I_z are the moments of inertia referred to the principal axes of rotation.

For a linear molecule—a molecule in which the atoms all lie along a common straight line—Eq. (4-10) can be further simplified. It is easy to recognize that diatomic molecules are always linear. There are, however, polyatomic molecules, such as CO_2, N_2O, acetylene, and many more that are also linear.

For the discussion of the rotation of linear molecules, we chose the simplest model of a diatomic molecule—the rigid rotator model. The two atoms in a rigid rotator are assumed to be mass-points and the distance between them—the bond length—is fixed. For such a molecule (see Figure 4-2), $I_x=I_z=I$, $I_y=0$,[3] and therefore Eq. (4-10) assumes the form

$$\varepsilon_{rot}=\tfrac{1}{2}I(\omega_x^2+\omega_z^2) \tag{4-11}$$

---

[2] $\varepsilon_{kin}=\tfrac{1}{2}mv^2$. The angular velocity is related to the linear velocity by $v=\omega r$, and therefore $\varepsilon_{kin}=\varepsilon_{rot}=\tfrac{1}{2}mr^2\omega^2=\tfrac{1}{2}I\omega^2$ where $I=mr^2$.

[3] Since the masses of atoms are concentrated in the nearly pointlike nuclei, which lie on the axis, it is presumed to be meaningless to consider rotations about the molecular axis itself.

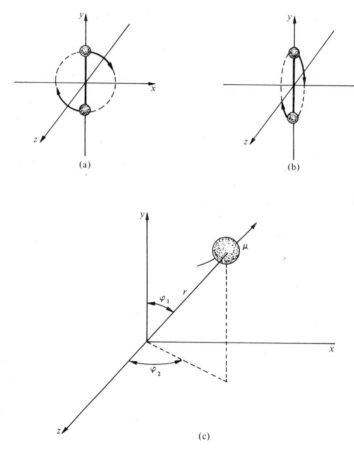

Figure 4-2 Rotational motions of a diatomic molecule. (a) Rotation about the z axis. (b) Rotation about the x axis. (c) Rotation of reduced mass, μ at a distance r from the origin.

The moment of inertia of a diatomic molecule can be written as

$$I = m_1 r_1^2 + m_2 r_2^2 \qquad (4\text{-}12)$$

r_1 and r_2, representing the distances of the two atoms from the center of gravity of the molecule, are

$$r_1 = \frac{m_1}{m_1 + m_2} r \qquad r_2 = \frac{m_1}{m_1 + m_2} r$$

Consequently

$$I = \frac{m_1 m_2}{m_1 + m_2} r^2 = \mu r^2 \qquad (4\text{-}13)$$

$\mu(\equiv m_1 m_2/(m_1 + m_2))$ is called the reduced mass. The rotational motion of a rigid rotator is equivalent to the rotation of a mass μ at a distance r from the origin. As Eq. (4-11) and Figure 4-2c indicate, only two coordinates, φ_1 and φ_2, respectively, are required to describe completely the rotational motion of a linear molecule. Thus, a linear molecule has two rotational degrees of freedom.

According to the equipartition principle, the energy corresponding to one degree of freedom is $\frac{1}{2}kT$. Consequently, the rotational energy is $\frac{3}{2}kT$ and $\frac{2}{2}kT$ for a nonlinear and linear molecule, respectively.

The remaining degrees of freedom, $(3n-6)$ and $(3n-5)$ for nonlinear and linear molecules respectively, belong to the vibrational motions.

It is experimentally proved that the distance between the centers of masses of the two atoms in a molecule—the bond length—is not fixed. The masses oscillate along the bond and, consequently, the bond length varies. The fixed distance in a rigid rotator must therefore be replaced by a stiff spring (see Figure 4-3). In such an oscillating system there is a continuous interchange between vibrational kinetic and potential energies. At the limits of maximum stretching and maximum compressing, the vibrational energy is all potential because the masses stop momentarily to reverse their direction. At the distance equal to the equilibrium position, r^*, the vibrational energy is all kinetic. For all the other distances, the molecule has both potential and kinetic energy

$$\varepsilon_{\text{vib}} = (\varepsilon_{\text{kin}})_{\text{vib}} + (\varepsilon_{\text{pot}})_{\text{vib}} \qquad \textbf{(4-14)}$$

The simplest model of a vibrating diatomic molecule is the harmonic oscillator.[4] Such an oscillator obeys Hooke's law, $f = -k(r-r^*)$, which states that, when the masses are displaced from their equilibrium position, the restoring force, f, is proportional to the displacement and has the opposite direction. The constant k, which is called the force constant, is a measure of the stiffness of the spring.

For a harmonic oscillator, Eq. (4-14) has the form

$$\varepsilon_{\text{vib}} = \frac{1}{2}\frac{p^2}{\mu} + \frac{1}{2}k(r-r^*)^2 \qquad \textbf{(4-15)}$$

in which μ denotes the reduced mass. The first quadratic term on the right-hand side of Eq. (4-15) represents the kinetic energy and is proportional to the square of momentum, p^2; the second term represents the potential energy and is proportional to the square of the displacement from equilibrium position, $(r-r^*)^2$. Since two

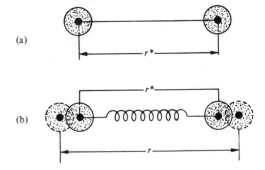

(a)

(b)

Figure 4-3 Vibrational motions in a diatomic molecule. (a) Fixed distance between the atoms. (b) Vibration about the equilibrium interatomic separation, r^*.

---

[4] It is physically unrealistic since it cannot account for the dissociation of the molecule; it can, however, account rather well for many vibrational transitions.

quadratic terms are required to express the vibrational energy of a harmonic oscillator, the energy per one mode of vibration is $2(\frac{1}{2})kT = kT$.

The vibrational motions in polyatomic molecules may be quite complicated, even for molecules, in which the displacement of the individual atoms follows Hooke's law. In addition to the stretching vibrations, polyatomic molecules may undergo other deformations, such as bending, torsion, and so on. Fortunately, it is always possible to replace the complex vibrational motions by a certain number of simple motions. These simple motions are called normal modes of vibration. In a normal mode of vibration each atom in the molecule vibrates with the same frequency. Since the energy per one vibrational degree of freedom (per one mode) is kT, the total vibrational energy is $(3n-5)kT$ for a linear and $(3n-6)kT$ for a nonlinear molecule, respectively.

According to the equipartition principle the total energy per one mole of a polyatomic substance is

$$\mathscr{E} = \begin{cases} (3n-3)\boldsymbol{N}\boldsymbol{k}T = (3n-3)\boldsymbol{R}T & \text{(nonlinear)} \\ (3n-\frac{5}{2})\boldsymbol{N}\boldsymbol{k}T = (3n-\frac{5}{2})\boldsymbol{R}T & \text{(linear)} \end{cases} \tag{4-16}$$

where $\boldsymbol{N}$ is Avogadro's number.

Following from this, the molar heat capacity at constant volume, as predicted by the equipartition principle, is

$$\mathscr{C}_V = \left(\frac{\partial \mathscr{E}}{\partial T}\right)_V = \begin{cases} (3n-3)\boldsymbol{R} & \text{(nonlinear)} \\ (3n-\frac{5}{2})\boldsymbol{R} & \text{(linear)} \end{cases} \tag{4-17}$$

In Table 4-1 we compare some experimentally found values of molar heat capacities with those calculated from the equipartition principle. As we can see, the equipartition principle fails badly for all but monatomic gases. This is explained by the fact that classical mechanics, from which the equipartition principle is derived, assumes that energy is stored in all the $3n$ degrees of freedom at any temperature. Such an assumption is incorrect. Although energy is put into the translational degrees of freedom even at very low temperatures, higher temperatures are required to excite a molecule rotationally and even higher temperatures must be reached before energy can be put into the vibrational degrees of freedom. This indicates that

Table 4-1 Molar heat capacities of some gases at 25°C

Gas	$\mathscr{C}_V/\boldsymbol{R}$ (experimental)	$\mathscr{C}_V/\boldsymbol{R}$ (equip. princ.)
He	1.501	1.5
H_2	2.468	3.5
O_2	3.152	3.5
I_2	3.45	3.5
CO_2	3.47	6.5
NH_3	3.31	9.0
C_2H_6	5.328	21.0

the variation of the molar heat capacities with temperature cannot be predicted by the classical equipartition principle. Eqs. (4-16) and (4-17) are therefore useless, except at very high temperatures when all the different kinds of molecular motions are fully excited and the results obtained from these equations are good.[5]

The variation of the molar heat capacities with temperature can be explained quantitatively only by quantum mechanics.

_____ *Problems*

1. If the distribution of molecular energies in an ideal gas is given approximately by

$$dN'_\varepsilon = \frac{N'}{kT} e^{-\varepsilon/kT} d\varepsilon$$

   calculate what percentage of molecules has an energy (a) higher than kT and (b) higher than $2kT$.

2. Make the change of variable $\varepsilon = \frac{1}{2}mc^2$ to derive an expression for the fraction of dN_ε/N of molecules having energy in two dimensions between ε and $\varepsilon + d\varepsilon$, as a function of ε. Calculate the fraction of molecules of a gas which have an energy equal to or greater than two times their average kinetic energy.

3. How many degrees would the temperature of 1 mole of ethyl alcohol be raised by the addition of an amount of energy equal to the translational kinetic energy at 25°C of 1 mole of alcohol vapor? $\mathscr{C}_P = 20$ cal mol^{-1} K^{-1}.

4. Sketch molar heat capacity at constant volume ($\mathscr{C}_V$) versus T curves for (a) Ne, (b) CO, (c) O_3, and (d) acetylene, from 200 to 800 K. Assume complete activation of the translational and rotational degrees of freedom at 200 K and make the wrong assumption about complete activation of the vibrational degrees of freedom at 750 K.

5. Making use of the equipartition principle, calculate the value of $\mathscr{C}_P/\mathscr{C}_V$ for propane. (a) All the degrees of freedom (except the electronic ones) are excited. (b) Only translational and rotational degrees of freedom are excited. Note: The heat capacity at constant pressure $\mathscr{C}_P = \mathscr{C}_V + R$.

6. The ratio $\mathscr{C}_V/R$ for ammonia, found experimentally at room temperature is 3.29. Find the ratio $\mathscr{C}_V/R$ from the equipartition principle. Discuss the discrepancy.

7. Discuss why the tetratomic molecule NH_3 has a lower value of $\mathscr{C}_V$ than the triatomic molecule SO_2 at 298 K.

8. Derive an equation for the average momentum in the positive x direction for a gas with a molecular mass m and temperature T.

9. Calculate the fraction of nitrogen molecules at 1 atm and 300 K whose (a) speeds are in the range $c_{mp} - 0.005c_{mp}$ to $c_{mp} + 0.005c_{mp}$ and (b) energies in the range $\varepsilon - 0.005\varepsilon$ and $\varepsilon + 0.005\varepsilon$.

---

[5] The classical equipartition principle predicts Dulong–Petit's law well: $\mathscr{C}_V = 3R$ for monatomic crystalline elements.

References and Recommended Reading

See the references and recommended reading at the end of Chapter 2.

Part II

Classical Thermodynamics

chapter 5

The First Law of Thermodynamics

5-1 INTRODUCTION

Thermodynamics is a branch of the natural sciences that deals with the interconversion of the different forms of energy, with the spontaneous direction of physical and chemical processes, and with equilibrium.

Thermodynamics is developed, principally, from four fundamental so-called laws of thermodynamics: the zeroth law, the first law, the second law, and the third law. These laws represent generalizations obtained on the basis of restricted experiments, and their truth is ascertained by inference; that is, from the fact that, of all the consequences derived from them, none have failed to be verified experimentally.

Classical thermodynamics cannot say anything about the properties of a certain substance. It can, however, forbid certain combinations of properties in a particular substance. In this manner, classical thermodynamics can provide tests for the consistency of measurements relating to the properties of a substance, as well as methods of evaluating some of these properties if some others have been measured. Classical thermodynamics is useful precisely because some quantities are easier to measure than others; it is a collection of useful relations between quantities [1, 2, 3, 4].

The properties of a substance can, however, in some simple cases, be calculated from the structure of matter. The sciences that make this calculation possible are called statistical thermodynamics and quantum mechanics.

It is possible to treat either classical or statistical thermodynamics without making use of the other, but such treatment has little to recommend it. This chapter deals mainly with classical thermodynamics.[1] Educationally it seems to be more logical to deal with statistical thermodynamics prior to classical thermodynamics. Statistical thermodynamics requires, however, some knowledge of structure of matter and therefore it will be discussed in Part IV after the reader has gained some knowledge about the basic principles of the structure of matter from quantum mechanics in Part III.

From an educational point of view it does not seem to be reasonable to present thermodynamics in a historical way. Such a method would require, for example, the introduction of the second law of thermodynamics before the first law was treated. Our presentation of the text is therefore logical rather than historical.

---

[1] Some recently revised and newly published textbooks of physical chemistry treat the two kinds of thermodynamics simultaneously. In our opinion, such a treatment, however logical, is not recommended from an educational point of view. The two disciplines use different approaches, and the frequent switching in the explanation from the macroscopic—classical—approach to the microscopic—statistical—approach and vice-versa may result in confusion and difficulties in understanding the subject.

5-2 BASIC DEFINITIONS

A **thermodynamic system** is any collection of material bodies isolated from the rest by a clearly defined boundary. The **boundary** may be either observable or drawn arbitrarily through space. The choice of a particular boundary for a particular system is dictated exclusively by convenience and by our intuitive grasp of the subject. The material objects not included within the boundary of the system constitute its **surroundings**. In the study of thermodynamics, it is impossible to leave the surroundings completely out of consideration when phenomena in the system are studied; the system invariably interacts with its surroundings, and therefore, it is affected by its surroundings.

A **closed system** is a system that exchanges only energy in the form of heat and work with its surroundings. It is important to recognize that a closed system is one of constant mass but that the converse is not necessarily the case. When a system loses matter through one channel and gains mass at exactly the same rate through another, its mass will remain constant; however, the system is open. In a true sense, there is no such thing as a closed system for, as is known, matter and energy are one and the same thing. Therefore, whenever there is a transfer of energy, there is also a transfer of mass. However, the magnitude of the energy transfers commonly discussed in chemical thermodynamics is such that the attendant change in mass is immeasurable; therefore, the abstraction can be made that a closed system is one whose mass is constant while it exchanges energy with its surroundings. An **open system** is a system that may exchange matter, as well as energy, with its surroundings. An **isolated system** is a system that does not exchange matter or energy in any form with its surroundings.

Having selected a system and its boundary, we shall proceed to describe it in terms of its properties. The system is said to be in a given state when all its properties—temperature, pressure, volume, and so on—have fixed values. It is known that the properties of a system do not change independently of one another and that a fixed number of properties determine all the others.[2] The simplest systems encountered in thermodynamics, namely, pure substances, are chacterized by having two independent properties for a fixed mass. Using the language of mathematics [5, 6], we say that any property, y, regarded as dependent, is a single-valued function of the n independent properties $x_1, x_2, \ldots, x_n$; therefore, we write

$$y = f(x_1, x_2, \ldots, x_n) \tag{5-1}$$

An equation of this form is known as an equation of state for a given system. It must be determined experimentally, except in the simple cases when it can also be derived theoretically by making use of statistical mechanics.

If we examine all the possible macroscopic thermodynamic properties of a system, we find it useful to distinguish between two classes. When the value of a property in a system is independent of its mass, the property is called **intensive**: pressure P, temperature T, and density ρ, are examples of intensive properties. In contrast, properties whose values are proportional to the mass of the system are called **extensive**: volume V, energy E, and mass m, are familiar examples of extensive

---

[2] See the equations of state in Part I.

properties. Extensive properties per any unit of quantity (gram, mole, and so on) become, however, intensive properties.

It will be remembered from mathematics [5, 6] that a function, f, of the independent variables $x_1, x_2, \ldots$ which has the property

$$f(\lambda x_1, \lambda x_2, \ldots) = \lambda^m f(x_1, x_2, \ldots) \tag{5-2}$$

where λ is an arbitrary constant, is called homogeneous of degree m. It is seen that, when all variables of a homogeneous function are multiplied by a constant, the function itself is multiplied by the same constant raised to a power which describes the degree of the function. It is easy to recognize that an equation of state for any extensive property of a system must be homogeneous of order $m = 1$ in terms of other extensive properties. Indeed, if the extensive properties that are considered independent are multiplied by a constant λ, the mass of the system becomes multiplied by it. The same applies to any dependent extensive property. An intensive dependent property must be homogeneous of degree $m = 0$ in terms of extensive properties because its value cannot change as the mass of the system is changed.

Temperature is one of the most important intensive properties of a system. The concept of temperature is linked in our minds with a definite intuitive idea based on our temperature sense, which enables us to distinguish hot from cold. This sense, however, is insufficient to establish a scientifically rigorous concept or a quantitative measure of temperature. The quantitative determination of temperature is based on empirical observations.

Consider, for example, two pieces of a metal. One is glowing red, that is, very hot, the other is taken from a refrigerator. If the two are clamped together, the red-hot piece of metal will cool while the refrigerated one will warm; and, if left together for a long period of time, both pieces will eventually have the same feeling of hotness or coldness. In other words, the two pieces will be in thermal equilibrium, and both will have the same temperature. Thus, temperature is defined as that property which determines whether or not a system is in thermal equilibrium with other systems or with its surroundings.

Now, suppose that when the two pieces of a metal were clamped together, another body was included that had some characteristic that varies with temperature in a well-known manner, and thus could be calibrated. The calibrated third body could then be a reference or a standard. This reference would be what is known as a thermometer.

The temperature is thus found indirectly by measuring the properties of the standard connected to the body of unknown temperature. This brings us to the definition of the so-called zeroth law of thermodynamics [7]. This law states that if a body A is in thermal equilibrium with another body B, and if body A is also in thermal equilibrium with a body C, then bodies B and C are also in thermal equilibrium. As a matter of fact, an extended form of the zeroth law of thermodynamics, which has been introduced in connection with temperature, has a much broader application. It applies to the measurement of any physical quantity. For example, when the lengths of two different waves are equal to the same multiple of a standard meter, they must also have equal length.[3]

---

[3] Because of its general validity within and outside thermodynamics, the denotation, zeroth law of thermodynamics, does not seem to be proper.

In order to establish a temperature scale—to calibrate our standard—we shall first choose some property of the standard—a thermometric property, a single-valued function of temperature—that varies continuously and reproducibly with the temperature. The most commonly used thermometric properties are the following:

1. The change in volume of solids, liquids, or gases with temperature at constant pressure.
2. The variation of the pressure of a gas with temperature at a constant volume.
3. The saturated vapor pressure of liquids as a function of temperature.
4. The electrical resistance as a function of temperature.
5. The thermoelectric force as a function of temperature.

Having chosen a thermometric property, we must determine its values at a certain number of fixed points, to choose a standard temperature interval and construct a temperature scale. Conventionally, the standard temperature interval is defined as the difference in the thermometric properties at the melting point of ice, the temperature at which solid and liquid water are in mutual equilibrium at a pressure of 1 atm, and the normal boiling point of water, the temperature at which liquid water is in equilibrium with its vapor at a pressure of 1 atm. The temperatures assigned to these two fixed points, on the centigrade scale, are 0°C and 100°C, respectively. The degree is then defined as 1/100 of the standard temperature interval.[4] The temperature, on the centigrade scale, may then be calculated from the formula

$$\Theta = \frac{z - z_0}{z_{100} - z_0} 100 \tag{5-3}$$

where Θ is the measured temperature, z is the thermometric property at this temperature, z_0 and z_{100} are the thermometric properties at the melting point of ice and normal boiling point of water, respectively. It is necessary to recognize that Eq. (5-3) does not supply an exact reading of the temperature. The rate of change of the thermometric property with temperature, $dz/d\Theta$, is not the same for different properties; thus, interpolation formula (5-3) would supply different readings for different thermometric properties. It is now necessary to decide which reading is correct.

In order to get a universal temperature scale, one should work with a thermometric property that is independent of the nature of the substance. Such a thermometric property is the volume of a gas at low pressures [8]. Writing Eq. (5-3) for the volume, we have

$$\Theta = \frac{V^+ - V_0^+}{V_{100}^+ - V_0^+} 100 = \frac{100 V^+}{V_{100}^+ - V_0^+} - \frac{100 V_0^+}{V_{100}^+ - V_0^+} \tag{5-4}$$

where $(+)$ denotes a gas at low pressures. The value of $(100 V_0^+ / V_{100}^+ - V_0^+)$ is

---

[4] The international practical temperature scale (IPTS) uses the triple-point of water, the temperature at which all the three phases of water are in mutual equilibrium, instead of the melting point of ice as a fixed point. The temperature of the triple-point is just slightly above 0°C. Consequently, the degree on the IPTS is slightly smaller than on the centigrade scale.

273.15^5 and is the same for all gases at low pressures—ideal behavior. Eq. (5-4) may be rewritten as

$$t = \Theta = 273.15 \frac{V^+}{V_0^+} - 273.15 \tag{5-5}$$

On substituting T for $t + 273.15$, we obtain the final equation:

$$T = 273.15 \frac{V^+}{V_0^+} \tag{5-6}$$

From the measured value of V^+, the temperature is found.

The temperature T is called the absolute temperature. It will be shown later that the ideal gas temperature scale, based on Eq. (5-6), is in full agreement with the so-called thermodynamic temperature scale, defined on the basis of the second law.

A gas thermometer is a very difficult instrument to operate and, because the thermodynamic temperature scale is not convenient for measuring temperatures, a new temperature scale, international practical temperature scale, has been adopted (in 1948, and then revised in 1954). This scale, defined on the basis of some easily reproducible fixed points and interpolation formulas, agrees, within the limits of experimental error, with the thermodynamic scale.

The definitions of pressure and volume are simpler than the definition of temperature. Pressure is the force acting per unit area,[6] and volume is the space occupied by a system.[7] Later, we shall discuss some other thermodynamic properties and will find that many useful thermodynamic properties are not directly measurable. Thermodynamics, however, establishes the necessary relationships between those that are and those that are not measurable.

5-3   EXACT DIFFERENTIALS

The properties of a system, called functions of state, depend only on the state in which the system is, and in no way on the path along which the system got into this state. Consequently, the change in any property of the system, Δz, when passing from state A to state B is given by the difference in values of z in the final state z_B, and original state z_A, respectively. Thus

$$\Delta z = z_B - z_A \tag{5-7}$$

Alternatively

$$\oint dz = 0 \tag{5-8}$$

The integral $\oint$ implies that the system is in the same state at the end of the path as it was at the beginning (closed path), and thus the property z is not affected by the change; dz is therefore an exact differential.

---

[5] The value of this number depends upon the arbitrary zero of the temperature scale and upon the arbitrary size of the centigrade degree.

[6] In the SI system of units N/m^2 ($\equiv kg\, m^{-1} s^{-2}$) = Pa, where N is the Newton.

[7] 1 liter = 1000 ml = 1000 cm^3 = 1 dm^3.

As an important corollary of the fact that dz is an exact differential, it can also be written as[8]

$$dz = \left(\frac{\partial z}{\partial x}\right)_y dx + \left(\frac{\partial z}{\partial y}\right)_x dy \qquad (5\text{-}9)$$

$$= M\, dx + N\, dy$$

where x and y are independent variables of the system and M and N are defined as

$$M = \left(\frac{\partial z}{\partial x}\right)_y \qquad N = \left(\frac{\partial z}{\partial y}\right)_x \qquad (5\text{-}10)$$

If z is a function of state—a property of the system—its change must be independent of the path, and therefore

$$\frac{\partial}{\partial y}\left(\frac{\partial z}{\partial x}\right)_y = \frac{\partial}{\partial x}\left(\frac{\partial z}{\partial y}\right)_x \qquad (5\text{-}11)$$

Consequently

$$\left(\frac{\partial M}{\partial y}\right)_x = \left(\frac{\partial N}{\partial x}\right)_y \qquad (5\text{-}12)$$

Eq. (5-12) is used very often in thermodynamics, and it is called the reciprocity relation.[9]

Another test for the exactness of a differential follows from Eq. (5-9)

$$\left(\frac{\partial z}{\partial x}\right)_y \left(\frac{\partial x}{\partial y}\right)_z \left(\frac{\partial y}{\partial z}\right)_x = -1 \qquad (5\text{-}13)$$

EXAMPLE 5-1
Given

$$dz = (68x^3y + 22y^5)\, dx + (17x^4 + 110xy^4)\, dy$$

prove that dz is an exact differential.
Solution: In this case,

$$M = 68x^3y + 22y^5 \qquad N = 17x^4 + 110xy^4$$

and thus

$$\left(\frac{\partial M}{\partial y}\right)_x = 68x^3 + 110y^4 \qquad \left(\frac{\partial N}{\partial x}\right)_y = 68x^3 + 110y^4$$

Consequently

$$\left(\frac{\partial M}{\partial y}\right)_x = \left(\frac{\partial N}{\partial x}\right)_y$$

and dz is an exact differential.   ●

---

[8] z is a function of x and y only.
[9] For more detailed discussion of functions of state and the mathematics dealing with them, see references [5] and [6] and also Appendix I.

EXAMPLE 5-2

The following equation is given for 1 mole of a gas

$$\mathcal{V} = f(P, T)$$

Derive the ideal gas equation.

Solution: If the molar volume is a property of the system, it has an exact differential:

$$d\mathcal{V} = \left(\frac{\partial \mathcal{V}}{\partial P}\right)_T dP + \left(\frac{\partial \mathcal{V}}{\partial T}\right)_P dT \tag{a}$$

Since we are considering an ideal gas, $(\partial \mathcal{V}/\partial P)_T$ and $(\partial \mathcal{V}/\partial T)_P$ may be evaluated from Boyle's and Gay-Lussac's laws, respectively, Eqs. (1-2) and (1-6).

$$\mathcal{V} = \frac{k}{P} \rightarrow \left(\frac{\partial \mathcal{V}}{\partial P}\right)_T = -\frac{k}{P^2} = -\frac{P\mathcal{V}}{P^2} = -\frac{\mathcal{V}}{P} \tag{b}$$

and

$$\mathcal{V} = k'T \rightarrow \left(\frac{\partial \mathcal{V}}{\partial T}\right)_P = k' = \frac{\mathcal{V}}{T} \tag{c}$$

Substituting from (b) and (c) into (a) and separating the unknowns, we have

$$\frac{d\mathcal{V}}{\mathcal{V}} + \frac{dP}{P} - \frac{dT}{T} = 0 \tag{d}$$

Integration results in

$$\ln\left(\frac{P\mathcal{V}}{T}\right) = \ln \boldsymbol{R} \tag{e}$$

or

$$\frac{P\mathcal{V}}{T} = \boldsymbol{R} \tag{f}$$

where $\boldsymbol{R}$ is the universal gas constant. The result of this derivation proves that the assumption that $d\mathcal{V}$ is an exact differential is correct. ●

EXAMPLE 5-3

One mole of an ideal gas is passed from state 1 (P_1, $\mathcal{V}_1$, T_1) to state 2 (P_2, $\mathcal{V}_2$, T_2). Such a change may be accomplished in an infinite number of ways. Two of the many possible paths are (see Figure 5-1)

> Path AC, T is a linear function of P.
> Path ABC, consisting of two partial steps: step AB, expressing isothermal compression and step BC representing an isobaric temperature change.

If $d\mathcal{V}$ is an exact differential, it must be the same for both paths. Prove that this is true.

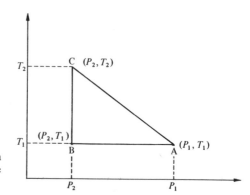

Figure 5-1 The process of passing of a system from initial state (A) to final state (C).

Solution: If $d\mathcal{V}$ is an exact differential, then

$$d\mathcal{V} = \left(\frac{\partial \mathcal{V}}{\partial T}\right)_P dT + \left(\frac{\partial \mathcal{V}}{\partial P}\right)_T dP \tag{a}$$

From the ideal gas equation,

$$\left(\frac{\partial \mathcal{V}}{\partial T}\right)_P = \frac{R}{P} \quad \text{and} \quad \left(\frac{\partial \mathcal{V}}{\partial P}\right)_T = -\frac{RT}{P^2}$$

and therefore

$$d\mathcal{V} = \frac{R}{P} dT - \frac{RT}{P^2} dP \tag{b}$$

This equation will now be applied to both paths.
For path AC, which is linear, it follows

$$T - T_1 = \frac{T_2 - T_1}{P_2 - P_1}(P - P_1) = \frac{\Delta T}{\Delta P}(P - P_1) \tag{c}$$

and for an infinitesimal change

$$dT = \frac{\Delta T}{\Delta P} dP \tag{d}$$

where $\Delta T / \Delta P$ is recognized to be the slope of the straight line.
Upon substituting dT from (d) and T from (c) into (b), we obtain

$$d\mathcal{V} = R\left(\frac{\Delta T}{\Delta P}P_1 - T_1\right)\frac{dP}{P^2} \tag{e}$$

Integration yields

$$\Delta \mathcal{V}_{AC} = R\left(\frac{\Delta T}{\Delta P}P_1 - T_1\right)\left(\frac{1}{P_1} - \frac{1}{P_2}\right)$$

$$= \frac{R(T_2 P_1 - T_1 P_2)}{P_1 P_2} \tag{f}$$

For path ABC, the integral is made up of two parts: the constant temperature part at T_1 and the constant pressure part at P_2. Thus

$$\Delta \mathcal{V}_{ABC} = -\int_{P_1}^{P_2} \frac{RT_1}{P^2} \, dP + \int_{T_1}^{T_2} \frac{R}{P_2} \, dT \tag{g}$$

$$= RT_1 \left(\frac{1}{P_2} - \frac{1}{P_1} \right) + \frac{R}{P_2} (T_2 - T_1)$$

$$= \frac{R(T_2 P_1 - T_1 P_2)}{P_1 P_2} \tag{h}$$

As seen from (f) and (h) $\Delta \mathcal{V}$ is the same for both paths; hence it is an exact differential. ●

5-4 MATHEMATICAL FORMULATION OF THE FIRST LAW OF THERMODYNAMICS

Instead of giving an exact definition of energy, we shall state that a system contains energy due to its position (potential energy), its motion as a whole (kinetic energy), and energy associated with the different molecular and internal motions, called internal energy.[10] A change in any one of these affects the total energy of the system. In order to achieve a change in the energy of the system, the system must interact with its surroundings.

In a closed system, the interaction is possible only in the form of heat (including radiation) and in the form of work. Heat and work are distinguishable from each other solely by the mode of transfer. Whereas work is defined as energy in transit across the boundary of a system by virtue of a force through a distance, heat is defined as energy in transit across the boundary of a system by virtue of a temperature difference between the system and its surroundings. It must be understood that neither work nor heat is contained in a system; each refers only to energy in transit. Heat and work cease to exist at the boundary of the system and become some form of the energy of the system. Heat and work, therefore, are not properties of the system; thus, they do not have exact differentials (see Example 5-4).

On the basis of what we have just said, the change in energy of a system for a process between states 1 and 2 is given by

$$E_2 - E_1 = \Delta E = \Delta U + \Delta E_p + \Delta E_k = q + w \tag{5-14}$$

where ΔU is the change in the internal energy of the system; ΔE_p and ΔE_k represent the change in the potential and kinetic energy, respectively, during the process; q and w denote the heat and work, respectively, exchanged by the system during the process. By convention, both are considered positive when added to the system. In most thermodynamics applications, the system as a whole does not change its

---

[10] Energy associated with the translational, rotational, vibrational, electronic, and nuclear motions (see Part III, Quantum Mechanics, Part IV, Statistical Thermodynamics, and Part I, States of Matter).

velocity and its position[11] with respect to some reference level during the process ($\Delta E_p = \Delta E_k = 0$), and thus Eq. (5-14) reduces to the form

$$\Delta E = \Delta U = q + w \tag{5-15}$$

It is understood that states 1 and 2 are equilibrium states.

If an infinitesimal quantity of heat and work is absorbed by the system, the infinitesimal increment in the internal energy is given by the differential of Eq. (5-15)

$$dU = dq + dw \tag{5-16}$$

where d is used rather than d to indicate inexact differentials. Since U is a function of state, dU may be integrated, and $U_2 - U_1$ therefore depends only on the initial and final states. However, dq and dw are not differentials of a function of state of the system. Therefore, their integrals, which are simply written q and w, respectively, rather than $q_2 - q_1$ and $w_2 - w_1$, depend on the particular path which is followed while the system passes from state 1 to state 2. For example, when a gas is allowed to expand, the amount of work obtained may vary from zero, if the gas is allowed to expand into a vacuum, to a maximum value which is obtained if the expansion is carried out reversibly as described in Section 5-5.

Eqs. (5-14), (5-15), and (5-16), are the mathematical expressions of the first law of thermodynamics.[12] It must be recognized that the first law provides a means for determining changes in internal energy but not the absolute value of the internal energy. According to the first law of thermodynamics, energy may be transformed from one form to another, but it cannot be created or destroyed; the total energy of an isolated system is constant.[13] It follows that, if there is no change in internal energy, the work performed by the system must be exactly equal to the heat absorbed by the system.

Nuclear reactions do not contradict the first law because, according to the theory of relativity, a certain mass is equivalent to a certain amount of energy [9] and the large amount of energy released by nuclear reactions is accompanied by an equivalent decrease in mass according to the equation

$$\Delta E = \Delta m c^2$$

At the end of the eighteenth and in the first half of the nineteenth century, Benjamin Thompson (1798), better known as Count Rumford, Humphrey Davy (1799), Julius R. Mayer (1842), and James P. Joule (1840) performed a number of experiments involving heat and work, that led to the postulation of the first law of thermodynamics [14, 18]. This law, following from human experience with nature, is really no more than an expression of the fact that energy is always conserved.

5-5 REVERSIBLE AND IRREVERSIBLE PROCESSES

It is a well-known fact from mechanics that a system may perform different kinds of work, depending on the kinds of forces acting on it [15]. We shall mention here

---

[11] The energy associated with the change in the kinetic and potential energies of the individual molecules is included in the internal energy.

[12] For the statistical interpretation, see page 584.

[13] Clausius (1865): "Die Energie der Welt ist konstant." The energy of the universe is constant.

only a few of these kinds of work: electrical work ($\Phi \, dQ$, where Φ is the potential and Q the charge), surface work ($\gamma \, dA$, where γ is the surface tension and A the area), $P\text{-}V$ work, and so on. Work is always composed of two terms: the work coefficient (an intensive property) and the work coordinate (an extensive property). The $P\text{-}V$ work is perhaps the most common form of work in chemical processes. For this reason, we shall separate it from all the other kinds of work, dw'', and will pay special attention to it. We therefore rewrite Eq. (5-16) as

$$dU = dq + dw' + dw'' \tag{5-17}$$

For the moment, we shall consider systems in which only $P\text{-}V$ work is performed. In such cases $dw'' = 0$ and therefore

$$dU = dq + dw' \tag{5-18}$$

To find a mathematical expression for the $P\text{-}V$ work, dw', let us consider a fluid contained in a cylinder, shown in Figure 5-2, of cross-sectional area A, fitted with a frictionless and weightless piston. Let the pressure on the piston be P_{op}.[14] Then, since the pressure is force per unit area, the total force acting on the piston is $f = P_{op}A$. If the piston is displaced a distance dl in a direction opposite to the direction of the force, the change in volume of the gas is $dV = -A \, dl$, and the work is

$$dw' = f \, dl = P_{op}A \, dl = -P_{op} \, dV \tag{5-19}$$

This is in accord with our sign convention. For compression, dV is negative, and thus the work done on the system is positive. On combining Eqs. (5-19) and (5-18), we have

$$dU = dq - P_{op} \, dV \tag{5-20}$$

As one can see, the magnitude of the work exchanged by the system and its surroundings depends on the opposing pressure. If $P_{op} = 0$, the force acting against the moving piston is zero, and thus, $w' = 0$. If P_{ext} is higher than the pressure of the fluid in the cylinder, P, compression will take place and work will be done on the system by its surroundings. In the opposite case, expansion will take place, and work will be done on the surroundings by the system.

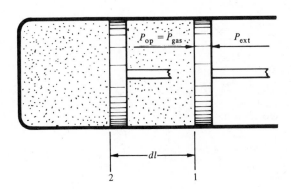

Figure 5-2 Cylinder with piston filled with a gas. Compression work.

[14] P_{op} is the opposing pressure; for expansion, it is the external pressure, P_{ext}, and, for compression, the pressure of the gas.

It is easy to recognize that, in the case when $P = P_{ext} + dP$, that is, when the pressure of the gas is just infinitesimally higher than the external pressure, the force acting on the moving piston will be at its maximum value (for expansion) and, therefore, the work done by the system will be the maximum work. If the force acting on the moving piston, and thus the external pressure, were increased to higher values, then $P_{ext} > P$, and instead of expansion, compression would take place. The maximum work of expansion is therefore performed in the case when the difference in the two pressures is infinitesimally small; thus, by a small change in pressure, the direction of the process can be reversed [11, 12, 13, 14]. Such a process is called **reversible**. For a reversible process, $P = P_{op} \pm dP$, and thus

$$dw' = -(P \mp dP)\, dV = -P\, dV \mp dP\, dV = -P\, dV \qquad (5\text{-}21)$$

and the opposing pressure may be replaced by the pressure of the gas in the cylinder for expansion as well as for compression since $dP\, dV$ is negligibly small.

Throughout the reversible process, the difference between the outside pressure and the gas pressure is infinitesimally small; the system is only infinitesimally removed from equilibrium. The process therefore requires an infinite amount of time for completion.[15] A reversible process can be reversed in every respect by an infinitesimal change in any of the intensive properties. During a reversible process the system and its surroundings can both be restored to their initial states without any changes whatsoever having taken place in the rest of the universe.

A thermodynamically reversible process is actually one for which the conditions are so strict that it cannot be completely realized in practice. All naturally occurring processes are therefore irreversible. Electricity tends to flow only from a point of higher electric potential to one of lower potential; water will flow by itself only from a higher to a lower level; diffusion will occur only from a point of higher concentration (density, pressure) to a point of lower concentration; chemical systems will tend to undergo reactions in a direction that will lead to the establishment of equilibrium; heat flows spontaneously from a point of higher temperature to a point of lower temperature. Life itself goes in one direction only and is an irreversible process.

EXAMPLE 5-4
Prove that heat is not a function of state.
Solution: Making use of Eq. (5-20), we may write

$$dq = dU + P\, dV$$

Since U is a function of state

$$dU = \left(\frac{\partial U}{\partial V}\right)_T dV + \left(\frac{\partial U}{\partial T}\right)_V dT$$

On combining these two equations, we obtain

$$dq = \left[\left(\frac{\partial U}{\partial V}\right)_T + P\right] dV + \left(\frac{\partial U}{\partial T}\right)_V dT$$

---

[15] Thermodynamics tells us nothing about how fast a process will occur; thermodynamically possible processes can occur at immeasurably slow rates.

If q were a function of state then, according to the reciprocity relation (5-12)

$$\left\{\frac{\partial}{\partial T}\left[\left(\frac{\partial U}{\partial V}\right)_T + P\right]\right\}_V = \left[\frac{\partial}{\partial V}\left(\frac{\partial U}{\partial T}\right)_V\right]_T$$

But this requires that

$$\left(\frac{\partial P}{\partial T}\right)_V = 0$$

which is never the case because the pressure changes with temperature at a constant volume. ●

5-6 CALORIMETRY

The first law of thermodynamics introduces a new extensive thermodynamic property: internal energy. It enables one to determine the change in the internal energy during a process, but it cannot say anything about the absolute value of the internal energy either in the original or final states of the system. As we will prove next, the change in the internal energy, under some restricted conditions, is equal to either the work or the heat that is added to or removed from the system.

Let us consider, first, a process taking place in a system which is separated from its surroundings by a boundary made of an insulating material. During such a process, which is called an **adiabatic process**, the system cannot exchange heat with its surroundings ($q = 0$), and therefore

$$\Delta U = w_{ad} \tag{5-22}$$

The work exchanged between the system and its surroundings during an adiabatic process, w_{ad}, is equal to the change in the internal energy of the system; adiabatic conditions can be approached by careful thermal insulation of the system.

If the adiabatically functioning boundary is now replaced by a diathermal one—a good heat conductor—in such a manner that no work may be exchanged between the system and its surroundings ($w = 0$), we obtain

$$\Delta U = q \tag{5-23}$$

When only P-V work is considered, the restriction $w' = 0$ requires that dV also equals 0, and therefore

$$\Delta U = q_V \tag{5-24}$$

During a process of constant volume—**isochoric process**—the change in the internal energy of a system is equal to the heat, q_V, exchanged by the system and its surroundings.

The quantitative determination of the amount of heat exchanged between a system and its surroundings during a process is performed in a calorimeter [18, 19]. A simple type of calorimeter is shown in Figure 5-3.

The calorimeter itself (A), made from polished metal sheets, is placed in a double-walled jacket filled with water and covered by a good insulator (B). Both the calorimeter and its jacket are equipped with covers in order to minimize heat loss.

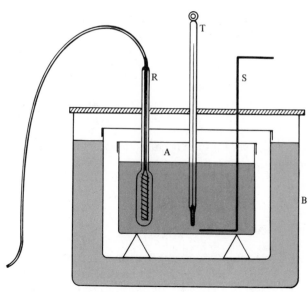

Figure 5-3 Simple calorimeter. S, stirrer; T, thermometer; R, electrical heater.

Other important parts of the calorimeter are an efficient stirrer and a sensitive, finely graded (to 0.01°), thermometer.

The process for which the thermal effect is to be studied takes place in the calorimeter, and the temperature change caused by the process is noted. The heat received by the calorimeter during the process is proportional to the change in temperature

$$q = \bar{C}(T_2 - T_1) \tag{5-25}$$

where T_1 is the temperature at the beginning, and T_2 the temperature at the end of the process. The proportionality constant, $\bar{C}$, is called the average heat capacity, or the water value of the calorimeter. This value is easily obtained from the measurement of the temperature change caused by the addition of a certain known amount of electrical work through an electrical heater to the system. As is obvious from Eq. (5-25), in calorimetry we measure temperature changes caused by adding a certain known amount of work (electric) or heat to the system; we do not measure heat directly, as the word calorimetry might indicate.

On the basis of the relation of the calorimeter to the surroundings, we classify calorimeters as adiabatic calorimeters, isothermal calorimeters, and some other special types. On the basis of the process taking place in the calorimeter, they are classified as combustion calorimeters, solution calorimeters, mixing calorimeters, and so on.

5-7 ENTHALPY

Constant pressure—**isobaric**—processes are more usual in chemistry than constant volume processes because most operations are carried out in open vessels. If the only work exchanged between the system and its surroundings is P-V work, then, according to Eq. (5-20), we may write for a constant pressure process

$$d\,q_P = dU + d(PV) = d(U + PV) = dH \tag{5-26}$$

since at constant pressure $P\,dV = d(PV)$. The term $d\,q_P$ is the infinitesimal amount of heat exchanged at constant pressure between the system and its surroundings; and H

$$H = U + PV \tag{5-27}$$

represents a new thermodynamic function called enthalpy. Since it is defined in terms of functions of state (U, P, V), enthalpy is itself a function of state. As seen from Eq. (5-26), the heat exchanged between the system and its surroundings at constant pressure is equal to the change in enthalpy if the only work done is P-V work. As in the case of internal energy, the first law allows the evaluation of the change in enthalpy during a process but not the absolute values of the enthalpy at the beginning and at the end of the process.

5-8 HEAT CAPACITY

It is a well-known experimental fact that whenever heat is added to or removed from a system, and no phase change takes place within the system, the temperature of the system changes. The amount of heat added to the system is related to the temperature increase by the equation

$$q = m\bar{c}(T_2 - T_1) \tag{5-28}$$

where m is the mass of the system. As is obvious from Eq. (5-29), the proportionality constant $\bar{c}$, called the average specific heat, is defined as the amount of heat required to increase the temperature of 1 g of a substance by 1 degree

$$\bar{c} = \frac{q}{m(T_2 - T_1)} \tag{5-29}$$

Generally, the heat capacities are temperature dependent; therefore $\bar{c}$ may be applied in the given temperature interval only.[16]
For the exact definition of the specific heat, one must consider an infinitesimal change in temperature; then

$$c = \lim_{(T_2 - T_1) \to 0} \frac{q}{m(T_2 - T_1)} = \frac{1}{m}\frac{d\,q}{dT} \tag{5-30}$$

The relationship between the mean and the exact specific heat capacities is the following:

$$\bar{c} = \frac{q}{m(T_2 - T_1)} = \frac{m \int_{T_1}^{T_2} c\,dT}{m(T_2 - T_1)} = \frac{\int_{T_1}^{T_2} c\,dT}{T_2 - T_1} \tag{5-31}$$

[16] For this reason, the calorie used to be defined as the amount of heat required to increase the temperature of 1 g of water from 14.5°C to 15.5°C; at present the thermochemical calorie, used most frequently in chemistry, including this text, is defined: $1\text{ cal} \equiv 4.184\text{ J}$. In the SI system of units joules ($\equiv$ Newton meter) are used instead of calories.

Heat capacity is an extensive property; when related to 1 g—**specific heat capacity**—or to 1 mole—**molar heat capacity**—it becomes an intensive property. In this text, we shall mostly use the molar heat capacities.

Since the amount of heat needed to increase the temperature by one degree is not the same at constant pressure as it is at constant volume, it is logical that the molar heat capacities will also differ at constant pressure and at constant volume. To differentiate between the constant pressure and constant volume values, we introduce $\mathscr{C}_P$ and $\mathscr{C}_V$:

$$\mathscr{C}_P = \frac{d\,q_\mathrm{p}}{dT} = \left(\frac{\partial \mathscr{H}}{\partial T}\right)_P \tag{5-32}$$

and

$$\mathscr{C}_V = \frac{d\,q_V}{dT} = \left(\frac{\partial \mathscr{U}}{\partial T}\right)_V \tag{5-33}$$

The equation for the difference in heat capacities, $\mathscr{C}_P - \mathscr{C}_V$, may be derived as follows: The molar enthalpy is given as

$$\mathscr{H} = \mathscr{U} + P\mathscr{V} \tag{5-27}$$

and

$$d\mathscr{H} = d\mathscr{U} + P\,d\mathscr{V} + \mathscr{V}\,dP \tag{5-34}$$

Since

$$d\mathscr{U} = \left(\frac{\partial \mathscr{U}}{\partial T}\right)_V dT + \left(\frac{\partial \mathscr{U}}{\partial \mathscr{V}}\right)_T d\mathscr{V} \tag{5-35}$$

we write at constant pressure

$$\left(\frac{\partial \mathscr{H}}{\partial T}\right)_P = \left(\frac{\partial \mathscr{U}}{\partial T}\right)_V + \left[\left(\frac{\partial \mathscr{U}}{\partial \mathscr{V}}\right)_T + P\right]\left(\frac{\partial \mathscr{V}}{\partial T}\right)_P \tag{5-36}$$

or

$$\mathscr{C}_P - \mathscr{C}_V = \left[\left(\frac{\partial \mathscr{U}}{\partial \mathscr{V}}\right)_T + P\right]\left(\frac{\partial \mathscr{V}}{\partial T}\right)_P \tag{5-37}$$

Since $(\partial \mathscr{V}/\partial T)_P > 0$, the heat capacity at constant pressure $\mathscr{C}_P$ is always larger than the heat capacity at constant volume $\mathscr{C}_V$.[17]

The difference between the two molar heat capacities represents not only the work done against the surroundings in isobaric heating, $P(\partial \mathscr{V}/\partial T)_P$, but also the work done against intermolecular attraction of the substance itself, $(\partial \mathscr{U}/\partial \mathscr{V})_T(\partial \mathscr{V}/\partial T)_P$.[18]

The variation of $\mathscr{C}_P$ with temperature has been determined for many substances, and is usually reported as an empirical power series in $T(K)$. For example

$$\mathscr{C}_P = a + bT + cT^2 + \cdots \tag{5-38a}$$

$$\mathscr{C}_P = a + bT + cT^{-2} \tag{5-38b}$$

$$\mathscr{C}_P = a + bT + cT^2 + dT^3 + eT^{-2} \tag{5-38c}$$

where a, b, c, and so on, are constants characteristic of the given substance (see Table 5-1).

---

[17] Liquid water behaves anomalously between 0°C and 4°C; the rate of the volume change with temperature, $(\partial V/\partial T)_P$, is negative and therefore $\mathscr{C}_P < \mathscr{C}_V$.

[18] This contribution may become significant especially in solid and liquid systems [16].

Table 5-1 Molar heat capacities at constant pressure[a]

$$\mathscr{C}_P = a + bT + cT^2 + dT^3 + eT^{-2} \text{ cal K}^{-1} \text{ mol}^{-1} \text{ (between 273 K and 2000 K)}$$

Substance	a	$b \times 10^3$	$c \times 10^6$	$d \times 10^9$	$e \times 10^{-5}$
$H_2(g)$	6.9469	−0.1999	0.4808		
$O_2(g)$	6.0954	3.2533	−1.0171		
$N_2(g)$	6.4492	1.4125	−0.0807		
$Cl_2(g)$	7.5755	2.4244	−0.9650		
C(s, graphite)	−1.265	14.008	−10.331	2.751	
CO(g)	6.3424	1.8363	−0.2801		
$CO_2(g)$	6.369	10.100	−3.405		
$H_2O(g)$	7.219	2.314	0.267		
$SO_2(g)$	6.147	13.844	−9.103		
$SO_3(g)$	13.70	6.42			−3.12
NO(g)	7.020	−0.370	2.546		
$NH_3(g)$	6.189	7.887	−0.728		
HCl(g)	6.7319	0.4325	0.3697		
CaO(s)	11.67	1.08			1.56
$CaCO_3$(s,calcite)	24.98	5.24			−6.20
$CH_4(g)$	4.171	14.450	0.267	−1.722	
$C_2H_6(g)$	1.074	43.561	−17.891	2.581	
$C_2H_4(g)$	1.003	36.948	−19.381	4.019	
$C_2H_2(g)$	12.184	3.879		−2.581	
$C_6H_6(g)$	−8.102	112.780	−71.306	16.930	

[a] Most of the data are taken from the compilations of H. M. Spencer and J. L. Justice: *J. Am. Chem. Soc.*, **56**:2311 (1934); H. M. Spencer and G. N. Flanagan: *J. Am. Chem. Soc.*, **64**:2511 (1942); H. M. Spencer: *Ind. Eng. Chem.*, **40**:2152 (1948).

EXAMPLE 5-5
Calculate the change in enthalpy involved in heating 1 kg of aluminum from 0°C to 800°C, if $t_{mp} = 658°C$, $\Delta H_{fusion} = 86.6 \text{ kcal kg}^{-1}$, $(C_P)_s = 0.218 + 0.48 \times 10^{-4} t$ $(\text{kcal °C}^{-1} \text{ kg}^{-1})$ for solid Al, and $(\bar{C}_P)_l = 0.259$ $(\text{kcal K}^{-1} \text{ kg}^{-1})$ for liquid Al (1 kcal = 4.184 kJ).

Solution: The total change in enthalpy is

$$\Delta H = \Delta H_1 + \Delta H_2 + \Delta H_3$$

$$= \int_{273}^{931} (C_P)_s \, dT + \Delta H_{fusion} + \int_{931}^{1073} (\bar{C}_P)_l \, dT$$

$$= \int_{273}^{931} [0.218 + 0.48 \times 10^{-4}(T - 273)] \, dT + 86.6 + \int_{931}^{1073} 0.259 \, dT$$

$$= (0.218 - 0.48 \times 10^{-4} \times 273)(931 - 273) + \frac{0.48 \times 10^{-4}}{2}(931^2 - 273^2)$$

$$+ 86.6 + 0.259(1073 - 931)$$

$$= 134.9 + 19.0 + 86.6 + 36.8 = 277.3 \text{ kcal} = 1160.2 \text{ kJ} \qquad \bullet$$

5-9 THE APPLICATION OF THE FIRST LAW OF THERMODYNAMICS TO CHEMICAL REACTIONS: THERMOCHEMISTRY

One of the most important aspects of a reaction is the energy change that accompanies it. We are often far more interested in the amount of energy that can be obtained from a reaction than in the products formed.

Almost without exception, chemical reactions, as they are ordinarily carried out in the laboratory, evolve or absorb energy in the form of heat. **Thermochemistry**, the study of the heat changes accompanying chemical reactions, will occupy our attention in this section.

From a thermochemical standpoint, it is possible to classify reactions into one of two categories. A reaction that proceeds with the evolution of heat is referred to as an **exothermic** reaction; one in which heat is absorbed is called **endothermic**. As the great majority of chemical reactions proceed at constant pressure, and, as the only work performed by chemical reactions is P-V work (in a cell, electrical work may also be performed, see page 900), the heat exchanged during the course of a reaction is equal to the change in enthalpy ($\Delta H = q_p$). For reactions proceeding at constant volume, the heat exchanged during the course of a reaction is equal to the change in the internal energy ($\Delta U = q_V$). ΔU and ΔH are related through the equation

$$\Delta U = \Delta H - \Delta(PV) \tag{5-39}$$

By $\Delta(PV)$ we mean the sum of the PV for each of the products of the reaction, minus the sum of the PV for each of the reactants. If all reactants and products are liquids or solids, $\Delta(PV)$ is usually so small compared to ΔH and ΔU that it may be neglected, and, in this case, $\Delta H \approx \Delta U$. For reactions in which gases occur, the value of $\Delta(PV)$ may become important, if the number of moles of gas changes as a result of the reaction. From the ideal gas equation, we obtain $\Delta(PV) = \Delta \nu RT$, and thus,

$$\Delta U = \Delta H - \Delta \nu RT \tag{5-40}$$

$\Delta \nu$ denotes the number of moles of gaseous products minus the number of moles of gaseous reactants.

EXAMPLE 5-6

During the burning of 1 g of naphthalene in a calorimetric bomb (constant volume process) at 18°C (the water produced by the burning condenses), 9621 cal are evolved (1 cal = 4.184 J). Calculate the enthalpy change for this process.

Solution: The burning proceeds according to the equation

$$C_{10}H_8(s) + 12 O_2(g) = 10 CO_2(g) + 4 H_2O(l)$$

$$\Delta \mathcal{U}_{comb} = -9621 \times 128 \times 4.184 = -5153 \text{ kJ mol}^{-1}$$

Since the naphthalene is in a solid state and water is in a liquid state, $\Delta \nu = 10 - 12 = -2$, and therefore

$$\Delta \mathcal{H}_{comb} = \Delta \mathcal{U}_{comb} + \Delta \nu RT = -5153 + (-2 \times 8.314 \times 10^{-3} \times 291)$$

$$= -5158 \text{ kJ mol}^{-1}$$

For 1 g of naphthalene $-5158/128 = -40.3$ kJ.   ●

It is particularly important that the physical states of reactants and products be specified in writing thermochemical equations [solid (s), liquid (l), gas (g)]. The magnitude of the heat effect is changed if the state of one of the products or reactants changes. For example, the formation of 1 mole of water according to the equation

$$H_2(g) + \tfrac{1}{2} O_2(g) = H_2O(g)$$

is accompanied by an enthalpy change of -241.83 kJ mol^{-1}, whereas according to the equation

$$H_2(g) + \tfrac{1}{2} O_2(g) = H_2O(l)$$

the change in enthalpy at 25°C is -285.84 kJ mol^{-1}. It is logical that the difference in the enthalpy change of the two reactions, 44.01 kJ, is the enthalpy of vaporization of 1 mole of water at 25°C.

It is also necessary to recognize the fact that the amount of heat evolved or absorbed in a reaction is directly proportional to the amount of reactants consumed or products formed. Since it is known that 241.83 kJ are evolved when 1 mole of water vapor is formed from its elements at 25°C, it follows that exactly twice as much heat, 483.66 kJ, is evolved when 2 moles of water vapor are formed.

Thermochemical equations resemble ordinary algebraic equations in some other ways, too:

1. Both sides of a chemical equation may be multiplied by the same factor without affecting its validity.
2. It is possible to add or to subtract the same quantity from both sides of a thermochemical equation without affecting its validity.
3. Two or more thermochemical equations may be added or subtracted to obtain a new, equally valid, equation.

To illustrate the use of this last principle, consider the following two equations:

$$Sn(s) + Cl_2(g) = SnCl_2(s) \qquad \Delta \mathcal{H}_{298} = -349.78 \text{ kJ mol}^{-1}$$

$$SnCl_2(s) + Cl_2(g) = SnCl_4(s) \qquad \Delta \mathcal{H}_{298} = -195.32 \text{ kJ mol}^{-1}$$

If we wish to obtain the enthalpy change evolved when 1 mole of SnCl$_4$ is formed from its elements, it is only necessary to add the last two equations:

$$Sn(s) + SnCl_2(s) + 2\, Cl_2(g) = SnCl_2(s) + SnCl_4(s)$$

$$\Delta \mathcal{H} = -349.78 - 195.32 = -545.10 \text{ kJ mol}^{-1}$$

$$Sn(s) + 2\, Cl_2(g) = SnCl_4(s) \qquad \Delta \mathcal{H} = -545.10 \text{ kJ mol}^{-1}$$

This is a consequence of the fact that the enthalpy change of reaction, ΔH, is a function of state and is therefore independent of the path, that is, of any intermediate reactions that may have occurred. This principle, first established experimentally by G. H. Hess in 1840, enables one to calculate the heat of a reaction from measurements made on quite different reactions.

EXAMPLE 5-7
Calculate the enthalpy change of the reaction

$$C_2H_4(g) + H_2(g) = C_2H_6(g)$$

Given the following thermochemical equations:

$$C_2H_4(g) + 3\,O_2(g) = 2\,CO_2(g) + 2\,H_2O(l) \qquad \Delta\mathcal{H}_{298} = -1411.26 \text{ kJ mol}^{-1} \qquad \textbf{(a)}$$

$$H_2(g) + \tfrac{1}{2}\,O_2(g) = H_2O(l) \qquad \Delta\mathcal{H}_{298} = -285.84 \text{ kJ mol}^{-1} \qquad \textbf{(b)}$$

$$C_2H_6(g) + \tfrac{7}{2}\,O_2(g) = 3\,H_2O(l) + 2\,CO_2(g) \qquad \Delta\mathcal{H}_{298} = -1559.80 \text{ kJ mol}^{-1} \qquad \textbf{(c)}$$

Solution: Add Eqs. (a) and (b), and subtract Eq. (c): The result is

$$C_2H_4(g) + H_2(g) = C_2H_6(g) \qquad \Delta\mathcal{H}_{298} = -137.30 \text{ kJ mol}^{-1} \qquad \bullet$$

5-10 THE STANDARD ENTHALPY OF FORMATION

There are many thousands of reactions for which enthalpy changes have been experimentally determined. It would require a volume considerably larger than this text to list all of the corresponding thermochemical equations. A more concise method of expressing thermochemical data involves tabulating quantities known as standard enthalpies of formation.

The standard enthalpy of formation, $(\Delta\mathcal{H}_f^\circ)_T$, is defined as the enthalpy change of a reaction by which a compound in its standard state is formed from its elements in their standard states. As a standard state, we choose the most stable modification at 25°C and 1 atm.[19] By convention, we assign to all elements in their standard state zero enthalpy. A few examples are

$$C(\text{graphite}) + O_2(g) = CO_2(g) \qquad (\Delta\mathcal{H}_f^\circ)_{298} = -393.51 \text{ kJ mol}^{-1}$$

$$S(\text{rhombic}) + O_2(g) = SO_2(g) \qquad (\Delta\mathcal{H}_f^\circ)_{298} = -296.90 \text{ kJ mol}^{-1}$$

$$3\,C(\text{graphite}) + 4\,H_2(g) = C_3H_8(g) \qquad (\Delta\mathcal{H}_f^\circ)_{298} = -103.81 \text{ kJ mol}^{-1}$$

The first law of thermodynamics (law of conservation of energy) permits the calculation of the enthalpy change of reaction from the standard enthalpies of formation. For this purpose the law may be expressed as

$$(\Delta H^\circ)_{298} = \sum_P \nu_P (\Delta\mathcal{H}_f^\circ)_{298} - \sum_R \nu_R (\Delta\mathcal{H}_f^\circ)_{298} \qquad (5\text{-}41)$$

where $\sum_P \nu_P (\Delta\mathcal{H}_f^\circ)_{298}$ and $\sum_R \nu_R (\Delta\mathcal{H}_f^\circ)_{298}$ denote the sum of the standard enthalpies of formation (multiplied by the number of moles in the balanced chemical equation) of the products and reactants.

An alternate method for the calculation of enthalpy changes of reactions follows from the equation

$$(\Delta\mathcal{H}^\circ)_{298} = \sum_R \nu_R (\Delta\mathcal{H}_{comb}^\circ)_{298} - \sum_P \nu_P (\Delta\mathcal{H}_{comb}^\circ)_{298} \qquad (5\text{-}42)$$

where $(\Delta\mathcal{H}_{comb}^\circ)_{298}$ is the standard enthalpy of combustion, defined as the change in

---

[19] More often the standard state is defined at the particular temperature of the system. Such a definition requires, however, the exact specification of the aggregate state that is the most stable at the temperature of the system. It also requires that the enthalpy of the elements be zero in the standard state regardless of temperature and the aggregate state. For the definition of other standard states see page 195.

enthalpy of a reaction in which 1 mole of a substance in its standard state reacts completely with pure oxygen. For example

$$C_2H_6(g) + \tfrac{7}{2}O_2(g) = 2\,CO_2(g) + 3\,H_2O(l) \qquad (\Delta\mathscr{H}^\circ_{comb})_{C_2H_6} = -1559.80 \text{ kJ mol}^{-1}$$

or

$$C(graphite) + O_2(g) = CO_2(g) \qquad (\Delta\mathscr{H}^\circ_{comb})_C = -393.51 \text{ kJ mol}^{-1}$$

Eqs. (5-41) and (5-42) respectively, represent the mathematical formulation of Hess's law. They are special forms of the law of conservation of energy.

Some values of standard enthalpies of formation are given in Table 5-2.[20]

Table 5-2 Standard enthalpies of formation[a] at 25°C

Compound	$\dfrac{\Delta\mathscr{H}^\circ_f}{\text{(cal mol}^{-1})}$	Compound	$\dfrac{\Delta\mathscr{H}^\circ_f}{\text{(cal mol}^{-1})}$
$CO(g)$	−26,416	$C_4H_{10}(g)$(iso-)	−32,150
$CO_2(g)$	−94,052	$C_5H_{12}(g)$	−35,000
$COCl_2(g)$	−53,300	$C_6H_6(g)$	19,820
$CaCO_3$(s,calcite)	−288,450	$C_6H_6(l)$	11,718
$CaO(s)$	−151,900	$C_6H_{12}(g)$(cyclohexane)	−29,430
$FeO(s)$	−63,700	$CH_4O(l)$(methyl alcohol)	−57,020
$H_2S(g)$	−4,815	$CH_2O(g)$(formaldehyde)	−27,700
$NH_3(g)$	−11,040	$CH_2O_2(l)$(formic acid)	−97,800
$SO_2(g)$	−70,960	$C_2H_4O_2(l)$(acetic acid)	−116,400
$SO_3(g)$	−94,450	$CCl_4(g)$	−25,500
$CH_4(g)$	−17,889	$CCl_4(l)$	−33,300
$C_2H_2(g)$	54,194	$CHCl_3(g)$	−24,000
$C_2H_4(g)$	12,496	$CH_2Cl_2(g)$	−21,000
$C_2H_6(g)$	−20,236	$CH_3Cl(g)$	−19,600
$C_3H_8(g)$	−24,820	$C_6H_5Cl(l)$	27,800
$C_4H_{10}(g)$	−30,150	$C_6H_6O(l)$(phenol)	−37,260

[a] *Nat. Bur. Stand. Circ.* 500, U.S.G.P.O., Washington, D.C., 1952; *Nat. Bur. Stand. Circ.* C461, U.S.G.P.O., Washington, D.C., 1947.

5-11 BOND ENERGIES

The enthalpy of formation, $(\Delta\mathscr{H}^\circ_f)_{298}$, represents an important thermodynamic property. Some effort has been made to relate this quantity to the structural formula of the compound.

The idea is simple: $(\Delta\mathscr{H}^\circ_f)_T$ is a function of state, and as such, its value is independent of the path of synthesis of the compound. The enthalpy of formation is the same whether the synthesis is carried out directly, or in an artificial way made up

---

[20] The values are given in calories rather than in joules as required by the SI system of units, since the great majority of thermochemical data is still expressed in calories rather than in joules.

of the two following steps:

1. First, the elements entering the reaction are transformed into their monatomic gaseous forms; energy is required to break the bonds in the molecule and to vaporize the liquid and solid elements, respectively.
2. Second, the monatomic gases thus formed combine to produce the desired compound; energy is released during the formation of the new bonds.

The enthalpy of formation is given by the algebraic sum of the enthalpy changes accompanying these two steps. Table 5-3 presents standard enthalpy changes for reactions in which isolated gaseous atoms are formed from their molecules, $(\Delta \mathcal{H}_{at}^{\circ})_T$; these values correspond to the first step and are obtained from spectroscopic data, enthalpies of dissociation, and enthalpies of vaporization.

Table 5-3 Standard enthalpies of atomization, $\Delta \mathcal{H}_{at}^{\circ}$, at 25°C (kcal mol^{-1})
(All elements in gaseous monatomic state.)

Element	$\Delta \mathcal{H}_{at}^{\circ}$	Element	$\Delta \mathcal{H}_{at}^{\circ}$	Element	$\Delta \mathcal{H}_{at}^{\circ}$
O	59.159	Ga	66.0	Zr	125
H	52.089	In	58.2	B	97.2
F	18.3	Tl	43.34	Al	75.0
Cl	29.012	Zn	31.19	Sc	93
Br	26.71	Cd	26.97	Ce	85
I	25.482	Hg	14.54	La	88
S	52.35	Cu	81.52	U	125
Se	48.37	Ag	69.12	Be	76.63
Te	47.6	Au	82.99	Mg	35.9
N	85.565	Pt	121.6	Ca	46.04
P	75.18	Ni	101.61	Ba	39.2
As	60.64	Pd	193	Li	37.07
Sb	60.8	Mn	68.34	Na	25.98
Bi	49.7	Cr	80.5	K	21.51
C	171.698	Mo	155.5	Rb	20.51
Si	88.04	W	201.6	Cs	18.83
Ge	78.44	V	120	Co	105
Sn	72	Ta	185	Fe	96.68
Pb	46.34	Ti	112		

The bond energy[21] is defined as the average dissociation energy of a certain bond present in different compounds. It is known for example that the dissociation energy of the C—H bond is not the same in the different hydrocarbon compounds; the bond energy is taken, therefore, as the average value over a large number of hydrocarbons

---

[21] More precisely the bond enthalpy.

Table 5-4 Bond energies at 25°C (kcal mol^{-1})

Bond	ε	Bond	ε
O—O	33.2	H—F	148.5
H—H	104.2	H—Cl	103.2
F—F	64.4	H—Br	87.5
Cl—Cl	58.0	H—I	71.4
Br—Br	46.1	C—O	82.3
I—I	36.1	C—F	115.8
S—S	50.2	C—Cl	78.3
N—N	20.2	C—Br	73.0
P—P	47.9	C—I	50.1
As—As	34.5	C—S	60.1
C—C	81.8	C—N	60.3
Si—Si	44.0	C—Si	143.2
Ge—Ge	39.2	Si—O	104.0
C—H	99.5	Si—F	146.7
Si—H	77.8	Si—Cl	87.4
N—H	84.3	Ge—Cl	102.6
P—H	76.4	C=C	146.6
As—H	58.6	C≡C	201.9
O—H	110.6	N=N[a]	225.1
S—H	81.1	C=O[a]	173
Se—H	66.0	C=O[a]	209
Te—H	57.4		

[a] These values are at 0 K

[17]. As is shown next, the bond energies listed in Table 5-4 may be calculated from the values of $(\Delta \mathcal{H}_{at}^\circ)_T$, given in Table 5-3 and from one piece of experimental data—enthalpy of formation. For example, the bond energy of the C—H bond, $\varepsilon_{C—H}$, at 25°C, may be calculated as follows:

Starting from the dissociation reaction of CH_4, we may write

$$CH_4(g) \rightarrow C(g) + 4\ H(g)$$

and from Eq. (5-41)

$$(\Delta \mathcal{H}_{dis}^\circ)_{CH_4} = (\Delta \mathcal{H}_{at}^\circ)_{C(g)} + 4\ (\Delta \mathcal{H}_{at}^\circ)_{H(g)} - (\Delta \mathcal{H}_f^\circ)_{CH_4(g)}$$

$$= 171.7 + 4(52.09) - (-17.89)$$

$$= 398 \text{ kcal mol}^{-1}$$

According to the definition of bond energy, the bond energy of the C—H bond, $\varepsilon_{C—H}$, is equal to one quarter the enthalpy of dissociation of CH_4. Thus, $\varepsilon_{C—H} = 398/4 = 99.5$ kcal mol^{-1} = 416.31 kJ mol^{-1}.

The bond energy of the C—C bond, $\varepsilon_{C—C}$, may be calculated on the basis of the following dissociation reaction of ethane:

$$C_2H_6(g) \rightarrow 2\ C(g) + 6\ H(g)$$

and

$$(\Delta \mathcal{H}^{\circ}_{\text{dis}})_{C_2H_6} = 2(\Delta \mathcal{H}^{\circ}_{\text{at}})_{C(g)} + 6(\Delta \mathcal{H}^{\circ}_{\text{at}})_{H(g)} - (\Delta \mathcal{H}^{\circ}_{f})_{C_2H_6(g)}$$
$$= 2 \times 171.7 + 6 \times 52.09 - (-20.24)$$
$$= 676.2 \text{ kcal mol}^{-1}$$

Since the dissociation represents the breakage of six C—H and one C—C bonds, we may write

$$(\Delta \mathcal{H}^{\circ}_{\text{dis}})_{C_2H_6} = 676.2 = \varepsilon_{C-C} + 6\varepsilon_{C-H} = 6 \times 99.5 + \varepsilon_{C-C}$$

and

$$\varepsilon_{C-C} = 676.2 - 597.0 = 79.2 \text{ kcal mol}^{-1} = 331.4 \text{ kJ mol}^{-1}$$

This value is slightly different from the value given in Table 5-4. The difference is explained by the fact that the value listed in Table 5-4 represents an average value calculated from more than one hydrocarbon, whereas the value 331.4 kJ is calculated only from ethane.

Generally, the enthalpy of formation may be calculated from the formula

$$(\Delta \mathcal{H}^{\circ}_{f})_T = \sum_{\text{react}} n_i (\Delta \mathcal{H}^{\circ}_{\text{at}})_T - \sum_{\text{prod}} n_j \varepsilon_j \qquad (5\text{-}43)$$

where n_i denotes the number of atoms of kind i in the reacting elements, and n_j the number of bonds of kind j, of energy ε_j, in the final product.

EXAMPLE 5-8

Calculate the enthalpy of formation of propyne at 25°C, using the data from Tables 5-3 and 5-4.

Solution: Making use of Eq. (5-43) for the reaction

$$3 \text{ C(graphite)} + 2 \text{ H}_2(g) \rightarrow C_3H_4(g) \rightarrow \overset{\displaystyle H}{\underset{\displaystyle H}{H-C\equiv C-\overset{|}{\underset{|}{C}}-H}}$$

we write

$$(\Delta \mathcal{H}^{\circ}_{f})_{C_3H_4(g)} = 3 (\Delta \mathcal{H}^{\circ}_{\text{at}})_{C(g)} + 4 (\Delta \mathcal{H}^{\circ}_{\text{at}})_{H(g)} - 4\varepsilon_{C-H} - \varepsilon_{C-C} - \varepsilon_{C\equiv C}$$
$$= 3 \times 171.7 + 4 \times 52.09 - 4 \times 99.5 - 81.8 - 201.9$$
$$= 41.76 \text{ kcal mol}^{-1} = 174.72 \text{ kJ mol}^{-1}$$

The calculated value is in reasonable agreement with the experimentally found value, 185.43 kJ mol^{-1}.   ●

The determination of bond energies is not always easy, and there are discrepancies among many of the values [17]. The three most commonly applied methods for the determination of bond energies, calorimetric, spectroscopic, and electron impact, furnish slightly different results. Even so, the difference between the calculated and experimentally found values of the enthalpy of formation is, in most cases, less than 10%.

The bond energy is a measure of the bonding forces in a molecule. This aspect of the bond energy is discussed in Part III, Quantum Mechanics, page 437.

5-12 VARIATION OF ENTHALPY CHANGE OF REACTION WITH TEMPERATURE

Up to this point, we have been concerned with the values of $\Delta H°$ for reactions at one temperature only, usually 298 K. It is conceivable that $\Delta H°$ for a reaction is a function of temperature. We shall next show how application of the fact that $\Delta H°$ is independent of the reaction path provides us with the temperature dependence of $\Delta H°$.

Let us consider a general reaction

$$a\,\text{A} + b\,\text{B} \rightarrow c\,\text{C} + d\,\text{D}$$

The conversion of reactants to products at a temperature T_1, may be carried out by either of two paths, as shown in Figure 5-4. Let us suppose we know $\Delta H°_{T_1}$, the enthalpy change when reactants and products are at temperature T_1. (This can be obtained from the tabulated values of enthalpies of formation at 298 K.) Referring to Figure 5-4, we see that, since $\Delta H°_{T_1}$ is independent of path, then

$$\Delta H°_{T_1} = \Delta H' + \Delta H°_{T_2} + \Delta H'' \tag{5-44}$$

In order to find $\Delta H°_{T_2}$, the enthalpy change for the reaction at temperature T_2, $\Delta H'$ and $\Delta H''$ must be evaluated first. $\Delta H'$ is the enthalpy change associated with changing the temperature of the reactants at constant pressure from T_1 to T_2, and $\Delta H''$ is the enthalpy change which results from changing the temperature of the products from T_2 to T_1 at constant pressure.

The total heat capacity of the reactants is

$$\sum_R C_P = a\mathscr{C}_P(\text{A}) + b\mathscr{C}_P(\text{B}) \tag{5-45}$$

so, for $\Delta H'$, we have

$$\Delta H' = \int_{T_1}^{T_2} \sum_R C_P \, dT \tag{5-46}$$

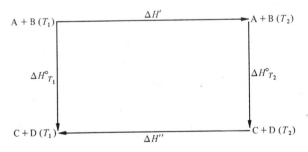

Figure 5-4   Graphical presentation of Kirchhoff's law.

Similårly, for the products

$$\sum_{P} C_P = c\mathscr{C}_P(C) + d\mathscr{C}_P(D) \tag{5-47}$$

and

$$\Delta H'' = \int_{T_2}^{T_1} \sum_{P} C_P\, dT \tag{5-48}$$

Combining Eqs. (5-44), (5-46), and (5-48) results in

$$\Delta H^\circ_{T_2} = \Delta H^\circ_{T_1} - \int_{T_1}^{T_2} \sum_{R} C_P\, dT - \int_{T_2}^{T_1} \sum_{P} C_P\, dT \tag{5-49}$$

We can change the sign of the second term on the right-hand side of this equation if we reverse the limits of integration. Thus

$$\Delta H^\circ_{T_2} = \Delta H^\circ_{T_1} + \int_{T_1}^{T_2} \left(\sum_{P} C_P - \sum_{R} C_P \right) dT$$

$$= \Delta H^\circ_{T_1} + \int_{T_1}^{T_2} \Delta C_P\, dT \tag{5-50}$$

where

$$\Delta C_P = \sum_{P} C_P - \sum_{R} C_P = [c\mathscr{C}_P(C) + d\mathscr{C}_P(D)] - [a\mathscr{C}_P(A) + b\mathscr{C}_P(B)] \tag{5-51}$$

Eq. (5-50) can also be written in a differential form

$$\left(\frac{\partial \Delta H^\circ}{\partial T} \right)_P = \Delta C_P \tag{5-52}$$

These equations were first obtained by G. R. Kirchhoff in 1858. They show that the rate of change of the enthalpy of reaction with the temperature is equal to the difference in heat capacities of products and reactants.

EXAMPLE 5-9
The literature gives the data shown in the following table for sulfur trioxide, sulfur dioxide, and oxygen.

Substance	$(\Delta\mathscr{H}^\circ_f)_{291}$ (cal mol^{-1})	$\mathscr{C}_P$ (cal K^{-1} mol^{-1})	Temperature Range (K)
$SO_3(g)$	−93,900	$12.65 + 6.40 \times 10^{-3}T$	(273–900)
$SO_2(g)$	−70,970	$11.40 + 1.714 \times 10^{-3}T$ $-2.045 \times 10^5 T^{-2}$	(273–2000)
$O_2(g)$		$7.52 + 0.81 \times 10^{-3}T$ $-9 \times 10^4 T^{-2}$	(273–2000)

Express the temperature dependence of the standard enthalpy change for the reaction

$$SO_2(g) + \tfrac{1}{2} O_2(g) \rightarrow SO_3(g)$$

and calculate ΔH° for the reaction at 600°C.

Solution: Using Hess's law (Eq. 5-41), we first calculate the enthalpy of the reaction at 291 K.

$$\Delta H^\circ_{291} = \sum_P \nu_P (\Delta \mathscr{H}^\circ_f)_{291} - \sum_R \nu_R (\mathscr{H}^\circ_f)_{291}$$
$$= -93,900 - (-70,970) = -22,930 \text{ cal}$$

Next, we evaluate ΔC_P

$$\Delta C_P = \mathscr{C}_{PSO_3} - \mathscr{C}_{PSO_2} - \tfrac{1}{2} \mathscr{C}_{PO_2}$$
$$= -2.51 + 4.281 \times 10^{-3} T + 2.495 \times 10^5 T^{-2}$$

Upon substituting for ΔC_P in Eq. (5-52) and integrating, we obtain

$$\Delta H^\circ_T = -2.51 T + 2.140 \times 10^{-3} T^2 - 2.495 \times 10^5 T^{-1} + I_H$$

The value of the integration constant, I_H, is calculated by substituting into the equation the known value of the standard enthalpy of reaction at 291 K.

$$I_H = -22,930 + 2.51 \times 291 - 2.140 \times 10^{-3} \times 291^2 + 2.495 \times 10^5 \times 291^{-1}$$
$$= -21,523.4 \text{ cal}$$

The equation giving the dependence of ΔH°_T on temperature thus has the form[22]

$$\Delta H^\circ_T = -21,523.4 - 2.51 T + 2.140 \times 10^{-3} T^2 - 2.495 \times 10^5 T^{-1}$$

From this equation, we get for ΔH° at a temperature of 600°C

$$\Delta H^\circ_{873} = -21,523.4 - 2.51 \times 873 + 2.140 \times 10^{-3} \times 873^2 - 2.495 \times 10^5 \times 873^{-1}$$
$$= -22,369.4 \text{ cal} = -93.59 \text{ kJ} \quad \bullet$$

Recently, rather extensive tables have become available which give the enthalpy function, $(\mathscr{H}^\circ_T - \mathscr{H}_0)$, as a function of T over a wide range of temperatures. The use of these tables makes the calculation much simpler and direct reference to the heat capacities unnecessary (see Section 9-11).

The calculation is more complicated when the temperatures of the reactants and products differ. For illustration let us consider the following reaction

$$a \, A_{T_1} + b \, B_{T_2} \rightarrow c \, C_{T_3} + d \, D_{T_4} \qquad \Delta H = x \qquad \text{(5-53)}$$

---

[22] The equation for the dependence of the standard enthalpy change of reaction on the temperature only applies strictly for the temperature range in which the equation $\Delta \mathscr{C}_P = f(T)$ is valid. In our case, this is the interval between 273 and 900 K.

Enthalpy is a function of state (its value is independent on the path) and therefore Eq. (5-53) may be written as a sum of three steps:

Step 1.   The reactants are transferred from their initial temperatures, T_1 and T_2, respectively, to the standard temperature of 298 K.

$$a \, A_{T_1} \rightarrow a \, A_{298}$$
$$b \, B_{T_2} \rightarrow b \, B_{298}$$

Step 2.   The reaction takes place at 298 K

$$a \, A_{298} + b \, B_{298} \rightarrow c \, C_{298} + d \, D_{298}$$

Step 3.   The products are transferred from 298 K to their final temperatures

$$c \, C_{298} \rightarrow c \, C_{T_3}$$
$$d \, D_{298} \rightarrow d \, D_{T_4}$$

The enthalpy changes corresponding to the three steps are the following:

$$\Delta H_1 = a \int_{T_1}^{298} \mathscr{C}_P(A) \, dT + b \int_{T_2}^{298} \mathscr{C}_P(B) \, dT \qquad (5\text{-}54)$$

$$\Delta H_2 = \Delta H_{298}^{\circ} \qquad (5\text{-}55)$$

$$\Delta H_3 = c \int_{298}^{T_3} \mathscr{C}_P(C) \, dT + d \int_{298}^{T_4} \mathscr{C}_P(D) \, dT \qquad (5\text{-}56)$$

The total enthalpy change of Eq. (5-53) is given as a sum of the three partial changes: Thus

$$\Delta H = \Delta H_1 + \Delta H_2 + \Delta H_3 = a \int_{T_1}^{298} \mathscr{C}_P(A) \, dT + b \int_{T_2}^{298} \mathscr{C}_P(B) \, dT$$

$$+ \Delta H_{298}^{\circ} + c \int_{298}^{T_3} \mathscr{C}_P(C) \, dT + d \int_{298}^{T_4} \mathscr{C}_P(D) \, dT \qquad (5\text{-}57)$$

Under adiabatic conditions,[23] the amount of heat exchanged between the reacting system and its surroundings is zero. Consequently $\Delta H = 0$. In such a case, all the heat produced by the reaction remains in the system; therefore, the temperature of the final products of an exothermic reaction will increase. The temperature of the products, called the final temperature of adiabatic reactions, may be evaluated from Eq. (5-57) by setting $\Delta H = 0$. Thus

$$c \int_{298}^{T_f} \mathscr{C}_P(C) \, dT + d \int_{298}^{T_f} \mathscr{C}_P(D) \, dT = -\Delta H_{298}^{\circ}$$

$$- a \int_{T_1}^{298} \mathscr{C}_P(A) \, dT - b \int_{T_2}^{298} \mathscr{C}_P(B) \, dT \qquad (5\text{-}58)$$

where T_f is the adiabatic temperature of the reaction.

[23] Fast reactions very often do not have enough time to exchange heat with their surroundings and may therefore be considered as adiabatic.

Due to the fact that the reaction is never completely adiabatic and that the final products may undergo decomposition at the relatively high final temperatures, the experimentally observed T_f is generally lower than T_f calculated from Eq. (5-58).

EXAMPLE 5-10

The following molar heat capacities[24] are given in calories per degree Kelvin per mole (cal K^{-1} mol^{-1}).

$$CH_4(g): \quad \mathscr{C}_P = 7.5 + 5 \times 10^{-3} T$$
$$O_2(g): \quad \mathscr{C}_P = 6.5 + 1 \times 10^{-3} T$$
$$N_2(g): \quad \mathscr{C}_P = 6.5 + 1 \times 10^{-3} T$$
$$H_2O(g): \quad \mathscr{C}_P = 8.15 + 5 \times 10^{-4} T$$
$$CO_2(g): \quad \mathscr{C}_P = 7.7 + 5.3 \times 10^{-3} T$$

Calculate the final adiabatic reaction temperature of the reaction

$$CH_4(g) + 2\,O_2(g) = CO_2(g) + 2\,H_2O(g) \qquad \Delta H^\circ_{298} = -165.2 \text{ kcal}$$

if the reactants, $CH_4(g)$ and air, respectively, are heated to 200°C before burning. Assume that 100% of the methane reacts.

Solution: Eq. (5-58) is applied to the calculation

$$n(CO_2) \int_{298}^{T_f} \mathscr{C}_P(CO_2)\,dT + n(H_2O) \int_{298}^{T_f} \mathscr{C}_P(H_2O)\,dT + n(N_2) \int_{298}^{T_f} \mathscr{C}_P(N_2)\,dT$$

$$= -\Delta H^\circ_{298} - n(CH_4) \int_{473}^{298} \mathscr{C}_P(CH_4)\,dT - n(O_2) \int_{473}^{298} \mathscr{C}_P(O_2)\,dT$$

$$- n(N_2) \int_{473}^{298} \mathscr{C}_P(N_2)\,dT \qquad \qquad \text{(a)}$$

Number of moles of reactants	Number of moles of products
$CH_4(g) = 1$	$CO_2(g) = 1$
$O_2(g) = 2$	$H_2O(g) = 2$
$N_2(g) = \dfrac{2 \times 0.79}{0.21} = 7.52$	$N_2(g) = 7.52$

Substituting these values and the respective molar heat capacities into Eq. (a) yields

$$\int_{298}^{T_f} (7.7 + 5.3 \times 10^{-3})\,dT + 2 \int_{298}^{T_f} (8.15 + 5 \times 10^{-4})\,dT$$

$$+ 7.52 \int_{298}^{T_f} (6.5 + 1 \times 10^{-3} T)\,dT$$

$$= -(-165,200) - \int_{473}^{298} (7.5 + 5 \times 10^{-3} T)\,dT$$

$$- 2 \int_{473}^{298} (6.5 + 1 \times 10^{-3} T)\,dT - 7.52 \int_{473}^{298} (6.5 + 1 \times 10^{-3} T)\,dT$$

[24] To make the calculation easier we use simpler equations than are given in Table 5-1.

Integration and rearrangement results in a quadratic equation of the form

$$6.96 \times 10^{-3} T_f^2 + 72.88 T_f - 200,957 = 0$$

The solution of this equation is $T_f = 2265 \text{ K}$.[25]   ●

5-13 APPLICATION OF THE FIRST LAW OF THERMODYNAMICS TO AN IDEAL GAS

In an ideal gas, which must obey the equation of state $PV = RT$, the molecules act independently and the potential energy arising from molecular interactions is therefore zero; the intermolecular forces have no effect on the thermodynamic properties of the gas. As a consequence of this, no work is required to separate the molecules during the course of expansion; and thus the change in volume at constant T leads to no change in the internal energy

$$\left(\frac{\partial U}{\partial V}\right)_T = 0 \qquad\qquad \textbf{(5-59)}$$

Direct attempts to study $(\partial U/\partial V)_T$ were first by Gay-Lussac in 1807 and Joule in 1844. Their experimental arrangement is shown in Figure 5-5. Two large vessels, A and B, are connected through a stopcock. In the initial state, A is filled with a gas at a pressure P, whereas B is evacuated. The apparatus is immersed in a large water bath and is allowed to equilibrate with the water at temperature T, which is read on the thermometer. The stopcock is opened and the gas expands to fill both containers

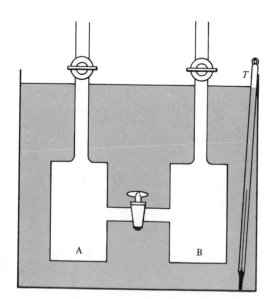

Figure 5-5 Experimental determination of the dependence of the internal energy on volume.

[25] Since the burning is incomplete and the products of burning CO_2 and H_2O, respectively, decompose to a certain extent at the final temperature, the actual temperature is lower by about 50–100°C.

uniformly. After equilibrium is established, the temperature of the water is read again. No temperature difference in the water before and after opening the stopcock was observed by Gay-Lussac or Joule.

The interpretation of this experiment is as follows: During the process of expansion, no work has been exchanged between the system and its surroundings, $dw' = 0$ (expansion into vacuum—free expansion). Since the temperature of the surroundings (the water) is unchanged, it follows that $dq = 0$, and thus $dU = dq + dw' = 0$. Consequently

$$dU = \left(\frac{\partial U}{\partial V}\right)_T dV + \left(\frac{\partial U}{\partial T}\right)_V dT = \left(\frac{\partial U}{\partial V}\right)_T dV = 0 \qquad \text{(5-60)}$$

Because $dV \neq 0$, it follows that

$$\left(\frac{\partial U}{\partial V}\right)_T = 0 \qquad \text{(5-61)}$$

However, Gay-Lussac's and Joule's device was not sensitive enough to detect small changes in temperature. A few years later (1852—1862), Joule used a more sensitive device and did observe some temperature changes. He came to the conclusion that the internal energy is volume dependent and that Eq. (5-61) represents only a limiting case, for pressures approaching zero. Thus, only the ideal gas obeys Joule's law exactly.[26]

On taking the differential of Eq. (5-27) with respect to V at constant temperature, it can be also proved that

$$\left(\frac{\partial H}{\partial V}\right)_T = \left(\frac{\partial U}{\partial V}\right)_T + \left[\frac{\partial (PV)}{\partial V}\right]_T$$
$$= \left(\frac{\partial U}{\partial V}\right)_T + \left[\frac{\partial (nRT)}{\partial V}\right]_T = 0 \qquad \text{(5-62)}$$

because, for an ideal gas, the first and second terms are equal to zero.

Similarly

$$\left(\frac{\partial U}{\partial P}\right)_T = \left(\frac{\partial H}{\partial P}\right)_T = 0 \qquad \text{(5-63)}$$

Thus for an ideal gas, the internal energy as well as the enthalpy is just a function of the temperature.

On taking the differential of the molar heat capacity at constant pressure with respect to the pressure at constant temperature, we have

$$\left(\frac{\partial \mathscr{C}_P}{\partial P}\right)_T = \frac{\partial}{\partial P}\left(\frac{\partial \mathscr{H}}{\partial T}\right)_P = \frac{\partial}{\partial T}\left(\frac{\partial \mathscr{H}}{\partial P}\right)_T = 0 \qquad \text{(5-64)}$$

since for an ideal gas $(\partial H/\partial P)_T = 0$. Similarly

$$\left(\frac{\partial \mathscr{C}_V}{\partial P}\right)_T = \left(\frac{\partial \mathscr{C}_P}{\partial V}\right)_T = \left(\frac{\partial \mathscr{C}_V}{\partial V}\right)_T = 0 \qquad \text{(5-65)}$$

[26] For quantitative proof, see page 147.

The molar heat capacities at constant pressure and at constant volume, respectively, of an ideal gas are also temperature dependent only.

According to Eq. (5-37), the difference in the molar heat capacities at constant pressure and at constant volume is

$$\mathscr{C}_P - \mathscr{C}_V = \left[\left(\frac{\partial \mathscr{U}}{\partial \mathscr{V}}\right)_T + P\right]\left(\frac{\partial \mathscr{V}}{\partial T}\right)_P \tag{5-37}$$

For an ideal gas, $(\partial \mathscr{U}/\partial \mathscr{V})_T = 0$, and $(\partial \mathscr{V}/\partial T)_P = \mathbf{R}/P$ and, thus, for an ideal gas Eq. (5-37) takes the form

$$\mathscr{C}_P - \mathscr{C}_V = \mathbf{R} \tag{5-66}$$

On the basis of what we have just said, we can discuss the work exchanged by an ideal gas with its surroundings during adiabatic and isothermal processes.

During an adiabatic process, the system does not exchange heat with its surroundings; $dq = 0$. Since the internal energy of an ideal gas depends on the temperature only, we may write for an adiabatic process

$$dU = \left(\frac{\partial U}{\partial T}\right)_V dT = n\mathscr{C}_V dT = dw' = -P_{\text{ext}} dV \tag{5-67}$$

If the expansion of the gas takes place against a constant external pressure (irreversible process), Eq. (5-67) takes the form

$$\Delta U = \int_{T_1}^{T_2} n\mathscr{C}_V dT = w' = -P_{\text{ext}}(V_2 - V_1) \tag{5-68}$$

If the expansion is reversible (at any time during the expansion $P_{\text{ext}} = P - dP$), then

$$\Delta U = \int_{T_1}^{T_2} n\mathscr{C}_V dT = w' = -\int_{V_1}^{V_2} P dV \tag{5-69a}$$

where the subscripts 1 and 2 indicate the initial and final states of the gas, respectively.

As is seen from the last two equations, work performed during an adiabatic process is at the expense of the internal energy of the system. A decrease in energy of an ideal gas is accompanied by a decrease in temperature; hence, as work is produced in an adiabatic process, the temperature of the system falls. It follows from Eqs. (5-68) and (5-69a), respectively, that, in order to find the work done during an adiabatic process, the initial and final properties of the system must be known.

First, the relationships between volume, temperature, and pressure for a reversible adiabatic process will be discussed.

Using the differential form of Eq. (5-69a)

$$n\mathscr{C}_V dT + P dV = 0 \tag{5-69b}$$

For the pressure, P, we may introduce $P = n\mathbf{R}T/V$, and consequently

$$n\mathscr{C}_V dT + \frac{n\mathbf{R}T}{V} dV = 0 \tag{5-70}$$

and, after separation of variables

$$\mathscr{C}_V \frac{dT}{T} + \mathbf{R} \frac{dV}{V} = 0 \tag{5-71}$$

Substituting $\mathscr{C}_P - \mathscr{C}_V$ for $\boldsymbol{R}$, and γ for $\mathscr{C}_P/\mathscr{C}_V$, we obtain

$$\frac{dT}{T} + (\gamma - 1)\frac{dV}{V} = 0 \tag{5-72}$$

Integration of (5-72) ($\mathscr{C}_V$ and hence γ is kept constant)[27] results in

$$\ln T + (\gamma - 1)\ln V = \ln \text{constant} \tag{5-73}$$

or

$$T_1 V_1^{\gamma-1} = T_2 V_2^{\gamma-1} = \text{constant} \tag{5-74}$$

γ is called Poisson's constant, and Eq. (5-74) represents one of the so-called Poisson's equations.

Using the ideal gas equation, $P\mathscr{V} = \boldsymbol{R}T$, we can transform Eq. (5-74), into the equivalent forms

$$T_1^{\gamma} P_1^{1-\gamma} = T_2^{\gamma} P_2^{1-\gamma} = \text{constant}' \tag{5-75}$$

$$P_1 V_1^{\gamma} = P_2 V_2^{\gamma} = \text{constant}'' \tag{5-76}$$

since

$$\frac{\mathscr{V}_1}{\mathscr{V}_2} = \frac{V_1/n}{V_2/n} = \frac{V_1}{V_2}$$

Eq. (5-76) may also be written as

$$P_1 \mathscr{V}_1^{\gamma} = P_2 \mathscr{V}_2^{\gamma} \tag{5-77}$$

In Figure 5-6b, Eq. (5-77) is compared with the well-known Boyle's isotherm, $PV = \text{constant}$. Since γ is always greater than unity ($\mathscr{C}_P > \mathscr{C}_V$), the two curves intersect. It follows from human experience with nature that an isotherm and an adiabatic may intersect only once (see page 134).

For an irreversible adiabatic process, Poisson's equations cannot be applied to evaluate the final properties of the system; the following procedure is recommended.

For a gas of constant $\mathscr{C}_V$, integration of Eq. (5-68) yields[28]

$$w' = n\mathscr{C}_V(T_2 - T_1) = P_2(V_1 - V_2)$$

$$= \frac{P_2 n\boldsymbol{R}T_1}{P_1} - \frac{n\boldsymbol{R}T_2 T_1}{T_1} \tag{5-78}$$

$$= n\boldsymbol{R}T_1\left[\frac{P_2}{P_1} - \frac{T_2}{T_1}\right]$$

Rearrangement of Eq. (5-78) results in an equation expressing the temperature at the end of an irreversible adiabatic expansion as a function of the initial temperature

$$T_2 = \left[1 - \frac{1 - P_2/P_1}{1 + \mathscr{C}_V/\boldsymbol{R}}\right]T_1 \tag{5-79}$$

---

[27] As a result of the fact that the number of degrees of freedom is affected by the temperature, $\mathscr{C}_V$ changes with temperature even in the case of an ideal gas (except monatomic). See page 590.

[28] For an irreversible process $P_2 = P_{\text{ext}}$.

Eq. (5-79) indicates that, as P_1 approaches infinity, T_2 approaches its lowest possible value

$$\lim_{P_1 \to \infty} T_2 = \left[1 - \frac{1}{1 + \mathscr{C}_V/\boldsymbol{R}} \right] T_1 = \frac{T_1}{\gamma} \qquad (5\text{-}80)$$

Consequently, the maximum adiabatic irreversible work is given as[29]

$$\lim_{P_1 \to \infty} w' = n\mathscr{C}_V(T_2 - T_1) = P_2(V_1 - V_2)$$

$$= -P_2 V_2 = -n\boldsymbol{R}T_2 \qquad (5\text{-}81)$$

since as $P_1 \to \infty$, $V_1 \to 0$.

Let us now replace the adiabatically functioning boundary between the system and its surroundings—the walls of the cylinder—with a boundary made from a good heat conducting material. Heat can, therefore, be exchanged between the gas in the cylinder and a constant heat bath, of high heat capacity, in which the cylinder containing the gas is immersed. The heat bath keeps the temperature constant during expansion and compression; heat is supplied to the gas during expansion and removed from the gas during compression. For such a process, $dT = 0$, and consequently

$$dU = n\mathscr{C}_V \, dT = dq + dw' = 0 \qquad (5\text{-}82)$$

Resulting from this

$$dq = -dw' = P_{\text{ext}} \, dV \qquad (5\text{-}83)$$

If P_{ext} is constant (irreversible process), integration results in[30]

$$q = -w' = P_{\text{ext}}(V_2 - V_1) \qquad (5\text{-}84)$$

If the process is carried out reversibly, ($P = P_{\text{ext}} + dP$ for expansion, and $P = P_{\text{ext}} - dP$ for compression), then, upon integration of Eq. (5-83), we obtain

$$q = -w' = \int_{V_1}^{V_2} (P \pm dP) \, dV = \int_{V_1}^{V_2} P \, dV \qquad (5\text{-}85)$$

since $dP \, dV$ is negligibly small.

Eq. (5-85) represents the maximum work of expansion and the minimum work of compression. It is worthwhile to mention that Eq. (5-85) is general and not restricted only to gases.

---

[29] The maximum adiabatic reversible work is (see Problem 33)

$$\lim_{P_1 \to \infty} w' = n\mathscr{C}_V T_1$$

[30] The maximum isothermal irreversible work is (see Problem 33)

$$\lim_{P_1 \to \infty} w' = n\boldsymbol{R}T$$

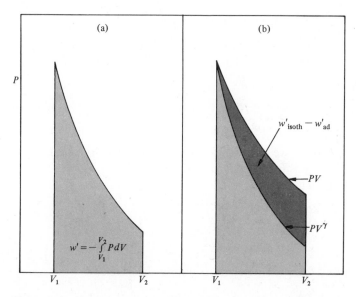

Figure 5-6 (a) Isothermal reversible P-V work. (b) Isothermal and adiabatic reversible P-V work.

In the case of an ideal gas, $P = n\mathbf{R}T/V$, and Eq. (5-85) takes the form

$$q = -w' = \int_{V_1}^{V_2} \frac{n\mathbf{R}T}{V} dV = n\mathbf{R}T \ln \frac{V_2}{V_1}$$

$$= n\mathbf{R}T \ln \frac{P_1}{P_2} \tag{5-86}$$

since, for an ideal gas at constant temperature[31] $V_2/V_1 = P_1/P_2$.

For compression, $V_2 < V_1$, $\ln(V_2/V_1)$ is negative, and the work w' is positive. In the case when $V_2 > V_1$ (expansion), work is done by the system, and w' is negative. This is in agreement with our convention.

The shaded area under the P-V isotherm in Figure 5-6a represents the maximum expansion work and the minimum compression work. Figure 5-6b represents the isothermal (the area under the PV curve) and adiabatic (the area under the PV^γ curve) reversible P-V work. As Figure 5-6b indicates, the work done by the system during the constant temperature process is greater than during the adiabatic process.

EXAMPLE 5-11

Two moles of hydrogen at standard conditions ($P_1 = 1$ atm, $t = 0°C$) are compressed adiabatically and reversibly to a volume of 10 liters. For hydrogen, $\gamma = 1.41$.

From the given data, calculate the pressure and temperature after the compression. In addition, calculate the work exchanged during the process.

---

[31] The maximum isothermal reversible work:

$$\lim_{P_1 \to \infty} w' = \infty$$

Solution: Making use of Eq. (5-77), we have, for the pressure after compression,

$$P_2 = P_1 \left(\frac{V_1}{V_2}\right)^\gamma = 1\left(\frac{44.8}{10}\right)^{1.41} = 8.30 \text{ atm}$$

Knowing P_2 and V_2, T_2 may be evaluated

$$T_2 = \frac{P_2 V_2}{nR} = \frac{8.30 \times 10}{2 \times 0.08206} = 505.7 \text{ K} = 232.5°C$$

Since $\mathscr{C}_V$ is not given directly,[32] we cannot use expression (5-69) to calculate the work. The following procedure is used instead:

  Differentiation of Eq. (5-77) results in

$$\gamma P V^{\gamma-1} \, dV + V^\gamma \, dP = 0$$
$$\gamma P \, dV + V \, dP = 0$$
$$V \, dP = -\gamma P \, dV$$

$V \, dP$ may also be expressed from the ideal gas equation, $PV = nRT$:

$$V \, dP = nR \, dT - P \, dV$$

On combining the last two equations, and by simple rearrangement, we obtain

$$-dw' = P \, dV = \frac{nR \, dT}{1 - \gamma}$$

and, upon integration

$$w' = -\frac{nR(T_2 - T_1)}{1 - \gamma} = -\frac{2 \times 1.987 \times (505.1 - 273.15)}{1 - 1.41}$$
$$= 2248 \text{ cal} = 9.406 \text{ kJ} \quad \bullet$$

--- *Problems*

1. Show that the expression $2xy \, dx + x^2 \, dy$ is an exact differential.
2. For an ideal gas the following equation may be written

$$dV = \frac{R}{P} \, dT - \frac{RT}{P^2} \, dP$$

  Prove that $P \, dV (\equiv -dw')$ is not a function of state.
3. Derive the equation

$$\left(\frac{\partial T}{\partial P}\right)_V \left(\frac{\partial P}{\partial V}\right)_T \left(\frac{\partial V}{\partial T}\right)_P = -1$$

[32] It can, however, be evaluated from the equipartition principle, see Section 4-2.

4. A gas system changes its volume by 1.2 liters at a constant external pressure of 30 atm. Find the work performed by the gas during expansion. Express this work in different energy units.

5. How much heat is evolved on the passage of 26.45 coulombs of electricity through a conductor, the potential drop being 2.432 V. (a) in joules (b) in ergs (c) in calories?

6. To heat 300 g of ethanol from 12.2°C to 18.9°C one needs 1.355 Whr. Calculate the mean specific and molar heat capacities of ethanol.

7. Calculate the change in internal energy when 20 g of ethanol are evaporated at the normal boiling point. The specific heat of vaporization of ethanol is 857.7 J g^{-1}, the specific volume of the ethanol vapor is 607 cm^3 g^{-1}. Neglect the volume of the liquid.

8. Two 10 g metal bars are made of lead and iron, respectively. Both are heated to 100°C and then placed in identical, insulated containers each containing 200 g of water at 25°C. In which container will there be a greater temperature rise? If it were desired to raise the temperature of the water in the two containers equally, by means of these two bars heated to the same temperature, what relative masses of water must be placed in the containers?

9. The specific heat capacity of carbon dioxide at constant pressure has the following values:

$t/°C$	0	100	200
c_P	0.8167	0.9075	0.9987 (J K^{-1} g^{-1})

Find (a) the equation giving the temperature dependence of the molar heat capacity at constant pressure, valid in the temperature range, 0–200°C; (b) the mean molar heat capacity at constant pressure in the temperature range 100–200°C; (c) the amount of heat taken from 1 mole of carbon dioxide when cooling from 200 to 0°C at constant pressure; and then (d) repeat the calculation by using the mean molar heat capacity. Estimate the error.

10. One mole of nitrogen is heated from 0 to 110°C under a pressure of 2 atm. Calculate: (a) the change in enthalpy; (b) the change in internal energy on the assumption that nitrogen behaves like an ideal gas. The mean specific heat of nitrogen at constant pressure is 1.021 J K^{-1} g^{-1}.

11. Derive the expression

$$\left(\frac{\partial \mathscr{U}}{\partial T}\right)_P = \mathscr{C}_P - P\left(\frac{\partial \mathscr{V}}{\partial T}\right)_P$$

12. Derive the expression

$$\mathscr{C}_P - \mathscr{C}_V = -\left(\frac{\partial P}{\partial T}\right)_{\mathscr{V}}\left[\left(\frac{\partial \mathscr{H}}{\partial P}\right)_T - \mathscr{V}\right]$$

13. Calculate the standard enthalpy of formation of gaseous sulfur trioxide from the standard enthalpy changes of the following reactions

$$PbO(s) + S(s) + \tfrac{3}{2}O_2(g) = PbSO_4(s) \qquad \Delta H° = -165,500 \text{ cal}$$
$$PbO(s) + H_2SO_4 \cdot 5\,H_2O(l) = PbSO_4(s) + 6\,H_2O(l) \qquad \Delta H° = -23,300 \text{ cal}$$
$$SO_3(g) + 6\,H_2O(l) = H_2SO_4 \cdot 5\,H_2O(l) \qquad \Delta H° = -41,100 \text{ cal}$$

14. The standard enthalpy of hydrogenation of gaseous propylene to propane is $-29{,}600$ cal and the standard enthalpy of combustion of propane is -530.6 kcal at 25°C. Calculate the standard enthalpy of combustion and standard enthalpy of formation of propylene if the following data are given:

$$CO_2(g): \quad (\Delta \mathcal{H}_f^\circ)_{298} = -94.03 \text{ kcal mol}^{-1}$$
$$H_2O(l): \quad (\Delta \mathcal{H}_f^\circ)_{298} = -68.32 \text{ kcal mol}^{-1}$$

15. Calculate $(\Delta H^\circ)_{298}$ for the reaction

$$C_2H_5OH(l) + CH_3COOH(l) = CH_3COOC_2H_5(l) + H_2O(l)$$

from the standard enthalpy of combustion of the constituents. Determine the accuracy of calculation if the enthalpies of combustion are measured with an accuracy of $\pm 0.1\%$. The enthalpies of combustion are as follows:

$$C_2H_5OH(l): \quad (\Delta \mathcal{H}_{comb}^\circ)_{298} = -326.66 \text{ kcal mol}^{-1}$$
$$CH_3COOH(l): \quad (\Delta \mathcal{H}_{comb}^\circ)_{298} = -208.00 \text{ kcal mol}^{-1}$$
$$CH_3COOC_2H_5(l): \quad (\Delta \mathcal{H}_{comb}^\circ)_{298} = -538.50 \text{ kcal mol}^{-1}$$

16. Calculate the heat evolved by burning 1 m^3 of methane at STP.

$$CH_4(g) + 2\,O_2(g) = CO_2(g) + 2\,H_2O(g)$$

Data:

$$CH_4(g): \quad (\Delta \mathcal{H}_f^\circ)_{298} = -17{,}889 \text{ cal mol}^{-1}$$
$$CO_2(g): \quad (\Delta \mathcal{H}_f^\circ)_{298} = -94{,}052 \text{ cal mol}^{-1}$$
$$H_2O(g): \quad (\Delta \mathcal{H}_f^\circ)_{298} = -57{,}798 \text{ cal mol}^{-1}$$

17. Magnetite (Fe_3O_4) is reduced to iron (a) by hydrogen, (b) by carbon monoxide. Calculate the change in enthalpy accompanying the reduction of 1 g of iron. Data:

$$Fe_3O_4(s): \quad (\Delta \mathcal{H}_f^\circ)_{298} = -267.00 \text{ kcal mol}^{-1}$$
$$CO(g): \quad (\Delta \mathcal{H}_f^\circ)_{298} = -26.417 \text{ kcal mol}^{-1}$$
$$H_2O(g): \quad (\Delta \mathcal{H}_f^\circ)_{298} = -57.798 \text{ kcal mol}^{-1}$$
$$CO_2(g): \quad (\Delta \mathcal{H}_f^\circ)_{298} = -94.052 \text{ kcal mol}^{-1}$$

18. Calculate the standard enthalpy of formation of ammonia from the bond energies and from the standard enthalpy change for the formation of elements in the gaseous state; given

$$(\Delta \mathcal{H}_{at}^\circ)_{N(g)} = 85.565 \text{ kcal g-atom}^{-1}$$
$$(\Delta \mathcal{H}_{at}^\circ)_{H(g)} = 52.089 \text{ kcal g-atom}^{-1}$$
$$\varepsilon_{N-H} = 83.3 \text{ kcal}$$

19. Calculate the standard enthalpy of formation of 2,2,3,3,-tetramethyl butane from the following data:

$$(\Delta\mathcal{H}^\circ_{at})_{H(g)} = 52.089 \text{ kcal g-atom}^{-1}$$

$$(\Delta\mathcal{H}^\circ_{at})_{C(g)} = 170.4 \text{ kcal g-atom}^{-1}$$

$$\varepsilon_{C-C} = 81.8 \text{ kcal}$$

$$\varepsilon_{C-H} = 99.5 \text{ kcal}$$

20. Prove that Eqs. (5-41) and (5-42) are equivalent.

21. Calculate the standard enthalpy change of reaction for the oxidation of ammonia at 427°C, which takes place according to the equation

$$4 \text{ NH}_3(g) + 5 \text{ O}_2(g) = 4 \text{ NO}(g) + 6 \text{ H}_2\text{O}(g)$$

if the following values are given at 293 K:

$$(\Delta\mathcal{H}^\circ_f)_{NH_3} = -11.00 \text{ kcal mol}^{-1}$$

$$(\Delta\mathcal{H}^\circ_f)_{NO} = -21.60 \text{ kcal mol}^{-1}$$

$$(\Delta\mathcal{H}^\circ_f)_{H_2O} = -57.85 \text{ kcal mol}^{-1}$$

$$(\mathscr{C}_P)_{O_2} = 6.26 + 2.746 \times 10^{-3} T - 0.770 \times 10^{-6} T^2 \text{ cal K}^{-1} \text{ mol}^{-1}$$

$$(\mathscr{C}_P)_{NO} = 6.21 + 2.436 \times 10^{-3} T - 0.612 \times 10^{-6} T^2 \text{ cal K}^{-1} \text{ mol}^{-1}$$

$$(\mathscr{C}_P)_{NH_3} = 5.92 + 8.963 \times 10^{-3} T - 1.764 \times 10^{-6} T^2 \text{ cal K}^{-1} \text{ mol}^{-1}$$

$$(\mathscr{C}_P)_{H_2O} = 6.89 + 3.283 \times 10^{-3} T - 0.343 \times 10^{-6} T^2 \text{ cal K}^{-1} \text{ mol}^{-1}$$

22. Express the standard enthalpy change as a function of the temperature for the reaction

$$CO_2(g) + H_2(g) = CO(g) + H_2O(g) \qquad (\Delta H^\circ)_{298} = 9814 \text{ cal}$$

Given:

$$(\mathscr{C}_P)_{CO_2} = 6.40 + 10.2 \times 10^{-3} T - 35.3 \times 10^{-7} T^2 \text{ cal K}^{-1} \text{ mol}^{-1}$$

$$(\mathscr{C}_P)_{H_2} = 6.95 - 0.2 \times 10^{-3} T + 4.8 \times 10^{-7} T^2 \text{ cal K}^{-1} \text{ mol}^{-1}$$

$$(\mathscr{C}_P)_{CO} = 6.34 + 1.8 \times 10^{-3} T - 2.8 \times 10^{-7} T^2 \text{ cal K}^{-1} \text{ mol}^{-1}$$

$$(\mathscr{C}_P)_{H_2O} = 7.19 + 2.4 \times 10^{-3} T + 2.1 \times 10^7 T^2 \text{ cal K}^{-1} \text{ mol}^{-1}$$

23. The standard enthalpy of formation of carbon dioxide is expressed as a function of the temperature as follows:

$$(\Delta\mathcal{H}^\circ_f)_T = -93,480 - 0.603 T - 0.675 \times 10^{-4} T^2 - 1.001 \times 10^5 T^{-1} \text{ cal mol}^{-1}$$

Find ΔC_P as a function of the temperature for this reaction.

24. The following are empirical equations giving the temperature dependence of the molar heat capacities:

$$(\mathscr{C}_P)_C = 1.1 + 4.80 \times 10^{-3} T - 1.20 \times 10^{-6} T^2 \text{ cal K}^{-1} \text{ mol}^{-1}$$

$$(\mathscr{C}_P)_{O_2} = 6.26 + 2.746 \times 10^{-3} T - 0.77 \times 10^{-6} T^2 \text{ cal K}^{-1} \text{ mol}^{-1}$$

$$(\mathscr{C}_P)_{CO} = 6.60 + 1.20 \times 10^{-3} T \text{cal K}^{-1} \text{ mol}^{-1}$$

Calculate the temperature at which the enthalpy of the reaction

$$C(s) + \tfrac{1}{2}O_2(g) = CO(g)$$

will be temperature independent.

25. Calculate the final adiabatic reaction temperature if methane is burned at constant pressure with the theoretical amount of air having an initial temperature 25°C. Assume that only 80% of the methane reacts. Use the data from Example 5-10.

26. The reaction

$$C(s) + \tfrac{1}{2}O_2(g) = CO(g)$$

is exothermic, and the reaction

$$C(s) + H_2O(g) = CO(g) + H_2(g)$$

is endothermic. A mixture of water vapor and air (20 mol % O_2 and 80 mol % N_2) can be forced over hot coke so that a constant temperature is maintained. Assuming that both reactions take place quantitatively, determine the ratio of the water vapor to the air if the mixture is heated to 100°C and if it is to maintain the temperature of the coke at 1000°C.
Given

$$CO(g): \quad (\Delta\mathcal{H}_f^\circ)_{291} = -26.62 \text{ kcal mol}^{-1}$$
$$H_2O(g): \quad (\Delta\mathcal{H}_f^\circ)_{291} = -58.00 \text{ kcal mol}^{-1}$$
$$(\mathscr{C}_P)_{N_2} = (\mathscr{C}_P)_{O_2} = (\mathscr{C}_P)_{CO} = 6.5 + 1.0 \times 10^{-3}T \text{ cal K}^{-1}\text{ mol}^{-1}$$
$$(\mathscr{C}_P)_{H_2O} = 8.15 + 5 \times 10^{-4}T \text{ cal K}^{-1}\text{ mol}^{-1}$$
$$(\mathscr{C}_P)_{H_2} = 6.62 + 0.81 \times 10^{-3}T \text{ cal K}^{-1}\text{ mol}^{-1}$$
$$(\mathscr{C}_P)_{C} = 1.1 + 4.8 \times 10^{-3}T \text{ cal K}^{-1}\text{ mol}^{-1}$$

27. In a pressure flask, there is an unknown gas. It is thought to be nitrogen or argon. A sample of the gas, taken at 25°C, was subjected to adiabatic expansion from 5 liters to 6 liters. The temperature decreased by 21°C. Which gas does the flask contain?

28. Air at a temperature of 25°C is adiabatically and reversibly compressed from a volume of 10 liters to 1 liter. Assuming ideal behavior and a value of 5 cal K^{-1} mol^{-1} for $\mathscr{C}_V$ for air, calculate the final temperature of the air.

29. Derive the following equations for the adiabatic reversible work done by an ideal gas

$$-w' = \frac{P_1 V_1}{\gamma - 1}\left[1 - \left(\frac{P_2}{P_1}\right)^{(\gamma-1)/\gamma}\right]$$
$$-w' = \frac{P_2 V_2}{\gamma - 1}\left[1 - \left(\frac{V_2}{V_1}\right)^{\gamma-1}\right]$$

30. Thirty kg of carbon dioxide were adiabatically compressed from 1 atm to 7 atm. Calculate: (a) the work of compression and (b) the final temperature, if the initial temperature was 15°C. For carbon dioxide $\gamma = 1.28$.

31. The limit of flammability of a certain mixture of gases was studied by means of adiabatic compression. During one experiment, an explosion occurred when the

volume was decreased from 377 to 30.2 cm^3. The initial conditions were 18°C and 1 atm. Find the temperature and pressure at the moment of explosion, $\gamma = 1.4$.

32. One mole of oxygen expands adiabatically against a constant external pressure of 1 atm (irreversible process) until both pressures are equal. Initial state: temperature 200°C, volume 20 liters. Calculate the work done during this process.

33. Derive equations for (a) the maximum reversible adiabatic work and (b) the maximum irreversible isothermal work.

34. Derive the following expression for the work done during an adiabatic process:

$$w' = \frac{P_2 V_2}{\gamma - 1}\left[1 - \frac{P_1 V_1}{P_2 V_2}\right]$$

35. The pressure of saturated water vapor at 25°C is 23.76 mm Hg. Assuming ideal behavior, calculate the work done during the reversible isothermal expansion of 100 g of water vapor to a pressure of 0.001 atm.

36. A steel bottle contains 20 liters of hydrogen at 27°C and a pressure of 50 atm. The gas expands isothermally and reversibly to 100 liters. Calculate: (a) how much heat had to be supplied; (b) the pressure of the hydrogen after expansion; (c) how much work was performed by the gas during expansion.

37. At a constant temperature of 100°C oxygen is transformed reversibly from a pressure of 340 mm Hg and volume of 2 liters to a pressure of 1 atm. Calculate the work of compression and the amount of heat evolved.

_____ *References and Recommended Reading*

[1] J. L. Dye: *J. Chem. Educ.*, **42**:193 (1965). Thermodynamics in the physical chemistry course.

[2] L. K. Nash: *J. Chem. Educ.*, **42**:64 (1965). Elementary chemical thermodynamics.

[3] Duncan MacRae: *J. Chem. Educ.*, **43**:586 (1966). The fundamental assumption of chemical thermodynamics.

[4] G. N. Lewis, M. Randall, K. S. Pitzer, and L. Brewer: *Thermodynamics.* McGraw-Hill, New York, 1961.

[5] S. M. Blinder: *J. Chem. Educ.*, **43**:85 (1966). Mathematical methods in elementary thermodynamics.

[6] A. J. Brainard: *J. Chem. Educ.*, **46**:104 (1969). The mathematical behavior of extensive and intensive properties of simple systems.

[7] Otto Redlich: *J. Chem. Educ.*, **47**:740 (1970). The so-called zeroth law of thermodynamics.

[8] M. L. Kremer: *J. Chem. Educ.*, **43**:583 (1966). The ideal gas equation and temperature scale.

[9] R. P. Bauman: *J. Chem. Educ.*, **43**:366 (1966). Can matter be converted to energy?

[10] M. L. McGlashan: *J. Chem. Educ.*, **43**:226 (1966). The use and misuse of the laws of thermodynamics.

[11] R. P. Bauman: *J. Chem. Educ.*, **41**:102 (1964). Work of compressing an ideal gas.

[12] S. D. Christian: *J. Chem. Educ.*, **42**:547 (1965). Reversible work.

[13] W. H. Eberhardt: *J. Chem. Educ.*, **41**:483 (1964). Reversible and irreversible work.

[14] Daniel Kivelson and Irwin Oppenheim: *J. Chem. Educ.*, **43**:233 (1966). Work in irreversible expansions.

[15] P. G. Wright: *J. Chem. Educ.*, **46**:380 (1969). Quantities of work in thermodynamic equations.

[16] N. O. Smith: *J. Chem. Educ.*, **42**:654 (1965). The difference between $\mathscr{C}_P$ and $\mathscr{C}_V$ for liquids and solids.

[17] B. E. Knox and H. B. Palmer: *J. Chem. Educ.*, **38**:292 (1961). The use and abuses of bond energies.

[18] M. W. Zemansky: *Heat and Thermodynamics.* McGraw-Hill Book Company, New York, 1957.

[19] G. T. Armstrong: *J. Chem. Educ.*, **41**:297 (1964). The calorimeter and its influence on the development of chemistry.

chapter 6

The Second Law of Thermodynamics

6-1 INTRODUCTION

The first law of thermodynamics is a necessary consequence of human experience with nature. There are, however, some natural phenomena for which the first law cannot account. Some of them were mentioned previously and some additional ones are discussed in this chapter.

A piece of paper can be burned in the presence of oxygen in a flame to form ashes and carbon dioxide. No one has ever succeeded in making a piece of paper and oxygen by mixing ashes and carbon dioxide in a flame. Eggs laid by a chicken can be converted with the help of necessary ingredients into a delicious omelet, but can you imagine any process that would reconvert an omelet into eggs capable of hatching into a chicken? It takes only a few seconds of shuffling to destroy the order of a new set of cards (as delivered by the manufacturers), however, the opposite process in which shuffling (even for a long time) would restore the complete order in the set of cards has very little chance.

Why natural processes proceed only in a certain direction is a question that cannot be answered by the first law. In 1878, M. Berthelot and J. Thomsen suggested that the enthalpy of reaction, ΔH, might be considered as a criterion for the spontaneous direction of chemical reactions. This principle, which would imply that no endothermic reaction can occur spontaneously, was rejected as incorrect [1, 2].

The question of why work can be completely transformed into heat but heat cannot be completely transformed into work[1] without additional changes in the universe also remains unanswered by the first law. A stirrer rotating in a water reservoir transforms electrical work into heat; the temperature of the reservoir rises as a result of friction. On the other hand, heat added to the reservoir does not cause the stirrer to rotate. When a baseball hit by a player collides with the ground and embeds itself in a patch of loose earth, its kinetic energy (as a whole) drops to zero. As a result of this the temperature of the ground and the temperature of the baseball (as a result of warming up, the molecules of the baseball will move faster) will increase. The temperature change is, however, very small because the heat capacity of the ground is very large. Thus, part of the work done by a batter on the baseball is converted eventually into heat. From our own experience we know that the reverse process is impossible: heating the baseball will not cause it to fly back to the batter. These examples prove that heat and work are not exactly equivalent forms of energy transfer.

---

[1] According to Eq. (5-86), the heat removed from a heat bath is completely transformed into work during an isothermal and reversible process. The transformation of heat into work stops, however, as soon as the pressure of the gas in the cylinder drops to the value of P_{ext}. In order to get more work out of the system, the gas must be returned to its original state by compression, which requires at least the same quantity of work as that gained during the expansion. The total work ($w'_{exp} + w'_{comp}$) is therefore zero.

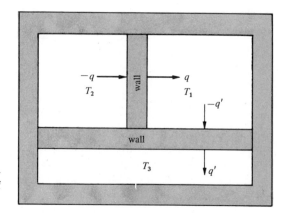

Figure 6-1 Heat flow in an isolated system. The principle of degradation of energy.

Let us now consider a completely isolated system divided by adiabatic walls into three parts (Figure 6-1). The temperatures in the three parts of the system, T_1, T_2 and T_3 are different; $T_2 > T_1 > T_3$. Heat cannot flow from one part of the system to the other parts because of the adiabatic walls separating them. By replacing the vertical adiabatic wall with a diathermal one, however, heat will be allowed to flow from the part having temperature T_2 to the part with temperature T_1. This heat flow can be partially transformed, by means of some type of device, into work. As soon as thermal equilibrium has been established between the two parts of the system ($T_2 = T_1$), the ability to do work disappears. Although the total energy of the system remains unchanged (isolated system), the system has lost its ability to do work; the energy of the system during the process of heat flow has therefore been degraded; its ability for doing work disappeared.

If, now, the third part of the isolated system, having a lower temperature than the already combined parts one and two, is brought in thermal contact with the combined parts, a further heat flow is observed in the system. The heat flow could again be transformed into work. This process would result in a further degradation of energy. The smaller the temperature differences between the different parts of the isolated system, the less the energy of the system is available for doing work. Consequently, at thermal equilibrium, when the temperature is the same in all parts of the system, no energy is available for doing work; the energy has been completely degraded.[2] The first law states the principle of conservation of energy without saying anything about the usefulness of the energy, and about its availability for doing work.

To answer the question about the amount of work a given system can do and the other questions raised in this chapter, it is necessary to go beyond the first law of thermodynamics. These questions are of such nature that an answer to any one of them would be an answer to all of them. Next, we shall discuss a very convenient way (however, not the most illuminating for a chemist) of answering these questions; we shall consider the transformation of heat into work in an ideal heat engine.

---

[2] Some scientists, beginning with Clausius, consider the universe as an isolated system and predict the so-called heat death of the universe. In this state the same amount of energy as ever would be present in the universe but, because no temperature differences will exist, none of the energy would be available for work.

6-2 CARNOT'S CYCLE

As has been mentioned before, there is a natural limitation to the convertibility of heat into work. In 1824, Sadi Carnot, a French engineer, recognized that in order to achieve maximum convertibility, the transformation of heat into work must be performed in a cyclic process, in which all the necessary intermediate steps are carried out reversibly. Such a reversibly functioning cycle is called the **Carnot cycle** [3, 4], and the engine operating on the basis of this hypothetical cycle is called **Carnot's heat engine**.

To derive an expression for the efficiency of Carnot's heat engine, we use the arrangement schematically shown in Figure 6-2. A cylinder equipped with a frictionless piston and filled with 1 mole of an ideal gas may be brought into contact with two heat reservoirs of constant temperatures, T_h and T_c respectively; $T_h > T_c$. The heat capacities of the two reservoirs must be very large, so that when heat is removed from or added to them by the system (the gas in the cylinder), the temperatures of the reservoirs remain unchanged. Carnot's engine performs in a cycle consisting of four

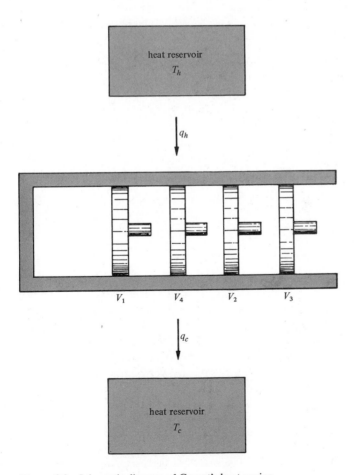

Figure 6-2 Schematic diagram of Carnot's heat engine.

reversible steps:

1. Isothermal expansion.
2. Adiabatic expansion.
3. Isothermal compression.
4. Adiabatic compression.

During the first step, the insulating layer is removed from the cylinder and the gas is brought into contact, through the diathermal walls of the cylinder, with the heat bath of higher temperature. According to Eq. (5-86) the gas removes heat, q_h, from the heat bath in order to compensate for the work done

$$q_h = -w_1' = RT_h \ln \frac{\mathcal{V}_2}{\mathcal{V}_1} \tag{6-1}$$

The cylinder is now covered by an insulating layer, and the gas is let to expand from $\mathcal{V}_2$ to $\mathcal{V}_3$ (see Figures 6-2 and 6-3). Following from Eq. (5-69a), we write for this reversible adiabatic process

$$w_2' = \int_{T_h}^{T_c} \mathcal{C}_V \, dT \tag{6-2}$$

The value of $\mathcal{V}_3$ is fixed by the requirement that during this adiabatic expansion the temperature of the gas in the cylinder drops from T_h to T_c.

The cylinder is now brought into contact with the lower temperature heat bath and the gas is compressed isothermally, at T_c, from $\mathcal{V}_3$ to a certain volume $\mathcal{V}_4$. This volume is not arbitrarily chosen; its value is fixed by the requirement that during the last step of the cycle—adiabatic compression—the gas in the cylinder must return to its initial state. The work added to the gas during the isothermal compression is immediately converted into heat; this heat, q_c, is given up to the lower temperature heat bath

$$q_c = -w_3' = RT_c \ln \frac{\mathcal{V}_4}{\mathcal{V}_3} \tag{6-3}$$

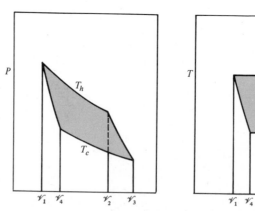

Figure 6-3 (a) Carnot's cycle in P–$\mathcal{V}$ coordinates. (b) Carnot's cycle in T–$\mathcal{V}$ coordinates.

Finally, after insulating the cylinder, the gas is brought back to its original state by a reversible adiabatic compression. The work done during this step is

$$w_4' = \int_{T_c}^{T_h} \mathscr{C}_V \, dT \tag{6-4}$$

According to the first law, the internal energy of a system remains unchanged during a cyclic process; thus

$$\oint d\mathscr{U} = \Delta \mathscr{U} = q + w'$$
$$= q_h + q_c + w_1' + w_2' + w_3' + w_4'$$
$$= q_h + q_c + w_1' + w_3'$$
$$= q_h + q_c + w' = 0 \tag{6-5}$$

since $w_2' = -w_4'$. Hence, $q_h + q_c = -w'$.

The efficiency of an engine is generally given as the ratio of the output to the input. Thus, for Carnot's engine, we have

$$\eta = -w'/q_h = \frac{q_h + q_c}{q_h}$$
$$= \frac{RT_h \ln (\mathscr{V}_2/\mathscr{V}_1) + RT_c \ln (\mathscr{V}_4/\mathscr{V}_3)}{RT_h \ln(\mathscr{V}_2/\mathscr{V}_1)} \tag{6-6}$$

This equation may be simplified considerably. Since points $(T_h, \mathscr{V}_2)$ and $(T_c, \mathscr{V}_3)$ lie on the same adiabatic, then, by Eq. (5-74)

$$T_h \mathscr{V}_2^{\gamma-1} = T_c \mathscr{V}_3^{\gamma-1} \tag{6-7}$$

Similarly

$$T_h \mathscr{V}_1^{\gamma-1} = T_c \mathscr{V}_4^{\gamma-1} \tag{6-8}$$

On dividing Eq. (6-7) by Eq. (6-8), we obtain $\mathscr{V}_2/\mathscr{V}_1 = \mathscr{V}_3/\mathscr{V}_4$, and consequently, $-\ln (\mathscr{V}_2/\mathscr{V}_1) = \ln (\mathscr{V}_4/\mathscr{V}_3)$. Substituting this result into Eq. (6-6), we finally obtain

$$\eta = \frac{q_h + q_c}{q_h} = \frac{RT_h \ln (\mathscr{V}_2/\mathscr{V}_1) - RT_c \ln (\mathscr{V}_2/\mathscr{V}_1)}{RT_h \ln (\mathscr{V}_2/\mathscr{V}_1)}$$
$$= (T_h - T_c)/T_h \tag{6-9}$$

As T_c can never be zero,[3] it is obvious that the efficiency of a heat engine is always less than 1; consequently, in a cyclic process, heat can never be completely transformed into work. Since Carnot's cycle is reversible, the work obtained is the maximum, and the efficiency is also a maximum. As no properties of the working fluid appear in Eq. (6-9), Carnot concluded that the efficiency of a reversibly working heat engine depends only on the temperature of the warmer and cooler heat baths, $\eta = \eta(T_h, T_c)$, and in no way on the properties of the working fluid.

Carnot's conclusion about the efficiency of a heat engine can be considered as the basic definition of the second law of thermodynamics [3, 6, 9, 10]. An equivalent definition was given in 1850 by W. Thomson (better known as Lord Kelvin): "It is

---

[3] For the unattainability of absolute zero, see Section 9-9.

impossible, by a cyclic process, to take heat from a reservoir and convert it into work without in the same operation, transferring some heat from a hot to a cold reservoir." In other words, it is impossible to construct any cyclic device that can extract work from an isothermal system. For this reason an adiabatic (PV^{γ}) and an isotherm (PV) may intersect only once (see page 119). If they intersected twice, a cycle would be formed, resulting in a heat engine working at one temperature only.

The warm upper layer and the cold deep water of the tropical oceans may serve as the two heat reservoirs of a heat engine. The power supplied by such a heat engine may be called "solar sea power," for the sun would rapidly restore to the upper layer the heat transferred to the cold deep water. Because solar sea power is essentially pollution free and is renewable, it has special relevance today. It will take some time for scientists to solve this very important problem satisfactorily. Perhaps, man will someday mine the ocean for heat to power his civilization, rather than mine the earth for fossil fuels [11].

6-3   THE THERMODYNAMIC TEMPERATURE SCALE

The temperature so far used in this text is based on the ideal gas temperature scale. We shall prove next that this temperature scale is identical to the so-called thermodynamic temperature scale, which is based on the second law of thermodynamics.[4]

According to the second law, the efficiency of a reversibly working heat engine is a function of the temperatures only, regardless of which temperature scale we use. As a consequence of this, we may rewrite Eq. (6-9)

$$\eta = \frac{q_h + q_c}{q_h} = \frac{T_h - T_c}{T_h} = \frac{\Theta_h - \Theta_c}{\Theta_h}$$

where Θ is the thermodynamic temperature.

Rearrangement yields

$$\frac{|q_h|}{|q_c|} = \frac{T_h}{T_c} = \frac{\Theta_h}{\Theta_c} \tag{6-10}$$

where $|q_h|$ is the absolute value of the heat exchanged between the working fluid and the hot reservoir, and $|q_c|$ is the absolute value of the heat exchanged between the working fluid and the cold reservoir.

Eq. (6-10) indicates that the temperature expressed on the ideal gas temperature scale, T, and the temperature expressed on the thermodynamic temperature scale, Θ, are related in some way to each other. The simplest relation may be written as[5]

$$T = k\Theta \tag{6-11}$$

---

[4] The thermodynamic temperature is an integration factor which, when divided into the inexact differential, dq, converts the latter to an exact differential of a state function, that is, dS.

[5] The final result is independent of the form of the relation applied.

Defining the standard temperature interval for the thermodynamic scale in the same way as for the ideal gas scale

$$\Theta_{100} - \Theta_0 = 100° \tag{6-12}$$

where Θ_{100} is the normal boiling point of water, and Θ_0 is the melting point of ice, both expressed on the thermodynamic temperature scale. We may set the proportionality constant, k, equal to unity, and thus

$$T = \Theta \tag{6-13}$$

The ideal gas temperature scale and the thermodynamic temperature scale are identical, which is what we set out to prove.

6-4 THE MATHEMATICAL FORMULATION OF THE SECOND LAW OF THERMODYNAMICS: ENTROPY

Although Eq. (6-10) has been derived for Carnot's cycle—a cycle consisting of two isothermal and two adiabatic steps—it also applies to any general cyclic process, as is evident from the following discussion.

A general cycle ABA, such as shown in Figure 6-4, may be divided by alternative isotherms and adiabatics into a great number of infinitesimally small Carnot cycles. The smaller the cycles, the more closely they follow the path of the general cycle. As seen from Figure 6-4 each isotherm and adiabatic inside the general cycle is shared by two small cycles, an expansion in one cycle and compression in the other one. All the work and heat terms arising from these small cycles cancel, and only the sum of q_i/T_i ($q_i = q_h$ or q_c of the ith cycle, see Eq. 6-9) arising from the outside (zigzag) isotherms must be considered. On the basis of this assumption, we may write for an

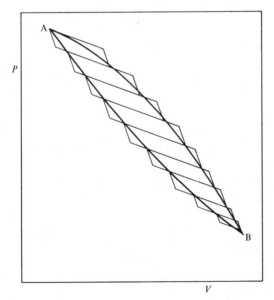

Figure 6-4 A general reversible cycle.

arbitrary cycle

$$q_1/T_1 + q_2/T_2 + q_3/T_3 + \cdots = \sum_i q_i/T_i = 0 \tag{6-14}$$

and in the limit of infinitesimally small reversible cycles

$$\sum_i \frac{d q_i}{T} = 0 \tag{6-15}$$

That the sums in Eqs. (6-14) and (6-15) are equal to zero follows from Eq. (6-10) (applicable to any small cycle), which can be rewritten as $q_h/T_h + q_c/T_c = 0$.

The final and original states in a cyclic process are identical, and thus, Eq. (6-15) may also be written in the form

$$\oint \frac{d q_{rev}}{T} = 0 \tag{6-16}$$

where $d q_{rev}$ indicates that Eq. (6-16) is applicable to reversible processes only. According to the criterion of Eq. (5-8), $d q_{rev}/T$ must be a function of state. Thus, Eq. (6-16) defines a new thermodynamic function

$$dS = \frac{d q_{rev}}{T} \tag{6–17}$$

This function was first introduced by Clausius in 1850 and called entropy.[6] Thus, the restriction put by nature on the convertibility of heat into work in a cyclic process leads to a new thermodynamic function, **entropy**.

In order to introduce entropy we have used the rather obscure irreversible flow of heat from a higher temperature heat bath to a lower temperature heat bath in an unrealistic operation with a heat engine. However, the reason for introducing entropy (and the second law of thermodynamics) is much more general that this spontaneous heat flow process indicates; it lies in the existence of irreversible (spontaneous) processes in nature. If such processes were absent from our everyday life, there would not be any need for entropy and the second of thermodynamics.

Eq. (6-17) represents the mathematical formulation of the second law of thermodynamics. It is worthwhile to observe that an exact differential like dS differs from a nonexact differential, dq, only by a multiplicative factor,[7] $1/T$. Temperature may therefore be defined as the reciprocal of the multiplicative factor converting reversibly exchanged heat to entropy.

6-5 ENTROPY AS A MEASURE OF SPONTANEITY OF PROCESSES

We shall now show that entropy is the function we are looking for to answer the questions raised earlier.

---

[6] The statistical treatment of the second law of thermodynamics is on page 585. A mathematically more impressive, but educationally and physically less illustrative way of introducing entropy is recommended by Caratheodory [10].

[7] This is also true for work

$$-(1/P) \, d w' = dV.$$

Let us consider the following process: A heat bath of temperature T_b, well but not perfectly insulated, loses heat slowly to its surroundings, of temperature T_s. During a certain (long) period of time, the bath loses a quantity of heat dq, whereas during the same period of time, the surroundings gain the quantity of heat dq. Although the total process is irreversible ($T_b > T_s$), one can assume that, because the exchange is so slow, the bath loses the heat reversibly and the surroundings gain the heat reversibly. Thus, the entropy changes in the bath and in the surroundings, respectively, may be expressed as (see Eq. 6-17)

$$dS_b = -dq_{rev}/T_b \qquad (6\text{-}18)$$

and

$$dS_s = dq_{rev}/T_s \qquad (6\text{-}19)$$

The total change in entropy of bath and surroundings is equal to the sum of the partial changes; thus

$$dS = dS_b + dS_s = -dq_{rev}/T_b + dq_{rev}/T_s$$
$$= dq_{rev}\left(\frac{1}{T_s} - \frac{1}{T_b}\right) > 0 \qquad (6\text{-}20)$$

since $T_b > T_s$, dS is positive.

During the irreversible process of heat flow, the total entropy of the system and its surroundings increases. This conclusion, drawn from the simple heat flow process, has a general validity. It applies to any irreversible process; only processes accompanied by an increase in entropy proceed spontaneously.[8]

By increasing the temperature of the surroundings slightly while keeping the temperature of the bath T_b constant, the entropy increase becomes smaller and smaller—the process is less and less irreversible—and, finally, in the limit when $T_b = T_s + dT$, the process becomes reversible, and

$$dS = 0 \qquad (6\text{-}21)$$

Generally, we write

$$dS = dS_{sys} + dS_{sur} \geqq 0 \qquad (6\text{-}22)$$

where >0 is valid for an irreversible—natural—process, whereas $=0$ is valid for a reversible process. Thus, entropy contains the answers to the questions why certain processes have a high chance to occur spontaneously, and why certain processes have practically no chance to occur.

The tendency of entropy to increase has also been expressed as a principle of the degradation of energy, by which energy becomes less available for doing work. During the spontaneous heat flow process in our isolated system (page 130), the energy has been conserved but degraded; consequently dS is positive.

According to the second law of thermodynamics a more organized system is less stable—less probable—than a less organized, more probable system. The direction of spontaneous processes is therefore toward greater stability of the system, Figure 6-5. Thus, the forward direction of time is unambiguously determined as the

---

[8] Clausius (1865): The entropy of the universe is increasing. "Die Entropie der Welt strebt einem Maximum zu."

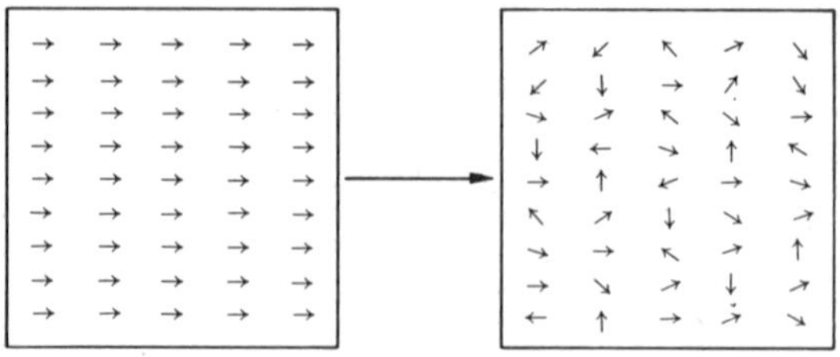

Figure 6-5 The process of spontaneous passage of a system from a more organized (less probable) state to a less organized (more probable) state.

direction in which order diminishes, stability increases and consequently entropy increases. For this reason, entropy can be described as "time's arrow" pointing from the past toward the future.

Entropy S and probability w,[9] (stability) are related by Boltzmann's equation

$$S = k \ln w \qquad (6\text{-}23)$$

where k is Boltzmann's constant. In this equation it is easy to find the answer to the question. Why is it so simple to destroy the order in a new set of cards? There are many ways of achieving a disorder in the cards (w is large), however, there is only one possibility ($w = l$) for restoring the cards into a completely ordered manner. The chaotic distribution of the cards is more probable than the orderly distribution. This is even more obvious when one deals with large systems such as one mole of gas ($N = 6 \times 10^{23}$ molecules). Entropy is a measure of the randomness of a system. Following from this, we may conclude that any naturally occurring process results in some permanent disorder in the universe, and no amount of effort can return it to its original state. Thus, the second law of thermodynamics predicts an inevitable state of complete disorder in the future, such as heat and death of the universe. There are, however, two questions to be answered before such a conclusion can be justified. (1) Is the universe an isolated system? (2) Can the second law be applied to the whole universe? In addition it must also be realized that the second law is statistical, stating merely the probabilities of various possible occurrences in nature.

6-6 THE COMBINED FIRST AND SECOND LAWS OF THERMODYNAMICS

For a closed system, on combining the first and second laws of thermodynamics (Eqs. 5-20 and 6-17), we obtain the following equation:

$$T \, dS \geq dq = dU + P \, dV \qquad (6\text{-}24)$$

Since $H = U + PV$ and $dH = dU + P \, dV + V \, dP$, Eq. (6-24) may also assume the form

$$T \, dS \geq dq = dH - V \, dP \qquad (6\text{-}25)$$

---

[9] This probability varies from one to infinity, in contrast to the mathematical probability, which varies only from zero to one. For more details see page 570.

Eqs. (6-24) and (6-25), respectively, represent the combined first and second laws of thermodynamics. Both equations, as they are presented here, apply for processes during which the only work exchanged between the system and its surroundings is P-V work.

6-7 THE HELMHOLTZ AND GIBBS FREE ENERGIES

According to Eq. (6-22), only processes accompanied by an increase in entropy can proceed spontaneously. The entropy increase referred to is that of the system plus its surroundings. As is evident from the equation $T\,dS \geq dU - dw$, this is equivalent to saying that the entropy of an isolated system (constant U and w) increases during an irreversible process and is unchanged in a reversible process. Thus

$$dS_{U,w} \geqq 0 \qquad\qquad (6\text{-}26)$$

where the subscript U,w refers to an isolated system.

Chemical processes are rarely studied under conditions of constant U and w. A criterion of spontaneity restricted to conditions of constant T,P, rather than to constant U,w, would be more preferable for chemists to work with. To derive such a criterion, let us start with the combined equation of the first and second laws of thermodynamics, written in the form

$$T\,dS \geq dq = dU - dw \qquad\qquad (6\text{-}24)$$

At constant T, $T\,dS = d(TS)$, and Eq. (6-24) may be rewritten as

$$-d(U - TS) \geqq -dw \qquad\qquad (6\text{-}27)$$

Since U, T, and S are functions of state, $U - TS$ defines a new thermodynamic function, called the Helmholtz free energy, A:

$$A = U - TS \qquad\qquad (6\text{-}28)$$

Hence, at constant temperature[10]

$$dA_T \leqq dw \qquad\qquad (6\text{-}29)$$

The total maximum work done on the system during an isothermal and reversible process equals the change in the Helmholtz free energy of the system. During a natural process (irreversible), $dA_T < dw$. When no work is exchanged between the system and its surroundings, $dw = 0$, and

$$dA_{T,w} \leqq 0 \qquad\qquad (6\text{-}30)$$

Since the only work we consider here is P-V work, Eq. (6-30) may also be written as

$$dA_{T,V} \leqq 0 \qquad\qquad (6\text{-}31)$$

As Eq. (6-31) indicates, only processes that are accompanied by a decrease in the Helmholtz free energy at constant T and V can proceed spontaneously. For a

---

[10] By multiplying the inequality by (-1), the sign greater than $>$ changes to the sign less than $<$.

reversible process under the same restrictions, the Helmholtz free energy does not change; $dA_{T,V} = 0$.

Let us now find the condition for spontaneity when temperature and pressure are held constant. This is the most important case, since many chemical processes are studied at constant T and P. Eq. (6-24) may also be written in the form

$$T\,dS \geq dq = dU - dw'' + P\,dV \qquad (6\text{-}24)$$

At constant temperature $T\,dS = d(TS)$, and at constant pressure $P\,dV = d(PV)$. Thus

$$-d(U + PV - TS) \geq -dw'' \qquad (6\text{-}32)$$

$$-d(H - TS) \geq -dw'' \qquad (6\text{-}33)$$

Since H, T, and S are functions of state, $H - TS$ defines a new thermodynamic function called the Gibbs free energy, G:

$$G = H - TS \qquad (6\text{-}34)$$

Hence, at constant T and P

$$-dG_{T,P} \geq -dw'' \qquad (6\text{-}35)$$

$$dG_{T,P} \leq dw'' \qquad (6\text{-}36)$$

The work[11] done on the system during a reversible process at constant temperature and pressure ($dw'' = dw + P\,dV$) equals the change in the Gibbs free energy. For an irreversible process, $dG < dw''$. When no such work is performed, then

$$dG_{T,P,w''} \leq 0 \qquad (6\text{-}37)$$

It follows that only processes of constant T and P accompanied by a decrease in the Gibbs free energy can proceed spontaneously. For a reversible process, $dG_{T,P,w''} = 0$.

In contrast to entropy, the Helmholtz free energy and the Gibbs free energy are criteria of spontaneity of processes that are independent of the surroundings.

On comparing Eq. (6-28) with Eq. (6-34), we find

$$G = A + PV \qquad (6\text{-}38)$$

$$\Delta G = \Delta A + \Delta(PV) \qquad (6\text{-}39)$$

It is also obvious that

$$H - U = G - A \qquad (6\text{-}40)$$

6-8 THE VARIATION OF THE HELMHOLTZ FREE ENERGY WITH VOLUME AND TEMPERATURE AND THE GIBBS FREE ENERGY WITH TEMPERATURE AND PRESSURE

For any substance of constant mass, A is most conveniently expressed in terms of the independent variables T and V. We thus have

$$dA = \left(\frac{\partial A}{\partial T}\right)_V dT + \left(\frac{\partial A}{\partial V}\right)_T dV \qquad (6\text{-}41)$$

Now, if we differentiate $A = U - TS$, and insert for dU from the first[12] and second laws, we get

---

[11] Work different from P-V work.

[12] Only P-V work is considered.

$$dA = dU - T\,dS - S\,dT = d\,q_{rev} - P\,dV - T\,dS - S\,dT$$
$$= T\,dS - P\,dV - T\,dS - S\,dT$$
$$= -P\,dV - S\,dT \tag{6-42}$$

On comparing Eq. (6-41) with Eq. (6-42), it follows that

$$\left(\frac{\partial A}{\partial T}\right)_V = -S \tag{6-43}$$

and

$$\left(\frac{\partial A}{\partial V}\right)_T = -P \tag{6-44}$$

An alternate equation for variation of A with T can be obtained as follows: differentiation of the ratio A/T with respect to T at constant V yields

$$\left[\frac{\partial(A/T)}{\partial T}\right]_V = \frac{T(\partial A/\partial T)_V - A}{T^2} = -\frac{(A + TS)}{T^2} = -\frac{U}{T^2} \tag{6-45}$$

since $(\partial A/\partial T)_V = -S$ and $A + TS = U$.

The Gibbs free energy is most conveniently expressed in terms of the independent variables T and P. In terms of these variables, we have for dG of any substance of constant mass

$$dG = \left(\frac{\partial G}{\partial T}\right)_P dT + \left(\frac{\partial G}{\partial P}\right)_T dP \tag{6-46}$$

Upon differentiation of Eq. (6-34) and introducing expressions for dH and dU from the first[13] and second laws, we obtain

$$dG = dH - T\,dS - S\,dT = dU + P\,dV + V\,dP - T\,dS - S\,dT$$
$$= T\,dS - P\,dV + P\,dV + V\,dP - T\,dS - S\,dT$$
$$= -S\,dT + V\,dP \tag{6-47}$$

From Eqs. (6-46) and (6-47) it follows that

$$\left(\frac{\partial G}{\partial T}\right)_P = -S \tag{6-48}$$

and

$$\left(\frac{\partial G}{\partial P}\right)_T = V \tag{6-49}$$

An alternate expression for the dependence of G on T follows from the differentiation of the quantity of G/T with respect to T at constant P

$$\left[\frac{\partial(G/T)}{\partial T}\right]_P = \frac{T(\partial G/\partial T)_P - G}{T^2} = -\frac{(G + TS)}{T^2} = -\frac{H}{T^2} \tag{6-50}$$

because $(\partial G/\partial T)_P = -S$ and $G + TS = H$.

---

[13] Again only P-V work is considered.

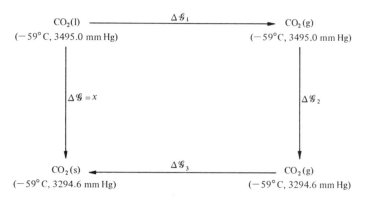

Figure 6-6 The isothermal change of $CO_2(l)$ from a pressure of 3495.0 mm Hg to $CO_2(s)$ at a pressure of 3294.6 mm Hg.

EXAMPLE 6-1

The saturated vapor pressure of supercooled CO_2 at $-59°C$ is 3495.0 mm Hg, and that of solid CO_2 at the same temperature is 3294.6 mm Hg. What is the change in the Gibbs free energy when 1 mole of supercooled liquid CO_2 solidifies at the given temperature? Assume ideal behavior of the vapor.

Solution: The Gibbs free energy is a function of state and its change thus depends only on the initial and final states of the system. We therefore replace the irreversible solidification of 1 mole of CO_2, under the given conditions, with three reversible partial processes that lead to the required final state (see Figure 6-6).

The first process represents the reversible vaporization of supercooled CO_2 at constant T and P. According to Eq. (6-37), $\Delta \mathcal{G}_1 = 0$.

The second step represents the expansion of gaseous CO_2 to a pressure equal to the vapor pressure of solid CO_2 at temperature $-59°C$. Thus

$$\Delta \mathcal{G}_2 = \int_{P_1}^{P_2} \mathcal{V} \, dP = \int_{P_1}^{P_2} \frac{RT}{P} \, dP$$

$$= RT \ln (P_2/P_1) = 1.987 \times 2.303 \times 214.2 \log \frac{3294.6}{3495.0}$$

$$= -25.13 \text{ cal} = -105.1 \text{ J.}$$

The third step expresses the reversible condensation of 1 mole of gaseous CO_2 at constant T and P. Following from Eq. (6-37), $\Delta \mathcal{G}_3 = 0$.

The total change, $\Delta \mathcal{G} = \Delta \mathcal{G}_1 + \Delta \mathcal{G}_2 + \Delta \mathcal{G}_3 = -25.13 \text{ cal} = -105.1 \text{ J}$, indicates that the process will take place spontaneously.   ●

6-9 EVALUATION OF ENTROPY CHANGES

Since entropy is a function of state, its value may be fixed if we specify any two convenient properties, such as T and V or T and P, for a homogeneous equilibrium system of fixed mass. Thus, for 1 mole, we write[14]

$$\mathscr{S} = f(T, \mathscr{V}) \tag{6-51}$$

and

$$d\mathscr{S} = \left(\frac{\partial \mathscr{S}}{\partial \mathscr{V}}\right)_T d\mathscr{V} + \left(\frac{\partial \mathscr{S}}{\partial T}\right)_V dT \tag{6-52}$$

Both coefficients, $(\partial \mathscr{S}/\partial \mathscr{V})_T$ as well as $(\partial \mathscr{S}/\partial T)_V$ may be expressed by easily measurable terms. Since, at constant volume

$$d q_{\text{rev}} = d\mathscr{U} = \mathscr{C}_V dT = T d\mathscr{S} \tag{6-53}$$

we have

$$\left(\frac{\partial \mathscr{S}}{\partial T}\right)_V = \frac{\mathscr{C}_V}{T} \tag{6-54}$$

On applying the reciprocity relationship (5-12) to Eq. (6-42), we obtain

$$\left(\frac{\partial \mathscr{S}}{\partial \mathscr{V}}\right)_T = \left(\frac{\partial P}{\partial T}\right)_V \tag{6-55}$$

Eq. (6-52) may thus be written in the form

$$d\mathscr{S} = \left(\frac{\partial P}{\partial T}\right)_V d\mathscr{V} + \frac{\mathscr{C}_V}{T} dT \tag{6-56}$$

When the heat capacity is known as a function of temperature and P-V-T data are available, entropy changes may be calculated. For example, for an ideal gas of constant $\mathscr{C}_V$, $(\partial P/\partial T)_V = R/\mathscr{V}$, and integration of Eq. (6-56) results in

$$\Delta \mathscr{S} = \mathscr{S}_2 - \mathscr{S}_1 = R \ln \frac{\mathscr{V}_2}{\mathscr{V}_1} + \mathscr{C}_V \ln \frac{T_2}{T_1} \tag{6-57}$$

The entropy may alternatively be expressed as a function of T and P, which is particularly convenient for constant pressure processes. In this case,

$$\mathscr{S} = f(T, P) \tag{6-58}$$

and

$$d\mathscr{S} = \left(\frac{\partial \mathscr{S}}{\partial T}\right)_P dT + \left(\frac{\partial \mathscr{S}}{\partial P}\right)_T dP \tag{6-59}$$

Since, at constant pressure, $d q_{\text{rev}} = d\mathscr{H} = \mathscr{C}_P dT = T d\mathscr{S}$, we have

$$\left(\frac{\partial \mathscr{S}}{\partial T}\right)_P = \frac{\mathscr{C}_P}{T} \tag{6-60}$$

On applying the reciprocity relationship to Eq. (6-47), we obtain

$$\left(\frac{\partial \mathscr{S}}{\partial P}\right)_T = -\left(\frac{\partial \mathscr{V}}{\partial T}\right)_P \tag{6-61}$$

[14] Since $\mathscr{S} = S/n$, $\mathscr{V} = V/n$ and $\mathscr{U} = U/n$, the derived equation may be easily modified from molar properties to system properties.

It then follows that

$$d\mathscr{S} = \frac{\mathscr{C}_P}{T}\, dT - \left(\frac{\partial \mathscr{V}}{\partial T}\right)_P dP \tag{6-62}$$

For an ideal gas of constant $\mathscr{C}_P$, integration yields

$$\Delta \mathscr{S} = \mathscr{S}_2 - \mathscr{S}_1 = \mathscr{C}_P \ln \frac{T_2}{P_1} - \boldsymbol{R} \ln \frac{P_2}{P_1} \tag{6-63}$$

Changes in entropy accompany not only variations in the temperature, pressure, or volume of a system but also physical transformations such as fusion, vaporization, or transitions from one crystalline form to another. All such processes may take place reversibly at constant temperature T and pressure P, and are accompanied by a change in enthalpy, ΔH. Their entropy changes are obtained from Eq. (6-17):

$$\int dS = \frac{1}{T}\int d\, q_{\mathrm{rev}} = \frac{1}{T}\int dH$$

and

$$\Delta S = \frac{\Delta H}{T} \tag{6-64}$$

Since the enthalpy of vaporization of water at the normal boiling point is 9717 cal mol^{-1}, the entropy accompanying the evaporation of 1 mole of water at $100°C$ and 1 atm is

$$\Delta \mathscr{S} = \mathscr{S}(g) - \mathscr{S}(l) = \frac{\Delta \mathscr{H}_{\mathrm{vap}}}{T_{\mathrm{nbp}}} = \frac{9717}{373.15}$$

$$= 26 \text{ cal K}^{-1} \text{ mol}^{-1} = 108.8 \text{ J K}^{-1} \text{ mol}^{-1}$$

where T_{nbp} is the normal boiling point of water.

Entropy changes may also be calculated for irreversible transitions, but the change is no longer given by the simple equation $dS = d\, q_{\mathrm{rev}}/T$. Nevertheless, entropy is a function of state, and, therefore, its change will depend only on the initial and final states of the system, regardless of whether the process is carried out reversibly or irreversibly. When dealing with an irreversible process, we break it up into a few reversible steps between the same original and final states, as can be seen from Example 6-2.

EXAMPLE 6-2

Calculate the entropy change at a constant pressure of 1 atm, when 2 moles of liquid ammonia at $-40°C$ are transformed to the gaseous state at $200°C$. The following data are given:

$$NH_3(l): \quad \mathscr{C}_P = 17.9 \text{ cal K}^{-1} \text{ mol}^{-1}$$
$$NH_3(g): \quad \mathscr{C}_P = 8.04 + 7.0 \times 10^{-4}T + 5.1 \times 10^{-5}T^2 \text{ cal K}^{-1} \text{ mol}^{-1}$$
$$(\Delta \mathscr{H}_{\mathrm{vap}})_{239.7} = 5560 \text{ cal mol}^{-1}$$
$$T_{\mathrm{nbp}}(NH_3) = 239.7 \text{ K}$$

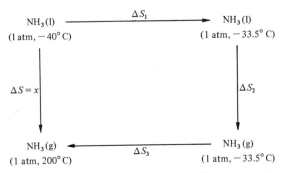

Figure 6-7 The change of liquid ammonia from a pressure of 1 atm and temperature of $-40°C$ to gaseous ammonia at a pressure of 1 atm and temperature of 200°C.

Solution: The process

$$NH_3(l) \ (1 \text{ atm}, -40°C) \rightarrow NH_3(g) \ (1 \text{ atm}, 200°C)$$

is irreversible. We therefore break it up into the following reversible steps (Figure 6-7).

1. ΔS_1 is the change in entropy during heating of 2 moles of liquid ammonia at constant pressure from -40 to $-33.5°C$

$$\Delta S_1 = 2 \int_{233.2}^{239.7} \frac{\mathscr{C}_P(l)}{T} dT = 2 \int_{233.2}^{239.7} 17.9 \frac{dT}{T}$$

$$= 2 \times 17.9 \times 2.303 \log \frac{239.7}{233.2} = 0.984 \text{ cal K}^{-1}$$

2. ΔS_2 is the change in entropy during evaporation of 2 moles of liquid ammonia at constant pressure and at constant temperature

$$\Delta S_2 = \frac{2(\Delta \mathscr{H}_{vap})}{T} = \frac{2(5560)}{239.7} = 46.391 \text{ cal K}^{-1}$$

3. ΔS_3 is the change in entropy accompanying the heating of 2 moles of gaseous ammonia from its boiling point to 200°C

$$\Delta S_3 = 2 \int_{239.7}^{473.2} \frac{\mathscr{C}_P(g)}{T} dT$$

$$= 2 \int_{239.7}^{473.2} \left(\frac{8.04 + 7 \times 10^{-4}T + 5.1 \times 10^{-5}T^2}{T} \right) dT$$

$$= 12.112 \text{ cal K}^{-1}$$

The total entropy change is

$$\Delta S = \Delta S_1 + \Delta S_2 + \Delta S_3 = 0.984 + 46.391 + 12.112$$
$$= 59.487 \text{ cal K}^{-1} = 248.894 \text{ J K}^{-1}$$

6-10 THE THERMODYNAMIC CRITERION OF EQUILIBRIUM

A spontaneous process taking place in an isolated system (constant U,w) is accompanied by an increase in entropy. In other words, the entropy of an isolated system continues to increase as long as spontaneous changes occur within it. When the changes cease, the system is in equilibrium and the entropy has reached a maximum value. Thus, the condition of equilibrium in an isolated system is that the entropy has a maximum value, and therefore

$$dS_{U,w} = 0 \tag{6-65}$$

Spontaneous processes of constant T and V are characterized by a decrease of the Helmholtz free energy. The Helmholtz free energy continues to decrease as long as spontaneous changes occur within the system. When the changes cease, the system is in equilibrium and the Helmholtz free energy has reached a minimum value; consequently

$$dA_{T,V} = 0 \tag{6-66}$$

Eq. (6-66) applies when only P-V work is considered.

Similarly, for constant T and P processes

$$dG_{T,P,w''} = 0 \tag{6-67}$$

At constant T and P, when only P-V work is considered, the condition of equilibrium is that G be a minimum and $dG_{P,T} = 0$.

6-11 SOME APPLICATIONS OF THE FIRST AND SECOND LAWS OF THERMODYNAMICS TO GASES

From Eq. (6-24), we derive

$$\left(\frac{\partial U}{\partial V}\right)_T = T\left(\frac{\partial S}{\partial V}\right)_T - P \tag{6-68}$$

According to Eq. (6-55), $(\partial S/\partial V)_T = (\partial P/\partial T)_V$, so that

$$\left(\frac{\partial U}{\partial V}\right)_T = T\left(\frac{\partial P}{\partial T}\right)_V - P \tag{6-69}$$

The property $(\partial U/\partial V)_T$, called the internal pressure, is very significant, especially for liquids and solids.[15]

Similarly, starting from Eq. (6-25) and making use of Eq. (6-61), we obtain

$$\left(\frac{\partial H}{\partial P}\right)_T = V - T\left(\frac{\partial V}{\partial T}\right)_P \tag{6-70}$$

Eqs. (6-69) and (6-70) are called the thermodynamic equations of state.

We are now in a position to prove quantitatively some of the conclusions made concerning the behavior of an ideal gas. For 1 mole of an ideal gas, $P\mathcal{V} = \mathbf{R}T$, and

---

[15] For a gas obeying the van der Waals equation it may be expressed as $a/\mathcal{V}^2$ (see page 45).

thus $(\partial P/\partial T)_V = R/\mathcal{V}$, and $(\partial\mathcal{V}/\partial T)_P = R/P$. Substitution of these expressions into Eqs. (6-69) and (6-70) yields the following relationships for an ideal gas

$$\left(\frac{\partial \mathcal{U}}{\partial \mathcal{V}}\right)_T = RT/\mathcal{V} - P = 0 \qquad (6\text{-}71)$$

$$\left(\frac{\partial \mathcal{H}}{\partial P}\right)_P = -RT/P + \mathcal{V} = 0 \qquad (6\text{-}72)$$

Thus, the internal pressure of an ideal gas is zero.[16]
The difference in the molar heat capacities at constant pressure and constant volume respectively is given by Eq. (5-37) as

$$\mathscr{C}_P - \mathscr{C}_V = \left[\left(\frac{\partial \mathcal{U}}{\partial \mathcal{V}}\right)_T + P\right]\left(\frac{\partial \mathcal{V}}{\partial T}\right)_P \qquad (5\text{-}37)$$

On substituting from Eq. (6-69) into Eq. (5-37), we obtain

$$\mathscr{C}_P - \mathscr{C}_V = T\left(\frac{\partial P}{\partial T}\right)_V \left(\frac{\partial \mathcal{V}}{\partial T}\right)_P \qquad (6\text{-}73)$$

and for 1 mole of an idea gas[17]

$$\mathscr{C}_P - \mathscr{C}_V = \frac{RT}{\mathcal{V}} \times \frac{R}{P} = \frac{RT}{P\mathcal{V}} R = R \qquad (6\text{-}74)$$

since $RT/P\mathcal{V} = 1$.
The variation of the molar heat capacity, $\mathscr{C}_P$, with pressure may also be obtained from the thermodynamic equation of state, Eq. (6-70)

$$\left(\frac{\partial \mathscr{C}_P}{\partial P}\right)_T = \frac{1}{\partial P}\left(\frac{\partial \mathcal{H}}{\partial T}\right)_P = \frac{1}{\partial T}\left(\frac{\partial \mathcal{H}}{\partial P}\right)_T$$

$$= \frac{1}{\partial T}\left[\mathcal{V} - T\left(\frac{\partial \mathcal{V}}{\partial T}\right)_P\right] = -T\left(\frac{\partial^2 \mathcal{V}}{\partial T^2}\right)_P \qquad (6\text{-}75)$$

For an ideal gas, $(\partial\mathcal{V}/\partial T)_P = R/P$, and

$$\left(\frac{\partial^2 \mathcal{V}}{\partial T^2}\right)_P = 0 \qquad (6\text{-}76)$$

Consequently[17]

$$\left(\frac{\partial \mathscr{C}_P}{\partial P}\right)_T = 0 \qquad (6\text{-}77)$$

Similarly, it can be shown that

$$\left(\frac{\partial \mathscr{C}_V}{\partial V}\right)_T = T\left(\frac{\partial^2 P}{\partial T^2}\right)_V \qquad (6\text{-}78)$$

---

[16] This is expected, since there are no intermolecular forces present in an ideal gas.
[17] For nonideal gases see Problems 13 and 14.

and again for an ideal gas

$$\left(\frac{\partial \mathscr{C}_V}{\partial V}\right)_T = 0 \tag{6-79}$$

Thus, the molar heat capacity of an ideal gas is temperature dependent only.

EXAMPLE 6-3

Prove that the ideal gas law may also be written in the form

$$\frac{\left(\dfrac{\partial \mathscr{U}}{\partial \mathscr{V}}\right)_S \left(\dfrac{\partial \mathscr{H}}{\partial P}\right)_S}{\left(\dfrac{\partial \mathscr{U}}{\partial \mathscr{S}}\right)_V} = -R \tag{a}$$

Solution: From the equations

$$d\mathscr{U} = T\,d\mathscr{S} - P\,d\mathscr{V}$$

and

$$d\mathscr{H} = T\,d\mathscr{S} + \mathscr{V}\,dP$$

it follows

$$\left(\frac{\partial \mathscr{U}}{\partial \mathscr{V}}\right)_S = -P \tag{b}$$

and

$$\left(\frac{\partial \mathscr{H}}{\partial P}\right)_S = \mathscr{V} \tag{c}$$

The denominator in (a) may be transformed to

$$\left(\frac{\partial \mathscr{U}}{\partial \mathscr{S}}\right)_V = \left(\frac{\partial \mathscr{U}}{\partial \mathscr{S}}\right)_V \left(\frac{\partial T}{\partial T}\right)_V = \left(\frac{\partial \mathscr{U}}{\partial T}\right)_V \left(\frac{\partial T}{\partial \mathscr{S}}\right)_V$$

$$= \mathscr{C}_V \frac{T}{\mathscr{C}_V} = T \tag{d}$$

since

$$\left(\frac{\partial \mathscr{U}}{\partial T}\right)_V = \mathscr{C}_V \qquad \text{and} \qquad \left(\frac{\partial T}{\partial \mathscr{S}}\right)_V = \frac{T}{\mathscr{C}_V}$$

Eqs. (b), (c), and (d) when substituted into Eq. (a) yield $P\mathscr{V}/T = R$.   ●

EXAMPLE 6-4

A certain gas behaves according to the equation

$$\frac{P\mathscr{V}}{RT} = 1 + \mathscr{B}P + \mathscr{C}P^2$$

where $\mathscr{B}$ and $\mathscr{C}$ are the second and third virial coefficients, respectively; these are functions of temperature only.

Evaluate $(\partial \mathscr{H}/\partial P)_T$ and $(\partial \mathscr{C}_P/\partial P)_T$.

Solution: On substituting the given equation of state into Eq. (6-70), we obtain

$$\left(\frac{\partial \mathcal{H}}{\partial P}\right)_T = \mathcal{V} - T\left(\frac{\partial \mathcal{V}}{\partial T}\right)_P$$

$$= RT/P + \mathcal{B}RT + \mathcal{C}RTP - T\frac{\partial}{\partial T}(RT/P + \mathcal{B}RT + \mathcal{C}RTP)$$

$$= RT/P + \mathcal{B}RT + \mathcal{C}RTP - RT/P - \mathcal{B}RT - RT^2\left(\frac{\partial \mathcal{B}}{\partial T}\right) - \mathcal{C}RTP$$

$$- RT^2 P\left(\frac{\partial \mathcal{C}}{\partial T}\right)$$

$$= -RT^2\left[\left(\frac{\partial \mathcal{B}}{\partial T}\right) + P\left(\frac{\partial \mathcal{C}}{\partial T}\right)\right]$$

According to Eq. (6-75)

$$\left(\frac{\partial \mathcal{C}_p}{\partial P}\right)_T = -T\left(\frac{\partial^2 \mathcal{V}}{\partial T^2}\right)_P = -2RT\left[\left(\frac{\partial \mathcal{B}}{\partial T}\right) + P\left(\frac{\partial \mathcal{C}}{\partial T}\right)\right] - RT^2\left[\left(\frac{\partial^2 \mathcal{B}}{\partial T^2}\right) + P\left(\frac{\partial^2 \mathcal{C}}{\partial T^2}\right)\right]$$

In order to evaluate the values of $(\partial \mathcal{H}/\partial P)_T$ and $(\partial \mathcal{C}_p/\partial P)_T$, the temperature dependence of both virial coefficients must be known.   ●

Important information about the thermodynamic properties of real gases may be obtained through the porous plug experiment of Joule and Thomson. In this experiment (Figure 6-8), a gas of volume V_1, temperature T_1, and pressure P_1, is forced by a piston through a porous barrier or plug D. During the irreversible process,[18] the conditions are kept as nearly adiabatic by perfect insulation as possible, so that $q = 0$.

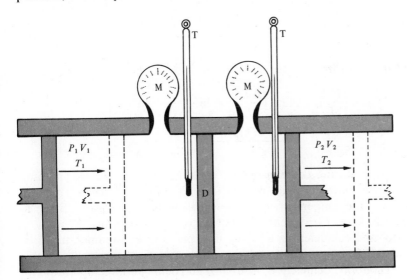

Figure 6-8 Schematic representation of Joule–Thomson experiment. M, manometer; T, thermometer; D, porous plug.

---

[18] Steady-state flow process.

If during the period of observation, 1 mole of gas is forced thorough the plug, the work done on the system by the left piston is $P_1 \mathcal{V}_1$, whereas that done by the system on the right piston is $P_2 \mathcal{V}_2$; thus, $w' = P_1 \mathcal{V}_1 - P_2 \mathcal{V}_2$. Since the process is adiabatic, we write

$$\Delta \mathcal{U} = \mathcal{U}_2 - \mathcal{U}_1 = w' = P_1 \mathcal{V}_1 - P_2 \mathcal{V}_2 \tag{6-80}$$

or

$$\mathcal{U}_2 + P_2 \mathcal{V}_2 = \mathcal{U}_1 + P_1 \mathcal{V}_1$$

$$\mathcal{H}_2 = \mathcal{H}_1 \tag{6-81}$$

Thus, the porous plug process is not only adiabatic but is also isenthalpic;[19] the enthalpy change is zero. Generally, the temperature of the gas changes during expansion through the porous plug, except at the inversion temperature. Since the total enthalpy change of the process is zero, the rate of the enthalpy change with temperature must be exactly counterbalanced by the rate of the enthalpy change with pressure. Thus

$$\left(\frac{\partial \mathcal{H}}{\partial P}\right)_T dP + \left(\frac{\partial \mathcal{H}}{\partial T}\right)_P dT = d\mathcal{H} = 0 \tag{6-82}$$

$$\mu \equiv \left(\frac{\partial T}{\partial P}\right)_H = -\frac{(\partial \mathcal{H}/\partial P)_T}{(\partial \mathcal{H}/\partial T)_P} = -\frac{(\partial \mathcal{H}/\partial P)_T}{\mathcal{C}_P} \tag{6-83}$$

The ratio, $(\partial T/\partial P)_H$, expressing the rate of temperature change with pressure at constant enthalpy, is called the Joule–Thomson coefficient, μ. It is a very important property of gases because it informs us whether a gas is cooled or heated up during expansion.

Substitution of Eq. (6-70) into Eq. (6-83) results in

$$\mu \equiv \left(\frac{\partial T}{\partial P}\right)_H = \frac{T(\partial \mathcal{V}/\partial T)_P - \mathcal{V}}{\mathcal{C}_P} \tag{6-84}$$

As is evident from Eq. (6-84), written in the form

$$\mu \equiv \left(\frac{\partial T}{\partial P}\right)_H = \lim_{\Delta P \to 0} \left(\frac{\Delta T}{\Delta P}\right)_H = \lim_{\Delta P \to 0} \left(\frac{T_2 - T_1}{P_2 - P_1}\right)$$

$$= \frac{T(\partial \mathcal{V}/\partial T)_P - \mathcal{V}}{\mathcal{C}_P} \tag{6-85}$$

the temperature rises during the expansion, $T_2 > T_1$, when the Joule–Thomson coefficient is negative, since P_2 is always less than P_1. When the Joule–Thomson coefficient is positive, the temperature drops during the expansion $T_1 > T_2$. Since $\mathcal{C}_P$ is always positive the sign of μ depends on the sign of the numerator.

The Joule–Thomson coefficient is a function of pressure and temperature (see Table 6-1 and also Figure 6-9). It also varies with the nature of the gas (see Table 6-2). The Joule–Thomson coefficient is positive at and below room temperature for all gases except hydrogen and helium, which have negative Joule–Thomson coeffi-

---

[19] This process is not a constant-pressure one, and therefore, it is not necessarily true that $dq = dH$.

Table 6-1 Joule–Thomson coefficients of nitrogen at different pressures and temperatures

P (atm)	$\dfrac{\mu}{(\text{K atm}^{-1})}$				
	$-150°C$	$-100°C$	$0°C$	$100°C$	$300°C$
1	1.266	0.6490	0.2656	0.1292	0.0139
20	1.125	0.5958	0.2494	0.1173	0.0096
33.5	0.1704	0.5494	0.2377	0.1100	0.0050
100	0.0202	0.2754	0.1679	0.0768	−0.0075
140	−0.0056	0.1373	0.1316	0.0582	−0.0129
200	−0.0284	0.0087	0.0891	0.0419	−0.0171

cients. The temperatures of these two gases rise during expansion at room temperature.

Every gas has a characteristic temperature at which its temperature does not change during expansion, $\mu = 0$. This temperaure is called the inversion temperature; it is characterized by the equation

$$T_{\text{inv}}\left(\frac{\partial \mathscr{V}}{\partial T}\right)_P = \mathscr{V} \tag{6-86}$$

and may be defined as the temperature at which the Joule–Thomson coefficient changes its sign.

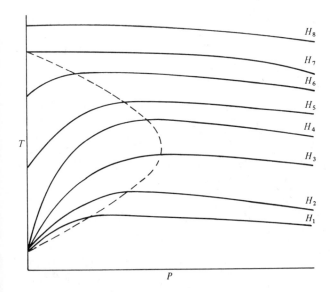

Figure 6-9 Isenthalps for a typical gas. The dashed line is the inversion curve $[\mu = (\partial T/\partial P)_H = 0]$.

Table 6-2 Joule–Thomson coefficients of helium, nitrogen, and carbon dioxide at 1 atm

$\dfrac{t}{(°C)}$	Helium $\dfrac{\mu}{(K\,atm^{-1})}$	Nitrogen $\dfrac{\mu}{(K\,atm^{-1})}$	Carbon Dioxide $\dfrac{\mu}{(K\,atm^{-1})}$
300	−0.0597	0.0139	0.2650
200	−0.0641	0.0558	0.3370
100	−0.0638	0.1291	0.6491
25	−0.0624	0.2216	1.0658
0	−0.0616	0.2655	1.2901
−100	−0.0584	0.6487	
−180	−0.0412	2.391	

The variation of temperature with pressure at constant enthalpies for a typical gas is shown in Figure 6-9. The solid curves represent isenthalps and the dashed line is the inversion temperature curve. From graphs similar to Figure 6-9, we note that the maximum inversion temperature of nitrogen is about 619 K, of hydrogen about 195 K, and of helium about 36 K. Pressures at these maxima are of the order of 400, 160, and 40 atm, respectively.

For an ideal gas, $(\partial \mathcal{V}/\partial T)_P = R/P$, and consequently

$$\mu = \frac{T(\partial \mathcal{V}/\partial T)_P - \mathcal{V}}{\mathscr{C}_P} = \frac{RT/P - \mathcal{V}}{\mathscr{C}_P} = 0 \qquad \textbf{(6-87)}$$

An ideal gas does not change its temperature during expansion.

The Joule–Thomson porous plug experiment has offered one of the most valuable methods for chilling gases to temperatures at which liquefaction occurs (below the critical temperature, see page 21). It is also utilized in many refrigeration devices.

EXAMPLE 6-5

Compare the experimentally found value of the Joule–Thomson coefficient of ammonia at 300°C and 40 atm, $\mu = 0.370$, with that calculated from the simplified van der Waals equation. *Given*: $\mathscr{C}_P = 11.0\,cal\,K^{-1}\,mol^{-1}$, and the constants of the van der Waals equation for ammonia are $a = 4.0\,liter^2 \cdot atm\,mol^{-2}$ and $b = 0.036\,liter\,mol^{-1}$.

Solution: For 1 mole of gas the van der Waals equation has the form

$$(P + a/\mathcal{V}^2)(\mathcal{V} - b) = RT$$

Multiplying out the terms in the parentheses, we obtain

$$P\mathcal{V} + a/\mathcal{V} - Pb - ab/\mathcal{V}^2 = RT$$

At moderate pressures, the last term on the left-hand side of the equation may be neglected (as $P \to 0$, $\mathcal{V} \to \infty$), and in the second term, $\mathcal{V}$ may be replaced by RT/P; thus

$$\mathcal{V} = RT/P - a/RT + b$$

and

$$\left(\frac{\partial \mathcal{V}}{\partial T}\right)_P \approx \frac{R}{P} + \frac{a}{RT^2}$$

Substituting this into Eq. (6-84), we obtain

$$\mu = \frac{T\left(\frac{\partial \mathcal{V}}{\partial T}\right)_P - \mathcal{V}}{\mathscr{C}_P} = \frac{RT/P + a/RT - RT/p + a/RT - b}{\mathscr{C}_P}$$

$$= \frac{2(a/RT) - b}{\mathscr{C}_P} = \frac{2[4.0/(0.082 \times 573)] - 0.036}{11.0 \times 0.0413} = 0.300$$

The calculated value differs appreciably from the experimentally found value. Using more accurate equations of state, one can expect better results. The value calculated from the experimentally found P-V-T data of ammonia is 0.338.

_____ *Problems*

1. Under a pressure of 35.4 atm water boils at a temperature of 244°C. Calculate the maximum fraction of heat that a steam engine can change into work if the steam in the boiler is under pressure of 35.4 atm and the temperature of the condenser is 50°C.
2. One mole of oxygen at an initial temperature of 100°C performs an ideal Carnot cycle. It expands isothermally to double its volume, and then expands adiabatically to triple the initial volume. It is then isothermally compressed to such a volume that, by the adiabatic compression which follows, it returns to the initial state. Calculate the work done by the gas in each part of the cycle, the total work done in the whole cycle, the efficiency of the cycle, and ΔU of the cycle ($\gamma = 1.4$).
3. Calculate the entropy change for the transformation of 1 g of water at 0°C into ice at the same temperature. The enthalpy of fusion of ice is 79.0 cal g^{-1}.
4. Calculate the change in entropy during the solidification of 1 mole of water (a) at 0°C, (b) at -20°C, given:

$$(\Delta \mathscr{H}_{\text{fusion}})_{273} = 1436.4 \text{ cal mol}^{-1}$$
$$H_2O(l): \mathscr{C}_P = 18 \text{ cal K}^{-1} \text{ mol}^{-1}$$
$$H_2O(s): \mathscr{C}_P = 8.60 \text{ cal K}^{-1} \text{ mol}^{-1}$$

5. The molar heat capacity of solid iodine in the temperature range 25–113.6°C (to the melting point of I_2) is given by the empirical equation

$$\mathscr{C}_P = 13.07 + 3.21 \times 10^{-4}(t - 25) \quad \text{cal } °C^{-1} \text{ mol}^{-1}$$

The enthalpy of fusion at the melting point is 3740 cal mol^{-1}. The molar heat capacity of liquid iodine is approximately constant, 19.5 cal K^{-1} mol^{-1}, and the enthalpy of vaporization at the boiling point (184°C) is 6100 cal mol^{-1}. What is the entropy change during the transformation of 1 mole of solid iodine from a temperature of 25°C to vapor at a temperature of 184°C and a pressure of 1 atm?

6. Calculate the entropy change involved in the isothermal reversible expansion of 5 moles of an ideal gas from a volume of 10 liters to a volume of 100 liters at 300 K.

7. The entropy of liquid ethanol is 38.4 cal K^{-1} mol^{-1} at 25°C. At this temperature, the pressure of its vapor is 59.0 mm Hg, and the enthalpy of vaporization is 10.19 kcal mol^{-1}. Calculate the entropy of ethanol vapor at a pressure of 1 atm and a temperature of 25°C. Assume ideal behavior of the ethanol vapor.

8. One mole of oxygen expands adiabatically against a constant external pressure of 1 atm until the pressures balance. The initial temperature is 200°C, the initial volume 20 liters. Calculate the change in entropy during this process, assuming that the oxygen behaves like an ideal gas.

9. One mole of an ideal gas at 300 K expands reversibly and isothermally from 20 liters to 40 liters. Calculate: (a) how much work (in calories) is performed by the gas; (b) how much the entropy of the gas changes. (c) Explain the fact that ΔS is not zero although this is a reversible process.

10. Evaluate the following coefficients for an ideal gas:

$$\left(\frac{\partial^2 P}{\partial T^2}\right)_V \qquad \left(\frac{\partial U}{\partial P}\right)_T \qquad \left(\frac{\partial P}{\partial V}\right)_S$$

11. Calculate the value of the Joule–Thomson coefficient for carbon monoxide at 25°C and 400 atm. The following are given: $(T/V)(\partial V/\partial T)_P = 0.984$, $\mathcal{V} = 76.25$ cm^3 mol^{-1}, and $\mathcal{C}_P = 8.91$ cal K^{-1} mol^{-1}.

12. Calculate the inversion temperature of nitrogen and methane. The van der Waals constants are

$$N_2: a = 1.390 \text{ liter}^2 \cdot \text{atm mol}^{-2}; b = 0.03914 \text{ liter mol}^{-1}$$
$$CH_4: a = 2.253 \text{ liter}^2 \cdot \text{atm mol}^{-2}; b = 0.04278 \text{ liter mol}^{-1}$$

13. The molar heat capacity of methane as a function of temperature at pressure of 1 atm is given by the equation

$$\mathcal{C}_P = 3.422 + 17.845 \times 10^{-3} T \text{ cal K}^{-1} \text{ mol}^{-1} (291\text{--}1500 \text{ K})$$

Using the equation

$$P\mathcal{V} = RT\left[1 + \frac{9}{128}\frac{PT_c}{P_c T}1 - 6\frac{T_c^2}{T^2}\right]$$

calculate the molar heat capacity of methane at a temperature of 100°C and a pressure of 100 atm. The critical temperature of methane, t_c, is −82.5°C, and the critical pressure, P_c, is 45.8 atm.

14. Using the equation from problem 13, calculate for methane the difference in the molar heat capacities at constant pressure and constant volume at a temperature of 100°C and pressure of 50 atm.

15. The molar volume of a gas as a function of P and T is given by the virial equation

$$\mathcal{V} = RT\left(\frac{1}{P} + \mathcal{B} + \mathcal{C}P\right)$$

Write an equation for the change of the Gibbs free energy when 1 mole of gas is compressed from 100 atm to 200 atm at constant temperature.

16. At a temperature of $-5°C$, the saturated vapor pressure of solid benzene is 17.1 mm Hg, and that of supercooled liquid benzene at the same temperature is 19.8 mm Hg. What is the change in the Gibbs free energy during the solidi-fication of 1 mole of supercooled benzene at the given temperature? Assume ideal behavior of the benzene vapors.

17. Prove that, for an ideal gas at constant temperature, the change in the Helmholtz free energy is equal to minus the change in the Gibbs free energy.

18. Calculate the change in the Helmholtz free energy for the evaporation of 1 mole of water at $100°C$. Assume ideal behavior of the water vapor and neglect the volume of the liquid phase.

19. One mole of water vapor is compressed reversibly to liquid water at the boiling point, $100°C$, The enthalpy of vaporization of water at $100°C$ and 1 atm is 539.7 cal g^{-1}. Calculate w', q, and each of the thermodynamic quantities: ΔH, ΔU, ΔG, ΔA, and ΔS.

20. Show that

$$\left(\frac{\partial A}{\partial V}\right)_S = S\left(\frac{\partial P}{\partial S}\right)_V - P$$

21. Given the equation of state $P\mathcal{V} = RT + \mathcal{B}P$, where $\mathcal{B}$ is temperature dependent, prove

$$\left(\frac{\partial \mathcal{U}}{\partial \mathcal{V}}\right)_T = \frac{RT^2}{(\mathcal{V}-\mathcal{B})^2}\frac{d\mathcal{B}}{dT}$$

Also obtain expressions for $(\partial \mathcal{S}/\partial \mathcal{V})_T$, $(\partial \mathcal{S}/\partial P)_T$ and $(\partial \mathcal{H}/\partial P)_T$.

22. Prove that for a van der Waals gas

$$\left(\frac{\partial \mathcal{U}}{\partial V}\right)_T = \frac{a}{\mathcal{V}^2}$$

23. Show that the molar heat capacities at constant pressure and constant volume, respectively, can be expressed as

$$\mathcal{C}_P = \frac{T\mathcal{V}\alpha^2\beta_S}{\beta_T(\beta_T-\beta_S)}$$

and

$$\mathcal{C}_V = \frac{T\mathcal{V}\alpha^2}{\beta_T-\beta_S}$$

where

$$\beta_S \equiv \left[-\left(\frac{\partial V}{\partial P}\right)_S\right]$$

is the adiabatic compressibility coefficient and

$$\beta_T \equiv \left[-\left(\frac{\partial V}{\partial P}\right)_T \right]$$

is the thermal compressibility coefficient and

$$\alpha \equiv \frac{1}{V}\left(\frac{\partial V}{\partial T}\right)_P$$

is the cubic expansion coefficient and S is the entropy.

———————————————— *References and Recommended Reading*

[1] G. W. J. Matthews: *J. Chem. Educ.*, **43**:476 (1966). Demonstrations of spontaneous endothermic reactions.

[2] G. E. MacWood and F. H. Verhoek: *J. Chem. Educ.*, **38**:334 (1961). How can you tell whether a reaction will occur?

[3] H. A. Bent: *J. Chem. Educ.*, **39**:491 (1962). The second law of thermodynamics.

[4] L. K. Nash: *J. Chem. Educ.*, **41**:368 (1964). The Carnot cycle and Maxwell's relation.

[5] F. J. Bockhoff: *J. Chem. Educ.*, **39**:340 (1962). A model for introducing the entropy concept.

[6] Milton Burton, *J. Chem. Educ.*, **39**:500 (1962). A second lecture on thermodynamics.

[7] A. H. Kalantar: *J. Chem. Educ.*, **43**:477 (1966). Nonideal gases and elementary thermodynamics.

[8] R. W. Hakala: *J. Chem. Educ.*, **41**:99 (1964). A method for relating thermodynamic first derivatives.

[9] H. A. Bent: *The Second Law.* Oxford Press, New York, 1965.

[10] H. Reiss: *Methods of Thermodynamics.* Blaisdell Publ., New York, 1965.

[11] C. Zener: *Physics Today*, **26**(1):48 (1973). Solar sea power.

chapter 7

Application of Thermodynamics to Phase Equilibria of Single Component Systems

7-1 INTRODUCTION

Before entering into a discussion of the application of thermodynamics to single and polycomponent multiphase systems, it is essential that the various terms employed be defined precisely.

A **phase** is a macroscopic homogeneous portion of a system. It may be a pure substance or a solution. A phase is a solid, liquid, or gas. For example, the three phases in which water exists at moderate and low pressures are ice, liquid water, and water vapor. Several pieces of ice of the same crystalline form are just one phase. Since gases are completely miscible[1] there can never be more than one gaseous phase present in the system; however, the system may contain a number of liquid and solid phases. Thermodynamics determines the conditions under which the possible phases of a system will be in equilibrium.

A phase may be composed of one or of several components. In thermodynamics, the number of components, c, is defined as the smallest number of distinct chemical substances required to form all of the phases present in the system at equilibrium. Algebraically the number of components, c, is given by $c = n - r$, where n is the total number of chemically distinct constituents in the system and r is equal to the number of independent chemical equations among the constituents. For example, in the aforementioned water system, $H_2O(s)-H_2O(l)-H_2O(g)$, each phase can be formed by using only one chemical substance, namely H_2O. Water is considered as a single component, even though species such as OH^-, H_3O^+, and hydrogen bonded polymers of varying complexity are undoubtedly present in water. In a nonreacting system, the number of components is always equal to the number of constituents present in the different phases of the system. As is obvious from the next two examples, the situation is more complicated in reacting systems.

If solid ammonium chloride is present in an evacuated vessel, the following reaction occurs:

$$NH_4Cl(s) = NH_3(g) + HCl(g)$$

Two phases exist in the reaction vessel: one gaseous (HCl and NH_3) and one solid (NH_4Cl). One component, NH_4Cl, is sufficient to form either of the two equilibrium phases. For an arbitrary nonstoichiometric amount of $NH_3 + HCl$ present in the vessel, two components are required to form the two phases because NH_4Cl alone can supply NH_3 and HCl in the stoichiometric ratio only.

For a similar reaction

$$CaCO_3(s) = CaO(s) + CO_2(g)$$

[1] Except at very high pressures when in some cases immiscibility is observed.

two components are required to form the three equilibrium phases: two solid ($CaCO_3$ and CaO) and one gaseous (CO_2) phase.

The **number of degrees of freedom** of a system, v, is the smallest number of intensive variables, such as pressure, temperature, concentration, and so on, that the experimentor can vary independently without changing the number of phases at equilibrium; expressed in a different way, v is the number of intensive variables that must be specified in order to define an equilibrium system completely. As long as pure water is present in only one phase, pressure and temperature may be changed independently without affecting the number of phases (see Figure 7-1). A single-component, one-phase system, has therefore two degrees of freedom. When two phases of a single component are present in the system at equilibrium, the number of degrees of freedom reduces to one. Water has, at any particular temperature, a specific vapor pressure. All three phases of a single component may exist in equilibrium only at a specific temperature and pressure, the triple point; the number of degrees of freedom at this point is zero.

The Gibbs phase rule provides a general relationship between the degrees of freedom of a system, v, the number of phases, p, and the number of components, c, and may be stated as

$$v = c + 2 - p \tag{7-1}$$

Eq. (7-1) is merely able to fix the number of variables involved, but it cannot give any quantitative relationship between these variables. For example, for pure liquid water in equilibrium with its vapor (two phases, one component), the number of degrees of freedom is one. Thus, either the pressure or the temperature, but never both, may vary independently; an arbitrary change in one of the intensive variables results in a certain, fixed change in the other. The quantitative relationship between these changes cannot be predicted from Eq. (7-1).

As will be seen further, Eq. (7-1) is valid only under the following assumptions (see page 188):

1. The system is in thermal and mechanical equilibrium. Consequently, the pressure and temperature are the same in all the phases of the system.
2. Surface contributions as well as contributions from any electric or magnetic field to the extensive properties of the system, are negligible.
3. Interphase surfaces are deformable and permeable to all components.

7-2 THE CONDITION OF EQUILIBRIUM IN SINGLE-COMPONENT MULTIPHASE SYSTEMS

Let us consider a closed system consisting of a substance in two phases. For example, a closed vessel partially filled with water; water is present in the vessel in two phases, the liquid and vapor phases. In this system a process may occur in which matter is transferred from one phase to the other. Each phase must therefore be considered as an open part of the system.

Since the Gibbs free energy is an extensive property, the Gibbs free energy of the system is the sum of the Gibbs free energies of the two phases, G' and G'':

$$dG = dG' + dG'' \tag{7-2}$$

The state of the individual phases may change with temperature, pressure, and number of moles, so that Eq. (7-2) can be written as[2]

$$dG = \left(\frac{\partial G'}{\partial T}\right)_{P,n'} dT + \left(\frac{\partial G'}{\partial P}\right)_{T,n'} dP + \left(\frac{\partial G'}{\partial n'}\right)_{T,P} dn'$$

$$+ \left(\frac{\partial G''}{\partial T}\right)_{P,n''} dT + \left(\frac{\partial G''}{\partial P}\right)_{T,n''} dP + \left(\frac{\partial G''}{\partial n''}\right)_{P,T} dn'' \qquad (7\text{-}3)$$

where n' and n'' are the number of moles in the respective phases. At a constant temperature and pressure, Eq. (7-3) reduces to the form

$$dG = \left(\frac{\partial G'}{\partial n'}\right)_{T,P} dn' + \left(\frac{\partial G''}{\partial n''}\right)_{T,P} dn'' \qquad (7\text{-}4)$$

The partial derivative, $(\partial G/\partial n)_{T,P}$, is the molar Gibbs free energy, $\mathcal{G}$, following from the fact that G is an extensive property; therefore, $G = n\mathcal{G}$. Thus

$$\mathcal{G}' = \left(\frac{\partial G'}{\partial n'}\right)_{T,P} \qquad (7\text{-}5)$$

$$\mathcal{G}'' = \left(\frac{\partial G''}{\partial n''}\right)_{T,P} \qquad (7\text{-}6)$$

Since we are dealing here with a closed system, the number of moles in both phases are related to each other by the equation

$$dn' = -dn'' \qquad (7\text{-}7)$$

On combining Eqs. (7-4), (7-5), (7-6), and (7-7) with the equilibrium condition, $dG = 0$ at constant T and P, (Eq. 6-67), we have

$$dG = (\mathcal{G}' - \mathcal{G}'')\, dn' = 0 \qquad (7\text{-}8)$$

dn' is not zero and therefore

$$\mathcal{G}' = \mathcal{G}'' \qquad (7\text{-}9)$$

Eq. (7-9) states that the system under consideration is in equilibrium if the molar Gibbs free energy of the substance constituting the system is the same in both phases. This conclusion can easily be extended to multiphase systems. The criterion of equilibrium represented by Eq. (7-9) has an advantage over Eq. (6-67) in that $\mathcal{G}$ is an intensive quantity like temperature and pressure and, as such, is independent of the size of the system or the quantity of substance present.

7-3 PHASE DIAGRAM OF A SINGLE-COMPONENT SYSTEM

In a single component system, the following phases may exist in mutual equilibrium:

    1.   solid $\underset{\text{freezing}}{\overset{\text{melting}}{\rightleftharpoons}}$ liquid

---

[2] Assuming that the phases are so large and of such shape that the surface contribution to the Gibbs free energy may be neglected in comparison with the bulk volume effects.

2. $\text{solid}(\alpha) \underset{}{\overset{\text{solid phase transition}}{\rightleftharpoons}} \text{solid}(\beta)$

3. $\text{liquid} \underset{\text{condensation}}{\overset{\text{evaporation}}{\rightleftharpoons}} \text{vapor}$

4. $\text{solid} \underset{\text{condensation}}{\overset{\text{sublimation}}{\rightleftharpoons}} \text{vapor}$

5. $\text{solid} \rightleftharpoons \text{liquid} \rightleftharpoons \text{vapor}$

These phase transitions, except 2, are shown for water in Figure 7-1.

    The labeled areas between the curves in Figure 7-1 denote conditions under which only one phase exists. Within these areas, the pressure and temperature can be independently varied without the appearance of a second phase; the limits of these variations being the boundaries of the labeled areas. In these areas, the system has two degrees of freedom.

    While two phases coexist in mutual equilibrium, the variation must be confined to the curves alone. A small independent change in one of the intensive properties, pressure or temperature, requires a certain, dependent, change in the other variable in order to maintain the two phases at equilibrium. Curves therefore represent systems with one degree of freedom only. Curve AC denotes equilibrium between the solid and vapor phases; it is the sublimation curve. Curve CD defines equilibrium between the liquid and vapor phases—the vaporization curve; this curve ends at the critical point (see page 22), which, for water, is at 374°C and 218 atm. At temperatures higher than that of point D, condensation does not occur at any

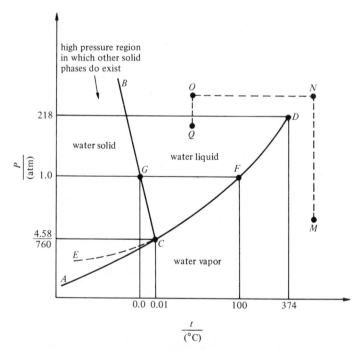

Figure 7-1   Phase diagram of the water system (not drawn to scale).

pressure. Point F represents the normal boiling point, the temperature at which the vapor pressure is equal to 760 mm Hg. The extension of curve CD down to E represents the metastable equilibrium between supercooled liquid water and its vapor. Finally, curve CB represents equilibrium between the solid and liquid phases—the melting curve. As is evident from this curve, the melting point of ice is lowered with increasing pressure. This behavior is anomalous for water; in the great majority of cases the melting point increases slightly with pressure.

At point C, the triple point, all three phases coexist in mutual equilibrium.[3] At the temperature of the triple point, ice and water exhibit the same vapor·pressure, 4.58 mm Hg. This point is a fundamental constant for the water system. The temperature of the triple point was set equal to 273.16 K and is now used as a reference point for the international practical temperature scale (see page 90). In Figure 7-1 the triple point temperature differs just slightly from the melting point of ice, G. This follows from the fact that melting points are only very slightly affected by the pressure.

It can be concluded from Figure 7-1 that it is possible to change one phase into another without crossing the equilibrium curves. By isothermal compression of the vapor from state M to state N, followed first by an isobaric lowering of the temperature to point O, and next, by an isothermal lowering of the pressure to point Q, we have succeeded in transforming the vapor into liquid without crossing the equilibrium curves.

At very high pressures, the diagram of the water system is more complex; ice may exist in different crystalline modifications not shown in Figure 7-1. Such a phenomenon, generally called **polymorphism**, is also observed in many other cases. There are two different kinds of polymorphism: enantiotropy and monotropy.

Figure 7-2 shows the phase diagram for sulfur. This graph indicates that rhombic sulfur is the variety which is stable at room temperature; slow heating converts it to monoclinic sulfur at 95.4°C. Above this temperature, monoclinic sulfur is stable up to 119.3°C, at which point it melts.

At both 95.4 and 119.3°C, there are three rather than two phases in mutual equilibrium. Correctly speaking these temperatures represent triple points rather than melting points. The normal boiling point of sulfur is 444.6°C.

Rapid heating allows rhombic sulfur to persist in metastable equilibrium up to its melting point of 113°C.[4] As seen from Figure 7-2, at the temperature of 95.4°C[4] both solid forms coexist in equilibrium with sulfur vapor. This temperature, called the transition temperature, lies below the melting points of both rhombic and monoclinic sulfur. Thus, the transition may go in either direction. This reversible type of phase transition is characteristic of the polymorphism called **enantiotropy**.

On the other hand, there are many substances where the transition is irreversible—possible in one direction only. The melting point of such a substance is below the transition temperature so that the substance melts before reaching the transition temperature. Phosphorus exhibits this type of polymorphism. The well-known crystalline forms of phosphorus are red and white phosphorus. The white

---

[3] Since ice can exist at high pressures in more than one crystalline form, the water system exhibits more than one triple point. Vapor, however, is present in only one of them.

[4] At these temperatures again, there are three rather than two phases in mutual equilibrium.

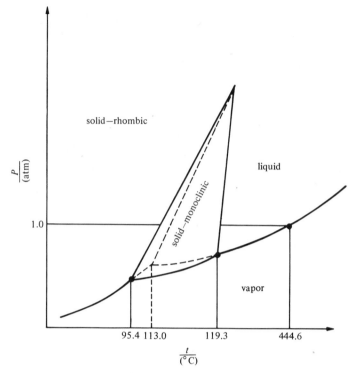

Figure 7-2   Phase diagram of the sulfur system (not drawn to scale).

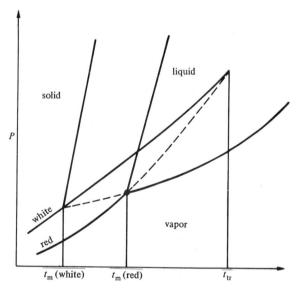

Figure 7-3   Schematic graphical representation of the phosphorus system: t_m (red), melting point of the red modification; t_m (white), melting point of the white modification; t_{tr}, hypothetical transition temperature.

modification is metastable at all pressures and temperatures; therefore it has a higher vapor pressure than red phosphorus. The higher the vapor pressure, the larger the Gibbs free energy and, thus, the less stable is the phase. As seen from Figure 7-3, the transition temperature lies above the melting point; therefore, the transition occurs in only one direction, from white to red phosphorus. This type of phase transition is characteristic of the polymorphism called **monotropy**.

7-4 THE CLAUSIUS–CLAPEYRON EQUATION

As previously stated, the phase rule is unable to give any information on the shape of the equilibrium curves in the phase diagrams. Only a more rigorous thermodynamic treatment can give quantitative expressions for the dependence of the equilibrium pressure on temperature for the different phases.

If considered independently of the other phases, the state of a single component depends on the temperature and pressure, which can change independently of each other. The equilibrium condition, that is, when both phases have equal molar Gibbs free energies, however, makes the pressure dependent on temperature. At a certain temperature T and pressure P, equilibrium is characterized by a certain value of G. If we now change the temperature by dT, from T to $T+dT$, then, if the system is to remain at equilibrium, the equilibrium pressure must also be changed by a certain value dP, from P to $P+dP$. In this new equilibrium state, the molar Gibbs free energies in both phases are still equal; however, they now have a value different from that in the preceding equilibrium state.

At T and P

$$\mathscr{G}' = \mathscr{G}'' \tag{7-9}$$

and at $(T+dT)$ and $(P+dP)$

$$\mathscr{G}' + d\mathscr{G}' = \mathscr{G}'' + d\mathscr{G}'' \tag{7-10}$$

On subtracting, we obtain

$$d\mathscr{G}' = d\mathscr{G}'' \tag{7-11}$$

Use of Eq. (6-47) results in

$$-\mathscr{S}'\,dT + \mathscr{V}'\,dP = -\mathscr{S}''\,dT + \mathscr{V}''\,dP \tag{7-12}$$

and upon simple rearrangement of terms, we obtain

$$\frac{dP}{dT} = \frac{\mathscr{S}'' - \mathscr{S}'}{\mathscr{V}'' - \mathscr{V}'} = \frac{\Delta\mathscr{S}}{\Delta\mathscr{V}} \tag{7-13}$$

where $\Delta\mathscr{S}$ and $\Delta\mathscr{V}$ denote the change in entropy and volume, respectively, when a certain quantity (in this case 1 mole) of a substance is transferred from one phase to another. As the transfer is an isothermal reversible process, the following relation may be used

$$\Delta\mathscr{S} = \frac{\Delta\mathscr{H}}{T} \tag{6-64}$$

where $\Delta\mathcal{H}$ is the corresponding molar enthalpy change; enthalpy of vaporization, enthalpy of sublimation, enthalpy of transition, or enthalpy of fusion.

On combining Eq. (7-13) with Eq. (6-64), we have

$$\frac{dP}{dT} = \frac{\Delta\mathcal{H}}{T\Delta\mathcal{V}} \tag{7-14}$$

Eq. (7-14), expressing the rate of change of the equilibrium pressure with temperature, is called the **Clapeyron equation**.

Since the foregoing derivation does not depend on any assumptions concerning the nature of the two coexisting phases, it may readily be used for any kind of phase equilibrium, except second and higher order phase transitions, in a single-component system (see Section 7-8).

The change of the molar volume at melting, $(\mathcal{V}(l) - \mathcal{V}(s))$, as well as at solid phase transitions, $(\mathcal{V}^\beta(s) - \mathcal{V}^\alpha(s))$, is small; therefore, the melting point and the solid phase transition temperature is only slightly dependent on pressure as seen from Eq. (7-14). For water, $\mathcal{V}(s) > \mathcal{V}(l)$, and thus dT/dP is negative.

For vaporization and sublimation, the molar volume of the vapor phase, $\mathcal{V}(g)$, can be expressed by any of the equations of state discussed previously (see Section 1-6). Using the virial equation (Eq. 1-24)

$$\mathcal{V}(g) = \boldsymbol{R}T/P + \mathcal{B}$$

Eq. (7-14) assumes the forms

$$\Delta\mathcal{H}_{vap} = [\boldsymbol{R}T + P(\mathcal{B} - \mathcal{V}(l))]T\frac{d\ln P}{dT} \tag{7-15}$$

$$\Delta\mathcal{H}_{subl} = [\boldsymbol{R}T + P_s(\mathcal{B} - \mathcal{V}(s))]T\frac{d\ln P_s}{dT} \tag{7-16}$$

where P_s denotes the sublimation pressure and $\mathcal{B}$ is the second virial coefficient of the vapor phase.

Note that the enthalpy of vaporization, $\Delta\mathcal{H}_{vap}$, enthalpy of sublimation, $\Delta\mathcal{H}_{subl}$, and enthalpy of fusion, $\Delta\mathcal{H}_{fus}$, are interrelated by the equation

$$\Delta\mathcal{H}_{subl} = \Delta\mathcal{H}_{fus} + \Delta\mathcal{H}_{vap} \tag{7-17}$$

Since enthalpy is a function of state, the change in enthalpy is independent of the path.

When the molar volumes of the condensed phases and the virial coefficient of the saturated vapor are not known, Eqs. (7-15) and (7-16) cannot be used. For vaporization and sublimation, however, $\mathcal{V}(g) \gg \mathcal{V}(l)$ and $\mathcal{V}(g) \gg \mathcal{V}(s)$, and so Eq. (7-14) may be simplified by neglecting the molar volume of the condensed phase $\mathcal{V}(l)$ and $\mathcal{V}(s)$, in comparison with the volume of the vapor phase. Using water as an example, $\mathcal{V}(g) = 30.2$ liter and $\mathcal{V}(l) = 0.0188$ liter at 100°C and 1 atm. Now, if the vapor phase behaves ideally, $\mathcal{V}(g) = \boldsymbol{R}T/P$, and Eq. (7-14) takes the form

$$d\ln P/dT = \Delta\mathcal{H}_{vap}/\boldsymbol{R}T^2 \tag{7-18}$$

Integration (assuming constant $\Delta\mathcal{H}_{vap}$) of Eq. (7-18), called the **Clausius–Clapeyron equation**, yields

$$\log P = -\frac{\Delta\mathcal{H}_{vap}}{2.303\boldsymbol{R}}\frac{1}{T} + C = \frac{A}{T} + C \tag{7-19}$$

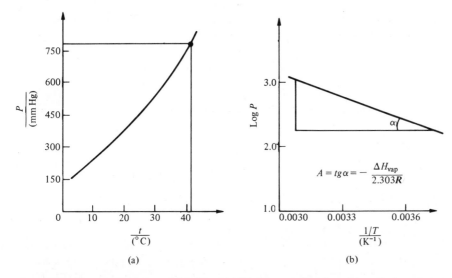

Figure 7-4 The vapor pressure of 3,3-dimethyl-1-butene as a function of temperature: $t_{nbp} = 41.3°C$. (a) P versus t; (b) log P versus $1/T$.

Thus, if $\Delta\mathcal{H}_{vap}$ is constant, a plot of log P versus $1/T$ should yield a straight line. The slope of this line, A, is equal to $-\Delta\mathcal{H}_{vap}/2.303\boldsymbol{R}$ (see Figure 7-4).

Expression (7-19) is limited to a narrow temperature range. In a more extended range of temperature, the dependence of $\Delta\mathcal{H}_{vap}$ on temperature must be considered. Making use of the relationship

$$\Delta\mathcal{H}_{vap} = a + bT + cT^2 \tag{7-20}$$

integration of Eq. (7-18) results in

$$\log P = -\frac{a}{2.303\boldsymbol{R}}\frac{1}{T} + \frac{b}{\boldsymbol{R}}\log T + \frac{c}{2.303\boldsymbol{R}}T + d$$
$$= A/T + B\log T + CT + D \tag{7-21}$$

where

$$A = -a/2.303\boldsymbol{R}, \ B = b/\boldsymbol{R}, \ C = c/2.303\boldsymbol{R}, \ D = d.$$

Eq. (7-21) yields good results over a wide temperature range; nevertheless, its use is not favored because of its complicated form. More frequently, the semiempirical Antoine equation

$$\log P = -A/(t + C) + B \tag{7-22}$$

is employed. This simple equation, with its three constants, A, B, and C (t is the temperature in °C), supplies excellent results in a wide temperature interval, except in the very low temperature range.

The values of the constants in the aforementioned equations are calculated from experimental vapor pressure data. Table 7-1 presents vapor pressure data for a few substances.

Table 7-1　Vapor pressure of some organic liquids as function of temperature

$\dfrac{t}{(^\circ C)}$	CCl$_4$	n-Octane	n-Hexane	Benzene	n-Propanol	H$_2$O
			Vapor Pressure (mm Hg)			
0.0	33.08	2.94	45.45	26.54	3.44	4.579
10.0	55.65	5.62	75.0	45.19	7.26	9.209
20.0	89.55	10.45	120.0	74.13	14.56	17.535
30.0	139.6	18.40	185.4	117.45	27.6	31.824
40.0	210.9	30.85	276.7	180.20	50.2	55.324
50.0	309.0	49.35	400.9	268.30	87.2	92.51
60.0	439.0	77.55	566.2	388.51	147.0	149.38
70.0	613.8	117.90	787.0	548.16	239.0	233.70
80.0		174.48	1062.0	755.0	376.0	355.10
90.0	1112.0	253.5	1407.0	1008.0	574.0	525.76
100.0	1457.0	353.6	1836.0	1335.0	842.5	760.00

Table 7-2　Entropy of vaporization at the normal boiling point

Substance	T_{nbp} (K)	$\Delta \mathcal{H}_{vap}$ (cal mol^{-1})	$\Delta \mathcal{S}_{vap}$ (cal mol^{-1} K^{-1}) from Eq. (7-23)	$\Delta \mathcal{S}_{vap}$ (cal mol^{-1} K^{-1}) from Eq. (7-24)
O$_2$(l)	90.18	1630	18.07	17.70
H$_2$O(l)	373.15	9717	26.04	20.52
HF(l)	293.05	1800	6.14	20.47
HCl(l)	188.10	3860	20.52	19.16
CH$_3$OH(l)	337.85	8430	24.95	20.32
CCl$_4$(l)	349.85	7170	20.49	20.34
CS$_2$(l)	319.40	6400	20.04	20.21
C$_2$H$_5$OH(l)	351.65	9220	26.22	20.40
methane(l)	111.66	1955	17.51	18.12
ethane(l)	184.52	3517	19.06	19.15
n-octane(l)	398.81	8360	20.96	20.65
benzene(l)	353.25	7353	20.81	20.41
toluene(l)	383.77	8001	20.84	20.57
cyclohexane(l)	353.89	5820	14.86	20.61
acetic acid(l)	391.45	5820	14.86	20.61

The Clapeyron as well as the Clausius–Clapeyron equation enable one to evaluate the enthalpy of vaporization from vapor pressure data. For excellent vapor pressure and enthalpy of vaporization correlation methods see Othmer et al [14, 15].

Approximate values of enthalpies of vaporization at the normal boiling points of liquids, T_{nbp}, can be obtained from Trouton's rule

$$\frac{\Delta \mathscr{H}_{vap}}{T_{nbp}} = \Delta \mathscr{S}_{vap} = 20\text{–}21 \text{ cal K}^{-1} \text{ mol}^{-1} \tag{7-23}$$

and Kistiakowsky's rule

$$\frac{\Delta \mathscr{H}_{vap}}{T_{nbp}} = 8.75 + 4.576 \log T_{nbp} \tag{7-24}$$

These rules indicate that at the normal boiling point liquids obey the principle of corresponding states; see Section 1-8.

Trouton's and Kistiakowsky's rules are obeyed fairly well by many nonpolar liquids, as seen from Table 7-2. Hildebrand showed that Trouton's rule gives more consistent results if the enthalpies of vaporization are calculated, not at the normal boiling points, but rather, at temperatures at which the vapors of the different liquids have the same molar concentration.

EXAMPLE 7-1
The melting point of monoclinic sulfur at a pressure of 1 atm is 119.3°C. The change in volume during fusion is 41 cm^3 kg^{-1} and $\Delta \mathscr{H}_{fus}$ is 422 cal mol^{-1}. Find the melting point of sulfur when the pressure is raised to 1000 atm.
Solution: One-component phase equilibria in condensed systems are represented by the Clapeyron equation, Eq. (7-14):

$$\frac{dP}{dT} = \frac{\Delta H}{T \Delta V}$$

If the ratio $\Delta H / \Delta V$ is assumed to be temperature and pressure independent, then integration yields

$$\int_{P_1}^{P_2} dP = \frac{\Delta H}{\Delta V} \int_{T_1}^{T_2} \frac{dT}{T}$$

$$P_2 - P_1 = \frac{\Delta H}{\Delta V} \ln \frac{T_2}{T_1}$$

After substituting the data, we have

$$1000 - 1 = \frac{1000 \times 422 \times 41.29}{32 \times 41} \times 2.303 \log \frac{T_2}{392.5}$$

where the value 41.29 is the conversion factor from calories to cubic centimeter-atmospheres which must be used if the units are to be consistent.

$$T_2 = 423.2 \text{ K} \quad \text{and} \quad t_2 = 150°C \quad \bullet$$

EXAMPLE 7-2

The vapor pressure of ethylene is given as a function of temperature by the equation

$$\log P/(\text{mm Hg}) = -\frac{834.13}{T} + 1.75 \log T - 8.375 \times 10^{-3} T + 5.32340$$

Calculate the enthalpy of vaporization of ethylene at its normal boiling point (−103.9°C).

Solution: On taking the derivative with respect to T, we transform the given equation to the differential form

$$\frac{d \ln P}{dT} = \frac{2.303 \times 834.13}{T^2} + \frac{1.75}{T} - 8.375 \times 10^{-3} \times 2.303$$

On comparing this with the Clausius–Clapeyron equation, (Eq. 7-18), it is obvious that

$$\frac{d \ln P}{dT} = \frac{\Delta \mathcal{H}_{\text{vap}}}{RT^2} = \frac{2.303 \times 834.13}{T^2} + \frac{1.75}{T} - 8.375 \times 10^{-3} \times 2.303$$

and

$$\Delta \mathcal{H}_{\text{vap}} = R \times 2.303 \times 834.13 + 1.75RT - 8.375 \times 10^{-3} \times 2.303 \times RT^2$$

On substituting $R = 8.314 \text{ J mol}^{-1} \text{ K}^{-1}$ and $T = 169.3$ K, we obtain

$$\Delta \mathcal{H}_{\text{vap}} = 13.84 \text{ kJ mol}^{-1}. \quad \bullet$$

7-5 VAPOR PRESSURE AS A FUNCTION OF THE EXTERNAL PRESSURE

The simple form of the Gibbs phase rule determines that the saturated vapor pressure of a pure liquid depends only on the temperature. When the liquid is under a pressure different from its saturated value at the given temperature, the saturated vapor pressure is, however, changed. The increased pressure over the liquid may be caused by addition of an inert gas—a gas insoluble in the liquid—to the vapor phase, or by means of a mechanical device, such as a piston.[5] The following equation may be written for such a system in equilibrium.

$$\bar{\mathcal{G}}(\text{g})(T, P) = \mathcal{G}(\text{l})(T, P_{\text{tot}}) \tag{7-25}$$

where $\bar{\mathcal{G}}(\text{g})$ and $\mathcal{G}(\text{l})$ is the partial molal Gibbs free energy of the substance in the vapor phase (since the vapor is a mixture of the substance plus inert gas, the molar properties are replaced by partial molal properties, see Section 8-2) and the molar Gibbs free energy of the liquid phase, respectively; P is the vapor pressure; and P_{tot} is the total pressure over the liquid. Eq. (7-25) may also be written in the differential form

$$d\bar{\mathcal{G}}(\text{g}) = d\mathcal{G}(\text{l}) \tag{7-26}$$

---

[5] The piston must be permeable to vapor but not to liquid.

Making use of Eq. (6-47) at constant temperature, we have

$$\bar{V}(g)\, dP = \bar{V}(l)\, dP_{tot} \tag{7-27}$$

or

$$\frac{\bar{V}(l)}{\bar{V}(g)} = \frac{dP}{dP_{tot}} \tag{7-28}$$

This expression, called the **Gibbs equation**, shows that the saturated vapor pressure is increased by the total pressure over the liquid; the rate of increase is, however, very small,[6] since $\bar{V}(l) \ll \bar{V}(g)$.

On assuming ideal behavior of the vapor phase, $\bar{V}(g) = \boldsymbol{R}T/P$, we obtain

$$d \ln P = \frac{\bar{V}(l)}{\boldsymbol{R}T}\, dP_{tot} \tag{7-29}$$

Since the molar volume of the liquid does not vary significantly with pressure, we may integrate this equation, assuming constant $\bar{V}(l)$, to obtain

$$\ln \frac{P}{P^*} = \frac{\bar{V}(l)}{\boldsymbol{R}T}(P_{tot} - P^*) \tag{7-30}$$

where P^* is the saturated vapor pressure of the pure liquid when no external pressure is present, and P is the saturated vapor pressure of the liquid under a pressure P_{tot}.

The fact that the vapor pressure of a pure liquid depends on the temperature as well as on the pressure is not in conflict with the phase rule. The phase rule, as given in the form of Eq. (7-1), is applicable only if the pressure in all phases is the same. In the case under discussion, the two phases exist under different pressures, and, thus, the phase rule must be written in a different form (see page 188).

$$v = c + 3 - p = 1 + 3 - 2 = 2$$

and such a system has two degrees of freedom.

7-6 THE EFFECT OF THE SIZE AND SHAPE OF THE SURFACE AREA ON VAPOR PRESSURE

Until now we have neglected the interface contribution to the Gibbs free energy. In some cases, however, the interface contribution to the Gibbs free energy is so large that it must be taken into consideration.

Molecules inside a liquid and those at the surface of a liquid have different environments. Molecules in the interior are surrounded by the same kind of molecules, whereas molecules at the surface are subject to different environments above and below the surface layer. As a result of the different environments, the surface assumes a certain characteristic shape[7]—the smallest possible area. Consequently, the Gibbs free energy, originally pressure and temperature dependent only,

---

[6] This is not true near the critical point.

[7] The shape depends on other factors also, for example, on the material from which the container holding the liquid is made.

will also depend on the surface area, A. Thus for a constant amount of liquid, we have

$$G = G(T, P, A) \qquad (7\text{-}31)$$

and

$$dG = \left(\frac{\partial G}{\partial P}\right)_{T,A} dP + \left(\frac{\partial G}{\partial T}\right)_{P,A} dT + \left(\frac{\partial G}{\partial A}\right)_{T,P} dA$$
$$= V\, dP - S\, dT + \gamma\, dA \qquad (7\text{-}32)$$

because, as seen from Eqs. (6-48) and (6-49) $(\partial G/\partial P)_{T,A} = V$, and $(\partial G/\partial T)_{P,A} = -S$. The partial derivative, $(\partial G/\partial A)_{T,P}$, expressing the Gibbs free energy change per unit area at constant T and P, is called the surface tension, γ. At constant T and P, Eq. (7-32) reduces to the form

$$dG = \gamma\, dA = d\, w'' \qquad (7\text{-}33)$$

As we can see from Eqs. (6-36) and (7-33), $\gamma\, dA$ may be interpreted as the work done on the interface by extending its area by dA. Since at constant temperature γ depends not only on the nature of the liquid itself but also on the surroundings in which the extension is done (air, vapor), the work will be different against different environments. The most commonly used units for surface tension are erg per square centimeter ($\equiv$ dyne per centimeter).[8]

The most widely used methods for measuring the surface tension are the capillary rise, maximum bubble pressure, and the Du Noüy tensiometer. Some data on surface tension are listed in Table 7-3.

Table 7-3 Surface tension of some organic liquids at 25°C

Substance	Surface Tension (erg/cm²)	Substance	Surface Tension (erg/cm²)
methyl alcohol	22.18	pyridine	36.48
ethyl alcohol	21.85	cyclohexane	24.40
benzene	28.25	chlorobenzene	32.70
toluene	27.96	n-hexane	17.90
carbon tetrachloride	26.12	n-pentane	15.50
1,2-dibromoethane	38.26	n-heptane	19.79
ethyl acetate	23.15	2,5-dimethylhexane	19.27
n-hexylacetate	25.74	n-octane	21.27
tetrachloroethylene	31.80	water	71.97

The surface tension is lowered with increasing temperature and becomes zero near the critical temperature. The relationship between γ and T may be written as

$$\frac{d[\gamma(M/\rho)^{2/3}]}{dT} = -k \qquad (7\text{-}34)$$

where (M/ρ) is the molar volume and $(M/\rho)^{2/3}$ is proportional to the surface area.

---

[8] In the SI system: Newton per meter ($N\,m^{-1}$).

The product $\gamma(M/\rho)^{2/3}$ is called the surface Gibbs free energy. Taking into consideration the fact that at T_c, $\gamma = 0$, integration of Eq. (7-34) yields

$$\gamma(M/\rho)^{2/3} = k(T_c - T) \qquad (7\text{-}35)$$

The surface tension vanishes roughly 6 degrees above the critical temperature rather than at the critical temperature T_c, and therefore Ramsay and Shield rewrote Eq. (7-35) into the form

$$\gamma(M/\rho)^{2/3} = k(T_c - T - 6) \qquad (7\text{-}36)$$

The value of k for normal liquids is found to be approximately 2.1 erg K^{-1}.

As a consequence of the surface contribution to the Gibbs free energy, vapor pressures over curved and flat surfaces are different. This difference becomes especially significant when the radius of the curvature is very small. Concave curvature toward the air (vapor) phase results in lowering the vapor pressure; convex curvature results in an increase of the vapor pressure. Consequently, small droplets, having higher vapor pressures, are less stable than large ones. Thus, small droplets will join by self-distillation and condensation into larger units.

The work done when the surface area of a droplet increases by an infinitesimal area dA, at constant T and P, is expressed by Eq. (7-33)

$$d w'' = dG = \gamma \, dA \qquad (7\text{-}33)$$

and the change in the surface area, corresponding to the droplet increase from a radius of r to a radius of $(r + dr)$ is: $dA = 4\pi(r + dr)^2 - 4\pi r^2 = 8\pi r \, dr.$[9] Consequently

$$d w'' = dG = 8\pi\gamma r \, dr \qquad (7\text{-}37)$$

During the increase, dn moles are transferred by distillation and condensation, isothermally and reversibly, from the smaller to the larger droplet; in the limit a large unit may be considered as having a flat surface. The work necessary for the transfer is $d w'' = dG = dn \, RT \ln P_d/P$ where P_d is the vapor pressure over the droplet, P is the vapor pressure over the flat surface and

$$dn = \frac{dm}{M} = \frac{\rho \, dV}{M} = \frac{\rho 4\pi r^2 \, dr}{M}$$

Equating this to the work in Eq. (7-37), we obtain

$$8\pi\gamma r \, dr = \frac{\rho 4\pi r^2 \, dr}{M} RT \ln \frac{P_d}{P} \qquad (7\text{-}38)$$

consequently

$$\ln \frac{P_d}{P} = \frac{2M\gamma}{\rho r RT} \qquad (7\text{-}39)$$

ρ, m and M denote here the density, mass and molecular mass, respectively.

Kelvin was the first to derive Eq. (7-39), showing quantitatively the effect of the drop size on the vapor pressure. Such an effect is not predicted by the simple Gibbs phase rule, Eq. (7-1), in which the surface contribution has not been considered.

[9]Neglecting the term $(dr)^2$.

The Kelvin equation explains simply why a vapor may remain supersaturated for an extended period of time. An analogous equation may be derived to explain the important fact that very small solid particles have a considerably greater solubility than large particles. The conclusions of the Kelvin equation have been experimentally verified. For water at 20°C ($P = 17.5$ mm Hg), the vapor pressure is raised by 1.1% when $r = 10^{-5}$ cm.

7-7 EXPERIMENTAL DETERMINATION OF THE VAPOR PRESSURE

In principle, there are two possible ways of determining the dependence of the saturated vapor pressure of volatile liquids on temperature: the dynamic method and the static method.

When using the dynamic method, we measure the temperature at which the liquid and vapor phases are in equilibrium under a certain constant pressure. In determining the vapor pressure by a static method, we measure the pressure of the saturated vapors that prevails over the liquid at constant temperature. These methods are equivalent and, when correctly used, give concordant results.

The most commonly used static method employs the isoteniscope, shown in Figure 7-5. The bulb and the attached U-tube of the isoteniscope are partially filled with the liquid under consideration; the liquid is allowed to boil vigorously until all the air has

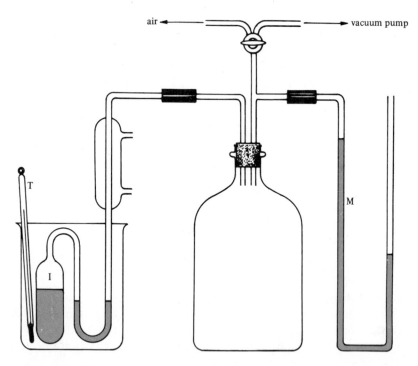

Figure 7-5 The static isoteniscope method for vapor pressure measurement: I, isoteniscope; T, thermometer; M, manometer.

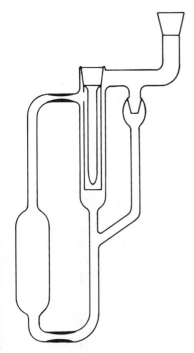

Figure 7-6   Ebulliometer.

been expelled from the sample. At each temperature, the external pressure is adjusted until the liquid in the arms of the U-tube is leveled, and the pressure and temperature are then recorded.

When using a dynamic method, the isoteniscope is replaced by a device called an ebulliometer (see Figure 7-6). The boiling point of the liquid is measured at different pressures.

A third method, the gas saturation method, is mostly used for measuring vapor pressure of less volatile liquids and solids. An inert gas is passed through the sample maintained at a constant temperature. The volume of the passing gas and the final vapor content are measured. From these data the vapor pressure of the sample may readily be calculated.

Knudsen's method for the determination of very low vapor pressures is briefly discussed on page 58.

7-8 SECOND-ORDER PHASE TRANSITIONS

In the phase transitions discussed so far, the molar Gibbs free energies of the two phases are equal at equilibrium, whereas there is a discontinuity observed in the volume, entropy, and enthalpy. Because volume, entropy, and enthalpy are first derivatives of the Gibbs free energy (see Eqs. 6-49, 6-48, and 6-50)

$$V = \left(\frac{\partial G}{\partial P}\right)_T \qquad S = -\left(\frac{\partial G}{\partial T}\right)_P \qquad H = -T^2\left[\frac{\partial(G/T)}{\partial T}\right]_P$$

these phase changes are called first-order phase transitions.

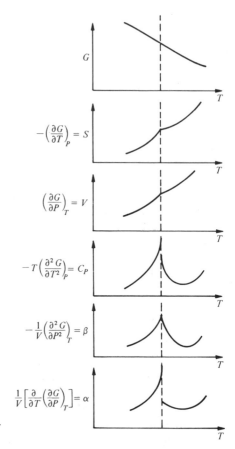

Figure 7-7 Second-order phase transitions.

In some cases, the discontinuity appears not in the volume, enthalpy, and entropy, but in the thermodynamic properties which are defined as second derivatives of the Gibbs free energy (see Figure 7-7). The best known properties of this type are the molar heat capacity, given by

$$\mathscr{C}_P = \left(\frac{\partial \mathscr{H}}{\partial T}\right)_P = -\frac{\partial\left[T^2 \frac{\partial(\mathscr{G}/T)}{\partial T}\right]_P}{\partial T} \tag{7-40}$$

consequently

$$\frac{\mathscr{C}_P}{T} = -\left(\frac{\partial^2 \mathscr{G}}{\partial T^2}\right)_P \tag{7-41}$$

compressibility, defined by

$$\beta = -\frac{1}{V}\left(\frac{\partial V}{\partial P}\right)_T = -\frac{1}{V}\left(\frac{\partial^2 G}{\partial P^2}\right)_T \tag{7-42}$$

and the cubic expansion coefficient given as

$$\alpha = \frac{1}{V}\left(\frac{\partial V}{\partial T}\right)_P = \frac{1}{V}\left[\frac{\partial}{\partial T}\left(\frac{\partial G}{\partial P}\right)_T\right]_P \tag{7-43}$$

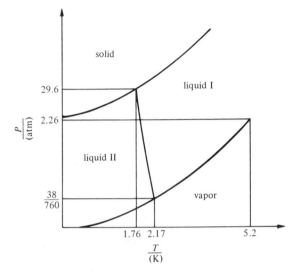

Figure 7-8 The phase diagram of the helium system (not drawn to scale).

Having discontinuous second derivatives of the Gibbs free energy, these changes are called second-order phase transitions.

Examples of second-order phase transitions are the changes of certain substances from ferromagnetic to paramagnetic solids at their Curie points, the transitions of some metals at low temperatures to a condition of electric superconductivity (zero resistance), some transitions in polymers (including biopolymers) and also the transition of helium from one liquid form to another.

Helium is so far the only known pure substance that can exist in two different isotropic liquid phases. There are many organic liquid substances capable of existing in two liquid phases; however, only one of the phases is isotropic, the other is anisotropic (liquid crystals). The phase diagram of helium is presented in Figure 7-8, and the heat capacity of helium as a function of temperature is seen in Figure 7-9.

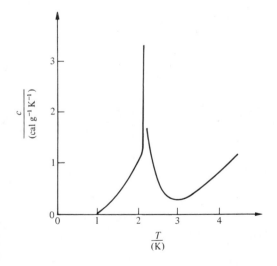

Figure 7-9 The heat capacity of liquid helium (not drawn exactly to scale).

The resemblance of the curve to the Greek letter lambda, λ, led to the name λ-point transition. Below the transition point, liquid helium exhibits peculiar flow properties that indicate the presence of a superfluid having no entropy or viscosity.

As ΔH and ΔV have practically zero values, the Clapeyron equation (7-14) cannot be used for second-order phase transitions.

Problems

1. The following data give the dependence of the melting point of water on the pressure and the corresponding volume changes upon melting

P	t	ΔV
(atm)	(°C)	(cm^3 kg^{-1})
1	0.0	−90.0
590	−5.0	−101.6
1090	−10.0	−112.2
1540	−15.0	−121.8
1910	−20.0	−131.3

   Calculate the enthalpy of fusion of water at −12°C.

2. The transition temperature between two solid phases of ammonium nitrate is 125.5°C at a pressure of 1 atm and 135°C at 1000 atm. The difference between the specific volumes of the high temperature and the low temperature phases is, on the average, 0.0126 cm^3 g^{-1}. Calculate the molar enthalpy of transition.

3. From the following data, roughly sketch the phase diagram for carbon dioxide: critical point at 31°C and 73 atm; triple point at −57°C and 5.3 atm; solid is denser than liquid at the triple point. Label all regions on the diagram.

4. Figure 7-10 represents a general phase diagram of a one-component system. Express the change in the Gibbs free energy and enthalpy, by general thermodynamic formulas only, when 1 mole of a substance is transferred from state M to state N.

5. Find the temperature and pressure corresponding to the triple point of arsenic if the temperature dependence of the pressure of the saturated vapors for liquid arsenic is given by

$$\log P/(\text{mm Hg}) = -\frac{2460}{T} + 6.69$$

   and for solid arsenic (sublimation equilibrium) by

$$\log P/(\text{mm Hg}) = -\frac{6947}{T} + 10.8$$

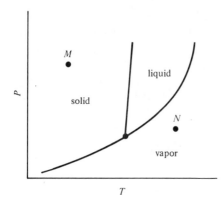

Figure 7-10 Single-component phase diagram.

6. At a temperature of 0°C the enthalpy of fusion of ice is 80 cal g^{-1} and the enthalpy of vaporization of water is 567 cal g^{-1}. (a) Calculate the molar enthalpy of sublimation of ice at this temperature. (b) If the triple point of water is at a temperature of 0.0098°C and a pressure of 4.6 mm Hg, find the change in pressure of the saturated vapors with the temperature (dP/dT) for ice and for liquid water at this point and compare the values of the two slopes.

7. The vapor pressure of ethyl ether as a function of temperature is given by the following values

$t/(°C)$	−49.0	34.5	121.0
$P/(\text{mm Hg})$	10	760	7600

Express this dependence analytically by an equation of the form

$$\log P/(\text{mm Hg}) = A + \frac{B}{T} + C \log T$$

8. The saturated vapor pressure of 2,2-dimethyl butanol is given by the equation

$$\log P/(\text{mm Hg}) = -\frac{4849.3}{T} - 14.701 \log T + 53.1187$$

(a) Find the temperature dependence of the molar enthalpy of vaporization.
(b) Calculate the molar enthalpy of vaporization at 25°C.

9. The saturated vapor pressure of ethyl ether at 0°C is 185 mm Hg, and the mean enthalpy of vaporization between 0 and 33°C is 92.5 cal g^{-1}. Calculate the saturated vapor pressure of ethyl ether at a temperature of 33°C.

10. The vapor pressure of thiophene as a function of temperature is given by the equation

$$\log P/(\text{mm Hg}) = -\frac{1246.038}{t + 221.354} + 6.95926$$

(a) Calculate the enthalpy of vaporization at 300 K.
(b) Compare this value with the enthalpy of vaporization calculated by assuming nonideal behavior of the vapor phase. The density of liquid thiophene at 300 K is 1.0520 g cm^{-3} and the second virial coefficient of the saturated vapor at 300 K is −1315 cm^3 mol^{-1}.

11. Making use of the data in Problem 10, (a) calculate the enthalpy of vaporization of thiophene at the normal boiling point and compare it with the values calculated from (b) Trouton's rule and (c) Kistiakowsky's rule.

12. In a determination of the vapor pressure of chloropicrin by the gas-saturation method, $2151 \, cm^3$ (STP) of air were passed through a saturator containing chloropicrin at 20°C and total pressure of 731.0 mm Hg. Chloropicrin lost a weight of 0.9200 g. Calculate the saturated vapor pressure of chloropicrin at the temperature of 35°C assuming that, under the given conditions, the air saturated with the vapors of the substance satisfies the equation of state of an ideal gas.

13. The temperature dependence of the surface tension and density of *n*-hexyl alcohol is given in the following table.

t (°C)	γ (dyne cm^{-1})	d (g cm^{-3})
85	21.04	0.77024
105	19.41	0.75341
125	17.75	0.73550
145	15.96	0.71618

Calculate the value of the constant k in the Ramsay–Shield equation (7-34).

References and Recommended Reading

[1] N. O. Smith: *J. Chem. Educ.*, **40**:317 (1963). The variation of vapor pressure with total pressure.

[2] R. P. Rastogi: *J. Chem. Educ.*, **41**:443 (1964). Thermodynamics of phase equilibria and phase diagrams.

[3] F. L. Swinton: *J. Chem. Educ.*, **44**:541 (1967). The triple point of water.

[4] A. Findlay, A. N. Campbell, and N. O. Smith: *The Phase Rule and its Applications.* Dover, New York, 1951.

[5] W. H. Keesom: *Helium.* Elsevier, Amsterdam, 1942.

[6] J. S. Marsh: *Principles of Phase Diagrams.* McGraw-Hill, New York, 1935.

[7] N. O. Smith: *J. Chem. Educ.*, **35**:125 (1958). Meaningful teaching of phase diagrams.

[8] J. E. Ricci: *The Phase Rule.* Van Nostrand, Princeton, N.J., 1951.

[9] S. T. Bowden: *The Phase Rule and Phase Reactions.* Macmillan, London, 1950.

[10] R. H. Petrucci: *J. Chem. Educ.*, **42**:323 (1965). Three dimensional models in phase studies.

[11] H. W. Salzberg, J. I. Morrow, S. R. Cohen, and M. E. Green: *A Modern Laboratory Course in Physical Chemistry.* Academic, New York, 1969.

[12] K. S. Pitzer and L. Brewer: *Thermodynamics* (2nd ed. of a book by G. N. Lewis and M. Randall), McGraw-Hill, New York, 1961.

[13] E. F. Hammel: *J. Chem. Educ.*, **35**:28 (1958). The vapor pressure of curved surfaces.

[14] D. F. Othmer and Erl-Sheen Yu: *Ind. Eng. Chem.*, **60**:23 (1968).

[15] D. F. Othmer and Hung-Tsung Chem: *Ind. Eng. Chem.*, **60**:39 (1968).

chapter 8

Application of Thermodynamics to Phase Equilibria of Multicomponent Systems

8-1 INTRODUCTION

Now that we have discussed the thermodynamics of single-component multiphase systems in detail, we shall turn to the discussion of thermodynamics of multi-component multiphase systems. In a multi-component system, the thermodynamic functions are dependent on the composition of the system as well as upon the parameters already discussed. The composition of a solution is usually expressed in one of the following ways:

Mole Fraction. Suppose a solution contains n_A moles of component A, n_B moles of component B, and n_C moles of component C. The mole fraction of A, N_A, is

$$N_A = \frac{n_A}{n_A + n_B + n_C} = \frac{n_A}{n} \tag{8-1a}$$

where $n = n_A + n_B + n_C$ is the total number of moles in the solution. Similarly

$$N_B = \frac{n_B}{n} \tag{8-1b}$$

and

$$N_C = \frac{n_C}{n} \tag{8-1c}$$

As is obvious

$$\sum N_i = 1 \tag{8-2}$$

In an n-component system, only $(n-1)$ mole fractions represent independent variables; the last mole fraction is given by Eq. (8-2).

Very often, especially when working with dilute solutions, the molarity and molality concentration scales are employed. The **molarity** of a solute B (commonly the minor constituent of the solution), c_B, is defined as the number of moles of B, n_B, dissolved in 1000 cm^3 of solution, or

$$c_B = \frac{n_B}{V} 1000 \tag{8-3}$$

where V is the volume of the solution expressed in cubic centimeters. The **molality** of a solute B, m_B, is defined as the number of moles of B, n_B, in 1000 g of solvent A (the major component of the solution):

$$m_B = \frac{n_B}{w_A} 1000 \tag{8-4}$$

where w_A is the weight of solvent A.

179

The relationship between the mole fraction and molality is very simple. For a two-component system,

$$N_B = \frac{m_B}{\left(\dfrac{1000}{M_A}\right) + m_B} = \frac{M_A m_B}{1000 + m_B M_A} \tag{8-5}$$

where M_A represents the molecular weight of the solvent. In a dilute solution $(1000 \gg m_B M_A)$ and the mole fraction becomes proportional to the molality $[N_B = (M_A m_B)/1000]$.

The relationship between the mole fraction and molarity is a little more complicated.

$$N_B = \frac{c_B}{\left(\dfrac{1000\rho - c_B M_B}{M_A}\right) + c_B} = \frac{c_B M_A}{1000\rho + c_B(M_A - M_B)} \tag{8-6}$$

where ρ denotes the density of the solution, and M_B the molecular weight of the solute. In a dilute solution ρ approaches the density of the pure solvent ρ_A, and $c_B(M_A - M_B)$ becomes much less than $1000\rho_A$, so that

$$N_B \approx \frac{c_B M_A}{1000\rho_A} \tag{8-7}$$

On comparison of Eq. (8-5) with Eq. (8-7), we have, for a dilute solution

$$m_B \rho_A \approx c_B \tag{8-8}$$

In dilute aqueous solutions, $\rho_A \approx 1 \text{ g cm}^{-3}$, and molarity becomes approximately equal to molality.

The mole fraction and molality are independent of temperature, $dN/dT = dm/dT = 0$, whereas molarity varies with temperature because volume is temperature dependent.

For nonaqueous solvents, as well as for mixed solvents, molality does not seem to be a proper concentration expression; 1000 g of different solvents contain different numbers of molecules, depending on the molecular weight of the solvent. Thus, in different solvents, molecules of the solute are surrounded, even at the same molality, by a different number of solvent molecules. This may significantly affect the properties of the solute in the solutions (solvent effect).[1]

8-2 PARTIAL MOLAL PROPERTIES

An arbitrary extensive property, Z, of a multicomponent system, is dependent on P, T, and also on the number of moles of each constituent of the system:

$$Z = Z(P, T, n_1, n_2, \ldots) \tag{8-9}$$

---

[1] This applies also to molarity.

Thus

$$dZ = \left(\frac{\partial Z}{\partial T}\right)_{P,n_1,n_2,\ldots} dT + \left(\frac{\partial Z}{\partial P}\right)_{T,n_1,n_2,\ldots} dP + \sum_i \left(\frac{\partial Z}{\partial n_i}\right)_{T,P,n_{j\neq i}} dn_i \qquad (8\text{-}10)$$

At constant T and P

$$dZ = \sum_i \left(\frac{\partial Z}{\partial n_i}\right)_{T,P,n_{j\neq i}} dn_i = \sum_i \bar{\mathcal{Z}}_i \, dn_i \qquad (8\text{-}11)$$

The partial derivative $(\partial Z/\partial n_i)_{T,P,n_{j\neq i}}$, called the partial molal quantity, is important in physical chemistry of solutions, and it is denoted by $\bar{\mathcal{Z}}_i$:

$$\left(\frac{\partial Z}{\partial n_i}\right)_{T,P,n_{j\neq i}} \equiv \bar{\mathcal{Z}}_i \qquad (8\text{-}12)$$

Since $N_i = n_i/n$, and $dn_i = n\,dN_i + N_i\,dn$, Eq. (8-11) may be rewritten as

$$dZ = \sum_i \bar{\mathcal{Z}}_i n \, dN_i + \sum_i \bar{\mathcal{Z}}_i N_i \, dn \qquad (8\text{-}13)$$

Now, if the restriction of constant concentration is imposed, dN_i becomes zero, and Eq. (8-13) reduces to

$$dZ = \sum_i \bar{\mathcal{Z}}_i N_i \, dn \qquad (8\text{-}14)$$

Eq. (8-14) expresses the change in the thermodynamic property Z with the addition of mass to the system in such a way that the mole fractions remain constant at constant T and P (mixing of two or more solutions of the same composition). Integration of Eq. (8-14) between states a and b results in

$$Z_b - Z_a = \sum_i \bar{\mathcal{Z}}_i N_i n_b - \sum_i \bar{\mathcal{Z}}_i N_i n_a \qquad (8\text{-}15)$$

and, in any arbitrary state

$$Z = \sum_i \bar{\mathcal{Z}}_i N_i n = \sum_i \bar{\mathcal{Z}}_i n_i \qquad (8\text{-}16)$$

since $N_i n = n_i$. Unlike Z, the partial molal quantity $\bar{\mathcal{Z}}_i$ does not depend on the total number of moles in the system; $\bar{\mathcal{Z}}_i$ depends on the concentration as well as on the pressure and temperature of the solution.

In the case of an ideal solution,[2] the partial molal properties derived from those extensive functions which are not defined in terms of entropy—$\bar{\mathcal{U}}_i$, $\bar{\mathcal{H}}_i$, $\bar{\mathcal{V}}_i$, $\bar{\mathcal{C}}_i$—are, however, concentration independent and are equal to the corresponding molar properties of the pure components; $\mathcal{U}_i$, $\mathcal{H}_i$, $\mathcal{V}_i$, $\mathcal{C}_i$. For example, the volume of a two-component ideal solution is

$$V = n_1 \mathcal{V}_1 + n_2 \mathcal{V}_2 \qquad (8\text{-}17)$$

Division of both sides by $(n_1 + n_2)$ yields

$$\frac{V}{n_1 + n_2} = \frac{n_1}{n_1 + n_2}\mathcal{V}_1 + \frac{n_2}{n_1 + n_2}\mathcal{V}_2 \qquad (8\text{-}18)$$

[2] For discussion see Section 8-9.

or

$$\mathcal{V} = N_1 \mathcal{V}_1 + N_2 \mathcal{V}_2 = \mathcal{V}_2 + N_1(\mathcal{V}_1 - \mathcal{V}_2) \qquad (8\text{-}19)$$

where $\mathcal{V}$ denotes the molar volume of the solution, and $\mathcal{V}_1$ and $\mathcal{V}_2$ the molar volumes of the pure components. The molar volume of an ideal solution is therefore a linear function of its composition. This also applies to the other functions whose definition does not require entropy but is invalid for the functions defined by the second law, even for an ideal solution.

The general thermodynamic theory of solutions is expressed in terms of these partial molal quantities and their derivatives, just as the theory of simple systems is based on the corresponding molar thermodynamic properties. For example, the Gibbs free energy is defined by the equation

$$G = H - TS \qquad (6\text{-}34)$$

and similarly, the partial molal Gibbs free energy is defined as

$$\bar{\mathcal{G}}_i = \bar{\mathcal{H}}_i - T\bar{\mathcal{S}}_i \qquad (8\text{-}20)$$

or

$$\left(\frac{\partial G}{\partial n_i}\right)_{T,P,n_{j\neq i}} = \left(\frac{\partial H}{\partial n_i}\right)_{T,P,n_{j\neq i}} - T\left(\frac{\partial S}{\partial n_i}\right)_{T,P,n_{j\neq i}} \qquad (8\text{-}21)$$

8-3 THE GIBBS–DUHEM EQUATION

On differentiation, Eq. (8-16) yields

$$dZ = \sum_i \bar{\mathcal{Z}}_i \, dn_i + \sum_i n_i \, d\bar{\mathcal{Z}}_i \qquad (8\text{-}22)$$

However, by comparison with Eq. (8-11), we find that

$$\sum_i n_i \, d\bar{\mathcal{Z}}_i = 0 \qquad (8\text{-}23)$$

Since $n \neq 0$, this equation may also be written in other forms:

$$\sum_i N_i \, d\bar{\mathcal{Z}}_i = 0 \qquad (8\text{-}24)$$

and

$$\sum_i N_i \left(\frac{\partial \bar{\mathcal{Z}}_i}{\partial N_j}\right)_{P,T} = 0 \qquad (8\text{-}25)$$

Relations (8-23), (8-24), and (8-25), written here at constant P and T, indicate that the partial molal properties of the constituents of a solution are not independent of each other; the variation of one partial molal quantity affects the others in the way prescribed by these equations. In fact, for a two-component system, if values of the partial molal quantities are known for one of the constituents, they can, by an appropriate integration of

$$d\bar{\mathscr{Z}}_2 = -\frac{n_1}{n_2} d\bar{\mathscr{Z}}_1 \qquad (8\text{-}23)$$

be obtained for the other constituents. Eqs. (8-23), (8-24), and (8-25), called the Gibbs–Duhem equations, are useful not only in the determination of the partial molal properties, but can also be successfully employed in proving the thermodynamic consistency of experimental data on partial molal properties and other data directly related to them.

8-4 DETERMINATION OF PARTIAL MOLAL PROPERTIES

Various methods are generally employed for the determination of the partial molal properties; just a few of them will be discussed here.

The definition of the partial molal properties (Eq. 8-12), is such that the number of moles of one component varies, while the amount of all the other constituents remains constant. Molality seems, therefore, to be a very suitable concentration expression in the determination of partial molal properties. The molality m, as defined on page 179, is equal to the number of moles of the varying substance (solute), n_2, in n_1 ($\equiv 1000/M_1$) moles of the solvent of molecular weight M_1. n_1 is constant and therefore $(\partial Z/\partial n_2)_{n_1} = \partial Z/\partial m$. Thus, the slope of a curve, expressing the experimental values of any extensive property of the system against molality (see Figure 8-1), taken at any concentration, represents the partial molal property of the solute at this concentration. This method directly determines only the partial molal property of the solute. The corresponding partial molal quantity of the solvent can be calculated from Eq. (8-16); it is

$$\bar{\mathscr{Z}}_1 = \frac{Z - n_2\bar{\mathscr{Z}}_2}{n_1}$$

An alternate method makes possible the direct determination of both partial molal properties. Dividing Eq. (8-16) by $(n_1 + n_2)$, and substituting $(1 - N_2)$ for N_1, we have

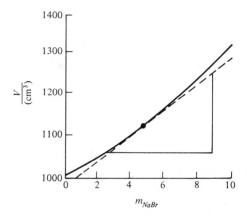

Figure 8-1 The variation of the volume with molality. System: water–sodium bromide at 25°C.

for the molar quantity $\mathscr{L}$

$$\mathscr{L} = \bar{\mathscr{L}}_1 + N_2(\bar{\mathscr{L}}_2 - \bar{\mathscr{L}}_1) \tag{8-26}$$

The derivative of Eq. (8-26) with respect to N_2 at constant T and P yields

$$\left(\frac{\partial \mathscr{L}}{\partial N_2}\right)_{T,P} = (1-N_2)\left(\frac{\partial \bar{\mathscr{L}}_1}{\partial N_2}\right)_{T,P} + N_2\left(\frac{\partial \bar{\mathscr{L}}_2}{\partial N_2}\right)_{T,P} + \bar{\mathscr{L}}_2 - \bar{\mathscr{L}}_1 \tag{8-27}$$

According to the Gibbs–Duhem equation (Eq. 8-25), the sum of the first two terms on the right-hand side of Eq. (8-27) is equal to zero. Consequently,

$$\left(\frac{\partial \mathscr{L}}{\partial N_2}\right)_{T,P} = \bar{\mathscr{L}}_2 - \bar{\mathscr{L}}_1 \tag{8-28}$$

On combining Eq. (8-26) with Eq. (8-28), we find

$$\bar{\mathscr{L}}_1 = \mathscr{L} - N_2\left(\frac{\partial \mathscr{L}}{\partial N_2}\right)_{T,P} \tag{8-29a}$$

$$\bar{\mathscr{L}}_2 = \mathscr{L} + N_1\left(\frac{\partial \mathscr{L}}{\partial N_2}\right)_{T,P} \tag{8-29b}$$

Both terms on the right-hand side of these equations, the molar property of the solution, $\mathscr{L}$, and its rate of change with N_2, $(\partial \mathscr{L}/\partial N_2)_{T,P}$, can be evaluated graphically or analytically from the experimental data.[3]

We shall introduce next the more accurate analytical method. The experimental data, expressing, let us say, the molar volume as a function of concentration at constant T and P, can be fitted to an equation of the form

$$\mathscr{V} = \mathscr{V}_1 + AN_2 + BN_2^2 + CN_2^3 + \cdots \tag{8-30}$$

where $\mathscr{V}_1$ is the molar volume of the pure solvent, and A, B, and C are constants whose values are to be determined from the experimental data. On taking the derivative of Eq. (8-30) with respect to N_2, and introducing the result into Eqs. (8-29a) and (8-29b), we obtain

$$\bar{\mathscr{V}}_1 = \mathscr{V}_1 - BN_2^2 - 2CN_2^3 \tag{8-31}$$

and

$$\begin{aligned} \bar{\mathscr{V}}_2 &= \mathscr{V}_1 + A + B + C - (B+3C)N_1^2 + 2CN_1^3 \\ &= \mathscr{V}_2 - (B+3C)N_1^2 + 2CN_1^3 \end{aligned} \tag{8-32}$$

where $\mathscr{V}_2 = \mathscr{V}_1 + A + B + C$. In the limit when $N_2 \to 0$, $\bar{\mathscr{V}}_1 = \mathscr{V}_1$; and when $N_1 \to 0$, $\bar{\mathscr{V}}_2 = \mathscr{V}_2$—the partial molal properties are equal to the molar properties of the pure substances.

EXAMPLE 8-1

The following expression gives the volume of the sodium bromide–water solution as function of molality at constant T and P:

$$V = 1002.93 + 23.189m + 2.197m^{3/2} - 0.178m^2$$

---

[3] Another method similar to this is discussed in Problem 3.

where V is the volume of a solution containing m moles of NaBr in 1000 g of water. Determine the partial molal volumes of NaBr and H_2O for solutions of $m = 0.25$ and $m = 0.50$.

Solution: To evaluate the partial molal volumes of NaBr in both solutions, we simply differentiate the given equation with respect to m. Thus

$$\bar{V}_{NaBr} = \left(\frac{\partial V}{\partial m}\right)_{T,P,n_1} = 23.189 + \frac{3}{2} \times 2.197 m^{1/2} - 2 \times 0.178 m$$

On substituting $m = 0.25$ and $m = 0.50$, respectively, we obtain

$$\bar{V}_{NaBr} = 24.748 \text{ cm}^3 \text{ mol}^{-1} \qquad \bar{V}_{NaBr} = 25.340 \text{ cm}^3 \text{ mol}^{-1}$$

From $\bar{V}_{NaBr}$ and V, the partial molal volume of water is calculated from Eq. (8-16)

$$\bar{V}_{H_2O} = \frac{V - m\bar{V}_{NaBr}}{n_{H_2O}}$$

$$\bar{V}_{H_2O} = 18.067 \text{ cm}^3 \text{ mol}^{-1} \qquad \bar{V}_{H_2O} = 18.045 \text{ cm}^3 \text{ mol}^{-1}. \qquad \bullet$$

8-5   PARTIAL MOLAL ENTHALPIES OF SOLUTION

The methods discussed in the preceding section are applicable to all the extensive properties of solutions: V, U, H, A, S, and G. Since the absolute values of these functions, except the volume and entropy[4] are not known, it is necessary to choose a certain standard state (page 195), and relate the partial molal quantities to this state.

We shall next discuss in detail the application of the partial molal enthalpies to the calculation of enthalpies of solution and enthalpies of dilution of two-component systems.

Consider a process by which a binary solution is formed from its constituents, A and B, at constant T and P. We may express this as

$$n_1 A + n_2 B = [n_1 A + n_2 B]$$

where n_1 denotes the number of moles of A, n_2 denotes the number of moles of B, and $[n_1 A + n_2 B]$ denotes the solution composed of n_1 moles of A and n_2 moles of B. The enthalpy change of this process of mixing, ΔH_{mix}, is given by the difference in the enthalpies between the final and initial states

$$\Delta H_{mix} = H_{sol} - n_1 \mathscr{H}_1 - n_2 \mathscr{H}_2 \qquad \text{(8-33)}$$

in which H_{sol} is the enthalpy of the solution, and $\mathscr{H}_1$ and $\mathscr{H}_2$ are the molar enthalpies of the pure components, A and B, at the given temperature and pressure. The enthalpy of the solution, however, may be expressed as the sum of the partial molal enthalpies of the two components, $\bar{\mathscr{H}}_1$ and $\bar{\mathscr{H}}_2$, respectively:

$$H_{sol} = n_1 \bar{\mathscr{H}}_1 + n_2 \bar{\mathscr{H}}_2 \qquad \text{(8-34)}$$

---

[4] See third law of thermodynamics, Section 9-7.

On combining the last two equations, we obtain

$$\Delta H_{\text{mix}} = n_1(\bar{\mathcal{H}}_1 - \mathcal{H}_1) + n_2(\bar{\mathcal{H}}_2 - \mathcal{H}_2) \tag{8-35}$$

The term $(\bar{\mathcal{H}}_i - \mathcal{H}_i)$, which relates the partial molal enthalpy of component i to its molar enthalpy in the standard state, is called the partial molal enthalpy of solution, $\bar{\mathcal{L}}_i$,

$$\bar{\mathcal{L}}_1 = \bar{\mathcal{H}}_1 - \mathcal{H}_1 \qquad \bar{\mathcal{L}}_2 = \bar{\mathcal{H}}_2 - \mathcal{H}_2 \tag{8-36}$$

Consequently

$$\Delta H_{\text{mix}} = n_1 \bar{\mathcal{L}}_1 + n_2 \bar{\mathcal{L}}_2 \tag{8-37}$$

$\bar{\mathcal{L}}_2$ and $\bar{\mathcal{L}}_1$ may be evaluated from Eqs. (8-29a), and (8-29b), respectively

$$\bar{\mathcal{L}}_1 = \Delta \mathcal{H}_{\text{mix}} - N_2 \left(\frac{\partial \Delta \mathcal{H}_{\text{mix}}}{\partial N_2} \right)_{T,P} \tag{8-29a}$$

and

$$\bar{\mathcal{L}}_2 = \Delta \mathcal{H}_{\text{mix}} + N_1 \left(\frac{\partial \Delta \mathcal{H}_{\text{mix}}}{\partial N_2} \right)_{T,P} \tag{8-29b}$$

From the experimental data, the rate of change $(\partial \Delta \mathcal{H}_{\text{mix}} / \partial N_2)_{T,P}$ is found graphically or analytically, and $\bar{\mathcal{L}}_1$ and $\bar{\mathcal{L}}_2$ is calculated.

The mixing of two solutions of the same components but of different compositions proceeds according to the equation

$$n[N_1 A + N_2 B] + n'[N_1' A + N_2' B] = (n + n')[N_1'' A + N_2'' B] \tag{8-38}$$

where n denotes the number of moles of the first solution of concentration N_1 and N_2, n' denotes the number of moles of the second solution of concentration N_1' and N_2', and $(n + n')$ gives the number of moles of the final solution of concentration N_1'' and N_2''.

The enthalpy change accompanying this process of mixing is given by the difference in enthalpies before and after mixing

$$\Delta \mathcal{H}_{\text{mix}} = (n + n')[N_1'' \bar{\mathcal{L}}_1'' + N_2'' \bar{\mathcal{L}}_2''] - n[N_1 \bar{\mathcal{L}}_1 + N_2 \bar{\mathcal{L}}_2] - n'[N_1' \bar{\mathcal{L}}_1' + N_2' \bar{\mathcal{L}}_2'] \tag{8-39}$$

The partial molal enthalpies, as well as the differential molal enthalpies of solution, depend on temperature, pressure, and especially on the concentration.

8-6   THE CHEMICAL POTENTIAL

The differential of the internal energy in a closed system, when only reversibly performed P-V work is considered, is given by the equation

$$dU = T\, dS - P\, dV \tag{6-24}$$

In a system consisting of two or more phases, matter can flow from one phase to another. Although the system as a whole is closed, each phase is open. The internal energy of such an open part of the system depends on the number of moles of the constituents $(n_1, n_2, \ldots, n_c)$ in the particular phase, as well as on the entropy and volume. We must, therefore write

$$dU = \left(\frac{\partial U}{\partial S}\right)_{V,n_1,n_2,...,n_c} dS + \left(\frac{\partial U}{\partial V}\right)_{S,n_1,n_2,...,n_c} dV + \left(\frac{\partial U}{\partial n_1}\right)_{S,V,n_2,...,n_c} dn_1$$

$$+ \left(\frac{\partial U}{\partial n_2}\right)_{S,V,n_1,...,n_c} dn_2 + \cdots + \left(\frac{\partial U}{\partial n_c}\right)_{S,V,n_1,n_2,...} dn_c$$

$$= T\,dS - P\,dV + \left(\frac{\partial U}{\partial n_1}\right)_{S,V,n_2,...,n_c} dn_1$$

$$+ \left(\frac{\partial U}{\partial n_2}\right)_{S,V,n_1,...,n_c} dn_2 + \cdots + \left(\frac{\partial U}{\partial n_c}\right)_{S,V,n_1,n_2,...} dn_c \qquad (8\text{-}40)$$

since as is obvious from Eq. (6-24)

$$\left(\frac{\partial U}{\partial S}\right)_{V,n_1,n_2,...,n_c} = T \quad \text{and} \quad \left(\frac{\partial U}{\partial V}\right)_{S,n_1,n_2,...,n_c} = -P$$

Substituting a new quantity, $(\mu_i)_{S,V}$, called the chemical potential, for the partial derivatives $(\partial U/\partial n_i)_{S,V,n_{j\neq i}}$, we obtain

$$dU = T\,dS - P\,dV + (\mu_1)_{S,V}\,dn_1 + (\mu_2)_{S,V}\,dn_2 + \cdots + (\mu_c)_{S,V}\,dn_c \qquad (8\text{-}41)$$

Relations analogous to Eq. (8-41) may also be derived for the enthalpy, Helmholtz free energy, and Gibbs free energy. By employing the equations

$$dH = dU + P\,dV + V\,dP \qquad (8\text{-}42)$$

$$dA = dU - T\,dS - S\,dT \qquad (8\text{-}43)$$

$$dG = dU + P\,dV + V\,dP - T\,dS - S\,dT \qquad (8\text{-}44)$$

and making use of Eq. (8-41), we have

$$dH = V\,dP + T\,dS + (\mu_1)_{S,V}\,dn_1 + (\mu_2)_{S,V}\,dn_2 + \cdots + (\mu_c)_{S,V}\,dn_c \qquad (8\text{-}45)$$

$$dA = -P\,dV - S\,dT + (\mu_1)_{S,V}\,dn_1 + (\mu_2)_{S,V}\,dn_2 + \cdots + (\mu_c)_{S,V}\,dn_c \qquad (8\text{-}46)$$

$$dG = V\,dP - S\,dT + (\mu_1)_{S,V}\,dn_1 + (\mu_2)_{S,V}\,dn_2 + \cdots + (\mu_c)_{S,V}\,dn_c \qquad (8\text{-}47)$$

From a comparison of the last three equations with Eq. (8-41), it follows that

$$(\mu_i)_{S,V} \equiv \left(\frac{\partial U}{\partial n_i}\right)_{S,V,n_{j\neq i}} = \left(\frac{\partial H}{\partial n_i}\right)_{S,P,n_{j\neq i}} = \left(\frac{\partial A}{\partial n_i}\right)_{T,V,n_{j\neq i}} = \left(\frac{\partial G}{\partial n_i}\right)_{T,P,n_{j\neq i}} \qquad (8\text{-}48)$$

or

$$(\mu_i)_{S,V} = (\mu_i)_{S,P} = (\mu_i)_{T,V} = (\mu_i)_{T,P} \qquad (8\text{-}49)$$

where

$$\left(\frac{\partial H}{\partial n_i}\right)_{S,P,n_{j\neq i}} \equiv (\mu_i)_{S,P} \quad \left(\frac{\partial A}{\partial n_i}\right)_{T,V,n_{j\neq i}} \equiv (\mu_i)_{T,V} \quad \left(\frac{\partial G}{\partial n_i}\right)_{T,P,n_{j\neq i}} \equiv (\mu_i)_{T,P} \qquad (8\text{-}50)$$

The subindices in μ_i may now be dropped (see also Problem 28, p. 253).

Next, the physical meaning of the chemical potential, μ_i, is described. The left-hand side of Eq. (8-41) is an energy differential; therefore, each term on the right-hand side must represent energy. Energy expressions consist of an extensive and an intensive (thermodynamic force) factor: dS, dV, and dn_i are the extensive factors, and T, P, and μ_i represent the intensive factors. The intensive factors determine when and in what direction a spontaneous process will proceed.

Temperature is defined as the rate of change of internal energy per unit change of entropy; $(\partial U/\partial S)_{V,n_1,n_2,\dots,n_c} = T$. Consequently, heat flows spontaneously from a region of higher $(\partial U/\partial S)_{V,n_1,n_2,\dots,n_c}$ to a region of lower $(\partial U/\partial S)_{V,n_1,n_2,\dots,n_c}$. At low temperatures a small change in the internal energy results in a large change in the entropy. Pressure is defined as the negative rate of change of internal energy per unit change of volume; $-(\partial U/\partial V)_{S,n_1,n_2,\dots,n_c} = P$. A fluid expands, therefore, from a region of higher $-(\partial U/\partial V)_{S,n_1,n_2,\dots,n_c}$ into a region of lower $-(\partial U/\partial V)_{S,n_1,n_2,\dots,n_c}$. The chemical potential is given as the rate of change of internal energy per unit change of mass (or 1 mole); $(\partial U/\partial n_i)_{S,V,n_{j\neq i}} = \mu_i$. Consequently, mass flows spontaneously from a region of higher $(\partial U/\partial n_i)_{S,V,n_{j\neq i}}$ to a region of lower $(\partial U/\partial n_i)_{S,V,n_{j\neq i}}$.

At equilibrium, no spontaneous process can occur in the system and, therefore, the temperature, pressure, and chemical potentials must be the same in all the phases of the system. Thus, for an equilibrium system consisting of p' phases, we write

$$T' = \left[\left(\frac{\partial U}{\partial S}\right)_{V,n_1,\dots}\right]' = \left[\left(\frac{\partial U}{\partial S}\right)_{V,n_1,\dots}\right]'' = \dots = \left[\left(\frac{\partial U}{\partial S}\right)_{V,n_1,\dots}\right]^{p'} = T^{p'} \qquad \text{(8-51)}$$

$$P' = -\left[\left(\frac{\partial U}{\partial V}\right)_{S,n_1,\dots}\right]' = -\left[\left(\frac{\partial U}{\partial V}\right)_{S,n_1,\dots}\right]'' = \dots = -\left[\left(\frac{\partial U}{\partial V}\right)_{S,n_1,\dots}\right]^{p'} = P^{p'} \qquad \text{(8-52)}$$

$$\mu_i' = \left[\left(\frac{\partial U}{\partial n_i}\right)_{S,V,n_{j\neq i}}\right]' = \left[\left(\frac{\partial U}{\partial n_i}\right)_{S,V,n_{j\neq i}}\right]'' = \dots = \left[\left(\frac{\partial U}{\partial n_i}\right)_{S,V,n_{j\neq i}}\right]^{p'} = \mu_i^{p'} \qquad \text{(8-53)}$$

It is important to distinguish between the partial molal quantities $(\partial Z/\partial n_i)_{T,P,n_{j\neq i}}$, which are always restricted to constant T and P conditions, and the chemical potentials which do not necessarily meet the constant T and P requirement. The only chemical potential that is defined at constant T and P is the partial molal Gibbs free energy

$$(\mu_i)_{T,P} = \bar{\mathcal{G}}_i = \left(\frac{\partial G}{\partial n_i}\right)_{T,P,n_{j\neq i}} = \mu_i \qquad \text{(8-54)}$$

For a single-component system, the molar Gibbs free energy defines the chemical potential

$$\mathcal{G}_i = \mu_i^*$$

where * denotes the pure component.

8-7 THE DERIVATION OF THE GIBBS PHASE RULE

On the previous pages we have restricted ourselves to stating the phase rule without giving any indication of how it is derived. Now, after introducing the equilibrium condition in multicomponent systems, we are in a position to give the derivation of this very important law.

Consider a closed system of p open phases composed of c components. The composition of each of the present phases is expressed by $(c-1)$ intensive variables of concentration. Only $(c-1)$ mole fractions are variables, since $\sum_i N_i = 1$; N_i are the various mole fractions. Thus, with respect to the composition, each phase possesses $(c-1)$ variables. Since there are p phases, the entire system has $p(c-1)$ concentration variables. Additional variables, such as the temperature and pressure of each phase, should also be given. When the temperature and pressure are the same in all phases, the total number of intensive variables is

$$p(c-1)+2$$

Not all the intensive variables are, however, independent. The equilibrium condition

$$\mu_i' = \mu_i'' = \cdots \mu_i^{p'} \qquad \text{where } i = 1, 2, \ldots, c \qquad \textbf{(8-53)}$$

reduces the number of the intensive variables by $(p-1)$ for each component, and by $c(p-1)$ for all the components in the system. The difference, v,

$$v = p(c-1)+2-c(p-1) = c+2-p \qquad \textbf{(7-1)}$$

is the number of the independent intensive variables. The number of these undefined intensive variables is called the degrees of freedom of the system. Values must be assigned to these variables before the system is completely defined.

Eq. (7-1) represents the famous Gibbs phase rule. This rule is not concerned with the quantity of material in each phase; rather, it is the intensive variables that are important. If, in addition to the uniform pressure and temperature, other intensive variables exist (for example, magnetic field, electric field, and so on) or if the pressure and temperature is not uniform throughout the system, the number 2 in Eq. (7-1) must be replaced by another number.

8-8 AUXILIARY THERMODYNAMIC FUNCTIONS

The chemical potential is a very important thermodynamic quantity; it enables one to specify the condition of equilibrium in a multiphase system and also allows the thermodynamic treatment of solutions. The use of chemical potentials in the thermodynamic analysis of solutions is inconvenient, however, because the chemical potential becomes negative without bound as the concentration or gas pressure approaches zero. In addition, the absolute value of the chemical potential is unknown and can never be computed. For these reasons it has been found convenient to introduce new auxiliary thermodynamic functions, such as fugacity and activity, which facilitate the analysis of many problems in chemical thermodynamics.

8-8.1 Fugacity Consider 1 mole of a pure gaseous substance at a pressure low enough that it obeys the ideal gas law $(P\mathcal{V} = \boldsymbol{R}T)$. The change of the chemical potential μ^* of this gas with pressure at constant temperature is, from Eq. (6-47)

$$d\mathcal{G} = d\mu_{\text{id}}^* = \mathcal{V}_{\text{id}}\, dP = \frac{RT}{P}\, dP = \boldsymbol{R}T\, d \ln P \qquad \textbf{(8-55a)}$$

The superscript $*$ denotes the pure component.

At higher pressures, when the ideal gas law fails, it is necessary to make substitutions for $\mathcal{V}$ from some of the more complicated equations of state, (see Section 1-6) and therefore, the relation between the chemical potential and pressure becomes more complicated. To make the simple form of Eq. (8-55a) applicable also to nonideal gases, G. N. Lewis introduced the fugacity, f.

$$d\mu^* = RT\,d\ln f \qquad\qquad (8\text{-}55\text{b})$$

or for a constituent of a mixture

$$d\mu_i = RT\,d\ln f_i \qquad\qquad (8\text{-}55\text{c})$$

Comparison with Eq. (8-55a) shows that the fugacity of a perfect gas must be proportional to its pressure; for convenience, the two are set equal to each other. The definition of fugacity for a real gas is then completed by setting the condition

$$\lim_{P\to 0} \frac{f}{P} = 1 \qquad\qquad (8\text{-}56\text{a})$$

and for a component in a mixture

$$\lim_{P\to 0} \frac{f_i}{p_i} = \lim_{P\to 0} \frac{f_i}{y_i P} = 1 \qquad\qquad (8\text{-}56\text{b})$$

where y_i is the mole fraction of component i in the gaseous phase, $p_i = y_i P$, and P is the total pressure. The fugacity of a real gas approaches its pressure (equal to that of an ideal gas) in the limit of zero pressure.

As is evident, the fugacity is related to the chemical potential of a real gas in the same way the pressure is related to the chemical potential of an ideal gas. Direct measurement of fugacity and chemical potential is impossible. To make them applicable, it is necessary to relate them to other directly measurable properties of the system.

At constant temperature, the chemical potential of a pure real gas is related to its pressure by the equation

$$d\mu^* = \mathcal{V}\,dP$$

and for a pure ideal gas

$$d\mu_{id}^* = \mathcal{V}_{id}\,dP \qquad\qquad (8\text{-}55\text{a})$$

On subtracting the last two equations from each other, we obtain

$$d\mu^* - d\mu_{id}^* = (\mathcal{V} - \mathcal{V}_{id})\,dP \qquad\qquad (8\text{-}57)$$

However, $d\mu^*$ and $d\mu_{id}^*$ are given by Eqs. (8-55a) and (8-55b), and thus

$$d\ln f - d\ln P = \frac{1}{RT}(\mathcal{V} - \mathcal{V}_{id})\,dP \qquad\qquad (8\text{-}58)$$

Upon integrating between the limits of pressure P and pressure approaching zero (P^+), and taking into consideration that at this low pressure the fugacity of any real gas approaches the ideal gas pressure, $f^+ = P^+$, we obtain

$$\ln \frac{f}{P} = \ln \nu = \frac{1}{RT}\int_0^P (\mathcal{V} - \mathcal{V}_{id})\,dP \qquad\qquad (8\text{-}59)$$

The ratio f/P, called the fugacity coefficient, ν, gives a direct measure of the extent to which any real gas deviates from ideality ($\nu = 1$) at a given temperature and pressure.

Eq. (8-59), which compares the properties of a nonideal system with those of an ideal one, provides a basic formula for the calculation of fugacities. Knowing $\mathscr{V}$ as a function of pressure, the quantity $[(\mathscr{V} - \mathscr{V}_{id})/RT]$ may be calculated and plotted into a graph against the pressure. The area under the curve from $P = 0$ to P is the value of the integral. This value can also be calculated analytically by expressing $\mathscr{V}$ as a function of P from any equation of state. Using the virial equation (see page 15)

$$\mathscr{V} = RT/P + \mathscr{B}$$

we have

$$\ln \nu = \mathscr{B}P/RT \tag{8-60a}$$

or

$$\nu = e^{\mathscr{B}P/RT} \tag{8-60b}$$

The use of the compressibility factor z (see page 13) results in

$$(\mathscr{V} - \mathscr{V}_{id}) = zRT/P - RT/P = (z-1)RT/P \tag{8-61}$$

Consequently

$$\ln \nu = \int_0^P (z-1)\, d \ln P \tag{8-62a}$$

As shown on page 26, $P = P_r P_c$, and Eq. (8-62a) can also be written as

$$\ln \nu = \int_0^{P_r} (z-1)\frac{P_c\, dP_r}{P_c P_r} = \int_0^{P_r} (z-1)\, d \ln P_r \tag{8-62b}$$

Thus, at the same reduced temperature and pressure all gases exhibit the same ν and thus the same deviations from ideal behavior.[5] A diagram obtained from Eq. (8-62b), presenting $\nu = f(P_r, T_r)$, is shown in Figure 8-2.

Fugacity seems to be even more helpful in dealing with gaseous mixtures, liquids, and solids, pure or mixed.

From the direct relation between the chemical potential and fugacity, Eqs. (8-55b,c), it also follows that, in an equilibrium system, the fugacity of any constituent is the same in all the phases of the system

$$(f_i)' = (f_1)'' = \cdots = (f_i)^{p'} \tag{8-63}$$

EXAMPLE 8-2
Carbon dioxide obeys the reduced Berthelot equation reasonably well

$$z = \frac{P\mathscr{V}}{RT} = 1 + \frac{9}{128}\frac{PT_c}{P_cT}\left(1 - 6\frac{T_c^2}{T^2}\right)$$

Given: $T_c = 304.3$ K and $P_c = 73.0$ atm.
(a) Calculate the fugacity of carbon dioxide at a temperature of 150°C and pressure of 50 atm. (b) Calculate also the fugacity by making use of Figure 8-2.

---

[5] Compare this with the compressibility factor on page 28.

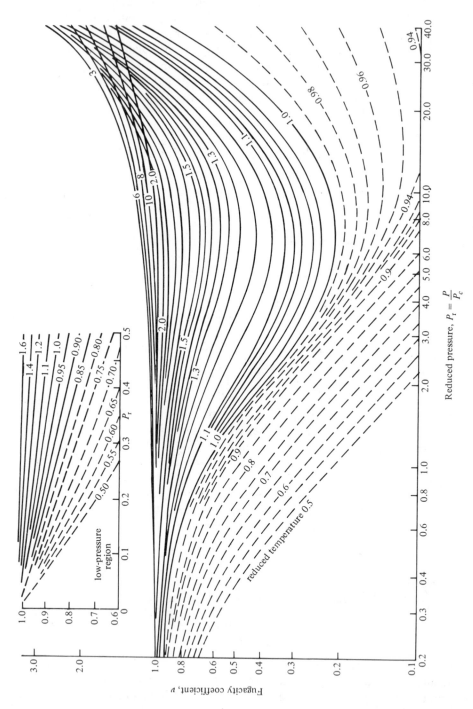

Figure 8-2   Generalized fugacity coefficient diagram.

Solution: (a) According to Eq. (8-58)

$$d \ln f = d \ln P + \frac{1}{RT}(\mathcal{V} - \mathcal{V}_{id}) \, dP$$

On substituting the solution for $\mathcal{V}$ from the above given equation and using $\mathcal{V}_{id} = RT/P$, we obtain

$$d \ln f = d \ln P + \frac{9T_c}{128 P_c T}\left(1 - 6\frac{T_c^2}{T^2}\right) dP$$

Integration, from a pressure approaching zero (at this pressure the fugacity can be put equal to the pressure) to pressure P, results in

$$\ln f = \ln P + \frac{9T_c}{128 P_c T}\left(1 - 6\frac{T_c^2}{T^2}\right) P$$

and after substituting the numerical data, we obtain

$$f = 46.4 \text{ atm.}$$

(b) The reduced pressure and temperature are

$$P_r = \frac{50}{73} = 0.68 \qquad T_r = \frac{423.15}{304.3} = 1.39$$

From Figure 8-2, we find $\nu = 0.93$. The fugacity is $f = P\nu = 50 \times 0.93 = 46.5$ atm. There is excellent agreement between the two methods. However, method (a) is more advisable, at least at moderate pressures. ●

The concentration dependence of fugacity may be obtained by substituting Eq. (8-55c) into the Gibbs–Duhem expression (Eq. 8-25)

$$N_1\left(\frac{\partial \ln f_1}{\partial N_i}\right)_{T,P} + N_2\left(\frac{\partial \ln f_2}{\partial N_i}\right)_{T,P} + \cdots + N_c\left(\frac{\partial \ln f_c}{\partial N_i}\right)_{T,P} = 0 \qquad \textbf{(8-64)}$$

In order to solve this differential equation, the partial derivatives must first be determined from experimental data. In a two-component system $dN_1 = -dN_2$ and thus

$$N_1\left(\frac{\partial \ln f_1}{\partial N_1}\right)_{T,P} = -N_2\left(\frac{\partial \ln f_2}{\partial N_1}\right)_{T,P} = N_2\left(\frac{\partial \ln f_2}{\partial N_2}\right)_{T,P} \qquad \textbf{(8-65)}$$

As seen, for a two-component system it is sufficient to know the variation of the fugacity with the mole fraction for one of the components; the variation for the other component may then be found through Eq. (8-65).

The fugacity also varies with temperature. In order to derive an expression relating the fugacity to temperature, let us proceed in the following manner: At a constant temperature, let 1 mole of a substance i be transferred from a state of pressure P to another state of pressure P^+. The change of the Gibbs free energy in this process is given by

$$\Delta \mathcal{G} = \mu_i^+ - \mu_i = RT \ln \frac{f_i^+}{f_i} \qquad \textbf{(8-66)}$$

or, rearranging terms,

$$\frac{\overset{+}{\mu_i}}{T} - \frac{\mu_i}{T} = R \ln f_i^+ - R \ln f_i \tag{8-67}$$

On partial differentiation with respect to T, we obtain

$$\left[\frac{\partial(\mu_i^+/T)}{\partial T}\right]_{P^+} - \left[\frac{\partial(\mu_i/T)}{\partial T}\right]_P = R\left[\left(\frac{\partial \ln f_i^+}{\partial T}\right)_{P^+} - \left(\frac{\partial \ln f_i}{\partial T}\right)_P\right] \tag{8-68}$$

The significance of the terms on the left-hand side of this equation may be seen from the following:

$$\left[\frac{\partial(\mu_i/T)}{\partial T}\right]_P = \frac{1}{T^2}\left[-\mu_i + T\left(\frac{\partial \mu_i}{\partial T}\right)\right]_P \tag{8-69}$$

$$= \frac{1}{T^2}(-\mu_i - T\bar{\mathscr{S}}_i)$$

$$= -\frac{\bar{\mathscr{H}}_i}{T^2} \tag{8-70}$$

because $(\partial \mu_i/\partial T)_P = -\bar{\mathscr{S}}_i$ and $\mu_i = \bar{\mathscr{H}}_i - T\bar{\mathscr{S}}_i$. Similarly

$$\left[\frac{\partial(\mu_i^+/T)}{\partial T}\right]_{P^+} = -\frac{\bar{\mathscr{H}}_i^+}{T^2} \tag{8-71}$$

Consequently, Eq. (8-68) assumes the form

$$\frac{-\bar{\mathscr{H}}_i^+ + \bar{\mathscr{H}}_i}{T^2} = R\left[\left(\frac{\partial \ln f_i^+}{\partial T}\right)_{P^+} - \left(\frac{\partial \ln f_i}{\partial T}\right)_P\right] \tag{8-72}$$

If the pressure P^+ is sufficiently low, then, $f_i^+ = P_i^+$, and $(\partial \ln f_i^+/\partial T)_{P^+} = 0$. Following from this

$$\left(\frac{\partial \ln f_i}{\partial T}\right)_P = \frac{\bar{\mathscr{H}}_i^+ - \bar{\mathscr{H}}_i}{RT^2} \tag{8-73}$$

$\bar{\mathscr{H}}_i^+$ being the partial molal enthalpy of constituent i at a very low pressure (at which the gaseous mixture behaves ideally), and hence $\bar{\mathscr{H}}_i^+ = \mathscr{H}_i$. The difference $(\mathscr{H}_i - \bar{\mathscr{H}}_i)$ is the enthalpy change for the expansion of the constituent from pressure P to pressure P^+.

8-8.2 Activity   When dealing with liquid or solid solutions, it is more convenient to use activity and activity coefficients instead of fugacity and fugacity coefficients.

Integration of Eq. (8-55c) between two states, at constant temperature, yields

$$(\mu_i)_2 - (\mu_i)_1 = RT \ln \frac{(f_i)_2}{(f_i)_1} \tag{8-74}$$

If state 1 is now, by definition, fixed and completely specified, then the difference in the chemical potentials will depend only on state 2. Let us denote the chemical potential and fugacity in this arbitrarily chosen fixed state by μ_i° and f_i°, respectively, so that

$$\mu_i - \mu_i^\circ = RT \ln \frac{f_i}{f_i^\circ} \tag{8-75}$$

We may substitute

$$\frac{f_i}{f_i^\circ} = \frac{a_i}{a_i^\circ} \qquad (8\text{-}76)$$

and obtain

$$\mu_i - \mu_i^\circ = RT \ln \frac{a_i}{a_i^\circ} \qquad (8\text{-}77)$$

where a_i and a_i° denote the activity in a given state and that in the fixed state, respectively. In the fixed state, which we shall call the standard state, the activity a_i° is always assigned a value of unity (in concentration units); consequently

$$\mu_i - \mu_i^\circ = RT \ln \frac{a_i}{1 \text{ (conc. unit)}} = RT \ln a_i \qquad (8\text{-}78)$$

From Eqs. (8-55) and (8-59) it follows that fugacity has the dimension of pressure, whereas the fugacity coefficient is dimensionless. Activity may be defined to be with or without dimension. In most physical chemistry texts, activity is treated as a dimensionless function. Since fugacity and activity are so closely related and, as shown below, by using the proper standard state they might even become equal, we prefer to assign dimensions to activity.

Just as fugacity is related to the pressure through the fugacity coefficient, activity is related to the concentration via the activity coefficient:

$$a_i = \gamma_i N_i \qquad (8\text{-}79)$$

$$a_i = {}^m\gamma_i m_i \qquad (8\text{-}80)$$

$$a_i = {}^c\gamma_i c_i \qquad (8\text{-}81)$$

where γ_i, ${}^m\gamma_i$, and ${}^c\gamma_i$ are the activity coefficients referred to the mole fraction, molality, and molarity, respectively. Since dimensions have been assigned to the activity, the activity coefficients are dimensionless.

The numerical value of the activity depends on the choice of the standard state, as well as on the concentration, temperature, and pressure; if f_i° is changed, a_i and γ_i is also changed (see Example 8-3). The activity is meaningless unless it is accompanied by a detailed description of the standard state to which it refers. Within certain limits, the standard state for a constituent may be arbitrarily selected. The primary limitation is that the temperature of the standard state always be that of the system.

The standard state of a gas at any given temperature is taken as the state in which the fugacity and pressure are equal to 1 atm, namely, $f^\circ = P^\circ = 1$ atm.[6] This definition requires that, at the pressure of 1 atm, the gas behave as if it were ideal. Since no real gas behaves ideally at $P = 1$ atm (see Figure 8-3), this standard state is not a real one; for convenience, hypothetical states are commonly used as standard states. Using this standard state, the activity of any gas becomes equal to its fugacity (see Eq. 8-76 and remember that $a_i^\circ = 1$)

$$\mu_i = \mu_i^\circ(T, 1 \text{ atm}) + RT \ln \frac{f_i}{1 \text{ (atm)}}$$

$$= \mu_i^\circ(T, 1 \text{ atm}) + RT \ln f_i \qquad (8\text{-}82)$$

---

[6] In the SI system the pressure unit is the pascal $(1 \text{ Pa} = 1 \text{ N m}^{-2}; 1 \text{ atm} = 101325 \text{ N m}^{-2})$; for convenience, however, we recommend that 1 atm be retained in the definition of the standard state.

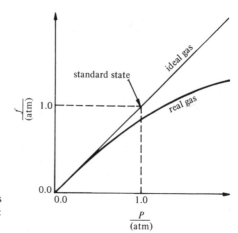

Figure 8-3 The standard state for gases in a hypothetical state of unit fugacity: $f^0 = P^0 = 1$ atm.

$\mu_i^\circ(T, 1 \text{ atm})$, the chemical potential in the standard state, is only temperature dependent.

For liquid and solid solutions, there are two commonly used standard states:

1. The first convention assumes as a standard state the pure liquid or solid constituent at the temperature and pressure of the system; sometimes, also at a pressure of 1 atm or at the vapor pressure of the pure constituent. This definition is based on the equations

$$\mu_i^*(T, P) = \lim_{N_i \to 1} (\mu_i - RT \ln N_i) \quad \text{where } i = 1, 2, \ldots, c \qquad \textbf{(8-83)}$$

Following from this all the constituents of the solutions behave ideally when their mole fractions approach unity:

$$\lim_{N_i \to 1} \frac{a_i}{N_i} = \lim_{N_i \to 1} \gamma_i = 1 \qquad \textbf{(8-84)}$$

2. The second convention used distinguishes between the solvent and solute constituents. The standard state of the solvent, the constituent present at highest concentration, is the same as in convention 1 (Eqs. 8-83 and 8-84). The standard state of the solutes, however, is defined as

$$\mu_i^\bullet(T, P) = \lim_{N_1 \to 1} (\mu_i - RT \ln N_i) \quad \text{where } i = 2, 3, \ldots, c \qquad \textbf{(8-85)}$$

Here N_1 is the mole fraction of the solvent and the superscript $\bullet$ denotes a standard state of component i at infinite dilution. As a result of this

$$\lim_{N_i \to 0} \frac{a_i}{N_i} = \lim_{N_i \to 0} \gamma_i^\bullet = 1 \qquad \textbf{(8-86)}$$

As is obvious, the solute standard state is again a hypothetical one. It is the state in which the pure solute $i(N_i = 1)$ has the properties it would have in an infinitely dilute solution (Figure 8-4). Such a state is commonly referred to as the infinitely dilute state.

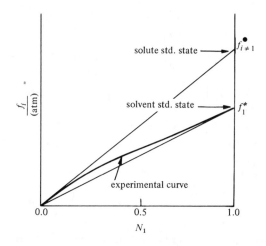

Figure 8-4   The standard states in a solution; mole fraction concentration scale.

On the molality concentration scale, the standard state of the solvent is again given by Eq. (8-83) and for the solutes

$$\mu_i^{\bullet}(T, P) = \lim_{N_1 \to 1} (\mu_i - RT \ln m_i) \quad \text{where } i = 2, 3, \ldots, c \tag{8-87}$$

$$\lim_{m_i \to 0} \frac{a_i}{m_i} = \lim_{m_i \to 0} {}^m\gamma_i^{\bullet} = 1 \tag{8-88}$$

where $\bullet$ denotes the standard state at infinite dilution. As is evident from Figure 8-5, this standard state is also hypothetical. It is a state in which the solute would exist at unit molality but would still have the environment typical of an extremely dilute solution.

It can be shown from the Gibbs–Duhem equation that for the region where ${}^m\gamma_i^{\bullet} = 1$, also $\gamma_i = 1$.

Similar expressions may be derived for the molarity concentration scale.

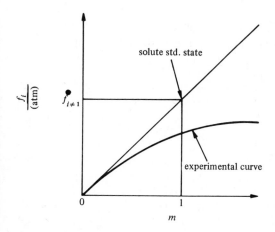

Figure 8-5   The standard state in a solution; molality concentration scale.

The pressure dependence of activity is given by the equation

$$\left(\frac{\partial \ln a_i}{\partial P}\right)_{T,N_i} = \frac{1}{RT}\left[\left(\frac{\partial \mu_i}{\partial P}\right)_{T,N_i} - \left(\frac{\partial \mu_i^\circ}{\partial P}\right)_{T}\right]$$

$$= \frac{1}{RT}[\bar{V}_i - V_i^\circ] \tag{8-89}$$

Pressure dependence may be ignored when dealing with liquid mixtures and solid solutions because the difference $(\bar{V}_i - V_i^\circ)$ is negligibly small. For gases, however, the change is significant and must be considered.

The change of the activity with temperature is represented by the expression

$$\left(\frac{\partial \ln a_i}{\partial T}\right)_{P,N_i} = -\frac{(\bar{\mathcal{H}}_i - \mathcal{H}_i^\circ)}{RT^2} \tag{8-90}$$

The activity decreases with temperature, when the difference in the partial molal enthalpy and molal enthalpy of the pure component i $(\bar{\mathcal{H}}_i - \mathcal{H}_i^\circ)$ is positive. It increases with temperature when $\bar{\mathcal{H}}_i - \mathcal{H}_i^\circ$ is negative.

The expressions for the dependence of the activity and activity coefficient on composition at constant T and P are based on the Gibbs–Duhem equation. On combining Eqs. (8-25) and (8-78), we obtain

$$N_1\left(\frac{\partial \ln a_1}{\partial N_i}\right)_{T,P} + N_2\left(\frac{\partial \ln a_2}{\partial N_i}\right)_{T,P} + \cdots + N_c\left(\frac{\partial \ln a_c}{\partial N_i}\right)_{T,P} = 0 \tag{8-91}$$

Since $a_i = N_i\gamma_i$, we may also write[7]

$$N_1\left(\frac{\partial \ln \gamma_1}{\partial N_i}\right)_{T,P} + N_2\left(\frac{\partial \ln \gamma_2}{\partial N_i}\right)_{T,P} + \cdots + N_c\left(\frac{\partial \ln \gamma_c}{\partial N_i}\right)_{T,P} = 0 \tag{8-92}$$

Just as in the case of the partial molal quantities, we must remember that, for the determination of the dependence of the activity and activity coefficient on composition, it is necessary to have experimental data.

The primary value of Eqs. (8-91) and (8-92) is that they allow the determination of this dependence from a minimum amount of experimental data and can also reveal inconsistencies and errors in the measured values (see Example 8-8 on page 227).

EXAMPLE 8-3

Eqs. (a) and (b) relate the logarithm of the activity coefficients of the two components of the ethyl alcohol(1)–toluene(2) mixture to the concentration of the solution

$$\log \gamma_1 = \frac{1.0074x_2^2}{(1.3343x_1 + x_2)^2} \tag{a}$$

and

$$\log \gamma_2 = \frac{0.7550x_1^2}{(0.7495x_2 + x_1)^2} \tag{b}$$

[7] $N_1\left(\dfrac{\partial \ln N_1}{\partial N_i}\right)_{P,T} + N_2\left(\dfrac{\partial \ln N_2}{\partial N_i}\right)_{P,T} + \cdots + N_c\left(\dfrac{\partial \ln N_c}{\partial N_i}\right)_{P,T} = 0$

The standard state to which these activity coefficients refer is the pure component, and the mole fraction concentration scale. Calculate: (a) the activity coefficients of both components referred to the standard state at infinite dilution and mole fraction concentration scale, $\gamma_1^\bullet$, $\gamma_2^\bullet$, (b) the activity coefficient of toluene(2) referred to the standard state at infinite dilution and molality concentration scale, $^m\gamma_2^\bullet$.

Solution: (a) According to Eq. (8-78) the chemical potential of component i in a nonideal solution is given by

$$\mu_i = \mu_i^\circ + RT \ln x_i \gamma_i \tag{c}$$

for the standard state referred to the pure component and mole fraction concentration scale, and

$$\mu_i = \mu_i^\bullet + RT \ln x_i \gamma_i^\bullet \tag{d}$$

for the standard state referred to infinite dilution and mole fraction concentration scale. The chemical potential, μ_i, is independent of the selection of the standard state and, therefore, for a certain concentration x_i we can write

$$\mu_i^\circ + RT \ln x_i \gamma_i = \mu_i^\bullet + RT \ln x_i \gamma_i^\bullet \tag{e}$$

For a binary solution, in the limit of $x_i \to 0$, Eq. (e) assumes the form

$$\mu_i^\circ + RT \ln \gamma_i^\infty = \mu_i^\bullet \tag{f}$$

since at infinite dilution $\gamma_i^\bullet = 1$ and $\gamma_i = \gamma_i^\infty$. Substitution of Eq. (f) into Eq. (e) yields

$$\mu_i^\circ + RT \ln \gamma_i = \mu_i^\circ + RT \ln \gamma_i^\infty + RT \ln \gamma_i^\bullet \tag{g}$$

and consequently

$$\gamma_i^\bullet = \frac{\gamma_i}{\gamma_i^\infty} \tag{h}$$

Thus, the activity coefficients referred to the standard state at infinite dilution are simply given as a ratio of the activity coefficients referred to the pure component standard state at a given concentration, γ_i, and at infinite dilution, γ_i^∞. The values of γ_1^∞ and γ_2^∞ obtained from Eqs. (a) and (b) are 10.172 and 5.688, respectively. The values of γ_1 and γ_2 calculated from (a) and (b) at various concentrations as well as $\gamma_1^\bullet$ and $\gamma_2^\bullet$ obtained from Eq. (h) are summarized in Table 8-1.

As expected from the definition of the infinite dilute standard state $\gamma_1^\bullet \to 1$ at $x_1 \to 0$ and $\gamma_2^\bullet \to 1$ at $x_2 \to 0$.

(b) Because the chemical potential is independent of the chosen standard state, we write for component 2

$$\mu_2 = \mu_2^{\bullet m} + RT \ln m\,^m\gamma_2^\bullet = \mu_2^\bullet + RT \ln x_2 \gamma_2^\bullet \tag{i}$$

Molality and mole fraction are related to each other by the equation

$$m = \frac{n_2}{w_1} 1000 = \frac{n_2}{M_1 n_1} 1000 = \frac{x_2}{x_1} \frac{1000}{M_1}$$

Table 8-1 Activity coefficients of ethyl alcohol (1) and toluene (2) in their respective mixtures referred to pure component and infinite dilute standard states

x_1	γ_1	γ_2	$\gamma_1^{\bullet}$	$\gamma_2^{\bullet}$	$^{m}\gamma_2^{\bullet}$
0.00	10.172	1.000			
0.10	5.808	1.029	0.5710	0.1810	0.0180
0.20	3.685	1.115	0.3623	0.1960	0.0392
0.30	2.557	1.259	0.2514	0.2213	0.0664
0.40	1.915	1.470	0.1883	0.2584	0.1034
0.50	1.531	1.765	0.1505	0.3113	0.1557
0.60	1.294	2.166	0.1272	0.3808	0.2284
0.70	1.147	2.688	0.1128	0.4725	0.3308
0.80	1.060	3.432	0.1042	0.6033	0.4826
0.90	1.014	4.399	0.0997	0.7734	0.6962
1.00	1.000	5.688			

and, therefore, Eq. (i) can be rewritten as

$$\mu_2^{\bullet m} + RT \ln m \, {}^{m}\gamma_2^{\bullet} = \mu_2^{\bullet} + RT \ln \frac{mx_1 M_1}{1000} \gamma_2^{\bullet}$$

$$= \mu_2^{\bullet} + RT \ln \frac{M_1}{1000} + RT \ln mx_1 \gamma_2^{\bullet} \qquad \text{(j)}$$

As is obvious

$$\mu_2^{\bullet m} = \mu_2^{\bullet} + RT \ln \frac{M_1}{1000} \qquad \text{(k)}$$

and

$$^{m}\gamma_2^{\bullet} = x_1 \gamma_2^{\bullet} \qquad \text{(l)}$$

In Eqs. (i) through (l) w_1 denotes the weight of the solvent, M_1 the molecular weight of the solvent and the superscript m refers to the molality concentration scale. It is assumed that component 2 is the solute and component 1 is the solvent in the entire concentration range. The values of the activity coefficients of toluene referred to the infinite dilute standard state and molality concentration scale, $^{m}\gamma_2^{\bullet}$, are given in the last column of Table 8-1. As you can see, toluene behaves less ideally when molality concentration scale is employed. ●

8-9 IDEAL MIXTURES

Just as the theory of gases is simplified by considering an idealized type of gas as a first approximation to the behavior of real gases, an idealized type of liquid mixture is also valuable in developing the theory of solutions. A perfect (ideal) mixture may be defined in a number of ways. A simple way is to state that a solution is ideal if the chemical potential of every constituent is a linear function of the logarithm of its mole fraction, N_i, according to the relation

$$\mu_i = \mu_i^*(P, T) + \boldsymbol{R}T \ln N_i \tag{8-93}$$

where N_i denotes a mole fraction in a gaseous, solid, or liquid state. It can be easily proved that this definition also requires the change, on mixing of the pure components, to be zero for the thermodynamic functions, U, H, C_P, C_V, V. We shall prove this for the volume and enthalpy.

On taking the derivative of Eq. (8-93) with respect to T at constant P and N_i, we have

$$\left[\frac{\partial(\mu_i/T)}{\partial T}\right]_{P,N_i} = \left[\frac{\partial(\mu_i^*/T)}{\partial T}\right]_P \tag{8-94}$$

But, according to Eq. (8-70)

$$\left[\frac{\partial(\mu_i/T)}{\partial T}\right]_{P,N_i} = -\frac{\bar{\mathcal{H}}_i}{T^2} \quad \text{and} \quad \left[\frac{\partial(\mu_i^*/T)}{\partial T}\right]_P = -\frac{\mathcal{H}_i^*}{T^2}$$

so that

$$\bar{\mathcal{H}}_i = \mathcal{H}_i^* \tag{8-95}$$

Consequently

$$\Delta H_{\text{mix}} = \sum_{i=1}^{i=c} n_i(\bar{\mathcal{H}}_i - \mathcal{H}_i^*) = 0 \tag{8-96}$$

The formation of an ideal solution from the pure components does not occur with the liberation or absorption of heat.

Let us now differentiate Eq. (8-93) with respect to the pressure at constant temperature and composition

$$\left(\frac{\partial\mu_i}{\partial P}\right)_{T,N_i} = \left(\frac{\partial\mu_i^*}{\partial P}\right)_T \tag{8-97}$$

Since

$$\left(\frac{\partial\mu_i}{\partial P}\right)_{T,N_i} = \bar{\mathcal{V}}_i \tag{8-98}$$

and

$$\left(\frac{\partial\mu_i^*}{\partial P}\right) = \mathcal{V}_i^* \tag{8-99}$$

we have

$$\bar{\mathcal{V}}_i = \mathcal{V}_i^* \tag{8-100}$$

Consequently

$$\Delta V_{\text{mix}} = \sum_{i=1}^{i=c} n_i(\bar{\mathcal{V}}_i - \mathcal{V}_i^*) = 0 \tag{8-101}$$

The formation of an ideal solution is not accompanied by a volumetric contraction or dilatation.

Mixing is a spontaneous process and, therefore, the entropy of mixing, and all thermodynamic functions related to the entropy, will be nonzero even for an ideal solution. For example, the Gibbs-free energy of mixing is given by the equation

$$\Delta G_{\text{mix}} = \Delta H_{\text{mix}} - T\,\Delta S_{\text{mix}} = \sum_{i=1}^{i=c} n_1(\mu_i - \mu_i^*)$$

$$= \sum_{i=1}^{i=c} n_i \boldsymbol{R} T \ln N_i \qquad (8\text{-}102)$$

For an ideal solution, $\Delta H_{\text{mix}} = 0$; thus

$$\Delta S_{\text{mix}} = - \sum_{i=1}^{i=c} n_i \boldsymbol{R} \ln N_i \qquad (8\text{-}103)$$

To understand the properties of a solution formed by mixing different constituents together, we must consider both the energy and the entropy changes involved. The energy change upon forming an ideal solution is simply zero; this is always the case when the three different kinds of molecular interaction energies (in two-component systems), ε_{11}, ε_{22}, and ε_{12}, are the same. The entropy change in forming an ideal solution is, however, nonzero, and always positive (see Eq. 8-103).

8-10 VAPOR–LIQUID EQUILIBRIUM IN AN IDEAL SOLUTION

Let us consider an isothermal system made up of two components and two phases: both phases, liquid and vapor, behave ideally. The chemical potential of any constituent in any phase will depend, at constant T, on the pressure and composition only.

$$d\mu_i = \left(\frac{\partial \mu_i}{\partial P}\right)_{T,N_i} dP + \left(\frac{\partial \mu_i}{\partial N_i}\right)_{T,P} dN_i \qquad (8\text{-}104)$$

N_i refers to the mole fraction in both phases. The first partial derivative on the right-hand side represents the molar volume, $\mathcal{V}_i$. The second derivative can be shown by simple differentiation of Eq. (8-93) to be

$$\left(\frac{\partial \mu_i}{\partial N_i}\right)_{T,P} = \frac{\boldsymbol{R}T}{N_i} \qquad (8\text{-}105)$$

and Eq. (8-104) takes the form

$$d\mu_i = \mathcal{V}_i\,dP + \frac{\boldsymbol{R}T}{N_i}\,dN_i \qquad (8\text{-}106)$$

Integration between the limits of $N_i = 1$ and N_i yields

$$\mu_i(P,\,T) = \mu_i^*(P_i^*\,T) + \int_{P_i^*}^{P} \mathcal{V}_i\,dP + \boldsymbol{R}T \ln N_i \qquad (8\text{-}107)$$

where P_i^* is the vapor pressure of pure component i and P is the pressure above the solution of composition N_i. On comparing Eq. (8-93) with Eq. (8-107) you can see that the standard state chemical potentials at the pressure of the system and the vapor pressure of the pure component, respectively, are related through the equation

$$\mu_i^*(P, T) = \mu_i^*(P_i^*, T) + \int_{P_i^*}^{P} \mathcal{V}_i \, dP \tag{8-108}$$

$\mu_i^*(P_i^*, T)$ is the standard chemical potential of the pure component at a pressure equal to its saturated vapor pressure at the temperature of the system, and $\mu_i^*(P, T)$ is the standard chemical potential of the pure component at the temperature and pressure of the system. After combining the equilibrium conditions

$$\mu_i(l, P, T) = \mu_i(g, P, T) \tag{8-109}$$

and

$$\mu_i^*(l, P_i^*, T) = \mu_i^*(g, P_i^*, T) \tag{8-110}$$

with Eq. (8-107), we have

$$\int_{P_i^*}^{P} \mathcal{V}_i(l) \, dP + RT \ln x_i = \int_{P_i^*}^{P} \mathcal{V}_i(g) \, dP + RT \ln y_i \tag{8-111}$$

where x_i and y_i are the mole fractions in the liquid (l) and vapor (g), respectively.

The molar volume of the component in the liquid phase, $\mathcal{V}_i(l)$, can be considered constant in the given pressure interval, and the molar volume in the vapor phase, $\mathcal{V}_i(g)$, may be expressed by the virial gas equation (see page 15)

$$\mathcal{V}_i(g) = \frac{RT}{P} + \mathcal{B}_{ii}$$

After evaluating the integral, Eq. (8-111) takes the form

$$\mathcal{V}_i(l)(P - P_i^*) + RT \ln x_i = \mathcal{B}_{ii}(P - P_i^*) + RT \ln \frac{P}{P_i^*} + RT \ln y_i \tag{8-112}$$

When we write Eq. (8-112) in an exponential form for a two-component system, we have

$$P = \frac{x_1 P_1^*}{y_1} \exp \left[\frac{(\mathcal{V}_1(l) - \mathcal{B}_{11})(P - P_1^*)}{RT} \right] \tag{8-113}$$

for component 1, and

$$P = \frac{x_2 P_2^*}{y_2} \exp \left[\frac{(\mathcal{V}_2(l) - \mathcal{B}_{22})(P - P_2^*)}{RT} \right] \tag{8-114}$$

for component 2. As $y_2 = 1 - y_1$, the last two equations may be combined. The result is

$$P = x_1 P_1^* \exp \left[\frac{(\mathcal{V}_1(l) - \mathcal{B}_{11})(P - P_1^*)}{RT} \right]$$
$$+ x_2 P_2^* \exp \left[\frac{(\mathcal{V}_2(l) - \mathcal{B}_{22})(P - P_2^*)}{RT} \right] \tag{8-115}$$

On comparing Eq. (8-115) with Dalton's law (see page 30)

$$P = p_1 + p_2$$

we find for the partial pressure

$$p_i = x_i P_i^* \exp \left[\frac{(\mathcal{V}_i(l) - \mathcal{B}_{ii})(P - P_i^*)}{RT} \right] \qquad \textbf{(8-116)}$$

The value of the exponent is in most cases close to zero, and therefore

$$p_i \approx P_i^* x_i \qquad \textbf{(8-117)}$$

Eq. (8-117) was found empirically by Raoult (1886). It states that the partial pressure of component i over an ideal solution, p_i, is equal to the product of the vapor pressure of the pure component i at the temperature of the solution, P_i^*, and the mole fraction of the component i in the liquid phase, x_i. Raoult's law does not define an ideal solution exactly. It follows from the exact definition, Eq. (8-93), only after making the assumption that $\exp[\ ] = 1$.

If we use Eq. (8-117), Dalton's law may be written in the form

$$P = x_1 P_1^* + x_2 P_2^* = x_1 P_1^* = P_2^*(1 - x_1) + x_1(P_1^* - P_2^*) + P_2^* \qquad \textbf{(8-118)}$$

For any given system and temperature, P_1^* and P_2^* are constant and, hence, the dependence of the total pressure on the composition of the liquid phase may be represented by a straight line (Figure 8-6, curve $P_2^* C P_1^*$).

To derive the relationship between the total pressure and the composition of the vapor phase, Dalton's law is used in the form of Eq. (1-65). According to this

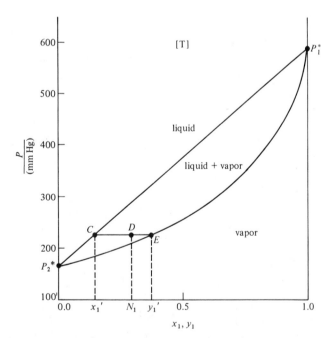

Figure 8-6 Vapor–liquid phase diagram of the nearly ideal system, carbon tetrachloride (1)–tetrachloroethylene (2), at 70°C.

equation, the partial pressure is proportional to the vapor phase composition, y_i, where the proportionality constant is the total pressure of the system

$$p_1 = Py_1 \qquad\qquad\qquad \textbf{(8-119a)}$$

$$p_2 = Py_2 = P(1 - y_1) \qquad\qquad \textbf{(8-119b)}$$

Combination of Eqs. (8-117) and (8-119) results in

$$P = \frac{P_1^* x_1}{y_1} \qquad\qquad\qquad \textbf{(8-120)}$$

On substituting the solution for x_1 from Eq. (8-118), we have

$$P = \frac{P_1^* P_2^*}{P_1^* - y_1(P_2^* - P_1^*)} \qquad\qquad \textbf{(8-121)}$$

It is obvious that the total pressure is not a linear function of the vapor phase composition. This dependence is shown in Figure 8-6, curve $P_2^* E P_1^*$.

Figure 8-6 presents the phase diagram of the nearly ideal system carbon tetrachloride–tetrachloroethylene at 70°C. Above the liquid phase composition line, $P_2^* C P_1^*$, only the liquid phase is stable; below the vapor phase composition curve, $P_2^* E P_1^*$, only the vapor phase can exist. In addition to the temperature, the system has two more degrees of freedom in these regions (see Eq. 7-1). Between the two curves, liquid and vapor coexist in equilibrium and the system has only one degree of freedom in addition to the temperature. Any system whose overall composition is between these two curves, for example D, will separate into two phases: liquid phase of composition x_1', and vapor phase of composition y_1'. The composition of the two phases depends on the pressure and is independent of the overall composition of the system.

Nevertheless as we show next, the overall composition affects the relative amounts of the two phases in equilibrium. From a mass balance, it follows that the total number of moles, n, is given as

$$n = n' + n'' \qquad\qquad\qquad \textbf{(8-122)}$$

where n' and n'' denote the number of moles in liquid and gas phases, respectively. A similar balance may be written for each component:

$$n_1 = n_1' + n_1'' \qquad\qquad\qquad \textbf{(8-123)}$$

Since $N_1 = n_1/n$, $x_1' = n_1'/n'$, and $y_1' = n_1''/n''$, we may rewrite Eq. (8-123) as

$$N_1 n = x_1' n' + y_1' n'' \qquad\qquad \textbf{(8-124)}$$

On substituting Eq. (8-122) into Eq. (8-124) and rearranging, we obtain

$$\frac{n'}{n''} = \frac{y_1' - N_1}{N_1 - x_1'} = \frac{\overline{DE}}{\overline{CD}} \qquad\qquad \textbf{(8-125)}$$

It is obvious from this equation that a change in N_1 (overall composition of component 1) results in a change of the ratio n'/n''. Relation (8-125) is called the lever rule. It applies to any type of phase equilibrium.

At a constant pressure, the boiling point of a solution depends on its composition; thus, the vapor pressures of the pure components, P_1^* and P_2^*, are no longer constant.

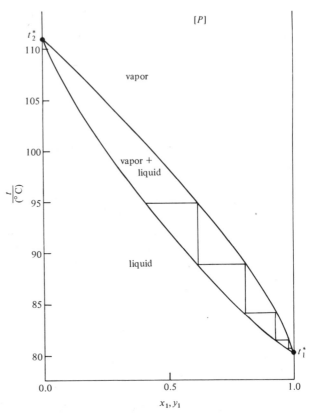

Figure 8-7 Vapor–liquid phase diagram of the nearly ideal system, benzene (1)–toluene (2), at 760 mm Hg.

Consequently, none of the curves $[T=f(x_1), \ T=f'(y_1)]$ will be represented by straight lines. The constant pressure vapor–liquid equilibrium diagram of the system benzene–toluene is shown in Figure 8-7. The upper curve represents the vapor composition as a function of temperature, and the lower curve gives the dependence of the liquid composition on the temperature. Benzene is the more volatile substance, and therefore, the vapor is richer in benzene throughout the entire concentration range than the liquid solution itself. It is obvious from Figure 8-7, that benzene may be separated from the other component by distillation (see Section 8-17).

EXAMPLE 8-4

The vapor pressures of ethylene glycol monomethyl ether and ethylene glycol monoethyl ether at 80°C are 154.4 mm Hg and 102.2 mm Hg, respectively. Assuming ideal behavior in the liquid as well as in the vapor phase, calculate the composition of the vapor that is in equilibrium with a liquid solution of composition $x_1 = 0.500$ at a temperature of 80°C.

Solution: The total vapor pressure above the solution is evaluated from Eq. (8-118)

$$P = x_1(P_1^* - P_2^*) + P_2^* = 0.5\,(154.4 - 102.2) + 102.2$$

$$= 128.3 \text{ mm Hg}$$

and the composition of the vapor phase is given by Eq. (8-120)

$$y_1 = \frac{P_1^* x_1}{P} = \frac{154.4 \times 0.500}{128.3} = 0.602 \quad \bullet$$

8-11   BEHAVIOR OF DILUTE SOLUTIONS

In this section we shall discuss the properties of solutions that are not ideal throughout the entire concentration range, but behave ideally when properly diluted.

8-11.1   The Solubility of Gases in Liquids

Let us consider a special type of vapor–liquid equilibrium, that in which the temperature and pressure of the system lie above the critical state of one of the components of the system. This component is a gas in its pure state and, therefore, this type of equilibrium is that of gases dissolved in liquids.

At a constant temperature, the chemical potential of a real gas, $\mu_2(g)$, is given by the equation

$$\mu_2(g) = \mu_2^\circ(g, T, 1 \text{ atm}) + RT \ln f_2 \qquad (8\text{-}82)$$

Similarly, the chemical potential of the gas dissolved in the dilute ideal solution, $\mu_2(l)$, is expressed by

$$\mu_2(l) = \mu_2^\bullet(l, T, P) + RT \ln x_2 \qquad (8\text{-}93)$$

where $\mu_2^\bullet(l, T, P)$ is the chemical potential referred to the infinitely dilute standard state, and $\mu_2^\circ(g)(T, 1 \text{ atm})$ is the chemical potential referred to the pure gas at the temperature of the system and a pressure of 1 atm (see Section 8-8.2). At equilibrium, $\mu_2(g) = \mu_2(l)$, and thus

$$\mu_2^\circ(g, T, 1 \text{ atm}) + RT \ln f_2 = \mu_2^\bullet(l, T, P) + RT \ln x_2 \qquad (8\text{-}126)$$

Consequently,

$$f_2 = x_2 \exp \left[\frac{\mu_2^\bullet(l, T, P) - \mu_2^\circ(g, T, 1 \text{ atm})}{RT} \right] \qquad (8\text{-}127)$$

and

$$f_2 = x_2 H \approx p_2 \qquad (8\text{-}128)$$

since, at pressures that are not very high, f_2 may be replaced by the partial pressure of the gas above the solution. As can be seen

$$H = \exp \left[\frac{\mu_2^\bullet(l, T, P) - \mu_2^\circ(g, T, 1 \text{ atm})}{RT} \right] \qquad (8\text{-}129)$$

Eq. (8-128) is the mathematical expression of Henry's law (1803). According to this law, the solubility of a gas in a dilute solution is directly proportional to its partial pressure above the solution. The proportionality constant is mainly temperature dependent; it also varies, however, with the total pressure, with the chosen standard state, and, as is evident from Table 8-2, with the nature of the solvent.

Table 8-2 Henry's law constant of some gases in water and benzene at 25°C

Substance	$\dfrac{H}{(\text{atm})}$	
	Water	Benzene
H_2	7.02×10^4	3.61×10^3
N_2	8.57×10^4	2.35×10^3
O_2	4.34×10^4	
CO	5.71×10^4	1.61×10^3
CO_2	1.64×10^3	1.12×10^2
CH_4	4.13×10^4	5.62×10^2
C_2H_2	1.33×10^3	
C_2H_4	1.14×10^4	
C_2H_6	3.03×10^4	

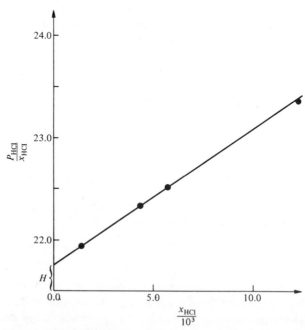

Figure 8-8   Determination of Henry's law constant: system, hydrogen chloride–toluene at 25°C.

Henry's law is not restricted to gas–liquid systems, but is followed by a wide variety of fairly dilute solutions and by all solutions in the limit of extreme dilution. However, corrections must be applied when the molecular structure of the solute undergoes a change in the solution (decomposition or association).

The value of Henry's law constant, H, may be obtained by plotting the value of p_2/x_2 versus x_2 and extrapolating to $x_2 = 0$. Such a plot is shown in Figure 8-8. The values of H for some gases are presented in Table 8-2. As seen from Eq. (8-128) the constant H cannot be identified with the vapor pressure of the pure solute P_2^*, because the law applies only to dilute solutions of component 2 and not to the pure substance. In addition, if component 2 is a gas at the temperature and pressure of the system, it is unrealistic to assign a vapor pressure value to it.

It may be recognized from the Gibbs–Duhem equation that, if Raoult's law holds for one of the components of a binary solution (solvent), then Henry's law must hold in the same region for the other component (solute).

If there is more than one gas present in the system, the solubility of each will depend on its individual partial pressure. Thus, the solubility of a certain gas in a mixture is not affected by the other gases; at low pressures H is independent of the composition of the gas.

The solubility of a gas in a liquid solvent is, however, affected significantly by the presence of other solutes in the solution, especially electrolytes. The solubility is usually reduced. The extent of this "salting out" varies considerably with different salts, but with a given salt the relative decrease in solubility is nearly the same for different gases.

On taking the partial derivative of Eq. (8-129) with respect to P at constant temperature, the pressure dependence of H is obtained.

$$\left(\frac{\partial \ln H}{\partial P}\right)_T = \frac{1}{RT}\left[\frac{\partial \mu_2^\bullet(l, T, P)}{\partial P} - \frac{\partial \mu_2^\circ(g, T, 1\text{ atm})}{\partial P}\right]$$

$$= \frac{\bar{V}_2^\bullet(l, P, T)}{RT} \tag{8-130}$$

The first term in the brackets is the partial molal volume of the dissolved gas in an infinitely dilute solution at the T and P of the system. The value of the second term in the brackets is zero. Because of the small compressibility of liquids, the volume dependence on pressure is very small, at least at low and moderate pressures, and thus the variation of H with pressure may be ignored.

The partial derivative of Eq. (8-129) with respect to T at constant pressure yields the temperature dependence of Henry's law constant.

$$\left(\frac{\partial \ln H}{\partial T}\right)_P = \frac{\mathcal{H}_2^\circ(g, T, 1\text{ atm}) - \bar{\mathcal{H}}_2^\bullet(l, T, P)}{RT^2}$$

$$= -\frac{\bar{\mathcal{L}}_2}{RT^2} \tag{8-131}$$

where $\bar{\mathcal{L}}_2 [\equiv \bar{\mathcal{H}}_2^\bullet(l, T, P) - \mathcal{H}_2^\circ(g, T, 1\text{ atm})]$ is the last differential heat of solution; that is, the change in enthalpy accompanying the transfer of 1 mole of gas into a practically saturated solution. For an exothermic reaction, the right-hand side of Eq. (8-131) is positive and thus H increases with increasing temperature. Consequently,

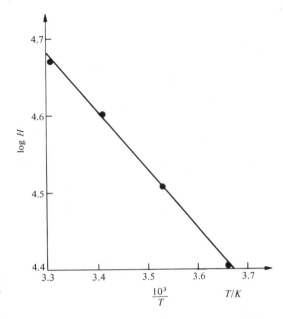

Figure 8-9 The logarithm of Henry's law constant versus $1/T$: system, oxygen–water.

the solubility is reduced. In the case of an endothermic reaction, the solubility increases with increasing temperature. Endothermic solubility is not very common, nevertheless, some systems exhibit such behavior.

In a narrow temperature interval $\bar{\mathscr{L}}_2$ may be taken as temperature independent and integration of Eq. (8-131) yields

$$\ln H = \frac{\bar{\mathscr{L}}_2}{R}\frac{1}{T} + I_H$$

$$= \frac{A}{T} + I_H$$

where I_H is an integration constant and $A = \bar{\mathscr{L}}_2/R$. Figure 8-9 presents the solubility of oxygen in water in the temperature interval between 0 and 30°C. Note that the relation is nearly linear.

EXAMPLE 8-5
Calculate the ratio of the mole fractions of oxygen to nitrogen dissolved in water at 0°C. *Given:* $H_{N_2} = 5.29 \times 10^4$ atm and $H_{O_2} = 2.55 \times 10^4$ atm.
Solution: From Eqs. (8-119) and (8-128), we obtain

$$p_{N_2} = Py_{N_2} = H_{N_2}x_{N_2}$$

and

$$p_{O_2} = Py_{O_2} = H_{O_2}x_{O_2}$$

Dividing the last two equations and rearranging terms results in[8]

$$\frac{x_{O_2}}{x_{N_2}} = \frac{y_{O_2}H_{N_2}}{y_{N_2}H_{O_2}} = \frac{0.21}{0.79} \times \frac{5.29}{2.55} = 0.55$$

This value is in excellent agreement with the ratio found experimentally: 0.54.

[8] Air is composed of 21% oxygen and 79% nitrogen.

Since the ratio of x_{O_2}/x_{N_2} in water is greater than the ratio of the mole fractions in the air (0.27), fractional solubility may result in a complete separation of oxygen from nitrogen in the air.   ●

EXAMPLE 8-6
The solubility of carbon monoxide in water, expressed in terms of reciprocal Henry's constant, is given in the following table.

Temperature (°C)	H (10^4 atm)
0	3.52
50	7.61
100	8.46

Express the solubility of carbon monoxide in water as function of temperature. Calculate also the last differential heat of solution at 50°C.
Solution: Let us consider, that the last differential heat of solution is a linear function of temperature: $\mathcal{L}_2 = a + bT$. On substituting this into Eq. (8-131) and integrating, we obtain

$$\ln H = \frac{a}{RT} - \frac{b}{R} \ln T + C = \frac{A}{T} + B \ln T + C$$

where $a/R = A$ and $-b/R = B$. The constants, A, B, and C are evaluated from the data given in the table by solving three equations with three unknowns:

$$\ln H = -\frac{7928.7}{T} - 22.126 \ln T + 163.627$$

and

$$a = AR = -7928.7 \times 1.987 = -15{,}754.3$$
$$b = -BR = -(-22.126) \times 1.987 = 43.96$$

The last differential heat of solution at 50°C is

$$\mathcal{L}_2 = a + bT = (-15{,}754.3 + 43.96 \times 323.15)\, 4.184$$
$$= -6.481 \text{ kJ mol}^{-1}   ●$$

8-11.2 Colligative Properties

A colligative property is any property that depends only on the number of particles of the solute in the solution and not in any way on the nature of the particles. The colligative properties discussed next are vapor pressure depression, boiling point elevation, freezing point depression, and osmotic pressure.

It is a well known experimental fact that addition of a nonvolatile solute to a solvent results in a solution whose vapor pressure, at a given temperature, is lower than that of the pure solvent.

According to Raoult's law, the vapor pressure of the solvent in a very dilute solution is given by

$$p_1 = P_1^* x_1 \qquad \text{(8-117)}$$

Since $x_1 = 1 - x_2$, Raoult's law may also be written in the form

$$x_2 = \frac{P_1^* - p_1}{P_1^*} = \frac{\Delta p}{P_1^*} \qquad \text{(8-133)}$$

in which P_1^* denotes the vapor pressure of the pure solvent, p_1 is the partial pressure of the solvent above the solution, both at the temperature of the solution, and x_2 is the mole fraction of the nonvolatile solute in the solution. The depression of the vapor pressure, $P_1^* - p_1 = \Delta p$, for dilute solutions of nonvolatile substances is thus seen to depend only on the mole fraction of the solute; it is, therefore, a colligative property.

In a dilute solution, however

$$x_2 = \frac{n_2}{n_1 + n_2} \approx \frac{n_2}{n_1} = \frac{w_2 M_1}{w_1 M_2} \qquad \text{(8-134)}$$

consequently

$$M_2 = \frac{w_2 M_1}{w_1} \frac{P_1^*}{\Delta p} \qquad \text{(8-135)}$$

where all the properties on the right-hand side of Eq. (8-135), that is, the weight of the solute, w_2, and solvent, w_1, and $P_1^*/\Delta p$, can be determined experimentally; M_1, the molecular weight of the solvent, is known and thus M_2, the molecular weight of the solute, may be calculated.

A direct consequence of the vapor pressure depression is an increase of the boiling point. It is evident from Figure 8-10 that the temperature at which the vapor pressure of a solution reaches atmospheric pressure is higher than for the pure solvent.

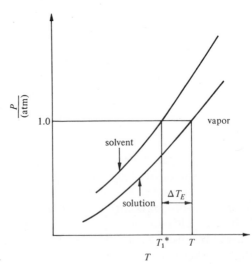

Figure 8-10 Boiling point elevation.

For a quantitative treatment of the boiling point elevation, let us consider the pure liquid solvent in equilibrium with its vapor. The condition for equilibrium is

$$\mu_1^*(g) = \mu_1^*(l) \tag{8-136}$$

consequently

$$d\mu_1^*(g) = d\mu_1^*(l) \tag{8-137}$$

If we now dissolve a small amount of a nonvolatile substance in the solvent, the boiling point will be increased. This will result in a change of the chemical potentials. Since the vapor phase is pure, the change of its chemical potential is given as

$$d\mu_1^*(g) = \left(\frac{\partial \mu_1^*(g)}{\partial T}\right)_P dT \tag{8-138}$$

The chemical potential of the solvent in the solution is, however, affected by the temperature change as well as by the concentration change, and thus we write

$$d\mu_1(l) = \left(\frac{\partial \mu_1(l)}{\partial T}\right)_{P,x_1} dT + \left(\frac{\partial \mu_1(l)}{\partial x_1}\right)_{T,P} dx_1 \tag{8-139}$$

The change of the chemical potential in the vapor phase with the temperature must be completely balanced by a change of the chemical potential in the liquid phase with the temperature and concentration if the system is to remain in equilibrium. Thus

$$\left(\frac{\partial \mu_1^*(g)}{\partial T}\right)_P dT = \left(\frac{\partial \mu_1(l)}{\partial T}\right)_{P,x_1} dT + \left(\frac{\partial \mu_1(l)}{\partial x_1}\right)_{T,P} dx_1 \tag{8-140}$$

Making use of earlier relationships

$$\left(\frac{\partial \mu_1^*(g)}{\partial T}\right)_P = -\mathscr{S}_1(g) \qquad \left(\frac{\partial \mu_1(l)}{\partial T}\right)_{P,x_1} = -\bar{\mathscr{S}}_1(l)$$

and

$$\left(\frac{\partial \mu_1}{\partial x_1}\right)_{T,P} = \frac{RT}{x_1}$$

we have

$$(\mathscr{S}_1(g) - \bar{\mathscr{S}}_1(l)) \, dT = -\frac{RT}{x_1} dx_1 \tag{8-141}$$

Because, in a very dilute solution, the partial molal entropy, $\bar{\mathscr{S}}_1(l)$, can be set equal to the molar entropy, $\mathscr{S}_1(l)$, we may write

$$\mathscr{S}_1(g) - \mathscr{S}_1(l) = \Delta \mathscr{S}_{1,\text{vap}} = \frac{\Delta \mathscr{H}_{1,\text{vap}}}{T} \tag{8-142}$$

Substituting Eq. (8-142) into Eq. (8-141) yields

$$-\frac{dx_1}{x_1} = -d \ln x_1 = \frac{\Delta \mathscr{H}_{1,\text{vap}}}{RT^2} dT \tag{8-143}$$

Over a small temperature interval, the enthalpy of vaporization of the pure solvent, $\Delta\mathcal{H}_{1,\text{vap}}$, may be considered independent of temperature, and upon integration, Eq. (8-143) becomes

$$-\ln x_1 = -\frac{\Delta\mathcal{H}_{1,\text{vap}}}{R}\left(\frac{1}{T} - \frac{1}{T_1^*}\right)$$

$$= -\frac{\Delta\mathcal{H}_{1,\text{vap}}}{R}\left(\frac{T_1^* - T}{TT_1^*}\right)$$

$$= \frac{\Delta\mathcal{H}_{1,\text{vap}}\,\Delta T_E}{RTT_1^*} \tag{8-144}$$

where T_1^* and T are the boiling points of the pure solvent and solution, respectively, and $\Delta T_E = T - T_1^*$ is the boiling point elevation. In a dilute solution, $T \approx T_1^*$, and $-\ln x_1 \approx x_2$.[9] Thus

$$x_2 = \frac{\Delta\mathcal{H}_{1,\text{vap}}}{RT_1^{*2}}\Delta T_E \tag{8-145}$$

and, on rearrangement of terms

$$\Delta T_E = \frac{RT_1^{*2}}{\Delta\mathcal{H}_{1,\text{vap}}}x_2 \tag{8-146}$$

On making use of the simple relation between the mole fraction, x_2, and molality, m

$$x_2 \approx \frac{mM_1}{1000}$$

we have

$$\Delta T_E = \frac{RT_1^{*2}}{\Delta\mathcal{H}_{1,\text{vap}}}\left(\frac{mM_1}{1000}\right) = \frac{RT_1^{*2}}{1000(\Delta\mathcal{H}_{1,\text{vap}}/M_1)}m = K_E m \tag{8-147}$$

where M_1 is the molecular weight of the solvent, $\Delta\mathcal{H}_{1,\text{vap}}/M_1 = l_{1,\text{vap}}$ is the specific enthalpy of vaporization of the solvent, and K_E represents the molal boiling point elevation. The value of K_E, called the ebulliometric constant, depends only on the properties of the pure solvent.

To calculate the molecular weight of the solute, M_2, Eq. (8-147) may be rewritten into the form

$$M_2 = K_E\frac{1000w_2}{\Delta T_E w_1} \tag{8-148}$$

in which the molality, m, has been expressed by

$$m = \frac{n_2}{w_1}1000 = \frac{w_2}{M_2 w_1}1000$$

---

[9]
$$\ln x_1 \approx (x_1 - 1) - \frac{(x_1 - 1)^2}{2} + \frac{(x_1 - 1)^3}{3}$$

Neglecting the higher terms

$$\ln x_1 \approx (x_1 - 1) \quad \text{or} \quad -\ln x_1 \approx (1 - x_1) = x_2$$

Here, n_2 and w_2 denote the number of moles and the weight of the solute, and w_1 is the weight of the solvent. Since the boiling point elevation is inversely proportional to the molecular weight, this method cannot be applied to the determination of high molecular weights.

Another consequence of the vapor pressure lowering is the freezing point depression. It should be apparent from Figure 8-11 that the temperature at which the solid and liquid phases are in equilibrium at a constant atmospheric pressure is lower for the solution than for the pure solvent. Here, we are concerned only with the simple case in which the solid that crystallizes out of the solution is the pure solvent.

Again, the condition for equilibrium between the pure solid and liquid solvent is

$$\mu_1^*(l) = \mu_1^*(s) \tag{8-149}$$

and

$$d\mu_1^*(l) = d\mu_1^*(s) \tag{8-150}$$

After dissolving a small amount of a nonvolatile substance, the freezing point will be lowered. As before, the change of the chemical potential of the pure solid phase with the temperature must be exactly balanced by a change in the chemical potential of the liquid phase with the temperature and concentration if the system is to remain in equilibrium. Thus

$$\left(\frac{\partial\mu_1^*(s)}{\partial T}\right)_P dT = \left(\frac{\partial\mu_1(l)}{\partial T}\right)_{P,x_1} dT + \left(\frac{\partial\mu_1(l)}{\partial x_1}\right)_{T,P} dx_1 \tag{8-151}$$

Proceeding as in the case of the boiling point elevation, we derive

$$\Delta T_F = \frac{RT_1^{*2}}{1000(\Delta\mathcal{H}_{1,\text{fus}}/M_1)}\left(\frac{1000w_2}{M_2 w_1}\right) = K_F\frac{1000w_2}{M_2 w_1} \tag{8-152}$$

and consequently

$$M_2 = K_F 1000\frac{w_2}{\Delta T_F w_1} \tag{8-153}$$

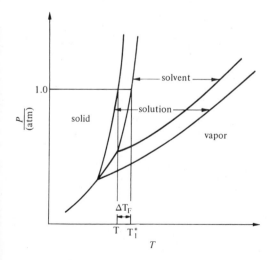

Figure 8-11 Freezing point depression.

Table 8-3 Ebulliometric constants of some solvents

Solvent	Normal Boiling Point (°C)	Ebulliometric Constant	
		Experimental	$K_E = \dfrac{RT_1^{*2}}{1000\, l_{1,vap}}$
water	100.00	0.52	0.52
methanol	64.51	0.81	0.83
ethanol	78.33	1.20	1.19
acetone	56.15	1.72	1.73
benzene	80.10	2.57	2.60
chloroform	61.20	3.78	3.85
carbon tetrachloride	76.72	5.01	5.02

The constant K_F represents the freezing point depression of a 1 molal solution, and its value depends only on the properties of the solvent. As in the case of the boiling point elevation, the freezing point depression is inversely proportional to the molecular weight. Because of the higher value of K_F (compare Tables 8-3 and 8-4), the freezing point depression method, **cryoscopy**, is more accurate than the boiling point elevation method, **ebulliometry**. Cryoscopy also has an additional advantage in that the freezing point is much less dependent on the pressure than the boiling point.

The colligative method discussed next deals with a process called **osmosis**. When a solution of a nonvolatile solute is separated from the pure solvent by a membrane permeable only to the solvent, it is observed that the solvent tends to pass through the membrane into the solution, thereby diluting it. This phenomenon, called osmosis, and first reported by Abbé Nollet in 1748, has a simple explanation in the equation

$$\mu_1 = \mu_1^*(T, P) + RT \ln x_1 \qquad (8\text{-}93)$$

Since x_1 in a solution is lower than unity, the chemical potential of component 1 (solvent) in the solution is lower than the chemical potential of the pure solvent at the

Table 8-4 Cryoscopic constants of some solvents

Solvent	Freezing Point (°C)	$K_F = \dfrac{RT_1^{*2}}{1000\, l_{1,fus}}$
water	0.00	1.86
acetic acid	16.60	3.90
benzene	5.533	5.10
nitrobenzene	5.70	6.9
naphthalene	80.25	7.0
bromoform	8.30	14.3
cyclohexane	6.554	20.2

same pressure and temperature (that is, $\mu_1 < \mu_1^*$). In order to equalize the chemical potentials and bring the system to equilibrium, solvent will flow through the membrane from the pure solvent (higher chemical potential) to the solution (lower chemical potential). When equilibrium is reached, $\mu_1^* = \mu_1$, and also $d\mu_1^* = d\mu_1$.

At constant temperature and pressure, the chemical potential of the pure solvent is a constant and therefore $d\mu_1^* = 0$. However, the chemical potential of the solvent in the solution is changed due to the presence of the solute. Applying the condition for equilibrium, the change in the chemical potential with the concentration has to be fully counterbalanced by some other change. At a constant temperature, the only remaining intensive variable is the pressure—the pressure over the solution must differ from the pressure over the pure solvent—and thus

$$d\mu_1^* = d\mu_1 = \left(\frac{\partial \mu_1}{\partial x_2}\right)_{T,P} dx_2 + \left(\frac{\partial \mu_1}{\partial P}\right)_{T,x_2} dP = 0 \qquad \text{(8-154)}$$

In a two-component system, $dx_2 = -dx_1$; consequently

$$\left(\frac{\partial \mu_1}{\partial x_2}\right)_{T,P} = -\left(\frac{\partial \mu_1}{\partial x_1}\right)_{T,P} \qquad \text{(8-155)}$$

and Eq. (8-154) can be written in the form

$$-\left(\frac{\partial \mu_1}{\partial x_1}\right)_{T,P}(-dx_1) = -\left(\frac{\partial \mu_1}{\partial P}\right)_{T,x_2} dP \qquad \text{(8-156)}$$

Since

$$\left(\frac{\partial \mu_1}{\partial x_1}\right)_{T,P} = \frac{RT}{x_1} \qquad \text{and} \qquad \left(\frac{\partial \mu_1}{\partial P}\right)_{T,x_2} = \bar{V}_1(l)$$

it follows that

$$-d \ln x_1 = \frac{\bar{V}_1(l)}{RT} dP \qquad \text{(8-157)}$$

Furthermore, if we assume $\bar{V}_1(l) = V_1(l)$ is independent of pressure and replace $-\ln x_1$ with x_2,[10] we obtain

$$x_2 = \frac{V_1(l)}{RT}(P - P_1^*) = \frac{V_1(l)}{RT}\pi \qquad \text{(8-158)}$$

where $V_1(l)$ is the molar volume of the pure solvent and $\pi = P - P_1^*$ is the pressure that must be applied to the solution in order to stop osmosis; that is, the osmotic pressure. The osmotic pressure may, therefore, be defined as the external pressure that must be applied to the solution in order to raise the chemical potential of the solvent in the solution to that of the pure solvent.

In a dilute solution, $x_2 = n_2/(n_1 + n_2) \approx n_2/n_1$, and therefore

$$\pi = \frac{n_2 RT}{n_1 V_1(l)} \qquad \text{(8-159)}$$

However, $n_1 V_1(l)$ is the total volume of the solvent containing n_2 moles of the solute

---

[10] We are dealing with a dilute solution, and therefore, $-\ln x_1 \approx x_2$ and $\bar{V}_1(l) \approx V_1(l)$.

which, for a dilute solution, is essentially the volume, V, of the solution. Consequently,

$$\pi = \frac{n_2}{V} RT = cRT = \frac{w_2}{M_2 V} RT \qquad (8\text{-}160)$$

or better

$$\lim_{c \to 0} \left(\frac{\pi}{c}\right) = RT \qquad (8\text{-}161)$$

where w_2 is the weight of the solute, M_2 its molecular weight, and c is the molarity.

The last equation, known as **van't Hoff's law** for a dilute solution, is used to find the molecular weight of solutes from measurements of osmotic pressure. Because of the difficulties in obtaining accurate osmotic pressures (molecules of solutes with lower molecular weights may diffuse through the membrane and produce an erratic value of the osmotic pressure), this method is used mostly in high polymer work. From the derivation of the osmotic pressure, it is obvious that the osmotic pressure is a property of the solution and is independent of the nature of the membrane. A simple device for measuring osmotic pressure is shown in Figure 8-12.

Osmosis is of great importance in biology. The walls of plant and animal cells act as semipermeable membranes. If the fluid surrounding the cells is more concentrated in solutes than the fluid within the cells (a hypertonic solution), the cells lose fluid by osmosis and shrink. If the external fluid is more dilute (hypotonic) than the internal fluid, it will diffuse into the cells; as a consequence of this diffusion, the cells may

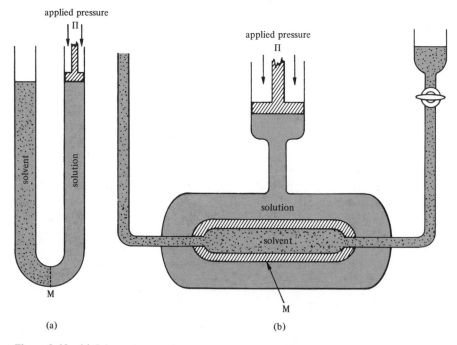

Figure 8-12 (a) Schematic osmotic pressure experiment. (b) Osmotic pressure apparatus of Berkeley and Hartley. M is the semipermeable membrane.

Table 8-5 Osmotic pressures of solutions of sucrose in water at 20°C

Concentration		Osmotic Pressure (atm)		
			Calculated	
Molality	Molarity	Experimental	from Eq. (8-159)	from Eq. (8-160)
0.1	0.098	2.59	2.40	2.36
0.2	0.192	5.06	4.81	4.63
0.3	0.282	7.61	7.21	6.80
0.4	0.370	10.14	9.62	8.90
0.5	0.453	12.75	12.0	10.9

burst. Hemolysis is the term used to describe the loss of red hemoglobin by the bursting of blood cells. Hemolysis may occur when the osmotic pressure of the blood drops very sharply. Cells will not be affected by osmosis in isotonic solutions (equal osmotic pressures), such as the physiological solution containing about 0.9% NaCl. Intravenous feeding must, therefore, be done with solutions isotonic with blood.

8-12 NONIDEAL SOLUTIONS

In a nonideal solution, the chemical potential is related to the activity rather than to the mole fraction

$$\mu_i = \mu_i^*(T, P) + RT \ln a_i \qquad (8\text{-}78)$$

It can be easily proved from this equation that neither ΔV_{mix} nor ΔH_{mix} is equal to zero in a nonideal solution. For the Gibbs free energy of mixing of a nonideal solution, we write

$$\Delta G_{mix} = \sum_{i=1}^{i=c} n_i[\mu_i - \mu_i^*(P, T)] = \Delta H_{mix} - T\Delta S_{mix}$$

$$= \sum_{i=1}^{i=c} n_i RT \ln a_i \qquad (8\text{-}162)$$

According to Eq. (8-79) $a_i = \gamma_i x_i$, and Eq. (8-162) assumes the form

$$\Delta G_{mix} = \sum_{i=1}^{i=c} n_i RT \ln x_i + \sum_{i=1}^{i=c} n_i RT \ln \gamma_i \qquad (8\text{-}163)$$

$$= \Delta G_{mix}^{id} + \Delta G^E \qquad (8\text{-}164)$$

Comparison with Eq. (8-102) shows that the first term on the right-hand side of Eq. (8-163) is identical with the Gibbs free energy of mixing of an ideal solution ($\gamma_i = 1$). The second term, ΔG^E, is responsible for the nonideal behavior of the solution and is called the excess Gibbs free energy. When $\Delta G^E > 0$ or $\gamma_i > 1$, the system is said to exhibit positive deviations from ideal behavior. In the opposite case, when $\Delta G^E < 0$, or $\gamma_i < 1$, the system exhibits negative deviations from ideal behavior. Finally, $\Delta G^E = 0$, $\gamma_i = 1$, defines an ideal solution.

8-13 VAPOR–LIQUID EQUILIBRIUM IN NONIDEAL SOLUTIONS

Let us consider a two-component nonideal liquid solution in equilibrium with an ideal vapor mixture of nonideal gases ($\delta_{12} = \mathcal{B}_{12} - 1/2\mathcal{B}_{11} - 1/2\mathcal{B}_{22} = 0$). If we follow the same procedure that we used for the ideal solution on page 202, Eq. (8-165) may be derived at constant temperature

$$P = x_1 P_1^* \gamma_1 \exp\left[\frac{(\mathcal{V}_1(l) - \mathcal{B}_{11})(P - P_1^*)}{RT}\right]$$

$$+ x_2 P_2^* \gamma_2 \exp\left[\frac{(\mathcal{V}_2(l) - \mathcal{B}_{22})(P - P_2^*)}{RT}\right] \tag{8-165}$$

On comparing this with Dalton's law, $P = p_1 + p_2$, where $p_1 = Py_1$ and $p_2 = Py_2$, we may find the partial pressure of component i

$$p_i = Py_i = x_i P_i^* \gamma_i \exp\left[\frac{(\mathcal{V}_i(l) - \mathcal{B}_{ii})(P - P_i^*)}{RT}\right] \tag{8-166}$$

The value of the exponent is, in most cases, close to zero, and thus

$$y_i = \frac{x_i P_i^* \gamma_i}{P} \tag{8-167}$$

Eq. (8-167) is the analog of Eq. (8-120), which describes the ideal solution; formally, they differ only in that Eq. (8-167) contains a correction factor, the activity coefficient γ_i. If we know the total pressure of the system P, the vapor pressure of the pure component P_i^*, and the appropriate activity coefficient, γ_i, we can easily calculate the composition of the vapor phase, y_i, in equilibrium with the liquid solution of composition, x_i.

The theoretical prediction of γ_i by means of classical and statistical thermodynamics is limited to simple systems. In the great majority of systems, the activity coefficients must be determined experimentally. For solutions of nonelectrolytes, vapor–liquid equilibrium data are often used for the evaluation of the activity coefficients.[11]

What we mean by direct experimental determination of vapor–liquid equilibrium is separation of samples of the liquid and vapor phases that are in mutual equilibrium, either at constant temperature or constant pressure, and the subsequent determination of the concentrations of both phases. A simple vapor–liquid equilibrium still is shown in Figure 8-13. The distillation flask, A, is filled with the liquid solution, which is brought to its boiling point with an electric heater. The evolved vapors and also

---

[11] Colligative properties as well as some other properties of solutions may also be used for the determination of the activity coefficients.

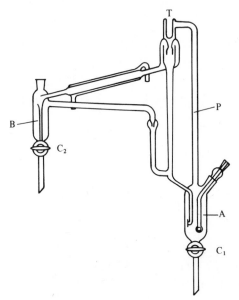

Figure 8-13 Vapor–liquid equilibrium still.

some liquid rise through the Cottrell pump, P, pass by the thermometer, T, condense in the cooler, and flow into the receiver, B. After a several-fold exchange of the contents of the receiver, a steady state is attained in which the composition of the liquid in the boiling flask and the composition of the distillate in the receiver no longer change. This state is indicated by the constancy of the boiling temperature. Samples of the liquid and condensed vapor are then withdrawn through the appropriate stopcocks C_1, C_2, into the sampling flasks for analysis. The distillation flask is then filled with a new solution, and the experiment is repeated.

Experimental vapor–liquid equilibrium data indicate that, for systems exhibiting complete miscibility in both the liquid and vapor phase, three different types of phase diagrams are observed (see Figure 8-14).

Type I. Systems whose total vapor pressure is intermediate between those of the pure components. The vapor in such systems is always richer in the more volatile component than the liquid. Examples: carbon tetrachloride–benzene, tetra-chloroethylene–benzene, methyl alcohol–water, and so on.

Type II. Systems exhibiting a maximum in the total vapor pressure curve or a minimum in the boiling point curve. To the left of the maximum, the vapor is richer in the more volatile substance than the liquid. Distillation with an efficient column would yield a distillate having the composition of the maximum in the curve, while the less volatile substance would remain at the bottom of the column as a residue. To the right of the maximum, the vapor is richer in the less volatile substance than the liquid; therefore, the more volatile substance will remain at the bottom of the column. At the maximum point, called the **azeotrope**—the constant boiling mixture—the liquid and vapor have the same composition. For this reason, azeotropic mixtures cannot be separated into their components by simple distillation; in fact, they were thought, at one time, to be real chemical compounds.

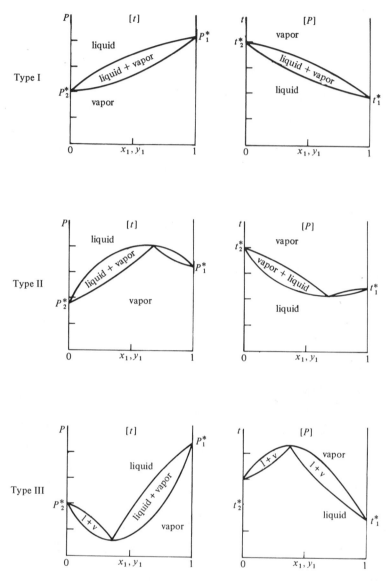

Figure 8-14   Different types of vapor–liquid phase diagrams.

In order to achieve separation of an azeotropic mixture into its components, the conditions of distillation, pressure, or temperature, must be changed, or a different component must be added to the system (azeotropic distillation). As an example, the well known azeotropic mixture ethyl alcohol–water is easily separated by the addition of benzene as a third component to the distillation mixture. The lowest boiling point is shown by the azeotrope benzene–water (a heterogeneous azeotrope), and therefore, all the water is distilled off with the benzene and removed from the mixture. The remaining solution of benzene and ethyl alcohol is then separated by distillation.

Systems with a maximum in the total vapor pressure curve exhibit positive deviations from ideal behavior, that is, $\gamma_i > 1$, and the excess Gibbs free energy is positive. Examples are carbon disulfide–acetone, benzene–ethyl alcohol, dioxane–water, and so on.

Type III. Systems exhibiting a minimum in the total vapor pressure curve or a maximum in the boiling point curve. Here, the azeotropic mixture has the lowest vapor pressure, and hence the highest boiling point. Systems to the left of the minimum will be richer in the less volatile component in the vapor phase than in the liquid phase. To the right of the minimum, just the opposite is true. At the minimum, the composition of both phases is the same. These systems exhibit negative deviations from ideal behavior, that is, $\gamma_i < 1$, and the excess Gibbs free energy is negative. Examples are chloroform–acetone, formic acid–water, water–nitric acid, and so on.

It can be generally stated that the vapor is always richer than the liquid in the component which, upon addition to the system, raises the vapor pressure (Konovalov's rule).

It can be proved mathematically that the composition of the two phases at the maximum or minimum point must be identical. For the sake of simplicity, let us assume that the vapor may be treated as an ideal gaseous mixture. The total pressure is therefore the sum of the two partial pressures (see page 30).

$$P = p_1 + p_2$$

At the maximum or minimum

$$\left(\frac{\partial P}{\partial x_1}\right)_T = \left(\frac{\partial p_1}{\partial x_1}\right)_T + \left(\frac{\partial p_2}{\partial x_1}\right)_T = 0 \tag{8-168}$$

The Gibbs–Duhem equation for an ideal vapor may be written as

$$x_1\left(\frac{\partial \ln p_1}{\partial x_1}\right)_T + x_2\left(\frac{\partial \ln p_2}{\partial x_1}\right)_T = \frac{x_1}{p_1}\left(\frac{\partial p_1}{\partial x_1}\right)_T + \frac{x_2}{p_2}\left(\frac{\partial p_2}{\partial x_1}\right)_T = 0 \tag{8-169}$$

Hence

$$\left(\frac{\partial p_2}{\partial x_1}\right)_T = -\frac{x_1 p_2}{p_1 x_2}\left(\frac{\partial p_1}{\partial x_1}\right)_T \tag{8-170}$$

and Eq. (8-168) assumes the form

$$\left(\frac{\partial p_1}{\partial x_1}\right)_T \left(1 - \frac{x_1 p_2}{x_2 p_1}\right) = 0 \tag{8-171}$$

Since $(\partial p_1/\partial x_1)_T$ is never zero, the quantity in the parenthesis must be zero at the azeotropic composition. Because $p_1 = P y_1$ and $p_2 = P y_2$, we may also write

$$\frac{p_1}{p_2} = \frac{y_1}{y_2} \tag{8-172}$$

which, upon substitution into Eq. (8-171) yields

$$1 - \frac{x_1 y_2}{x_2 y_1} = 0 \tag{8-173}$$

One of the conditions required by this equation is $x_1 = y_1$ and $x_2 = y_2$.

Figure 8-14 illustrates that the vapor–liquid equilibrium diagrams of two-component systems are, in general, not very complicated. Nevertheless, their experimental determination may be, in some cases, tedious and time consuming. Thermodynamics furnishes us with some relations that we can use to reduce to a minimum the number of experiments that must be performed.

The most important relation in this connection is the Gibbs–Duhem equation, (8-25):

$$x_1\left(\frac{\partial \ln \gamma_1}{\partial x_1}\right)_{T,P} = x_2\left(\frac{\partial \ln \gamma_2}{\partial x_2}\right)_{T,P} \tag{8-174}$$

As is evident from this equation, if the dependence of the activity coefficient on composition is known for one of the components, the dependence for the other component may be evaluated by integration.

A different procedure applies the Margules expansions:

$$\bar{\mathcal{G}}_1^E = \mu_1^E = RT \ln \gamma_1 = A_1 x_2 + B_1 x_2^2 + C_1 x_2^3 + \cdots \tag{8-175a}$$

and

$$\bar{\mathcal{G}}_2^E = \mu_2^E = RT \ln \gamma_2 = A_2 x_1 + B_2 x_1^2 + C_2 x_1^3 + \cdots \tag{8-175b}$$

μ_1^E and μ_2^E are the partial molal excess Gibbs free energies of components 1 and 2 respectively, and A_1, A_2, B_1, B_2, C_1, C_2, ... are constants, which are pressure and temperature dependent only. There are no constant terms in the power series because $\lim_{x_2 \to 0} \gamma_1 = 1$ and $\lim_{x_2 \to 1} \gamma_2 = 1$.[12]

When the Gibbs–Duhem equation is written in the form

$$RT(x_1 \, d \ln \gamma_1 + x_2 \, d \ln \gamma_2) = 0 \tag{8-176}$$

the constants of the two series (8-175a) and (8-175b) can be proved interrelated. The Gibbs–Duhem equation also requires that $A_1 = A_2 = 0$ (see Problem 19).

If the activity coefficients of both components are known in a wide range of concentration, the Gibbs–Duhem equation may serve as a test for the thermodynamic consistency of the experimental vapor–liquid equilibrium data, from which the activity coefficients were calculated. The consistency test is derived as follows.

Since the activity coefficient is related to the partial molal excess Gibbs free energy by the equation

$$\bar{\mathcal{G}}_i^E = \mu_i^E = RT \ln \gamma_1 \tag{8-177}$$

the Gibbs–Duhem equation (8-25) may also be expressed as

$$x_1\left(\frac{\partial \mu_1^E}{\partial x_1}\right)_{T,P} dx_1 = x_2\left(\frac{\partial \mu_2^E}{\partial x_2}\right)_{T,P} dx_2 \tag{8-178}$$

Upon integrating within the limits $x_1 = 0$, $x_1 = 1$, and $x_2 = 0$, $x_2 = 1$, we obtain

$$[x_1 \mu_1^E]_{x_1=0}^{x_1=1} - \int_{x_1=0}^{x_1=1} \mu_1^E \, dx_1 = [x_2 \mu_2^E]_{x_2=0}^{x_2=1} - \int_{x_2=0}^{x_2=1} \mu_2^E \, dx_2 \tag{8-179}$$

---

[12] The standard state is defined here as the pure component at the temperature and pressure of the system.

Since μ_1^E is finite at $x_1 = 0$ and zero at $x_1 = 1$, and μ_2^E is finite at $x_2 = 0$ and zero at $x_2 = 1$, Eq. (8-179) reduces to the form (the convention represented by Eq. 8-83 has been used for the definition of the standard state)

$$\int_{x_1=0}^{x_1=1} (\mu_1^E - \mu_2^E)\, dx_1 = 0 \qquad (8\text{-}180)$$

Consequently

$$\int_{x_1=0}^{x_1=1} \ln\,(\gamma_1/\gamma_2)\, dx_1 = 0 \qquad (8\text{-}181)$$

which implies that, for thermodynamic consistency, the value of the integral in the given limits is equal to zero at constant T and P. Application of Eq. (8-181) to the system, of pyridine–tetrachloroethylene at 60°C, is shown in Figure 8-15.

It can be readily seen from Eq. (8-180) that the deviations from ideality of the two components cannot be of opposite sign throughout the entire concentration range. If this were so $\mu_1^E - \mu_2^E$ would have the same sign over the whole range of concentration, and the necessary condition set by Eq. (8-180) could not be fulfilled. A common misconception is that, for a given x_1, the sign of the deviations from ideal behavior must always be the same for both components. Although this is the most common observation of the behavior of solutions of nonelectrolytes, it is by no means the only kind permitted by the Gibbs–Duhem equation.

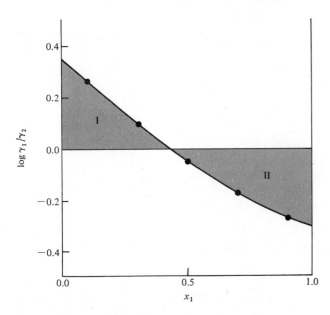

Figure 8-15 Thermodynamic consistency test method: system, pyridine (1)–tetrachloroethylene (2) at 60°C.

EXAMPLE 8-7

The vapor–liquid equilibrium data of the system chloroform–ethanol at 45°C are listed in the following table.

x_1	y_1	$\dfrac{P}{(\text{mm Hg})}$	$\dfrac{p_1}{(\text{mm Hg})}$	$\dfrac{p_2}{(\text{mm Hg})}$
0.1260	0.3974	249.92	99.30	150.62
0.2569	0.6060	329.62	199.60	130.02
0.4015	0.7143	391.51	279.50	112.01
0.6283	0.7954	438.89	349.00	89.89
0.8206	0.8516	455.56	388.60	66.96
0.9557	0.9319	448.49	418.00	30.49

Assuming ideal behavior of the vapor phase, prove the thermodynamic consistency of the experimental data.

Solution: In the case of an ideal vapor phase ($f_i = p_i$), the Gibbs–Duhem equation (8-25) may be written as:[13]

$$\Delta = \frac{x_1}{p_1}\frac{dp_1}{dx_1} - \frac{x_2}{p_2}\frac{dp_2}{dx_2}$$

The smaller Δ is the more consistent of the experimental data.

   In order to prove the thermodynamic consistency of the data, the slopes dp_1/dx_1 and dp_2/dx_2 as well as p_1 and p_2, respectively, at the corresponding mole fractions, have to be known. A plot of the partial pressures, p_1 and p_2, respectively, as function of the mole fraction, can supply the necessary information (Figure 8-16). The values of the slopes computed for the values of x_1 in the preceding table, and the calculated deviations are given in the following table.

$\dfrac{dp_1}{dx_1}$	$\dfrac{dp_2}{dx_2}$	Δ
$\dfrac{\text{mm Hg}}{\text{mole fraction}}$		
792	−176	−0.016
724	−140	+0.132
416	−118	−0.033
229	−96	+0.015
176	−153	−0.038
290	−472	−0.023

As seen from the small values of Δ the data are thermodynamically consistent.

---

[13] In this form the equation is known as the Duhem–Margules equation.

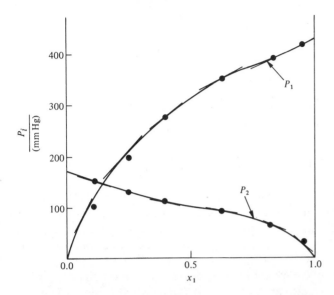

Figure 8-16 Variation of the partial pressure with composition: system, chloroform–ethanol at 45°C.   ●

EXAMPLE 8-8

For an acetone(1)–methanol(2) system, the following boiling point and equilibrium composition of the liquid and vapor·phase were measured at a pressure of 760 mm Hg:

x_1	y_1	$t/(°C)$	$P_1^*/\text{mm Hg}$	$P_2^*/\text{mm Hg}$
0.280	0.420	58.3	819	579

Calculate the activity coefficients of both components in a solution of $x_1 = 0.676$. The boiling point of this solution is 55.70°C at 760 mm Hg.

Solution: It can be shown that the activity coefficients as functions of the mole fraction are given by (see Problem 20)

$$\ln \gamma_1 = \frac{B_1}{RT}x_2^2 + \frac{C_1}{RT}x_2^3 \qquad \text{(a)}$$

$$\ln \gamma_2 = \frac{(B_1 + \frac{3}{2}C_1)}{RT}x_1^2 - \frac{C_1}{RT}x_1^3 \qquad \text{(b)}$$

The constants B_1 and C_1 respectively, may be determined from the given experimental data. Making use of Eq. (8-167), we have

$$\gamma_1 = \frac{y_1 P}{x_1 P_1^*} = \frac{0.420 \times 760}{0.280 \times 819} = 1.3920 \qquad \ln \gamma_1 = 0.3307$$

$$\gamma_2 = \frac{y_2 P}{x_2 P_2^*} = \frac{0.580 \times 760}{0.720 \times 579} = 1.0574 \qquad \ln \gamma_2 = 0.0558$$

On substituting these values into Eqs. (a) and (b) we obtain, for B_1 and C_1, respectively

$$B_1 = 349.9 \qquad C_1 = 97.6$$

These constants, when substituted into (a) and (b), enable one to calculate the activity coefficients in a solution of $x_1 = 0.676$. The results are:

$$\gamma_1 = 1.063 \qquad \gamma_2 = 1.352 \qquad \bullet$$

8-14   REGULAR SOLUTIONS

Some real solutions behave like ideal solutions in certain respects. For example, Hildebrand [11] found experimentally that a large group of nonideal solutions exhibits the same entropy of mixing as an ideal solution. Such solutions are called **regular solutions**. We shall next discuss some of the properties of these regular solutions.

The partial molal entropy of component i in an ideal solution is obtained from Eq. (8-93) by taking the partial derivative with respect to T at constant P and x_1. Thus

$$\left(\frac{\partial \mu_i}{\partial T}\right)_{P,x_i} = \left(\frac{\partial \mu_i^*(\mathrm{l})}{\partial T}\right)_P + R \ln x_i$$

or

$$\bar{\mathscr{S}}_i = \mathscr{S}_i(\mathrm{l}) - R \ln x_i \tag{8-182}$$

Due to the fact that $\Delta S_{\mathrm{mix}}^{\mathrm{id}} = \Delta S_{\mathrm{mix}}^{\mathrm{reg}}$, Eq. (8-182) also relates the partial molal entropy to the mole fraction in a regular solution.

For pure component i

$$\mu_i^*(\mathrm{l}) = \mathscr{H}_i(\mathrm{l}) - T\mathscr{S}_i(\mathrm{l}) \tag{8-183}$$

and for component i in the solution

$$\mu_i = \bar{\mathscr{H}}_i - T\bar{\mathscr{S}}_i \tag{8-184}$$

On subtracting Eq. (8-183) from Eq. (8-184), we obtain

$$\mu_i - \mu_i^*(\mathrm{l}) = \bar{\mathscr{H}}_i - \mathscr{H}_i(\mathrm{l}) - T(\bar{\mathscr{S}}_i - \mathscr{S}_i(\mathrm{l})) \tag{8-185}$$

Substitution of Eq. (8-182) into Eq. (8-185) results in

$$\mu_i - \mu_i^*(\mathrm{l}) = \bar{\mathscr{H}}_i - \mathscr{H}_i(\mathrm{l}) + RT \ln x_i \tag{8-186}$$

On comparison of Eq. (8-186) with Eq. (8-163), we find that

$$RT \ln \gamma_i = \bar{\mathscr{H}}_i - \mathscr{H}_i(\mathrm{l}) \tag{8-187}$$

Since in a regular solution, the activity coefficient is different from 1, $\bar{\mathscr{H}}_i \neq \mathscr{H}_i(\mathrm{l})$; thus the partial molal enthalpy of a component in a regular solution is different from the molal enthalpy of the pure component.

Differentiation of Eq. (8-186) with respect to T at constant P and x_i yields

$$\bar{\mathscr{S}}_i = \mathscr{S}_i(\mathrm{l}) - R \ln x_i + \mathscr{C}_{Pi}(\mathrm{l}) - \bar{\mathscr{C}}_{Pi} \tag{8-188}$$

However, in a regular solution, $\bar{\mathscr{S}}_i = \mathscr{S}_i(\mathrm{l}) - R \ln x_i$, and therefore

$$\bar{\mathscr{C}}_{Pi} = \mathscr{C}_{Pi}(\mathrm{l}) \tag{8-189}$$

Thus the partial molal heat capacity at constant pressure of component i in a regular solution is the same as is the molar heat capacity of the pure liquid component i.

Differentiation of (8-186) with respect to P at constant T and x_i results in another interesting consequence for a regular solution

$$\bar{V}_i - V_i(l) = \left(\frac{\partial \bar{\mathcal{H}}_i}{\partial P}\right)_{T,x_i} - \left(\frac{\partial \mathcal{H}_i(l)}{\partial P}\right)_T \tag{8-190}$$

Since

$$\left(\frac{\partial \bar{\mathcal{H}}_i}{\partial P}\right)_{T,x_i} = \bar{V}_i - T\left(\frac{\partial \bar{V}_i}{\partial T}\right)_{P,x_i}$$

and

$$\left(\frac{\partial \mathcal{H}_i(l)}{\partial P}\right)_T = V_i(l) - T\left(\frac{\partial V_i(l)}{\partial T}\right)$$

it follows for a regular solution

$$\left(\frac{\partial \bar{V}_i}{\partial T}\right)_{P,x_i} = \left(\frac{\partial V_i(l)}{\partial T}\right)_P \tag{8-191}$$

Eqs. (8-189) and (8-191) represent two aspects in which the ideal and regular solutions behave in the same way. It should be understood that Eq. (8-191) does not necessarily require that $\bar{V}_i = V_i(l)$.

It has been proved experimentally that, in some simple cases of regular solutions, the following equation may be applied:

$$\bar{\mathcal{H}}_1 - \mathcal{H}_1(l) = Bx_2^2 \tag{8-192}$$

Comparison of Eq. (8-192) with Eq. (8-187) shows that

$$\ln \gamma_1 = \frac{Bx_2^2}{RT} \tag{8-193}$$

Such systems are called symmetric solutions.

8-15 SOLUTIONS OF PARTIALLY MISCIBLE LIQUIDS

If the positive deviations from ideal behavior are large enough ($\Delta G^E \gg 0$), immiscibility may result. Thus, systems that separate into two liquid phases are composed of constituents for which the activity coefficients are appreciably greater than unity and ΔG^E is large and positive.

As mixing is an irreversible process, the Gibbs free energy of actual mixing of pure components, at constant T and P, must always be negative, regardless of whether the system is completely miscible or not; that is, $\Delta G_{mix} < 0$. A system will separate into two solution phases if the formation of the two phases results in even a more negative value of the Gibbs free energy than if the system formed only a single phase.

An example of this kind of behavior is presented in Figure 8-17, where the molar Gibbs free energy of mixing, $\Delta \mathcal{G}_{mix}/RT$, is plotted against x_1 at constant T and P. The

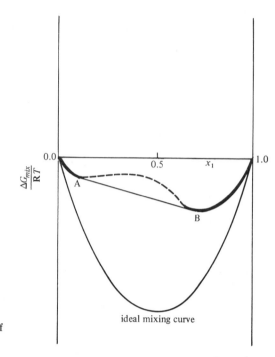

Figure 8-17   The Gibbs free energy of
mixing of liquids with limited solubility.

heavy solid curves at both ends and the center portion—shown dotted—give
$\Delta \mathcal{G}_{\text{mix}}/RT$ as a function of x_1 for the binary system, assuming complete miscibility.
For a system that separates into two liquid phases, the actual $\Delta \mathcal{G}_{\text{mix}}/RT$ for the
incomplete miscibility region is given by a straight line connecting the solutions
representing the two existing phases (conjugate phases) in mutual equilibrium. Such
a line must lie below the dotted curve which would result from complete miscibility.

The linear relationship between $\Delta \mathcal{G}_{\text{mix}}/RT$ and x_1 in the immiscible region results
from the fact that the composition of the two conjugate liquid phases, although
different from each other, remains constant as long as the two phases are present at
the given temperature and pressure. Addition of either of the two components
merely changes the relative masses of the two layers, not their composition. Due to
the constancy of the composition, the activities also remain unchanged, and the
following equation may be written:

$$\frac{\Delta \mathcal{G}_{\text{mix}}}{RT} = x_1 \ln a_1 + (1 - x_1) \ln a_2$$

$$= x_1(\ln a_1 - \ln a_2) + \ln a_2 \tag{8-194}$$

which is the equation of a straight line.

Different solubility diagrams are shown in Figures 8-18, 8-19, and 8-20. Figure
8-18 represents a solution with an upper critical solution temperature; that is, above
this temperature, the system is completely miscible. Any solution within the curve
will separate into two liquid phases. The line A′B′, joining the two conjugate
solutions, is called the tie line. Some examples of solutions with upper critical
solution temperatures are water–aniline and water–phenol. Solutions also exist, for

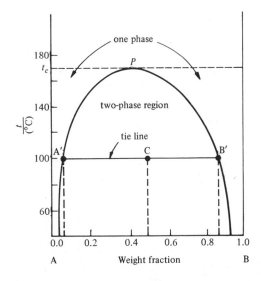

Figure 8-18 Liquid–liquid phase diagram: system, water(A)–aniline(B).

example, triethylamine–water, that exhibit a lower critical solution temperature (Figure 8-19). There are even solutions having upper and lower critical solution temperatures (Figure 8-20), for example, water–nicotine.

A distillation diagram of a partially miscible liquid solution at a constant pressure is shown in Figure 8-21. Below the boiling point tie line, BHC, only two liquid phases exist in mutual equilibrium. At the boiling point, however, vapor is formed above the liquid solution. Since any liquid solution, whose overall composition lies between B and C, will separate into two liquid phases of constant composition, x_B and x_C, the composition of the equilibrium vapor phase will also be constant and is denoted by y_H. Anywhere on line $\overline{BC}$ there are three phases in equilibrium, two of liquid and one of vapor; the system, therefore, has only one degree of freedom. At a given pressure, the boiling point and the composition of the three phases are therefore fixed.

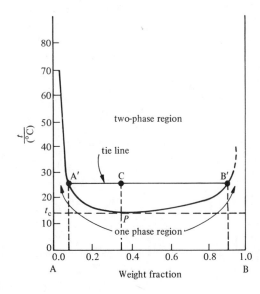

Figure 8-19 Liquid–liquid phase diagram: system, water(A)–triethylamine (B).

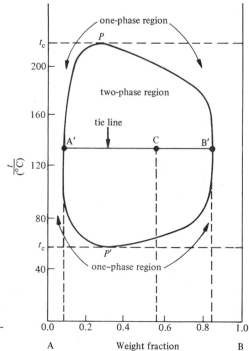

Figure 8-20 Liquid–liquid phase diagram: system, water(A)–nicotine(B).

A liquid solution whose overall composition lies between B and H will yield a vapor richer than the liquid in the less volatile component. Upon distillation, the overall composition of the liquid will shift to the left. On reaching point B, the liquid phase of composition x_C disappears, and the distillation continues as in the case of a homogeneous solution. On the other hand, any solution whose overall composition lies between H and C will yield a vapor phase richer in the more volatile component.

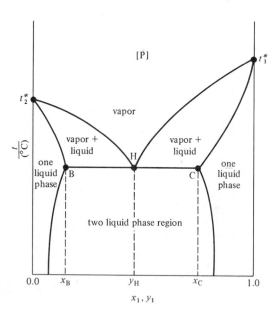

Figure 8-21 Vapor–liquid phase diagram of a system with limited solubility in the liquid phase.

The composition of the liquid will, therefore, shift to the right upon distillation, At C, the liquid phase of composition x_B disappears, and the distillation continues again as in the case of a homogeneous liquid solution.

8-16 VAPOR PRESSURE OF IMMISCIBLE LIQUIDS

A limiting case of partial miscibility is that of practically complete immiscibility. Since immiscible liquids are mutually insoluble, the addition of one liquid to the other does not affect the properties of either liquid. Hence, each will behave as if the other were not present. The total vapor pressure of the system P, at any temperature, is merely the sum of the saturated vapor pressures of the pure components, P_1^* and P_2^*, or

$$P = P_1^* + P_2^* \tag{8-195}$$

The boiling point of any system is the temperature at which the total vapor pressure is equal to the prevailing external pressure. Since, by Eq. (8-195), the two immiscible liquids together can reach any given total pressure at a lower temperature than either liquid alone, it must follow that any mixture of two immiscible liquids must boil at a temperature lower than the boiling point of either of the liquids. Figure 8-22 shows how the vapor pressures add at various temperatures. Curve a gives the vapor pressure as a function of the temperature of water (not drawn to scale), curve b shows an organic substance immiscible with water, and curve c is their sum. It is evident that their sum becomes 1 atm somewhat below 100°C (boiling point of water).

Experimentally, we take advantage of this behavior in a process called steam distillation. Suppose, for example, that some substance decomposes at its normal boiling point. If it is immiscible with water and stable in steam, it can be safely steam

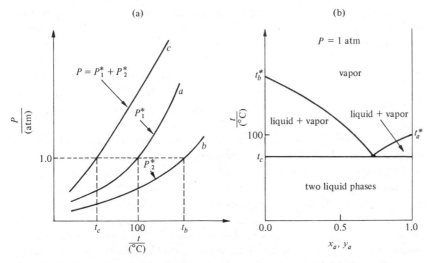

Figure 8-22 Vapor–liquid phase diagram of a system of two immiscible liquid constituents.

distilled at a lower temperature than 100°C at 1 atm pressure, because steam distillation, like vacuum distillation, occurs at a temperature below the normal boiling point of the substance.

The ratio of the weights of the two components in the vapor phase may be calculated from the equations

$$p_1 = P_1^* = Py_1 \tag{8-196a}$$

$$p_2 = P_2^* = Py_2 \tag{8-196b}$$

On division of the last two equations, we obtain

$$\frac{P_1^*}{P_2^*} = \frac{y_1}{y_2} = \frac{n_1/n}{n_2/n} = \frac{n_1}{n_2} = \frac{w_1 M_2}{w_2 M_1} \tag{8-197}$$

Consequently

$$\frac{w_1}{w_2} = \frac{P_1^* M_1}{P_2^* M_2} \tag{8-198}$$

Eq. (8-198) relates directly the ratio of masses of the two components, w_1/w_2, in the distillate to the molecular weights M_1 and M_2 and to the vapor pressures of the pure components, P_1^* and P_2^*.

8-17 FRACTIONAL DISTILLATION

As we have seen from the vapor–liquid equilibrium diagrams, the composition of the equilibrium liquid and vapor phases is different, except in the cases of pure components and azeotropic mixtures. By a simple distillation, it is therefore possible to obtain a liquid and vapor phase of different composition. If the distillation is repeated, a new vapor phase is obtained which differs even more in its composition from the original liquid. By several repetitive vaporizations and condensations, it is possible to separate the constituents of the solution from each other.

The process of separation of mixtures by distillation would be extremely complicated and tedious if it had to be performed by repeated evaporations and condensations in such a simple discontinuous manner. Instead, the separation may be performed in a continuous process, known as fractional distillation, by using a distillation apparatus called a fractionating column, shown in Figure 8-23. It consists essentially of three parts: a boiling flask A, equipped with a heater B, a column D, composed of a series of plates whose construction is shown in the figure, and a condenser F.

On the plates, the liquid flowing downward comes into contact with the vapor moving upward through the bubble caps. The vapor must bubble through the layer of liquid on each plate before it goes to the upper plates. In doing so, part of the less volatile constituent condenses. The heat released during the process of condensation evaporates part of the more volatile constituent of the liquid on the plate. The vapor advancing to the next higher plate is therefore richer in the more volatile constituent than the vapor approaching the plate from below, whereas the liquid flowing back to the next lower plate is richer in the less volatile constituent than the liquid reaching the plate from the next higher one.

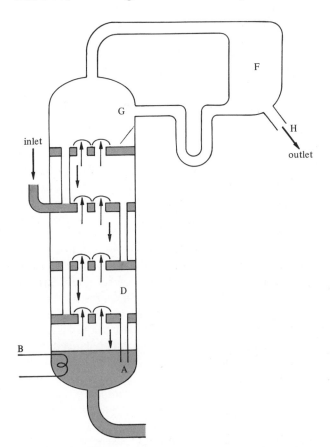

Figure 8-23   Schematic diagram of a fractionating column.

The net result of the interaction between vapor and liquid at the plate is, therefore, a redistribution in favor of the more volatile constituent in the vapor phase and the less volatile constituent in the liquid phase. Since this process repeats itself at each plate, it is possible, with a sufficient number of plates, to separate the mixture into two final fractions, a residue of the less volatile fraction in the distillation flask A and a vapor passing from the top of the column containing essentially the more volatile fraction. (An azeotropic mixture cannot be separated by such a distillation.) After liquefying the vapor in the condenser F, part of it is drawn off through H and part is returned to the column through G in order to maintain the stock of essentially pure distillate on the upper plates. The ratio of the part of the condensed vapor returned to the column to the part drawn off from the column head, called reflux, is a factor that may affect the quality of the distillation product. The lower this ratio, the better the quality of the product obtained in distillation.

8-18   PHASE DIAGRAMS OF CONDENSED SYSTEMS

In our discussion of phase diagrams we dealt mostly with cases in which the vapor phase was an important part of the system. Nevertheless, there are many systems in which the vapor phase may be completely ignored. Some of them, like the solid–liquid equilibrium or solid–solid equilibrium of single-component systems as well as the mutual solubility of two partially miscible liquids, have already been discussed. Systems in which the vapor phase may be ignored are called **condensed systems**. As liquids and solids are practically incompressible, the properties of the condensed systems are only slightly pressure dependent except at very high pressures.

8-18.1 Solid–Liquid Diagrams   As has been stated before, a single-component, single-phase system has two degrees of freedom. When a second phase appears in such a system, the number of degrees of freedom reduces to one. As a result of this, the freezing point of a pure component will depend only on the pressure (and only slightly). Fixing the pressure to a certain value also fixes the freezing point; the system is univariant.

When a pure liquid is cooled at constant pressure, the plot of temperature versus time has a nearly constant nonzero slope. As the temperature drops to the freezing point, the cooling curve becomes horizontal, if the process of cooling has been carried out slowly enough. The halt in the cooling curve results from the appearance of the second—the solid—phase; the system loses its single degree of freedom and, therefore, the temperature has to remain unchanged until the entire liquid is frozen. The heat evolved during this liquid–solid phase transition is exactly balanced by the heat lost due to cooling. As soon as all the liquid is frozen, a further drop in the temperature is observed. This is shown by the cooling curves *a* and *f* in Figure 8-24a. As is evident, the slope of the curve in the liquid region is different from that in the solid region. This is explained by the difference in the heat capacities of the two phases.

The cooling process is different in a two-component one-phase system. Such a system has two degrees of freedom, besides the one fixed by the pressure. Consequently, its cooling curves (*b, c, d, g,* and *e* in Figure 8-24a) look different from those of the pure components. At the appearance of the solid phase, composed, in the case shown in Figure 8-24, of one of the pure components, the system still possesses one more degree of freedom. Therefore, instead of getting a halt in the cooling curve, the temperature drops continuously during crystallization. The concentration change in the liquid as well as the heat evolved during the process of crystallization cause a change of the slope of the cooling curve. At a certain composition of the liquid, the other component begins to crystallize also, and the whole liquid freezes. Curves *b, c, d, g,* and *e* are the cooling curves of liquid solutions of different composition.

The temperature at the onset of crystallization (marked by small circles) is different for the different compositions, whereas the temperature at which both components crystallize simultaneously is the same, regardless of the composition of the original liquid. This temperature is called the **eutectic temperature**. Curve *c* is of special interest in that it completely resembles a cooling curve for a single component (*a, f*). Both solid phases, pure A and pure B, appear at the same temperature, the eutectic temperature. Such a mixture is called a **eutectic mixture**. The two solid phases and one liquid phase present in the system at the eutectic temperature at constant pressure make the system invariant.

The phase diagram constructed from the different cooling curves is shown on the right in Figure 8-24b. The points at the onset of crystallization lie on the curves *CE*

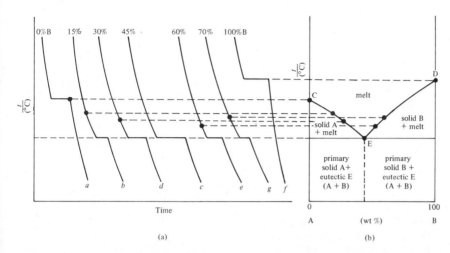

Figure 8-24 (a) Cooling curves. (b) Solid–melt phase diagram (immiscibility in the solid phase).

and *ED*. To the left of *E*, the liquid is always in equilibrium with pure solid A, whereas to the right of *E*, pure solid B crystallizes from the liquid. At *E*, both solids, pure A and pure B, crystallize from the liquid. As is evident from Figure 8-24b, the eutectic temperature is the last temperature at which a liquid phase can still exist. Below this temperature, only the two solid phases coexist in mutual equilibrium. (Since the process of diffusion in solids is very slow, in practice these phases will never reach the equilibrium composition.) Looking at the different cooling curves, it is also apparent that the horizontal halt, called the **eutectic halt**, diminishes as the composition of the liquid phase departs from the eutectic composition.

It is worthwhile to mention that, in an ideal case, the curves *CE* and *ED* can be approximated by the equation

$$\ln x_i = -\frac{\Delta \mathcal{H}_{i,\text{fus}}}{R}\left(\frac{1}{T} - \frac{1}{T_i^*}\right) \tag{8-199}$$

where x_i is the solubility of component i in the liquid phase at temperature T, and T_i^* and $\Delta \mathcal{H}_{i,\text{fus}}$ are the freezing point and molar enthalpy of fusion, respectively, of pure component i.

The type of diagram discussed in the preceding paragraphs represents systems in which the mutual solubility of the two components in the solid phase is negligibly small. When a limited mutual solubility is to be considered, the diagram looks like that shown in Figure 8-25 (compare with Figure 8-24b). To the right of *E*, the liquid is in equilibrium with a homogeneous solid mixture of A in B, whereas to the left of *E*, the liquid is in equilibrium with a solid solution of B in A. Let us now consider the cooling of a liquid solution of composition x_1. At temperature t_1, a solid solution of composition x_s appears. As the solid solution is richer in B than the liquid, the composition of the liquid is shifted towards A as the temperature drops. Upon reaching point *E*, solid solution of B in A crystallizes, as does the solid solution of A

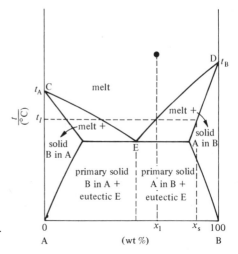

Figure 8-25 Solid–melt phase diagram.
(Limited solubility in the solid phase.)

in B. The temperature at which this happens in the eutectic temperature. Below this temperature, the two solid solutions are in mutual equilibrium; their composition depends on the temperature. Examples of this type of behavior may be found in phase diagrams of Al–CuAl$_2$ and Cu–Zn systems.

The diagram of a system with complete miscibility in both liquid and solid phases is shown in Figure 8-26. The upper curve represents the liquid solution and the lower curve represents the solid solution. The interpretation of this diagram is similar to that of a liquid–vapor phase diagram (see Figure 8-7). At temperature t_1, the liquid of composition x_1 is in equilibrium with a solid solution of composition x_s. During the

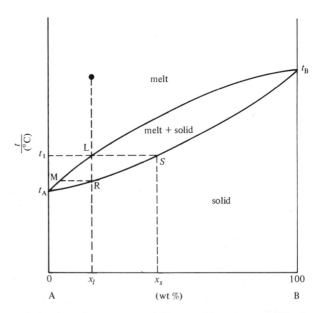

Figure 8-26 Solid–melt phase diagram. (Complete solubility in the melt and solid phase.)

process of cooling, the composition of the liquid is changing along the LM curve, while the composition of the solid solution is moving along the SR line. At point M, the liquid phase disappears. The composition of the solid phase at this point is the same as the composition of the liquid phase at the beginning of cooling, assuming that during the entire process of cooling the two phases were in mutual equilibrium. The Co–Ni, Au–Ag, naphthalene-β-naphthol and sodium chloride–silver chloride systems have phase diagrams of this type.

There are also binary systems that are completely miscible in both liquid and solid phases and exhibit either a maximum (comparatively rare) or a minimum in the melting point. Some examples are the Cu–Au, Mn–Ni, As–Sb and KCl–KBr systems.

The increase in the melting point of A by addition of B, as indicated in Figure 8-26, does not seem to be in agreement with the results derived for the freezing point depression on page 215. We will now prove that, in the case of solid solutions, the freezing point of a substance may also be increased by adding a foreign substance.

From the equilibrium condition, $\mu_i(l) = \mu_i(s)$, the following equation may be written for an ideal liquid mixture in mutual equilibrium with an ideal solid solution

$$\mu_i^*(s) + RT \ln x_i(s) = \mu_i^*(l) + RT \ln x_i(l) \tag{8-200}$$

Rearranging, results in

$$\frac{\mu_i^*(l) - \mu_i^*(s)}{RT} = \frac{\Delta \mathcal{G}_{i,\text{fus}}}{RT} = \ln \frac{x_i(s)}{x_i(l)}$$

$$= \frac{\Delta \mathcal{H}_{i,\text{fus}}}{RT} - \frac{\Delta \mathcal{S}_{i,\text{fus}}}{R} \tag{8-201}$$

where T is the freezing point of the solution. For a reversible process at constant temperature, however, $\Delta \mathcal{H}_{i,\text{fus}}/T^* = \Delta \mathcal{S}_{i,\text{fus}}$, and therefore

$$\ln \frac{x_i(s)}{x_i(l)} = \frac{\Delta \mathcal{H}_{i,\text{fus}}}{R} \left(\frac{1}{T} - \frac{1}{T_i^*} \right) \tag{8-202}$$

where T_i^* is the freezing point of pure component i. On solving for T, we obtain

$$T = T_i^* \left[\frac{\Delta \mathcal{H}_{i,\text{fus}}}{\Delta \mathcal{H}_{i,\text{fus}} + RT_i^* \ln \frac{x_i(s)}{x_i(l)}} \right] \tag{8-203}$$

When the mole fraction in the solid solution is greater than the mole fraction in the liquid phase, that is, $x_i(s)/x_i(l) > 1$, the freezing point T is less than T_i^* and depression is observed. This is also the case when the pure solid component i crystallizes from the solution, that is, $x_i(s) = 1$. In the opposite case, when $x_i(s)/x_i(l) < 1$, an increase in the freezing point is observed. The higher the ratio $x_i(s)/x_i(l)$, the greater the difference between T and T_i^*.

8-18.2 Phase Diagrams of Reacting Components

In some cases the two components react together to form a compound that is stable up to its melting point; thus, at the melting point, the composition of the liquid phase is the same as the composition of the solid compound. It is then said that the compound has a **congruent melting point**. The cooling curve of a liquid of composition equal to the composition of the

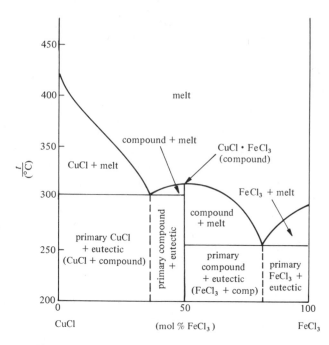

Figure 8-27   Solid–melt phase diagram: system, CuCl–FeCl$_3$.

compound looks, in every respect, like the cooling curve of a single component. Systems such as aniline–phenol, cuprous chloride–ferric chloride, gold–tellurium, aluminum–selenium, and so on, form compounds with congruent melting points.

The phase diagram of the CuCl–FeCl$_3$ system is shown in Figure 8-27. The maximum on the curve corresponds to the compound of composition, CuCl·FeCl$_3$. This diagram can be considered as composed of two simple eutectic diagrams attached to each other.[14] One of the diagrams represents the system CuCl–CuCl·FeCl$_3$, and the other represents the system CuCl·FeCl$_3$–FeCl$_3$. In the composite diagram there are two eutectics, one of CuCl–CuCl·FeCl$_3$, the other of CuCl·FeCl$_3$–FeCl$_3$. The phases corresponding to the various regions of the diagram are labeled.

In some cases, the solid compound does not exhibit stability up to its melting point. When such a compound is heated, it decomposes before it reaches its melting point to yield a new solid phase and a liquid phase of different composition from that of the original solid phase. The compound is said to have an **incongruent melting point**, and the process of decomposition is called a **peritectic** or phase reaction. Since, during the peritectic reaction, three phases—two solid and one liquid—are present at equilibrium, the system is invariant (the pressure is fixed), and hence the temperature, as well as the composition of all phases, is fixed. Some examples of this type of behavior are the systems Na$_2$SO$_4$–H$_2$O, SiO$_2$–Al$_2$O$_3$, CaF$_2$–CaCl$_2$, Na–K, and so on.

---

[14] This is not exactly true; the slope of the curve at the maximum—corresponding to the melting point of the compound—is horizontal, whereas the slopes at the two ends of the diagram—corresponding to the melting points of the pure components—are different from zero.

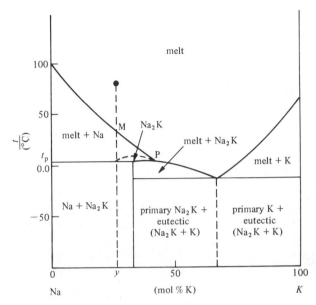

Figure 8-28   Solid–melt phase diagram: system, Na–K.

The Na–K system is shown in Figure 8-28. After heating the compound Na_2K to the peritectic temperature, t_P, the compound decomposes according to the equation

$$Na_2K(s) \rightleftharpoons Na(s) + \text{liquid phase}$$

Above this temperature, the liquid phase is in equilibrium with solid Na only. Upon cooling a liquid of composition y after the point M is reached, solid Na begins to separate from the liquid, whose composition therefore becomes richer in K, falling along the line MP. At point P, solid sodium reacts with the liquid phase, and the compound Na_2K is formed.

When the phase diagrams of condensed systems are examined, several features become obvious: eutectic points, solid solutions, compound formation, peritectic reactions, and so on. Any more complicated phase diagrams may be readily interpreted in terms of these features.

One of the most important phase diagrams in the steel industry is the iron–carbon phase diagram. Its most interesting portion is shown in Figure 8-29. Pure iron crystallizes in a few modifications: α-iron, γ-iron, and δ-iron. The temperature at which the body centered cubic structure, α-iron, is transformed into the face centered structure, γ-iron, is 910°C (point G). At a temperature of 1401°C, (point (N)), γ-iron is transformed back into a new body centered cubic structure, called δ-iron. This structure is stable up to the melting point of iron, 1539°C. The solid solution of carbon in α-iron, existing in the region limited by the curves QP and PG, is called α-ferrite. The solid solution of carbon in γ-iron, called austenite, exists in the region $GSEJN$. The curves GS and GP denote the compositions at which α-ferrite and austenite are in equilibrium. At the point S, three solid phases are in mutual equilibrium: α-ferrite of composition P, austenite of composition S, and a compound of iron and carbon, Fe_3C, called cementite. Point S has, therefore, the property of a eutectic (the liquid phase present at the eutectic temperature is

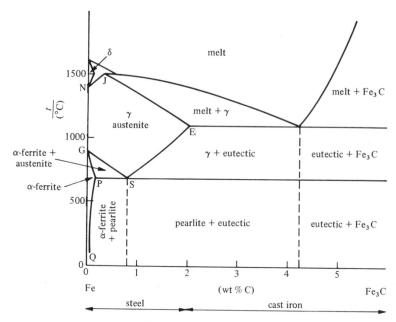

Figure 8-29   Solid–melt phase diagram: system, Fe–C.

replaced here by a third solid phase), and is called a **eutectoid**. A mixture of α-ferrite and cementite, of composition given by S, is called pearlite.

From Figure 8-29 the different behavior of steel and cast iron is evident. A system of iron and carbon (0.008–2%), known as steel, can be heated until the homogeneous solid solution, austenite, is obtained. In this condition, the alloy is readily hot rolled or subjected to other forming processes. A system having more than 2% carbon, called cast iron, cannot be brought into a homogeneous solid form by heating and therefore cannot be conveniently worked by mechanical means.

8-18.3   Phase Equilibrium in Three-Component Systems

According to the phase rule, a three-component, one-phase system, has four degrees of freedom: $v = c + 2 - p = 3 + 2 - 1 = 4$. To illustrate the behavior of such a system by graphical means, a four-dimensional diagram is required. Most commonly, the temperature and pressure are kept constant, so that two-dimensional diagrams can be used.

The two degrees of freedom that remain under constant pressure and temperature conditions are two composition variables. These two independent variables may conveniently be thought of as two of the three mole fractions, where the third is given by

$$\sum_{i=1}^{i=3} N_i = 1$$

Several methods have been proposed for plotting two-dimensional equilibrium diagrams for ternary systems. The equilateral triangle method, suggested by Stokes and Roozeboom, seems to be the most convenient. This method is based upon the fact that, in an equilateral triangle, the sum of the three distances, drawn from a point

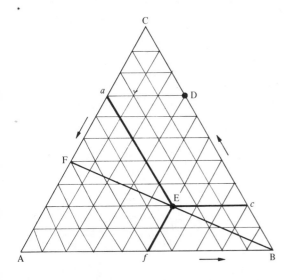

Figure 8-30 Graphical representation of a three-component system at a constant temperature.

within the triangle parallel to each of the three sides, is equal to the side of the triangle. Thus, the condition

$$\sum_{i=1}^{i=3} N_i = 1$$

is automatically satisfied, and plotting becomes very simple. Each apex of the triangle is taken as 100% of the component with which it is identified. The sides of the triangle represent binary systems. For example, point D in Figure 8-30 represents a binary solution of composition 30% B and 70% C. Any three-component mixture must lie within the triangle. A point such as E represents a system containing 30% A (line Ca), 50% B (line Af), and 20% C (line Bc). Any line going from one apex of the triangle to a point on the opposite side, such as BEF in Figure 8-30 represents the composition of all possible mixtures that have the same relative amount of the other two components.

We shall consider only certain types of the many possible ternary systems; those formed by three liquids and those formed by two solids and a liquid. Systems composed of three liquid constituents are presented in Figures 8-31 and 8-32. In the simplest case, where only one pair of the components is of limited solubility (Figure 8-31), the phase diagram is divided by the solubility curve DNPLE into two regions: upper homogeneous and lower heterogeneous. Any ternary mixture within the heterogeneous part of the diagram, for example a mixture of composition M, separates at the given temperature and pressure into two equilibrium phases, whose compositions are given by points L and N on the solubility curve. The line connecting the compositions of these phases in mutual equilibrium, $\overline{NML}$, is called a tie line. All mixtures whose overall composition lies on this tie line separate into two phases of composition N and L.

The relative masses of the two equilibrium phases depend on the overall composition of the heterogeneous mixture. For mixture M, the relative masses of the two phases are given by the ratio $\overline{LM}/\overline{NM}$, the ratio being greater than unity. As is evident from Figure 8-31, when we add component C to a binary solution of

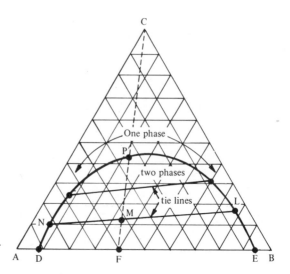

Figure 8-31 Liquid–liquid phase diagram. One partially miscible pair, A–B.

composition F, the concentrations of the two equilibrium phases approach each other, the tie lines become shorter, and finally, at point P, they coincide. This point is not necessarily the maximum point on the solubility curve. There are many systems with this relatively simple behavior. Some examples are H_2O–$CH_3COOC_2H_5$–C_2H_5OH, H_2O–C_6H_6–C_2H_5OH, H_2O–$CHCl_3$–CH_3COOH.

Examples of systems with more than one pair of liquids with limited solubility are shown in Figure 8-32. In part (a), there are two separated heterogeneous regions. In (b) the two regions overlap. Part (c) illustrates a system in which all three heterogeneous regions are separated from each other. Part (d) represents a system with three two-phase regions (areas crossed by the tie lines) and one three-phase region DEF. Any mixture within the three-phase region separates into three phases of composition D, E, and F.

Since the solubility is affected by the temperature, the shape of the solubility curve changes with temperature. Very often systems exhibiting limited solubility at one temperature may become homogeneous at other temperatures. The solubility curves for the system water–aniline–phenol at various temperatures are shown in Figure 8-33. At lower temperatures, water–aniline and water–phenol exhibit a limited solubility. With increasing temperature, the solubility increases; and, as is evident from Figure 8-33, at some higher temperature (approximately 180°C) the system becomes homogeneous.

8-18.4 The Distribution of a Solute Between Two Mutually Insoluble Solvents

Let us now consider a three-component system in which two of the components, 1 and 3, are insoluble or slightly soluble in each other. The third component, 2, will distribute itself between the two insoluble liquid components according to the thermodynamic requirement

$$\mu_{2,1} = \mu_{2,3} \tag{8-204}$$

On combining Eq. (8-204) with Eq. (8-78), we obtain

$$\mu_{2,1}^{\bullet} + RT \ln a_{2,1} = \mu_{2,3}^{\bullet} + RT \ln a_{2,3} \tag{8-205}$$

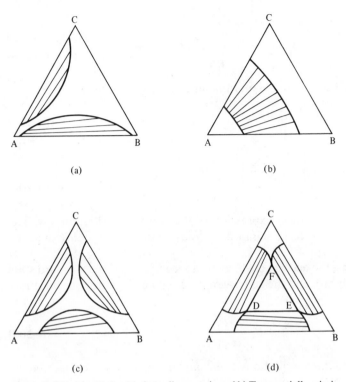

(a)   (b)

(c)   (d)

Figure 8-32 Liquid–liquid phase diagram. (a and b) Two partially miscible pairs; (c and d) Three partially miscible pairs.

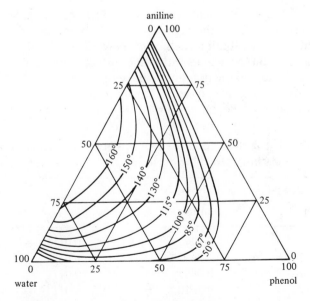

Figure 8-33 Liquid–liquid phase diagram: system, water–
phenol–aniline at various temperatures.

Consequently

$$\ln \frac{a_{2,3}}{a_{2,1}} = \frac{\overset{\bullet}{\mu}_{2,1} - \overset{\bullet}{\mu}_{2,3}}{RT} \tag{8-206}$$

Since both $\overset{\bullet}{\mu}_{2,1}$ and $\overset{\bullet}{\mu}_{2,3}$ are independent of composition, it follows that

$$\ln \frac{a_{2,3}}{a_{2,1}} = \text{constant} \tag{8-207}$$

or

$$K = \frac{a_{2,3}}{a_{2,1}} \tag{8-208}$$

Eq. (8-208) is the mathematical expression of the Nernst distribution law. The temperature dependence of K is significant; the pressure dependence may be ignored.

At low concentrations, the activity coefficients approach unity (the standard state is chosen at infinite dilution), and the activity may be replaced by the concentration. Using molarity, we have

$$K_c \approx \frac{c_{2,3}}{c_{2,1}} \tag{8-209}$$

The constant K_c is called the distribution coefficient and, in general, depends on the concentration as well as on the temperature. At low concentrations, the change of K_c with concentration is negligible, as can be seen from Table 8-6. When the solute undergoes a change, such as association or dissociation in one of the solvents, the simple form of the distribution law, Eq. (8-209), does not give good results, even in very dilute solutions.

Extraction with immiscible solvents finds many practical applications. Organic compounds can be easily transferred from their aqueous solutions to organic solvents. Substances having different distribution coefficients may be separated by simple extraction. When the distribution coefficients of the different substances differ only slightly, successive or multistep extraction must be employed. Partition chromatography is a special type of successive extraction. One of the liquid phases is held by a finely divided solid in a column—the stationary phase—and the other liquid

Table 8-6   **The distribution coefficient of I_2 between water and carbon tetrachloride at 25°C**

c_{H_2O} (mol liter^{-1})	c_{CCl_4} (mol liter^{-1})	$K_c = \dfrac{c_{H_2O}}{c_{CCl_4}}$
0.000322	0.02745	0.0117
0.000503	0.0429	0.0117
0.000763	0.0654	0.0117
0.00115	0.1010	0.0114
0.00134	0.1196	0.0112

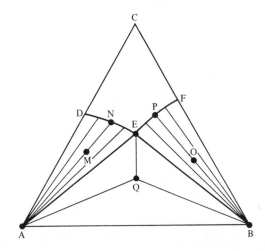

Figure 8-34 Phase diagram of a three-component system, A and B solid and C liquid.

phase flows continuously through the column. Distribution of the solutes (contained in either of the phases) between the two phases occurs according to their distribution coefficients. Biological products that are difficult to separate by other processes have been separated successfully by this method.

8-18.5 Equilibrium Between Solid and Liquid Phases in Three-Component Systems

When the pure components A, B, and C are in different states of aggregation, for example, when A and B are pure solids and C is a liquid, new kinds of phase diagrams are obtained. When the three components are nonreactive, the diagram shown in Figure 8-34 is applicable.

Point D represents the saturated solution of A in C in the absence of B. Similarly, F represents the solubility of B in C in the absence of A. The presence of B changes the solubility of A in C along the curve DE, whereas the presence of A changes the solubility of B in C along the line FE. Point E represents a solution that is saturated with respect to both A and B. In the region CDEFC, only homogeneous solutions exist. A mixture within the region ADE, for example M, consists of the pure solid A and the saturated solution N. A mixture within the area BFE, for example O, consists of the pure solid B and the saturated solution P. A mixture within the triangle AEB, Q for example, separates into three phases: pure solid A, pure solid B, and a saturated solution, E.

Such diagrams have great interest from a practical standpoint, crystallization, for example; it allows us to estimate the amount of solvent that must be added to a solid mixture of A and B in order to separate them from each other.

8-19 THERMODYNAMICS OF SURFACES

An equation relating the saturated vapor pressure of pure liquids to surface phenomena has been discussed in Section 7-6. Here, we shall briefly discuss the behavior of solutions in the vicinity of a surface. We are interested especially in the

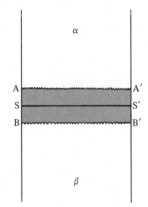

Figure 8-35 Two bulk liquid solutions separated by an interfacial region.

effect of concentration on the surface tension of solutions and in the consequences resulting from this effect.

The treatment presented next is essentially that of Gibbs. The system to be investigated, seen in Figure 8-35, is composed of two phases, α and β, and of an interfacial region positioned between planes AA' and BB'. The two phases are arbitrarily separated by a planar surface SS' drawn within the interfacial region.[15] It is assumed that within the interfacial region the properties of the system vary continuously from purely α at AA' to purely β at BB'; the properties of both bulk phases remain unchanged up to the boundaries of the interfacial region.

The extensive properties of the system, such as the internal energy, entropy, and Gibbs free energy, for example, can be expressed as

$$U = U^\alpha + U^\beta + U^s \tag{8-210}$$

$$S = S^\alpha + S^\beta + S^s \tag{8-211}$$

$$G = G^\alpha + G^\beta + G^s \tag{8-212}$$

We may also write for the number of moles

$$n_i = n_i^\alpha + n_i^\beta + n_i^s \tag{8-213}$$

It is understood that at equilibrium the chemical potentials of constituent i are the same in all the parts of the system. Thus

$$\mu_i^\alpha = \mu_i^\beta = \mu_i^s \tag{8-214}$$

For a small reversible change in the surface phase (designated by the superscript s) the Gibbs free energy change is equal to[16]

$$dG^s = -S^s \, dT + \gamma \, dA + \sum_i \mu_i \, dn_i^s \tag{8-215}$$

where γ is the surface tension and A area. Integration of Eq. (8-215), at constant T,

---

[15] Since the surface is planar, an arbitrary location will not have any influence on the results of the derivation.

[16] It is assumed that the total volume of the solution is shared by the two bulk phases only: $V = V^\alpha + V^\beta$; consequently $V^s = 0$.

γ, and μ_i, yields

$$G^s = \gamma A + \sum_i \mu_i n_i^s \qquad (8\text{-}216)$$

Differentiation of Eq. (8-216) and comparing the result with Eq. (8-215) at constant temperature indicates that

$$\sum_i n_i^s \, d\mu_i + A \, d\gamma = 0 \qquad (8\text{-}217)$$

When expressed per unit area Eq. (8-217) becomes

$$\sum_i \frac{n_i^s}{A} \, d\mu_i + d\gamma = 0 \qquad (8\text{-}218)$$

Introducing the surface concentration, $\Gamma_i = n_i^s/A$, Eq. (8-218) for a two-component solution assumes the form

$$d\gamma = -\Gamma_1 \, d\mu_1 - \Gamma_2 \, d\mu_2 \qquad (8\text{-}219)$$

The exact location of the planar surface SS' is not specified. The values of Γ_1 and Γ_2 will vary with the location. If the position of the surface is such that $\Gamma_1 = 0$, then Eq. (8-219) reduces to

$$\Gamma_2^1 = -\left(\frac{\partial \gamma}{\partial \mu_2}\right)_T \qquad (8\text{-}220)$$

This equation is called the **Gibbs adsorption isotherm**. The chemical potential may be expressed in terms of activity (Eq. 8-78). Thus

$$\Gamma_2^1 = -\frac{1}{RT} \frac{d\gamma}{d \ln a_2} \qquad (8\text{-}221)$$

For dilute solutions, $a_2 = c$, and consequently

$$\Gamma_2^1 = -\frac{c}{RT} \frac{d\gamma}{dc} \qquad (8\text{-}222)$$

If the change of the surface tension with concentration is known, Γ_2^1 can be evaluated. As seen from Eq. (8-222), whenever a solute lowers the surface tension, $d\gamma/dc < 0$, the surface phase is enriched in that solute, $\Gamma_2^1 > 0$. In electrolytic solutions $d\gamma/dc > 0$ and therefore $\Gamma_2^1 < 0$; the surface concentration of the electrolyte is less than it is in the bulk solution.

For a regular solution the surface tension is given by the simple equation

$$\gamma = x_1 \gamma_1 + x_2 \gamma_2 - A x_1 x_2 \qquad (8\text{-}223)$$

Such solutions may exhibit both positive and negative adsorption, depending on the value of A. Commonly, however, $\partial\gamma/\partial x < 0$.

In a dilute solution

$$\gamma = \gamma_0 - Bc \qquad (8\text{-}224)$$

where γ_0 is the surface tension of the pure solvent. Following from this

$$-c \frac{d\gamma}{dc} = \gamma_0 - \gamma = cB \qquad (8\text{-}225)$$

so that Eq. (8-222) assumes the form

$$\Gamma_2^1 = \frac{\gamma_0 - \gamma}{RT} = \frac{cB}{RT} \tag{8-226}$$

Substitution of σ for $1/\Gamma_2^1$ and π for $\gamma_0 - \gamma$ yields

$$\pi\sigma = RT \tag{8-227}$$

π is called the surface pressure and σ is the surface area per mole. The surface pressure is an important property of monolayers—films. Some insoluble substances spread over a surface of liquid to form films exactly one molecule thick; oil over water for example. Detergents are chemicals that strongly affect the surface tension of water.

Problems

1. At 18°C, 100 g of water dissolves 53.846 g of anhydrous calcium chloride giving a saturated solution. The density of the solution is 1.3420 g cm^{-3}. Express the concentration of the solution in (a) weight per cent, (b) molarity, (c) molality, (d) mole fraction, and (e) mol %.

2. The molal enthalpies of mixing of a water–glycol solution at 20°C, are given in the following table:

H_2O (mol %)	$\Delta\mathcal{H}_{mix}$ (cal mol^{-1})	H_2O (mol %)	$\Delta\mathcal{H}_{mix}$ (cal mol^{-1})
90	−2781	40	−6736
80	−4554	30	−4929
70	−5541	20	−4441
60	−6563	10	−2402
50	−6965		

Determine the partial molal enthalpy of mixing of both components for a solution having the following composition: (a) 25 mol % water; (b) 70 mol % water.

3. Derive the following two equations for the partial molal properties:

$$\bar{\mathcal{Z}}_1 = -N_2^2 \left[\frac{\partial(\mathcal{Z}/N_2)}{\partial N_2}\right]_{P,T} \quad ; \quad \bar{\mathcal{Z}}_2 = -N_1^2 \left[\frac{\partial(\mathcal{Z}/N_1)}{\partial N_1}\right]_{P,T}$$

4. Express the concentration dependence of the partial molal volume for both components of the ethanol(1)–water(2) solution at 20°C and 1 atm. Given:

$$\mathcal{V}/(\text{cm}^3\,\text{mol}^{-1}) = 58.36 - 32.64N_2 - 42.98N_2^2 + 58.77N_2^3 - 23.45N_2^4$$

5. Calculate the fugacity of carbon dioxide at a temperature of 50°C and a pressure of 100 atm assuming that under these conditions carbon dioxide follows the van der Waals equation. *Given:*

$$a = 3.59 \times 10^6 \ (\text{cm}^3)^2 \ \text{atm mol}^{-2} \qquad b = 42.7 \ \text{cm}^3 \ \text{mol}^{-1}$$

6. Calculate the fugacity of ammonia at a temperature of 200°C and a pressure of 500 atm. Apply the Berthelot equation of state. *Given:*
$$t_c = 132.4°C \qquad P_c = 111.5 \ \text{atm}$$

7. At a temperature of 200°C the following data were measured for ammonia

$P/(\text{atm})$	20	60	100
$V/(\text{cm}^3 \ \text{mol}^{-1})$	1866	570.8	310.9

Find the fugacity of ammonia at the given temperature and at a pressure of 100 atm.

8. In a two-component solution the fugacity of one of the components is given by the expression

$$\log \frac{f_1}{f_1^\circ} = \log x_1 + B(1-x_1)^2$$

where B is a constant. Find an expression for $\log (f_2/f_2^\circ)$.

9. Assuming the ideal behavior of the system benzene–toluene, derive the following equations at 50°C:

$$P/(\text{mm Hg}) = 179.2x_1 + 92.1$$

$$P/(\text{mm Hg}) = \frac{24.986.7}{271.3 - 179.2y_1}$$

$$y_1 = \frac{2.946(x_1/x_2)}{1 + 2.946(x_1/x_2)}$$

where P is the total vapor pressure, x_1 and y_1 are the mole fractions of benzene in the liquid and vapor phase, respectively. *Given:*

$$P_1^* = 271.3 \ \text{mm Hg} \qquad P_2^* = 92.1 \ \text{mm Hg}.$$

10. The following is a table of the vapor pressures of pure benzene(1) and chlorobenzene(2)

$t(°C)$	90	100	110	120	132
$P_1^*/(\text{mm Hg})$	1013	1340	1744	2235	2965
$P_2^*/(\text{mm Hg})$	208	293	403	542	760

Assuming ideal behavior of the vapor and liquid phases, respectively, determine the boiling point of a mixture containing 40 mol % of benzene and 60 mol % of chlorobenzene at a pressure of 1000 mm Hg.

11. How does the entropy change during the mixing of 0.8 moles of nitrogen with 0.2 moles of oxygen? Assume that both gases behave ideally.

12. How many parts by weight of hydrogen at 20°C dissolve in 100 parts by weight of water from a gaseous mixture in which the partial pressure of hydrogen is 200 mm Hg? Henry's law constant is 6.83×10^4 atm at a temperature of 20°C.

13. The vapor pressure of carbon tetrachloride at 19°C is 85.513 mm Hg. The vapor pressure above a solution containing 0.5455 g of a substance in 25 g of carbon tetrachloride at the same temperature is 83.923 mm Hg. Calculate the molecular weight of this substance and determine its empirical formula if the results of the elementary analysis are as follows: C, 94.34%; H, 5.66%.

14. Ten grams of glucose, dissolved in 400 g of ethanol, elevated the boiling point of the latter by 0.1428°C; 2 g of an organic substance, dissolved in 100 g of ethanol, elevated the boiling point by 0.1250°C. What is the molecular weight of the unknown substance in this solution?

15. If 2.441 g of benzoic acid in 250 g of benzene lowers its melting point by 0.2048°C, what is the molecular formula of benzoic acid in this solvent; K_F of benzene is 5.12.

16. What is the osmotic pressure of a solution of 4.0 g of polyvinyl chloride in 1 liter of dioxane at a temperature of 27°C, if the mean molecular weight of this polymer is 1.5×10^5?

17. The saturated vapor pressure of ethanol and water at 78.8°C is 773.4 mm Hg and 338.3 mm Hg, respectively. Calculate the activity coefficients of both components in a solution of ethanol(1)–water(2) having the composition in the liquid phase $x_1 = 0.663$ and in the vapor phase $y_1 = 0.733$ at a total pressure of 760.00 mm Hg.

18. Making use of the data from the previous problem, find an expression giving the dependence of the activity coefficients on concentration for the system ethanol(1)–water(2) at 78.8°C.

19. The following two series are given:

$$\mu_1^E = RT \ln \gamma_1 = A_1 x_2 + B_1 x_2^2 + C_1 x_2^3$$

and

$$\mu_2^E = RT \ln \gamma_2 = A_2 x_1 + B_2 x_1^2 + C_2 x_1^3$$

Derive the relations

$$A_1 = A_2 = 0 \qquad B_2 = B_1 + \tfrac{3}{2}C_1 \qquad C_2 = -C_1$$

20. Prove that equations

$$RT \ln \gamma_1 = B_1 x_2^2 + C_1 x_2^3 + \cdots$$

$$RT \ln \gamma_2 = (B_1 + \tfrac{3}{2}C_1)x_1^2 - C_1 x_1^3$$

are consistent with the Gibbs–Duhem equation.

21. A phenol–water system separates into two liquid phases at a temperature of 60°C; the first contains 16.8 wt % of phenol, the second 44.9 wt % of water. If the system contains 90 g of water and 60 g of phenol, what is the weight of each phase?

22. How much water must be added to a 100 g solution containing 80 wt % of phenol for the solution to become turbid at 60°C. Use the data from Problem 21.

23. The vapor pressures of bromobenzene and water at 80°C are 66.2 mm Hg and 355.1 mm Hg, respectively. The normal boiling point of bromobenzene is 156°C. Calculate: (a) the temperature at which bromobenzene steam distills in a laboratory where the barometric pressure is 760 mm Hg; (b) the weight percentage of bromobenzene in the vapors in such a steam distillation.

24. The melting point of mercury is −39°C and that of thallium is 303°C. The compound Tl_2Hg_5 has a melting point of 15°C; 8% thallium lowers the melting point of mercury to the minimum, −60°C. The eutectic temperature for thallium and the compound Tl_2Hg_5 is 0.4°C, and the corresponding composition of the eutectic mixture is 41% of thallium. Sketch a phase diagram and determine the maximum amount of thallium obtainable from 10 kg of thallium amalgam containing 80% thallium.

25. Determine how many degrees of freedom there are in a system composed of a saturated solution of sodium chloride, water vapor, and sodium chloride crystals.

26. Mercuric chloride is distributed between water and benzene so that the ratio of the molarity of the aqueous solution to the benzene solution is 12 : 1. How many grams of mercuric chloride are extracted from a solution containing 0.2 g of mercuric chloride in 600 cm³ of benzene, if the solution is extracted twice, each time with 200 cm³ of pure water.

27. The partial pressures and surface tensions of water(1)—ethanol(2) solutions at different concentrations are listed below:

x_2	P_1 (mm Hg)	P_2 (mm Hg)	γ (dyne cm^{-1})
0.2	20.4	26.8	29.7
0.4	18.35	34.2	26.35
0.6	15.8	40.1	24.6
0.8	10.0	48.3	23.2

Calculate the surface area occupied by 1 mole of ethanol.

28. Show that the chemical potential can also be expressed as

$$\mu_i = -T\left(\frac{\partial S}{\partial n_i}\right)_{U,V,n_{j\neq i}}$$

_____ *References and Recommended Reading*

[1] D. W. Rogers: *J. Chem. Educ.*, **39**:527 (1962). Deriving the Gibbs-Duhem equation.
[2] P. J. Robinson: *J. Chem. Educ.*, **41**:654 (1964). Chemical potentials in an ideal mixture of ideal gases.

[3] M. L. McGlashan: *J. Chem. Educ.*, **40**:516 (1963). Deviations from Raoult's law.

[4] A. G. Williamson: *J. Chem. Educ.*, **43**:211 (1966). Raoult's law and the thermodynamic definition of ideal mixing.

[5] V. Fried: *J. Chem. Educ.*, **45**:720 (1968). Some comments on thermodynamics of ideal solutions.

[6] A. Ben-Naim: *J. Chem. Educ.*, **39**:242 (1962). The activity functions of nonideal gases and solutions.

[7] S. D. Christian: *J. Chem. Educ.*, **39**:521 (1962). Gibbs-Duhem equation and free energy calculations.

[8] A. F. Berndt: *J. Chem. Educ.*, **46**:594 (1969). Binary phase diagrams.

[9] K. J. Mysels: *J. Chem. Educ.*, **32**:179 (1955). The mechanism of vapor pressure lowering.

[10] G. N. Copley: *J. Chem. Educ.*, **36**:596 (1959). Eutectics.

[11] J. H. Hildebrand and R. L. Scott: *Regular Solutions.* Prentice-Hall, Inc., Englewood Cliffs, N.J., 1962.

[12] E. Hála, J. Pick, V. Fried, and O. Vilím: *Vapor–Liquid Equilibrium.* Pergamon Press, London, 1958, 1965.

[13] F. P. Chinard: *J. Chem. Educ.*, **32**:377 (1955). Colligative properties.

[14] G. W. Castellan: *J. Chem. Educ.*, **32**:424 (1955). Phase equilibria and the chemical potential.

Application of the First and Second Laws of Thermodynamics to Chemical Reactions

9-1   INTRODUCTION

The previous section dealt with systems in which chemical reactions were not present or, at least, were not of great importance. In this section we shall discuss systems in which chemical reactions are of prime interest. Thermodynamics can supply exact information on the direction in which chemical changes may proceed spontaneously. It also permits the evaluation of the maximum equilibrium yield of a reaction. It can, however, say nothing about the rate of chemical reactions. Chemical changes favored by thermodynamics sometimes proceed at such a low rate, for example, $H_2(g) + \frac{1}{2} O_2(g) = H_2O(g)$, that within a certain reasonable period of time no change is observed in the system. Chemical reactions, like stones that roll downhill, may sometimes require a "nudge" to get started, but the barrier to be surmounted is not in any way the concern of thermodynamics.

9-2   EQUILIBRIUM CONSTANTS OF CHEMICAL REACTIONS

Consider the chemical reaction

$$aA + bB \rightarrow cC + dD$$

Let this reaction proceed in a closed system. Despite the fact that the total mass of the system is unchanged, the number of moles of each reactant and product changes as the reaction proceeds. Now, if the so called progress variable or "extent of the reaction," ζ, increases by an infinitesimal amount $d\zeta$, the amount of A and B decreases by $a\, d\zeta$ and $b\, d\zeta$ and that of C and D increases by $c\, d\zeta$ and $d\, d\zeta$. Thus,

$$dn_A = -a\, d\zeta \qquad dn_B = -b\, d\zeta$$

and

$$dn_C = c\, d\zeta \qquad dn_D = d\, d\zeta$$

If the system in which the reaction takes place is constrained so that it may perform only P-V work, the change in the Gibbs free energy of the reaction, at constant T and P, is given by

$$d\Delta G' = \sum_i \mu_i\, dn_i$$

$$= (d\mu_D + c\mu_C - a\mu_A - b\mu_B)\, d\zeta \qquad \text{(9-1)}$$

255

Consequently

$$\left(\frac{\partial \Delta G'}{\partial \zeta}\right)_{P,T} = d\mu_D + c\mu_C - a\mu_A - b\mu_B$$

$$= \sum_i \nu_i \mu_i \equiv \Delta G \tag{9-2}$$

where ν_i is the stoichiometric coefficient taken with a positive sign for the products (C, D) and a negative sign for the reactants (A, B). The term $(\partial \Delta G'/\partial \zeta)_{P,T}$, introduced by DeDonder in 1922, expresses the change of the Gibbs free energy per unit change in the extent of reaction, ΔG. If this derivative is negative, the reaction proceeds spontaneously in the direction given by the stoichiometric equation—from the left to the right. A positive value of the derivative means that the reverse reaction proceeds spontaneously. When $(\partial \Delta G'/\partial \zeta)_{P,T}$ is zero, the Gibbs free energy has a minimum value and the system is in equilibrium.

Making use of the equation, $\mu_i = \mu_i^\circ + RT \ln a_i$, Eq. (9-2) may be transformed into

$$\Delta G = \sum_i \nu_i \mu_i = \sum_i \nu_i (\mu_i^\circ + RT \ln a_i)$$

$$= c\mu_C^\circ + d\mu_D^\circ - a\mu_A^\circ - b\mu_B^\circ$$
$$+ cRT \ln a_C + dRT \ln a_D - aRT \ln a_A - bRT \ln a_B$$

$$= \Delta G^\circ + RT \ln \left(\frac{a_C^c a_D^d}{a_A^a a_B^b}\right)_{ne} \tag{9-3}$$

where $\Delta G^\circ (\equiv \sum_i \nu_i \mu_i^\circ)$ denotes the change in the standard Gibbs free energy of the reaction, and the subscript ne means nonequilibrium activities.

In equilibrium, at constant T and P, $\Delta G = 0$ and consequently

$$\Delta G^\circ = -RT \ln \left(\frac{a_C^c a_D^d}{a_A^a a_B^b}\right)_e = -RT \ln K \tag{9-4}$$

The subscript e indicates that these are the equilibrium activities; the ratio of the product of the equilibrium activities of the products to the product of the equilibrium activities of the reactants, each raised to its stoichiometric number, defines the equilibrium constant, K:

$$K = \frac{a_C^c a_D^d}{a_A^a a_B^b} \tag{9-5}$$

The thermodynamic equilibrium constant defined by Eq. (9-5) is concentration independent; it varies, however, with the temperature and may vary also with the pressure, depending on the chosen standard state.

9-2.1 Pressure Dependence of the Equilibrium Constant
In order to investigate the dependence of K on pressure, it is convenient to rewrite the combined Eqs. (9-2) and (9-4) in the form

$$\ln K = -\frac{\Delta G^\circ}{RT} = -\frac{\sum_i \nu_i \mu_i^\circ}{RT} \tag{9-6}$$

Upon differentiation with respect to the pressure at constant T, we obtain

$$\left(\frac{\partial \ln K}{\partial P}\right)_T = -\frac{\sum_i \nu_i (\partial \mu_i^\circ / \partial P)_T}{RT}$$

$$= -\frac{\sum_i \nu_i \mathcal{V}_i^\circ}{RT} = -\frac{\left(\sum_P \nu_i \mathcal{V}_i^\circ - \sum_R \nu_i \mathcal{V}_i^\circ\right)}{RT} = -\frac{\Delta V^\circ}{RT} \qquad (9\text{-}7)$$

where $\mathcal{V}_i^\circ$ is substituted for $(\partial \mu_i^\circ / \partial P)_T$ and represents the standard molar volume of component i, ΔV° is the change of the standard volume of the reaction, and $\sum_P$ and $\sum_R$ represent the sum over the products and reactants, respectively.

Since ΔV° is very small for solid and liquid phase reactions, the dependence of K on P for this type of reaction is negligibly small. For gaseous reactions, ΔV° can be significant, and the pressure dependence of the equilibrium constant is to be considered. When $V_i^\circ = \nu_i z_i (RT/P^\circ)$, $\Delta V^\circ = (RT/P^\circ)\Delta(\nu_i z_i)$ and Eq. (9-7) assumes the form

$$d \ln K = -\Delta(\nu_i z_i) \, d \ln P^\circ \qquad (9\text{-}8)$$

To make the integration of Eq. (9-8) simple, we shall neglect the pressure dependence of the $\Delta(\nu_i z_i)$ term. The stoichiometric numbers, ν_i, are always pressure independent, the compressibility factors, z_i, however, can be strongly pressure dependent. Thus

$$\ln K = -\Delta(\nu_i z_i) \ln P^\circ + C \qquad (9\text{-}9)$$

As seen, when the standard state pressure, P°, is 1 atm or any other fixed value the equilibrium constant is independent of pressure. In any other case, except when $\Delta(\nu_i z_i) = 0$, the equilibrium constant is pressure dependent. When $\Delta(\nu_i z_i)$ is negative, K increases with pressure, whereas when $\Delta(\nu_i z_i)$ is positive, K decreases with pressure. At low and moderate pressures, the compressibility factors differ only slightly from unity, and thus $\Delta(\nu_i z_i) = \Delta \nu_i$. Eq. (9-9) is the mathematical expression of one aspect of the Le Chatelier principle.

9-2.2 Temperature Dependence of the Equilibrium Constant

The derivative of Eq. (9-6) with respect to the temperature at constant P leads to

$$\left(\frac{\partial \ln K}{\partial T}\right)_P = -\frac{\sum_i \nu_i [\partial(\mu_i^\circ / T)/\partial T]_P}{R} \qquad (9\text{-}10)$$

If we make use of Eq. (8-70)

$$\left[\frac{\partial(\mu_i^\circ / T)}{\partial T}\right]_P = -\frac{(\Delta \mathcal{H}_f^\circ)_i}{T^2}$$

we can write

$$\left(\frac{\partial \ln K}{\partial T}\right)_P = \frac{\sum_i \nu_i (\Delta \mathcal{H}_f^\circ)_i}{RT^2} = \frac{\Delta H^\circ}{RT^2} \qquad (9\text{-}11)$$

where $(\Delta \mathcal{H}_f^\circ)_i$ is the standard enthalpy of formation of component i, and $\Delta H^\circ [\equiv \sum_i \nu_i (\Delta \mathcal{H}_f^\circ)_i]$ is the standard enthalpy change of the reaction. Eq. (9-11), known as the **van't Hoff equation**, is just another mathematical expression of the Le Chatelier principle. If ΔH° is negative, the equilibrium constant decreases with increasing temperature, whereas, when ΔH° is positive, K increases with temperature.

In order to find the temperature change of the equilibrium constant quantitatively, Eq. (9-11) must be integrated. Over a small temperature range, ΔH° may be considered constant, and integration results in

$$\ln K = -\frac{\Delta H^\circ}{R}\frac{1}{T} + I_K \tag{9-12}$$

In such a case, a straight line is obtained by plotting $\ln K$ against $1/T$. The integration constant I_K is simply related to the standard entropy change of the reaction as can be seen from the following thermodynamic relation

$$-\frac{\Delta G^\circ}{RT} = \ln K = -\frac{\Delta H^\circ}{R}\frac{1}{T} + \frac{\Delta S^\circ}{R} \tag{9-13}$$

The Gibbs free energy of a reaction is always composed of two terms: the enthalpy term and the entropy term. When ΔH° dominates the value of ΔG°, the reaction is said to be enthalpically driven; when the entropy term, ΔS°, contributes extensively to ΔG°, the reaction is said to be entropically driven.

Over a wider temperature interval, the dependence of ΔH° upon temperature must be considered. Using Kirchhoff's equation (5-50) and Eq. (5-38a), we have

$$\ln K = -\frac{I_H}{RT} + \frac{\Delta a}{R}\ln T + \frac{\Delta b}{2R}T + \frac{\Delta c}{6R}T^2 + I_K \tag{9-14}$$

Δa, Δb, and Δc are constants whose values can be easily calculated from the molar heat capacities of the products and reactants. To evaluate the integration constant, I_H, one value of the standard enthalpy change of the reaction must be known, as is shown in Example 5-9. The second integration constant, I_K, can be evaluated either from a known value of K or from other thermodynamic properties directly related to K, such as the molar Gibbs free energies of formation or standard enthalpies of formation and standard entropies, respectively.

The saturated vapor pressure of pure liquids can be treated as an equilibrium constant of the process

$$A(l) \rightarrow A(g)$$

and

$$K = \frac{a(g)}{a(l)} = f(g) \approx P^*$$

The standard state in the liquid phase is chosen so that $a(l) = 1$ and the standard state in the vapor phase is $f^\circ(g) = 1$ atm.

Consequently, all the equations derived in Chapter 7 resemble the equations derived for K in Chapter 9.

EXAMPLE 9-1

H. Zeisse [12] gives the following data for the reaction

$$H_2(g) + \tfrac{1}{2} O_2(g) \rightleftharpoons H_2O(g)$$

$\log K = 10.059$ at 1000 K

$\Delta \mathcal{H}^\circ_{298} = -57{,}798$ cal mol^{-1}

$O_2(g): \mathcal{C}_P = 8.27 + 0.258 \times 10^{-3}T - 1.877 \times 10^5 T^{-2}$ cal K^{-1} mol^{-1}

$H_2(g): \mathcal{C}_P = 6.65 + 0.69 \times 10^{-3}T$ cal K^{-1} mol^{-1}

$H_2O(g): \mathcal{C}_P = 7.20 + 2.70 \times 10^{-3}T$ cal K^{-1} mol^{-1}

Determine the temperature dependence of the equilibrium constant of the above reaction.

Solution: Eq. (9-11) may also be written as

$$\frac{d \ln K}{dT} = \frac{I_H + \int \Delta C_P \, dT}{RT^2}$$

Since ΔC_P for the above reaction is given by

$$\Delta C_P = (\mathcal{C}_P)_{H_2O} - (\mathcal{C}_P)_{H_2} - \tfrac{1}{2}(\mathcal{C}_P)_{O_2}$$
$$= -3.585 + 1.881 \times 10^{-3}T + 9.385 \times 10^4 T^{-2}$$

the temperature dependence of the equilibrium constant is expressed by

$$\ln K = \frac{1}{R}\left[-\frac{I_H}{T} - 3.585 \ln T + 0.9405 \times 10^{-3}T + \frac{4.692 \times 10^4}{T^2} \right] + I_K$$

The integration constant, I_H, is evaluated (see Example 5-9) from Kirchhoff's law by making use of the value of the enthalpy change of the reaction at 298 K; $I_H = -56{,}499$. By inserting the given value of $\log K$ at 1000 K and the value of $I_H = -56{,}499$, the value of the integration constant, I_K, is obtained: $I_K = 6.693$. The final expression for $\ln K$ as function of temperature is therefore

$$\ln K = \frac{28{,}434}{T} - 1.804 \ln T + 0.473 \times 10^{-3}T + \frac{2.361 \times 10^4}{T^2} + 6.693$$

In Example 9-4, we will show a more convenient and more reliable method of expressing K as a function of T.   ●

9-3   STANDARD GIBBS FREE ENERGIES OF FORMATION

Eq. (9-4) represents one of the most interesting relations in thermodynamics. It relates an equilibrium property, the equilibrium constant K, to a nonequilibrium property, the standard Gibbs free energy change of a reaction, ΔG°. If ΔG° is known, the equilibrium constant can be calculated.

The third law of thermodynamics (Section 9-7) permits the evaluation of the standard entropy, $\mathcal{S}^\circ$, which in combination with the standard enthalpy of formation,

$\Delta \mathcal{H}_f^\circ$, provides a simple way to determine ΔG° and thus K:

$$\Delta G^\circ = -RT \ln K = \sum_P \nu_i(\Delta \mathcal{H}_f^\circ)_i - \sum_R \nu_i(\Delta \mathcal{H}_f^\circ)_i - T\left(\sum_P \nu_i \mathcal{S}_i^\circ - \sum_R \nu_i \mathcal{S}_i^\circ\right) \quad \text{(9-15)}$$

where $\sum_P$ and $\sum_R$ represents the sum over the products and reactants, respectively.

As seen from Eq. (9-6), the standard Gibbs free energy of a reaction may also be calculated from the following equation:

$$\Delta G^\circ = \sum_P \nu_i \mu_i^\circ - \sum_R \nu_i \mu_i^\circ \quad \text{(9-16)}$$

where μ_i° is the standard molar Gibbs free energy of formation $[\equiv (\Delta \mathcal{G}_f^\circ)_i]$ of component i. It is defined as the change of the Gibbs free energy of a reaction in which 1 mole of a compound in its standard state is formed from its elements in their standard states. By convention, we assign a zero value for the Gibbs free energy to the elements in their most stable form at 1 atm and at the temperature of interest; for example, $\mu_{O_2}^\circ(g) = 0$, $\mu_{Hg}^\circ(l) = 0$, $\mu_{I_2}^\circ(s) = 0$ at 25°C. Physically, both $\mathcal{H}^\circ$ and μ° of elements cannot be equal to zero. This violates the equation, $\mu^\circ = \mathcal{H}^\circ - T\mathcal{S}^\circ$, and consequently the third law of thermodynamics ($S^\circ \neq 0$). This convention is, however, widely applied and it is without an effect on thermodynamic calculations.

Table 9-1 The standard Gibbs free energies of formation at 25°C

Substance	$(\Delta \mathcal{G}_f^\circ)_{298}$ (kcal mol^{-1})	Substance	$(\Delta \mathcal{G}_f^\circ)_{298}$ (kcal mol^{-1})
AgBr(s)	−22.930	Fe$_3$O$_4$(s)	−242.4
AgCl(s)	−26.224	H$_2$O(g)	−54.635
AgI(s)	−15.852	H$_2$O(l)	−56.690
CaSO$_4$(s)	−315.560	H$_2$O$_2$(g)	−24.736
CH$_4$(g)	−12.140	H$_2$O$_2$(l)	−28.23
C$_2$H$_2$(g)	50.0	D$_2$O(l)	−58.206
C$_2$H$_4$(g)	16.282	D$_2$O(g)	−56.067
C$_2$H$_6$(g)	−7.860	H$_2$S(g)	−7.892
C$_6$H$_6$(g)	30.989	H$_2$SO$_4$(aq)	−177.34
C$_6$H$_6$(l)	29.756	NaCl(s)	−91.785
CH$_3$OH(l)	−39.73	NH$_3$(g)	−3.976
CH$_3$CH$_2$OH(l)	−41.77	H$_2$O(g)	24.76
CCl$_4$(l)	−16.4	NO(g)	20.719
CO(g)	−32.8079	NO$_2$(g)	12.390
CO$_2$(g)	−94.2598	N$_2$O$_4$(g)	23.44
CuO(s)	−30.40	SO$_2$(g)	−71.79
Cu$_2$O(s)	−34.98	SO$_3$(g)	−88.52
CuBr$_2$(s)	−30.3	UF$_6$(s)	−486
Fe$_2$O$_3$(s)	−177.1	UF$_6$(g)	−485

Source: Rossini, F. D., et al.: Selected values of chemical thermodynamic properties. *N.B.S. Circular 500.* Washington, D.C., 1952; Rossini, F. D., et al.: Selected values of properties of hydrocarbons. *N.B.S. Circular C461.* Washington, D.C., 1947.

Some values of the standard Gibbs free energies of formation at 25°C are given in Table 9-1.

9-4 RELATIONSHIPS AMONG EQUILIBRIUM CONSTANTS

Eq. (9-5) defines the thermodynamic equilibrium constant: the equilibrium constant expressed in terms of activities. The existing relationships between the activity, fugacity, and concentration permit the expression of the equilibrium constant in other concentration terms.

For a gas phase reaction, $f_i^\circ = 1$ atm, and

$$K = K_f = \frac{f_C^c f_D^d}{f_A^a f_B^b} \tag{9-17}$$

For a gaseous mixture of imperfect gases, $f_i = \nu_i p_i$, and Eq. (9-17) takes the form

$$K_f = \frac{\nu_C^c \nu_D^d}{\nu_A^a \nu_B^b} \frac{p_C^c p_D^d}{p_A^a p_B^b} = K_\nu K_p \tag{9-18}$$

where ν_i denotes the fugacity coefficient of component i and p_i its partial pressure.

If the components of the gas phase behave ideally, then all $\nu_i = 1$, the fugacity equals the partial pressure, and Eq. (9-18) reduces to

$$K_p = \frac{p_C^c p_D^d}{p_A^a p_B^b} \tag{9-19}$$

For ideal gases the partial pressure, p_i, is related to the molarity, c_i, by

$$p_i = \frac{n_i}{V} RT = c_i RT \tag{9-20}$$

and thus we may write

$$K_p = \frac{c_C^c c_D^d}{c_A^a c_B^b} (RT)^{\Delta\nu} = K_c (RT)^{\Delta\nu} \tag{9-21}$$

On expressing the partial pressures by Dalton's law, $p_i = P y_i$, Eq. (9-19) assumes the form

$$K_p = \frac{y_C^c y_D^d}{y_A^a y_B^b} P^{\Delta\nu} = K_y P^{\Delta\nu} \tag{9-22}$$

Since the mole fraction is given by $y_i = n_i/n$, Eq. (9-22) may also be written as

$$K_p = \frac{n_C^c n_D^d}{n_A^a n_B^b} \left(\frac{P}{n}\right)^{\Delta\nu} = K_n \left(\frac{P}{n}\right)^{\Delta\nu} \tag{9-23}$$

In all of the above equations, P denotes the total pressure, p_i the partial pressure of component i, n_i the number of moles of component i, n the total number of moles, c_i the molar concentration of i, $\Delta\nu$ the change in the stoichiometric number of moles of gaseous species for the reaction considered ($\Delta\nu = c + d - a - b$) and ν_i is the fugacity coefficient of component i.

Table 9-2 The influence of pressure on the equilibrium of the reaction $\frac{1}{2} N_2(g) + \frac{3}{2} H_2(g) \rightleftharpoons NH_3(g)$ at 723 K, $\Delta\nu = -1$

$\dfrac{\text{Pressure}}{\text{(atm)}}$	$\dfrac{\text{Amount } NH_3 \text{ at equilibrium}}{(\%)}$	$K_p \times 10^3$	K_ν	$K_f \times 10^3$
10	2.04	6.59	0.995	6.55
50	9.17	6.90	0.945	6.50
100	16.36	7.25	0.880	6.36
600	53.6	12.94	0.497	6.42
2000	89.8	133.7	0.342	45.81

It is worthwhile to mention that, although the value of K and K_f is independent of concentration, the equilibrium constants K_p, K_c, K_y, and K_n may be strongly concentration dependent. As shown in Table 9-2, K_p varies strongly with pressure, whereas K_f remains constant except at very high pressures, when the ideal mixture assumption is not justified any more.

For reactions taking place in a liquid phase, it is more convenient to use as a standard state the pure liquid components at the temperature and pressure of the system. In this case, $a_i = \gamma_i x_i$ and

$$K = \frac{x_C^c x_D^d}{x_A^a x_B^b} \frac{\gamma_C^c \gamma_D^d}{\gamma_A^a \gamma_B^b} = K_x K_\gamma \tag{9-24}$$

γ_i is the activity coefficient. K_γ is a very complicated function of concentration that must be determined experimentally. In solutions approaching ideal behavior, however, $K_\gamma \to 1$, and K_x expresses the equilibrium behavior of the reaction system fairly well, as seen from Table 9-3. The difference in K_x and K_c is also evident from Table 9-3. Remember that K_x and K_c are based on different standard states.

9-5 THE EFFECT OF PRESSURE, INERT GAS, AND TEMPERATURE UPON THE YIELD OF A GASEOUS REACTION

The effect of pressure on the yield of a reaction may be discussed on the basis of Eq. (9-18)

$$K_f = K_\nu K_p = K_\nu \frac{n_C^c n_D^d}{n_A^a n_B^b} \left(\frac{P}{n}\right)^{\Delta\nu} \tag{9-18}$$

We will assume that $K_\nu \approx 1$. If $\Delta\nu$ is positive, for example, in the reaction $2 H_2O(g) \rightleftharpoons 2 H_2(g) + O_2(g)$ where $\Delta\nu = 3 - 2 = 1$, an increase in the pressure will decrease the yield of the reaction. If $\Delta\nu$ is negative, as, for example, in the reaction $\frac{1}{2} N_2(g) + \frac{3}{2} H_2(g) \rightleftharpoons NH_3(g)$ where $\Delta\nu = 1 - 2 = -1$, the yield increases with increasing pressure. Finally, when $\Delta\nu$ is zero, the yield of the reaction is independent of the pressure.

The number of moles of the inert gas—the constituent that does not take part in the reaction—is included in the total number of moles, n, and will therefore affect the yield in a manner opposite to that of pressure.

Table 9-3 The dissociation constant of the reaction $N_2O_4(g) \rightleftharpoons 2\,NO_2(g)$ in chloroform solutions at 8.2°C

Mole Fractions		Molarities		
$x_{N_2O_4}$	x_{NO_2}	$K_x = \dfrac{x_{NO_2}^2}{x_{N_2O_4}}$	$c_{N_2O_4}$	c_{NO_2}	$K_c = \dfrac{c_{NO_2}^2}{c_{N_2O_4}}$ (mol liter^{-1})
1.03×10^{-2}	0.93×10^{-6}	8.57×10^{-11}	0.129	1.17×10^{-3}	1.07×10^{-5}
2.48×10^{-2}	1.47×10^{-6}	8.70×10^{-11}	0.324	1.85×10^{-3}	1.05×10^{-5}
6.10×10^{-2}	2.26×10^{-6}	8.35×10^{-11}	0.778	2.84×10^{-3}	1.04×10^{-5}
		Av.: 8.54×10^{-11}			Av.: 1.05×10^{-5} (mol liter^{-1})

The temperature effect on the yield is given by Eq. (9-11). For an exothermic reaction, K, and consequently the yield of the reaction, decreases with increasing temperature. For an endothermic reaction, just the opposite is true.

The effect of variables, such as pressure, temperature, and number of moles of inert gas, on the position of chemical equilibrium is summarily given by the Le Chatelier principle (1888). According to this principle, the equilibrium will always be displaced in such a manner as to oppose the applied change. For example, when the temperature is raised, the equilibrium shifts in the direction that causes an absorption of heat.

EXAMPLE 9-2

The equilibrium of the reaction

$$\tfrac{1}{2}N_2(g) + \tfrac{3}{2}H_2(g) = NH_3(g)$$

was investigated experimentally at high pressures and temperatures by A. T. Larson [13]. At 450°C and 300 atm he found 35.82 mol % of ammonia in the equilibrium mixture. The initial mixture in this experiment contained hydrogen and nitrogen in the ratio 3 : 1. Determine the value of the equilibrium constant at a temperature of 450°C and pressure of 300 atm.

Given: The fugacity coefficients of the gases present in the equilibrium mixture at 450°C and 300 atm: $\nu_{H_2} = 1.08$, $\nu_{N_2} = 1.14$, $\nu_{NH_3} = 0.91$.

Solution: The equilibrium constant is given by relation (9-18)

$$K = K_f = \frac{f_{NH_3}}{f_{N_2}^{0.5}\, f_{H_2}^{1.5}} = \frac{\nu_{NH_3}}{\nu_{N_2}^{0.5}\, \nu_{H_2}^{1.5}}\, \frac{n_{NH_3}}{n_{N_2}^{0.5}\, n_{H_2}^{1.5}} \left(\frac{P}{n}\right)^{-1}$$

The number of moles of each component at the beginning of the reaction: $n_{N_2} = 0.5$, $n_{H_2} = 1.5$, $n_{NH_3} = 0$. After equilibrium is attained the number of moles are as follows: $n_{N_2} = 0.5\,(1-x)$, $n_{H_2} = 1.5\,(1-x)$, $n_{NH_3} = x$. x is the number of moles of ammonia formed by the reaction. The total number of moles in the equilibrium mixture then is $n = 0.5\,(1-x) + 1.5\,(1-x) + x = 2 - x$. Substituting these values into the preceding equation results in

$$K = K_f = \frac{0.91}{1.14^{0.5} \times 1.08^{1.5}}\, \frac{x}{[0.5(1-x)]^{0.5}[1.5(1-x)]^{1.5}} \left(\frac{300}{2-x}\right)^{-1}$$

The equilibrium mixture was found to contain 35.82 mol % of ammonia; from this we get for the equilibrium conversion, x.

$$\frac{n_{NH_3}}{n} 100 = \left(\frac{x}{2-x}\right) 100 = 35.82 \qquad x = 0.5275$$

and thus

$$K = K_f = 6.79 \times 10^{-3}$$

This value is in good agreement with the constant obtained by extrapolation from F. Haber's [14] low pressure measurements: $K = 5.92 \times 10^{-3}$. ●

9-6 CHEMICAL EQUILIBRIUM IN HETEROGENEOUS REACTIONS

If the substances participating in the chemical reaction consist of more than one phase, the reaction is heterogeneous.

In the reaction

$$NH_4Cl(s) \rightleftharpoons NH_3(g) + HCl(g)$$

there are two phases present: one solid, $NH_4Cl(s)$, and one gaseous, $NH_3(g) + HCl(g)$. The equilibrium constant is simply given as

$$K = \frac{a_{NH_3(g)} \; a_{HCl(g)}}{a_{NH_4Cl(s)}} \qquad (9\text{-}25)$$

Using the pure solid component at T and P of the system as the standard state for $NH_4Cl(s)$, $a_{NH_4Cl(s)} = 1$, and $f^\circ_{NH_3(g)} = f^\circ_{HCl(g)} = 1$ atm for the constituents of the gaseous phase, we have

$$K_f = f_{NH_3(g)} \; f_{HCl(g)} \qquad (9\text{-}26)$$

and in the case of ideal behavior

$$K_p = p_{NH_3(g)} \; p_{HCl(g)} \qquad (9\text{-}27)$$

Thus, the equilibrium constant contains only the partial pressures of the gaseous components. If the only starting material is $NH_4Cl(s)$, it follows from the stoichiometric condition that $p_{NH_3(g)} = p_{HCl(g)}$ and

$$K_p = p^2_{NH_3(g)} = p^2_{HCl(g)} \qquad (9\text{-}28)$$

Another example of a heterogeneous reaction is that of solid carbon with oxygen to form carbon monoxide

$$C(s) + \tfrac{1}{2} O_2(g) \rightleftharpoons CO(g)$$

K_p of this reaction is

$$K_p = \frac{p_{CO}}{p^{1/2}_{O_2}} \qquad (9\text{-}29)$$

Whenever oxygen and carbon monoxide are at equilibrium in the presence of solid carbon, the ratio of their partial pressures is determined by the value of K_p.

The decomposition of hydrates belongs to this group of chemical reactions. For example the equilibrium constant for the reactions

$$CuSO_4 \cdot 5 \, H_2O(s) \rightleftharpoons CuSO_4 \cdot 3 \, H_2O(s) + 2 \, H_2O(g)$$
$$CuSO_4 \cdot 3 \, H_2O(s) \rightleftharpoons CuSO_4 \cdot H_2O(s) + 2 \, H_2O(g)$$

is simply given as

$$K_p = p^2_{H_2O} \qquad (9\text{-}30)$$

9-7 THE THIRD LAW OF THERMODYNAMICS

As we have stated on page 258, the evaluation of the temperature dependence of the equilibrium constant requires knowledge of the following data: Δa, Δb, Δc, I_H and I_K. Although Δa, Δb, Δc, and I_H are easily obtained from published thermal data ($\mathscr{C}_P$, $\Delta \mathscr{H}_f^\circ$), the evaluation of I_K requires additional data. These may be direct experimental data on the composition of the equilibrium mixture; however, knowledge of either the standard Gibbs free energy of formation or the standard entropy of the reactants and products, respectively, is also sufficient for the evaluation of the integration constant, I_K.

As is shown next, the third law of thermodynamics permits the determination of the standard entropies and, thus, of I_K. As with the first two laws of thermodynamics, the third law is also an expression of human experience with nature.

The differential change in entropy of a reversible process is given by the equation

$$dS = \frac{d\,q_{\text{rev}}}{T} \tag{9-31}$$

Integration between T and the absolute zero yields

$$S_T = S_0 + \int_0^T \frac{d\,q_{\text{rev}}}{T} \tag{9-32}$$

where S_0 is the entropy of the given substance at absolute zero.

In practice, measurements of $\mathscr{C}_P$ are made to some low temperature T^*, below which a theoretical extrapolation is made. For this reason it seems to be appropriate to rewrite Eq. (9-32) as

$$S_T = S_0 + \left(\int_0^{T^*} \frac{d\,q_{\text{rev}}}{T} \right)_{\text{extr}} + \left(\int_{T^*}^T \frac{d\,q_{\text{rev}}}{T} \right)_{\text{exp}} \tag{9-33}$$

The first integral is usually evaluated with the aid of Debye's theory of the heat capacity of crystalline substances (see page 608). This theory gives, at low temperatures, the following equation for the molar heat capacity:

$$\mathscr{C}_P \approx \mathscr{C}_V \approx aT^3 \tag{9-34}$$

where a is an empirical constant, evaluated from the properties of the substance (see Eq. 6-11) or, better, by fitting the extrapolated curve of the molar heat capacity to the experimental one. The second integral is evaluated from experimental measurements of heat capacities and enthalpies of phase transitions. Thus, for 1 mole of a gas at temperature T, the general expression for the entropy becomes

$$\mathscr{S}_T = \mathscr{S}_0 + \int_0^{T^*} \frac{aT^3}{T}\, dT + \int_{T^*}^{T_{\text{tr}}} \frac{\mathscr{C}_P(s, \alpha)}{T}\, dT$$

$$+ \frac{\Delta \mathscr{H}_{\text{tr}}(\alpha \to \beta)}{T_{\text{tr}}} + \int_{T_{\text{tr}}}^{T_{\text{fus}}} \frac{\mathscr{C}_P(s, \beta)}{T}\, dT + \frac{\Delta \mathscr{H}_{\text{fus}}}{T_{\text{fus}}}$$

$$+ \int_{T_{\text{fus}}}^{T_b} \frac{\mathscr{C}_P(l)}{T}\, dT + \frac{\Delta \mathscr{H}_{\text{vap}}}{T_b} + \int_{T_b}^T \frac{\mathscr{C}_P(g)}{T}\, dT \tag{9-35}$$

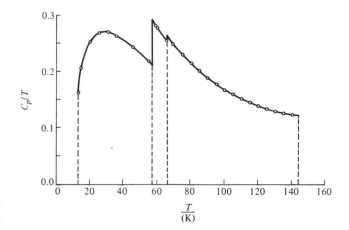

Figure 9-1 The evaluation of the third law entropy of nitrogen fluoride.

where T_{tr}, T_{fus}, T_b are the phase transition temperatures. The integrations are usually performed graphically from a plot of either $\mathscr{C}_P/T$ versus T or $\mathscr{C}_P$ versus $\ln T$. In either treatment, the area under the curve between the two temperatures gives the entropy increment of that temperature range. The method is illustrated in Example 9-3, and in Figure 9-1. In order to obtain the value of the entropy at a certain temperature, $\mathscr{S}_0$ must be known.

In 1902, T. W. Richards found experimentally that the Gibbs free energy change of a reaction approaches the enthalpy change asymptotically as the temperature is decreased. From a study of Richard's data, Nernst in 1906 suggested that the entropy change for all reactions approaches zero as the temperature approaches zero. This idea has, on the whole, proved valid, provided it is restricted to reactions between pure crystalline substances. The statement

$$\lim_{T \to 0} \Delta S_T = 0 \qquad (9\text{-}36)$$

is generally known as the third law of thermodynamics, or as the Nernst heat theorem. It must be understood that Eq. (9-36) requires only

$$\lim_{T \to 0} \Delta S_T = \lim_{T \to 0} \sum_i \nu_i (\mathscr{S}_i)_T = 0 \qquad (9\text{-}37)$$

and does not imply that

$$\lim_{T \to 0} \mathscr{S}_T = \mathscr{S}_0 = 0 \qquad (9\text{-}38)$$

The third law expresses only the fact that, as the temperature of a system tends to absolute zero, its entropy tends to a constant value $\mathscr{S}_0$. This value may, however, more or less, deviate from zero.

The nuclear spin entropy arising from the multiplicity of the ground state (see page 600), and the entropy of mixing arising from the isotopic constitution of elements make the value of $\mathscr{S}_0$ deviate from zero.

Since nuclei and isotopic constitution of elements remain unchanged in chemical reactions, the nuclear spin entropy and the entropy of mixing, respectively, do not contribute to the entropy change of a reaction, ΔS_T (they appear on both sides of the chemical reaction and cancel out). Thus, for practical purposes we may put $\mathcal{S}_0$ equal to zero (by this the entropy of any substance becomes normalized in an absolute way). Consequently, the third law entropies used for the calculation of the entropy change of chemical reaction do not include the nuclear spin and isotopic mixing contributions.

Third law entropies, like enthalpies of formation and Gibbs free energies of formation, are usually calculated for 25°C and unit activity—standard entropies (see Table 9-4). All the terms in Eq. (9-35) that do not deal with solids must be neglected when calculating the standard entropy of solids. When dealing with liquids, only the terms involving solids and liquids have to be considered; for gases, the entire equation must be considered. The standard entropies of gaseous substances are given for ideal behavior.

Usually, gases at 25°C and 1 atm do not behave ideally; therefore, a correction must be applied to the real gas to convert it to an ideal gas at the pressure of 1 atm. The correction is usually made by use of the low pressure form of the Berthelot equation (Eq. 1-56), and it is given by

$$\mathcal{S}^\circ_{id} = \mathcal{S}^\circ_{real} + \frac{27}{32} \frac{RT^3_c P}{T^3 P_c} \tag{9-39}$$

where T_c and P_c are the critical temperature and pressure of the substance and $\mathcal{S}^\circ_{real}$ is the standard entropy of the real gas at pressure P.

The importance of the third law of thermodynamics lies in the fact that it permits the evaluation of the equilibrium constant and its temperature dependence from calorimetric data alone. What a triumph of thermodynamics!

Table 9-4 Third law entropies of some basic substances

Substance	$\dfrac{\mathcal{S}^\circ_{298}}{(\text{cal K}^{-1}\,\text{mol}^{-1})}$	Substance	$\dfrac{\mathcal{S}^\circ_{298}}{(\text{cal K}^{-1}\,\text{mol}^{-1})}$
$O_2(g)$	49.003	$H_2S(g)$	49.15
$H_2(g)$	31.211	$NO(g)$	50.339
$N_2(g)$	45.766	$NO_2(g)$	57.47
$C(s)(graphite)$	1.3609	$NH_3(g)$	46.01
$Cl_2(g)$	53.286	$CO(g)$	47.301
$Br_2(g)$	58.639	$CO_2(g)$	51.061
$I_2(s)$	27.9	$CH_4(g)$	44.50
$S(s)(rhombic)$	7.62	$C_2H_6(g)$	54.85
$H_2O(g)$	45.106	$C_3H_8(g)$	64.51
$H_2O(s)$	16.716	$C_4H_{10}(g)$	74.10
$HCl(g)$	44.617	$C_5H_{12}(g)$	83.27
$HBr(g)$	47.437	$C_6H_{14}(g)$	92.45
$HI(g)$	49.314	$C_6H_6(g)$	64.34
$SO_2(g)$	59.40	$C_2H_4(g)$	52.45
$SO_3(g)$	61.24	$C_2H_2(g)$	47.997

SOURCE: Rossini, F. D., et. al.: Selected values of chemical thermodynamic properties. *N.B.S. Circular 500.* Washington, D.C., 1952; Rossini, F. D., et al.: Selected values of properties of hydrocarbons. *N.B.S. Circular C461.* Washington, D.C., 1947.

9-8 SOME CONSEQUENCES OF THE THIRD LAW OF THERMODYNAMICS

From the definition of the molar Gibbs free energy it follows that

$$\lim_{T\to 0} \mathscr{S}_T = \lim_{T\to 0} \left(\frac{\mathscr{H} - \mathscr{G}}{T}\right)_T = \mathscr{S}_0 \qquad (9\text{-}40)$$

Making use of the l'Hopital rule, we have

$$\lim_{T\to 0} \mathscr{S}_T = \lim_{T\to 0} \left[\frac{(\partial H/\partial T)_P - (\partial G/\partial T)_P}{1}\right]_T = \lim_{T\to 0} (\mathscr{C}_P + \mathscr{S}_T) = \mathscr{S}_0 \qquad (9\text{-}41)$$

and thus

$$\lim_{T\to 0} \mathscr{C}_P = 0 \qquad (9\text{-}42)$$

All measurements indicate that $\mathscr{C}_P$ does approach zero as the temperature approaches absolute zero (see also Eq. 9-34).

If the Nernst heat theorem is accepted, it can also be proved that

$$\lim_{T\to 0} \left(\frac{\partial S}{\partial V}\right)_T \to 0 \qquad (9\text{-}43)$$

$$\lim_{T\to 0} \left(\frac{\partial S}{\partial P}\right)_T \to 0 \qquad (9\text{-}44)$$

and consequently,

$$\lim_{T\to 0} \left(\frac{\partial P}{\partial T}\right)_V \to 0 \qquad (9\text{-}45)$$

$$\lim_{T\to 0} \left(\frac{\partial V}{\partial T}\right)_P \to 0 \qquad (9\text{-}46)$$

EXAMPLE 9-3

Low temperature calorimetric data on nitrogen fluoride, NF_3, are reported by L. Pierce and F. L. Pace [15]. They measured the heat capacity of nitrogen fluoride in a temperature range of 12–144 K and determined the transition point, melting point, boiling point, and the corresponding enthalpy changes of these phase transitions.

Given:

T (K)	$\mathscr{C}_P$ (cal K^{-1} mol^{-1})	T (K)	$\mathscr{C}_P$ (cal K^{-1} mol^{-1})	T (K)	$\mathscr{C}_P$ (cal K^{-1} mol^{-1})
12.95	2.154	58.41	16.52	100	16.75
15	3.077	60	16.53	105	16.71
20	5.031	65	16.62	110	16.69
25	6.719	70	17.27	115	16.70
30	8.100	75	17.17	120	16.74
35	9.193	80	17.07	125	16.81
40	10.12	85	16.98	130	16.93
45	10.99	90	16.90	135	17.08
55	11.95	95	16.82	140	17.26

The transition temperature is 56.62 K and the enthalpy of transition is 361.8 cal mol^{-1}. The melting point is 66.37 K and the enthalpy of fusion is 95.10 cal mol^{-1}. Finally, the boiling point is reported as 144.15 K and the enthalpy of vaporization is 2769 cal mol^{-1}. The Debye characteristic temperature Θ_D (see page 610) of NF_3 is 120.8 K.

Making use of the data given, calculate the standard third law entropy of NF_3 vapor at its boiling point, 144.15 K.

Solution: Substituting for the constant a into Debye's cubic law (see page 611), we write

$$\mathscr{C}_P \approx \mathscr{C}_V = \frac{12}{5} R\pi^4 \left(\frac{T}{\Theta_D}\right)^3 = aT^3$$

From this equation, we can evaluate the entropy change between 0 and 13 K.

$$\Delta \mathscr{S}_1 = \int_0^{13} \frac{\mathscr{C}_P}{T}\, dT = \frac{12 \times 1.987 \times 3.14^4 \times 13^3}{5 \times 120.8^3 \times 3}$$
$$= 0.193 \text{ cal K}^{-1} \text{ mol}^{-1}$$

The entropy change between 13 K and 56.62 K is given by

$$\Delta \mathscr{S}_2 = \int_{13}^{T_{tr}} \frac{\mathscr{C}_P(\alpha)\, dT}{T}$$

graphical integration (see Figure 9-1) results in $\Delta S_2 = 10.702$ cal K^{-1} mol^{-1}.

The entropy change corresponding to the solid phase transition is simply

$$\Delta \mathscr{S}_3 = \frac{\Delta \mathscr{H}_{tr}}{T_{tr}} = \frac{361.8}{56.62} = 6.390 \text{ cal K}^{-1} \text{ mol}^{-1}$$

The entropy change between the transition temperature and the melting point is

$$\Delta \mathscr{S}_4 = \int_{56.62}^{66.37} \frac{\mathscr{C}_P(\beta)\, dT}{T}$$

By graphical integration, (see Figure 9-1) we obtain $\Delta \mathscr{S}_4 = 2.633$ cal K^{-1} mol^{-1}.
For fusion

$$\Delta \mathscr{S}_5 = \frac{\Delta \mathscr{H}_{fus}}{T_{fus}} = \frac{95.10}{66.37} = 1.433 \text{ cal K}^{-1} \text{ mol}^{-1}$$

The heating of liquid NF_3 from its melting point (66.37 K) to the boiling point (144.15 K) is accompanied by a change in entropy

$$\Delta \mathscr{S}_6 = \int_{66.37}^{144.15} \frac{\mathscr{C}_P(l)\, dT}{T}$$

Graphical integration (see Figure 9-1) gives $\Delta S_6 = 13.126$ cal K^{-1} mol^{-1}.
Finally the change in entropy accompanying the evaporation at the boiling point is

$$\Delta \mathscr{S}_7 = \frac{\Delta \mathscr{H}_{vap}}{T_{bp}} = \frac{2769}{144.15} = 19.21 \text{ cal K}^{-1} \text{ mol}^{-1}$$

The third law entropy of gaseous NF_3 at its boiling point and a pressure of 1 atm is given by the sum of all the entropy contributions: $\mathscr{S}° = 53.69$ cal K^{-1} mol^{-1}.

This value differs significantly from that determined spectroscopically, 54.61 cal K^{-1} mol^{-1}. The deviation is partly due to neglecting the deviation from ideal behavior, partly to the approximate graphical integration, and partly to the approximate nature of Debye's cubic law. ●

9-9 THE PRINCIPLE OF UNATTAINABILITY OF ABSOLUTE ZERO

Since the third law is inextricably bound to phenomena that take place at low temperatures, it is appropriate here to make some remarks about how these low temperatures may be achieved.

The simplest method for cooling a system is to place it in an environment with a lower temperature than its own. If no such environment is available, some means will have to be used that involves the behavior of the system itself, that is, evaporation, Joule–Thomson expansion, demagnetization, and so on. Generally, any process that operates between states with different entropies can be used for cooling.

We will now prove that an adiabatic and reversible cooling process is the most effective. The proof will be demonstrated for the expansion of a gas, but the result can be generalized to any cooling process.

Consider 1 mole of gas constrained so that it may only perform P-V work. If V and S are chosen as the independent variables of the system, the differential change in temperature during such an expansion may be expressed by

$$dT = \left(\frac{\partial T}{\partial \mathscr{S}}\right)_V dS + \left(\frac{\partial T}{\partial V}\right)_S dV \tag{9-47}$$

Since T and $\mathscr{C}_V$ are always positive, the first term in Eq. (9-47) is also positive, as seen from Eq. (6-54)

$$\left(\frac{\partial T}{\partial \mathscr{S}}\right)_V = \frac{T}{\mathscr{C}_V}$$

This means that, if cooling is to take place, the second term $(\partial T/\partial V)_S$ must be negative. For a certain change in volume, dV, cooling is therefore maximized when the first term vanishes. But, dS is zero if expansion is conducted reversibly and adiabatically; therefore, the greatest cooling effect is achieved during an adiabatic and reversible process.

In 1926, P. Debye and W. F. Giauque independently proposed a very effective cooling method: the adiabatic demagnetization of paramagnetic substances. A paramagnetic crystalline substance—a substance having unpaired electrons—is cooled to the temperature of liquid helium in the presence of a magnetic field. Once the crystal has been cooled and magnetized, it is adiabatically isolated and the magnetic field is turned off. The crystal slowly loses its magnetism. The electron spins, lined up in the presence of the magnetic field, take up random orientation as they proceed to a more probable arrangement. As a consequence of this step, the entropy of the spins is increased. The heat necessary for this increase is taken from the thermal vibrational energy of the crystal lattice and its temperature is therefore lowered. Temperatures within 10^{-3} degree of absolute zero have been obtained by this method.[1] It appears, however, that in spite of this remarkably close approximation to absolute zero, absolute zero cannot be reached.

To prove this, let us consider an arbitrary cooling process in an isolated system. If we denote the entropy in the original state by $\mathscr{S}_1$ and after cooling by $\mathscr{S}_2$, we have for the spontaneous cooling process

$$\mathscr{S}_1 \leqq \mathscr{S}_2 \qquad (9\text{-}48)$$

Consequently

$$\mathscr{S}_0 + \int_0^{T_1} \frac{\mathscr{C}_P}{T} dT \leqq \mathscr{S}_0 + \int_0^{T_2} \frac{\mathscr{C}_P}{T} dT \qquad (9\text{-}49)$$

If the final cooling temperature T_2 is to be zero, we must have

$$\int_0^{T_1} \frac{\mathscr{C}_P}{T} dT \leqq 0 \qquad (9\text{-}50)$$

However, this is impossible because $\mathscr{C}_P$ is positive for all temperatures.

The third law may therefore be stated also as follows: It is impossible for any process involving a finite number of steps to reduce the temperature of any system to absolute zero. This is the principle of the unattainability of absolute zero.

9-10 THE EXPERIMENTAL VERIFICATION OF THE THIRD LAW OF THERMODYNAMICS

Substances that exist in two different crystalline forms may serve to verify the third law. For the reversible, isothermal transition, $(\beta \to \alpha)$, we may write

$$\Delta \mathscr{S} = \mathscr{S}_\alpha - \mathscr{S}_\beta = \Delta \mathscr{H}_{tr} / T_{tr} \qquad (9\text{-}51)$$

where $\Delta \mathscr{H}_{tr}$ and T_{tr} are the experimentally found enthalpy of transition and temperature of transition, respectively. By making use of Eq. (9-35), relation (9-51) can be expressed as

$$\Delta \mathscr{S} = \mathscr{S}_{0,\alpha} + \int_0^{T_{tr}} \frac{\mathscr{C}_P(\text{s}, \alpha)}{T} dT - \mathscr{S}_{0,\beta} - \int_0^{T_{tr}} \frac{\mathscr{C}_P(\text{s}, \beta)}{T} dT = \frac{\Delta \mathscr{H}_{tr}}{T_{tr}} \qquad (9\text{-}52)$$

Now, if an experiment proves that

$$\int_0^{T_{tr}} \frac{\mathscr{C}_P(\text{s}, \alpha)}{T} dT - \int_0^{T_{tr}} \frac{\mathscr{C}_P(\text{s}, \beta)}{T} dT = \frac{\Delta \mathscr{H}_{tr}}{T_{tr}} \qquad (9\text{-}53)$$

it also proves that

$$\mathscr{S}_{0,\alpha} = \mathscr{S}_{0,\beta} \qquad (9\text{-}54)$$

Hence, both crystalline modifications have equal entropies at absolute zero, as required by the third law of thermodynamics.

---

[1] A similar method when applied to nuclear magnetic moments resulted in a temperature of about 10^{-5} K.

Experiments performed on systems such as sulfur, tin, and phosphine have proved the validity of the third law. The following experimental results have been obtained for phosphine:

$$\Delta \mathscr{S}_{tr} = \frac{\Delta \mathscr{H}_{tr}}{T_{tr}} = \frac{185.7}{49.43} = 3.76 \text{ cal K}^{-1} \text{ mol}^{-1}$$

The difference of the two integrals in Eq. (9-53) determined experimentally is

$$\int_0^{T_{tr}} \frac{\mathscr{C}_P(s, \alpha)}{T} dT - \int_0^{T_{tr}} \frac{\mathscr{C}_P(s, \beta)}{T} dT = 8.13 - 4.38$$

$$= 3.75 \text{ cal K}^{-1} \text{ mol}^{-1}$$

On comparison of the two results, it is obvious that, within the limits of experimental error, the experimental data on phosphine prove the validity of the third law.

9-11 ENTHALPY FUNCTIONS $(\mathscr{H}_T^\circ - \mathscr{H}_0^\circ)/T$ AND GIBBS FREE ENERGY FUNCTIONS $(\mathscr{G}_T^\circ - \mathscr{H}_0^\circ)/T$

The method of evaluation of the temperature dependence of the equilibrium constant as illustrated in Example 9-1 requires tedious and complicated numerical calculations. In addition, even with three constants the empirical equations for the molar heat capacities used in the calculation are only of limited accuracy. The reliability of the method is therefore questionable.

The third law of thermodynamics furnishes us with a more convenient and also more reliable method of evaluating the dependence of K on T. The fact that, at absolute zero, $\Delta S_0^\circ = 0$ allows us to write

$$\mathscr{U}_0^\circ = \mathscr{H}_0^\circ = \mathscr{A}_0^\circ = \mathscr{G}_0^\circ$$
$$\Delta U_0^\circ = \Delta H_0^\circ = \Delta A_0^\circ = \Delta G_0^\circ \qquad \text{(9-55)}$$

Spectroscopy and low temperature calorimetry permit the evaluation of the so-called **enthalpy function** $(\mathscr{H}_T^\circ - \mathscr{H}_0^\circ)/T$ and **Gibbs free energy function** $(\mathscr{G}_T^\circ - \mathscr{H}_0^\circ)/T$ at any temperature. If, in addition, the standard enthalpies of formation of the products and reactants are known at absolute zero, $(\Delta \mathscr{H}_f^\circ)_0$, the standard enthalpy change of the reaction as well as the standard Gibbs free energy change of the reaction and, consequently, K may be easily obtained at any temperature using the following procedure:

Enthalpy and Gibbs free energy are functions of state and, therefore, we may write for a pure substance at any temperature

$$(\Delta \mathscr{H}_f^\circ)_T = (\Delta \mathscr{H}_f^\circ)_0 + (\mathscr{H}_T^\circ - \mathscr{H}_0^\circ) \qquad \text{(9-56)}$$

and assuming that $\Delta \mathscr{S}_0^\circ = 0$

$$(\Delta \mathscr{G}_f^\circ)_T = (\Delta \mathscr{H}_f^\circ)_0 + (\mathscr{G}_T^\circ - \mathscr{H}_0^\circ) \qquad \text{(9-57)}$$

where the terms $(\mathscr{H}_T^\circ - \mathscr{H}_0^\circ)$ and $(\mathscr{G}_T^\circ - \mathscr{H}_0^\circ)$ represent the enthalpy and Gibbs free energy changes associated with the change in the temperature from absolute zero to

T. Consequently, for the enthalpy change of a reaction we may write

$$\frac{(\Delta H^\circ)_T}{T} = \sum_P \nu_i \frac{(\Delta \mathcal{H}^\circ_{f,i})_T}{T} - \sum_R \nu_i \frac{(\Delta \mathcal{H}^\circ_{f,i})_T}{T}$$

$$= \sum_P \nu_i \left[\frac{(\Delta \mathcal{H}^\circ_{f,i})_0}{T} + \frac{(\mathcal{H}^\circ_T - \mathcal{H}^\circ_0)_i}{T} \right] - \sum_R \nu_i \left[\frac{(\Delta \mathcal{H}^\circ_{f,i})_0}{T} + \frac{(\mathcal{H}^\circ_T - \mathcal{H}^\circ_0)_i}{T} \right] \qquad (9\text{-}58)$$

and for the equilibrium constant

$$-\boldsymbol{R} \ln K = \frac{(\Delta G^\circ)_T}{T} = \sum_P \nu_i \frac{(\Delta \mathcal{G}^\circ_{f,i})_T}{T} - \sum_R \nu_i \frac{(\Delta \mathcal{G}^\circ_{f,i})_T}{T}$$

$$= \sum_P \nu_i \left[\frac{(\Delta \mathcal{H}^\circ_{f,i})_0}{T} + \frac{(\mathcal{G}^\circ_T - \mathcal{H}^\circ_0)_i}{T} \right] - \sum_R \nu_i \left[\frac{(\Delta \mathcal{H}^\circ_{f,i})_0}{T} + \frac{(\mathcal{G}^\circ_T - \mathcal{H}^\circ_0)_i}{T} \right] \qquad (9\text{-}59)$$

where $\sum_P$ and $\sum_R$ denote the sum over the products and reactants, respectively, ν_i denotes the stoichiometric coefficients.

Values of enthalpy functions and Gibbs free energy functions, for some substances, are presented in Tables 9-5 and 9-6. Accurate reaction enthalpies and equilibrium constants at intermediate temperatures can be calculated from functions obtained by a linear interpolation between the tabulated functions. The majority of the data are taken from reference [16]. A very large compilation of this kind of data is found in JANAF Thermochemical Tables (1965) and in the Addenda (1966, 1967, 1968) prepared by the Thermal Research Laboratory of the Dow Chemical Company under the direction of D. R. Stull.

EXAMPLE 9-4

Styrene is produced by catalytic dehydrogenation of ethylbenzene according to the equation

$$C_6H_5 \cdot CH_2 \cdot CH_3(g) \rightleftharpoons C_6H_5CH{=}CH_2(g) + H_2(g)$$

Determine the equilibrium constant of this reaction at 500 K.

Given: At $T = 500$ K, $(\mathcal{G}^\circ - \mathcal{H}^\circ_0)/T$ is -27.947 cal K^{-1} mol^{-1} for hydrogen, -74.44 cal K^{-1} mol^{-1} for styrene, and -79.64 cal K^{-1} mol^{-1} for ethylbenzene. $(\Delta \mathcal{H}^\circ_f)_0$ of styrene is 40,340 cal mol^{-1} and of ethylbenzene 13,917 cal mol^{-1}.

Solution: The equilibrium constant is calculated from Eq. (9-59).

$$\frac{\Delta G^\circ}{T} = -\boldsymbol{R} \ln K = \sum_P \left[\frac{\mathcal{G}^\circ - \mathcal{H}^\circ_0}{T} + \frac{(\Delta \mathcal{H}^\circ_f)_0}{T} \right] - \sum_R \left[\frac{\mathcal{G}^\circ - \mathcal{H}^\circ_0}{T} + \frac{(\Delta \mathcal{H}^\circ_f)_0}{T} \right]$$

The indices P and R denote the products and reactants, respectively.

On substituting the given values, we obtain

$$\ln K = -\frac{1}{1.987} \left[\left(-74.44 + \frac{40,340}{500} - 27.947 \right) - \left(-79.64 + \frac{13,917}{500} \right) \right]$$

$$= -15.15$$

and $K = 2.638 \times 10^{-7}$.

With increasing temperature the equilibrium constant increases significantly, and at a temperature of 1000 K it reaches a value of 2.005. ●

Table 9-5 Thermodynamic functions $(\mathcal{H}_T^\circ - \mathcal{H}_0^\circ)/T$ and $(\Delta\mathcal{H}_f^\circ)_0$ of some basic substances at different temperatures

Substance	$\dfrac{(\Delta\mathcal{H}_f^\circ)_0}{(\text{kcal mol}^{-1})}$	$\dfrac{(\mathcal{H}_T^\circ - \mathcal{H}_0^\circ)/T}{(\text{cal K}^{-1}\text{mol}^{-1})}$ Temperature (K)						
		298.16	400	500	600	800	1000	1500
methane	−15.987	8.039	8.307	8.73	9.249	10.401	11.56	14.09
ethane	−16.517	9.578	10.74	12.02	13.36	15.95	18.28	23.00
ethylene	14.522	8.47	9.28	10.23	11.22	13.10	14.76	18.07
propane	−19.482	11.78	13.89	16.08	18.22	22.20	25.67	32.43
propylene	8.468	10.86	12.48	14.15	15.82	18.94	21.69	27.05
n-butane	−23.67	15.58	18.35	21.19	23.96	29.08	33.54	42.18
butene(l)	4.96	13.79	16.21	18.68	21.08	25.46	29.25	36.56
Isobutane	−25.30	14.34	17.41	20.50	23.45	28.76	33.31	42.03
Isobutene	−0.98	13.69	16.30	18.83	21.25	25.62	29.37	36.67
n-pentane	−27.23	18.88	22.38	25.94	29.38	35.71	41.19	51.75
n-hexane	−30.91	22.21	26.45	30.72	34.82	42.35	48.85	61.34
acetylene	54.329	8.021	8.853	9.582	10.212	11.249	12.090	13.694
graphite	0	0.8437	1.257	1.642	1.997	2.602	3.075	3.876
$H_2(g)$	0	6.7877	6.8275	6.8590	6.8825	6.9218	6.9658	7.1295
$H_2O(g)$	−57.107	7.941	7.985	8.051	8.137	8.362	8.608	9.232
$CO(g)$	−27.2019	6.951	6.959	6.980	7.016	7.125	7.257	7.572
$CO_2(g)$	−93.9686	7.506	7.987	8.446	8.871	9.612	10.222	11.336
$N_2(g)$	0	6.9502	6.9559	6.9701	6.9967	7.0857	7.2025	7.5024
$O_2(g)$	0	6.942	6.981	7.048	7.132	7.320	7.497	7.851

Table 9-6 Thermodynamic function $-[(\mathscr{S}_T^\circ - \mathscr{H}_0^\circ/T)]$ of some basic substances at different temperatures/(cal K^{-1} mol^{-1})

Substance	Temperature (K)						
	298.16	400	500	600	800	1000	1500
methane	36.46	38.86	40.75	42.39	45.21	47.65	52.84
ethane	45.27	48.24	50.77	53.08	57.29	61.11	69.46
ethylene	43.98	46.61	48.74	50.70	54.19	57.29	63.94
propane	52.73	56.48	59.81	62.93	68.74	74.10	85.86
propylene	52.95	56.39	59.32	62.05	67.04	71.57	81.43
n-butane	58.54	63.51	67.91	72.01	79.63	86.60	101.95
butene (l)	59.25	63.64	67.52	71.14	77.82	83.93	97.27
Isobutane	56.08	60.72	64.95	68.95	76.45	83.38	98.64
Isobutene	56.47	60.90	64.77	68.42	75.15	81.29	94.66
n-pentane	64.52	70.57	75.94	80.96	90.31	98.87	117.72
n-hexane	70.62	77.75	84.11	90.06	101.14	111.31	133.64
acetylene	39.976	42.451	44.508	46.313	49.400	52.005	57.231
graphite	0.5172	0.824	1.146	1.477	2.138	2.771	4.181
H$_2$	24.423	26.422	27.947	29.203	31.186	32.738	35.590
H$_2$O(g)	37.165	39.505	41.293	42.766	45.128	47.010	50.598
CO(g)	40.350	42.393	43.947	45.222	47.254	48.860	51.864
CO$_2$(g)	43.555	45.828	47.667	49.238	51.895	54.109	58.481
N$_2$(g)	38.817	40.861	42.415	43.688	45.711	47.306	50.284
O$_2$(g)	42.061	44.112	45.675	46.968	49.044	50.697	53.808

1. For the reaction

$$CO(g) + Cl_2(g) \rightleftharpoons COCl_2(g)$$

G. N. Lewis and M. Randall [17] give the following value for the standard Gibbs free energy at 25°C: $\Delta G^\circ_{298} = -48.77$ kcal. Calculate the equilibrium constant at this temperature.

2. In which direction can the following reactions take place spontaneously at a temperature of 25°C and pressure of 1 atm

$$CO(g) + H_2O(g) \rightleftharpoons CO_2(g) + H_2(g) \qquad \text{(a)}$$

$$CH_3CH_2CH_2CH_3(g) \rightleftharpoons CH_2{=}CHCH{=}CH_2(g) + 2\,H_2(g) \qquad \text{(b)}$$

The values of the standard Gibbs free energies of formation of the components are as follows (in kcal mol^{-1}):

$CO(g)$	-32.807	$n\text{-}C_4H_{10}(g)$	-4.100
$CO_2(g)$	-94.260	$C_4H_6(g)$	36.010
$H_2O(g)$	-54.635		

3. Calculate the standard Gibbs free energy of formation of gaseous water at a temperature of 25°C if the following data are given:

$$H_2(g) + \tfrac{1}{2}O_2(g) \rightleftharpoons H_2O(g) \qquad (\Delta \mathcal{H}^\circ_f)_{298} = -57,800 \text{ cal } mol^{-1}$$

Substance	$\mathcal{S}^\circ_{298}$ (cal $K^{-1}\,mol^{-1}$)
$H_2(g)$	31.23
$H_2O(g)$	45.13
$O_2(g)$	49.056

4. An initial mixture, consisting of 3 parts of hydrogen and 1 part of nitrogen, contained 15.3 mol % of ammonia at equilibrium at a temperature of 200°C and pressure of 1 atm. Assuming ideal behavior, calculate the equilibrium constant for this reaction.

5. The equilibrium constant of the reaction

$$\tfrac{3}{2}H_2(g) + \tfrac{1}{2}N_2(g) \rightleftharpoons NH_3(g)$$

at 450°C and 1000 atm is $K_f = 1.010 \times 10^{-2}$. Assuming ideal behavior, the equilibrium constant is $K_p = 2.328 \times 10^{-2}$. Calculate the error introduced in the equilibrium constant by neglecting the nonideal behavior of the equilibrium mixture.

6. For the reaction

$$CO(g) + H_2O(g) \rightleftharpoons CO_2(g) + H_2(g)$$

L. S. Kassel [18] gives the value of the equilibrium constant as $K_p = 1.39$ for a temperature of 1000 K. Assuming ideal behavior, calculate the equilibrium composition of the reacting mixture in mol % after water has been removed from it by quenching the mixture. The reactants were in stoichiometric ratio and the total pressure was 1 atm.

7. H. F. Giauque and R. Overstreet [19] give the value $K_p = 2.45 \times 10^{-7}$ for the reaction

$$Cl_2(g) \rightleftharpoons 2\ Cl(g)$$

at a temperature of 1000 K. Assuming ideal behavior, calculate the degree of dissociation of chlorine at this temperature.

8. For the hydration of ethylene

$$C_2H_4(g) + H_2O(g) \rightleftharpoons C_2H_5OH(g)$$

H. M. Stanley, J. E. Youell, and J. B. Dymock [20] found experimentally the following values of the equilibrium constants

$\dfrac{t}{(°C)}$	K_p
145	6.8×10^{-2}
175	3.60×10^{-2}
200	1.65×10^{-2}
225	1.07×10^{-2}
250	6.73×10^{-3}

The value of K_ν at 200 atm and 300°C is 0.860. Assuming that the enthalpy change of the reaction is temperature independent, determine (a) the temperature dependence of the equilibrium constant, (b) the enthalpy change of the reaction and (c) the equilibrium conversion, if the initial mixture is composed of an equimolar mixture of ethylene and water and if the hydration is carried out at 300°C and a pressure of 200 atm.

9. For the reaction

$$C_2H_6(g) \rightleftharpoons C_2H_4(g) + H_2(g)$$

A. V. Frost [21] gives the equilibrium conversion of C_2H_6 as $x = 0.485\%$ at 1000 K and a pressure of 1 atm. Assuming ideal behavior, calculate the equilibrium constant at this temperature and pressure.

10. For the reaction

$$CO_2(g) \rightleftharpoons CO(g) + \tfrac{1}{2}O_2(g) \tag{a}$$

at 1000 K and a total pressure of 1 atm, L. S. Kassel [18] gives the equilibrium conversion of CO_2 as $x = 2.0 \times 10^{-7}\%$.

Determine the equilibrium constant of the reaction and also evaluate the equilibrium constant of the same reaction but written in a double molar ratio.

$$2\ CO_2(g) \rightleftharpoons 2\ CO(g) + O_2(g) \tag{b}$$

11. Cyclohexane is produced by the hydrogenation of benzene according to the equation

$$C_6H_6(g) + 3\ H_2(g) \rightleftharpoons C_6H_{12}(g)$$

Find the expression for the temperature dependence of the standard Gibbs free energy of the reaction and calculate the standard Gibbs free energy at a temperature of 400 K. *Given*:

	$(\Delta \mathcal{H}^\circ_f)_{298}$	$\mathcal{S}^\circ_{298}$	$\mathcal{C}_P = a + bT + cT^2/(\text{cal K}^{-1}\text{ mol}^{-1})$ [1]		
	(kcal mol^{-1})	(cal K^{-1} mol^{-1})	a	$b \times 10^3$	$c \times 10^6$
$C_6H_6(g)$	19.82	64.3	−3.54	90.37	−36.67
$C_6H_{12}(g)$	−29.43	71.3	−12.46	143.2	−54.94
$H_2(g)$		31.23	6.50	0.90	

12. Using the data from Example 9-1 calculate the equilibrium constant of the reaction

$$H_2(g) + \tfrac{1}{2} O_2(g) \rightleftharpoons H_2O(g)$$

at 2000 K.

13. During the calcination of sodium bicarbonate, the following reaction takes place

$$2\ NaHCO_3(s) \rightleftharpoons Na_2CO_3(s) + H_2O(g) + CO_2(g)$$

R. M. Caren and H. J. S. Sand [22] measured the equilibrium pressure of bicarbonate as a function of the temperature. The data obtained are given in the table that follows:

$t/(°C)$	$P/(\text{mm Hg})$
30	6.2
50	30.0
70	120.4
90	414.3
100	731.1
110	1252.6

Determine (a) the temperature dependence of the equilibrium constant of the reaction, (b) the decomposition temperature of sodium bicarbonate,[2] and (c) the enthalpy change of the reaction.

---

[2] The temperature at which the decomposition pressure equals 1 atm.

14. Cyclohexanone is produced by the hydrogenation of phenol and the dehydrogenation of the cyclohexanol thereby produced

$$C_6H_5OH(g) + 3 H_2(g) \rightleftharpoons C_6H_{11}OH(g) \tag{a}$$
$$C_6H_{11}OH(g) \rightleftharpoons C_6H_{10}O(g) + H_2(g) \tag{b}$$

These two reactions give rise to the question whether phenol could not be hydrogenated to cyclohexanone directly by 2 moles of hydrogen. This problem has been dealt with by a number of workers. A. P. Terenter and A. N. Guseva [23] judged from the results of their experiments that cyclohexanol, produced by reaction (a), reacts with the phenol as shown in Eq. (c)

$$C_6H_5OH(g) + 2 C_6H_{11}OH(g) \rightleftharpoons 3 C_6H_{10}O(g)$$

and that, by choosing suitable conditions, the synthesis of cyclohexanone by the dehydrogenating reaction (b) can be replaced by reaction (c).

On the basis of the data given below, calculate the equilibrium yield of reaction (c) at a temperature of 500 K and a pressure of 1 atm, if the initial mixture is phenol with cyclohexanol in stoichiometric ratio (1 : 2). *Given*:

$$\Delta G = I_H + aT \log T + bT^2 + cT^3 + I_G T$$

Reaction	I_H	a	$b \times 10^3$	$c \times 10^6$	I_G
(1)	$-44,428$	48.374	-16.920	1.442	-43.44
(2)	13,279	-19.387	7.411	0.972	24.27

15. D. P. Stevenson and J. H. Morgan [24] expressed the temperature dependence of the equilibrium constant of the reaction (isomerization of cyclohexane to methylcyclopentane)

$$C_6H_{12}(g) \rightleftharpoons C_5H_9CH_3(g)$$

by the relation

$$\ln K = 4.814 - \frac{2059}{T}$$

Calculate the isomerization entropy change at a temperature of 25°C.

16. HNO_3 in the gaseous phase may decompose according to the equation

$$4 HNO_3(g) \rightleftharpoons 4 NO_2(g) + 2 H_2O(g) + O_2(g)$$

If the reactions starts with pure HNO_3 prove that

$$K_p = \frac{1024 \, P_{O_2}^7}{(P - 7P_{O_2})^4}$$

17. The temperature dependence of the standard Gibbs free energy for the reaction

$$Sb_2O_3(s) + 3 CO(g) \rightleftharpoons 2 Sb(l) + 3 CO_2(g)$$

is given by the relation

$$\Delta G° = -33,461 + 34.286T \log T - 11.1 \times 10^{-3} T^2 + 0.96 \times 10^{-6} T^3 - 88.65 T$$

Calculate ΔG°_{1000} for the reaction

$$Sb_2O_3(s) = 2\,Sb(l) + \tfrac{3}{2}\,O_2(g)$$

if the Gibbs free energies of formation of $CO_2(g)$ and $CO(g)$ at 1000 K are

$(\Delta \mathscr{G}^{\circ}_f)_{CO_2} = -94.610$ cal mol^{-1} and $(\Delta \mathscr{G}^{\circ}_f)_{CO} = -47.942$ cal mol^{-1}

18. A mixture of hydrogen and carbon monoxide for synthetic purposes can be manufactured catalytically by the action of water vapor on methane at high temperatures. Assuming that only two reactions take place simultaneously, that is

$$CH_4(g) + H_2O(g) \rightleftharpoons CO(g) + 3\,H_2(g) \tag{a}$$
$$CO(g) + H_2O(g) \rightleftharpoons CO_2(g) + H_2(g) \tag{b}$$

calculate the composition of the equilibrium mixture (in mol %) if the reaction is carried out at a temperature of 600°C and a pressure of 1 atm and the initial mixture contains 5 moles of water vapor to 1 mole of methane. The equilibrium constants of the above reactions under the given conditions are $K_{(a)} = 0.574$, $K_{(b)} = 2.21$.

19. Calculate the standard Gibbs free energy change for the reaction

$$H_2(g) + \tfrac{1}{2}\,O_2(g) \rightleftharpoons H_2O(g)$$

at a temperature of 25°C if the following data are given

	$H_2(g)$	$O_2(g)$	$H_2O(g)$
$\left(\dfrac{\mathscr{G}^{\circ}_T - \mathscr{H}^{\circ}_0}{T}\right)_{298}$ (cal mol^{-1} K^{-1})	-24.423	-42.061	-37.165
$(\mathscr{H}^{\circ}_T - \mathscr{H}^{\circ}_0)_{298}$ (kcal mol^{-1})	2.024	2.070	2.365
$(\Delta \mathscr{H}^{\circ}_f)_{298}$ (cal mol^{-1})			-57.800

20. From the data given in the table that follows, calculate $(\Delta \mathscr{H}^{\circ}_f)_{298}$ and $(\Delta \mathscr{G}^{\circ}_f / T)_{298}$ of isobutane.

	iso-$C_4H_{10}(g)$	$H_2(g)$	$C(s)$
$\left(\dfrac{\mathscr{G}^{\circ} - \mathscr{H}^{\circ}_0}{T}\right)_{298}$ (cal K^{-1} mol^{-1})	-56.08	-24.423	-0.5172
$(\Delta \mathscr{H}^{\circ}_f)_0$ (kcal mol^{-1})	-25.30		
$(\mathscr{H}^{\circ}_T - \mathscr{H}^{\circ}_0)_{298}$ (kcal mol^{-1})	4.275	2.024	0.251

21. The Landolt–Börnstein tables give the following values for the temperature dependence of the specific heat capacity of nickel

T (K)	c_P (cal K^{-1} g^{-1})	T (K)	c_P (cal K^{-1} g^{-1})
15.05	7.92×10^{-4}	133.4	7.28×10^{-2}
25.20	2.44×10^{-3}	204.05	9.25×10^{-2}
47.10	1.438×10^{-2}	256.5	1.010×10^{-1}
67.13	3.11×10^{-2}	283.0	1.062×10^{-1}
82.11	4.11×10^{-2}		

Determine the third law entropy of nickel at 25°C by graphical integration.

22. W. F. Giauque and R. W. Blue [25] studied the thermodynamic properties of hydrogen sulphide at low temperatures and obtained the following values:

first transition point: 103.5 K; $\Delta \mathcal{H}_{tr} = 368$ cal mol^{-1}
second transition point: 126.2 K; $\Delta \mathcal{H}_{tr} = 121$ cal mol^{-1}
melting point: 187.6 K; $\Delta \mathcal{H}_{fus} = 568$ cal mol^{-1}
boiling point: 212.8 K; $\Delta \mathcal{H}_{vap} = 4463$ cal mol^{-1}

T (K)	$\mathcal{C}_P$ (cal K^{-1} mol^{-1})	T (K)	$\mathcal{C}_P$ (cal K^{-1} mol^{-1})	T (K)	$\mathcal{C}_P$ (cal K^{-1} mol^{-1})
20	1.25	100	9.36	160	13.65
30	2.48	105	11.25	170	13.92
40	3.56	110	11.79	180	14.26
50	4.56	120	13.27	185	14.53
60	5.51	125	15.02	liquid hydrogen sulphide	
70	6.43	130	13.25	190	16.21
80	7.31	140	13.33	200	16.26
90	8.26	150	13.45	210	16.31

These authors found that at the lowest temperatures the molar heat capacity of solid hydrogen sulphide follows the Debye law, $\mathcal{C}_P = aT^3$, and they give the value of the characteristic temperature $\Theta_D = 136$ K. By means of the Berthelot equation they calculated the entropy correction to nonideal behavior of the hydrogen sulphide vapors at normal boiling point as 0.1 cal K^{-1} mol^{-1}.

Using these data calculate the third law entropy of hydrogen sulphide in the state of an ideal gas at normal boiling point, $T_{nbp} = 212.8$ K.

23. Calculate the entropy correction for gas imperfection of nitrogen at 77.32 K and 1.00 atm if the critical temperature of nitrogen is 126 K and its critical pressure is 33.5 atm.

24. Derive the following equation for a reversible adiabatic demagnetization process.

$$\left(\frac{\partial T}{\partial \kappa}\right)_{P,S} = -\frac{T}{\mathscr{C}_{P,\kappa}}\left(\frac{\partial M}{\partial T}\right)_{P,\kappa}$$

where κ is the magnetic field strength, M is the magnetization and $\mathscr{C}_{P,\kappa} \equiv (dq/\partial T)_{P,\kappa}$ [Hint: $dw_{magn} = \kappa dM$ and $M = k\kappa/T^2$, Curie's law].

_____ *References and Recommended Reading*

[1] J. deHeer: *J. Chem. Educ.*, **35**:133 (1958). Le Chatelier, scientific principle or "sacred cow"?

[2] J. deHeer: *J. Chem. Educ.*, **34**:375 (1957). The principle of Le Chatelier and Braun.

[3] K. Mendelsohn: *The Quest for Absolute Zero*. McGraw-Hill, New York, 1966.

[4] P. A. Rock: *J. Chem. Educ.*, **44**:104 (1967). Fixed pressure standard states in thermodynamics and kinetics.

[5] D. Chandler and I. Oppenheim: *J. Chem. Educ.*, **43**:535 (1966). Some comments on the second and third laws of thermodynamics.

[6] C. J. G. Raw: *J. Chem. Educ.*, **38**:140 (1961). Chemical equilibrium in imperfect gases.

[7] E. F. Westrum Jr.: *J. Chem. Educ.*, **39**:443 (1962). Cryogenic calorimetric contributions to chemical thermodynamics.

[8] B. H. Mahan: *J. Chem. Educ.*, **40**:293 (1963). Temperature dependence of equilibrium.

[9] E. M. Loebl: *J. Chem. Educ.*, **37**:361 (1960). The third law of thermodynamics, the unattainability of absolute zero, and quantum mechanics.

[10] M. W. Lindauer: *J. Chem. Educ.*, **39**:384 (1962). The evolution of the concept of chemical equilibrium from 1775 to 1923.

[11] J. G. Eberhart and J. E. McDonald: *J. Chem. Educ.*, **42**:601 (1965). Graphical estimation of thermodynamic properties.

[12] H. Zeisse: *Z. Elektrochem.*, **43**:706 (1937).

[13] A. T. Larson: *J. Am. Chem. Soc.*, **46**:367 (1914).

[14] F. Haber: *Z. Elektrochem.*, **20**:597 (1914).

[15] L. Pierce and F. L. Pace: *J. Chem. Phys.*, **23**:531 (1955).

[16] G. N. Lewis, M. Randall, K. S. Pitzer, and L. Brewer: *Thermodynamics*. McGraw-Hill, New York, 1961.

[17] G. N. Lewis and M. Randall: *J. Am. Chem. Soc.*, **38**:2348 (1916).

[18] L. S. Kassel: *J. Am. Chem. Soc.*, **56**:1838 (1934).

[19] H. F. Giauque and R. Overstreet: *J. Am. Chem. Soc.*, **54**:1731 (1932).

[20] H. M. Stanley, J. E. Youell, and J. B. Dymock: *J. Soc. Chem. Ind.*, **53**:205T (1934).

[21] A. V. Frost: *Compt. rend acad. sci. USSR*, **1933**:158.

[22] R. M. Caren and H. J. S. Sand: *J. Chem. Soc.*, **99**:1359 (1911).

[23] A. P. Terenter and A. N. Guseva: *Proc. Acad. Sci. USSR*, **52**:135 (1946).

[24] D. P. Stevenson and J. H. Morgan: *J. Am. Chem. Soc.*, **70**:2773 (1948).

[25] W. F. Giauque and R. W. Blue: *J. Am. Chem. Soc.*, **58**:831 (1936).

Part III

Quantum Mechanics

chapter 10

The Principles of Quantum Mechanics

10-1 INTRODUCTION

The theory of valence is concerned with the problem of predicting the properties of a molecule by calculating the motion of its nuclei and its electrons. It is well known that the motion of macroscopic systems may be described by means of classical or Newtonian mechanics. The term "macroscopic system" means here the objects that we deal with in everyday life, such as cars, trains, ships, baseballs, and so on, or much larger objects such as stars or planets. However, if we study the motion of an electron in a molecule, then we deal with a particle that has a mass of the order of 10^{-27} g and moves around in an orbit of the order of 10^{-8} cm. For such a small system, classical mechanics is no longer valid. Instead we should use a new type of mechanics, which was developed in the period between 1900 and 1930 and is known as quantum mechanics or wave mechanics.

Many people think that quantum mechanics is much more difficult to learn than classical mechanics because the mathematical techniques in quantum mechanics are more difficult. Actually, the mathematical techniques in both types of mechanics are about equally complicated. The reason why classical mechanics seems easier is that most people have an intuitive feeling for the principles of classical mechanics from their experience in everyday life. For example, driving a car, hitting a baseball, or throwing a football all involve classical mechanics. Most people are familiar with the physical quantities that play a role in classical mechanics even though they may not even be aware of it. It is convenient to begin our study of mechanics by discussing these quantities.

The simplest situation that we can consider in classical mechanics is the motion of a point particle in one dimension. An example would be a car moving along a straight highway, if we ignore the dimensions of the car. The motion of the particle (or of the car) is completely determined if we know its position as a function of time. If we use the symbol x to describe the position we can write this as $x(t)$.

The main purpose of classical mechanics is to determine how the position x depends on the time t. This is usually achieved by solving the equation of motion, which is a differential equation. In order to find the solution of this equation it is necessary to know both the position and the velocity of the particle at a given time, say $t = 0$. In this procedure it is assumed that $x(0)$ and $v(0)$, the position and the velocity at $t = 0$, are known exactly. In that case the function $x(t)$ determines exactly where the particle is at any given time t.

If we consider the same problem, namely, the motion of a point particle in one dimension in quantum mechanics, then we must solve a different equation of motion, which is again a differential equation. However, at this stage we should not worry about differences in mathematical techniques because the difference between quantum mechanics and classical mechanics is much more basic than just these mathematical details. We should realize that the essential concepts in the two kinds of mechanics are fundamentally different. We have already mentioned that the main purpose of classical mechanics is to determine the position x of the particle as a function of the time t, that is to derive the function $x(t)$.

In principle, classical mechanics leads to exact predictions about the function $x(t)$, which is called the orbit of the particle. In practice, however, this is usually not true. Even if we consider the simple situation that we discussed previously, we have to measure $x(0)$ and $v(0)$, the position x and the velocity v of the particle at time $t = 0$, before we can calculate $x(t)$. The observations of $x(0)$ and $v(0)$ are necessarily subject to the possible inaccuracies of our measurements, and our prediction of $x(t)$ must allow also for these possible deviations. One of the basic differences between classical mechanics and quantum mechanics is in the approach to the treatment of these possible experimental errors.

In classical mechanics it is assumed that there are no limits to the accuracy of our observations and that it is, in principle, possible to improve our measuring techniques to the extent where experimental errors are arbitrarily small. In quantum mechanics, on the other hand, we have to recognize that there is a finite limit to the accuracy of our measurements.

Let us illustrate this difference in view-point by considering a simple and familiar situation, namely the motion of an ocean liner. If it is a calm day in midocean and the weather forecast predicts no change, it may be assumed that the ship will move with a constant velocity v. A measurement of v shows that this velocity is 16 nautical miles per hour. On most ocean liners the purser runs a betting pool every day where the passengers can predict the distance that the ship will cover from noon on one day until noon the next day. The mileage is rounded off to the closest integer, and the passenger who predicts the right number (or comes closest to it) wins the pot. Obviously, the calculation of the ship's daily course is a useful application of classical mechanics to anyone who happens to be a passenger on the ship.

Actually, we do not need to know much classical mechanics to calculate the ship's course over a 24 hr period. Since the velocity v is 16 miles hr^{-1} the result is simply $24 \times 16 = 384$ nautical miles. According to classical mechanics this result should be rigorous; that is, if we give this number to the purser we should have a 100% chance to win the pool. We know that in real life we are usually not that lucky. There are many unforeseeable factors that might cause the ship to speed up or slow down during the next 24 hr, the measurement of the velocity v is subject to possible errors, the weather might change and, as a consequence, it is easily possible that the ship's course might be different from 384 miles. An experienced bettor does not believe that the prediction of 384 miles is 100% accurate, and he will attempt instead to make a guess about the accuracy of his number.

Now, in the situation that we have described there is not enough information available to make such a guess, but for the sake of the argument, we assume that we know the probability pattern for the prediction and that the probability is given by Table 10-1, in pattern A. We should imagine that this probability pattern contains all the uncertainties in the weather, the measurements, possible engine trouble, and so on. We may then conclude from probability pattern A, that there is only a 22% chance that the prediction of 384 miles will turn out to be exact, that there is a 36% chance that the prediction is 1 mile off, and so on. We may also imagine a different situation, with fewer uncertainties, which is described by probability pattern B in the

Table 10-1 Probability patterns A and B for the prediction of the ship's course

A		B	
Course	Probability/(%)	Course	Probability/(%)
<380	1		
380	2		
381	6	<382	1
382	12	382	4
383	18	383	25
384	22	384	40
385	18	385	25
386	12	386	4
387	6	>386	1
388	2		
>388	1		

table; here the chance for a correct prediction is 40%, the chance for a 1-mile error is 50%, and so on.

We see that classical mechanics is actually an idealization of everyday life situations, since we assume that in classical mechanics everything can be measured with 100% accuracy and that there are no unforeseeable changes in our system. The result of these assumptions is that in classical mechanics the predictions about the future behavior of our system should be rigorously accurate. We have seen in the example concerning the ship that in real life these assumptions are not always satisfied and that predictions about the future positions of our system involve probability patterns rather than exact predictions. Such probability patterns play a prominent role in the formalism of quantum mechanics.

The examples we have given are somewhat misleading because we know that in classical mechanics it is in principle possible to improve our measuring techniques to any extent that we desire. However, if we now consider the motion of an electron in a molecule, the situation becomes completely different. Here it is not only impossible in practice to measure the position and velocity of the electron with unlimited accuracy but it is not even possible in principle to perform such measurements. The only way to obtain information about the electron is by letting it interact with radiation or with other particles. Such interactions necessarily affect the motion of the electron.

Consequently, we have to choose between two unattractive alternatives in planning experiments on the motion of the electron. In the first alternative we use as little radiation as possible so that the motion of the electron will not be seriously affected by our measurement; however, this technique gives us only a very crude idea of the electron's position. In the second experiment we use enough radiation to obtain an accurate result for the electron's position; but in this case the electron will be subject to a drastic change in velocity as a result of our measurement. It follows that we cannot measure both the position and the velocity of the electron with unlimited accuracy. As soon as we try to make the result for one of the two quantities more accurate, the other becomes less accurate. In order to put this limitation into mathematical form, we introduce Δx as the possible error in our determination of the electron's position and Δv as the possible error for the velocity. It follows that there is a lower limit for the product of the uncertainties, and we can write this as

$$\Delta x \cdot \Delta v > A \qquad (10\text{-}1)$$

This result was first obtained by Heisenberg in 1927. We should point out that Heisenberg arrived at this conclusion by following a different line of reasoning and also that his argument was based on the momentum p of the particle rather than the velocity v. The momentum p is simply defined as the velocity v multiplied by the mass m of the particle

$$p = mv \qquad (10\text{-}2)$$

Heisenberg derived the condition

$$\Delta x \cdot \Delta p > h \qquad (10\text{-}3)$$

where Δp is the uncertainty in the observed momentum and h is Planck's constant, which has the value 6.62×10^{-27} erg $\cdot$ s. Eq. (10-3) is known as **Heisenberg's uncertainty principle**. In three dimensions the position of the particle is given by the three Cartesian coordinates x, y, and z and the momentum has also three components p_x, p_y, and p_z. The Heisenberg uncertainty relations refer to each pair of coordinates, and momentum components that fall along the same axis

$$\Delta x \cdot \Delta p_x > h$$
$$\Delta y \cdot \Delta p_y > h \qquad (10\text{-}4)$$
$$\Delta z \cdot \Delta p_z > h$$

According to Heisenberg's uncertainty relations, the product of the uncertainties Δx and Δp_x must be larger than h, but the product of, for example Δx and Δp_z, can be arbitrarily small.

It follows from the Heisenberg uncertainty relations that we cannot describe the motion of an electron in terms of the functions $x(t)$, $y(t)$, and $z(t)$ and that classical mechanics is not applicable to electronic motion. Instead, we must recognize the uncertainties of our observations as an essential part of the theory, and we should describe the motion of the particle by means of probability patterns, like those in Table 10-1 that we discussed. In the following section we will discuss the mathematical description of such probability patterns in quantum mechanics.

10-2 THE WAVE FUNCTION

Let us start out by trying to give a more precise mathematical description of the probability patterns that we introduced previously as examples. For instance, take pattern A of Table 10-1. We should realize that the numbers we have listed are all rounded off to the nearest integer so that there is a 22% probability that x lies between 383.5 and 384.5, an 18% possibility that x lies between 384.5 and 385.5, and so on. We can give a graphical representation of this probability pattern by

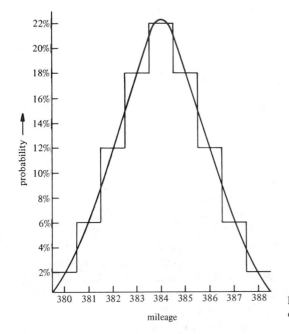

probability →

mileage

Figure 10-1 Continuous and discontinuous probability functions.

means of a step function, as shown in Figure 10-1. We have divided the possible values of x into 1-mile intervals and we are interested only in the probability distribution over these relatively large intervals.

It is obvious that we would have obtained a more precise probability pattern if we had used smaller intervals. For example, if we take 0.2-mile intervals, we get a finer-grained step function and, consequently, a more accurate probability pattern. Clearly, if we let the intervals get smaller and smaller we should get a continuous function in the limit. This is the probability distribution function $\rho(x)$ that we have drawn in Figure 10-1.

By definition, the probability of finding x between the two values x_1 and x_2 is given by

$$P(x_1, x_2) = \int_{x_1}^{x_2} \rho(x)\, dx \qquad (10\text{-}5)$$

It follows that

$$\int_{382.5}^{383.5} \rho(x)\, dx = (18/100)$$

$$\int_{383.5}^{384.5} \rho(x)\, dx = (22/100) \qquad \text{etc.} \qquad (10\text{-}6)$$

If we take the interval between x_1 and x_2 very small, for example

$$x_1 = x_0 - \tfrac{1}{2}\Delta x$$

$$x_2 = x_0 + \tfrac{1}{2}\Delta x \qquad (10\text{-}7)$$

then the probability becomes

$$P(x_0, \Delta x) = \int_{x_0-(1/2)\Delta x}^{x_0+(1/2)\Delta x} \rho(x)\, dx \tag{10-8}$$

If Δx becomes small enough this may be written as

$$P(x_0, \Delta x) = \rho(x_0) \cdot \Delta x \tag{10-9}$$

Since the sum of all probabilities must be equal to unity we have

$$\int_{-\infty}^{\infty} \rho(x)\, dx = 1 \tag{10-10}$$

In general, we may expect that our predictions for x depend also on the time that has elapsed so that our forecasts for x are represented by a probability function $\rho(x, t)$, which depends also on the time. However, for the sake of simplicity we have disregarded this time dependence in the above discussion.

In order to avoid confusion it may be useful to discuss the certainties and uncertainties of probability theory. Let us consider some examples from everyday life. If we flip pennies we know that there is a 50% chance to get heads and a 50% chance to get tails. These probabilities are exact, and this can be verified by flipping a large number of pennies. On the other hand, if we flip a penny only once we have absolutely no idea what will turn up, the only thing that we know is that the chances for heads or tails are exactly equal. Another example is the life expectancy tables that the life insurance companies use to determine their premiums. Nobody knows exactly when an individual death will occur, but the life expectancy tables are quite accurate if we consider a large group of people; if this were not the case, life insurance companies would soon be bankrupt.

Similarly, if we predict the future motion of a particle by means of a probability function $\rho(x, t)$, we have no exact knowledge about the positions of the particle and we can predict only the relative probabilities. On the other hand, the probability function is quite rigorous and exact. If we could consider a large number of identical systems and take the average over all systems, then we would find that these averages agree exactly with the predictions from the probability density function.

Let us now return to quantum mechanics. One of the basic concepts of quantum mechanics is that the motion of a particle, that is, its position as a function of space and time, is exactly described by means of a probability density function $\rho(x, y, z; t)$. By definition, the probability of finding the particle at a time t_0 in a volume element dv, surrounding the point (x_0, y_0, z_0), is given by

$$P = \rho(x_0, y_0, z_0; t_0)\, dv \tag{10-11}$$

In quantum mechanics the probability density function ρ is derived from another function ψ, which is known as the wave function. The relation between ρ and ψ is given by

$$\rho(x, y, z; t) = \psi(x, y, z; t)\psi^*(x, y, z; t) \tag{10-12}$$

This definition allows for the possibility that the wave function ψ is a complex function, which may contain a real part ψ_1 and an imaginary part ψ_2,

$$\psi(x, y, z; t) = \psi_1(x, y, z; t) + i\psi_2(x, y, z; t) \tag{10-13}$$

The symbol ψ^* stands for the complex conjugate of ψ, which is defined as

$$\psi^*(x, y, z; t) = \psi_1(x, y, z; t) - i\psi_2(x, y, z; t) \tag{10-14}$$

if ψ is given by Eq. (10-13). It follows that

$$\psi\psi^* = (\psi_1 + i\psi_2)(\psi_1 - i\psi_2) = \psi_1^2 - i^2\psi_2^2 = \psi_1^2 + \psi_2^2 \tag{10-15}$$

because $i^2 = -1$.

We should point out that it is always possible to write a molecular or atomic wave function as a real function. In that case the probability density function ρ is given simply by

$$\rho(x, y, z; t) = [\psi(x, y, z; t)]^2 \tag{10-16}$$

and there is no need to consider the complications that result from ψ being complex.

The wave function is obtained as the solution of a differential equation that is known as the **Schrödinger equation**. For a particle of mass m, which moves in three dimensions, the Schrödinger equation has the form

$$-\frac{\hbar^2}{2m}\left(\frac{\partial^2\psi}{\partial x^2} + \frac{\partial^2\psi}{\partial y^2} + \frac{\partial^2\psi}{\partial z^2}\right) + V(x, y, z)\psi = E\psi \tag{10-17}$$

Let us now discuss the various aspects of this equation. In order to do this, it is useful to return to classical mechanics.

In classical mechanics the motion of a particle is determined by the forces that act on it. In three-dimensional space the total force that acts on a particle can be written as a sum of three components, namely the forces in the X, Y, and Z directions F_x, F_y, and F_z, respectively. In general these force components are functions of the coordinates x, y, and z of the particle and of the time. However, we will consider only situations where the forces are not directly dependent on the time and where they can all be represented as the derivative of a function $V(x, y, z)$

$$F_x = -\frac{\partial V}{\partial x} \qquad F_y = -\frac{\partial V}{\partial y} \qquad F_z = -\frac{\partial V}{\partial z} \tag{10-18}$$

The function $V(x, y, z)$ is known as the potential energy of the particle.

The total energy E of the particle is the sum of the potential energy V and its kinetic energy T

$$E = T + V \tag{10-19}$$

The kinetic energy is given by

$$T = \tfrac{1}{2}mv^2 \tag{10-20}$$

or by

$$T = \frac{p^2}{2m} \tag{10-21}$$

if we use the momentum p instead of the velocity v.

It may be interesting to give the classical equations of motion for the particle. They are

$$m\frac{d^2x}{dt^2} = -\frac{\partial}{\partial x}V(x, y, z)$$

$$m\frac{d^2y}{dt^2} = -\frac{\partial}{\partial y}V(x, y, z) \qquad (10\text{-}22)$$

$$m\frac{d^2z}{dt^2} = -\frac{\partial}{\partial z}V(x, y, z)$$

If we are able to solve them we obtain x, y, and z as functions of time.

Since V depends on the coordinates, the potential energy depends implicitly on the time and the kinetic energy T is also time dependent. However, it can be shown that the total energy E is time independent in the situation described. It follows then that E remains constant during the motion of the particle and that the total energy is also independent of the position coordinates. Hence, it is called a constant of motion.

Both in classical mechanics and in quantum mechanics we are mainly interested in stationary states, that is, states where the energy remains constant during the motion. This means that in the Schrödinger equation (10-17) the symbol E may be considered a constant.

Let us now return to the Schrödinger equation (10-17). It contains the mass m of the particle and the Dirac constant $\hbar$, which is defined as

$$\hbar = (h/2\pi) = 1.05443 \times 10^{-27} \text{ erg} \cdot \text{s} \qquad (10\text{-}23)$$

Here, h is Planck's constant. In quantum mechanics we often find Planck's constant divided by (2π) and as a result a separate symbol $\hbar$ was introduced.

In atoms and molecules the forces between the particles are all electrostatic, apart from small perturbations that we neglect for the time being. The electrostatic interaction between two particles with charges Z_1 and Z_2 is given by

$$V = Z_1 Z_2 / r \qquad (10\text{-}24)$$

if the charges have the same sign and by

$$V = -Z_1 Z_2 / r \qquad (10\text{-}25)$$

if they have different signs. Here r is the distance between the particles.

To give an example, the Schrödinger equation for the hydrogen atom is

$$-\frac{\hbar^2}{2m}\left(\frac{\partial^2}{\partial x^2}+\frac{\partial^2}{\partial y^2}+\frac{\partial^2}{\partial z^2}\right)\psi(x, y, z)-\frac{e^2\psi(x, y, z)}{r}=E\psi(x, y, z) \qquad (10\text{-}26)$$

if we take the position of the nucleus fixed, or

$$-\frac{\hbar^2}{2m}\left(\frac{\partial^2}{\partial x^2}+\frac{\partial^2}{\partial y^2}+\frac{\partial^2}{\partial z^2}\right)\psi(x, y, z; X, Y, Z)$$

$$-\frac{\hbar^2}{2M}\left(\frac{\partial^2}{\partial X^2}+\frac{\partial^2}{\partial Y^2}+\frac{\partial^2}{\partial Z^2}\right)\psi(x, y, z; X, Y, Z)$$

$$-\frac{e^2\psi(x, y, z; X, Y, Z)}{r}=E\psi(x, y, z; X, Y, Z) \qquad (10\text{-}27)$$

if we are more precise. Here m is the mass of the electron, M is the mass of the proton, (x, y, z) are the coordinates of the electron, (X, Y, Z) are the coordinates of the proton, and r is the distance between the electron and the proton. We will discuss how this equation is solved in Chapter 11.

It should be noted that the Schrödinger equation (10-17) is suitable only for deriving the wave functions of stationary states that satisfy the condition that the energy of the system is a constant. The equation is known as the time-independent Schrödinger equation, and its solutions are time independent also. For the time being we are interested only in stationary states, but at a later stage in the book we will wish to consider the quantum theoretical aspects of spectroscopy and magnetic resonance. At that stage we have to use the time-dependent Schrödinger equation, which is suitable for nonstationary states.

10-3 THE BOHR ATOM

The quantum mechanics of bound particles is in many respects dependent on the Bohr theory of the atom. We will give a brief outline of the experimental results of atomic spectroscopy, followed by a discussion of the Bohr theory.

In studying the spectroscopic properties of a given sample, we measure either the absorption spectrum or the emission spectrum. Atomic spectra are usually measured in emission, where we subject the sample to a very high temperature by placing it in a high voltage discharge and then we measure the spectroscopic distribution of the light that is emitted. The result of the experiment is often a photographic plate from which we derive the intensity of the emitted light as a function of the wavelength. It has been known for many years that atomic spectra are usually line spectra, a typical example may be seen in Figure 10-2. Here light is emitted only in very narrow wavelength ranges, in a typical case the range might be of the order of 0.001 Å for a spectral line with a wavelength of about 4000 Å. The wavelength of a spectral line in atomic spectra is thus measured with an accuracy of $1 : 10^7$ or even better.

In reporting spectroscopic measurements, various sets of units may be used. Wavelengths are usually reported in terms of Ångstrom units ($1 \text{ Å} = 10^{-8}$ cm). It is possible also to describe a spectrum in terms of the frequency of the light. The frequency and the wavelength are related by means of

$$c = \lambda \nu \qquad (10\text{-}28)$$

where c is the velocity of light, which is approximately equal to $3 \cdot 10^{10}$ cm s^{-1}. Obviously a wavelength of 5000 Å corresponds to a frequency of $6 \cdot 10^{14}$ s^{-1}. It is

Figure 10-2 Line spectrum of the He atom. Wavelengths are in terms of Ångstroms.

often advantageous to use the wave number σ instead of the wavelength λ for the description of a spectrum. The wave number σ is defined simply as the reciprocal of the wavelength

$$\sigma = 1/\lambda \qquad (10\text{-}29)$$

and it is expressed in terms of cm^{-1}.

The final result of an atomic spectrum measurement consists of a very large collection of numbers, each of which represents a spectroscopic line. Around the turn of the century some semiempirical rules were discovered that led to a more concise description of these numbers. The most important of these rules was discovered by Ritz in 1908. Ritz described the atomic spectra in terms of wave numbers instead of wavelengths, and he noted that, if wave numbers are used, it is possible to construct a set of terms T_n so that each spectral line is the difference of a pair of terms,

$$T_{n,m} = |T_n - T_m| \qquad (10\text{-}30)$$

The spectrum of the hydrogen atom can be represented in a very simple form because in this spectrum the terms are given by

$$T_n = \boldsymbol{R}_H/n^2 \qquad (10\text{-}31)$$

where n is a positive integer ($n = 1, 2, 3, 4, \ldots$). All spectral lines in the hydrogen spectrum may be derived from the equation

$$T_{n,m} = \boldsymbol{R}_H\left(\frac{1}{n^2} - \frac{1}{m^2}\right) \quad \text{where } m, n = 1, 2, 3, \ldots \text{ and } n < m \qquad (10\text{-}32)$$

The constant $\boldsymbol{R}_H$ is known as the Rydberg constant for the hydrogen atom, and its value is $\boldsymbol{R}_H = 109{,}677.501 \ cm^{-1}$. It should be noted that Eq. (10-32) exactly represents the value of each hydrogen line to within the accuracy that these lines are measured.

Some attempts were made to explain the preceding results in terms of classical mechanics, but these attempts met with very serious difficulties. If we accept the Rutherford model of the hydrogen atom, where the electron moves in a closed orbit around the proton, it is very difficult to understand even why an atomic spectrum consists of discrete lines.

According to classical mechanics we may represent the motion of the electron as an oscillating charge, or an oscillating electric dipole, and we may derive from electromagnetic theory that such a dipole emits electromagnetic radiation and, consequently, produces light. However, such a system should emit light of a broad range of frequencies; thus, the classical theory predicts a band spectrum instead of a line spectrum. Furthermore, it follows from the principle of conservation of energy that the emission of light should cause a decrease in the energy of the atom. As more and more light is emitted, the atom should lose more and more energy, and the electron should orbit in a smaller orbit, closer to the nucleus. Eventually the electron should fall down on the nucleus, and the atom should collapse. It follows that classical theory predicts that a hydrogen atom will emit a short burst of light with a continuous spectroscopic distribution and, subsequently, collapse. It is easy to see that this prediction cannot be true because it does not agree with the experimental facts.

In order to avoid the difficulties just described Bohr introduced a set of new fundamental assumptions that must be used in the theoretical description of atomic structure. This quantum theory of atomic structure was published in a series of papers before the beginning of the first world war.

The first Bohr assumption states that an atom (or molecule) can only be in certain stable (or stationary) states with a fixed and well-defined energy. The atom does not absorb or emit radiation as long as it remains in one of these stationary states.

The second Bohr assumption states that an atomic spectral line corresponds to a transition from one stationary atomic state to another. Bohr made use of the ideas of Planck and Einstein, and he proposed that a spectral transition between two stationary states with energies E_n and E_m has a frequency given by

$$h\nu_{nm} = |E_n - E_m| \tag{10-33}$$

We see that the two Bohr assumptions account for those features of atomic spectra that cannot be explained classically; the first assumption avoids the prediction of atomic collapse and the second assumption leads to a discrete line spectrum. Bohr and Sommerfeld proposed a third assumption, which enabled them to make theoretical predictions of the energies of the stationary states of the hydrogen atom. In this way some useful theoretical results were derived.

On the other hand, the Bohr quantum theory by itself does not offer a really fundamental explanation of the structure of an atom even though it was one of the major scientific breakthroughs of this century. We know that Bohr's assumptions are basically correct, but we want to know why an atom has stationary states and how we can explain these assumptions from a more fundamental theoretical description. In the following section we will show, therefore, how the Bohr theory can be derived in a simple way from the Schrödinger equation.

10-4   STATIONARY STATES AND THE PARTICLE IN A BOX

It follows from the Bohr model of atomic structure that in an atomic system only specific, discrete values of the energy are permissible. This cannot be explained from classical mechanics. We wish to show now how the existence of discrete energy states is derived from quantum theory.

Let us again review the procedure that we must follow in order to derive the quantum mechanical description of the behavior of a particle moving in three-dimensional space. First, we must solve the Schrödinger equation

$$\frac{-\hbar^2}{2m}\left(\frac{\partial^2\psi}{\partial x^2} + \frac{\partial^2\psi}{\partial y^2} + \frac{\partial^2\psi}{\partial z^2}\right) + V\psi = E\psi \tag{10-34}$$

Next we derive the probability density function ρ from the solution ψ of Eq. (10-34)

$$\rho = \psi^2 \tag{10-35}$$

Finally, we use the function ρ for the physical interpretation of the motion of the particle, following the concepts of probability theory that we outlined in Section 10-1.

The first stage of this procedure is purely mathematical. Let us imagine that we are able to solve Eq. (10-34). Then, the solution is dependent on the value of E, and for every value of E we obtain a different solution. In other words, the solution may be written as $\psi(x, y, z; E)$ and it contains E as a parameter. If we then proceed to the next stage, we find that in most cases the general solution is not suitable for the purpose of physical interpretation. Then we must impose the condition that our solution make sense from a physical point of view, and usually this condition leads to restrictions on the energy values that may be used.

It is difficult to put these conditions in an exact mathematical form that has general validity because the conditions that we impose depend on the circumstances, and they vary from one problem to the next. In general the conditions are determined by simple common sense and, for a particular problem, it is usually quite easy to see what should be done. Let us consider specific examples.

It is easily seen that the probability density function should satisfy the condition

$$\iiint \rho(x, y, z; E)\, dx\, dy\, dz = 1 \qquad (10\text{-}36)$$

because the integral represents the total probability of finding the particle anywhere in space. The solution of a differential equation may always be multiplied by an arbitrary constant. Therefore, if we find that

$$\iiint \psi^2(x, y, z; E)\, dx\, dy\, dz = C \qquad (10\text{-}37)$$

where C is a finite number, we can make sure that the condition of Eq. (10-36) is satisfied by multiplying by a suitable constant. However, if the integral of Eq. (10-37) is infinite, this procedure does not work and we find that the corresponding wave function is not suitable. Those wave functions for which the integral of Eq. (10-37) is finite are called normalizable. It follows that one of the conditions we have to impose on the wave function is that it is normalized.

A second condition for the wave function is that it should be continuous in every point. This means that, if we consider the value of the wave function at a given point P (given by x_0, y_0, z_0), then the function should change only by a small amount ε if we move from the point P by a small distance δ. In the limit where δ approaches zero, ε should also approach zero. The wave function should be continuous in every point; if it is not continuous in even one single point, it becomes difficult to interpret the probability patterns. Also, if the function is not continuous, its derivative is not defined and it cannot satisfy the Schrödinger equation too well.

There may be other conditions that the wave function must satisfy, depending on the nature of the system. All these conditions may be summarized in a simple way: the wave function should give a probability density function that makes sense from a physical point of view.

As an illustration we will solve the Schrödinger equation for the simplest possible system in quantum mechanics, namely the one-dimensional particle in a box. So far we have mentioned only the three-dimensional Schrödinger equation, but it is easily seen that if we restrict the motion of the particle to only one dimension the Schrödinger equation takes the form

$$\frac{-\hbar^2}{2m}\frac{d^2\psi}{dx^2} + V(x)\psi = E\psi \qquad (10\text{-}38)$$

A one-dimensional box is defined by a potential which is zero for $0 \le x \le a$ and which

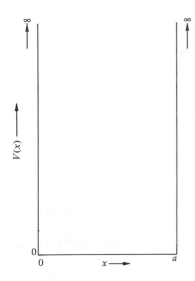

Figure 10-3 Potential function of the particle in a box.

is infinite elsewhere (see Figure 10-3). Algebraically, the potential is given by

$$V(x) = \infty \qquad x < 0$$
$$V(x) = 0 \qquad 0 \leq x \leq a \qquad \textbf{(10-39)}$$
$$V(x) = \infty \qquad a < x$$

It is clear that the particle cannot move outside the box and we have, therefore

$$\psi(x) = 0 \qquad x < 0 \quad \text{and} \quad x > a \qquad \textbf{(10-40)}$$

Inside the box the Schrödinger equation reduces to

$$\frac{-\hbar^2}{2m} \frac{d^2\psi}{dx^2} = E\psi \qquad \textbf{(10-41)}$$

where E must be real and positive. We may introduce a new parameter λ by means of

$$\frac{2mE}{\hbar^2} = \lambda^2 \qquad \textbf{(10-42)}$$

The Schrödinger equation then becomes

$$\frac{d^2\psi}{dx^2} = -\lambda^2\psi \qquad \textbf{(10-43)}$$

It is easily verified that this differential equation has two solutions, namely

$$\psi_1(x) = \sin \lambda x \qquad \textbf{(10-44)}$$

and

$$\psi_2(x) = \cos \lambda x \qquad \textbf{(10-45)}$$

The general solution of Eq. (10-43) is a linear combination of the two solutions of Eqs. (10-44) and (10-45) and it is represented as

$$\psi(x) = A \sin \lambda x + B \cos \lambda x \qquad \textbf{(10-46)}$$

This expression contains three parameters, namely A, B, and λ and it satisfies the Schrödinger equation (10-43) for any values of these parameters.

Let us now consider the various conditions that the wave function must satisfy. The function of Eq. (10-46) is normalizable because it is nonzero only in the finite interval $0 \leq x \leq a$ and it is finite everywhere within the interval. The important condition for the wave function is the continuity condition. If we summarize the results that we obtained for the wave function we have

$$
\begin{aligned}
\psi(x) &= 0 & x &< 0 \\
\psi(x) &= A \sin \lambda x + B \cos \lambda x & 0 &\leq x \leq a & \textbf{(10-47)} \\
\psi(x) &= 0 & a &< x
\end{aligned}
$$

This function is continuous within any of the three intervals, but it is not necessarily continuous if we go from one interval to the other, that is, at the points $x = 0$ and $x = a$. We must impose the conditions that the function be continuous at these points; this leads to the equations

$$\lim_{x \to 0} (A \sin \lambda x + B \cos \lambda x) = B = 0 \qquad \textbf{(10-48)}$$

and

$$\lim_{x \to a} (A \sin \lambda x + B \cos \lambda x) = A \sin \lambda a + B \cos \lambda a = 0 \qquad \textbf{(10-49)}$$

By making use of Eq. (10-48) we find that Eq. (10-49) reduces to

$$A \sin \lambda a = 0 \qquad \textbf{(10-50)}$$

It is easily seen that the parameter A cannot be zero. If it were zero the wave function would be zero everywhere and the total probability of finding the particle would be zero also; this does not make any sense so we do not allow for it. Consequently, we have the condition

$$\sin \lambda a = 0 \qquad \textbf{(10-51)}$$

This equation has an infinite number of solutions, namely

$$\lambda a = n\pi \qquad \text{where } n = 0, \pm 1, \pm 2, \pm 3, \ldots \qquad \textbf{(10-52)}$$

Some of these solutions may be eliminated. If $n = 0$ the wave function would again be zero everywhere, and this is not allowed. Also, we should recognize that the two solutions, $n = \nu$ and $n = -\nu$, are really only one solution because they lead to the same probability density function. Consequently, the possible values of n are

$$n = 1, 2, 3, 4, \ldots \qquad \textbf{(10-53)}$$

We should remember that λ is related to the energy E by means of Eq. (10-42). If λ must satisfy Eq. (10-52), the possible values of E are given by

$$E_n = \frac{n^2 \hbar^2 \pi^2}{2ma^2} \qquad \textbf{(10-54)}$$

where the possible values of n are listed in Eq. (10-53).

The preceding considerations indicate how the use of quantum mechanics leads to the existence of stationary states because only the energy values that are given by Eq.

(10-54) are allowed. All other energy values are not allowed because they would lead to results that do not make any sense from a physical point of view.

Each of the energy values E_n has a corresponding wave function ψ_n that is obtained by substituting Eq. (10-52) into Eq. (10-46),

$$\psi_n = A_n \sin \frac{n\pi x}{a} = \sqrt{\frac{a}{2}} \sin \frac{n\pi x}{a} \qquad \text{(10-55)}$$

Here we have used the result $B = 0$ and we have substituted the value for the normalization constant A_n.

EXAMPLE 10-1

The constant A_n is determined from the normalization condition

$$\int_0^a [\psi_n(x)]^2 \, dx = A_n^2 \int_0^a \sin^2 \frac{n\pi x}{a} \, dx = 1 \qquad \text{(a)}$$

The integral is evaluated by substituting

$$\frac{n\pi x}{a} = t \qquad x = \frac{at}{n\pi} \qquad dx = \frac{a}{n\pi} \, dt \qquad \text{(b)}$$

This gives

$$\int_0^a \sin^2 \frac{n\pi x}{a} \, dx = \frac{a}{n\pi} \int_0^{n\pi} \sin^2 t \, dt = \frac{a}{\pi} \int_0^\pi \sin^2 t \, dt = \frac{a}{2} \qquad \text{(c)}$$

Consequently

$$A_n^2 = \frac{a}{2} \qquad A_n = \sqrt{\frac{a}{2}} \qquad \text{(d)}$$

which is the value we have substituted into Eq. (10-55). ●

In Figure 10-4 we show the behavior of some of the eigenfunctions. This type of representation is fairly common. First we draw the potential function as a function of x, just as we did in Figure 10-3 and we draw horizontal lines at the positions of the eigenvalues; we show only the lowest eigenvalues E_1, E_2, E_3, and E_4. Then we plot the eigenfunction $\psi_1(x)$ as a function of x with respect to the horizontal line that corresponds to E_1; the eigenfunction $\psi_2(x)$ is plotted with respect to the line belonging to E_2; and so on. In this way we can see the behavior of both the eigenvalues and eigenfunctions in a convenient way.

Now that we have seen at least one example of solving the Schrödinger equation we can present the problem from a different point of view. It is possible to rewrite Eq. (10-34) as

$$\left[\frac{-\hbar^2}{2m} \left(\frac{\partial^2}{\partial x^2} + \frac{\partial^2}{\partial y^2} + \frac{\partial^2}{\partial z^2} \right) + V(x, y, z) \right] \psi(x, y, z) = E\psi(x, y, z) \qquad \text{(10-56)}$$

This may be written in a much more compact form, namely as

$$\mathfrak{H}_{op}\psi = E\psi \qquad \text{(10-57)}$$

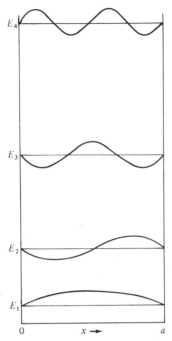

Figure 10-4   Eigenvalues and eigenfunctions of the particle in a box.

Here the symbol $\mathfrak{H}_{op}$ stands in place of

$$\mathfrak{H}_{op} = -\frac{\hbar^2}{2m}\left(\frac{\partial^2}{\partial x^2}+\frac{\partial^2}{\partial y^2}+\frac{\partial^2}{\partial z^2}\right)+V(x,\,y,\,z) \tag{10-58}$$

and it is known as the **Hamiltonian operator**.

EXAMPLE 10-2

Let us define what an operator is. Any procedure that transforms an arbitrary function f into another function g may be written in the form

$$g = \Lambda f \tag{a}$$

where Λ is defined as the operator that represents this procedure. The most common examples of operators are multiplicative operators and differential operators. If g is obtained by multiplying f by a given function we may write

$$g = hf \tag{b}$$

In this notation h is an operator and it is known as a multiplicative operator. If g is obtained by differentiating f,

$$g = \frac{df}{dx} = \frac{d}{dx}f \tag{c}$$

then the operator is (d/dx) and it is known as a differential operator. Another well known operator is

$$\Delta = \frac{\partial^2}{\partial x^2}+\frac{\partial^2}{\partial y^2}+\frac{\partial^2}{\partial z^2} \tag{d}$$

which occurs in Eq. (10-56). It is known as the **Laplace operator**.   ●

We rewrite the problem of the particle in a box in terms of the operator language. The differential equation is

$$\mathfrak{H}_{op}(x)\psi(x) = E\psi(x) \qquad 0 \leq x \leq a$$

where

$$\mathfrak{H}_{op}(x) = -\frac{\hbar^2}{2m}\frac{d^2}{dx^2} \qquad\qquad \textbf{(10-59)}$$

and the solutions must satisfy the conditions

$$\psi(0) = \psi(a) = 0 \qquad\qquad \textbf{(10-60)}$$

We have seen that Eqs. (10-59) and (10-60) are satisfied for a certain set of energy values E_n ($n = 1, 2, 3, \ldots$). The corresponding solutions are the functions $\psi_n(x)$ and they satisfy the equations

$$\mathfrak{H}_{op}(x)\psi_n(x) = E_n\psi_n(x)$$
$$\psi_n(0) = \psi_n(a) = 0 \qquad\qquad \textbf{(10-61)}$$

In mathematical language the conditions of Eq. (10-60) are called the boundary conditions of the differential equation (10-59). The differential equation is defined by the operator $\mathfrak{H}$. The set of numbers E_n are called the eigenvalues of the operator $\mathfrak{H}$ and the corresponding functions ψ_n are the eigenfunctions of $\mathfrak{H}$.

In the example that we have discussed, namely the particle in a box, there is exactly one eigenfunction belonging to each eigenvalue. In that case we call the eigenvalue nondegenerate. In general, it is possible that more than one eigenfunction belongs to the same eigenvalue. If that is the case we call the eigenvalue degenerate. We speak of a twofold degenerate eigenvalue if it has two different eigenfunctions, a threefold degenerate eigenvalue if there are three different eigenfunctions, and so on.

Let us consider a twofold degenerate eigenvalue λ_n of an operator Λ. We have the two equations

$$\Lambda\phi_{n,1} = \lambda_n\phi_{n,1}$$
$$\Lambda\phi_{n,2} = \lambda_n\phi_{n,2} \qquad\qquad \textbf{(10-62)}$$

where $\phi_{n,1}$ and $\phi_{n,2}$ are the two eigenfunctions belonging to λ_n. It follows from Eq. (10-62) that we have also

$$\Lambda(a\phi_{n,1} + b\phi_{n,2}) = \lambda_n(a\phi_{n,1} + b\phi_{n,2}) \qquad\qquad \textbf{(10-63)}$$

where a and b are arbitrary parameters. In other words, if a degenerate eigenvalue has a set of different eigenfunctions, then any linear combination of those eigenfunctions is also an eigenfunction belonging to that eigenvalue.

We have seen that we can make probability predictions about the particle from the wave function. In particular we discussed the predictions about the position of the particle. By using probability theory, we can more or less predict what the most probable position of the particle is. In fact, the different definitions give somewhat different results, and probability theory is somewhat ambiguous in this respect. If the probability density function is symmetric, then the various definitions of the most probable position all give the same result. But if the probability function is not symmetric, as in Figure 10-5, then the result depends on the definition.

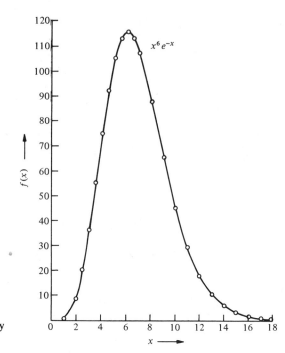

Figure 10-5 An asymmetric probability
function $f(x) = x^6 e^{-x}$.

EXAMPLE 10-3

If we take a probability density function that is given by

$$P(x) = x^n e^{-\alpha x} \qquad x \geq 0$$
$$P(x) = 0 \qquad\qquad x < 0 \tag{a}$$

then this function has a maximum value x_m for

$$\left(\frac{dP}{dx}\right)_{x_m} = (nx_m^{n-1} - \alpha x_m^n) e^{-\alpha x_m} = 0 \tag{b}$$

or

$$x_m = \frac{n}{\alpha} \tag{c}$$

This value is known in statistics as the most probable value.

An alternative definition is by means of

$$\bar{x}_0 = \frac{\int xP(x)\,dx}{\int P(x)\,dx} \tag{d}$$

Here every value of x is multiplied by a certain weight factor, which is given by $P(x)$, and we integrate over x. Naturally, the probability density function should then be normalized to unity, and we must divide by the second integral. It is easily derived from Eq. (d) that x_0 is given by

$$\bar{x}_0 = \frac{\int_0^\infty x^{n+1} e^{-\alpha x}\,dx}{\int_0^\infty x^n e^{-\alpha x}\,dx} = \frac{(n+1)!/\alpha^{n+2}}{n!/\alpha^{n+1}} = \frac{n+1}{\alpha} \tag{e}$$

This value is known as the average value in statistics. ●

It has been found that in quantum mechanics we must predict the values of experimental quantities by means of an equation of the type of Eq. (d) of Example 10-3. We predict the position of a particle in one dimension by means of the integral

$$\bar{x} = \frac{\int \Psi^*(x) x \Psi(x)\, dx}{\int \Psi^*(x) \Psi(x)\, dx} \tag{10-64}$$

The result is called the expectation value of the coordinate x. In three dimensions the expectation value of the position is given by the three integrals

$$\bar{x} = \iiint \Psi^*(x, y, z) x \Psi(x, y, z)\, dx\, dy\, dz$$

$$\bar{y} = \iiint \Psi^*(x, y, z) y \Psi(x, y, z)\, dx\, dy\, dz \tag{10-65}$$

$$\bar{z} = \iiint \Psi^*(x, y, z) z \Psi(x, y, z)\, dx\, dy\, dz$$

Here it is assumed that the wave function Ψ is normalized to unity

$$\int \Psi^*(x, y, z) \Psi(x, y, z)\, dx\, dy\, dz = 1 \tag{10-66}$$

By definition, the expectation value is the number that we should measure if we perform an experiment. It follows then that the position of the particle should be the vector $\bar{\mathbf{r}}$ with $\bar{x}$, $\bar{y}$, and $\bar{z}$, defined by Eq. (10-65). This is the most probable value as predicted by quantum mechanics.

If we wish to find the expectation value for any other physical observable we should know the operator that is representative for this observable. For example, the energy E is represented by the Hamiltonian operator $\mathfrak{H}_{op}$ and its expectation value $\bar{E}$ is given by

$$\bar{E} = \iiint \Psi^*(x, y, z) \mathfrak{H}_{op} \Psi(x, y, z)\, dx\, dy\, dz \tag{10-67}$$

In the case where the wave function is an eigenfunction Ψ_n of the Hamiltonian operator we have

$$\bar{E} = \iiint \Psi_n^*(x, y, z) \mathfrak{H}_{op} \Psi_n(x, y, z)\, dx\, dy\, dz = E_n \tag{10-68}$$

The expectation value of the energy is the eigenvalue E_n of the operator.

It is obvious that the result of a physical measurement cannot be an imaginary number. It follows that a physical observable should always be real. Its expectation value should also be real, otherwise the above definitions would not make much sense. If we consider a physical observable L, which is represented by an operator Λ, then the expectation value $\bar{L}$ is given by

$$\bar{L} = \int \Psi^* \Lambda \Psi\, dv \tag{10-69}$$

The complex conjugate of Eq. (10-69) is

$$(\bar{L})^* = \int \Psi \Lambda^* \Psi^*\, dv \tag{10-70}$$

The condition that L is a real quantity is equivalent with the condition

$$\int \Psi^* \Lambda \Psi \, dv = \int \Psi \Lambda^* \Psi^* \, dv \tag{10-71}$$

In mathematical language this means that the operator Λ must be Hermitian because a **Hermitian operator** is defined as an operator that satisfies the condition

$$\int f \Lambda g \, dv = \int g \Lambda^* f \, dv \tag{10-72}$$

for any pair of functions f and g. In summary, we have the condition that any operator that represents a physical observable must be Hermitian.

EXAMPLE 10-4

Hermitian operators have some useful properties. First, the eigenvalues of a Hermitian operator are all real and, second, the eigenfunctions of a Hermitian operator are orthogonal to each other. We will prove both properties and discuss what they mean.

Let H be a Hermitian operator and ε_n and ϕ_n one of its eigenvalues and the corresponding eigenfunction, respectively. We have then

$$H\phi_n = \varepsilon_n \phi_n \tag{a}$$

If we multiply both sides of this equation on the left by ϕ_n^* and then integrate we obtain

$$\int \phi_n^* H \phi_n \, dv = \varepsilon_n \int \phi_n^* \phi_n \, dv \tag{b}$$

The complex conjugate of this equation must also be valid, it is

$$\int \phi_n H^* \phi_n^* \, dv = \varepsilon_n^* \int \phi_n^* \phi_n \, dv \tag{c}$$

The left hand side of Eq. (b) is equal to the left hand side of Eq. (c) because the operator is Hermitian. Therefore, if we subtract the two equations we obtain

$$(\varepsilon_n - \varepsilon_n^*) \int \phi_n^* \phi_n \, dv = 0 \tag{d}$$

The integral must be nonzero and positive so that

$$\varepsilon_n = \varepsilon_n^* \tag{e}$$

In other words, the eigenvalue ε_n is real.

By definition, two functions Ψ and χ are orthogonal to each other if they satisfy the condition

$$\int \chi^* \Psi \, dv = 0 \tag{f}$$

We wish to prove that two eigenfunctions ϕ_n and ϕ_m, belonging to different eigenvalues ε_n and ε_m of the Hermitian operator H are always orthogonal. The function ϕ_m satisfies the equation

$$H\phi_m = \varepsilon_m \phi_m \tag{g}$$

If we multiply this equation on the left by ϕ_n^* and integrate we obtain

$$\int \phi_n^* H \phi_m \, dv = \varepsilon_m \int \phi_n^* \phi_m \, dv \qquad \text{(h)}$$

The function ϕ_n must satisfy the complex conjugate of Eq. (a), which is

$$H^* \phi_n^* = \varepsilon_n \phi_n^* \qquad \text{(i)}$$

Remember that ε_n is real. When we multiply Eq. (i) on the left by ϕ_m and integrate, the result is

$$\int \phi_m H^* \phi_n^* \, dv = \varepsilon_n \int \phi_n^* \phi_m \, dv \qquad \text{(j)}$$

Again, the left hand side of Eq. (h) is equal to the left hand side of Eq. (j) because the operator is Hermitian. Consequently, subtraction of the two equations gives

$$(\varepsilon_m - \varepsilon_n) \int \phi_n^* \phi_m \, dv = 0 \qquad \text{(k)}$$

We have assumed that the two eigenvalues are different, hence

$$\int \phi_n^* \phi_m \, dv = 0 \qquad \text{(l)}$$

This means that the two eigenfunctions ϕ_n and ϕ_m are orthogonal to each other.

The two theorems about Hermitian operators that we have proved in this example will be very useful later on. ●

Finally, we wish to point out that in the theory of valence the molecular eigenfunctions may always be taken as real since in atoms and molecules the molecular Hamiltonians are real. This is easily proved by considering the case where an eigenfunction Ψ_n would be complex, in that case Ψ_n may be written as

$$\Psi_n = \Psi_{n,1} + i \Psi_{n,2} \qquad \textbf{(10-73)}$$

where $\Psi_{n,1}$ and $\Psi_{n,2}$ are real functions. We have then

$$\mathfrak{H}(\Psi_{n,1} + i \Psi_{n,2}) - \varepsilon_n (\Psi_{n,1} + i \Psi_{n,2}) = 0 \qquad \textbf{(10-74)}$$

The real and imaginary parts of this equation should each be zero, therefore $\Psi_{n,1}$ and $\Psi_{n,2}$ should satisfy the equations

$$\begin{aligned} \mathfrak{H}\Psi_{n,1} &= \varepsilon_n \Psi_{n,1} \\ \mathfrak{H}\Psi_{n,2} &= \varepsilon_n \Psi_{n,2} \end{aligned} \qquad \textbf{(10-75)}$$

If ε_n is nondegenerate, then $\Psi_{n,1}$ and $\Psi_{n,2}$ should be proportional to each other and we can take $\Psi_{n,1}$ as the eigenfunction. If ε_n is degenerate, we can take the real functions $\Psi_{n,1}$ and $\Psi_{n,2}$ as the eigenfunctions rather than the complex function Ψ_n. It is often convenient to take all eigenfunctions real, and it is useful to know that it is always possible to do so.

10-5 THE COLOR OF CONJUGATED ORGANIC MOLECULES

The particle in a box has served as a starting point for a number of simple theories in chemistry. We think that it may be interesting to discuss one of them here in order to show that even simple calculations in quantum mechanics can have some practical use. We have to make use of some concepts that we have not yet discussed, but we do not believe that this will cause any difficulties.

In Kuhn's theory of the color of conjugated organic molecules, it is assumed that the color is mostly due to the π electrons in the molecule. It is further assumed that these electrons move freely along the molecular bonds and that the particle in a box offers a reasonable approximation to their quantum-mechanical description.

The energy levels of the π electrons are derived in this model from Eq. (10-54) since this represents the eigenvalues of the particle in a box. Let us consider the conjugated hydrocarbons and take butadiene to start with. This molecule has three carbon–carbon bonds; if we take the average length of these bonds as 1.38 Å, the total length of the carbon–carbon bonds is 4.14 Å. We assume that the π electrons may move slightly beyond the carbon skeleton, and we allow them half a bond length, or 0.69 Å at each end of the molecule. The length of the box is therefore 4×1.38 Å.

We evaluate the energy levels of the butadiene π electrons from Eq. (10-54). We substitute $\hbar = 1.0544 \times 10^{-27}$ erg $\cdot$ s, $m = 9.107 \times 10^{-28}$ g and $a = 4 \times 1.38 \times 10^{-8}$ cm. The result is

$$E_n = \frac{(1.0544)^2 \times 10^{-54} \times (3.14159)^2}{2 \times 9.107 \times 10^{-28} \times (1.38)^2 \times 10^{-16}} \frac{n^2}{4^2} \text{ erg}$$

$$= (n^2/4^2) \times 3.163 \times 10^{-11} \text{ erg}$$

$$= (n^2/4^2) \times 159,300 \text{ cm}^{-1} \tag{10-76}$$

if we remember that 1 erg $= 5.0358 \times 10^{15}$ cm^{-1}.

Butadiene has four π electrons, and we have to consider how they should be distributed over the energy levels. According to the Pauli exclusion principle, which will be discussed at a later stage in the book, each nondegenerate energy level can accommodate no more than two electrons. It follows that the lowest molecular energy is obtained if two electrons occupy energy level E_1 and two electrons occupy

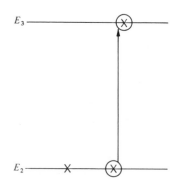

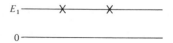

Figure 10-6 Distribution of electrons over the energy eigenstates in butadiene.

energy level E_2 (see Figure 10-6). The lowest excited state of the molecule is obtained by transferring one electron from energy level E_2 to energy level E_3. The lowest excitation energy of the molecule is therefore given by

$$E_3 - E_2 = \tfrac{5}{16} \times 159,300 \text{ cm}^{-1}$$
$$= 49,780 \text{ cm}^{-1} \tag{10-77}$$

according to Eq. (10-76). It follows that the first absorption band occurs at about 2000 Å and, because this is in the ultraviolet, that the molecule is colorless.

Let us now investigate how many carbon atoms a conjugated hydrocarbon must have so that the compound is colored. An arbitrary conjugated hydrocarbon has $2N$ carbon atoms and $(2N-1)$ carbon–carbon bonds. Its energy levels $E_n(N)$ may be derived from Eq. (10-76) by replacing the number 4 in this equation by $2N$.

$$E_n(N) = (n^2/4N^2) \times 159,300 \text{ cm}^{-1} \tag{10-78}$$

The molecule has $2N$ π electrons, and its lowest excitation energy is obtained as the difference between E_{N+1} and E_N

$$\Delta E_N = E_{N+1}(N) - E_N(N)$$
$$= [(2N+1)/4N^2] \times 159,300 \text{ cm}^{-1} \tag{10-79}$$

The wavelength of the first absorption band, λ_N, is given by the inverse of Eq. (10-79)

$$\lambda_N = \left(1 - \frac{1}{2N+1}\right) N \times 1255 \text{ Å} \tag{10-80}$$

In Table 10-2 we have listed the values of λ_N and the colors for the first four conjugated hydrocarbons. It should be noted that the colors that we see are complementary to the colors that are prepresented by λ_N. For instance, in the

Table 10-2 Absorption bands and colors of the conjugated hydrocarbons $C_{2N}H_{2N+2}$

n	$\lambda_N/(\text{Å})$	Color
2	2008	colorless
3	3270	colorless
4	4462	yellow
5	5704	violet

compound C_8H_{10} light is absorbed in the blue-violet region and the compound will look yellow to us because yellow is complementary to blue-violet. We must be careful if we want to extend our predictions beyond the compound $C_{10}H_{12}$ since we may have to consider more than one absorption band in the visible region. It is a useful exercise to look up the colors of the compounds that we have listed in Table 10-2 to verify how accurate our theoretical predictions are.

_____ *Problems*

1. Look up in the literature the colors of the four conjugated hydrocarbons C_4H_6 (butadiene), C_6H_8 (hexatriene), C_8H_{10} (octatetraene), and $C_{10}H_{12}$ (decapentaene) and verify the accuracy of the theoretical results of Table 10-2.

2. Calculate the first excitation energy $(E_2 - E_1)$ in terms of electron volts (eV) for an electron and for a proton in a one-dimensional box of 1 Å.

3. Write down the Hamiltonian operator and the Schrödinger equation for the electron in the hydrogen molecular ion H_2^+, for the He atom, and for the hydrogen molecule.

4. Calculate the wavelength of the absorption line that corresponds to an energy difference of 4 eV and also for an energy difference of 80 kcal mol^{-1}. (Look up the values of the physical constants in Appendix IX.)

5. By making use of the Heisenberg uncertainty principle, we can predict a lower limit for Δp, Δv and for the kinetic energy of a particle in a box since Δx is equal to the length a of the box. Show that these predictions are consistent with the result we obtained for the lowest energy eigenvalue by solving the Schrödinger equation.

6. Calculate the lower limit $\Delta v \cdot \Delta x$ for a car (mass 100 kg), a bullet (mass 10 g), and for a hydrogen molecule by making use of the Heisenberg uncertainty principle. Determine Δx for the car if it moves at 60 mile hr^{-1}, for the bullet if it moves at 2000 ft s^{-1}, and for the hydrogen molecule if it moves at 20 km s^{-1}.

7. How long should a linear conjugated hydrocarbon molecule be in order to absorb light between 8000 and 9000 Å in an electronic transition?

8. We know that one throw of a die will give equal possibilities for throwing 1, 2, 3, 4, 5, or 6 so that the probability density function is a straight line. Derive the probability density functions for the average result of two throws of the die and of three throws of the die.

9. Show explicitly that for a particle in a box the eigenfunctions belonging to the eigenvalues E_1 and E_3 are orthogonal.

10. It is easily shown that the multiplicative operator x is Hermitian. What about the operators (d/dx) and (d^2/dx^2)?

11. If we take it that visible light is light with a wavelength between 4000 and 7000 Å, which spectral transitions of the hydrogen atom, given by Eq. (10-32), are observed in the visible?

12. If we have an electron in a box of 10 Å long, what is the wavelength of the light that is absorbed in a transition from the ground state to the first excited state?

13. Let the probability density function of a particle be given by $(x^2 - 4x + 5)^{-1}$. Determine the expectation value x_0 for this probability density function. Show that x_0 is equal to the point where the probability density function has its maximum. Determine also the probability for finding the particle between the two points $x_0 - 1$ and $x_0 + 1$ and between the two points $x_0 - \frac{1}{2}$ and $x_0 + \frac{1}{2}$.

14. Consider the probability density function

$$P(x) = x^6 e^{-x} \qquad x \geq 0$$
$$P(x) = 0 \qquad x < 0$$

Determine the position x_m of the maximum of this function and also the expectation value x_0 with respect to this function. According to quantum mechanics where are we supposed to find the particle?

_____ *References and Recommended Reading*

In Chapter 10 we gave a brief outline of quantum theory; for a more complete treatment of the subject, we refer the reader to any of the quantum mechanics texts listed at the end of Chapter 11. We feel that it may be of interest to read the historical development of quantum theory, and we recommend reading the two Bohr biographies that we have listed [1, 2] and the two histories [A1, A2] listed in Chapter 11. In addition, we have listed here a description of the old quantum theory [3] and the original Bohr papers [4, 5].

[1] R. Moore: *Niels Bohr: The Man, His Science and the World They Changed.* Alfred A. Knopf, New York, 1966.

[2] S. Rozental: *Niels Bohr, His Life and Work as Seen by His Friends and Colleagues.* North-Holland, Amsterdam, 1967.

[3] D. ter Haar: *The Old Quantum Theory.* Pergamon Press, Oxford, 1967.

[4] N. Bohr: *Kgl. Danske Vid. Selsk. Skr., Nat.-Math. Afd.,* **8**(Raekke IV):1, Part I (April 1918). On the quantum theory of line-spectra.

[5] N. Bohr: *Phil. Mag.,* **26**:1 (1913); **26**:875 (1913); **27**:506 (1914); **29**:332 (1915); **30**:394 (1915).

Solutions of the Schrödinger Equation

11-1 INTRODUCTION

In 1926 Erwin Schrödinger published a series of three papers on quantum theory; at the time he was 39 years old and a professor of physics in Vienna. The first paper contains the differential equation that we discussed in Chapter 10. It was named after Schrödinger and it is one of the fundamental equations in quantum theory. In the second and third papers Schrödinger gave the exact solution of this equation for three systems that play an essential role in the theory of atomic structure; these systems are the hydrogen atom, the harmonic oscillator, and the rigid rotor. It was known in classical theory that in a molecule the rotational motion, the vibrational motion, and the electronic motion may, to a first approximation, be considered separately. We shall see that the vibrational motion may be approximated by a model that is known as the harmonic oscillator. For a diatomic molecule the rotational motion is described by a model that is known as the rigid rotor.

It is somewhat ironical that there are very few systems for which the Schrödinger equation can be solved exactly beyond the three systems that were treated by Schrödinger. The most complicated system for which the Schrödinger equation was solved is the hydrogen molecular ion H_2^+ (one electron moving in the field of two protons), but this required some fairly sophisticated mathematics and it cannot be discussed in an elementary (or even advanced) textbook.

In the last paper of the series Schrödinger introduced perturbation theory, which is an approximate method for treating systems that cannot be solved exactly. It is not much of an exaggeration to say that Schrödinger's three papers covered all important features of nonrelativistic quantum mechanics. On the other hand, all subsequent work on the theory of valence has also played an important role in our understanding of chemistry because it was necessary to extend Schrödinger's principles to large molecules. This involves the use of suitable approximate methods, theoretical explanations of important chemical features and experimental results, numerical evaluations, and so on.

In this chapter we shall discuss the solution of the Schrödinger equation for the three systems mentioned, namely, the harmonic oscillator, the hydrogen atom, and the rigid rotor. In order to understand valence theory, it is necessary to know the energy eigenvalues of these systems and to have some knowledge of the properties of the eigenvalues and eigenfunctions. We will not give complete mathematical derivations for solving the differential equations and introducing the boundary conditions because we feel that this might be too complicated for a beginning student in physical chemistry. Instead we will introduce certain functions and show how these functions satisfy the equation. This procedure is more or less a compromise between undertaking a complete mathematical derivation and omitting the mathematical derivation altogether.

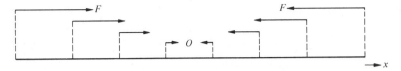

Figure 11-1 The force acting on a harmonic oscillator. The force F tends to move the particle back to the origin.

11-2 THE HARMONIC OSCILLATOR

Around the turn of the century it was not known how the electrons in an atom moved, but it was fairly obvious that their motion must satisfy two conditions. First, the electrons cannot stand still but must move in some fashion and, second, the electrons must stay within the vicinity of the nucleus. A simple one-dimensional model that satisfies both conditions is the harmonic oscillator.[1] This is a particle moving in the x direction, subject to a force F that is given by

$$F = -kx \qquad \text{(11-1)}$$

It may be seen that the force always tends to move the particle to the origin (see Figure 11-1). The further the particle gets away from the origin, the larger the force that tends to move it back. The potential energy that corresponds to this force is determined by the equation

$$F = -\frac{dV}{dx} \qquad \text{(11-2)}$$

It follows easily that V is given by

$$V = \tfrac{1}{2}kx^2 \qquad \text{(11-3)}$$

You can see that in classical mechanics the particle will oscillate back and forth around the origin. If it moves to the right, the force will tend to slow it down until the particle is turned around; then the force will accelerate it towards the origin. As soon as it has passed the origin, moving to the left, the force will slow it down again, and so on.

It is relatively easy to solve the classical equation of motion for the harmonic oscillator, and it may then be shown that the particle behaves exactly as we expect. We have

$$ma = m\frac{d^2x}{dt^2} = -kx \qquad \text{(11-4)}$$

where the acceleration of the particle a, multiplied by its mass m must be equal to the force F that acts on it. F is given by Eq. (11-1). It is convenient to introduce the angular frequency ω of the harmonic oscillator by defining it as

$$\omega^2 = \frac{k}{m} \qquad \text{(11-5)}$$

[1] See also pages 81, 351.

If we substitute ω^2 into Eq. (11-4) we obtain

$$\frac{d^2x}{dt^2} = -\omega^2 x \qquad (11\text{-}6)$$

This equation has two solutions

$$x = \sin \omega t \qquad x = \cos \omega t \qquad (11\text{-}7)$$

and the general solution is

$$x = A \cos \omega t + B \sin \omega t \qquad (11\text{-}8)$$

where A and B are arbitrary parameters.

We can describe the classical behavior of the harmonic oscillator by means of the simple model of Figure 11-2. Here we consider a particle that moves with a constant velocity along the circumference of a circle with radius R. The position of the particle is determined by the polar angle ϕ because the radius R is constant. The dependence of the angle ϕ on the time t must be given by the linear relationship

$$\phi = \omega t \qquad (11\text{-}9)$$

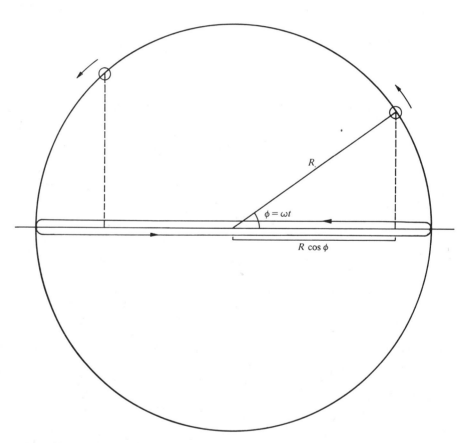

Figure 11-2   The harmonic oscillator moves as the projection of the motion of the particle around the circle.

because the particle has a constant angular velocity ω. The position of the projection is then given by

$$x(t) = R \cos \phi = R \cos \omega t \tag{11-10}$$

which is the first of the two solutions of Eq. (11.8).

Let us now derive an expression for the energy of the harmonic oscillator in classical mechanics. We find the velocity $v(t)$ of the particle by differentiating Eq. (11-10) with respect to the time,

$$v(t) = -R\omega \sin \omega t \tag{11-11}$$

The total energy is obtained as the sum of the kinetic energy T and the potential energy V,

$$E = T + V = \tfrac{1}{2}mv^2 + \tfrac{1}{2}kx^2 \tag{11-12}$$

By substituting Eqs. (11-10) and (11-11) into Eq. (11-12) we obtain

$$E = \tfrac{1}{2}mR^2\omega^2 \sin^2 \omega t + \tfrac{1}{2}kR^2 \cos^2 \omega t = \tfrac{1}{2}kR^2 = \tfrac{1}{2}m\omega^2 R^2 \tag{11-13}$$

We shall now consider the description of the harmonic oscillator in quantum mechanics. Note that the Hamiltonian is given by

$$\mathfrak{H} = \frac{p^2}{2m} + \frac{kx^2}{2} \tag{11-14}$$

so that the Schrödinger equation is

$$-\frac{\hbar^2}{2m}\frac{d^2\psi}{dx^2} + \frac{kx^2}{2}\psi = E\psi \tag{11-15}$$

We must solve this differential equation first; then we must determine for which values of the energy parameter E these solutions are acceptable from a physical point of view. These values, E_n, are then eigenvalues of the Schrödinger Eq. (11-15), which belong to the stationary states. The corresponding solutions $\psi_n(x)$ are the eigenfunctions, which describe the motion of the particle when it is in the stationary state n.

We will show that the energy eigenfunctions of the harmonic oscillator are given by

$$E_n = (n + \tfrac{1}{2})\hbar\omega \qquad n = 0, 1, 2, 3, 4, \ldots, \text{etc.} \tag{11-16}$$

where ω is equal to the classical frequency that we defined in Eq. (11-5).

In Figure 11-3 we have given the customary representation of the quantum mechanical eigenstates of the harmonic oscillator. We have plotted the potential energy V of Eq. (11-3) as a function of x and we have drawn horizontal lines at the positions of the energy eigenvalues E_n. It is obvious from Eq. (11-16) that these lines must be equidistant; that is, they are all a distance $\hbar\omega$ apart.

Before we consider the eigenfunctions of the various eigenstates let us first determine the points of intersection between the horizontal line belonging to the eigenstate E_0 and the parabolic curve of the potential curve $V(x)$; these are the points $\pm x_0$ in Figure 11-3. These are determined from the condition

$$E_0 = \tfrac{1}{2}\hbar\omega = V(x_0) = \tfrac{1}{2}kx_0^2 \tag{11-17}$$

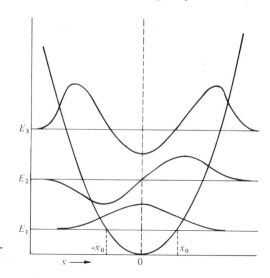

Figure 11-3 Eigenvalues and eigenfunctions of the harmonic oscillator.

It follows easily that

$$x_0^2 = \frac{\hbar\omega}{k} = \frac{\hbar}{\sqrt{km}} \tag{11-18}$$

or

$$x_0 = \sqrt{\hbar}/(km)^{1/4} \tag{11-19}$$

It seems logical to use the quantity x_0 as the unit of length in the eigenfunctions of the harmonic oscillator. This means that we introduce the new variable y

$$y = \frac{x}{x_0} \tag{11-20}$$

This is a dimensionless quantity that represents the position x of the particle in terms of x_0 as its unit of length.

The eigenfunctions of the harmonic oscillator are usually expressed in terms of y; they are given by

$$\psi_n(y) = H_n(y) \, e^{-(y^2/2)} \qquad n = 0, 1, 2, 3, \ldots \tag{11-21}$$

Here the functions $H_n(y)$ are known as Hermite polynomials; they are polynomials of the nth degree in the variable y. The first few Hermite polynomials are given by

$$\begin{aligned}
H_0(y) &= 1 \\
H_1(y) &= 2y \\
H_2(y) &= 4y^2 - 2 \\
H_3(y) &= 8y^3 - 12y
\end{aligned} \tag{11-22}$$

We show the corresponding eigenfunctions $\psi_0(x), \ldots, \psi_3(x)$ in Figure 11-3.

In Appendix A of this chapter we will prove that the preceding Hermite polynomials are indeed the eigenfunctions of the harmonic oscillator and we will discuss also some of the properties of the polynomials. Here, we just mention that the general expression for $H_n(y)$, the nth Hermite polynomial, is given by

$$H_n(y) = e^{y^2} \frac{d^n}{dy^n}(e^{-y^2}) \tag{11-23}$$

It is easily verified that the expressions of Eq. (11-22) may be derived from the general definition, Eq. (11-23).

It may be instructive to consider the probability density of the lowest eigenstate. We have plotted the normalized probability density

$$\rho_0(y) = (1/\sqrt{\pi})\, e^{-y^2} \tag{11-24}$$

in Figure 11-4. We notice two things. First, the probability density is largest for $y = 0$. Second, there is a finite probability for $y > y_0$, that is to say, beyond the two points where the energy line intersects with the potential energy curve. This second feature is quite remarkable. We should remember that the points $y = \pm y_0$ represent the two points where the total energy E is equal to the potential energy V, so that the kinetic energy T is zero. In the outside regions E is smaller than V and in our classical picture the kinetic energy would be negative. We find therefore that there is a finite probability of finding the particle in a region where its kinetic energy is negative. This type of behavior is fairly common in quantum mechanics, but it is not possible in classical mechanics where the kinetic energy must always be positive.

In a diatomic molecule the distance R between the two nuclei is not a fixed quantity. The two nuclei oscillate around the equilibrium position R_0 (see Figure 11-5) and this oscillation, which is called the molecular vibration, may be represented as a harmonic oscillator with the distance $q = R - R_0$ as its coordinate. We will try to derive the magnitude of the nuclear motion from the experimental data by making use of the theoretical results of the harmonic oscillator.

The vibrational motion is described by the Hamiltonian

$$\mathfrak{H}_{\text{vib}} = -\frac{\hbar^2}{2M}\frac{d^2}{dq^2} + \tfrac{1}{2}kq^2 \tag{11-25}$$

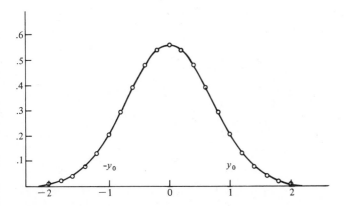

Figure 11-4 Probability density in the ground state of the harmonic oscillator.

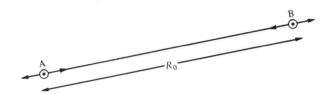

Figure 11-5 Vibrational motion of the nuclei in a diatomic molecule AB.

where μ is the reduced mass of the nuclei with masses M_A and M_B,

$$\frac{1}{\mu} = \frac{1}{M_A} + \frac{1}{M_B} \tag{11-26}$$

and k is the force constant of the harmonic oscillator.

We wish to find out how much the nuclei move away from their equilibrium position due to the vibrational motion when the molecule is in its vibrational ground state. In quantum mechanical terms this means that we wish to calculate the expectation value of the quantity q^2. We can define the extent of the vibrational motion then in terms of a quantity Δq, which is related to this expectation value by means of

$$(\Delta q)^2 = \langle \psi_0 | (R - R_0)^2 | \psi_0 \rangle = \langle \psi_0 | q^2 | \psi_0 \rangle \tag{11-27}$$

Here ψ_0 is the normalized eigenfunction of the ground state of the harmonic oscillator.

We have seen that the ground state eigenfunction ψ_0 of the harmonic oscillator is given by Eqs. (11-21) and (11-22) as

$$\psi_0(y) = \exp\left(-\tfrac{1}{2}y^2\right) \tag{11-28}$$

where the variable y is given by Eq. (11-20) as

$$y = q/q_0 \tag{11-29}$$

In our problem the variable is q and q_0 is our unit of length which is given by Eq. (11-20). It should be noted that the function (11-28) is not normalized, if we wish to use it for the evaluation of $(\Delta q)^2$ we must use the expression

$$(\Delta q)^2 = \frac{\int q^2 \exp\left(-q^2/q_0^2\right) dq}{\int \exp\left(-q^2/q_0^2\right) dq} \tag{11-30}$$

The calculation of the integral (11-30) is discussed in Appendix B of this chapter and from the result we derive that

$$(\Delta q)^2 = \tfrac{1}{2}q_0^2 \tag{11-31}$$

or

$$\Delta q = \tfrac{1}{2}q_0\sqrt{2} \tag{11-32}$$

It is clear that the calculation of Δq boils down to the calculation of the unit of length q_0. This quantity is defined as the point of intersection between the ground state energy $\frac{1}{2}\hbar\omega$ of the harmonic oscillator and the potential function $V(q)$, it is given by Eq. (11-19)

$$\hbar\omega = kq_0^2 \tag{11-33}$$

EXAMPLE 11-1

Let us now consider the specific case of the CO molecule. Here, it is known from the infrared spectrum that the transition from the vibrational ground state to the first excited state has a wave number σ of 2170 cm^{-1}. This corresponds to a frequency ν_0 which is given by

$$\nu_0 = c/\lambda = 2.998 \times 10^{10} \times 2170 \text{ s}^{-1} = 6.506 \times 10^{13} \text{ s}^{-1} \tag{a}$$

The energy difference ΔE between the two states is given by

$$\Delta E = \hbar\omega = (h\omega/2\pi) \tag{b}$$

and from the relation

$$\Delta E = h\omega/2\pi = h\nu_0 \tag{c}$$

we derive that

$$\omega = 2\pi\nu_0 = 4.088 \times 10^{14} \text{ s}^{-1} \tag{d}$$

We obtain the reduced mass μ from Eq. (11-26) by substituting

$$M_C = 12.01 \times 1.6748 \times 10^{-24} \text{ g} \quad \text{and} \quad M_O = 16.00 \times 1.6748 \times 10^{-24} \text{ g}$$

$$\mu = \frac{12.01 \times 16.00 \times 1.6748 \times 10^{-24}}{12.01 + 16.00} = 1.149 \times 10^{-23} \text{ g} \tag{e}$$

We must determine q_0 from Eq. (11-33). We do not know the force constant k but we can express k in terms of ω and μ by using the definition of Eq. (11-5).

$$k = \omega^2 \mu \tag{f}$$

If we substitute this in Eq. (11-33), we obtain

$$\hbar\omega = \omega^2 \mu q_0^2 \tag{g}$$

or

$$q_0^2 = (\hbar/\omega\mu) = \frac{1.054 \times 10^{-27}}{4.088 \times 10^{14} \times 1.149 \times 10^{-23}} = 0.2244 \times 10^{-18} \text{ cm}^2 \tag{h}$$

It follows that

$$q_0 = 0.474 \times 10^{-9} \text{ cm} = 0.0474 \text{ Å} \tag{i}$$

The average deviation Δq of the nuclei with respect to their equilibrium position is now obtained by substituting the result in Eq. (i) for q_0 into Eq. (11-32). The result is

$$\Delta q = 0.0335 \text{ Å} \tag{j}$$

This is about the magnitude that we would expect. ●

11-3   THE RIGID ROTOR

In Section 11-2 we discussed the vibrational motion of a diatomic molecule and its relation to the harmonic oscillator. In this section we shall discuss the rotational motion of a diatomic molecule; this is related to the quantum mechanical model of the rigid rotor.[2]

Let us first consider the rotational motion of a diatomic molecule from a classical point of view. We can show that the rotational motion of the two nuclei A and B may be considered separately and that it may be represented by the model where each nucleus rotates around the center of gravity. During this motion, the distance between the nuclei remains fixed and it is as if the two nuclei were connected by an invisible rigid rod of length R. We have sketched this situation in Figure 11-6. Realize that during this motion the center of gravity C of the two nuclei (with masses M_A and M_B) remains fixed and we have taken this center of gravity as the origin in both Figures 11-6a and 11-6b. In Figure 11-6a we represent the situation where the two masses M_A and M_B are equal to one another; in that case the center of gravity C is exactly at the middle of the rod. In Figure 11-6b we have assumed that one of the two nuclei, A, is much heavier than the other, or that M_A is much larger than M_B. In that case the center of gravity C is still on the line between the two nuclei, but it is much closer to the heavier nucleus A than to B. The exact ratio between the two distances R_A and R_B is given by

$$\frac{R_A}{R_B} = \frac{M_B}{M_A} \tag{11-34}$$

We see that if M_A is much larger than M_B, the distance R_A becomes smaller; this means that the center of gravity is closer to the heavier nucleus. We may express the two distances R_A and R_B in terms of the distance R between the two nuclei. Since

$$R = R_A + R_B \tag{11-35}$$

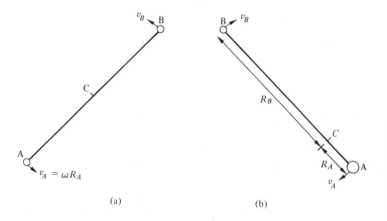

(a)   (b)

Figure 11-6   (a) The masses M_A and M_B are equal and the center of gravity C is in the middle. (b) M_A is larger and C is closer to A.

[2] See pages 79, 355.

it is easily derived from Eq. (11-34) that

$$R_A = \frac{M_B}{M_A + M_B} R \qquad R_B = \frac{M_A}{M_A + M_B} R \qquad \text{(11-36)}$$

The position of the point A is given by the vector $\mathbf{R}_A$ and it moves with a velocity $\mathbf{v}_A$ which is perpendicular to $\mathbf{R}_A$. The magnitude v_A may be written as the product of the length R_A and the angular velocity ω_A which describes the rate at which the orientation of the vector $\mathbf{R}_A$ changes with time

$$v_A = \omega_A R_A \qquad \text{(11-37)}$$

Similarly we write the velocity v_B of the point B as

$$v_B = \omega_B R_B \qquad \text{(11-38)}$$

and we note again that $\mathbf{v}_B$ must be perpendicular to $\mathbf{R}_B$.

It is obvious that the two angular velocities ω_A and ω_B must be equal to each other

$$\omega_A = \omega_B = \omega \qquad \text{(11-39)}$$

otherwise the rod between the two nuclei would bend while they are moving.

Let us now consider the energy of the rigid rotor. Since there is no potential energy, the energy is simply equal to the sum of the kinetic energies of the two nuclei

$$E = \tfrac{1}{2}M_A v_A^2 + \tfrac{1}{2}M_B v_B^2 \qquad \text{(11-40)}$$

It is convenient to rewrite Eq. (11-40) in another form. Since the center of gravity of the system is fixed, the position of the rigid rotor is completely determined by the vector $\mathbf{R}$, which is defined as the line segment from the point A to the point B. The magnitude of $\mathbf{R}$ is then given by Eq. (11-35); it is the sum of R_A and R_B. We define the velocity $\mathbf{v}$ now as

$$\mathbf{v} = \frac{d\mathbf{R}}{dt} ; \quad v = \omega R \qquad \text{(11-41)}$$

The kinetic energy E of Eq. (11-40) may then be written as

$$E = \tfrac{1}{2}\mu v^2 \qquad \text{(11-42)}$$

where

$$\frac{1}{\mu} = \frac{1}{M_A} + \frac{1}{M_B} \qquad \mu = \frac{M_A M_B}{M_A + M_B} \qquad \text{(11-43)}$$

We see that if we define the momentum $\mathbf{p}$ as

$$\mathbf{p} = \mu \mathbf{v} \qquad \text{(11-44)}$$

then the energy of Eq. (11-42) may also be written as

$$E = \frac{p^2}{2\mu} \qquad \text{(11-45)}$$

In this way the motion of the rigid rotor is represented by the motion of a particle of mass μ over the surface of a sphere with radius R.

EXAMPLE 11-2

It is fairly easy to prove that Eq. (11-40) is equivalent to Eq. (11-42) by rewriting it in terms of the angular velocity ω. If we substitute Eqs. (11-37) and (11-38) into Eq. (11-40) we obtain

$$E = \tfrac{1}{2}\omega^2(M_A R_A^2 + M_B R_B^2) \tag{a}$$

where we have used Eq. (11-39). The two distances R_A and R_B are expressed in terms of R by means of Eq. (11-36). By substituting this into Eq. (a) we obtain

$$E = \frac{\omega^2}{2}\left(\frac{M_A M_B^2}{(M_A+M_B)^2}R^2 + \frac{M_B M_A^2}{(M_A+M_B)^2}R^2\right) = \frac{M_A M_B}{M_A+M_B}\frac{\omega^2 R^2}{2} \tag{b}$$

The first term of this equation is equal to the reduced mass μ, defined by Eq. (11-43) and the second term may be expressed in terms of the velocity v, defined by Eq. (11-41). It follows then that

$$E = \tfrac{1}{2}\mu v^2 \tag{c}$$

which is equal to Eq. (11-42). ●

The quantum mechanical Hamiltonian corresponding to Eq. (11-45) is given by

$$\mathfrak{H}_{op} = \frac{p^2}{2\mu} = -\frac{\hbar^2}{2\mu}\left(\frac{\partial^2}{\partial X^2} + \frac{\partial^2}{\partial Y^2} + \frac{\partial^2}{\partial Z^2}\right) \tag{11-46}$$

where X, Y, and Z are the components of the vector $\mathbf{R}$. The Schrödinger equation for the rigid rotor is

$$-\frac{\hbar^2}{2\mu}\left(\frac{\partial^2}{\partial X^2} + \frac{\partial^2}{\partial Y^2} + \frac{\partial^2}{\partial Z^2}\right)\Psi = E\Psi \tag{11-47}$$

Remember though that the particle is constrained to move over the surface of the sphere of radius R. This means that we must impose the condition

$$(X^2 + Y^2 + Z^2)^{1/2} = R \tag{11-48}$$

where R is constant. It means also that the eigenfunctions of the rigid rotor depend only on the position of the particle on the sphere so that the eigenfunctions are not dependent on the variable R. Consequently, these eigenfunctions Ψ must be functions of the variables

$$x_0 = \frac{X}{R} \qquad y_0 = \frac{Y}{R} \qquad z_0 = \frac{Z}{R} \tag{11-49}$$

which satisfy the relation

$$x_0^2 + y_0^2 + z_0^2 = 1 \tag{11-50}$$

The variables (x_0, y_0, z_0) are the direction cosines that determine the direction of the vector $\mathbf{R}$. They form the components of a unit vector pointing in the same direction as $\mathbf{R}$. The position of the particle is completely determined by the unit vector (x_0, y_0, z_0) since it is known that the distance from the particle to the origin must be equal to R.

The eigenvalues and eigenfunctions of the rigid rotor are usually denoted by a quantum number J. We show in Appendix C of this chapter that the eigenvalues are

given by

$$E_J = \frac{\hbar^2}{2\mu R^2} J(J+1) \qquad J = 0, 1, 2, 3, \ldots \qquad \textbf{(11-51)}$$

In molecular spectroscopy this is often written as

$$E_J = BJ(J+1) \qquad \textbf{(11-52)}$$

where B is known as the rotational constant of the diatomic molecule. By comparing the two equations (11-51) and (11-52) we find that B is given by

$$B = \frac{\hbar^2}{2\mu R^2} \qquad \textbf{(11-53)}$$

In Example 11-3 we calculate the value of B for some typical diatomic molecules and in Table 14-1 of Chapter 14 we list the rotational values for about 20 molecules. It may be seen that B varies from about 0.04 cm^{-1} for I_2 to about 61 cm^{-1} for H_2.

EXAMPLE 11-3
We want to calculate the rotational constant B, as given by Eq. (11-53) for the three molecules H_2, $H^{79}Br$, and $^{35}Cl_2$. If we express all quantities on the right hand side in cgs units, we obtain B in ergs. We should remember then that $1\,\text{erg} = 5.0358 \times 10^{15}$ cm^{-1} if we want to obtain B in terms of cm^{-1}.

The internuclear distance R is usually close to an Ångstrom unit (10^{-8} cm) and it is helpful to write R as

$$R = 10^{-8}\rho \qquad \textbf{(a)}$$

where ρ is expressed in Ångstrom units. For the same reason we express the mass μ in terms of the proton mass M_p by substituting

$$\mu = M \cdot M_p = 1.6724 \times 10^{-24} M \qquad \textbf{(b)}$$

where M is the ratio between the reduced mass and the proton mass. If we substitute all this into Eq. (11-53) we obtain

$$B = \frac{\hbar^2}{2\mu R^2} = \frac{\hbar^2}{2M_p \times 10^{-16}} \times \frac{1}{M\rho^2} \text{erg}$$

$$= \frac{(1.05443 \times 10^{-27})^2}{2 \times 1.6724 \times 10^{-24} \times 10^{-16}} \times \frac{1}{M\rho^2} \times 5.0358 \times 10^{15} \text{ cm}^{-1}$$

$$= \frac{16.714}{M\rho^2} \text{ cm}^{-1} \qquad \textbf{(c)}$$

Let us now consider the three molecules in which we are interested. In the case of H_2 we have $\rho = 0.7416$ Å and $\mu = \frac{1}{2}$ so that

$$B(H_2) = \frac{16.714}{\frac{1}{2} \times (0.7416)^2} = 60.9 \text{ cm}^{-1} \qquad \textbf{(d)}$$

For HBr we have $\rho = 1.414$ Å and

$$\mu = \frac{M_H M_{Br}}{M_H + M_{Br}} = \frac{78.918}{79.926} = 0.98738 \qquad \textbf{(e)}$$

so that

$$B(H^{79}Br) = \frac{16.714}{0.98738 \times (1.414)^2} = 8.47 \text{ cm}^{-1} \qquad \text{(f)}$$

Finally, for $^{35}Cl_2$ we have $\rho = 1.988$ Å and $M = \frac{1}{2} \times 34.969$ so that

$$B(^{35}Cl_2) = \frac{16.714}{\frac{1}{2} \times 34.969 \times (1.988)^2} = 0.242 \text{ cm}^{-1} \qquad \bullet \qquad \text{(g)}$$

We also show in Example 11-4 that each eigenvalue E_J is $(2J+1)$-fold degenerate. This means that there are $2J+1$ different, linearly independent eigenfunctions belonging to the same eigenvalue E_J. In our derivation in Appendix C of Chapter 11 we find that the eigenfunction Ψ_J contains $2J+1$ arbitrary parameters, and this result is consistent with the $(2J+1)$-fold degeneracy because our wave function Ψ_J is obtained as a linear combination of $2J+1$ different functions.

In our derivation in Appendix C we also show that the eigenfunction Ψ_J is an Euler polynomial of the Jth degree in the three variables x_0, y_0, and z_0. We write such an Euler polynomial as $F_J(x_0, y_0, z_0)$, so that

$$\Psi_J = F_J(x_0, y_0, z_0) \qquad \text{(11-54)}$$

We define Euler polynomials and we discuss their properties in Appendix C, but we will summarize the definition here and we will also list the first few.

First we define a homogeneous polynomial $f(x, y, z)$ of the three variables x, y, and z. A homogeneous polynomial is a linear combination of terms of the type $x^k y^l z^m$ with the condition that for each term the sum of the three powers must have the same value. If that value is n, we have a homogeneous polynomial of the nth degree. Such a polynomial is a sum of terms of the type

$$a_{k,l,m} x^k y^l z^m \qquad \text{(11-55)}$$

with the condition that

$$k + l + m = n \qquad \text{(11-56)}$$

We can write the polynomial as

$$f_n(x, y, z) = \sum_k \sum_l \sum_m a_{k,l,m} x^k y^l z^m \quad (k + l + m = n) \qquad \text{(11-57)}$$

or as

$$f_n(x, y, z) = \sum_{k=0}^{n} \sum_{l=0}^{n-k} a_{k,l,n-k-l} x^k y^l z^{n-k-l} \qquad \text{(11-58)}$$

It may be helpful if we list the general expressions of the first few homogeneous polynomials

$$f_0(x, y, z) = a_{0,0,0}$$

$$f_1(x, y, z) = a_{0,0,1}z + a_{0,1,0}y + a_{1,0,0}x$$

$$f_2(x, y, z) = a_{0,0,2}z^2 + a_{0,1,1}yz + a_{0,2,0}y^2 + a_{1,0,1}xz + a_{1,1,0}xy + a_{2,0,0}x^2 \qquad \text{(11-59)}$$

$$f_3(x, y, z) = a_{0,0,3}z^3 + a_{0,1,2}yz^2 + a_{0,2,1}y^2z + a_{0,3,0}y^3 + a_{1,0,2}xz^2 + a_{1,1,1}xyz$$
$$+ a_{1,2,0}xy^2 + a_{2,0,1}x^2z + a_{2,1,0}x^2y + a_{3,0,0}x^3$$

An Euler polynomial of the nth degree $F_n(x, y, z)$ is now defined as a homogeneous polynomial $f_n(x, y, z)$ that satisfies the condition

$$\left(\frac{\partial^2}{\partial x^2} + \frac{\partial^2}{\partial y^2} + \frac{\partial^2}{\partial z^2}\right) f_n(x, y, z) = 0 \tag{11-60}$$

An Euler polynomial is thus a homogeneous polynomial where the coefficients are
subject to certain restrictions that are determined by the condition (11-60). We may
derive the restrictions for the four polynomials that are listed in Eq. (11-59). By
straightforward differentiation we find

$$\left(\frac{\partial^2}{\partial x^2} + \frac{\partial^2}{\partial y^2} + \frac{\partial^2}{\partial z^2}\right) f_0 = 0$$

$$\left(\frac{\partial^2}{\partial x^2} + \frac{\partial^2}{\partial y^2} + \frac{\partial^2}{\partial z^2}\right) f_1 = 0 \tag{11-61}$$

$$\left(\frac{\partial^2}{\partial x^2} + \frac{\partial^2}{\partial y^2} + \frac{\partial^2}{\partial z^2}\right) f_2 = 2(a_{0,0,2} + a_{0,2,0} + a_{2,0,0}) = 0$$

The first two homogeneous polynomials f_0 and f_1 satisfy the Euler condition so that
they are Euler polynomials. Clearly, the first two Euler polynomials F_0 and F_1 are
given by

$$F_0(x, y, z) = f_0(x, y, z) = a$$
$$F_1(x, y, z) = f_1(x, y, z) = ax + by + cz \tag{11-62}$$

The polynomial f_2 is an Euler polynomial only if it satisfies the condition (11-60). It
follows from Eq. (11-61) that it becomes an Euler polynomial if

$$a_{0,0,2} + a_{0,2,0} + a_{2,0,0} = 0 \tag{11-63}$$

or

$$a_{0,0,2} = -(a_{0,2,0} + a_{2,0\,0}) \tag{11-64}$$

It follows that F_2 is given by

$$F_2(x, y, z) = axy + bxz + cyz + p(x^2 - z^2) + q(y^2 - z^2) \tag{11-65}$$

The eigenfunctions of the rigid rotor are Euler polynomials so that we have

$$\mathfrak{H}_{op} F_J(x_0, y_0, z_0) = E_J F_J(x_0, y_0, z_0) \tag{11-66}$$

where E_J is given by Eq. (11-51) and where F_J is the Euler polynomial of the Jth
degree as previously defined. The Euler polynomials F_0 and F_1 are given by Eq.
(11-62) and F_2 is given by Eq. (11-65). We show in Example 11-4 that E_J is
$(2J+1)$-fold degenerate and we show in Appendix C of this chapter that the
polynomials F_J are eigenfunctions of the rigid rotor.

The properties of the eigenvalues and eigenfunctions of the rigid rotor may be
better understood if we relate them to the properties of the angular momentum,
which are discussed in Section 11-4.

EXAMPLE 11-4
We can prove that the eigenvalue E_J is $(2J+1)$-fold degenerate by showing that
the Euler polynomial F_J has $2J+1$ arbitrary parameters.

First we calculate the number of parameters in a homogeneous polynomial f_n as defined by Eq. (11-58). The number of parameters is equal to the number of terms. It may be seen from Eq. (11-58) that for a given value of k, the subscript l has $n - k + 1$ possible values. So, if $k = 0$, l has $n + 1$ possible values; if $k = 1$, l has n possible values; and if $k = n$, there is only one value for l. The number of terms, q_n, is thus given by

$$q_n = (n + 1) + n + (n - 1) + (n - 2) + \cdots + 1 \tag{a}$$

This sum is equal to

$$q_n = \tfrac{1}{2}(n + 1)(n + 2) \tag{b}$$

If we now calculate

$$\frac{\partial^2 f_n}{\partial x^2} + \frac{\partial^2 f_n}{\partial y^2} + \frac{\partial^2 f_n}{\partial z^2} \tag{c}$$

we obtain a homogeneous polynomial of degree $n - 2$, which, according to Eq. (b), has

$$q_{n-2} = \tfrac{1}{2}n(n - 1) \tag{d}$$

terms.

Let us now consider the Euler polynomial F_n. This is a homogeneous polynomial f_n that satisfies the condition that Eq. (c) must be identically zero. This means that the coefficient of every term in expression (c) must be zero. Because there are $\tfrac{1}{2}n(n - 1)$ coefficients, there must be $\tfrac{1}{2}n(n - 1)$ equations for the parameters of the original polynomial f_n. We conclude that the Euler polynomial has

$$q_n - q_{n-2} = \tfrac{1}{2}(n + 1)(n + 2) - \tfrac{1}{2}n(n - 1) = 2n + 1 \tag{e}$$

arbitrary parameters.

Hence, the Euler polynomial F_J has $2J + 1$ arbitrary parameters and the eigenvalue E_J is $(2J + 1)$-fold degenerate. ●

11-4 ANGULAR MOMENTUM

We mentioned in Chapter 10 that in quantum mechanics we use the momentum p of a particle rather than its velocity v. The momentum is a useful quantity for describing the motion of a particle that moves, more or less, in a straight line. However, if the particle moves around in a closed orbit, it is more useful to have another quantity that represents the extent of the rotational motion. For this purpose the concept of angular momentum was introduced into classical mechanics. In the case of a circular motion where the velocity $\mathbf{v}$ is always perpendicular to the position vector $\mathbf{r}$ of the particle, the magnitude of the angular momentum M is defined as

$$M = pr = mvr \tag{11-67}$$

In other cases, where $\mathbf{v}$ and $\mathbf{r}$ are not perpendicular to each other, the magnitude of the angular momentum is given by

$$M = pr \sin \theta = mvr \sin \theta \tag{11-68}$$

where θ is the angle between $\mathbf{v}$ and $\mathbf{r}$.

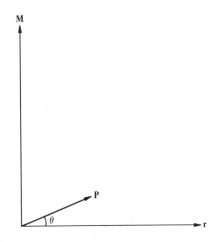

Figure 11-7 Definition of the angular momentum vector **M**.

If we want to describe a rotational motion we must know the rotation axis in addition to the magnitude M. Consequently, angular momentum is defined as a vector quantity **M**. Its magnitude is given by Eq. (11-68) and its direction is along the rotation axis. To be more precise, the direction of **M** is taken perpendicular to the plane through the vectors **r** and **p** (see Figure 11-7). The sign of the direction is given by the corkscrew rule: If we rotate a cork-screw from **r** to **p** over the smallest angle θ, then **M** points in the direction that the corkscrew moves. A more complete definition of the angular momentum, in terms of classical mechanics, is given in Appendix D of this chapter.

Since the rigid rotor has a purely rotational motion, there should be a close correlation between its eigenfunctions and eigenvalues and the properties of the angular momentum. It is useful to discuss this correlation because it helps to interpret the eigenvalues and eigenfunctions of the rigid rotor, and it also helps us understand the quantum mechanical description of the angular momentum.

In classical mechanics the energy E is a constant of the motion; that means that, if there are no outside influences on a system, its energy does not change in time. We can write this in mathematical form as

$$\frac{dE}{dt} = 0 \tag{11-69}$$

In quantum mechanics we represent the energy by an operator, the Hamiltonian operator $\mathfrak{H}_{op}$ and we look for the eigenvalues and eigenfunctions of this operator

$$\mathfrak{H}_{op} \Psi_n = E_n \Psi_n \tag{11-70}$$

Each stationary state n of the system is thus characterized by a constant specific energy value E_n.

We see that the two results in quantum mechanics and in classical mechanics are consistent with one another. Since the energy is a constant of the motion in classical mechanics, each stationary state in quantum mechanics is characterized by an energy eigenvalue E_n and its wave function is an eigenfunction of the energy operator $\mathfrak{H}_{op}$.

We show in Appendix D of this chapter that the angular momentum vector **M** is a constant of the motion for both the rigid rotor and the hydrogen atom, or, in other words

$$\frac{d\mathbf{M}}{dt} = 0 \tag{11-71}$$

This leads us to believe that each stationary state of the two systems is characterized also by a specific value for the angular momentum. We will investigate this question for the rigid rotor because here the problem is relatively simple.

In the case of the rigid rotor, the angular momentum is the sum of the two contributions from each nucleus. According to the definition (11-67), it is given by

$$M = M_A v_A R_A + M_B v_B R_B \tag{11-72}$$

We showed in Section 11-3 that the rigid rotor may also be represented as a particle with mass μ moving over the surface of a sphere with radius R. Its velocity $\mathbf{v}$ is then given by Eq. (11-41). In this model the angular momentum is given by

$$M = \mu v R \tag{11-73}$$

It may be shown that the two equations (11-72) and (11-73) are the same.

Recall that the energy E of the rigid rotor is given by

$$E = \tfrac{1}{2}\mu v^2 \tag{11-74}$$

It follows then from Eqs. (11-73) and (11-74) that the energy may also be written as

$$E = \frac{M^2}{2\mu R^2} \tag{11-75}$$

where μ and R are constants. If we represent both E and M^2 by operators, namely $\mathfrak{H}_{op}$ and $(M^2)_{op}$ respectively, then it follows from Eq. (11-75) that

$$\mathfrak{H}_{op} = (2\mu R^2)^{-1}(M^2)_{op}$$
$$(M^2)_{op} = (2\mu R^2)\mathfrak{H}_{op} \tag{11-76}$$

Since the two operators are proportional, they must have the same eigenfunctions. We have seen that

$$\mathfrak{H}_{op}\Psi_J = E_J\Psi_J = \frac{\hbar^2}{2\mu R^2}J(J+1)\Psi_J \qquad J = 0, 1, 2, 3, \ldots \tag{11-77}$$

Obviously

$$(M^2)_{op}\Psi_J = 2\mu R^2 E_J\Psi_J = J(J+1)\hbar^2\Psi_J \qquad J = 0, 1, 2, 3, \ldots \tag{11-78}$$

We see that in the quantum state J, the quantum number J describes the magnitude of the angular momentum vector.

The results of the quantum mechanical description of the angular momentum may be summarized in the form of a simple nonmathematical model. This is the old vector model, which was used by spectroscopists to explain and interpret the results of atomic spectroscopy. In this model it is assumed that the angular momentum is quantized and that it is determined by an integral quantum number, in our case that is the quantum number J, which has the values $J = 0, 1, 2, 3, \ldots$ and so on. Initially it was assumed that the magnitude of the angular momentum was equal to $\hbar J$, but it soon became apparent that this was not true. The magnitude $|M|$ of the angular momentum must be slightly larger than $\hbar J$, and it is given by

$$|M| = \hbar\sqrt{J(J+1)} \qquad (11\text{-}79)$$

This is consistent with the result that we derived from the Schrödinger equation in Eq. (11-78).

The vector model also makes predictions about the possible orientations of the angular momentum vector. We show in Figure 11-8 how this works by considering the possible values of the projection of the angular momentum vector along a given direction in space. Physically speaking, all directions in space are equivalent so that we can take any direction we want. It turns out, however, that it is convenient to take

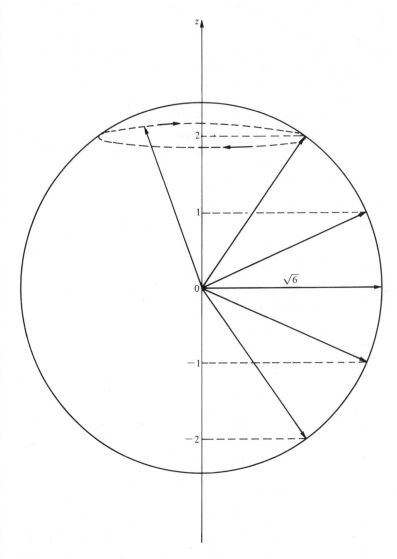

Figure 11-8   Quantization of the angular momentum along the z axis for $J=2$.

this direction as the z axis. According to the vector model the projection M_z must have an integer value

$$M_z = m\hbar \qquad m = 0, \pm 1, \pm 2, \ldots, \pm J \qquad (11\text{-}80)$$

if the system is in a stationary state. The quantum number m has the possible values

$$m = -J, -J+1, -J+2, \ldots, J-1, J \qquad (11\text{-}81)$$

In Figure 11-8 we have drawn the situation for $J = 2$. Here the possible values for the projection are $-2, -1, 0, 1$, and 2. In one of the stationary states, say $m = 1$, we know only the value of the projection of the vector **M** with respect to the z axis. In our model we imagine that the vector rotates around the z axis at a given angle in such a way that its projection on the z axis is always one unit of $\hbar$. In this way we can visualize the quantization of the angular momentum; it is convenient to do this even if it may not be quite the truth.

Note that there is no stationary state where the angular momentum vector is directed exactly along the z axis. The maximum value for the projection is J and, because the length of the vector is slightly larger (see Eq. 11-79) the vector **M** rotates around the z axis in this stationary state also. This aspect of the quantization of the angular momentum vector can be explained from the uncertainty principle.

Let us consider finally the mathematical description of the vector model. We have said that for a given value for the total angular momentum or for a given value of J, the possible values of the projection M_z must be given by Eq. (11-80). In quantum mechanical language this means that the operator $(M_z)_{op}$ representing M_z must have a set of eigenvalues and eigenfunctions that are given by

$$(M_z)_{op}\Psi_{J,m} = m\hbar\Psi_{J,m} \qquad m = 0, \pm 1, \pm 2, \ldots, \pm J \qquad (11\text{-}82)$$

Each eigenfunction $\Psi_{J,m}$ must be an eigenfunction also of the operator $(M^2)_{op}$ and it must satisfy the equation

$$(M^2)_{op}\Psi_{J,m} = J(J+1)\hbar^2\Psi_{J,m} \qquad J = 0, 1, 2, 3, \ldots \qquad (11\text{-}83)$$

In the case where $J = 0$ the quantum number m can only be equal to zero. The corresponding eigenfunction is given in Eq. (11-62)

$$\Psi_{0,0} = F_0(x_0, y_0, z_0) = a \qquad (11\text{-}84)$$

where a is a constant.

In the case where $J = 1$ the quantum number m can have the three values $m = 0, \pm 1$ and there are three eigenfunctions $\Psi_{1,0}$, $\Psi_{1,1}$, and $\Psi_{1,-1}$. We have seen in Section 11-3, Eqs. (11-62) and (11-66), that

$$\Psi_1 = F_1(x_0, y_0, z_0) = ax_0 + by_0 + cz_0 \qquad (11\text{-}85)$$

Each of the three eigenfunctions $\Psi_{1,0}$, $\Psi_{1,1}$, and $\Psi_{1,-1}$ must be an eigenfunction of $(M^2)_{op}$ belonging to $J = 1$. Each of the three functions must therefore be of the form of Eq. (11-85), and they are obtained by substituting suitable values of the three parameters a, b, and z. The three eigenfunctions $\Psi_{1,m}$ are given by

$$\Psi_{1,1} = x_0 + iy_0$$

$$\Psi_{1,0} = z_0 \qquad (11\text{-}86)$$

$$\Psi_{1,-1} = x_0 - iy_0$$

The first function $\Psi_{1,1}$ represents the situation where the projection M_z has the value $\hbar$; $\Psi_{1,0}$ corresponds to $M_z = 0$; and $\Psi_{1,-1}$ corresponds to $M_z = -\hbar$. We prove this in Example 11-5.

EXAMPLE 11-5
The value of the angular momentum M_z is defined in classical mechanics as

$$M_z = xp_y - yp_x \tag{a}$$

according to Appendix D of this chapter. The corresponding quantum mechanical operator is given by

$$(M_z)_{op} = \frac{\hbar}{i}\left(x\frac{\partial}{\partial y} - y\frac{\partial}{\partial x}\right) \tag{b}$$

In order to find the eigenvalues and eigenfunctions of this operator we introduce the polar coordinates

$$x = r\sin\theta\cos\phi$$

$$y = r\sin\theta\sin\phi \tag{c}$$

$$z = r\cos\theta$$

The partial derivative with respect to the polar angle ϕ is then given by

$$\frac{\partial}{\partial\phi} = \frac{\partial x}{\partial\phi}\frac{\partial}{\partial x} + \frac{\partial y}{\partial\phi}\frac{\partial}{\partial y} + \frac{\partial z}{\partial\phi}\frac{\partial}{\partial z}$$

$$= -r\sin\theta\sin\phi\frac{\partial}{\partial x} + r\sin\theta\cos\phi = x\frac{\partial}{\partial y} - y\frac{\partial}{\partial x} \tag{d}$$

It follows from Eqs. (b) and (d) that

$$(M_z)_{op} = \frac{\hbar}{i}\frac{\partial}{\partial\phi} \tag{e}$$

We see that $(M_z)_{op}$ depends only on the polar angle ϕ. Consequently, its eigenfunctions are functions of the angle ϕ only, and we may write the eigenvalue problem as

$$\frac{\hbar}{i}\frac{\partial}{\partial\phi}\chi(\phi) = \lambda\chi(\phi) \tag{f}$$

or

$$\frac{\partial\chi}{\partial\phi} = \frac{i\lambda}{\hbar}\chi \tag{g}$$

The solutions are

$$\chi(\phi) = \exp(i\lambda\phi/\hbar) \tag{h}$$

This function should be single-valued and the eigenvalues are derived from the boundary condition

$$\chi(\phi) = \chi(\phi + 2\pi) \tag{i}$$

or

$$\frac{\lambda}{\hbar} = m \qquad m = 0, \pm 1, \pm 2, \ldots \tag{j}$$

It follows that the eigenvalues are given by

$$\lambda = m\hbar \tag{k}$$

and the corresponding eigenfunctions are

$$\chi(\phi) = e^{im\phi} \tag{l}$$

In other words

$$(M_z)_{op} e^{im\phi} = m\hbar \, e^{im\phi} \qquad m = 0, \pm 1, \pm 2, \ldots \tag{m}$$

defines the eigenvalues and eigenfunctions of $(M_z)_{op}$.

Realize that Eq. (m) determines only the ϕ dependence of the eigenfunctions. If we want to find the three eigenfunctions that belong to the state $J = 1$ they should be linear combinations of the three functions x_0, y_0, and z_0, according to Eq. (11-85). In terms of polar coordinates these functions are

$$x_0 = \sin \theta \cos \phi$$
$$y_0 = \sin \theta \sin \phi \tag{n}$$
$$z_0 = \cos \theta$$

The three eigenfunctions of Eq. (11-86) are

$$x_0 + iy_0 = \sin \theta(\cos \phi + i \sin \phi) = \sin \theta \, e^{i\phi}$$
$$z_0 = \cos \theta \tag{o}$$
$$x_0 - iy_0 = \sin \theta(\cos \phi + i \sin \phi) = \sin \theta \, e^{-i\phi}$$

It follows that these three functions are eigenfunctions of M_z belonging to the three eigenvalues $\hbar$, 0 and $-\hbar$, respectively.   ●

11-5   THE HYDROGEN ATOM

It is important to know the eigenvalues and eigenfunctions of the hydrogen atom. The wave functions of the ground state and the lower excited states are used as the basis for almost all valence theories. In the previous sections we have considered only one-dimensional problems and the rigid rotor, which is a two-dimensional problem. The hydrogen atom is a three-dimensional problem; that means that the Schrödinger equation contains three variables, and we expect that such an equation may be somewhat difficult to solve. Fortunately we can simplify the equation and it can then be solved exactly. We would prefer not to give a purely mathematical analysis of this problem; instead we will consider some of the more physical aspects that are related to the angular momentum. In this way it is possible to get a better understanding of the general features of the quantum mechanical description.

The hydrogen atom consists of a proton with mass M and a positive charge e and an electron with mass m and a negative charge $-e$.

The proton mass is almost 2000 times larger than the electron mass. The potential energy of this system is the Coulomb attraction between the two particles which is

given by

$$V = -\frac{e^2}{r} \tag{11-87}$$

where r is the distance between the electron and the proton. Since the proton is so much heavier than the electron, it is a good approximation to assume that the proton remains at a fixed position and the electron moves around it. The Schrödinger equation of the hydrogen atom is then

$$\left(-\frac{\hbar^2}{2m}\Delta - \frac{e^2}{r}\right)\psi(x, y, z) = E\psi(x, y, z) \tag{11-88}$$

with

$$\Delta = \frac{\partial^2}{\partial x^2} + \frac{\partial^2}{\partial y^2} + \frac{\partial^2}{\partial z^2} \tag{11-89}$$

Here, the vector $\mathbf{r}$ which runs from the nucleus to the electron has the components (x, y, z). If we want to be more precise we have to realize that the proton does not remain at a fixed position. Instead, both proton and electron move around the center of gravity which remains at a fixed position. The net effect of this correction is that in Eq. (11-88) we must replace the electron mass m by the reduced mass μ

$$\frac{1}{\mu} = \frac{1}{M} + \frac{1}{m} \tag{11-90}$$

which is slightly different from the electron mass. The Schrödinger equation then takes the form

$$\left(-\frac{\hbar^2}{2\mu}\Delta - \frac{e^2}{r}\right)\psi(x, y, z) = E\psi(x, y, z) \tag{11-91}$$

In classical mechanics the motion of the electron around the nucleus is in every respect the same as the motion of a satellite around the earth (or the earth around the sun). In all these cases there is a force that tends to pull the orbiting particle towards the center. In our case the force is electrostatic, and in astronomy the force is gravitational; but both forces are given by the same type of mathematical expression (Eq. 11-87). We know that the earth describes an orbit around the sun that is elliptical, and this type of behavior may in general be derived from classical mechanics; the moving particle describes either a circular or an elliptical orbit around the center as long as its energy is below a certain limit (otherwise it is not in a closed orbit).

In order to understand the behavior of the moving particle it is useful to consider the two quantities that are constants of the motion, namely, the energy and the angular momentum. The fact that

$$\frac{dE}{dt} = 0 \tag{11-92}$$

follows from the conservation of energy principle. We show in Appendix D of this chapter that the angular momentum vector $\mathbf{M}$ must also stay the same as a function of time

$$\frac{d\mathbf{M}}{dt} = \mathbf{0} \tag{11-93}$$

It follows from Eqs. (11-92) and (11-93) that a stationary state of the hydrogen atom in classical mechanics is characterized by a constant, time-independent, value of the energy E and the angular momentum vector $\mathbf{M}$. Conversely if E and $\mathbf{M}$ are given, it is possible to calculate the classical orbit of the electron. It is then found that the electron moves in an elliptical orbit around the origin in such a way that the angular momentum remains constant.

Naturally, the quantum mechanical description of the hydrogen atom is quite different from the classical description. However, the two theories have one thing in common; in both cases the energy E and the angular momentum $\mathbf{M}$ are constants of motion. We have seen what this means in classical mechanics. In quantum mechanics the consequences are a little different. Here, a stationary state is represented by a wave function, which is an eigenfunction Ψ_n of the Hamiltonian operator $\mathfrak{H}_{\mathrm{op}}$

$$\mathfrak{H}_{\mathrm{op}}\Psi_n = E_n\Psi_n \tag{11-94}$$

It should be noted that the wave function of a stationary state must be an eigenfunction of $\mathfrak{H}_{\mathrm{op}}$ because $\mathfrak{H}_{\mathrm{op}}$ represents the energy, which is a constant of the motion. It follows then that the wave functions of the stationary states must also be eigenfunctions of the angular momentum operators, because the angular momentum is also a constant of the motion.

In Section 11-4 we saw that the angular momentum is represented in quantum mechanics by the two operators $(M^2)_{\mathrm{op}}$ and $(M_z)_{\mathrm{op}}$. The first operator $(M^2)_{\mathrm{op}}$ represents the magnitude of the angular momentum vector (to be precise, the square of the magnitude); the second operator $(M_z)_{\mathrm{op}}$ represents the projection of $\mathbf{M}$ on the z axis, and it describes the orientation of the angular momentum vector. It follows that each stationary state of the hydrogen atom is characterized by three quantum numbers, n, l, and m. The first quantum number n labels the value of the energy E_n; the second quantum number l gives us the magnitude of the angular momentum vector; and the third number m describes the orientation of the angular momentum vector.

We can also express this statement in terms of mathematics. It means that the wave function $\Phi_{n,l,m}$ of the stationary state (n,l,m) must be an eigenfunction of the three operators $\mathfrak{H}_{\mathrm{op}}$, $(M^2)_{\mathrm{op}}$ and $(M_z)_{\mathrm{op}}$

$$\mathfrak{H}_{\mathrm{op}}\Phi_{n,l,m} = E_n\Phi_{n,l,m}$$

$$(M^2)_{\mathrm{op}}\Phi_{n,l,m} = (M^2)_l\Phi_{n,l,m} \tag{11-95}$$

$$(M_z)_{\mathrm{op}}\Phi_{n,l,m} = (M_z)_m\Phi_{n,l,m}$$

Thus, the value E_n of the energy depends on the quantum number n, the value $(M^2)_l$ of the magnitude of the angular momentum depends on l, and the value $(M_z)_m$ is determined by m.

Once we understand the general behavior of the eigenvalues and eigenfunctions of the hydrogen atom, we can derive their specific forms without too much difficulty. Remember that we derived the eigenvalues and eigenfunctions of the angular momentum in Section 11-4, and we can also use these results for the hydrogen atom by making a few minor changes.

In the description of the hydrogen atom it is helpful to make use of the polar coordinates that we show in Figure 11-9. The variable r is the distance OP, the angle θ is the angle between the vector $\mathbf{OP}$ and the z axis and ϕ is the angle between $\mathbf{OP'}$

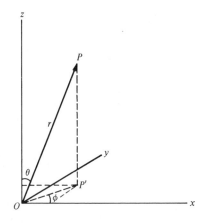

Figure 11-9 Definition of polar coordinates. P' is the projection of P in the xy plane.

and the x axis. The relation between the Cartesian coordinates (x, y, z) and the polar coordinates (r, θ, ϕ) is given by the equations

$$x = r \sin \theta \cos \phi$$
$$y = r \sin \theta \sin \phi \qquad \text{(11-96)}$$
$$z = r \cos \theta$$

We saw in Section 11-4, Eq. (e) of Example 11-5, that the operator $(M_z)_{op}$ is a function only of the polar angle ϕ. It is given by

$$(M_z)_{op} = \frac{\hbar}{i} \frac{\partial}{\partial \phi} \qquad \text{(11-97)}$$

Its eigenvalues and eigenfunctions were obtained in Example 11-5. They are given by

$$(M_z)_{op} \, e^{im\phi} = m\hbar \, e^{im\phi} \qquad \text{(11-98)}$$

Realize that these eigenfunctions may be multiplied by an arbitrary function ψ of the other polar coordinates r and θ. We have, namely,

$$(M_z)_{op}[\psi(r, \theta) \, e^{im\phi}] = m\hbar[\psi(r, \theta) \, e^{im\phi}] \qquad \text{(11-99)}$$

because the operator does not contain the two variables r and θ. This makes it possible to find a set of functions that are eigenfunctions of both $(M^2)_{op}$ and $(M_z)_{op}$. For example, in Section (11-4), Eq. (11-86), we mentioned that the set of functions

$$\Psi_{1,1} = x_0 + iy_0 = \sin \theta \, e^{i\phi}$$
$$\Psi_{1,0} = z_0 = \cos \theta \qquad \text{(11-100)}$$
$$\Psi_{1,-1} = x_0 - iy_0 = \sin \theta \, e^{-i\phi}$$

are eigenfunctions of both operators $(M^2)_{op}$ and $(M_z)_{op}$.

The same argument can be applied to the eigenfunctions of the hydrogen atom. The operator $(M^2)_{op}$ depends only on the two angles θ and ϕ and its eigenfunctions,

given by

$$(M^2)_{op} F_l(x_0, y_0, z_0) = l(l+1)\hbar^2 F_l(x_0, y_0, z_0)$$

$$x_0 = \frac{x}{r} \qquad y_0 = \frac{y}{r} \qquad z_0 = \frac{z}{r}$$

(11-101)

depend only on the two polar angles θ and ϕ. Of course, we may multiply its eigenfunctions by an arbitrary function of the variable r, and they should remain eigenfunctions of $(M^2)_{op}$. Then

$$(M^2)_{op}[g_l(r) F_l(x_0, y_0, z_0)] = l(l+1)\hbar^2[g_l F_l(x_0, y_0, z_0)]$$

(11-102)

It follows then that the eigenfunctions of the operator $\mathfrak{H}_{op}$ should have the form

$$\Phi_{n,l}(r, \theta, \phi) = g_{n,l}(r) F_l(x, y, z)$$

(11-103)

because they must also be eigenfunctions of the operator $(M^2)_{op}$.

We show in Appendix E of this chapter that it is helpful to introduce new units of length and energy in describing the hydrogen atom. The new unit of length is

$$a_0 = \frac{\hbar^2}{\mu e^2}$$

(11-104)

and it is approximately equal to 0.53 Å. The new unit of energy is

$$\varepsilon_0 = \frac{e^2}{a_0}$$

(11-105)

and is approximately equal to 27.2 eV. We use these units throughout our discussion of the hydrogen atom.

The energy eigenvalues of the hydrogen atom are then given by

$$\varepsilon_n = -\frac{1}{2n^2}$$

(11-106)

where the quantum number n can take the values

$$n = l+1, l+2, l+3, \ldots$$

The corresponding eigenfunctions have the form of Eq. (11-103). The radial functions $g_{n,l}(r)$ depend on both quantum numbers n and l. We have listed some of them in Table 11-1. It may be seen that for $n = l+1$ the function $g_{n,l}$ is simply an

Table 11-1 Radial eigenfunctions $g_{n,l}(r)$ for various values of the quantum numbers n and l

n	$g_{n,l}(r)$
$l+1$	$\exp(-r/n)$
$l+2$	$\left[\dfrac{r}{n} - (l+1)\right] \exp\left(-\dfrac{r}{n}\right)$
$l+3$	$\left[\left(\dfrac{r}{n}\right)^2 - (2l+3)\dfrac{r}{n} + \tfrac{1}{2}(l+1)(2l+3)\right] \exp\left(-\dfrac{r}{n}\right)$

exponential. For $n = l + 2$, it is the exponential multiplied by a polynomial of the first degree; for $n = l + 3$, it is a polynomial of the second degree, and so on. In principle it is possible to find all the eigenfunctions of the hydrogen atom by means of the method of Appendix E in this chapter.

Before we consider the detailed form of the eigenfunctions, it may be useful to look into the bookkeeping in classifying them. We have already mentioned that each stationary state of the hydrogen atom is characterized by the quantum numbers, n, which refers to the energy, l, which refers to the magnitude of the angular momentum, and m, which refers to the orientation of the angular momentum. The possible values of these quantum numbers are

$$l = 0, 1, 2, 3, 4, \ldots$$

$$m = 0, \pm 1, \pm 2, \pm 3, \ldots, \pm l \qquad \textbf{(11-107)}$$

$$n = l + 1, l + 2, l + 3, \ldots$$

We see that the possible values of n and l must satisfy the condition

$$n \geq l + 1 \qquad \textbf{(11-108)}$$

In Eqs. (11-107) we let the quantum number l take all possible values and then limit the possible values of n to those allowed by the condition (11-108). However, we classify the quantum states in a different way; we let n take all possible values and restrict the values of l. In this way we have

$$n = 1, 2, 3, 4, \ldots$$

$$l = 0, 1, 2, \ldots, n - 1 \qquad \textbf{(11-109)}$$

$$m = 0, \pm 1, \pm 2, \ldots, \pm l$$

This is the customary way to classify the quantum states of the hydrogen atom.

The symbols that are used to describe these various quantum states date from before quantum mechanics. They are derived from the symbols that the experimental spectroscopists used to describe the various terms in an atomic spectrum. In this terminology the value of the quantum number l, the magnitude of the angular momentum is given by a letter. States with $l = 0$ are described by a letter s, states with $l = 1$ with a letter p, $l = 2$ is d, $l = 3$ is f, $l = 4$ is g, and so on. The value of the quantum number n is then put in front of this letter.

Since the energy eigenvalues are given by

$$\varepsilon_n = -\frac{1}{2n^2} \qquad \textbf{(11-110)}$$

we see that the lowest energy occurs for the value $n = 1$ of the quantum number n. It follows from Eq. (11-109) that the quantum number l must then have the value $l = 0$, and this is called the $1s$ state of the hydrogen atom. The quantum number m can have only one value $m = 0$, and it follows that the $1s$ state is nondegenerate. The eigenfunction of the $1s$ state ($n = 1$, $l = 0$, $m = 0$) follows from Eq. (11-103) and Table 11-1.

$$\phi(1s) = F_0(x, y, z) g_{1,0}(r) = \exp(-r) \qquad \textbf{(11-111)}$$

We obtain the first excited state of the hydrogen atom if we take $n = 2$. Now it follows from Eq. (11-109) that the quantum number l can have two possible values, namely, $l=0$ and $l=1$. Both states have the same energy but different values of the angular momentum. The eigenfunctions of these states are again derived from Eq. (11-103) and Table 11-1.

$$\phi(2s) = F_0(x, y, z)g_{2,0}(r) = (\tfrac{1}{2}r - 1) \exp(-r/2)$$

$$\phi(2p) = F_1(x, y, z)g_{2,1}(r) = (ax + by + cz) \exp(-r/2)$$

(11-112)

The energy eigenvalue $n=2$ is fourfold degenerate because the function $\phi(2p)$ contains three parameters so that there are four functions that belong to the same energy eigenvalue. We could have predicted this degeneracy by considering the possible values of the quantum numbers l and m

$$l = 0 \qquad m = 0$$
$$l = 1 \qquad m = 0, \pm 1$$

Since there are four possibilities the energy eigenvalue $n=2$ must be fourfold degenerate.

The next higher energy eigenvalue occurs for $n = 3$. The quantum numbers l and m can have the values

$$l = 0 \qquad m = 0$$
$$l = 1 \qquad m = 0, \pm 1$$
$$l = 2 \qquad m = 0, \pm 1, \pm 2$$

and the energy eigenvalue is ninefold degenerate. We see that we have three different states if we just look at the quantum number l, namely, a $3s$ state for $l=0$, a $3p$ state for $l=1$ and a $3d$ state for $l=2$. The corresponding eigenfunctions follow again from Eq. (11-103) and Table 11-1.

$$\phi(3s) = F_0(x, y, z)g_{3,0}(r) = [(r^2/9) - r + \tfrac{3}{2}] \exp(-r/3)$$
$$= \tfrac{1}{18}(2r^2 - 18r + 27) \exp(-r/3)$$
$$\phi(3p) = F_1(x, y, z)g_{3,1}(r) = (ax + by + cz)[(r/3) - 2] \exp(-r/3)$$
$$= \tfrac{1}{3}(r - 6)(ax + by + cz) \exp(-r/3)$$
$$\phi(3d) = F_2(x, y, z)g_{3,2}(r) = [c_1(x^2 - z^2) + c_2(y^2 - z^2) + c_3xy$$
$$+ c_4xz + c_5yz] \exp(-r/3)$$

(11-113)

The $3s$ function is nondegenerate; the $3p$ function is threefold degenerate because it contains three parameters; and the $3d$ function is fivefold degenerate because it contains five parameters. We can predict these degeneracies from the angular momentum. If $l=0$ (in the $3s$ state), there is no degeneracy; if $l=1$, the projections of **M** can have the three values $0, \pm 1$, so there is a threefold degeneracy for the $3p$ state; and, if $l=2$ (in a d state), the projection has the five values $0, \pm 1, \pm 2$ so there must be a fivefold degeneracy. (See Figure 11-8.)

The final question that we want to consider is how to select the basis functions that constitute a p or a d state. There is no clearcut answer to this question because we can make a number of different choices, and the answer depends on what we want to do. We discuss the two most common procedures.

The first possible choice is the one we discussed in the beginning of this section. Here we chose the wave functions so that they were eigenfunctions of the three operators $\mathfrak{H}_{op}$, $(M^2)_{op}$ and $(M_z)_{op}$. We noted that the homogeneous polynomials were all eigenfunctions of the operator $(M^2)_{op}$. For p and d states we have

$$(M^2)_{op}F_1(x, y, z) = 2\hbar^2 F_1(x, y, z) \qquad (l=1)$$
$$(M^2)_{op}F_2(x, y, z) = 6\hbar^2 F_1(x, y, z) \qquad (l=2) \tag{11-114}$$

and the same relationship holds for the eigenfunctions $\phi(np)$ and $\phi(nd)$

$$(M^2)_{op}\phi(np) = 2\hbar^2 \phi(np)$$
$$(M^2)_{op}\phi(nd) = 6\hbar^2 \phi(nd) \tag{11-115}$$

We have seen in Section 11-4 that each function $F_1(x, y, z)$ may be decomposed into a set of functions $\Psi_{l,m}(x, y, z)$ so that

$$(M^2)_{op}\Psi_{l,m}(x, y, z) = l(l+1)\hbar^2\Psi_{l,m}(x, y, z) \tag{11-116}$$
$$(M_z)_{op}\Psi_{l,m}(x, y, z) = m\hbar\Psi_{l,m}(x, y, z) \qquad m = 0, \pm 1, \ldots, \pm l$$

We listed the specific forms of these functions for $l = 1$

$$\Psi_{1,1}(x, y, z) = x + iy$$
$$\Psi_{1,0}(x, y, z) = z \tag{11-117}$$
$$\Psi_{1,-1}(x, y, z) = x - iy$$

(see Eq. 11-86). Let us now consider the $2p$ wave function, as defined by Eq. (11-112).

$$\phi(2p) = (ax + by + cz)\exp(-r/2) \tag{11-118}$$

It follows from Eq. (11-117) that the $2p$ wave function can be decomposed into the following three functions

$$\phi(2p_1) = (x + iy)\exp(-r/2)$$
$$\phi(2p_0) = z \exp(-r/2) \tag{11-119}$$
$$\phi(2p_{-1}) = (x - iy)\exp(-r/2)$$

These are eigenfunctions of the three operators $\mathfrak{H}_{op}$, $(M^2)_{op}$ and $(M_z)_{op}$. In this way we have implemented the plan we discussed at the beginning of this section, that is, to classify the eigenstates by means of the quantum numbers n, l, and m, which refer to the eigenvalues of $\mathfrak{H}_{op}$, $(M^2)_{op}$, and $(M_z)_{op}$, respectively. This procedure is useful if there is a physical reason to single out the z axis as the quantization axis for the angular momentum axis, for example, if we have an electric or a magnetic field along this axis.

Ordinarily, there is no physical reason to prefer one specific direction in space over any other direction, and we do not necessarily have to use the classification of the eigenfunctions given by Eqs. (11-119). It may be seen that the functions of Eqs. (11-119) are complex functions, and in many cases this gives a mathematical complexity that can easily be avoided. Instead of Eqs. (11-119) we can break down the $2p$ functions into

$$\phi(2p_x) = x \exp(-r/2)$$
$$\phi(2p_y) = y \exp(-r/2) \tag{11-120}$$
$$\phi(2p_z) = z \exp(-r/2)$$

which are easier to work with than the functions (11-119) because they are all real. It should be noted that the functions (11-120) are not eigenfunctions of the operator $(M_z)_{op}$ (except for the $2p_z$ function), but that does not really matter. The various quantum mechanical theories of the chemical bond customarily use the functions (11-120) as a basis.

It is easily seen that the $3p$ functions can be classified by analogy with Eqs. (11-120) as

$$\phi(3p_x) = x(r-6) \exp(-r/3)$$
$$\phi(3p_y) = y(r-6) \exp(-r/3) \tag{11-121}$$
$$\phi(3p_z) = z(r-6) \exp(-r/3)$$

Here we have used the expression (11-113) for the $3p$ wave function.

The classification of the various $3d$ functions (defined by Eq. 11-113) is somewhat more complicated. In Eq. (11-113) we have the three functions

$$\phi(3d_{xy}) = xy \exp(-r/3)$$
$$\phi(3d_{xz}) = xz \exp(-r/3) \tag{11-122}$$
$$\phi(3d_{yz}) = yz \exp(-r/3)$$

There are three more functions, namely

$$\phi(3d_1) = (x^2 - y^2) \exp(-r/3)$$
$$\phi(3d_2) = (y^2 - z^2) \exp(-r/3) \tag{11-123}$$
$$\phi(3d_3) = (z^2 - x^2) \exp(-r/3)$$

Two of these functions occur in Eq. (11-113), and we have added the third one by cyclic substitution. The two sets of equations (11-122) and (11-123) contain six $3d$ functions; we know, however, that there are only five, so that we have one function too many. This discrepancy is easily explained if we note that the functions (11-123) are linearly dependent. It may be seen that

$$\phi(3d_1) + \phi(3d_2) + \phi(3d_3) = 0 \tag{11-124}$$

Actually, Eq. (11-123) contains only two linearly independent functions, and it is customary to select the functions

$$\phi(3d_0) = (x^2 - y^2) \exp(-r/3)$$
$$\phi(3d_{zz}) = (r^2 - 3z^2) \exp(-r/3) \tag{11-125}$$

as the basis functions. We have listed all the hydrogen atom eigenfunctions with quantum numbers $n = 1, 2,$ and 3 in Table 11-2.

Let us finally try to visualize the various functions that we have listed in Table 11-2. Here it is useful to consider the s, p, and d functions separately. The s functions

Table 11-2 Eigenfunctions of the hydrogen atom. All lengths are expressed in terms of atomic units and the eigenfunctions are normalized to unity

State	
1s	$\phi(1s) = (1/\sqrt{\pi})e^{-r}$
2s	$\phi(2s) = (1/4\sqrt{\pi})(r-2)\,e^{-(r/2)}$
2p	$\phi(2p_x) = (1/4\sqrt{2\pi})x\,e^{-(r/2)}$
	$\phi(2p_y) = (1/4\sqrt{2\pi})y\,e^{-(r/2)}$
	$\phi(2p_z) = (1/4\sqrt{2\pi})z\,e^{-(r/2)}$
3s	$\phi(3s) = (1/81\sqrt{3\pi})(2r^2 - 18r + 27)\,e^{-(r/3)}$
3p	$\phi(3p_x) = (\sqrt{2}/81\sqrt{\pi})(r-6)x\,e^{-(r/3)}$
	$\phi(3p_y) = (\sqrt{2}/81\sqrt{\pi})(r-6)y\,e^{-(r/3)}$
	$\phi(3p_z) = (\sqrt{2}/81\sqrt{\pi})(r-6)z\,e^{-(r/3)}$
3d	$\phi(3d_{xy}) = (\sqrt{2}/81\sqrt{\pi})xy\,e^{-(r/3)}$
	$\phi(3d_{yz}) = (\sqrt{2}/81\sqrt{\pi})yz\,e^{-(r/3)}$
	$\phi(3d_{xz}) = (\sqrt{2}/81\sqrt{\pi})xz\,e^{-(r/3)}$
	$\phi(3d_{zz}) = (1/81\sqrt{6\pi})(r^2 - 3z^2)\,e^{-(r/3)}$
	$\phi(3d_0) = (1/81\sqrt{2\pi})(x^2 - y^2)\,e^{-(r/3)}$

depend only on the variable r and in an s state the probability density function is radially symmetric. The probability that the electron is found between the distances r and $r + dr$ from the origin is given by

$$P(r) = 4\pi r^2 [\phi(ns)]^2\,dr \tag{11-126}$$

and we write this as

$$P(r) = R(ns; r)\,dr \tag{11-127}$$

where $R(ns)$ is known as the radial distribution function. It follows easily from Eqs. (11-126) and (11-127) that the radial distribution function is given by

$$R(ns) = 4\pi r^2 [\phi(ns)]^2 \tag{11-128}$$

We show the radial distribution functions for the $1s$ and $2s$ states in Figure 11-10. It may be seen that these functions have fairly sharp maxima, and we can visualize an ns state by a probability density that looks roughly like an orange peel. The radius of this orange peel is approximately given by the expectation value of the variable r,

$$\bar{r}(ns) = \int_0^\infty rR(ns; r)\,dr \tag{11-129}$$

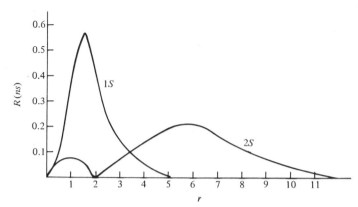

Figure 11-10   Radial distribution functions $R(ns)$ for the $1s$ and $2s$ states of the hydrogen atom.

It is then found that

$$\bar{r}(1s) = 1.5 \qquad \bar{r}(2s) = 6 \qquad \bar{r}(3s) = 13.5$$

In general

$$\bar{r}(ns) = (3n^2/2) \tag{11-130}$$

and the radius of the charge shell increases quadratically with the quantum number n.

It is a bit more difficult to visualize the behavior of the $\phi(np)$ functions because these functions depend both on the variable r and on the polar angles. Let us consider, for example, the function $\phi(2p_z)$, which is given by

$$\phi(2p_z) = \left(\frac{1}{4\sqrt{2\pi}}\right) z\, e^{-(r/2)} \tag{11-131}$$

If we write this in terms of polar coordinates, the probability density function is given by

$$[\phi(2p_z)]^2 = \left(\frac{1}{32\pi}\right) r^2 \cos^2 \theta\, e^{-r} \tag{11-132}$$

We can represent this function by means of two figures, Figures 11-11a and 11-11b. For a given value of r, the probability density function depends on the polar angle θ, which is $0°$ or $180°$ along the z axis and $90°$ in the xy plane. We have listed the values of the angular function $\cos^2 \theta$ at some points along the circle. The radial dependence of the function may again be given by means of the radial distribution function $R(np)$, and we show these functions in Figure 11-11b for the $2p$ and $3p$ states. These functions again have fairly well defined maxima. The values of the expectation values $\bar{r}(np)$ indicate where these maxima are located. They are

$$\bar{r}(2p) = 5 \qquad \text{and} \qquad \bar{r}(3p) = 12.5$$

The functions $\phi(np_z)$ are zero when $z = 0$, this means that the functions are zero in the xy plane. Such a plane, where the wave function is zero, is called a **nodal plane**.

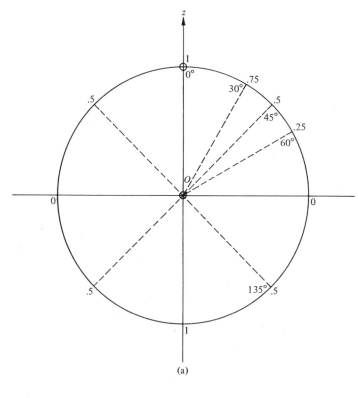

(a)

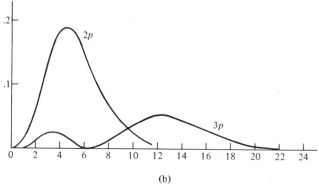

(b)

Figure 11-11   (a) Angular dependence of probability density in 2p function. (b) Radial distribution functions $R(np)$ for the 2p and 3p states of the hydrogen atom.

If we could take a picture of the charge density, we would see something like Figure 11-12. We have marked in the figure the two maxima of the probability density that are situated on the z axis. We observe two charge blobs around these two points.

The other p functions np_x and np_y have the same behavior as the function $2p_z$; the only difference is that the charge density is now along the x axis or the y axis, respectively.

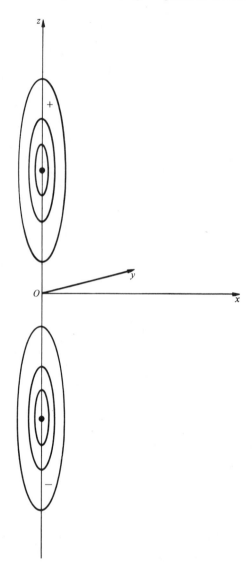

Figure 11-12 Representation of the hydrogen $2p_z$ wave function.

The behavior of the $3d$ functions is best understood if we concentrate on the angular dependence of these functions. The expectation value $\bar{r}(3d)$ is given by

$$\bar{r}(3d) = 10.5$$

and the probability density is mostly located on or near the surface of a sphere with radius 10.5.

Let us consider first the function $\phi(3d_{xy})$, which is given by

$$\phi(3d_{xy}) = \frac{\sqrt{2}}{81\sqrt{\pi}} xy \exp \frac{-r}{3} \qquad (11\text{-}133)$$

This function is zero in the two nodal planes $x = 0$ and $y = 0$, and the z axis is the intersection between these two planes. The wave function has its maximum values in the xy plane, and in Figure 11-13a we have sketched the behavior of the wave function in the xy plane. We can see that the maximum values of the wave function are at the points A and B and the points C and D. All these points are on the two lines that bisect the x and the y axes, at a distance of about 10.5 from the origin. At A and C the function is positive; at B and D the function is negative. The probability density of the function $\phi(3d_{xy})$ consists of four charge blobs around these four points. They are separated by the nodal planes $x = 0$ and $y = 0$.

It is interesting to compare this behavior with the behavior of the function

$$\phi(3d_0) = \frac{1}{81\sqrt{\pi}}(x^2 - y^2)\exp\frac{-r}{3} \tag{11-134}$$

which we have sketched in Figure 11-13b. Here the nodal planes are $x = \pm y$ and the maximum values of the function occur at the points A', B', C', and D', which are located on the x axis and the y axis. It follows that the function $\phi(3d_0)$ may be

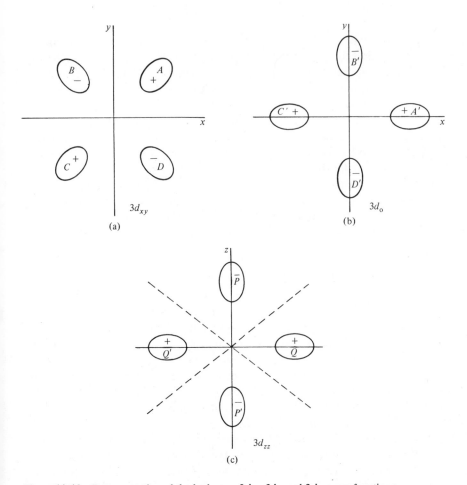

Figure 11-13 Representation of the hydrogen $3d_{xy}$, $3d_0$, and $3d_{zz}$ wave functions.

obtained from $\phi(3d_{xy})$ by rotation around the z axis through an angle of $45°$. After this rotation, the two functions are identical.

It is obvious that the functions $\phi(3d_{xz})$ and $\phi(3d_{yz})$ have the same behavior as $\phi(3d_{xy})$. They may be derived from each other by renaming the coordinate axes.

The most complicated of the $3d$ functions is $\phi(3d_{zz})$, which is given by

$$\phi(3d_{zz}) = \frac{1}{81\sqrt{6\pi}}(r^2 - 3z^2) \exp\frac{-r}{3} \tag{11-135}$$

This function may be expressed in terms of r and the polar angle θ

$$\phi(3d_{zz}) = \frac{1}{81\sqrt{6\pi}}r^2(1 - 3\cos^2\theta) \exp\frac{-r}{3} \tag{11-136}$$

This function has cylindrical symmetry around the z axis and it has the same behavior in every plane that contains the z axis. We have sketched one of these planes in Figure 11-13c.

It follows from Eq. (11-136) that the function $\phi(3d_{zz})$ is zero when

$$1 - 3\cos^2\theta = 0 \qquad \cos^2\theta = \tfrac{1}{3} \tag{11-137}$$

As a result, the nodal surface for this function is a cone around the z axis with an angle $\theta_0 = 54° \, 45'$. This is the angle for which $\cos^2\theta_0 = \tfrac{1}{3}$. In the cross section of Figure 11-13c, we have drawn two dashed lines where this cone intersects with the plane. We have also indicated the points where the wave function has its maximum values, the two points P and P' on the z axis and the two points Q and Q' in the xy plane. All these points are at a distance of about 10.5 from the origin. From Figure 11-13c we can visualize the three-dimensional charge density. It consists of two charge blobs around the points P and P' on the z axis and a ring of charge in the xy plane at a distance of about 10.5 from the z axis.

We have limited this discussion to the states $n = 1, 2$, and 3 of the hydrogen atom. The higher states can be discussed in a similar way, but their behavior is more complicated and we feel that there is no need to discuss them.

11-6 APPROXIMATE METHODS

In Section 11-5 we discussed the exact solution of the Schrödinger equation for some simple systems, but it is obvious that exact solutions can be obtained only for such very simple systems. The largest molecule that can be described exactly is the hydrogen molecular ion, H_2^+, and here the solution is so complicated that it is of little practical use. Thus, it is clear that, if we want to apply quantum mechanics to problems that are of chemical interest, we must resort to approximate methods. We will discuss the two methods of approximation that are most widely used in quantum chemistry, namely, the variation principle and perturbation theory.

We know that it is not possible to find the exact eigenfunctions of a molecule such as benzene or even a simple molecule such as H_2. However, by making use of our intuition and knowledge of chemistry, we may write down approximate expressions for the eigenfunctions. The variation principle enables us to find out how good our approximate functions are. Also, if our approximate expressions contain any parameters, we can determine the most suitable values for these parameters by making use of the variational principle.

The variation principle states that, if we have an approximate wave function ψ for a molecule and if we use this wave function to calculate the molecular energy, the result must always be higher than the true molecular ground state energy E_0. In mathematical language, if the atom or molecule is represented by a Hamiltonian $\mathfrak{H}$, then we have

$$E(\Psi) = \frac{\langle \Psi | \mathscr{H} | \Psi \rangle}{\langle \Psi | \Psi \rangle} \geqq E_0 \tag{11-138}$$

where E_0 is the smallest of the set of eigenvalues E_n of $\mathfrak{H}$,

$$\mathfrak{H}\Phi_n = E_n \Phi_n \qquad E_0 \leqq E_1 \leqq E_2 \cdots \tag{11-139}$$

The expectation value $E(\Psi)$ is always larger than the exact eigenvalue E_0 except when Ψ is equal to the eigenfunction Φ_0, in that case the expectation value $E(\Phi_0)$ is equal to E_0.

The theorem of Eq. (11-138) can be proved fairly easily, but before we do that let us see first how we make use of it. The approximate energy $E(\Psi)$ is equal to the exact eigenvalue E_0 only when Ψ is equal to Φ_0. This means that the lower the energy $E(\Psi)$, the closer it is to E_0 and the more the function Ψ resembles the function Φ_0. In practice, if we have a number of different approximate functions Ψ that contain some parameters, we should take the values of the parameters which give the lowest energy, those parameter values then give the best form of the wave function.

As an illustration, let us consider the hydrogen molecule (see Figure 11-14). We can construct an approximate eigenfunction for its ground state by assuming that the first electron 1 is in the vicinity of nucleus A and the second electron 2 is in the vicinity of nucleus B. This situation may be approximated by an eigenfunction

$$\Psi_{\mathrm{I}} = s_A(1)s_B(2) \tag{11-140}$$

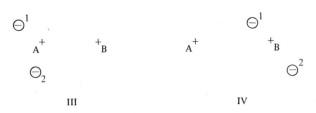

Figure 11-14 Possible electronic structures of the hydrogen molecule.

where s_A is a $1s$ hydrogen eigenfunction centered on nucleus A and s_B is a $1s$ hydrogen eigenfunction centered on nucleus B. Obviously, it is just as likely that electron 1 is on nucleus B and electron 2 is on nucleus A; this situation is represented by

$$\Psi_{II} = s_A(2)s_B(1) \tag{11-141}$$

There is also a possibility that the two electrons are centered on the same nucleus, either on A, which is described by

$$\Psi_{III} = s_A(1)s_A(2) \tag{11-142}$$

or on B, where we have

$$\Psi_{IV} = s_B(1)s_B(2) \tag{11-143}$$

We have sketched the four situations in Figure 11-14. It may be seen that I and II describe an ordinary chemical bond composed of two electrons, and that III and IV represent ionic structures. The total wave function may now be written as a superposition of the preceding four functions.

$$\Psi = \lambda(\Psi_I + \Psi_{II}) + \mu(\Psi_{III} + \Psi_{IV}) \tag{11-144}$$

It may be derived from symmetry considerations that Ψ_I and Ψ_{II} must have the same coefficient, and Ψ_{III} and Ψ_{IV} also.

If we want to make use of the variation principle we calculate the expectation value $E(\Psi)$ from the function (11-144). From the result we may then derive the probability for ionic structures in the H_2 molecule. This probability is approximately given by $\mu^2/(\lambda^2 + \mu^2)$.

In the preceding example, we have used the principle of superposition of states, which has been widely applied in the theory of the chemical bond. In order to describe a given molecule, we write down a number of chemical structures for it. Each of these structures is then represented by a wave function $\Phi_I, \Phi_{II}, \Phi_{III}, \ldots$, and so on. The total molecular wave function is now written as a linear combination of these separate wave functions.

$$\Psi = \sum_N a_N \Phi_N \tag{11-145}$$

We determine the values of the coefficients by using the variation principle, that is, by minimizing the energy. The relative probability of each structure is then given by the square of its coefficient.

EXAMPLE 11-6

The principle of superposition is a special case of a mathematical theorem which states that any function Ψ may be expanded in terms of a complete set of functions. The theorem states also that the set of eigenfunctions of a Hamiltonian $\mathfrak{H}$ forms such a complete set. Let us define the complete set of eigenfunctions of $\mathfrak{H}$ by means of

$$\mathfrak{H}\Phi_n = E_n\Phi_n \tag{a}$$

Then any arbitrary function Ψ may be expressed as a linear combination of the functions Φ_n according to this theorem,

$$\Psi = \sum_n a_n \Phi_n \tag{b}$$

It should be recalled from Example 10-4 that the eigenfunctions Φ_n form an orthonormal set of functions. They satisfy the relations

$$\langle \Phi_n | \Phi_m \rangle = \delta_{n,m} = 1 \qquad \text{if } n = m$$
$$= 0 \qquad \text{if } n \neq m \tag{c}$$

The variation principle, Eq. (11-138), is easily derived from Eqs. (a), (b), and (c). We take E_0 as the smallest eigenvalue of the operator $\mathfrak{H}$ and we determine

$$(\mathfrak{H} - E_0)\Psi = ? \tag{d}$$

where Ψ is an arbitrary function. We know that Ψ may be expressed in terms of the expansion (b). We do not know the values of the coefficients a_n but that does not matter; we just substitute Eq. (b) into Eq. (d) and we find

$$(\mathfrak{H} - E_0)\Psi = (\mathfrak{H} - E_0) \sum_{n=0}^{\infty} a_n \Phi_n = \sum_{n=0}^{\infty} a_n (E_n - E_0)\Phi_n = \sum_{n=1}^{\infty} a_n (E_n - E_0)\Phi_n \tag{e}$$

If we multiply this equation on the left by Ψ^* and then integrate, we obtain

$$\langle \Psi | \mathfrak{H} - E_0 | \Psi \rangle = \left\langle \Psi \Big| \sum_{n=1}^{\infty} a_n (E_n - E_0)\Phi_n \right\rangle$$
$$= \sum_{n=1}^{\infty} a_n (E_n - E_0)\langle \Psi | \Phi_n \rangle = \sum_{n=1}^{\infty} a_n (E_n - E_0) \left\langle \sum_{m=0}^{\infty} a_m \Phi_m \Big| \Phi_n \right\rangle$$
$$= \sum_{n=1}^{\infty} a_n a_n^* (E_n - E_0) \tag{f}$$

In the last step of Eq. (f) we have made use of the orthogonality relation (c).

It is easily seen that the last term of Eq. (f) must be positive because each term in the sum must be positive. We find thus that

$$\langle \Psi | \mathfrak{H} - E_0 | \Psi \rangle \geq 0 \tag{g}$$

or

$$\langle \Psi | \mathfrak{H} | \Psi \rangle - E_0 \langle \Psi | \Psi \rangle \geq 0 \tag{h}$$

This may be rewritten as

$$\langle \Psi | \mathfrak{H} | \Psi \rangle \geq E_0 \langle \Psi | \Psi \rangle \tag{i}$$

or

$$\frac{\langle \Psi | \mathfrak{H} | \Psi \rangle}{\langle \Psi | \Psi \rangle} \geq E_0 \tag{j}$$

which is the variation principle of Eq. (11-138).   ●

The second method of approximation in quantum mechanics is the perturbation theory. This method is particularly suitable to study the effects of electric or magnetic fields on atoms or molecules. It plays a prominent role in the theories of magnetic resonance and spectroscopy.

The detailed formulation of perturbation theory is all mathematical, but we will attempt to outline it in general terms. Again, we consider an atom or molecule that is represented by a Hamiltonian, which we call now $\mathfrak{H}_0$. This Hamiltonian has a set of eigenvalues ε_n and eigenfunctions Φ_n which are defined by the relation

$$\mathfrak{H}_0 \Phi_n = \varepsilon_n \Phi_n \tag{11-146}$$

If we place this molecule in an electric or magnetic field then the system is described by a slightly different Hamiltonian $\mathfrak{H}$, and the new eigenfunctions and eigenvalues are Ψ_n and E_n. They are defined by

$$\mathfrak{H} \Psi_n = E_n \Psi_n \tag{11-147}$$

It may be seen that if the Hamiltonian $\mathfrak{H}$ differs only slightly from $\mathfrak{H}_0$,

$$\mathfrak{H} \approx \mathfrak{H}_0 \tag{11-148}$$

then the sets of eigenvalues and eigenfunctions must be similar also; that is

$$E_n \approx \varepsilon_n \quad \text{and} \quad \Psi_n \approx \Phi_n \tag{11-149}$$

Of course, we assume here that the two sets of eigenvalues and eigenfunctions are numbered the same way.

The main problem of perturbation theory is to formulate the expressions (11-149) in a more precise mathematical form. This can be done only if we first write Eq. (11-148) in a more precise form. If $\mathfrak{H}$ and $\mathfrak{H}_0$ differ only slightly, we can write

$$\mathfrak{H} = \mathfrak{H}_0 + \mathfrak{H}' \quad \text{with} \quad \mathfrak{H}' \ll \mathfrak{H}_0 \tag{11-150}$$

However, the precise formulation of perturbation theory makes use of power series expansions, and we write instead

$$\mathfrak{H} = \mathfrak{H}_0 + \lambda V \quad \text{with} \quad \lambda \ll 1 \tag{11-151}$$

Here λ is a somewhat artificial quantity that is known as a scaling parameter. Since it is much smaller than unity we may assume that power series expansions in terms of λ are rapidly convergent.

If we introduce the scaling parameter λ, we expand the eigenvalues and eigenfunctions of $\mathfrak{H}$ as power series in λ and they take the form

$$\begin{aligned} E_n &= \varepsilon_n + \lambda E'_n + \lambda^2 E''_n + \cdots \\ \Psi_n &= \Phi_n + \lambda \Psi'_n + \lambda^2 \Psi''_n + \cdots \end{aligned} \tag{11-152}$$

Since λ is supposed to be small, it is assumed that each successive term in these expansions is an order of magnitude smaller than its preceding term; an order of magnitude is thus defined here as a factor λ.

It is customary to call the term E'_n the first order perturbation to the energy; E''_n is the second order perturbation to the energy, and so on. Similarly, Ψ'_n is the first order perturbation to the eigenfunction Φ_n, and so on. The goal of perturbation theory is to express the various perturbation terms $E'_n, E''_n, \Psi'_n, \ldots$, and so on, in terms of the eigenvalues and eigenfunctions ε_n and Φ_n of the unperturbed Hamiltonian $\mathfrak{H}_0$.

We will just mention some of the results of perturbation theory and not give the mathematical derivations because these are fairly elaborate. Also, we limit ourselves to perturbations of nondegenerate energy levels because the perturbation theory of degenerate states is much more involved.

If the state n is nondegenerate, then the first-order perturbation to the energy, E_n' is given by

$$E_n' = \langle \Phi_n | V | \Phi_n \rangle = V_{n,n} \qquad \textbf{(11-153)}$$

This is simply the expectation value of the perturbation term V. The second-order perturbation E_n'' is given by an infinite sum

$$E_n'' = - \sum_{m \neq n} \frac{\langle \Phi_n | V | \Phi_m \rangle \langle \Phi_m | V | \Phi_n \rangle}{\varepsilon_m - \varepsilon_n} = - \sum_{m \neq n} \frac{V_{n,m} V_{m,n}}{\varepsilon_m - \varepsilon_n} \qquad \textbf{(11-154)}$$

The perturbation terms of the eigenfunction Ψ_n are usually expressed as series expansions in terms of the set of functions Φ_n. The first order perturbation Ψ_n' is given by

$$\Psi_n' = - \sum_{m \neq n} \frac{\langle \Phi_m | V | \Phi_n \rangle \Phi_m}{\varepsilon_m - \varepsilon_n} = - \sum_{m \neq n} \frac{V_{m,n}}{\varepsilon_m - \varepsilon_n} \Phi_m \qquad \textbf{(11-155)}$$

In magnetic resonance and in spectroscopy the value of the scaling parameter λ is usually very small, between 10^{-3} and 10^{-7}. The perturbation expansions, then, converge very rapidly, and there is no need to consider the terms beyond second order in the energy and beyond the first order in the wave function. Perturbation theory has also been applied in situations where λ is much larger, say between 0.1 and 0.5, but the method then becomes less satisfactory.

Finally, we should point out that we have oversimplified perturbation theory in our discussion. There is much that we have not discussed, but we decided to limit ourselves to the main features of the theory only.

_____ *Appendixes*

APPENDIX A EIGENFUNCTIONS OF THE HARMONIC OSCILLATOR

The Schrödinger equation for the harmonic oscillator is

$$-\frac{\hbar^2}{2m} \frac{d^2\psi}{dx^2} + \frac{kx^2}{2} \psi = E\psi \qquad \textbf{(A-1)}$$

Before we try to solve this equation we want to eliminate the various constants. This is done by introducing suitable new units of length and energy. We take the quantity x_0 of Eq. (11-19) as the new unit of length. It should be recalled that x_0 is the point of intersection between the energy of the ground state and the potential function. We found that

$$x_0^2 = \frac{\hbar}{\sqrt{km}} \qquad mx_0^2 = \hbar\sqrt{\frac{m}{k}} = \frac{\hbar}{\omega} \qquad \textbf{(A-2)}$$

Consequently, we introduce the new variable y

$$y = \frac{x}{x_0} \qquad \text{(A-3)}$$

into Eq. (A-1), which then becomes

$$-\frac{\hbar^2}{2mx_0^2}\frac{d^2\psi}{dy^2} + \frac{kx_0^2}{2}y^2\psi = E\psi \qquad \text{(A-4)}$$

By using Eq. (A-2), Eq. (A-4) is reduced to

$$\left(\frac{d^2}{dy^2} - y^2\right)\psi = -\frac{2E}{\hbar\omega}\psi \qquad \text{(A-5)}$$

Now we introduce $\hbar\omega$ as the new unit of energy. This means that we define a new energy parameter ε as

$$\varepsilon = \frac{E}{\hbar\omega} \qquad \text{(A-6)}$$

Substitution of ε into Eq. (A-5) gives

$$\left(\frac{d^2}{dy^2} - y^2\right)\psi = -2\varepsilon\psi \qquad \text{(A-7)}$$

We solve this equation by introducing two new operators D_+ and D_- which we define as

$$D_+ = \frac{d}{dy} - y \qquad D_- = \frac{d}{dy} + y \qquad \text{(A-8)}$$

The two products D_+D_- and D_-D_+ are different. It is easily verified that

$$D_+D_-\Phi = \left(\frac{d}{dy} - y\right)\left(\frac{d\Phi}{dy} + y\Phi\right) = \left(\frac{d^2}{dy^2} - y^2 + 1\right)\Phi$$

$$D_-D_+\Phi = \left(\frac{d}{dy} + y\right)\left(\frac{d\Phi}{dy} - y\Phi\right) = \left(\frac{d^2}{dy^2} - y^2 - 1\right)\Phi$$

$$\tfrac{1}{2}(D_+D_- + D_-D_+) = \frac{d^2}{dy^2} - y^2 \qquad \text{(A-9)}$$

We may write the Schrödinger equation (A-7) in the form

$$\tfrac{1}{2}(D_+D_- + D_-D_+)\psi = -2\varepsilon\psi \qquad \text{(A-10)}$$

We wish to show now that the eigenfunctions of this Schrödinger equation are the functions ψ_n, defined by Eqs. (11-21) and (11-23)

$$\psi_n(y) = e^{-(1/2)y^2}H_n(y)$$

$$H_n(y) = e^{y^2}\frac{d^n e^{-y^2}}{dy^n} \qquad \text{(A-11)}$$

We note that the functions $H_n(y)$ are polynomials of order n in the variable y and that the functions ψ_n tend to zero when y tends to plus or minus infinity. These functions can be normalized to unity and are acceptable from a physical viewpoint.

We have to derive the effect of the operators D_+ and D_- on the functions $\psi_n(y)$. For this purpose we write them as

$$\psi_n(y) = e^{(1/2)y^2} g_n(y)$$

$$g_n(y) = \frac{d^n e^{-y^2}}{dy^n} \tag{A-12}$$

It follows from the definition that

$$g_{n+1}(y) = \frac{d^{n+1} e^{-y^2}}{dy^{n+1}} = \frac{d^n}{dy^n} \frac{d e^{-y^2}}{dy} = \frac{d^n}{dy^n}(-2y e^{-y^2}) \tag{A-13}$$

This last derivative may be reduced to

$$\frac{d^n}{dy^n}(-2y e^{-y^2}) = -2y g_n(y) - 2n g_{n-1}(y) \tag{A-14}$$

We find thus that

$$g_{n+1}(y) = -2y g_n(y) - 2n g_{n-1}(y) \tag{A-15}$$

or

$$y g_n(y) = -n g_{n-1}(y) - \tfrac{1}{2} g_{n+1}(y) \tag{A-16}$$

It also follows from the definition that

$$\frac{d}{dy} g_n(y) = \frac{d^{n+1} e^{-y^2}}{dy^{n+1}} = g_{n+1}(y) \tag{A-17}$$

Let us now consider the functions $\psi_n(y)$. We see from Eq. (A-12) that

$$\frac{d}{dy} \psi_n(y) = \frac{d}{dy}(e^{(1/2)y^2} g_n(y)) = y e^{(1/2)y^2} g_n(y) + e^{(1/2)y^2} \frac{d g_n(y)}{dy}$$

$$= e^{(1/2)y^2}\left(y g_n(y) + \frac{d g_n}{dy}\right) \tag{A-18}$$

It follows then that

$$D_+\psi_n = \left(\frac{d g_n}{dy} + y g_n - y g_n\right) e^{(1/2)y^2} = g_{n+1} e^{(1/2)y^2} = \psi_{n+1} \tag{A-19}$$

and

$$D_-\psi_n = \left(\frac{d g_n}{dy} + y g_n + y g_n\right) e^{(1/2)y^2}$$

$$= (g_{n+1} - 2n g_{n-1} - g_{n+1}) e^{(1/2)y^2} = -2n\psi_{n-1} \tag{A-20}$$

We can see now why the operators D_+ and D_- are also known as ladder operators. It follows from Eq. (A-19) that D_+ transforms a function ψ_n into a function ψ_{n+1} so that it has the effect of increasing the subscript by one. The operator D_+ is therefore called the step-up operator. It follows in the same way from Eq. (A-20) that D_- lowers the subscript by one; it is therefore known as the step-down operator.

It is easily derived from Eqs. (A-19) and (A-20) that

$$D_+D_-\psi_n = D_+(D_-\psi_n) = D_+(-2n\psi_{n-1}) = -2n\psi_n \tag{A-21}$$

and

$$D_-D_+\psi_n = D_-(D_+\psi_n) = D_-(\psi_{n+1}) = -(2n+2)\psi_n \qquad \text{(A-22)}$$

This is easily understood. If one of the operators lowers the subscript by one and the other increases it by one, we must get the original function back if both operators work on the function. Or, in the language of ladder operators, if we go up the ladder one step and then down one step we are back exactly where we started.

Obviously, the functions ψ_n are eigenfunctions of both D_+D_- and D_-D_+ and they are also eigenfunctions of the sum of these two operators. By adding the Eqs. (A-21) and (A-22) we obtain

$$\tfrac{1}{2}(D_+D_- + D_-D_+)\psi_n = -2(n+\tfrac{1}{2})\psi_n \qquad \text{(A-23)}$$

If we compare this equation with the Schrödinger equation (A-10)

$$\tfrac{1}{2}(D_+D_- + D_-D_+)\psi = -2\varepsilon\psi$$

it is obvious that the functions ψ_n are eigenfunctions of the harmonic oscillator, belonging to the eigenvalues

$$\varepsilon_n = n + \tfrac{1}{2} \qquad \text{(A-24)}$$

Since the energies ε_n are expressed in terms of the units $\hbar\omega$ according to Eq. (A-5), the eigenvalues E_n of the harmonic oscillator are given by

$$E_n = (n+\tfrac{1}{2})\hbar\omega \qquad \text{(A-25)}$$

The corresponding eigenfunctions are the functions ψ_n defined by Eq. (A-11). If we express these functions of the variable y in terms of the original variable x by using Eq. (A-3), we obtain

$$\psi_n(x) = H_n(x/x_0)\exp(-x^2/2x_0^2) \qquad \text{(A-26)}$$

where the Hermite polynomials are defined by Eq. (A-11).

APPENDIX B   THE RATIO OF TWO INTEGRALS

It is somewhat difficult to evaluate the two integrals

$$I(a) = \int_{-\infty}^{\infty} e^{-ax^2}\, dx$$

$$J(a) = \int_{-\infty}^{\infty} x^2 e^{-ax^2}\, dx \qquad \text{(B-1)}$$

but the evaluation of their ratio

$$R(a) = \frac{J(a)}{I(a)} = \frac{\int_{-\infty}^{\infty} x^2 e^{-ax^2}\, dx}{\int_{-\infty}^{\infty} e^{-ax^2}\, dx} \qquad \text{(B-2)}$$

is much easier.

If we differentiate the integral $I(a)$ with respect to a, we obtain

$$\frac{dI(a)}{da} = -\int_{-\infty}^{\infty} x^2 e^{-ax^2}\, dx = -J(a) \qquad \text{(B-3)}$$

We do not need to know the exact value of the integral $I(a)$ in order to calculate $R(a)$. Instead the dependence of $I(a)$ on the variable a is sufficient. This dependence is easy to find. If we substitute

$$x\sqrt{a} = y \tag{B-4}$$

into the integral, we find

$$I(a) = \int_{-\infty}^{\infty} e^{-ax^2}\, dx = (1/\sqrt{a}) \int_{-\infty}^{\infty} e^{-y^2}\, dy \tag{B-5}$$

The second integral in Eq. (B-5) is independent of a so that we may set it equal to a constant C,

$$I(a) = C/\sqrt{a} \tag{B-6}$$

It follows now from Eq. (B-3) that

$$J(a) = -\frac{dI(a)}{da} = \frac{C}{2a\sqrt{a}} \tag{B-7}$$

By substituting Eq. (B-6) and (B-7) into Eq. (B-2) we obtain

$$R(a) = \frac{C/2a\sqrt{a}}{C/\sqrt{a}} = \frac{1}{2a} \tag{B-8}$$

APPENDIX C   EIGENFUNCTIONS OF THE RIGID ROTOR

A homogeneous polynomial of the nth degree is defined by Eq. (11-58) as

$$f_n(x, y, z) = \sum_{k=0}^{n} \sum_{l=0}^{n-k} a_{k,l,n-k-l}\, x^k y^l z^{n-k-l} \tag{C-1}$$

It is a linear combination of terms of the type $x^k y^l z^m$ with the condition that the sum of the powers of each term must be the same for all terms.

$$k + l + m = n \tag{C-2}$$

An Euler polynomial of the nth degree $F_n(x, y, z)$ is defined as a homogeneous polynomial $f_n(x, y, z)$ which satisfies the condition

$$\frac{\partial^2 f_n}{\partial x^2} + \frac{\partial^2 f_n}{\partial y^2} + \frac{\partial^2 f_n}{\partial z^2} = 0 \tag{C-3}$$

We showed in Example 11-4 that an Euler polynomial $F_n(x, y, z)$ contains $2n+1$ arbitrary parameters.

In Section 11-3 we argued that the eigenfunctions Ψ_J of the rigid rotor have the form

$$\Psi_J = F_J(x_0, y_0, z_0) \tag{C-4}$$

where the variables are given by

$$x_0 = \frac{X}{R} \qquad y_0 = \frac{Y}{R} \qquad z_0 = \frac{Z}{R} \tag{C-5}$$

and the variable R is given by

$$R = (X^2 + Y^2 + Z^2)^{1/2} \tag{C-6}$$

In order to prove this we note first that

$$\Psi_J = F_J(x_0, y_0, z_0) = \frac{1}{R^J} F_J(X, Y, Z) \tag{C-7}$$

This follows from the definition (C-1) of homogeneous polynomials.

We differentiate the function (C-7) with respect to the variable X. The result is

$$\frac{\partial \Psi_J}{\partial X} = \frac{\partial}{\partial X}\left(\frac{F_J(X, Y, Z)}{R^J}\right) = \frac{1}{R^J}\frac{\partial F_J}{\partial X} + F_J\left(\frac{\partial}{\partial X}\frac{1}{R^J}\right)$$

$$= \frac{1}{R^J}\frac{\partial F_J}{\partial X} - \frac{JX}{R^{J+2}}F_J \tag{C-8}$$

A second differentiation with respect to X gives

$$\frac{\partial^2 \Psi_J}{\partial X^2} = \frac{1}{R_J}\frac{\partial^2 F_J}{\partial X^2} - \frac{2JX}{R^{J+2}}\frac{\partial F_J}{\partial X} - \frac{J}{R^{J+2}}F_J + \frac{J(J+2)X^2}{R^{J+4}}F_J \tag{C-9}$$

The differentiations with respect to Y and Z give analogous results, and if we add them up we obtain

$$\frac{\partial^2 \Psi_J}{\partial X^2} + \frac{\partial^2 \Psi_J}{\partial Y^2} + \frac{\partial^2 \Psi_J}{\partial Z^2} = \frac{1}{R^J}\left(\frac{\partial^2}{\partial X^2} + \frac{\partial^2}{\partial Y^2} + \frac{\partial^2}{\partial Z^2}\right)F_J$$

$$- \frac{2J}{R^{J+2}}\left(X\frac{\partial F_J}{\partial X} + Y\frac{\partial F_J}{\partial Y} + Z\frac{\partial F_J}{\partial Z}\right) - \frac{3J}{R^{J+2}}F_J + \frac{J(J+2)}{R^{J+2}}F_J \tag{C-10}$$

The first term of Eq. (C-10) is zero because of the definition of the Euler polynomials, Eq. (C-3). The second term of Eq. (C-10) may be reduced to a simple form. We note that

$$\left(X\frac{\partial}{\partial X} + Y\frac{\partial}{\partial Y} + Z\frac{\partial}{\partial Z}\right)(X^k Y^l Z^m) = (k + l + m)(X^k Y^l Z^m) \tag{C-11}$$

In a homogeneous polynomial we have for each term

$$k + l + m = J \tag{C-12}$$

so that the effect of the operator (C-11) is to multiply each term in F_J by the same factor J. Hence

$$\left(X\frac{\partial}{\partial X} + Y\frac{\partial}{\partial Y} + Z\frac{\partial}{\partial Z}\right)F_J = JF_J \tag{C-13}$$

If we substitute this result into Eq. (C-10) we obtain

$$\left(\frac{\partial^2}{\partial X^2} + \frac{\partial^2}{\partial Y^2} + \frac{\partial^2}{\partial Z^2}\right)\Psi_J = -\frac{2J^2}{R^{J+2}}F_J - \frac{3J}{R^{J+2}}F_J + \frac{J(J+2)}{R^{J+2}}F_J$$

$$= \frac{-J^2 - J}{R^{J+2}}F_J = -\frac{J(J+1)}{R^2}\frac{F_J}{R^J} = -\frac{J(J+1)}{R^2}\Psi_J \tag{C-14}$$

The Hamiltonian $\mathfrak{H}_{op}$ of the rigid rotor is defined in Eq. (11-46) as

$$\mathfrak{H}_{op} = -\frac{\hbar^2}{2\mu}\left(\frac{\partial^2}{\partial X^2} + \frac{\partial^2}{\partial Y^2} + \frac{\partial^2}{\partial Z^2}\right) \qquad \text{(C-15)}$$

It is easily derived from Eq. (C-14) that

$$\mathfrak{H}_{op}\Psi_J = J(J+1)\cdot\frac{\hbar^2}{2\mu R^2}\Psi_J \qquad \text{(C-16)}$$

Consequently Ψ_J is an eigenfunction of the rigid rotor belonging to the eigenvalue

$$E_J = \frac{\hbar^2}{2\mu R^2}J(J+1) \qquad \text{(C-17)}$$

The eigenvalue is $(2J+1)$-fold degenerate because the function Ψ_J contains $2J+1$ arbitrary parameters.

APPENDIX D   ANGULAR MOMENTUM

We have seen that the angular momentum M of a particle that moves in a closed circular orbit is given by

$$M = mvr = pr \qquad \text{(D-1)}$$

In this case the velocity $\mathbf{v}$ is always perpendicular to the position vector $\mathbf{r}$. In other cases, where $\mathbf{v}$ and $\mathbf{r}$ are not perpendicular, the angular momentum is given by

$$M = pr \sin\theta = mvr \sin\theta \qquad \text{(D-2)}$$

where θ is the angle between $\mathbf{v}$ and $\mathbf{r}$. The magnitude of M then describes the extent of the rotational motion.

In general, it is useful to know not only the extent of the rotational motion but also its direction, which is described by the axis of rotation. The plane of the rotational motion is perpendicular to this axis of rotation. We can also describe this direction by means of the angular momentum, then we must define the angular momentum $\mathbf{M}$ as a vector. Its magnitude is still given by Eq. (D-2), and its direction is taken as the axis of rotation, which is the line perpendicular to the plane of the two vectors $\mathbf{r}$ and $\mathbf{p}$ (see Figure 11-7). The way we define $\mathbf{M}$, then, is known in vector analysis as the cross product of the two vectors $\mathbf{r}$ and $\mathbf{p}$, and it is written as

$$\mathbf{M} = \mathbf{r} \times \mathbf{p} \qquad \text{(D-3)}$$

Again, the magnitude of $\mathbf{M}$ as defined by Eq. (D-3) is given by Eq. (D-2), and the direction of $\mathbf{M}$ is perpendicular to the plane of the two vectors $\mathbf{r}$ and $\mathbf{p}$. The sign of the direction is determined by the corkscrew rule: If we rotate a corkscrew from $\mathbf{r}$ to $\mathbf{p}$ over the smallest angle, then $\mathbf{M}$ points in the direction that the corkscrew moves (see Figure 11-7).

There is also an algebraic definition of the vector $\mathbf{M}$. Let us first consider the simple case where the two vectors $\mathbf{p}$ and $\mathbf{r}$ are both located in the xy plane. Thus, $\mathbf{r}$ has the components x and y, and $\mathbf{p}$ has the components p_x and p_y (see Figure 11-15). The

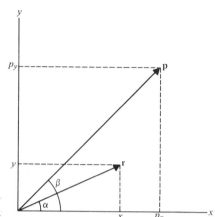

Figure 11-15 Definition of the angular momentum M_z if **r** and **p** are in the xy plane.

vector **M** is then directed along the z axis, and its magnitude M_z may be defined as

$$M_z = xp_y - yp_x \qquad \text{(D-4)}$$

It is easily shown that this definition gives the same result as Eq. (D-2). If α is the polar angle for the vector **r** and β the angle for the vector **p** (see Figure 11-15), we have

$$x = r \cos \alpha \qquad p_x = p \cos \beta$$

$$y = r \sin \alpha \qquad p_y = p \sin \beta \qquad \text{(D-5)}$$

Consequently,

$$M_z = pr(\cos \alpha \sin \beta - \sin \alpha \cos \beta) = pr \sin (\beta - \alpha) \qquad \text{(D-6)}$$

This is equivalent to Eq. (D-2) because $(\beta - \alpha)$ is the angle between the two vectors. Eq. (D-6) automatically gives the right sign for M_z. In Figure 11-15, we have sketched the situation where $\beta > \alpha$. In that case it follows from the corkscrew rule that **M** must point upward, in the positive z direction. Eq. (D-6) gives the same result.

In the general case, the vector **M** is not necessarily oriented along the z axis but has three components M_x, M_y, and M_z. These vector components are defined similarly to Eq. (D-4), namely

$$M_x = yp_z - zp_y$$

$$M_y = zp_x - xp_z \qquad \text{(D-7)}$$

$$M_z = xp_y - yp_x$$

Definition (D-7) of the vector **M** is again equivalent to definition (D-2).

Finally, we wish to show that in many systems, such as the hydrogen atom, the rigid rotor, and many others, the angular momentum is a constant of the motion. This means that

$$\frac{dM_x}{dt} = \frac{dM_y}{dt} = \frac{dM_z}{dt} = 0 \qquad \text{(D-8)}$$

The systems mentioned all fall in the category of a particle in a central force field. This means that, at any point **r**, the force acting on the particle is directed towards the origin so that the force **F** has the same direction as the vector **r**. We can write

$$F_x : F_y : F_z : F = x : y : z : r \qquad \text{(D-9)}$$

The three components F_x, F_y, and F_z may then be represented as

$$F_x = \frac{xF}{r} \qquad F_y = \frac{yF}{r} \qquad F_z = \frac{zF}{r} \qquad \text{(D-10)}$$

Since the acceleration **a** is proportional to **F**,

$$\mathbf{F} = m\mathbf{a} \qquad \text{(D-11)}$$

the same relationship as (D-10) holds for a

$$a_x = \frac{xa}{r} \qquad a_y = \frac{ya}{r} \qquad a_z = \frac{za}{r} \qquad \text{(D-12)}$$

Let us now consider M_x, as defined by Eq. (D-7). If we differentiate M_x with respect to t we obtain

$$\frac{dM_x}{dt} = \frac{dy}{dt}p_z + y\frac{dp_z}{dt} - \frac{dz}{dt}p_y - z\frac{dp_y}{dt}$$

$$= m(v_y v_z - v_z v_y) + m(ya_z - za_y) = 0 \qquad \text{(D-13)}$$

We may prove in the same way that the time derivatives of M_y and M_z are zero so that the vector **M** remains at a constant value at all times. This means that the angular momentum is a constant of the motion, just like the energy.

APPENDIX E   EIGENFUNCTIONS OF THE HYDROGEN ATOM

The Schrödinger equation of the hydrogen atom is given by

$$-\frac{\hbar^2}{2\mu}\left(\frac{\partial^2}{\partial x^2}+\frac{\partial^2}{\partial y^2}+\frac{\partial^2}{\partial z^2}\right)\psi - \frac{e^2}{r}\psi = E\psi \qquad \text{(E-1)}$$

In order to simplify the equation we first introduce a new unit of length a_0 and a new unit of energy $\varepsilon_0 = e^2/a_0$. The equation then becomes

$$-\frac{\hbar^2}{2\mu a_0^2}\left(\frac{\partial^2}{\partial x^2}+\frac{\partial^2}{\partial y^2}+\frac{\partial^2}{\partial z^2}\right)\psi - \frac{e^2}{a_0 r}\psi = \varepsilon\frac{e^2}{a_0}\psi \qquad \text{(E-2)}$$

If we divide by ε_0 and take

$$\frac{\hbar^2}{\mu e^2 a_0} = 1 \qquad a_0 = \frac{\hbar^2}{\mu e^2} \qquad \text{(E-3)}$$

all constants cancel, and the equation takes the simple form

$$-\tfrac{1}{2}\left(\frac{\partial^2\psi}{\partial x^2}+\frac{\partial^2\psi}{\partial y^2}+\frac{\partial^2\psi}{\partial z^2}\right) - \frac{1}{r}\psi = \varepsilon\psi \qquad \text{(E-4)}$$

We mentioned in Eq. (11-102) that the eigenfunction ψ should also be an eigenfunction of $(M^2)_{op}$; thus, it should have the form

$$\psi = \chi_l(r)F_l(x_0, y_0, z_0) \tag{E-5}$$

Here F_l is an Euler polynomial of the three variables

$$x_0 = \frac{x}{r} \qquad y_0 = \frac{y}{r} \qquad z_0 = \frac{z}{r} \tag{E-6}$$

(see Appendix C). Because χ_l is an arbitrary function of r we can write this also as

$$\psi = G_l(r)F_l(x, y, z) \tag{E-7}$$

Let us now derive the first and second derivatives with respect to x of this function. We have

$$\frac{\partial \psi}{\partial x} = G_l \frac{\partial F_l}{\partial x} + \frac{x}{r}\frac{dG_l}{dr}F_l \tag{E-8}$$

and

$$\frac{\partial^2 \psi}{\partial x^2} + \frac{\partial^2 \psi}{\partial y^2} + \frac{\partial^2 \psi}{\partial z^2} = G_l\left(\frac{\partial^2 F_l}{\partial x^2} + \frac{\partial^2 F_l}{\partial y^2} + \frac{\partial^2 F_l}{\partial z^2}\right) + \frac{2}{r}F_l\frac{dG_l}{dr} + F_l\frac{d^2 G_l}{dr^2}$$

$$+ \frac{2}{r}\frac{dG_l}{dr}\left(x\frac{\partial F_l}{\partial x} + y\frac{\partial F_l}{\partial y} + z\frac{\partial F_l}{\partial z}\right) \tag{E-9}$$

We have

$$x\frac{\partial F_l}{\partial x} + y\frac{\partial F_l}{\partial y} + z\frac{\partial F_l}{\partial z} = lF_l \tag{E-10}$$

and

$$\frac{\partial^2 F_l}{\partial x^2} + \frac{\partial^2 F_l}{\partial y^2} + \frac{\partial^2 F_l}{\partial z^2} = 0$$

because F_l is an Euler polynomial. If we substitute this into Eq. (E-9) we obtain

$$\frac{\partial^2 \psi}{\partial x^2} + \frac{\partial^2 \psi}{\partial y^2} + \frac{\partial^2 \psi}{\partial z^2} = F_l\left(\frac{d^2 G_l}{dr^2} + \frac{2l+2}{r}\frac{dG_l}{dr}\right) \tag{E-11}$$

Substitution of Eq. (E-11) into the Schrödinger equation (E-4) gives

$$-\frac{1}{2}\frac{d^2 G_l}{dr^2} - \frac{l+1}{r}\frac{dG_l}{dr} - \frac{1}{r}G_l = \varepsilon G_l \tag{E-12}$$

This equation can be solved by the standard methods of differential equations, and we would get all the eigenvalues and eigenfunctions. There is a simpler way, however, to find some of the eigenvalues. If we substitute, namely,

$$G_l = \exp\left(-\frac{r}{\rho}\right) \tag{E-13}$$

we obtain

$$-\frac{1}{2\rho^2} + \frac{l+1}{\rho r} - \frac{1}{r} = \varepsilon \tag{E-14}$$

It follows that the function (E-13) is an eigenfunction if we take

$$\varepsilon = -\frac{1}{2\rho^2} \quad \text{and} \quad \rho = l+1 \tag{E-15}$$

The next set of eigenvalues and eigenfunctions are obtained by substituting

$$G_l = (ar+b)\exp.(-r/\rho) \tag{E-16}$$

This is a solution if we take

$$\varepsilon = -\frac{1}{2\rho^2} \quad \rho = l+2 \quad G_l = \left(\frac{r}{l+2}-l+1\right)\exp\left(-\frac{r}{l+2}\right) \tag{E-17}$$

It may be derived that the general eigenvalue is given by

$$\varepsilon = -\frac{1}{2n^2} \quad n = l+1, l+2, l+3, \dots, \text{ etc.} \tag{E-18}$$

The corresponding eigenfunctions $G_l(r)$ are listed in Table 11-1.

_____ *Problems*

1. The vibrational transitions for the molecules H_2 and HBr are 4395 cm^{-1} and 2650 cm^{-1}, respectively. Determine the force constants for the harmonic oscillators that represent these vibrations in terms of dynes per centimeter.

2. Show that the two functions $\psi_0 = \exp(-\alpha x^2)$ and $\psi_1 = x\exp(-\alpha x^2)$ are eigenfunctions for the harmonic oscillator if we take the proper value for the parameter α. What is the value of α for which these two functions are eigenfunctions and what are the corresponding eigenvalues?

3. The vibrational transition in H_2 occurs at 4395 cm^{-1}. If we assume that the vibrational motion in D_2 has the same force constant as in H_2, where does the vibrational transition in D_2 occur?

4. Assuming that the two molecules HF and DF have the same force constant for their vibrational motion, determine the wave number of the vibrational transition in DF from the value 4139 cm^{-1} for the wave number of the vibrational transition in HF.

5. Calculate the expectation values of x^2 and p^2 for the ground state of a harmonic oscillator (x is the coordinate and p is the momentum). Determine the product of the square roots of x^2 and p^2 and relate the result to Heisenberg's uncertainty principle.

6. Calculate the expectation value of $(\Delta q)^2$, the square of the displacement coordinate for the ground state of a harmonic oscillator. What are the numerical values for $(\Delta q)^2$ and Δq for the HF molecule where the vibrational transition is 4139 cm^{-1}?

7. In the ground state of the $^1H^{35}Cl$ molecule, the equilibrium internuclear distance R_0 is 1.2746 Å. What is the position of the center of gravity?

8. The internuclear distance for H_2 is 0.7416 Å and its rotational constant is 60.809 cm^{-1}. The rotational constants for HD and D_2 are 45.655 cm^{-1} and 30.429 cm^{-1}, respectively. What are the internuclear distances for these two molecules? Do they differ to an appreciable extent from H_2?

9. Calculate the rotational constant B for the HB molecule where the internuclear distance is 1.2325 Å.

10. The rotational constant for the ground state of the BF molecule ($^{11}B^{19}F$) is 1.518 cm^{-1}. Determine the internuclear distance R_0 for the ground state. For the first excited electronic state the rotational constant is 1.423 cm^{-1}. Determine the internuclear distance R_1 for this excited state.

11. Determine the general form of the Euler polynomial $F_3(x, y, z)$ from the homogeneous polynomial $f_3(x, y, z)$ of Eq. (11-59).

12. Consider an electron that moves with a constant angular velocity around a circle with radius 0.5292 Å (the atomic unit of length for the hydrogen atom). If the angular momentum of this electron is equal to Dirac's constant $\hbar$ what is the velocity of the electron and what is its momentum? What is its kinetic energy and how does this compare with the ground state energy (for the state $n = 1$) of the hydrogen atom?

13. Show explicitly that the three functions $x \pm iy$ and z are eigenfunctions of $(M_z)_{op}$ and derive the corresponding eigenvalues.

14. Calculate the rotational constant for the motion of an electron on a sphere of radius 0.5292 Å. Is the rotational excitation energy for this system going from $J = 0$ to $J = 1$, comparable to the energy of the hydrogen atom $1s$ state?

15. The eigenvalues and eigenfunctions of the He^+ ion may be derived in the same way as for the H atom. What are the values for the reduced masses μ for these two systems? What are the values for the units of length and energy a_0 and ε_0 that we use in the two systems?

16. What is the value of the Coulomb attraction between an electron and a proton that are a distance 0.529 Å apart?

17. Before the discovery of quantum mechanics the energy levels of the hydrogen atom were found to be $E_n = \boldsymbol{R}_H/n^2$ where the Rydberg constant $\boldsymbol{R}_H$ is expressed in terms of cm^{-1}. Determine the value of $\boldsymbol{R}_H$ from the results of the quantum mechanical calculation.

18. Show explicitly that the eigenfunctions for the $1s$ and the $2s$ states of the hydrogen atom are orthogonal.

19. List the various eigenfunctions that belong to the eigenvalue E_4 of the hydrogen atom. What is the total degeneracy of E_4?

20. What is the value of the angular momentum for the $1s$ state of the hydrogen atom according to quantum mechanics? Can this result be explained by means of classical mechanics?

21. If we use atomic units, the Hamiltonian of the hydrogen atom is $-\frac{1}{2}\Delta - (1/r)$ and its lowest energy eigenvalue is -0.5. Calculate the expectation value of the energy from the variational function $\psi = \exp[-(1+\delta)r]$. Show that this energy has a minimum of -0.5 for $\delta = 0$.

22. Calculate the expectation value of the energy of the hydrogen atom from the Hamiltonian of Problem 21 by using the variational function $\psi = \exp(-\alpha r^2)$. Determine the energy minimum and the corresponding value of α and compare it with the exact value -0.5.

References and Recommended Reading

In Chapters 10 and 11 we have given a brief outline of the general theory of quantum mechanics and of some of its applications. For a more detailed treatment of the subject we refer the student to any of the numerous textbooks on quantum mechanics that are now available. Our bibliography gives a selected representation of these books. We have divided them into five categories: A are historical descriptions of the developments of the basic ideas underlying quantum mechanics; B are introductory treatments for beginning students; C are "classics," that is, the fundamental texts that were written in the 1930s; D are texts that were written specifically for chemists; and E are the texts that are most commonly used in physics courses. Very likely we have omitted some good books on the subject, but we do not claim that our list is complete.

[A1] M. Jammer: _The Conceptual Development of Quantum Mechanics._ McGraw-Hill, New York, 1966.

[A2] E. Whittaker: _A History of the Theories of Aether and Electricity,_ Vols. I and II. Harper & Row, New York, 1960.

[B1] W. Heitler: _Elementary Wave Mechanics._ Oxford University Press, London, 1945.

[B2] V. Rojanski: _Introductory Quantum Mechanics._ Prentice-Hall, Englewood Cliffs, N.J., 1938.

[B3] A. Sommerfeld: _Wave Mechanics._ Methuen, London, 1930.

[C1] P. A. M. Dirac: _The Principles of Quantum Mechanics,_ 4th ed. Oxford University Press, London, 1958.

[C2] E. C. Kemble: _The Fundamental Principles of Quantum Mechanics._ McGraw-Hill, New York, 1937.

[C3] H. A. Kramers: _Quantum Mechanics._ North-Holland, Amsterdam, 1958.

[D1] H. F. Hameka: _Introduction to Quantum Theory._ Harper & Row, New York, 1967.

[D2] L. Pauling and E. B. Wilson, Jr.: _Introduction to Quantum Mechanics._ McGraw-Hill, New York, 1935.

[E1] A. S. Davydov: _Quantum Mechanics._ Addison-Wesley, Reading, Mass., 1965.

[E2] R. H. Dicke and J. P. Wittke: _Introduction to Quantum Mechanics._ Addison-Wesley, Reading, Mass., 1960.

[E3] P. Fong: _Elementary Quantum Mechanics._ Addison-Wesley, Reading, Mass., 1962.

[E4] L. D. Landau and E. M. Lifshitz: _Quantum Mechanics._ Pergamon Press, Oxford, 1958.

[E5] J. L. Powell and B. Crasemann: _Quantum Mechanics._ Addison-Wesley, Reading, Mass., 1961.

[E6] L. Schiff: _Quantum Mechanics,_ 3rd ed. McGraw-Hill, New York, 1968.

chapter 12

Atomic Structure

12-1 INTRODUCTION

In the previous chapters there was no particular need to consider experimental information. We only discussed simple systems for which the Schrödinger equation could be solved exactly, and our mathematical derivations provided a complete description for every system that we considered. Now that we wish to proceed to more complicated systems, it becomes desirable to include the experimental results in our discussion. We should realize that the Schrödinger equation cannot be solved for any atom other than the hydrogen atom. This means that a quantum mechanical discussion of atomic structure must necessarily be based on approximations. Especially for the more complicated atoms, we need some guidelines for devising suitable approximate methods and the obvious place to look for such guidelines is the experimental information.

The bulk of experimental information on atomic structure is derived from atomic spectra. Some aspects of atomic spectra had already been interpreted by means of classical, semiclassical, or semiempirical arguments before the introduction of the Schrödinger equation. All these semiempirical descriptions are based on two broad assumptions: (1) in each atomic eigenstate it is possible to recognize the individual electrons, and (2) each individual electron may be identified by a set of hydrogen-type quantum numbers. The validity of these assumptions is justified by the agreement between the semiempirical theories and the experiments. By using them it is possible to explain the general features of most atomic spectra. A more detailed interpretation of the atomic spectra is based on what is known as the vector model of the atom. Here the concept of angular momentum plays an important role.

We must also discuss some additional postulates of quantum theory. The first deals with the electron spin. Even though it applies also to one-electron systems, it is particularly relevant to many-electron systems. The second is known as the Pauli exclusion principle, and it applies only to many-particle systems. Again, the concept of angular momentum plays an important role in the theory of electron spin.

In order to illustrate these assumptions we consider the helium atom. This is the simplest atom with more than one electron, and it is the easiest system to understand. The Hamiltonian of the helium atom is given by

$$\mathfrak{H} = -\frac{\hbar^2}{2m}\Delta_1 - \frac{\hbar^2}{2m}\Delta_2 - \frac{Ze^2}{r_1} - \frac{Ze^2}{r_2} + \frac{e^2}{r_{12}} \tag{12-1}$$

The first two terms are the kinetic energies for electrons 1 and 2, respectively; the third term is the Coulomb attraction between electron 1 and the nucleus (with charge $Z = 2$); the next term is the Coulomb attraction between the second electron and the nucleus; and the last term is the Coulomb repulsion between the two electrons. If we did not have the last term, we could solve the corresponding Schrödinger equation;

however, because we do have the last term, we have a problem that cannot be solved exactly. The only thing we can say is that the exact eigenfunctions of the helium atom must be rather involved functions of the six variables that denote the positions of the two electrons.

To find an approximate description for the helium atom, we consider only the behavior of one of the two electrons, say electron 1. Electron 1 moves in a potential field that is the sum of the Coulomb attraction of the nucleus and the Coulomb repulsion of the other electron. We assume now that the position of the second electron is described by a probability density function $\rho(\mathbf{r}_2)$, and we assume further (and this is the important assumption) that this probability density function $\rho(\mathbf{r}_2)$ is independent of the position of the first electron. We then have the situation that is sketched in Figure 12-1, where electron 1 moves through the charge cloud that represents the probability density of the second electron. At this point we do not know the exact form of $\rho(\mathbf{r}_2)$, but, since the total probability of finding the second electron must be unity, we know that

$$\int \rho(\mathbf{r}_2) d\mathbf{r}_2 = 1 \tag{12-2}$$

Our next assumption is that the charge cloud of electron 2 is spherically symmetric; this means that $\rho(\mathbf{r}_2)$ is a function of the radius r_2 only and is independent of the angles. Such an assumption simplifies the situation considerably because we can now use a property of electrostatics shown in Figure 12-2. If we have a sphere of radius R, covered with a homogeneous layer of charge, total charge q, then the electric field strength is zero inside the sphere and it is equal to

$$\mathbf{F} = \frac{q\mathbf{r}}{r^3} \tag{12-3}$$

outside the sphere. Thus, outside the sphere the field strength is the same as the field strength due to a charge q, located at the origin (the center of the sphere), and inside the sphere the field strength is zero.

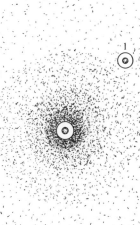

Figure 12-1 Sketch of the potential field acting on one of the two electrons, electron 1, in the helium atom. Electron 1 experiences the field due to the nucleus and the field due to the charge cloud of the second electron.

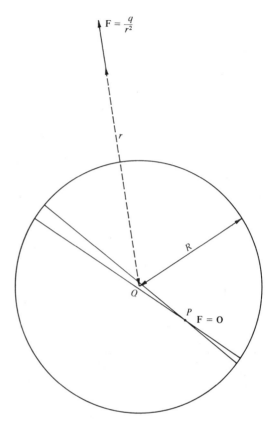

Figure 12-2 Field strength due to a sphere of radius R covered with a homogeneous layer of charge. The total charge is q. Inside the sphere the field strength is zero because the effects of opposite surface elements cancel. Outside the sphere the field strength is equal to the field strength due to a charge q located at the center O of the sphere.

If we apply this to the helium atom we see that the part δ of the charge cloud that is between electron 1 and the nucleus acts on electron 1 as if the whole charge δ were concentrated at the origin. Mathematically δ is given by

$$\delta(r_1) = 4\pi \int_0^{r_1} \rho(r_2) r_2^2 \, dr_2 \qquad (12\text{-}4)$$

The rest of the charge cloud, which is $1 - \delta(r_1)$ has no effect on electron 1. It is easily seen that $\delta(r_1)$ is zero when r_1 is zero, and δ increases with increasing values of r_1 until it approaches the value 1 asymptotically for large values of r_1.

It follows thus that electron 1 moves in an effective potential field $V_{\text{eff}}(r_1)$ which is given by

$$V_{\text{eff}}(r_1) = -\frac{Z - \delta(r_1)}{r_1} \qquad (12\text{-}5)$$

with

$$Z = 2 \qquad 0 < \delta(r_1) < 1 \qquad (12\text{-}6)$$

The corresponding Schrödinger equation for electron 1 is then

$$\left[-\frac{\hbar^2}{2m} \Delta_1 + V_{\text{eff}}(r_1) \right] \psi(1) = \varepsilon \psi(1) \qquad (12\text{-}7)$$

We say that the quantity $\delta(r_1)$ describes the amount of shielding of the nuclear charge by the second electron. It varies from zero near the nucleus, where there is no shielding, to unity far away, where there is complete shielding by the second electron. As a final approximation we can now replace the function $\delta(r_1)$ by an effective average value δ_{eff}, which must be somewhere between zero and unity. If we substitute this into the Schrödinger equation (12-7) we obtain

$$\left[-\frac{\hbar^2}{2m}\Delta_1 - \frac{Z_{\text{eff}}e^2}{r_1}\right]\psi(1) = \varepsilon\psi(1) \tag{12-8}$$

with

$$Z_{\text{eff}} = Z - \delta_{\text{eff}} \tag{12-9}$$

Here Z_{eff} is called the effective nuclear charge. It is equal to 2 minus the effective shielding constant due to the other electron.

The Schrödinger equation (12-8) is very similar to the Schrödinger equation of the hydrogen atom; in fact, it can be transformed to the same equation by choosing suitable units of length and energy. This is easily shown. We take a_Z as our unit of length and

$$\varepsilon_Z = \frac{Z_{\text{eff}}e^2}{a_Z} \tag{12-10}$$

as our unit of energy. Then Eq. (12-8) becomes

$$\left[-\frac{\hbar^2}{2ma_Z^2}\Delta_1 - \frac{Z_{\text{eff}}e^2}{a_Zr_1}\right]\psi(1) = \frac{Z_{\text{eff}}e^2}{a_Z}\varepsilon\psi(1) \tag{12-11}$$

We can transform this to

$$\left[-\frac{\hbar^2}{2a_Zme^2Z_{\text{eff}}}\Delta_1 - \frac{1}{r_1}\right]\psi(1) = \varepsilon\psi(1) \tag{12-12}$$

and if we take

$$a_Z = \frac{\hbar^2}{me^2Z_{\text{eff}}} \tag{12-13}$$

the equation becomes

$$\left[-\tfrac{1}{2}\Delta_1 - \frac{1}{r_1}\right]\psi(1) = \varepsilon\psi(1) \tag{12-14}$$

which is the same as the hydrogen atom Schrödinger equation of the previous chapter.

Let us now consider the eigenvalues and eigenfunctions of the Schrödinger equation (12-11), and compare them with the eigenvalues and eigenfunctions of the hydrogen atom. The difference between the two sets is mostly a matter of units. We should recall that, in the case of the hydrogen atom, our unit of length was

$$a_H = \frac{\hbar^2}{\mu e^2} \tag{12-15}$$

where μ is the reduced mass of the electron mass m and the proton mass M, it is given by

$$\mu = \frac{mM}{m+M} = m\left(1+\frac{m}{M}\right)^{-1} \tag{12-16}$$

The general unit of length, which is used in atomic calculations, is the Bohr radius a_0, which is defined by

$$a_0 = \frac{\hbar^2}{me^2} \tag{12-17}$$

We see that a_H differs slightly from a_0 because

$$\frac{a_H}{a_0} = \frac{m}{\mu} = 1 + \frac{m}{M} \tag{12-18}$$

and m/M is about 0.5×10^{-3}. It is obvious why a_0 has been chosen as a general unit. It applies to all atoms and molecules, whereas a_H is strictly limited to the hydrogen atom.

Let us now consider the eigenvalues and eigenfunctions of the original Schrödinger equation (12-8). It follows from Eq. (12-13) and (12-17) that the unit of length a_Z that we introduced in Eq. (12-13) is given by

$$a_Z = \frac{a_0}{Z_{\text{eff}}} \tag{12-19}$$

Our unit of energy ε_Z is then given by

$$\varepsilon_Z = \frac{Z_{\text{eff}}e^2}{a_Z} = Z_{\text{eff}}^2 \frac{e^2}{a_0} = Z_{\text{eff}}^2 \varepsilon_0 \tag{12-20}$$

The energy eigenvalues of Eq. (12-14) are the same as for the hydrogen atom, namely

$$\varepsilon_n = -\frac{1}{2n^2} \tag{12-21}$$

but realize that they are now expressed in terms of the energy unit ε_Z of Eq. (12-20). This means that the energy eigenvalues of the original equation (12-8) are given by

$$\varepsilon_n = -\frac{Z_{\text{eff}}^2}{2n^2}\varepsilon_0 \tag{12-22}$$

It follows that, for a hydrogenlike system with nuclear charge Ze, the eigenvalues are equal to the hydrogen atom eigenvalues if they are all multiplied by a factor Z^2.

The eigenfunctions of Eq. (12-8) are also the same as the hydrogen atom eigenfunctions, but now they are expressed in terms of (a_0/Z) instead of a_H. For example, the $1s$ function is now

$$\psi(1s) = \exp(-Zr/a_0) \tag{12-23}$$

and it is contracted by a factor Z as compared to the hydrogen $1s$ function. Similarly, the $2p$ functions are

$$\psi(2p_x) = x \exp(-Zr/2a_0) \qquad \text{etc.} \tag{12-24}$$

They are also contracted by a factor Z, just as all the other eigenfunctions.

It follows that the solution of the Schrödinger equation (12-8), which is the equation for electron 1 moving in an effective field of nuclear charge Z_{eff}, is a hydrogen atom eigenfunction, contracted by a factor Z. The lowest eigenstate is a $1s$ state, which is described by the $1s$ function of Eq. (12-23). In order to find the total wave function of the helium atom we note that electron 2 moves in the effective field of the nucleus, shielded by electron 1 and, as we shall see later, the total ground state eigenfunction should be symmetric in the two electrons, 1 and 2. Our approximate treatment leads thus to the following expression for the ground state wave function of the helium atom

$$\Psi(1, 2) = \exp\left(- Zr_1/a_0\right) \exp\left(- Zr_2/a_0\right) \qquad (12\text{-}25)$$

which is the product of the $1s$ function (12-23) for electron 1 and the same function for electron 2.

We still do not know the value of the effective nuclear charge Z, but it is possible to determine Z by making use of the variation principle. In this method the function Ψ is used to calculate the expectation value of the energy, using the helium atom Hamiltonian (12-1), then the value of Z for which this energy value has a minimum is determined. It turns out that this value is $Z = 1.6875$ so that the effective nuclear charge acting on each electron is 1.6875.

In the preceding discussion we have used quite a few approximations to obtain a wave function of the helium atom. But some of our conclusions are valid, even if we make fewer approximations. The first conclusion is that we can still recognize the individual electrons, even in a complicated molecule.[1] The second conclusion is that each electron can be identified by means of the same set of quantum numbers (n, l, m) that are used for the hydrogen atom. We can use these two conclusions to describe the ground state and also the various excited states of the helium atom.

The lowest eigenstate of the helium atom is the state where both the electrons are in a $1s$ state. The customary notation for this atomic configuration is $1s^2$. The excited states are obtained by exciting one of the two electrons to a higher orbital.[2] These are atomic configurations such as $1s2s$, $1s2p$, $1s3s$, and so on. The excited states, where both electrons are in higher orbitals, are not ordinarily observed in atomic spectra. In Figure 12-3 we show the energy levels of the helium atom that have been obtained from the atomic spectrum. In this level diagram we can recognize the configurations that we mentioned. We note that the eigenstate $1s2s$ has a slightly lower energy than $1s2p$. We can understand this if we realize that the effective nuclear charge Z_{eff} should be a little larger for the $2s$ than for the $2p$ state because, in the $2s$ orbital, the electron is on the average closer to the nucleus. In helium an ns state has a lower energy than an np state, which has a lower energy than an nd state, and so on, even though in hydrogen all these states have the same energy.

If we proceed to larger atoms, then, according to the preceding argument their ground states would be obtained by placing all electrons in $1s$ orbitals since that gives the lowest energies. However, this is inconsistent with the experimental information. In order to resolve this contradiction, Pauli introduced in 1924 the exclusion principle, where he imposes the condition that in an atomic eigenstate there cannot be more than two electrons with the same set of quantum numbers n, l, and m.

---

[1] This statement should be qualified so that it is compatible with the exclusion principle, but we want to gloss over this shortcoming at this point.

[2] Orbital is defined as a one-electron wave function.

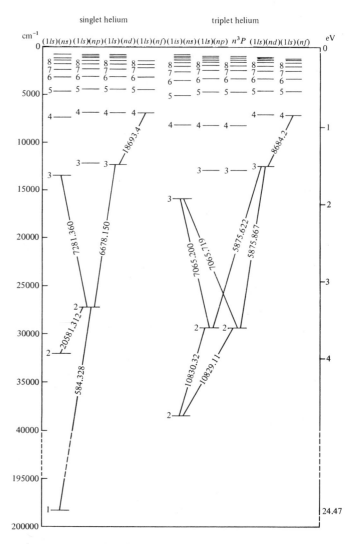

Figure 12-3 Energy level diagram of the helium atom.

According to the exclusion principle the ground state of lithium must have a configuration $1s^2\,2s$ and the ground state of beryllium has a configuration $1s^2\,2s^2$. In the $2p$ state the quantum number m can have three possible values, namely, $m = \pm 1, 0$. According to the exclusion principle we can place six electrons in the $2p$ orbitals before they are filled up. Hence, the ground state configuration of boron is $1s^2\,2s^2\,2p$, carbon is $1s^2\,2s^2\,2p^2$, and so on, until we reach neon, where we have $1s^2\,2s^2\,2p^6$, and the $2p$ orbitals are all filled.

We have listed all atomic configurations in Table 12-1, and it may be seen that with increasing atomic number we add an electron every time we proceed to the next element. The configuration of argon is $1s^2\,2s^2\,2p^6\,3s^2\,3p^6$, and it may be seen from the table that the next element, potassium, has the ground state configuration $1s^2\,2s^2\,2p^6\,3s^2\,3p^6\,4s$. It follows that in potassium the $4s$ orbital has a lower energy

Table 12-1 Atomic ground-state configurations

Atomic Number	Symbol	Configuration	Atomic Number	Symbol	Configuration
1	H	$1s$	42	Mo	$[Kr]+4d^5\,5s$
2	He	$1s^2$	43	Tc	$[Kr]+4d^5\,5s^2$
			44	Ru	$[Kr]+4d^7\,5s$
3	Li	$[He]+2s$	45	Rh	$[Kr]+4d^8\,5s$
4	Be	$[He]+2s^2$	46	Pd	$[Kr]+4d^{10}$
5	B	$[He]+2s^2\,2p$	47	Ag	$[Kr]+4d^{10}\,5s$
6	C	$[He]+2s^2\,2p^2$	48	Cd	$[Kr]+4d^{10}\,5s^2$
7	N	$[He]+2s^2\,2p^3$	49	In	$[Kr]+4d^{10}\,5s^2\,5p$
8	O	$[He]+2s^2\,2p^4$	50	Sn	$[Kr]+4d^{10}\,5s^2\,5p^2$
9	F	$[He]+2s^2\,2p^5$	51	Sb	$[Kr]+4d^{10}\,5s^2\,5p^3$
10	Ne	$[He]+2s^2\,2p^6$	52	Te	$[Kr]+4d^{10}\,5s^2\,5p^4$
11	Na	$[Ne]+3s$	53	I	$[Kr]+4d^{10}\,5s^2\,5p^5$
12	Mg	$[Ne]+3s^2$	54	Xe	$[Kr]+4d^{10}\,5s^2\,5p^6$
13	Al	$[Ne]+3s^2\,3p$	55	Cs	$[Xe]+6s$
14	Si	$[Ne]+3s^2\,3p^2$	56	Ba	$[Xe]+6s^2$
15	P	$[Ne]+3s^2\,3p^3$	57	La	$[Xe]+5d\,6s^2$
16	S	$[Ne]+3s^2\,3p^4$	58	Ce	$[Xe]+4f\,5d\,6s^2$
17	Cl	$[Ne]+3s^2\,3p^5$	...	...	...
18	Ar	$[Ne]+3s^2\,3p^6$			
19	K	$[Ar]+4s$	70	Yb	$[Xe]+4f^{13}\,5d\,6s^2$
20	Ca	$[Ar]+4s^2$	71	Lu	$[Xe]+4f^{14}\,5d\,6s^2$
21	Sc	$[Ar]+3d\,4s^2$	72	Hf	$[Xe]+4f^{14}\,5d^2\,6s^2$
22	Ti	$[Ar]+3d^2\,4s^2$	73	Ta	$[Xe]+4f^{14}\,5d^3\,6s^2$
23	V	$[Ar]+3d^3\,4s^2$	74	W	$[Xe]+4f^{14}\,5d^4\,6s^2$
24	Cr	$[Ar]+3d^5\,4s$	75	Re	$[Xe]+4f^{14}\,5d^5\,6s^2$
25	Mn	$[Ar]+3d^5\,4s^2$	76	Os	$[Xe]+4f^{14}\,5d^6\,6s^2$
26	Fe	$[Ar]+3d^6\,4s^2$	77	Ir	$[Xe]+4f^{14}\,5d^7\,6s^2$
27	Co	$[Ar]+3d^7\,4s^2$	78	Pt	$[Xe]+4f^{14}\,5d^9\,6s$
28	Ni	$[Ar]+3d^8\,4s^2$	79	Au	$[Xe]+4f^{14}\,5d^{10}\,6s$
29	Cu	$[Ar]+3d^{10}\,4s$	80	Hg	$[Xe]+4f^{14}\,5d^{10}\,6s^2$
30	Zn	$[Ar]+3d^{10}\,4s^2$	81	Tl	$[Xe]+4f^{14}\,5d^{10}\,6s^2\,6p$
31	Ga	$[Ar]+3d^{10}\,4s^2\,4p$	82	Pb	$[Xe]+4f^{14}\,5d^{10}\,6s^2\,6p^2$
32	Ge	$[Ar]+3d^{10}\,4s^2\,4p^2$	83	Bi	$[Xe]+4f^{14}\,5d^{10}\,6s^2\,6p^3$
33	As	$[Ar]+3d^{10}\,4s^2\,4p^3$	84	Po	$[Xe]+4f^{14}\,5d^{10}\,6s^2\,6p^4$
34	Se	$[Ar]+3d^{10}\,4s^2\,4p^4$	85	At	$[Xe]+4f^{14}\,5d^{10}\,6s^2\,6p^5$
35	Br	$[Ar]+3d^{10}\,4s^2\,4p^5$	86	Rn	$[Xe]+4f^{14}\,5d^{10}\,6s^2\,6p^6$
36	Kr	$[Ar]+3d^{10}\,4s^2\,4p^6$	87	Fr	$[Rn]+7s$
37	Rb	$[Kr]+5s$	88	Ra	$[Rn]+7s^2$
38	Sr	$[Kr]+5s^2$	89	Ac	$[Rn]+6d\,7s^2$
39	Y	$[Kr]+4d\,5s^2$	90	Th	$[Rn]+6d^2\,7s^2$
40	Zr	$[Kr]+4d^2\,5s^2$	91	Pa	$[Rn]+6d^3\,7s^2$
41	Nb	$[Kr]+4d^4\,5s$	92	U	$[Rn]+6d^4\,7s^2$

than the $3d$ orbital. Apparently the differences in energy between orbitals with different l values become larger with increasing quantum numbers, and in potassium these energy differences become larger than the differences in energy between states with different values of the quantum number n. In calcium the $4s$ orbital is filled and, then, moving from scandium to nickel, the $3d$ orbital is filled before we can fill up the $4p$ orbital. It follows from Table 12-1 that the order in which the orbitals are filled up is

$$1s \rightarrow 2s \rightarrow 2p \rightarrow 3s \rightarrow 3p \rightarrow 4s \rightarrow 3d \rightarrow 4p \rightarrow 5s \rightarrow 4d$$
$$\rightarrow 5p \rightarrow 6s \rightarrow 4f \rightarrow 5d \rightarrow 6p \rightarrow 7s \rightarrow 6d \qquad \textbf{(12-26)}$$

In general, a completely filled atomic orbital has relatively a lower energy, and for that reason there are some minor exceptions to the above rules. For example, in copper it is possible to fill the $5d$ orbital with ten electrons if we take one of the electrons out of the $4s$ orbital and place it in the $5d$ orbital, and it appears that this is the ground state configuration of copper.

The preceding formulation of the exclusion principle is not too accurate because it depends on the assumption that atomic configurations can be constructed from atomic orbitals. In addition it is limited to atoms only. A more rigorous formulation of the exclusion principle will be given in the next section, after we have discussed the electron spin.

12-2   THE ELECTRON SPIN

It is customary to explain the structure of an atom by comparing it with the solar system. The motion of the various electrons around the nucleus is then compared with the motion of the planets around the sun. We know that the earth describes an elliptic orbit around the sun with a time period of a year. In addition, the earth rotates around its axis, and it completes a full rotation in one day. The orbit of the electron around the nucleus may be compared with the yearly motion of the earth around the sun. Once we begin to think about all these analogies, we may wonder whether there is a motion of the electron that is comparable with the daily rotation of the earth around its axis. In fact, such a motion exists, and it is called the electron spin.

The concept of the electron spin was introduced in 1925 by Goudsmit and Uhlenbeck in order to explain the fine structure that occurs in some atomic spectra. This early theoretical description of the electron spin relied heavily on the vector model of the atom, and it contained some general assumptions about the properties of the spin. Very likely these assumptions were introduced mainly for the purpose of explaining the splittings in experimental atomic spectra. At the time some of these assumptions seemed somewhat arbitrary, but they were all explained satisfactorily in later years by making use of relativity theory.

Let us summarize the vector model for the orbital angular momentum $\mathbf{M}$ of an atom (see Figure 12-4). The magnitude of $\mathbf{M}$ is determined from a quantum number l that must have a nonnegative integer value. The possible projections of $\mathbf{M}$ in a given direction (say, the z axis) are $m\hbar$, where $m = 0, \pm 1, \pm 2, \pm, \ldots, \pm l$. In each stationary state $\mathbf{M}$ can be represented as a vector that rotates around the z axis so that its projection is constant, and the length of this vector is equal to $\hbar\sqrt{l(l+1)}$.

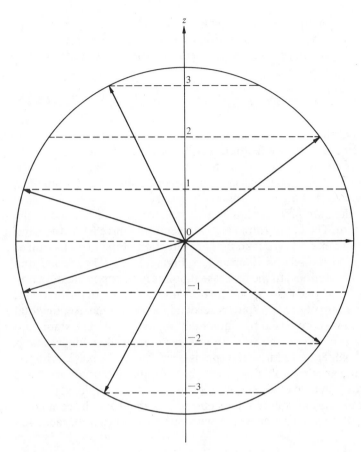

Figure 12-4 Quantization of the angular momentum vector **M** along the z axis for $l = 3$.

An electron that moves in a closed orbit can be considered as a ring current and, as such, it creates a magnetic field. The magnetic field is identical with the magnetic field due to a magnetic dipole **μ**. The ratio between the magnetic moment **μ** and the angular momentum **M** is easily derived from classical electromagnetic theory. It is given by

$$\boldsymbol{\mu} = -\frac{e}{2mc}\mathbf{M} \tag{12-27}$$

According to Goudsmit and Uhlenbeck the spin angular momentum **S** may be represented in a similar fashion by a vector. For the electron spin the quantum number s, which plays the same role as the quantum number l for orbital angular momentum, must then be taken equal to $\frac{1}{2}$. Consequently, the length S of the spin vector is given by

$$S = \hbar\sqrt{\tfrac{1}{2}(\tfrac{1}{2}+1)} = \tfrac{1}{2}\hbar\sqrt{3} \tag{12-28}$$

The projection of the spin vector along a given direction can have two possible values, namely $\pm\frac{1}{2}\hbar$. If we take the z axis as the axis of quantization, the possible eigenvalues of S_z are $\pm\frac{1}{2}\hbar$ and the possible values of the spin quantum number m_s are $\pm\frac{1}{2}$. The spinning motion of the electron gives rise to a magnetic moment $\boldsymbol{\mu}_s$ which is also related to the spin angular momentum $\mathbf{S}$. The relation between $\boldsymbol{\mu}_s$ and $\mathbf{S}$ is given by

$$\boldsymbol{\mu}_s = -\frac{e}{mc}\mathbf{S} \qquad (12\text{-}29)$$

It should be noted that this ratio is twice as large as in the case of orbital angular momentum (compare Eq. 12-27). The relation (12-29) was introduced because it led to a correct quantitative interpretation of the fine structure of atomic spectra. It is difficult to justify Eq. (12-29) from classical electromagnetic theory, but eventually it was shown to be correct by making use of relativity theory.

It may be useful to illustrate how the simple vector model can be applied to the theory of atomic structure. The helium atom has two electrons, and each of them has a spin with quantum number $s = \frac{1}{2}$. We take the spin direction of the first electron as the axis of quantization of the second electron (see Figure 12-5). The second spin then has two possible orientations. In the first case it points in the opposite direction of the first spin; and in the second case it points in the same direction. In the first case, the total angular momentum of the two spins is zero; and in the other case, the total spin angular momentum is represented by a quantum number $s = 1$. The stationary states of the helium atom, where the total spin is zero, are called singlet states because they are nondegenerate as far as the spin is concerned. The states with the total spin s equal to unity are threefold degenerate because the spin vector can have three possible projections in a given direction; they are known as triplet states.

Let us now consider the various states of the helium atom that have the configuration $1s2p$. Altogether there are 12 eigenstates with this configuration because the

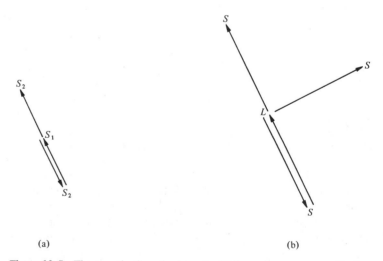

(a)   (b)

Figure 12-5   The quantization of spin and orbital angular momentum for the $(1s)(2p)$ configuration of helium. (a) Represents the quantization of the spin angular momentum. There are two cases with $\mathbf{S} = 1$ and $\mathbf{S} = 0$ respectively. The $\mathbf{S} = 1$ state combines with $\mathbf{L} = 1$ in (b) to give three states with total angular momentum $\mathbf{J} = 2$, 1, or 0 respectively.

($2p$) orbital is threefold degenerate and because there are four possible spin states. We have seen in the previous paragraph that the two electron spins can be combined to form either a nondegenerate singlet state with $s = 0$ or a threefold degenerate triplet state with $s = 1$. We may now use the vector model to derive the possible values of the total angular momentum **J** of the helium atom by combining the spin angular momentum **S** with the orbital angular momentum **L**. For a triplet ($2p$) state the quantum numbers l and s are each equal to unity. If we take the vector **L** as the axis of quantization for **S** then it is easily seen that the quantum number j can have the values 2, 1, or 0. These three stationary states belonging to the different j values have slightly different energies and they are fivefold degenerate, threefold degenerate, and nondegenerate, respectively. The singlet $1s2p$ state is threefold degenerate, and should have an energy that differs considerably from the energies of the triplet states. It may be seen in Figure 12-6 that these predictions agree with the experimental information.

The mathematical description of the electron spin can be derived by drawing an analogy with the angular momentum. We take the z axis as the axis of quantization; there are two possible stationary states of the spin. The first has a projection $\frac{1}{2}\hbar$, and the second has a projection $-\frac{1}{2}\hbar$ along the z axis. Both these stationary states are represented by an eigenfunction. We denote the eigenfunction of the state with projection $\frac{1}{2}\hbar$ by α and the eigenfunction of the state with projection $-\frac{1}{2}\hbar$ by β. Obviously, we must then have

$$S_z\alpha = \tfrac{1}{2}\hbar\alpha \qquad S_z\beta = -\tfrac{1}{2}\hbar\beta \tag{12-30}$$

where S_z is the operator that represents the z component of the electron spin S. The spin functions α and β are taken to be orthonormal; that means that they are normalized to unity and orthogonal to one another

$$\langle\alpha|\alpha\rangle = \langle\beta|\beta\rangle = 1 \qquad \langle\alpha|\beta\rangle = \langle\beta|\alpha\rangle = 0 \tag{12-31}$$

The two functions α and β must be eigenfunctions also of the operator S^2 and it follows from Eq. (12-28) and by drawing an analogy with the angular momentum that

$$S^2\alpha = \tfrac{1}{2}(\tfrac{1}{2}+1)\hbar^2\alpha = \tfrac{3}{4}\hbar^2\alpha$$
$$S^2\beta = \tfrac{1}{2}(\tfrac{1}{2}+1)\hbar^2\beta = \tfrac{3}{4}\hbar^2\beta \tag{12-32}$$

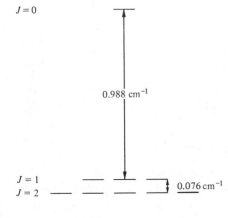

$J = 0$

0.988 cm^{-1}

$J = 1$
$J = 2$
0.076 cm^{-1}

Figure 12-6 Fine structure of the triplet ($1s$)($2p$) state of the helium atom.

The functions α and β are not eigenfunctions of the operators S_x and S_y. The effect of these two operators is described by the equations

$$S_x\alpha = \tfrac{1}{2}\hbar\beta \qquad S_y\alpha = \tfrac{1}{2}i\hbar\beta$$
$$S_x\beta = \tfrac{1}{2}\hbar\alpha \qquad S_y\beta = -\tfrac{1}{2}i\hbar\beta \tag{12-33}$$

If we consider the spin of an electron in addition to its orbital motion, we must write the electron eigenfunctions as

$$\psi_n(x, y, z; s) = \phi_n(x, y, z)\{C_1\alpha + C_2\beta\} \tag{12-34}$$

The two possible spin states of the electron are described by the functions α and β and, in the general case where we have no information as regard to the spin, we must write the spin part of the eigenfunction as an arbitrary linear combination of α and β. The probability of finding the spin parallel to the z axis is given by $C_1 C_1^*$, and the probability of finding an antiparallel spin is given by $C_2 C_2^*$. Obviously, the parameters must satisfy the normalization condition

$$C_1 C_1^* + C_2 C_2^* = 1 \tag{12-35}$$

Symbolically, we use the spin variable s to indicate the dependence of the eigenfunction on the electron spin.

Let us now proceed to the case of two electrons. We begin by considering the spin functions only. Since each of the two electron spins can have the possible projections $\pm\tfrac{1}{2}\hbar$ along the z axis, the four spin functions of a two-electron system are

$$\phi_1 = \alpha_1\alpha_2$$
$$\phi_2 = \beta_1\beta_2$$
$$\phi_3 = \alpha_1\beta_2 \tag{12-36}$$
$$\phi_4 = \beta_1\alpha_2$$

The total spin of the two electrons is represented by an operator $\mathbf{S}$, which is defined by

$$\mathbf{S} = \mathbf{S}_1 + \mathbf{S}_2 \tag{12-37}$$

or

$$S_x = S_{1x} + S_{2x}$$
$$S_y = S_{1y} + S_{2y} \tag{12-38}$$
$$S_z = S_{1z} + S_{2z}$$

The operator S^2 is defined as

$$S^2 = S_x^2 + S_y^2 + S_z^2 \tag{12-39}$$

It is now possible to construct a set of functions ζ from the four functions ϕ_k of Eq. (12-36) that are eigenfunctions of both operators S^2 and S_z. We write this as

$$(S^2)^1\zeta(1, 2) = 0$$
$$(S^2)^3\zeta_j(1, 2) = (2\hbar^2)^3\zeta_j(1, 2) \tag{12-40}$$

The first function $^1\zeta(1, 2)$ represents the situation where the total spin is zero; obviously this state is nondegenerate. This state is called a singlet state, and it is denoted by a superscript 1 on the left. The second function ${}^{3}\zeta_{j}$ represents the situation where the total spin quantum number s is unity. This state is threefold degenerate; it is called a triplet state and denoted by a superscript 3. We can prove that the singlet spin function is given by

$$
{}^{1}\zeta(1, 2) = \tfrac{1}{2}\sqrt{2}(\phi_{3} - \phi_{4}) = \tfrac{1}{2}\sqrt{2}(\alpha_{1}\beta_{2} - \beta_{1}\alpha_{2}) \tag{12-41}
$$

This result may be understood on the basis of the vector model shown in Figure 12-5a. If we quantize the second spin with respect to the first spin, we find that the total spin can be equal to zero when the two spins are antiparallel or equal to unity when they are parallel. The first case is the singlet state described by the function ${}^{1}\zeta(1, 2)$.

In the second case, where the total spin is unity, its possible projections along the z axis are $\hbar$, 0, and $-\hbar$. These three situations are described by the eigenfunctions ${}^{3}\zeta_{1}$, ${}^{3}\zeta_{0}$ and ${}^{3}\zeta_{-1}$, respectively. These functions must satisfy the equations

$$
\begin{aligned}
S_{z}\,{}^{3}\zeta_{1}(1, 2) &= \hbar\,{}^{3}\zeta_{1}(1, 2) \\
S_{z}\,{}^{3}\zeta_{0}(1, 2) &= 0 \\
S_{z}\,{}^{3}\zeta_{-1}(1, 2) &= -\hbar\,{}^{3}\zeta_{-1}(1, 2)
\end{aligned} \tag{12-42}
$$

Every one of these functions corresponds to the case where the total spin is unity so they must all satisfy the equation

$$
(S^{2})\,{}^{3}\zeta_{k}(1, 2) = 1(1+1)\hbar^{2}\,{}^{3}\zeta_{k}(1, 2) = (2\hbar^{2})\,{}^{3}\zeta_{k}(1, 2) \tag{12-43}
$$

It may be verified that the various spin functions are given by

$$
\begin{aligned}
{}^{3}\zeta_{1}(1, 2) &= \alpha_{1}\alpha_{2} \\
{}^{3}\zeta_{0}(1, 2) &= \tfrac{1}{2}\sqrt{2}(\alpha_{1}\beta_{2} + \beta_{1}\alpha_{2}) \\
{}^{3}\zeta_{-1}(1, 2) &= \beta_{1}\beta_{2}
\end{aligned} \tag{12-44}
$$

EXAMPLE 12-1

The easiest way to prove Eqs. (12-41) and (12-44) is first to derive the effect of the operator S_{z} on the four spin functions of Eq. (12.36). We have

$$
\begin{aligned}
S_{z}\phi_{1} &= (S_{1z} + S_{2z})\alpha_{1}\alpha_{2} = \hbar\alpha_{1}\alpha_{2} = \hbar\phi_{1} \\
S_{z}\phi_{2} &= (S_{1z} + S_{2z})\beta_{1}\beta_{2} = -\hbar\beta_{1}\beta_{2} = -\hbar\phi_{2} \\
S_{z}\phi_{3} &= (S_{1z} + S_{2z})\alpha_{1}\beta_{2} = 0 \\
S_{z}\phi_{4} &= (S_{1z} + S_{2z})\beta_{1}\alpha_{2} = 0
\end{aligned} \tag{a}
$$

It follows immediately that the function ϕ_{1} must be the function ${}^{3}\zeta_{1}$ since it describes the situation where the projection S_{z} is $\hbar$. The spin can only have a projection $\hbar$ if the total spin is at least unity, so ϕ_{1} must be a triplet function. By means of the same argument we conclude that ϕ_{2} must be the function ${}^{3}\zeta_{-1}$.

We cannot apply this argument to the functions ϕ_{3} and ϕ_{4}. The situation where the projection S_{z} is zero corresponds either to a total spin unity perpendicular to the z axis or to a total spin 0. We must therefore determine the effect of the operator S^{2} on these two functions. It follows from Eqs. (12-38) and (12-39) that S^{2} may also be written as

$$
S^{2} = S_{1}^{2} + S_{2}^{2} + 2(S_{1x}S_{2x} + S_{1y}S_{2y} + S_{1z}S_{2z}) \tag{b}
$$

By making use of Eqs. (12-30), (12-32), and (12-33) we find

$$S^2\phi_3 = S^2(\alpha_1\beta_2) = \hbar^2(\alpha_1\beta_2 + \beta_1\alpha_2) = \hbar^2(\phi_3 + \phi_4)$$
$$S^2\phi_4 = S^2(\beta_1\alpha_2) = \hbar^2(\alpha_1\beta_2 + \beta_1\alpha_2) = \hbar^2(\phi_3 + \phi_4)$$

(c)

Consequently

$$S^2(\phi_3 + \phi_4) = 2\hbar^2(\phi_3 + \phi_4)$$
$$S^2(\phi_3 - \phi_4) = 0$$

(d)

or

$$^1\zeta(1, 2) = \phi_3 - \phi_4 = \tfrac{1}{2}\sqrt{2}(\alpha_1\beta_2 - \beta_1\alpha_2)$$
$$^3\zeta_0(1, 2) = \phi_3 + \phi_4 = \tfrac{1}{2}\sqrt{2}(\alpha_1\beta_2 + \beta_1\alpha_2) \quad \bullet$$

(e)

12-3 EXCLUSION PRINCIPLE AND SPIN

Our formulation of the exclusion principle in Section 12-1 was not very precise because it was limited to atoms only and it was based on the assumption that each electron could be identified by means of a set of hydrogenlike quantum numbers. The introduction of the electron spin allows a much more precise formulation of the exclusion principle, namely, every eigenfunction of a many-electron system must satisfy the requirement that it is antisymmetric with respect to permutations of the electron spin and space coordinates. In order to discuss what this means we first consider a two-electron system.

A two-electron function is called symmetric with respect to permutations if it remains the same when we permute the two electrons; that is

$$\Psi_s(1, 2) = \Psi_s(2, 1) \tag{12-45}$$

If the function changes sign when we permute the two electrons, or

$$\Psi_a(1, 2) = -\Psi_a(2, 1) \tag{12-46}$$

then it is called antisymmetric.

EXAMPLE 12-2

Realize that an arbitrary two-electron function may be neither symmetric nor antisymmetric. For example, the function

$$\Psi(1, 2) = \exp(-r_1) \cdot z_2 \exp(-\tfrac{1}{2}r_2) \tag{a}$$

is nonsymmetric. However, it is possible to construct both a symmetric and an antisymmetric function, starting from Eq. (a). The symmetric function is

$$\Psi_s = \Psi(1, 2) + \Psi(2, 1) \tag{b}$$

and the antisymmetric function is

$$\Psi_a = \Psi(1, 2) - \Psi(2, 1) \quad \bullet \tag{c}$$

According to the exclusion principle the total wave function $\Psi(1, 2)$ of a two-electron system must be antisymmetric. It must therefore satisfy the condition

$$\Psi(1, 2) = -\Psi(2, 1) \tag{12-47}$$

Note that the function $\Psi(1, 2)$ must include both the space and spin coordinates of the two electrons.

We concluded in Section 12-2 that the most general wave function of a two-electron system can be written in the form

$$\Psi(1, 2) = {}^3\psi_1(\mathbf{r}_1, \mathbf{r}_2)\,{}^3\zeta_1(1, 2) + {}^3\psi_0(\mathbf{r}_1, \mathbf{r}_2)\,{}^3\zeta_0(1, 2)$$
$$+ {}^3\psi_{-1}(\mathbf{r}_1, \mathbf{r}_2)\,{}^3\zeta_{-1}(1, 2) + {}^1\psi(\mathbf{r}_1, \mathbf{r}_2)\,{}^1\zeta(1, 2) \tag{12-48}$$

namely as a linear combination of all possible spin states. If we permute the two electrons we obtain

$$\Psi(2, 1) = {}^3\psi_1(\mathbf{r}_2, \mathbf{r}_1)\,{}^3\zeta_1(2, 1) + {}^3\psi_0(\mathbf{r}_2, \mathbf{r}_1)\,{}^3\zeta_0(2, 1)$$
$$+ {}^3\psi_{-1}(\mathbf{r}_2, \mathbf{r}_1)\,{}^3\zeta_{-1}(2, 1) + {}^1\psi(\mathbf{r}_2, \mathbf{r}_1)\,{}^1\zeta(2, 1) \tag{12-49}$$

It follows from Eqs. (12-41) and (12-44) that the singlet spin function is antisymmetric and that the triplet spin functions are all symmetric

$$\begin{aligned} {}^1\zeta(1, 2) &= -{}^1\zeta(2, 1) \\ {}^3\zeta_k(1, 2) &= {}^3\zeta_k(2, 1) \end{aligned} \tag{12-50}$$

The wave function (12-48) satisfies the antisymmetry condition (12-47) only if

$$\begin{aligned} {}^3\psi_k(\mathbf{r}_1, \mathbf{r}_2) &= -{}^3\psi_k(\mathbf{r}_2, \mathbf{r}_1) \\ {}^1\psi(\mathbf{r}_1, \mathbf{r}_2) &= {}^1\psi(\mathbf{r}_2, \mathbf{r}_1) \end{aligned} \tag{12-51}$$

In other words, the space part of a triplet function is antisymmetric, and the space part of a singlet function is symmetric.

The orbital functions ${}^3\psi_k$ or ${}^1\psi$ must be eigenfunctions of the Schrödinger equation

$$\mathfrak{H}(1, 2)\psi(1, 2) = \varepsilon\psi(1, 2) \tag{12-52}$$

According to the exclusion principle the eigenfunction $\psi(1, 2)$ must be either symmetric or antisymmetric with respect to permutation of the two electrons. If it is symmetric, it is part of a singlet wave function; if it is antisymmetric, it is part of a triplet wave function.

The definition of antisymmetry with respect to permutations is a little more complicated for many-electron systems. Here we call a wave function antisymmetric with respect to permutations if it changes sign when we permute a pair of electron coordinates. Let us illustrate this definition for a three-electron system where the wave function is written as $\Psi(1, 2, 3)$. There are six possible ways in which we can rank-order three objects, and there must be five permuted functions in addition to $\Psi(1, 2, 3)$: namely, $\Psi(1, 3, 2)$, $\Psi(2, 1, 3)$, $\Psi(2, 3, 1)$, $\Psi(3, 1, 2)$, and $\Psi(3, 2, 1)$. Only three of these functions, $\Psi(1, 3, 2)$, $\Psi(2, 1, 3)$, and $\Psi(3, 2, 1)$, may be obtained from the original function by a pairwise exchange of electron coordinates. Hence, if the function $\Psi(1, 2, 3)$ is antisymmetric with respect to permutations it must satisfy the conditions

$$\Psi(1, 3, 2) = -\Psi(1, 2, 3)$$

$$\Psi(2, 1, 3) = -\Psi(1, 2, 3) \qquad \text{(12-53)}$$

$$\Psi(3, 2, 1) = -\Psi(1, 2, 3)$$

The remaining permuted functions, $\Psi(2, 3, 1)$ and $\Psi(3, 1, 2)$, may be obtained by a pairwise exchange of electron coordinates from one of the functions on the left hand side of Eq. (12-53). They must satisfy the conditions

$$\Psi(2, 3, 1) = -\Psi(2, 1, 3) = \Psi(1, 2, 3)$$

$$\Psi(3, 1, 2) = -\Psi(3, 2, 1) = \Psi(1, 2, 3) \qquad \text{(12-54)}$$

If the function $\Psi(1, 2, 3)$ is not antisymmetric to start with, it may be used as a basis for the construction of an antisymmetric function $\Psi_a(1, 2, 3)$, namely

$$\Psi_a(1, 2, 3) = \Psi(1, 2, 3) + \Psi(2, 3, 1) + \Psi(3, 1, 2)$$

$$-\Psi(1, 3, 2) - \Psi(2, 1, 3) - \Psi(3, 2, 1) \qquad \text{(12-55)}$$

It is easily verified that the function Ψ_a of Eq. (12-55) is antisymmetric with respect to permutations.

You can understand from the preceding considerations that it is useful to divide permutations into even and odd permutations. A permutation is defined as odd if it can be derived from the original sequence by means of an odd number of pairwise exchanges; similarly, a permutation is even if it is derived by means of an even number of pairwise exchanges. A permutation may be represented by a permutation operator, which is usually denoted by the symbol P. The symbol δ_P represents unity if P is even, and minus unity if P is odd. If we use this notation we can abbreviate Eq. (12-55) as

$$\Psi_a(1, 2, 3) = \sum_P P\delta_P \Psi(1, 2, 3) \qquad \text{(12-56)}$$

Eq. (12-56) is easily generalized to an N-electron system. Let $\Psi(1, 2, 3, \ldots, N)$ be an arbitrary function of N electron coordinates. Then the function

$$\Psi_a(1, 2, 3, \ldots, N) = \sum_P P\delta_P \Psi(1, 2, 3, \ldots, N) \qquad \text{(12-57)}$$

is antisymmetric with respect to permutation of the electrons.

Let us now consider the question of how many possible permutations of N electrons exist, or, the same question, in how many different ways we can rank-order N objects. We call this number S_N, and we have seen already that $S_1 = 1$ and $S_2 = 2$. If we add an object, the number S_3 is three times larger than S_2 because each permutation of two objects gives rise to three permutations of three objects

$$S_3 = 3 \cdot S_2 = 1 \cdot 2 \cdot 3 = 3! \qquad \text{(12-58)}$$

This is easily generalized to the relation

$$S_N = N \cdot S_{N-1} = 1 \cdot 2 \cdot 3 \cdots\cdots N = N! \qquad \text{(12-59)}$$

We have mentioned that it is not possible to derive the exact solutions of the Schrödinger equations fo. many-electron systems. It is necessary, therefore, to introduce approximations to represent atomic and molecular wave functions. In the most customary approximation, the atomic wave functions are derived from products of one-electron wave functions. These one-electron wave functions are known as orbitals. For example, in the helium atom we consider the situation where the two electrons are either in the same orbital, where the total wave function is derived from the product

$$\Psi = \phi(\mathbf{r}_1)\phi(\mathbf{r}_2), \tag{12-60}$$

or in different orbitals, where the total wave function is derived from the function

$$\Psi = \phi(\mathbf{r}_1)\chi(\mathbf{r}_2) \tag{12-61}$$

It is important to note that in the case of Eq. (12-60) the atomic wave function must be a singlet function

$$\Phi(1, 2) = \phi(\mathbf{r}_1)\phi(\mathbf{r}_2) \, {}^1\zeta(1, 2) \tag{12-62}$$

Because the orbital part of the wave function is symmetric with respect to permutations, the spin part must be antisymmetric. We find therefore that if we have two electrons in the same orbital they must have antiparallel spins. We can derive either a singlet or a triplet eigenfunction from the product of Eq. (12-61)

$$^1\Phi(1, 2) = (1/\sqrt{2})[\phi(\mathbf{r}_1)\chi(\mathbf{r}_2) + \chi(\mathbf{r}_1)\phi(\mathbf{r}_2)] \, {}^1\zeta(1, 2)$$

$$^3\Phi(1, 2) = (1/\sqrt{2})[\phi(\mathbf{r}_1)\chi(\mathbf{r}_2) - \chi(\mathbf{r}_1)\phi(\mathbf{r}_2)] \, {}^3\zeta(1, 2) \tag{12-63}$$

Let us now consider a four-electron system. According to the exclusion principle it is not allowed to have two electrons with identical spatial orbitals and spin functions. In that situation, namely the wave function must be zero because it must be antisymmetric with respect to permutation of these two electrons with identical space and spin functions and it can be antisymmetric only if the total wave function is zero. According to the exclusion principle there cannot be more than two electrons in the same spatial orbital and there can be two electrons in the same spatial orbital only if they have different spin functions (or antiparallel spins). If we assign an energy ε_i to every orbital χ_i we can have only two electrons in the orbital χ_1 with the lowest energy ε_1. The atomic ground state corresponds to the configuration with two electrons each in the orbitals χ_1 and χ_2 with the lowest energy ε_1 and the next lowest energy ε_2. The atomic eigenfunction is then obtained as

$$\Psi = \sum_P P\delta_P[\chi_1(\mathbf{r}_1)\alpha(1)\chi_1(\mathbf{r}_2)\beta(2)\chi_2(\mathbf{r}_3)\alpha(3)\chi_2(\mathbf{r}_4)\beta(4)] \tag{12-64}$$

This is called a closed-shell state. It should be noted that the total spin is zero because it consists of pairs of electrons that have antiparallel spins so that each such electron pair has spin zero, and the sum of these spins is also zero.

It may be seen that the early formulation of the exclusion principle discussed in Section 12-1 is a special case of the general definition. If we derive the total wave function from a product of one-electron orbitals, then we are allowed to have no more than two electrons in any one orbital. If there are two electrons in one orbital, they must have antiparallel spins. In atomic systems it is usually possible to identify the orbitals by means of hydrogenlike quantum numbers. If we do that, we are back at the original formulation of the exclusion principle as discussed in Section 12-1.

The lower excited states of the four-electron system may be derived from Eq. (12-64) by taking one of the electrons out of the orbital χ_2 and by placing it in an excited orbital χ_3. The configuration of this state is $(\chi_1)^2(\chi_2)(\chi_3)$. The two electrons in the orbitals χ_2 and χ_3 have either parallel or antiparallel spins. In the first case the total antisymmetric eigenfunction is

$$^3\Psi = \sum_P P\delta_P[\chi_1(\mathbf{r}_1)\alpha(1)\chi_1(\mathbf{r}_2)\beta(2)\chi_2(\mathbf{r}_3)\chi_3(\mathbf{r}_4)\,^3\zeta(3,4)] \tag{12-65}$$

The total spin of the system is unity, and we have a triplet state. If the two spins are antiparallel, the total wave function is

$$^1\Psi = (1/\sqrt{2})\sum_P P\delta_P[\chi_1(\mathbf{r}_1)\alpha(1)\chi_1(\mathbf{r}_2)\beta(2)\{\chi_2(\mathbf{r}_3)\chi_3(\mathbf{r}_4)$$
$$+\chi_3(\mathbf{r}_3)\chi_2(\mathbf{r}_4)\}\alpha(3)\beta(4)] \tag{12-66}$$

and we have a singlet state. Both wave functions of Eqs. (12-65) and (12-66) satisfy the exclusion principle because they are antisymmetric with respect to all possible electron permutations.

Systems with more than two electrons may be described in a similar manner to the four-electron systems. In writing the antisymmetrized total wave functions, it is advantageous to take the one-electron orbitals orthonormal

$$\langle \chi_i | \chi_j \rangle = \delta_{ij} \tag{12-67}$$

It is not strictly necessary that the preceding orthogonality condition be satisfied, but it may be shown that the nonorthogonal parts of the orbitals cancel out anyway in the antisymmetrized wave function and these parts do not contribute to the expectation values of Hermitian operators. For that reason we may just as well take the orbitals orthonormal because this simplifies the mathematics considerably. If the orbitals are orthonormal, it is easily shown that the eigenfunctions in Eqs. (12-64), (12-65), and (12-66) should all be multiplied by a factor $(1/\sqrt{24})$ in order to be normalized to unity. For an N-electron system the normalization factor is $(1/N!)^{1/2}$ because here there are $N!$ terms in the antisymmetrized wave function.

12-4   ATOMIC ORBITALS

It is an approximation to write an atomic wave function as a product (or an antisymmetrized product) of one-electron orbitals, but is a very useful method in the theory of atomic structure. The approximation is fairly satisfactory, and it gives us a good overall understanding of the structure of an atom. Besides, if we want to make use of more precise theoretical descriptions, the theory becomes so complicated that it is difficult to appreciate what the results mean.

Once we have made the approximation of writing an atomic wave function as a product of one-electron orbitals, we are still left with the problem of determining the specific form of these orbitals. Actually, we have already discussed some aspects of this problem in Section 12-1, where we considered the helium atom. It may be recalled that we derived an approximate expression for the atomic wave function. Let us summarize this discussion here, as an introduction to the theory for more complex atoms.

The Hamiltonian of the helium atom was given in Eq. (12-1). We found later that if we use the atomic units of length and energy a_0 and ε_0 that we define in Eqs. (12-17) and (12-20), then this Hamiltonian may be written in the simple form

$$\mathfrak{H} = -\tfrac{1}{2}\Delta_1 - \tfrac{1}{2}\Delta_2 - \frac{2}{r_1} - \frac{2}{r_2} + \frac{1}{r_{12}} \tag{12-68}$$

In the ground state of the helium atom the atomic configuration is $(1s)^2$. This means that both electrons are in the same $(1s)$ orbital and that the helium atom must have a singlet spin function. Consequently, the atomic eigenfunction of the helium ground state must have the form

$$\Psi_0 = s(r_1)s(r_2)[\alpha(1)\beta(2) - \beta(1)\alpha(2)](1/\sqrt{2}) \tag{12-69}$$

We can show that the orbitals $s(r_1)$ and $s(r_2)$ are functions only of the distance r_i between the electron and the nucleus. The orbitals do not depend on the angles because the total angular momentum of the helium atom is zero for the atomic ground state.

In Section 12-1 we made some additional approximations beyond the approximation of Eq. (12-69) and we finally obtained the orbital $s(r)$ in the form

$$s(r) = \exp(-Zr) \tag{12-70}$$

where Z is the effective nuclear charge acting on the electron. From the variation principle we may then derive the value of Z, which is 1.6875.

We must point out now that it is possible to derive a more precise form of the atomic orbitals than expression (12-70). In fact, once we have made the approximation (12-69) of expressing the atomic wave function in terms of atomic orbitals, it is possible to do the rest of the calculation exactly without making use of any additional approximations. The orbitals that are derived in this way are the most precise atomic orbitals that can be obtained. They are known as Hartree–Fock orbitals or SCF orbitals (self-consistent field), and they have been tabulated in numerical form for the majority of atomic ground states.

The general derivation of the Hartree–Fock method is fairly complicated, but it is possible to get a general understanding of the method by considering again the ground state of the helium atom.

We follow the same approach as in Section 12-1 where we considered one of the two electrons, electron 1, and where we tried to determine the effective potential field $V_{\text{eff}}(\mathbf{r})$ in which the electron moves. If we know this effective potential field then $s(r_1)$ should be the solution of the Schrödinger equation

$$[-\tfrac{1}{2}\Delta_1 - V_{\text{eff}}(\mathbf{r}_1)]s(r_1) = \varepsilon s(r_1) \tag{12-71}$$

To be more precise, $s(r_1)$ should be the eigenfunction belonging to the lowest eigenvalue of this equation.

In Section 12-1, we used an approximate expression for the effective potential V_{eff}, but there is no need to resort to approximations because there is an exact expression for the potential. We again consider the situation sketched in Figure 12-1, where electron 1 moves in the charge cloud of the second electron. The behavior of electron 2 is characterized by the orbital $s(r_2)$ so that its probability density is given by the expression

$$\rho(r_2) = [s(r_2)]^2 \tag{12-72}$$

The total interaction between electron 1 and the charge cloud of electron 2 is given by

$$V_{int}(r_1) = \int \frac{\rho(r_2)}{r_{12}} d\mathbf{r}_2 = \int \frac{[s(r_2)]^2}{r_{12}} d\mathbf{r}_2 \tag{12-73}$$

This is a function of the coordinate r_1 only because the charge cloud of the second electron is spherically symmetric. It follows now easily that the effective potential $V_{eff}(r_1)$ is the sum of the Coulomb attraction of the nucleus and the interaction term of Eq. (12-73),

$$V_{eff}(r_1) = -\frac{2}{r_1} + V_{int}(r_1) = -\frac{2}{r_1} + \int \frac{[s(r_2)]^2}{r_{12}} d\mathbf{r}_2 \tag{12-74}$$

This is an exact expression for the effective potential, and if we substitute it into the Schrödinger equation (12-71) for the atomic orbital $s(r_1)$ we can attempt to solve this equation. Unfortunately, there is one difficulty. The function $s(r)$ that we are trying to find occurs in the potential function V_{eff}, which we have to know in order to solve the equation. Due to this difficulty we cannot solve the equation in a straightforward manner. However, the equation can be solved by means of the self-consistent field method. Here, we start with an approximate form $s_0(r)$ of $s(r)$ and we substitute this into Eq. (12.74) for V_{eff}. Now we can solve the Schrödinger equation for this approximate expression for V_{eff}. The solution that we obtain in this way, $s_1(r)$, should be a better approximation than $s_0(r)$. And, if we substitute this new expression into Eq. (12-74), we should obtain a more accurate expression for V_{eff} so that we can solve the Schrödinger equation again. The new solution, $s_2(r)$, should be better than the previous one, and if we substitute it then we should obtain a better approximation $s_3(r)$. We keep repeating this procedure, which is known as an iterative method, until the function $s_n(r_1)$ that we obtain is identical with the function $s_{n-1}(r_1)$ that we substitute into V_{eff}. At that point we have obtained self-consistency of the equation, and the function $s_n(r_1)$ is the solution. The method is known either as the Hartree–Fock method, named after the two scientists who proposed it, or as the self-consistent field method or SCF method.

It is clear that the above method is fairly involved and laborious and that it does not lead to simple analytical expressions for the orbitals. On the other hand, it is feasible to do the calculation if we make use of an electronic computer. Hartree calculated the SCF orbitals for a large number of molecules long before electronic computers were even invented. He made use of analog computers. These are electronic devices that were specially designed to deal with one specific problem. These analog computers may be considered predecessors of the present-day electronic computers.

It should be realized that a one-dimensional differential equation, no matter how complicated, can usually be solved numerically if we use an analog computer or an electronic computer. The solutions that were derived by Hartree are all in numerical form. They come in the form of a table where the values of the orbital for various values of the variable are listed.

The Hartree–Fock method can also be applied to the excited state of the helium atom and to the ground and excited states of more complicated atoms. However, we must know the configuration of the state we are dealing with to know enough about the orbitals so that they can be written as functions of one variable only. For example, the $1s^2 2s^2$ state of beryllium can be studied in this way. Here we have to find two different orbitals for the $1s$ and $2s$ states, and each of the orbitals is a function of the variable r only. In the case of neon, the ground state configuration is $1s^2 2s^2 2p^6$. The various p orbitals are then written as

$$p_x = xp(r) \qquad p_y = yp(r) \qquad p_z = zp(r) \tag{12-75}$$

and we have to determine the three functions $s_1(r)$, $s_2(r)$, and $p(r)$, corresponding to the $1s$, $2s$, and $2p$ orbitals. In all these cases the Hartree–Fock orbitals are known, and they can be found in the literature in numerical form.

Even though the exact Hartree–Fock orbitals are available for most atoms, the form in which they are given is not too convenient. Also, it is not easy to relate a numerical table to the chemical properties of an atom. Consequently, there have been various attempts to construct simple analytical expressions for atomic orbitals, which would form reasonable approximations to the exact Hartree–Fock orbitals. The best known of these are the Slater orbitals, which were derived by making use of the concept of the effective nuclear charge Z_{eff}. In fact, the helium orbitals that we discussed in Section 12-1 are the Slater $1s$ orbitals for the helium atom. It may be recalled that we wrote the $1s$ orbital of the helium atom as

$$s(r) = e^{-Zr} \tag{12-76}$$

where Z is the effective nuclear charge. It follows from the variation principle that Z should be taken as $Z = 1.6875$.

We argued that each of the two electrons in the helium atom experiences the Coulomb attraction of the nucleus with charge 2, shielded by the charge cloud of the other electrons δ. In the case of helium we have

$$Z = 1.6875 \qquad \text{and} \qquad \delta = 0.3125 \tag{12-77}$$

and this value of δ represents the amount of shielding.

It is possible to perform similar calculations for all other atoms. For example, in the case of neon we may introduce a variational function of the form

$$\Psi = (10!)^{-1/2} \sum_P P\delta_P[(1s)_1\alpha_1(1s)_2\beta_2(2s)_3\alpha_3(2s)_4\beta_4(2p_x)_5\alpha_5$$

$$\times (2p_x)_6\beta_6(2p_y)_7\alpha_7(2p_y)_8\beta_8(2p_z)_9\alpha_9(2p_z)_{10}\beta_{10}] \tag{12-78}$$

where we introduce an effective nuclear charge Z_1 for the $(1s)$ electrons, a charge Z_2 for the $(2s)$ electrons, and a charge Z_3 for the $(2p)$ electrons. The values for the quantities Z_i are obtained, again, by minimizing the energy that is derived from the function Ψ. It turns out that the various effective nuclear charges for the electrons in different atoms may be estimated fairly accurately by making use of a set of simple semiempirical rules. These rules were first proposed by Slater and the resulting orbitals are called Slater orbitals.

In order to find the effective nuclear charge Z' for an electron in an atom we write it as

$$Z' = Z - \sigma \tag{12-79}$$

where Z is the charge of the nucleus and σ is the amount of shielding of the nuclear potential due to the other electrons in the atom. It is assumed now that the contributions to the shielding from the other electrons are additive and that it is possible to assign a constant amount of shielding to each type of electron. These amounts are determined from the following rules:

1. Nothing from any electron that has a principal quantum number n higher than the electron we are considering.
2. An amount 0.35 from each electron that has the same principal quantum number n as the electron we are considering; except that, when we consider a $1s$ electron, the contribution from the other $1s$ is 0.30.
3. An amount 0.85 from each electron whose principal quantum number n is 1 less than the quantum number of the electron that we are considering if the latter is an s or a p electron; an amount 1.00 from each electron whose principal quantum number is 1 less than the electron we are considering if the latter is in a d, f, or g state.
4. An amount 1.00 from each electron with a principal number that is 2 or more less than the quantum number of the electron we are considering.

According to the preceding rules, the effective nuclear charge for the $1s^2$ state in helium is 1.70; this is fairly close to the value $Z = 1.6875$ that follows from a variational calculation.

EXAMPLE 12-3

Let us illustrate the Slater rules by considering the $1s^2 \, 2s^2 \, 2p^2$ configuration of the carbon atom. The nuclear charge is 6 and the amount of shielding is 0.30 for each $1s$ electron and $2 \times 0.85 + 3 \times 0.35 = 2.75$ for the $2s$ and $2p$ electrons. It follows that the effective nuclear charge is 5.70 for the $1s$ electrons and 3.25 for the other electrons.●

There are some differences between the Slater orbitals and the hydrogen atom eigenfunctions, and these are related to the orthogonality properties of these functions. We saw in Chapter 11 that all the hydrogen atom orbitals are orthogonal to each other, but we should realize that in the hydrogen atom the nuclear charge is the same for each eigenstate. If, for example, we substitute different nuclear charges Z_1 and Z_2 in the $1s$ and $2s$ eigenfunctions, these functions are no longer orthogonal. Since the orthogonality is destroyed anyway the Slater $2s$ orbital is defined as

$$\Phi(2s) = r \exp\left(-\tfrac{1}{2}Z_2 r\right) \tag{12-80}$$

which is different from the hydrogen-type $2s$ eigenfunction

$$\Psi(2s) = (Z_2 r - 2) \exp\left(-\tfrac{1}{2}Z_2 r\right) \tag{12-81}$$

In performing calculations we should impose the condition that the Slater $2s$ orbital is orthogonal to the Slater $1s$ orbital

$$\Phi(1s) = \exp(-Z_1 r) \tag{12-82}$$

This is achieved by replacing $\Phi(2s)$ of Eq. (12-80) by the function

$$\Phi'(2s) = \Phi(2s) - a\Phi(1s) \tag{12-83}$$

where a is chosen in such a way that

$$\langle \Phi'(2s)|\Phi(1s)\rangle = 0 \tag{12-84}$$

Accordingly, the Slater ns orbitals are defined as

$$\Phi(ns) = r^{n-1} \exp(-Z_n r/n) \tag{12-85}$$

and the Slater np orbitals are defined as

$$\Phi(np_x) = xr^{n-2} \exp(-Z_n r/n) \tag{12-86}$$

and so on.

The Slater orbitals are very useful for quick, order-of-magnitude calculations. They can also be used as starting orbitals for the Hartree–Fock procedures. These Slater orbitals give a fairly good approximation of the overall charge density of an atom, and they are used for molecular calculations also.

_____ *Problems*

1. Use the vector model of the atom to discuss the various states and their degeneracies that may be derived from the $1s3d$ configuration of the helium atom.
2. Consider the configuration $1s^2 2s^2 2p^2$ of the carbon atom. There are a number of different atomic states that have this configuration. By using the vector model, determine the possible values of the orbital angular momentum $\mathbf{M}$ that these states can have and derive the possible values of the spin angular momentum $\mathbf{S}$ for each of these states. Finally, give the possible values of the total angular momentum.
3. Determine the Slater orbitals for the $1s$ electrons in the helium, oxygen, and iodine atoms.
4. We have seen that the energy of a $1s$ electron is approximately equal to $-\frac{1}{2}Z_{\text{eff}}^2 \varepsilon_0$. Relativistic effects may be neglected if the energy of an electron is small with respect to mc^2, where m is the mass of the electron and c is the velocity of light. Investigate whether or not relativistic effects may be neglected for the $1s$ electron in helium, oxygen, and iodine. Use the values of Z_{eff} that were derived in Problem 3.
5. We know that for one electron the eigenfunctions of the spin operator S_z are the functions α and β. Derive the eigenfunctions of the operator S_x.
6. Show explicitly that the triplet spin function $\alpha_1 \alpha_2$ is an eigenfunction of the operator S^2 and determine the eigenvalue.

7. Show explicitly that $S^2 \alpha_1 \beta_2 = \hbar^2(\alpha_1 \beta_2 + \beta_1 \alpha_2)$.

8. Write down all possible 24 permutations of four numbers 1, 2, 3, 4. Determine which of these permutations are even and which are odd.

9. Construct an antisymmetric function from the two orbitals $\phi(1s) = e^{-pr}$ and $\phi(2s) = r e^{-qr}$. (Do not include the spins.) Now construct an antisymmetric function from the two orbitals $\psi(1s) = e^{-pr}$ and $\psi(2s) = r e^{-qr} - \lambda e^{-pr}$. Show that the two antisymmetric functions are the same.

10. Derive the Slater 1s and 2s orbitals for the 1s2s configuration of the helium atom. Normalize these two functions and calculate their overlap integral $\langle 1s | 2s \rangle$. *Hint:* $\int_0^\infty r^n e^{-\alpha r} dr = (n!/\alpha^{n+1})$.

11. Determine the Slater orbitals for the configuration $1s^2 2s^2$ of beryllium.

12. Derive the Slater orbitals for the ground state configuration $1s^2 2s^2 2p^6 3s$ of sodium and for the ground state configuration $1s^2 2s^2 2p^6 3s^2 3p^6 4s$ of potassium.

13. Write the complete antisymmetrized eigenfunctions, including spin, for the various eigenstates of the lithium atom that have the configuration $1s^2 2s$.

14. What are the possible values of the orbital angular momentum in the $1s^2 2s 2p^3$ configuration of the carbon atom.

15. Determine the Slater coefficients of the various orbitals in the $1s^2 2s^2 2p^6$ configuration of the Na^+ ion. Determine the expectation value of the coordinate r with respect to the $2p$ orbital. *Hint:* $\int_{0.}^\infty r^n e^{-\alpha r} = (n! \alpha^{n+1})$.

16. Prove explicitly that inside the charged sphere of Figure 12-2 the field strength is zero.

_____ *References and Recommended Reading*

The best-known books on atomic spectra and atomic structure are the book by Herzberg [1] and the book by Condon and Shortley [2]. Herzberg gives an excellent account of atomic spectra, based on physical arguments, and it is easy to read because it avoids complicated mathematics. Condon and Shortley's book is written on a more sophisticated level, but is the best-known advanced treatment of atomic structure. We also list some other books on the subject.

[1] G. Herzberg: *Atomic Spectra and Atomic Structure*. Prentice-Hall, Englewood Cliffs, N.J., 1937.

[2] E. U. Condon and G. H. Shortley: *The Theory of Atomic Spectra*. Cambridge University Press, London, 1935.

[3] H. G. Kuhn: *Atomic Spectra*. Academic Press, New York, 1962.

[4] H. E. White: *Introduction to Atomic Spectra*. McGraw-Hill, New York, 1934.

chapter 13

Light and Spectroscopy

13-1 THE NATURE OF LIGHT

Scientists and philosophers have expressed their ideas on the nature of light since the time of Aristotle, who lived during the fourth century before Christ. The various theories on the nature of light make fascinating reading. We should realize that some of the early descriptions of light were consistent with what was known at that time, even though later developments showed that they were either partially or totally wrong. Also, a comprehensive theory had to explain more and more different features, such as the various colors, the diffraction of light through a prism, and, at a later time, interference phenomena, the existence of polarized light, optical activity, and so on.

For example, during the seventeenth century Newton described a beam of light as a stream of fast-moving particles of various kinds where each color is represented by a particular kind of particle. At the same time the Dutch scientist Huygens proposed an ondulatory description of light, and he showed how his description was also consistent with straight-line propagation. At the time it was hard to say which of the two theories was right, but the various interference experiments on light during the beginning of the nineteenth century showed that Huygens' ideas were preferable to Newton's.

Nowadays the established view point is that light is a form of electromagnetic radiation. All types of electromagnetic radiation are mathematically described by the four Maxwell equations; to be exact, the radiation field is the solution of these equations in the vacuum. We will just outline the results here without going into the mathematical derivations.

The properties of any electromagnetic field are described by two vector quantities, the electric field strength $\mathbf{E}$ and the magnetic field strength $\mathbf{H}$. These two field strengths are usually related to one another, and the relation is described by the Maxwell equations.

A radiation field is a superposition of plane waves. We have sketched such a plane wave in Figure 13-1. For simplicity, we have taken the z axis as the direction of propagation of the wave. Then at each point on the z axis there is an oscillating electric field vector in the y direction; the magnitude is given by

$$E_y = A \cos\left[\frac{2\pi}{\lambda}(z - ct)\right] \tag{13-1}$$

At the same time there is an oscillating magnetic field vector in the x direction, with a magnitude

$$H_x = A \cos\left[\frac{2\pi}{\lambda}(z - ct)\right] \tag{13-2}$$

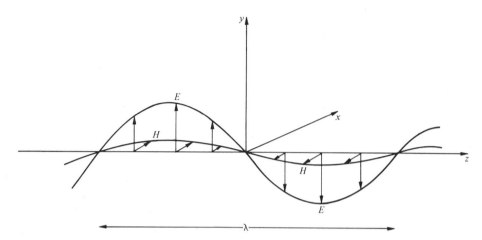

Figure 13-1   Representation of a plane electromagnetic wave.

In Eqs. (13-1) and (13-2) c is the velocity with which the waves propagate in the z direction, t is time, and λ is the wavelength of the wave. It follows from Maxwell's equations that in vacuum all electromagnetic waves have the same velocity, c, which is equal to the velocity of light.

We know now that a variety of seemingly different kinds of radiation are all electromagnetic radiation. The wavelength λ determines the kind of radiation. For example, the waves that we receive in an ordinary AM radio have wavelengths that vary between 100 and 500 m. Shortwave radio receivers are capable of receiving transatlantic signals with wavelengths varying from 5 to 100 m. Radar waves have wavelengths of the order of 1 cm. The wavelengths of visible light vary between 4000 and 7000 Å; ultraviolet light may be taken as having a wavelength between 1000 and 4000 Å; and infrared light between 7000 and 10,000 Å. (These figures are not precise, but all we want to do is to give a rough idea.) If we proceed to shorter wavelengths we find X rays with a wavelength of the order of 1 Å and, beyond that, γ rays which have a wavelength of 0.01 Å and lower.

Let us now return to Eqs. (13-1) and (13-2). It is well known that the velocity of a wave, its wavelength λ, and its frequency ν are related to one another as follows:

$$c = \lambda \nu \qquad\qquad (13\text{-}3)$$

The frequency ν is defined as the inverse of the period T of the wave; T is the amount of time it takes to complete one oscillation.

$$\nu = \frac{1}{T} = \frac{c}{\lambda} \qquad\qquad (13\text{-}4)$$

Similarly, the wave number σ is defined as the inverse of the wave length λ

$$\sigma = 1/\lambda \qquad\qquad (13\text{-}5)$$

By making use of the above relations we can rewrite Eqs. (13-1) and (13-2) as

$$E_y = A \cos\left[2\pi(\sigma z - \nu t)\right]$$
$$H_x = A \cos\left[2\pi(\sigma z - \nu t)\right] \qquad\qquad (13\text{-}6)$$

We should realize that an electromagnetic wave is a plane wave in three dimensions. In Figure 13-1 we have simplified the situation to make it a little easier to understand, but in reality the electric and magnetic field strengths that are given by Eq. (13-6) are oscillating in every point in space. We see that the functions in Eq. (13-6) depend only on the variable z. This means that for a constant z the oscillations are in phase. The direction of propagation is perpendicular to the planes of constant phase and it is therefore along the z axis.

By rotating the coordinate system we can transform the plane waves of Eq. (13-6) into waves with arbitrary directions of propagation. The general expression for such waves is

$$E_\alpha = E_0 \cos\left[2\pi(\sigma_x x + \sigma_y y + \sigma_z z - vt)\right]$$
$$H_\beta = H_0 \cos\left[2\pi(\sigma_x x + \sigma_y y + \sigma_z z - vt)\right] \tag{13-7}$$

The planes of constant phase are found from the condition that E_α and H_β must be constant in such a phase; hence, these planes are determined by the condition

$$\sigma_x x + \sigma_y y + \sigma_z z = \text{constant} \tag{13-8}$$

The vector

$$\boldsymbol{\sigma} = (\sigma_x, \sigma_y, \sigma_z) \tag{13-9}$$

is perpendicular to all planes of equal phase, consequently $\boldsymbol{\sigma}$ represents the direction of propagation of the wave. The magnitude of $\boldsymbol{\sigma}$ is still given by Eq. (13-5) because it is not affected by a rotation of the coordinate system. Hence

$$\sigma = (\sigma_x^2 + \sigma_y^2 + \sigma_z^2)^{1/2} = 1/\lambda = v/c \tag{13-10}$$

The vectors $\mathbf{E}$ and $\mathbf{H}$ must be perpendicular to each other and both of them must be perpendicular to $\boldsymbol{\sigma}$. Hence E_α and H_β are the components of the electric and the magnetic field strengths in the directions perpendicular to $\boldsymbol{\sigma}$.

The direction of polarization of the light coincides with the direction of the vector $\mathbf{E}$. Eq. (13-6) represents a light wave that is polarized in the y direction. It is easily seen that there must be a second light wave of the type of Eq. (13-6) that is polarized in the x direction. It is described by the equations

$$\mathbf{E} = (0, E_x, 0) \qquad E_x = A \cos\left[2\pi(\sigma z - vt)\right]$$
$$\mathbf{H} = (H_y, 0, 0) \qquad H_y = A \cos\left[2\pi(\sigma z - vt)\right] \tag{13-11}$$

This wave also propagates in the z direction because the vector $\boldsymbol{\sigma} = (0, 0, \sigma)$ points in the z direction. The vector $\mathbf{E}$ may have any arbitrary direction in the xy plane (it must be perpendicular to $\boldsymbol{\sigma}$), but any vector $\mathbf{E}_\alpha$ in the xy plane may be written as a linear combination of a vector in the x direction and a vector in the y direction. Therefore, there are two possible directions of polarization for a light wave once $\boldsymbol{\sigma}$ is specified.

In the general case, where $\boldsymbol{\sigma}$ has an arbitrary direction, it is more difficult to give a unique definition of the two basic directions of polarization. In principle we may define two directions perpendicular to $\boldsymbol{\sigma}$ and take these as the basis for defining the possible values of $\mathbf{E}$. We see then that a light wave is determined by the value of $\boldsymbol{\sigma}$, which determines the direction of propagation, and by an additional symbol that tells us which of the two possible directions of polarization the light wave has. The general plane waves have the form

$$\mathbf{E}_1(\boldsymbol{\sigma}) = \mathbf{A}_1 \cos\left[2\pi(\sigma_x x + \sigma_y y + \sigma_z z - \nu t)\right]$$

$$\mathbf{H}_1(\boldsymbol{\sigma}) = \mathbf{A}'_1 \cos\left[2\pi(\sigma_x x + \sigma_y y + \sigma_z z - \nu t)\right]$$

$$\mathbf{E}_2(\boldsymbol{\sigma}) = \mathbf{A}_2 \cos\left[2\pi(\sigma_x x + \sigma_y y + \sigma_z z - \nu t)\right] \qquad \textbf{(13-12)}$$

$$\mathbf{H}_2(\boldsymbol{\sigma}) = \mathbf{A}'_2 \cos\left[2\pi(\sigma_x x + \sigma_y y + \sigma_z z - \nu t)\right]$$

with

$$\mathbf{A}_1 \perp \boldsymbol{\sigma} \qquad \mathbf{A}_2 \perp \boldsymbol{\sigma} \qquad \mathbf{A}_1 \perp \mathbf{A}_2 \qquad\qquad \textbf{(13-13)}$$

The radiation field is always a superposition of plane waves of the type (13-12) because it is not possible to produce a radiation field that consists of only one wave. The simplest case that we can have is a beam of polarized light. Here the direction of propagation $\boldsymbol{\sigma}$ must be the same for all components. We can write the corresponding radiation field as

$$E_x = \sum_j A_j \cos\left[2\pi(\sigma_j z - \nu_j t)\right]$$

$$\qquad\qquad\qquad\qquad\qquad\qquad\qquad\qquad \textbf{(13-14)}$$

$$H_y = \sum_j A'_j \cos\left[2\pi(\sigma_j z - \nu_j t)\right]$$

if we take the vector $\boldsymbol{\sigma}$ along the z axis. Each component plane wave is determined by the value of σ, since the possible values of σ form a continuum the most general superposition takes the form of an integral

$$E_x = \int A(\sigma) \cos\left[2\pi(\sigma z - \nu t)\right] d\sigma$$

$$\qquad\qquad\qquad\qquad\qquad\qquad\qquad \textbf{(13-15)}$$

$$H_y = \int A'(\sigma) \cos\left[2\pi(\sigma z - \nu t)\right] d\sigma$$

Realize that it is not possible to have a purely monochromatic beam of light. At best we can have the situation where the function $A(\sigma)$ has a sharp maximum around a given value σ, but this maximum must have a finite line width. Hence, the term monochromatic is somewhat misleading because purely monochromatic light cannot be produced.

The most general expression for the radiation field is a superposition of all possible plane waves of the type (13-12). It is given by

$$\mathbf{E} = \int \left[\mathbf{A}_1(\boldsymbol{\sigma}) + \mathbf{A}_2(\boldsymbol{\sigma})\right] \cos\left[2\pi(\boldsymbol{\sigma} \cdot \mathbf{r} - \nu t)\right] d\boldsymbol{\sigma}$$

$$\qquad\qquad\qquad\qquad\qquad\qquad\qquad \textbf{(13-16)}$$

$$\mathbf{H} = \int \left[\mathbf{A}'_1(\boldsymbol{\sigma}) + \mathbf{A}'_2(\boldsymbol{\sigma})\right] \cos\left[2\pi(\boldsymbol{\sigma} \cdot \mathbf{r} - \nu t)\right] d\boldsymbol{\sigma}$$

This represents a radiation field that contains light waves with all possible frequencies, directions of propagation, and directions of polarization.

We stated at the beginning of this section that light is a form of electromagnetic radiation. Its mathematical description may be derived from the Maxwell equations and leads to Eq. (13-16). We should mention some of the more recent developments in the theory of the radiation field, such as Planck's description of black body radiation and Einstein's theory of the photoelectric effect, because these two theories played a very important role in the discovery of quantum mechanics. Ultimately, their approach led to the quantum mechanical description of the radiation field, which is known as quantum electrodynamics. We are somewhat hesitant to discuss this description of the radiation field because, at first sight, some of the basic concepts of this model seem to be inconsistent with what we discussed previously. However, a more careful analysis shows that the two theoretical models are in accord with one another.

Let us first consider Planck's work on black body radiation. This type of radiation can be observed by anyone who has an electric heater or an electric kitchen range. If the heating coils are turned on they begin to have a reddish glow when they warm up, and as they get hotter the color becomes more intense and changes over to yellow and white. We can measure the spectral distribution of this glow, that is, the intensity of the light as a function of the frequency. If we were to measure the radiation from the kitchen stove, we should observe the sum of the light that is emitted by the stove and of the light that is reflected. By definition, a black body is an object that does not reflect any light at all, and black body radiation is observed experimentally by eliminating all the reflected light. In practice, a black body is a hollow sphere with a little hole in it. The geometrical arrangement is such that the reflected light is dissipated within the sphere and the black body radiation is measured as the light that comes out of the hole when the sphere is heated up. Black body radiation depends on the temperature but it is independent of the material of which the sphere is made.

The properties of black body radiation were studied by measuring the spectral distribution of the light at various temperatures. Here the light is passed through a spectrograph and its intensity is plotted versus the wavelength. If we repeat this procedure at various temperatures, we obtain a set of curves. We have sketched two of these curves in Figure 13-2, where the temperature T_2 is supposed to be larger than T_1. We see that both curves in Figure 13-2 have a maximum, and if we look at the light we will see the color that corresponds to the wavelength of this maximum. When the temperature T increases, the total intensity increases (the light becomes brighter) and at the same time the maximum shifts to the left. This is why we see the color change from red to yellow to white as the temperature of the radiating body goes up.

We see that the black body radiation is a function of both the wavelength λ of the light and of the temperature T. After the experimental curves became available, several attempts were made to find analytical expressions that would represent the experimental results. Planck finally showed that the correct expression, which is consistent with all experimental data, is given by

$$\rho(v, T) = \frac{8\pi h v^3}{c^3} \frac{1}{e^{hv/kT} - 1} \tag{13-17}$$

Here $\rho(v, T)$ is the energy density of the radiation per unit volume and per unit frequency. This means that the energy of the radiation with frequencies between v and $v + dv$ is given by

$$dE = \rho(v, T) \, dv \tag{13-18}$$

Planck showed that the radiation formula (13-17) could be derived only if it is assumed that the radiation is quantized.

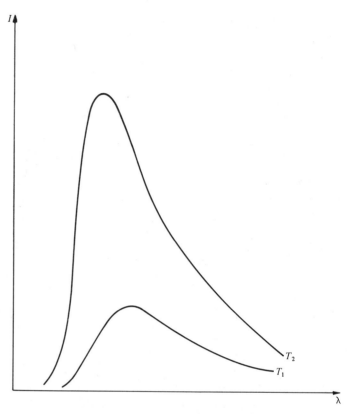

Figure 13-2 Two typical spectral distribution curves for black body radiation at different temperatures, T_2 is larger than T_1.

In order to understand this we should realize that in Planck's model, just as in the classical model, it is first assumed that the radiation field is a superposition of plane waves of the type that we described by Eq. (13-12). The total energy of the radiation field E may then be written as the sum of the partial energies E_λ of the various components,

$$E = \sum_\lambda E_\lambda \qquad (13\text{-}19)$$

Here we use the subscript λ to label the various plane waves; λ is a shorthand notation for the value of the vector $\boldsymbol{\sigma}$ and for the direction of polarization. Planck's quantization rule requires that each energy E_λ can have only certain discrete energy values $E_{n\lambda}$, which are given by

$$E_{n\lambda} = n_\lambda h\nu_\lambda \qquad (13\text{-}20)$$

where ν_λ is the frequency of the plane wave λ and n_λ must be an integer

$$n_\lambda = 0, 1, 2, 3, \ldots \qquad (13\text{-}21)$$

According to Planck's quantization rule the total energy of the radiation field is then given by

$$E = \sum_\lambda E_{n\lambda} = \sum_\lambda n_\lambda h\nu_\lambda \qquad (13\text{-}22)$$

The energy E is thus determined by the infinite set of quantum numbers $(n_1, n_2, \ldots, n_\lambda, \ldots)$ corresponding to the various plane waves λ that constitute the radiation field.

It is important to note that Planck's quantization rule refers only to the energy of the radiation field. It assumes that each component λ of the radiation field has an energy E_λ, which is composed of energy parcels $h\nu_\lambda$ that are indivisible in the same way a chemical substance is composed of indivisible atoms. Much later the term "photon" was introduced by G. N. Lewis, and today it is commonly used for describing these energy packages.

We mentioned before that the quantum theory of the radiation field has certain features that might be confusing. What we had in mind is that, in this approach, photons are often treated as particles. It is then said that the radiation field consists of a large number of particles, the photons, and that its total energy is the sum of the energies of all photons that are present. What we should realize is that this terminology is permissible as long as we consider the energy only. The energy of the radiation field is given by Eq. (13-22) as the sum of the photon energies, and in this equation it does not make any difference whether we think of the photons as energy parcels or as particles. It is often convenient to visualize the photons as particles, but whatever we like to think does not alter the fact that the radiation field is a superposition of electromagnetic plane waves.

Einstein made use of Planck's quantization rules and the photon concept to explain certain features of the photoelectric effect that could not be understood on the basis of the classical electromagnetic theory. The photoelectric effect is measured by placing a piece of metal in a beam of light. The light ejects electrons out of the metal, and these electrons are attracted by an electrode which is placed a little distance away and has a different potential from the metal. The number of ejected electrons is derived from measuring the electron current between the metal and the electrode. A surprising feature of the photoelectric effect is that no current is observed if the frequency of the light is below a certain frequency ν_0, no matter how intense the light is.

Einstein offered a simple explanation of the photoelectric effect. Let W be the potential barrier that the electron must pass in order to be able to leave the metal. An electron can be ejected only if it collides with a photon and absorbs the photon energy. It follows, then, that the photon must have an energy of at least eW in order to eject an electron. Therefore, if the frequency of the light is below a frequency ν_0, which is given by

$$h\nu_0 = eW \tag{13-23}$$

the photoelectric effect is not observed. If the light has a frequency ν that is higher than ν_0, the electrons are ejected and their kinetic energies are given by

$$\tfrac{1}{2}mv^2 = h(\nu - \nu_0) \tag{13-24}$$

From this relation Einstein calculated photoelectric currents, and he obtained satisfactory agreement with the experimental results.

Planck's work on radiation is important from a historical point of view because it was the first time that the concept of quantization was introduced in theoretical physics. The validity of Planck's assumptions was strongly supported by Einstein's work on the photoelectric effect and by subsequent studies on the specific heat in solids by Debye. No doubt it helped Bohr in formulating his quantum theory of atomic structure and, thus, led ultimately to the formulation of quantum mechanics as we know it today.

13-2 TRANSITION PROBABILITIES

In atomic and molecular spectroscopy we measure the intensity distribution of the light that comes out of the sample that we wish to study. If we wish to measure an emission spectrum, we must treat the sample in such a way that a large fraction of atoms or molecules are produced in an excited state. A common procedure is to place a gaseous, low pressure sample in a high voltage electric discharge. When atoms return to their ground states they emit light, and we measure its spectral distribution with a photographic plate or by other means. This is called an emission spectrum. In order to measure an absorption spectrum, we must have a light source that has a continuous intensity distribution over a sufficiently large range. We let the light pass through our sample, and we measure the intensity distribution of the light that has passed through the sample. In phosphorescence or fluorescence spectra the experimental arrangement is the same as in absorption spectra, but here the light is first absorbed by the sample and then reemitted. The frequency of the emitted light may be different from the frequency of the light that has been absorbed.

The simplest way to measure the intensity distribution of a beam of light is to let it pass through a prism and then focus it on a photographic plate. However, this is fairly time-consuming, especially if we wish our results to be reasonably accurate. In more sophisticated and more rapid experiments, we make use of a monochromator. This is an arrangement of prisms, lenses, and mirrors, through which only light within a narrow frequency range can pass. At the exit of the monochromator, there is a photomultiplier cell with an amplifier. In this way we can immediately measure the intensity of the light that comes out of the monochromator. By turning the prisms we can vary the frequency of the light which the monochromator lets through and thus we can cover the whole frequency range of the spectrum that we wish to measure. Nowadays the apparatus is often automated, and the absorption or emission spectrum that we wish to measure is recorded on tracing paper.

Einstein gave a formal analysis of the various absorption and emission processes that can occur. In order to understand this analysis we first consider an absorption process where a molecule is excited from a lower state k to a higher state l by absorbing light of frequency ν_{kl}. Because of the conservation of energy, we must have

$$h\nu_{kl} = E_l - E_k \qquad (13\text{-}25)$$

In Einstein's description the absorption process is represented by a quantity $B_{k \to l}$ which is known as the Einstein transition probability of absorption. In order to define it we consider a sample of molecules, and we assume that the sample contains N_k molecules in the lower state k. The number of transitions per unit time $N_{k \to l}$ from a lower state k to a higher state l is then given by

$$N_{k \to l} = N_k B_{k \to l} \rho(\nu_{kl}) \qquad (13\text{-}26)$$

where $\rho(\nu)$ is the energy of the radiation field.

Let us now consider the emission of light. Einstein was the first to point out that in this case we must consider two processes instead of the single process that we have in the case of absorption. The first emission process is described by a quantity $B_{l \to k}$, and it is in every respect analogous to the absorption process that we have described. It may be shown in general that, if the light causes a transition from a stationary state k to a stationary state l with a probability $w_{k \to l}$, then the light also induces a transition from state l to state k with a probability $w_{l \to k}$ which is equal to $w_{k \to l}$. Hence, if we define the quantities $B_{k \to l}$ and $B_{l \to k}$ in the same way, we must have

$$B_{l \to k} = B_{k \to l} \tag{13-27}$$

The quantity $B_{l \to k}$ is called the Einstein coefficient of stimulated emission.

It is easy to understand the preceding mechanism from a theoretical point of view, but it is not so easy to appreciate it if we think about the experimental arrangements for measuring emission spectra. We know that, if we have a sufficiently large number N_l of molecules in the excited state l, these molecules will return to the lower state k while emitting light. This process occurs in the absence of radiation. In fact, in measuring an emission spectrum we usually measure this spontaneous emission and not the stimulated emission that we discussed previously.

It follows that there must be two possible emission mechanisms, namely, the spontaneous emission, which we usually measure, and the stimulated emission, which we predict from theoretical considerations. In the most general case, where we have a sample in the presence of radiation the number of transitions per unit time $N_{l \to k}$ from a higher state l to a lower state k is given by

$$N_{l \to k} = N_l [B_{l \to k} \rho(\nu_{kl}) + A_{l \to k}] \tag{13-28}$$

Here $B_{l \to k}$ is the Einstein coefficient of stimulated emission and $A_{l \to k}$ is the Einstein coefficient of spontaneous emission. The energy density $\rho(\nu_{kl})$ of the radiation field and its intensity are related by means of

$$I = c\rho \tag{13-29}$$

It is possible to derive two relationships between the three Einstein coefficients. We consider a transition between a lower state o and a higher state n of a set of identical molecules, and we take N_o as the number of molecules per cubic centimeter in the lower state and N_n as the same number for the higher state. The emission of light is described by Eq. (13-28), and it follows that

$$N_{n \to o} = N_n [B_{n \to o} \rho(\nu_{on}) + A_{n \to o}] \tag{13-30}$$

By analogy, it follows from Eq. (13-26) that

$$N_{o \to n} = N_o B_{o \to n} \rho(\nu_{on}) \tag{13-31}$$

If our system is in equilibrium, the rates of change due to the two processes (13-30) and (13-31) must be equal to one another, and we have

$$N_n [B_{n \to o} \rho(\nu_{on}) + A_{n \to o}] = N_o B_{o \to n} \rho(\nu_{on}) \tag{13-32}$$

We know that in equilibrium the ratio between the two numbers N_n and N_o must be given by the Boltzmann distribution (see Chapter 19)

$$\frac{N_n}{N_o} = \exp\left(-\frac{h\nu_{on}}{kT}\right) \tag{13-33}$$

It follows from Eqs. (13-33) and (13-32) that

$$\frac{N_o}{N_n} = \exp\left(\frac{h\nu_{on}}{kT}\right) = \frac{B_{n \to o}\rho(\nu_{on}) + A_{n \to o}}{B_{o \to n}\rho(\nu_{on})} \tag{13-34}$$

or

$$\rho(\nu_{on}) = \frac{A_{n \to o}}{B_{o \to n} \exp(h\nu_{on}/kT) - B_{n \to o}} \tag{13-35}$$

Eq. (13-35) expresses the energy density of the radiation field (in thermal equilibrium) in terms of the three Einstein coefficients of absorption and emission. In Eq. (13-17) we mentioned that the energy density ρ is also given by the expression that was derived by Planck, namely

$$\rho(\nu_{on}) = \frac{8\pi h\nu_{on}^3}{c^3} \frac{1}{\exp(h\nu_{on}/kT) - 1} \tag{13-36}$$

By comparing Eqs. (13-35) and (13-36) we find that

$$A_{n \to o} = \frac{8\pi h\nu_{on}^3}{c^3} B_{n \to o} \qquad B_{n \to o} = B_{o \to n} \tag{13-37}$$

As we expect, the two coefficients $B_{n \to o}$ and $B_{o \to n}$ are equal. It is important to note that the ratio between the two coefficients $A_{n \to o}$ and $B_{n \to o}$ depends very strongly on the frequency. The coefficient $A_{n \to o}$ is known as the coefficient for spontaneous emission, and $B_{n \to o}$ is the coefficient for induced or stimulated emission. For optical transitions, the spontaneous emission is usually much larger than the stimulated emission due to light of customary intensity (this excludes laser light). In magnetic resonance the frequencies are much smaller than in optical transitions, namely, by a factor of the order of 10^4; here the situation is just the reverse, the stimulated emission being predominant.

13-3   TRANSITION PROBABILITIES AND QUANTUM THEORY

In order to understand how the properties of spectroscopic transitions can be explained in terms of quantum theory, it is helpful to recall Bohr's theory of atomic structure that we discussed in Chapter 10. The immediate purpose of this theory was to explain, firstly, how an atom could remain in a stable, stationary state when it is not subject to outside perturbations and, second, why atomic spectra consist of discrete lines. To explain the first difficulty, Bohr assumed that an atom could only be in certain stationary states, each having a fixed energy and angular momentum, and that the atom does not emit or absorb radiation as long as the atom remains in one of those stationary states. A spectroscopic transition occurs when the atom jumps from one stationary state k to another stationary state l. If the energy E_k of state k is higher than the energy E_l of state l then this jump is accompanied by the emission of a photon with an energy

$$h\nu_{kl} = E_k - E_l \tag{13-38}$$

and we observe an emission line of frequency ν_{kl}. If E_k is smaller than E_l, the atom absorbs a photon of energy

$$h\nu_{kl} = E_l - E_k \qquad (13\text{-}39)$$

and we observe an absorption line of frequency ν_{kl}.

We have seen that the stationary states of an atom are obtained as the eigenvalues and eigenfunctions of the Schrödinger equation

$$\mathfrak{H}_{op}\Phi_n = E_n\Phi_n \qquad (13\text{-}40)$$

Here the atom is represented by the Hamiltonian $\mathfrak{H}_{op}$; the energy of stationary state n is given by the eigenvalue E_n, and the behavior of the electrons in this stationary state is described by the eigenfunction Φ_n. It is obvious that Eq. (13-40) describes the properties of stationary states, but it does not account for transitions between different stationary states. If we wish to study such transitions, we should realize that the time independent Schrödinger equation (13-40) is a special case of a more general time dependent Schrödinger equation, which has the form

$$\mathfrak{H}_{op}\Psi = i\hbar\frac{\partial\Psi}{\partial t} \qquad (13\text{-}41)$$

Here the function Ψ depends both on the electron coordinates, which we denote symbolically by x, and on the time t.

The two Schrödinger equations (13-40) and (13-41) are equivalent if we consider a stationary state and if the Hamiltonian $\mathfrak{H}_{op}$ is time independent (otherwise we could not have stationary states). In the time independent formalism, the stationary state n is described by the function Φ_n. In the time dependent formalism the same stationary state n is described by the function $\Psi_n(x; t)$ which is defined as

$$\Psi_n(x; t) = \Phi_n(x)\exp\left(-iE_nt/\hbar\right) \qquad (13\text{-}42)$$

We can easily verify that substitution of the time dependent function $\Psi_n(x; t)$ into the time dependent Schrödinger equation (13-41) leads to the time independent equation (13-40) since the energy E_n is a constant for the stationary state n.

The advantage of using the time dependent Schrödinger equation is that it also describes the situation where the system is not in a stationary state. Supposedly the system can be in any one of an infinite number of stationary states n. Each of these states is described by a time dependent function $\Psi_n(x; t)$, defined by Eqs. (13-40) and (13-42), which satisfies the equation

$$\left(\mathfrak{H}_{op} + \frac{\hbar}{i}\frac{\partial}{\partial t}\right)\Psi_n(x; t) = 0 \qquad (13\text{-}43)$$

Obviously an arbitrary linear combination

$$\Psi(x; t) = \sum_n b_n\Psi_n(x; t) = \sum_n b_n\Phi_n(x)\exp\left(-iE_nt/\hbar\right) \qquad (13\text{-}44)$$

also satisfies the time dependent Schrödinger equations (13-41) and (13-43). It follows that function (13-44) is the most general solution of the time dependent Schrödinger equation and that the general behavior of our system, when it is not in any given stationary state, is described by function (13-44).

If our system is described by the time dependent wave function (13-44), we can ask what is the probability P_m of finding the system in a given stationary state m. By definition this probability is given by

$$P_m = b_m b_m^*$$ (13-45)

Obviously, we must have

$$\sum_m P_m = \sum_m b_m b_m^* = 1$$ (13-46)

Let us now proceed to the problem in which we are interested, namely, how to describe spectroscopic transitions. The quantum mechanical analysis of this problem is quite complicated because there are a number of different approaches involving various approximations and it is not easy to appreciate the consequences of these approximations. In addition, even the simplest approach is quite involved mathematically. Our discussion is therefore limited to a very broad outline of the problem, but we will give the results.

In order to study spectroscopic transitions we consider the case of a molecule in the presence of a radiation field. The molecule is described by a time independent Hamiltonian $\mathfrak{H}_{mol}$ and the interaction between the molecule and the radiation field is described by a time dependent Hamiltonian $\mathfrak{H}_{int}$. The total system is thus described by a Hamiltonian

$$\mathfrak{H}_{op}(x; t) = \mathfrak{H}_{mol}(x) + \mathfrak{H}_{int}(x; t)$$ (13-47)

Strictly speaking, this Hamiltonian does not have stationary states because the Hamiltonian contains the time, but if its time dependent part is much smaller than its time independent part

$$\mathfrak{H}_{int}(x; t) \ll \mathfrak{H}_{mol}(x)$$ (13-48)

then we may express the behavior of the system in terms of the stationary states of the operator $\mathfrak{H}_{mol}$.

In mathematical terms, the behavior of the total system is described by a time dependent function $\Psi(x; t)$, which must be a solution of the Schrödinger equation

$$\mathfrak{H}_{op}(x; t)\Psi(x; t) = -\frac{\hbar}{i}\frac{\partial\Psi}{\partial t}$$ (13-49)

This function may be expressed in the form

$$\Psi(x; t) = \sum_n b_n(t)\exp(-i\varepsilon_n t/\hbar)\Phi_n(x)$$ (13-50)

where ε_n and Φ_n are the eigenvalues and eigenfunctions of the molecular Hamiltonian $\mathfrak{H}_{mol}$.

We can show that expansion (13-50) is permissible as long as the coefficients $b_n(t)$ are dependent on the time t. On the other hand, the expansion makes sense only if the time dependent part of the operator is small and satisfies condition (13-48). In that case, namely, the coefficients $b_n(t)$ are slowly varying functions of the time, they may be calculated by means of perturbation theory. Also, the behavior of the total system may be described in terms of the coefficients $b_n(t)$. The probability of finding the molecule in one of its stationary states n is given by

$$P_n(t) = b_n(t)b_n^*(t)$$ (13-51)

just as in Eq. (13-45). The time dependence of the probabilities $P_n(t)$ describes how the probability distribution over the various molecular stationary states changes as a function of time, and it may be related to the three Einstein coefficients of absorption and emission.

We have seen in Eq. (13-37) that there are two relations between the three Einstein coefficients. This is fortunate because the simplest approach for calculating the probabilities $P_n(t)$ by means of perturbation theory gives a result only for the Einstein coefficient of absorption $B_{o \to n}$. The values of the other two coefficients of emission, $B_{n \to o}$ and $A_{n \to o}$, may then be derived by using Eq. (13-37).

The theoretical result for the Einstein coefficient $B_{o \to n}$ is

$$B_{o \to n} = (2\pi/3\hbar^2)(\mu_{n,o} \cdot \mu_{o,n}) \tag{13-52}$$

The vector $\mu_{n,o}$ is known as the transition moment between the states n and o. It is the matrix element over the dipole moment operator

$$\mu = e \sum_j \mathbf{r}_j \tag{13-53}$$

between the two states

$$\mu_{n,o} = \langle \Phi_n | \mu | \Phi_o \rangle$$
$$\mu_{o,n} = \langle \Phi_o | \mu | \Phi_n \rangle \tag{13-54}$$

In most cases we are not so much interested in the exact value of a transition probability as in its order of magnitude. The experimental spectroscopists speak of allowed and forbidden transitions. An allowed transition has a large transition probability, and a forbidden transition has a small transition probability. The nature of the transition probabilities can often be derived from a set of rules that are known as selection rules. Ideally, these rules can tell us immediately which transitions are allowed and which are forbidden.

We like to define a forbidden transition as a transition for which the transition moment, as defined by Eqs. (13-53) and (13-54), is zero. Realize, however, that we have used a number of approximations in deriving our expression for the transition probability. Therefore, even if the transition moment $\mu_{k,o}$ is zero, the transition may still be observable; however, such a transition is caused by higher order effects, and its probability is generally much smaller than that of an allowed transition.

If we use our definition of a forbidden transition, a selection rule is simply a general rule that tells us when the transition moment $\mu_{k,o}$ is zero and when it is nonzero. For example, the operator in equation (13-54) is spin independent, and the transition moment $\mu_{k,o}$ is zero if the states k and o have different spin multiplicity because then the spin functions of the states k and o are orthogonal to one another. This is known as the spin selection rule $\Delta S = 0$. Other selection rules may be derived from the symmetry of the atom or molecule. For example, it is easily seen that the transition between the $(1s)$ and the $(2s)$ states of the hydrogen atom must be forbidden because the transition moment of that transition is zero. Of course, forbidden transitions may still be observed because they are caused by mechanisms that were neglected in the derivation of the result (13-52) for the transition probability. Some symmetry forbidden transitions can be observed because of these mechanisms. In a similar fashion, spin forbidden transitions are nonzero if we take the interactions between the electron spins into account.

The probability of an ordinary allowed transition is of the order of $10^{10}\,\text{s}^{-1}$. This means that a molecule in an excited state returns to the ground state in a time of about $10^{-10}\,\text{s}$ if the transition is allowed. Forbidden transition probabilities range all the way from 10^{-1} to $10^{8}\,\text{s}^{-1}$. Some transitions are both spin forbidden and symmetry forbidden, and they have very low probabilities. In such cases the life times of molecular excited states can be as high as $10\,\text{s}$. We postpone a more detailed discussion of the selection rules until later.

13-4   THE MASER AND THE LASER

The word "maser" is an acronym for "microwave amplification by stimulated emission of radiation" and the word "laser" is derived from "light amplification by stimulated emission of radiation." The maser is a device that amplifies microwaves. It was invented around 1950, and it was used primarily as a noise-free amplifier. The first lasers were constructed around 1960, and they were used to produce beams of light with very high intensities in a very narrow frequency range. The invention of the laser has led to some interesting experimental discoveries, which are known as nonlinear optical phenomena or multiple-photon phenomena. The result (13-52) of the transition probability represents processes where only one photon at a time is absorbed or emitted. In principle there are multiple-photon absorption or emission processes, but these are negligible in ordinary light. However, if we have light beams of very high intensity, these higher order effects may become observable. Recent experiments have uncovered a large variety of such multiple-photon processes that might have useful practical applications.

The first maser that was constructed made use of ammonia. We know that the molecule NH_3 has a pyramidal structure (see Figure 13-3) but remember that there

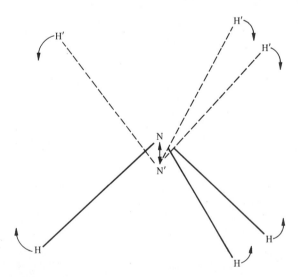

Figure 13-3 The inversion motion of the ammonia molecule.

are two equilibrium structures, with the nitrogen atom either above or below the plane. The molecule can rock back and forth between these two structures, and this motion can be roughly represented by a harmonic oscillator with a frequency ν that corresponds to a wavelength of 1.26 cm. Under ordinary circumstances a sample of NH_3 molecules has N_0 molecules in the vibrational ground state and N_1 in the first excited state and the ratio between N_0 and N_1 is given by the Boltzmann distribution

$$(N_1/N_0) = e^{-h\nu/kT} \tag{13-55}$$

Obviously N_1 is smaller than N_0.

The operation of the maser is based on population inversion. If some way can be found to make N_1 larger than N_0, that is, to produce an excess of molecules in the excited state, then we can use stimulated emission to liberate the excess energy and, in this way, produce an intense flash of radiation. It is now possible to produce such a population inversion in the case of ammonia. If we let a beam of ammonia pass through an inhomogeneous electric field, the beam splits into two because the electric field interacts differently with the molecules in the excited vibrational state and the molecules in the ground state. One of the two beams has a large fraction of molecules N_1 in the excited state, and in this way we can obtain a sample of ammonia molecules with N_1 larger than N_0.

We can see what happens when a photon of energy $h\nu$ enters this sample. As soon as it encounters an excited molecule, it gives rise to stimulated emission where the molecule returns to its ground state; as a result we have two photons. These two photons produce more photons, and so on. In the most efficient device, the ammonia sample is in a resonance cavity where the photons move back and forth through the sample and keep producing more and more photons. Such an ammonia maser can amplify an incoming signal by about a factor 100 to 1 in energy. What is more important, it is a practically noise free amplifier. The ammonia maser is obviously too expensive for use in ordinary radios or televisions, but it was used fairly extensively in astronomy for the detection of microwave radiation from outer space.

It became clear that a maser or a laser can operate effectively if two conditions are satisfied. First, we must have a population inversion where there are more molecules in a higher excited state than in a lower state. Second, we must have a resonance condition where the radiation can travel back and forth through the sample and where it is amplified with each passage through. In order to construct a laser, we must find a system where these two conditions can be satisfied; moreover, it is clear that the first condition is the crucial one. The first two lasers that were constructed in 1960 were the ruby laser and the helium-neon laser. Other kinds of lasers were constructed later, but we will limit our discussion to the two lasers mentioned.

Ruby consists of aluminum oxide with small amounts of chromium. The chromium atoms replace some of the aluminum atoms in the lattice. In the ruby laser we use a cylinder of about 1 cm in diameter and about 5 cm long that contains 0.05% chromium. In Figure 13-4 we give the energy level diagram for the chromium atoms embedded in the lattice. Level 3 is not a single energy level but consists of a very large number of closely spaced near-degenerate levels; in solid state physics this is called an energy band. If we irradiate the ruby with green light, we obtain excitation into the energy band. Some of the excited chromium atoms drop back to the nondegenerate level 2 through spontaneous emission. In ruby, level 2 is metastable; this means that the probability of the transition from level 2 to level 1 is quite small. Consequently,

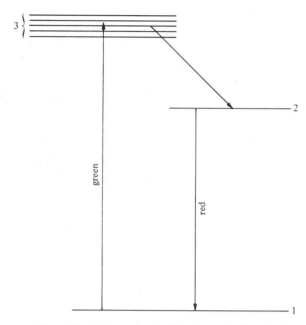

Figure 13-4   Energy levels of the chromium ions in ruby.

the chromium atoms in the excited state 2 have a relatively long lifetime, and it is possible in this way to obtain a fairly large number of chromium atoms in the excited state 2. This procedure is known as optical pumping. We see that we can obtain a significant population inversion between levels 1 and 2 by means of optical pumping via the energy band 3.

Resonance is obtained in a ruby laser by placing a mirror at one end of the cylinder and a partially transparent mirror at the other end. This forces the bulk of the light to travel back and forth between the two mirrors and acquire a greater and greater intensity. Eventually all of this light passes through the partially transparent mirror in the form of the laser beam. In the ruby laser, all the emitted photons have a frequency that is very close to the energy difference between levels 2 and 1. In fact, the line width of the frequency distribution is equal to the natural line width of the emission line, which is of the order of 0.01 Å. Thus, the energy of the emitted light is concentrated within a very narrow frequency range, and its energy density is very high. The laser can be operated so that it emits pulses of light or can be constructed so that it has a continuous light output. In the latter case, the optical pumping must also be continuous.

A completely different type of laser is the gas laser; the helium–neon (He–Ne) laser is the best known example of this group. A gas laser consists of a mixture of two gases and population inversion in the one gas is obtained by means of energy transfer from the second gas. In the He–Ne laser the He atoms are excited to a higher state, and they transfer their excess energy to the Ne atoms through collisions. The Ne atoms lose their energy through a series of downward transitions from a high-excited state via the intermediate lower excited states to the ground state. It is clear that, if one or more of these intermediate levels is metastable, the concentration of Ne atoms in this metastable excited state will increase. In this way we can obtain a significant population inversion over certain energy levels in the Ne atoms.

Optimum efficiency for the He–Ne level occurs when we take five parts of He to one part of Ne. The lasing material in the gas mixture is usually the smaller fraction in the mixture. Gas lasers are built in the form of a cylindrical glass tube. At the beginning two plane mirrors were placed within the glass tube at both ends, but it soon became apparent that it is more convenient to make use of spherical reflectors that are placed outside the tube. The final construction of the laser makes use of more sophisticated engineering devices, but we do not wish to go into these. Actually, the engineering skills that are needed to construct a gas laser are much less sophisticated than for solid state lasers. The difficult problem is to know the energy levels and their lifetimes in the gases involved and to have a good understanding of the energy transfer processes.

_____ *Problems*

1. What is the representation for a plane light wave that propagates in the x direction and is polarized in the y direction?
2. If an electromagnetic wave has a direction of propagation that bisects the positive x and y axes, what are its possible directions of polarization? Give the corresponding equations.
3. The work function W is the potential barrier that an electron must pass in order to be able to leave a metal. If W is 2 eV, what is the value of the frequency ν_0 and wavelength λ of the light that we must shine on the metal to begin to observe a photoelectric effect?
4. If the work function of a metal is 1 eV and we illuminate with light of 5000 Å, what is the velocity v of the electrons that are ejected from the metal?
5. What is the ratio between the coefficients of spontaneous and stimulated emission for light with a wavelength of 6000 Å and for microwave radiation with a wavelength of 3 cm?
6. What is the transition moment between a $1s$ and $2p_z$ state of the hydrogen atom? Determine the corresponding Einstein coefficient of absorption.
7. Derive the value of the Einstein coefficient of spontaneous emission from the $2p_z$ state to the $1s$ state of the hydrogen atom.
8. Is the transition from a $1s$ state to a $3d$ state in the hydrogen atom allowed or forbidden? Why?
9. If we want to construct a laser, which two conditions must be satisfied?
10. How does the He–Ne laser operate?

_____ *References and Recommended Reading*

The historical development of the various theories of light and radiation is treated very extensively in reference [1], which makes very interesting reading. The quantum theory of the radiation field is discussed in a number of books written for physicists, but we find that the majority of the modern books are very abstract and sophisticated and do not discuss any applications. The only exception is Heitler's book [2]. Hameka's book [3] was written specifically to deal with molecular applications of quantum electrodynamics. The classical theory of the radiation field is discussed in most physics textbooks, but we recommend particularly Feynman's book [4] because it gives such an excellent account of the subject. Time-dependent perturbation theory is discussed in [2] and [3] and in most of the quantum mechanics texts that we have listed at the end of Chapter 11. Finally, we list some books that deal with lasers.

[1] E. Whittaker: *A History of the Theories of Aether and Electricity.* Harper & Row, New York, 1960.

[2] W. Heitler: *The Quantum Theory of Radiation.* Oxford University Press, London, 1957.

[3] H. F. Hameka. *Advanced Quantum Chemistry.* Addison-Wesley, Reading, Mass., 1965.

[4] R. P. Feynman, R. B. Leighton, and M. Sands: *The Feynman Lectures on Physics.* Addison-Wesley, Reading, Mass., 1963.

[5] B. A. Lengyel: *Introduction to Laser Physics.* Wiley, New York, 1966.

[6] A. L. Bloom: *Gas Lasers.* Wiley, New York, 1968.

chapter 14

The Spectra of Diatomic Molecules

14-1 EXPERIMENTAL INFORMATION

The bulk of the available experimental information on diatomic molecules has been derived from molecular spectra. These are not just the spectra in the visible and ultraviolet regions (with wavelengths ranging between 2000 and 7000 Å and wave numbers between 15,000 and 50,000 cm^{-1}) but also the spectra in the infrared region (wave numbers between 100 and 5000 cm^{-1}) and in the microwave region (wave numbers between 0.1 and 100 cm^{-1}). It is necessary to consider all these spectra because a diatomic molecule has a very large number of energy levels, and the spectra in the visible region do not always give us enough information to identify and classify all the energy levels.

Naturally, the classification of the energy levels was originally derived from the experimental data, but in our discussion we will reverse the historical order of the events. We feel that it is easier if we first outline some of the general features of molecular spectra and use these to discuss some of the experimental results.

It has been well established that the energy eigenvalues of a diatomic molecule can be represented as a sum of three terms

$$E_{mol} = E_{el} + E_{vib} + E_{rot} \tag{14-1}$$

The first term is an energy eigenvalue of the electronic motion for a specific fixed, nuclear configuration; the second term is the vibrational energy, where the two nuclei move relative to one another so that the internuclear distance R oscillates around an equilibrium distance R_0; and the third term represents the rotational motion of the whole molecule. It should be noted that Eq. (14-1) is only an approximation, but it is a fairly good approximation, which agrees quite well with experiment.

The rotational energy levels may be obtained by assuming that the molecule behaves like a rigid rotor with a distance R, which is equal to the internuclear distance, and with a mass μ_{AB}, which is the reduced mass of the two nuclear masses M_A and M_B,

$$\frac{1}{\mu_{AB}} = \frac{1}{M_A} + \frac{1}{M_B} \tag{14-2}$$

We have seen in Chapter 11 that the eigenvalues of the rigid rotor are given by

$$E_J = \frac{\hbar^2}{2\mu R^2} J(J+1) \tag{14-3}$$

Consequently, we represent the rotational energy levels of a diatomic molecule as

$$E_J = BJ(J+1) \tag{14-4}$$

where B is known as the rotational constant. It should be equal to

$$B = \frac{\hbar^2}{2\mu_{AB}R^2} \qquad (14\text{-}5)$$

In Table 14-1 we have listed some experimental values for rotational constants. Note that the energy differences between the rotational levels fall in the microwave region.

Table 14-1 Rotational constants $B_{0,0}$ for the electronic and vibrational ground states ($n = 0$, $v = 0$) of some diatomic molecules and the equilibrium distances R_0 for those states

Molecule	$B_{0,0}/(\text{cm}^{-1})$	$R_0/(\text{Å})$
Cl_2	0.2348	1.988
CO	1.9314	1.1282
H_2	60.809	0.7417
HD	45.655	0.7414
D_2	30.429	0.7416
HBr	8.473	1.414
HCl	10.591	1.275
HF	20.939	0.9171
HI	6.551	1.604
I_2	0.03736	2.667
ICl	0.11416	2.3207
Li_2	0.6727	2.6725
LiH	7.5131	1.595
N_2	2.010	1.094
O_2	1.4457	1.2074
S_2	0.2956	1.889

The vibrational energy levels are roughly approximated by the energy levels of a harmonic oscillator,

$$E_v = (v + \tfrac{1}{2})h\nu \qquad (14\text{-}6)$$

This approximation is quite satisfactory for the lower vibrational levels, but it becomes less satisfactory for the higher states. In Table 14-2 we have listed the values of ν for the electronic ground states of some molecules. It may be seen that these energies correspond to photons in the far infrared.

Finally, the eigenvalues of the operator E_{el}, which we denote by ε_n have the same order of magnitude as atomic energy eigenvalues. The differences between different electronic energies ε_n and ε_m are of the order of 1–$10\,\text{eV}$, and the corresponding spectral transitions range from the infrared to the far ultraviolet region.

It follows that each stationary state of a diatomic molecule may be represented by a set of three quantum numbers, n, v, and J. The first quantum number, n, refers to the electronic motion, the second quantum number, v, describes the vibrational state,

Table 14-2 Vibrational frequencies $\nu_{0,1}$ for the electronic ground states of some diatomic molecules, obtained as the energy differences between the lowest and the first excited vibrational levels

Molecule	$\nu_{0,1}/(\text{cm}^{-1})$
Cl_2	564.9
CO	2170.2
H_2	4395.2
HD	3817.1
D_2	3118.5
HBr	2649.7
HCl	2989.7
HF	4138.5
HI	2309.5
I_2	214.6
ICl	384.2
Li_2	351.4
LiH	1405.6
N_2	2359.6
O_2	1580.4
S_2	725.7

and the third quantum number J represents the rotational state. Realize that the vibrational frequency depends on the electronic structure of the molecule and, consequently, it depends on the quantum number n. The same is true for the rotational constant B, because the equilibrium distance R is usually different for different values of n. This means that we ought to write the vibrational and rotational energies as

$$E_v^{(n)} = (v + \tfrac{1}{2})h\nu_n$$
$$E_J^{(n)} = B_n J(J+1) \tag{14-7}$$

The total energy $E_{n,v,J}$ of the molecular eigenstate (n, v, J) is then approximated as

$$E_{n,v,J} = \varepsilon_n + E_v^{(n)} + E_J^{(n)}$$
$$= \varepsilon_n + (v + \tfrac{1}{2})h\nu_n + B_n J(J+1) \tag{14-8}$$

We will use this expression for discussing the experimental spectra.

A spectroscopic transition between two stationary states (n, v, J) and (n', v', J') has a frequency ν that is given by

$$h\nu = E_{n',v',J'} - E_{n,v,J}$$
$$= \varepsilon_{n'} - \varepsilon_n + h\nu_{n'}(v' + \tfrac{1}{2}) - h\nu_n(v + \tfrac{1}{2}) + B_{n'}J'(J'+1) - B_n J(J+1) \tag{14-9}$$

if the state (n', v', J') has the higher energy. If the quantum numbers n' and n are

different, we observe a transition between different electronic states and we expect
the corresponding light to be in the visible or ultraviolet region. Realize that Eq.
(14-9) describes a very large number of transitions because the quantum numbers v
and v' and the quantum numbers J and J' may have a large number of possible
values. If n and n' have given specific values, we observe a specific electronic
transition; this transition, however, contains a large number of different vibrational
and rotational transitions with frequencies that are very close together. In many
instances the various lines are so close together that they overlap, and we observe a
continuous absorption band. The resulting spectrum is known as a band spectrum, a
frequency range of continuous absorption, extending over 100–1000 cm^{-1}, and it is
called an absorption or emission band. We mentioned in Chapter 12 that an atomic
spectrum is usually a line spectrum because it consists of discrete lines.

It may be helpful to discuss briefly the experimental width of spectral lines.
Ordinarily they are determined by experimental factors, namely, the width of the slit
in the spectrograph. In the visible region, then, the widths are of the order of 5 cm^{-1}.
If we use more sophisticated equipment the widths of the spectral lines are smaller
and they are determined by the collisions between the molecules, the Doppler effect,
and collisions with the wall. They are of the order of 0.1–1 cm^{-1}. We can show that
even in ideal circumstances a spectral line of one isolated molecule has a finite line
width; this is known as the natural line width, and it is of the order of 0.001–
0.01 cm^{-1}. We should emphasize that all the numbers given refer to transitions in the
visible. In other spectral regions the orders of magnitude may be different.

It is now obvious that for heavy molecules, where the values of B are small, it is not
possible to detect the rotational structure in the electronic spectra. For lighter
molecules, where B is larger, it is in principle possible to find the rotational fine
structure in the electronic absorption bands if the experiments are done carefully so
that the spectrograph has a high enough resolution.

The molecular spectra in the infrared are usually vibration spectra. In this case, we
observe transitions between different vibrational levels within the same electronic
state. Theoretically, if the vibrational motion is exactly described by a harmonic
oscillator, we should see only one absorption line at the vibrational frequency ν_n. In
practice, we observe more than one line; the additional lines have frequencies $2\nu_n$,
$3\nu_n$, and so on, because the vibrational motion is not exactly harmonic. Again, there
is rotational fine structure in the spectrum due to the fact that the rotational constant
B_n depends slightly on the vibrational quantum number; thus, the rotational
constants for different vibrational levels are slightly different.

Transitions between different rotational levels within the same electronic and
vibrational states fall in the microwave region, as may be seen from Table 14-1. The
microwave spectra give us quite accurate values for the rotational constants B and
also for the internuclear distance R. As a result, these distances are quite well known
for diatomic molecules.

In Section 14-2 we wish to discuss how the approximations we have sketched can
be understood from theoretical arguments. We should mention once again that the
validity of the approximations can be justified from the experimental results, because
the use of the approximations leads to a very satisfactory interpretation of most of the
experimental results. Hence, we do not have to prove the validity of Eq. (14-1);
rather, we wish to explain how it is compatible with the quantum mechanical
description of diatomic molecules.

14-2 THE MOLECULAR WAVE FUNCTION

We mentioned in Section 14-1 that the energy levels of a diatomic molecule may be written as the sum of an electronic energy, a vibrational energy, and a rotational energy. Strictly speaking, there is a fourth contribution to the total molecular energy, namely, the kinetic energy of the overall translational motion of the total molecule. However, this translational energy does not play an important role in the molecular spectrum so that we may safely ignore it, even though we should remember that this motion is always present and that it is important in the kinetic theory of gases.

Recall that the molecular energy levels E_{mol} were written in the form

$$E_{mol} = E_{el} + E_{vib} + E_{rot} + E_{tr} \tag{14-10}$$

For completeness sake we have added the fourth energy contribution E_{tr} here, even though we are not particularly interested in this term.

The molecular wave function Ψ may be written as a product of four factors

$$\Psi = \psi_{el}\psi_{vib}\psi_{rot}\psi_{tr} \tag{14-11}$$

The first factor represents the electronic motion, the second the vibrational motion, the third the rotational motion, and the fourth the translational motion of the molecule.

At first sight Eq. (14-11) seems quite logical because it is consistent with Eq. (14-10). Also, it is well known in quantum mechanics that, if the Hamiltonian of a system is separable, its energy eigenvalues may be written as a sum of different contributions and its eigenfunctions are products of different terms. However, in a molecule the situation is not all that simple; as we will see, the molecular Hamiltonian is not a simple sum of four terms representing the four different motions of the molecule. The mathematical analysis leading to Eqs. (14-10) and (14-11) is fairly complex and involves a number of approximations. We do not plan to discuss this mathematical analysis in detail but we would like to get a general understanding of the form of the wave function. For this purpose we must at least outline the most important features of the mathematical analysis.

EXAMPLE 14-1

We consider a Hamiltonian $\mathfrak{H}(x, y)$ which depends on two sets of coordinates x and y, and we assume that it may be separated as

$$\mathfrak{H}(x, y) = \mathfrak{H}_1(x) + \mathfrak{H}_2(y) \tag{a}$$

The first term $\mathfrak{H}_1$ depends only on the coordinates x, and the second term $\mathfrak{H}_2$ depends only on the coordinates y. Let the eigenvalues and eigenfunctions of these Hamiltonians be given by

$$\mathfrak{H}_1(x)\phi_n(x) = \lambda_n\phi_n(x)$$
$$\mathfrak{H}_2(y)\chi_m(y) = \omega_m\chi_m(y) \tag{b}$$

It follows then that

$$\mathfrak{H}(x, y)\phi_n(x)\chi_m(y) = \mathfrak{H}_1(x)\phi_n(x)\chi_m(y) + \mathfrak{H}_2(y)\phi_n(x)\chi_m(y)$$
$$= (\lambda_n + \omega_m)\phi_n(x)\chi_m(y) \tag{c}$$

The eigenvalues of $\mathfrak{H}$ are the sums of the eigenvalues of $\mathfrak{H}_1$ and $\mathfrak{H}_2$, and the eigenfunctions are products of the eigenfunctions of $\mathfrak{H}_1$ and $\mathfrak{H}_2$. ●

Let us consider a diatomic molecule AB. The situation is easier to understand for a diatomic than for a polyatomic molecule, even though there are no basic differences between them. The total molecular Hamiltonian $\mathfrak{H}_{AB}$ for a diatomic molecule may be written as

$$\mathfrak{H}_{AB} = T_A + T_B + (e^2 Z_A Z_B / R_{AB}) + \mathfrak{H}_{el} \tag{14-12}$$

Here T_A and T_B represent the kinetic energies of nuclei A and B; the next term is the Coulomb repulsion between the two nuclei; and $\mathfrak{H}_{el}$ represents the electronic energy.

The electronic Hamiltonian is given by

$$\mathfrak{H}_{el} = -\frac{\hbar^2}{2m} \sum_j \Delta_j - \sum_j \frac{Z_A e^2}{r_{Aj}} - \sum_j \frac{Z_B e^2}{r_{Bj}} + \sum_{j>k} \frac{e^2}{r_{jk}} \tag{14-13}$$

The electron coordinates are defined with respect to a coordinate system that is attached to the nuclear framework. The origin is the center of gravity of the molecule, which is situated on the nuclear axis, and the z axis is taken along the nuclear axis AB.

The electronic energies E_n and eigenfunctions F_n are the solutions of the Schrödinger equation

$$\mathfrak{H}_{el} F_n = E_n F_n \tag{14-14}$$

The problem of how to solve this Schrödinger equation is discussed in Chapter 15. Here we are concerned only with the question how E_n and F_n depend on the nuclear configuration. We should realize that the second term of $\mathfrak{H}_{el}$ in Eq. (14-13), that is the Coulomb potential for the electrons, depends indirectly on the internuclear distance R_{AB}. As a result, both the eigenvalues E_n and the eigenfunctions F_n must depend on R_{AB}. The eigenfunctions F_n depend of course on the electron coordinates, which we denote symbolically by r_{el}. It follows that the Schrödinger equation (14-14) should be written in the form

$$\mathfrak{H}_{el}(r_{el}; R_{AB}) F_n(r_{el}; R_{AB}) = E_n(R_{AB}) F_n(r_{el}; R_{AB}) \tag{14-15}$$

The energy eigenvalues E_n depend only on the distance R_{AB} and not on any other nuclear coordinates. The coordinate system for the electronic coordinates depends on the orientation of the molecule, but the specific form of the eigenfunctions F_n depends only on r_{el} and R_{AB} and not on any other nuclear coordinates.

In order to arrive at the form (14-11) of the molecular eigenfunction Ψ_{AB} we make use of the Born–Oppenheimer approximation. According to this approximation we may write Ψ_{AB} as the product of one of the eigenfunctions F_n and of a corresponding function f_n that depends on the nuclear coordinates r_{nuc} only

$$\Psi_{AB} = f_n(r_{nuc}) F_n(r_{el}; R_{AB}) \tag{14-16}$$

The function F_n is an eigenfunction of the Schrödinger equation (14-15), and the function f_n must be a solution of the Schrödinger equation

$$[T_A + T_B + (e^2 Z_A Z_B / R_{AB}) + E_n(R_{AB})] f_n(r_{nuc}) = E_{AB} f_n(r_{nuc}) \tag{14-17}$$

EXAMPLE 14-2

The molecular eigenfunctions Ψ_{AB} must be solutions of the Schrödinger equation

$$\mathfrak{H}_{AB} \Psi_{AB} = E \Psi_{AB} \tag{a}$$

where the Hamiltonian is defined in Eq. (14-12). Any molecular function Ψ_{AB} may be expanded in terms of the complete set of functions F_n as long as the expansion coefficients depend on the nuclear coordinates,

$$\Psi_{AB} = \sum_n f_n(r_{nuc}) F_n(r_{el}; R_{AB}) \tag{b}$$

According to the Born–Oppenheimer approximation, if the separation between the vibrational energy levels in a molecule is much smaller that the separation between the electronic energy levels E_n, then an eigenfunction $\Psi_{AB}(n)$ that belongs to the electronic eigenstate n may be approximated as

$$\Psi_{AB}(n) \approx f_n(r_{nuc}) F_n(r_{el}; R_{AB}) \tag{c}$$

The validity of this approximation may be proved by using perturbation theory.

In order to arrive at Eq. (14-17) we need a second approximation, which is also part of the Born–Oppenheimer approximation. We note that the whole vibrational motion of a nucleus extends over a relatively small area. The extent of this vibrational motion is of the order of 0.1 Å or less. The electronic wave function F_n varies only slightly over this range, and it seems reasonable to assume that the derivative of F_n with respect to an arbitrary nuclear coordinate X is much smaller than the corresponding derivative of f_n,

$$\frac{\partial F_n}{\partial X} \ll \frac{\partial f_n}{\partial X} \tag{d}$$

If this is true, then we have

$$(T_A + T_B) f_n(r_{nuc}) F_n(r_{el}; R_{AB}) \approx F_n(r_{el}; R_{AB})(T_A + T_B) f_n(r_{nuc}) \tag{e}$$

It follows that

$$\begin{aligned}
\mathfrak{H}_{AB}(f_n F_n) &= [T_A + T_B + e^2 Z_A Z_B/R_{AB} + \mathfrak{H}_{el}](f_n F_n) \\
&= F_n(T_A + T_B) f_n + f_n \mathfrak{H}_{el} F_n + F_n(e^2 Z_A Z_B/R_{AB}) f_n \\
&= F_n[T_A + T_B + (e^2 Z_A Z_B/R_{AB}) + E_n(R_{AB})] f_n \\
&= E_{AB} F_n f_n
\end{aligned} \tag{f}$$

We may divide this equation by F_n and we find that f_n must satisfy the equation

$$[T_A + T_B + (e^2 Z_A Z_B/R_{AB}) + E_n(R_{AB})] f_n = E_{AB} f_n \tag{g}$$

which is the same as Eq. (14-17). ●

The nuclei move in a potential field which is the sum of the energy eigenvalue E_n of Eq. (14-15) and the Coulomb repulsion term of Eq. (14-12). This potential field is thus described by a potential function $U_n(R_{AB})$, which is defined as

$$U_n(R_{AB}) = E_n(R_{AB}) + (e^2 Z_A Z_B/R_{AB}) \tag{14-18}$$

Naturally, the form of the potential function depends on the electronic state of the molecule. Its form can be derived either theoretically by calculating the function $E_n(R_{AB})$ from approximate quantum mechanical methods or experimentally from the distribution of the vibrational energy levels in the molecule.

In Figure 14-1 we have sketched some typical potential curves for an imaginary molecule AB. Even though the exact forms of the potential curves in molecules are

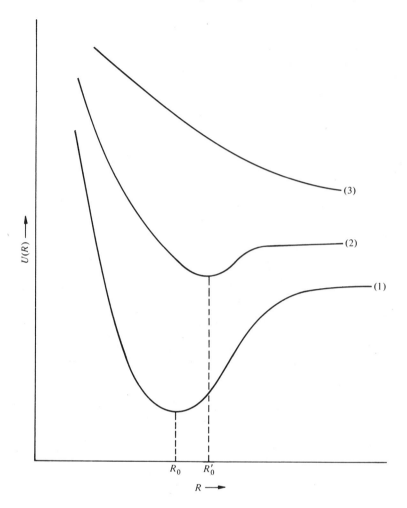

Figure 14-1   Potential curves of a molecule AB.

not always known, their general behavior can easily be understood. First, we note that every potential curve must tend to infinity when $R \rightarrow 0$ because the Coulomb repulsion between the two nuclei tends to infinity for $R \rightarrow 0$. Also, the curve tends asymptotically to a constant value when R tends to infinity; this value is the sum of the energies of the two atoms A and B. If the molecule is stable, there must be a value R_0 for which the potential curve has a minimum, as in curves 1 and 2 of Figure 14-1. If the curve does not have a minimum, there is no value of R for which the molecule has a stable configuration and the molecule will dissociate.

We mentioned already that we wish to separate Eq. (14-17) into three different equations, namely, the equations for the overall translational motion of the molecule, for the rotational motion, and for the vibrational motion,

$$\mathfrak{H}_{tr}\Phi_{tr} = E_{tr}\Phi_{tr}$$
$$\mathfrak{H}_{rot}\Phi_{rot} = E_{rot}\Phi_{rot} \qquad\qquad \textbf{(14-19)}$$
$$\mathfrak{H}_{vib}\Phi_{vib} = E_{vib}\Phi_{rot}$$

It may be helpful if we consider these various motions from a classical point of view.

Figure 14-2 shows how, in the classical approach, the overall motion of the molecule may be decomposed into three different motions. Figure 14-2a illustrates the translational motion of the total molecule, which may be described by the motion of a particle with mass

$$M_C = M_A + M_B \tag{14-20}$$

located at the center of gravity of the molecule. This motion is described by the Cartesian coordinates of the center of gravity

$$\mathbf{R}_C = (x_C, y_C, z_C) \tag{14-21}$$

which are defined in Appendix A of this chapter.

The internal motion of the molecule is characterized by the reduced mass μ

$$\frac{1}{\mu} = \frac{1}{M_A} + \frac{1}{M_B} \tag{14-22}$$

and by the internal coordinates $\mathbf{R} = (x, y, z)$, where

$$x = x_B - x_A \qquad y = y_B - y_A \qquad z = z_B - z_A \tag{14-23}$$

This motion may be decomposed into the rotational motion of the molecule (Figure 14-2b), which is characterized by the orientation of the vector $\mathbf{R}$, and the vibrational motion (Figure 14-2c), which is characterized by changes in the internuclear distance $R = R_{AB}$.

The separation of the nuclear motion into the translational motion of Figure 14-2a and into the internal motion is quite straightforward. We show in Appendix A of this chapter that

$$T_A + T_B = T_C + T \tag{14-24}$$

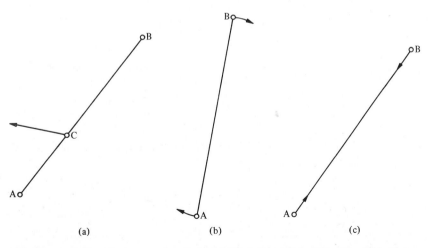

(a)   (b)   (c)

Figure 14-2   Schematic presentation of the (a) translational, (b) rotational and (c) vibrational motion of a molecule.

where T_C is the kinetic energy of the center of gravity and T is given by

$$T = -\frac{\hbar^2}{2\mu}\left(\frac{\partial^2}{\partial x^2}+\frac{\partial^2}{\partial y^2}+\frac{\partial^2}{\partial z^2}\right) \tag{14-25}$$

If we substitute this into the Schrödinger equation (14-17) we obtain

$$[T_C + T + U_n(R)]f_n(r_{\text{nuc}}) = E_{\text{AB}}f_n(r_{\text{nuc}}) \tag{14-26}$$

Since only the kinetic energy term T_C depends on the coordinates of the center of gravity, this equation may be separated into

$$\begin{aligned}\mathfrak{H}_{\text{tr}}\Phi_{\text{tr}} &= E_{\text{tr}}\Phi_{\text{tr}} \\ \mathfrak{H}_{\text{int}}\Phi_{\text{int}} &= E_{\text{int}}\Phi_{\text{int}}\end{aligned} \tag{14-27}$$

with

$$\mathfrak{H}_{\text{int}} = T + U(R) \tag{14-28}$$

The translational motion is of little interest because it is not observable in the spectrum. We will not discuss it any further.

In order to separate $\mathfrak{H}_{\text{int}}$ into a rotational and a vibrational part, we change from the set of Cartesian coordinates x, y, z to the set of polar coordinates R, θ, ϕ that are defined as

$$\begin{aligned}x &= R \sin \theta \cos \phi \\ y &= R \sin \theta \sin \phi \\ z &= R \cos \theta\end{aligned} \tag{14-29}$$

It is then possible to separate the kinetic energy operator T of Eq. (14-25) into a part that depends on R and represents the vibrational motion and a part that depends on the polar angles and represents the rotational motion.

We have seen in Chapter 11 that the rotational Hamiltonian is given by

$$\mathfrak{H}_{\text{rot}} = \frac{(M^2)_{\text{op}}}{2\mu R^2} \tag{14-30}$$

where $(M^2)_{\text{op}}$ stands for the square of the angular momentum. It may be shown that

$$\mathfrak{H}_{\text{int}} = T + U(R) = -\frac{\hbar^2}{2\mu}\left(\frac{\partial^2}{\partial R^2}+\frac{2}{R}\frac{\partial}{\partial R}\right)+\frac{(M^2)_{\text{op}}}{2\mu R^2}+U(R) \tag{14-31}$$

The solutions of the rigid rotor, represented by $\mathfrak{H}_{\text{rot}}$ have been discussed extensively in Chapter 11. They are given by

$$\mathfrak{H}_{\text{rot}}\Psi_J = \frac{\hbar^2}{2\mu R^2}J(J+1)\Psi_J \tag{14-32}$$

We should realize that the eigenfunctions Ψ_J are functions of the polar angles θ and ϕ only because the rotational motion depends only on the orientation of the molecular axis.

We can separate the rotational and vibrational motions of the molecule by writing the function Φ_{int} of Eq. (14-27) as

$$\Phi_{int} = \Phi_{vib}(R)\Psi_J(\theta, \phi) \tag{14-33}$$

If we substitute this into the Schrödinger equation

$$\mathfrak{H}_{int}\Phi_{int} = E_{int}\Phi_{int} \tag{14-34}$$

we can separate this equation into the two parts

$$\mathfrak{H}_{rot}\Psi_J = \frac{(M^2)_{op}}{2\mu R^2}\Psi_J = E_J\Psi_J \tag{14-35}$$

for the rotational motion and

$$\mathfrak{H}_{vib}\Phi_{vib} = \left[-\frac{\hbar^2}{2\mu}\left(\frac{\partial^2}{\partial R^2} + \frac{2}{R}\frac{\partial}{\partial R} - \frac{J(J+1)}{R^2}\right) + U(R)\right]\Phi_{vib} = E_{vib}\Phi_{vib} \tag{14-36}$$

for the vibrational motion. The rotational equation is the equation for the rigid rotor that we discussed in Chapter 11. The vibrational motion may be approximated by the harmonic oscillator, but we have to discuss the approximations involved.

In order to solve the vibrational equation of motion we must know the form of the vibrational function $U_n(R)$ for the electronic state n, and in most cases we do not know the exact form of this function. However, we can find an approximate solution by considering the general behavior of the potential curves shown in Figure 14-1. We consider the ground state configuration of a stable molecule. The potential curve $U_0(R)$ should then have a minimum at a point R_0, which we call the equilibrium nuclear distance. In the vicinity of this minimum, we may expand the function $U_0(R)$ as power series in $(R - R_0)$

$$U_0(R) = U_0(R_0) + \left(\frac{\partial U_0}{\partial R}\right)_{R_0}(R - R_0) + \frac{1}{2}\left(\frac{\partial^2 U_0}{\partial R^2}\right)_{R_0}(R - R_0)^2 + \cdots \tag{14-37}$$

Since $U_0(R)$ has a minimum at the point R_0, its derivative at that point is zero. Furthermore, if we neglect the higher terms in the power series expansion, we may approximate $U_0(R)$ as

$$U_0(R) = U_0(R_0) + \frac{1}{2}k(R - R_0)^2 \tag{14-38}$$

with

$$k = \left(\frac{\partial^2 U_0}{\partial R^2}\right)_{R_0} \tag{14-39}$$

In this way we have approximated the potential function $U_0(R)$ by a parabola, and it may be seen in Figure 14-3 how good (or bad) this approximation is. Obviously, it is a very poor approximation for large values of R, and it even has the wrong asymptotic behavior for $R \to \infty$. On the other hand, it seems a fairly good approximation in the vicinity of the point R_0. This means, in the quantum mechanical description, that the lower vibrational eigenstates are adequately described by approximation (14-38) but the approximation breaks down for the higher vibrational levels. We should thus limit our theoretical description to the situation where $R \approx R_0$.

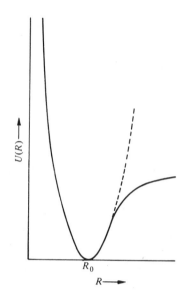

Figure 14-3   A potential function $U(R)$ approximated by a parabola near the minimum.

Let us now substitute approximation (14-38) into the Schrödinger equation (14-36), bearing in mind that $R \approx R_0$

$$\left[-\frac{\hbar^2}{2\mu}\left(\frac{d^2}{dR^2} + \frac{2}{R}\frac{d}{dR} - \frac{J(J+1)}{dR} \right) + U_0(R_0) + \tfrac{1}{2}k(R-R_0)^2 \right]\Phi_{vib} = E_{vib}\Phi_{vib} \quad \textbf{(14-40)}$$

We introduce the displacement coordinate

$$q = R - R_0 \qquad\qquad\qquad \textbf{(14-41)}$$

which describes the vibration of the nuclei with respect to their equilibrium positions. The vibrational equation (14-40) may then be reduced to the harmonic oscillator equation

$$-\frac{\hbar^2}{2\mu}\frac{\partial^2 \psi}{\partial q^2} + \tfrac{1}{2}kq^2\psi = \varepsilon\psi \qquad\qquad \textbf{(14-42)}$$

EXAMPLE 14-3

It may not be obvious how we arrive from Eq. (14-40) to Eq. (14-42). First we make the substitution

$$\Phi_{vib} = \frac{\psi}{R} \qquad \psi = R\Phi_{vib} \qquad\qquad \textbf{(a)}$$

and we note that

$$\frac{d^2\psi}{dR^2} = R\frac{d^2\Phi_{vib}}{dR^2} + 2\frac{d\Phi_{vib}}{dR} \qquad\qquad \textbf{(b)}$$

Substitution into the Schrödinger equation (14-40) gives

$$-\frac{\hbar^2}{2\mu}\frac{\partial^2\psi}{\partial R^2} + \tfrac{1}{2}k(R-R_0)^2\psi = \left(E_{vib} - U_0(R_0) - \frac{\hbar^2 J(J+1)}{2\mu R^2} \right)\psi \qquad \textbf{(c)}$$

Since the amplitude of the vibration is small, we may replace R in the rotational energy expression by its equilibrium value R_0

$$\frac{J(J+1)}{R^2} \approx \frac{J(J+1)}{R_0^2} \tag{d}$$

It is then possible to substitute

$$\varepsilon = E_{vib} - U_0(R_0) - \frac{\hbar^2 J(J+1)}{2\mu R_0^2} \tag{e}$$

because the two terms on the right of Eq. (e) are both constants. If we finally introduce the displacement coordinate of Eq. (14-41) we obtain

$$-\frac{\hbar^2}{2\mu}\frac{d^2\psi}{dq^2} + \tfrac{1}{2}kq^2\psi = \varepsilon\psi \tag{f}$$

which is the harmonic oscillator equation. ●

In Chapter 11 we saw that the eigenvalues of Eq. (14-42) are given by

$$E_n = (n+\tfrac{1}{2})\hbar\omega_0 = (n+\tfrac{1}{2})h\nu_0 \tag{14-43}$$

with

$$\begin{aligned}\omega_0 &= (k/\mu)^{1/2}\\ \nu_0 &= \omega_0/2\pi\end{aligned} \tag{14-44}$$

The corresponding eigenfunctions are the functions ψ_n that we defined in Eq. (11-21).

The approximation (14-38), where we replaced the true potential function $U_n(R)$ by a parabola, is known as the harmonic approximation because it reduces the vibrational motion to a harmonic oscillator.

If we want to consider an excited electronic state of the molecule, we can follow the same procedure as for the ground state but we should realize that there are some differences. For one thing, the potential curve $U_k(R)$ that belongs to the electronic state k is different from the curve $U_0(R)$ for the ground state (see Figure 14-1). We can only observe vibrational energy levels if the potential curve has a minimum, and the minimum for the curve $U_k(R)$ is usually located at a different equilibrium value R_k than the ground state equilibrium distance R_0. It is an empirical fact that the equilibrium distances R_k are usually larger than the ground state value R_0. This means that the nuclei are further apart in excited electronic states. The frequency ω_k for the upper vibrational level is also different from the ground state frequency ω_0 and, in general, we find that ω_k is much smaller than ω_0 because the upper potential curve is flatter than the lower one (see Figure 14-1).

It may be useful to summarize the results for the molecular eigenvalues and eigenfunctions that are obtained from the various approximations that we have used. The molecular energy levels $E_{n,v,J}$ are obtained in the form

$$E_{n,v,J} = U_n(R_n) + (v+\tfrac{1}{2})h\nu_n + B_{n,v}J(J+1) \tag{14-45}$$

The first term represents the electronic energy, the second term the vibrational energy, and the third term the rotational energy. We have already mentioned that the vibrational frequency ν_n depends on the electronic quantum number n. The rotational constant $B_{n,v}$ is given by

$$B_{n,v} = (\hbar^2/2\mu R_n^2) \tag{14-46}$$

and it depends both on the electronic quantum number n and on the vibrational quantum number v. It is obvious why $B_{n,v}$ depends on n; it contains the equilibrium distance R_n. Realize that $B_{n,v}$ also depends on the quantum number v because we made the approximation of replacing the actual internuclear distance R by the equilibrium distance R_n, and this approximation is less accurate for the higher vibrational states.

The eigenfunctions that correspond to the eigenvalues $E_{n,v,J}$ of Eq. (14-45) have the form

$$\Psi_{n,v,J}(r; R, \theta, \phi) = F_n(r; R)f_v^n(R)\psi_J(\theta, \phi) \tag{14-47}$$

They may be written as a product of an electronic eigenfunction $F_n(r; R)$, a vibrational function $f_v^n(R)$, which also depends on the quantum number n, and a rotational function $\psi_J(\theta, \phi)$. Remember that the specific forms of the vibrational and rotational eigenfunctions were discussed in Chapter 11.

14-3 MOLECULAR SYMMETRY

In Section 14-2 we considered only the frequencies of the various spectral lines. It is clear that additional information may be derived by also considering the intensities. In general, it is not easy to calculate the absolute intensity of a given transition, but it is often possible to predict the relative intensities of a progression of lines within an absorption or emission band. The most useful tool for correlating the intensities of spectroscopic intensities with the molecular structure is the selection rules. These are rules that predict whether a given transition is allowed or forbidden. Remember that we define a forbidden transition as a transition for which the electric dipole transition moment is zero.

Most selection rules are related to the symmetry properties of the eigenstates that are involved in the transition. We realize that a rigorous treatment of the symmetry properties of molecular eigenfunctions should be based on the branch of mathematics that is known as group theory. However, we feel that we can discuss the few theorems that are needed for our purpose without making use of group theory.

All diatomic molecules have cylindrical symmetry. In terms of mathematics this means that the potential function for the electrons (or the electronic Hamiltonian) remains the same if we rotate the molecule around its axis. It may then be derived that, if we rotate the molecule around its internuclear axis by an angle ϕ, each electronic eigenfunction changes by an amount

$$e^{im\phi}$$

where m is an integer

$$m = 0, \pm 1, \pm 2, \ldots$$

Each electronic eigenstate is thus characterized by a value of the quantum number m, and in describing the eigenstate the value of m is denoted by a symbol. For example, states with $m = 0$ are called Σ states; states with $m = \pm 1$ are Π states; states with $m = \pm 2$ are known as Δ states; and so on.

From a physical point of view, the projection M_z of the total electronic angular momentum is a constant of the motion, and its magnitude is described by the quantum number m. The magnitude of M_z is usually denoted by the symbol Λ, which is the absolute value of the quantum number m

$$|M_z| = \Lambda \hbar \qquad \Lambda = 0, 1, 2, 3, \ldots \qquad \textbf{(14-48)}$$

In a Σ state $\Lambda = 0$; in a Π state $\Lambda = 1$; and so on. This nomenclature is similar to the hydrogen atom, where the letters s, p, d, and so on, were used to denote the magnitude of the angular momentum.

A homonuclear diatomic molecule A_2 has, in addition to cylindrical symmetry, a center of inversion that is the center of gravity or the midpoint of the internuclear axis. As a result, it may be shown that each molecular eigenfunction ψ_n must be either symmetric or antisymmetric with respect to inversion. A symmetric eigenfunction obeys the relation

$$\psi_n(\mathbf{r}) = \psi_n(-\mathbf{r}) \qquad \textbf{(14-49)}$$

Here $\psi_n(-\mathbf{r})$ is obtained by replacing every electron coordinate $\mathbf{r}_i$, defined with respect to the center of inversion as origin, by its negative value $-\mathbf{r}_i$. Similarly an antisymmetric eigenfunction must obey the relation

$$\psi_n(\mathbf{r}) = -\psi_n(-\mathbf{r}) \qquad \textbf{(14-50)}$$

EXAMPLE 14-4

If a molecule has a center of inversion, its Hamiltonian must be symmetric with respect to inversion

$$\mathfrak{H}(\mathbf{r}) = \mathfrak{H}(-\mathbf{r}) \qquad \textbf{(a)}$$

Every eigenfunction $\psi_n(\mathbf{r})$ must be a solution of the Schrödinger equation

$$\mathfrak{H}(\mathbf{r})\psi_n(\mathbf{r}) = E_n\psi_n(\mathbf{r}) \qquad \textbf{(b)}$$

and it must also obey the equation

$$\mathfrak{H}(-\mathbf{r})\psi_n(-\mathbf{r}) = E_n\psi_n(-\mathbf{r}) \qquad \textbf{(c)}$$

or

$$\mathfrak{H}(\mathbf{r})\psi_n(-\mathbf{r}) = E_n\psi_n(-\mathbf{r}) \qquad \textbf{(d)}$$

If the eigenvalue E_n is nondegenerate, there is only one eigenfunction, the two functions $\psi_n(\mathbf{r})$ and $\psi_n(-\mathbf{r})$ must then be proportional to one another

$$\psi_n(\mathbf{r}) = \delta\psi_n(-\mathbf{r}) \qquad \textbf{(e)}$$

If we use this equation twice we find that

$$\delta^2 = 1 \qquad \delta = \pm 1 \qquad \textbf{(f)}$$

This means that $\psi_n(\mathbf{r})$ is either symmetric or antisymmetric with respect to inversion.

If the eigenvalue E_n is degenerate, its eigenfunctions may be divided into a set of symmetric and a set of antisymmetric eigenfunctions and we may also say that every eigenfunction ψ_n is either symmetric or antisymmetric. ●

If the eigenfunction of a molecular eigenstate is symmetric with respect to inversion then we indicate this by a subscript g, this is the first letter of the German word "gerade," meaning "even." If the eigenfunction is antisymmetric with respect to inversion then we denote this by a subscript u, this is derived from the German work "ungerade," meaning "odd."

Finally, the spin multiplicity of a molecular eigenstate is usually described by a superscript which has the value $2S + 1$. In this way, a superscript 1 refers to a singlet state, 2 is a doublet, 3 a triplet, and so on.

The lowest electronic eigenstates of the oxygen molecule O_2 are $^3\Sigma_g$, $^1\Delta_g$, $^1\Sigma_g$ and $^3\Sigma_u$. We see that the groundstate is a triplet with spin $S = 1$, its eigenfunction is symmetric with respect to inversion, and it does not change if we rotate around the internuclear axis. The next two states are singlets, both gerade; the first is a Δ state so that its wave function changes by a factor $e^{2i\phi}$ if we rotate around the molecular axis by an angle ϕ, the second is a Σ state and its wave function is not affected by rotations. The next excited state is again a triplet; it is ungerade and its wave function has cylindrical symmetry.

In the subsequent discussions we will use the symmetry properties of molecular eigenfunctions for describing some of the spectroscopic selection rules.

14-4 SELECTION RULES AND SPECTRAL INTENSITIES IN ELECTRONIC BANDS

We have seen in Section 14-2 that an eigenstate of a diatomic molecule is characterized by three quantum numbers, n, which describes the electronic state, v, which describes the vibrational state, and J, which refers to the rotational state. In general, a spectroscopic transition may occur between two states (n, v, J) and (n', v', J'), but in calculating transition probabilities it is not practical to try to derive a general expression for the transition probabilities that applies to all situations. Instead we will divide the general problem into a number of special cases, discuss some selection rules, and make some predictions about relative intensities.

In this section we consider the situation where the two quantum numbers n and n' differ from one another. This is an electronic transition, and in most cases the light that is absorbed or emitted will be in the visible or the ultraviolet region of the spectrum. Experimentally we observe an absorption or an emission band composed of a large number of separate lines due to the various values that the quantum numbers v, v', J and J' may assume. In Chapter 13 we discussed how the intensity of each individual line is derived from the square of the transition moment, which is defined as

$$\mathbf{P}(n, v, J; n', v', J') = \langle \Psi_{n,v,J} | \mathbf{\mu}_{op} | \Psi_{n',v',J'} \rangle \tag{14-51}$$

Here $\mathbf{\mu}_{op}$ is defined as the electric dipole moment operator

$$\mathbf{\mu}_{op} = e \sum_j \mathbf{r}_j \tag{14-52}$$

where we have to sum over all electrons in the molecule. For the eigenfunction Ψ we substitute the approximate expression of Eq. (14-47)

$$\Psi_{n,v,J}(r; R, \theta, \phi) = F_n(r; R)f_v^n(R)\Psi_J(\theta, \phi) \tag{14-53}$$

where r stands symbolically for all the electron coordinates.

The first term represents the electronic part of the wave function; the second part represents the vibrational part; and the third part represents the rotational part. If we substitute this function (14-53) into Eq. (14-51) for the transition moment, we obtain

$$\mathbf{P}(n, v, J; n', v', J') = \langle F_n(r; R)f_v^n(R)\Psi_J(\theta, \phi)|\boldsymbol{\mu}_{\text{op}}(r)|F_{n'}(r; R)f_{v'}^{n'}(R)\Psi_{J'}(\theta, \phi)\rangle \tag{14-54}$$

for the transition moment of the transition $(n, v, J) \rightarrow (n', v', J')$. Here we must integrate over the electron coordinates, the vibrational coordinates, and the rotational coordinates.

In order to understand the general properties of the transition moment it may be helpful to look at the situation illustrated in Figure 14-4. Here we have assumed that we have polarized light. This may be assumed without loss of generality, and it makes the situation a lot easier. We take the z axis along the direction of polarization of the light and the origin as the center of gravity of the molecule. The molecular axis makes an angle θ with the z axis, and the angle θ together with the second polar angle ϕ describe the orientation of the molecular axis. The two angles θ and ϕ also describe the rotational motion of the molecule. The vibrational motion of the molecule is described by the distance AB, which we denote by R.

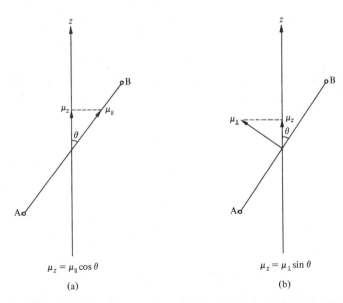

$$\mu_z = \mu_{\parallel} \cos \theta \qquad\qquad \mu_z = \mu_{\perp} \sin \theta$$

(a)   (b)

Figure 14-4 Electronic transition moments. The light is polarized along the z axis. (a) The electronic transition moment $\mu_{\parallel}$ is along the molecular axis and $\mu_z = \mu_{\parallel} \cos \theta$. (b) The electronic transition moment $\mu_{\perp}$ is perpendicular to the molecular axis and $\mu_z = \mu_{\perp} \sin \theta$.

The electronic transition moment $\mu_{n,n'}(R)$ is now defined as

$$\mu_{n,n'}(R) = \int F_n^*(r, R)\mu_{op}F_{n'}(r, R)\, dr \qquad (14\text{-}55)$$

where we integrate over the electron coordinates r. These electron coordinates are defined with respect to a coordinate system that is attached to the molecular axis. It is important to realize that this coordinate system is different from the space-fixed coordinate system that we use to describe the radiation field.

There are three sets of selection rules for electronic transitions. The first selection rule states that the electronic transition moment is different from zero only if one of the two states, n or n', is gerade and the other is ungerade. According to this selection rule the only allowed transitions are

$$g \rightarrow u \qquad u \rightarrow g \qquad\qquad (14\text{-}56)$$

The other transitions

$$g \rightarrow g \qquad u \rightarrow u \qquad\qquad (14\text{-}57)$$

are forbidden.

The second selection rule

$$\Delta S = 0 \qquad\qquad (14\text{-}58)$$

requires that the two electronic states have the same spin functions. If the spins of the two states are different from each other, the molecular eigenfunctions have orthogonal spin parts and the corresponding transition moment is zero.

The third selection rule is related to the quantum number Λ that we described in Eq. (14-48), this quantum number represents the magnitude of M_z. It may be shown that the electronic transition moment $\mu_{n,n'}$ of Eq. (14-55) must be either directed along the molecular axis or perpendicular to the molecular axis. In the first case we call it $\mu_\parallel$ (see Figure 14-4a), and this transition is allowed only if

$$\Delta\Lambda = 0 \qquad\qquad (14\text{-}59)$$

This means that the values of the quantum numbers Λ for the two states n and n' must be the same. In the second case it is called $\mu_\perp$ (see Figure 14-4b) and this is nonzero only if the values of Λ for the two states differ by unity

$$\Delta\Lambda = \pm 1 \qquad\qquad (14\text{-}60)$$

It follows that the selection rule for the quantum numbers Λ is given by

$$\Delta\Lambda = 0, \pm 1 \qquad\qquad (14\text{-}61)$$

To be more specific, the selection rule is

$$\begin{array}{lll}
\Delta\Lambda = 0 & \mu_\parallel \neq 0 & \mu_\perp = 0 \\
\Delta\Lambda = \pm 1 & \mu_\parallel = 0 & \mu_\perp \neq 0
\end{array} \qquad (14\text{-}62)$$

EXAMPLE 14-5

Let us apply the above selection rules to some specific cases. A $^1\Sigma_g$ state combines with a $^1\Sigma_u$ state ($\Delta\Lambda = 0$) to give $\mu_\parallel$ and with a $^1\Pi_u$ state ($\Delta\Lambda = 1$) to give $\mu_\perp$. A $^1\Pi_u$ state combines with a $^1\Pi_g$ state to give $\mu_\parallel$ and with a $^1\Sigma_u$ and a $^1\Delta_u$ state to give $\mu_\perp$. Other transitions are not allowed.   ●

Now we want to consider the intensities or transition moments of transitions between different electronic states. We should realize that absolute values of these transition probabilities are hard to determine, both theoretically and experimentally, and for that reason we are not too interested in them. The only question that we want to answer is whether a certain transition is allowed or forbidden, and that answer is derived from the preceding selection rules.

Even though we may not know the absolute value of a certain electronic transition, we can make quite a few predictions about the relative intensities of the many separate transitions that occur in the same electronic band. These separate transitions occur between the various vibrational and rotational levels that belong to the same two electronic states n and n'. For all these separate transitions the electronic transition moment $\mu_{n,n'}$ has the same value, and it may be considered as a parameter that describes the net intensity of each line. We will see that we can predict the relative intensities of the various lines in a specific electronic band quite accurately, even though we do not know too much about the absolute intensity. Our discussion is therefore concerned with relative intensities only.

Let us now return to the situation illustrated in Figure 14-4, where we have a beam of light that is polarized in the z direction. According to Eq. (14-54) the transition moment between the two states (n, v, J) and (n', v', J') is then given by

$$\mathbf{P}(n, v, J; n', v', J') = \langle F_n(r; R) f_v^n(R) \Psi_J(\theta, \phi) | \mu_z(r) | F_{n'}(r; R) f_{v'}^{n'}(R) \Psi_{J'}(\theta, \phi) \rangle \tag{14-63}$$

This expression must be integrated successively over the electronic coordinates r, the vibrational coordinate R, and the rotational coordinates (θ, ϕ). We may substitute expression (14-55) for the electronic transition moment, but we should take its projection along the z axis. As we show in Figure 14-4, in the first case where the electronic transition moment is parallel to the molecular axis we should substitute

$$\mu_z(R) = \cos \theta \mu_\parallel(R) = \cos \theta \mu_{n,n'}(R) \tag{14-64}$$

into Eq. (14-63). The result is

$$P(n, v, J; n', v', J') = \langle f_v^n(R) \Psi_J(\theta, \phi) | \cos \theta \mu_\parallel(R) | f_{v'}^{n'}(R) \Psi_{J'}(\theta, \phi) \rangle$$
$$= \langle f_v^n(R) | \mu_\parallel(R) | f_{v'}^{n'}(R) \rangle \langle \Psi_J(\theta, \phi) | \cos \theta | \Psi_{J'}(\theta, \phi) \rangle \tag{14-65}$$

In the second case we should substitute

$$\mu_z(R) = \sin \theta \mu_\perp(R) = \sin \theta \mu_{n,n'}(R) \tag{14-66}$$

and the result is

$$P(n, v, J; n', v', J') = \langle f_v^n(R) | \mu_\perp(R) | f_{v'}^{n'}(R) \rangle \langle \Psi_J(\theta, \phi) | \sin \theta | \Psi_{J'}(\theta, \phi) \rangle \tag{14-67}$$

It follows that the transition moment is the product of a vibrational and a rotational part and that these parts may be considered separately. The rotational part depends on the orientation of the electronic transition moment. We consider the vibrational part first. This part behaves differently for absorption and emission so that we must also consider these two situations separately.

In an electronic absorption band the electronic quantum number n of the initial state must be zero, representing the electronic ground state. In addition the vibrational quantum number in the initial state must also be zero. It may be seen from

Table 14-2 that the vibrational frequencies of diatomic molecules are fairly large; thus, at ordinary temperatures only a very small fraction of molecules are in excited vibrational states due to the thermal distribution over these states. We may assume that in a typical sample practically all molecules will be in their vibrational ground states. In an absorption band we measure transitions from an initial state $(0, 0, J)$ to a set of excited states (k, v, J').

According to either Eq. (14-65) or (14-67) the vibrational transition moments in an electronic absorption band are given by

$$P(0, 0; k, v) = \langle f_0^0(R)|\mu_{0,k}(R)|f_v^k(R)\rangle_R$$
$$\mu_{0,k}(R) = \langle F_0(r)|\mu_\alpha(r)|F_k(r)\rangle_r \tag{14-68}$$

Here the subscripts R and r indicate that in the first integral we integrate over the vibrational coordinate R and in the second integral we integrate over the electronic coordinates r. The subscript α is either $\parallel$ or $\perp$, but the two situations are treated in the same way.

The electronic transition moment $\mu_{0,k}(R)$ of Eq. (14-68) is a function of the internuclear distance R but it may be assumed that it varies very slowly over the range that is covered by the molecular vibrations. This means that in the integral (14-68) it may be approximated as

$$\mu_{0,k}(R) \approx \mu_{0,k}(R_0) \tag{14-69}$$

where $\mu_{0,k}(R_0)$ is the value at the equilibrium distance of the electronic ground state. If we substitute approximation (14-69) into integral (14-68) then $\mu_{0,k}(R_0)$ is a constant so that we obtain

$$P(0, 0; k, v) = \mu_{0,k}(R_0)\langle f_0^0(R)|f_v^k(R)\rangle = \mu_{0,k}(R_0)I_{0,0;k,v} \tag{14-70}$$

The integrals $I_{0,0;k,v}$ are called overlap integrals. They contain the product of the vibrational eigenfunctions of the electronic ground state and of the excited electronic state k. They depend very strongly on the relative positions of the potential curves $U_0(R)$ and $U_k(R)$ for the two electronic states.

In Figure 14-5 we have drawn the two curves $U_0(R)$ and $U_k(R)$ for a typical situation in a diatomic molecule where the electronic state k has a stable configuration. It is known experimentally that the equilibrium distances R_k for excited states are generally larger than the equilibrium distance R_0 for the corresponding ground state, the differences are of the order of 0.1 Å. Also, the curvature of the upper curve is usually smaller than for the lower curve. This means that the force constant and the frequency ω for the higher electronic state are smaller than the corresponding quantities for the electronic ground state. This difference may be quite large; the frequencies may differ by a factor of 2 or more.

We consider the various transitions between the vibrational levels $v = 0, 1, 2, 3, \ldots$, of the electronic state k and the lowest vibrational level $v = 0$ of the electronic ground state 0 because these are the transitions that are observed in the spectrum. The term **vibronic** is used to denote an electronic–vibrational energy level. Accordingly we consider transitions between the vibronic states (k, v) and the vibronic state $(0, 0)$. The situation is illustrated in Figure 14-5. It may be seen here that the energy differences between the vibronic states are given by

$$\Delta\varepsilon_{k,v} = U_k(R_k) + (v + \tfrac{1}{2})h\nu_k - U_0(R_0) - \tfrac{1}{2}h\nu_0$$
$$= [U_k(R_k) - U_0(R_0) + \tfrac{1}{2}h(\nu_k - \nu_0)] + vh\nu_k \qquad v = 0, 1, 2, 3, \ldots \tag{14-71}$$

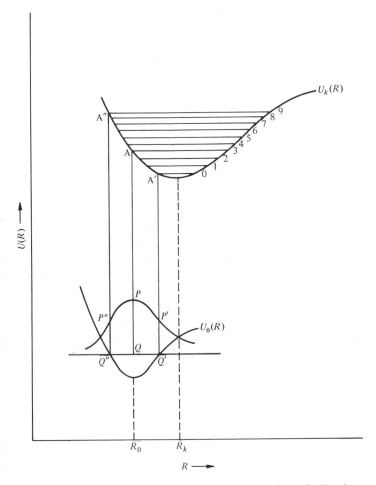

Figure 14-5   Vibrational fine structure of an electronic absorption band.

Here ν_k is the vibrational frequency of the electronic state k, and ν_0 is the vibrational frequency of the electronic ground state 0. We use the harmonic approximation for simplicity's sake even though its accuracy may leave something to be desired for the higher vibrational levels.

According to Eq. (14-71) the spectrum consists of a number of equidistant lines, separated by the vibrational frequency ν_k. It is important to note that in an absorption band the vibrational separation of the lines is the frequency ν_k belonging to the excited state curve $U_k(R)$.

Let us now consider the relative intensities of these lines. According to Eq. (14-70) the transition moments are proportional to the overlap integrals $I_{0,0;k,v}$ so that the intensity of the transition $(0, 0) \rightarrow (k, v)$ is proportional to the square of this overlap integral

$$w_v = (\langle f_0^0(R)| f_v^k(R)\rangle)^2 \tag{14-72}$$

The relative intensities of the various lines are thus given by Eq. (14-72).

The overlap integrals of Eq. (14-72) are called Franck–Condon factors. They may be calculated exactly from the harmonic oscillator functions of the two electronic states, but their magnitudes may also be estimated by making use of the Franck–Condon approximation. We also show this in Figure 14-5. Here we have sketched the ground state vibrational function of the vibronic state $(0, 0)$. This is a Gaussian of the form

$$\exp\left[-\lambda(R-R_0)^2\right] \tag{14-73}$$

Let us now consider, for example, the transition $(k, 9) \rightarrow (0, 0)$. Here we draw a line for the energy level $v = 9$ and we look to see where this intersects with the curve $U_k(R_k)$ on the left. If we then draw a vertical line from this intersection point A'' the overlap integral is about equal to the length of line $P''Q''$, which is the value of the ground state eigenfunction (14-73) at the intersection point. The relative intensity is then the square of the length of line $P''Q''$.

In Figure 14-6 we show the relative intensity curves for a few typical situations. The overall intensity distribution always has the form of a Gaussian because the square of a Gaussian function (14-73) is also a Gaussian. The first curve (a) represents the situation of Figure 14-5. Here it may be seen that the $0 \rightarrow 3$ transition is the one with the highest intensity. Towards the left, with decreasing intensity we find the transitions $0 \rightarrow 2$, $0 \rightarrow 1$ and $0 \rightarrow 0$. The latter is the cutoff because there are no transitions to the left of it. The result is a Gaussian which is truncated on the left. Curve (b) represents the situation where $R_k - R_0$ is larger; that is, the two potential curves are shifted further with respect to one another. Here the transition $0 \rightarrow 6$ has the highest intensity and there is practically no cutoff on the left; this is a situation where $R_k - R_0$ is larger than in Figure 14-5. Finally, in curve (c) $R_k - R_0$ is small, and the cutoff point is closer to the maximum. It follows that we can derive the relative shift of the two potential curves from the cutoff point of the vibrational intensity distribution curve.

The Franck–Condon approximation may be explained either by classical or by quantum mechanics. In the classical explanation we note that the nuclei move much slower than the electrons so that, during an electronic transition, the nuclei remain in the same configuration. The quantum mechanical explanation follows from the properties of the harmonic oscillator wave functions, and it leads to the same results.

We must recognize that the potential curve is not really a parabola but that it has the form of Figure 14-1, for large values of R it approaches a constant value $U_k(\infty)$ and the harmonic approximation is no longer valid. There are only discrete vibra-

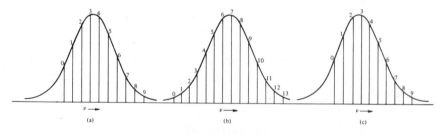

Figure 14-6 Relative intensities of vibrational lines. (a) Corresponds to Figure 14-5. (b) The difference $R_k - R_0$ is larger than in Figure 14-5. (c) The difference $R_k - R_0$ is smaller.

tional levels below energy $U_k(\infty)$; for higher energies the energy levels form a continuum. Therefore, if the frequency of the absorbed light becomes larger than $U_k(\infty) - U_0(R_0)$, the molecule is excited into a continuum of states and there is no structure left in the absorption band. In addition, we excite to a molecular state that is not stable, and the molecule will dissociate. We see that for this particular electronic transition the dissociation energy is given by

$$E_{\text{dis}} = U_k(\infty) - U_0(R_0) - \tfrac{1}{2}h\nu_0 \tag{14-74}$$

An electronic emission band may be treated the same way as an absorption band, but the result is different. In the emission the molecules are initially in the state $(k, 0)$, and we observe transitions to the various vibronic levels $(0, v)$ of the electronic ground state (see Figure 14-7). The energy differences for these transitions are given by

$$\Delta\varepsilon_{0,v} = U_k(R_k) + \tfrac{1}{2}h\nu_k - U_0(R_0) - (v + \tfrac{1}{2})h\nu_0$$
$$= [U_k(R_k) - U_0(R_0) + \tfrac{1}{2}h(\nu_k - \nu_0)] - vh\nu_0 \qquad v = 0, 1, 2, 3, \ldots \tag{14-75}$$

If we compare Eq. (14-75) with Eq. (14-71) for an absorption band, we see that the

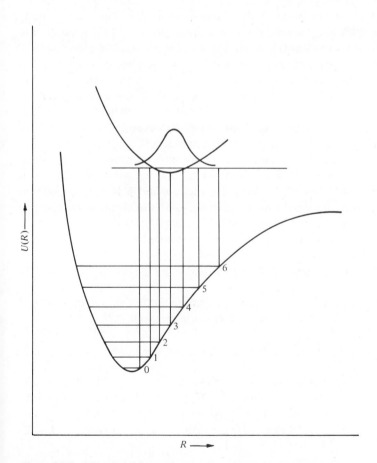

Figure 14-7   Vibrational fine structure of an electronic emission band.

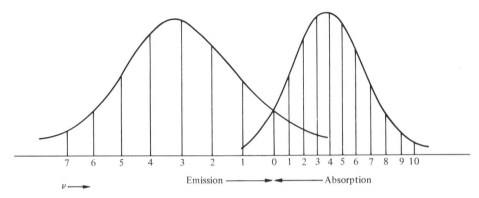

$v \longrightarrow$

Emission ————————►◄———————— Absorption

Figure 14-8 Combined absorption and emission spectra, the vibrational fine structure of both.

two expressions are equal for $v = 0$. In emission we have, in addition, a set of equidistant lines at lower frequencies separated by an amount ν_0. In absorption we have a set of equidistant lines at higher frequencies separated by an amount ν_k. We show both spectra in Figure 14-8; at the right side we have the absorption spectrum and at the left the emission spectrum, both spectra have the $0 \to 0$ line in common.

In some cases the probability for transitions into the continuum of the ground state potential curve becomes significant. If such a transition occurs, the molecule will dissociate. There are, thus, two possible mechanisms for the photodissociation of a diatomic molecule. In the first mechanism we excite the molecule directly into the vibrational continuum of an excited electronic state. In the second mechanism we excite the molecule first into a discrete vibrational level of the excited state (k). The molecule then falls back into one of the continuum vibrational states of the electronic state (0), either directly or via intermediate vibrational levels.

In order to discuss the rotational structure within an electronic band we must return to the general expressions (14-65) and (14-67) for the rotational transition moment, that is the part that depends on the polar angles θ and ϕ. It may be seen that the selection rules for the rotational transitions depend on the symmetry of the two electronic states involved. If the electronic transition moment is directed along the molecular axis, the rotational transition moments are given by

$$P_{J,J'} = \int \int \Psi_J^*(\theta, \phi) e^{i\phi} \sin \theta \Psi_{J'}(\theta, \phi) \sin \theta \, d\theta \, d\phi \qquad (14\text{-}77)$$

If the electronic transition moment is directed perpendicular to the molecular axis, it follows from Eq. (14-67) that the rotational transition moment is

$$P_{J,J'} = \int \int \Psi_J^*(\theta, \phi) e^{i\phi} \sin \theta \Psi_{J'}(\theta, \phi) \sin \theta \, d\theta \, d\phi \qquad (14\text{-}77)$$

The first integral (14-76) is always zero unless $\Delta J = J - J' = \pm 1$. The second integral (14-77) is nonzero if $\Delta J = 0$ or $\Delta J = \pm 1$. The result is, thus, that we have the selection rule

$$\Delta J = \pm 1 \qquad \text{if } \Delta\Lambda = 0 \qquad (14\text{-}78)$$

and

$$\Delta J = 0, \pm 1 \qquad \text{if } \Delta\Lambda = 1 \qquad (14\text{-}79)$$

The specific values of the transition moments $P_{J,J'}$ of Eqs. (14-76) and (14-77) may also be derived from the rotational eigenfunctions, but these expressions are fairly complicated so that we will not discuss them here.

Let us now consider the frequencies of the rotational transitions that are allowed for given values of the vibrational and electronic quantum numbers of the upper and the lower states. It follows from Eq. (14-45) that the energies of the upper and lower states are given by

$$E_{k,w,J'} = U_k(R_k) + (w + \tfrac{1}{2})h\nu_k + B_{k,w}J'(J'+1)$$
$$E_{0,v,J} = U_0(R_0) + (v + \tfrac{1}{2})h\nu_0 + B_{0,v}J(J+1)$$

(14-80)

The difference between the upper and lower states may be written as

$$\Delta E_{J,J'} = h\nu_{J,J'} = \Delta\varepsilon_0 + B'J'(J'+1) - BJ(J+1) \qquad (14\text{-}81)$$

where the definition of the various quantities is self-explanatory. If we substitute the possible values $J' = J$ and $J' = J \pm 1$ we obtain three possible expressions for the rotational lines in the spectrum

$$R(J) = \Delta E_{J,J+1} = \Delta\varepsilon_0 + B'(J+1)(J+2) - BJ(J+1) = \Delta\varepsilon_0 + 2B'$$
$$\qquad + (3B' - B)J + (B' - B)J^2$$

$$Q(J) = \Delta E_{J,J} = \Delta\varepsilon_0 + B'J(J+1) - BJ(J+1) = \Delta\varepsilon_0 + (B' - B)J + (B' - B)J^2 \qquad (14\text{-}82)$$

$$P(J) = \Delta E_{J,J-1} = \Delta\varepsilon_0 + B'J(J-1) - BJ(J+1) = \Delta\varepsilon_0 - (B' + B)J + (B' - B)J^2$$

These expressions give the frequencies of the spectral lines as a function of an integer J which can in principle have the values $0, 1, 2, \ldots$. The distribution depends on the selection rule. It is customary to denote the lines that belong to transitions $J' = J + 1$ by the R branch of the spectrum, the lines that are due to $J' = J$ are called the Q branch and the lines that are due to $J' = J - 1$ are known as the P branch. The P and the R branch may both be described by one equation

$$\Delta E_n = \Delta\varepsilon_0 + (B' + B)n + (B' \overset{\backslash}{-} B)n^2 \qquad (14\text{-}83)$$

For negative values of n, $n = -J$, Eq. (14-83) represents the P branch, and for positive values of n, $n = J + 1$, Eq. (14-83) describes the R branch.

Usually, the equilibrium distance R_k is larger in the upper electronic state than it is in the lower state; consequently, B' is usually smaller than B. We see that Eq. (14-83) is a quadratic expression in n, and, since the coefficient of n^2 is negative, it will have a maximum for some positive value of n. The P and R branches may thus be derived from a parabola, as we have sketched in Figure 14-9; they have a band head at the higher frequency end. The Q branch behaves differently.

So far we have discussed only the frequencies of the rotational lines. The relative intensities can also be calculated, but this is fairly complicated so that we will limit ourselves to an outline of the theory. We must first consider the thermal population over the various rotational levels, either in the lower electronic state in an absorption band or in the higher electronic state in an emission band. It may be seen that for most molecules the rotational constant is quite small so that even the higher rotational levels are populated to an appreciable extent. For a heavy molecule such as O_2 the rotational levels are almost equally populated at room temperature. In most cases the thermal distribution does not impose any limitation on the observation of the rotational lines, even for high values of the quantum number J. The exact

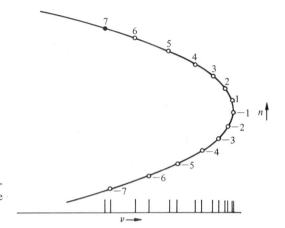

Figure 14-9 Representation of the rotational lines in the P and R branches. The lines are all derived from a parabola.

values of the rotational transition moments may be derived from the properties of the rotational eigenfunctions. The relative intensities of the rotational lines can be calculated quite accurately. The results contain the exponential factors that represent the thermal distribution over the various levels, and, from a comparison with the experimental intensities, it is even possible to determine the temperature of the sample.

In the preceding treatment we have neglected some small perturbations, in particular the interaction between the electronic and the rotational motions and also the spin-orbit coupling. Even though these perturbations are quite small, there are situations where they lead to observable effects, especially when we consider states where the quantum number Λ is not zero. However, we feel that it is beyond the scope of this book to discuss these various effects.

14-5   INTENSITIES IN THE INFRARED AND MICROWAVE REGIONS

The transitions that fall in the infrared usually occur between states that differ only in their vibrational and rotational quantum numbers. The theory of infrared intensities is based on the calculation of transition moments between states of the type $(0, 0, J)$ and $(0, v, J')$. We consider only absorption spectra, and we may assume that in the initial state all molecules are in their electronic and vibrational ground states. The transition moment is, again, given by Eq. (14-54), which now has the form

$$\mathbf{P}(0, 0, J; 0, v, J') = \langle F_0(r; R)f_0^0(R)\Psi_J(\theta, \phi)|\boldsymbol{\mu}_{\mathrm{op}}(r)|F_0(r; R)f_v^0(R)\Psi_{J'}(\theta, \phi)\rangle$$

$$(14\text{-}84)$$

The electronic part of the transition moment is equal to the electric dipole moment of the molecule as a function of the distance R

$$\boldsymbol{\mu}_0(R) = \langle F_0(r; R)|\boldsymbol{\mu}_{\mathrm{op}}(r)|F_0(r; R)\rangle_r \qquad (14\text{-}85)$$

This dipole moment is rigorously zero for homonuclear molecules, and it is to be expected that the infrared transitions are forbidden for such molecules. For heteronuclear molecules, the electric dipole moment is directed along the molecular axis and, if we make use of Eq. (14-65), we find that the transition moment reduces to

$$P(0, 0, J; 0, v, J') = \langle f_0^0(R)\Psi_J(\theta, \phi)| \cos \theta \mu_0(R)|f_v^0(R)\Psi_{J'}(\theta, \phi)\rangle \quad \textbf{(14-86)}$$

In order to calculate the transition moment we assume again that the dipole moment $\mu_0(R)$ varies only slightly in the vicinity of the equilibrium distance R_0 so that we may expand its magnitude as

$$\mu_0(R) \approx \mu_0(R_0) + \left(\frac{\partial\mu_0}{\partial R}\right)_{R_0} (R - R_0) = \mu_0(R_0) + \mu_0'(R_0)(R - R_0) \quad \textbf{(14-87)}$$

We first consider a vibrational transition where the quantum number v of the second state is different from zero. The vibrational transition moment is then given by

$$\begin{aligned} P_{0,v} &= \langle f_0^0(R)|\mu_0(R)|f_v^0(R)\rangle \\ &= \mu_0(R_0)\langle f_0^0(R)|f_v^0(R)\rangle + \mu_0'(R_0)\langle f_0^0(R)|R - R_0|f_v^0(R)\rangle \end{aligned} \quad \textbf{(14-88)}$$

The first term of Eq. (14-88) is rigorously zero because the vibrational eigenfunctions are orthogonal. The second term is, in principle, nonzero. It may be derived from the properties of the harmonic oscillator eigenfunctions that, in the harmonic approximation, this second term is nonzero only if $v = 1$.

If we use the harmonic approximation, the theory predicts only one line in the vibrational absorption spectrum, namely, the line from $v = 0$ to $v = 1$, which has the vibrational frequency ν_0. This prediction is in reasonable agreement with the experimental results. Even though vibrational transitions to states $v = 2, 3, 4,$ and 5 have been observed, these lines are extremely weak. The finite intensity of these lines is due to anharmonicity effects or to higher terms in the expansion of Eq. (14-87) that we have neglected, and a measurement of the intensities gives us some information about the magnitude of the anharmonicity effects.

From the vibrational intensity we can derive the magnitude of the quantity $\mu_0'(R_0)$. This is the derivative of the electronic dipole moment at the point R_0 with respect to the internuclear distance R.

Each vibrational line has a rotational fine structure that is easier to measure than the rotational fine structure in an electronic band because the spectral resolution is higher in the infrared. The theoretical description of the rotational fine structure is identical with the theory for electronic bands that we discussed in the previous section. The only difference is that the electronic dipole moment $\mu(R)$ is always directed along the molecular axis and, according to Eq. (14-78), the rotational selection rule must be $\Delta J = \pm 1$. This means that in vibrational transitions there is no Q branch in the rotational fine structure. It is possible to derive the values of the rotational constants $B_{0,0}$ and $B_{0,v}$ from the rotational structure and we can see how the rotational constants depend on the vibrational quantum numbers.

Pure rotational transitions between states $(0, 0, J)$ and $(0, 0, J')$ fall within the microwave region of the spectrum because the energies range between 0.01 and 50 cm^{-1}. They can be observed only if the molecule has a nonzero dipole moment, and even then they are difficult to measure because the experimental techniques are fairly difficult. The selection rules for these transitions are again $\Delta J = \pm 1$. The results lead to quite accurate values of the rotational constant B_0. The microwave spectra have been measured only for a limited number of molecules because the measurements are much more difficult than in the optical regions of the spectrum.

Appendix

APPENDIX A

The translational and internal motion of a diatomic molecule may be separated by a simple coordinate transformation. The nuclear kinetic energy is the sum of the two kinetic energy operators T_A and T_B which are given by

$$T_A = -\frac{\hbar^2}{2M_A}\left(\frac{\partial^2}{\partial x_A^2} + \frac{\partial^2}{\partial y_A^2} + \frac{\partial^2}{\partial z_A^2}\right)$$

$$T_B = -\frac{\hbar^2}{2M_B}\left(\frac{\partial^2}{\partial x_B^2} + \frac{\partial^2}{\partial y_B^2} + \frac{\partial^2}{\partial z_B^2}\right) \tag{A-1}$$

We introduce a new set of coordinates, namely the coordinates for the center of gravity

$$x_C = \frac{M_A x_A + M_B x_B}{M_A + M_B}$$

$$y_C = \frac{M_A y_A + M_B y_B}{M_A + M_B} \tag{A-2}$$

$$z_C = \frac{M_A z_A + M_B z_B}{M_A + M_B}$$

and the internal coordinates

$$x = x_B - x_A$$

$$y = y_B - y_A \tag{A-3}$$

$$z = z_B - z_A$$

The vector $\mathbf{R} = (x, y, z)$ runs from nucleus A to nucleus B.

It is easily found that

$$\frac{\partial^2}{\partial x_A^2} = \frac{\partial^2}{\partial x^2} - \frac{2M_A}{M_A + M_B}\frac{\partial^2}{\partial x \partial x_C} + \frac{M_A^2}{(M_A + M_B)^2}\frac{\partial^2}{\partial x_C^2}$$

$$\frac{\partial^2}{\partial x_B^2} = \frac{\partial^2}{\partial x^2} + \frac{2M_B}{M_A + M_B}\frac{\partial^2}{\partial x \partial x_C} + \frac{M_B^2}{(M_A + M_B)^2}\frac{\partial^2}{\partial x_C^2} \tag{A-4}$$

or

$$\frac{1}{M_A}\frac{\partial^2}{\partial x_A^2} + \frac{1}{M_B}\frac{\partial}{\partial x_B^2} = \left(\frac{1}{M_A} + \frac{1}{M_B}\right)\frac{\partial^2}{\partial x^2} + \frac{1}{M_A + M_B}\frac{\partial^2}{\partial x_C^2} \tag{A-5}$$

We introduce the reduced mass μ

$$\frac{1}{\mu} = \frac{1}{M_A} + \frac{1}{M_B} \tag{A-6}$$

and the total mass

$$M_C = M_A + M_B \tag{A-7}$$

and we obtain

$$\frac{1}{M_A}\frac{\partial^2}{\partial x_A^2}+\frac{1}{M_B}\frac{\partial^2}{\partial x_B^2}=\frac{1}{M_C}\frac{\partial^2}{\partial x_C^2}+\frac{1}{\mu}\frac{\partial^2}{\partial x^2} \qquad \text{(A-8)}$$

Clearly

$$T_A+T_B=-\frac{\hbar^2}{2M_C}\left(\frac{\partial^2}{\partial x_C^2}+\frac{\partial^2}{\partial y_C^2}+\frac{\partial^2}{\partial z_C^2}\right)-\frac{\hbar^2}{2\mu}\left(\frac{\partial^2}{\partial x^2}+\frac{\partial^2}{\partial y^2}+\frac{\partial^2}{\partial z^2}\right) \qquad \text{(A-9)}$$

_____ *Problems*

1. Calculate the reduced nuclear mass for the molecules H_2, HCl, HI, and N_2.
2. In the electronic ground state of HCl the rotational constant $B_{0,0}$ for the lowest vibrational level is 10.4400 cm^{-1} and the rotational constant $B_{0,1}$ for the first excited vibrational level is 10.1366 cm^{-1}. Calculate the effective distance R_0 between the two nuclei for both cases from the rotational constants.
3. The electronic ground state of N_2 is $^1\Sigma_g$. The lowest excited states have $^1\Pi_g$, $^1\Sigma_u$, $^1\Pi_u$, $^1\Sigma_u$ and $^1\Pi_u$ symmetry. Which electronic transitions between the ground state and these excited states are allowed? What can you say about the direction of the electronic transition moment?
4. In the ground state of CO ($C=12$, $O=16$) the rotational constant $B_{0,0}$ is 1.9314 cm^{-1}. What is the value of the equilibrium internuclear distance R_0? In the first excited electronic state $B_{1,0}$ has the value 1.6116 cm^{-1}. What is the value of the equilibrium internuclear distance R_1?
5. According to the Boltzmann distribution the relative population of an energy level E_k is given by $g_k \exp(-E_k/kT)$, where g_k is the degeneracy of level k. Calculate the relative populations for the levels $J=0$, $J=2$, and $J=10$ for the molecules H_2 and N_2 at liquid helium temperature ($T=4$ K) and at room temperature ($T=300$ K). The rotational constants are given in Table 14-1.
6. The vibrational transition in $H^{35}Cl$ is at 2989.74 cm^{-1}. Calculate the force constant of the vibrational motion on the assumption that it is a harmonic oscillator. If we now assume that the vibrational motion in $D^{35}Cl$ has the same force constant as in $H^{35}Cl$, where does the vibrational transition occur in $D^{35}Cl$?
7. Explain the Born–Oppenheimer approximation in simple nonmathematical terms.
8. Why does the rotational constant B of a diatomic molecule depend on the electronic state the molecule is in?
9. If we want to determine the electric dipole moment of a diatomic molecule from molecular spectra which spectroscopic experiment should we perform?
10. It may be shown that the rotational selection rules of Eqs. (14-78) and (14-79) should be supplemented by the selection rule that the rotational transition from the state $J=0$ to the state $J'=0$ is always forbidden. How does this affect the P, Q and R branches?

11. The rotational constant B for a specific electronic state of a diatomic molecule depends slightly on the vibrational quantum number v. Explain why.

12. The potential function $U_0(R)$ of a diatomic molecule is sometimes approximated by the Morse potential $U_0(R) = D[1 - \exp\{-b(R - R_0)\}]^2$ where R_0 is the equilibrium distance and D the dissociation energy. If the harmonic force constant of the motion is k what should we take as the value of the parameter b?

13. The rotational constants for the ground states of H_2 and D_2 are 60.809 cm^{-1} and 30.429 cm^{-1} respectively. Are the equilibrium internuclear distances R_0 for the two molecules different from one another or are they basically the same?

14. In an excited electronic state the vibrational frequency is usually much smaller than in the ground state. Explain why this is from the behavior of the potential curves $U(R)$.

15. Explain why you would not expect to observe an infrared spectrum of the N_2 molecule.

16. The electronic ground state of H_2 has $^1\Sigma_g$ symmetry and the first excited state has $^1\Sigma_u$ symmetry. Do you expect the rotational fine structure of the electronic transition between these two states to have a P, Q, and an R branch? Explain your answer.

17. In the infrared spectrum of a heteronuclear diatomic molecule we observe the transition $v = 0 \rightarrow v = 1$ with a large intensity but we also observe the transition $v = 0 \rightarrow v = 2$ with a much smaller intensity. Why can this second transition be observed?

-- *References and Recommended Reading*

The best books on molecular spectra are the three books written by Herzberg [1, 2, 3], which contain everything there is to be known on molecular spectra. As a textbook, we recommend Herzberg's recent book on free radicals [4], which gives a simplified and more concise version of the theory of molecular spectra and is easier to read than the more complete books [1, 2, 3]. In our discussion we have avoided the use of group theory. For those students who are interested, we list a few good books on the subject; Hall's book [5] is more concerned with chemical questions than is Tinkham's.

[1] G. Herzberg: *Molecular Spectra and Molecular Structure, I. Spectra of Diatomic Molecules*. Van Nostrand Reinhold, New York, 1950.

[2] G. Herzberg: *Molecular Spectra and Molecular Structure, II. Infrared and Raman Spectra*. Van Nostrand Reinhold, New York, 1945.

[3] G. Herzberg: *Molecular Spectra and Molecular Structure, III. Electronic Spectra and Electronic Structure of Polyatomic Molecules*. Van Nostrand Reinhold, New York, 1966.

[4] G. Herzberg: *The Spectra and Structures of Simple Free Radicals*. Cornell University Press, Ithaca, N.Y., 1971.

[5] L. H. Hall: *Group Theory and Symmetry in Chemistry*. McGraw-Hill, New York, 1969.

[6] M. Tinkham: *Group Theory and Quantum Mechanics*. McGraw-Hill, New York, 1964.

chapter 15

The Chemical Bond

15-1 INTRODUCTION

From a theoretical point of view the chemical bond in a diatomic molecule is characterized by its potential curve, that is, the ground state eigenvalue $E_0(R)$ as a function of the internuclear distance R, and by the eigenfunction Ψ_0 of the electronic ground state. From an empirical point of view the bond is characterized by quantities that we can measure such as the bond length, the bond energy, the force constant of the vibration, the electric dipole moment, and so on. Remember that the Schrödinger equation cannot be solved exactly for any molecule other than the hydrogen molecular ion, and even in that case the solutions are too complicated to have much practical use. Consequently, the theory of valence is based on approximate methods, that is to say, on the variation principle.

In the early papers on valence theory the variational functions were usually constructed by making use of chemical ideas and, sometimes, of experimental information that was available. The variational function could contain some parameters that were determined by minimizing the energy. Nowadays electronic computers are widely used in quantum chemical calculations. Thus, it is possible to do calculations with variational functions that have a very large number of parameters. This has the advantage that the accuracy of the more recent calculations is much higher than the results that were obtained some thirty or forty years ago. On the other hand, the early work was more closely related to chemical ideas, and in many cases it gave a good understanding of the general nature of the chemical bond. In this chapter the emphasis will be on the early work in quantum chemistry because we are basically interested in the general ideas on the chemical bond.

We have seen in the previous chapter that for a diatomic molecule the translational and rotational motions of the nuclei may be considered separately. If we then introduce the Born–Oppenheimer approximation, we may calculate the electronic wave function on the assumption that the nuclei have fixed positions in space. The solutions contain the internuclear distance R as a parameter because the electronic potential function contains R implicitly, but the kinetic energy of the nuclear motion may be disregarded while solving the electronic Schrödinger equation. The situation is basically the same for polyatomic molecules, even though it is more complicated. Again, the translational and rotational motions of the nuclei may be separated off, but the theory of rotational motion is more complicated than for diatomic molecules. The rotational motion is described by the moments of inertia with respect to three mutually orthogonal axes of symmetry. We speak of a spherical top if the three moments of inertia are the same, of a symmetric top if two moments are equal and the third is different, and of an asymmetric top if all three are different. The symmetric top has more or less the same eigenvalues as the rigid rotor, and the eigenvalues of the asymmetric top are derived from the symmetric top by means of perturbation theory.

It is somewhat more difficult to formulate the Born–Oppenheimer approximation for a polyatomic molecule than for a diatomic molecule, but it may be concluded from general considerations that the approximation should be equally valid in both cases. If we use the Born–Oppenheimer approximation, we find that the electronic eigenvalues depend on a number of parameters that represent the relative distances of the nuclei, rather than just the one parameter for a diatomic molecule. Within the harmonic approximation these parameters may be combined in certain linear combinations of displacement coordinates, which are known as normal coordinates. Each normal coordinate vibrates as a harmonic oscillator with a given frequency and its vibration, which is called a normal mode, does not interact with any of the other normal modes. In this model the vibrational motion of a polyatomic molecule is a superposition of noninteracting normal modes, each with its own frequency. The theoretical description of one normal mode is identical with the theoretical description of the vibration in a diatomic molecule.

The electronic eigenfunctions of a polyatomic molecule are again derived on the assumption that the nuclei are at rest, but the eigenvalues and eigenfunctions now contain more than one parameter. For example, in the case of the water molecule these parameters are the O—H bond distances and the H—O—H bond angle. In large molecules the number of parameters may be quite large. Usually the electronic Schrödinger equation is solved only for the equilibrium configuration of the nuclei, but, if we wish to make predictions about the nuclear configuration or about changes in nuclear configuration, we must solve the Schrödinger equation for different bond lengths and bond angles.

It is clear that the accuracy that can be achieved in calculations on small molecules is much higher than for large molecules. The various methods for deriving electronic eigenfunctions may be divided into two categories: First we have methods that are quite accurate, but they are so laborious that they can be applied only to small molecules. At the other extreme we have methods that can be applied to a large number of molecules but have only limited accuracy. Such methods are often semiempirical since they make use of experimental information. The various methods are usually tested on some small molecules, in particular the hydrogen molecule and the hydrogen molecular ion, because here we can verify the accuracy of every approximation that we wish to introduce and we obtain some understanding of the reliability of the method. The literature contains many calculations on the hydrogen molecule, and some of these are very accurate; the results agree with the experimental data to within the experimental error. Unfortunately, many of the methods that work so well for the hydrogen molecule cannot be applied to larger molecules, and their practical use is therefore limited.

In this chapter we first discuss the hydrogen molecular ion and the hydrogen molecule and we proceed then to larger molecules.

15-2 THE HYDROGEN MOLECULAR ION

In valence theory the concept of resonance has played a useful role. This concept is nothing more than an application of the variation principle that we discussed in Section 11-6.

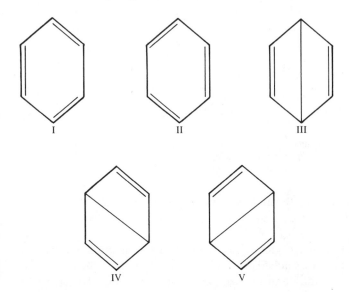

Figure 15-1   The two Kékulé structures I and II and the three Dewar structures III, IV, and V of the benzene molecule.

The benzene molecule is a good example of applying the resonance principle. It is well known in organic chemistry that we can write two possible structures of benzene, namely the two Kékulé structures I and II of Figure 15-1. However, it follows from experiments that the true structure of benzene is neither I nor II because the experimental bond lengths in the benzene molecule are all equal to one another; whereas, in either one of the structures I or II, the double bonds should be shorter than the single bonds. It was proposed, therefore, that the benzene molecule oscillates between structures I and II and that these oscillations are so rapid that we observe some kind of an average of structures I and II. This is called **resonance** between the two structures I and II.

In quantum mechanics we may, in principle, represent each of the structures I or II by means of a ground state eigenfunction Ψ_I and Ψ_{II}, respectively. The ground state wave function Ψ of benzene is then represented as

$$\Psi = a_I \Psi_I + a_{II} \Psi_{II} \qquad (15\text{-}1)$$

according to the resonance principle. It is easily seen that this is a special case of the variation principle where we expand the molecular eigenfunction in terms of the wave functions of the separate chemical structures. In principle we obtain a more accurate approximation to the benzene eigenfunction if we include more terms in the expansion. In Figure 15-1 we have also sketched the three Dewar structures III, IV and V, in addition to the two Kékulé structures I and II. If we represent the Dewar structures by wave functions Ψ_{III}, Ψ_{IV} and Ψ_V, respectively, we may represent the benzene wave function also as

$$\Psi = a_I \Psi_I + a_{II} \Psi_{II} + a_{III} \Psi_{III} + a_{IV} \Psi_{IV} + a_V \Psi_V \qquad (15\text{-}2)$$

It is to be expected that the energy that we derive from function (15-1) is lower than the energy of structure I, and that the energy that we derive from function (15-2) is

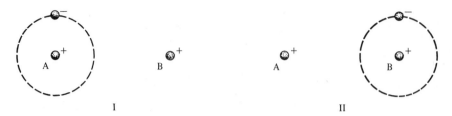

Figure 15-2 The two resonance structures of the hydrogen molecular ion H_2^+. In structure I the electron is on atom A and in structure II the electron is on atom B.

lower yet. This difference between the molecular energy and the energy of structure I is called the resonance energy and it supplies an additional stabilization for the benzene molecule.

Let us now apply the resonance principle to the hydrogen molecular ion. The two resonance structures I and II are sketched in Figure 15-2. In structure I the electron is located in a $1s$ orbital on atom A and in structure II the electron occupies the $1s$ orbital of atom B. The two resonance structures are described by the functions

$$\Psi_I = s_A = (1/\sqrt{\pi})\, e^{-r_A}$$
$$\Psi_{II} = s_B = (1/\sqrt{\pi})\, e^{-r_B}$$
(15-3)

The molecular wave function is then represented as

$$\Psi = a_I \Psi_I + a_{II} \Psi_{II} = a_I s_A + a_{II} s_B \tag{15-4}$$

Since the molecule has a center of inversion the eigenfunctions must be either symmetric or antisymmetric with respect to inversion; this means that the possible wave functions are

$$\Psi_1 = s_A + s_B$$
$$\Psi_2 = s_A - s_B$$
(15-5)

The expectation values of the energies E_1 and E_2 with respect to the functions Ψ_1 and Ψ_2 are given by

$$E_1 = \frac{\langle \Psi_1 | \mathfrak{H} | \Psi_1 \rangle}{\langle \Psi_1 | \Psi_1 \rangle}$$
$$E_2 = \frac{\langle \Psi_2 | \mathfrak{H} | \Psi_2 \rangle}{\langle \Psi_2 | \Psi_2 \rangle}$$
(15-6)

EXAMPLE 15-1

The energies E_1 and E_2 in Eq. (15-6) can be calculated in closed form. The Hamiltonian of the hydrogen molecular ion is given by

$$\mathfrak{H} = -\tfrac{1}{2}\Delta - \frac{1}{r_A} - \frac{1}{r_B} \tag{a}$$

if we use atomic units of length and energy. Since s_A is an eigenfunction of hydrogen atom A with eigenvalue $-\frac{1}{2}$ we have

$$\mathfrak{H} s_A = \left(-\tfrac{1}{2}\Delta - \frac{1}{r_A} \right) s_A - \frac{1}{r_B} s_A = -\tfrac{1}{2} s_A - \frac{1}{r_B} s_A \qquad \textbf{(b)}$$

It may be shown now that E_1 and E_2 may be expressed in terms of the three integrals

$$I = \langle s_A | r_B^{-1} | s_A \rangle = \langle s_B | r_A^{-1} | s_B \rangle$$

$$J = \langle s_A | r_A^{-1} | s_B \rangle = \langle s_B | r_B^{-1} | s_A \rangle \qquad \textbf{(c)}$$

$$S = \langle s_A | s_B \rangle$$

From Eq. (b) we may easily derive

$$\mathfrak{H}_{AA} = \langle s_A | \mathfrak{H} | s_A \rangle = -\tfrac{1}{2} - I$$

$$\mathfrak{H}_{AB} = \langle s_B | \mathfrak{H} | s_A \rangle = -\tfrac{1}{2}S - J \qquad \textbf{(d)}$$

Furthermore, we have

$$E_1 = \frac{\langle \Psi_1 | \mathfrak{H} | \Psi_1 \rangle}{\langle \Psi_1 | \Psi_1 \rangle} = \frac{\mathfrak{H}_{AA} + \mathfrak{H}_{AB}}{1 + S} = -\tfrac{1}{2} - \frac{I + J}{1 + S}$$

$$E_2 = \frac{\langle \Psi_2 | \mathfrak{H} | \Psi_2 \rangle}{\langle \Psi_2 | \Psi_2 \rangle} = \frac{\mathfrak{H}_{AA} - \mathfrak{H}_{AB}}{1 - S} = -\tfrac{1}{2} - \frac{I - J}{1 - S} \qquad \textbf{(e)}$$

The three integrals I, J, and S may be calculated by making use of elliptical coordinates, which are discussed in Appendix A of this chapter. The results are

$$S = e^{-R}[1 + R + (R^2/3)]$$

$$I = R^{-1}[1 - e^{-2R}(1 + R)] \qquad \textbf{(f)}$$

$$J = e^{-R}(1 + R)$$

where R is the internuclear distance. If we substitute the results of Eqs. (f) into Eqs. (e) and add the Coulomb repulsion term $(1/R)$ between the nuclei, we obtain the potential curves $U_1(R)$ and $U_2(R)$ for the two states shown in Figure 15-3.   ●

The behavior of the two potential functions $U_1(R)$ and $U_2(R)$ belonging to the two eigenstates defined by Ψ_1 and Ψ_2 is sketched in Figure 15-3. We have used atomic units in our calculations, and we find that the lower curve has a minimum for $R = 2.5$ and the energy value at the minimum is -0.5648. Thus the equilibrium distance is $R_0 = 2.5 \times 0.529$ Å $= 1.32$ Å and the dissociation energy is 0.0648 au $= 1.76$ eV. The experimental values are 1.06 Å for R_0 and 2.79 eV for the dissociation energy. The agreement with our calculation is not too bad.

It may be seen that $U_2(R)$ does not have a minimum. There is no stable molecular configuration in this electronic state. Therefore, if the molecule is excited from the state E_1 to the state E_2 it dissociates.

By definition, a one-electron wave function is called an orbital. In an atomic orbital the electron moves in the vicinity of one particular atomic nucleus. The functions Ψ_1 and Ψ_2 are called molecular orbitals because they represent situations where the electron moves through the whole molecule.

More specifically they are molecular orbitals, which are linear combinations of atomic orbitals, or LCAO for short. It is customary to call the function Ψ_1 a bonding

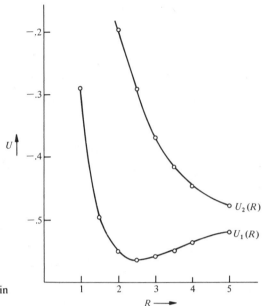

Figure 15-3 Potential curves for H_2^+ in atomic units.

orbital because it represents a situation where there is a stable chemical bond between the two atoms. Similarly, the function Ψ_2 is known as an antibonding orbital.

It is possible to obtain better energies for the ground state of the hydrogen molecular ion by starting from more elaborate variational functions. We may argue that the electron, even when it is in the vicinity of atom A is subject to the Coulomb attraction of both nuclei. To a first approximation we may account for this effect by writing the wave function as

$$\Psi = e^{-\rho r_A} + e^{-\rho r_B} \tag{15-7}$$

where ρ is a variational parameter which represents the effective nuclear charge. A variational calculation based on this wave function gives an energy expression that has a minimum for $\rho = 1.23$ and $R = 1.06$ Å; the corresponding dissociation energy is 2.25 eV.

The next effect that we may consider is the polarization of the atomic orbitals. If the electron is on nucleus A then the Coulomb attraction of the other nucleus will pull the electron cloud slightly towards nucleus B. We can represent this effect by writing the wave function in the form

$$\Psi = A\, e^{-\rho r_A} + \lambda z_A\, e^{-\rho' r_A} + e^{-\rho r_B} - \lambda z_B\, e^{-\rho' r_B} \tag{15-8}$$

The first two terms of this equation represent an atomic orbital on nucleus A that is pulled off-center towards nucleus B; the magnitude of the effect depends on the value of the parameter λ. A variational calculation gives

$$\rho = 1.247 \qquad \rho' = 0.934 \qquad \lambda = 0.145 \tag{15-9}$$

The equilibrium distance is 1.06 Å and the dissociation energy is 2.71 eV.

The most efficient approximation to the ground state eigenfunction of H_2^+ makes use of elliptical coordinates (see Appendix A of this chapter). For example, the function

$$\Psi = e^{-\delta\mu} \tag{15-10}$$

gives a dissociation energy of 2.17 eV and an equilibrium distance R_0 of 1.06 Å. The improved function

$$\Psi = e^{-\delta\mu}(1 + a\nu^2) \tag{15-11}$$

with $\delta = 1.3540$ and $a = 0.4480$ gives $R_0 = 2$ au $= 1.06$ Å and a dissociation energy of 0.102386 au $= 2.78538$ eV. These values agree with the experimental data to within the possible experimental errors. We have combined the various results that are derived from the preceding variational functions in Table 15-1.

Table 15-1 Values of dissociation energy and equilibrium distance for the hydrogen molecular ion derived from various trial functions by means of the variational principle

Wave Function	$\dfrac{R_0}{(\text{Å})}$	Dissociation Energy (eV)
Eq. (15-5)	1.32	1.76
Eq. (15-7)	1.06	2.25
Eq. (15-10)	1.06	2.17
Eq. (15-11)	1.06	2.785
Eq. (15-8)	1.06	2.71
experimental	1.06	2.79

It may be seen that the expansion of Ψ in terms of elliptical coordinates is the most efficient approach for calculating the dissociation energy and the equilibrium distance. However, this method is not easily extended to more complex molecules and, also, it does not give a clear picture of the behavior of the electrons in relation to chemical ideas. Therefore, it is generally preferred to represent molecular wave functions by expansion in terms of atomic orbitals even though the convergence of the expansion may be slower.

If we use atomic orbitals we may try to relate the energy of a chemical bond to the charge distribution of the electron. Let us consider, for example, the simple molecular orbital of Eq. (15-5). The corresponding charge density function (normalized to unity) is given by

$$P = (2 + 2S)^{-1}[(s_A)^2 + 2(s_A)(s_B) + (s_B)^2] \tag{15-12}$$

It may be said that the first term here represents the probability that the electron is associated with nucleus A, the last term the probability that it is associated with nucleus B, and the middle term, which is called the overlap charge, is the probability that the electron may be found between the two nuclei. The numerical values of the three possibilities are obtained by integrating the corresponding terms in Eq. (15-12)

$$P_A = P_B = 1/(2 + 2S)$$

$$P_{AB} = 2S/(2 + 2S) \tag{15-13}$$

We may derive from Eq. (15-13) that for $R = 2$ we have $P_A = P_B = 31.5\%$, and $P_{AB} = 37\%$. It is interesting to note that the same calculation for the antibonding function Ψ_2 of Eq. (15-5) gives the strange results $P_A = P_B = 1.209$ and $P_{AB} = -1.418$. Here the overlap charge appears to be negative.

It seems to be an empirical fact that a chemical bond is stronger, that is it has a higher dissociation energy, if the overlap charge is larger. Usually in antibonding orbitals such as Ψ_2 the overlap charge is negative with respect to the electronic charge densities at the atoms. However, all semiempirical theories should be used with caution, and the various theories that attempt to relate the bond strengths with overlap charges are not always reliable.

15-3 THE HYDROGEN MOLECULE

A chemical bond usually contains a pair of electrons and the properties of the bond are determined not only by the interaction between each of the electrons and the nuclei but also by the interaction between the electrons. The hydrogen molecule is the simplest molecule that contains at least one pair of electrons, and it is therefore a useful system to illustrate some of the general features of the chemical bond.

Just as in the previous section, we may approximate the molecular eigenfunction by writing it as a linear combination of functions that represent various resonance structures of the molecule. Figure 15-4 shows the two most obvious resonance structures of the hydrogen molecule. In structure I electron 1 is centered on nucleus A and electron 2 is centered on nucleus B. Obviously this situation may be represented by a function Ψ_I which is defined as

$$\Psi_I = s_A(1) s_B(2) \tag{15-14}$$

where the atomic orbitals are defined in Eq. (15-3). In structure II electron 1 is centered on B and electron 2 is centered on A. The corresponding function Ψ_{II} is given by

$$\Psi_{II} = s_B(1) s_A(2) \tag{15-15}$$

The molecular wave function is then a linear combination of functions Ψ_I and Ψ_{II}.

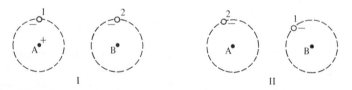

Figure 15-4   Resonance structures I and II for the hydrogen molecule.

It is easily seen that we may use only the sum and the difference of the two functions because of the exclusion principle

$$\Psi_0 = \tfrac{1}{2}[s_A(1)s_B(2) + s_B(1)s_A(2)][\alpha(1)\beta(2) - \beta(1)\alpha(2)] \tag{15-16}$$

and

$$^3\Psi_1 = \tfrac{1}{2}\sqrt{2}[s_A(1)s_B(2) - s_B(1)s_A(2)]^3\zeta(1, 2) \tag{15-17}$$

The total molecular wave function must be antisymmetric with respect to permutations of the electrons. We can easily verify that the sum of Ψ_I and Ψ_{II} is symmetric with respect to permutations. It must therefore be combined with an antisymmetric spin function, that is, a singlet spin function. By the same argument, the difference of Ψ_I and Ψ_{II} is antisymmetric with respect to permutations, and it must be combined with a triplet spin function so that the total wave function is antisymmetric. The function Ψ_0 belongs to the ground state of the hydrogen molecule, and the function $^3\Psi_1$ belongs to an excited state.

It is again possible to determine the expectation values $E(R)$ of the energy with respect to the wave functions Ψ_0 and Ψ_1. This leads to a theoretical prediction for the potential curves $U(R)$ for the two states of the hydrogen molecule. The calculation is much more involved than in the analogous case for the hydrogen molecular ion that we discussed in Example 15-1. The difficulty is due to the Coulomb repulsion term between the two electrons. As a result it is no longer possible to derive analytical expressions for $E(R)$. Nowadays, calculations of this type are usually performed by electronic computers. If we use a computer, the problem of treating the hydrogen molecule is trivial and the whole calculation takes less than a second. The result is that function (15-16) leads to a dissociation energy of 3.14 eV and an equilibrium distance of 0.869 Å. The experimental values are 4.747 eV and 0.741 Å, respectively.

The preceding procedure, where the molecular wave function is approximated as a superposition of resonance structures, is called the **valence bond method**. Function (15-16) is called a valence bond function or a Heitler–London function, after the two scientists who first proposed this method. We have seen that the valence bond method is closely related to chemical ideas, and its use leads to a general understanding of how the various resonance structures contribute to the chemical bond. However, the valence bond method is not appropriate for making numerical predictions for large molecules because the calculations become very tedious. Therefore, an alternative method has been developed; this approach is known as the **molecular orbital theory**.

In the molecular orbital theory, the molecular wave function is written as an antisymmetrized product of one-electron functions. We have already mentioned that these functions are called molecular orbitals. It is customary to make the additional assumption that each molecular orbital may be approximated as a linear combination of atomic orbitals; this procedure is abbreviated in the literature as the **LCAO MO** method.

We shall illustrate the LCAO MO method for the H_2 molecule. To a first approximation we may expand the molecular orbitals in terms of the atomic $1s$ orbitals s_A and s_B only. It follows from symmetry considerations that the possible linear combinations are

$$\phi_1 = s_A + s_B$$
$$\phi_2 = s_A - s_B \tag{15-18}$$

where ϕ_1 has the lower energy. The molecular ground state is thus obtained by placing two electrons in the orbital ϕ_1. This state must be a singlet state because of the Pauli exclusion principle.

$$\Psi = \phi_1(1)\phi_1(2)[\alpha(1)\beta(2) - \beta(1)\alpha(2)] \tag{15-19}$$

This function gives a dissociation energy of 2.70 eV and an equilibrium distance of 0.85 Å, which is slightly worse than the results from the valence bond method (we have collected the various results in Table 15-2).

Table 15-2 Values of dissociation energy and equilibrium distance for hydrogen molecule, derived from different trial functions by means of the variational principle

Wave Function	$R_0/(\text{Å})$	Dissociation Energy (eV)
Eq. (15-16)	0.869	3.14
Eq. (15-20)	0.85	2.70
Eqs. (15-21), (23)	0.743	3.78
Eqs. (15-20), (23)	0.732	3.49
Eqs. (15-22), (23)	0.749	4.02
Hartree–Fock	0.74	3.63
Eq. (15-24)	0.71	4.11
James–Coolidge	0.740	4.72
Kolos–Roothaan	0.741	4.7467
experimental	0.741	4.747

It is interesting to substitute expansion (15-18) into Eq. (15-19) and write it all out. The result is

$$\Psi_{MO} = s_A(1)s_A(2) + s_A(1)s_B(2) + s_B(1)s_A(2) + s_B(1)s_B(2) \tag{15-20}$$

where we have omitted the spin function. It is easily seen that this function contains two more terms than the corresponding valence bond function

$$\Psi_{VB} = s_A(1)s_B(2) + s_B(1)s_A(2) \tag{15-21}$$

We may write Eq. (15-20) as a superposition of the four resonance structures shown in Figure 15-5. The first two of these structures are identical with those of Figure 15-4, but here in addition we have structure III, where both electrons are centered on nucleus A, and structure IV, where both electrons are on nucleus B. These two resonance structures are known as ionic structures. Thus, in the valence bond method the ionic structures are completely neglected, whereas in the molecular

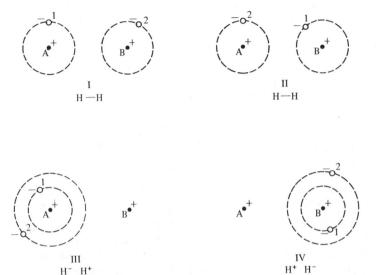

Figure 15-5 Covalent (I, II) and ionic (III, IV) resonance structures of the hydrogen molecule.

orbital method they are treated on the same basis as the nonionic structures. It may be seen that a more realistic representation would be

$$\Psi = \rho[s_A(1)s_B(2) + s_B(1)s_A(2)] + \sigma[s_A(1)s_A(2) + s_B(1)s_B(2)] \tag{15-22}$$

where ρ and σ indicate the relative weights of the nonionic and the ionic structures, respectively. The parameters may be determined from the variational principle, and it turns out that the ionic character of the bond, σ^2, is about 5%.

The variational calculations may also be performed for the three functions (15-20), (15-21), and (15-22) by substituting

$$s_A = e^{-Zr_A} \qquad s_B = e^{-Zr_B} \tag{15-23}$$

for the $1s$ orbitals. Here, the effective nuclear charge Z is determined variationally. It is also called a screening constant, and its value is around 1.17. The results of the three variational calculations with the atomic orbitals (15-23) are 3.78 eV for the valence bond function (15-21), 3.49 eV for the molecular orbital function (15-20), and 4.02 eV for the intermediate function (15-22).

There are a number of ways to improve the wave functions by adding parameters and by using the variational principle. Some of the results are particularly interesting because they give us some indication of the accuracy we might expect in calculations on larger molecules. Once we write the molecular wave function in the form of Eq. (15-19) as a product of molecular orbitals, we can determine the best possible form of the function ϕ_1 by expanding it in a very large number of terms. This is known as the Hartree–Fock limit, and it means that we do not make any approximations in the wave function beyond approximation (15-19). The method leads to a dissociation energy of 3.63 eV.

In the Hartree–Fock approach we do not consider the electron repulsion in constructing the wave function so that the wave function does not contain the electron correlation. The repulsion between the electrons is represented in the Hamiltonian, and the energy expression contains the expectation value of the Coulomb repulsion. We note that the discrepancy between the energy derived from the best possible molecular orbital function and the experimental energy is a little more than 1 eV. This difference is known as the correlation energy because the correlation energy is defined as the difference between the exact energy and the energy of the Hartree–Fock limit. We may conclude that for larger molecules molecular orbital wave functions should give energies that have errors of 1 eV or more; thus, we should not expect too high an accuracy from LCAO MO calculations.

Let us now consider what methods we have available to improve upon the accuracy of the Hartree–Fock method. In quantum chemistry this is known as the problem of finding the correlation energy because this energy is defined as the difference between the Hartree–Fock energy and the true energy. It is clear that we must incorporate the electron correlation somehow in the wave function to improve upon the Hartree–Fock method. This may be done either directly or indirectly.

The most efficient method is to put the electron correlation directly into the eigenfunction. For example, instead of the molecular orbital function (15-19) we take

$$\Psi(1, 2) = \phi(\mathbf{r}_1)\phi(\mathbf{r}_2)[1 + ar_{12}] \tag{15-24}$$

where the parameter a is determined from the variation principle. If we simply take the molecular orbital ϕ as a linear combination of $1s$ functions, the function Ψ of Eq. (15-24) gives an energy of 4.11 eV, which is significantly better than the Hartree–Fock energy. In the extreme case the function Ψ is expanded in terms of the elliptical coordinates of the two electrons and the electron distance r_{12}. Such a calculation was performed in 1933 by James and Coolidge, and they obtained a dissociation energy of 4.72 eV, very close to the experimental value. More recently, Kolos and Roothaan did a similar calculation on an electronic computer. Since they included many more terms in their expansion (about 50), the result of this calculation, a dissociation energy of 4.7467 eV, is quite accurate. It agrees with the experimental value to within the experimental error, and it is generally considered to be even more accurate than the experimental dissociation energy.

The disadvantage of the calculations we have discussed is that the methods cannot be applied to molecules other than hydrogen; it is not practical to include the electron distances r_{ij} in the wave functions of systems with more than two electrons. For that reason we must look for other methods, even though these methods may be less effective for small molecules. We have already mentioned a wave function that gives a better result than the Hartree–Fock energy, namely, function (15-22), which is a valence bond function including ionic terms. It is worth noting that this function can also be represented within the framework of the molecular orbital method. In Eq. (15-18), we saw that two molecular orbitals can be derived from the two $1s$ orbitals, namely

$$\phi_1 = s_A + s_B \qquad \phi_2 = s_A - s_B \tag{15-25}$$

Here ϕ_1 has a lower energy than ϕ_2. In the molecular orbital description we place two electrons in orbital ϕ_1 (see Eq. 15-19). We can construct a more elaborate variational function by also considering the excited states

$$\Psi = \lambda\phi_1(1)\phi_1(2) + \mu\phi_2(1)\phi_2(2) \tag{15-26}$$

Note that we may not mix the configuration $(\phi_1\phi_2)$ because its function has a different symmetry than the ground state wave function. It is easily seen that Eq. (15-26) may also be written as

$$\Psi = (\lambda - \mu)[s_A(1)s_B(2) + s_B(1)s_A(2)] + (\lambda + \mu)[s_A(1)s_A(2) + s_B(1)s_B(2)] \quad \textbf{(15-27)}$$

which is the same as function (15-22). We say that function (15-26) is constructed by using configuration interaction; in addition to the lowest configuration that may be derived from the molecular orbitals, it also contains excited configurations. It may be seen that the configuration interaction method is an indirect way of including electron correlation into the wave function. It is not as efficient as the direct method of Eq. (15-24), but it is at least possible to use the configuration interaction method (CI for short) in larger molecules.

Finally, we should point out that we can construct only a finite number of molecular configurations from a finite number of atomic orbitals. These configurations may either be constructed by means of the valence bond method or by means of the molecular orbital method. In order to obtain all possible molecular configurations we must include the ionic states in the valence bond method and we must use configuration interaction in the molecular orbital method. We see that the two methods may differ when we consider only the lowest possible molecular orbital configuration and only a few resonance structures in the valence bond method. Naturally, the two methods should give the same result if we were to use the complete function containing all possible molecular configurations, but in practice this is never done. Initially, the two methods were about equally popular, but it became clear that the molecular orbital method is easier to use for numerical calculations on large molecules, and this method has become much more popular than the valence bond method.

15-4 DIATOMIC MOLECULES

The customary approach for dealing with diatomic molecules is by means of the molecular orbital method. In the early 1930s the derivation of approximate molecular eigenfunctions relied heavily on experimental information, mainly derived from the molecular spectra, and on chemical concepts. At that time it was not possible to perform elaborate numerical calculations, and it was not really possible to derive reliable numerical results by means of ab initio calculations, that is, calculations starting from first principles. This state of affairs was changed drastically by the invention of the electronic computer. During the 1950s a group of scientists at the University of Chicago, Mulliken, Roothaan, Ruedenberg and others, recognized how the electronic computer might be used for performing molecular calculations. Even so, it took them many years to develop a method of calculation that may be applied to all diatomic molecules and gives reasonably accurate results. The method is known as the Roothaan SCF method because Roothaan developed the main ideas. Rather than discuss Roothaan's method, we plan to treat the chemical bond and its quantum mechanical description in a more qualitative manner. However, it is useful to have the results of the Chicago group because it enables us to verify the accuracy of our qualitative arguments.

It is customary now to describe the electronic structure of a diatomic molecule by means of the molecular orbital method. It is then assumed that the molecular wave function Ψ may be expressed in terms of one-electron functions, the molecular orbitals χ_i. There is an orbital energy ε_i associated with each molecular orbital χ_i. If we number these energies in order of increasing energy,

$$\varepsilon_1 \leqq \varepsilon_2 \leqq \varepsilon_3 \ldots \tag{15-28}$$

then the ground state of the molecule is obtained by placing two electrons in the orbital χ_1, two in the orbital χ_2, and so on, until we have accommodated all the electrons. Remember that we cannot place more than two electrons in any one orbital because of the Pauli exclusion principle. If the molecule has $2N$ electrons, its ground state configuration is described by $(\chi_1)^2(\chi_2)^2(\chi_3)^2 \ldots (\chi_N)^2$.

EXAMPLE 15-2

The molecular wave function should contain the electron spins and should be antisymmetric with respect to permutations of the electrons. It may be seen that for a two-electron system the molecular wave function Ψ should be written as

$$\Psi = (2)^{-1/2}[\chi_1(1)\chi_1(2)\alpha(1)\beta(2) - \chi_1(2)\chi_1(1)\alpha(2)\beta(1)]$$

$$= 2^{-1/2} \sum_P P \, \delta_P[\chi_1(1)\chi_1(2)\alpha(1)\beta(2)] \tag{a}$$

Similarly the antisymmetrized wave function for a $2N$-electron system is given by

$$\Psi = [(2N)!]^{-1/2} \sum_P P \, \delta_P[\chi_1(1)\alpha(1)\chi_1(2)\beta(2)\chi_2(3)\alpha(3)\chi_2(4)\beta(4) \ldots$$

$$\ldots \chi_N(2N-1)\alpha(2N-1)\chi_N(2N)\beta(2N)] \quad \bullet \tag{b}$$

It is also customary to write the molecular orbitals χ_i as linear combinations of atomic orbitals ϕ_k,

$$\chi_i = \sum_k c_{i,k}\phi_k \tag{15-29}$$

The values of the expansion coefficients $c_{i,k}$ and the specific form of the atomic orbitals ϕ_k can be determined by means of the Roothaan method, which is based on the variational principle. However, it is also possible to make reasonable estimates of the molecular orbitals from general chemical considerations. It seems most practical to illustrate this approach by considering a specific diatomic molecule, and for this purpose we choose the nitrogen molecule N_2.

In N_2 (and in molecules such as O_2, CO, F_2, and so on) we write the molecular orbitals as linear combinations of the atomic $1s$, $2s$, and $2p$ orbitals. Each atom has one $1s$ orbital, one $2s$ orbital, and three $2p$ orbitals, so that there is a total of ten atomic orbitals. From this basis set we can construct ten molecular orbitals. Of course, we would get more accurate results if we expanded our basis set to include $3s$, $3p$, and $3d$ orbitals, but this makes the theory much more complicated and the results from a more limited basis set are not bad.

The ten molecular orbitals in N_2 may be classified according to symmetry. First we can distinguish the σ orbitals, for which the component M_z of the angular momentum along the molecular axis is zero, and the π orbitals, for which $M_z = \pm\hbar$. The σ orbitals are linear combinations of the $1s$, $2s$, and $2p_z$ atomic orbitals and the π orbitals are linear combinations of either the $2p_x$ or the $2p_y$ orbitals. Each of the molecular orbitals must be either gerade or ungerade. It is easily verified that there are three σ_g and three σ_u molecular orbitals. These are usually denoted by $1\sigma_g$, $2\sigma_g$, $3\sigma_g$ and $1\sigma_u$, $2\sigma_u$, and $3\sigma_u$. The other symmetry species, π_{xg}, π_{yg}, π_{xu} and π_{yu} have one orbital each.

Let us now see if we can make any additional predictions about the molecular orbitals and their energies.

The energy of the $1s$ atomic orbital is considerably less than that of other atomic orbitals. It may be expected, therefore that the molecular orbitals with the lowest energies consist almost exclusively of the atomic $1s$ orbitals. For symmetry reasons we must take the sum and the difference of the $1s$ orbitals. In this way we obtain the two orbitals, $1\sigma_g$ and $1\sigma_u$, which have energies that are far below the other energies.

The nature of the other σ orbitals may be understood by introducing the concept of hybridized atomic orbitals. In Figure 15-6 we have drawn a contour map of a $2s$

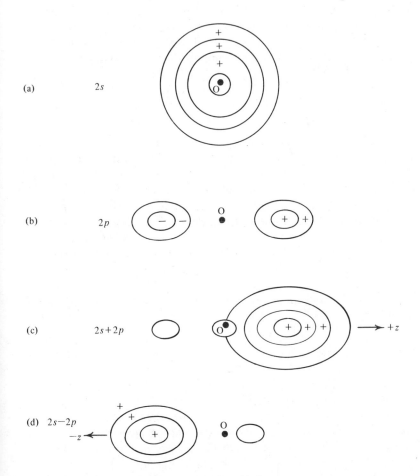

Figure 15-6 Hybridization of a $2s$ and a $2p$ orbital.

orbital and a $2p_z$ orbital. It may be seen that in a linear combination of the type

$$\phi_r = \sqrt{1-\rho^2}\, s + \rho p_z \tag{15-30}$$

for positive ρ the two functions s and p reinforce one another for positive values of z, and they partially cancel for negative values of z. The result is the type of function illustrated in Figure 15-6c. In the hybridized function the charge cloud is shifted along the positive z axis, and we may even say that it points in the positive z direction. Naturally, for negative values of ρ we have a similar situation, but now the charge cloud is displaced in the negative z direction and we have a hybridized orbital that points to the left, as shown in Figure 15-6d.

If in a molecule such as N_2 we have two hybridized orbitals, centered on the nuclei A and B and pointing towards each other, it is clear that the overlap integral between these two orbitals may be quite large. The overlap integrals have been calculated for a variety of situations and values as high as 0.6 or 0.7 are not uncommon. Since the degree of bonding may be associated with the magnitude of the overlap integral, it may be expected that hybridized orbitals give much stronger bonding than nonhybridized orbitals.

Let us now return to our discussion of the nitrogen molecule. On each atom we may combine the $2s$ and $2p_z$ functions into two hybridized orbitals, ϕ_r, which points to the other atom, and ϕ_{lp}, which points away from the other atom. The two orbitals $\phi_{r,A}$ and $\phi_{r,B}$ may be combined into the symmetric and antisymmetric linear combinations.

$$\phi_{\text{bond}} = \phi_{r,A} + \phi_{r,B}$$
$$\phi_{\text{anti}} = \phi_{r,A} - \phi_{r,B} \tag{15-31}$$

We expect that the molecular orbital ϕ_{bond} has by far the lowest energy of the four remaining σ orbitals, and since it is gerade we take it as the $2\sigma_g$ orbital. Similarly, the antibonding orbital ϕ_{anti} must have the highest energy of the molecular orbitals, and since it is ungerade we take it as the $3\sigma_u$ molecular orbital. The two orbitals ϕ_{lp} point away from the other atom and they do not really participate in the bonding process. Such orbitals are known as lone pair orbitals. Because of symmetry requirements the two atomic lone pair orbitals must be combined to give a symmetric and an antisymmetric σ orbital. These orbitals are neither strongly bonding nor strongly antibonding, they are known as nonbonding orbitals. These are the nonbonding molecular orbitals $2\sigma_u$ and $3\sigma_g$.

The two atomic $2p_x$ orbitals may be combined to form a bonding and an antibonding π orbital

$$1\pi_u = 2p_{x,A} + 2p_{x,B}$$
$$1\pi_g = 2p_{x,A} - 2p_{x,B} \tag{15-32}$$

The $2p_y$ orbitals may be treated in the same way, and it is easily seen that the orbitals $1\pi_u$ and $1\pi_u'$ (constructed from $2p_y$ orbitals) are degenerate with one another. The overlap integral between the $2p_x$ orbitals is much smaller than for the hybridized σ orbitals and we may conclude that the π orbitals are not as strongly bonding and antibonding as the σ orbitals. We expect that the energy of the orbital $1\pi_u$ lies between the bonding and the nonbonding σ orbitals and that the energy of the orbital $1\pi_g$ lies between the nonbonding and the antibonding σ orbitals.

If we combine the preceding general arguments, we find that the rank order of the molecular orbitals in terms of their energies is

$$(1s)_{\text{sym}} \rightarrow (1s)_{\text{antisym}} \rightarrow \sigma_{\text{bond}} \rightarrow \pi_{\text{bond}} \rightarrow \text{lone pair } 1 \rightarrow \text{lone pair } 2 \rightarrow \pi_{\text{antibond}} \rightarrow \sigma_{\text{antibond}}$$

$$(15\text{-}33)$$

or

$$1\sigma_g \rightarrow 1\sigma_u \rightarrow 2\sigma_g \rightarrow 1\pi_u \rightarrow 2\sigma_u \rightarrow 3\sigma_g \rightarrow 1\pi_g \rightarrow 3\sigma_u \qquad (15\text{-}34)$$

We can compare this with the results of the Roothaan calculation. Here the rank-order of the molecular orbitals according to their energies is

$$1\sigma_g \rightarrow 1\sigma_u \rightarrow 2\sigma_g \rightarrow 2\sigma_u \rightarrow 3\sigma_g \rightarrow 1\pi_u \rightarrow 1\pi_g \rightarrow 3\sigma_u \qquad (15\text{-}35)$$

Apparently, in our general arguments we made the wrong prediction about the relative magnitudes of the energies of the lone pair σ electrons and the electrons in the π bond, but otherwise our general considerations gave the correct relative magnitudes of the energy levels.

The N_2 molecule has 14 electrons so that its ground state configuration is given by

$$(1\sigma_g)^2 (1\sigma_u)^2 (2\sigma_g)^2 (2\sigma_u)^2 (3\sigma_g)^2 (1\pi_u)^4 \qquad (15\text{-}36)$$

We must remember that the π orbitals are twofold degenerate so that we can accommodate four electrons in them.

It is interesting to consider some other diatomic molecules. We should realize that the rank order (15-35) is approximately valid for all diatomic molecules. If we move from N_2 to O_2 we have two additional electrons that should be placed in the orbital $1\pi_g$ because this is the next orbital. However, the $1\pi_g$ state is degenerate, the two orbitals are

$$1\pi_g = 2p_{x,A} - 2p_{x,B}$$
$$1\pi_g' = 2p_{y,A} - 2p_{y,B} \qquad (15\text{-}37)$$

There are three different ways to distribute two electrons over two orbitals, corresponding to the configurations

$$x^2 = (1\pi_g)^2 \qquad y^2 = (1\pi_g')^2 \qquad xy = (1\pi_g)(1\pi_g') \qquad (15\text{-}38)$$

The first two configurations must have antiparallel spins because of the exclusion principle, and they must therefore be singlet states. In the configuration xy the electron spins may be either parallel or antiparallel, and we expect to have a singlet xy state and a triplet xy state. It follows that the ground state configuration of the O_2 molecule gives rise to four different eigenstates. The lowest state is the xy triplet state, and the other three are the xy, x^2, and y^2 singlet states. It may be shown that the singlet states give rise to two molecular states, one with Σ symmetry and the other one with Δ symmetry. The O_2 molecule is one of the few molecules that has a triplet ground state, and we have seen now how this may be understood on the basis of molecular orbital theory.

If we proceed to the next homonuclear molecule, F_2, the situation becomes more straightforward. Here we have to accommodate four more electrons than in the N_2 molecule, and these can all be placed in the twofold degenerate $1\pi_g$ orbital. The ground state configuration of F_2 is

$$(1\sigma_g)^2(1\sigma_u)^2(2\sigma_g)^2(2\sigma_u)^2(3\sigma_g)^2(1\pi_u)^4(1\pi_g)^4 \qquad \textbf{(15-39)}$$

So far we have considered only homonuclear diatomic molecules where the molecular orbitals must be symmetric with respect to the coordinate z (the origin is taken at the midpoint of the internuclear axis). In the case of heteronuclear molecules, such as CO or HF, we cannot make use of this symmetry with respect to z and the molecular orbitals contain an additional unknown variable. For example, if we know the form of the hybridized atomic orbitals t_C and t_O on the carbon and on the oxygen atoms, respectively, then the molecular bonding orbital must be represented as

$$\chi_{bond} = t_C + \lambda t_O \qquad \textbf{(15-40)}$$

This represents a charge cloud that is not symmetric. It is easily seen that the charge cloud is pulled towards the oxygen atom if λ is larger than unity, and that it is displaced towards the nitrogen if λ is smaller than unity. If we want to make qualitative predictions about the asymmetry of the charges in heteronuclear chemical bonds, we should have some feeling for the relative power of the atoms in pulling electrons towards themselves. This question is related to the concept of electronegativity, which we shall discuss in Section 15-5. We should point out that the electronegativity rules are useful only if we wish to make qualitative predictions about the general form of the charge distribution because they are not accurate enough to make precise qualitative predictions.

15-5 ELECTRONEGATIVITY

The electronegativity of an atom may be defined as its capability of pulling electrons towards itself in a chemical bond. This definition is somewhat vague, but then, the whole concept of electronegativity is somewhat imprecise. In a diatomic molecule AB it is possible to measure only the total charge distribution of all the electrons; there is no unique way in which we can assign part of this charge cloud to atom A and part of it to atom B. It is very difficult to define electronegativity in terms of molecular orbital theory, but it is possible to propose a definition by making use of valence bond theory. We can imagine that a molecule AB is a superposition of the three resonance structures AB, A^+B^- and A^-B^+, and we can imagine also that each of these structures may be represented by a wave function, $\Psi(AB)$, $\Psi(A^+B^-)$ and $\Psi(A^-B^+)$, respectively. The total molecular wave function may then be written as a linear combination

$$\Psi = \Psi(AB) + \rho_1\Psi(A^+B^-) + \rho_2\Psi(A^-B^+) \qquad \textbf{(15-41)}$$

If ρ_2 is larger than ρ_1 we may say that atom A is more electronegative than atom B, and the difference $(\rho_2 - \rho_1)$ or the ratio (ρ_2/ρ_1) is a measure of the difference in electronegativity between the two atoms.

In polyatomic molecules there is ample evidence to support the idea that the various bonds in the molecule may be treated as separate, well-defined entities. For one thing, bond energies seem to be additive, and we can assign a fairly accurate value to the energy of a bond AB, independent of the rest of the molecule. As a result, the concept of electronegativity can also be used to describe chemical bonds in polyatomic molecules.

The goal of the various theories on electronegativity is the derivation of an electronegativity scale that applies to all atoms. This means that an electronegativity number x_A is assigned to each atom A in such a way that the relative electronegativity between two atoms A and B in a chemical bond AB is related to the difference $x_A - x_B$. The best known electronegativity scales are those by Pauling and by Mulliken. Both of these were derived from semiempirical arguments, and neither one of them has been justified by means of rigorous theoretical derivations. However, the two electronegativity scales seem to be consistent with one another and their predictions seem to be quite reasonable. Also, they have been useful in giving us a qualitative understanding of the chemical bond between different atoms.

Pauling noted that the dissociation energy of a bond AB between two different atoms is generally larger than the arithmetic average of the dissociation energies of the bonds AA and BB. Consequently, we can define a positive quantity Δ_{AB} as

$$\Delta_{AB} = D_{AB} - \tfrac{1}{2}(D_{AA} + D_{BB}) \tag{15-42}$$

where D_{AB} stands for the dissociation energy. It is customary to express all quantities in the equation in terms of kilocalories. Pauling now defines the electronegativities x_A and x_B of the two atoms by means of

$$|x_A - x_B| = 0.208\sqrt{\Delta_{AB}} \tag{15-43}$$

where Δ is expressed in terms of kilocalories.

Unfortunately, the bond energies D_{AA} are available only for a limited number of elements so that Eq. (15-42) can be used only for a few atoms. In order to derive the electronegativity constants for the other elements, Pauling had to use more complicated thermochemical arguments involving the heats of formation of polyatomic molecules. In this way he derived the electronegativity values that we have listed in Table 15-3. Note that the method gives only differences in electronegativity values. In order to choose a reference value, Pauling assigned the value $x_H = 2.1$ to the electronegativity value of the hydrogen atom; the electronegativities of the other elements are then defined uniquely.

The second electronegativity scale was derived by Mulliken. Here a quantity χ_M is defined as

$$\chi_M(A) = \tfrac{1}{2}(I_A + E_A) \tag{15-44}$$

where I_A is the ionization potential and E_A is the electron affinity of atom A in its valence state. The valence state of an atom is its electron configuration when it participates in a chemical bond. This may be different from the configuration of the isolated atom. It is customary to express I_A, E_A, and χ_A in terms of electron volts. The relation between Mulliken's quantities χ_M and Pauling's electronegativities x is then given by

$$\chi_M(A) - \chi_M(B) = 2.78|x_A - x_B| \tag{15-45}$$

It seems that the Mulliken electronegativity scale is more precisely defined than the Pauling scale because the ionization potentials and the electron affinities are usually better known than the bond energies that were used by Pauling. Also, the

Table 15-3 Electronegativity values

Atom	Pauling[a]	Mulliken[b]	Best[c]
Ag	—	1.36	1.7
Al	1.5	1.81	1.5
As	2.0	1.75	2.0
B	2.0	2.01	2.0
Ba	0.9	—	0.9
Be	1.5	1.46	1.5
Br	2.8	2.76	2.8
C	2.5	2.63	2.6
Ca	1.0	—	1.0
Cl	3.0	3.00	3.0
Cs	0.7	—	0.7
F	4.0	3.91	4.0
H	2.1	2.28	2.1
I	2.5	2.56	2.5
K	0.8	0.80	0.8
Li	1.0	0.94	1.0
Mg	1.2	1.32	1.3
N	3.0	2.33	3.0
Na	0.9	0.93	0.9
O	3.5	3.17	3.5
Rb	0.8	—	0.8
S	2.5	2.41	2.5
Si	1.8	2.44	1.9

[a] Pauling's values for the electronegativity.
[b] Mulliken's values. These values were derived from Eq. (15-44) and converted to the Pauling scale by making use of Eq. (15-45).
[c] This column contains the values used at present, which are considered the most suitable.

Mulliken scale is easier to understand because I_A is the energy of an electron in the highest filled molecular orbital and E_A is the energy of an electron in the lowest unfilled molecular orbital so that χ is roughly equal to the energy of a bonding electron when it is located on atom A. Clearly, the higher this energy, the greater the chance that the electron is on atom A so that χ_M should be a measure of the electronegativity of atom A. The Mulliken and Pauling scales are derived from different concepts, but the resulting sets of electronegativity values are very similar if we use the conversion of Eq. (15-45). This may be verified from Table 15-3 where we have listed both sets of values.

Even though the electronegativity scale consists of a set of numbers, its practical use is more of a qualitative than a quantitative nature. We should mention that several relations have been proposed between the electronegativity difference of the atoms in a bond and the parameters that occur in valence bond theory, but none of these relationships have been adequately justified. In addition the valence bond parameters are poorly defined quantities that cannot be measured.

One quantity that is well defined and easily measured is the electric dipole moment μ_0 of a molecule. We mentioned already that the dipole moment may be derived from the rotational spectrum of a molecule, but most experimental dipole moments are obtained by measuring the dielectric constant. In quantum mechanics the theoretical dipole moment is defined as

$$\mu_0 = \langle \Psi_0 | \mu_{op} | \Psi_0 \rangle$$

$$\mu_{op} = e \sum_n Z_n \mathbf{R}_n - e \sum_j \mathbf{r}_j$$

(15-46)

Here Ψ_0 is the wave function of the molecular ground state. In the expression for the dipole moment operator, the first summation should be extended over all nuclei and the second summation over all electrons in the molecule.

A dipole moment has the dimension of charge multiplied by length. If we express the charge in terms of electrostatic units and the length in terms of centimeters, the customary unit for dipole moments is 10^{-18} esu $\cdot$ cm, which is known as a Debye unit. Remember that the charge of an electron is 4.8×10^{-10} esu and that an Ångstrom unit is 10^{-8} cm so that the dipole moment of two electronic charges with opposite signs, separated by 1 Å is 4.8 Debye units. Most experimental dipole moments range from 0.1 to 10 Debye units.

It may be seen from Eq. (15-46) that the dipole moment is a vector, but experimentally we can determine only the magnitude of μ_0 and not its direction. In most cases we have a fairly good idea what the direction of the dipole moment should be, mainly by considering the difference in electronegativity and by making a comparison with similar molecules. However, for some molecules it is difficult to say what the direction of the dipole moment should be. A well known example is CO, which has a small dipole moment of 0.11 D.

It is remarkable that in the case of the hydrogen halides the dipole moments in terms of Debye units are almost equal to the electronegativity differences $x_X - x_H$, as may be seen from Table 15-4. However, this excellent agreement is nothing more than a fortuitous coincidence because there are many reasons to believe that any simple relation between experimental dipole moments and electronegativity values

Table 15-4 A comparison between the electric dipole moments μ of hydrogen halides HX and the electronegativity differences $\Delta_{HX} = x_X - x_H$

Molecule	$\dfrac{\mu}{\text{(Debye)}}$	Δ
HF	1.9	1.9
HCl	1.07	0.9
HBr	0.8	0.7
HI	0.38	0.4

cannot really be justified theoretically. From the various ab initio calculations, that is, the fairly accurate SCF calculations of the Roothaan group and from subsequent more precise work, it may be seen that the theoretical dipole moments are quite sensitive to small variations in the eigenfunctions. Even these highly accurate calculations give relatively poor predictions about dipole moments. Also, the atomic charge clouds are not spherically symmetric so that we must consider atomic dipole moments in addition to the effects due to differences in electronegativity. Altogether there are so many different effects that contribute to molecular dipole moments that simple rules for predicting their values generally do not work.

15-6 HYBRIDIZATION

In Section 15-4 on diatomic molecules we discussed some general features of the chemical bond. In this section we use these general ideas to treat the structures of some polyatomic molecules, such as, water, ammonia, and methane. First we consider the methane molecule.

It has been shown experimentally that bond energies are nearly additive. This means that we can assign a bond energy to a given bond in a saturated molecule. In Table 5-4 we have listed some of these bond energies. The total dissociation energy of methane is approximately equal to four times the C—H bond energy (the difference from the experimental value is 4 kcal mol^{-1}). The dissociation energy of ethane is six C—H bond energies plus one C—C bond energy, and so on. For the higher alkanes the sums of the separate bond energies agree with experimental energies of formation to within less than a kcal mol^{-1}. We may conclude therefore that the additivity rules for bond energies are surprisingly accurate for saturated molecules.

Let us now consider the quantum mechanical description of the methane molecule. We assume that the four C—H bonds may be treated as separate entities because of the additivity rules for the bond energies. This means that each C—H bond is represented by a molecular orbital χ_i and that the bond consists of two electrons with opposite spins which are both placed in the molecular orbital χ_i. The methane molecule has a total of ten electrons. Two of these are in the carbon $1s$ orbital, which we denote by k_C, and the other eight electrons form four pairs, each of which occupies a bonding orbital χ_i. The total, antisymmetrized, molecular wave function is then obtained as

$$\Psi_0 = (10!)^{-1/2} \sum_P P \, \delta_P [k_C(1)\alpha(1)k_C(2)\beta(2)\chi_1(3)\alpha(3)\chi_1(4)\beta(4)$$

$$\times \; \chi_2(5)\alpha(5)\chi_2(6)\beta(6)\chi_3(7)\alpha(7)\chi_3(8)\beta(8)\chi_4(9)\alpha(9)\chi_4(10)\beta(10)] \quad \textbf{(15-47)}$$

The ground state of the carbon atom has the configuration $1s^2 2s^2 2p^2$. It seems reasonable to assume that the bonding orbitals χ_i may be represented as linear combinations of the $2s$ and $2p$ orbitals on the carbon and of the $1s$ orbital of the hydrogen atom. We write this as

$$\chi_i = t_i + \rho s_H \quad \textbf{(15-48)}$$

where t_i is a linear combination of the carbon $2s$ and $2p$ orbitals and ρ is a parameter which depends on the electronegativities of carbon and hydrogen. It may be seen in Table 15-3 that $x_C = 2.5$ and $x_H = 2.1$. It seems therefore that ρ must be smaller than unity, but that it is only slightly smaller than unity because the difference $x_C - x_H$ is only 0.4.

We mentioned in Section 15-4 that the strongest bonds are obtained if the overlap between the atomic orbitals that form the bond is large. The most favorable situation, that is, the largest amount of overlap, occurs when the atomic orbital is hybridized. This means that the atomic orbital should be a linear combination

$$t_i = \alpha s_C + \sqrt{1-\alpha^2}\, p_{C,i} \tag{15-49}$$

where $p_{C,i}$ is the linear combination of $2p$ orbitals which points in the direction of the ith hydrogen atom. Naturally, the magnitude of the overlap integral depends on the value of the parameter α.

It is customary to impose the condition that the four orbitals t_i are orthogonal to one another. It may be shown that we can always choose a set of orbitals that satisfy this condition, and, since the theoretical description then becomes much simpler, there is every reason to take the orbitals orthogonal. In our present discussion on methane we choose the hybridized atomic orbitals t_i orthogonal to one another rather than the molecular orbitals χ_i because this is more convenient in the present case.

In the case of methane we know that the four C—H bonds are equivalent and that they form a tetrahedral structure. The same must be true for the hybridized atomic orbitals t_i. Furthermore, we impose the condition that the orbitals t_i are orthogonal to one another. The derivation of the orbitals is illustrated in Example 15-3.

The specific form of the hybridized orbitals depends on how we choose our coordinate system. One set of hybridized orbitals have the form

$$t_1 = \tfrac{1}{2}s_C + \tfrac{1}{2}\sqrt{3}\, p_z$$

$$t_2 = \tfrac{1}{2}s_C + \tfrac{1}{3}\sqrt{6}\, p_x - \tfrac{1}{6}\sqrt{3}\, p_z$$

$$t_3 = \tfrac{1}{2}s_C - \tfrac{1}{6}\sqrt{6}\, p_x + \tfrac{1}{2}\sqrt{2}\, p_y - \tfrac{1}{6}\sqrt{3}\, p_z \tag{15-50}$$

$$t_4 = \tfrac{1}{2}s_C - \tfrac{1}{6}\sqrt{6}\, p_x - \tfrac{1}{2}\sqrt{2}\, p_y - \tfrac{1}{6}\sqrt{3}\, p_z$$

It is easily verified that these four orbitals are orthogonal to one another. They are also equivalent because the hybridization coefficients of the orbital s_C are the same for all four orbitals. We have chosen our coordinate system in such a way that the orbital t_1 points in the z direction and the orbital t_2 is directed in the xz plane. It may be seen that the directions of the four orbitals may be described by four unit vectors $\mathbf{e}_i$ which are given by

$$\mathbf{e}_1 = (0, 0, 1) \qquad \mathbf{e}_3 = \left(-\frac{\sqrt{2}}{3}, \frac{\sqrt{2}}{\sqrt{3}}, -\frac{1}{3}\right)$$

$$\tag{15-51}$$

$$\mathbf{e}_2 = \left(\frac{2\sqrt{2}}{3}, 0, -\frac{1}{3}\right) \qquad \mathbf{e}_4 = \left(-\frac{\sqrt{2}}{3}, -\frac{\sqrt{2}}{\sqrt{3}}, -\frac{1}{3}\right)$$

Another possible set of hybridized orbitals is given by

$$t_1 = \tfrac{1}{2}(s_C + p_x + p_y + p_z)$$
$$t_2 = \tfrac{1}{2}(s_C + p_x - p_y - p_z)$$
$$t_3 = \tfrac{1}{2}(s_C - p_x + p_y - p_z)$$
$$t_4 = \tfrac{1}{2}(s_C - p_x - p_y + p_z)$$

(15-52)

These orbitals are again orthogonal and equivalent. They point in four different directions that are given by the three vectors $\mathbf{e}_i$,

$$\mathbf{e}_1 = (1, 1, 1) \qquad \mathbf{e}_3 = (-1, 1, -1)$$
$$\mathbf{e}_2 = (1, -1, -1) \qquad \mathbf{e}_4 = (-1, -1, 1)$$

(15-53)

The set of orbitals (15-52) is equivalent with the set (15-50). The only difference is that we have rotated the coordinate axes. Both sets form a tetrahedron and the angle θ between any pair of orbitals (or vectors $\mathbf{e}_i$) is equal to the tetrahedron angle. We can easily calculate this angle, which should be equal to the H—C—H bond angle in methane. For example, if we take the unit vectors $\mathbf{e}_1$ and $\mathbf{e}_2$ of Eq. (15-51) we have

$$\cos \theta = \mathbf{e}_1 \cdot \mathbf{e}_2 = -\tfrac{1}{3}$$

(15-54)

It follows that $\theta = 109°28'$.

EXAMPLE 15-3

The specific form of the orbitals (15-50) may be derived from the orthonormality conditions between the orbitals and from the requirement that they are equivalent.

We take the first orbital along the z axis, its form is then

$$t_1 = as + bp_z$$

(a)

The second orbital is in the xz plane, it must have the form

$$t_2 = as + cp_x - dp_z$$

(b)

because the coefficients of the s orbitals must be the same in t_1 and t_2. The remaining two orbitals must have the form

$$t_3 = as - fp_x + gp_y - dp_z$$
$$t_4 = as - fp_x - gp_y - dp_z$$

(c)

because t_2, t_3, and t_4 must all make the same angle with t_1, and the two orbitals t_3 and t_4 make the same angle with t_2.

The normalization conditions for the four orbitals are

$$a^2 + b^2 = 1$$

(d)

$$a^2 + c^2 + d^2 = 1$$

(e)

$$a^2 + f^2 + g^2 + d^2 = 1$$

(f)

The orthogonality conditions are

$$a^2 - bd = 0$$

(g)

$$a^2 - cf + d^2 = 0$$

(h)

$$a^2 + f^2 - g^2 + d^2 = 0$$

(i)

It follows immediately from Eqs. (f) and (i) that $g = \frac{1}{2}\sqrt{2}$. Then from Eqs. (e) and (h), we find the values of c and f, and from the remaining equations we derive a, b, and d. In this way we obtain the coefficients that are listed in Eq. (15-50). ●

We may look upon the formation of the methane molecule as a two-step procedure. First we "prepare" the carbon atom for bonding by rearranging its atomic structure so that we have four hybridized atomic orbitals pointing in the four tetrahedron directions and each of them with one electron. This atomic configuration may be written as $(1s)^2(t_1)(t_2)(t_3)(t_4)$; it is known as the valence state of the carbon atom. The methane molecule is then formed by placing the four hydrogen atoms in the proper positions. Each hydrogen atom contributes one electron, which combines with the electron in the corresponding orbital t_i to form a two-electron bond.

In each of the one-electron states the probability of finding the electron in the $2s$ state is given by a^2, which is 0.25. In the valence state there is on the average $4 \times 0.25 = 1$ electron in the $2s$ state and $4 \times 0.75 = 3$ electrons in the $2p$ state, and we can write the atomic configuration of the atomic valence state also as $1s^2 2s 2p^3$. We have already mentioned that the configuration of the carbon atom ground state is $1s^2 2s^2 2p^2$ so that the valence state is derived from the ground state by transferring (or promoting) an electron from the $2s$ to the $2p$ state. It may be shown that the valence state has a higher energy than the atomic ground state, the energy difference, known as the promotion energy, is 96 kcal mol^{-1}.

We may explain the situation by assuming that the valence state of carbon is much more suitable for bond formation than the original atomic ground state. Apparently, it pays to excite the atom first to its valence state. This costs an amount of energy of 96 kcal mol^{-1}, which is 24 kcal mol^{-1} per bond. We have listed the C—H bond energy in Table 5-4. It is about 100 kcal mol^{-1}, and it seems that the increase in bond energy due to suitable hybridization more than compensates for the excitation energy that is needed to bring the carbon atom to its valence state.

Let us now consider the ammonia molecule NH_3. It is interesting to discuss its electronic structure by making a comparison with methane, and we have sketched the two molecules in Figure 15-7. Remember that in the valence state the configuration of the carbon atom is $(1s)^2(t_1)(t_2)(t_3)(t_4)$, because the s and p orbitals combine to form the four hybridized valence orbitals t_i. Each valence orbital t_i combines with the

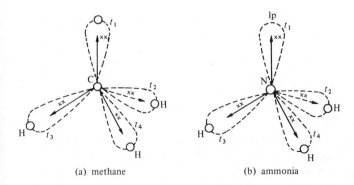

(a) methane   (b) ammonia

Figure 15-7 Electronic structures of (a) methane and (b) ammonia.

corresponding hydrogen orbital h_i to form a bond orbital χ_i, and the configuration of CH_4 is $(1s)^2(\chi_1)^2(\chi_2)^2(\chi_3)^2(\chi_4)^2$.

In the ammonia molecule the situation is slightly different because the nitrogen atom has one more electron than carbon. Here three of the four valence orbitals t_2, t_3, and t_4 combine with the corresponding hydrogen orbitals to form the bonding orbitals χ_2, χ_3, and χ_4. The remaining two electrons are accommodated in the valence orbital t_1, and this pair of electrons does not participate in the bonding. It is called a lone pair of electrons. The configuration of NH_3 is thus $(1s)^2(t_1)^2(\chi_2)^2(\chi_3)^2(\chi_4)^2$, where the orbital t_1 accommodates the lone pair of electrons and the other orbitals χ_2, χ_3, and χ_4 represent the three N—H bonds.

It may be seen now what the difference is between methane and ammonia. In methane the four hybridized orbitals are all equivalent, and for that reason they must have the form of Eq. (15-50). In the case of ammonia the three orbitals t_2, t_3, and t_4 that form the N—H bonds are equivalent, but the remaining lone pair orbital t_1 may be different. Consequently, if we try to determine the form of the four ammonia orbitals by the same method as in Example 15-3 we cannot determine all the coefficients in the orbitals because we have more parameters than equations. It may be shown, however, that the four nitrogen hybridized orbitals should have the form

$$t_1 = \rho s_N + \sqrt{1-\rho^2}\, p_z$$

$$t_2 = \tfrac{1}{3}\sqrt{3}\,\sqrt{1-\rho^2}\, s_N - \tfrac{1}{3}\sqrt{3}\,\rho p_z + \tfrac{1}{3}\sqrt{6}\, p_x$$

$$t_3 = \tfrac{1}{3}\sqrt{3}\,\sqrt{1-\rho^2}\, s_N - \tfrac{1}{3}\sqrt{3}\,\rho p_z - \tfrac{1}{6}\sqrt{6}\, p_x + \tfrac{1}{2}\sqrt{2}\, p_y \qquad \textbf{(15-55)}$$

$$t_4 = \tfrac{1}{3}\sqrt{3}\,\sqrt{1-\rho^2}\, s_N - \tfrac{1}{3}\sqrt{3}\,\rho p_z - \tfrac{1}{6}\sqrt{6}\, p_x - \tfrac{1}{2}\sqrt{2}\, p_y$$

where ρ is an unknown parameter at this point.

The value of ρ may be derived from the experimental H—N—H bond angle, which is 108°. We note that the coefficients of the p orbitals represent vectors that point in the direction of the bond. The direction of t_2 is given by the vector $(\tfrac{1}{3}\sqrt{6}, 0, -\tfrac{1}{3}\rho\sqrt{3})$ and the direction of t_3 is given by the vector $(-\tfrac{1}{6}\sqrt{6}, \tfrac{1}{2}\sqrt{2}, -\tfrac{1}{3}\rho\sqrt{3})$. The angle θ between two vectors $\mathbf{u}$ and $\mathbf{v}$ may be derived from their inner product

$$\mathbf{u} \cdot \mathbf{v} = uv \cos\theta \qquad \textbf{(15-56)}$$

In our case this gives

$$-\tfrac{1}{3} + \tfrac{1}{3}\rho^2 = (\tfrac{2}{3} + \tfrac{1}{3}\rho^2)\cos 108° \qquad \textbf{(15-57)}$$

or

$$\rho^2 = \frac{1 - 2\sin 18°}{1 + \sin 18°} \qquad \rho = 0.540 \qquad \textbf{(15-58)}$$

We see that the coefficient of the orbitals in the lone pair orbital t_1 differs only slightly from the value 0.5 in carbon.

The molecular wave function is easily obtained from the hybridized orbitals t_i. Each of the bonding orbitals t_2, t_3, and t_4 combines with the $1s$ orbital s_i of the corresponding hydrogen atom to form a molecular orbital χ_i,

$$\chi_i = t_i + \lambda s_i \qquad \textbf{(15-59)}$$

It may be seen in Table 15-3 that the electronegativity values of nitrogen and hydrogen are $x_N = 3.0$ and $x_H = 2.1$ so that λ is somewhat smaller than unity. The ammonia molecule has ten electrons and its configuration is $(k_N)^2(\chi_2)^2(\chi_3)^2(\chi_4)^2(t_1)^2$, with a pair of electrons in each of the bonding orbitals χ_i and the lone pair of electrons in the orbital t_i.

We may take the point of view that the ammonia molecule is a slightly perturbed tetrahedral structure. If we start off with four tetrahedral atomic orbitals (as in methane), then we place two electrons in the lone pair orbital t_1 and we use the other three to form the N—H bond. It may be seen that the charge density of the lone pair in the vicinity of the nitrogen nucleus is somewhat higher than in the other orbitals, and this causes the Coulomb repulsion between the lone pair and a bond to be somewhat larger than the Coulomb repulsion between two bonds. The result is that the bond orbitals t_2, t_3, and t_4 are pushed away from the lone pair orbital t_1, and the angle between a pair of bonding orbitals should be a little smaller than the tetrahedral angle of $109°28'$. The experimental H—N—H bond angle is $108°$, and this agrees with our model of the ammonia molecule as a perturbed tetrahedral structure.

It may be seen in Figure 15-6 that for a hybridized atomic orbital the electrons are displaced in the direction of the hybridization; this means that a lone pair electron contributes to the molecular dipole moment. It is easily derived that the dipole moment of a pair of electrons in the orbital t_1 is given by

$$\mu_1 = 4\rho\sqrt{1-\rho^2}\langle s_N|ez|p_z\rangle \tag{15-60}$$

If we evaluate this number we obtain a surprisingly large value, namely, $\mu_1 = 3.42$ D. The experimental dipole moment of the ammonia molecule is only 1.47 D. If we assume that the total molecular dipole moment is the vector sum of the lone pair moment $\boldsymbol{\mu}_1$ and of the three N—H bond moments $\boldsymbol{\mu}_2$, $\boldsymbol{\mu}_3$, and $\boldsymbol{\mu}_4$, we find that each N—H bond moment must be 1.9 D (N^+—H^-) in order to reproduce the experimental value of the total dipole moment. This may seem a large value, but it is not too different from what we calculate from the molecular orbitals χ_i. It may be seen that in the hybridized orbital t_i the electronic charge is shifted towards the proton (compare Eq. 15-60), whereas the hydrogen orbital s_i is spherically symmetric.

It follows that the bond moment N^+—H^- may be fairly large as a result of these atomic contributions even though the electronegativity scale predicts an excess of electronic charge on the nitrogen. We might add that we have little confidence in the various theoretical descriptions of electric dipole moments that are based on electronegativity differences only because they disregard the existence of the lone pair moments and of the atomic dipole moments that we have mentioned.

Finally, let us discuss the water molecule. Again, we treat the molecule as a perturbed tetrahedral structure. The four hybridized orbitals t_i consist of two equivalent pairs. The orbitals t_1 and t_2 participate in the two O—H bonds, and the orbitals t_3 and t_4 are the lone pair orbitals. We have sketched the molecule in Figure 15-8. It may be seen that we take the bonding orbitals t_1 and t_2 in the xy plane, symmetric with respect to the x axis. They are then obtained as

$$t_1 = as_0 + bp_x + cp_y$$
$$t_2 = as_0 + bp_x - cp_y \tag{15-61}$$

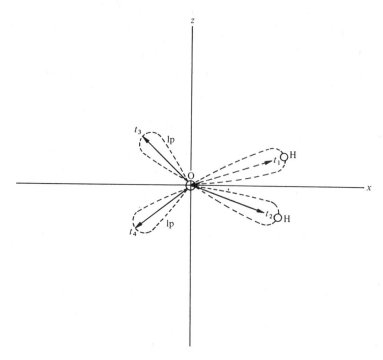

Figure 15-8   Electronic structure of the water molecule.

Because of the molecular symmetry the lone pair orbitals t_3 and t_4 must be in the xz plane, symmetric with respect to the x axis. They may be written as

$$t_3 = ds_0 - fp_x + gp_z$$
$$t_4 = ds_0 - fp_x - gp_z \tag{15-62}$$

From the orthonormality relations between the orbitals we may easily derive their form.

$$t_1 = \tfrac{1}{2}\sqrt{2}(\rho s_0 + \sqrt{1-\rho^2}\, p_x + p_y)$$

$$t_2 = \tfrac{1}{2}\sqrt{2}(\rho s_0 + \sqrt{1-\rho^2}\, p_x - p_y)$$

$$t_3 = \tfrac{1}{2}\sqrt{2}(\sqrt{1-\rho^2}\, s_0 - \rho p_x + p_z) \tag{15-63}$$

$$t_4 = \tfrac{1}{2}\sqrt{2}(\sqrt{1-\rho^2}\, s_0 - \rho p_x - p_z)$$

If the H—O—H bond angle is θ, the parameter ρ is determined from the relation

$$\cot \tfrac{1}{2}\theta = \sqrt{1-\rho^2} \tag{15-64}$$

The bond angle θ is 105° and $\rho = 0.4865$.

The molecular configuration is $(k_0)^2(\chi_1)^2(\chi_2)^2(t_3)^2(t_4)^2$ where

$$\chi_i = t_i + \lambda s_i \tag{15-65}$$

The electronegativity difference between oxygen and hydrogen is 1.4, and we expect that λ in Eq. (15-65) is considerably smaller than the corresponding quantities for the N—H and C—H bonds.

Again, we have two electrons in each lone pair orbital, and the electronic charge density in the lone pair orbitals is considerably higher than in the bonding orbitals in the vicinity of the oxygen molecule. Because of the Coulomb repulsions between the four orbitals (and some additional effects), the two bonding orbitals are slightly squeezed together as compared with the tetrahedral structure. This explains why the H—O—H bond angle of 105° is a bit smaller than the tetrahedral angle $109\frac{1}{2}°$.

The experimental dipole moment of the water molecule is 1.82 D. This is mainly due to the dipole moment of the lone pair electrons. If we calculate the lone pair moments from the orbitals t_3 and t_4 of Eq. (15-63), we obtain a value that is considerably larger than the experimental moment 1.82 D. Again, we are forced to conclude that the O—H bond moments are of the type O^+—H^-, in spite of the fact that the electronegativity values would predict just the opposite sign.

15-7 UNSATURATED MOLECULES

The chemical bonds that we discussed in Section 15-6 are all known as single bonds. We have seen that each of them is formed by one pair of electrons with opposite spins, which occupy a localized molecular orbital. We know that there are molecules with double, or even triple bonds, and it should be expected that these bonds differ in some respects from single bonds. We shall discuss the properties of multiple bonds by considering the hydrocarbons, ethane, ethylene, and acetylene. Table 15-5 lists the distances and energies of the various bonds in these molecules,

Table 15-5 Bond lengths and bond energies in some hydrocarbons

Molecule	Bond	Length/(Å)	Energy (kcal mol^{-1})
CH_4	CH	1.090	103
C_2H_6	CH	1.10	99
C_2H_4	CH	1.069	106
C_2H_2	CH	1.060	121
C_2H_6	CC	1.54	83
C_2H_4	CC	1.34	146
C_2H_2	CC	1.205	201

and you can see that these quantities may vary considerably from one molecule to another. The structures of the three molecules are also quite different. The ethane molecule consists of two tetrahedral CH_3 groups, linked together by a C—C bond. The structure of ethane is thus quite similar to methane. The ethylene molecule is planar with an H—C—C bond angle of 120°. Finally, the acetylene molecule is linear.

The electronic structure of the ethane molecule is readily understood by drawing an analogy with the methane molecule. We first construct two sets of tetrahedral hybridized orbitals u_i and t_i on the two carbon atoms, just as in methane, and we see to it that two of the orbitals on different carbon atoms, say t_4 and u_4 point towards each other. The C—C bond is then represented by an orbital

$$\chi_{CC} = t_4 + u_4 \tag{15-66}$$

and the C—H bonds by orbitals

$$\chi_i = t_i + \lambda h_i$$
$$\chi_i' = u_i + \lambda h_i' \tag{15-67}$$

The molecular configuration is then $(k_C)^2(k_C')^2(\chi_1)^2(\chi_2)^2(\chi_3)^2(\chi_1')^2(\chi_2')^2(\chi_3')^2(\chi_{CC})^2$, where the orbitals k_C and k_C' are $1s$ orbitals on the two carbons. The structures of propane and all the other saturated hydrocarbons may be derived in the same way, and all their bonds have the same orbitals as in ethane or methane.

Let us now consider the ethylene molecule. Its structure is quite different from ethane because ethylene is planar. Also the energies and distances of its bonds differ significantly from ethane, as may be seen from Table 15-5. If we wish to construct the various bonds from hybridized carbon orbitals, these orbitals must all have directions that are in the plane of the molecule. We use the coordinate system of Figure 15-9. The hybridized orbitals t_i on the first carbon atom are then given by

$$t_1 = \tfrac{1}{3}\sqrt{3}\, s_C + \tfrac{1}{3}\sqrt{6}\, p_x$$
$$t_2 = \tfrac{1}{3}\sqrt{3}\, s_C - \tfrac{1}{6}\sqrt{6}\, p_x + \tfrac{1}{2}\sqrt{2}\, p_y \tag{15-68}$$
$$t_3 = \tfrac{1}{3}\sqrt{3}\, s_C - \tfrac{1}{6}\sqrt{6}\, p_x - \tfrac{1}{2}\sqrt{2}\, p_y$$

The values of these coefficients are easily derived from the orthonormality relations between the orbitals and from the conditions that they are equivalent and in the xy plane.

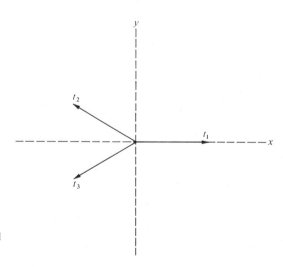

Figure 15-9 The three sp^2 hybridized orbitals on a carbon atom in ethylene.

It is useful to note the differences between these orbitals and the methane orbitals of Eq. (15-50). In the present case we have only three hybridized orbitals, which are constructed from the $2s$ orbital and from two $2p$ orbitals, the $2p_z$ orbital has not been used in the construction of the hybridized orbitals. In the case of methane we have four hybridized orbitals, which are constructed from the $2s$ orbital and from all three $2p$ orbitals. The methane orbitals are known as sp^3 hybridized orbitals, and the ethylene orbitals are called sp^2 hybridized orbitals, for obvious reasons.

The various bonds in the ethylene molecule may now be constructed from the sp^2 type hybridized orbitals of Eq. (15-68) and from the corresponding set of hybridized orbitals u_i that belong to the other carbon atom. The C—C bond is represented by a molecular orbital

$$\chi_{CC} = t_1 + u_1 \tag{15-69}$$

and the C—H bonds are represented by the orbitals

$$\chi_1 = t_2 + \lambda h_1 \qquad \chi_3 = u_2 + \lambda h_3$$
$$\chi_2 = t_3 + \lambda h_2 \qquad \chi_4 = u_3 + \lambda h_4 \tag{15-70}$$

We may accommodate ten electrons in these bonding orbitals and four electrons in the $1s$ orbitals of the two carbon atoms, a total of 14 electrons. The ethylene molecule has 16 electrons so that we have two electrons left that are not accounted for. Remember also that we have not yet considered the two $2p_z$ orbitals of the two carbon atoms because they do not occur in the bonding orbitals of Eqs. (15-69) and (15-70). The two p_z orbitals (p_z and p'_z) may be combined to form a bonding and an antibonding molecular orbital,

$$\pi_1 = p_z + p'_z$$
$$\pi_2 = p_z - p'_z \tag{15-71}$$

The molecular configuration is then obtained as $(k_C)^2(k'_C)^2(\chi_1)^2(\chi_2)^2(\chi_3)^2$ $\cdot (\chi_4)^2(\chi_{CC})^2(\pi_1)^2$.

If we compare the two molecules C_2H_6 (ethane) and C_2H_4 (ethylene) we see that the greatest difference between the molecules occurs in the C—C bond. In the case of ethane, the C—C bond is formed by one electron pair in an orbital χ_{CC}; in the case of ethylene, the C—C bond consists of two electron pairs, one in an orbital χ_{CC} and one in an orbital π_1, which is composed of the two p_z orbitals. It is customary in conjugated molecules to speak of σ orbitals, which are symmetric with respect to the plane of the molecule, and of π orbitals, which are antisymmetric with respect to the molecular plane. The double C—C bond in ethylene consists thus of a σ bond and a π bond. Table 15-6 shows that the C—C bond distances are 1.54 Å in ethane and 1.34 Å in ethylene, and the bond energies are 83 and 146 kcal mol^{-1}, respectively. The overlap integral between two $2p_z$ orbitals is much smaller than the overlap integral between two hybridized σ orbitals. The first integral is roughly 0.25, and the second is about 0.65. We can then expect that the π bond energy is smaller than the σ bond energy, which is in agreement with the experimental values 83 and 146. It is not really permissible to say that the bond energy in ethylene is the sum of a 83 kcal mol^{-1} σ bond energy and a 63 kcal mol^{-1} π bond energy because the C—C distance in the two molecules is different also, but it seems safe to conclude that the π bond energy is smaller than the σ bond energy.

Table 15-5 also shows that there are differences in the C—H bond lengths and energies between the two molecules. These differences may be ascribed to the different hybridization parameters in the carbon orbitals from which the bonds are constructed. We may recall that an sp^3 hybridized orbital has the form

$$t(sp^3) = \tfrac{1}{2}s_C + \tfrac{1}{2}\sqrt{3}\,p_\sigma \tag{15-72}$$

and that an sp^2 hybridized orbital is given by

$$t(sp^2) = \tfrac{1}{3}\sqrt{3}\,s_C + \tfrac{1}{3}\sqrt{6}\,p_\sigma \tag{15-73}$$

The two orbitals in Eqs. (15-72) and (15-73) should have somewhat different overlap integrals with the hydrogen $1s$ orbitals; thus, we expect that the C—H bonds that result from these orbitals should differ somewhat in their lengths and bond energies. It follows from Table 15-5 that the C—H bond is a little stronger in ethylene than it is in ethane, and this seems to indicate that the sp^2 hybridized orbitals have the larger overlap integrals. An exact calculation of the overlap integral agrees with this prediction.

Finally, let us consider the acetylene molecule, C_2H_2 (see Figure 15-10). Since the molecule is linear, the bonds must all be formed by hybridized orbitals that point in the x direction; hence, they must be linear combinations of the carbon $2s$ and $2p_x$ orbitals only. It is easily derived that the proper hydrized orbitals are

$$t_1 = \tfrac{1}{2}\sqrt{2}\,(s_C + p_x)$$
$$t_2 = \tfrac{1}{2}\sqrt{2}(s_C - p_x) \tag{15-74}$$

on the first carbon and

$$u_1 = \tfrac{1}{2}\sqrt{2}\,(s'_C + p'_x)$$
$$u_2 = \tfrac{1}{2}\sqrt{2}\,(s'_C - p'_x) \tag{15-75}$$

on the second carbon. The two C—H bonds are then represented by the molecular orbitals

$$\chi_1 = t_2 + \lambda h_1$$
$$\chi_2 = u_2 + \lambda h_2 \tag{15-76}$$

and the C—C bond by an orbital

$$\chi_{CC} = t_1 + u_1 \tag{15-77}$$

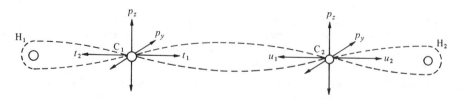

Figure 15-10 Electronic structure of acetylene, C_2H_2.

(Note that the coordinate systems on the individual atoms are always chosen in such a way that the positive coordinate direction points towards the other atom. In this case that means that the coordinate directions for the x axes on the two atoms have opposite directions.) The acetylene molecule has two π bonds; the first is represented by a linear combination of $2p_z$ orbitals

$$\pi_1 = p_z + p'_z \tag{15-78}$$

and the second by a linear combination of $2p_y$ orbitals,

$$\pi'_1 = p_y + p'_y \tag{15-79}$$

The molecular configuration is then $(k_C)^2(k'_C)^2(\chi_1)^2(\chi_2)^2(\chi_{CC})^2(\pi_1)^2(\pi'_1)^2$. We see that the C—C bond is now formed by three electron pairs; the first pair occupies a σ orbital and the other two pairs occupy π orbitals.

Table 15-5 shows that the C—C bond in acetylene has a larger bond energy and a shorter bond length than the C—C bond in ethylene. This is consistent with the description of the acetylene bond as a superposition of a σ and two π bonds. The second π bond accounts for the difference of 55 kcal mol^{-1} between the C—C bonds in the two molecules. In acetylene the hybridized orbitals are linear combinations of one s and one p orbital, and they are called sp hybridized orbitals. These orbitals have different overlap integrals, bond energies, and bond lengths from the corresponding quantities in ethylene and ethane. It may be seen in Table 15-5 that the experiments bear this out; the C—H bond energy in acetylene is 121 kcal mol^{-1}, which is quite a bit higher than the 106 kcal mol^{-1} in ethylene.

The concept of separate σ and π bonds and of σ and π orbitals is quite important in organic molecules. In general, if a molecule has a plane of symmetry we may divide the molecular orbitals into σ orbitals, which are symmetric with respect to the molecular plane, and π orbitals, which are antisymmetric. It is customary then to make use of the Σ—Π approximation, where it is assumed that the σ and π orbitals may be treated separately as if there is no interaction between them. In the examples that we have discussed in this section, all the bonds and all the orbitals are localized; however, molecules exist where the π orbitals are not localized. We shall discuss these molecules in the following section.

15-8 CONJUGATED AND AROMATIC MOLECULES

We mentioned the benzene molecule in Section 15-2, and we used its valence bond description as an example for discussing the resonance principle. In Figure 15-1 we illustrated some of the possible valence bond structures of benzene, and we concluded that the benzene molecule is a superposition of these various structures. Experimentally, the benzene molecule is a regular hexagon with a C—C bond length of 1.397 Å. Note that all the bond lengths are the same and lie somewhere between the value of 1.54 Å for the single C—C bond in ethane and 1.34 Å for the double C—C bond in ethylene. You can see that any theoretical description of benzene based on three localized single C—C bonds and three localized double C—C bonds is not consistent with the experimental information.

It was proposed in the early thirties that in aromatic molecules the π electrons are basically delocalized. In the molecular orbital description we may treat the σ electrons in the same way as in the ethylene molecule. This means that we construct three sp^2 hybridized orbitals for each carbon atom as linear combinations of the $2s$, the $2p_x$, and the $2p_y$ orbitals on the atom. (We take the z axis perpendicular to the plane of the molecule.) From the sp^2 hybridized orbitals, we can then construct the bonding molecular orbitals that form the six C—H bonds and the six σ C—C bonds. At this point we have six electrons not accounted for (one for each carbon atom) and we have six p_z orbitals that we have not yet considered in deriving the total molecular wave function. Now we assume that each of the six π electrons can move through the whole molecule. In the molecular orbital description this means that a π electron is represented by a molecular orbital ϕ which is a linear combination of the six $2p_z$ orbitals,

$$\phi = \sum_j a_j p_{z,j} \tag{15-80}$$

The numbering of the carbon atoms is shown in Figure 15-11, $p_{z,j}$ is the $2p_z$ orbital of the jth carbon atom.

The molecular orbitals for the π electrons are derived by means of the Hückel molecular orbital theory. This theory is based on a number of rather drastic approximations, and the method cannot really be justified from theoretical principles. On the other hand, the results of the Hückel theory have been quite useful, and they have been used to interpret a large number of experimental phenomena. In short, the method has worked much better than might have been expected. We will discuss it briefly and illustrate it for the benzene molecule.

First we assume that the molecular orbitals for the π electrons are eigenfunctions of some sort of effective Hartree–Fock Hamiltonian $\mathfrak{H}_{\text{eff}}$

$$\mathfrak{H}_{\text{eff}}\phi = \varepsilon\phi \tag{15-81}$$

We substitute expansion (15-80) into this equation

$$\sum_j a_j \mathfrak{H}_{\text{eff}} p_{z,j} = \varepsilon \sum_j a_j p_{z,j} \tag{15-82}$$

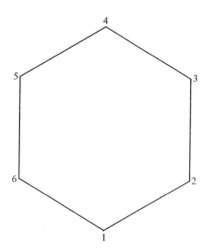

Figure 15-11 Numbering of atoms in molecular orbital calculations of benzene.

If we multiply this equation on the left by one of the atomic orbitals $p_{z,l}$ and integrate, we obtain a set of linear equations for the coefficients a_j

$$\sum_{j=1}^{6} (H_{l,j} - \varepsilon S_{l,j})a_j = 0 \qquad l = 1, 2, \ldots, 6 \qquad \text{(15-83)}$$

where

$$H_{l,j} = \langle p_{z,l} | \hat{\mathfrak{H}}_{\text{eff}} | p_{z,j} \rangle$$
$$S_{l,j} = \langle p_{z,l} | p_{z,j} \rangle \qquad \text{(15-84)}$$

We can calculate the overlap integrals $S_{l,j}$, and we find that they are equal to unity if $l = j$; they are about 0.25 if l and j denote neighboring atoms; and they are 0.01 or less in all other cases. In the Hückel method it is assumed that $S_{l,j}$ may be set equal to zero if $l \neq j$ and is unity if $l = j$. It is difficult to calculate the matrix elements $H_{l,j}$ because we do not really know the form of the Hamiltonian. We do know that the matrix elements depend only on the distance between the atoms l and j and not on their absolute positions. We may therefore define two parameters α and β as

$$\alpha = H_{l,l}$$
$$\beta = H_{l,l+1} = H_{l+1,l} \qquad \text{(15-85)}$$

The second parameter is the matrix element between two neighboring carbon atoms. In the Hückel theory all other matrix elements $H_{l,j}$ are set equal to zero.

If we make use of the preceding approximations, we may write the set of equations (15-83) as

$$\begin{aligned}
(\alpha - \varepsilon)a_1 + \beta a_2 + \beta a_6 &= 0 \\
\beta a_1 + (\alpha - \varepsilon)a_2 + \beta a_3 &= 0 \\
\beta a_2 + (\alpha - \varepsilon)a_3 + \beta a_4 &= 0 \\
\beta a_3 + (\alpha - \varepsilon)a_4 + \beta a_5 &= 0 \\
\beta a_4 + (\alpha - \varepsilon)a_5 + \beta a_6 &= 0 \\
\beta a_1 + \beta a_5 + (\alpha - \varepsilon)a_6 &= 0
\end{aligned} \qquad \text{(15-86)}$$

We do not know the values of the parameters α and β, but they are treated as adjustable parameters in Hückel theory. Besides, they may be eliminated from Eqs. (15-86). If we divide each equation by β and substitute

$$\frac{\alpha - \varepsilon}{\beta} = -x \qquad \varepsilon = \alpha + \beta x \qquad \text{(15-87)}$$

we obtain

$$\begin{aligned}
-xa_1 + a_2 + a_6 &= 0 \\
a_1 - xa_2 + a_3 &= 0 \\
a_2 - xa_3 + a_4 &= 0 \\
a_3 - xa_4 + a_5 &= 0 \\
a_4 - xa_5 + a_6 &= 0 \\
a_1 + a_5 - xa_6 &= 0
\end{aligned} \qquad \text{(15-88)}$$

The set of equations (15-88) form a set of six homogeneous equations in six unknowns. They have a solution only if the determinant of the coefficients is zero, and this condition gives an equation in the variable x from which the eigenvalues may be derived. This may be a fairly laborious procedure for large molecules, but nowadays it is a routine problem for an electronic computer. Most Hückel calculations on organic molecules are now easily performed by using computers.

Benzene happens to be one of the few molecules where the Hückel equations can be solved analytically without even using the theory of determinants. We show this calculation in Appendix B of this chapter. The results are given by Eqs. (B-12) and (B-13). In the case of benzene $N = 6$ and the eigenvalues, as given by Eq. (B-12), are

$$x_n = 2 \cos \frac{2\pi n}{6} \tag{15-89}$$

It follows from Eq. (15-87) that the energy eigenvalues ε_n are given by

$$\varepsilon_n = \alpha + \beta x_n = \alpha + 2\beta \cos \frac{2\pi n}{6} \tag{15-90}$$

The corresponding eigenfunctions may be written as

$$\phi_n = \sum_k a_k^{(n)} p_{z,k} \tag{15-91}$$

according to Eq. (15-80). The coefficients are given by Eq. (B-13). We may write them as

$$a_k^{(n)} = \exp \left(\frac{2\pi i k n}{6} \right) \qquad n = 0, \pm 1, \pm 2, 3 \tag{15-92}$$

The energy eigenvalues ε_0 and ε_3 are nondegenerate, but the other energies ε_1 and ε_2 are twofold degenerate. The different eigenfunctions are obtained by taking $n = 1$, $n = -1$ for ε_1 and $n = 2$, $n = -2$ for ε_2. The results are all listed in Table 15-6.

We may choose different basis sets for the eigenfunctions that belong to degenerate eigenvalues. It is thus possible to select real eigenfunctions. These are the functions that we have listed in the lower half of Table 15-6. They are sums and differences of the complex functions.

Realize that α and β must be negative quantities. The lowest eigenvalue is therefore $\alpha + 2\beta$, and the next lowest is the twofold degenerate eigenvalue $\alpha + \beta$. In the ground state we have two electrons in the state $k = 0$ and four electrons in the states $k = \pm 1$; the ground state energy of the π electrons in benzene is $6\alpha + 8\beta$.

The value of the parameter β may be related to a quantity known as the resonance energy and defined as the difference between the energies of the actual structure of a conjugated molecule and the structure of one of the resonance forms. The resonance energy can both be measured and calculated in terms of β, and by comparing the experimental and theoretical values we can derive the value of β.

The calculation of the resonance energy is quite simple. The energy of a localized bonding π orbital (as in ethylene) is $\alpha + \beta$ so that the energy of an electron pair in a π bond is $2\alpha + 2\beta$. In the resonance structure (see Figure 15-1) benzene has three π bonds so that the energy of the π electrons in the resonance structure is $6\alpha + 6\beta$. It follows then that the resonance energy of benzene is given by

$$E_{\text{res}} = (6\alpha + 8\beta) - (6\alpha + 6\beta) = 2\beta \tag{15-93}$$

Table 15-6   Eigenvalues and eigenfunctions of the bezene molecule

$$(\gamma = 2\pi/6)$$

k	ε	
0	$\alpha + 2\beta$	$p_{z,1} + p_{z,2} + p_{z,3} + p_{z,4} + p_{z,5} + p_{z,6}$
1	$\alpha + \beta$	$e^{i\gamma}p_{z,1} + e^{2i\gamma}p_{z,2} - p_{z,3} - e^{i\gamma}p_{z,4} - e^{2i\gamma}p_{z,5} + p_{z,6}$
−1	$\alpha + \beta$	$e^{-i\gamma}p_{z,1} + e^{-2i\gamma}p_{z,2} - p_{z,3} - e^{-i\gamma}p_{z,4} - e^{-2i\gamma}p_{z,5} + p_{z,6}$
2	$\alpha - \beta$	$e^{2i\gamma}p_{z,1} + e^{4i\gamma}p_{z,2} + p_{z,3} + e^{2i\gamma}p_{z,4} + e^{4i\gamma}p_{z,5} + p_{z,6}$
−2	$\alpha - \beta$	$e^{-2i\gamma}p_{z,1} + e^{-4i\gamma}p_{z,2} + p_{z,3} + e^{-2i\gamma}p_{z,4} + e^{-4i\gamma}p_{z,5} + p_{z,6}$
3	$\alpha - 2\beta$	$-p_{z,1} + p_{z,2} - p_{z,3} + p_{z,4} - p_{z,5} + p_{z,6}$
	$\alpha + 2\beta$	$(1/\sqrt{6})(p_{z,1} + p_{z,2} + p_{z,3} + p_{z,4} + p_{z,5} + p_{z,6})$
	$\alpha + \beta$	$\begin{cases} (\sqrt{3}/6)(p_{z,1} - p_{z,2} - 2p_{z,3} - p_{z,4} + p_{z,5} + 2p_{z,6}) \\ \frac{1}{2}(p_{z,1} + p_{z,2} - p_{z,4} - p_{z,5}) \end{cases}$
	$\alpha - \beta$	$\begin{cases} (\sqrt{3}/6)(-p_{z,1} - p_{z,2} + 2p_{z,3} - p_{z,4} - p_{z,5} + 2p_{z,6}) \\ \frac{1}{2}(p_{z,1} - p_{z,2} + p_{z,4} - p_{z,5}) \end{cases}$
	$\alpha - 2\beta$	$(1/\sqrt{6})(-p_{z,1} + p_{z,2} - p_{z,3} + p_{z,4} - p_{z,5} + p_{z,6})$

The resonance energy can be derived experimentally by measuring the heat of hydrogenation of the molecule and by making a comparison with the experimental bond energy of a localized π bond. In this way we obtain a value of 37 kcal mol^{-1}. In Table 15-7 we have listed the theoretical and experimental resonance energies for some other aromatic molecules. It follows that β is roughly equal to 20 kcal mol^{-1}.

Table 15-7   Resonance energies for some aromatic molecules

Molecule	Theoretical	Experimental (kcal mol^{-1})	β
benzene	2.00β	37	18.5
naphthalene	3.68β	75	20.4
anthracene	5.32β	105	19.7
phenanthrene	5.45β	110	20.2

It may be interesting to illustrate the Hückel theory with a second example, namely, the butadiene molecule. Here the normal structure is given by Figure 15-12, but there are various indications that the π electrons are not strictly localized in this

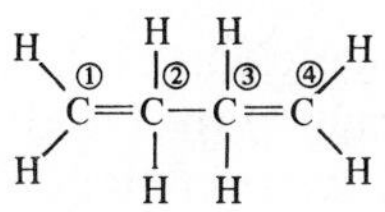

Figure 15-12 Normal structure of butadiene and numbering of carbon atoms.

molecule and we may attempt to settle this question by performing the Hückel calculation. We write the molecular orbitals for the π electrons again as

$$\phi = \sum a_n p_{z,n} \tag{15-94}$$

The coefficients must satisfy the equations

$$(\alpha - \varepsilon)a_1 + \beta a_2 = 0$$
$$\beta a_1 + (\alpha - \varepsilon)a_2 + \beta a_3 = 0$$
$$\beta a_2 + (\alpha - \varepsilon)a_3 + \beta a_4 = 0 \tag{15-95}$$
$$\beta a_3 + (\alpha - \varepsilon)a_4 = 0$$

or

$$-xa_1 + a_2 = 0$$
$$a_1 - xa_2 + a_3 = 0$$
$$a_2 - xa_3 + a_4 = 0 \tag{15-96}$$
$$a_3 - xa_4 = 0$$

if we substitute Eq. (15-87).

The solution of these equations is also discussed in Appendix B of this chapter, and the results are reported in Eqs. (B-21) and (B-22). In the case of butadiene $N = 4$ and the eigenvalues are given by

$$E_1 = \alpha + 2\beta \cos\ 36° = \alpha + 1.6180\beta$$
$$E_2 = \alpha + 2\beta \cos\ 72° = \alpha + 0.6180\beta$$
$$E_3 = \alpha + 2\beta \cos 108° = \alpha - 0.6180\beta \tag{15-97}$$
$$E_4 = \alpha + 2\beta \cos 144° = \alpha - 1.6180\beta$$

The corresponding eigenfunctions may be discussed by substituting $N = 4$ into Eq. (B-22). In the ground state there are two electrons in the state E_1 and two electrons in the state E_2, and the energy of these electrons is

$$E_{\text{tot}} = 4\alpha + 4.4720\beta \tag{15-98}$$

This energy is lower than that of the resonance structure with two localized π bonds because the resonance state has an energy $\alpha + 4\beta$. The resonance energy of the butadiene molecules is thus 0.47β.

The Hückel molecular orbital theory has also been used to make theoretical predictions about the C—C bond lengths in conjugated and aromatic molecules. The bond length in a molecule may be related to a quantity that is known as the bond order, which was introduced by Coulson. Before we give its formal definition, let us consider the case of the ethylene molecule. In the Hückel approximation the normalized bonding π orbital is given by

$$\phi = \tfrac{1}{2}\sqrt{2}\ p_{z,1} + \tfrac{1}{2}\sqrt{2}\ p_{z,2} \tag{15-99}$$

The bond order of the orbital is now defined as the product of the coefficients of the two orbitals, which is equal to one half. The total bond order is the sum of the bond orders for the individual electrons in their orbitals. Since we have two electrons in the orbital ϕ, the total bond order is two times one half, which is equal to unity. The C—C bond order in ethylene is thus one and this agrees with the fact that there is one π bond in the molecule.

In general, the π bond order between two adjacent carbon atoms k and l is defined as

$$P_{k,l} = 2 \sum_n c_{n,k} c_{n,l} \tag{15-100}$$

where the sum is to be extended over all occupied orbitals ϕ_n and where the orbitals are defined as

$$\phi_n = \sum_k c_{n,k} p_{z,k} \tag{15-101}$$

In Eq. (15-100) we have assumed that the coefficients are all real; otherwise the bond order should be defined as

$$P_{k,l} = \sum_n (c_{n,k} c_{n,l}^* + c_{n,k}^* c_{n,l}) \tag{15-102}$$

In Table 15-8 we have calculated the bond orders in the benzene molecule from the real orbitals of Table 15-6. (Obviously, the complex orbitals give the same result.) We see that the π—π bond orders in benzene are all equal to $\frac{2}{3}$.

Table 15-8   Bond orders for benzene and butadiene

			Benzene			
bond	1–2	2–3	3–4	4–5	5–6	6–1
orb 1	$\frac{1}{6}$	$\frac{1}{6}$	$\frac{1}{6}$	$\frac{1}{6}$	$\frac{1}{6}$	$\frac{1}{6}$
orb 2	$-\frac{1}{12}$	$\frac{1}{6}$	$\frac{1}{6}$	$-\frac{1}{12}$	$\frac{1}{6}$	$\frac{1}{6}$
orb 3	$\frac{1}{4}$	0	0	$\frac{1}{4}$	0	0
total	$\frac{1}{3}$	$\frac{1}{3}$	$\frac{1}{3}$	$\frac{1}{3}$	$\frac{1}{3}$	$\frac{1}{3}$
bond order	$\frac{2}{3}$	$\frac{2}{3}$	$\frac{2}{3}$	$\frac{2}{3}$	$\frac{2}{3}$	$\frac{2}{3}$

		Butadiene	
bond	1–2	2–3	3–4
orb 1	0.2236	0.3618	0.2236
orb 2	0.2236	−0.1382	0.2236
total	0.4472	0.2236	0.4472
bond order	0.8944	0.4472	0.8944

In butadiene the molecular orbitals that belong to the eigenvalues of Eq. (15-97) are easily derived from equation (B-22). The normalized orbitals are

$$\phi_1 = 0.3717 p_{z,1} + 0.6015 p_{z,2} + 0.6015 p_{z,3} + 0.3717 p_{z,4}$$
$$\phi_2 = 0.6015 p_{z,1} + 0.3717 p_{z,2} - 0.3717 p_{z,3} - 0.6015 p_{z,4}$$
$$\phi_3 = 0.6015 p_{z,1} - 0.3717 p_{z,2} - 0.3717 p_{z,3} + 0.6015 p_{z,4}$$
$$\phi_4 = 0.3717 p_{z,1} - 0.6015 p_{z,2} + 0.6015 p_{z,3} - 0.3717 p_{z,4}$$

$$\tag{15-103}$$

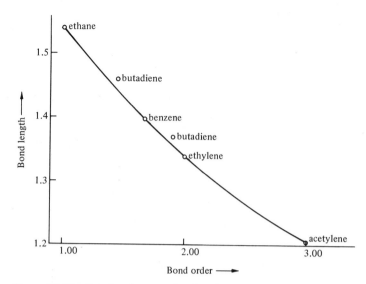

Figure 15-13 Relation between bond order and bond length in some conjugated molecules.

It has been found empirically that the theoretical bond orders may be related to the experimental bond lengths. We know the bond orders and the C—C bond lengths for ethane, ethylene, acetylene, benzene, and butadiene, and in Figure 15-13 we have plotted the bond lengths versus the bond orders. We see that all the points that we have are situated on a curve, and we feel confident that we may use this curve for predicting the bond lengths of conjugated molecules from the calculated bond orders.

Note that the 1–2 bond length in butadiene is 1.37 Å, which is somewhat longer than the bond length 1.34 in ethylene, and that the 2–3 bond length is 1.46, which is shorter than the bond length of 1.54 Å in ethane. We can see immediately that the 1–2 bond is not a pure π bond so that it must have a bond order smaller than unity. We also see that the 2–3 bond must have some π character. It follows from the bond lengths in butadiene that the simple valence bond structure of Figure 15-12 does not give an adequate representation of the true butadiene structure. If we wish to use valence bond theory we must consider resonance with the structure of Figure 15-14 in order to explain the bond lengths in the molecule.

The Hückel molecular orbital method has been widely used in organic chemistry to interpret a variety of experimental results. We have already seen how the method can be used to calculate resonance energies and bond lengths. Its most important applications are the various ways it provides for making predictions about chemical reactions. The method is easily modified to include molecules that contain heteroatoms, such as nitrogen or oxygen, in the π electron system, and it can be applied to a wide variety of organic molecules.

Figure 15-14 Ionic resonance structure of butadiene.

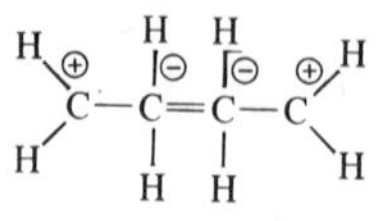

Appendices

APPENDIX A EVALUATION OF INTEGRALS *I, J,* AND *S* BY USE OF ELLIPTICAL COORDINATES

We want to calculate the following three integrals

$$S = \langle s_a | s_b \rangle = (1/\pi) \int e^{-r_a} e^{-r_b} \, d\mathbf{r}$$

$$I = \langle s_a | r_b^{-1} | s_a \rangle = (1/\pi) \int e^{-r_a} (1/r_b) \, e^{-r_a} \, d\mathbf{r} \qquad \textbf{(A-1)}$$

$$J = \langle s_b | r_b^{-1} | s_a \rangle = (1/\pi) \int e^{-r_b} (1/r_b) \, e^{-r_a} \, d\mathbf{r}$$

The three integrals *I, J,* and *S* are calculated by using elliptical coordinates. We feel that it may be useful to discuss these coordinates because they are often used in quantum mechanical calculations.

We consider the point *P* in Figure 15-15, which is described by the vector **r**. We take two points *a* and *b* on the *z* axis. Each point is at a distance $\frac{1}{2}R$ from the origin so that point *a* has the coordinates $(0, 0, -\frac{1}{2}R)$ and point *b* has the coordinates $(0, 0, \frac{1}{2}R)$. We define r_a as the distance from *a* to *P* and r_b as the distance from *b* to *P*. We know then that

$$r_a + r_b \geq R \geq |r_a - r_b| \qquad \textbf{(A-2)}$$

Hence

$$\frac{r_a + r_b}{R} \geq 1 \qquad -1 \leq \frac{r_a - r_b}{R} \leq 1 \qquad \textbf{(A-3)}$$

Point *P* is determined by r_a, r_b, and the angle ϕ between the plane *aPb* and the *x* axis. The elliptical coordinates of the point *P* are now defined as μ, ν and ϕ, where

$$\mu = \frac{r_a + r_b}{R} \qquad \nu = \frac{r_a - r_b}{R} \qquad \textbf{(A-4)}$$

The range of values of the coordinates follows from Eq. (A-3),

$$1 \leq \mu \leq \infty$$
$$-1 \leq \nu \leq 1 \qquad \textbf{(A-5)}$$
$$0 \leq \phi \leq 2\pi$$

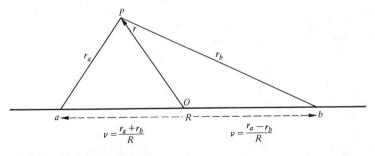

Figure 15-15 Elliptical coordinates.

It is obvious that

$$r_a = \tfrac{1}{2}R(\mu + \nu) \qquad r_b = \tfrac{1}{2}R(\mu - \nu) \tag{A-6}$$

The transformation between the Cartesian coordinates (x, y, z) and the elliptical coordinates (μ, ν, ϕ) is given by

$$
\begin{aligned}
x &= \tfrac{1}{2}R \cos \phi [(\mu^2 - 1)(1 - \nu^2)]^{1/2} \\
y &= \tfrac{1}{2}R \sin \phi [(\mu^2 - 1)(1 - \nu^2)]^{1/2} \\
z &= \tfrac{1}{2}R\mu\nu
\end{aligned} \tag{A-7}
$$

Three-dimensional integrations are transformed as follows

$$\int_{-\infty}^{\infty} \int_{-\infty}^{\infty} \int_{-\infty}^{\infty} f(x, y, z)\, dx\, dy\, dz = (R^3/8) \int_0^{2\pi} d\phi \int_{-1}^{1} d\nu \int_1^{\infty} f(\mu, \nu, \phi)(\mu^2 - \nu^2)\, d\mu \tag{A-8}$$

Let us now consider the three integrals S, I, and J. We have

$$
\begin{aligned}
S &= (1/\pi) \iiint \exp(-r_a) \exp(-r_b)\, dx\, dy\, dz \\
&= (R^3/8\pi) \int_0^{2\pi} d\phi \int_{-1}^{1} d\nu \int_1^{\infty} (\mu^2 - \nu^2)\, e^{-\mu R}\, d\mu \\
I &= (R^2/4\pi) \int_0^{2\pi} d\phi \int_{-1}^{-1} d\nu \int_1^{\infty} (\mu + \nu)\, e^{-R(\mu+\nu)}\, d\mu \\
J &= (R^2/4\pi) \int_0^{2\pi} d\phi \int_{-1}^{1} d\nu \int_1^{\infty} (\mu + \nu)\, e^{-\mu R}\, d\mu
\end{aligned} \tag{A-9}
$$

The various integrals can be calculated exactly by standard methods. The results are

$$
\begin{aligned}
S &= e^{-R}[1 + R + (R^2/3)] \\
I &= (1/R)[1 - e^{-2R}(1 + R)] \\
J &= e^{-R}(1 + R)
\end{aligned} \tag{A-10}
$$

APPENDIX B ANALYTIC SOLUTIONS OF THE HÜCKEL EQUATIONS

There are two systems for which the Hückel equations can be solved analytically: the first system is a ring of N carbon atoms and the second system is a chain of N carbon atoms. The benzene molecule is a ring system with $N = 6$ and butadiene is a chain with $N = 4$. We consider the ring system first.

It may be seen from Eq. (15-88) that for a ring system the first and the last (or Nth) equations are given by

$$
\begin{aligned}
-xa_1 + a_2 + a_N &= 0 \\
a_1 + a_{N-1} - xa_N &= 0
\end{aligned} \tag{B-1}
$$

All the other equations may be written as

$$a_{k-1} - xa_k + a_{k+1} = 0 \qquad (k = 2, 3, \ldots, N-1) \tag{B-2}$$

We can solve the set of equations (B-2) by substituting

$$a_k = e^{ik\rho} \tag{B-3}$$

This gives

$$e^{i(k-1)\rho} - x e^{ik\rho} + e^{i(k+1)\rho} = 0 \tag{B-4}$$

or

$$x = e^{i\rho} + e^{-i\rho} = 2 \cos \rho \tag{B-5}$$

We can also solve the equations (B-2) by substituting

$$a_k = e^{-ik\rho} \tag{B-6}$$

Substitution into Eq. (B-2) leads to the same value of x as in Eq. (B-5) because

$$e^{-i(k-1)\rho} - x e^{-ik\rho} + e^{-i(k+1)\rho} = 0 \tag{B-7}$$

It follows that the general solution of the set of equations (B-2) is given by

$$a_k = A e^{ik\rho} + B e^{-ik\rho}$$
$$x = 2 \cos \rho = e^{i\rho} + e^{-i\rho} \tag{B-8}$$

where A and B are arbitrary parameters.

We still must solve the first and the last equations (B-1) of the total set. We substitute the results (B-8) into the two equations (B-1). The result is

$$-(e^{i\rho} + e^{-i\rho})(A e^{i\rho} + B e^{-i\rho}) + (A e^{2i\rho} + B e^{-2i\rho}) + (A e^{Ni\rho} + B^{-Ni\rho}) = 0 \tag{B-9}$$
$$(A e^{i\rho} + B e^{-i\rho}) + (A e^{(N-1)i\rho} + B e^{-(N-1)i\rho}) - (e^{i\rho} + e^{-i\rho})(A e^{Ni\rho} + B e^{-Ni\rho}) = 0$$

or

$$A(-1 + e^{Ni\rho}) + B(-1 + e^{-Ni\rho}) = 0$$
$$A e^{i\rho}(1 - e^{Ni\rho}) + B e^{i\rho}(1 - e^{-Ni\rho}) = 0 \tag{B-10}$$

It follows that both equations are solved for all values of A and B if we take

$$e^{iN\rho} = 1$$

or

$$\rho = \frac{n \cdot 2\pi}{N} \qquad n = 0, 1, 2, 3, \ldots \tag{B-11}$$

It may be useful to summarize this result. The possible values of x (the energy eigenvalues) are given by

$$x_n = 2 \cos \frac{2\pi n}{N} \tag{B-12}$$

and the corresponding solutions are

$$a_k^{(n)} = A \exp(2\pi i k n/N) + B \exp(-2\pi i k n/N) \tag{B-13}$$

Each eigenvalue is twofold degenerate except when $n = 0$ and when $n = \frac{1}{2}N$. The possible values of n are thus

$$n = 0, 1, 2, \ldots, \tfrac{1}{2}N \tag{B-14}$$

if N is even and

$$n = 0, 1, 2, \ldots, \tfrac{1}{2}(N-1) \tag{B-15}$$

if N is odd.

The corresponding solutions for benzene, where $N = 6$, are listed in Table 15-6.

Let us now consider the solution of the Hückel equations for a chain of N carbon atoms. This situation is similar to the ring system that we just discussed. For a chain the first and the last equations are given by

$$-xa_1 + a_2 = 0$$
$$a_{N-1} - xa_N = 0 \tag{B-16}$$

instead of by Eq. (B-1), but the other equations are the same as Eqs. (B-2) for the ring. This means that the solution of the chain is again obtained in the form (B-8), but now we must substitute this result into Eq. (B-16) in order to determine A, B and ρ. The result is

$$-(e^{i\rho} + e^{-i\rho})(A\, e^{i\rho} + B\, e^{-i\rho}) + (A\, e^{2i\rho} + B\, e^{-2i\rho}) = 0$$
$$[A\, e^{(N-1)i\rho} + B\, e^{-(N-1)i\rho}] - (e^{i\rho} + e^{-i\rho})(A\, e^{Ni\rho} + B\, e^{-Ni\rho}) = 0 \tag{B-17}$$

These two equations reduce to

$$A + B = 0$$
$$A\, e^{i(N+1)\rho} + B\, e^{-i(N+1)\rho} = 0 \tag{B-18}$$

The condition for ρ is

$$e^{2i(N+1)\rho} = 1 \tag{B-19}$$

or

$$\rho = \frac{n\pi}{N+1} \qquad n = 1, 2, 3, \ldots, N \tag{B-20}$$

Let us summarize this result also. The possible values for x are

$$x_n = 2 \cos \frac{n\pi}{N+1} \tag{B-21}$$

and the corresponding solutions are

$$a_k^{(n)} = \sin \frac{nk\pi}{N+1} \tag{B-22}$$

_____ *Problems*

1. In Eq. (15-11) we take the variational function for the hydrogen molecular ion as $e^{-\delta\mu}(1 + a\nu^2)$. Why didn't we take the variational function as $e^{-\delta\mu}(1 + a\nu + b\nu^2)$?

2. Show that the probabilities P_{AA}, P_{AB} and P_{BB} for the hydrogen molecular ion, as defined by Eq. (15-13) have the values that we report in the text for $R = 2$ au. Calculate the values also for $R = 2.5$ au and compare the two sets of values.

3. Which method gives the best result for the bond energy of the hydrogen molecule, the valence bond method or the molecular orbital method? Can you explain this result?

4. Which method do you expect to give better results for the hydrogen molecule, the valence bond method with the inclusion of ionic structures or the molecular orbital method with the inclusion of configuration interaction?

5. What is the value of the overlap integral $\langle s_A | s_B \rangle$ at the internuclear distance $R = 2$ au of the hydrogen molecular ion? What is the value of the same integral for $R = 1$ au $R = 4$ au and $R = 10$ au?

6. Determine the values of the normalized wave function $s_A + s_B$ for the hydrogen molecular ion at the position of one of the protons and at the center of the molecule. Where does the wave function have a larger value?

7. Why does the wave function of Eq. (15-22) give a better energy for the hydrogen molecule than the much more complicated Hartree–Fock function?

8. The oxygen molecule O_2 is one of the few molecules whose ground state is a triplet state. Can you explain why this is so by making use of Eq. (15-35)?

9. Until recently it was believed that the CH_2 radical has a linear configuration in its ground state. Explain why it may be expected that in this configuration the ground state of the molecule is a triplet state.

10. It has been found recently that the CH_2 radical is slightly bent, and it was also found that its ground state is a triplet state. Can you explain this by means of arguments similar to Problem 9?

11. The electric dipole moment of the NH_3 molecule is 1.47 and the dipole moment of the NF_3 molecule is 0.21 D. Explain the relative magnitudes of these numbers and, in particular, why the NF_3 moment is so small in spite of the large difference in electronegativity constants between N and F.

12. Why is the H—N—H bond angle of $108°$ in ammonia smaller than the tetrahedral angle of $109\frac{1}{2}°$?

13. If you have to do some numerical calculations with the set of sp^3 hybridized orbitals of carbon of Eq. (15-52) what would you substitute for the atomic $2s$ and $2p$ orbitals? Give the specific form of the orbitals and justify your choice.

14. Calculate the expectation value of the coordinate z with respect to the lone pair orbital of Eq. (15-55) of the nitrogen atom. Choose the atomic orbitals yourself and justify your choice. Use your result to calculate the dipole moment in Debye units due to the lone pair of electrons in ammonia.

15. What are the relative magnitudes of the C—H, N—H, and O—H bond moments in methane, ammonia, and water (qualitative, not quantitative)? Justify your answer.

16. Why are the C—H bond energies different in ethane, ethylene, and acetylene?

17. Explain the electronic structure of the HCN molecule in terms of σ and π orbitals.

18. The triphenyl methyl radical has a planar structure; it consists of three phenyl rings attached to a carbon atom. How do you explain the relative stability of this radical as compared with the nonplanar triphenyl methane molecule?

19. In the aromatic molecule pyridine, C_5H_5N, the C—N—C bond angle is smaller than $120°$. (Pyridine is benzene with one C—H group in the ring replaced by a nitrogen atom.) How do you explain this in terms of the sp^2 hybridized σ orbitals of the nitrogen atom?

20. Derive the Hückel energy eigenvalues for the conjugated molecule hexatriene

$$H_2C=C-C=C-C=CH_2$$

$$\overset{\displaystyle H}{|} \quad \overset{\displaystyle H}{|} \quad \overset{\displaystyle H}{|} \quad \overset{\displaystyle H}{|}$$

from the results in Appendix B. Derive the resonance energy of this molecule in terms of the Hückel parameter β.

21. Write down the equations for the molecular orbitals in the Hückel calculation of the naphthalene molecule. [Use a form similar to the benzene equations (15-88).] There is no simple method for solving these equations and we are not asking you to solve them.

22. The bond order for the central bond in naphthalene is 1.518. Predict the length of this bond from the curve in Figure 15-13 and compare the result with the experimental value 1.418 Å.

23. Write down the Hückel equations for cyclopentadiene, C_5H_6. (Consider carefully the electronic structure of the σ electrons before doing this.) Give the solutions of these equations. List the energy eigenvalues and the corresponding molecular orbitals.

_____ *References and Recommended Reading*

In this chapter we have given a brief discussion of the quantum theory of the chemical bond. This subject is also known under the name quantum chemistry, and it is treated in a number of books at different levels and from different points of view. In the following list the books have been divided into different categories. Category A contains some of the older books. The review article by Van Vleck and Sherman [A1] may be old, but it still gives an excellent brief review of the subject. The book by Hellmann [A2] is no longer available and it is written in German, but it is an excellent book, especially considering its age. Coulson's book [A3] is very well written and gives a very good insight into the subject without using very much mathematics. The book by Eyring, Walter, and Kimball [A4] contains a lot of material, but it is more suitable as a reference book than as a text. Kauzmann's book [A5] is just the opposite of Coulson's book; it is highly mathematical and it discusses general theory but not too many chemical applications. In category B some recent texts on quantum theory are listed. These are all good books and it is difficult to make a comparison of their relative quality, so they are just listed in alphabetical order. Category C includes books that are concerned mainly with molecular orbital theory and its applications to organic chemistry of conjugated molecules.

[A1] J. H. Van Vleck and A. Sherman: *Rev. Mod. Phys.* **7**:167 (1935). The quantum theory of valence.

[A2] H. Hellmann: *Einführung in die Quantenchemie.* Deuticke, Leipzig, 1937.

[A3] C. A. Coulson: *Valence,* 2nd ed. Oxford University Press, London, 1961.

[A4] H. Eyring, J. Walter, and G. E. Kimball: *Quantum Chemistry.* Wiley, New York, 1944.

[A5] W. Kauzmann: *Quantum Chemistry.* Academic Press, New York, 1957.

[B1] J. M. Anderson: *Introduction to Quantum Chemistry.* W. A. Benjamin, Menlo Park, Calif., 1969.

[B2] M. Karplus and R. N. Porter: *Atoms and Molecules.* W. A. Benjamin, Menlo Park, Calif., 1970.

[B3] I. N. Levine: *Quantum Chemistry*, Vols. I and II. Allyn & Bacon, Boston, 1970.

[B4] P. O'D. Offenhartz: *Atomic and Molecular Orbital Theory.* McGraw-Hill, New York, 1970.

[B5] R. G. Parr: *The Quantum Theory of Molecular Electronic Structure.* W. A. Benjamin, Menlo Park, Calif., 1964.

[B6] F. L. Pilar: *Elementary Quantum Chemistry.* McGraw-Hill, New York, 1968.

[B7] J. C. Slater: *Quantum Theory of Molecules and Solids*, Vol I. McGraw-Hill, New York, 1963.

[C1] R. Daudel, R. Lefebvre, and C. Moser: *Quantum Chemistry.* Wiley-Interscience, New York, 1959.

[C2] A. Streitwieser: *Molecular Orbital Theory for Organic Chemists.* Wiley, New York, 1961.

[C3] M. J. S. Dewar: *The Molecular Orbital Theory of Organic Chemistry.* McGraw-Hill, New York, 1969.

chapter 16

Magnetic Resonance

16-1 INTRODUCTION

Magnetic resonance concerns itself with the study of transitions between the various spin states within an atom or a molecule. In Chapter 12 we discussed the properties of the electron spin. We mentioned that the electron spin has a constant magnitude that is characterized by the quantum number $s = \frac{1}{2}$. Most nuclei also possess a spin angular momentum and magnetic moment, characterized by a spin quantum number I. We will show that the formal theoretical description of the nuclear spin is very similar to the theory of the electron spin.

For magnetic resonance we can either study the transitions between different electronic spin states or between different nuclear spin states. In the first case we speak of electron spin resonance (ESR) and in the second case we speak of nuclear magnetic resonance (NMR). Practically every atom has at least one isotope that has a nonzero nuclear spin, and NMR is in principle applicable to any of the nuclei in a molecule. In practice, though, the bulk of the work in NMR has been on protons. An ESR experiment can be performed only on a molecule with a nonzero total electronic spin. We have seen that practically all stable molecules have singlet ground states, and it follows that they cannot be studied by means of ESR. Initially, ESR experiments were confined to the study of free radicals, which have an unpaired electron spin. Many free radicals are intermediate products in chemical reactions, and ESR has been used extensively to study chemical reactions. More recently it has become possible also to measure ESR of molecules in their triplet states.

The possible observation of magnetic resonance was predicted as early as 1936, but the early attempts to measure the phenomenon all failed. We will discuss later why the experiment is not quite as easy to do as it may seem at first sight. Magnetic resonance was discovered in 1945–46. The discovery may be attributed to three independent and almost simultaneous achievements: (1) the first ESR measurement by Zavoiski in the USSR, (2) the nuclear resonance absorption measurement by Purcell, Torrey, and Pound at Harvard and, (3) the nuclear induction work by Bloch, Hansen, and Packard at Stanford. We should realize that the electromagnetic radiation that is absorbed in electron spin resonance has a wave length of the order of 1 cm; this is in the range of radar waves. The advances in radar technology that were made during the war were essential in developing the experimental techniques that were needed for measuring magnetic resonance.

In Chapter 12 we saw that a spinning electron has a magnetic moment $\boldsymbol{\mu}_s$ which is given by

$$\boldsymbol{\mu}_s = -\frac{e}{mc}\mathbf{S} \tag{16-1}$$

where $\mathbf{S}$ is the spin operator (see Eq. 12-29). If we place a magnetic moment $\boldsymbol{\mu}$ in a homogeneous magnetic field **H** its energy is given by

$$E = -\boldsymbol{\mu}_s \cdot \mathbf{H} = -(\mu_x H_x + \mu_y H_y + \mu_z H_z) \tag{16-2}$$

The interaction between the electron spin and a homogeneous magnetic field **H** is therefore represented by a Hamiltonian

$$\mathfrak{H} = -\boldsymbol{\mu}_s \cdot \mathbf{H} = \frac{e}{mc}(\mathbf{H} \cdot \mathbf{S}) \tag{16-3}$$

It is convenient to choose the z axis in the direction of the magnetic field **H**, because the Hamiltonian reduces then to

$$\mathfrak{H} = \frac{e}{mc} H S_z \tag{16-4}$$

The electron spin can have two possible orientations with respect to the z axis. The first, where the spin is parallel to the z axis and where it points in the positive z direction is represented by a spin function α. In the second, the spin points in the negative z direction; this is represented by a spin function β. The spin functions α and β are eigenfunctions of the operator S_z as we have seen in Eq. (12-30), and we have

$$S_z \alpha = \frac{\hbar}{2}\alpha \qquad S_z \beta = -\frac{\hbar}{2}\beta \tag{16-5}$$

Obviously, the functions α and β are also eigenfunctions of the Hamiltonian (16-4), because

$$\mathfrak{H}\alpha = \frac{e\hbar H}{2mc}\alpha \qquad \mathfrak{H}\beta = -\frac{e\hbar H}{2mc}\beta \tag{16-6}$$

It is convenient to make use of the constant μ_0, the Bohr magneton, which is given by

$$\mu_0 = \frac{e\hbar}{2mc} = 0.9273 \times 10^{-20} \text{ erg gauss}^{-1} = 4.670 \times 10^{-5} \text{ cm}^{-1} \text{ gauss}^{-1} \tag{16-7}$$

1 erg gauss$^{-1} = 5.034 \times 10^{15}$ cm^{-1} gauss^{-1}.

It follows easily from Eq. (16-6) that the energy difference ΔE between the two eigenstates of the electron spin is given by

$$\Delta E = 2\mu_0 H \tag{16-8}$$

The magnetic fields that are customarily used for magnetic resonance experiments vary between 3000 and 10,000 gauss; in electron spin resonance the field is usually around 3400 gauss. In that case the energy difference is

$$\Delta E = 2\mu_0 H = 2(4.670 \times 10^{-5})(3.4 \times 10^3) \text{ cm}^{-1} = 0.318 \text{ cm}^{-1} \tag{16-9}$$

The resonance radiation that corresponds to ΔE has a wavelength of about 3.15 cm, which is in the radar region.

In magnetic resonance theory the spin operators are usually defined somewhat differently from the preceding treatment, namely as

$$S_z \alpha = \tfrac{1}{2}\alpha \qquad S_z \beta = -\tfrac{1}{2}\beta \tag{16-10}$$

instead of by Eq. (16-5). This means that the factor $\hbar$ is included in the Hamiltonian, which takes the form

$$\mathfrak{H} = \frac{e\hbar}{mc}(\mathbf{H} \cdot \mathbf{S}) = 2\mu_0(\mathbf{H} \cdot \mathbf{S}) \qquad (16\text{-}11)$$

instead of Eq. (16-4). Customarily, this is written as

$$\mathfrak{H} = g\mu_0(\mathbf{H} \cdot \mathbf{S}) \qquad (16\text{-}12)$$

By comparing Eqs. (16-11) and (16-12) you can see that $g = 2$ but it has been found experimentally that g differs slightly from this value. The experimental value for a free electron spin is $g = 2.002292$. This discrepancy in the g factor is due to a number of small corrections that are not considered in the simple classical theory that we have discussed.

The quantum theory of nuclear spins is very similar to the theory of the electron spin, especially if we use the formulation of Eqs. (16-10), (16-11), and (16-12). The interaction between a nuclear spin $\mathbf{I}$ and a homogeneous magnetic field $\mathbf{H}$ is represented by a Hamiltonian

$$\mathfrak{H} = -g_N\mu_N(\mathbf{H} \cdot \mathbf{I}) \qquad (16\text{-}13)$$

Here μ_N is the nuclear magneton, which is defined as

$$\mu_N = \frac{e\hbar}{2Mc} \qquad (16\text{-}14)$$

where M is the mass of the proton. The spin operator $\mathbf{I}$ plays the same role as the spin operator $\mathbf{S}$ for electron spin, but there is one important difference between the two cases. We have seen that the total electron spin has a value $\frac{1}{2}$ and, consequently, the electron spin can only have two possible projections $\pm\frac{1}{2}$ in a given direction. The spin functions α and β are defined with respect to the z direction; consequently, they are eigenfunctions of the operator S_z with the eigenvalues given by

$$S_z\alpha = \tfrac{1}{2}\alpha \qquad S_z\beta = -\tfrac{1}{2}\beta \qquad (16\text{-}15)$$

Certain nuclei (some are listed in Table 16-1) also have spin $\frac{1}{2}$. Here the situation is

Table 16-1 Nuclei with spin $\frac{1}{2}$ and their g values

	g
^{1}H	5.5854
^{3}H	5.9576
^{13}C	1.4043
^{15}N	−0.5661
^{19}F	5.2546
^{29}Si	−1.1095
^{31}P	2.2610
^{195}Pt	1.2008
^{199}Hg	0.9986

identical with the electron. The nuclear spin has two possible projections with respect to the z axis, and the two situations may be represented by the spin functions $\zeta_{1/2}$ and $\zeta_{-1/2}$. The eigenfunctions are given by

$$I_z\zeta_{1/2} = \tfrac{1}{2}\zeta_{1/2} \qquad I_z\zeta_{-1/2} = -\tfrac{1}{2}\zeta_{-1/2} \tag{16-16}$$

There is no simple way to calculate the gyromagnetic ratio g_N for a nucleus and the g factors have to be determined experimentally. We have listed some of them in Table 16-1.

An important difference between the nuclear and the electronic spins is that the spin quantum number I for a nucleus can have a value different from $\tfrac{1}{2}$. For example, the nuclear spins of deuterium and of ^{14}N both have a magnitude $\hbar$ (see Table 16-2).

Table 16-2 Nuclei with spin 1 and their g values

	g
D	0.85738
^{14}N	0.40357

In this case the projections of the spin in a given direction can have the values 1, 0, and −1. The operator I_z has three eigenfunctions, which we denote by ζ_1, ζ_0, and ζ_{-1} and the eigenvalues are given by

$$I_z\zeta_1 = \zeta_1 \qquad I_z\zeta_0 = 0 \qquad I_z\zeta_{-1} = -\zeta_{-1} \tag{16-17}$$

In general, if the spin quantum number of a nucleus X has the value I_X, the projection in the z direction can have the values $I_X, I_X - 1, I_X - 2, \ldots, -I_X + 1, -I_X$ (see Figure 16-1). The corresponding eigenfunctions and eigenvalues are given by

$$I_z\zeta_m = m\zeta_m \qquad m = I_X, I_X - 1, \ldots, -I_X + 1, -I_X \tag{16-18}$$

The nucleus with the largest spin is ^{50}V, which has spin 6. Other values are $\tfrac{3}{2}$ for ^{7}Li, ^{9}Be and ^{11}B, $\tfrac{5}{2}$ for ^{17}O, ^{25}Mg and ^{127}I, $\tfrac{7}{2}$ for ^{133}Cs, and so on. Some nuclei, such as ^{12}C and ^{16}O have no spin, and their spin quantum number is zero.

The rest of this chapter is mainly concerned with the magnetic resonance of proton spins, and it may be useful to calculate some of the values that are relevant for proton magnetic resonance. The nuclear magneton μ_N has the value

$$\mu_N = \frac{e\hbar}{2Mc} = 2.5428 \times 10^{-8} \text{ cm}^{-1} \text{ gauss}^{-1} \tag{16-19}$$

In a homogeneous magnetic field H the energy difference between the two levels of the proton spin is

$$\Delta E = g_N\mu_N H = 1.4202 \times 10^{-7} H \text{ cm}^{-1} \text{ gauss}^{-1} \tag{16-20}$$

Customarily, the magnetic field in nuclear magnetic resonance is 10,000 gauss; the resonance radiation in such a field has a wave number of $1.4202 \times 10^{-3} \text{ cm}^{-1}$.

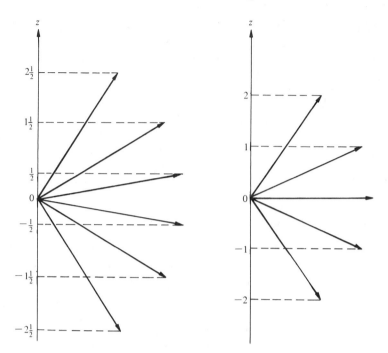

Figure 16-1   Quantization of the nuclear spin along the z axis for the two cases $I = \frac{5}{2}$ and $I = 2$.

It is obvious from our discussion in Chapter 13 that radiation will only be absorbed by a spin system if its frequency ν satisfies the resonance condition

$$h\nu = \Delta E = g_N \mu_N H \qquad (16\text{-}21)$$

In optical spectroscopy the energy difference is a fixed quantity, and we can observe absorption or emission of light by varying the frequency of the light. In magnetic resonance we can vary the energy difference ΔE by varying the magnitude of the magnetic field H so that we have two ways of satisfying the resonance condition (16-21). We can either vary the frequency ν of the radiation or the magnitude of the magnetic field. In most magnetic resonance experiments it is not possible to change the frequency beyond a very narrow range, so the resonance condition must be satisfied by varying the magnetic field.

The value of H in Eq. (16-21) is the magnitude of the magnetic field at the position of the nucleus. This is slightly different from the exterior field H_{ext} that we apply because it is the sum of H_{ext} and the small magnetic fields that are due to the motion of the electrons in the vicinity and to the interaction with other nuclei. Magnetic resonance experiments are accurate enough so that these small magnetic fields can be measured, and this is the reason why magnetic resonance is of such general interest to the chemists.

The small magnetic fields H_{ind} that are due to the motion of the electrons in the vicinity of a proton are proportional to the exterior field H_{ext}, and the proportionality factor is of the order of 10^{-5}. The magnitude of H_{ind} is slightly different for different protons; for example, in the ethyl alcohol molecule, there are three different types of protons and each of them will have a slightly different resonance line, due to differences in the induced field H_{ind}. The interactions between the magnetic moments of the different protons, the spin-spin interactions, also affect the resonance frequencies. As a result of these various effects the proton magnetic resonance spectrum of the ethyl alcohol molecule contains a number of different lines. In this chapter we shall first discuss the experimental conditions under which magnetic resonance may be observed and then we shall discuss the various types of induced magnetic fields and how they affect the magnetic resonance spectra.

16-2 RELAXATION PHENOMENA

If it were not for relaxation, it would never be possible to measure magnetic resonance. Realize that in magnetic resonance we want to observe transitions between energy levels that are very close together—the separation is 0.3 cm^{-1} in ESR and $1.4 \times 10^{-3} \text{ cm}^{-1}$ in proton magnetic resonance. As a result, there are two important differences between magnetic resonance spectroscopy and optical spectroscopy. First, in magnetic resonance the probability of spontaneous emission is negligible because of the low frequency of the radiation. This means that the probability of emission is due to induced emission only, and it is therefore exactly equal to the probability of absorption. Second, all the energy levels are about equally populated at ordinary temperatures. That means that, if we have a large number N of spins, there will be equal numbers of spins N_l and N_u in the lower and the upper states. Since N_l and N_u are equal to one another and since the two transition probabilities W_{abs} and W_{em} are equal also, the total number of absorption transitions is equal to the total number of emission transitions. Consequently, there is no net transfer of energy between the spin system and the radiation field, and we do not observe any magnetic resonance for the system just described.

It may be helpful to use a situation in everyday life as a model for the behavior of the spin system. Imagine that I start out with a number of cigarettes N_1 and that I have a friend who has a number N_2 of the same brand of cigarettes. We make a deal where I give him 10% of my cigarettes every day and he gives me 10% of his cigarettes at the same time. Clearly, after a certain number of days, each of us ends up with the same number

$$n = \tfrac{1}{2}(N_1 + N_2) \tag{16-22}$$

of cigarettes. Any further exchange of cigarettes is somewhat futile because we just pass the same number of cigarettes back and forth, and the net effect of all this trading is zero. We have assumed that neither I or my friend smokes any of the cigarettes we have. Obviously, the situation changes if my friend starts smoking, either a few cigarettes a day or a few packs. In that case I will also begin to lose cigarettes as a result of the exchange since I will get fewer cigarettes back each day than I am giving away. In fact, through the continual exchange I am partially financing his smoking.

A magnetic resonance experiment works in a very similar way as the cigarette exchange. If we have just a spin system, we have equal amounts of upward and downward transitions and we do not observe any transfer of energy between the radiation field and the spin system. What we need is some kind of an energy leak through which the energy can escape from the system. In practice this is easily accomplished because in most cases the spin system is in thermal equilibrium with its surroundings. In a solid, this is the set of lattice vibrations; in a gas, this may be the translational motion of the molecules. Formally, we can say that the spin system is in thermal contact with a reservoir and that it will try to maintain thermal equilibrium with this reservoir. This causes an energy transfer between the spin system and the reservoir and, consequently, an energy transfer out of the radiation field.

Let us try and give a mathematical description of this situation. We consider a system of N spins of spin quantum number $\frac{1}{2}$. At a given time t there are $N_+(t)$ spins in the lower state $\zeta_{1/2}$ and $N_-(t)$ spins in the upper state $\zeta_{-1/2}$. The probability for a downward transition is W_- and for an upward transition it is W_+ (see Figure 16-2). The changes in time of $N_+(t)$ and $N_-(t)$ are then given by

$$\frac{dN_+(t)}{dt} = W_- N_-(t) - W_+ N_+(t)$$

$$\frac{dN_-(t)}{dt} = W_+ N_+(t) - W_- N_-(t) \tag{16-23}$$

The two transition probabilities W_- and W_+ due to the radiation field are equal to one another because we can neglect the spontaneous emission

$$W_+ = W_- = W \tag{16-24}$$

If we define the difference in population $n(t)$ as

$$n(t) = N_+(t) - N_-(t) \tag{16-25}$$

it follows from Eq. (16-23) that

$$\frac{dn(t)}{dt} = -2Wn(t) \tag{16-26}$$

This is a simple differential equation that has the solution

$$n(t) = n(0)\,e^{-2Wt} \tag{16-27}$$

We see that in the stationary state (for $t \to \infty$) the population difference is zero and the total number of upward transitions is equal to the total number of downward transitions.

Let us now consider the same spin system without the radiation field and let us try to analyze the interactions with the reservoir. We know that in the equilibrium

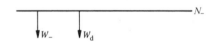

Figure 16-2 Transitions between a two-level spin system that are due to the radiation field (W_- and W_+) and due to thermal transitions (W_d and W_u).

situation the spin system is in thermal equilibrium with the reservoir. The ratio between N_+ and N_- is then determined by Maxwell's equation

$$\frac{N_+(\infty)}{N_-(\infty)} = \exp\left(\frac{\hbar\omega}{kT}\right) \tag{16-28}$$

where $\hbar\omega$ is the energy difference between the two levels and T is the temperature of the reservoir. If at an earlier time t the spin system is not in equilibrium with the reservoir, then there must be a mechanism through which thermal equilibrium is attained. We may represent this mechanism by means of a downward transition probability W_d and an upward transition probability W_u; both these transition probabilities cause changes in the spin level population which approach the thermal equilibrium. We have again

$$\frac{dN_+}{dt} = W_d N_-(t) - W_u N_+(t)$$

$$\frac{dN_-}{dt} = W_u N_+(t) - W_d N_-(t) \tag{16-29}$$

It is easily shown that the two transition probabilities, W_d and W_u, must be different because in the stationary state we have

$$\frac{dN_+}{dt} = 0 = W_d N_-(\infty) - W_u N_+(\infty) \tag{16-30}$$

or

$$\frac{W_d}{W_u} = \frac{N_+(\infty)}{N_-(\infty)} = \exp\left(\frac{\hbar\omega}{kT}\right) \tag{16-31}$$

We solve the differential equations (16-29) by introducing the population difference $n(t)$,

$$n(t) = N_+(t) - N_-(t) \tag{16-32}$$

and the equilibrium population $n(\infty)$,

$$n(\infty) = \frac{W_d - W_u}{W_d + W_u}[N_+(t) + N_-(t)] \tag{16-33}$$

Eq. (16-29) may then be written as

$$\frac{dn(t)}{dt} = 2 W_d N_-(t) - 2 W_u N_+(t) = (W_d - W_u)N - (W_d + W_u)n(t)$$

$$= (W_d + W_u)[n(\infty) - n(t)] = \frac{n(\infty) - n(t)}{T_1} \tag{16-34}$$

Here we have substituted

$$T_1 = (W_d + W_u)^{-1} \tag{16-35}$$

and we call T_1 the relaxation time. The solution of Eq. (16-34) is

$$n(t) = n(\infty) + \alpha e^{-t/T_1} \tag{16-36}$$

where α is an integration constant. Clearly, the population difference $n(t)$ approaches its stationary state value $n(\infty)$ exponentially.

In the preceding treatment we have considered the two mechanisms through which transitions may occur. First, we considered the transitions that are due to the radiation field only, and then we considered the transitions that represent the approach to thermal equilibrium. In reality, of course, the two mechanisms occur at the same time. It is assumed now that this situation may be described theoretically by adding the two separate transition probabilities (see Figure 16-2). We obtain, then, from Eqs. (16-26) and (16-34)

$$\frac{dn(t)}{dt} = -2Wn(t) + \frac{n' - n(t)}{T_1} \tag{16-37}$$

Here we have written n' rather than $n(\infty)$ because n' is no longer the stationary state value.

There is no need to solve the differential equation (16-37) because we only wish to know the value $n(\infty)$ in the stationary state. This is obtained by setting

$$\left(\frac{dn}{dt}\right)_{\infty} = 0 = -2Wn(\infty) + \frac{n' - n(\infty)}{T_1} \tag{16-38}$$

or

$$n(\infty) = \frac{n'}{1 + 2WT_1} \tag{16-39}$$

It is helpful to compare this result with Eq. (16-27) where we found that, without the thermal transition probabilities, $n(\infty)$ would be zero. Because of the thermal relaxation the population difference $n(t)$ has become different from zero in the stationary state.

Let us now return to our original question, namely, what is the energy transfer from the radiation field to the spin system? Every time we have a transition from a lower spin state $(+)$ to a higher spin state $(-)$ the field must supply an amount of energy $\hbar\omega$ and, vice versa, for every transition from a state $(-)$ to a state $(+)$ the field receives an amount of energy $\hbar\omega$. Per second there are WN_+ transitions from a lower to a higher state and WN_- transitions from a higher to a lower state. It is easily seen that per unit of time the field supplies an amount of energy

$$\frac{dE}{dt} = N_+ W\hbar\omega - N_- W\hbar\omega = nW\hbar\omega \tag{16-40}$$

In the stationary state this becomes

$$\left(\frac{dE}{dt}\right)_{\infty} = n(\infty)W\hbar\omega = \frac{n'W\hbar\omega}{1 + 2WT_1} \tag{16-41}$$

according to Eq. (16-39).

We see that it is possible to have a net transfer of energy from the radiation field to the spin system if there is thermal relaxation. In that case, magnetic resonance is, in principle, observable. It follows from Eq. (16-41) that the energy transfer becomes very small if T_1 becomes large. This is understandable because a large T_1 value means that the energy transfer from the spin system to the reservoir is a very slow process so that it becomes negligible. We then again approach the situation where there is no reservoir, and we have seen that in that case magnetic resonance cannot be observed.

EXAMPLE 16-1

Magnetic resonance of a given sample (either a solid, a liquid, or a gas) is formally described by the Bloch equations. These were obtained from the classical equations of motion for the total magnetic moment **M** of the sample by adding the thermal relaxation effects. The classical equations of motion are

$$\frac{dM_x(t)}{dt} = \gamma_n [H_z M_y(t) - H_y M_z(t)]$$

$$\frac{dM_y(t)}{dt} = \gamma_n [H_x M_z(t) - H_z M_x(t)] \qquad \text{(a)}$$

$$\frac{dM_z(t)}{dt} = \gamma_n [H_y M_x(t) - H_x M_y(t)]$$

where γ_n is the ratio between the magnetic moment **μ** of a spin and its angular momentum **J**.

In a magnetic resonance experiment the magnetic field **H**(t) is always the sum of a constant homogeneous magnetic field **H**$_0$, which causes the splitting of the spin energy levels, and a small time dependent field **H**$'(t)$, which acts as the radiation field with which we measure the absorption (Figure 16-3). If we take the homogeneous field **H** as the z axis, the Bloch equations are

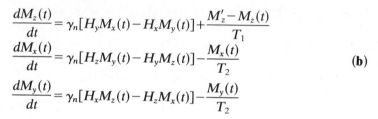

$$\frac{dM_z(t)}{dt} = \gamma_n [H_y M_x(t) - H_x M_y(t)] + \frac{M'_z - M_z(t)}{T_1}$$

$$\frac{dM_x(t)}{dt} = \gamma_n [H_z M_y(t) - H_y M_z(t)] - \frac{M_x(t)}{T_2} \qquad \text{(b)}$$

$$\frac{dM_y(t)}{dt} = \gamma_n [H_x M_z(t) - H_z M_x(t)] - \frac{M_y(t)}{T_2}$$

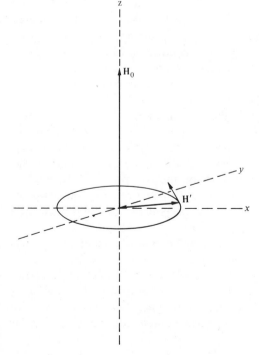

Figure 16-3 A magnetic resonance experiment is performed by applying a large constant magnetic field **H**$_0$ to split the spin levels and a rotating field **H**$'$ to cause transitions.

It is important to note that there are two different relaxation times, T_1 and T_2. In solids the relaxation time T_1 corresponds generally to the energy transfer between the spin system and the lattice, and T_1 is then called the spin-lattice relaxation time. The transverse relaxation time T_2 may be due to a variety of mechanisms, but in many cases it may be calculated from the dipole-dipole interactions between different spins; it is often referred to as the spin-spin relaxation time.

By solving the Bloch equations we find that magnetic resonance may be observed only if the two relaxation times T_1 and T_2 have suitable values. We have already seen in our model calculation that the effect cannot be observed if the two relaxation times are too long; in that case, the intensity of the signal is too small. It may also be shown that magnetic resonance cannot be observed if T_2 is too small because the resonance line becomes too broad. The line shape of the absorption line has the general form

$$I(\omega) = \frac{A}{1 + T_2^2(\omega - \omega_0)^2 + B} \qquad \textbf{(c)}$$

This is known as a Lorentzian function with a sharp maximum for $\omega = \omega_0$. The resonance frequency ω_0 is given by the resonance condition of Eq. (16-21)

$$\hbar\omega_0 = h\nu = g_N\mu_N H \qquad \textbf{(d)}$$

for nuclear magnetic resonance. It is clear now why the early attempts at measuring magnetic resonance failed. The samples that had been chosen for the experiments had unsuitable relaxation times so that the resonance signals could not be observed.

Given the Bloch equations (Eqs. b), derive expression (c) for the shape of a magnetic resonance absorption line.

To understand the derivation we must refer to Figure 16-3 to describe the mechanism of absorption as it is reflected in the Bloch equations. Visualize the spin system that is doing the absorbing as a set of magnetic moments (small magnets), each of which is parallel in direction and proportional in magnitude to the angular momentum of the spin. When put in an external magnetic field H_0, which is parallel to z axis, the xy plane component of such a magnetic moment will precess about H_0 just as the horizontal angular momentum component of a spinning toy top precesses about the gravitational field (vertical) direction.

In either case the precessional change of the angular momentum is caused by a torque that points in a direction perpendicular to the plane defined by the field and the angular momentum vectors. For the magnetic moment of a spin the torque (represented by the two terms in Eqs. (b) that contain $H_0 = H_z$) is caused by the force-couple which the magnetic field H_0 exerts on the two poles of the magnet. Without angular momentum this torque would simply line up the magnetic moment with the field (for example, a compass needle). If the magnetic moment is associated with an angular momentum, the result is precession.

At equilibrium in a constant field $H_z = H_0$, the total magnetic moment **M**, which is proportional to the average moment of a spin system, will point in the $+z$-direction with a small positive value because of the slightly lower energy of magnets that point toward the positive z axis (see Figure 16-2). To cause absorption of energy, it is necessary to reverse the z components of the magnetic moments. This can be done by superimposing a second precession, this time in a plane that *includes* the z axis, upon the precession of the horizontal component. The second precession is caused by a magnetic field H' which rotates in the xy plane (Figure 16-3). If the field rotates and the xy component of the magnetic moment of a spin precesses with the same frequency, then a torque, constant in time, will tend to precess the spin in a plane perpendicular to H'. The Bloch equations for this case simplify considerably if expressed in an XYZ-coordinate system that rotates with the field H'. As can be seen from the accompanying figure, the relationships between the components of the

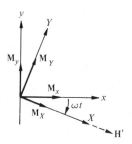

magnetic field and the magnetic moment are

$$H_x = H' \cos \omega t \tag{e}$$

$$H_y = -H' \sin \omega t \tag{f}$$

$$M_x = M_X \cos \omega t + M_Y \sin \omega t \tag{g}$$

$$M_y = -M_X \sin \omega t + M_Y \cos \omega t \tag{h}$$

Of course, $H_z = H_Z$, $M_z' = M_Z'$, and $M_z = M_Z$.

Substitution of Eqs. (e) through (h) into Eqs. (b) yields

$$\frac{dM_z}{dt} = \frac{dM_Z}{dt} = -\gamma_n H' M_Y + (M_Z' - M_Z)/T_1 \tag{i}$$

$$\frac{dM_x}{dt} = \frac{dM_X}{dt} \cos \omega t - M_X \omega \sin \omega t + \frac{dM_Y}{dt} \sin \omega t + M_Y \omega \cos \omega t$$

$$= \gamma_n H' M_Z \sin \omega t - M_X (\gamma_n H_0 \sin \omega t + \frac{1}{T_2} \cos \omega t) \tag{j}$$

$$+ M_Y (\gamma_n H_0 \cos \omega t - \frac{1}{T_2} \sin \omega t)$$

$$\frac{dM_y}{dt} = \frac{dM_X}{dt} \sin \omega t - M_X \omega \cos \omega t + \frac{dM_Y}{dt} \cos \omega t - M_Y \omega \sin \omega t$$

$$= \gamma_n H' M_Z \cos \omega t - M_X (\gamma_n H_0 \cos \omega t - \frac{1}{T_2} \sin \omega t) \tag{k}$$

$$- M_Y (\gamma_n H_0 \sin \omega t - \frac{1}{T_2} \cos \omega t)$$

In order to solve for dM_Y/dt Eqs. (j) and (k) are multiplied by $\sin \omega t$ and $\cos \omega t$, respectively. Addition of the resulting equations leads (after cancellation of the terms containing $\sin \omega t \cdot \cos \omega t$, and after the introduction of the identity $\cos^2 \omega t + \sin^2 \omega t = 1$) to

$$\frac{dM_Y}{dt} = \gamma_n H' M_Z + (\omega - \omega_0) M_X - M_Y / T_2 \tag{I}$$

where $\omega_0 = \gamma_n H_0$. A similar equation may be derived for dM_X/dt.

$$\frac{dM_X}{dt} = \frac{M_X}{T_2} - (\omega - \omega_0) M_Y \tag{m}$$

If the magnetic field H_0 is changed very slowly (as is done in the typical "slow passage" NMR experiment), the magnetic moment **M** can adjust readily to the field. At all times steady state conditions exist: $dM_Z/dt \approx dM_X/dt \approx dM_Y/dt \approx 0$. Then Eqs. (i), (l), and (m) may be combined to eliminate M_Z, M_X. The resulting equation for M_Y is

$$M_Y = \frac{\gamma_n H' M_Z' T_2}{1 + (\omega - \omega_0)^2 T_2^2 + T_1 T_2 \gamma_n^2 (H')^2} \tag{n}$$

Eq. (h) has the same form as Eq. (c) with B from Eq. (c) equal to $T_1 T_2 \gamma_n^2 (H')^2$, $I(\omega) \propto M_Y$, $A \propto \gamma_n H' M_Z T_2$. The reason why the absorption intensity $I(\omega)$ is proportional to M_Y is clear: the magnitude of the torque that changes M_Z and causes absorption of energy from the field H' is proportional to M_Y (see Eq. i).

From the classical Bloch equations, we have derived a result that applies to a fundamentally quantum mechanical phenomenon: the transitions between two energy levels. An alternate quantum mechanical derivation is possible by methods that are related to those described in Chapter 13. In this derivation the classical precision frequency ω_0 would be reinterpreted as a resonance frequency (see Eq. d).

In the majority of routine magnetic resonance experiments we just wish to determine the resonance frequency ω_0, and we are only interested in the relaxation times to the extent that the resonance signal should be observable. In other experiments, however, the main purpose is to measure the relaxation times in order to obtain some information about the various relaxation mechanisms. In the latter situation we would measure the intensities and the line widths of the resonance signals as a function of the temperature and the concentrations of the various components in the sample. The relaxation mechanisms relate mostly to intermolecular phenomena, such as interactions between different molecules, diffusion, molecular rotations, and so on. However, in the remainder of this chapter we shall confine our discussion to intramolecular phenomena, that is, the magnetic interactions within a single isolated molecule at rest. Thus, we shall not discuss the relaxation mechanisms any further.

16-3 CHEMICAL SHIFTS

If we place a molecule in a homogeneous magnetic field **H**, the field causes small changes in the motion of the electrons. These changes give rise to a small magnetic field called the induced field. A general law of physics, Lenz's law, states that the induced magnetic field must always be in the opposite direction of the original field **H**.

The induced magnetic fields can be measured in magnetic resonance. Realize that in the resonance equation (Eq. d of Example 16-1) we ought to substitute the field H_{loc} at the position of the nucleus for the magnetic field H. For a particular nucleus A we have

$$H_{loc}^A = H_{ext} - H_{ind}^A \qquad (16\text{-}42)$$

where H_{ext} is the homogeneous field that we apply and H_{ind}^A is the value of the induced magnetic field at the positions of nucleus A. It may be shown that the induced field H_{ind} is proportional to the exterior field H_{ext} and we may write

$$H_{ind}^A = \sigma_A H_{ext} \qquad (16\text{-}43)$$

The constant σ_A depends on the electronic environment of nucleus A. By substituting Eqs. (16-42) and (16-43) into the resonance equation (16-21), we find that the resonance frequency for nucleus A is given by

$$h\nu_A = g_N\mu_N H_{loc}^A = g_N\mu_N H_{ext}(1 - \sigma_A) \qquad (16\text{-}44)$$

The constant σ_A depends on the chemical environment of the nucleus that we measure, and it follows that the resonance frequency of the NMR signal depends on the type of molecule in which the nucleus is located and on the way it is bonded to the rest of the molecule. For example, in the ethyl alcohol molecule CH_3-CH_2-OH we have three different kinds of protons: the three protons on the methyl group, the two protons on the central carbon atom, and the proton on the hydroxyl group. These different types of protons have different values for the constant σ_A; therefore we expect that the NMR spectrum of C_2H_6O consists of three lines at different frequencies and that the intensities of these lines will have a ratio $3:2:1$.

We have mentioned that the majority of the nuclear magnetic resonance experiments are concerned with protons. For protons the value of σ varies between 2×10^{-5} and 4×10^{-5}, depending on the electronic environment. A resonance frequency in proton magnetic resonance can be measured with an accuracy of $1 : 10^8$ to $1 : 10^9$ so that we can derive the values of the constants σ_A with an accuracy of $1 : 10^3$ to $1 : 10^4$.

In the initial magnetic resonance experiments, the physicists were interested mainly in determining the magnetic moment $g_N\mu_N$ of the proton. It was an unpleasant surprise for them to find that the magnetic moment depends on the chemical environment when the accuracy of their measurements surpassed $1 : 10^5$. The effect was called "chemical shift" and the constants σ_A are sometimes known as chemical shift constants. A more precise name for σ_A is the proton magnetic shielding constant.

The quantum mechanical theory of the chemical shift is fairly complicated and the numerical results for chemical shifts that have been derived from calculations are not particularly accurate. It may be useful to outline the theoretical problem.

We consider the situation where we have a very small magnetic dipole $\boldsymbol{\mu}_A$ at the position of one of the nuclei A in a molecule and where we have placed the molecule in a homogeneous magnetic field **H**. We have seen already that in zeroth approximation the interaction between the dipole and the field is given by

$$E_{\text{int}} = -\boldsymbol{\mu}_A \cdot \mathbf{H} \tag{16-45}$$

The electrons in the molecule interact both with the dipole and the magnetic field, and the total Hamiltonian $\mathfrak{H}$ for the electrons in the molecule contains both **H** and $\boldsymbol{\mu}$.

$$\mathfrak{H} = \mathfrak{H}_{0,0} + \mathfrak{H}_{1,0}H + \mathfrak{H}_{0,1}\mu + \mathfrak{H}_{1,1}\boldsymbol{\mu} \cdot \mathbf{H} + \cdots \tag{16-46}$$

Here $\mathfrak{H}_{0,0}$ is the Hamiltonian of the molecule in the absence of $\boldsymbol{\mu}$ and **H**, and its lowest eigenvalue is E_0. The lowest eigenvalue ε_0 of the full Hamiltonian $\mathfrak{H}$ may be derived by means of perturbation theory. It is again obtained as a power series in μ and H,

$$\varepsilon_0 = E_0 + HE_{1,0} + \mu H_{0,1} + (\boldsymbol{\mu} \cdot \mathbf{H})E_{1,1} + \cdots \tag{16-47}$$

It is easily seen that the perturbation term $E_{1,1}$ relates to the shielding constant σ_A for nucleus A. The value of $E_{1,1}$ can be calculated by using second order perturbation theory, but the expression that is obtained in this way contains an infinite sum of terms that depend on the eigenfunctions of all excited states of the molecule. In practice, it is difficult to obtain reliable numbers from this expression.

The only systems for which it is relatively easy to calculate the shielding constants are atomic S states (states with zero angular momentum). Here the theoretical expression for σ reduces to

$$\sigma = \frac{e^2}{3mc^2a_0}\left\langle \Psi_0 \left| \sum_j \frac{1}{r_j} \right| \Psi_0 \right\rangle \tag{16-48}$$

In the case of atomic hydrogen $\sigma = 1.8 \times 10^{-5}$. Also, a number of fairly accurate calculations have been performed for molecular hydrogen. The best calculations give results that are in the vicinity of $\sigma = 2.70 \times 10^{-5} (\pm 0.05 \times 10^{-5})$ and we take this as the best value for H_2.

All these theoretical results are obtained with the bare proton as the reference point. Unfortunately, it has not been possible to measure the proton magnetic resonance frequency for the bare proton. This means that we have no experimental values for the absolute magnitudes of the proton shielding constants; instead we have results for the differences in proton shielding constants for different molecules. Also, we cannot measure the magnetic moment of the proton with a greater accuracy than $1 : 10^5$. The most accurate value of the proton magnetic moment is obtained by measuring the proton resonance frequency for the hydrogen molecule and by correcting this with the theoretical value of the proton shielding constant for molecular hydrogen.

The relative values of the proton magnetic shielding constants in different molecules can be measured fairly accurately. It is not practical to report the absolute values of the chemical shift constants because we cannot measure the resonance frequency of the bare proton. We must therefore choose a reference compound which we take as a "zero-point" for reporting the σ values. In Table 16-3 we have reported the chemical shifts of some molecules with respect to methane; that is, we have chosen the relative constant in σ in such a way that the methane value is equal to zero.

Table 16-3 Values for proton shielding constants
σ in some selected molecules[a]

Molecule	$\sigma_1 \times 10^5$	$\sigma_2 \times 10^5$
H_2	-0.422	2.70
H_2O	-0.060	3.062
HF	-0.251	2.871
HCl	0.045	3.167
HBr	0.435	3.557
HI	1.325	4.447
NH_3	-0.008	3.114
CH_4	0	3.122
C_2H_6	-0.075	3.047
C_2H_4	-0.518	2.604
C_2H_2	-0.135	2.987
C_6H_6	-0.713	2.409
C_6H_{12}	-0.129	2.993

[a] In the second column, σ_1 represents the experimental values taken with respect to CH_4. In the third column, σ_2 represents the absolute values, which were obtained by taking $\sigma = 2.70 \times 10^{-5}$ for the hydrogen molecule.

The chemical shifts are easy to measure, and if we could find some simple rules to correlate the value of σ with the molecular structure this would be very useful. Numerous attempts have been made to find such correlations, but the situation is not simple. The theoretical expression for σ consists of a term equal to Eq. (16-48) and, in addition, of an infinite sum over all excited states of the molecule. If the infinite sum could be neglected, we could try and relate the value of σ_A to the electronic charge density in the vicinity of atom A. It may be seen in Table 16-3 that such a correlation exists for the series H_2, HF, HCl, HBr, and HI. However, there are no general simple rules to relate the constant σ to the electronegativity constants of the neighboring atoms. It may be seen in Table 16-3 that in the series CH_4, NH_3, H_2O the proton shielding constants are almost the same in spite of the differences in structure.

Organic chemists are particularly interested in possible applications of NMR spectroscopy because the structure of an organic molecule can be quite complicated and any information that may be derived from an NMR spectrum is welcome. It may be seen from Table 16-3 that the σ values for the protons in saturated hydrocarbons are all fairly close together. The protons in benzene have a completely different σ value, and it is usually possible to recognize the resonance frequencies of protons in saturated C—H bonds and the signals from protons attached to aromatic rings. For example, in toluene the σ value for the CH_3 protons is 0.480×10^{-5} higher than the σ value for the other five protons.

We cannot even begin to describe the many semiempirical theories for the values of proton shielding constants in organic molecules because there are just too many of them. The only straightforward rule links the intensity of the signal to the number of equivalent protons. For example in toluene there are three protons in the CH_3 group and five protons in the phenyl group and we should observe two absorption lines with an intensity ratio of 5 : 3. None of the other theories on proton shielding constants is very reliable, but we should realize that any semiempirical rule that works can save organic chemists an enormous amount of work. Therefore, it is well worthwhile to look for such rules even if they cannot be fully justified by theoretical arguments.

16-4 SPIN–SPIN COUPLING

We mentioned several times that the theoretical description of an NMR spectrum may be derived from the resonance equation (16-21) as long as we realize that we must take the magnetic field H in the equation as the magnetic field at the position of the nucleus. This field is the sum of the field H_{ext}, which we apply, and the magnetic fields that are due to the interactions between the nuclear spin $\mathbf{I}_A$ and its environment. In Section 16-3 we discussed the chemical shifts, which represent the interaction between the nuclear magnetic dipole and the motion of the electrons in the molecule. In this section we shall discuss the spin–spin interactions, which represent the interactions between the magnetic moment $\mathbf{\mu}_A$ and the magnetic moments due to the other spins.

In classical electromagnetic theory the interaction energy of two magnetic dipoles $\mathbf{\mu}_A$ and $\mathbf{\mu}_B$ is given by

$$E_{int} = \frac{R^2{}_{A,B}(\mathbf{\mu}_A \cdot \mathbf{\mu}_B) - 3(\mathbf{R}_{A,B} \cdot \mathbf{\mu}_A)(\mathbf{R}_{A,B} \cdot \mathbf{\mu}_B)}{R^5_{A,B}} \qquad (16\text{-}49)$$

where $\mathbf{R}_{A,B}$ is the distance from A to B.

If we have two nuclei with spin $\frac{1}{2}$ and we place them in a homogeneous magnetic field, each of the two spins can be either parallel or antiparallel to the field. The magnetic moment of the second spin gives rise to a magnetic field at the position of the first spin, and this magnetic field depends on the orientation of the second spin. In principle, this magnetic field can be in the same direction as H_{ext} or it can have the opposite direction, depending on the orientation of the second spin. This effect causes a splitting of the energy levels of the first spin, as we have shown in Figure 16-4. This argument is somewhat of an oversimplification of the situation, but in general it may be said that the interactions between the spins give rise to splittings in the energy levels causing a fine structure in the magnetic resonance spectrum.

The magnitude of the spin–spin interactions may be determined from the fine structure in the magnetic resonance spectrum. However, this is not such a simple procedure. We must start with the Hamiltonian for the spin system, which contains as parameters the various magnetic shielding constants and the terms that represent the spin–spin interactions. Then we must find the eigenvalues of this Hamiltonian and we must determine the selection rules for the transitions between the energy levels. By comparing this calculated spectrum with the experimental absorption lines, we can obtain the values of the parameters. It was found that in this method the spin–spin interaction between two nuclei i and j may be represented by a term

$$\mathfrak{H}_{i,j} = J_{i,j}(\mathbf{I}_i \cdot \mathbf{I}_j) \qquad (16\text{-}50)$$

Here $\mathbf{I}_i$ and $\mathbf{I}_j$ are the spin operators for nuclei i and j and the parameter $J_{i,j}$

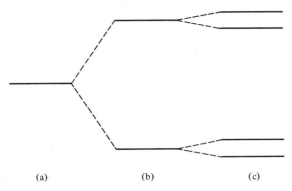

(a)   (b)   (c)

Figure 16-4   A simple picture of the effect of spin–spin coupling for two protons. In (b) we show the splitting of the first
proton due to the magnetic field **H**. The second proton is
parallel or antiparallel to **H** and it either helps or opposes **H**. As
a result each level in (b) is split further in (c) into two different
levels.

represents the magnitude of the spin–spin interaction between the two nuclei. The
total Hamiltonian $\mathfrak{H}$ for the nuclear spin system is then obtained by combining Eq.
(16-50) with Eq. (16-44)

$$\mathfrak{H} = \sum_i (1 - \sigma_i) g_i \mu_i (\mathbf{I}_i \cdot \mathbf{H}) + \sum_{j>i} J_{i,j} (\mathbf{I}_i \cdot \mathbf{I}_j) \tag{16-51}$$

It may be useful to illustrate how the eigenvalues and eigenfunctions of $\mathfrak{H}$ are
obtained for a simple system. We take a system of two protons and we take **H** along
the z axis. The Hamiltonian then reduces to

$$\mathfrak{H} = (1 - \sigma_1) g_p \mu_p I_{1z} H + (1 - \sigma_2) g_p \mu_p I_{2z} H + J(\mathbf{I}_1 \cdot \mathbf{I}_2) \tag{16-52}$$

In Appendix A of this chapter we illustrate how the eigenvalues and eigenfunctions
of this Hamiltonian are obtained for a simple spin system, consisting of two protons.

It may be seen from the treatment in Appendix A that the relation between the
experimental NMR frequencies and the various parameters may be fairly complicated, even for a simple case of only two spins. It is interesting to note that the
coupling constant between two equivalent protons cannot be derived from the NMR
spectrum of the two-proton system. This is the reason why the H—H coupling
constant in the H_2 molecule has not been observed experimentally. The coupling
constant has been calculated theoretically, but it has not been possible to verify the
accuracy of the theoretical result by comparing it with the experimental value.

A number of typical spin systems (such as three or four equivalent spins, and so on)
have been analyzed theoretically in the same way as the two-spin system in Appendix
A. The theoretical results for the relations between the various parameters in the
Hamiltonian (16-51) and the experimental frequencies are available in the literature,
and they can be used for the interpretation of NMR spectra. This material is
discussed in most textbooks on the interpretation of NMR spectra. If we want to
interpret an unknown NMR spectrum we can often tell from the number of lines and
their relative positions and intensities what the nature of the spin system is; it is then
relatively easy to derive the values of the chemical shift constants and of the coupling

constants from the NMR spectrum. Note also that the effects of magnetic shielding are proportional to the applied magnetic field H whereas the splittings due to the spin–spin coupling are field independent. In complex situations we can derive the values of the chemical shift constants by performing an NMR experiment at a high magnetic field, and we can determine the coupling constants by repeating the experiment at low magnetic field strengths.

In the customary definition of the coupling constants by means of Eqs. (16-50) and (16-51) the constants J have the dimension of energy and their values are usually listed in terms of cycles per second (Hz). The coupling constant for H_2 has been calculated to be about 250 Hz, the proton–proton coupling constant in methane is 12.4 Hz, and the coupling constants in the ethylene molecule are 19.1, 11.6, and 2.5 Hz.

Up to this point we have discussed only how the values of the coupling constants J_{ij} are obtained experimentally from the NMR spectra. Let us now consider how they can be evaluated theoretically from the molecular eigenfunctions. Again, the complete theory of coupling constants is fairly complicated, and we do not intend to discuss it in detail. We just want to outline the various mechanisms that are responsible for the coupling between the nuclear spins.

At first glance it seems that the spin–spin coupling is easily calculated from Eq. (16-49), which represents the classical electromagnetic interaction between two magnetic dipoles. In this calculation we do not even have to know the molecular eigenfunctions; all we have to do is to substitute the distance between the two nuclei into the equation. However, it turns out that the values obtained in this way are much smaller than the experimental coupling constants, and it follows that the direct dipole–dipole interaction between the nuclear spins cannot be responsible for the experimentally observed spin–spin coupling.

It is generally believed now that the main contribution to the nuclear spin interactions is the indirect coupling between the nuclei by means of the electron nuclear interactions. In order to understand this mechanism let us look at the situation illustrated in Figure 16-5. Here we have two protons A and B, and two electrons 1 and 2. At a given moment one of the electrons, electron 1, is in the vicinity of nucleus A and the other electron, electron 2, is close to nucleus B. Since the distance between proton A and electron 1 is so small, there is a fairly large interaction between the proton spin and the electron spin and the two spins will have a tendency to be parallel.[1] This means that, if proton spin A points upwards as in Figure 16-5, the

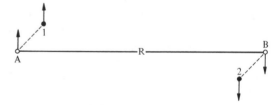

Figure 16-5 Representation of electron-coupled proton–proton interaction. The two electron spins must be antiparallel because of the Pauli exclusion principle.

[1] Two magnetic dipoles have the lowest energy if they are antiparallel. Since the proton spin has the same direction as the proton magnetic moment and since the electron spin has the opposite direction as the electron magnetic moment, the two spins have the lowest energy if they are parallel.

spin of electron 1 also has the tendency to point upwards. According to the Pauli principle the two electron spins must be antiparallel if they are in the same orbital. Consequently, if the two electrons form a chemical bond, the spin of electron 2 must point down if the spin of electron 1 points upward. Now, electron 2 is quite close to proton B and, as a result, there is a fairly strong interaction between the spins of electron 2 and proton B. This means that the two spins have a tendency to point in the same direction and that the spin of proton B is likely to point downward.

You can see, in the preceding model, that the interaction between the two nuclear spins A and B is due to the two interactions between spin A and spin 1 and between spin B and spin 2. These two interactions are much larger than the direct interaction between spins A and B because the distances A → 1 and B → 2 are so much smaller than the distance A → B. It does not matter that the distance between the two electrons is fairly large because their coupling is due to the Pauli principle, and they must be antiparallel at any distance.

The formal quantum mechanical theory of electron-coupled spin–spin interactions is fairly complicated, but we feel that it may be useful if we at least outline the theory. First we must know the interaction between an electron spin i and a nuclear spin N. This interaction is represented by a Hamiltonian

$$\mathfrak{H}_{i,N} = -2\mu_0 g_N \mu_N \left[\frac{(\mathbf{S}_i \cdot \mathbf{I}_N)(\mathbf{r}_{N,i} \cdot \mathbf{r}_{N,i}) - 3(\mathbf{S}_i \cdot \mathbf{r}_{N,i})(\mathbf{I}_N \cdot \mathbf{r}_{N,i})}{r_{N,i}^5} - \frac{8\pi}{3}(\mathbf{S}_i \cdot \mathbf{I}_N)\delta(\mathbf{r}_{N,i}) \right]$$

(16-53)

The first term in this Hamiltonian is the classical dipole–dipole interaction between the magnetic moments of the electron and the nucleus. It has the same form as the nuclear spin–spin interaction of Eq. (16-49). The second term in Eq. (16-53) cannot be visualized so easily. It is known as the Fermi contact potential, and its existence can be shown from relativistic arguments or, in an easier way, by taking the finite dimensions of the nucleus into account. The Fermi contact potential is quite important in the theory of coupling constants because some model calculations show that the contribution of the Fermi potential to the electron-nuclear spin interaction is usually much larger than the contribution from the classical dipole–dipole interaction.

EXAMPLE 16-2
In Eq. (16-53) we formulated the Fermi contact potential in terms of the three-dimensional δ function. We have sketched the one-dimensional δ function $\delta(x - x_0)$ in Figure 16-6. It is defined as a function that is very large in the vicinity of the point $x = x_0$ and practically zero when x differs from x_0. The function also satisfies the condition

$$\int_{-\infty}^{\infty} \delta(x - x_0)\, dx = 1 \qquad \textbf{(a)}$$

Clearly, if we have a function $f(x)$ that is finite and continuous at the point $x = x_0$ we must have

$$\int_{-\infty}^{\infty} f(x)\delta(x - x_0)\, dx = f(x_0) \qquad \textbf{(b)}$$

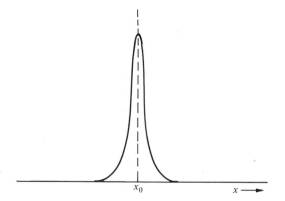

Figure 16-6 Representation of the one-dimensional δ function $\delta(x-x_0)$.

A three-dimensional δ function is a product of three one-dimensional δ functions

$$\delta(\mathbf{r}-\mathbf{r}_0) = \delta(x-x_0)\delta(y-y_0)\delta(z-z_0) \tag{c}$$

It follows from Eq. (b) that it has the property

$$\int f(\mathbf{r})\delta(\mathbf{r}-\mathbf{r}_0)\,d\mathbf{r} = \iiint f(x,y,z)\delta(x-x_0)\delta(y-y_0)\delta(z-z_0)\,dx\,dy\,dz$$

$$= f(x_0,y_0,z_0) = f(\mathbf{r}_0) \tag{d}$$

Thus, you can see that it is always convenient to have a δ function occur in an integral. In that case the integration is easy to do and the result is usually quite simple. ●

The quantum mechanical theory of the nuclear spin–spin interactions is fairly complicated, and it is not easy to evaluate the spin–spin coupling constants with high accuracy. To do the calculation we must treat the Hamiltonian

$$\mathfrak{H}' = \sum_i \sum_N \mathfrak{H}_{i,N} \tag{16-54}$$

as a perturbation, and we must calculate the various perturbations E_0', E_0'', ... to the unperturbed ground state energy E_0. In Eq. (16-54) the Hamiltonian $\mathfrak{H}'$ is derived from the Hamiltonian (16-53), which represents the interaction between the spin of electron i and the spin of nucleus N by summing over all electrons and all nuclei in the molecule. It follows from the definition (16-50) of the coupling constants that they are obtained as the coefficients of terms of the type $(\mathbf{I}_N \cdot \mathbf{I}_M)$ in the energy expression and that they are derived from the second order perturbation E_0'' to the energy. We have seen in our discussion on perturbation theory that this energy perturbation is given by

$$E_0'' = \sum_k \frac{\langle \Psi_0|\mathfrak{H}'|\Psi_k\rangle\langle \Psi_k|\mathfrak{H}'|\Psi_0\rangle}{E_k - E_0} \tag{16-55}$$

where E_k and Ψ_k are the eigenvalues and eigenfunctions of the unperturbed molecular Hamiltonian.

In general, the eigenfunctions of the excited molecular eigenstates are not known too accurately so that it is difficult to perform reliable calculations of the nuclear coupling constants. However, some calculations were performed on the hydrogen molecule in order to find out the relative magnitudes of the contributions due to the different terms in the Hamiltonian (16-53). It turned out that the main contribution is due to the Fermi contact term

$$\mathcal{H}'_{i,N} = \frac{16\pi\mu_0 g_N \mu_N}{3}(\mathbf{S}_i \cdot \mathbf{I}_N)\delta(\mathbf{r}_{N,i}) \tag{16-56}$$

If we assume that this is true in general, and not just for the hydrogen molecule, it might be possible to relate the coupling constants between two nuclear spins to the electron densities on the two nuclei. Thus, we might attempt to construct semiempirical theories of the values of the various spin–spin coupling constants. However, none of these semiempirical theories are very reliable because they rest on too many approximations.

It may be seen from the definition that the coupling constants $J_{i,j}$ have the dimension of energy and they are usually expressed in terms of cycles per second (Hz). From Eq. (16-20) we found that proton resonance occurs at 42.577×10^6 Hz, the various proton–proton coupling constants range from 1 to 100 Hz, and their effect is comparable to the effect of the chemical shift. The largest proton–proton coupling occurs in the H_2 molecule where it is about 300 Hz. Note that the H_2 coupling constant cannot be measured, as we have shown in Table 16-A. However, it can be calculated and it can also be estimated from the coupling constant of HD, which has been measured.

Note that appreciable coupling can also occur between protons that are not directly bonded to one another. For example, the H—C—H proton–proton coupling constant is around 12 Hz; the exact value varies for different molecules. Coupling has been observed between protons that are separated by several bonds; the record seems to be a coupling constant of the order of 1 Hz between two protons that are separated by eight bonds.

An interesting situation arises in the ethylene molecule. Here the three coupling constants are $J_{1,2} = 2.5$, $J_{2,3} = 11.6$, and $J_{1,3} = 19.1$ Hz (the numbering of the protons is illustrated in Figure 16-7). Here the largest coupling occurs for the two protons that are the farthest apart, whereas the H—C—H coupling constant is only 2.5 Hz. The reader may notice that this number is much smaller than the 12 Hz that we quoted in the previous paragraph, but this difference is due to the presence of π electrons, in the previous paragraph we were talking about saturated molecules.

From the various numbers we quoted it may be seen that the coupling constants can be quite sensitive to small changes in chemical structure. For that reason the experimental hyperfine splittings are often used by organic chemists to help them in determining the structure of a molecule. If nothing else, it is often possible to find out how many groups of equivalent protons there are just by counting the number of lines in the NMR spectrum. Additional information may be derived from the

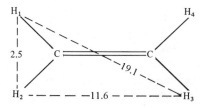

Figure 16-7 Proton–proton coupling constants in ethylene.

magnitudes of the various experimental coupling constants. Someone who knows the values of the many experimental coupling constants that have been measured and who has experience in interpreting NMR spectra can make fairly good predictions about the structure of a molecule. As a result, NMR techniques have become very important in determining molecular structures.

16-5 ELECTRON-SPIN RESONANCE

Electron-spin resonance can be used only for molecules that have a nonzero net electronic spin. Most molecules have singlet ground states (one of the few exceptions is the oxygen molecule) so that ESR cannot be applied to them. In general, ESR may be applied only to free radicals, which have an odd number of electrons so that there is one unpaired electron spin, and to molecules in their triplet state. In this section we will limit ourselves to free radicals only.

A free radical has the configuration $(\chi_1)^2(\chi_2)^2(\chi_3)^2 \cdots (\chi_N)^2(\chi_{N+1})$ if we use simple molecular orbital theory to represent its wave function. Here the functions χ_i are the molecular orbitals. Each of the first N molecular orbitals $\chi_1, \chi_2, \cdots, \chi_N$ contains a pair of electrons with antiparallel spins. The next highest molecular orbital χ_{N+1} contains a single electron that can have either spin α or spin β. The complete molecular eigenfunctions $\Psi_{1/2}$ and $\Psi_{-1/2}$ are given by

$$\Psi_{1/2} = [(2N+1)!]^{-1/2} \sum_P P\delta_P\{\chi_1(1)\alpha(1)\chi_1(2)\beta(2)\chi_2(3)\alpha(3)\chi_2(4)\beta(4) \cdots$$

$$\cdots \chi_N(2N-1)\alpha(2N-1)\chi_N(2N)\beta(2N)\chi_{N+1}(2N+1)\alpha(2N+1)\}$$

(16-57)

$$\Psi_{-1/2} = [(2N+1)!]^{-1/2} \sum_P P\delta_P\{\chi_1(1)\alpha(1)\chi_1(2)\beta(2)\chi_2(3)\alpha(3)\chi_2(4)\beta(4) \cdots$$

$$\cdots \chi_N(2N-1)\alpha(2N-1)\chi_N(2N)\beta(2N)\chi_{N+1}(2N+1)\beta(2N+1)\}$$

We should realize that in ESR of a free radical we measure the resonance radiation between the two spin states of the unpaired electron, that is, the electron in the molecular orbital χ_{N+1}. We have seen in Section 16-1, Eqs. (16-8) and (16-12) that the resonance radiation of a free electron is given by

$$h\nu = g\mu_0 H \tag{16-58}$$

where the g factor is very close to 2 and where H is the applied homogeneous magnetic field that splits the two spin states. In a free radical we must substitute the magnetic field H_{loc} at the position of the unpaired electron. Just as in magnetic resonance, the field H_{loc} consists of the applied field H_{ext} and of the small fields due to the environment of the electron. These small fields are the sum of the induced fields due to the motion of the other electrons and of the magnetic fields due to the nuclear spins. This means that the spin Hamiltonian for the electron should be written as

$$\mathfrak{H} = g\mu_0 H(1-\sigma)S_z + \sum_n a_n \mathbf{S} \cdot \mathbf{I}_n \tag{16-59}$$

instead of the Hamiltonian (16-12) for the free electron.

It may be seen that the parameter σ, which is the analog of the chemical shift constant in NMR, causes a displacement of the resonance line and that the coupling constants a_n cause a splitting of the resonance line into a number of separate lines. It is customary to incorporate the effect of σ into the g factor. Since we consider the coupling between the electron and the nuclear spins, we must also consider the interaction between the nuclear spins and the magnetic field H. Consequently, the ESR spectrum should be described by using the Hamiltonian

$$\mathfrak{H} = g\mu_0 H S_z - \sum_n g_n \mu_n H I_{nz} + \sum_n a_n \mathbf{S} \cdot \mathbf{I}_n \tag{16-60}$$

If we neglect higher-order effects we may approximate this Hamiltonian as

$$\mathfrak{H} = g\mu_0 H S_z - \sum_n g_n \mu_n H I_{nz} + S_z \cdot \sum_n a_n I_{nz} \tag{16-61}$$

Recall that the nuclear spin–spin coupling we discussed in the Section 16-4 was a second-order effect and, as a result, it was relatively difficult to interpret the spectra and to calculate the coupling constants. The electron-nuclear spin–spin coupling is a first order effect, and its description is therefore much simpler.

In order to illustrate the situation let us consider a system consisting of one electron and two protons. This system has eight possible spin functions, which we write as

$$
\begin{array}{ll}
\zeta_{1/2,1} = \alpha_{\mathrm{el}}\alpha_1\alpha_2 & \zeta_{-1/2,1} = \beta_{\mathrm{el}}\alpha_1\alpha_2 \\
\zeta_{1/2,2} = \alpha_{\mathrm{el}}\alpha_1\beta_2 & \zeta_{-1/2,2} = \beta_{\mathrm{el}}\alpha_1\beta_2 \\
\zeta_{1/2,3} = \alpha_{\mathrm{el}}\beta_1\alpha_2 & \zeta_{-1/2,3} = \beta_{\mathrm{el}}\beta_1\alpha_2 \\
\zeta_{1/2,4} = \alpha_{\mathrm{el}}\beta_1\beta_2 & \zeta_{-1/2,4} = \beta_{\mathrm{el}}\beta_1\beta_2
\end{array}
\tag{16-62}
$$

where the subscripts 1 and 2 refer to the two protons. These functions are all eigenfunctions of the spin operator (16-61). It is easily verified that

$$
\begin{aligned}
\mathfrak{H}\zeta_{1/2,1} &= [(\tfrac{1}{2}g\mu_0 - \tfrac{1}{2}g_1\mu_1 - \tfrac{1}{2}g_2\mu_2)H + \tfrac{1}{4}(a_1 + a_2)]\zeta_{1/2,1} \\
\mathfrak{H}\zeta_{-1/2,1} &= [(-\tfrac{1}{2}g\mu_0 - \tfrac{1}{2}g_1\mu_1 - \tfrac{1}{2}g_2\mu_2)H - \tfrac{1}{4}(a_1 + a_2)]\zeta_{-1/2,1} \\
\mathfrak{H}\zeta_{1/2,2} &= [(\tfrac{1}{2}g\mu_0 - \tfrac{1}{2}g_1\mu_1 + \tfrac{1}{2}g_2\mu_2)H + \tfrac{1}{4}(a_1 - a_2)]\zeta_{1/2,2} \\
\mathfrak{H}\zeta_{-1/2,2} &= [(-\tfrac{1}{2}g\mu_0 - \tfrac{1}{2}g_1\mu_1 + \tfrac{1}{2}g_2\mu_2)H - \tfrac{1}{4}(a_1 - a_2)]\zeta_{-1/2,2} \\
\mathfrak{H}\zeta_{1/2,3} &= [(\tfrac{1}{2}g\mu_0 + \tfrac{1}{2}g_1\mu_1 - \tfrac{1}{2}g_2\mu_2)H + \tfrac{1}{4}(-a_1 + a_2)]\zeta_{1/2,3} \\
\mathfrak{H}\zeta_{-1/2,3} &= [(-\tfrac{1}{2}g\mu_0 + \tfrac{1}{2}g_1\mu_1 - \tfrac{1}{2}g_2\mu_2)H - \tfrac{1}{4}(-a_1 + a_2)]\zeta_{-1/2,3} \\
\mathfrak{H}\zeta_{1/2,4} &= [(\tfrac{1}{2}g\mu_0 + \tfrac{1}{2}g_1\mu_1 + \tfrac{1}{2}g_2\mu_2)H - \tfrac{1}{4}(a_1 + a_2)]\zeta_{1/2,4} \\
\mathfrak{H}\zeta_{-1/2,4} &= [(-\tfrac{1}{2}g\mu_0 + \tfrac{1}{2}g_1\mu_1 + \tfrac{1}{2}g_2\mu_2)H + \tfrac{1}{4}(a_1 + a_2)]\zeta_{-1/2,4}
\end{aligned}
\tag{16-63}
$$

The selection rules for ESR transitions are

$$\Delta S_{\mathrm{el}} = 1 \qquad \Delta S_{\mathrm{nuc}} = 0 \tag{16-64}$$

This means that we observe transitions only between states with identical nuclear spin functions. In the preceding case these are the transitions $(\tfrac{1}{2}, i) \to (-\tfrac{1}{2}, i)$ with the frequencies

$$h\nu_i = E(\tfrac{1}{2}, i) - E(-\tfrac{1}{2}, i) \tag{16-65}$$

It follows from Eqs. (16-63) that these frequencies are given by

$$
\begin{aligned}
h\nu_1 &= g\mu_0 H + \tfrac{1}{2}(a_1 + a_2) \\
h\nu_2 &= g\mu_0 H + \tfrac{1}{2}(a_1 - a_2) \\
h\nu_3 &= g\mu_0 H - \tfrac{1}{2}(a_1 - a_2) \\
h\nu_4 &= g\mu_0 H - \tfrac{1}{2}(a_1 + a_2)
\end{aligned}
\tag{16-66}
$$

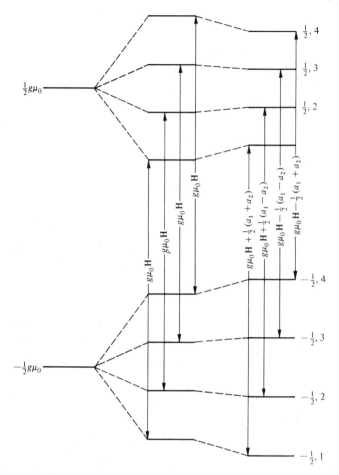

Figure 16-8   Energy level diagram for a system of one electron and two protons.

We have illustrated the energy levels of Eqs. (16-63) in Figure 16-8. It follows that the relation between the experimental splittings and the electron-nuclear spin coupling constants is quite simple. The splittings are even symmetric with respect to the central line. The four frequencies in Eqs. (16-66) may be written as

$$h\nu = g\mu_0 H + \tfrac{1}{2}(\pm a_1 \pm a_2) \qquad (16\text{-}67)$$

This expression is easily generalized to a system of one electron and N protons, here it becomes

$$h\nu = g\mu_0 H + \tfrac{1}{2}(\pm a_1 \pm a_2 \pm a_3 \cdots \pm a_N) \qquad (16\text{-}68)$$

The maximum number of lines is 2^N. It is clear that some of the lines will coincide if some of the coupling constants are equal. In the extreme case where all coupling constants are equal to one another the lines are given by

$$h\nu = g\mu_0 H + \tfrac{1}{2}a(\pm 1 \pm 1 \pm 1 \pm 1 \cdots \pm 1) \qquad (16\text{-}69)$$

EXAMPLE 16-3

In the case where we have N equivalent protons (that is, all the coupling constants are equal), there are $N + 1$ different resonance lines. Their relative intensities may be derived from the Pascal triangle

$$
\begin{array}{ccccccccccc}
 & & & & 1 & & 1 & & & & \\
 & & & 1 & & 2 & & 1 & & & \\
 & & 1 & & 3 & & 3 & & 1 & & \\
 & 1 & & 4 & & 6 & & 4 & & 1 & \\
1 & & 5 & & 10 & & 10 & & 5 & & 1 \\
\end{array}
$$

. . . .

Here each term is the sum of the two next terms in the line above. Obviously, in the case of four equivalent protons we have five ESR lines with their relative intensities given as $1 : 4 : 6 : 4 : 1$.   ●

The theoretical calculation of the coupling constants a_A corresponding to nucleus A is also much simpler than the evaluation of the nuclear spin–spin coupling. The interaction between the electron and the nuclear spins may be represented by the Hamiltonian (16-56). You can see that the interaction between the electronic spins and the spin of proton A may be written as

$$\mathfrak{H}_A = C \sum_i (\mathbf{S}_i \cdot \mathbf{I}_A) \delta(\mathbf{r}_{A,i}) \tag{16-70}$$

where C is a constant that may be derived from Eq. (16-56) and $\mathbf{r}_{A,i}$ is the distance from proton A to electron i; the summation should be performed over all electrons in the molecule. The coupling constant a_A may now simply be derived from first order perturbation theory; it is given by

$$a_A(\mathbf{S} \cdot \mathbf{I}_A) = \langle \Psi_{(1/2)} | \mathfrak{H}_A | \Psi_{(1/2)} \rangle \tag{16-71}$$

Here the molecular eigenfunction is given by Eqs. (16-57).

It is easily seen that Eq. (16-71) reduces to

$$a_A = C \langle \chi_{N+1} | \delta(\mathbf{r}_{A,i}) | \chi_{N+1} \rangle = C \chi_{N+1}^*(\mathbf{r}_A) \chi_{N+1}(\mathbf{r}_A), \tag{16-72}$$

which is the probability density at the position of nucleus A of the unpaired electron in the molecular orbital χ_{N+1}. It follows thus that the coupling constant is simply related to the value of the electronic molecular orbital at the position of the nucleus.

The electron-nuclear spin–spin coupling constants again have the dimension of energy, but they are usually expressed in terms of gauss. It is customary to perform an ESR experiment for a fixed wavelength of resonance radiation, namely 3.15 cm (see Section 16-1). The various absorption lines are then observed by varying the value of the homogeneous magnetic field, H. In this way we can vary the differences between the energy levels and we observe resonance absorption whenever the difference between two energy levels corresponds to the radiation of 3.15 cm. For a free electron spin, we observe resonance for a field of 3400 gauss, as we have shown in Section 16-1.

The coupling between the electron and the proton in a hydrogen atom is easily calculated from Eq. (16-72); it is around 500 gauss. This is about the order of magnitude for the first order electron-nuclear spin–spin coupling.

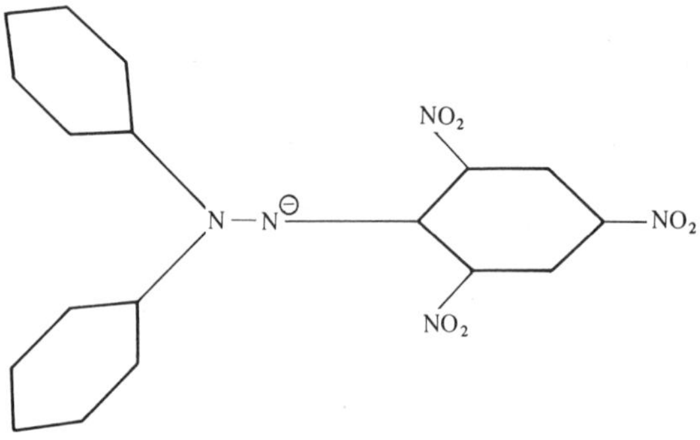

Figure 16-9 Diphenyl picryl hydrazine
(DPPH).

Unfortunately, the simple theory that we have outlined does not cover most of the situations that we must deal with in practice. Most of the molecules that are measured in ESR spectroscopy are fairly large aromatic free radicals; a typical example is diphenyl picryl hydrazine (see Figure 16-9), which is sometimes used for testing the equipment. In these aromatic free radicals the unpaired electron is usually in a π molecular orbital, which is zero in the plane of the molecule. According to Eq. (16-72) the coupling constants between the electron spin and the various protons should all be zero because the protons are all located in the molecular plane where the molecular orbital is zero. In practice, we measure finite values for the electron–proton coupling constants in these aromatic molecules, even though their values are much smaller than the effect we calculated for the hydrogen atom. It is interesting that some of the experimental electron–proton coupling constants are negative; this, of course, is not consistent with Eq. (16-72).

It is clear that we need a more precise theory in order to describe these small coupling constants in aromatic free radicals. In Eqs. (16-57) we wrote the molecular eigenfunction as a product of one-electron functions, and this is too crude an approximation if we wish to account for the aromatic coupling constants. A model calculation was performed on the C—H free radical, and it was shown here that we obtain a finite coupling constant between the π electron and the proton if we introduce configuration interaction. It was shown that in the C—H radical the π electron has a coupling constant of about -20 gauss with the proton spin. This number agrees quite satisfactorily with the experimental coupling value of -23 gauss for the methyl radical. It may be argued that in larger free radicals, where the π electron is in a delocalized molecular orbital, the coupling constant with a proton should be roughly proportional to the charge density of the π electron on the carbon atom to which the proton is bonded. This π electron atom charge density on the jth carbon atom, ρ_i is easily derived from simple Hückel molecular orbital theory. Accordingly, it might be expected that the coupling constant $a_{H,i}$ for the corresponding proton can be written as

$$a_{H,i} = \phi \rho_i \tag{16-73}$$

This rule seems to work reasonably well, and the empirical values of ϕ range between -22 and -28 gauss.

Most of the applications of ESR have been in organic chemistry. ESR has been used for structure determinations, for identifying intermediate products in reactions, and for explaining reaction mechanisms. Especially in the area of reaction mechanisms in organic chemistry ESR has been widely used.

_____ *Appendix*

APPENDIX A   TWO-PROTON SPIN SYSTEM

We wish to derive the eigenvalues and eigenfunctions of the Hamiltonian

$$\mathfrak{H} = \sum_i (1-\sigma_i)g_i\mu_i(\mathbf{I}_i \cdot \mathbf{H}) + \sum_{j>i} (\mathbf{I}_i \cdot \mathbf{I}_j) \tag{A-1}$$

of Eq. (16-51) for the simplest possible situation, that is, for a spin system consisting of two protons.

We take a system of two protons in the presence of a magnetic field $\mathbf{H}$ and we take $\mathbf{H}$ along the z axis. The Hamiltonian $\mathfrak{H}$ of Eq. (A-1) then reduces to

$$\mathfrak{H} = (1-\sigma_1)g_p\mu_p I_{1z}H + (1-\sigma_2)g_p\mu_p I_{2z}H + J(\mathbf{I}_1 \cdot \mathbf{I}_2) \tag{A-2}$$

We write this as

$$\mathfrak{H} = \gamma_1 H I_{1z} + \gamma_2 \dot{H} I_{2z} + J(I_{1x}I_{2x} + I_{1y}I_{2y} + I_{1z}I_{2z}) \tag{A-3}$$

The spin of the first proton can be either parallel or antiparallel to the z axis. The first situation is represented by a spin function α_1 and the second situation by a spin function β_1. The two possible orientations of the second spin are represented by the spin functions α_2 and β_2. It follows that there are four possible spin functions for the total spin system, namely

$$\begin{aligned}\zeta_1 = \alpha_1\alpha_2 \qquad \zeta_3 = \beta_1\alpha_2 \\ \zeta_2 = \alpha_1\beta_2 \qquad \zeta_4 = \beta_1\beta_2\end{aligned} \tag{A-4}$$

The effect of the Hamiltonian $\mathfrak{H}$ on this spin function may be derived from the properties of the spin functions that we discussed in Chapter 12, Eqs. (12-30) and (12-33). We have

$$\begin{aligned} I_x\alpha = \tfrac{1}{2}\beta \qquad & I_x\beta = \tfrac{1}{2}\alpha \\ I_y\alpha = \tfrac{1}{2}i\beta \qquad & I_y\beta = -\tfrac{1}{2}i\alpha \\ I_z\alpha = \tfrac{1}{2}\alpha \qquad & I_z\beta = -\tfrac{1}{2}\beta \end{aligned} \tag{A-5}$$

and

$$\begin{aligned}
(\mathbf{I}_1 \cdot \mathbf{I}_2)\zeta_1 &= (I_{1x}I_{2x} + I_{1y}I_{2y} + I_{1z}I_{2z})(\alpha_1\alpha_2) = \tfrac{1}{4}(\alpha_1\alpha_2) = \tfrac{1}{4}\zeta_1 \\
(\mathbf{I}_1 \cdot \mathbf{I}_2)\zeta_2 &= (I_{1x}I_{2x} + I_{1y}I_{2y} + I_{1z}I_{2z})(\alpha_1\beta_2) \\
&= \tfrac{1}{4}\beta_1\alpha_2 + \tfrac{1}{4}\beta_1\alpha_2 - \tfrac{1}{4}\alpha_1\beta_2 = -\tfrac{1}{4}\zeta_2 + \tfrac{1}{2}\zeta_3 \\
(\mathbf{I}_1 \cdot \mathbf{I}_2)\zeta_3 &= (I_{1x}I_{2x} + I_{1y}I_{2y} + I_{1z}I_{2z})(\beta_1\alpha_2) \\
&= \tfrac{1}{4}\alpha_1\beta_2 + \tfrac{1}{4}\alpha_1\beta_2 - \tfrac{1}{4}\beta_1\alpha_2 = -\tfrac{1}{4}\zeta_3 + \tfrac{1}{2}\zeta_2 \\
(\mathbf{I}_1 \cdot \mathbf{I}_2)\zeta_4 &= (I_{1x}I_{2x} + I_{1y}I_{2y} + I_{1z}I_{2z})(\beta_1\beta_2) = \tfrac{1}{4}(\beta_1\beta_2) = \tfrac{1}{4}\zeta_4
\end{aligned} \tag{A-6}$$

Also,

$$
\begin{aligned}
(\gamma_1 I_{1z} + \gamma_2 I_{2z})\zeta_1 &= (\tfrac{1}{2}\gamma_1 + \tfrac{1}{2}\gamma_2)\zeta_1 \\
(\gamma_1 I_{1z} + \gamma_2 I_{2z})\zeta_2 &= (\tfrac{1}{2}\gamma_1 - \tfrac{1}{2}\gamma_2)\zeta_2 \\
(\gamma_1 I_{1z} + \gamma_2 I_{2z})\zeta_3 &= (-\tfrac{1}{2}\gamma_1 + \tfrac{1}{2}\gamma_2)\zeta_3 \\
(\gamma_1 I_{1z} + \gamma_2 I_{2z})\zeta_4 &= (-\tfrac{1}{2}\gamma_1 - \tfrac{1}{2}\gamma_2)\zeta_4
\end{aligned}
\tag{A-7}
$$

It follows easily that the functions ζ_1 and ζ_4 are eigenfunctions of the Hamiltonian (A-3) because

$$
\begin{aligned}
\mathfrak{H}\zeta_1 &= [\tfrac{1}{2}H(\gamma_1 + \gamma_2) + \tfrac{1}{4}J]\zeta_1 \\
\mathfrak{H}\zeta_4 &= [-\tfrac{1}{2}H(\gamma_1 + \gamma_2) + \tfrac{1}{4}J]\zeta_4
\end{aligned}
\tag{A-8}
$$

Because there are only four eigenfunctions of the operator $\mathfrak{H}$ and two of the eigenfunctions are given by Eq. (A-8), the remaining two eigenfunctions must be linear combinations of the spin functions ζ_2 and ζ_3. It follows from Eqs. (A-6) and (A-7) that

$$
\begin{aligned}
\mathfrak{H}\zeta_2 &= [-\tfrac{1}{4}J + \tfrac{1}{2}H(\gamma_1 - \gamma_2)]\zeta_2 + \tfrac{1}{2}J\zeta_3 \\
\mathfrak{H}\zeta_3 &= \tfrac{1}{2}J\zeta_2 + [-\tfrac{1}{4}J - \tfrac{1}{2}H(\gamma_1 - \gamma_2)]\zeta_3
\end{aligned}
\tag{A-9}
$$

or

$$
\begin{aligned}
\mathfrak{H}(\zeta_2 + \zeta_3) &= \tfrac{1}{4}J(\zeta_2 + \zeta_3) + \tfrac{1}{2}H(\gamma_1 - \gamma_2)(\zeta_2 - \zeta_3) \\
\mathfrak{H}(\zeta_2 - \zeta_3) &= \tfrac{1}{2}H(\gamma_1 - \gamma_2)(\zeta_2 + \zeta_3) - \tfrac{3}{4}J(\zeta_2 - \zeta_3)
\end{aligned}
\tag{A-10}
$$

We define the eigenfunctions by means of the equation

$$
(\mathfrak{H} - \lambda)\chi = 0
$$
$$
\chi = a\chi_1 + b\chi_2 = a(\zeta_2 + \zeta_3) + b(\zeta_2 - \zeta_3)
\tag{A-11}
$$

It follows then easily that the eigenvalue λ and the coefficients a and b must satisfy

$$
\begin{aligned}
[\tfrac{1}{4}J - \lambda]a + \tfrac{1}{2}H(\gamma_1 - \gamma_2)b &= 0 \\
\tfrac{1}{2}H(\gamma_1 - \gamma_2)a + [-\tfrac{3}{4}J - \lambda]b &= 0
\end{aligned}
\tag{A-12}
$$

The equation for the eigenvalue λ is then

$$
(-\lambda + \tfrac{1}{4}J)(-\lambda - \tfrac{3}{4}J) - \tfrac{1}{4}H^2(\gamma_1 - \gamma_2)^2 = 0
\tag{A-13}
$$

which has two solutions

$$
\lambda = -\tfrac{1}{4}J \pm \tfrac{1}{2}\sqrt{J^2 + H^2(\gamma_1 - \gamma_2)^2}
\tag{A-14}
$$

We have assumed here that $\gamma_1 \neq \gamma_2$.

It may be useful to summarize the four eigenvalues that we have listed in Eqs. (A-8) and (A-14)

$$
\begin{aligned}
\lambda_1 &= \tfrac{1}{4}J + \tfrac{1}{2}H(\gamma_1 + \gamma_2) \\[4pt]
\lambda_2 &= -\tfrac{1}{4}J + \tfrac{1}{2}\sqrt{J^2 + H^2(\gamma_1 - \gamma_2)^2} \\[4pt]
\lambda_3 &= -\tfrac{1}{4}J - \tfrac{1}{2}\sqrt{J^2 + H^2(\gamma_1 - \gamma_2)^2} \\[4pt]
\lambda_4 &= \tfrac{1}{4}J - \tfrac{1}{2}H(\gamma_1 + \gamma_2)
\end{aligned}
\tag{A-15}
$$

From the selection rules, we may derive that the transition between the states 1 and 4 is forbidden so that we observe only the four transitions $1 \rightarrow 2$, $1 \rightarrow 3$, $4 \rightarrow 2$, and $4 \rightarrow 3$. The frequencies of these four transitions are derived by taking the corresponding energy differences in Eq. (A-15),

$$\nu_{1,2} = \tfrac{1}{2}J + \tfrac{1}{2}H(\gamma_1 + \gamma_2) - \tfrac{1}{2}\sqrt{J^2 + H^2(\gamma_1 - \gamma_2)^2}$$

$$\nu_{1,3} = \tfrac{1}{2}J + \tfrac{1}{2}H(\gamma_1 + \gamma_2) + \tfrac{1}{2}\sqrt{J^2 + H^2(\gamma_1 - \gamma_2)^2}$$

$$\nu_{2,4} = \tfrac{1}{2}J - \tfrac{1}{2}H(\gamma_1 + \gamma_2) - \tfrac{1}{2}\sqrt{J^2 + H^2(\gamma_1 - \gamma_2)^2} \qquad \text{(A-16)}$$

$$\nu_{3,4} = \tfrac{1}{2}J - \tfrac{1}{2}H(\gamma_1 + \gamma_2) + \tfrac{1}{2}\sqrt{J^2 + H^2(\gamma_1 - \gamma_2)^2}$$

We observe four lines in the proton magnetic resonance spectrum, and we can derive the values of J and the chemical shift constants from the spectrum.

Note that we have assumed the two constants σ_1 and σ_2 to be different in the preceding discussion. This means that the discussion applies only to a pair of nonequivalent protons. If the two protons are equivalent, the two constants γ_1 and γ_2 are equal to each other and Eq. (A-10) reduces to

$$\mathfrak{H}\chi_1 = \tfrac{1}{4}J\chi_1 \qquad \chi_1 = \zeta_2 + \zeta_3$$
$$\mathfrak{H}\chi_2 = -\tfrac{3}{4}J\chi_2 \qquad \chi_2 = \zeta_2 - \zeta_3 \qquad \text{(A-17)}$$

This means that the spin functions χ_1 and χ_2 are eigenfunctions of the Hamiltonian. The eigenvalues and corresponding eigenfunctions of the Hamiltonian $\mathfrak{H}$ for that particular case are listed in Table 16-A. The only allowed transitions are between the

Table 16-A Eigenvalues and eigenfunctions for a system of two equivalent spins

Eigenvalue	Eigenfunction
$\tfrac{1}{4}J + \gamma H$	$\alpha_1\alpha_2$
$\tfrac{1}{4}J$	$2^{-1/2}(\alpha_1\beta_2 + \beta_1\alpha_2)$
$\tfrac{1}{4}J - \gamma H$	$\beta_1\beta_2$
$-\tfrac{3}{4}J$	$2^{-1/2}(\alpha_1\beta_2 - \beta_1\alpha_2)$

levels 1 and 2 and between levels 2 and 3 of Table 16-A. Both of these transitions have the same frequency, $h\nu = \gamma H$, which is independent of J. We must conclude, therefore, that it is not possible to measure the coupling constant between two equivalent protons. As a result, the H—H coupling constant in the H_2 molecule is not known, and the simplest system for which an experimental value is available is HD.

_____ *Problems*

1. Why did the early attempts in 1936 to measure magnetic resonance fail?
2. ESR is performed with radiation of 3.15 cm wavelength. At which magnetic field does resonance occur for a free electron?

3. Calculate the wavelength of the resonance radiation for ^{13}C in a magnetic field of 10,000 gauss.

4. Calculate the magnetic field that produces resonance radiation with a wavelength of 50 m for ^{19}F.

5. If we have 3×10^{22} protons in a homogeneous magnetic field of 10,000 gauss, what is the difference in population between the two spin levels at a temperature of 27°C (300 K) and at a temperature of 20 K?

6. Why are the upward and downward transition probabilities W_d and W_u for establishing thermal equilibrium in a spin system different from each other?

7. Find the solution of the population n as a function of the time t by solving the differential equation (16-37). How does your result relate to the equilibrium population difference $n(\infty)$?

8. What are the relaxation times T_1 and T_2 that occur in the Bloch equations called? Why are they different, and to which physical processes are they related?

9. How is the chemical shift constant σ defined? Explain what it is caused by and why it shifts the resonance to lower frequencies (as compared with the bare nucleus).

10. Are the absolute values of proton shielding constants σ available from experiments? If not, how are the values of Table 16-3 obtained?

11. If we perform proton magnetic resonance with a fixed homogeneous magnetic field of 10,000 gauss, how far are the resonance lines in HF and HI apart in terms of megaHertz (MHz)?

12. The value of the proton shielding constant σ increases in the sequence of molecules HF, HCl, HBr, HI. Can you explain this in terms of general chemical considerations, guided by Eq. (16-48) for atoms?

13. Is it possible to explain nuclear spin–spin coupling constants in terms of the direct dipole–dipole interactions between the spin magnetic moments? If not, how should the magnitude of the coupling constants be interpreted?

14. Is it possible to measure the spin–spin coupling constant of the H_2 molecule? We mentioned in Section 16-4 that the value of the coupling constant is 300 Hz. How can this number be obtained?

15. Do the relative magnitudes of the splittings due to chemical shifts and to spin–spin coupling constants depend on the magnitude of the magnetic field **H** in which the NMR experiment is performed? If so, which of the two effects is predominant at high fields? Which at low fields?

16. Find out what the magnitude of the earth magnetic field is where you are and calculate the wavelength of proton resonance radiation in that magnetic field.

17. Calculate the ESR spectrum of the negative benzene ion, assuming that the six protons are equivalent and that the six coupling constants between the unpaired electron and the six spins are all the same. Determine the number of lines in the spectrum and their relative intensities.

18. Which quantum mechanical quantity do we derive by measuring the ESR spectrum of a free radical with one unpaired electron?

19. Why is the spin–spin coupling between the unpaired electron and the protons in a planar aromatic free radical non-zero in spite of the result we derived in Eq. (16-72)?

20. In ammonia, NH_3, there are three equivalent protons, which have the same chemical shift constants. The three proton–proton coupling constants are also

equal to one another. Write the spin Hamiltonian for the three-proton system and express it in terms of the spin operators I_1^2, I_2^2, and I_3^2 of the three protons and of the operator I, where I is the total proton spin ($\mathbf{I} = \mathbf{I}_1 + \mathbf{I}_2 + \mathbf{I}_3$).

References and Recommended Reading

We have listed first a few general textbooks on magnetic resonance from a physicist's point of view [1, 2] and then a few books that deal with applications of magnetic resonance to chemical problems. Memory's book [5] lists the various theories for calculating the parameters that occur in NMR. As an example, we also list a table of available experimental NMR data to show how they can be used for the purpose of identification [6].

[1] A. Abragam: *The Principles of Nuclear Magnetism*. Oxford University Press, London, 1961.

[2] C. P. Slichter: *Principles of Magnetic Resonance*. Harper & Row, New York, 1963.

[3] J. A. Pople, W. G. Schneider, and H. J. Bernstein: *High-Resolution Nuclear Magnetic Resonance*. McGraw-Hill, New York, 1959.

[4] A. Carrington and A. D. McLachlan: *Introduction to Magnetic Resonance*. Harper & Row, New York, 1967.

[5] J. D. Memory: *Quantum Theory of Magnetic Resonance Parameters*. McGraw-Hill, New York, 1968.

[6] F. A. Bovey: *NMR Data Tables for Organic Compounds*. Wiley-Interscience, New York, 1967.

chapter 17

The Solid State

17-1 CRYSTAL STRUCTURES

Any solid consists of atoms or molecules. Usually these are arranged in space in an orderly fashion, which we call a crystal. By definition, a crystal is characterized by translational symmetry. In order to illustrate what this means we have sketched a one-dimensional crystal in Figure 17-1, consisting of two kinds of atoms, A and B. The atoms A are all a distance a apart and so are the atoms B. You can see that if we start from the position of an atom A and we move a distance na, where n is an integer, we will end up on another atom A, that is an equivalent position. The same happens when we move a distance na away from an atom B. In order to describe the crystal structure we need to consider only the elementary cell that we have drawn in Figure 17-1 because the whole crystal consists of an infinite succession of identical elementary cells of length a. The elementary cell contains one atom A and one atom B. It may be seen that there are two atoms A at the boundaries of the elementary cell, but each of the two atoms is shared with the adjacent cells so that we count them as half; this gives a total of one atom A in our cell.

In a two-dimensional crystal the translational symmetry is characterized by two vectors **a** and **b**. Here we should move from a particular atom to an equivalent atom if we move over a distance

$$\mathbf{r} = p\mathbf{a} + q\mathbf{b} \tag{17-1}$$

where p and q are integers. We illustrate this in Figure 17-2, where we have drawn the structure of graphite. The graphite crystal has a layered structure, and each layer looks like a giant benzene molecule that may be considered a two-dimensional crystal. We have drawn the elementary cell in the middle of Figure 17-2; it is a parallelogram, determined by the two vectors **a** and **b**. It may be seen that **a** and **b** have the same length and make an angle of 60° with one another.

In three dimensions the crystal structure is determined by three vectors **a**, **b**, and **c**, and the elementary cell is the space spanned by these three vectors. The crystal is an infinite three-dimensional array of identical elementary cells.

The type of crystal structure is determined by the shape of the elementary cell. There are six types of elementary cells illustrated in Figure 17-3. In the first three

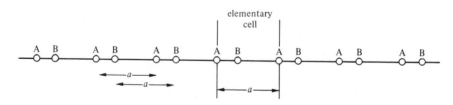

Figure 17-1   One-dimensional crystal.

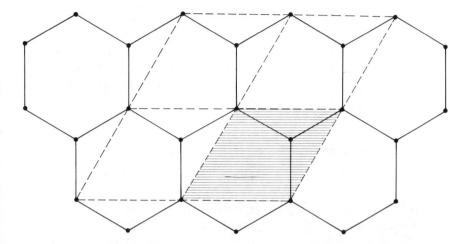

Figure 17-2 Two-dimensional graphite crystal. The elementary cell is the shaded area.

structures, which are known as cubic, tetragonal, and orthorhombic, the three vectors **a**, **b**, and **c** are all perpendicular to one another. In the cubic structure the three vectors also have the same length, and the elementary cell is shaped like a cube. In the tetragonal structure two of the vectors have the same length, and the third one is different. In the orthorhombic structure the three vectors all have different lengths. In the monoclinic structure two of the vectors, say **a** and **b**, make an arbitrary angle with each other, whereas the third vector **c** is perpendicular to the plane of the other

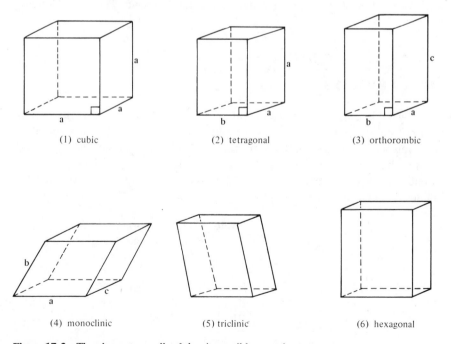

(1) cubic   (2) tetragonal   (3) orthorombic

(4) monoclinic   (5) triclinic   (6) hexagonal

Figure 17-3 The elementary cells of the six possible crystal structures.

two. In the triclinic structure the three vectors make arbitrary angles (different from 90°) with one another. The sixth crystal structure, hexagonal, stands apart from the above classification. This structure consists of parallel layers of the two-dimensional crystal structure shown in Figure 17-2. The base of the elementary cell is identical with the elementary cell of Figure 17-2; the two vectors **a** and **b** make an angle of 60° with one another, and the third vector **c** is then perpendicular to the plane of the other two. The length of **c** is usually different from the lengths of the other two vectors.

The preceding classification of crystals is concerned only with the geometry. A more chemical classification depends on the types of bonds between the atoms in the crystal. According to this point of view we distinguish four different types of crystals.

1. Ionic crystals, which are composed of positive and negative ions, a typical example is NaCl. Here the forces between the different ions are mainly electrostatic.
2. Metals, where each atom contributes one or more valence electrons that can then move freely through the whole crystal. A metal consists of an array of positive ions and a sea of almost-free electrons.
3. Covalent crystals or semiconductors, such as diamond, graphite, germanium. These are basically giant molecules where the atoms are bonded together through covalent bonds.
4. Molecular crystals, such as benzene, anthracene, and so on. Here the molecules maintain their separate identity and the interactions between different molecules are much smaller than the interactions within a specific molecule.

We shall discuss each of the four types of crystals separately in the following sections.

17-2 IONIC CRYSTALS

An ionic crystal is composed of positive and negative ions and the forces between the ions are of a purely electrostatic nature. This crystal structure occurs when the atoms involved have very different electronegativities. For example, in Table 17-1 we have listed the electronegativities for the alkali metals and for the halogens; since the differences between the two groups are quite large, it is understandable that the alkali halides all occur as ionic crystals.

Table 17-1 Electronegativities of alkali metals and halogens

Li	1.0	F	4.0
Na	0.9	Cl	3.0
K	0.8	Br	2.8
Rb	0.8	I	2.5
Cs	0.7		

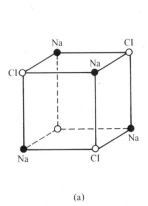

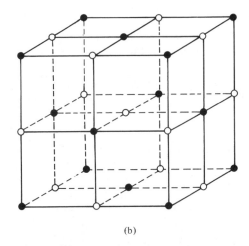

Figure 17-4 Sketch (a) and elementary cell (b) of the NaCl structure.

Let us consider two typical structures of alkali halide crystals, namely NaCl and CsCl. The NaCl crystal is composed of cubes of the type of Figure 17-4a. There are Na^+ and Cl^- ions at the corners of the cube arranged in such a way that the nearest neighbors of one species all belong to the other ion species. It is clear that the cube in Figure 17-4a is not the elementary cell because the adjacent cubes have a different structure. The elementary cell is obtained by placing eight cubes of the Figure 17-4a type together; the resulting elementary cell is drawn in Figure 17-4b. This structure is known as face-centered cubic because it contains ions at the centers of the six faces. The length a of the side of the cube is 5.63 Å. The ion in the center of the cube, which we take as Cl^-, has six nearest neighbors, which are Na^+ ions, at a distance of 2.81 Å.

We saw in Chapter 12 that in an atom (or an ion) the charge density is a continuous function and that there is no sharply defined boundary for the dimensions of an atom. However, in ionic crystals it is assumed that the charge density of each individual ion drops off sharply at a certain distance and that we may define an ionic radius which represents the effective dimension of the charge cloud. In this model we approximate the charge cloud of the ion as a hard sphere and the ionic radius is the radius of the sphere. According to this model, the shortest possible distance between two ions A^+ and B^- in an ionic crystal must be the sum of the ionic radii $r(A^+)$ and $r(B^-)$. In Table 17-2 we have listed the atomic radii for the alkali metal ions and the halogen ions. It

Table 17-2 Ionic radii of the alkali metal ions and of the halogen ions

Ion	Radius/(Å)	Ion	Radius/(Å)
Li^+	0.60	F^-	1.36
Na^+	0.95	Cl^-	1.81
K^+	1.33	Br^-	1.95
Rb^+	1.48	I^-	2.16
Cs^+	1.69		

may be seen that the sum of the atomic radii for Na^+ and Cl^- is 2.76 Å, which is fairly close to the experimental value of 2.81 Å. A number of ionic crystals have the same lattice structure as NaCl, namely LiH (lattice constant $a = 4.08$ Å), KBr ($a = 6.59$ Å), and RbI ($a = 7.33$ Å).

Another frequently occurring crystal structure is the body-centered cubic lattice illustrated in Figure 17-5; the CsCl crystal has this structure. Here we have a Cl^- ion in the center of the cube and Cs^+ ions at the eight corners. The lattice constant $a = 4.11$ Å. The shortest distance between a Cs^+ and a Cl^- ion is $\frac{1}{2}a\sqrt{3} = 3.56$ Å. It may be seen from Table 17-2 that the sum of the atomic radii of Cs^+ and Cl^- is 3.50 Å, which is again fairly close to the experimental value.

If we assume that every ion in the crystal is a hard sphere with a spherical charge density, we may calculate the total electrostatic energy of the crystal by assuming that we deal with a set of point charges at the various lattice points. Let us first consider the NaCl structure of Figure 17-4 and calculate the electrostatic energy of the central Cl^- ion. If we take R as the shortest $Na^+ - Cl^-$ distance, then there are 6 Na^+ ions at a distance R, 12 Cl^- ions at a distance $R\sqrt{2}$, 8 Na^+ ions at a distance $R\sqrt{3}$, and so on. The electrostatic energy $E(Cl)$ of the central Cl^- ion is therefore given by the series

$$E(Cl) = -e^2\left\{\frac{6}{R} - \frac{12}{R\sqrt{2}} + \frac{8}{R\sqrt{3}} \cdots\right\} \tag{17-2}$$

We can write this as

$$E(Cl) = -\frac{Ae^2}{R} \tag{17-3}$$

where the constant A is derived by summing the infinite series in Eq. (17-2). The constant is known as the Madelung constant, and it depends on the crystal structure. In general, we may write the electrostatic energy for any ion in any crystal structure in the form of Eq. (17-3) where A is the Madelung constant and R is the closest distance between two ions in the crystal structure. The sum of Eq. (17-2) is difficult to evaluate, but the result has been derived for a variety of crystal structures. For the face-centered NaCl structure, the Madelung constant is 1.748; and for the body-centered cubic CsCl structure, the constant is 1.763. It follows that the body-centered cubic structure has a more favorable electrostatic energy. In general, it may

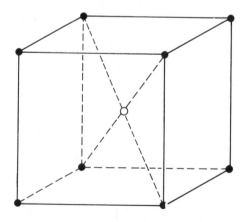

Figure 17-5   The elementary cell of the body-centered crystal structure of CsCl.

be predicted that an alkali halide crystal has the CsCl structure if the atomic radii of the positive and negative ions are comparable in magnitude. If the radii are widely different, the crystal structure is more likely to be face-centered cubic.

Apart from some small corrections, the lattice energy $\mathcal{U}_0$ per mol of crystal is derived by multiplying Eq. (17-3) by Avogadro's number N,

$$\mathcal{U}_0 = -NE(\mathrm{Cl}) = \frac{NAe^2}{R} \qquad (17\text{-}4)$$

This energy is defined as the energy of formation of 1 mol of Na^+ ions and 1 mol of Cl^- ions. Obviously, this energy cannot be measured directly because the solid NaCl is obtained experimentally from solid Na metal and gaseous Cl_2 molecules. It can be determined indirectly by means of the Born–Haber cycle, which gives the result

$$\Delta\mathcal{E} = \mathcal{U}_0 + \tfrac{1}{2}\mathcal{D}_0 + \mathcal{E}_s + \mathcal{I}(M) - \mathcal{E}(X) \qquad (17\text{-}5)$$

Here $\Delta\mathcal{E}$ is the energy of formation per 1 mol crystal MX, starting with solid metal M and gaseous X_2, $\mathcal{D}_0$ is the dissociation energy of 1 mol of X_2 into atoms, $\mathcal{E}(X)$ the electron affinity, $\mathcal{E}_s$ is the sublimation energy per 1 mol of atoms and $\mathcal{I}(M)$ its ionization energy. It is easily seen that the right hand side of Eq. (17-5) represents the formation of ionic M^+ and X^- and, if we add the unknown lattice energy $\mathcal{U}_0$, the total should be equal to the experimental heat of formation $\Delta\mathcal{E}$. In this fashion the lattice energy of the NaCl crystal may be derived; it is 7.98 eV per NaCl unit.

From Eq. (17-4) we find that $\mathcal{U}_0$ should be

$$\mathcal{U}_0 = \frac{1.748 \times 27.205 \times 0.5292}{2.815} = 8.94 \text{ eV per unit} \qquad (17\text{-}6)$$

The difference between the theoretical and experimental values is due to a number of secondary effects, such as the polarization of the ions, the repulsive forces between the charge clouds, and so on. These effects can all be calculated, and they account almost exactly for the difference of 0.96 eV between the experimental value of $\mathcal{U}_0$ and the value derived in Eq. (17-6). It may therefore be concluded that the electrostatic model gives an excellent account of the properties of ionic crystals.

17-3 METALS AND SEMICONDUCTORS

The division of the chemical elements into metals and nonmetals is discussed in most freshman chemistry courses. Some of the characteristic features of metals are their high electrical and thermal conductivity. It is generally assumed that in a solid metal one or more valence electrons can move freely through the whole crystal and that this accounts for most of the characteristic properties of metals. The crystal structures of metals are such that each ion is surrounded by a large number of neighbors. As a result of this most metals have either a body-centered cubic, a face-centered cubic, or a hexagonal close-packed structure; the latter is sketched in Figure 17-6.

Nonmetals usually crystallize in the form of covalent crystals. Typical examples are diamond, graphite, and germanium where the crystal may be considered a giant

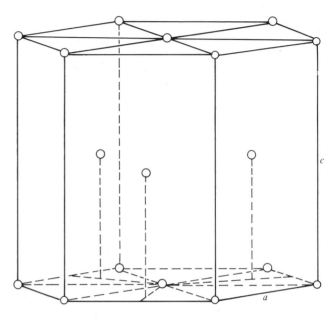

Figure 17-6 The elementary cell of a hexagonal close-packed structure.

molecule in which the atoms are connected by covalent bonds. In diamond the bonds on each carbon form a tetrahedron, and they are bonded to the set of four nearest neighbors. Graphite is not a typical covalent crystal; it consists of separate layers of giant benzene-shaped molecules (without the hydrogens of course).

In spite of the large differences between metals and covalent crystals, their theoretical descriptions are fairly similar. Thus, we can present a general quantum mechanical description of both types of crystals. It turns out that the decisive factor determining whether a crystal is a metal or an insulator is the form of the eigenvalue spectrum and the distribution of the electrons over the energy levels.

To understand the structure of a metal we shall consider Li, which crystallizes in the body-centered cubic structure illustrated in Figure 17-5. A Li atom has the configuration $1s^2 2s$, and in the crystal the $1s$ electrons stay attached to the nucleus while the outer valence electron, the $2s$ electron, moves freely through the whole crystal. The crystal consists of Li^+ ions at the lattice points and a set of $2s$ electrons that move freely through the whole crystal. There are different ways to describe the behavior of the $2s$ electrons, but one of the simpler approaches is to represent each electron by a one-electron orbital which may then be written as a linear combination of atomic obitals.

$$\phi_k = \sum_k a_{k,n}(2s)_n \qquad (17\text{-}7)$$

Here $(2s)_n$ is the atomic orbital associated with the nth Li atom, and the summation is to be performed over all Li atoms in the crystal. In this model it is assumed that each electron moves in an effective potential field V_{eff} so that its behavior is represented by an effective Hamiltonian

$$\mathfrak{H}_{\text{eff}} = (-\hbar^2/2m)\Delta + V_{\text{eff}} \qquad (17\text{-}8)$$

The orbitals ϕ_k should then be solutions of the Schrödinger equation

$$\mathfrak{H}_{\text{eff}}\phi_k = \varepsilon_k \phi_k \tag{17-9}$$

We can obtain some insight into the properties of the eigenvalues and eigenfunctions of this equation by considering a linear array of Li atoms (see Figure 17-7); in other words we consider a one-dimensional lattice. The eigenvalue problem (17-9) then becomes equivalent with the eigenvalue problem that we discussed in Chapter 15, Section 15-8, or with the chain or ring systems that we discussed in Appendix B of Chapter 15. The matrix elements of the effective Hamiltonian $\mathfrak{H}_{\text{eff}}$ depend only on the relative positions of the atomic orbitals so that we have

$$\mathfrak{H}_{n,n} = \alpha$$
$$\mathfrak{H}_{n,n+1} = \mathfrak{H}_{n+1,n} = \beta \tag{17-10}$$

where

$$\mathfrak{H}_{n,m} = \langle (2s)_n | \mathfrak{H}_{\text{eff}} | (2s)_m \rangle \tag{17-11}$$

It is customary to treat a one-dimensional lattice as an infinite ring of N atoms. Its eigenvalues and eigenfunctions may then be obtained from Appendix B, Eq. (B-12), of Chapter 15.

$$\varepsilon_m = \alpha + 2\beta \cos \frac{2\pi m}{N} \tag{17-12}$$

$$\phi_m = \sum_n (2s)_n \exp (2\pi i n m/N)$$

with

$$m = 0, \pm 1, \pm 2, \pm 3, \ldots \tag{17-13}$$

In Figure 17-7 we have sketched the distribution of the energy levels for the four cases $N = 2$, $N = 6$, $N = 12$, and $N = 30$. You can see that the energy levels must always be between the two limits

$$\alpha + 2\beta \leqq \varepsilon_m \leqq \alpha - 2\beta \tag{17-14}$$

If N becomes very large, the energy levels are very close together and, in the limit where N tends to infinity, we obtain a continuum of energy levels situated within the limits of Eq. (17-14). Such a set of energy levels is called an energy band; in the present case the band has a width 4β.

If we deal with crystals, we write the eigenvalues and eigenfunctions of Eq. (17-9) in a somewhat different form, namely

$$\phi_k = \sum_n (2s)_n \exp (ikr_n) \tag{17-15}$$

where r_n denotes the position of the nth Li atom. If the distance between two adjacent Li atoms is a, then

$$r_n = na \tag{17-16}$$

and by comparing this with Eq. (17-12), we find that k can assume the values

$$k = \frac{2\pi}{a} \frac{m}{N} \qquad m = 0, \pm 1, \pm 2, \ldots, \pm \tfrac{1}{2} N \tag{17-17}$$

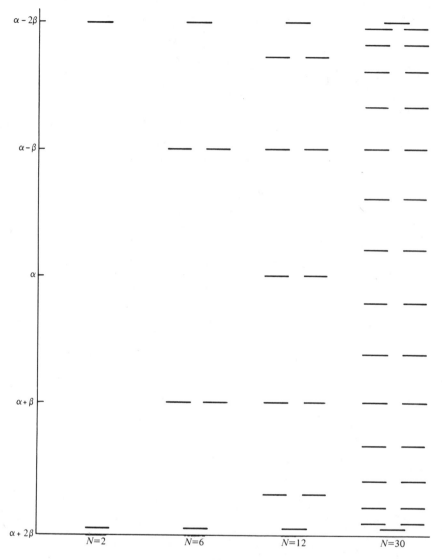

Figure 17-7 Energy levels of a ring of *N* atoms.

If *N* is very large *k* can take all values between the limits

$$-\frac{\pi}{a} \leq k \leq \frac{\pi}{a} \tag{17-18}$$

The energy values may also be rewritten in terms of *k*,

$$\varepsilon(k) = \alpha + 2\beta \cos ak \tag{17-19}$$

The most important conclusions that we draw from this treatment are that the energy levels form a band and that the energy eigenvalues and the eigenfunctions are both continuous functions of a parameter *k*. The eigenfunctions are given by Eq.

(17-15); these functions are known as Bloch functions. Note that the form (17-15) of the eigenfunctions may be derived from the crystal symmetry because these are the eigenfunctions that describe a system with translational symmetry.

Obviously, the atomic $2p$ functions also combine in the form of Bloch functions

$$\phi_k^\sigma = \sum_n (p_\sigma)_n \exp{(ikr_n)}$$

$$\phi_k^\pi = \sum_n (p_\pi)_n \exp{(ikr_n)}$$ (17-20)

At first sight it may seem that we have three different energy bands of the type of Eq. (17-19) namely a band for the $2s$ energy levels, one for the $2p_\sigma$ levels and one for the $2p_\pi$ levels. We have sketched this situation in Figure 17-8a and it is obvious that this arrangement of energy bands is not likely to occur. Usually the $2s$ and $2p_\sigma$ bands overlap, and we have the situations of either Figure 17-8b or 17-8c. In Figure 17-8b we have one broad energy band containing the $2s$ and $2p_\sigma$ levels. In Figure 17-8c we have two energy bands separated by an energy gap. Each of the two bands consists of $2s$ and $2p$ levels but the lower band has lower energies. In an energy band the one-electron orbitals take the form

$$\psi_k = a_k \phi_k + b_k \phi_k^\sigma$$ (17-21)

and the values of a_k and b_k depend on the values of k. The corresponding energies are also functions of k.

The preceding description may be generalized to three dimensions. Imagine that we have a monoclinic structure and that the elementary cell is defined by the three vectors $\mathbf{a}$, $\mathbf{b}$ and $\mathbf{c}$ and that we have one atom per elementary cell. The positions of the

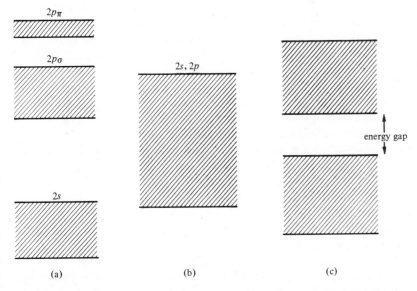

Figure 17-8 Possible arrangement of energy bands that may be constructed from $2s$ and $2p$ atomic orbitals.

atoms are then given by

$$\mathbf{r}_n = \mathbf{r}_0 + n_a \mathbf{a} + n_b \mathbf{b} + n_c \mathbf{c} \tag{17-22}$$

where the three numbers n_a, n_b, and n_c denote the elementary cell. The Bloch functions are then defined as

$$\phi(\mathbf{k}) = \sum_n \phi_n \exp[i\mathbf{k} \cdot \mathbf{r}_n]$$

$$= \sum_n \phi_n \exp[ik_x n_a a + ik_y n_b b + ik_z n_c c + i\mathbf{k} \cdot \mathbf{r}_0] \tag{17-23}$$

where ϕ_n is the atomic orbital belonging to the atom in cell n. The components k_x, k_y and k_z of the vector $\mathbf{k}$ must satisfy the conditions

$$-\frac{\pi}{a} \leq k_x \leq \frac{\pi}{a}$$

$$-\frac{\pi}{b} \leq k_y \leq \frac{\pi}{b} \tag{17-24}$$

$$-\frac{\pi}{c} \leq k_z \leq \frac{\pi}{c}$$

by analogy with Eq. (17-18). Again, the energy levels form an energy band but the energy is now a function of the three variables k_x, k_y, and k_z.

It is important to note that the theoretical description of a covalent crystal, such as diamond, is basically the same as the treatment of a metal discussed previously. Again, we must consider the various $2s$ and $2p$ functions of each carbon atom in the elementary cell and combine them in the form of Bloch functions. It is then possible to derive the form of the energy bands, that is, the energy as a function of the vector $\mathbf{k}$ from the set of Bloch functions. The fact that diamond is an insulator and that a metal is a conductor is due to the form of the energy bands and the distribution of the electrons over these bands, as we will discuss in the rest of this section.

We mentioned before that diamond is a giant molecule, and at first sight it may seem that we can describe such a system by assuming localized bonds between neighboring carbon atoms and by placing a pair of electrons in each localized bond orbital. However, you must realize that in such a model each bond orbital has the same energy so that this model would lead to a degeneracy of very large order. Due to the interactions between electrons in adjacent bonds, this degenerate energy level broadens into an energy band. If we take these effects into account, we might as well start off with Bloch orbitals in the first place because this approach is simpler and it leads to the same answer.

Usually the form of the function $E(\mathbf{k})$ is fairly complicated, but fortunately we can draw some conclusions from the general behavior of the energy bands without having to know their specific forms.

We note first that the energy levels occur in pairs, corresponding to positive and negative values of the vector $\mathbf{k}$. One level represents an electron traveling in direction $\mathbf{k}$, and the other level represents an electron traveling in the opposite direction. Such a pair of electrons cannot give rise to any net transport of electrons in any direction, and it does not contribute to the electrical conductivity. It follows that we must have unpaired electrons in order to observe electrical conductivity.

Let us now consider how the available electrons are distributed over the available energy levels. At zero temperature all electrons are in the lowest energy levels available; that is, we place the first pair of electrons in the lowest energy levels, the next pair in the next lowest, and so on, until all electrons are accommodated. In Figure 17-9 we have sketched the two possible situations that may occur. In Figure 17-9a the energy band is completely filled by all the available electrons, and the next higher energy band, which is empty at zero temperature, is separated from the filled band by a fairly large energy gap. In Figure 17-9b the energy band is only partially filled by electrons.

It may be seen that Figure 17-9b represents the situation that exists in a metal. At finite temperatures the electrons in the higher occupied levels will move to the lower of the empty levels because of thermal excitation. Since there is no separation between the filled and the empty energy levels (they form a continuum), a substantial fraction of the electrons will be in singly occupied energy levels even at low temperatures. All the unpaired electrons contribute to the electrical conductivity, and it follows that we have a conductor.

The situation in Figure 17-9a is quite different. Here the highest occupied level and the lowest empty level are separated by a fairly large energy gap ΔE. The fraction of electrons that is excited into the empty energy band is given by $\exp(-\Delta E/kT)$, and it follows that we have a very small electrical conductivity unless the temperature T is high enough so that kT becomes comparable with ΔE. It follows that Figure 17-9a represents a semiconductor. Its electrical conductivity is quite small, but, if the temperature becomes high enough so that some of the electrons are thermally excited into the empty band, we have a measurable electrical conductivity. Of course, if we shine light on a semiconductor it may be possible to excite the electrons optically from the filled to the empty energy band; in that case we have photoconductivity.

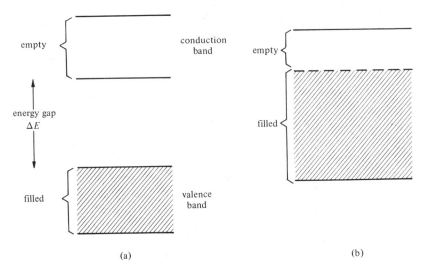

Figure 17-9 The distribution of electrons in the energy bands in (a) a semiconductor and (b) a metal.

The electrical conductivity of a semiconductor is determined mostly by the magnitude of the energy gap ΔE. A typical example is germanium where ΔE is approximately 0.7 eV. This corresponds to a temperature of about 8000 K. At room temperature $\exp(-\Delta E/kT)$ is approximately 10^{-11} so that the conductivity is quite small. In the case of diamond the energy gap is 6 eV so that for all practical purposes this may be considered an insulator. Gray tin has a fairly small energy gap, namely 0.08 eV; this corresponds to a temperature of 900 K. It follows that gray tin has a sizable conductivity at room temperature.

We mentioned that the calculation of the energy bands of a solid, that is, the calculation of E as a function of the three variables k_x, k_y and k_z, is a fairly complicated problem. However, some of the properties of metals and semiconductors may be derived from fairly simple models. We will illustrate this by deriving the Fermi energy and the electronic heat capacity of a metal.

We have mentioned that in a metal (or semiconductor) we place the available electrons in the lowest possible energy levels at zero temperature. We start with a pair of electrons in the lowest level, then we place a pair in the next lowest, and we keep doing this until we have accommodated all the available electrons. The highest occupied energy level at zero temperature is known as the Fermi level of the metal, and the corresponding energy is called the **Fermi energy** $E_F(0)$. The effective Fermi temperature T_F is then defined by the relation (see also page 618)

$$kT_F = E_F(0) \qquad\qquad \textbf{(17-25)}$$

The Fermi energy of a metal may be calculated by assuming that the electrons can move freely through the metal so that their behavior may be represented by the simple model of a particle in a box, which we discussed in Chapter 10, Section 10-4.

We found that in a one-dimensional box of length L the energy levels are given by

$$E_n = \frac{n^2\hbar^2\pi^2}{2mL^2} = \frac{n^2h^2}{8mL^2} \qquad n = 1, 2, \ldots \qquad \textbf{(17-26)}$$

where m is the mass of the particle. In a three-dimensional cubic box of length L the energy levels are given by

$$E_n = \frac{\pi^2\hbar^2}{2mL^2}(n_x^2 + n_y^2 + n_z^2) = \frac{\pi^2\hbar^2 n^2}{2mL^2} \qquad \textbf{(17-27)}$$

with

$$
\begin{aligned}
n_x &= 1, 2, 3, \ldots \\
n_y &= 1, 2, 3, \ldots \\
n_z &= 1, 2, 3, \ldots \\
n &= (n_x^2 + n_y^2 + n_z^2)^{1/2}
\end{aligned}
\qquad \textbf{(17-28)}
$$

We have N electrons in the box and we fill the energy levels by placing a pair of electrons in each energy level until we have accommodated all N electrons. The highest occupied level is characterized by a value n_{max} and it follows from the definition that

$$E_F(0) = E(n_{max}) = \frac{\pi^2\hbar^2 n_{max}^2}{2mL^2} \qquad \textbf{(17-29)}$$

We must determine first how many energy levels there are below the energy value $E(n_{max})$. If we consider a three-dimensional space and we represent each set of values (n_x, n_y, n_z) by a point in space with the Cartesian coordinates (n_x, n_y, n_z), then a sphere of radius n_{max} contains

$$S(n_{max}) = \frac{4\pi(n_{max})^3}{3} \tag{17-30}$$

points because there is one point per unit volume. The values of (n_x, n_y, n_z) must be positive so that only one octant of the sphere is filled; consequently, the number of energy levels S_0 below n_{max} is given by

$$S_0 = (\pi/6)(n_{max})^3 \tag{17-31}$$

The relation between the number of electrons N and n_{max} is thus given by

$$\tfrac{1}{2}N = S_0 = (\pi/6)(n_{max})^3 \tag{17-32}$$

or

$$n_{max} = (3N/\pi)^{1/3} \tag{17-33}$$

By substituting this value into Eq. (17-29), we find that the Fermi energy is given by

$$E_F(0) = \frac{\pi^2\hbar^2(3N/\pi)^{2/3}}{2mL^2} = \frac{\hbar^2}{2m}\left(\frac{3\pi^2N}{L^3}\right)^{2/3} \tag{17-34}$$

If we define

$$N_0 = N/L^3 = \frac{N}{V} \tag{17-35}$$

as the number of particles per unit volume, Eq. (17-34) reduces to

$$E_F(0) = (\hbar^2/2m)(3\pi^2N_0)^{2/3} \tag{17-36}$$

In the case of Li there are approximately 5×10^{22} free electrons per cubic centimeter, and the Fermi energy is about 5 eV. In general, the Fermi energies of most metals derived from Eq. (17-36) range from about 1 to about 6 eV. This is the range of the energy band part that is filled with electrons at zero temperature. The corresponding Fermi temperatures T_F vary from about 10,000 K to about 60,000 K.

Let us now derive an expression for the specific heat of a metal. If we consider a metal at zero temperature and if we raise the temperature to a value T, then only the electrons at the upper part of the filled band will be excited. The fraction of electrons that is excited thermally is roughly equal to T/T_F. Each of these electrons will gain an amount of energy kT. It follows that the electronic energy per mol is then

$$\mathscr{E}_{el}(T) = (NTkT/T_F) = NkT^2/T_F = RT^2/T_F \tag{17-37}$$

where N is Avogadro's number of electrons. The electronic heat capacity is then

$$\mathscr{C}_V(el) = RT/T_F \tag{17-38}$$

This is in agreement with the experimental results, where it is found that the electronic specific heat is indeed proportional to the absolute temperature.

The magnetic susceptibility of a metal may be discussed in a similar fashion. If we consider a free electron with spin magnetic moment μ, its net magnetic moment may be derived from Boltzmann's distribution law as

$$\mu(T) = \frac{\mu\, e^{\mu H/kT} - \mu\, e^{-\mu H/kT}}{e^{\mu H/kT} + e^{-\mu H/kT}} = \frac{\mu^2 H}{kT} \qquad (17\text{-}39)$$

If we have N free electrons then the total magnetization would be

$$\mu_{\text{tot}} = N\mu H^2 / kT \qquad (17\text{-}40)$$

However, remember that in a metal most of the electrons are in doubly occupied orbitals with their spins paired. Only the small fraction NT/T_F that are thermally excited are free so that the magnetic moment of the electrons in a metal is given by

$$\mu_{\text{tot}} = N\mu H^2 / kT_F \qquad (17\text{-}41)$$

instead of by Eq. (17-40). This too is in agreement with experimental evidence which shows that the magnetic susceptibility of a metal is small and temperature independent, as follows from Eq. (17-41).

Of course, a complete and accurate description of the properties of a metal can be derived only if the exact form of the energy band is known, but the preceding examples illustrate how some of the properties of a metal may be derived by means of very simple arguments.

17-4 MOLECULAR CRYSTALS

The metallic and covalent crystals that we discussed in Section 17-3 are characterized by the fact that the forces between adjacent atoms are all of the same order of magnitude. In a molecular crystal we can recognize the individual molecules because the forces between different molecules are much smaller than the forces between the atoms within one molecule. We may assume that there is no exchange of electrons between different molecules and that the forces between different molecules are the relatively weak van der Waals forces. As a result, molecular crystals should have fairly low melting points. The best known class of molecular crystals are the aromatic hydrocarbons, such as benzene, naphthalene, anthracene, and so on. Another category are the crystals of gases such as hydrogen, argon, nitrogen, and so on. Obviously these gases solidify only at very low temperatures, below 100 K, but they have been studied experimentally and they fall definitely in the category of molecular crystals. In the present discussion we are mainly interested in the aromatic hydrocarbons.

There is ample experimental evidence that in molecular crystals the molecules have practically the same structures and configurations as in the gaseous states. The internuclear distances measured in crystals are almost identical with the distances for the free molecules, and the molecular configurations in the crystalline and gaseous states are identical in most cases (there are of course some exceptions). The infrared and electronic spectra in the two phases are very similar. Each vibronic transition in the free molecule is easily recognized in the crystal spectrum. To a first approximation, the vapor spectrum is identical with the crystal spectrum. However, there are some small differences between the two spectra, and, even though they are very small, they are important because they serve to give us some understanding of the type of interactions between the molecules in the crystal.

Table 17-3 Some typical vibronic transition lines for naphthalene (cm^{-1}). The first column gives the values for naphthalene vapor and the second and third columns give the values for two different crystal directions

Vapor	Direction b	Direction a	Splitting	Shift
35,905	31,063	31,050	13	4849
36,398	31,620	31,474	146	4851
37,302	32,255	32,227	28	5061

In Table 17-3 we have listed some vibronic transitions in the naphthalene molecule as they occur in the free molecule (vapor) and in a molecular crystal. Note first that the crystal transitions are shifted by an amount of about $5000 \ cm^{-1}$ as compared to the free molecule. If we measure the crystal absorption spectra with polarized light, the crystal also exhibits some anisotropy, the frequencies are slightly different, depending on whether the light is polarized along different crystal axes. At sufficiently low temperatures this same effect may also be observed as a splitting of the spectral line into two different lines. Table 17-3 shows that the distance between the two lines can be quite small; it varies from $13 \ cm^{-1}$ for one doublet to $146 \ cm^{-1}$ for the other. However, such splittings have been clearly observed for the majority of vibronic transitions in molecular crystals. They are known as Davydov splittings, and they play an important role in the description of the intermolecular interactions in molecular crystals.

In order to understand the nature of the intermolecular interactions in a molecular crystal, we shall consider a typical case where there are two molecules per elementary cell; naphthalene falls into this category. Within the elementary cell, the first molecule is located at position $\mathbf{r}_1$ and the second molecule is located at position $\mathbf{r}_2$. The position of any molecule in the crystal is then given by

$$\mathbf{r}_\mathbf{n}^{(1)} = \mathbf{r}_1 + n_x \mathbf{a} + n_y \mathbf{b} + n_z \mathbf{c}$$
$$\mathbf{r}_\mathbf{n}^{(2)} = \mathbf{r}_2 + n_x \mathbf{a} + n_y \mathbf{b} + n_z \mathbf{c} \qquad (17\text{-}42)$$

We can label the molecules by means of $(\mathbf{n}, s)$ where $\mathbf{n}$ denotes the elementary cell and s is either 1 or 2 and denotes either one of the two molecules in the cell.

Each molecule can be in its ground state, characterized by an eigenfunction ψ_0, or in one of its excited states, characterized by eigenfunctions ψ_m; the corresponding energies are ε_0 and ε_m, respectively. It is clear that in the ground state of the crystal all molecules are in their ground states. The corresponding wave function for the crystal is the product of all molecular ground state eigenfunctions

$$\Psi_0 = \Pi \psi_0(\mathbf{n}, s) \qquad (17\text{-}43)$$

where $\psi_0(\mathbf{n}, s)$ is the ground state eigenfunction of molecule $(\mathbf{n}, s)$. Let us now consider the situation where one of the molecules is in an excited state, described by an eigenfunction $\psi'(\mathbf{n}, s)$ and an eigenvalue ε'. If the excited molecule is located at the position $(\mathbf{n}, s)$, the corresponding molecular wave function is given by

$$\Psi(\mathbf{n}, s) = \frac{\Psi_0 \psi'(\mathbf{n}, s)}{\psi_0(\mathbf{n}, s)} \qquad (17\text{-}44)$$

In this situation the excitation is localized at the position $(\mathbf{n}, s)$, and we speak of a **localized exciton**.

Realize that the situation where one molecule is excited, described by the wave function (17-44), corresponds to a degenerate eigenvalue with a very large degree of degeneracy. In fact, the degeneracy is N-fold, where N is the total number of molecules in the crystal, since all functions $\Psi(\mathbf{n}, s)$ of Eq. (17-44) for all possible values of $\mathbf{n}$ and s correspond to the same energy.

As a result of the small interactions between adjacent molecules, the degeneracy is lifted and we again obtain an energy band, just as in the case of semiconductors. Then the localized exciton wave functions of Eq. (17-44) combine to form Bloch functions of the form

$$\Phi_1(\mathbf{k}) = \sum_{\mathbf{n}} \Psi(\mathbf{n}, 1) \exp(i\mathbf{k} \cdot \mathbf{r}_{\mathbf{n}}^{(1)})$$

$$\Phi_2(\mathbf{k}) = \sum_{\mathbf{n}} \Psi(\mathbf{n}, 2) \exp(i\mathbf{k} \cdot \mathbf{r}_{\mathbf{n}}^{(2)})$$

(17-45)

The crystal energies form an energy band with the energies given as a function of the three variables k_x, k_y, and k_z. The corresponding eigenfunctions are obtained in the form

$$\Phi(\mathbf{k}) = \alpha(\mathbf{k})\Phi_1(\mathbf{k}) + \beta(\mathbf{k})\Phi_2(\mathbf{k})$$

(17-46)

where the coefficients α and β are also functions of $\mathbf{k}$. They must be determined from the intermolecular interactions and from the positions of the molecules in the elementary cell.

Let us now first consider the question of why a crystal spectrum consists of sharp lines. In fact, at sufficiently low temperatures the lines of the crystal spectrum may even be sharper than the lines of the vapor spectrum. It may seem at first sight that the width of a crystal spectrum line should be much larger than the corresponding width in the vapor spectrum because of the presence of an energy band. If there were a finite transition probability for all transitions from the crystal ground state to every energy level in the band, the width of the crystal spectral line would be approximately equal to the width of the energy band; but this is contrary to what we observe experimentally.

If we consider the selection rules for transitions between the ground state and an exciton state characterized by a vector $\mathbf{k}$, it can be shown fairly easily that the only allowed transitions are those for which

$$\mathbf{k} \approx 0$$

(17-47)

This means that we observe only transitions between the ground state and the part of the exciton band where $\mathbf{k} = \mathbf{0}$; usually this is at the bottom of the exciton band. This is why spectroscopic transitions in a molecular crystal give lines that are just as sharp as in the vapor.

EXAMPLE 17-1

To prove the selection rule of Eq. (17-47) let us imagine that the transition is induced by a light wave of the form $\exp(i\boldsymbol{\sigma} \cdot \mathbf{r})$. The transition between the ground state and the exciton state $\mathbf{k}$ is then derived from the expression

$$\langle \Phi(\mathbf{k}) | \exp(i\boldsymbol{\sigma} \cdot \mathbf{r}) | \Psi_0 \rangle = \int \Phi^*(\mathbf{k}) \exp(i\boldsymbol{\sigma} \cdot \mathbf{r}) \Psi_0 \, d\mathbf{r}$$

(a)

It easily follows from the translational symmetry of the lattice that this expression is zero unless

$$-\mathbf{k} + \boldsymbol{\sigma} = 0 \tag{b}$$

It should be noted that the vector $\boldsymbol{\sigma}$, which characterizes the light wave is much smaller than the ordinary values of the vector $\mathbf{k}$. The wavelength λ of the light, which is $(2\pi/\sigma)$ is of the order of 5000 Å, whereas k is of the order of a, the dimension of the elementary cell, which is around 10 Å. For that reason we may neglect $\boldsymbol{\sigma}$ in Eq. (b), which then reduces to the selection rule of Eq. (17-47).

In the same way it may be argued that the probability for a transition between two exciton states $\mathbf{k}$ and $\mathbf{k}'$ is derived from the expression

$$\langle \Phi(\mathbf{k}) | \exp(i\boldsymbol{\sigma} \cdot \mathbf{r}) | \Phi(\mathbf{k}') \rangle = \int \Phi^*(\mathbf{k}) \exp(i\boldsymbol{\sigma} \cdot \mathbf{r}) \Phi(\mathbf{k}') \, d\mathbf{r} \tag{c}$$

It follows again from the translational symmetry of the crystal that this expression is zero unless

$$\mathbf{k}' + \boldsymbol{\sigma} - \mathbf{k} = 0$$
$$\mathbf{k}' \approx \mathbf{k} \tag{d}$$

Again we may neglect $\boldsymbol{\sigma}$ for the reasons stated previously. ●

Let us now again consider the naphthalene crystal, having two molecules per unit cell. We have seen in Eq. (17-46) that for a given value of $\mathbf{k}$ the exciton eigenfunctions are given by

$$\Phi(\mathbf{k}) = \alpha(\mathbf{k})\Phi_1(\mathbf{k}) + \beta(\mathbf{k})\Phi_2(\mathbf{k}) \tag{17-48}$$

where $\Phi_1(\mathbf{k})$ and $\Phi_2(\mathbf{k})$ are Bloch functions for the first and second molecule in the elementary cell, respectively. It is somewhat difficult to determine the coefficients $\alpha(\mathbf{k})$ and $\beta(\mathbf{k})$ for arbitrary values of the vector $\mathbf{k}$. However, when $\mathbf{k} = 0$ it follows easily from symmetry considerations that there are two exciton eigenstates, characterized by the eigenfunctions

$$\Phi_+(0) = \Phi_1(0) + \Phi_2(0)$$
$$\Phi_-(0) = \Phi_1(0) - \Phi_2(0) \tag{17-49}$$

The corresponding energies are $E_+(0)$ and $E_-(0)$, respectively. Because of the selection rule (Eq. b of Example 17-1), we observe two spectroscopic transitions with frequencies

$$h\nu_+ = E_+(0) - E_0$$
$$h\nu_- = E_-(0) - E_0 \tag{17-50}$$

where E_0 is the energy of the crystal ground state.

Since the exciton band is derived from an excited molecular eigenstate with energy ε', the transition in the isolated molecule corresponds to a frequency ν' given by

$$h\nu' = \varepsilon' - \varepsilon_0 \tag{17-51}$$

We see that this transition in the isolated molecule with frequency ν' gives rise to two transitions with frequencies ν_+ and ν_- in the molecular crystal.

If the transition moment for the transition $(\varepsilon_0 \to \varepsilon')$ in molecule 1 is the vector $\boldsymbol{\mu}_1$ and the same transition moment for molecule 2 is $\boldsymbol{\mu}_2$, then the transition moment for the transition ν_+ is given by

$$\boldsymbol{\mu}_+ = \boldsymbol{\mu}_1 + \boldsymbol{\mu}_2 \qquad (17\text{-}52)$$

and the transition moment for the transition ν_- is given by

$$\boldsymbol{\mu}_- = \boldsymbol{\mu}_1 - \boldsymbol{\mu}_2 \qquad (17\text{-}53)$$

It may be seen that if we use polarized light there are certain directions in the crystal where only the first transition may be observed and other directions where only the other transition is observed. In most cases these directions coincide with the crystal axes.

It is even possible to calculate the magnitude of the shifts and splittings of the spectral lines because it turns out that they may be derived from the dipole–dipole interactions between the transition moments on the different molecules. It follows then that for very intense transitions these quantities are large, and for weak lines the splittings become very small. The various situations that have been calculated all give excellent agreement with the experimental values.

It has now become customary to talk about excitons as if they were particles in the same way as photons are often treated as particles. Absorption of light is then considered the absorption of a photon and the simultaneous creation of an exciton. Such excitons may either be totally delocalized as we have described them or they may be localized, due to interactions with the environment. The whole area of molecular crystal phenomena is a fairly active research field at the moment. Even though the main properties of molecular crystals are now well understood, there are many problems that can be studied and both experimental and theoretical studies of molecular crystals are of interest to many scientists.

_____ *Problems*

1. How many carbon atoms are there per unit cell in the two-dimensional graphite structure of Figure 17-2?
2. How many Na^+ and Cl^- ions are there per unit cell in the NaCl crystal?
3. The density of a NaCl crystal in the form of rocksalt is 2.18 g cm^{-3}. Use this number to derive the lattice constant a of the NaCl crystal. (This is the length of the lattice vector **a** that forms the elementary cell. The crystal structure of NaCl is cubic as may be seen from Figure 17-4.)
4. The elementary cell of a hexagonal structure is determined by the length a of the side of the hexagon and by the height c of the elementary cell (see Figure 17-6). Derive the distance between one of the particles on the hexagon points and one of the particles inside the structure of Figure 17-6 in terms of a and c.
5. According to the chemical classification, what type of crystal is solid argon? What type of crystal is SiC? What type is LiH? Sometimes a crystal is an

intermediate between two categories; for example, how would you classify ice and the SnS crystal?

6. The electrical conductivity of a semiconductor at room temperature depends on the value of the energy gap between the valence and the conduction bands. The values are 6 eV for diamond, 1.10 eV for Si, and 0.7 eV for Ge. Which of these compounds has the highest conductivity and why?

7. The lattice constant of NaCl is 5.63 Å. What is the shortest distance R between a Na^+ ion and a Cl^- ion? What is the lattice energy of NaCl as derived from the Madelung constant?

8. The KBr crystal has the same crystal structure as NaCl. Its lattice constant is 6.59 Å. What is the lattice energy of the KBr crystal?

9. The CsCl crystal has a lattice constant of 4.11 Å. What is the shortest distance R between a Cs^+ and a Cl^- ion in the crystal? What is the lattice energy of CsCl as derived from purely electrostatic interactions?

10. Derive the energy levels of a ring of eight Li atoms by considering only the $2s$ electrons in molecular orbital theory. The energy levels should be expressed in terms of the parameters α and β.

11. If we assume that in metallic Li each Li atom contributes one free electron, how many free electrons are there per cubic centimeter of Li? The density of Li is $0.534 \, g \, cm^{-3}$.

12. Metallic Na has 2.5×10^{-22} free electrons per cubic centimeter. Derive the Fermi energy of metallic Na in terms of electron volts.

13. Ordinarily the paramagnetic susceptibility of a paramagnetic substance (such as oxygen) is proportional to the inverse of the absolute temperature. However, the paramagnetic susceptibility of a metal is temperature independent and much smaller than for nonmetallic paramagnetic substances. Explain why this is true.

14. Explain why a vibronic transition in an isolated naphthalene molecule gives rise to two spectral lines in the molecular crystal of naphthalene.

15. What is the Davydov splitting? Explain it.

References and Recommended Reading

The bulk of the material in this chapter falls in the category of solid state physics. We list two books on this subject, the book by Kittel [1], which contains a lot of up-to-date applications, and the book by Peierls [2], which does not contain too many applications but gives an elegant treatment of the general quantum theory of solids. The third book we list deals with the theory of molecular crystals.

[1] C. Kittel: *Quantum Theory of Solids.* Wiley, New York, 1963.
[2] R. E. Peierls: *Quantum Theory of Solids.* Oxford University Press, London, 1955.
[3] A. S. Davydov: *Theory of Molecular Excitons.* McGraw-Hill, New York, 1962.

chapter 18

Scattered Radiation and Molecular Geometry

18-1 INTRODUCTION

A proper study for physical chemistry is the properties of molecules. Among these, the most important—geometries, vibrations, rotations, even reactivity—are determined by the potential energy of interaction between the atoms of the molecule.

Potential-energy-determined molecular properties may be calculated theoretically (ab initio) from the Schrödinger equation or semiempirically by experimental determination of parameters in formulas derived from quantum mechanics (as in the Hückel method, Chapter 15).

Experimentally, molecular properties are measured directly from the effects on a type of radiation—electromagnetic or particle (electron or neutron, usually)—that can interact with molecules. In one general class of methods, a sample of molecules is energized and the emitted radiation is observed. In a second and more important class of methods the effect on a radiation probe incident upon the sample (Section 18-2) is observed.

Since a molecule is a collection of electrical charges, by far the most important probes are the electric fields of electromagnetic radiation. Stationary electric fields are also used. Alternating and stationary magnetic fields, apart from their exclusive interactions with electron and nuclear spins (Chapter 16), have been of less importance in the study of molecular properties. Other than electromagnetic radiation, useful probes are electron and neutron beams. The latter interact with the nuclei of atoms via the strong nuclear force and with unpaired electron spins via the magnetic dipole of neutrons. The other two known forces of nature—the gravitational and the weak nuclear forces—are much less important in the study of molecular properties.

In this chapter we continue the study of radiation-matter interaction that was begun in Chapter 13 with a description of the scattering of radiation (Sections 18-2 through 18-6) and the relevant classical and quantum mechanical theories. The emphasis is on the uses of scattered radiation for the determination of molecular geometry, which is probably the most extensively studied molecular property. Finally, in Section 18-7, we give an overview of molecular geometry determination—the methods and the results.

18-2 INTERACTIONS OF RADIATION AND MATTER

Here we shall discuss scattered radiation in relation to absorption and will extend the discussion to radiations other than electromagnetic, that is, electrons and neutrons. The reason for the extension is the importance of electron and neutron scattering in the study of molecular properties.

To begin, we consider a monochromatic beam of unpolarized electromagnetic radiation incident upon a sample made up of molecules whose properties we wish to measure (Figure 18-1). The incident beam decreases in intensity because the sample absorbs and scatters it. The loss of intensity of the incident beam is frequently called **extinction.** A measure of the amount of extinction is the **extinction coefficient** α

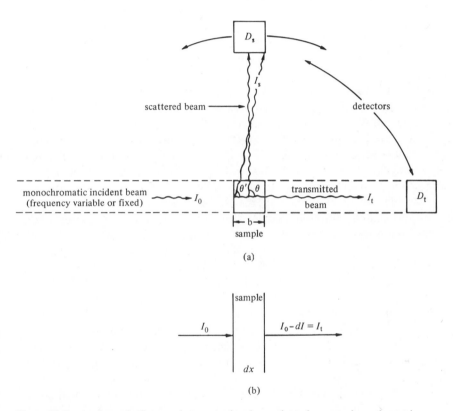

Figure 18-1 A schematic diagram that comprehends a variety of commonly used experimental arrangements for measuring the absorption and/or scattering of electromagnetic radiation by matter.

I, intensity of radiation (units: energy/time/area on which radiation falls). θ, the experimental scattering angle, that is, the angle between the direction of the incident beam and the average direction of the scattered rays detected by detector D_s. Note that, due to finite sample size and detector aperture width, the scattered intensity measured by D_s set at θ is an average over different scattering angles such as the angle θ' shown. [The situation would be more complex if repeatedly scattered light were to contribute to I_s appreciably. Fortunately, the interaction between the matter and electromagnetic radiation is weak enough to allow a practical sample size (sample volume × sample density) to be used without causing multiple scattering.]

Some experimental methods that use apparatus corresponding to the schematic diagram: (1) NMR, ESR, microwave, infrared, visible, ultraviolet, and x-ray absorption spectroscopies as well as light extinction experiments measure I_t with the detector D_t; (2) x-ray diffraction and light scattering measures I_s (with the detector D_s) as a function of the scattering angle θ; (3) Raman and fluorescence spectroscopies measure I_s as a function of frequency (D_s is a spectrograph) at a fixed scattering angle $\theta = 90°$. (If the sample fluoresces at a single frequency the spectrograph is not necessary.) A detector that includes a polarizing filter may be used for all of the methods to detect the polarization of transmitted or scattered radiation.

defined by

$$\frac{-dI}{I} = \alpha \, dx \qquad (18\text{-}1a)$$

where symbols are defined and explained in Figure 18-1 and its legend.

Thus the extinction coefficient is the fractional change of incident intensity I that occurs in a sample of infinitesimal width dx. For a sample of finite width b, Eq. (18-1a) integrates to

$$\ln (I_0/I_t) = \alpha b = (\alpha_a + \alpha_s)b \qquad (18\text{-}1b)$$

The extinction coefficient is a sum of the absorption coefficient, $\alpha_a(\nu)$, and the scattering coefficient, $\alpha_s(\nu)$. The three coefficients are functions of the radiation frequency ν. Whereas $\alpha_a(\nu)$ typically varies rather rapidly with frequency, showing strong maxima at some frequencies and near-zero values over extended frequency ranges, $\alpha_s(\nu)$ generally changes much more slowly.

The different frequency dependence of the absorption and the scattering coefficients permits the separation of the two spectra, in principle. In practice, it is more frequent, however, to measure scattering away from strong absorption regions and to determine absorption with reference to a "blank" which has the same scattering (and possibly unwanted absorption) characteristics as the sample.

The uses and analyses of the molecular absorption frequencies and intensities to yield some of the more important molecular properties have been described in Chapters 13, 14, and 16. Emission spectra (of electromagnetic radiation or electrons) from molecules that have been excited by electromagnetic radiation (emission, so obtained, is frequently called "fluorescence"), by high temperatures, by electric discharges, or by other means usefully supplement absorption spectra. Whereas the usual absorption spectrum yields properties of the ground and other low-lying quantum states of a stable molecule, the emission spectrum will contain information about higher lying states of stable and unstable molecules as well. The analysis of emission spectra depends on the same fundamental principles as that of absorption spectra.

The scattered radiation shown in Figure 18-1 may be of two kinds. If the wave scattered by a *single scatterer* (electron, nucleus, atom, ion, molecule, or group of molecules) has the same frequency as the incident wave, we have **coherently** scattered radiation. If the incident and scattered frequencies are unequal the radiation is **incoherently** scattered. Coherently scattered radiation has the property that phases of the electric and magnetic fields of incident and scattered radiation are in a definite relation to each other.

An alternate description of scattering, most useful for high-energy probes, is possible from a particle viewpoint. The possibility arises from the fact that every wave is associated with some elementary particle (for example, electron, proton, neutron, or photon). The relation between the momentum p of the particle and the wavelength λ of the corresponding wave was given by de Broglie in 1923

$$p = h/\lambda \qquad (18\text{-}2)$$

The kinetic energy (which, in the case of photons, is also the total energy) of three different types of particles is given by

$$KE = h\nu = pc = mc^2 \qquad (18\text{-}3)$$

for a photon in vacuum, by

$$\text{KE} = m_0 c^2 [1/(1 - v^2/c^2)^{1/2} - 1] = (m_0^2 c^4 + p^2 c^2)^{1/2} - m_0 c^2 \qquad \textbf{(18-4a)}$$

for a relativistic particle (rest mass m_0 and velocity $v \approx c$), and by

$$\text{KE} \approx \tfrac{1}{2} m_0 v^2 = p^2/2m_0 \qquad \textbf{(18-4b)}$$

for a nonrelativistic particle ($v \ll c$).

In mechanics, scattering in which a particle rebounds with unchanged kinetic energy is called **elastic.** From Eq. (18-3) we see that an elastically scattered particle is equivalent to a coherently scattered wave. Correspondingly, an **inelastically** scattered particle is equivalent to an incoherently scattered wave.

18-3 INTENSITY AND AMPLITUDE

Intensity—the amount of energy[1] falling on unit area per unit time—is proportional to the square of the average amplitude and to the velocity of the radiation. The amplitude of an electromagnetic wave is the electrical or the magnetic field strength (see Chapter 13) of the wave. The amplitude of particle radiation—electrons or neutrons, for example—is the solution of the Schrödinger equation for a free particle (if the particle is in the incident beam) or for the particle plus scatterer (if the particle is in the scattered beam). If the solution—the wave function—is complex, then the intensity $\propto |\psi\psi^*|$. As an example of the intensity–amplitude relations we discuss a monochromatic wave of wavelength λ, frequency ν, circular frequency $\omega = 2\pi\nu$ and period T. Applications to electromagnetic radiation are used to illustrate the relations.

The intensity of electromagnetic radiation, which is the amount of electromagnetic energy falling upon unit area per unit time, is given for SI units by

$$I = \varepsilon_0 c \langle E^2 \rangle_{\text{av}} \qquad \textbf{(18-5)}$$

where E (in newtons per coulomb) is the electric field strength, $\varepsilon_0 = 8.85418 \times 10^{-12}$ $C^2\,N^{-1}\,m^{-2}$, and c is the speed of light in vacuum (in meters per second). Hence the intensity I has the units joules per square meter per second ($J\,m^{-2}\,s^{-1}$). For monochromatic radiation the intensity may be readily expressed as the number of photons per area per time by dividing Eq. (18-5) by $h\nu$.

The averaging of E^2 indicated in Eq. (18-5) is over the time interval ($\gg T = 1/\nu$) during which I is measured. However, for a sinusoidal wave, such as that described in Eq. (13-1), the average over the time interval of measurement is equal to the average over one period T of the wave. When Eq. (13-1) is substituted into Eq. (18-5) and averaged over one period (see Problem 3), the result is

$$I = \varepsilon_0 c (\tfrac{1}{2} E_m^2) \qquad \textbf{(18-6)}$$

where E_m is the amplitude of the wave.

In most scattering (and absorption) experiments relative, rather than absolute, intensities need to be measured. If so, then the constants in Eqs. (18-5) and (18-6) may be (and frequently are) dropped.

---

[1] The number of particles—photons, electrons, neutrons—is also used.

We now consider the total scattered intensity from many scatterers. If the radiation entering the detector is incoherent, that is, the different waves have random phases (because, for example, during one period the scatterers move at random relative to each other over distances that are of the order of one wavelength), we simply add the scattered intensities. For n identical scatterers, each of which scatters an intensity I_s, the total intensity is

$$I_n = nI_s \tag{18-7}$$

For example, two incoherent sources, such as two identical lamps (in equivalent relative positions), will illuminate a surface with double the intensity of a single lamp.

18-3.1 Addition of Amplitudes If the detected radiation is coherent, we must first add the amplitudes, rather than the intensities, of the waves from each source. The student has encountered the amplitude addition (or superposition) problem before—in his studies of interference and diffraction and in Chapter 13. In the following discussion we will review the problem and set it up in a way useful for the study of molecular properties by scattered radiation.

For monochromatic radiation, the instantaneous displacement (or field) S at a time t and a specific point in space is

$$S = \sum_k A_k \cos{(\omega t + \phi_k)} \tag{18-8}$$

where A_k, ϕ_k are the amplitude and phase shift, respectively, of the kth scattered wave. Eq. (18-8) shows that the superposition problem has two parts: computation of the phase shifts ϕ_k and addition of the instantaneous displacements of all scattered waves to yield S.

The method of computing phase shifts depends upon the physical cause of the shift. The causes are of three kinds: the individual waves have different wavelengths, the scatterer shifts the phase of the scattered wave relative to the incident wave, and the paths traveled by waves incident upon and scattered by various scatterers differ. For monochromatic radiation and identical scatterers (for example, atoms) the first two causes need not be considered.[2] Hence, we need to compute only the phase shifts due to the third cause.

Figure 18-2 shows the path differences of waves associated with two different scatterers and defines the symbols necessary for solving the path difference–phase shift problem. The general relation between the phase shift of a wave (relative to a reference wave) and the path difference is

$$\frac{\phi}{2\pi} = \frac{\text{phase shift}}{2\pi} = \frac{\text{path difference}}{\lambda} = \frac{d}{\lambda} \tag{18-9}$$

If the phase shifts of the scattered waves at P are referred to the phase of the zeroth wave (scattered from O in Figure 18-2; $\phi_0 = 0$) then the phase shift ϕ of the wave

---

[2] All identical scatterers will shift the phase of the scattered wave relative to the incident wave by an equal amount. Hence, the phases of the scattered waves relative to each other will remain unchanged. (The phase shifts due to variation of wavelengths are given in Eq. 13-6.)

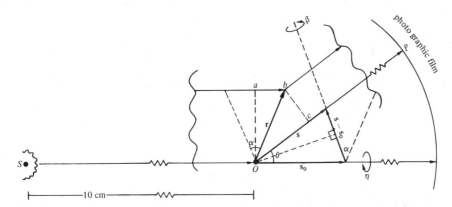

Figure 18-2 Scattering of radiation incident from source on left upon two scatterers at O and b (for example, the two atoms of a diatomic molecule). The scattered radiation falls upon the photographic film at P. Note that both the incident and scattered waves are effectively plane waves due to the very small size of the scattering system compared to $\overline{OP}$ and $\overline{SO}$ ($|\mathbf{r}| \approx 100$ pm, $\overline{OP} \approx 10$ cm).

The vectors $\mathbf{s_0}$, $\mathbf{s}$ are unit vectors ($|\mathbf{s_0}| = |\mathbf{s}| = 1$) that define the propagation directions of the incident and scattered rays, respectively. The scattering angles θ, η specify the orientation of $\mathbf{s}$ relative to $\mathbf{s_0}$. The angles α, β specify the orientation of the vector $\mathbf{r}$ relative to the vector $\mathbf{s} - \mathbf{s_0}$. *Note*: In the literature on diffraction by crystals, the Bragg angle θ', rather than the scattering angle θ, is frequently used ($\theta' = \frac{1}{2}\theta$).

(from b) is computed as follows: The path difference d is given[3] from Figure 18-2 as

$$d = \overline{Oc} - \overline{ab} = \mathbf{r} \cdot \mathbf{s} - \mathbf{r} \cdot \mathbf{s_0}$$
$$= \mathbf{r} \cdot (\mathbf{s} - \mathbf{s_0}) \qquad (18\text{-}10)$$

The magnitudes of the vectors $\mathbf{r}$ and $\mathbf{s} - \mathbf{s_0}$ are $r = |\mathbf{r}|$ and $2 \sin \frac{1}{2}\theta$, respectively. Therefore, from Eqs. (18-9), (18-10), and Figure 18-2 it follows that

$$\phi = 2\pi \frac{d}{\lambda} = 2\pi \frac{\mathbf{r} \cdot (\mathbf{s} - \mathbf{s_0})}{\lambda} = 2\pi \frac{2r \sin \frac{1}{2}\theta \cos \alpha}{\lambda} \qquad (18\text{-}11)$$

An application of Eq. (18-11) is shown in Example 18-1.

Eq. (18-11) shows that the phase shift depends on the distance between two scatterers. The scatterers in diffraction experiments may be, as we shall see later, electrons, nuclei, or atoms. Therefore, phase shifts contain information about the distribution of electrons and about internuclear distances (molecular structure). If it were possible to measure phase shifts of the waves from each scatterer, electron distribution and molecular structure could be readily determined. However, only intensity as a function of the scattering angle can be measured. The intensity results from superposition of amplitudes of waves such as the wave shown in Figure 18-3a.

The superposition of amplitudes is simpler if we describe the wave in complex number notation (see Appendix IV). The justification for the changed notation follows.

---

[3] The scalar product of two vectors, $\mathbf{A}$, $\mathbf{B}$ is $\mathbf{A} \cdot \mathbf{B} = |\mathbf{A}||\mathbf{B}| \cos \eta$ where η is the angle between the two vectors. If $\mathbf{B}$ is a unit vector ($|\mathbf{B}| = 1$) then the scalar product is equal to the projection of vector $\mathbf{A}$ on $\mathbf{B}$.

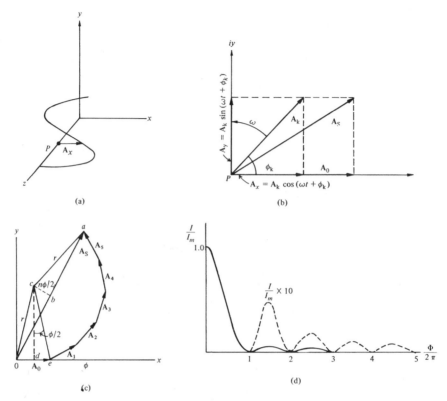

Figure 18-3 Addition of amplitudes. (a) The instantaneous displacement $\mathbf{A}_x$ of a single wave propagating along z. (b) Same wave viewed as a vector that rotates in the complex xy plane. Also, the vector addition of amplitudes $\mathbf{A}_k$ and $A_0(\phi_0 = 0)$ to yield total amplitude A_S. (c) Calculation of A_S given by Eq. (18-16). The circle with radius r that circumscribes the polygon formed by the arrows has its center at c. (d) Intensity dependence on the phase angle, from Eq. (18-19).

Although the oscillating field at P in Figure 18-3a

$$A_x(t) = A_k \cos (\omega t + \phi_k) \tag{18-12}$$

is caused by the propagation of the wave along z, Eq. (18-12) also describes the x component of a vector $\mathbf{A}_k$ which rotates about P in the xy-plane (Figure 18-3b). We signify the fact that the y component $A_y = A_k \sin (\omega t + \phi_k)$ has no physical significance in the present context by making y the imaginary axis in a complex number plane. Hence, from Eq. (IV-b) in Appendix IV, the complex field in our "vector rotation" problem is

$$\mathbf{A}_k^+(t) = A_k[\cos (\omega t + \phi_k) + i \sin (\omega t + \phi_k)]$$
$$= A_k\, e^{i(\omega t + \phi_k)} = A_k\, e^{i\phi_k}\, e^{i\omega t} = \mathbf{A}_k^+\, e^{i\omega t} \tag{18-13}$$

The last equality shows that the instantaneous electric field of the wave is the product of a time-dependent and a time-independent part. The complex number equivalent of Eq. (18-8) is obtained by summing over the amplitudes of all waves that pass through P along the z axis.

$$\mathbf{S}^+ = \sum_k A_k\, e^{i(\omega t + \phi_k)} = \left(\sum_k \mathbf{A}_k^+\right) e^{i\omega t} = \mathbf{A}_S^+\, e^{i\omega t} \qquad (18\text{-}14)$$

Eq. (18-14) shows that the superposition of many waves with the same frequency results in a new sinusoidal wave with amplitude

$$\mathbf{A}_S^+ = \sum_k A_k\, e^{i\phi_k} = \sum_k A_k(\cos \phi_k + i \sin \phi_k) \qquad (18\text{-}15)$$

The actual amplitude along the x axis is the real part of this expression. Eq. (18-15) allows the computation of the amplitude of any total wave, given the amplitudes and phase angles of the constituent waves. Figure 18-3b shows that Eq. (18-15) can be interpreted as the resultant of vectors representing the individual waves with magnitudes $\mathbf{A}_k$ and relative orientations determined by the phase shifts ϕ_k.

In Figure 18-3c vector addition is used to compute A_S (the real part of $\mathbf{A}_S^+$) for an illustrative case: n identical scatterers (or, more generally, sources of radiation) with equal phase shift ϕ between successive scatterers, that is

$$A_S = A\left(\sum_{k=0}^{n-1} \cos k\phi\right) \qquad (18\text{-}16)$$

From triangle cde in Figure 18-3c we have $A = A_0 = 2r \sin \phi/2$, whereas triangle abc gives $A_S = 2r \sin n\phi/2$. Elimination of r yields

$$A_S = A\frac{\sin n\phi/2}{\sin \phi/2} \qquad (18\text{-}17)$$

Eq. (18-17) applies to n sources equidistantly spaced along a line. A single narrow slit (width $\gtrsim \lambda$) may be viewed as an infinite number of sources placed across the slit with neighboring sources an infinitesimal distance apart. The resultant amplitude is obtained by taking the limit of Eq. (18-17) when $n \to \infty$ and $\phi \to 0$ while $\Phi = n\phi = $ constant. (Φ is the phase shift between the waves at the two edges of the slit.) The denominator of Eq. (18-17) is very small in the limit; that is, $\sin \phi/2 = \sin \Phi/2n \approx \Phi/2n$. Hence, Eq. (18-17) becomes

$$A_S = nA\frac{\sin \Phi/2}{\Phi/2} = A_m\frac{\sin \Phi/2}{\Phi/2} \qquad (18\text{-}18)$$

where we have recognized nA as the maximum amplitude A_m (for all oscillators in phase). The corresponding intensity (Figure 18-3d) is

$$I = I_m\frac{\sin^2 \Phi/2}{(\Phi/2)^2} \qquad (18\text{-}19)$$

EXAMPLE 18-1

Show that, for a freely rotating (gas phase) diatomic molecule, the average scattered amplitude (see Eq. 18-15) is

$$\langle A \rangle_{av} \propto \langle \cos \phi \rangle_{av} = \frac{\sin \mu r}{\mu r} \qquad \textbf{(a)}$$

where ϕ is the phase shift between the two atoms, $\mu = (4\pi/\lambda) \sin \tfrac{1}{2}\theta$, and the average is over all possible orientations of the molecule.

Refer to Figure 18-2. The possible orientations of the internuclear distance vector **r** with respect to vector $\mathbf{s} - \mathbf{s}_0$ may be specified by the angles α, β. The probability that **r** simultaneously points between the angles α and $\alpha + d\alpha$, β and $\beta + d\beta$ is $(\sin \alpha \, d\alpha \, d\beta)/4\pi$ (since the sum of all possible orientations is equal to $\int_0^{2\pi} \int_0^{\pi} \sin \alpha \, d\alpha \, d\beta = 4\pi$). Therefore the required average is (from Eq. 18-11)

$$\langle \cos \phi \rangle_{av} = \frac{1}{4\pi} \int_0^{2\pi} \int_0^{\pi} \cos (\mu r \cos \alpha) \sin \alpha \, d\alpha \, d\beta \tag{b}$$

To integrate Eq. (b) we change the variable to $x = \mu r \cos \alpha$. Therefore, $dx = -\mu r \sin \alpha \, d\alpha$, and the integration limits change from $\alpha = 0$ to $x = \mu r$ and from $\alpha = \pi$ to $x = -\mu r$. The right-hand side of Eq. (b) becomes

$$\frac{1}{4\pi} \cdot \frac{2\pi}{\mu r} \int_{-\mu r}^{\mu r} \cos x \, dx = \frac{\sin \mu r}{\mu r} \tag{c}$$

This result shows that the average amplitude scattered by a rotating molecule exhibits μ (or θ) and r dependent minima and maxima which possibly could serve to determine the internuclear distance r.

18-4 SCATTERING BY ATOMS

In the most important chemical application of scattering—that of determination of molecular structure from diffraction patterns—the atom is regarded as a fundamental scattering unit. Therefore, to understand molecular diffraction we must learn about the scattering properties of atoms. We shall discuss both coherent and incoherent scattering of x rays, electrons, and neutrons by atoms.

Coherently scattered radiation in the x ray region (see Table 18-1) is utilized for the **x-ray diffraction** method, a very important tool for determining molecular structures as well as other molecular properties. A common model that explains the relation between x-ray diffraction patterns and molecular structure is based on Maxwell's theory of electromagnetic radiation (see Chapter 13). Maxwell's theory predicts that the force f due to the electric field E_0 of electromagnetic radiation will accelerate particles (scattering centers) of charge q and mass m. If the particle is free (unbound), then from Newton's second law

$$f = qE_0 = m\dot{v} \tag{18-20}$$

Maxwell's theory further predicts that an accelerated charge will create electromagnetic radiation, that is, electrical and magnetic fields that propagate at the speed of light away from the charge. In 1903, J. J. Thomson showed that, at distances which are far from the charge q compared to the distance over which the charge moves, the time-dependent component of the electrical field is given by

$$E_q = \frac{q\dot{v} \sin \gamma}{4\pi\varepsilon_0 Rc^2} \tag{18-21}$$

where the new symbols are defined in Figure 18-4. Acceleration $\dot{v}$ may be eliminated

Table 18-1 A comparison of three probes—x rays, electrons, and neutrons—used in diffraction studies

	X Rays	Electrons	Neutrons
Source	Radiation from an energetic electron beam stopped by metallic target (in x-ray tube)	Electrons accelerated through 20–60 kV in high vacuum (in electron gun)	Thermalized fluxes from nuclear fission reactors
wavelength	0.05–0.25 nm	0.005–0.008 nm	0.05–0.4 nm
kinetic energy	25–5 keV	60–20 keV	0.3–0.0051 eV
detection	fluorescence (scintillation counter, etc.), photography	photography	reactions with ^{6}Li or ^{10}Be
coherently scattered by	(mostly) core electrons	Coulomb potential from nuclei and electrons	nuclei or unpaired electron spins
atomic scattering lengths			
expression[a]	$r_e \dfrac{(1+\cos^2\theta)f}{2}$	$\dfrac{2}{a_0}\cdot\dfrac{(Z-f)^2}{\mu^2}$	constant[b]
magnitude (m)	10^{-14}	10^{-10}	10^{-14}
major application to structure determination of	crystals	gases	H-atom posns., alloys, magnetic materials

[a] $r_e = e^2/(4\pi\varepsilon_0 mc^2)$ is the classical electron radius; $a_0 = \mathbf{h}^2\varepsilon_0/(\pi me^2)$ is the Bohr radius; Z = atomic number; $f(\mu)$ is the atomic form factor for x rays (Eq. 18-26); $\mu = 4\pi/\lambda \cdot \sin\frac{1}{2}\theta$; for θ, see Figures 18-1 and 18-2.

[b] Applies to nuclear scattering length; magnetic scattering length depends on μ through a form factor.

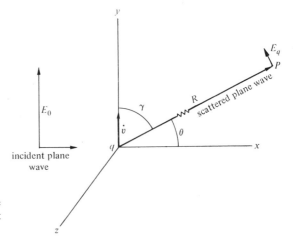

Figure 18-4 The scattering of a plane wave propagating along x by a charge q at the origin.

between Eqs. (18-20) and (18-21) to yield

$$E_q = \frac{q^2 E_0 \sin \gamma}{4\pi\varepsilon_0 mRc^2} = \frac{a}{R} \cdot E_0 \tag{18-22}$$

where a is called the **scattering length.** For application to x-ray diffraction we identify E_0 and E_q with the oscillating electrical fields of the incident and scattered plane waves, respectively (see Figure 18-2). Since the ratio q^2/m of an electron is much larger than that of any nucleus, it is clear that in x-ray diffraction the electrons, rather than nuclei, scatter the radiation.

18-4.1 X-ray Scattering by Electrons   The ratio of the intensity, I_e, scattered by a single electron of charge e to the incident intensity, I_0, is given by (from Eqs. 18-5, 18-6 and 18-22)

$$\frac{I_e}{I_0} = \frac{\langle E_e^2 \rangle_{av}}{\langle E_0^2 \rangle_{av}} = \frac{e^4 \sin^2 \gamma}{16\pi^2 \varepsilon_0^2 m^2 c^4 R^2} \tag{18-23}$$

For unpolarized incident radiation (such is commonly used in x-ray diffraction work) the intensity ratio must be averaged over all possible orientations of E_0. The result of the averaging (see Problem 4) is

$$\frac{I_e}{I_0} = \frac{e^4}{16\pi^2 \varepsilon_0^2 m^2 R^2 c^4} \cdot \frac{(1 + \cos^2 \theta)}{2} = \frac{a^2}{R^2} \tag{18-24}$$

where a is now the electron scattering length for unpolarized radiation. We note that I_e has maxima at $0°$ (forward scattering) and $180°$ (backward scattering). Furthermore, since for an electron the quantity $r_e = e^2/(4\pi\varepsilon_0 mc^2)$, known as the classical electron radius $(\approx a)$, is 2.83×10^{-15} m, the experimental ratio of the scattered to the incident intensity I_e/I_0 will be very small at all conceivable distances R. This implies long exposure times (of the order of hours for gases) to gather measurable scattered intensity.

Next, we consider scattering of a plane wave by a group of free electrons, n in number. If the distances between all pairs of electrons within the group are negligibly small compared to the wavelength of the incident radiation (see Eq. 18-11), then all electrons will scatter in phase (that is, constructive interference at all scattering angles). Therefore, the amplitude of the total scattered wave, E_a, is n times the amplitude scattered by a single electron (given in Eq. 18-22)

$$E_a = nE = \frac{naE_0}{R} \tag{18-25}$$

Eq. 18-25 is not valid, except at $\gamma = 90°$, for scattering of a plane wave by the electrons of an atom. The reason is that the wavelength, λ, of a typical x ray, that is, 100 pm, is comparable to the size of an atom. Therefore, destructive interference may occur between the scattered waves arriving from different parts of the atom. The total amplitude scattered from an atom with atomic number Z must be *less* than (or at most equal to) that given by Eq. (18-25) with $n = Z$.

18-4.2 Coherent Scattering of X rays by Atoms To calculate E_a for an atom we proceed as follows. We define the **atomic form factor** f as the ratio of the total amplitude scattered by the atom to the scattered amplitude due to a single free electron

$$f = E_a / E_e \tag{18-26}$$

For a spherical atom the form factor is given by

$$f(\mu) = \int_0^\infty dr \left(\frac{4\pi r^2 \rho(r) \sin \mu r}{\mu r} \right) \tag{18-27}$$

where $\rho(r)$ is the spherically symmetric electron density of the atom, r is the radial distance, and $\mu = (4\pi \sin \frac{1}{2}\theta)/\lambda$. The expression under the integral sign in Eq. (18-27) is the ratio of the scattered amplitude due to the electrons in the spherical shell dr $(4\pi r^2)$ to the scattered amplitude due to a single electron at $r = 0$. Hence integration over all spherical shells will yield an atomic form factor that is in accordance with the definition in Eq. (18-26). The expression assumes that the atom does not absorb significantly at the wavelength of the scattered x rays.

The dependence of the atomic form factor on the wavelength, the scattering angle θ, and the spatial distribution of the electron density $\rho(r)$ is shown in Figure 18-5. We see that the factor decreases toward zero as the wavelength decreases or as the scattering angle increases. Short wavelengths and large scattering angles are just the conditions for which the phases of the waves scattered from the different parts of the atom will differ, thus leading to destructive interference. In addition, we see that, for $\mu = $ constant, the atoms with higher atomic numbers (that is, those with higher average $\rho(r)$) are more effective scatterers (that is, have less destructive interference). Indeed, it can be seen in Figure 18-5 that for $\mu \gtrless 60$ nm^{-1} only the high-density core electrons contribute appreciably to the scattering.

The scattered intensity due to an atom is given (from Eqs. 18-6, 18-24, 18-26) by

$$I_a = f(\mu)^2 \cdot I_e = \frac{I_0}{R^2} \cdot \left[\frac{e^2}{4\pi\varepsilon_0 mc^2} \right]^2 \cdot \frac{1+\cos^2\theta}{2} \cdot f(\mu)^2 = \frac{I_0}{R^2} \cdot a_x^2(\mu) \tag{18-28}$$

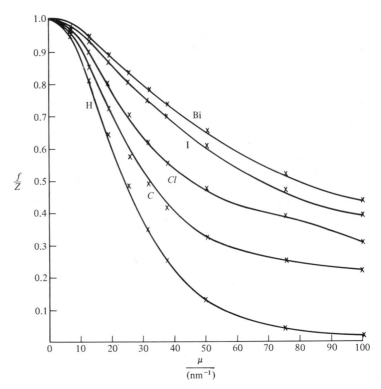

Figure 18-5 A comparison of the atomic form factors per electron, f/Z, for the scattering of x rays by H, C, Cl, I, and Bi atoms. [From data in K. Lonsdale (ed.): *International Tables of X-ray Crystallography*, Vol. III, Knyoch Press, Birmingham, England, 1962.]

for unpolarized radiation. The quantity $a_x(\mu)$, of magnitude similar to a (see Eq. 18-24), is the x-ray scattering length of an atom. Eq. 18-28 applies to the scattering of x rays by monatomic gases. In Figure 18-6a we show the experimental intensity for helium and in Figure 18-6b an $f(\mu)^2$ determined from a quantum mechanically computed electron density (curve - - -). We see that for the larger μ there is a discrepancy between experiment and theory. The reason for the discrepancy is that the x-ray detector measures the total intensity—coherently plus incoherently scattered—whereas the theoretical treatment has ignored incoherent scattering.

18-4.3   Incoherent Scattering of X rays by Atoms   Incoherent scattering of x rays is known as the **Compton effect.** In 1923 Compton studied the scattering of x rays by graphite. He found that the scattered radiation contained a component for which the wavelength increases with increasing scattering angle. A formula that explains the observation may be derived from a model which assumes that x-ray photons are being scattered inelastically by "free" electrons in graphite. (X-ray photons have energies of the order of 50 keV. This is much greater than the ionization potentials of most or all electrons in atom. Hence, most or all electrons will respond to an x-ray photon as if they were unbound or "free.") The formula is

$$\lambda = \lambda_0 + (2\boldsymbol{h}/\boldsymbol{mc})\sin^2 \tfrac{1}{2}\theta \qquad\qquad (18\text{-}29)$$

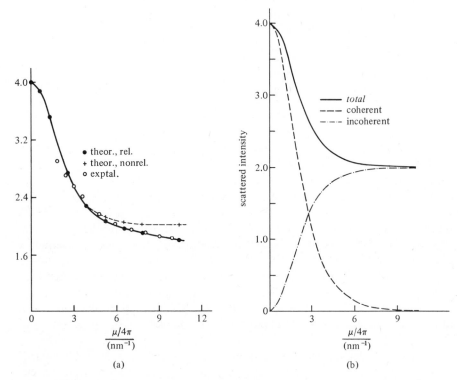

Figure 18-6 X-ray scattering by the helium atom. (a) Total scattered intensity: ○, experimental; ●, theoretical—relativistic (equivalent to Eq. 18-30 at large μ); +, theoretical—nonrelativistic (at large μ $I_{tot}/I_e \approx I_{a,incoh}/I_e \approx Z$). (b) Decomposition of the theoretical nonrelativistic intensity (——) into the coherent (– – – –) and the incoherent (– · – · –) components. [G. Herzog: Z. *Physik*, **70**:590 (1931).]

where m is the electron rest mass. The quantity $2h/mc = 4.84$ pm. Therefore incident waves with $\lambda_0 \gg 4.84$ pm will not be scattered incoherently to a measurable extent.

The intensity of radiation incoherently scattered by helium gas is shown in Figure 18-6b. A formula for the intensity that is valid in the limit of large μ (that is, large wavelength or momentum changes) may be derived by treating the electrons as "free." The formula for the ratio of incoherently scattered light intensity $I_{a,incoh}$ to I_e (see Eq. 18-24) for Z free, independently scattering electrons is

$$I_{a,incoh}/I_e = Z\left(\frac{\nu}{\nu_0}\right)^3 = Z\left(\frac{\lambda_0}{\lambda}\right)^3$$

$$= Z\left(1 + \frac{2h}{\lambda_0 mc}\sin^2\tfrac{1}{2}\theta\right)^{-3}$$

$$= Z\left(1 + \frac{\lambda - \lambda_0}{\lambda_0}\right)^{-3}$$

$$= Z\left(1 + \frac{2h\lambda_0\mu^2}{16\pi^2 mc}\right)^{-3} \tag{18-30}$$

where the second and third lines follow upon substitution of λ from Eq. (18-29) and the last line introduces the definition $\mu = (4\pi \sin \frac{1}{2}\theta)/\lambda_0$.

The total scattered intensity (Figure 18-6) is

$$I_{\text{tot}} = I_a + I_{a,\text{incoh}} \tag{18-31}$$

18-4.4 Electron and Neutron Scattering

We turn now to electron and neutron scattering by atoms and a comparison (see Table 18-1) of the two with x-ray scattering. The differences among the formulas for the scattered intensity of the three kinds of probes may have two explanations: (1) the scattering potentials are different, (2) the model used to derive the intensity formulas is different.

For comparison purposes we describe the physical features of the model that leads to the x-ray scattering formula (18-28). The most important feature of the model is that x-ray interaction with the atom is weak. [The resulting simplification of the scattering equations is known as the (first) Born approximation.] As a consequence the polarization and excitation of the atom by the wave may be neglected, and the scattered amplitude is a small fraction of the incident amplitude. A second feature is that all scatterers sampled by the probe "see" the same incident amplitude E_0; that is, the incident photon flux is not appreciably diminished by the thickness of the sample. (The model just described is known as the kinematic model, especially with reference to electron scattering.)

The kinematic model applies, besides x rays, to high energy (≈ 50 keV) electron scattering by thin samples. The high energy, compared to typical excitation energies of an atom, insures a weak (when averaged over the very short interaction time) electron–atom interaction. The interaction of a probe electron is with the effective potential due to the nucleus and the electrons combined. This effective potential leads to the intensity, I_a, scattered by an atom

$$I_a = \frac{I_0}{R^2}\left[\frac{2\pi m e^2}{h^2 \varepsilon_0}\right]^2 \left[\frac{Z - f(\mu)}{\mu^2}\right]^2 = \frac{I_0}{R^2}\, a_e^2(\mu) \tag{18-32}$$

where $a_e(\mu)$ is the electron scattering length of an atom. The other symbols have been defined previously. The last square brackets contain the atomic form factor for electron scattering.

We can account for most differences between x ray and electron scattering (Figure 18-7) on the basis of the greatly differing scattered intensities. Consider the case when $f(\mu)^2 \approx (Z - f(\mu))^2$ and incident intensities are equal. Note that the quantity in the first brackets of Eq. (18-32) is $2/a_0 = 2/(0.53 \times 10^{-10}\,\text{m}) \approx 4 \times 10^{10}\,\text{m}^{-1}$ where a_0 is the Bohr radius (see Eq. 11-104). Therefore, from Eqs. (18-28) and (18-32) and the value of r_e calculated earlier, we obtain

$$\frac{I_a\,(\text{x ray})}{I_a\,(\text{electron})} = \frac{a_x^2(\mu)}{a_e^2(\mu)} \approx \frac{r_e^2}{(2/a_0)^2 \mu^4} = (5.6 \times 10^{-51}\,\text{m}^4)\cdot\mu^4 \tag{18-33}$$

For $\mu = 10\text{–}100\,\text{nm}^{-1} = 10^{10}\text{–}10^{11}\,\text{m}^{-1}$ the intensity ratio is about $10^{-10}\text{–}10^{-6}$. As a consequence electron scattering needs much shorter exposure times (seconds compared with hours) and thinner samples (maximum thickness of about 100 nm compared with 1 mm for solid samples that obey the kinematic model) than x-ray scattering.

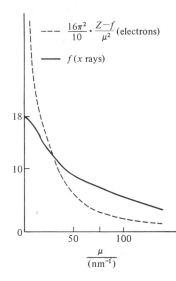

Figure 18-7 Atomic form factors for x ray and for electron scattering. [From M. H. Pirenne: *J. Chem. Phys.*, **7**:144 (1939).]

Inelastic scattering of high energy electrons, analogous to Compton scattering, obeys the approximate expression

$$I_{a,\text{incoh}} = \frac{I_0}{R^2} \cdot \left[\frac{2}{a_0 \mu^2} \right]^2 \cdot Zq \tag{18-34}$$

where $q = 0$ when $\mu = 0$ and $q \approx 1$ for large μ.

Neutron scattering by atoms occurs because (1) the neutron interacts via the strong force with the nucleus of the atom; (2) (only for paramagnetic atoms) the magnetic moments of the neutron and the atom (due to unpaired electron spins) interact. The most important application of scattering due to the second interaction is to the study of the distribution of unpaired electron spins in paramagnetic, ferromagnetic, and antiferromagnetic materials. We will not discuss magnetic scattering further.

The intensity of thermal (Table 18-1) neutrons scattered due to nuclear forces is

$$I_a = \frac{I_0}{R^2} \cdot a_n^2 \tag{18-35}$$

where the neutron scattering length a_n is nearly independent of both the scattering angle θ and the electron distribution. Experiment shows that a_n varies irregularly from nucleus to nucleus within a factor of two about the value 0.6×10^{-14} m. This value is comparable to a_x ($\approx a$). Therefore, the scattered intensities of neutrons and x rays are comparable. The exceptions are hydrogen and helium which, not having core electrons, scatter x rays very weakly although they scatter neutrons normally. Hence, neutron scattering may be used to detect the position of hydrogen atoms.

A second application of neutron scattering is to distinguish between atoms whose atomic numbers are very similar. The scattering of x rays and electrons by such atoms is indistinguishable. Their neutron scattering lengths, on the other hand, may be quite different. For example, the a_n of ^{59}Co and ^{58}Ni differ by a factor of three.

The mechanism of inelastic scattering of neutrons is different from that of high energy x rays and electrons. High neutron scattering intensities are found at frequencies equal to $\nu_0 \pm \nu_{\text{vib}}$ where ν_0 is the frequency (proportional to the kinetic energy) of the incident neutrons and ν_{vib} is a frequency characteristic of a vibrational transition in the sample. The scattered neutron frequencies result when the neutron excites ($\nu_0 - \nu_{vib}$) or deexcites ($\nu_0 + \nu_{vib}$) a vibrational mode of a molecule or crystal in the sample. For thermal neutrons with energies comparable to vibrational energies, such energy exchanges have a large and readily observable effect on the neutron frequency (or wavelength). Furthermore, the comparable energies together with the comparable masses of neutrons and nuclei make it easy to conserve momentum during the exchange. The measurement of vibrational frequencies from neutron scattering complements (primarily because of different selection rules) the data from infrared spectroscopy.

The inelastic scattering of the photons of visible light, analogous to the inelastic scattering of neutrons, is known as the **Raman effect.** The spectra caused by the Raman effect can be used to study molecular rotational and vibrational or even electronic energy levels with a resolution somewhat higher than that possible for neutron spectra.

18-5 SCATTERING CROSS SECTIONS

Up to now we have described the results of scattering in terms of the scattered intensity. An alternate description, much used in descriptions of particle scattering, is possible with the help of a new concept—scattering cross section.

Consider a scattered wave with amplitude given by

$$A = \frac{a}{R} A_0 \tag{18-36}$$

where a is a scattering length whose particular form (see Table 18-1) depends upon the probe-scatterer interaction potential. (Eq. 18-22 is an example of the general Eq. 18-36.) If $I(R, \theta, \eta)$ (see Figure 18-2) is the corresponding intensity received by area $R^2 \sin \theta \, d\theta \, d\eta$ from a single scatterer, then the power ($I \times$ area) falling on the area, from Eqs. (18-6) and (18-36), is

$$dP = I \cdot R^2 \sin \theta \, d\theta \, d\eta = I_0 a^2 \sin \theta \, d\theta \, d\eta = I_0 a^2 \, d\omega \tag{18-37}$$

where we have introduced the abbreviated notation $d\omega$ for the differential $\sin \theta \, d\theta \, d\eta$. ($\omega$ is called the **solid angle.** It subtends an area on the surface of a sphere. Its magnitude, measured in steradians, is proportional to the subtended area. Thus, $\omega = 4\pi$ steradians for total surface area of a sphere, and $\omega = 2\pi$ steradians for the surface area of a hemisphere.)

The quantity

$$\frac{d\sigma}{d\omega} = a^2(\theta, \eta) = I_0^{-1} \cdot \frac{dP}{d\omega} = \frac{IR^2}{I_0} \tag{18-38}$$

is known as the **differential cross section.** It is the fraction scattered into $d\omega$ at θ, η of the power (or energy or number of particles) incident upon a unit area. It may be visualized as an imaginary target of area a^2 (placed within the unit area) that scatters all radiation falling upon it into $d\omega$. The scattering length a then becomes the length of the edge of a square "target."

The **total** cross section σ is obtained by integrating Eq. (18-38) over all solid angles

$$\sigma = \int_0^{4\pi} \left(\frac{d\sigma}{d\omega}\right) d\omega = \int_0^{2\pi} \int_0^{\pi} a^2(\theta, \eta) \sin\theta \, d\theta \, d\eta \qquad \textbf{(18-39)}$$

The total cross section is the scattered fraction, into all angles, of the power incident on a unit area. For the scattering of neutrons by nuclei, for which the scattering length $a = a_n$ is independent of the angles (see Table 18-1), Eq. (18-39) integrates to $\sigma = 4\pi a_n^2$.

Cross sections depend only upon probe-scatterer interactions. In this respect they are preferable to intensities I which also depend on apparatus parameters—the scatterer-detector distance R, and the incident intensity I_0.

18-6 DIFFRACTION BY MOLECULES

Diffraction patterns (Figure 18-8) obtained by cameras such as those diagrammed in Figure 18-9, are important sources of information about molecular structure. The description of diffraction patterns of molecules in gas, solid, and liquid phases follows easily from the fundamental expressions in Sections 18-3 and 18-5 and from the results on scattering by atoms in Section 18-4.

For the simplest molecular scattering model we assume that the molecule is rigid and that the scattering length of the bonded atoms is the same as for free atoms. Hence, an expression for the diffracted intensity results from the superposition of the complex scattered amplitudes A^+ of all atoms in the molecule. From Eqs. (18-13) and (18-36) the amplitude of the nth atom is

$$A_n^+ = A_0 \frac{a_n}{R} \cdot e^{i\phi_n} \qquad \textbf{(18-40)}$$

Taking a diatomic molecule with atoms n and m as an example, we write for the total amplitude of the scattered radiation (from Eq. 18-15)

$$A_s^+ = A_0 \cdot R^{-1}(a_n e^{i\phi_n} + a_m e^{i\phi_m}) \qquad \textbf{(18-41)}$$

18-6.1 Diffraction by Gases We now apply Eq. (18-41) to diffraction in the gas phase. The ratio of the diffracted to incident intensities for a particular orientation of the molecule is

$$I/I_0 = A_s^+ A_s^{+*}/A_0^2 = R^{-2}[a_n^2 + a_m^2 + a_n a_m(e^{i\phi_{nm}} + e^{-i\phi_{nm}})]$$

$$= R^{-2}(a_n^2 + a_m^2 + a_n a_m \cos\phi_{nm}) \qquad \textbf{(18-42)}$$

where $\phi_{nm} = \phi_n - \phi_m$. The second line follows from the identity (IV-5) in Appendix IV. The random motion and large separations of gas phase molecules insure that waves scattered by two or more different molecules are incoherent. Therefore, the observed intensity is simply the average of Eq. (18-42) over all possible orientations of a molecule, multiplied by the total number of molecules n in the sample. Since the only orientation-dependent quantity in Eq. (18-42) is the phase shift ϕ_{nm}, the

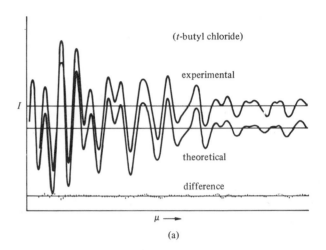

(a)

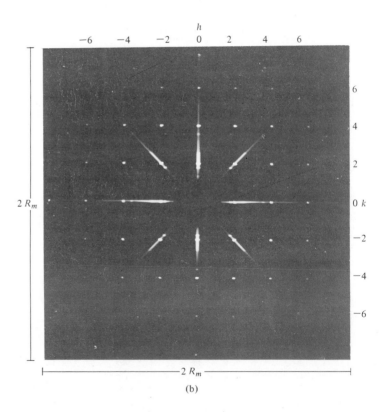

(b)

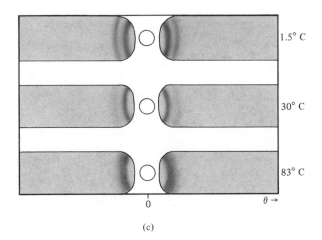

1.5° C

30° C

83° C

0                    $\theta \rightarrow$

(c)

Figure 18-8 Diffraction patterns. (a) Gas-phase electron diffraction curves of t-butyl chloride. The structure-dependent fraction of total diffracted intensity I is shown in arbitrary units. The zero intensity line of the experimental curve has been displaced with respect to the theoretical zero line. Difference curve = experimental − theoretical curve. Theoretical curve has been calculated from a molecular structure that gives optimal agreement with the experimental diffraction curve. [From R. L. Hilderbrandt and J. D. Wieser: *J. Chem. Phys.*, **55**:4648 (1971).]

(b) X-ray diffraction pattern of an NaCl crystal (face-centered cubic lattice) taken with a precession camera (see Figure 18-9a) using continuous radiation from molybdenum. The pairs of intense spots are due to the Mo K(shell)α and Mo $K\beta$ emissions (superimposed on the continuum) with wavelengths 71.07 (the outside spot in a pair) and 63.23 pm, respectively. The camera is set up so that the crystal-to-film distance M (Figure 18-9a) equals the maximum recordable center-of-film-to-diffraction-spot distance R_m. Then the distance between the crystal planes that have caused the diffraction spot at R is given by $d = \lambda R_m / R$. The diffraction spots are labelled with the indices h, k. (There are no spots for odd h, k because, due to symmetry, each elementary cell scatters with zero amplitude in directions that correspond to odd-numbered indices.) The index $l = 0$. For the spot with $\lambda = 71.07$ pm, $k = 2$, $h = 0$ it can be determined from the figure that $R_m/R = 3.97$. Hence $d = 282$ pm. The elementary cell dimension $a = \sqrt{h^2 + k^2 + l^2}\, d$ for a cubic lattice and $a = 564$ pm. [Courtesy of Professor Benjamin Post, Polytechnic Institute of New York.]

(c) Liquid-phase x-ray diffraction patterns of water at three different temperatures. The entire pattern is due to scattering by O atoms. Note the broad and indistinct diffraction rings (due to averaging over a range of O–O distances) and the absence of diffraction rings at large angles (due to the short range of molecular order in liquids). [From J. Morgan and B. E. Warren: *J. Chem. Phys.*, **6**:666 (1938).]

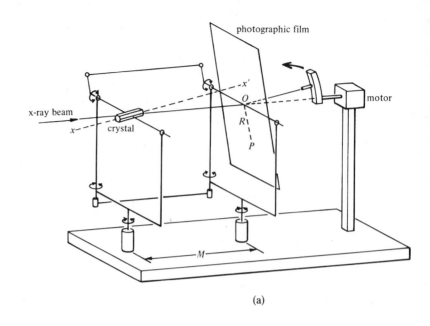

photographic film

x-ray beam

crystal

x

x'

O

R

P

motor

M

(a)

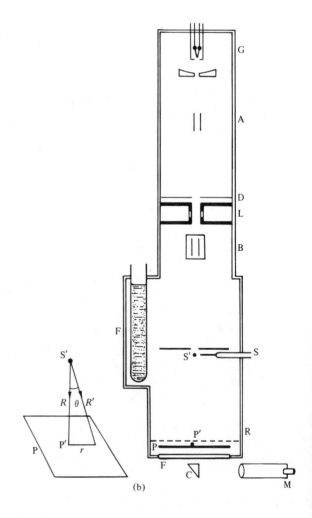

(b)

observed intensity is obtained readily from Eq. (a) of Example 18-1 and Eq. 18-42

$$I_{obs}/I_0 = \frac{n}{R^2}\left(a_n^2 + a_m^2 + a_n a_m \frac{\sin \mu r_{nm}}{\mu r_{nm}}\right) \tag{18-43}$$

where r_{nm} is the internuclear distance.

Eq. (18-43) applies to diffraction of any probe by gas phase diatomic molecules, provided the appropriate atomic scattering lengths are used. For x-ray, electron, and neutron probes, the expressions for the scattering lengths are given in Table 18-1. Typical angles of diffraction minima for the three probes are compared in Example 18-2. Eq. (18-43) may readily be extended to polyatomic molecules. An application of the extended equation to determine a low-precision structure of $SiCl_4$ from electron diffraction patterns is shown in Example 18-3. Some results from more refined methods of analysis are presented in the next section.

EXAMPLE 18-2

The bond length of the molecule Cl_2 is 200 pm. For 100 pm x rays or neutrons and for 5 pm electrons, predict from Eq. (18-43) (a) the first, (b) the fourth maxima, and (c) the first minimum of the diffraction pattern in the gas phase.

Solution: The maxima and minima are those in $\sin \mu r_{mn}$ and not necessarily in the total scattered intensity. This caution applies with particular force to electron diffraction where the total intensity contains a large component (see Eq. 18-32) which dominates the intensity-dependence on μ (or θ) and contains no information on molecular structure. This component must be eliminated in order to identify the scattering angles at which $\sin \mu r_{mn}$ has a maximum or minimum.

For the maxima we have $\sin \mu r_{mn} = 1$, which occurs for all μr_{mn} obeying the condition

$$\mu r_{mn} = (4n + 1)\pi/2 \tag{a}$$

where n is an integer. Eq. (a) combined with $\mu = (4\pi/\lambda) \cdot \sin \frac{1}{2}\theta$ yields

$$\sin \frac{1}{2}\theta = (4n + 1)\lambda/8r_{mn} \tag{b}$$

Figure 18-9 (a) A precession camera that records x-ray diffraction patterns of single crystals. The xx' axis of the crystal and the normal to the film plane precess identically about the incident x-ray beam direction. Therefore the film plane always remains perpendicular to the xx' axis.

The xx' axis is parallel with one of the three vectors defining the elementary cell, for example, **c**. A screen (not shown) with a hole in it passes diffracted beams with a single chosen value of l (see Eq. 18-46), for example, $l = 0$. The recorded diffraction spots correspond to different indices h, k. For a diffraction spot with indices $h, k, 0$ the relation between the crystal plane spacing d (Eq. 18-47), the wavelength λ, the radial distance $\overline{OP} = R$, and the crystal-to-film distance M is $d/\lambda = M/R$.

(b) A simple electron diffraction camera for gaseous samples. The chamber of the camera is evacuated. G, electron gun; A, deflector plates to switch the beam on (when it passes through the hole in diaphragm, D) and off; L, magnetic focussing lens; B, deflector plates to center the beam; S, sample injection nozzle; F, cold finger to trap injected sample; R, rotating sector for removing (approximately) the μ^{-4} (Eq. 18-32) dependence of the scattered intensity so that the $\sin \mu r$-dependence (Eq. 18-43) becomes more prominent; P, photographic plate for recording the diffraction pattern; f, fluorescent screen for visual observation of beam via prism C and microscope M. Inset defines the observables r, R from which the scattering angle θ is calculated (see Example 18-3). [From T. B. Rymer: *Electron Diffraction*. Methuen & Co., Ltd., London, 1970.]

(a) $n = 0$. For x rays or neutrons

$$\sin \tfrac{1}{2}\theta = \lambda/8r_{mn} = 100/8 \cdot 200 = 1/16$$
$$\tfrac{1}{2}\theta = 3.6° \quad \text{or} \quad \theta = 7.2°$$

Similarly for electrons

$$\theta = 0.36°$$

(b) $n = 3$. For x rays or neutrons

$$\theta = 108.7°$$

For electrons

$$\theta = 4.7°$$

(c)   At minima $\sin \mu r_{mn} = -1$. This occurs at all μr_{mn} obeying the condition

$$\mu r_{mn} = (4n + 3)\pi/2 \tag{c}$$

which, solved for $\sin \tfrac{1}{2}\theta$, yields

$$\sin \tfrac{1}{2}\theta = (4n + 3)\lambda/8r_{mn} \tag{d}$$

The first minima ($n = 0$) for x rays or neutrons are

$$\theta = 21.6°$$

For electrons

$$\theta = 1.07°$$

Thus we see that for x-ray and neutron diffraction the minima and maxima are at wide angles and widely spaced, whereas the opposite is true for electron diffraction. This difference is reflected in the design of the respective diffraction cameras (Figure 18-9).   ●

EXAMPLE 18-3

Determine the Si–Cl bond length from the following electron diffraction data for $SiCl_4$ (computed by Rymer [7, p. 130] from results of Brockaway and Wall [11]).

The diffraction pattern of $SiCl_4$ vapor due to electrons with wavelength $\lambda = 6.06$ pm, in a diffraction camera with $R = 121.9$ mm (see Figure 18-9b), consists of bright and dark rings with these radii:

bright (r mm^{-1})	2.8	**5.0**	7.4	9.5	**11.6**	14.0	**16.3**	18.7
dark (r mm^{-1})		3.9	6.3	8.5	10.6	12.9	—	17.5

The radii in boldface type are for the rings with highest intensities.
Solution: We extend Eq. (18-43) to five atoms and rewrite it with help of Eq. (18-38) to get the differential cross section for elastic scattering

$$\frac{d\sigma}{d\omega} = \frac{IR^2}{I_0} = \sum_{n=1}^{5} a_n^2 + \sum_{n=1}^{5} \sum_{m>n}^{5} a_n a_m \frac{\sin \mu r_{nm}}{\mu r_{nm}} \tag{a}$$

We assume that the scattering lengths may be approximated by their values in the limit of large μ (see Figure 18-5 and Eq. 18-32), that is, $a_n \approx 2Z_n a_0^{-1} \mu^{-2}$.

Therefore,

$$\frac{d\sigma}{d\omega} = \frac{4}{a_0^2 \mu^4}\left(\sum Z_n^2 + \sum\sum Z_n Z_m \frac{\sin \mu r_{nm}}{\mu r_{nm}}\right) \tag{b}$$

If we assume further, in analogy with CCl_4, that $SiCl_4$ is tetrahedral, the molecule has only two unequal interatomic distances—r_{SiCl} and r_{ClCl}. There are four r_{SiCl} and six r_{ClCl} distances. Hence, Eq. (b) becomes

$$\frac{d\sigma}{d\omega} = \frac{4}{a_0^2 \mu^4}\left[Z_{Si}^2 + 4Z_{Cl}^2 + 4Z_{Si}Z_{Cl}\frac{\sin \mu r_{SiCl}}{\mu r_{SiCl}} + 6Z_{Cl}^2\frac{\sin \mu r_{ClCl}}{\mu r_{ClCl}}\right] \tag{c}$$

Eq. (c) shows that the intensity ratio of the last two terms is, approximately

$$\frac{4Z_{Si}Z_{Cl}}{6Z_{Cl}^2} = \frac{4\times14\times17}{6\times17\times17} = 0\cdot55$$

The Cl–Cl interference terms may be expected to dominate the interference pattern.

To calculate r_{ClCl} from the bright rings, we solve for r from formula (b), Example 18-2, substitute the given values of λ, R and get

$$r_{nm} = (4n+1)\frac{\pi}{2\mu} = (4n+1)\frac{\lambda}{8\sin\frac{1}{2}\theta} \approx (4n+1)\frac{\lambda R}{4r} = (4n+1)\frac{184.7}{(r\ \text{mm}^{-1})}\text{pm} \tag{d}$$

where we have assumed $\frac{1}{2}\theta$ to be small, thus $R \approx R'$ in Figure 18-9b and $\sin\frac{1}{2}\theta \approx \frac{1}{2}r/R$.

Eq. (d) is our working equation for the bright rings. For the first ring we calculate $r_{ClCl} = (4n+1)184.7/2.8 = (4n+1)\ 66.0$ pm. If $n = 0$ we would get an unreasonably short value for a nonbonded (or even a bonded) interatomic distance. With $n = 1$, $r_{ClCl} = 330$ pm which is reasonable. If we assume $n = 2$ through 8 for the remaining successive bright rings we get for r_{ClCl} a self-consistent set of values: 332, 325, 330, 334, 330, 329, 326.

From Eq. (d) Example 18-2, we can get the working equation for the dark rings

$$r_{nm} = (4n+3)\frac{184.7}{(r\ \text{mm}^{-1})} \tag{e}$$

from which we can calculate from the successive dark rings, with assumed $n = 1$ through 5 and 7 for the Cl–Cl distances, the values 332, 322, 326, 331, 332, 327 pm. The average of all values yields the best estimate for the Cl–Cl distance: 329 pm with a standard deviation of 3 pm. The consistency of the results supports our interpretation of the pattern.

The relationship between r_{ClCl} and r_{SiCl} is, from the geometry of a tetrahedron

$$r_{SiCl} = r_{ClCl}/2\sin\frac{1}{2}\alpha \tag{f}$$

where $\alpha = 109.5°$ is the tetrahedral angle. From Eq. (f) and $r_{ClCl} = 329$ pm, we get for r_{SiCl} a reasonable value, that is, 201 pm.

To find the expected position of the bright rings due to Si–Cl interferences we use our r_{SiCl} value in Eq. (d) to calculate r. The results for $n = 1, 3, 4$ are $r = 4.6, 11.9, 15.6$ mm. These distances are very close to the radii of the three brightest rings. Hence the result of the Si–Cl interference maxima (bright rings) is to enhance the intensity of the Cl–Cl maxima if the two happen to fall close together and, possibly, to shift the positions of the dark rings slightly. ●

18-6.2   Diffraction by Solids   We now briefly consider diffraction from crystalline solids. To describe the diffraction pattern (Figure 18-8b) of a particular solid crystal in detail is a technical problem whose solution is determined by the shape (Figure 17-3) and atomic contents of the crystal's elementary cell. However, (as for gases) the solution still follows from the principle of adding the scattered amplitudes of all atoms in the sample. In the first stage, analogous to computing the amplitude for a single gas molecule, we might add the amplitudes scattered from all atoms in a single elementary cell. In the next stage we then add the amplitudes from all elementary cells in the sample crystal. (For gases we added intensities at this stage because the radiation scattered from different gas molecules is not coherent.) Since there are many elementary cells in a crystal the resultant amplitudes (and hence intensities) can reach very high values at favored scattering angles θ, η (see Figure 18-2).

The favored scattering angles θ, η are those at which all elementary cells in a sample crystal scatter in phase. If the points 0, b in Figure 18-2 represent two symmetrically equivalent positions in two different elementary cells, then from Eq. (18-11) the condition for in-phase scattering by the two points is

$$\mathbf{r} \cdot (\mathbf{s} - \mathbf{s}_0) = m\lambda \qquad\qquad \textbf{(18-44)}$$

where m is an integer. The distance vector between the two points is given by an extension of Eq. (17-1) to three dimensions, that is,

$$\mathbf{r} = p\mathbf{a} + q\mathbf{b} + s\mathbf{c} \qquad\qquad \textbf{(18-45)}$$

where p, q, s are integers and $\mathbf{a}$, $\mathbf{b}$, $\mathbf{c}$ are the vectors that specify the shape of the elementary cell. Substitution of Eq. (18-45) into (18-44) shows that, if Eq. (18-44) is to be valid simultaneously between all pairs of the equivalently located points (that is, for any p, q, s), then we must have

$$\mathbf{a} \cdot (\mathbf{s} - \mathbf{s}_0) = h\lambda \qquad \mathbf{b} \cdot (\mathbf{s} - \mathbf{s}_0) = k\lambda \qquad \mathbf{c} \cdot (\mathbf{s} - \mathbf{s}_0) = l\lambda \qquad \textbf{(18-46)}$$

where h, k, l are integers.[4] Eqs. (18-46) are the **Laue equations** which specify the possible orientations of the elementary cell, relative to $\mathbf{s} - \mathbf{s}_0$, necessary for a strong diffracted beam (Figure 18-10).

An alternate way to specify the necessary condition for a strong diffracted beam is as follows: consider the case $p = q = s = 1$. In this case the lower set of points × and the upper set of points × in Figure 18-10 define two adjacent parallel planes normal to $\mathbf{s} - \mathbf{s}_0$. The distance d between the two planes is the projection of $\mathbf{r}$ on $\mathbf{s} - \mathbf{s}_0$. Since the magnitude of $\mathbf{s} - \mathbf{s}_0$ is (from Figure 18-2) $2 \sin \frac{1}{2}\theta$ we get, from Eqs. (18-44) through (18-46)

$$2d \sin \tfrac{1}{2}\theta = 2d \sin \theta' = m\lambda = (h + k + l)\lambda \qquad \textbf{(18-47)}$$

where we have defined the Bragg angle θ'. **Bragg's equation,** Eq. (18-47), shows (Figure 18-10) that strong diffracted beams may be expected at angles of reflection θ' from sets of parallel planes which obey the condition (18-47). (Note from Figure 18-10 that the angle η specifies the condition that the reflection planes be normal to the plane of the paper.)

Eqs. (18-46) and (18-47) are necessary, but not sufficient for a strong diffracted beam. It may happen, due to the arrangement of atoms within the elementary cell,

---

[4] These integers are the **Miller indices,** see page 731.

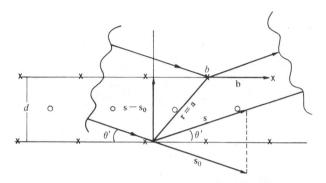

Figure 18-10 Scattering of radiation from a crystal. [This drawing is a rotated (about an axis normal to the plane of paper) and otherwise modified version of Figure 18-2.] Vectors **a**, **b**, **c** (**c** not shown because it is normal to plane of the paper) define the elementary cell of a monoclinic crystal (Figure 17-3). Symbols: d, the distance between two planes (normal to plane of paper) made up of identical atoms, ×. The plane of different atoms (○) would scatter out of phase with the × planes if, in Eq. 18-47, m = odd for the × planes. If the two kinds of atoms × and ○ have nearly the same atomic number the total scattered intensity for m = odd will be nearly zero. For definition of **s**, **s**$_0$, **r**, and η see Figure 18-2.

that the cell scatters zero amplitude at the Bragg angle. In that case, the diffracted intensity will be zero as well (see the example in Figure 18-10).

In favorable cases, diffraction patterns of liquids (Figure 18-8c) show interference of waves scattered by nearest and next-nearest neighbor molecules. These interferences are not as strong as in the case of solids, indicating that ordering of molecules in liquids extends over a short range compared to solids. Since the intra- and intermolecular features of the diffraction pattern cannot be separated without additional information, diffraction by liquids cannot serve to determine molecular structure.

18-7 MOLECULAR GEOMETRY

In this section we shall list and compare the major experimental and theoretical sources of molecular geometry data and discuss the effect of molecular vibrations on structure determinations.

18-7.1 Experimental Geometries and Molecular Vibrations The major experimental sources of molecular bond lengths and angles are rotational constants (from microwave, infrared, Raman, and optical spectroscopy) and electron diffraction patterns for gaseous molecules and x ray (also neutron) diffraction patterns for solid crystals. Less importantly, for liquids one can use NMR to obtain ratios of interatomic distances within a molecule. If one distance can be assigned from other information, the absolute values of the others can be obtained.

Table 18-2 Comparison of molecular geometries on the rigid molecule model

Molecule	Structural Parameters		
		Experimental Values $(R \text{ pm}^{-1})$	Experimental Method[a]
H_2	R_{HH}	75.11	visible–UV
		75.66	ED
OCS	R_{OC}, R_{CS}	116.47, 155.76	MW ($^{16}O^{12}C^{32}S$, $^{16}O^{12}C^{34}S$)
		115.52, 156.53	MW ($^{16}O^{12}C^{32}S$, $^{18}O^{12}C^{32}S$)
SO_2	R_{SO}, ∢OSO	143.21, 119.53°	MW
		143.51, 119.22°	ED
C_6H_6	R_{CC}, R_{CH}	139.7, 108.4	IR and Raman
		(139.8),[b] 114.2	NMR
		140.1, 111.6	ED
NH_4F	R_{NH}, R_{NF}	92, 270.8	x rays
(in crystal)	∢HNH	108.9°	
		104.0, 270.7	neutrons
		109.6°	

[a] ED, electron diffraction; MW, microwave.

[b] Assumed value.

DATA SOURCES: H_2 [G. Herzberg: *Molecular Spectra and Molecular Structure*. Van Nostrand, New York, 1950, vol. I. S. H. Bauer in H. Eyring et al. (eds.): *Physical Chemistry*. Academic Press, New York, 1967, 11 vols.]. OCS [C. H. Townes, A. N. Holden, and F. R. Merritt: *Phys. Rev.*, **74**:1113 (1948)]. SO_2 [V. W. Laurie and D. R. Herschbach: *J. Chem. Phys.*, **37**:1687 (1962); K. Hedberg in D. R. Lide, Jr., and M. A. Paul (eds.): *Critical Evaluation of Chemical and Physical Structural Information*. National Academy of Sciences, Washington, D.C., 1974]. C_6H_6 [L. C. Snyder and S. Meibom in D. R. Lide, Jr., and M. A. Paul (as given)]. NH_4F [S. H. Bauer in Eyring et al. (eds.)]

In Table 18-2 we compare the data on molecular geometries obtained from the various experimental sources. In all cases the interatomic distances have been calculated from a rigid molecule model. The inconsistencies between the data from different sources are beyond experimental error. The major reason for the greatly different N—H distances in NH_4F is the use of free atom form factors in the analysis of the x-ray diffraction data. Since an appreciable fraction of the H atom charge has shifted into the NH bond, the center of this charge is closer to the N atom than the position of the proton. The remaining differences between structural data from the various sources are due to neglect of molecular vibrations.

We shall now discuss the effect of molecular vibrations on experimental molecular structures. Consider a diatomic molecule with a potential energy curve such as the one shown in Figure 14-3. The equilibrium interatomic distance R_e (R_0 in Chapter 14) of the molecule is at the minimum of the curve. The instantaneous distance R is

$$R = R_e\left(1 + \frac{R - R_e}{R_e}\right) = R_e(1 + \xi) \tag{18-48}$$

As shown in Table 18-3 the different experimental methods measure the average of

Table 18-3 Diatomic molecules: a comparison of variously defined interatomic distances

Interatomic Distance	Expression[a]	Major Source
equilibrium, R_e		ab initio calculations
average, $\langle R \rangle_v$	$R_e(1 + \langle \xi \rangle_v)$	calculated from experimental data[b]
effective, R_v	$R_e(1 + \langle \xi \rangle_v - \frac{3}{2}\langle \xi^2 \rangle_v)$	rotational constants[c]
inverse cube, $\langle R^{-3} \rangle_v^{-1/3}$	$R_e(1 + \langle \xi \rangle_v - 2\langle \xi^2 \rangle_v)$	NMR

[a] See text for definitions of symbols.
[b] Diffraction patterns give distances (R_a) which usually are within 0.1 pm of the average distance (R_g) in all molecules in the sample. At 0 K $R_g = \langle R \rangle_0$.
[c] For polyatomic molecules, effective interatomic distances are not unique; they depend upon the particular isotopic species chosen (see OCS data in Table 18-2).

the nth power of R or a closely related quantity. For example, a rotational constant determined with the help of the rigid rotor formula Eq. (11-52), from measured frequencies is not proportional to R_e^{-2}, but rather to $\langle R^{-2} \rangle_v$ where the average is over the vth vibrational state. Therefore, the distance determined from the experimental rotational constant and Eq. (11-53) is an **effective** interatomic distance

$$R_v = \langle R^{-2} \rangle_v^{-1/2} \qquad (18\text{-}49)$$

We wish to relate the theoretically fundamental interatomic distance R_e to experimentally determined quantities of the form $\langle R^n \rangle^{1/n}$ such as R_v. Assuming that $\xi \ll 1$, we get from Eq. (18-48) by means of a binomial expansion (see Appendix III)

$$\langle R^n \rangle = R_e^n \langle (1 + \xi)^n \rangle = R_e^n (1 + n\langle \xi \rangle + \tfrac{1}{2}n(n-1)\langle \xi^2 \rangle + \cdots) \qquad (18\text{-}50)$$

Therefore, by a second binomial expansion

$$\langle R^n \rangle^{1/n} = R_e(1 + \langle \xi \rangle + \tfrac{1}{2}(n-1)\langle \xi^2 \rangle + \cdots) \qquad (18\text{-}51)$$

To check Eq. (18-51) we may observe that it reduces to the exact expression for the average distance ($n = 1$)

$$\langle R \rangle = R_e(1 + \langle \xi \rangle) \qquad (18\text{-}52)$$

The averages in the vth vibrational state of the fractional displacement ξ and of its square are given to a good approximation by

$$\langle \xi \rangle_v = \left(\frac{\alpha_e}{2B_e} + \frac{3B_e}{\nu_e} \right)(v + \tfrac{1}{2}) = -a_1(3B_e/\nu_e)(v + \tfrac{1}{2}) \qquad (18\text{-}53)$$

and

$$\langle \xi^2 \rangle_v = (2B_e/\nu_e)(v + \tfrac{1}{2}) \qquad (18\text{-}54)$$

where B_e is the rigid rotor rotational constant (calculated from R_e), ν_e is the harmonic vibrational frequency, α_e is the rotation-vibration interaction constant, and a_1 is an anharmonic constant, proportional to $(d^3U/dR^3)_{R_e}$ in the Taylor expansion of the vibrational potential energy, Eq. (14-37). The preceding constants appear in an approximate expression (compare with Eq. 14-45) for the rotation-vibration energy of a nonrigid anharmonic diatomic molecule:

$$E_{vJ} = \boldsymbol{h}\nu_e(v+\tfrac{1}{2}) + (B_e - \alpha_e(v+\tfrac{1}{2}))J(J+1) - \boldsymbol{h}\nu_e x_e(v+\tfrac{1}{2})^2 \qquad (18\text{-}55)$$

Thus, given a sufficient number of accurate transition frequencies between rotation-vibration levels, the constants can be measured experimentally.

The significance of Eqs. (18-53), and (18-54) for structure determination is as follows. The average displacement, $\langle \xi \rangle$, is zero if the anharmonic constant a_1 is zero. That is, there is no harmonic contribution to $\langle \xi \rangle$ because the displacements of a harmonic oscillator are symmetric about R_e (see Figure 14-3). On the other hand, Eq. (18-54) has a harmonic contribution only, the anharmonic one being negligibly small. Since these results apply to polyatomic molecules as well (for which anharmonic constants are difficult to measure although vibrational frequencies and rotational constants are available), the average structure recommends itself as one to which data from various sources could be reduced to achieve consistency. The different structures are compared in Table 18-4.

Table 18-4 Comparison of various molecular geometries

Molecule		Structural Parameters (R pm^{-1})		
		Equilibrium	Average	Effective
HF		91.70	93.36	92.57
DF		91.71	92.84	92.34
TF		91.77	92.72	92.30
H$_2$O	R_{OH}, $\sphericalangle$HOH	95.72, 104.5°	97.35, 104.3°	95.79, 104.9°
D$_2$O			96.83, 104.4°	95.71, 104.9°
CH$_4$	R_{CH}	108.5	109.9[a]	109.4
CD$_4$		108.6	109.6[a]	109.4
^{12}C^{16}O		112.82	113.23	113.09
^{12}C^{18}O		112.82	113.22	113.08
SO$_2$	R_{SO}, $\sphericalangle$OSO	143.08, 119.32°	143.49, 119.21°	143.36, 119.42°

[a] Electron diffraction R_g distances (defined in footnote b, Table 18-3) are 110.7 pm for CH$_4$, and 110.3 pm for CD$_4$ (S. H. Bauer in H. Eyring et al. (eds.): *Physical Chemistry*. Academic Press, 1967–1975, vol. IV).
DATA SOURCE: V. W. Laurie and D. R. Herschbach: *J. Chem. Phys.*, **37**:1687 (1962).

The following observations on the results in Table 18-4 are generally valid. Equilibrium parameters are the same, within experimental error, for different isotopic species of a molecule. This is to be expected since all species have the same potential energy surface. The average interatomic distances are invariably larger than the equilibrium ones since the potential energy curve is less steep for $R > R_e$ than for $R < R_e$. Among various isotopic species of a molecule, the heavier ones have smaller $\langle R \rangle$ since they have smaller zero point energies and therefore are "deeper down" in the potential energy well where oscillation amplitudes are smaller. For a similar reason, apparent hydrogen–other atom bond lengths exhibit large variations from molecule to molecule compared to bond lengths not involving H atoms.

18-7.2 Theoretical Geometries Theoretical (ab initio) prediction of molecular geometries has reached an acceptable level of accuracy in very recent times. Such predictions will be very useful for the study of the geometries of reactive species— reaction intermediates such as CH_2 (page 698), excited states (Table 18-5), and others which are difficult or impossible to study experimentally.

Table 18-5 Ab initio molecular geometries

Molecule	Equilibrium Structural Parameters (R pm^{-1})	
	Ab Initio	Experimental
Extended basis set-configuration interaction calculations		
HF	92.0	91.70
F_2	141	141.7
CH $\begin{cases} X^2\Pi \text{ (ground state)} \\ A^2\Delta \\ B^2\Sigma^- \\ C^2\Sigma^+ \end{cases}$	111.8	112.0
	110.2	110.2
	117.3	116.4 (CD: 117.3)
	111.1	111.4
DZ-self-consistent field calculations		
CH_3NC' R_{CH}, $\sphericalangle HCN$	108.1, 110.0°	110.1, 109.1°[a]
R_{CN}, $R_{NC'}$	196.7, 116.7	196.2, 116.6[a]

[a] These are the so-called (isotopic) substitution parameters. Their values fall between the values of effective and equilibrium parameters.
DATA SOURCE: H. F. Schaefer III in D. R. Lide, Jr., and M. A. Paul (eds.): *Critical Evaluation of Chemical and Physical Structural Information*. National Academy of Sciences, Washington, D.C., 1974.

Ab initio geometries are calculated with the help of the electronic energy from the Schrödinger equation and from a knowledge of the number, charges, and masses of electrons and nuclei in a molecule. The electronic energy of a state (n) plus the nuclear repulsion energy is, within the Born–Oppenheimer approximation, the potential energy U_n for the nuclear motion (see Eq. 14-18). U_n is a minimum at the equilibrium (R_e) molecular geometry, which is specified by $3N-6$ interatomic distances (or equivalent parameters) for a molecule with N atoms. Thus, the calculation of a molecular geometry is a trial-and-error search along $3N-6$ (or less if the molecule is known to have symmetry) independent coordinates for the minimum in the potential energy surface U_n.

This very complex and computer-time consuming search has been made feasible by the discovery that approximate potential energy surfaces exist with minima which coincide to a high degree of approximation with those of the exact surface. First, identical errors in approximately computed molecule and constituent atom energies will cancel if we write

$$\Delta U_n = U_n - U_a = U_n - \sum_A U_A \qquad (18\text{-}56)$$

where U_a is the sum of the energies of the atoms into which the nth state of the molecule dissociates. Furthermore, the potential energy ΔU_n may be decomposed

$$\Delta U_n = \Delta U_{HF} + \Delta U_{corr} + \Delta U_{rel} \qquad (18\text{-}57)$$

where the three terms on the right-hand side denote the differences between the molecule and the constituent atoms of their Hartree–Fock energies (pages 383, 447), the correlation energies (page 448), and the relativistic correction (to the non-relativistic Schrödinger equation) energies. $\Delta U_{rel} \approx 0$ since the relativistic correction energy is appreciable only for core electrons which are not affected by bond formation. Thus we see that the first two right-hand side terms of Eq. (18-57) should closely approximate the minima of U_n. The two terms can be calculated rather accurately for small molecules by means of configuration interaction (CI, page 449) if the wave functions of the atoms and the molecule are expanded in terms of the same large set of functions (termed an extended basis set—EBS). Some equilibrium geometries from such calculations are listed in Table 18-5.

Interatomic distances accurate to 5 pm or better can be obtained from the ΔU_{HF} term alone for molecules which are formed without the pairing of electron spins, for example ionic molecules such as NaCl (formed from Na^+ and Cl^-), also HCl^+, Cl_2^-, KrF. An additional fact of importance is that Hartree–Fock distances are, in general, the lower limits to the true equilibrium distances.

For the remaining molecule types it is found empirically that better than Hartree–Fock geometries are obtained with an even more approximate potential energy—ΔU_{DZ}. This energy is calculated by a self-consistent field (SCF) method from the molecular and atomic wave functions expanded in terms of the so-called double zeta (ζ) basis set, denoted DZ. The quantity ζ is the exponential coefficient of the radial part of the Slater function, page 385. As an example, for the carbon atom, a minimum basis set contains three different radial functions or zeta's: one each for the $1s$, $2s$, $2p$ orbitals. A DZ set of carbon would contain six zeta's. An example of a DZ–SCF structure is shown in Table 18-5.

_____ *Problems*

1. Compute for an x ray photon, electron, and neutron
   (a) the range of momenta
   (b) the range of velocities
   (c) the range of wavelengths corresponding to the ranges of kinetic energy listed in Table 18-1.
2. The power of the visible light emitted by a ruby laser is about 10^7 W; that of the x-ray radiation from an x-ray tube about 1 W. If in each case all radiation falls on a surface 10 cm from the source through a solid angle equal to $\frac{1}{2}\pi$, what is
   (a) the intensity
   (b) the maximum electric field at the surface in each case?

3. Derive the intensity formula, Eq. (18-6) for a sine wave from Eq. (18-5).
4. Derive the expression for the average intensity of unpolarized x-ray radiation scattered by an electron (Eq. 18-24) from Eq. (18-22).
5. What is the ratio of the intensities of scattered and incident unpolarized x-ray radiation at scattering angles $0°$, $90°$ if the scatterer is an electron? Assume that scattered intensity is measured at a distance of 10 cm from the electron.
6. Show, from Eq. (18-27) that the x-ray form factor of the H atom is $f = (1 + \frac{1}{4}\mu^2 a_0^2)^{-2}$ where a_0 is the Bohr radius (see Figure 18-5). *Hint*: First evaluate the integral equivalent to Eq. (18-27) in which $\sin \mu r$ has been replaced by $\exp(i\mu r)$.
7. Plot I_a/I_0 versus θ (for $R = 10$ cm) and I_a/I_e versus μ for 1-Å x rays scattered by an H atom (see problem 6 for the form factor of H).
8. Plot I_a/I_0 versus θ (assume $R = 10$ cm) for
   (a) 0.005-nm electrons scattered by an H atom,
   (b) 0.1-nm neutrons scattered by an H and by a D atom.
   (The neutron scattering lengths are 4 fm for the H atom, 6 fm for the D atom.)
9. What is the wavelength of x rays scattered incoherently at $0°$, $90°$, $180°$ if the wavelength of incident rays is 5 pm?
10. Derive expressions for the differential and total scattering cross sections for unpolarized electromagnetic radiation scattered by an electron. What is the magnitude of the total cross section?
11. (a) Compute and plot the theoretical electron diffraction curve of $y = (a_0^2\mu^4/4)(d\sigma/d\omega)$ versus $x = \mu r_{SnCl}$. Approximate the atomic scattering length a_n by $2Z_n a_0^{-1}\mu^{-2}$ (see Example 18-3) and disregard the x-independent terms in the expression for the differential cross section. The resulting curve approximates the experimental intensity curve obtained with the rotating sector (Figure 18-9b). What is the scattering mechanism that corresponds to the approximation for a_n?
   (b) Determine the SnCl bond length from the following data on the electron diffraction pattern of $SnCl_4$ [11] due to electrons with wavelength $\lambda = 6.04$ pm with $R = 121.9$ mm (see Figure 18-9b). Radii of rings

bright (r mm^{-1})	4.30		6.59		10.3		13.7		16.7
dark (r mm^{-1})		5.55		9.12		11.8		15.3	

12. Calculate the Na^+–Cl^- distance in the NaCl crystal from the Mo $K\beta$ (0.63225 Å) radiation spots in Figure 18-8b.
13. Compute the quantities $\langle \xi \rangle_v$, $\langle \xi^2 \rangle_v^{1/2}$, $\langle R \rangle_v$, R_v, $\langle R^{-3} \rangle_v^{1/3}$ for $v = 0$, 1 of HCl $(\nu_e = 2989.74$ cm^{-1}, $B_e = 10.5909$ cm^{-1}, $\alpha_e = 0.3019$ cm^{-1}, $R_e = 127.460$ pm) and $Cl_2(\nu_e = 564.9$ cm^{-1}, $B_e = 0.2438$ cm^{-1}, $\alpha_e = 0.0017$ cm^{-1}, $R_e = 198.8$ pm).

References and Recommended Reading

[1] J. G. Brown: *X-rays and Their Applications*. Plenum Press, New York, 1966.
[2] M. Davies: *Electrical and Optical Aspects of Molecular Behaviour*. Pergamon Press, Oxford, 1965. Supplements material in this chapter. Determination of molecular properties from bulk measurements of the index of refraction, the dielectric constant, etc.

[3] J. A. Kapecki: *J. Chem. Educ.*, **49**:231 (1972). An introduction to x-ray structure determination. Introductory discussion of x-ray diffraction as a means of structure determination.

[4] D. R. Lide, Jr. and M. A. Paul (eds.): *Critical Evaluation of Chemical and Physical Structural Information.* National Academy of Sciences, Washington, D.C., 1974.

[5] W. M. McIntyre: *J. Chem. Educ.*, **41**:526 (1964). X-ray crystallography as a tool for structural chemists.

[6] M. H. Pirenne: *The Diffraction of X-rays and Electrons by Free Molecules.* Cambridge University Press, London, 1946.

[7] T. B. Rymer: *Electron Diffraction.* Methuen & Co., London, 1970.

[8] H. F. Schaefer III: *The Electronic Structure of Atoms and Molecules. A Survey of Rigorous Quantum Mechanical Results.* Addison-Wesley, Reading, Pa., 1972.

[9] J. Waser: *J. Chem. Educ.*, **45**:446 (1968). Pictorial representation of the Fourier method of x-ray crystallography. Uses diagrams to explain the Fourier method.

[10] L. J. Wood: *J. Chem. Educ.*, **8**:952 (1931). Construction and operation of a simple x-ray spectrograph.

[11] L. O. Brockway and F. T. Wall: *J. Am. Chem. Soc.*, **56**:2373 (1934).

Part IV

Statistical Thermodynamics

chapter 19

Statistical Thermodynamics

19-1 INTRODUCTION

As we have seen in Part II, classical thermodynamics deals with macroscopic systems only. It begins with some basic postulates called the laws of thermodynamics, and, from these and few experimental facts, derives relationships between the observed macroscopic properties of a system. It is recognized, however, that thermodynamics cannot explain why a system has the properties it does.

An explanation of this can be found in statistical mechanics which applies the result of quantum mechanics to the evaluation of the properties of a macroscopic system from the properties of its constituent particles (atoms, molecules, and so on). Statistical mechanics enables us to calculate the bulk properties of a system from fundamental constants and microscopic properties. Since the number of particles is extremely large, it is understood that the evaluation must be of a statistical nature; all results represent statistical averages only. Such averages are, however, very precise.

Quantum mechanics associates a wave function with each particle or, better, with each microscopic system. This function, described by a set of quantum numbers, specifies exactly the state of the microscopic system, that is, the quantum state. Different combinations of quantum numbers, each combination designating a quantum state, can lead to the same energy. The quantum states having the same energy may be grouped together into an energy level. The number of quantum states belonging to the same energy level is called the degeneracy or statistical weight of that level, g.

A basic assumption of quantum statistics is that all quantum states are equally probable. Due to the different degeneracies of the levels, however, the energy levels are not equally probable.

Since the particles of a closed system are to be found in one or another energy level, it follows that

$$N = \sum_j n_j \qquad \textbf{(19-1)}$$

and

$$U = \sum_j \varepsilon_j n_j \qquad \textbf{(19-2)}$$

where N and U denote the total number of particles and the total energy of the system, respectively, n_j and ε_j are the number of particles and energy associated with level j.

The number of ways of distributing the N particles among the many different levels, satisfying Eqs. (19-1) and (19-2), is extremely large. As is shown next, however, not all the possible distributions are equally probable.

571

Let us consider the fictitious case of a three-particle system sharing the energy equal to three energy units. Let us also assume that the spacing between levels is equal to one energy unit; the levels are not degenerate; $g_j = 1$; and there is no physical limitation to the number of particles in a certain state. All possible distributions are given in Table 19-1.

Table 19-1 Distribution of three energy units among three particles

Energy Level	Number of Particles in Each Energy Level		
	Distribution 1	Distribution 2	Distribution 3
3	1	0	0
2	0	1	0
1	0	1	3
0	2	1	0

If the particles were distinguishable, it is obvious that distribution 1 could occur in three different ways, distribution 2 in six different ways, and distribution 3 in one way. The number of ways in which a certain distribution can occur—the number of possible microstates per macrostate—is called the thermodynamic probability, w. It may vary between one and infinity. Thermodynamic probability is different from the mathematical probability, which varies between zero and one. Thus $w_1 = 3$, $w_2 = 6$, and $w_3 = 1$; distribution 2 is therefore twice as probable as distribution 1 and six times as probable as distribution 3. Distribution 2 represents the most probable—the equilibrium—distribution; w corresponding to this distribution, denoted as w_{mp}, represents $\frac{6}{10}$ of the total w. The difference between w_{mp} and all the other distributions becomes more noticeable in systems containing large numbers of particles. In systems in which the number of particles approaches Avogadro's number ($N = 6 \times 10^{23}$), w_{mp} becomes so dominant that all the other w's can be ignored when compared with it (see Example 19-2); thus

$$\frac{w_{mp}}{\Sigma w} \approx 1$$

When the number of particles is very large, the evaluation of w for the different distributions is not as simple and the following formula is applied

$$w = \frac{N!}{n_0! n_1! n_2! n_3! \cdots} = \frac{N!}{\prod_j n_j!} \tag{19-3}$$

where n_j is the number of particles in level j and $\prod_j$ represents the product over all the levels. From mathematics, $x! = x(x-1)(x-2)(x-3) \cdots (x-x+2)(x-x+1)$ and as is obvious $1! = 1$. By definition also $0! \equiv 1$.

When the levels are degenerate, $g_j \neq 1$, and Eq. (19-3) must be modified. In such a case the jth level has g_j states and thus each particle has g_j different possibilities of entering the level; the total number of different arrangements is $g_j^{n_j}$ and the modified formula is

$$w = N! \prod_j \left(\frac{g_j^{n_j}}{n_j!} \right) \tag{19-4}$$

Distinguishable particles with no limitation on their number in a state, called **boltzmanons** or **maxwellons**, obey Eq. (19-4) exactly. The statistics dealing with such particles is the Maxwell–Boltzmann statistics.

Real physical particles are, however, indistinguishable and may belong to one of the following two groups:

1. No physical limitation is put upon the number of particles in a state. Using the language of quantum mechanics, we say that any number of such particles may have the same wave function. These particles are called **bosons** and the statistics dealing with them is the Bose–Einstein statistics. Composite particles built up from an even number of fundamental particles (protons, electrons, neutrons) behave like bosons, for example, deuterium nuclei, ^{4}He nuclei, and so on.

2. No more than one particle may exist in a single quantum state; thus the quantum state is either empty or occupied by one particle. Such particles are called **fermions** and the statistics dealing with them is the Fermi–Dirac statistics. The fundamental particles such as electrons, neutrons, and protons as well as composite particles built up from an odd number of fundamental particles belong to this group. Recall Pauli's exclusion principle, which states that no two electrons in the same atom can have the same set of four quantum numbers. Consequently, the electrons must be in different quantum states.

The number of ways of distributing n_j bosons among the g_j states of level j is

$$\frac{(g_j + n_j - 1)!}{(g_j - 1)! n_j!}$$

The product over all the levels is the thermodynamic probability

$$w = \prod_j \left[\frac{(g_j + n_j - 1)!}{(g_j - 1)! n_j!} \right] \tag{19-5}$$

For fermions, $g_j \geq n_j$, and the number of ways of distributing n_j particles among the g_j single states of level j is

$$\frac{g_j!}{(g_j - n_j)! n_j!}$$

and the thermodynamic probability of a certain distribution is

$$w = \prod_j \left[\frac{g_j!}{n_j!(g_j - n_j)!} \right] \tag{19-6}$$

19-2 THE DISTRIBUTION LAWS

For a system containing a constant number of particles, N, boltzmanons, bosons, or fermions, and having a constant internal energy, U, and volume, V, we may write

$$N = \sum_j n_j$$

and

$$dN = \sum_j dn_j = 0 \tag{19-7}$$

From the condition that

$$U = \sum_j n_j \varepsilon_j$$

is constant it follows that

$$dU = \sum_j \varepsilon_j \, dn_j = 0 \tag{19-8}$$

since ε_j, the energy of level j, is constant. As we can prove from the Schrödinger equation, the magnitude of the energy of a certain level, ε_j, is only a function of the boundary parameters of the system; see the solution of the Schrödinger equation in Chapters 10 and 11. For the translational energy, for example, the boundary parameter is the volume or length. If these parameters are fixed, the energy ε_j remains constant.

If the particles behave like boltzmanons, the thermodynamic probability, defined also as the number of microstates per macrostate, is given by Eq. (19-4). Taking the logarithm of this equation, we have[1]

$$\ln w = N \ln N - N + \sum_j [(n_j \ln g_j) - (n_j \ln n_j) + n_j] \tag{19-9}$$

since the logarithm of a product is equal to the sum of logarithms. For an infinitesimal change in $\ln w$ the result is

$$d \ln w = \sum_j (\ln g_j - \ln n_j) \, dn_j = \sum_j [\ln (g_j/n_j)] \, dn_j \tag{19-10}$$

Similar relations may be derived for bosons and fermions from Eqs. (19-5) and (19-6) respectively.

$$d \ln w = \sum_j \left[\ln \left(\frac{g_j + n_j - 1}{n_j} \right) \right] dn_j \tag{19-11}$$

$$d \ln w = \sum_j \left[\ln \left(\frac{g_j - n_j}{n_j} \right) \right] dn_j \tag{19-12}$$

---

[1] Stirling's formula, $\ln x! = x \ln x - x$, has been used for the evaluation of $\ln N!$ and $\ln n_j!$. For large values of x, this formula represents a very good approximation (see Appendix VII).

The number 1 in Eq. (19-11) may be neglected when compared with the large values of $(g_j + n_j)$, and the last three differential equations may be written in the general form

$$d \ln w = \sum_j \left[\ln \left(\frac{g_j + cn_j}{n_j} \right) \right] dn_j \qquad (19\text{-}13)$$

The value of c is $+1$ and -1 for bosons and fermions respectively, and 0 for boltzmanons.

For the most probable distribution (w and therefore $\ln w$ is a maximum), we may conclude that

$$\sum_j \left[\ln \left(\frac{g_j + cn_j}{n_j} \right) \right] dn_j = 0 \qquad (19\text{-}14)$$

The summation must be taken over all the energy levels: $j = 1, 2, 3, \ldots, (k-1), k$. However, because of the existing restrictions, $\sum_j dn_j = 0$ and $\sum \varepsilon_j dn_j = 0$, only $(k-2)$ n_j's are independent.

If the undetermined multipliers for the two restrictions are $-\alpha$ and $-\beta$ then (see Appendix VI for Lagrange's multipliers)

$$-\sum_j \alpha \, dn_j = 0 \qquad (19\text{-}15)$$

$$-\sum_j \beta \varepsilon_j \, dn_j = 0 \qquad (19\text{-}16)$$

Addition of Eqs. (19-14), (19-15), and (19-16) yields

$$\sum_j \left[\ln \left(\frac{g_j + cn_j}{n_j} \right) - \alpha - \beta \varepsilon_j \right] dn_j = 0 \qquad (19\text{-}17)$$

Now, each term in the summation is independent and therefore must be individually zero. This can be true only if the coefficients of every dn_j are zero; thus

$$\ln \left(\frac{g_j + cn_j}{n_j} \right) = \alpha + \beta \varepsilon_j \qquad (19\text{-}18)$$

Following from this, we have, for boltzmanons

$$(n_j)_{\text{MB}} = \frac{g_j}{e^\alpha e^{\beta \varepsilon_j}} \qquad (19\text{-}19)$$

for bosons

$$(n_j)_{\text{BE}} = \frac{g_j}{e^\alpha e^{\beta \varepsilon_j} - 1} \qquad (19\text{-}20)$$

and for fermions

$$(n_j)_{\text{FD}} = \frac{g_j}{e^\alpha e^{\beta \varepsilon_j} + 1} \qquad (19\text{-}21)$$

In the derivation of the last three equations we have assumed that all levels are very richly populated and Stirling's formula may be applied to all of them. This is however, very often not the case. As is shown in Example 19-1 for the Bose–Einstein distribution, Eqs. (19-19), (19-20), and (19-21) may be derived without making any use of Stirling's formula [15, 16].

Eqs. (19-19), (19-20), and (19-21) represent the mathematical expressions of the Maxwell–Boltzmann, Bose–Einstein, and Fermi–Dirac distribution laws. As is obvious from these equations, there are more bosons than boltzmanons and more boltzmanons than fermions per single quantum state

$$\frac{(n_j)_{BE}}{g_j} > \frac{(n_j)_{MB}}{g_j} > \frac{(n_j)_{FD}}{g_j} \tag{19-22}$$

When $n_j/g_j \ll 1$, $e^\alpha e^{\beta \varepsilon_j} \gg 1$, the three different laws predict essentially the same distribution. This is exactly the case of gases at moderate and low densities. In this textbook we are going to deal mostly with gases under such conditions, and therefore only the Maxwell–Boltzmann distribution law will be considered from here on.

The values of α and β in the distribution laws remain to be determined; α may be evaluated from the condition $\sum_j n_j = N$. The summation of Eq. (19-19) over all the levels results in

$$\sum_j n_j = N = e^{-\alpha} \sum_j g_j e^{-\beta \varepsilon_j} \tag{19-23}$$

and

$$e^{-\alpha} = \frac{N}{\sum_j g_j e^{-\beta \varepsilon_j}} \tag{19-24}$$

The final form of the Maxwell–Boltzmann distribution law is then

$$P_j = \frac{n_j}{N} = \frac{g_j e^{-\beta \varepsilon_j}}{\sum_j g_j e^{-\beta \varepsilon_j}} = \frac{g_j e^{-\beta \varepsilon_j}}{q} \tag{19-25}$$

where P_j is the mathematical probability of finding a particle in level j. As is shown in Problem 8, Eq. (19-25) is equivalent to the distribution law derived in Part I, page 1, for the distribution of the kinetic energy.

The summation over all the levels appears so frequently in statistical mechanics that it has been given a special name—the molecular partition function—and symbol q;

$$q = \sum_j g_j e^{-\beta \varepsilon_j} = \sum_i e^{-\beta \varepsilon_i} \tag{19-26}$$

Since a level is a group of states of equal energies, the summation may also be written over all the states, $\sum_i$. The partition function is a measure of the extent of energy distribution among the various energy levels of the system.

For the population ratio of two states it follows that

$$\frac{n_{j+1}/g_{j+1}}{n_j/g_j} = e^{-\beta(\varepsilon_{j+1} - \varepsilon_j)} < 1 \tag{19-27}$$

because $\varepsilon_{j+1} > \varepsilon_j$ and β is always positive. Thus, the higher the energy of the state, the less it is populated.

As seen from Eq. (19-27) a decrease in the value of β is accompanied by an increase in the ratio $[(n_{j+1}/g_{j+1})]/n_j/g_j$. The lower β the greater is the ratio; more particles are shifted toward the higher energy states. (For an equilibrium distribution however, the ratio must always remain less than unity.) Consequently, the energy and, hence, the temperature of the system is increased. β is therefore related to the temperature of the system; the higher the temperature the lower is β.

To derive a quantitative relationship between T and β we shall use a simple classical approach rather than a rigorous but more complicated quantum statistical procedure $g_{j=1}$ [5]. The average kinetic energy per molecule, $\bar{\varepsilon}$, is

$$\bar{\varepsilon} = \frac{\sum_j \varepsilon_j n_j}{N} \tag{19-28}$$

Substitution from the Maxwell–Boltzmann distribution law into (19-28) results in

$$\bar{\varepsilon} = \frac{\sum_j \varepsilon_j e^{-\beta \varepsilon_j}}{\sum_j e^{-\beta \varepsilon_j}} \tag{19-29}$$

If the energy is purely kinetic and, hence, expressible in terms of momentum, we write for the x component of the momentum, p_{xj}

$$\varepsilon_{xj} = \frac{1}{2m} p_{xj}^2 \tag{19-30}$$

and consequently

$$\bar{\varepsilon}_x = \frac{\sum_j (p_{xj}^2/2m) e^{-\beta p_{xj}^2/2m}}{\sum_j e^{-\beta p_{xj}^2/2m}} \tag{19-31}$$

In the classical approach the summation is replaced by an integral and therefore

$$\bar{\varepsilon}_x = \frac{\dfrac{1}{2m} \int_{-\infty}^{+\infty} p_x^2 e^{-\beta p_x^2/2m} \, dp_x}{\int_{-\infty}^{+\infty} e^{-\beta p_x^2/2m} \, dp_x} \tag{19-32}$$

If we substitute $\beta/2m \equiv a$, the value of the numerator becomes $\frac{1}{2}(\pi/a^3)^{1/2}$ and the value of the denominator is $(\pi/a)^{1/2}$ (see Appendix II, page 934), and thus the result of the integral is

$$\bar{\varepsilon}_x = \frac{1}{2\beta} \tag{19-33}$$

Since the energy per one degree of freedom is $\frac{1}{2}kT$ (see Part I, Eq. (4-9)), we obtain for β

$$\beta = \frac{1}{kT} \tag{19-34}$$

where k is Boltzmann's constant.

19-3 NEGATIVE ABSOLUTE TEMPERATURES

As we have stated before, the higher the temperature the closer the population ratio is to unity. An "infinite temperature" is required to achieve a population ratio equal to one. To reverse the population ratio and have the higher levels more populated than the lower ones (such a distribution, however, would not be an equilibrium distribution any more), temperatures even higher than "infinity" are needed. If the distribution laws (19-19), (19-20), (19-21) were also applicable to nonequilibrium distributions, a population ratio larger than one would require that $T < 0$. This is why temperatures at which the population ratio is larger than one are called "negative absolute temperatures." Such temperatures must be however, beyond "infinity" [6].

Are there real physical systems in which the population ratio can be reversed? Yes, in systems having a limited number of energy levels that can be separated from the remaining levels, the population ratio can be reversed. In certain nuclear spin systems it is possible, by rapid reversal of the magnetic field direction, to achieve a population inversion of the spin distribution [8]. Lasers are systems in which the population ratio can be reversed (see page 402).

By "infinite temperature" we mean a temperature at which the population of the levels is equalized. It is obvious that such a definition has only a relative meaning. Systems having only a few levels available may reach the state of equal population quite easily. The more energy levels there are available to the system, the more difficult it is to reach "the infinite temperature." In the majority of systems it is practically impossible to separate the few levels from the remaining levels and consequently the population ratio in these systems can never, under any circumstances, be reversed.

For better understanding of the terms "infinite temperature" and "negative absolute temperature," let us assume a system of N particles being able to exist in two levels only, zero level and the first level. The curve illustrating the relation between the entropy S and energy U for such a system is shown in Figure 19-1. At zero energy, all N atoms are in the zero energy level, which is a state of minimum disorder, or zero entropy. When the two levels are equally populated ($N/2$), the

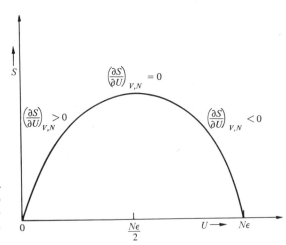

Figure 19-1 A schematic diagram representing the energy-entropy relationship in a simple system composed of N particles and having only two energy levels. The concept of negative temperatures.

energy of the system is $U = N\varepsilon/2$ and there is a maximum disorder—the most probable state—and hence maximum entropy. When all N particles are in the upper energy level, $U = N\varepsilon$, we have a state of maximum energy but of zero entropy—minimum disorder. The left half of the curve has a positive slope, and T is positive because $(\partial S/\partial U)_{V,N} = 1/T$. The right half, with a negative slope, is the region of negative temperatures. Since the maximum on the curve corresponds to a zero slope and thus to an infinite temperature, the region of negative absolute temperatures is beyond infinity, as stated before.

19-4 ADDITIONAL DISCUSSION OF THE PARTITION FUNCTION

The numerical value of the partition function, q, depends on the zero (ground state) energy from which the values of ε_j are measured. If the energy in the lowest state is set equal to zero, then

$$q = \sum_j g_j e^{-\varepsilon_j/kT} = \sum_i e^{-\varepsilon_i/kT} = e^{-\varepsilon_0/kT} + e^{-\varepsilon_1/kT} + e^{-\varepsilon_2/kT} + \cdots$$

$$= 1 + e^{-\varepsilon_1/kT} + e^{-\varepsilon_2/kT} + \cdots \tag{19-35}$$

Since $\varepsilon_1 > \varepsilon_0$, $\varepsilon_2 > \varepsilon_1$, and so on, each of the successive terms is less than unity. With increasing values of ε_i, the values of the successive terms diminish, and when $\varepsilon_i \gg kT$ the values are so small that further terms need not be included.

This is well illustrated in Table 19-2 where the distribution of N boltzmanons in the lowest ten energy levels is presented. More than 99% of the molecules are in the six lowest levels, and the sum over these levels amounts to 1.5783, which represents

Table 19-2 The distribution of N boltzmanons among ten equally spaced energy levels

Energy level	$\dfrac{\varepsilon_j}{kT}$	$e^{-\varepsilon_j/kT}$	$\dfrac{n_j}{N}$
0	0	1.0000	0.6317
1	1	0.3679	0.2323
2	2	0.1353	0.0855
3	3	0.0501	0.0314
4	4	0.0182	0.0115
5	5	0.0069	0.0043
6	6	0.0024	0.0016
7	7	0.0009	0.0006
8	8	0.0003	0.0002
9	9	0.0001	0.0001
10	10	0.0000	0.0001

$$q = \sum_j e^{-\varepsilon_j/kT} = 1.5821$$

over 99% of the total value of the partition function. For larger values of ε_j/kT this becomes even more obvious. For low values of ε_j/kT the molecules are spread out over more levels and consequently the partition function becomes larger. These conclusions are not restricted to equally spaced energy levels only.

Although we usually take the lowest state at zero energy, sometimes we shall wish to use a partition function in which the states are measured from a zero that lies below the ground state. Let us now take as an example a zero that lies below the ground state by an amount η. We may denote the new energies of the states by ε_0', ε_1', ε_2', $\ldots$, where ε_i' is equal to $(\varepsilon_i + \eta)$. The new partition function is

$$q' = \sum_i e^{-\varepsilon'/kT} = e^{-(1/kT)(\varepsilon_0+\eta)} + e^{-(1/kT)(\varepsilon_1+\eta)} + e^{-(1/kT)(\varepsilon_2+\eta)} + \cdots + \qquad \text{(19-36)}$$

This, however, is equal to

$$q' = e^{-\eta/kT} q \qquad \text{(19-37)}$$

As you can see the value of the partition function depends on the state from which the energy levels are measured.

EXAMPLE 19-1

Derive the Bose–Einstein distribution law without making use of the Stirling's approximation formula and the Lagrange multipliers.

Solution: The thermodynamic probability for the Bose–Einstein distribution (see Eq. 19-5) has the form

$$w = \prod_j \left[\frac{(g_j + n_j - 1)!}{n_j!(g_j - 1)!} \right]$$

$$= \frac{(g_1 + n_1 - 1)!}{n_1!(g_1 - 1)!} \cdots \frac{(g_r + n_r - 1)!}{n_r!(g_r - 1)!} \cdots \frac{(g_s + n_s - 1)!}{n_s!(g_s - 1)!} \cdots \frac{(g_N + n_N - 1)!}{n_N!(g_N - 1)!}$$

By adding a small quantity of heat to the system at constant volume (all the other general displacements are also kept constant) a particle may rise from level r to level s. The change of energy for this process is

$$\Delta U = \varepsilon_s - \varepsilon_r$$

where ε_s and ε_r are the energies of the particle in levels s and r, respectively. The thermodynamic probability of the new equilibrium state is

$$w' = \frac{(g_1 + n_1 - 1)!}{n_1!(g_1 - 1)!} \cdots \frac{(g_r + n_r - 2)!}{(n_r - 1)!(g_r - 1)!} \cdots \frac{(g_s + n_s)!}{(n_s + 1)!(g_s - 1)!} \cdots \frac{(g_N + n_N - 1)!}{n_N!(g_N - 1)!}$$

Making use of the Boltzmann relation (see page 138, and also page 586), the entropy change of the system due to the rise of the particle from level r to level s can be evaluated from

$$\Delta S = k \ln \frac{w'}{w}$$

Consequently

$$\Delta S = k \ln \frac{(g_s + n_s)/(n_s + 1)}{(g_r + n_r - 1)/n_r}$$

$$= k \ln \frac{g_s[(n_s/g_s) + 1]/(n_s + 1)}{g_r[(n_r/g_r) + 1 - (1/g_r)]/n_r}$$

If level s is highly populated, $n_s \gg 1$, and the degeneracy of each level is much greater than unity, $g_j \gg 1$, then

$$\Delta S = k \ln \left[\frac{1 + (g_s/n_s)}{1 + (g_r/n_r)} \right]$$

The changes in energy and entropy are nearly continuous when a single quantum of energy is added to the system and therefore

$$T = \left(\frac{\partial U}{\partial S} \right)_V = \left(\frac{\Delta U}{\Delta S} \right)_V$$

Following from this

$$\Delta U = \varepsilon_s - \varepsilon_r = T\Delta S = kT \ln \left[\frac{1 + (g_s/n_s)}{1 + (g_r/n_r)} \right]$$

or

$$\frac{1 + (g_s/n_s)}{1 + (g_r/n_r)} = \frac{e^{\varepsilon_s/kT}}{e^{\varepsilon_r/kT}}$$

For any general level j it follows that

$$1 + (g_j/n_j) = A \, e^{\varepsilon_j/kT}$$

where A is a proportionality constant. Written in a reciprocal form, we have

$$n_j = \frac{g_j}{A \, e^{\varepsilon_j/kT} - 1}$$

This is the Bose–Einstein distribution law. (Compare it to Eq. 19-20). This derivation requires only $g_j \gg 1$ and that at least one of the energy levels of the system is highly populated, $n_s \gg 1$.

This derivation is taken from Turoff [15]. Another approach is discussed by Burns and Brown [16]. ●

EXAMPLE 19-2

Prove mathematically that, in a system containing a large number of particles, the most probable distribution is so dominant that all the other distributions may be neglected when compared with it.

Solution: The method presented next was described by Keil and Nash [7].

Let us consider only two of the many possible distributions in an isolated system containing a large number of distinguishable molecules, N:

1. The predominant distribution for which the number of associated microstates has a maximum value, w_{mp}.

2. A distribution that is only infinitesimally shifted from the most probable distribution; let the number of microstates associated with this distribution be w.

According to Eq. (19-3) the most probable distribution for a nondegenerate system is given as

$$w_{mp} = \frac{N!}{\prod_j n_j!} \tag{a}$$

where n_j represent the population of level j and it can be expressed as

$$n_j = n_0 e^{-\beta \varepsilon_j} \tag{b}$$

Similar equations can also be written for the slightly shifted distribution.

Let us now introduce a new variable, α_j, relating the populations of level j for the two distributions:

$$\alpha_j \equiv \frac{n_j' - n_j}{n_j} \tag{c}$$

the prime relates to the shifted distribution. Due to the fact that the two distributions differ only slightly from each other, $|\alpha| \ll 1$ in all cases.

In an isolated system the total number of particles as well as the total energy remain constant. Following from this

$$\Delta N = (n_0' - n_0) + (n_1' - n_1) + \cdots + (n_k' - n_k) = \sum_j \alpha_j n_j = 0 \tag{d}$$

Similarly

$$\Delta E = \varepsilon_0(n_0' - n_0) + \varepsilon_1(n_1' - n_1) + \cdots + \varepsilon_k(n_k' - n_k) = \sum_j \varepsilon_j \alpha_j n_j = 0 \tag{e}$$

since as seen from the definition of α, $n_j' - n_j = \alpha_j n_j$.

The ratio of the probabilities for the two distributions is

$$\frac{w_{mp}}{w} = \frac{N!/\prod_j n_j!}{N!/\prod_j n_j'!} = \frac{\prod_j (n_j + \alpha_j n_j)!}{\prod_j n_j!} \tag{f}$$

because $n_j' = n_j + \alpha_j n_j$. Written in a logarithmic form

$$\ln(w_{mp}/w) = \sum_j \ln(n_j + \alpha_j n_j)! - \sum_j \ln n_j! \tag{g}$$

Applying Stirling's formula and realizing that $\alpha_j \ll 1$, we derive

$$\ln(w_{mp}/w) = \sum_j \alpha_j n_j \ln n_j + \sum_j n_j \alpha_j^2 \tag{h}$$

Substitution of the value of n_j from (b) into (h) yields

$$\ln(w_{mp}/w) = \ln n_0 \sum_j \alpha_j n_j - \beta \sum_j \varepsilon_j \alpha_j n_j + \sum_j n_j \alpha_j^2 \tag{i}$$

Due to the fact that in an isolated system $\sum_j \alpha_j n_j = 0$ and $\sum_j \varepsilon_j \alpha_j n_j = 0$, the last equation assumes the form

$$\ln (w_{mp}/w) = \sum_j n_j \alpha_j^2 \qquad\qquad \text{(j)}$$

The product $n_j \alpha_j^2$ varies from level to level, and it seems therefore more convenient to introduce an average value of α defined by the equation

$$\bar{\alpha} = \left(\frac{\sum_j n_j \alpha_j^2}{N} \right)^{1/2} \qquad \text{or} \qquad N\bar{\alpha}^2 = \sum_j n_j \alpha_j^2 \qquad\qquad \text{(k)}$$

which, after substitution into (j) yields

$$\ln (w_{mp}/w) = N\bar{\alpha}^2 \qquad\qquad \text{(l)}$$

or

$$w_{mp}/w = e^{N\bar{\alpha}^2} \qquad\qquad \text{(m)}$$

For $N = 6 \times 10^{23}$ and an assumed value of $\bar{\alpha} = 10^{-10}$ the result is

$$\frac{w_{mp}}{w} = e^{(6 \times 10^{23})(10^{-20})} = e^{6000}$$

This tremendously large number indicates that the contribution arising from the shifted distribution is very small when compared to w_{mp} and may be neglected completely.   ●

19-5 THERMODYNAMIC FUNCTIONS IN TERMS OF THE PARTITION FUNCTION

As we shall show next, if the partition function is known, the thermodynamic properties of a system can be evaluated. Since thermodynamic quantities are discussed on a molar rather than a molecular basis, the molar partition function is more appropriate for the statistical definition of thermodynamic functions.

When rearrangement of independent particles among the energy states gives rise to different quantum states, the molecular, q, and molar partition functions, Q, are related to each other through the equation

$$Q = q^N \qquad\qquad \text{(19-38)}$$

Consequently

$$\ln Q = N \ln q \qquad\qquad \text{(19-39)}$$

Such behavior is characteristic of an ideal crystal, where the different oscillators (vibrating particles, see page 610) occupy distinct positions at different locations in the crystal lattice and an interchange of two oscillators in the lattice creates a new quantum state.

In the case of an ideal gas, each independent particle is free to move throughout the whole system, and an interchange of coordinates between particles does not create a new quantum state. Therefore, the particles of an ideal gas loose their distinguishability. Consequently, quantum states in the gaseous phase that differ merely by the interchange of two particles are indistinguishable and should be counted only once. If each level contains only one particle (a good approximation for a gaseous system at moderate and low pressures, when the number of available levels is much greater than the number of particles in the system—a dilute system), the number of permutations of the particles among the levels is $N!$, and the relation becomes

$$Q = \frac{1}{N!} q^N \tag{19-40}$$

Utilizing Stirling's formula (the use is not completely justified here; nevertheless it has no effect on the final result), we have

$$\ln Q = -\ln N! + N \ln q$$
$$= -N \ln N + N + N \ln q$$
$$= N \left(\ln (q/N) + 1 \right) = N \left(\ln (q/N) + \ln e \right) \tag{19-41}$$
$$= N \ln (qe/N)$$

Eq. (19-41) differs significantly from Eq. (19-39). This difference, when not understood properly, may lead to inaccuracies in the statistical definition of entropy.

The first thermodynamic property to be discussed is the internal energy, $\mathscr{U}$. Upon substituting n_j from Eq. (19-25) into Eq. (19-2), we obtain

$$\mathscr{U} = \sum_j n_j \varepsilon_j = \frac{N}{q} \sum_j g_j \varepsilon_j \, e^{-\varepsilon_j/kT} \tag{19-42}$$

From the definition of the partition function it follows that

$$\sum_j g_j \varepsilon_j \, e^{-\varepsilon_j/kT} = kT^2 \left(\frac{\partial q}{\partial T} \right)_{V,N} \tag{19-43}$$

and the equation for the molar internal energy assumes the form

$$\mathscr{U} = \frac{N}{q} kT^2 \left(\frac{\partial q}{\partial T} \right)_{V,N} = kT^2 N \left(\frac{\partial \ln q}{\partial T} \right)_{V,N}$$
$$= kT^2 \left(\frac{\partial \ln Q}{\partial T} \right)_{V,N} \tag{19-44}$$

The restriction of constant volume is introduced, so that ε_j remains a constant.

The derivative of the internal energy with respect to T at constant V and N is the molar heat capacity

$$\mathscr{C}_V = k \frac{\partial}{\partial T} \left[T^2 \left(\frac{\partial \ln Q}{\partial T} \right)_{V,N} \right]$$
$$= k \left[T^2 \left(\frac{\partial^2 \ln Q}{\partial T^2} \right)_{V,N} + 2T \left(\frac{\partial \ln Q}{\partial T} \right)_{V,N} \right] \tag{19-45}$$

To obtain a relation between the partition function and entropy, we combine the classical relation (6-54) with (19-45). The result is

$$\int_0^T d\mathscr{S} = \int_0^T \frac{\mathscr{C}_V}{T} \, dT = k \int_0^T \left[T\left(\frac{\partial^2 \ln Q}{\partial T^2}\right)_{V,N} + 2\left(\frac{\partial \ln Q}{\partial T}\right)_{V,N} \right] dT \quad \textbf{(19-46)}$$

The integration of the first term on the right hand side of the equality sign can be carried out by parts.

$$\int v \, du = uv \Big]_0^v - \int u \, dv$$

If we set $T = v$ and $u = (\partial \ln Q / \partial T)_{V,N}$, we have

$$\mathscr{S} - \mathscr{S}_0 = kT\left(\frac{\partial \ln Q}{\partial T}\right)_{V,N} + k \ln Q - k \ln Q_0 \quad \textbf{(19-47)}$$

In this equation only $\mathscr{S}_0$ and $k \ln Q_0$ are temperature independent. Consequently,

$$\mathscr{S}_0 = k \ln Q_0 \quad \textbf{(19-48)}$$

and

$$\mathscr{S} = kT\left(\frac{\partial \ln Q}{\partial T}\right)_{V,N} + k \ln Q = \frac{\mathscr{U}}{T} + k \ln Q \quad \textbf{(19-49)}$$

For a system of independent and indistinguishable particles, $\ln Q = N \ln (qe/N)$, and therefore

$$\mathscr{S} = \frac{\mathscr{U}}{T} + R \ln \left(\frac{qe}{N}\right) \quad \textbf{(19-50)}$$

The Helmholtz free energy, $\mathscr{A}$, is equal to $\mathscr{U} - T\mathscr{S}$. Consequently

$$\mathscr{A} = -kT \ln Q \quad \textbf{(19-51)}$$

and again, for independent and indistinguishable particles, we obtain

$$\mathscr{A} = -RT\left[\ln \left(\frac{qe}{N}\right)\right] \quad \textbf{(19-52)}$$

For the pressure, we write

$$P = -\left(\frac{\partial \mathscr{A}}{\partial V}\right)_{T,N} = kT\left(\frac{\partial \ln Q}{\partial V}\right)_{T,N} = RT\left(\frac{\partial \ln q}{\partial V}\right)_{T,N} \quad \textbf{(19-53)}$$

Since enthalpy is defined as $\mathscr{H} = \mathscr{U} + P\mathscr{V}$, we have for 1 mole of an ideal gas

$$\mathscr{H} = kT^2\left(\frac{\partial \ln Q}{\partial T}\right)_{V,N} + RT$$

$$= RT\left[T\left(\frac{\partial \ln q}{\partial T}\right)_{VN} + 1\right] \quad \textbf{(19-54)}$$

Finally, the result for the Gibbs free energy is

$$\mathscr{G} = \mathscr{A} + P\mathscr{V} = -kT \ln Q + RT$$

$$= -RT \ln (q/N) \quad \textbf{(19-55)}$$

Statistical mechanics gives identical results for $\mathscr{U}$, $\mathscr{C}_V$ and $\mathscr{H}$ of localized and nonlocalized particle systems. This is a consequence of the simple fact that $d \ln \mathbf{N}! = 0$. For the second law functions ($\mathscr{S}$, $\mathscr{A}$, and $\mathscr{G}$), however, the results are different for systems of the two different kinds of particles.

19-6 THE MOLECULAR INTERPRETATION OF THE BASIC LAWS OF THERMODYNAMICS

For an infinitesimal reversible change taking place in a closed system of constant mass, the first law of thermodynamics states that

$$d\mathscr{U} = d\,q_{\text{rev}} + d\,w_{\text{rev}} \tag{19-56}$$

For the same infinitesimal change, the statistical equation (19-2) may be written as

$$d\mathscr{U} = \sum_j \varepsilon_j\, dn_j + \sum_j n_j\, d\varepsilon_j \tag{19-57}$$

It follows from classical thermodynamics that the internal energy of a system may only be changed by adding or removing heat or work to or from it. According to Eq. (19-57) the internal energy can be altered either by changing the distribution of particles among the levels, $\sum_j \varepsilon_j\, dn_j$, or by shifting levels up and down, $\sum_j n_j\, d\varepsilon_j$. Thus, one of these statistical terms must be associated with heat and the other with work. Since a change in the energy of a level may be accomplished only by a change in the boundary parameters, it is obvious that the term $\sum_j n_j\, d\varepsilon_j$ must represent work. When the only boundary parameter is the volume, we have

$$\sum_j n_j\, d\varepsilon_j = \sum_j n_j \left(\frac{\partial \varepsilon_j}{\partial \mathscr{V}} \right)_{T,N} d\mathscr{V}$$

$$= -\sum_j n_j kT \left(\frac{\partial \ln q}{\partial \mathscr{V}} \right)_{T,N} d\mathscr{V} = -\sum_j n_j p_j\, d\mathscr{V}$$

$$= -P\, d\mathscr{V} = d\,w_{\text{rev}} \tag{19-58}$$

since as can be proven from the distribution law

$$\left(\frac{\partial \varepsilon_j}{\partial \mathscr{V}} \right)_{T,N} = -kT \left(\frac{\partial \ln q}{\partial \mathscr{V}} \right)_{T,N} = -p_j \tag{19-59}$$

where p_j is the pressure exhibited by a particle in level j. The quantity $\sum_j n_j\, d\varepsilon_j$ is thus interpreted as the work that must be done in order to change the energy levels from ε_j to $\varepsilon_j + d\varepsilon_j$ while keeping the distribution of particles over the levels fixed. The other term, $\sum_j \varepsilon_j\, dn_j$, must therefore be equal to the reversibly exchanged heat between the system and its surroundings:

$$d\,q_{\text{rev}} = \sum_j \varepsilon_j\, dn_j = -kT \sum_j \left[\ln\left(\frac{n_j}{Ng_j} \right) + \ln q \right] dn_j \tag{19-60}$$

where the second equality follows from the basic relations—Eqs. 19-19, 19-24, 19-26, and 19-34.

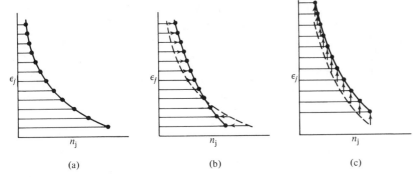

Figure 19-2 The statistical interpretation of the transfer of heat and work in a reversible process. (a) original distribution; (b) after adding heat; (c) after adding work.

Thus, in statistical mechanics, heat is identified with that part of the internal energy change of a system which results from the redistribution of particles among the existing energy levels. The statistical difference between reversible heat and work is obvious from Figure 19-2. The population of the levels is indicated by the length of the horizontal lines. As you can see, when heat is added to the system reversibly, the population of the lower levels is reduced, while the population of the higher levels is increased; the energy levels remain, however, unchanged. When work is added to the system, the energy levels are shifted while their population remains unaffected.

Upon dividing Eq. (19-60) by T, the statistical equivalent of the second law of thermodynamics is obtained.

$$d\mathscr{S} = \frac{d\,q_{\text{rev}}}{T} = \frac{1}{T} \sum_j \varepsilon_j \, dn_j$$

$$= -k \sum_j \left[\ln \left(\frac{n_j}{Ng_j} \right) + \ln q \right] dn_j \tag{19-61a}$$

which is equivalent to

$$d\mathscr{S} = -kd\left(\sum_j P_j \ln P_j \right) \tag{19-61b}$$

where P_j is the mathematical probability of finding a molecule in level j. This equation is identical with Eq. (6-13) on page 139.

Eq. (19-50) can be used for the statistical interpretation of the third law of thermodynamics. At low temperatures the population of the higher energy levels in an ideal crystal may be ignored and Eq. (19-50) may be rewritten by means of Eqs. 19-42, 19-39, and 19-26 as

$$\mathscr{S}/N = \frac{1}{T} \frac{g_0\varepsilon_0 \, e^{-\varepsilon_0/kT} + g_1\varepsilon_1 \, e^{-\varepsilon_1/kT} + \cdots}{g_0 \, e^{-\varepsilon_0/kT} + g_1 \, e^{-\varepsilon_1/kT} + g_2 \, e^{-\varepsilon_2/kT}} + k \ln \left(g_0 \, e^{-\varepsilon_0/kT} + g_1 \, e^{-\varepsilon_1/kT} \right) \tag{19-62}$$

In the limit $T \to 0$, $g_1 \, e^{-\varepsilon_1/kT} \ll g_0 \, e^{-\varepsilon_0/kT}$ and consequently

$$\mathscr{S}_{T\to 0}/N = \mathscr{S}_0/N = \frac{\varepsilon_0}{T} + \frac{\varepsilon_1}{T} \frac{g_1}{g_0} \, e^{-(\varepsilon_1-\varepsilon_0)/kT} + k \ln g_0 - \frac{\varepsilon_0}{T} + k\frac{g_1}{g_0} \, e^{-(\varepsilon_1-\varepsilon_0)/kT} = k \ln g_0 \tag{19-63}$$

because the exponential terms decrease more rapidly than T. If the ground energy level is nondegenerate, $g_0 = 1$, and $\mathscr{S}_0 = 0$.

At absolute zero all particles are in the ground level; thus, $w_0 = 1$. There is only one way of putting all the particles into one level—and Boltzmann's relation (see page 138) leads at absolute zero to the same conclusion

$$\mathscr{S}_0 = k \ln w_0 = 0 \qquad \qquad \textbf{(19-64)}$$

Eqs. (19-63) and (19-64) are considered as the statistical definitions of the third law of thermodynamics.

If a system is able to exist in more than one quantum state at absolute zero, its entropy is not zero. As we stated earlier, the different spin orientations, the isotopic constitution of elements, and the two or more possible geometrical arrangements of particles in the crystal lattice may cause $\mathscr{S}_0$ to differ from zero. For example, in nitrous oxide, two adjacent molecules can be oriented either as ONN—NNO or NNO—ONN. The entropy corresponding to the different arrangements—configurational entropy—cannot be predicted from the temperature dependence of the molar heat capacity. If we assume that half of the molecules are oriented in one direction and half in the other, the residual entropy is[2]

$$\mathscr{S}_0 = k \ln w = kN \ln 2$$
$$= R \ln 2 = 8.314 \times 0.6931 = 5.76 \text{ J K}^{-1} \text{ mol}^{-1}$$

In Table 19-3 values of entropy obtained calorimetrically, from $\mathscr{C}_P$ measurements, and statistically, from spectroscopic measurements (the method of calculation will be

Table 19-3 Comparison of the calorimetric and statistical values of entropy at 298 K and 1 atm

Substance	$\mathscr{S}^\circ_{cal}$ (cal K^{-1} mol^{-1})	$\mathscr{S}^\circ_{stat}$	Substance	$\mathscr{S}^\circ_{cal}$ (cal K^{-1} mol^{-1})	$\mathscr{S}^\circ_{stat}$
Ne	35.01	34.95	H$_2$O*	44.28	45.104
Kr	39.18	39.20	PH$_3$	50.35	50.36
Zn	38.4	38.5	CH$_3$Cl	55.94	55.98
O$_2$	49.09	49.04	CH$_3$D*	36.72	39.49
CO*	46.22	47.31	D$_2$O*	45.89	46.66
H$_2$S	49.10	49.17	C$_6$H$_6$	64.45	64.36
CO$_2$	51.11	51.07	NO*	49.7	50.43
N$_2$O*	51.43	52.58	C$_2$H$_4$	52.50	52.47
SO$_2$	59.24	59.27	toluene	76.77	76.69
AsF$_3$	69.05	69.08	CH$_4$	44.46	44.47
Cl$_2$	53.31	53.31	HCl	44.47	44.66
N$_2$	45.76	45.77	NH$_3$	45.96	45.98

* Molecules with frozen-in configurational entropy.

[2] $w = N!/(N/2)!(N/2)!$; using Stirling's formula, we obtain; $\ln w = N \ln 2$.

discussed further), are compared. The values agree reasonably well except for molecules with frozen-in configurational entropy, denoted by an asterisk.

19-7 EVALUATION OF THE PARTITION FUNCTION

An ideal gas molecule possesses translational, rotational, vibrational, nuclear, and electronic degrees of freedom. Its energy is given by (see also Eq. 14-10)

$$\varepsilon = \varepsilon_{tr} + \varepsilon_{rot} + \varepsilon_{vib} + \varepsilon_{el} + \varepsilon_{nuc} \tag{19-65}$$

and consequently

$$q = \sum_j g_j\, e^{-\varepsilon_j/kT} = q_{tr}q_{rot}q_{vib}q_{el}q_{nuc}$$

$$= q_{tr}q_{int}q_{el}q_{nuc} \tag{19-66}$$

where $q_{int} = q_{rot}q_{vib}$. The principle of separation of energies (19-65) is not exactly valid; in many cases, however, it is a good approximation.

Our immediate objective is the evaluation of each of the particle partition functions.

19-7.1 Translational Partition Function
Since we are dealing with an ideal gas—independent particles—the result of the Schrödinger equation for a particle in a three dimensional box may be applied here (see page 300)

$$\varepsilon_{tr} = \frac{h^2}{8m}\left(\frac{n_x^2}{a^2} + \frac{n_y^2}{b^2} + \frac{n_z^2}{c^2}\right)$$

where h is Planck's constant, m is the mass of the particle, n_x, n_y, and n_z are integral numbers called quantum numbers, and a, b, and c denote the dimensions of the box. On substituting this solution into the particle partition function, we obtain

$$q_{tr} = \sum_i e^{-\varepsilon_{i,tr}/kT}$$

$$= \sum_{n_x=0}^{\infty} \sum_{n_y=0}^{\infty} \sum_{n_z=0}^{\infty} e^{-\frac{h^2}{8mkT}\left(\frac{n_x^2}{a^2}+\frac{n_y^2}{b^2}+\frac{n_z^2}{c^2}\right)} \tag{19-67}$$

The spacing between the translational levels is so small that it can be considered as continuous, and thus, the sums may be replaced by integrals.

$$q_{tr} = \int_0^\infty \int_0^\infty \int_0^\infty e^{-\frac{h^2}{8mkT}\left(\frac{n_x^2}{a^2}+\frac{n_y^2}{b^2}+\frac{n_z^2}{c^2}\right)}\, dn_x\, dn_y\, dn_z \tag{19-68}$$

Upon making the substitutions

$$\frac{n_x^2 h^2}{8ma^2 kT} = x^2 \qquad dx = \frac{h}{a}\left(\frac{1}{8mkT}\right)^{1/2} dn_x \tag{19-69a}$$

$$\frac{n_y^2 h^2}{8mb^2 kT} = y^2 \qquad dy = \frac{h}{b}\left(\frac{1}{8mkT}\right)^{1/2} dn_y \tag{19-69b}$$

and

$$\frac{n_z^2 h^2}{8mc^2 kT} = z^2 \qquad dz = \frac{h}{c}\left(\frac{1}{8mkT}\right)^{1/2} dn_z \qquad \text{(19-69c)}$$

we obtain

$$q_{tr} = \frac{abc}{h^3}(8mkT)^{3/2}\int_0^\infty\int_0^\infty\int_0^\infty e^{-(x^2+y^2+z^2)}\,dx\,dy\,dz$$

$$= \frac{V(2\pi mkT)^{3/2}}{h^3} \qquad \text{(19-70)}$$

The triple integral reduces to a product of three identical single integrals. The value of the triple integral is $(\pi)^{3/2}/8$ (see Appendix II) and V denotes the volume of the box ($\equiv abc$).

For a monatomic gas, $q_{rot} = q_{vib} = 1$. At low and moderate temperatures, except for some species as, for example, fluorine and chlorine atoms, q_{el} is also equal to 1. If q_{nuc} can be neglected (see page 602), the partition function of a monatomic gas is given merely by its translational part. Thus, for 1 mole of a gas it follows

$$Q_{tr} = \left(\frac{1}{N!}\right)q_{tr}^N = \left(\frac{1}{N!}\right)\left[\frac{V(2\pi mkT)^{3/2}}{h^3}\right]^N \qquad \text{(19-71)}$$

and

$$\ln Q_{tr} = -N\ln N + N\ln e + N\ln q_{tr}$$
$$= N\ln\left[(Ve/Nh^3)(2\pi mkT)^{3/2}\right] \qquad \text{(19-72)}$$

The translational contributions to the thermodynamic functions of an ideal gas follow.

1. The molar internal energy (see Eq. 19-44)

$$\mathcal{U}_{tr} = kT^2\left(\frac{\partial\ln Q_{tr}}{\partial T}\right)_{V,N} = \frac{3}{2}NkT^2\left(\frac{\partial\ln T}{\partial T}\right)_{V,N} = \frac{3}{2}RT \qquad \text{(19-73)}$$

2. The molar heat capacity at constant volume

$$(\mathcal{C}_V)_{tr} = \left(\frac{\partial\mathcal{U}_{tr}}{\partial T}\right)_{V,N} = \frac{\partial}{\partial T}\left[kT^2\left(\frac{\partial\ln Q_{tr}}{\partial T}\right)_{V,N}\right]$$

$$= \frac{\partial}{\partial T}\left(\frac{3}{2}RT\right) = \frac{3}{2}R \qquad \text{(19-74)}$$

These, of course, are the equipartition values for the translational energy and the translational molar heat capacity, which are just the results obtained in Part I, Chapter 4 for an ideal gas on a purely classical basis. This is, however, to be expected since, by replacing the summation with an integral, quantum statistics become identical with classical statistics. Such an approximation is always justified if the spacing between successive energy levels is so small that the condition $\Delta\varepsilon \ll kT$ is met.

3. The molar Helmholtz free energy is given as

$$\mathcal{A}_{tr} = -kT\ln Q_{tr} = -RT\ln\left[\frac{Ve}{Nh^3}(2\pi mkT)^{3/2}\right] \qquad \text{(19-75)}$$

4. Pressure

$$P_{tr} = -\left(\frac{\partial \mathscr{A}_{tr}}{\partial \mathscr{V}}\right)_{T,N} = kT\left(\frac{\partial \ln Q_{tr}}{\partial \mathscr{V}}\right)_{T,N} = \frac{NkT}{\mathscr{V}} = \frac{RT}{\mathscr{V}} \qquad (19\text{-}76)$$

As seen from Eq. (19-76), the pressure of an ideal gas is independent of the rotational, vibrational, nuclear, and electronic degrees of freedom.

5. The molar Gibbs free energy

$$\mathscr{G}_{tr} = -RT \ln\left(\frac{q_{tr}}{N}\right)$$

$$= -RT \ln\left[\frac{\mathscr{V}(2\pi mkT)^{3/2}}{Nh^3}\right] \qquad (19\text{-}77)$$

6. The molar entropy

$$\mathscr{S}_{tr} = \frac{\mathscr{U}_{tr}}{T} + R \ln\left(\frac{eq_{tr}}{N}\right)$$

$$= \frac{3}{2}R + R \ln\left[\frac{e\mathscr{V}(2\pi mkT)^{3/2}}{Nh^3}\right]$$

$$= R \ln\left[\frac{e^{5/2}\mathscr{V}}{Nh^3}(2\pi mkT)^{3/2}\right] \qquad (19\text{-}78)$$

Since for an ideal gas $\mathscr{V} = RT/P$, Eq. (19-78) assumes the form

$$\mathscr{S}_{tr} = R[-\ln P + \tfrac{5}{2}\ln T + \tfrac{3}{2}\ln M + a] \qquad (19\text{-}79)$$

The value of the constant a is the same for all gases and depends only on the units in which k, m, h, and P are expressed. At standard conditions, $P = 1$ atm, $T = 298$ K, Eq. (19-79) reduces to

$$\mathscr{S}_{tr}^{\circ} = b + c \ln M \qquad (19\text{-}80)$$

where b and c are constants; their value depend again on the units only.

With increasing molecular weight, M, and temperature, T, the entropy increases, whereas an increase in pressure, P, lowers the entropy. This is the conclusion that Sackur and Tetrode reached in 1913 in a completely different way.

Monatomic molecules do not possess rotational and vibrational degrees of freedom. Electronic excitations, except in a few cases, may also be neglected at not very high temperatures, and therefore, the translational contributions to the thermodynamic functions represent the total thermodynamic property.

The measured values of the molar heat capacities of nitrogen are plotted in Figure 19-3. The agreement between the observed and calculated values of $\mathscr{C}_V$($\mathscr{C}_V = \tfrac{3}{2}R = 12.47$ J K^{-1} mol^{-1}) is excellent for the monatomic gases. Since N_2 possesses rotational and vibrational degrees of freedom in addition to the translational, its $\mathscr{C}_V$ changes with temperature.

19-7.2 Rotational Partition Function Polyatomic gas molecules possess rotational and vibrational degrees of freedom in addition to the translational and electronic ones. The details are complicated for the rotation of polyatomic

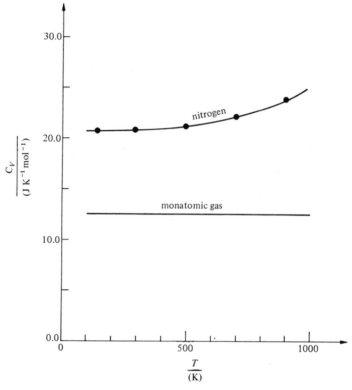

Figure 19-3 The variation of the molar heat capacity with temperature.

molecules, arising from both physical and mathematical complexities. For this reason, we shall consider here in detail only the rotation of linear molecules (all diatomic molecules must be linear) with fixed separation of the atoms in the molecule—molecules without vibration. This kind of a molecule is called a rigid rotor (see also Sections 4-2 and 11-3).

The indistinguishability of gaseous particles has already been taken into account in the translational partition function; therefore, for the rotation, we simply write

$$Q_{rot} = q_{rot}^N \qquad (19\text{-}81)$$

Some additional indistinguishability, arising from the rotation of symmetrical molecules, will be introduced later into the partition function.

The Schrödinger equation gives the following solution for the energy levels of a rigid rotor (see page 323)

$$\varepsilon_{rot} = \frac{l(l+1)h^2}{8\pi^2 I}$$

where I is the moment of inertia defined on page 79 and l is the rotational quantum number (J is used on page 323). For a certain value of l, there are $2l + 1$ quantum states associated with the same energy; the degeneracy number, g_j, is therefore

$2l+1$, and the rotational partition function becomes

$$q_{rot} = \sum_j g_j \, e^{-\varepsilon_{j,rot}/kT} = \sum_{l=0}^{\infty} (2l+1) \, e^{-\frac{l(l+1)h^2}{8\pi^2 IkT}}$$

$$= \sum_{l=0}^{\infty} (2l+1) \, e^{-l(l+1)\Theta_{rot}/T} \tag{19-82}$$

where

$$\Theta_{rot} = \frac{h^2}{8\pi^2 Ik}$$

is the characteristic temperature for rotation. If $T \gg \Theta_{rot}$, the separation between successive levels relative to kT will be very small and the sum in Eq. (19-82) may be replaced by an integral. Then the classical or high temperature limit of q_{rot} becomes

$$q_{rot} = \int_0^{\infty} (2l+1) \, e^{-l(l+1)\Theta_{rot}/T} \, dl \tag{19-83}$$

On substitution of z for $l(l+1)$ and dz for $(2l+1) \, dl$, we obtain

$$q_{rot} = \int_0^{\infty} e^{-z\Theta_{rot}/T} \, dz \tag{19-84}$$

Since at constant temperature Θ_{rot}/T is a constant, integration results in

$$q_{rot} = \int_0^{\infty} e^{-bz} \, dz = \frac{1}{b} = \frac{T}{\Theta_{rot}} = \frac{8\pi^2 IkT}{h^2} \tag{19-85}$$

As seen from the values of Θ_{rot} for different gases in Table 19-4, the replacement of the sum with an integral is justified for N_2, O_2, CO, NO, and so on. It gives, however, only approximate results for HF, HCl, HBr, HI, and so on. For these gases

Table 19-4 Rotational and vibrational characteristic temperatures of some gases

Substance	$\dfrac{\Theta_{rot}}{(K)}$	$\dfrac{\Theta_{vib}}{(K)}$
H_2	85.4	6210
N_2	2.86	3340
O_2	2.07	2230
CO	2.77	3070
NO	2.42	2690
HCl	15.2	4140
HBr	12.1	3700
HI	9.0	3200
Cl_2	0.346	810
Br_2	0.116	470
I_2	0.054	310

a term-by-term summation will result in much better values of q_{rot}. The high values of Θ_{rot} for H_2, D_2, and HD (low molecular weight gases) are responsible for the unusual behavior of these gases (quantum effect).

When T is not high enough, the following equation must be applied

$$q_{rot} = \frac{T}{\Theta_{rot}}\left(1 + \frac{\Theta_{rot}}{3T} + \cdots\right) \tag{19-86}$$

Formula (19-85) is valid as it stands for heteronuclear diatomic molecules such as HCl, NO, and so on. For homonuclear molecules such as N_2, Cl_2, and others (which have identical nuclei, such as $^{35}Cl^{35}Cl$) symmetry rules require the partition function be divided by the symmetry number, σ

$$q_{rot} = \frac{8\pi^2 I k T}{\sigma h^2} \tag{19-87}$$

σ is equal to the number of indistinguishable positions into which the molecule can be turned by simple rigid rotation. It may be interpreted as additional indistinguishability of the particles in the system. For any symmetrical linear molecule $\sigma = 2$.

The classical limit of the rotational energy, $\mathcal{U}_{rot}$, may be easily obtained from Eqs. (19-44) and (19-87)

$$\mathcal{U}_{rot} = NkT^2\left(\frac{\partial \ln q_{rot}}{\partial T}\right)_N$$

$$= RT^2\left\{\left[\frac{\partial \ln (8\pi^2 I k/\sigma h^2)}{\partial T}\right] + \frac{\partial \ln T}{\partial T}\right\}$$

$$= RT^2\frac{1}{T} = RT \tag{19-88}$$

The rotational contribution to the molar heat capacity is

$$(\mathcal{C}_V)_{rot} = \left(\frac{\partial (RT)}{\partial T}\right)_V = R \tag{19-89}$$

which again is in agreement with classical calculations.

Except for hydrogen, all gaseous diatomic molecules have classical rotational energies at and above their boiling points. Polyatomic molecules in the liquid state, however, have very little rotational excitation at low temperatures, and higher temperatures are needed to excite them rotationally. As the temperature increases, the molecules are getting excited more and more rotationally and the rotational contributions to the thermodynamic functions are approaching their classical values.

The molar rotational entropy is

$$\mathcal{S}_{rot} = k \ln Q_{rot} + kT\left(\frac{\partial \ln Q_{rot}}{\partial T}\right)_N$$

$$= R \ln\left(\frac{kT8\pi^2 I}{\sigma h^2}\right) + R$$

$$= R \ln\left(\frac{ekT8\pi^2 I}{\sigma h^2}\right) \tag{19-90}$$

This equation is rather cumbersome, but we can modify it by collecting the various constants into a single term. Then we obtain

$$\mathscr{S}_{rot} = R[\ln I + \ln T - \ln \sigma + a']$$ (19-91)

The value of the constant a' depends on the units in which k, I and h are expressed. The rotational entropy increases with increasing moment of inertia (increasing mass of the molecule), with increasing temperature, and decreases with increasing symmetry number.

To the same approximation, the partition function of a nonlinear polyatomic molecule is

$$q_{rot} = \frac{8\pi^2(2\pi kT)^{3/2}(I_A I_B I_C)^{1/2}}{\sigma h^3}$$ (19-92)

In this equation, I_A, I_B, and I_C are the principal moments of inertia along the three respective axes. The symmetry number of water is 2, of ammonia is 3, for CH_4 is 12, and for benzene also 12. The symmetry number does not affect the thermodynamic properties in which the derivatives of the logarithm of the partition function appears; since $\ln \sigma$ is constant, $(d \ln \sigma)/dx = 0$.

The moments of inertia can be calculated if the interatomic distance is known (see page 81). This distance is often obtained from electron diffraction measurements of the gas, or x-ray diffraction measurements of the solid state. Microwave spectra and spectra taken in the near-infrared region also make possible the evaluation of the moment of inertia.

19-7.3 Vibrational Partition Function

In evaluating the vibrational partition function, it is often sufficient to use the harmonic oscillator approximation (see page 315). A harmonic oscillator represents the simplest system for which one may write a partition function because the energy levels are equally spaced (see Figure 19-4).

$$\varepsilon_{vib} = (v + \tfrac{1}{2})h\nu$$ (19-93)

where ν is the fundamental vibration frequency of the oscillator and v is the vibrational quantum number. The vibrational partition function is thus given by

$$q_{vib} = \sum_{v=0}^{\infty} e^{-(v+\frac{1}{2})(h\nu/kT)}$$

$$= e^{-\frac{1}{2}(h\nu/kT)}\left(\sum_{v=0}^{\infty} e^{-(vh\nu/kT)}\right)$$ (19-94)

The zero energy is taken here at the bottom of the parabolic energy well (see Figure 19-4) rather than at the lowest energy level, where the vibrational quantum number, v, equals zero. The latter is $\frac{1}{2}h\nu$ above the bottom of the well. If we write $e^{-(h\nu/kT)} = x$, the expression within the parenthesis is simply $x^0 + x + x^2 + x^3$. For $|x| < 1$, this sum is equal to $1/(1-x)$, and Eq. (19-94) becomes

$$q_{vib} = \frac{e^{-\frac{1}{2}(h\nu/kT)}}{1 - e^{-(h\nu/kT)}} = \frac{e^{-\Theta_{vib}/2T}}{1 - e^{-\Theta_{vib}/T}}$$ (19-95)

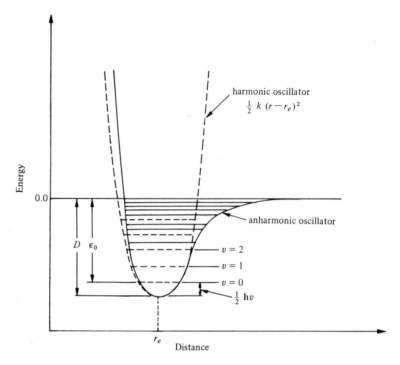

Figure 19-4 The potential energy function of a harmonic (dashed curve, equal spacing between levels) and anharmonic oscillator (solid curve, nonequal spacing between levels; the spacing becomes practically continuous at the higher levels). v, vibrational quantum number; D, dissociation energy of the molecule; ε_0, dissociation energy of the molecule related to $v = 0$.

When the zero of energy is at $v = 0$, we write

$$q_{vib} = \frac{1}{1 - e^{-h\nu/kT}} = \frac{1}{1 - e^{-\Theta_{vib}/T}} \tag{19-96}$$

where $\Theta_{vib}(\equiv h\nu/k)$ is the characteristic temperature of vibration.

Θ_{vib} is a very important property of a substance. The higher it is, the lower the fraction of molecules in the excited states ($v > 0$). Carbon monoxide has a high value of Θ_{vib}, 3070 K; the ratio of Θ_{vib}/T at 300 K is 10.2. Consequently, the fraction of molecules in the excited vibrational states at 300 K is

$$e^{-\Theta_{vib}/T} = e^{-10.2} = 3.7 \times 10^{-5}$$

This fraction is very small and can frequently be neglected. Such conclusions can be drawn for the majority of diatomic molecules. Iodine (solid at room temperature) belongs among the few molecules having loose vibrations at room temperature. Its value of Θ_{vib} is 310 K and the ratio of Θ_{vib}/T at 300 K is 1.03. Thus, the fraction of excited iodine molecules is

$$e^{-1.03} = 0.357$$

At low temperatures, $\Theta_{vib}/T \gg 1$, and $e^{-\Theta_{vib}/T}$ is negligible when compared with unity in the denominator of Eq. (19-95). Thus, the equation reduces to the form

$$q_{vib} = e^{-\Theta_{vib}/2T} \tag{19-97}$$

At high temperatures, $\Theta_{vib}/T \ll 1$, and the exponential in the denominator can be written as $e^{-\Theta_{vib}/T} \approx 1 - \Theta_{vib}/T$; this yields

$$q_{vib} = \frac{T}{\Theta_{vib}} e^{-\Theta_{vib}/2T} \tag{19-98}$$

In the case of a harmonic oscillator the summation and integral give different results for the partition function. Due to the large spacing between the vibrational energy levels, only the summation supplies reliable values for q_{vib}, except at high temperatures and for low frequency oscillators, when the summation and integral give identical results.

For a system containing N independent harmonic oscillators, the partition function and the vibrational contributions to the thermodynamic functions are given by Eqs. (19-99) through (19-107).

$$\ln Q_{vib} = N \ln q_{vib} = -\frac{N\Theta_{vib}}{2T} - N \ln (1 - e^{-\Theta_{vib}/T}) \tag{19-99}$$

and consequently

$$\mathcal{U}_{vib} - \mathcal{U}_{0,vib} = kT^2 \left(\frac{\partial \ln Q_{vib}}{\partial T} \right)_{N,V}$$

$$= RT \frac{\Theta_{vib}/T}{e^{\Theta_{vib}/T} - 1} \tag{19-100}$$

in which, $\mathcal{U}_{0,vib} \equiv Nh\nu/2 = R\Theta_{vib}/2$, is the zero-point energy.

The molar heat capacity is

$$(\mathscr{C}_V)_{vib} = \left[\frac{\partial(\mathcal{U}_{vib} - \mathcal{U}_{0,vib})}{\partial T} \right]_{N,V}$$

$$= R \frac{(\Theta_{vib}/T)^2 \, e^{\Theta_{vib}/T}}{(e^{\Theta_{vib}/T} - 1)^2} \tag{19-101}$$

In Figure 19-5 we have compared the classical and quantum values of $(\mathscr{C}_V)_{vib}$ of a single mode of vibration—one harmonic oscillator. As seen, the quantum value approaches the classical one only at high temperatures. It follows from Eq. (19-101) that at the same T/Θ_{vib} all gases following the harmonic oscillator model have the same molar heat capacities; that is, at the same T/Θ_{vib} the gases are in corresponding states. Consequently, gases having large Θ_{vib} must be raised to higher temperatures in order to have the same $(\mathscr{C}_V)_{vib}$.

The expression for the molar vibrational entropy is

$$\mathscr{S}_{vib} = R \left[\frac{\Theta_{vib}/T}{e^{\Theta_{vib}/T} - 1} - \ln (1 - e^{-\Theta_{vib}/T}) \right] \tag{19-102}$$

The molar vibrational Gibbs free energy is

$$\left(\frac{\mathscr{G} - \mathscr{G}_0}{T} \right)_{vib} = R \ln (1 - e^{-\Theta_{vib}/T}) \tag{19-103}$$

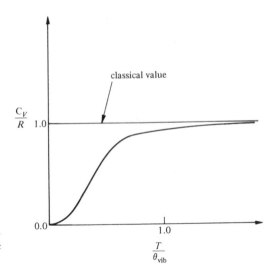

Figure 19-5 Vibrational contribution to the molar heat capacity of a harmonic oscillator.

As seen from (19-95), q_{vib} is neither a function of volume nor pressure. Consequently

$$\mathscr{H}_{vib} = \mathscr{U}_{vib} \tag{19-104}$$

$$(\mathscr{C}_P)_{vib} = (\mathscr{C}_V)_{vib} = \mathscr{C}_{vib} \tag{19-105}$$

and

$$\left(\frac{\mathscr{G} - \mathscr{G}_0}{T}\right)_{vib} = \left(\frac{\mathscr{A} - \mathscr{A}_0}{T}\right)_{vib} \tag{19-106}$$

It is also obvious that in the limit when $T \to 0$

$$\mathscr{U}_0 = \mathscr{H}_0 = \mathscr{A}_0 = \mathscr{G}_0 = \frac{Nh\nu}{2} = \frac{R\Theta_{vib}}{2} \tag{19-107}$$

Linear polyatomic molecules containing n atoms have $3n - 5$ normal modes of vibration, whereas nonlinear molecules have $3n - 6$ modes. The normal vibration frequencies are usually determined by an analysis of the Raman and infrared spectra (see page 542 and Chapter 14). In dealing with these, the common practice is to apply the preceding equations to each vibrational mode separately and then to add the separate contributions to obtain the total vibrational contribution to a given thermodynamic property. This procedure is illustrated in the following example.

Suppose we want to evaluate S_{vib} for $CO_{2(g)}$ at 25°C. Carbon dioxide is a linear molecule and therefore has four normal modes of vibration ($3n - 5 = 9 - 5 = 4$), which are shown in Figure 19-6. Modes 1 and 2 represent stretching vibrations (1 symmetric and 2 antisymmetric), whereas 3 and 4 are bending vibrations. The latter differ only in that the motion takes place in different directions; the bending vibrations are therefore degenerate. The wave numbers ($\bar{\nu} = 1/\lambda$, $\nu = c\bar{\nu}$) of these four normal modes of vibration are

$$\bar{\nu}_1 = 1351 \text{ cm}^{-1} \qquad \bar{\nu}_2 = 2396 \text{ cm}^{-1} \qquad \bar{\nu}_3 = \bar{\nu}_4 = 672 \text{ cm}^{-1}$$

$\bar{\nu}_1$ is not active in the infrared; it is taken from the Raman spectrum.

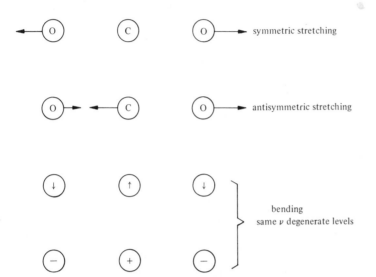

symmetric stretching

antisymmetric stretching

bending
same ν degenerate levels

Figure 19-6   The four normal modes of vibration of carbon dioxide. (The + and − signs refer to displacement perpendicular to the paper.)

Making use of Eqs. (19-100), (19-102), and (19-103), we obtain

$\bar{\nu}$ (cm^{-1})	$\dfrac{h\bar{\nu}c\,\Theta_{vib}}{kT\,298}$	$\left(\dfrac{\mathcal{U}-\mathcal{U}_0}{T}\right)_{vib}$	$\left(\dfrac{\mathcal{A}-\mathcal{A}_0}{T}\right)_{vib}$	$\mathcal{S}_{vib}$
1351	6.52	0.0192	−0.0029	0.092
2396	11.56	0.0002	0.0000	0.000
672	3.24	0.2618	−0.0791	1.423
672	3.24	0.2618	−0.0791	1.423
	$\mathcal{S}_{vib} = 2.938$ J K^{-1} mol^{-1}			

Hence, the vibrational contribution to the entropy of $CO_{2(g)}$ at 25°C is 2.94 J K^{-1} mol^{-1} from a total of 213.6 J K^{-1} mol^{-1}.

Note that the strict separation of the internal degrees of freedom into rotation and vibration can be applied only to some molecules. There are many molecules, especially hydrocarbons, for which the separation is unjustified and leads to erratic results. For better understanding of this complex problem, let us observe the two simple molecules, ethylene and ethane. The stronger double bond in the ethylene molecule

$$
\begin{array}{ccc}
\text{H} & & \text{H} \\
| & & | \\
\text{C} & = & \text{C} \\
| & & | \\
\text{H} & & \text{H}
\end{array}
$$

permits only torsional vibration of the two methylene groups about the bond. In ethane

$$\begin{array}{ccc} & \text{H} & \text{H} \\ & | & | \\ \text{H} & - \text{C} - \text{C} - \text{H} \\ & | & | \\ & \text{H} & \text{H} \end{array}$$

however, the weaker single (C—C) bond enables the rotation of one methyl group relative to the other. Thus, one of the 18 modes of vibration in ethane is replaced by one mode of rotation—internal rotation. The theoretical treatment of internal rotations is quite difficult, due to the fact that in most cases the rotation is hindered; in other words, the rotation is not completely free. There are potential energy barriers that must be overcome before rotation can occur. The internal rotation is neither Raman nor infrared active; microwave spectroscopy, however, has supplied major information for the study of internal rotation of polar molecules.

A very good example of a compound with free internal rotation is the organometallic compound, cadmium dimethyl

$$\begin{array}{ccc} \text{H} & & \text{H} \\ | & & | \\ \text{H} - \text{C} - \text{Cd} - \text{C} - \text{H} \\ | & & | \\ \text{H} & & \text{H} \end{array}$$

The different contributions to the entropy of this compound are as follows: translation and rotation, $253.80 \text{ J K}^{-1} \text{ mol}^{-1}$; vibration, $36.65 \text{ J K}^{-1} \text{ mol}^{-1}$; and free internal rotation, $12.26 \text{ J K}^{-1} \text{ mol}^{-1}$. The total entropy is $302.71 \text{ J K}^{-1} \text{ mol}^{-1}$. This statistical value is in excellent agreement with the calorimetrically obtained value of $302.92 \text{ J K}^{-1} \text{ mol}^{-1}$.

19-7.4 Electronic Partition Function

The electronic partition function of a molecule (or atom) is given by the relation

$$q_{el} = \sum_l g_j e^{-\varepsilon_{el,j}/kT} = g_0 + g_1 e^{-\varepsilon_{el,1}/kT} + g_2 e^{-\varepsilon_{el,2}/kT} + \cdots \qquad (19\text{-}108)$$

where $\varepsilon_{el,j}$ is the electronic energy of level j taken relative to the ground level. Since the first excited electronic state often lies about 400 kJ mol^{-1} above the ground state, the temperature must be very high before there is a significant fraction of the molecules in the electronically excited state. Near room temperature, most molecules are in the ground electronic state. Thus, to a first approximation, the q_{el} can be simplified to

$$q_{el} \approx g_0 \qquad (19\text{-}109)$$

and

$$\ln Q_{el} = N \ln q_{el} = N \ln g_0 = \text{constant} \qquad (19\text{-}110)$$

Hence

$$\mathcal{U}_{el} = \mathcal{H}_{el} = (\mathcal{C}_P)_{el} = (\mathcal{C}_V)_{el} = 0 \qquad (19\text{-}111)$$

The entropy, Gibbs free energy, and Helmholtz free energy are given by the simple expressions

$$\mathcal{S}_{el} = R \ln g_0 \qquad (19\text{-}112)$$

and

$$\mathscr{A}_{el} = \mathscr{G}_{el} = -RT \ln g_0 \qquad (19\text{-}113)$$

Note that $\mathscr{S}_{el}$, $\mathscr{A}_{el}$, and $\mathscr{G}_{el}$ are zero only if the ground electronic level degeneracy, g_0, is unity.

Even some very simple molecules, like monatomic fluorine, for example, can be electronically excited at lower temperatures (see Example 19-3). In such a case the partition function cannot be approximated by Eq. (19-109) and the electronic contribution to the thermodynamic functions must be considered.

EXAMPLE 19-3

The following data are given for monatomic fluorine:

Term	$\bar{\nu} = \varepsilon/hc/(\text{cm}^{-1})$	g_{el}
$P_{3/2}$	0.0	4
$P_{1/2}$	404.0	2
$P_{5/2}$	102,406.5	6

Determine the fraction of fluorine atoms in each of the first three electronic levels at 1000 K.[3]

Solution: First we shall evaluate the electronic partition function

$$\begin{aligned}
q_{el} &= g_0 \, e^{-\varepsilon_{0,el}/kT} + g_1 \, e^{-\varepsilon_{1,el}/kT} + g_2 \, e^{-\varepsilon_{2,el}/kT} + \cdots \\
&= 4 \, e^{-(\varepsilon_{0,el}/hc)(hc/kT)} + 2 \, e^{-(\varepsilon_{1,el}/hc)(hc/kT)} + 6 \, e^{-(\varepsilon_{2,el}/hc)(hc/kT)} + \cdots \\
&= 4 \, e^0 + 2 \, e^{-0.5813} + 6 \, e^{-147.4} \\
&= 5.118
\end{aligned}$$

The fraction of fluorine atoms in the different levels is calculated from the distribution law

$$\frac{n_j}{N} = \frac{g_{el,j} \, e^{-\varepsilon_{el,j}/kT}}{q_{el}}$$

In the ground level

$$\frac{n_0}{N} = \frac{g_{el,0} \times 1}{q_{el}} = \frac{4}{5.118} = 0.782$$

In the first level

$$\frac{n_1}{N} = \frac{g_{el,1} \, e^{-\varepsilon_{el,1}/kT}}{q_{el}} = \frac{2 \times e^{-0.5813}}{5.118} = 0.218$$

For the second excited level

$$\frac{n_2}{N} = \frac{g_{el,2} \, e^{-\varepsilon_{el,2}/kT}}{q_{el}} = \frac{6 \times e^{-147.4}}{5.118} = 0$$

[3] Given the term symbol for the electronic state of an atom one can readily compute the electronic degeneracy, g_{el}, as $2j + 1$, where j is the subscript on the term symbol.

The fraction of fluorine atoms in the second and higher excited levels is negligibly small and may be ignored. However, since the first level is heavily populated, it must be considered in the calculation of the thermodynamic functions (see also Problem 21).

As the data for the other halogen atoms at 1000 K indicate, the population of the first level decreases with the increasing atomic weight of the halogen: $(n_1/N)_{Cl} = 0.12$, $(n_1/N)_{Br} = 2.4 \times 10^{-3}$ and $(n_1/N)_I = 9 \times 10^{-6}$.   ●

19-7.5 Nuclear Partition Function   Owing to the very large energy separation between the two lowest nuclear energy levels at ordinary temperatures, the nucleus is essentially certain to be in its ground state. For a nucleus having a spin quantum number, i, the degeneracy of the ground nuclear energy level is $(2i + 1)$, and therefore, the nuclear partition function can be written as

$$q_{nuc} = \sum_j g_j\, e^{-\varepsilon_j/kT} = g_0 = (2i + 1) \tag{19-114}$$

For a molecule containing r atoms, we write

$$q_{nuc} = \prod_r (2i_r + 1) \tag{19-115}$$

The nuclear partition function gives rise to a nuclear spin entropy

$$\mathscr{S}_{nuc} = k \ln Q_{nuc} = R \ln q_{nuc}$$

$$= R \ln \prod_r (2i_r + 1) \tag{19-116}$$

As mentioned before, this nuclear spin entropy is not included in the thermodynamic calculation of the third law entropy, because the nuclear spin entropy is not ordinarily removed at the lowest temperatures of the calorimetric measurements.

EXAMPLE 19-4
K. Clusius and R. Riccoboni [18] determined the entropy of xenon from calorimetric data and found the value $\mathscr{S}_{298} = 170.29 \pm 1.2 \text{ J K}^{-1}\text{ mol}^{-1}$.

Using the methods of statistical thermodynamics, calculate the entropy of xenon at 298 K and compare it with the experimental value. Find also the values of the thermodynamic functions $\mathscr{C}_P^\circ$, $(\mathscr{H}^\circ - \mathscr{H}_0^\circ)/T$ and $(\mathscr{G}^\circ - \mathscr{H}_0^\circ)/T$, respectively.
Solution: The thermodynamic properties of xenon at 298 K (monatomic gas) result from translational motion only

$$\ln Q_{tr} = N \ln \left[\frac{\mathscr{V}e}{Nh^3} (2\pi mkT)^{3/2} \right]$$

Substituting RT/P for $\mathscr{V}$ and taking the derivative with respect to T, we obtain

$$\left[\frac{\partial \ln Q_{tr}}{\partial T} \right]_P = \frac{5}{2} \frac{1}{T}$$

and therefore

$$\mathscr{H}^\circ - \mathscr{H}_0^\circ = RT^2 \left(\frac{\partial \ln Q_{tr}}{\partial T} \right)_P = RT^2 \frac{5}{2} \frac{1}{T}$$

$$= \frac{5}{2} RT = 6193.9 \text{ J mol}^{-1}$$

Consequently

$$\frac{\mathscr{H}^\circ - \mathscr{H}_0^\circ}{T} = \frac{5}{2} R = 20.78 \text{ J K}^{-1} \text{ mol}^{-1}$$

The molar heat capacity at constant pressure

$$\mathscr{C}_P^\circ = \left(\frac{\partial (\mathscr{H}^\circ - \mathscr{H}_0^\circ)}{\partial T} \right)_P = \frac{\partial}{\partial T} \left(\frac{5}{2} RT \right) = \frac{5}{2} R$$

$$= 20.78 \text{ J K}^{-1} \text{ mol}^{-1}$$

The value of the entropy at $P = 1$ atm and $T = 298$ K is

$$\mathscr{S}_{tr}^\circ = R \ln \left(\frac{V e^{5/2}}{h^3 N} (2\pi m k T)^{3/2} \right) = 169.58 \text{ J K}^{-1} \text{ mol}^{-1}$$

in good agreement with the calorimetric value.
Finally

$$-\left(\frac{\mathscr{G}^\circ - \mathscr{H}_0^\circ}{T} \right)_{tr} = -\left(\frac{\mathscr{H}^\circ - \mathscr{H}_0^\circ}{T} \right)_{tr} + \mathscr{S}_{tr}^\circ$$

$$= -\frac{5}{2} R + 169.58$$

$$= 148.80 \text{ J K}^{-1} \text{ mol}^{-1}$$

EXAMPLE 19-5.
The following data are given for the $^{35}Cl_2$ molecule. The equilibrium internuclear distance is 1.988×10^{-8} cm; the reduced mass is 17.4894; and the fundamental vibrational frequency is 1.6947×10^{13} s^{-1}.
Calculate the molar heat capacity at constant pressure, the enthalpy function, the Gibbs free energy function and entropy of $^{35}Cl_2$ at 25°C and 1 atm. Neglect the electronic contribution.
Solution: The translational contributions:
Internal energy

$$\left(\frac{\mathscr{U}^\circ - \mathscr{U}_0^\circ}{T} \right)_{tr} = kT \left(\frac{\partial \ln Q_{tr}}{\partial T} \right)_{V,N} = \frac{3}{2} R$$

The enthalpy function and the molar heat capacity

$$\left(\frac{\mathscr{H}^\circ - \mathscr{H}_0^\circ}{T} \right)_{tr} = (\mathscr{C}_P^\circ)_{tr} = \frac{3}{2} R + R = 20.78 \text{ J K}^{-1} \text{ mol}^{-1}$$

The Gibbs free energy function

$$\left(\frac{\mathscr{G}^\circ - \mathscr{H}_0^\circ}{T} \right)_{tr} = -R \ln \left(\frac{q_{tr}}{N} \right) = -141.00 \text{ J K}^{-1} \text{ mol}^{-1}$$

The rotational contributions are

$$\left(\frac{\mathcal{H}° - \mathcal{H}_0°}{T}\right)_{rot} = kT\left(\frac{\partial \ln Q_{tr}}{\partial T}\right)_N = (\mathcal{C}_P°)_{rot}$$

$$= R = 8.314 \text{ J K}^{-1} \text{ mol}^{-1}$$

The moment of inertia, $I = \mu r^2 = 17.4894 \times (1.988 \times 10^{-8})^2 \text{ g cm}^2 \text{ mol}^{-1}$, and thus the rotational contribution to the Gibbs free energy function is

$$\left(\frac{\mathcal{G}° - \mathcal{H}_0°}{T}\right)_{rot} = -R \ln\left(\frac{q_{rot}}{N}\right) = -50.33 \text{ J K}^{-1} \text{ mol}^{-1}$$

The vibrational contributions are

$$\left(\frac{\mathcal{H}° - \mathcal{H}_0°}{T}\right)_{vib} = R\frac{\Theta_{vib}/T}{(e^{\Theta_{vib}/T} - 1)} = 1.573 \text{ J K}^{-1} \text{ mol}^{-1}$$

$$\mathcal{C}_{vib}° = R\frac{(\Theta_{vib}/T)^2 \, e^{\Theta_{vib}/T}}{(e^{\Theta_{vib}/T} - 1)^2} = 4.607 \text{ J K}^{-1} \text{ mol}^{-1}$$

and

$$\left(\frac{\mathcal{G}° - \mathcal{H}_0°}{T}\right)_{vib} = R \ln(1 - e^{-\Theta_{vib}/T}) = -0.556 \text{ J K}^{-1} \text{ mol}^{-1}$$

In all these equations $\Theta_{vib}/T = 2.74$

The total values of the thermodynamic functions are

$$\mathcal{C}_P° = (\mathcal{C}_P°)_{tr} + (\mathcal{C}_P°)_{rot} + (\mathcal{C}_P°)_{vib}$$

$$= 20.78 + 8.31 + 4.60 = 33.69 \text{ J K}^{-1} \text{ mol}^{-1}$$

$$\left(\frac{\mathcal{H}° - \mathcal{H}_0°}{T}\right) = 20.78 + 8.31 + 1.57 = 30.66 \text{ J K}^{-1} \text{ mol}^{-1}$$

$$\left(\frac{\mathcal{G}° - \mathcal{H}_0°}{T}\right) = -141.00 - 50.33 - 0.56 = -191.89 \text{ J K}^{-1} \text{ mol}^{-1}$$

Finally, the entropy is given by

$$\mathcal{S}° = \left(\frac{\mathcal{H}° - \mathcal{H}_0°}{T}\right) - \left(\frac{\mathcal{G}° - \mathcal{G}_0°}{T}\right) = 30.66 + 191.89$$

$$= 222.55 \text{ J K}^{-1} \text{ mol}^{-1}$$

The agreement with the calorimetric values given in Table 19-3 is impressive. ●

19-8 STATISTICAL THERMODYNAMICS OF GASEOUS MIXTURES

Let us consider a gaseous mixture of volume V, containing N_A and N_B independent particles of species A and B, respectively. The particles are independent and therefore the distribution of molecules A among the energy levels available to them will not be affected by the molecules of B, and vice versa. Consequently, the partition function of the binary mixture, Q_m, is given by

$$Q_m = \frac{1}{N_A! N_B!} q_A^{N_A} q_B^{N_B}$$

$$= \frac{1}{N_A! N_B!} (q_A' \, e^{-\varepsilon_{0,A}/kT})^{N_A} (q_B' \, e^{-\varepsilon_{0,B}/kT})^{N_B} \qquad (19\text{-}117)$$

Applying Stirling's formula, we obtain for $\ln Q_m$

$$\ln Q_m = - N_A \ln N_A - N_B \ln N_B - \frac{N_A \varepsilon_{0,A}}{kT} - \frac{N_B \varepsilon_{0,B}}{kT}$$

$$+ N_A \ln q_A' + N_B \ln q_B' + N_A + N_B \qquad (19\text{-}118)$$

Thus, the Helmholtz free energy of the mixture assumes the form

$$A_m = - kT \ln Q_m$$
$$= kTN_A \ln N_A + kTN_B \ln N_B + N_A \varepsilon_{0,A} + N_B \varepsilon_{0,B}$$
$$- N_A kT \ln q_A' - N_B kT \ln q_B' - N_A kT - N_B kT \qquad (19\text{-}119)$$

Using molar rather than molecular energies, $\mathscr{U}_{0,A} = N \varepsilon_{0,A}$ and $\mathscr{U}_{0,B} = N \varepsilon_{0,B}$, and defining the molar energy of the mixture at absolute zero as

$$\mathscr{U}_{0,m} = N_A \varepsilon_{0,A} + N_B \varepsilon_{0,B} = x_A \mathscr{U}_{0,A} + x_B \mathscr{U}_{0,B} \qquad (19\text{-}120)$$

Eq. (19-119) can be modified into

$$\mathscr{A}_m - \mathscr{U}_{0,m} = N_A kT \ln \frac{N_A}{N} + N_B kT \ln \frac{N_B}{N} - N_A kT (1 - \ln N + \ln q_A')$$

$$- N_B kT (1 - \ln N + \ln q_B')$$
$$= NkT(x_A \ln x_A + x_B \ln x_B) + x_A (\mathscr{A}_A - \mathscr{U}_{0,A}) + x_B (\mathscr{A}_B - \mathscr{U}_{0,B}) \qquad (19\text{-}121)$$

x_A and x_B denote the mole fractions of component A and B, respectively. The equality

$$(1 - \ln N + \ln q_A') = \mathscr{A}_A - \mathscr{U}_{0,A}$$

has been used in the derivation of Eq. (19-121) (see Problem 30).

Taking the derivative of Eq. (19-118) with respect to T, at constant volume and composition, results in an expression for the energy of the mixture

$$\mathscr{U}_m - \mathscr{U}_{0,m} = x_A (\mathscr{U}_A - \mathscr{U}_{0,A}) + x_B (\mathscr{U}_B - \mathscr{U}_{0,B}) \qquad (19\text{-}122)$$

The entropy of the mixture is

$$\mathscr{S}_m = - R(x_A \ln x_A + x_B \ln x_B) + x_A \mathscr{S}_A + x_B \mathscr{S}_B \qquad (19\text{-}123)$$

and finally for the pressure, we write

$$P = - \left(\frac{\partial A_m}{\partial V} \right)_{T, N_A, N_B} = - \frac{\partial}{\partial V} (- N_A kT \ln q_A' - N_B kT \ln q_B')$$

$$= \frac{N_A kT}{V} + \frac{N_B kT}{V} = p_A + p_B \qquad (19\text{-}124)$$

since at constant T, N_A, and N_B, the partition function depends only on the volume. Eq. (19-124) represents Dalton's law.

From the previously derived equations, we may further conclude that

$$\Delta \mathscr{S}_{mix} = -R(x_A \ln x_A + x_B \ln x_B) \tag{19-125}$$

$$\Delta \mathscr{H}_{mix} = 0 \tag{19-126}$$

$$\Delta \mathscr{V}_{mix} = 0 \tag{19-127}$$

where $\Delta \mathscr{S}_{mix}$, $\Delta \mathscr{V}_{mix}$ and $\Delta \mathscr{H}_{mix}$ are the entropy, volume, and enthalpy of mixing per 1 mole of mixture.

These equations are identical with the relations derived earlier for an ideal solution. This is not surprising since, in our discussion, we have considered a system of noninteracting particles.

19-9 STATISTICAL INTERPRETATION OF THE EQUILIBRIUM CONSTANT

As we have seen, the partition function of a mixture of two noninteracting gaseous species A and B can be written as

$$Q = Q_A Q_B = \frac{q_A^{N_A}}{N_A!} \times \frac{q_B^{N_B}}{N_B!} \tag{19-128}$$

Application of Stirling's formula results in

$$\ln Q = \ln Q_A + \ln Q_B = N_A \ln q_A + N_B \ln q_B$$
$$- N_A \ln N_A - N_B \ln N_B + N_A + N_B \tag{19-129}$$

On taking the derivative of (19-129) with respect to N_A, at constant T, V, and N_B, we obtain

$$\left(\frac{\partial \ln Q}{\partial N_A}\right)_{T,V,N_B} = \ln q_A - \ln N_A = \ln\left(\frac{q_A}{N_A}\right) \tag{19-130}$$

We are now in a position to express the chemical potential in terms of the partition function. The chemical potential per molecule is defined as

$$\mu_A' = \left(\frac{\partial A}{\partial N_A}\right)_{V,T,N_B}$$

and, since the Helmholtz free energy, A, is related to the partition function by the equation $A = -kT \ln Q$, we may also conclude that

$$\mu_A' = -kT\left(\frac{\partial \ln Q}{\partial N_A}\right)_{V,T,N_B} = -kT \ln\left(\frac{q_A}{N_A}\right)_{V,T,N_B} \tag{19-131}$$

and for 1 mole

$$\mu_A = -RT \ln\left(\frac{q_A}{N_A}\right)_{V,T,N_B} \tag{19-132}$$

A similar equation may be derived for μ_B.

We shall next derive the relationship between the equilibrium constant and the partition function. For this purpose, the partition functions of the reactants and products, respectively, must refer to the same ground state. In order to meet this condition, we use the relation

$$q = q' \, e^{-\varepsilon_0/kT} \tag{19-37}$$

Realizing that, the nuclear contribution does not have any influence on the equilibrium constant, we write

$$q' = q_{tr}q_{int}q_{el}$$
$$= \frac{(2\pi mkT)^{3/2}}{h^3} \frac{RT}{P} q_{int}q_{el} \tag{19-133}$$

where the volume in q_{tr} has been expressed as RT/P and $q_{int} = q_{rot}q_{vib}$.

For the partition function in the standard state ($P = 1$ atm), we may write similarly

$$q'^{\circ} = \frac{(2\pi mkT)^{3/2}}{h^3} RT q_{int}q_{el} \tag{19-134}$$

Obviously

$$q' = q'^{\circ}/P \tag{19-135}$$

and Eq. (19-37) assumes the form

$$q = \frac{q'^{\circ}}{P} e^{-\varepsilon_0/kT} \tag{19-136}$$

Hence, the chemical potential becomes

$$\mu = -RT \ln \frac{q'^{\circ} e^{-\varepsilon_0/kT}}{PN}$$
$$= -RT \ln \frac{q'^{\circ} e^{-\varepsilon_0/kT}}{N} + RT \ln P \tag{19-137}$$

On comparing Eq. (19-137) with the classical definition of the chemical potential of an ideal gas, $\mu = \mu^{\circ} + RT \ln P$, it is obvious that

$$\mu^{\circ} = -RT \ln \frac{q'^{\circ} e^{-\varepsilon_0/kT}}{N} \tag{19-138}$$

This represents the statistical definition of the standard chemical potential.

The equilibrium constant for an ideal homogeneous gaseous reaction can now be written in the following way:

$$\Delta G^{\circ} = \sum_i \nu_i \mu_i^{\circ} = -RT \sum_i \nu_i \ln \left(\frac{q_i'^{\circ}}{N}\right) e^{-\varepsilon_{0,i}/kT}$$
$$= -RT \ln K_P \tag{19-139}$$

where ν_i is the stoichiometric coefficient of component i, taken as positive for the products and as negative for the reactants. Eq. (19-139) is also equal to

$$K_P = \prod_i \left(\frac{q_i'^{\circ}}{N}\right)^{\nu_i} e^{-\Delta\mathcal{E}_0/RT} \tag{19-140}$$

where $\Delta\mathcal{E}_0$ is the change in the energy for the reaction at 0 K($\equiv \sum_P \nu_i\varepsilon_{0,i} - \sum_R \nu_i\varepsilon_{0,i}$).

The equilibrium constant expressed in molecules per cubic centimeter is

$$K' = \prod_i (q_i^\circ)^{\nu_i} e^{-\Delta \mathscr{E}_0/RT} \tag{19-141}$$

where $q_i^\circ = q_i'^\circ / 1000RT$.

EXAMPLE 19-6

Figure 19-7 illustrates the potential energy curve for the dissociation reaction

$$N_2(g) \to 2\,N(g)$$

Assuming the rigid rotor and the harmonic oscillator approximation, find the equilibrium constant for the dissociation of molecular nitrogen into monatomic nitrogen at 5000 K.

The degeneracy numbers of the electronic ground levels are four and one for monatomic and molecular nitrogen respectively.

Solution: For the dissociation of $N_2(g)$ into $2\,N(g)$, the equilibrium constant expressed by Eq. (19-140) has the form

$$K_P = \left[\left(\frac{q'^\circ q_{el}}{N} \right)^2_{N(g)} \middle/ \left(\frac{q'^\circ q_{el}}{N} \right)_{N_2(g)} \right] e^{-\Delta \mathscr{E}_0/RT}$$

$$= \frac{\left[\frac{(2\pi mkT)^{3/2} RT}{Nh^3 P^\circ} \right]^2_{N(g)}}{\left[\frac{(2\pi mkT)^{3/2} RT}{Nh^3 P^\circ} \right]_{N_2(g)}} \left(\frac{h^2 \sigma}{8\pi^2 IkT} \right)_{N_2(g)} (1 - e^{-\Theta_{vib}/T})_{N_2(g)} (q_{el})^2_{N(g)} \, e^{-\Delta \mathscr{E}_0/RT}$$

$$\tag{a}$$

$$\downarrow \qquad\qquad \downarrow \qquad\qquad \downarrow \qquad\qquad \downarrow \qquad\quad \downarrow$$
$$K_{tr} \qquad\qquad K_{rot} \qquad\qquad K_{vib} \qquad\qquad K_{el} \qquad\quad K_0$$

Making use of the basic data

$$k = 1.3804 \times 10^{-16} \text{ erg K}^{-1} \text{ molecule}^{-1}$$
$$h = 6.625 \times 10^{-27} \text{ erg sec}$$
$$N = 6.023 \times 10^{23} \text{ molecules mol}^{-1}$$
$$P^\circ = 1 \text{ atm} = 1.0133 \times 10^6 \text{ dyne cm}^{-2}$$

and the values shown in Figure 19-7 we get

$$I = \mu r_e^2 = \left(\frac{m_1 m_2}{m_1 + m_2} \right) r_e^2 = \frac{\left(\frac{14.008}{6.023 \times 10^{23}} \right)^2}{\left(\frac{28.016}{6.023 \times 10^{23}} \right)} \times (1.098 \times 10^{-8})^2$$

$$= 1.402 \times 10^{-39} \text{ g cm}^2$$

$$\frac{\Theta_{vib}}{T} = \frac{h\nu}{kT} = \frac{hc\bar{\nu}}{kT} = \frac{(6.625 \times 10^{-27})(3 \times 10^{10})2357.6}{(1.3804 \times 10^{-16})(5 \times 10^3)}$$

$$= 0.678$$

$$\Delta \mathscr{E}_0 = N\varepsilon_0 = \frac{6.023 \times 10^{23} \times 9.751 \times 1.602 \times 10^{-19}}{4.184}$$

$$= 225,023 \text{ cal mol}^{-1}$$

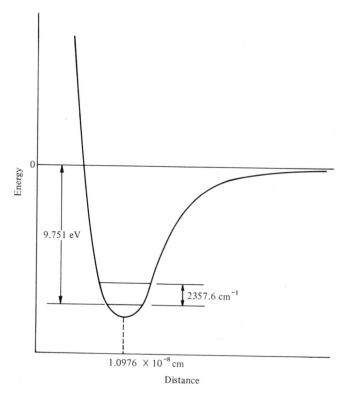

Figure 19-7 The potential energy function of nitrogen.

Substitution of these values into (a) yields

$$K_P = (8.393 \times 10^8)(1.150 \times 10^{-3}) \times 0.4923 \times 16 \times (1.457 \times 10^{-10})$$

$$\downarrow \qquad\qquad \downarrow \qquad\qquad\qquad \downarrow \quad \downarrow \quad \downarrow$$

$$K_{tr} \qquad\qquad K_{rot} \qquad\qquad K_{vib} \;\; K_{el} \;\; K_0$$

$$= 1.108 \times 10^{-3} \text{ atm.}$$

The value of K_P is largely determined by two competitive terms, K_{tr} and K_0, respectively. Nevertheless, K_{rot}, K_{vib}, and K_{el} also contribute significantly to the equilibrium constant.

In an isotopic exchange reaction, for example

$$^{16}O_2 + {}^{18}O_2 \rightarrow 2\ {}^{16}O^{18}O$$

the greatest contribution to the ratio of partition functions, and thus to the equilibrium constant, comes from the rotational partition function of the different isotopes, and especially from the difference in their symmetry numbers. The different contributions to the equilibrium constant of the oxygen isotope exchange reaction are the following: $K_{tr} = 1.005$, $K_{rot} = 3.970$, $K_{vib} = 1.000$, and the difference in the zero-level energy of the different isotopes contributes about 0.971 to the equilibrium constant. Thus

$$K = (1.005)(3.970)(1.000)(0.971) = 3.87 \qquad \bullet$$

19-10 STATISTICAL THERMODYNAMICS OF AN IDEAL CRYSTAL

According to Einstein (1907), 1 mole of an ideal crystal can be considered as a system of N noninteracting particles. Because of their thermal energy, the particles oscillate about their equilibrium positions in the crystal lattice. Due to the fact that each particle vibrates independently of the others and has three degrees of freedom, a crystal may be treated as a system of $3N$ independent and distinguishable harmonic oscillators, and its partition function may be written in the form[4]

$$\ln Q_{vib} = 3N \ln q_{vib} = -\frac{3N\Theta_E}{2T} - 3N \ln (1 - e^{-\Theta_E/T}) \qquad (19\text{-}142)$$

where Θ_E is the characteristic Einstein temperature of vibration ($\equiv h\nu/k$).

Using this equation to calculate the internal energy of an ideal Einstein crystal, we obtain

$$\mathcal{U} = kT^2 \left(\frac{\partial \ln Q_{vib}}{\partial T} \right)_{V,N} = \frac{3}{2} Nh\nu + \frac{3Nk\Theta_E}{e^{\Theta_E/T} - 1} \qquad (19\text{-}143a)$$

or

$$\mathcal{U} - \mathcal{U}_0 = 3 RT \frac{\Theta_E/T}{e^{\Theta_E/T} - 1} \qquad (19\text{-}143b)$$

$\mathcal{U}_0$ is the zero-point energy ($\equiv \frac{3}{2}Nh\nu = \frac{3}{2}R\Theta_E$).

The molar heat capacity is expressed as

$$\mathcal{C}_V = 3R \frac{(\Theta_E/T)^2 e^{\Theta_E/T}}{(e^{\Theta_E/T} - 1)^2} \qquad (19\text{-}144)$$

As can be seen, the zero-point energy does not contribute to the molar heat capacity.

For the entropy of an ideal Einstein crystal, we have

$$\mathcal{S} = \frac{\mathcal{U} - \mathcal{U}_0}{T} + k \ln Q = 3R \left[\frac{\theta_E/T}{e^{\Theta_E/T} - 1} - \ln (1 - e^{-\Theta_E/T}) \right] \qquad (19\text{-}145)$$

The Gibbs free energy of vibration is equal to the Helmholtz free energy of vibration

$$\mathcal{G} = \mathcal{A} = -kT \ln Q = \frac{3}{2} R\Theta_E + 3RT \ln (1 - e^{-\Theta_E/T}) \qquad (19\text{-}146)$$

and

$$\mathcal{G} - \mathcal{U}_0 = \mathcal{A} - \mathcal{U}_0 = 3RT \ln (1 - e^{-\Theta_E/T}) \qquad (19\text{-}147)$$

Of all the thermodynamic functions, the molar heat capacity is the easiest to measure. For this reason, we shall now compare the experimentally determined values of the molar heat capacities with those predicted by the Einstein theory. It is found experimentally that, at temperatures approaching zero, the molar heat capacity approaches zero; in the limit of high temperatures, $\mathcal{C}_V$ approaches Dulong–Petit's value of $3R$ [3].

---

[4] The vibration frequencies of atomic crystals are much lower than those of diatomic molecules. Thus the vibrations of most atomic crystals are completely activated at temperatures far below those required to activate vibrations of diatomic molecules.

As can be proved, Einstein's theory predicts these limiting values of $\mathscr{C}_V$ very well. For the limit $T \to \infty$, $e^{\Theta_E/T} \approx 1 + (\Theta_E/T)$ and therefore

$$\lim_{T \to \infty} \mathscr{C}_V \approx 3R \frac{(\Theta_E/T)^2 \, e^{\Theta_E/T}}{(1 + \Theta_E/T - 1)^2} \approx 3R \qquad (19\text{-}148)$$

and for the limit $T \to 0$, $e^{\Theta_E/T} - 1 \approx e^{\Theta_E/T}$. Consequently

$$\lim_{T \to \infty} \mathscr{C}_V \approx 3R \left(\frac{\Theta_E}{T}\right)^2 e^{-\Theta_E/T} \approx 0 \qquad (19\text{-}149)$$

since the exponential is limiting very fast to zero as the temperature approaches zero.

In the intermediate and low ranges of temperature, however, the Einstein theory predicts $\mathscr{C}_V$ values that are lower than those actually observed (see Figure 19-8). The most interesting feature of this theory is that, for all monatomic solids, it predicts equal values of $\mathscr{C}_V$ at the same (Θ_E/T). The value of Θ_E, which varies from crystal to crystal, is chosen empirically to give the best experimental fit.

A more successful theory of heat capacities of solids was formulated by Debye (1912). Debye's model, like Einstein's, pictures a crystal of N atoms as a collection of $3N$ harmonic oscillators. Debye's oscillators, however, are not single atoms, but vibrations of the entire crystal; the crystal is considered to be a huge molecule.

The distribution of frequencies among the different oscillators may be calculated from the classical theory of standing waves in a continuous elastic medium. This theory, however, provides no limit to the number of oscillators. Debye assumed that there is an upper limit for the allowed vibrations, which is characterized by ν_D and is determined by the requirement that the total number of normal vibrations be equal to $3N$.

$$\int_0^{\nu_D} g(\nu) \, d\nu = 3N \qquad (19\text{-}150)$$

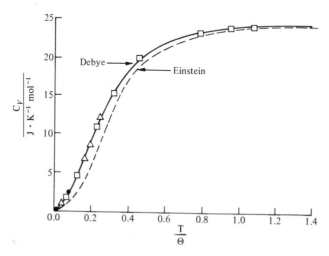

Figure 19-8 The heat capacity of solid crystalline silver, graphite, and aluminum as a function of T/Θ. ●, silver; △, graphite; and □, aluminum.

The vibrations are harmonic, and as N is very large, they form a continuous spectrum. The vibrations are characterized by a distribution function $g(\nu)$, which is defined so that the expression $g(\nu)\, d\nu$ gives the number of oscillators vibrating in the frequency interval between ν and $\nu + d\nu$.

The partition function of such a system is given by the equation

$$Q = q(\nu_1)^{g(\nu_1)} \cdot q(\nu_2)^{g(\nu_2)} \cdots q(\nu_D)^{g(\nu_D)} \tag{19-151}$$

and

$$\ln Q = \sum_{i=1}^{i=D} g(\nu_i) \ln q(\nu_i) \tag{19-152}$$

Because of the continuity, the summation may be replaced by an integral; thus

$$\ln Q = \int_0^{\nu_D} g(\nu) \ln q(\nu)\, d\nu \tag{19-153}$$

where $q(\nu)$ is the partition function of a harmonic oscillator, defined by Eq. (19-95).

In his derivation Debye applied the Rayleigh–Jeans' relation for the distribution function:

$$g(\nu)\, d\nu = c\nu^2\, d\nu \tag{19-154}$$

where the constant c is evaluated from the restricting conditions (19-150)

$$\int_0^{\nu_D} g(\nu)\, d\nu = c \int_0^{\nu_D} \nu^2\, d\nu = \tfrac{1}{3} c\nu_D^3 = 3N$$

Following from this, we have

$$c = \frac{9N}{\nu_D^3} \tag{19-155}$$

Eq. (19-154) can thus be written as

$$g(\nu)\, d\nu = \frac{9N}{\nu_D^3} \nu^2\, d\nu \tag{19-156}$$

and the partition function of a Debye crystal takes the form

$$\ln Q = -\frac{9N}{\nu_D^3} \int_0^{\nu_D} \left[\frac{h\nu}{2kT} + \ln\left(1 - e^{-h\nu/kT}\right) \right] \nu^2\, d\nu \tag{19-157}$$

from which we obtain

$$\mathcal{U} = \frac{9NkT}{\nu_D^3} \int_0^{\nu_D} \left(\frac{h\nu/kT}{e^{h\nu/kT} - 1} + \frac{h\nu}{2kT} \right) \nu^2\, d\nu \tag{19-158}$$

or

$$\mathcal{U} = \frac{9Nh\nu_D}{8} + \frac{9NkT}{\nu_D^3} \int_0^{\nu_D} \left(\frac{h\nu/kT}{e^{h\nu/kT} - 1} \right) \nu^2\, d\nu \tag{19-159a}$$

If we let $\Theta = h\nu/k$ and $\Theta_D = h\nu_D/k$, where Θ_D is known as the characteristic Debye temperature of the solid, we obtain

$$\mathscr{U} = \frac{9R\Theta_D}{8} + 9RT\left(\frac{T}{\Theta_D}\right)^3 \int_0^{\Theta_D/T} \frac{(\Theta/T)^3}{e^{\Theta/T}-1} \, d\left(\frac{\Theta}{T}\right) \qquad \textbf{(18-159b)}$$

The integral can be evaluated only numerically. Tables of the Debye function, $D(\Theta/T)$

$$D\left(\frac{\Theta}{T}\right) = 3\left(\frac{T}{\Theta_D}\right)^3 \int_0^{\Theta_D/T} \frac{(\Theta/T)^3}{e^{\Theta/T}-1} \, d\left(\frac{\Theta}{T}\right) \qquad \textbf{(19-160)}$$

are available in the literature [4]. Combination of Eqs. (19-159b) and (19-160) results in

$$\mathscr{U} = \frac{9R\Theta_D}{8} + 3RTD\left(\frac{\Theta}{T}\right) \qquad \textbf{(19-161)}$$

Differentiation of $D(\Theta/T)$ followed by integration by parts yields

$$\mathscr{C}_V = 3RD\left(\frac{\Theta}{T}\right) + 3RT\frac{\partial}{\partial T}\left[D\left(\frac{\Theta}{T}\right)\right]$$

$$= 3R\left[4D\left(\frac{\Theta}{T}\right) - \frac{3(\Theta_D/T)}{e^{\Theta_D/T}-1}\right] \qquad \textbf{(19-162)}$$

For the high temperature value of $\mathscr{C}_V$ we have

$$e^{\Theta_D/T} - 1 \approx 1 + \frac{\Theta_D}{T} - 1 \approx \frac{\Theta_D}{T}$$

and

$$\lim_{T\to\infty} D\left(\frac{\Theta}{T}\right) \approx 1$$

Consequently

$$\lim_{T\to\infty} \mathscr{C}_V \approx 3R \qquad \textbf{(19-163)}$$

For the low temperature limit

$$\lim_{T\to 0} D\left(\frac{\Theta}{T}\right) \approx \frac{\pi^4}{5}\left(\frac{T}{\Theta_D}\right)^3$$

and

$$\lim_{T\to 0} \mathscr{C}_V \approx \frac{12}{5}\pi^4 R\left(\frac{T}{\Theta_D}\right)^3 \approx aT^3 \qquad \textbf{(19-164)}$$

The Debye equation fits the experimental $\mathscr{C}_V$ versus T curve much better than the Einstein equation, especially at low and moderate temperatures (see Figure 19-8). The value of Θ_D is chosen again to give the best overall experimental fit. Values for the parameter Θ_D for several substances are listed in Table 19-5. Anharmonic vibrations in the crystal may result in a volume dependence of $\Theta_D(d\Theta_D/dV \neq 0)$.

It follows from the given theory that the smaller Θ_D, the lower is the temperature at which the molar heat capacity reaches the classical $3R$ value.

Table 19-5 The Debye characteristic temperatures

Substance	$\dfrac{\Theta_D}{(K)}$	Substance	$\dfrac{\Theta_D}{(K)}$
Al	389	Fe	453
Ag	215	Hg	90
Au	169	K	99
Be	980	Na	160
Bi	111	Pb	90
C (diamond)	1855	Pt	225
Ca	230	Sn	165
Cu	315	Zn	240

19-11 IDEAL LATTICE GAS

Let us consider a simple process in which molecules of a gaseous species are bound by physical forces on a two-dimensional lattice of sites on the surface of a crystal; sites are locations on the surface of minimum potential energy for an adsorbed molecule. Such a process is called **adsorption**, and the system of molecules adsorbed on the sites is sometimes referred to as a lattice gas.

We restrict the derivation presented by T. I. Hill [17] to a model in which all the sites are distinguishable, independent, and in which no more than one molecule is attached to one site. The adsorbed molecules are far apart so that no interactions exist between them. The adsorbed molecules have three vibrational degrees of freedom and may be considered as three-dimensional oscillators with energy levels $\varepsilon_1, \varepsilon_2, \ldots, \varepsilon_i$. The partition function for one molecule at a site is then $q = \sum_i e^{-\varepsilon_i/kT}$, and for N molecules $Q = q^N$, or $Q = q^{3N}$ for a set of $3N$ one-dimensional oscillators, just as in the case of an Einstein model for a crystal. Generally, the number of sites M is larger than the number of molecules N, $(M > N)$, and therefore $(M - N)$ sites remain unoccupied. Consequently, a configurational degeneracy term, not present in the Einstein crystal—it would be present if the crystal had lattice vacancies—appears in the expression for the partition function, Q. This degeneracy term may be defined as the number of ways in which N indistinguishable molecules can be distributed among M equivalent but distinguishable sites and is equal to

$$\frac{M!}{N!(M-N)!} \tag{19-165}$$

Following from this, the partition function assumes the form

$$Q = \frac{M!}{N!(M-N)!} q^N \tag{19-166}$$

Expressing $\ln N!$, $\ln M!$ and $\ln (M-N)!$ from Stirling's formula, Eq. (19-166) can be rewritten into

$$\ln Q = M \ln M - N \ln N - (M-N) \ln (M-N) + N \ln q \tag{19-167}$$

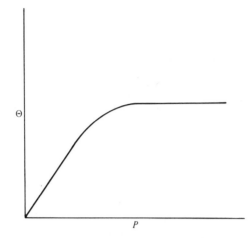

Figure 19-9 The Langmuir adsorption isotherm.

On taking the derivative of Eq. (19-167) with respect to N, at constant T and M, we obtain

$$-\left(\frac{\partial \ln Q}{\partial N}\right)_{M,T} = \ln \frac{\Theta}{(1-\Theta)q} = \frac{\mu}{kT} \tag{19-168}$$

where μ is the chemical potential and Θ denotes the fraction of occupied sites, (N/M).

At equilibrium, the chemical potentials of the gas adsorbed on the surface, $\mu_{(ad)}$, and the gas in the bulk phase, $\mu_{(g)}$, must be equal. Consequently

$$\ln \frac{\Theta}{(1-\Theta)q} = \frac{\mu_{(ad)}}{kT} = \frac{\mu_{(g)}}{kT} = \frac{\mu^\circ(T)}{kT} + \ln P \tag{19-169}$$

which can be modified into

$$\Theta = \frac{N}{M} = \frac{Pq\, e^{\mu^\circ(T)/kT}}{1 + Pq\, e^{\mu^\circ(T)/kT}} = \frac{K(T)P}{1 + K(T)P} \tag{19-170}$$

where $K(T) = q\, e^{\mu^\circ(T)/kT}$. Eq. (19-170) is the well-known Langmuir adsorption isotherm (also see page 713).

As is obvious at small P, $K(T)P \ll 1$, and Θ is linearly dependent on P; at large values of P, $K(T)P \gg 1$, and $\Theta = 1$. Thus, the adsorption isotherm must have the simple form shown in Figure 19-9. This isotherm may be applied in dealing with problems of adsorption of species from liquid as well as from gaseous mixtures. Its main use is, however, in the field of heterogeneous catalysis.

19-12 STATISTICAL DERIVATION OF THE EQUATION OF STATE FOR NONIDEAL FLUIDS

The equation relating the pressure to the partition function

$$P = kT\left(\frac{\partial \ln Q}{\partial V}\right)_{T,N} \tag{19-171}$$

is the basic statistical equation of state for any system. Once the partition function of the system is known, its equation of state can be derived.

The characteristic feature of dense fluids is the existence of strong intermolecular forces. The energy resulting from these forces makes the evaluation of the partition function and, consequently, the derivation of an equation of state difficult. The great majority of theories of dense fluids follows from the laws of classical mechanics and from the fundamental theorems of statistical mechanics by means of few well-defined physical and mathematical approximations.

Classical mechanics gives the following mathematical definition for the partition function:

$$Q = \frac{1}{N!} \frac{1}{h^{3N}} \int \overset{6N}{\cdots} \int e^{-(1/kT)\mathfrak{H}(q_1 \cdots p_{3N})} dq_1 \cdots dp_{3N} \qquad \textbf{(19-172)}$$

where $\mathfrak{H}(q_1 \cdots p_{3N})$ is the Hamiltonian of the system, and obviously, it is a function of the momentum (p) as well as of the position (q) coordinates. The integral is taken over the coordinates of all molecules of the system. Since each molecule has three position and three momentum coordinates and there are N molecules in the system, the integral in Eq. (19-172) must be a $6N$-fold integral.

A rigorous solution of the integral, known as the phase integral, is seldom possible; various approximations, depending on the kind of molecular interactions, must be assumed before a solution can be found. The accuracy of the equation of state obtained depends primarily on the approximations made.

Next we intend to derive an equation of state for a simple monatomic gas, for which the interactions between molecules, although not negligible, are small—a dilute gas. For such a gas the Hamiltonian has the form

$$\mathfrak{H} = \sum_{i=1}^{3N} \frac{1}{2m} p_i^2 + U(q_1 \cdots q_{3N}) \qquad \textbf{(19-173)}$$

where U is the interaction potential energy. Substitution in Eq. (19-172) shows that the $6N$-dimensional phase integral separates into two independent partition functions:

$$Q = Q_{(p)} Q_{(q)} \qquad \textbf{(19-174)}$$

where

$$Q_{(p)} = \frac{1}{N!} \frac{1}{h^{3N}} \int_{-\infty}^{+\infty} \overset{3N}{\cdots} \int_{-\infty}^{+\infty} e^{-(1/kT) \sum_i (p_i^2/2m)} dp_1 \cdots dp_{3N} \qquad \textbf{(19-175)}$$

is only momentum coordinate dependent, and

$$Q_{(q)} = \int \overset{3N}{\cdots} \int e^{-U/kT} dq_1 \cdots dq_{3N} \qquad \textbf{(19-176)}$$

is only position coordinate dependent. Eq. (19-176) is often called the configuration integral.

In a dilute gas the kinetic energy is the same as for an ideal gas; consequently, the p's are all independent and $Q_{(p)}$ may be expressed as a product of $3N$ integrals, all of the same form

$$\int_{-\infty}^{+\infty} e^{-p^2/kT2m} dp = (2\pi mkT)^{1/2} \qquad \textbf{(19-177)}$$

Thus

$$Q_{(p)} = \frac{1}{N!} \frac{(2\pi m k T)^{3N/2}}{h^{3N}} \tag{19-178}$$

As seen the evaluation of $Q_{(p)}$ is quite simple; it is the same as for an ideal gas. The difficulty arises with the evaluation of the configuration integral. The evaluation of $Q_{(q)}$ requires the knowledge of the potential energy function. Since such a function is not available in general, $Q_{(q)}$ cannot be evaluated in closed terms. It may, however, be approximated in various ways, each of which leads to a different equation of state.

In our discussion we will consider a gas for which the potential energy can be approximated as a sum of a set of terms, each describing the potential energy of a pair of molecules as a function of their separation, r_{ij}. Thus

$$U = \sum_{i>j}^{N} \sum_{j=1}^{N-1} u(r_{ij}) \tag{19-179}$$

The limits of the summations result from the simple fact that there are N ways available of selecting the first molecule of the pair and $(N-1)$ ways for the second. The order of chosing the molecules is immaterial; therefore, the number of terms involved in the summation is $N(N-1)/2$. The terms that Eq. (19-179) introduces into the exponent in Eq. (19-176) are such that we may conclude

$$e^{-U/kT} = \prod_{N \geq i \geq j \geq 1} e^{-u(r_{ij})/kT} \tag{19-180}$$

The product $\prod$ includes again $N(N-1)$ terms, each of which approaches the value of one as r_{ij} becomes large and $u(r_{ij}) \to 0$. If $u(r_{ij})/kT \ll 1$, then the exponential may be expanded into a series of the form

$$e^{-u(r_{ij})/kT} \approx 1 + f_{ij} \tag{19-181}$$

As is obvious, $f_{ij} \to 0$ as r_{ij} becomes very large. Introducing this into Eq. (19-180) results in

$$e^{-U/kT} \approx 1 + \sum f_{ij} + \sum\sum f_{ij}f_{kl} + \cdots \tag{19-182}$$

and the configuration integral takes the form

$$Q_{(q)} = \int \overset{3N}{\cdots} \int (1 + \sum f_{ij} + \sum\sum f_{ij}f_{kl}) \, dq_1 \cdots dq_{3N} \tag{19-183}$$

Integration of Eq. (19-183) results in a series of terms; the term arising from unity is equal to

$$\int \overset{3N}{\cdots} \int dq_1 \cdots dq_{3N} = \mathcal{V}^N \tag{19-184}$$

We see that integration over the three position coordinates of a molecule yields the volume accessible to the molecule, so that Eq. (19-183) is equivalent to

$$Q_{(q)} = \mathcal{V}^N + \int \overset{3N}{\cdots} \int (\sum f_{ij} + \sum\sum f_{ij}f_{kl}) \, dq_1 \cdots dq_{3N} \tag{19-185}$$

The integrals over the f_{ij} terms, the $f_{ij}f_{kl}$ terms, and so on, are known as cluster integrals. For a dilute gas it is sufficient to retain only the f_{ij} terms; each of these terms can be replaced by $e^{-u_{ij}/kT} - 1$, and $Q_{(q)}$ may be written as

$$Q_{(q)} = \mathcal{V}^N + \sum_i \sum_j \int \cdots \int (e^{-u_{ij}/kT} - 1)\, dq_1 \cdots dq_{3N} \qquad (19\text{-}186)$$

The double sum contains $N(N-1)/2$ integrals, all of the same value; each represents an integration over all the possible coordinates. If the value of one integral is I

$$I = \int \cdots \int (e^{-u_{ij}/kT} - 1)\, dq_1 \cdots dq_{3N} \qquad (19\text{-}187)$$

then

$$Q_{(q)} = \mathcal{V}^N + \frac{N^2}{2} I \qquad (19\text{-}188)$$

N is large and therefore $N(N-1)/2 \approx N^2/2$; the term $(N^2/2)I$ is responsible for the non-ideal behavior of the gas.

The term u_{ij} depends only on the coordinates of molecules i and j respectively; therefore, the integration can be performed over all the other coordinates. The result is $\mathcal{V}^{N-2}$. Consequently

$$I = \mathcal{V}^{N-2} \iint (e^{-u_{ij}/kT} - 1)\, dV_i\, dV_j \qquad (19\text{-}189)$$

where $dV_i = dq_i\, dq_{i+1}\, dq_{i+2}$ and $dV_j = dq_j\, dq_{j+1}\, dq_{j+2}$. If the coordinates of molecules i and j are replaced by their relative coordinates and the integration is taken over the relative coordinates, we have

$$I = \mathcal{V}^{N-2}\mathcal{V} \int (e^{-u_{ij}/kT} - 1)\, dV$$

$$= \mathcal{V}^{N-1} \int (e^{-u_{ij}/kT} - 1)\, dV \qquad (19\text{-}190)$$

Finally, the configuration integral becomes

$$Q_{(q)} = \mathcal{V}^N + \frac{N^2 \mathcal{V}^{N-1}}{2} \int (e^{-u_{ij}/kT} - 1)\, dV$$

$$= \mathcal{V}^N \left(1 - \frac{N^2 \mathcal{B}(T)}{\mathcal{V}}\right) \qquad (19\text{-}191)$$

where $\mathcal{B}(T) \equiv \frac{1}{2} \int (1 - e^{-u_{ij}/kT})$ is the second virial coefficient. The equation of state is

$$P = kT\left(\frac{\partial \ln Q_{(q)}}{\partial V}\right)_{T,N} = kTN\left(\frac{\partial \ln \mathcal{V}}{\partial \mathcal{V}}\right)_{T,N} + kT\left[\frac{\partial \ln\left(1 - \dfrac{N^2 \mathcal{B}(T)}{\mathcal{V}}\right)}{\partial \mathcal{V}}\right]_{T,N} \qquad (19\text{-}192)$$

and therefore

$$P = \frac{NkT}{\mathcal{V}}\left(1 + \frac{N\mathcal{B}(T)}{\mathcal{V}}\right) \qquad (19\text{-}193)$$

Eq. (19-193) applies to dilute gases only. For dense gases, a series development of $Q_{(q)}$ in terms of cluster integrals is possible. This cannot be applied to liquids because the series does not converge. Recently, some progress has been made in the development of theories of liquids; nevertheless, the problem of theory of liquids is still far from being solved.

Thus, we shall discuss a rather simple, if also rather inadequate, partition function of liquids. No attempt will be made to reproduce the detailed mathematics of the approach to the problem. Let us assume that the whole liquid is subdivided into cells of very small volume, v. A molecule may move freely within the cell. The boundary of the cell is set by the nearest neighbors of the molecule. Inside the cell, the interaction energy has a uniform value, $-U$, where U is a function only of the volume of the cell. Outside the cell the potential energy is infinite. Thus, the molecule behaves like a gas within the small volume of the cell, known as free volume. If we assume further that the translational degrees of freedom are essentially classical and that the internal degrees of freedom are the same in the liquid as in the gas (temperature dependent only), then the molar partition function can be written as

$$Q = q_{int}^N \left(\frac{2\pi m k T}{h^2}\right)^{3N/2} [eve^{-(-U/kT)}]^N \tag{19-194}$$

where the term $(eve^{U/kT})^N$ is essentially a measure of the possible arrangement of the N particles in space and is recognized as the configurational partition function.

For the Helmholtz free energy of 1 mole of a liquid that obeys the smoothed potential free-volume model, we have

$$\mathscr{A} = -kT \ln Q = -[RT \ln q_{int} + RT$$
$$+ RT \ln \left(\frac{2\pi m k T}{h^2}\right)^{3/2} + RT \ln v + NU] \tag{19-195}$$

If the free volume and U is a function of the volume only, then

$$P = -\left(\frac{\partial A}{\partial V}\right)_{T,N} = -\left[\frac{\partial(NU)}{\partial V}\right]_{T,N} + RT\left(\frac{\partial \ln v}{\partial V}\right)_{T,N} \tag{19-196}$$

In order to get reasonable results, the free volume is to be adjusted to agree with experiment.

As is obvious, in some aspects the theory presented resembles the solid state theories (the organized structure of the nearest neighbors forming the boundary of the free volume), and in some aspects it resembles the gaseous state theories (translational and internal degrees of freedom are the same as for a gas). Many statistical theories treat the liquid state in a similar way. Near the melting point, there is still a great part of organized structure left in the liquid; therefore the solidlike approximations are more justified. However, near the critical temperature, there is practically no organized structure left in the liquid; therefore, the gaslike assumptions seem to be more justified. Between these two extreme cases, a proper potential energy function of liquids must be derived (so far not available) before a general theory of liquids can be postulated.

19-13 ELECTRON GAS

A solid metal may be considered as a system of vibrating particles (positive ions) bound to the lattice sites of a crystal and of freely mobile electrons—an electron gas—which are responsible for the high electrical conductance of metals. The metal structure is regarded merely as a home for electrons where they can move freely in a field of uniform potential energy. The term "electron gas," even when not fully justified, is often used because the free electrons—either one electron per atom or all the normal valence electrons—confined in their motion to the volume occupied by the metal, exhibit some kinetic properties similar to those exhibited by monatomic gas molecules.

Debye, in his heat capacity theory, completely ignored the contribution to the heat capacity of solids arising from the electron gas. In spite of this the molar heat capacities of solids calculated from the Debye theory and those observed experimentally were in close agreement. If the electron gas truly behaved like a monatomic gas, the experimentally found molar heat capacities would differ by $\frac{3}{2}R$ from the Debye value. This, however, has not been observed. Thus, the free electrons, unlike the atoms of a monatomic gas, contribute very little to the heat capacity of metals.

To find an explanation of this phenomenon one must turn to statistics. Maxwell–Boltzmann statistics cannot be applied here, since this would require that at absolute zero temperature $n_j = 0$ for all levels for which $j > 0$ [all the electrons occupy the same (ground) energy level] in contradiction to Pauli's principle. The electrons must be spread out over all the allowed translational states of the system (like a monatomic gas, free electrons possess translational degrees of freedom only) and their distribution can be described by Fermi–Dirac statistics only.

The Fermi–Dirac distribution law has the form

$$n_j = \frac{g_j}{A\, e^{\varepsilon_j/kT} + 1} = \frac{g_j}{e^{(\varepsilon_j - \varepsilon_F)/kT} + 1} \qquad (19\text{-}21)$$

where A is replaced by $e^{-\varepsilon_F/kT}$; ε_F is the so called Fermi-level energy (see also Section 17-3). It can be identified with the chemical potential of the electron gas at absolute zero; it varies between 1–5 eV. n_j and g_j are the occupancy number and degeneracy number of level j, respectively.

At absolute zero the value of $1/kT$ is infinite, and the following conclusions can be drawn from the distribution law. For n_j to be greater than zero it is necessary that $\varepsilon_F > \varepsilon_j$. Consequently, only levels below the Fermi level can be occupied by electrons; all the levels having $\varepsilon_j > \varepsilon_F$ must be empty. There are as many energy states at or below ε_F as there are free electrons.

A few Fermi–Dirac distribution curves are plotted in Figure 19-10. As is obvious the shape of the curves is affected by the temperature; at very high temperatures the Fermi–Dirac and the Maxwell–Boltzmann distribution curves become equal. At all temperatures, states with energy equal to ε_F are half occupied: $n_j = g_j/2$.

As a first approximation it can be shown that the Fermi-level energy is proportional to the free electron density, $N/\mathscr{V}$ [see Eqs. (17-34) and (17-36)]

$$\varepsilon_F \approx \frac{h^2}{8m_e}\left(\frac{3N}{\pi\mathscr{V}}\right)^{2/3} \qquad (19\text{-}197)$$

where h is Planck's constant, m_e the mass of an electron and $\mathscr{V}$ the volume of the metal containing Avogadro's number, N, electrons. For a typical metal with one free electron, $\varepsilon_F/k\ (\equiv T_F)$ is approximately 6×10^4 K. This means that at room temperature almost all electrons are in the lowest states and those above ε_F are empty.

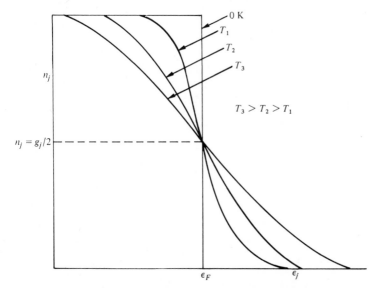

Figure 19-10   Fermi–Dirac distribution curves at different temperatures.

It is now easy to explain why an electron gas does not contribute appreciably to the heat capacity of metals. The only way an electron could gain energy when it is heated would be to move into a higher energy level. This is, however, impossible, since almost all the higher levels near by are already filled with electrons.

The approximation (see Eq. 17-38)

$$\mathscr{C}_{V,e} = \frac{R\pi^2 k}{2\varepsilon_F} T = \frac{R\pi^2}{2} \cdot \frac{T}{T_F} \qquad (19\text{-}198)$$

indicates that the molar heat capacity of the electron gas becomes significant at very high temperatures only. For a metal with one valence electron $\mathscr{C}_{V,e} = 7 \times 10^{-4}\, T$ J K^{-1} mol^{-1}. The electronic term, $\mathscr{C}_{V,e}$, is linear in T and therefore at very low temperatures it may contribute more than the Debye term $(\mathscr{C}_V = aT^3)$ to the heat capacity of the crystal.

The Fermi–Dirac statistics and the electron gas concept are used to explain the characteristic behavior of metallic conductors, semiconductors, insulators, and the band theory of crystals (see also Section 17-3).

_____ *Problems*

1. In how many ways may ten distinguishable particles be distributed among the given energy levels in such a manner that four particles be in the zero level, three particles in the first level, two particles in the second level, and one particle in the third level. Let all the remaining levels be empty.

2. Using Stirling's formula evaluate ln $N!$ for $N = 10$ and $N = 12$. What is the per cent error introduced by Stirling's approximation?

3. As proved in the text, the three different statistics predict essentially the same distribution when $n_j/g_j \ll 1$. Prove that the thermodynamic probability for any given distribution is different by $N!$ for the Boltzmann statistics [11]:

$$w_{BE} \approx w_{FD} \approx \left(\frac{w}{N!}\right)_{MB}$$

4. Using Boltzmann's distribution law, calculate the population ratio at 25°C of two energy levels separated from each other by (a) 2 kcal mol^{-1}, (b) 100 kcal mol^{-1}.

5. Prove that

$$\left(\frac{\partial \ln Q}{\partial \beta}\right)_{V,N} = -\mathcal{U}$$

6. Calculate the translational partition function, q_{tr}, for hydrogen in a volume of $1 \, cm^3$ and $T = 298$ K.

7. Calculate the partition function, Q, for hydrogen at 298 K and 1 atm.

8. Prove that the Maxwell–Boltzmann distribution law (Eq. 19-25) is equivalent to the energy distribution law derived on page 77.

9. Discuss the difference between the following two mathematical expressions:

$$e^{-(\sum_i \varepsilon_i/kT)} \qquad \text{and} \qquad \sum_i e^{-\varepsilon_i/kT}$$

10. Calculate the absolute entropy of 1 mole of helium gas in the ideal gas state at 25°C and 1 atm.

11. Determine the thermodynamic functions of neon, argon, and krypton at 25°C and 1 atm.

12. Calculate the ratio of the number of HBr molecules in state $v = 2$, $l = 5$ to the number in state $v = 1$, $l = 2$ at 1000 K. Assume that all the molecules are in their electronic ground states.

13. Calculate the rotational partition function for $^{14}N^{14}N$ and $^{14}N^{15}N$ molecules at 25°C.

14. To what temperature must water vapor be raised to excite 1.00% of the molecules to the first excited vibrational energy level. The wave numbers of the three normal modes of vibration for water vapor are 1595, 3652, and 3756 cm^{-1}.

15. Find the temperature at which 10% of the molecules will be in the first excited electronic state if this state is 100 kcal mol^{-1} above the ground level.

16. Making use of the molar partition function show that for a monatomic gas $\mathcal{U} = \frac{3}{2}NkT$ and that $P = NkT/\mathcal{V}$.

17. Calculate the rotational contribution to the thermodynamic functions of SO_2 at 298 K and 1 atm. The moments of inertia are $I_A = 1.386 \times 10^{-39}$ g cm^2, $I_B = 8.143 \times 10^{-39}$ g cm^2, and $I_C = 9.529 \times 10^{-39}$ g cm^2. The symmetry number of SO_2 is 2.

18. Calculate the thermodynamic functions of SO_2 at 25°C and 1 atm given: $M_{SO_2} = 64.063$, $\bar{\nu}_1 = 1151.4 \, cm^{-1}$, $\bar{\nu}_2 = 517.7 \, cm^{-1}$, $\bar{\nu}_3 = 1361.8 \, cm^{-1}$. For the

rotational contribution use data from Problem 17. Neglect the electronic and nuclear contribution.

19. Evaluate the rotational contribution to the thermodynamic functions of carbon monoxide at 298 K and 1 atm. The internuclear equilibrium distance in CO is 1.13 Å, the atomic masses of carbon and oxygen respectively are 12.0115 and 15.9994.

20. (a) Determine the vibrational contribution to the thermodynamic functions of Cl_2, if its normal mode of vibration $\bar{\nu} = 565 \text{ cm}^{-1}$. (b) If the ground electronic state of Cl is a doublet with a separation of 881 cm^{-1} calculate the electronic contributions to $\mathscr{S}$ and $\mathscr{C}_V$ for Cl at 500 K.

21. From the data given in Example 19-3, calculate the thermodynamic properties of fluorine.

22. Calculate the equilibrium constant for the isotope exchange reaction

$$D + H_2 = H + DH$$

at 25°C. Assume that the equilibrium distance and the force constant of H_2 and DH are the same.

23. The energy of dissociation of Na_2 molecules was measured spectroscopically as $\Delta E_0 = 0.73 \text{ eV}$. The fundamental vibration frequency of Na_2 is 159.23 cm^{-1} and the internuclear distance is 3.078 Å. Calculate the equilibrium constant, K_P, of the reaction

$$Na_2(g) \rightarrow 2 \text{ Na}(g)$$

at 1000 K. (Neglect the degeneracy of levels: $g_j = 1$).

24. The following data are given for H_2, D_2, and HD at 300 K:

	H_2	HD	D_2
$\bar{\nu}/(\text{cm}^{-1})$	4,371	3,786	3,092
$I \times 10^{40}/(\text{g cm}^2)$	0.458	0.613	0.919

Calculate the equilibrium constant of the reaction

$$H_2(g) + D_2(g) \rightarrow 2 \text{ HD}(g)$$

at 300 K. Compare the answer with the result of Problem 22.

25. By multiplying the zero-point energy of a single oscillator by $g(\nu) \, d\nu$ and integrating, show that the zero-point energy of a Debye crystal is $\frac{9}{8} Nk\Theta_D$.

26. Use L'Hopital's rule to show that the limiting values of the molar heat capacities of solid substances are zero and $3R$ respectively.

27. Derive the following equation for the entropy of a Debye crystal

$$\mathscr{S} = \frac{4}{3} \frac{\mathscr{U}}{T} - 3R \ln (1 - e^{-\Theta_D/T})$$

28. Calculate the molar heat capacity of aluminum at 100 K. Given $\Theta_D = 398$ K and $D(\Theta/T) = 0.1837$ at 100 K.

29. Recently measured values of specific heat capacities of copper are listed below:

T (K)	c (J K^{-1} g^{-1})
4	0.060
60	0.137
100	0.254
180	0.346
300	0.386

Find values of the characteristic Debye and Einstein Temperatures from each of these data. Which theory results in a better correlation of the data?

30. Prove that for pure component A the following equation can be derived

$$1 - \ln N + \ln q'_A = \mathscr{A}_A - \mathscr{U}_{0,A}$$

_____ *References and Recommended Reading*

[1] N. Davidson: *Statistical Mechanics*. McGraw-Hill, New York, 1962.

[2] D. D. Fitts and J. F. Mucci, *J. Chem. Educ.*, **39**:515 (1962). The Boltzmann H-function.

[3] R. K. Fitzgerald and F. H. Verhoek: *J. Chem. Educ.*, **37**:545 (1960). The law of Dulong and Petit.

[4] G. T. Furukawa and T. B. Douglas, in D. E. Gray (ed.): *American Institute of Physics Handbook*. McGraw-Hill, New York, 1957, pp. 4-44 to 4-46.

[5] R. W. Hakala: *J. Chem. Educ.*, **44**:436 (1967). Evaluation of the exponent in Boltzmann's distribution law.

[6] C. E. Hecht: *J. Chem. Educ.*, **44**:124 (1967). Negative absolute temperatures.

[7] R. G. Keil and L. K. Nash: *J. Chem. Educ.*, **48**:601 (1971). The most probable distribution in statistical thermodynamics.

[8] J. Lee: *J. Chem. Educ.*, **42**:340 (1965). The entropy of a nuclear spin system in a magnetic field.

[9] J. W. Lorimer: *J. Chem. Educ.*, **43**:39 (1966). Elementary statistical mechanics without Lagrange multipliers.

[10] P. E. Rock: *Classical Thermodynamics*. Macmillan, New York, 1969.

[11] P. Van Rysselberghe: *J. Chem. Educ.*, **39**:511 (1962). The term $k \ln N!$ in the derivation of the Sackur Tetrode equation.

[12] R. E. Sonntag and G. I. Van Wylen: *Fundamentals of Statistical Mechanics*. Wiley, New York, 1966.

[13] P. A. H. Wyatt: *J. Chem. Educ.*, **39**:27 (1962). Elementary statistical mechanics without Stirling's approximation.

[14] J. L. Hollenberg: *J. Chem. Educ.*, **47**:2 (1970). Resource papers VIII. Energy states of molecules.

[15] R. D. Turoff: *Am. J. Phys.*, **38**(3):387 (1970). Avoiding Stirling's approximation and Lagrange multipliers in introductory statistical mechanics.

[16] M. L. Burns and R. A. Brown: *Am. J. Phys.*, **39**(7):802 (1971). Becker averaging technique for obtaining distribution function in statistical mechanics.

[17] T. I. Hill: *Introduction to Statistical Thermodynamics.* Addison-Wesley, Reading, Pa., 1962.

[18] K. Clusius and R. Riccoboni: *Z. Phys. Chem.*, **B38**:81 (1937).

Part V

Chemical Kinetics

chapter 20

Rates and Mechanisms of Reactions

20-1 INTRODUCTION

Chemical kinetics is about time and chemistry: How fast do reactants change into products? How does the rate of the reaction depend upon the properties of the reacting system?

To date most (and the best) answers to the first question have been determined experimentally (Chapter 20). In regard to the second question, the known relations between the reaction rate and such properties as concentrations of the components or the temperature of the reacting system are largely empirical (Section 20-2).

There are four major theoretical activities within chemical kinetics: (1) reaction mechanisms to explain rate-concentration relations are proposed (Section 20-3); (2) relations to thermodynamics are explored (Section 20-4); (3) rates are predicted from statistical mechanical models (Chapter 21); and (4) classical or quantum mechanics are used to explain the reactions on a molecular level (Chapters 21 and 22). The theoretical activities have stimulated the development of new and specialized experimental methods, for example, for the measurement of fast reaction rates (Section 20-6) and for the study of reactive intermediates (Section 20-7) and molecular dynamics (Section 21-3 and Chapter 22).

Our exposition of chemical kinetics starts with its most important dimension—time. What is the shortest possible time in which chemical species can react? A very simple reaction involving the lightest chemical species, that is, an electron, is $Na + Cl \rightarrow Na^+ + Cl^-$. The minimum time for Na and Cl to react is the time it takes for the electron to travel from the Na to the Cl atom. This time from elementary considerations is

$$t = d/v = d/(2E/m)^{1/2} \qquad (20\text{-}1)$$

where d is the distance between the two atoms; v is the velocity, E the kinetic energy, and m the mass of the electron. Taking typical values—the interatomic distance $d = 5 \text{ Å} = 0.5 \text{ nm}$, the kinetic energy $E = 400 \text{ kJ mol}^{-1} \approx 7 \times 10^{-19} \text{ J}$ per electron (typical of the energies released when chemical bonds are formed)—we get $t \approx 5 \times 10^{-16}$ s. For the transfer of a proton, for example, $C + HF \rightarrow CH + F$, with the above d and E, $t \approx 2 \times 10^{-14}$ s.

Electron transfer is about the fastest possible chemical process. Therefore, the time 5×10^{-16} s is the lower limit of the experimental ability needed in chemistry to measure time intervals. It is not, however, the shortest meaningful time interval known to all of science. Figure 20-1 compares the time to complete the fastest chemical process to the completion times of other natural processes.

In the organic chemistry laboratory the student has had experience with reactions that may take minutes if not hours to complete. Indeed, most (if not all) reactions in a flask take much longer to completion than 5×10^{-16} s or even 5×10^{-14} s. The long

629

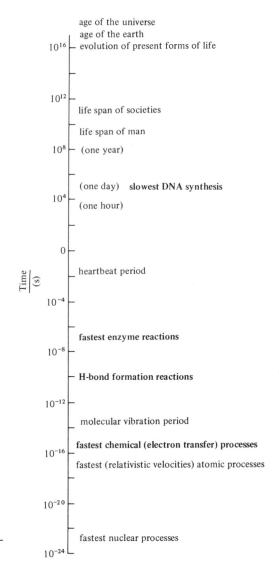

Figure 20-1 A comparison of the duration of natural processes and structures.

completion times are evidence that most reactions proceed, on the average, by processes more complex than envisualized in our simple models of transfer reactions.

Is there a slowest possible reaction? The answer is "No." Among the slowest reactions ever measured is the isomerization (at ambient temperatures) of L-aspartic acid in fossil bones. Approximately one half of the L-aspartic acid molecules in a sample isomerize to D-aspartic acid in 10^5 years [1].

In our presentation of chemical kinetics we will emphasize recent developments more than in the first four parts of the book for two reasons: (1) chemical kinetics is a much less developed field than, for example, thermodynamics. Therefore the chemical kinetics material traditionally presented in physical chemistry texts is much less fundamental and more readily dispensable than that on thermodynamics. (2) Chemical kinetics is a very rapidly developing field (see especially Sections 20-6 through 20-8 and Chapter 22) where important advances have been made quite recently. To omit the recent developments would mean that the student would be learning the kind of chemical kinetics that is no longer practised.

20-2 RATES OF HOMOGENEOUS REACTIONS

To obtain the simplest macroscopic description of a reacting homogeneous system we extend the description of the corresponding equilibrium system in a straightforward manner. For example, a gaseous system in which two isomers A and B are in chemical equilibrium is completely specified (page 189) by two variables, usually pressure and temperature. To specify the related reacting system at a given time, the partial pressure of A or B would have to be given in addition to total pressure and temperature. The values of the three variables, listed as a function of time, would describe the system completely. In practice, one usually deals with reactions for which a more compact but equivalent description is possible in an analytic form—the **rate law**.

The above description may become invalid for very fast reactions because the variables, pressure and temperature, are defined only at equilibrium—for the Boltzmann distribution (see Part IV). Only if the processes that lead to temperature, pressure, and similar physical equilibria are faster than the chemical reactions to be studied can we assume that temperature and pressure are well defined. This condition is more readily achieved if the system is stirred. Reactions in solids, even homogeneous ones, must be approached with care because strains, distribution of imperfections, and so on, may reach equilibrium very slowly. Note further that the physical equilibria need only be local—they can be confined to the particular region where the reaction is being observed.

As an example of rate law description of reactions, let us consider the rate of the reaction

$$H_2O_2 + 2\ HI \rightarrow 2\ H_2O + I_2 \qquad (20\text{-}2)$$

This reaction was one of the first on which accurate determinations were made (by Harcourt and Esson in [4]). If the concentration of HI is held constant by regenerating it from the product I_2 with added $NaHSO_3$, the H_2O_2 concentration C varies with time t as shown by curve a in Figure 20-2a. Curve a is exponential.

$$C = C_0\, e^{-kt} \qquad (20\text{-}3)$$

where $C_0 \equiv C_{t=0}$ and k is the rate constant. Eq. (20-3) when differentiated and simplified yields the reaction rate

$$r = -\frac{dC}{dt} = kC \qquad (20\text{-}4)$$

This equation is the **rate law** for reaction (20-2). Eq. (20-4), unlike Eq. (20-3), does not contain any quantities, for example, C_0, determined by the initial conditions. Therefore, general theories of reaction kinetics concentrate on explaining differential rate equations similar to Eq. (20-4) rather than the corresponding integral forms. The reaction rate r is thus an important quantity. Its various definitions must be considered.

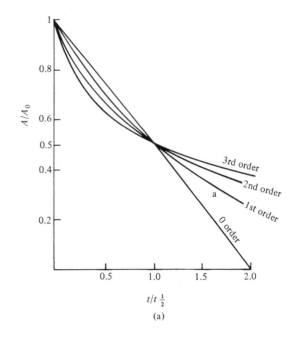

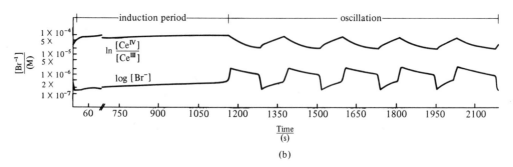

Figure 20-2 Time-dependence of concentration in a homogeneous closed reacting system.
(a) Some curves of the concentration variable (A/A_0) versus reduced time $(t/t_{\frac{1}{2}})$ for systems that obey the simple rate law $r = kA^\alpha$ (see Eq. 20-8). The initial concentration variable is A_0. The half-life $t_{\frac{1}{2}}$ is defined in the text. Sample reactions follow.

$$\text{camphene hydrochloride } (C_{10}H_{17}Cl) \rightarrow \text{ isobornyl chloride}$$
$$(A)$$

In nitrobenzene solution at 20°C this isomerization obeys the rate law $r = kA^{3/2}$. (The A/A_0 versus $t/t_{\frac{1}{2}}$ curve for this rate law would fall between first- and second-order curves.) When camphene is added the reaction obeys a second-order rate law [56].

$$2\,NO + O_2 \rightarrow 2\,NO_2$$
$$(A)$$

In the temperature range 0–65°C, $P_{tot} \approx 1$–20 torr, $P_{NO_2} < 6$ torr the rate law is $r = k\,(NO)^{2.0}(O_2)^{1.0}$. If $P_{NO} = 2\,P_{O_2}$ the reaction will follow the third-order curve [42].

(b) Time-dependence of some concentrations in a complex reacting system. Initial concentrations were: malonic acid, $CH_2(CO_2H)_2 = 0.013\,M$; $KBrO_3 = 0.063\,M$; $Ce(NH_4)_2(NO_3)_5 = 0.001\,M$; and $H_2SO_4 = 0.8\,M$ in water. The net reaction, averaged over the oscillations of the concentrations, is

$$3\,H^+ + 3\,BrO_3^- + 5\,CH_2(CO_2H)_2 \xrightarrow{Ce^{3+}} 3\,BrCH(CO_2H)_2 + 2\,HCO_2H + 4\,CO_2 + 5\,H_2O$$

20-2.1 Reaction Rate Although in a commercial production context a total reaction rate (an extensive quantity) would be useful, in fundamental reaction kinetics we shall need intensive reaction rate only. This reaction rate may be defined most generally as a partial derivative

$$r = -\left(\frac{\partial R}{\partial t}\right)_v \qquad \text{or} \qquad r = \left(\frac{\partial P}{\partial t}\right)_v \qquad \textbf{(20-5)}$$

where R and P are the concentrations[1] of a reactant and a product, respectively. The minus sign in the first definition is included so that the rate is positive under most conditions. The subscript v stands for variables that may affect concentration through means other than destruction of reactant or creation of product. In closed systems volume V is the most important variable of this kind.

Although in most reactions the volume change which accompanies a reaction is negligible, there are some, for example, polymerization reactions, where the volume may change by 10% or more at constant pressure [2]. Then we have

$$\frac{dR}{dt} = \left(\frac{\partial R}{\partial t}\right)_V + \left(\frac{\partial R}{\partial V}\right)_t \cdot \frac{dV}{dt} = -r - \frac{n_R}{V^2} \cdot \frac{dV}{dt} = -r - R \cdot \frac{d\ln V}{dt} \qquad \textbf{(20-6)}$$

where n_R is the moles of the reactant R. In order to get the true rate r of a reaction, we must determine the variation of both R and V with respect to t.

In open systems, a change in the net flow rate through the system or diffusion across a concentration gradient may affect concentrations at the point of measurement. Rather than correcting the experimental results for these effects, it is usually easier to carry out the experiment so that the effects of flow rate changes and concentration gradients on R are negligible, that is

$$\frac{dR}{dt} \approx \left(\frac{\partial R}{\partial t}\right)_v = -r$$

The general definition of the reaction rate r in Eq. (20-5) has one disadvantage: it depends upon a particular reactant or product. If a single stoichiometric equation can describe the reacting system at all times, this dependence may be eliminated. Suppose

$$a\mathrm{A} + b\mathrm{B} \rightarrow c\mathrm{C} + d\mathrm{D}$$

---

[1] In Part V, the concentration or partial pressure (only for gases) of a chemical species is denoted by the symbol of that chemical species printed in italic (*sloping*) type.

At any instant during the reaction any one (or any two) of the following reactions may prevail:

$$\mathrm{BrO_3^-} + 4\,\mathrm{Ce^{3+}} + \mathrm{CH_2(COOH)_2} + 5\,\mathrm{H^+} \rightarrow \mathrm{BrCH(COOH)_2} + 4\,\mathrm{Ce^{4+}} + 3\,\mathrm{H_2O}$$

$$4\,\mathrm{Ce^{4+}} + \mathrm{BrCH(COOH)_2} + 2\,\mathrm{H_2O} \rightarrow \mathrm{Br^-} + 4\,\mathrm{Ce^{3+}} + \mathrm{HCOOH} + 2\,\mathrm{CO_2} + 5\,\mathrm{H^+}$$

$$\mathrm{BrO_3^-} + 2\,\mathrm{Br^-} + 3\,\mathrm{CH_2(COOH)_2} + 3\,\mathrm{H^+} \rightarrow 3\,\mathrm{BrCH(COOH)_2} + 3\,\mathrm{H_2O}$$

A single rate law of the form of Eqs. (20-11) and (20-12) valid for all stages of the reaction, has not been written for this system. [Part (b) reprinted with permission from R. J. Field, E. Körös, and R. M. Noyes; *J. Am. Chem. Soc.*, **94**:8649 (1972). Copyright by the American Chemical Society.]

where the stoichiometric coefficients a, b, c, d are the smallest possible integers. Then the reaction rate can be defined unambiguously using the extent of the reaction, ζ (page 255). Introducing the concentration $\mu = \zeta / V$ we have

$$r = \left(\frac{\partial \mu}{\partial t}\right)_v = -\frac{1}{a}\left(\frac{\partial A}{\partial t}\right)_v = -\frac{1}{b}\left(\frac{\partial B}{\partial t}\right)_v = \frac{1}{c}\left(\frac{\partial C}{\partial t}\right)_v = \frac{1}{d}\left(\frac{\partial D}{\partial t}\right)_v \qquad (20\text{-}7)$$

In every use of r one should establish unambiguously how it is defined. In Part V, unless stated otherwise, the definition is as in Eq. (20-7).

20-2.2 Determination of Rate Law

We now turn to the determination of the rate law. We assume that the reacting system has two reactants A and B and a substance C that might be a catalyst for the reaction or a product. The rate law, which expresses the concentration dependence of r, very often can be fitted to the following form:

$$r = k(T) \cdot A^{\alpha} \cdot B^{\beta} \cdot C^{\gamma} \qquad (20\text{-}8)$$

where α, β, γ are most commonly equal to 1 or 2; less frequently they equal other integers and half-integral fractions in the range 0 to 3. α, β, and γ are the orders of the reaction with respect to species A, B, and C. In **mixed order** reactions more than one species has nonzero order. $\alpha + \beta + \gamma$ is the overall order of the reaction. The rate law for a reaction is "determined" when k, α, β, and γ are known.

The units for all definitions of r may be moles per cubic decimeter per second $(\text{mol dm}^{-3}\,\text{s}^{-1})$. (Occasionally the non-SI units—minutes or hours—are used.) This choice of units for r means that the stoichiometric coefficients a, b, c, and d are unitless (see Eq. 20-7). The units of k must be $(\text{mol dm}^{-3})^{-(\alpha+\beta+\gamma-1)}\text{s}^{-1}$ from Eq. (20-8) (if mole A = mole B = mole C). In gas reactions the concentrations in Eq. (20-8) are sometimes replaced by pressures in atmospheres or Newtons per square meter (N m^{-2}). Our choice of units (and all such choices) must obey the tests of convenience and self-consistency but is otherwise arbitrary. The conveniences deriving from our choice are consistency with traditional and SI usage and invariance with change in the definition of r.

We will discuss only the more generally applicable mathematical and experimental tools and strategies for determining the rate law. First, let us consider concentration versus time data for this purpose. Data of this type require that Eq. (20-8) be integrated to get $\mu = f(t)$. Different assumptions are made about α, β, and γ; the integrations are performed; and the different functions f are compared with the experimental data to find the function f which fits best.

Consider the case when $r = (\partial \mu / \partial t)_v = d\mu/dt$. If in addition $\beta = \gamma = 0$, the integration can be done in a straightforward manner by separation of variables. The results are listed in Table 20-1, column 3. Column 4 lists the results in another form, as the half-life, $t_{\frac{1}{2}}$, of the reaction. The half-life is defined as the time required for the concentration of A to reach the value $\frac{1}{2}A_0$.

If the rate law has the form

$$r = kAB \qquad (20\text{-}9)$$

then, for integration to be possible, time-independent stoichiometric relations between the concentrations A and B must exist and be known: $a\text{A} + b\text{B} \rightarrow$ products. $A - A_0 = -a\mu$ and $B - B_0 = -b\mu$. Substitution of the preceding relations in Eq.

Table 20-1 Expressions used to determine reaction order and rate constants for the reaction $aA+bB \rightarrow$ products

Reaction Order	Differential Equation	Rate Constant, k	Half-life, $t_{1/2}$
0	$r = \left(\dfrac{d\mu}{dt}\right)_v = k$	$\dfrac{A_0 - A}{at}$	$\dfrac{A_0}{2ak}$
1	$r = \left(\dfrac{d\mu}{dt}\right)_v = kA$	$\dfrac{1}{at} \ln \dfrac{A_0}{A}$	$\dfrac{1}{ak} \ln 2$
2	$r = \left(\dfrac{d\mu}{dt}\right)_v = kA^2$	$\dfrac{1}{at}\left(\dfrac{1}{A} - \dfrac{1}{A_0}\right)$	$\dfrac{1}{akA_0}$
$n\,(n \neq 1)$	$r = \left(\dfrac{d\mu}{dt}\right)_v = kA^n$	$\dfrac{1}{a(n-1)t}\left(\dfrac{1}{A^{n-1}} - \dfrac{1}{A_0^{n-1}}\right)$	$\dfrac{(2^{n-1} - 1)}{akA_0^{n-1}(n-1)}$
2, mixed	$r = \left(\dfrac{d\mu}{dt}\right)_v = kAB$	$\dfrac{\ln[(A/A_0)(B_0/B)]}{t(bA_0 - aB_0)}$	$\dfrac{1}{k(aB_0 - bA_0)} \ln\left(2 - \dfrac{bA_0}{aB_0}\right)$

(20-9) and separation of variables yields

$$\int_0^\mu \frac{d\mu}{(A_0 - a\mu)(B_0 - b\mu)} = k \int_0^t dt$$

The left-hand side may be integrated by breaking it into partial fractions. The result is listed on the last line of Table 20-1. If $bA_0 \approx aB_0$, the expressions listed are indeterminate or nearly so, and different expressions must be used [3, p. 20]. The method just illustrated may be used to integrate even more complicated rate laws of the form defined in Eq. (20-8).

Once the integrations have been performed, it is possible to determine the reaction orders (Example 20-1). This may be done analytically or graphically. Analytically, one may compute k according to the various expressions in column 3 of Table 20-1 from the experimental data and determine which expression yields a constant value k (except for random error). Graphically, the concentration functions in column 3 [that is, $A_0—A$ for the zero order expression, $\ln A_0/A$ for the first order and so on] are plotted versus t; a straight line indicates the correct order. Alternatively, plot the half-lives, $t_{1/2}$, against the various initial concentration functions in column 4 of Table 20-1; a straight line again indicates the correct order.

If the assumed rate law is too difficult to integrate or the integrated expressions derived from various assumed rate laws are experimentally indistinguishable or too unwieldy, then certain experimental strategies may be brought into play. The most generally used strategy is the **isolation** or **flooding method**. In this method we choose the initial conditions so that the reactant A is present initially in a concentration much lower than all other reactants, that is, we "isolate" A or "flood" the system with the other reactants. The result is that only the concentration of A changes appreciably during the run, and the rate law in Eq. (20-8) takes the simplified form

$$r = kA^\alpha B_0^\beta C_0^\gamma = k'A^\alpha$$

α is determined as described previously. By varying the initial concentration $[B_0]$ in two runs (numbered 1 and 2) we obtain

$$k_1'/k_2' = \left(\frac{B_{01}}{B_{02}}\right)^\beta$$

The only unknown in this expression is β which therefore can be determined (Example 20-1).

Reactions in dilute solution are always flooded with respect to the solvent. It is not usually practical to vary the solvent concentration enough to determine the order with respect to the solvent.

Another general strategy for determining the rate law is the **initial rate method**. In this method we determine the initial change in concentration, $\Delta\mu$, such that $\Delta\mu \ll A_0/a, B_0/b$. Then

$$r \approx \frac{\Delta\mu}{\Delta t} = kA_0^\alpha B_0^\beta \qquad\qquad (20\text{-}10)$$

By varying the initial concentrations one at a time, the orders with respect to A and B may be determined. The method avoids integration of the rate law all together, an especially important advantage before computers made it possible to perform repeated numerical integrations and to compute many values of complicated expressions. Another simplifying aspect of the initial rate method is that the reaction system contains only reactants. Therefore, complications due to the presence of reaction products cannot occur. Of course, if desired, the initial reaction mixture may contain a known (nonzero) concentration of one or more products.

EXAMPLE 20-1

A reaction of the type $A + 2B \rightarrow P$ was investigated by determining the rate of disappearance of B. Two typical runs gave the following results:

t	$A_{init} = 0.500$ M	$A_{init} = 0.250$ M
(min)	B	B
0	1.00×10^{-3} M	1.00×10^{-3} M
4	0.75	0.87
8	0.56	0.75
12	0.42	0.65
16	0.32	0.56

Determine the rate law for this reaction.

Solution: The rate, from Eq. (20-7), would be

$$r = -\frac{dA}{dt} = -\frac{1}{2}\frac{dB}{dt} = \frac{dP}{dt}$$

assuming that the concentrations do not depend on volume. The rate law is assumed to be $r = f(A, B) = kA^{\alpha}B^{\beta} \approx k'B^{\beta}$ since the system is flooded with respect to A.

The value of β is calculated first. Four methods for doing this are illustrated. The nature of the data and the calculator's personal preferences will determine the method of choice.

1. *The half-life method.* Using two convenient concentrations of B from the first run, for example, $B_0 = 1.0 \times 10^{-3}$ M and $B_4 = 0.75 \times 10^{-3}$ M, we see that when either $B = \frac{1}{2}B_0$ or $B = \frac{1}{2}B_4$ the time $t_{1/2} \approx 10$ min. As seen from Table 20-1 the half-life is independent of the initial concentration for a first-order rate law. Therefore $\beta = 1$.

How reliable is this conclusion? Let us consider two other possible choices $\beta = \frac{1}{2}$ and $\beta = 2$. If $\beta = \frac{1}{2}$ then from Table 20-1

$$\frac{t_{1/2,0}}{t_{1/2,4}} = \left(\frac{B_0}{B_4}\right)^{1/2} = \left(\frac{1.00}{0.75}\right)^{1/2} = 1.15$$

Thus, for $t_{1/2,4} = 10$ min, $t_{1/2,0}$ would have to be 11.5 min. If $\beta = 2$, then $t_{1/2,0}/t_{1/2,4} = (B_0/B_4)^{-1} = 0.75$; or, if $t_{1/2,4} = 10$ min, then $t_{1/2,0} = 7.5$ min. Clearly, $\beta = 1$ yields the best agreement with the data for run 1. However, confirmation of the agreement by methods 3 or 4 would be desirable.

2. *The initial-rates method.* We take as two "initial rates" the average rates during the first two intervals of run 1:

$$r_1 = -\frac{1}{2}\left(\frac{B_4 - B_0}{4-0}\right) = -\frac{1}{2}\left(\frac{0.75-1.00}{(4-0)}\right) \times 10^{-3} = 0.031 \times 10^{-3} \text{ mole liter}^{-1} \text{ min}^{-1}$$

$$r_2 = -\frac{1}{2}\left(\frac{0.56-0.75}{8-4}\right) \times 10^{-3} = 0.024 \times 10^{-3} \text{ mole liter}^{-1} \text{ min}^{-1}$$

$$\frac{r_1}{r_2} = \frac{0.031}{0.024} = 1.29 \approx \left(\frac{B_0}{B_4}\right)^{\beta} = \left(\frac{1.00}{0.75}\right)^{\beta} = (1.33)^{\beta}$$

hence $\beta \approx 1$.

3. *Analytic method.* This method utilizes the integrated rate laws from Table 20-1. For a first-order rate law with $a = 2$ we have

$$k' = \frac{\ln (B_1/B_2)}{2\Delta t}$$

This yields, for the first two points of the first run

$$k' = \frac{\ln (1.0/0.75)}{2(4-0)} = 0.036 \text{ min}^{-1}$$

The results for all successive pairs of points in both runs are

t_1, t_2 (min)	k' (min^{-1}), run 1	k' (min^{-1}), run 2
0, 4	0.036	0.017
4, 8	0.036	0.019
8, 12	0.036	0.018
12, 16	0.034	0.019
average	0.036	0.018

Since the variation of the rate constants is within experimental error and since it does not display any trends with time, $\beta = 1$. (If $\beta < 1$ we expect an increase with time for the effective k' calculated from a first-order rate law; if $\beta > 1$ a decrease is expected). An alternative method for data collected at equal time intervals would be to check the difference $f(B_2) - f(B_1)$ for constancy with time (f is the function of concentrations in column 3 of Table 20-1, for example, $f(B) = \ln B$ for a first-order rate law).

4. *Graphical method.* This method also utilizes the integrated rate laws from Table 20-1. When $\log B$ is plotted versus t, the result is a straight line, hence $\beta = 1$.

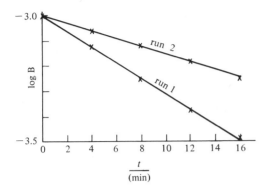

The order with respect to A, is best determined from k' for both runs.

$$\frac{k'_{run\,1}}{k'_{run\,2}} = \frac{0.036}{0.018} \approx \left(\frac{A_{run\,1}}{A_{run\,2}}\right)^\alpha = \left(\frac{0.500}{0.250}\right)^\alpha = (2.00)^\alpha$$

hence $\alpha \approx 1$. ●

Next, we consider the calculation of the rate constant k. Although the rate constant may be and often is calculated as a by-product of the determination of the reaction orders, this is not always the best procedure. The reason is that for determining the reaction orders α, β, γ we need data merely to discriminate between about half-a-dozen possible values, whereas we wish to compute the rate constant with the minimum error possible. Therefore, the decision on what kind of data to collect should be governed by considerations about the precision and accuracy of the rate constant. Precision, being a measure of reproducibility of data, should be determined by repeating runs under conditions as identical as possible. Accuracy, being a measure of agreement with the correct value, may be tested to a limited extent by varying the initial concentrations. A graphic determination of the rate constant for the anionic polymerization of isoprene in benzene at 35.1°C is shown in Figure 20-3.

What if, unlike the isoprene case, the rate constant systematically varies with the initial conditions or, even worse, cannot be fitted to an expression of the form given in Eq. (20-8)? Then a more complex empirical form than that of Eq. (20-8) must be considered, most commonly

$$r = \sum_i k_i f_i(A, B, C, D) - \sum_i k'_i f'_i(A, B, C, D) \qquad (20\text{-}11)$$

or

$$r = \frac{k f(A, B, C, D)}{1 \pm k' f'(A, B, C, D)} \qquad (20\text{-}12)$$

The functions f, f' have the general form

$$A^\alpha B^\beta C^\gamma D^\delta$$

where D is a product and α, β, γ, δ are integers or half-integral fractions $(n/2)$ in the range -3 to $+3$. There are no general strategies for determining complex rate laws.

There are clues other than those already mentioned that indicate when complex forms of rate laws should be considered. For example, checks on stoichiometry in the

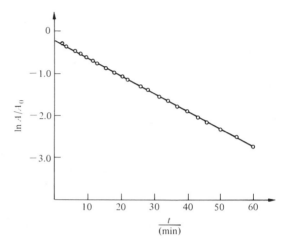

Figure 20-3 A graphical determination of the rate constant at 35.1°C for the first order reaction between polyisoprenyl lithium and isoprene

$$R(C_5H_8)_i CH_2\text{---}CH\text{=}\underset{\underset{CH_3}{|}}{C}\text{---}CH_2Li + CH_2\text{=}CH\text{---}\underset{\underset{CH_3}{|}}{C}\text{=}CH_2 \rightarrow$$

$$R(C_5H_8)_{i+1}CH_2\text{---}CH\text{=}\underset{\underset{CH_3}{|}}{C}\text{---}CH_2Li$$

The concentration A is that of isoprene; the polyisoprene lithium concentration is constant. Note that the line typically does not pass through the origin. This is caused by the fact that the initiation procedure—mixing of reactants and thermostating—is not instantaneous on the time scale of the plot.

The rate constant k is the slope of the line (see Table 20-1). What is its value as determined from the graph? For comparison we cite the analytical least squares results (see Appendix VIII): k for above run is $7.097 \times 10^4 \, s^{-1}$; for five other runs the results ranged from 6.944 to $7.275 \times 10^{-4} \, s^{-1}$ with a mean value and its standard error $(7.139 \pm 0.069) \times 10^{-4} \, s^{-1}$. The results were independent of the initial concentration of isoprene. It was concluded that the major source of error is the inconstancy of the concentration of polyisoprenyl concentration from run to run. [From C. H. Bamford and C. F. H. Tipper (eds.): *Comprehensive Chemical Kinetics*, Vol. I. Elsevier, New York, 1969, p. 412.]

reacting system may indicate that the stoichiometry varies with time, that is, that no single equation $a\,A - b\,B \rightarrow d\,D + e\,E$ with constant coefficients a, b, d, e, can describe the system, or the rate depends upon product concentration, or it increases rather than decreases with time, or the rate constants of ostensibly the same reaction from different experiments do not agree. For an example see Figure 20-2b.

20-2.3 Temperature dependence of rate constant Finally, we shall discuss the functional dependence of the rate constant upon the macroscopic variables: concentration, pressure, temperature, magnetic and electrical fields. Theoretical considerations (Section 20-4) indicate that the rate constant should depend upon the concentrations of the reactants. However, with the exception of reactions between ions (page 884), this dependence is so slight that it is difficult to observe. The same may be said about the effect of magnetic and weak electrical fields. (The effect of electrical fields on heterogeneous reactions—electrode reactions—can be readily observed, see page 894). The effects of pressure on the rate constants of solution reactions can be studied at pressures of several thousand atmospheres [3, p. 510].

However, by far the most dramatic effect on rate constants is caused by temperature changes: an increase of a few kelvins may easily double the rate constant. Among the first to attempt an empirical fit of the dependence of the rate constant upon temperature was Hood in 1878. His equation was an approximation to the form

$$k = A\, e^{-\mathscr{E}_a/RT} \tag{20-13}$$

Because Arrhenius in 1889 gave a theoretical justification (reprinted in [4], p. 31) for this equation, it became generally accepted and known as the Arrhenius equation. A and $\mathscr{E}_a$ are empirical parameters. A is called the **Arrhenius**, or **preexponential**, or **frequency parameter** or factor. $\mathscr{E}_a$ is the **activation energy**.

Eq. (20-13) may also be written as

$$\ln k = \ln A - \mathscr{E}_a/RT \tag{20-14}$$

Eq. (20-14) has the same form as Eqs. (7-19) and (9-12) for the temperature dependence of the vapor pressure of a pure liquid and the equilibrium constant, respectively. Therefore A and $\mathscr{E}_a$ may be determined using the same graphical (Figure 20-4) or analytical (least squares usually) methods as are used to obtain the corresponding quantities in Eqs. (7-19) and (9-12). Sample values of A and $\mathscr{E}_a$ are shown in Table 20-2.

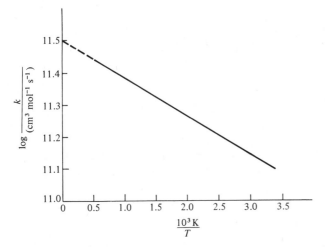

Figure 20-4 Arrhenius plot for the reaction $CO + OH \rightarrow CO_2 + H$. Data from Table 20-2.

Table 20-2 Gas phase rate constants

Reaction	Order (α)	k^a	A^a	$\mathscr{E}_a/R$ (K)	T range (K)
$N_2O \rightarrow N_2 + O$	1^b		1.4×10^{11}	30,000	800–2100
$OH + OH \rightarrow H_2O + O^c$	2	$\left.\begin{array}{l} 1.55 \pm 0.15 \\ 0.50 \pm 0.16 \end{array}\right\} \times 10^{12}$			300 300
$CO + OH \rightarrow CO_2 + H^c$	2	$\left.\begin{array}{l} 1.15 \pm 0.05 \\ 0.51 \pm 0.2 \end{array}\right\} \times 10^{10}$			300 300
			3.1×10^{11}	300	300–2000
$SO_2 + O + Ar \rightarrow$ $SO_3 + Ar$	3		4×10^{13}	500	250–1000

[a] Units: $cm^{3(\alpha-1)} mol^{-(\alpha-1)} s^{-1}$.

[b] High pressure limit.

[c] The lower value for the rate constant is thought to be due to omission of terms in the rate law representing reactions in addition to the reaction named.

DATA FROM: K. Scholfield: *J. Phys. Chem. Ref. Data*, **2**:25 (1973), (first and last reactions); W. E. Wilson, Jr.: *J. Phys. Chem. Ref. Data*, **1**:535 (1972), (second and third reactions).

Table 20-2 shows that two different reactions may have very different activation energies. What effect does a given difference in $\mathscr{E}_a$ have on the ratio of their rate constants? The effect is readily computed from a modified version of Eq. (20-14).

$$\log k = \log A - 0.053 \cdot \mathscr{E}_a/T \qquad \textbf{(20-15)}$$

where $\mathscr{E}_a$ and T are in kilojoules per mole (kJ mol^{-1}) and kilokelvins (kK), respectively. If $\mathscr{E}_a/T$ changes by slightly less than 20, k will change by a factor of 10. When $T = 0.1$, 0.3, 1 kK, differences in $\mathscr{E}_a$ of about 2, 6, and 20 kJ mol^{-1}, respectively, will cause the rate constants to differ by a factor of 10. Eq. (20-15) may also be used to estimate the effect of a temperature change on the rate constant, given the activation energy (Figure 20-5).

How generally valid is the Arrhenius equation? The great sensitivity of rate constants to changes in temperature means that it is difficult to answer this question experimentally. Small deviations are unmeasurable because temperature of the reacting system would need to be controlled to an impossible precision. Although deviations could be increased by extending the observations over a greater temperature range, this would require very different experimental methods to measure slow and fast rates (Section 20-6) and to overcome interference from side reactions at high temperatures. Therefore, it is not surprising that the large majority of reactions appear to obey the Arrhenius equation within the narrow temperature ranges in which they usually have been studied.

An expression used to fit data that do not obey the Arrhenius equation is

$$k = BT^{\mathscr{C}/R} \cdot e^{-\mathscr{E}'_a/RT} \qquad \textbf{(20-16)}$$

where B, $\mathscr{C}$, $\mathscr{E}'_a$ are empirical constants. Various theories (Chapter 21) suggest that $\mathscr{C}/R$ should be an integer or half-integer in the range -10 to 1.

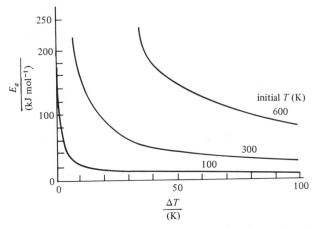

Figure 20-5 Plots of activation energy, E_a, versus change of temperature, ΔT, at three different initial temperatures T. All pairs of E_a, ΔT values read from the curves will cause the rate constant k to change tenfold.

Over a narrow temperature range Eqs. (20-13) and (20-16) would fit the data equally well. To convert the activation energy of one equation to that of the other, the following expressions are used.

$$\left(\frac{d \ln k}{dT}\right)_{\text{Eq. (20-13)}} = \mathscr{E}_a/\boldsymbol{R}T^2 = \left(\frac{d \ln k}{dT}\right)_{\text{Eq. (20-16)}} = \mathscr{E}'_a/\boldsymbol{R}T^2 + \mathscr{C}/\boldsymbol{R}T$$

or

$$\mathscr{E}_a = \mathscr{E}'_a + \mathscr{C}T \tag{20-17}$$

To get an expression for the Arrhenius parameter we equate the rate constants from Eqs. (20-13) and (20-16) at the average T for the temperature range and substitute for $\mathscr{E}_a$ from Eq. (20-17). This yields

$$A = B(eT)^{\mathscr{C}/\boldsymbol{R}} \tag{20-18}$$

In the following sections we shall present answers to some questions about the rate law and the rate constants at different temperatures, such as: What does the form of the rate law tell us about the reaction? Why do we restrict reaction orders to small integers and half-integers? Why are reaction rates unusually (compared to many equilibrium properties) sensitive to temperature changes? What other ways of describing reacting systems are there?

20-3 MECHANISMS[2] AND ELEMENTARY STEPS

Assuming that we have determined the form of the rate law, what does it tell us about the reaction?

[2] The word "chemism," from the German "Chemismus," has been proposed as a preferable alternative [5].

The rate law for the gas reaction (Table 20-2)

$$OH + OH \rightarrow H_2O + O \qquad (20\text{-}19)$$

is

$$r = k \cdot OH^2 \qquad (20\text{-}20)$$

According to the kinetic molecular theory of ideal gases the number of collisions between OH molecules is proportional to the OH concentration squared: OH^2 (page 60). Therefore the appearance of the OH^2 factor in the rate law, Eq. (20-20), is generally interpreted as implying that the reaction proceeds via collisions between two OH molecules at a time. We say that Eq. (20-19) describes the **mechanism** of the reaction. All reactions that proceed via a collision of two reactant molecules A and B are called **bimolecular**. Their stoichiometric equation and rate law is

$$A + B \rightarrow \text{products} \qquad r = kAB \qquad (20\text{-}21)$$

Termolecular reactions proceed via a collision of three molecules A, B, C

$$A + B + C \rightarrow \text{products} \qquad r = kABC \qquad (20\text{-}22)$$

Unimolecular reactions involve a single molecule

$$A \rightarrow \text{products} \qquad r = kA \qquad (20\text{-}23)$$

The three reaction types are the only ones recognized as **elementary**. Their role is fundamental. The mechanism of all **complex** reactions consists of two or more consecutive or parallel elementary reactions.

The definitions of the three types of elementary reactions can be generalized to include reactions in liquids and solids. A molecule of A in a condensed medium is in a "cage" of other (solvent) molecules. It is continually "bumping" into the molecules forming the walls of the cage. A clear distinction between bimolecular, termolecular, and so on, collisions is no longer possible. Rather the significant event is the **diffusion** (jumping) of the molecule A from one cage to the next. There it may encounter molecule B, and a reaction may occur. The rate of **encounters** has the same concentration-dependence as the rate of collisions in gas reactions. We may generalize the earlier definitions by replacing the word "collisions" with "encounters." The latter, more general, term includes both collisions and those jumps from one cage to the next that result in the meeting of two or more molecules. The role of diffusion with respect to reactions in condensed media is discussed quantitatively in Section 21-2.

Two concepts often used to describe the progress of an elementary reaction are the reaction coordinate and the activated complex (or transition state). The **reaction coordinate** is a function (often not explicitly defined) of the relative position coordinates in space of all atoms participating in the reaction. Its value (again, often not explicitly given) defines the progress of the reaction along the path of least possible potential energy at any moment beginning with the reactant molecules and ending with the product molecules (Figure 20-6). The **activated complex** is located at the point where the reaction coordinate has the highest potential energy. At this point the atoms of the reactants are in a configuration of maximum mutual repulsion. The difference between the energy of this configuration and that of the reactants may be taken as approximately equal to the empirical Arrhenius activation energy $\mathscr{E}_a$ which was defined in Section 20-2.3. (For exact relations see Section 21-1).

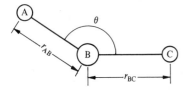

Figure 20-6 Reaction coordinate (x) for a hypothetical reaction

$$A + BC \rightarrow AB + C.$$

$x = f(r_{AB}, r_{BC}, \theta)$. Start of reaction: $r_{AB} = \infty$, $r_{BC} = r_{BC,eq}$. End of reaction: $r_{AB} = r_{AB,eq}$, $r_{BC} = \infty$; $0° \leqq \theta \leqq 180°$.

Of the three types of elementary reactions, bimolecular reactions are by far the most common. They may be transfer reactions (Eq. 20-19), exchange reactions ($H_2 + D_2 \rightarrow HD + DH$), redox reactions ($Na + Cl \rightarrow Na^+ + Cl^-$), and others. Unimolecular elementary reactions are decomposition, isomerization, and photochemical reactions. Clear-cut examples of termolecular reactions are not known.

20-3.1 Complex Reactions We now turn to mechanisms of complex reactions. A complex mechanism consisting of *consecutive* steps may be depicted schematically thus

$$R \xrightarrow{k_0} I_1 \xrightarrow{k_1} I_2 \xrightarrow{k_2} \cdots \xrightarrow{k_{n-1}} I_n \xrightarrow{k_n} P \qquad (20\text{-}24)$$

where R, I_i, and P are one or more reactant, intermediate, and product molecules, respectively. Some molecules represented by R, I_i, or P may be spectators in a particular step. The whole sequence may be described by a single reaction coordinate made up of the coordinates of the individual steps (Figure 20-7). The intermediates are more or less stable species: the deeper an energy valley in Figure 20-7, the more stable the corresponding species.

A *parallel* mechanism may consist of two or more sequences similar to that in Eq. (20-24) proceeding along two or more different reaction coordinates simultaneously. However, it is more common to have a combination of parallel and sequential steps, where the parallel steps are along the same reaction coordinate in the reverse direction (Eq. 20-26 is a particular example).

How do we know that the reaction is complex? In some (but not all) cases the answer is unequivocal. If the coefficients of the assumed stoichiometric equation vary with time or if the coefficients of the stoichiometric equation when written with smallest possible values of integers do not agree with the respective reaction orders (note the agreement in Eqs. 20-21 to 20-23), then the reaction must be complex. If the stoichiometric coefficients and the corresponding reaction orders do agree, the reaction may still be complex. For many years the reaction

$$H_2 + I_2 \rightarrow 2\,HI \qquad r = k \cdot H_2 \cdot I_2 \qquad (20\text{-}25)$$

was a textbook example of an elementary bimolecular reaction. It is now known that the reaction is complex (see Eq. 20-27).

If the reaction is known to be, or is suspected to be, complex then the problem is, what is its mechanism? The usual procedure is to propose a mechanism based on all available information and then see whether it is consistent with the kinetic data. In this section we shall discuss the methods used to derive the rate law from the proposed mechanism for the purpose of testing its consistency with the experimentally determined form of the rate law.

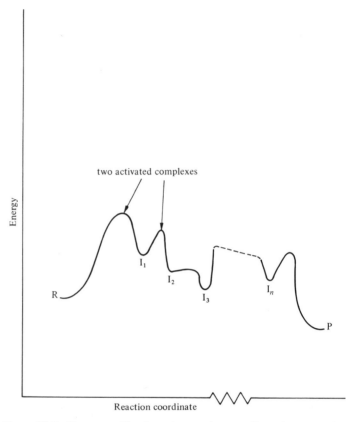

Figure 20-7 Energy profile along the reaction coordinate for a complex reaction (consecutive steps).

Usually, the rate law is derived under simplifying assumptions. One often used assumption is that a rate-determining step exists. The **rate-determining step** is the slowest step, by far, in a sequence of steps. A consequence of this bottleneck situation is that the steps preceding the rate-determining (nth) step are at equilibrium.

$$R \underset{}{\overset{K_0}{\rightleftharpoons}} I_1 \underset{}{\overset{K_1}{\rightleftharpoons}} \cdots \underset{}{\overset{K_{n-1}}{\rightleftharpoons}} I_n \overset{k_n}{\longrightarrow} I_{n+1} \rightarrow \cdots \qquad \textbf{(20-26)}$$

where the K_i are equilibrium constants. For example, a comparison of photochemical and thermal results for the stoichiometric reaction (20-25) has lead to the conclusion that the thermal mechanism is complex, possibly [6]

$$H_2 + I_2 \overset{K_0}{\rightleftharpoons} 2\,I + H_2 \overset{k_1}{\longrightarrow} 2\,HI$$

The overall reaction rate is equal to that of the rate-determining sep

$$r = k_1 \cdot H_2 \cdot I^2$$
$$= k_1 \cdot H_2 \cdot K_0 \cdot I_2$$
$$= k \cdot H_2 \cdot I_2 \qquad \textbf{(20-27)}$$

where we have used the equilibrium relation $I^2 = K_0 \cdot I_2$ and where $k = k_1 K_0$. We have shown that the complex mechanism is consistent with the experimentally determined rate law (Eq. 20-25).

A second more complicated example—a reaction in aqueous solution—is

$$H_3AsO_3 + I_3^- + H_2O \rightarrow H_3AsO_4 + 3\,I^- + 2\,H^+$$

The empirically determined rate law is

$$r = \frac{k \cdot H_3AsO_3 \cdot I_3^-}{(I^-)^2 \cdot H^+} \tag{20-28}$$

Two slightly different mechanisms have been proposed [7]. Both mechanisms start with the same two parallel steps in equilibrium and continue with identical steps. They differ in the assumptions about the rate-determining step

$$H_3AsO_3 \underset{}{\overset{K_0}{\rightleftharpoons}} H_2AsO_3^- + H^+$$

$$H_2O + I_3^- \underset{}{\overset{K_0'}{\rightleftharpoons}} H_2OI^+ + 2\,I^- \tag{20-29}$$

$$H_2AsO_3^- + H_2OI^+ \underset{(\overline{K_1})}{\overset{k_1}{\rightleftharpoons}} H_2AsO_3I + H_2O$$

$$H_2AsO_3I \underset{(\overline{k_2})}{\rightrightarrows} \text{products}$$

The constants for the second mechanism are enclosed in parentheses.

The rate law for the first mechanism whose rate-determining step has the label "1" may be derived in a manner entirely analogous to derivation (20-27)

$$r = k_1 \cdot H_2AsO_3^- \cdot H_2OI^+$$

$$= k_1 \cdot K_0 \frac{H_3AsO_3}{H^+} \cdot K_0' \frac{I_3^-}{(I^-)^2}$$

$$= k \frac{H_3AsO_3 \cdot I_3^-}{H^+ \cdot (I^-)^2} \tag{20-30}$$

where $k = k_1 K_0 K_0'$.

The second mechanism, with the step labelled "2" being rate-determining, will lead to a rate law of the same form (see Problem 10). This result illustrates a general principle about the relation between rate laws and mechanisms: many mechanisms may be compatible with a given rate law.

However, the rate laws do impose certain limitations upon the mechanisms. For example, all mechanisms with a single rate-determining step will lead to a rate law with a single term. Rate laws of the form of Eq. (20-11) (with the sum containing more than one term) or Eq. (20-12), are impossible for these mechanisms. Furthermore, the example of the aqueous reaction of H_3AsO_3 with I_3^- shows that inverse orders arise if some products of the equilibrium steps do not participate in the rate-determining step. Half-integral orders arise in a similar way (Problem 13). Parallel mechanisms, each with its own rate-determining step, would lead to a sum of two or more terms in a rate law of the form of Eq. (20-11) (Problem 15).

The molecular formula of the activated complex for the rate-determining step may be determined from the rate law. The number of atoms of each kind in the complex is given by the sum

$$\sum_i \alpha_i m_i \qquad\qquad (20\text{-}31)$$

where i designates a species in the rate law expression and the sum is over all species; α_i is the reaction order with respect to the ith species; m_i is the number of atoms of a kind in the ith species. The net charge of the complex can be computed by the same formula if m_i is now the charge of the ith species. Knowledge of the formula for the activated complex can suggest what the reactants of the rate-determining step are. From the rate law in Eq. (20-28) the formula of the activated complex can be shown (see Example 20-2) to be $AsO_2I \cdot nH_2O$. This suggests the two rate-determining steps shown in Eq. (20-29) with the reactants being $H_2AsO_3^- + H_2OI^+$ ($n = 2$) or H_2AsO_3I alone ($n = 1$).

EXAMPLE 20-2
What is the formula of the activated complex for the rate law in Eq. (20-28)?
Solution: Applying Eq. (20-31) to the rate law we get for the number of I atoms: $1\cdot 0 + 1\cdot 3 + (-2)\cdot 1 + (-1)\cdot 0 = 1$; similarly, there are two H, one As, and three O atoms. For the net charge Eq. (20-31) yields: $1\cdot 0 + 1\cdot(-1) + (-2)\cdot(-1) + (-1)\cdot 1 = 0$. The activated complex is H_2AsO_3I. However, since the dependence of the rate upon the concentration of H_2O has not been determined (why? See page 636) the number of water molecules participating in the formation of the activated complex is uncertain. Therefore the formula is more correctly $AsO_2I \cdot nH_2O$ with n a small but otherwise arbitrary positive integer.   ●

Frequently reactions are carried out under conditions where a rate-determining (that is, very slow) step does not exist. If so, a more general method for deriving rate laws from reaction mechanisms must be used. The most common method of this kind is based on the **steady state approximation**. The simplest mechanism to which it may be applied is the sequence

$$R \rightarrow I \qquad r_0 = k_0 R \qquad\qquad (20\text{-}32a)$$

$$I \rightarrow P \qquad r_1 = k_1 I \qquad\qquad (20\text{-}32b)$$

The net rate for the intermediate I is

$$r_1 = \frac{dI}{dt} = r_0 - r_1 = k_0 R - k_1 I \approx 0 \qquad\qquad (20\text{-}33)$$

The steady state approximation assumes that the concentration of I is "steady" during the time interval of interest, that is, $r_1 \approx 0$.

The rate of product formation, r_1, in terms of the concentration R is determined from Eqs. (20-32b) and (20-33).

$$r_1 \approx k_0 R \qquad\qquad (20\text{-}34)$$

Eq. (20-34) is valid whenever $r_0 \approx r_1 \gg r_1$. Except when the concentration of I is changing very rapidly r_1 must be small compared to r_0 if I is small compared to R. Since $r_1/r_0 = k_1 I/k_0 R \approx 1$ (from Eq. 20-33) we see that $I/R \ll 1$ implies $k_1/k_0 \gg 1$. If

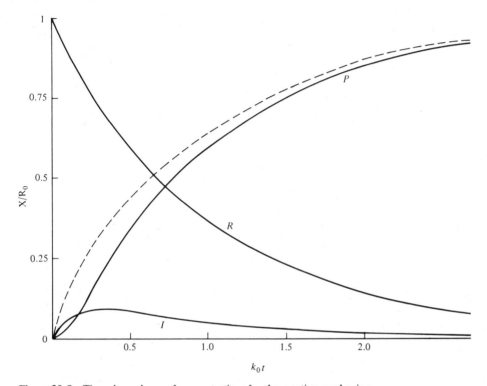

Figure 20-8   Time-dependence of concentrations for the reaction mechanism

$$A \xrightarrow{k_0} I \xrightarrow{k_1} P.$$

X is the concentration of R, I, or P. R_0 is the initial concentration of R. $k_1 = 10k_0$. The solid lines are the exact concentrations. The dashed line is from the steady state approximation, for $P/R_0 = 1 - \exp(-k_0 t)$ [obtained by integrating Eq. (20-34)]. Note the induction period (from $k_0 t = 0$ to $k_0 t \approx 0.2$) when $I > P$ and hence the average $r_I > r_1$. During the induction period the steady state approximation is not valid.

the intermediate I is reactive, k_1 will be large and the steady state approximation may be applied to I. Hence, the steady state approximation is usually valid for reactive species—most atoms, free radicals, electronically excited species, gaseous ions, and so on. For further detail see Figure 20-8.

The mechanism shown in Eq. (20-32) is only one of many compatible with the form of the rate law in Eq. (20-34). Other mechanisms with rate laws of the same form are

$$R \xrightarrow{k_0} I_1 \xrightarrow{k_1} I_2 \xrightarrow{k_2} \cdots \rightarrow I_n \xrightarrow{k_n} P \qquad (20\text{-}35)$$

when steady state is assumed for all intermediates (Problem 11). Furthermore, a mechanism in which step n is rate-determining will also lead to the same form.

20-3.2   Rice–Herzfeld Chain Reactions   Next, we shall illustrate the application of the steady state approximation to specific reactions. The selected reactions belong to the rather large and important class of **chain reactions**. Among the processes which proceed via chain reactions are gas phase pyrolysis (high temperature decomposition) of hydrocarbons and other organic compounds, explosions, reactions of hydrogen with halogens and oxygen, and polymerizations.

Rice and Herzfeld proposed in 1934 (see reprint in [4], p. 154) that gas phase pyrolysis of organic compounds proceeds via a chain mechanism involving free radicals. One example of a Rice–Herzfeld mechanism is the pyrolysis of acetaldehyde

$$CH_3CHO \rightarrow CH_4 + CO \qquad r = kR^{3/2} \tag{20-36}$$
$$(R)$$

(Deviations from this rate law are observed at late stages of the decomposition.)

The accepted mechanism for the decomposition is

$$CH_3CHO \xrightarrow{k_0} CH_3 + CHO^3 \qquad \text{initiation}$$
$$(R) \qquad\quad (C_1)$$

$$CH_3 + CH_3CHO \xrightarrow{k_1} CH_4 + CH_3CO \qquad \text{propagation} \tag{20-37}$$
$$(C_2) \qquad\qquad \text{(products formed)}$$

$$CH_3CO \xrightarrow{k_2} CH_3 + CO$$

$$2\,CH_3 \xrightarrow{k_3} C_2H_6 \qquad\qquad \text{termination}$$

The chain is initiated by the first step. The two propagation steps together constitute one link of the chain. New links of the chain are formed by repeating the propagation steps in sequence until, finally, two chains are simultaneously terminated by the last step. The species C_1 and C_2, responsible for propagating the chain, are the **chain centers** (or **carriers**).

The rate law is derived by applying the steady state approximation to the chain centers.

$$r_{C_1} = r_0 - r_1 + r_2 - 2r_3 = k_0 R - k_1 C_1 R + k_2 C_2 - 2k_3 C_1^2 \approx 0$$
$$r_{C_2} = \quad r_1 - r_2 \quad = \quad k_1 C_1 R - k_2 C_2 \quad\quad \approx 0 \tag{20-38}$$
$$r_{C_1} + r_{C_2} = r_0 \qquad -2r_3 = k_0 R \qquad\qquad -2k_3 C_1^2 \approx 0$$

Therefore

$$C_1 = (k_0/2k_3)^{1/2} R^{1/2}$$

and

$$r_{CH_4} = r_1 = k_1 C_1 R = k_1 (k_0/2k_3)^{1/2} R^{3/2} \tag{20-39}$$

This expression has the same form as the empirical rate law (Eq. 20-36). There is additional firm evidence proving that CH_3CHO pyrolyses via a chain mechanism, although the exact identity of some individual steps may be questionable.

---

[3] Leads to about 1% H_2 via $CHO + CH_3CHO \rightarrow H_2 + CO + CH_3CO$.

A measure of the effectiveness of the chain mechanism is the average chain length, defined as

$$\frac{\text{number of chain links}}{\text{number of chains initiated}} = \frac{r_{CH_4}}{r_0} = k_1(2k_0k_3)^{-1/2}R^{1/2} \qquad (20\text{-}40)$$

For acetaldehyde each chain has about 100 links at temperatures of about 800 K and atmospheric pressure.

Some work has been done on classifying the order n in the general Rice–Herzfeld rate law, $r = kR^n$, according to the particular mechanism. The two factors that determine n are (1) the order with respect to R of the initiation step, and (2) the nature of the reactants in the termination step. There are three kinds of terminating reactants: (1) radicals that participate in bimolecular propagation steps, for example C_1 in mechanism (20-37), (2) radicals that decompose unimolecularly, for example, C_2 in mechanism (20-37); (3) inert (third) bodies M that deactivate the radicals. The results for the possible combinations of initiation and termination steps are given in Table 20-3. In the language of the table, mechanism (20-37) has first-order initiation and C_1C_1 termination; therefore $n = \frac{3}{2}$

Table 20-3 Rice–Herzfeld mechanisms: over-all orders of reaction for various types of initiation and termination reactions

First-order Initiation		Second-order Initiation		Over-all Order
Simple Termination	Third-body Termination	Simple Termination	Third-body Termination	
		C_1C_1		2
C_1C_1		C_1C_2	C_1C_1M	$\frac{3}{2}$
C_1C_2	C_1C_1M	C_2C_2	C_1C_2M	1
C_2C_2	C_1C_2		C_2C_2M	$\frac{1}{2}$
	C_2C_2M			0

20-3.3 Explosions: $H_2 + O_2$ reaction　As a final example of the uses of the steady state approximation we apply it to the mechanism for the gas phase reaction

$$H_2 + \tfrac{1}{2}O_2 \rightarrow H_2O \qquad (20\text{-}41)$$

Every high school chemistry student knows that a mixture of H_2 and O_2 or air will explode upon application of a glowing splint. Much less known is the rather complex behavior of $H_2 + O_2$ mixtures with respect to spontaneous explosion (Figure 20-9). It has been found that in the region bound (approximately) by the temperatures 350 and 600°C and the pressures 1 and 10^4 torr, a sharp limit exists beyond which explosion does not occur. Instead a steady reaction is observed. The limit varies in a complicated but regular way with temperature, pressure, composition, and the properties of the reaction vessel. It is remarkable that a theory exists which can explain, almost quantitatively, these complicated variations [8]. The basics of the theory were first given by Semenov and by Hinshelwood. For this and other work

Figure 20-9 Explosion limits of a stoichiometric mixture of $H_2 + O_2 (p_{H_2} = 2p_{O_2}; P = 3p_{O_2})$. Curve R, explosion limits in vessels with reactive walls (KCl coated). Curve I, same for vessels with inert walls (boric acid coated). The data in the figure are for spherical reaction vessels about 5 cm in diameter. (The limits, particularly first and third depend upon size and shape of vessel.) Addition of inert gases or change in the proportion of H_2 to O_2 also affects the limits (not shown).

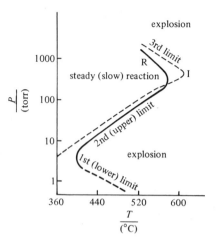

in gas kinetics N. N. Semenov and C. N. Hinshelwood shared the Nobel prize in 1956. (Semenov's first article on the subject is reprinted in [4].)

We shall introduce the theory by considering only the first explosion limit (Figure 20-9). The theory is based on a chain mechanism with one new feature: the *branched* chain replaces the linear chain of Eq. (20-37). A branching step is one that produces more chain centers than it consumes.

At the low pressures of the first limit the accepted mechanism is

$$H_2 + O_2 \xrightarrow{k_0} HO_2 + \underline{H} \qquad \text{initiation}^4 \qquad \textbf{(20-42a)}$$

$$\underline{H} + O_2 \xrightarrow{k_1} \underline{OH} + \underline{O} \qquad \text{branching} \qquad \textbf{(20-42b)}$$

$$\underline{O} + H_2 \xrightarrow{k_2} \underline{OH} + \underline{H} \qquad \text{branching} \qquad \textbf{(20-42c)}$$

$$\underline{OH} + H_2 \xrightarrow{k_3} H_2O + \underline{H} \qquad \text{propagation} \qquad \textbf{(20-42d)}$$

$$\underline{H} + \text{wall} \xrightarrow{k_4} \text{destruction} \qquad \text{termination} \qquad \textbf{(20-42e)}$$

The chain centers have been underlined. The last step is an adsorption on the wall of the vessel that removes H from the chain (for more detail see Section 20-8.2).

In order to avoid unnecessary complexity we assume (1) the concentrations of the reactants H_2 and O_2 are constant (hence our results will be applicable to the initial stage of the reaction only) and (2) steady states for all chain centers (hence our results are valid only in the steady reaction region).

With these assumptions the steady state equations are

$$
\begin{aligned}
r_H &= r_0 - r_1 + r_2 + r_3 - r_4 \\
&= k_0 \cdot H_2 \cdot O_2 - k_1 \cdot O_2 \cdot H + k_2 \cdot H_2 \cdot O + k_3 \cdot H_2 \cdot OH - k_4 \cdot H \approx 0
\end{aligned}
\qquad \textbf{(20-43a)}
$$

---

[4] The identity of the actual initiation step is unknown. Any other initiation step which yields H would lead to very similar results.

$$r_O = \quad r_1 - r_2$$

$$= \quad k_1 \cdot O_2 \cdot H - k_2 \cdot H_2 \cdot O \qquad \approx 0 \qquad \text{(20-43b)}$$

$$r_{OH} = \quad r_1 + r_2 - r_3$$

$$= \quad k_1 \cdot O_2 \cdot H + k_2 \cdot H_2 \cdot O - k_3 \cdot H_2 \cdot OH \qquad \approx 0 \qquad \text{(20-43c)}$$

$$r_H + r_O = r_0 \qquad + r_3 - r_4$$

$$= k_0 \cdot H_2 \cdot O_2 \qquad\qquad + k_3 \cdot H_2 \cdot OH - k_4 \cdot H \approx 0 \qquad \text{(20-43d)}$$

$$r_O + r_{OH} = \quad 2r_1 \qquad - r_3$$

$$= \quad 2k_1 \cdot O_2 \cdot H \qquad - k_3 \cdot H_2 \cdot OH \qquad \approx 0 \qquad \text{(20-43e)}$$

Elimination of concentration of H between the last two equations gives the OH concentration.

$$OH = \frac{(k_0 H_2 \cdot O_2) \cdot (2k_1 O_2)}{k_3 H_2 (k_4 - 2k_1 O_2)} \qquad \text{(20-44)}$$

Therefore, H_2O is produced at the rate

$$r_{H_2O} = r_3 = k_3 H_2 \cdot OH = \frac{(k_0 \cdot H_2 \cdot O_2) \cdot (2k_1 \cdot O_2)}{k_4 - 2k_1 O_2}$$

$$= \frac{r_0}{\delta - \beta} \cdot \beta = \frac{r_0}{\nu} \cdot \beta \qquad \text{(20-45)}$$

where we have defined the parameters; r_0, the rate of intiation; δ, the chain center destruction factor; β, the branching factor; and the *net* destruction factor $\nu = \delta - \beta$.

Eq. (20-45)—in terms of the factors r_0, β, and ν—can explain the existence of more than one explosion limit (as we shall show later) and many aspects of explosions as well.

The explosion limit is defined from Eq. (20-45)

$$\nu = \nu(P, T) = 0$$

$$= k_4(T) - 2k_1(T) \cdot (3RT)^{-1} P \qquad \text{(20-46)}$$

since for a stoichiometric ($p_{H_2} = 2p_{O_2}$) mixture of ideal gases the concentration $O_2 = n_{O_2}/V = n_{tot}/3\ V = P/3RT$ (see Figure 20-9). We see from Eq. (20-45) that when ν is zero H_2O is being produced at an infinite rate; that is, the mixture is exploding. When ν is positive H_2O is produced at a finite and steady rate. When ν becomes negative Eq. (20-45) is meaningless: the steady state assumptions on which it was based no longer apply. Thus we can understand an explosion limit as the condition when the rate at which new chains are being started via branching escapes the control of the chain termination rate. In vessels with walls that are inert toward H atoms (Figure 20-9) an efficient termination process does not exist, and neither does a lower explosion limit.

In order to explain the second limit as well as its dependence upon the properties of the reaction system additional reactions must be considered. Some of them are

$$\underline{H} + O_2 + M \xrightarrow{k_5} \underline{HO_2} + M \qquad \text{(20-47a)}$$

$$\underline{HO_2} + H_2 \xrightarrow{k_6} \underline{H_2O_2} + H \qquad \text{branching} \qquad \text{(20-47b)}$$

$$\underline{H_2O_2} + M \xrightarrow{k_7} 2\underline{OH} + M' \tag{20-47c}$$

$$\underline{H} + \underline{H_2O_2} \xrightarrow{k_8} H_2O + \underline{OH} \qquad \text{termination-propagation} \tag{20-47d}$$

H_2O_2 is regarded as two chain centers because it can form two OH radicals. Reaction (20-47d) is a termination step since it replaces three chain centers by one. The second limit, which is not very sensitive to the size and wall properties of the reaction vessel, is thought to arise because, with increasing pressure, reaction (20-47d) gains control of the branching reactions. Mathematically, the bimolecular reaction (20-47d) adds another term, proportional to P^2, to ν. Therefore Eq. (20-46) becomes quadratic in P and has two roots for at least some T. The two roots correspond to the two limits. Additional wall termination reactions—for O, OH, HO_2, and H_2O_2—are found to modify the second limit but slightly.

The third limit introduces a new (thermal) factor which adds to the kinetic effect of branching. The thermal factor is self-heating: the exothermic system begins to generate heat more rapidly than it can loose it. The reactions in the system speed up as a result. The theory handles the new factor by adding an energy conservation equation to the kinetic steady state equations. The modified theory predicts, in agreement with experiment, that the third limit is in part thermal and in part kinetic in origin.

We will now summarize the results of this section. The fundamental assumption we have made—that any reaction may be viewed as a collection of one or more elementary reactions—is seen to be adequate for explaining the functional form of the empirical rate laws described in the previous section. For example, the allowed values of the reaction orders derive from the limitation of the elementary reactions to uni-, bi-, or termolecular. Parallel reactions, along two independent reaction coordinates, lead to two terms in the rate law (as in Eq. 20-11). If the reaction occurs along the same reaction coordinate part of the way but along two different ones for another part, we get rate laws of the form shown in Eq. (20-12). An example of this is the rate law for the formation of H_2O (Eq. 20-45). Its form derives from two competing reactions—termination and branching—of the H atom which share the same initiation reaction (reaction 20-42a).

The power of the reaction mechanism approach to rate laws is demonstrated by our successful explanation of the complicated behavior of the hydrogen-oxygen system. A description of reacting systems different from the rate law description may also be interpreted successfully by a very similar approach (Section 20-5).

Nevertheless, this section has posed a question far more important than the answers it has provided. The question is, what is *the* reaction mechanism of a given reaction? It is clear that, so far, we do not know how to answer this question. We have demonstrated that, in general, the rate law is compatible with more than one mechanism. As a consequence of the approximations made in deriving the rate law, detailed information on the concentrations of intermediates as a function of time has been lost. Our search for answers to the question posed will lead to methods for direct measurement of elementary reaction rates and the concentrations of intermediates (Sections 20-6 and 20-7). First, in Section 20-4, we will introduce the principle of detailed balancing which is occasionally useful for eliminating some of several possible mechanisms, and in Section 20-5 we will explore some further difficulties with the study of mechanisms.

20-4 DETAILED BALANCING: KINETICS AND THERMODYNAMICS

We have seen that many different mechanisms can conform with the empirical rate law of a reacting system. This fact makes it important to consider any general principles that might guide us in selecting the actual mechanism or, at least, help us eliminate some mechanisms from consideration.

Besides well-known thermodynamic laws there is one general principle of frequent application which helps in elimination if we are dealing with reacting systems at or near chemical equilibrium. This is the principle of **detailed balancing**. It asserts that *at equilibrium* the rate of every elementary reaction in the system must equal the rate of the reverse elementary reaction, that is

if

$$A + B \rightarrow C + D \qquad r_+ = k_+ A \cdot B$$

and

$$C + D \rightarrow A + B \qquad r_- = k_- C \cdot D \qquad \text{(20-48)}$$

then $r_+ = r_-$. Detailed balancing may be proved from the quantum mechanical principle of microscopic reversibility.[5] However, we shall deduce it from a thermodynamic reductio ad absurdum argument.

Suppose we have a system where equilibrium is maintained, in violation of the principle of detailed balancing, by two reactions:

$$\text{heat} + X_2 \rightarrow 2X \qquad \text{(20-49a)}$$

$$2X + C \rightarrow X_2 + C + \text{heat} \qquad \text{(20-49b)}$$

with the two reverse reactions being negligibly slow. X_2 and X are gases; C is a solid catalyst. With this system it is possible to construct the nearly isothermal perpetual motion machine shown in Figure 20-10 by making the system oscillate about the equilibrium. As explained in the caption of Figure 20-10, the net effect of the machine would be to convert heat at the ambient temperature to work. This is a violation of the second law of thermodynamics. In systems similar to the one described by Eq. (20-49), it will always be possible to stop or slow down one of the two reactions without affecting the other. Whenever this is done, the system moves away from equilibrium. A violation of the second law results. Therefore, we may conclude that the principle of detailed balancing is a necessary condition for all reacting systems to obey the second law.

The principle has two qualitative consequences. The first one is that for systems near equilibrium one should not propose mechanisms that do not include the reverse steps of every elementary reaction. For example, the mechanism in Eq. (20-29) should not be applied if the system is near equilibrium because the reverse of the rate-determining step, as well as the steps beyond it, is not included. The second consequence is that in near-equilibrium systems one should not propose elementary

---

[5] This principle of quantum mechanics asserts that the probabilities of the forward and reverse transitions between two quantum states are equal. When this result is averaged over all molecular states present in a system at equilibrium, we get the principle of detailed balancing [9]. Many texts on reaction kinetics use the term "microscopic reversibility" to mean "detailed balancing."

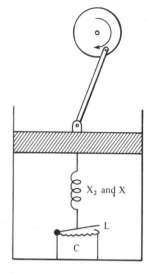

Figure 20-10 A perpetual motion machine based on the hypothetical reactions in Eq. (20-49). During the up-stroke the latch, L, keeps the catalyst box locked; only reaction (20-49a) occurs; $P_{gas} > P_{ext}$. During down-stroke the box is open and both reactions occur; $P_{ext} > P_{gas}$. T is held nearly constant at the ambient value. Net work done during a cycle: $w = q_{\text{in during up}} - q_{\text{out during down}}$.

reactions whose reverses are improbable. The bimolecular reaction

$$NH_3 + NH_3 \rightarrow N_2 + 3\,H_2$$

is unacceptable because its reverse is an unlikely tetramolecular reaction.

Next, let us consider the quantitative relations between the rate constants of the forward and reverse reactions and the forms of the two rate laws that result from the principle of detailed balancing. We start with an elementary reaction and then proceed to complex reactions.

Consider the elementary reactions (20-48) at equilibrium:

$$r_+ = k_+ A \cdot B = r_- = k_- C \cdot D$$

or

$$\frac{C \cdot D}{A \cdot B} = \frac{k_+}{k_-} \tag{20-50}$$

Applying the definition of the equilibrium constant K, Eq. (9-5), and of the activities in terms of concentrations, that is, $a_A = \gamma_A \cdot A$, and so on we get

$$Q_c = \frac{C \cdot D}{A \cdot B} = K\left[\frac{\gamma_C \gamma_D}{\gamma_A \gamma_B}\right]^{-1} = K/Q_\gamma \tag{20-51}$$

where we have defined the quotients of concentrations and of activity coefficients, Q_c and Q_γ, respectively. Equating Eqs. (20-50) and (20-51) yields

$$Q_c = k_+/k_- = K/Q_\gamma \tag{20-52}$$

If the system is ideal, $Q_\gamma = 1$, and we recover a familiar relation $K = k_+/k_-$.

From Eq. (20-52) we get a general expression for the dependence of the rate constants upon the concentrations of the species A, B, C, D. The deviation of γ_A from 1 is a measure of the influence of the concentrations A, B, C, D upon the properties of the species A; $\gamma_A = \gamma_A(A, B, C, D)$. Similarly for the other species. If the properties are the mutual reactivities of A and B or C and D (measured by k_+

and k_-, respectively), Eq. (20-52) suggests the following physically reasonable result:

$$K = \frac{k_+}{k_-} \cdot Q_\gamma = \frac{k_+/\gamma_A \gamma_B}{k_-/\gamma_C \gamma_D} \cdot \frac{\gamma_a(A, B, C, D)}{\gamma_a(A, B, C, D)} \cdot \frac{f}{f}$$

$$= \frac{k_+ \gamma_a / \gamma_A \gamma_B f}{k_- \gamma_a / \gamma_C \gamma_D f} = \frac{k_+^\ominus / f}{k_-^\ominus / f} \tag{20-53}$$

The factor $\gamma_a/\gamma_a = 1$ was introduced to account for the possibility that both rate constants may be equally influenced by the concentrations of A, B, C, D. The factor $f/f = 1$ accounts for similar influences that are independent of concentration variables. One theory of reaction rates (Chapter 21) identifies γ_a and f, respectively, as the activity coefficient of the activated complex and as a frequency with which the reactant or product molecules pass through the activated complex configuration. As is obvious from Eq. (20-53) both γ_a and f cancel out of the equilibrium constant; that is, they are typically kinetic quantities no matter what their theoretical interpretation may be. The rate constants $k_+^\ominus$, $k_-^\ominus$ refer to reactant and product standard states where all activity coefficients are equal to 1. We see that the equilibrium constant for an elementary reaction is exactly defined as the ratio of rate constants that are determined in a system where reactants and products are in their standard states. The elementary rate constant for any state is, taking k_+ as an example,

$$k_+ = k_+^\ominus \gamma_A \gamma_B / \gamma_a \tag{20-54}$$

The precision of rate constant measurements is seldom high enough to make the difference between k_+ and $k_+^\ominus$ observable. One exception is ionic reactions (page 884). The extension of Eqs. (20-53) and (20-54) to uni- and termolecular elementary reactions is straightforward.

As an example of the application of Eq. (20-52) consider the reaction in the gas phase at 296 K [10].

$$CO_2H^+ + CH_4 \underset{k_-}{\overset{k_+}{\rightleftharpoons}} CH_5^+ + CO_2$$

for which the rate constants have been measured in both directions under non-equilibrium conditions:

$$k_+ = (7.8 \pm 0.2) \times 10^{-10} \text{ cm}^3 \text{ mol}^{-1} \text{ s}^{-1}$$
$$k_- = (3.2 \pm 0.2) \times 10^{-11} \text{ cm}^3 \text{ mol}^{-1} \text{ s}^{-1}.$$

Hence $k_+/k_- = 24 \pm 1$. An independently measured value of the equilibrium constant K is 25 ± 1. The agreement implies that the rate constants away from equilibrium are the same as at equilibrium in this case, and that the determination of the rate constants has correctly taken into account side reactions.

The rate constants $k_+^\ominus$ and $k_-^\ominus$ can be related to the standard Gibbs free energy, $\Delta G^\ominus$, of the elementary forward and reverse reactions

$$k_+^\ominus / k_-^\ominus = \frac{f}{f} K = \frac{f}{f} e^{-(\Delta G^\ominus / RT)} \tag{20-55}$$

In order to get an expression for, say $k_+^\ominus$, we partition $\Delta G^\ominus$ (and $\Delta S^\ominus$, $\Delta H^\ominus$ as well) as follows

$$\Delta G^\ominus = (G_P^\ominus - G_a^\ominus) - (G_R^\ominus - G_a^\ominus)$$

$$= \Delta H^\ominus - T\Delta S^\ominus \qquad\qquad\qquad (20\text{-}56)$$

$$\Delta G^\ominus = (H_P^\ominus - H_a^\ominus) - (H_R^\ominus - H_a^\ominus) - T[(S_P^\ominus - S_a^\ominus) - (S_R^\ominus - S_a^\ominus)] \qquad (20\text{-}57)$$

where the subscripts P, R, a indicate all products, all reactants, and the activated complex, respectively.

Eq. (20-57) allows separation of Eq. (20-55) into parts corresponding to $k_+^\ominus$ and $k_-^\ominus$. For $k_+^\ominus$ we have

$$k_+^\ominus = f e^{(S_a^\ominus - S_R^\ominus)/R} \cdot e^{-(H_a^\ominus - H_R^\ominus)/RT} \qquad\qquad (20\text{-}58)$$

and similarly for $k_-^\ominus$. If f, $S_a^\ominus - S_R^\ominus$, and $H_a^\ominus - H_R^\ominus$ are assumed to be independent of temperature, Eq. (20-58) has the form of the Arrhenius equation (20-13) with

$$A = f \cdot e^{(S_a^\ominus - S_R^\ominus)/R} \qquad\qquad\qquad (20\text{-}59)$$

and

$$E_a = H_a^\ominus - H_R^\ominus \qquad\qquad\qquad (20\text{-}60)$$

For a non-Arrhenius temperature dependence, Eq. (20-58) will correspond in form to Eq. (20-16). Eqs. (20-59) and (20-60) apply to reactions studied under conditions of constant pressure. Similar equations may be derived for constant volume conditions.

Eq. (20-60) together with Eq. (20-57) puts a constraint on the value of E_a permitted for an endothermic elementary reaction. From the two equations we get (see Figure 20-7 for a graphical statement of the following expression)

$$E_{a+} = \Delta H^\ominus + E_{a-} \qquad\qquad\qquad (20\text{-}61)$$

where the E_{a+}, E_{a-} are the activation energies of the forward and reverse reactions, respectively. Since $\Delta H^\ominus$ is positive and $E_{a-} \geq 0$, we have $E_{a+} \geq \Delta H^\ominus$. In practice it is found that activation energies determined away from equilibrium occasionally violate this condition to a small degree. For example, the E_a for dissociation of diatomic molecules is frequently found to be a few kilocalories per mole less than $\Delta H_{dis}^\ominus$.

Let us now proceed to apply the results obtained for elementary reactions to complex ones. Consider a near-equilibrium complex reaction of ideal gases whose stoichiometric equation and equilibrium constant are

$$a A + b B \rightleftharpoons c C + d D \qquad K = \frac{C^c D^d}{A^a B^b} \qquad (20\text{-}62)$$

We assume that its mechanism consists of a sequence of steps along a single reaction coordinate. Using the formalism of Eq. (20-24) and applying the principle of detailed balancing we have

$$a A + b B \underset{k_{-0}}{\overset{k_0}{\rightleftharpoons}} I_1 \underset{k_{-1}}{\overset{k_1}{\rightleftharpoons}} \cdots \underset{k_{-(n-1)}}{\overset{k_{n-1}}{\rightleftharpoons}} I_n \underset{k_{-n}}{\overset{k_n}{\rightleftharpoons}} c C + d D \qquad (20\text{-}63)$$

where a, b, c, d are integers.

Not all molecules A and B need to participate in the zeroth elementary step; some may be mere "spectators." Similarly I_i may represent reactants for the ith step, products from step $i-1$, and spectator molecules. Since we are dealing with ideal gases, $k_i = k_i^{\ominus}$ and therefore from Eq. (20-52) we get

$$K_i = \frac{k_i}{k_{-i}} \qquad (20\text{-}64)$$

where K_i is the equilibrium constant for the ith elementary reaction. A relation between the overall equilibrium constant K and the K_i follows readily since the stoichiometric Eq. (20-62) is just the sum of the elementary reactions in Eq. (20-63). Therefore, we multiply the K_i together to get the constant for the overall reaction. The result is

$$K = \prod_{i=0}^{n} (K_i)^{\alpha_i} = \prod_{i=0}^{n} \left(\frac{k_i}{k_{-i}}\right)^{\alpha_i} \qquad (20\text{-}65)$$

where the $\alpha_i (= 1, 2, \text{ or } 3 \text{ at most})$ have been included to account for the possibility (not shown explicitly in Eq. 20-63) that some of the elementary reactions may have to occur more than once in order to provide a sufficient number of reactants for further steps. The stoichiometric Eq. (20-62) is determined by the mechanism, Eq. (20-63). The coefficients, a, b, c, d must therefore be integers, because a mechanism cannot involve fractional molecules, but they need not necessarily be the smallest possible integers (although usually they will be). The equilibrium constant K of Eqs. (20-62) and (20-65) must then be the one which corresponds to the stoichiometric equation determined by the mechanism.

These results and considerations do not apply for mechanisms far from equilibrium where reverse reactions may be insignificant. Such mechanisms are often proposed without regard for the principle of detailed balancing. Even correct stoichiometry may be disregarded if only the initiation of the reaction is to be observed. Examples are found most readily among mechanisms based on the steady state assumption (see the mechanism in Eq. 20-42).

Results similar to those for the mechanism in Eq. (20-63) may be obtained for mechanisms that involve two or more parallel reaction coordinates. Consider either mechanism in Eq. (20-29). Each may be described as a reactant branching mechanism; the reactants start to react along two different branches (or reaction coordinates) which then both join to form a trunk (a single reaction coordinate). Equation (20-65) applies unchanged both to a reactant branching mechanism and to a product branching mechanism.

If a mechanism contains cycles, for example

$$aA + bB \underset{k_{-0}}{\overset{k_0}{\rightleftharpoons}} I_1 \qquad cC + dD \qquad (20\text{-}66)$$

the overall stoichiometric equation may be obtained by adding the elementary reactions along one of the two reaction coordinates. Therefore we get two expressions for the overall equilibrium constant of the form of Eq. (20-65)—one with the unprimed and the other with the primed rate constants. Physically this means that equilibrium is maintained along either path independently of the other path.

EXAMPLE 20-3

Let us apply the results of this section to the elucidation of the reaction mechanism for the gas phase reaction

$$H_2 + I_2 = 2\,HI \tag{a}$$

The rate of reaction becomes measurable near 600 K.

The early workers—Bodenstein (1894–99) and Taylor and Crist (1941)—found that the net rate of the reaction fits the rate law

$$r = k'_+ \cdot H_2 \cdot I_2 - k'_- \cdot HI^2 \tag{b}$$

Typical values of the rate constants obtained with rate law (b) were $k'_+ = 6.4 \times 10^{-2}$, and $k'_- = 2.3 \times 10^{-3}\,dm^3\,mol^{-1}\,s^{-1}$ at 700 K; this yields, for the equilibrium constant of reaction (a)

$$K = k'_+ / k'_- = (6.4 \times 10^{-2}) / (2.3 \times 10^{-3}) = 28$$

This value was in agreement, within experimental error, with independently determined values for the equilibrium constant, thus supporting the assumption of a one-step bimolecular mechanism for both forward and reverse reactions.

However, in 1959 Sullivan showed that the agreement was the result of a fortuitous cancellation (under the particular experimental conditions employed by the early workers) of factors in the experimental rate constants k'_+, k'_-. He established that a chain mechanism (mechanism B in Problem 15) becomes increasingly important at higher temperatures and that the more exact rate law has the form shown in Problem 15. In this case the attempt to check the rate constants for compatibility with thermodynamic data led the scientists up a false path.

In 1958 Semenov suggested, with better luck, that mechanism A (Problem 15) is compatible with the experimental rate constant and activation energy determined on the basis of rate law (b), $k'_+ \approx 10^{12} \exp\left[-39{,}000\,mole/(kcal \cdot \boldsymbol{R}T)\right]$. His argument was based on Eq. (20-65) applied to mechanism A:

$$K = k'_+ / k'_- = K_0 \cdot (k_1 / k_{-1})$$

Therefore, using Eq. (9-12) to eliminate K_0, we have

$$k'_+ = K_0 \cdot k_1 = A' \exp\left(-\Delta H^\ominus / \boldsymbol{R}T\right) \exp\left(-E_{a1} / \boldsymbol{R}T\right)$$

where $\Delta H^\ominus$ is the enthalpy of dissociation for $I_2(g)$ and E_{a1} is the activation energy for reaction $2\,H + I_2 \rightarrow 2\,HI$. A comparison with the experimental k'_+ yields the expression

$$39{,}000 = \Delta H^\ominus + E_{a1}$$

Since $\Delta H^\ominus$ is known to equal $35{,}500\,cal\,mol^{-1}$, $E_{a1} = 4500\,cal\,mol^{-1}$. This rather low activation energy, is reasonable for a reaction involving the reactive I atoms. If the *experimental* activation energy had been less than $\Delta H^\ominus$, the thermodynamic evidence would have argued against mechanism A.

In 1967 Sullivan was able to show that the mechanism suggested by Semenov is very probably correct. Sullivan independently determined k_1 to be $(6.7 \pm 0.5) \times 10^7 \exp[-(5310 \pm 85) \; \text{mol}/(\text{cal} \cdot RT)] \, \text{dm}^6 \, \text{mol}^{-2} \, \text{s}^{-1}$, in approximate agreement with Semenov's argument.   ●

In this section we have seen that both thermodynamics and the principle of detailed balancing place certain constraints upon the forward and reverse rate constants of reactions and upon the activation energies as well. These constraints are, however, merely necessary conditions that the correct mechanism must obey; they are by no means sufficient to guarantee that a proposed mechanism is the correct one. Further constraints are discussed in Chapter 21.

20-5 ACTIVATOR-INITIATED REACTIONS

There is a large class of systems that our experience has taught us to regard as stable. Pure water belongs in this class. Yet, when water is irradiated with 4.8 MeV α particles (from the decay of ^{234}U nuclei), H_2 and H_2O_2 are produced. Irradiation of all stable solids, liquids, and gases produces a great variety of substances ranging from simple molecules to polymers.

Irradiation with α particles, electrons, γ rays, even neutrons of a given energy forms the same products from water, although not necessarily in the same proportions (Table 20-4). This result leads us to conclude that the most important

Table 20-4 The effect of different radiations on the primary product yields[b] from acidic (pH = 0.5) dilute aqueous solutions

Radiation	$g_{H_2O_2}$	g_{H_2}	g_{OH}	$(g_H + g_{e^-})$	g_{HO_2}	g_{-H_2O}
	Number of Species per 100 eV					
1.25 MeV (^{60}Co)γ rays(0.3)[a]	0.8	0.45	2.95	3.65	0.008	4.55
18 keV (^{3}H)β rays(3.2)[a]	1.0	0.6	2.1	2.9		4.1
18 MeV D$^+$	1.03	0.7	1.75	2.4		3.85
8 MeV D$^+$	1.2	1.05	1.45	1.7		3.85
32 MeV α rays(<134)[a]	1.25	1.15	1.05	1.3		3.55
5.5 MeV α rays(134)[a]	1.34	1.57	0.5	0.6	0.11	3.5

[a] LET in keV μm^{-1}, see page 672.
[b] The symbol g_X distinguishes a primary yield from the final molecular yield (G_X), see Section 20-5.4. The distinction is not relevant to the discussion in Section 20-5.1.

characteristic of the radiation that causes chemical reaction is its energy. The radiation deposits the energy necessary to activate the system so that reactions can occur. Although the deposition of energy may increase the temperature of the system, the temperature change is not sufficient to create reactive species—electronically excited molecules and unstable ions. Rather, the reactive species are created by energy directly absorbed by their parent molecules. We shall call all radiations that act in this manner **activators**. Other common activators are visible and ultraviolet (UV) light, and x rays. Activators need not be, though they may be, reactants. In this respect they are similar to catalysts, although the two are opposites in their modes of action. Activators cause reaction by supplying energy; catalysts do the same by lowering energy requirements.

Two basic experimental arrangements for depositing energy are shown in Figure 20-11a and 20-11b. In order to keep an already complex reaction system as simple as possible, it is customary to stop irradiation before no more than a small fraction of the reactants has disappeared. The different products can be analyzed quantitatively either directly or after chromatographic separation. The amounts thus determined are combined with measurements of adsorbed intensity to calculate the yields of the various products.

Traditionally the reaction kinetics discussed here belong to several distinct areas of chemistry. Reactions initiated by visible and UV radiation belong to **photochemistry**. Radiations that emanate from nuclear reactions or from high energy accelerators initiate reactions studied in **radiation** (or **radio-**) **chemistry**. Low energy ($<100\,\text{eV}$) electron initiated reactions are most often investigated by means of **mass spectrometry**. Each of these areas has its own terminology and definitions of units. The underlying concepts that are used to describe reactive systems are the same, however. We shall define these concepts and then describe the modifications necessary to apply them to the two most important areas—photo- and radiochemistry.

The number of products of an activator-initiated reaction is large. A single stoichiometric equation therefore does not usually describe the overall reaction. Even more importantly these reactions are fast; they are over in less than 1 s. On the other hand, typical irradiation intensities may be so low that product concentrations become measurable only after irradiation times much longer than 1 s. For these reasons it may be impractical to describe these reacting systems by an overall rate law (see the thermal homogeneous systems discussed in Section 20-2). A different method of description is necessary.

20-5.1 The Yield An important quantity that describes a reaction in an activator-initiated system is the yield.

$$\text{yield} = \frac{\text{amount of product formed (or reactant destroyed)}}{\text{amount (or energy) of activator absorbed}} \qquad \textbf{(20-67)}$$

If amounts are used in both the numerator and denominator the units are either the number or the moles of species (chemical species or photons[6]). These are the usual units of yield in photochemistry. Photochemists call yield the "quantum yield" and use the symbol Φ for it. In radiochemistry, energy is used in the denominator—the common units being the number of species for amount and $100\,\text{eV}$ for energy. The yield is then referred to as the "G-value," with symbol G.

---

[6] A common non-SI name for a mole of photons is "einstein."

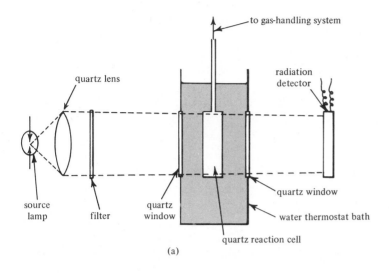

(a)

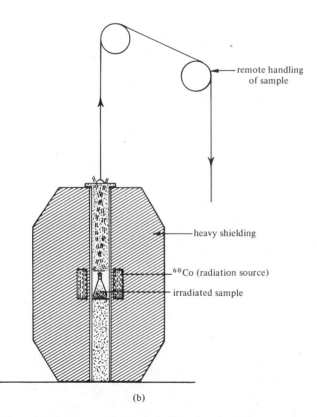

(b)

Figure 20-11   Two typical basic facilities for the initiation of reactions via activators. The products of the reactions are analyzed in separate facilities. (a) Near-UV apparatus. Quartz is used in place of glass because the latter absorbs UV radiation. Radiation detector is a photomultiplier or thermopile-galvanometer. (b) ^{60}Co γ-ray apparatus. [From R. P. Wayne: *Photochemistry*. Butterworth, London, 1970, p. 196.]

The numerator of Eq. (20-67) is determined, after irradiation has ended, from direct or indirect chemical analysis or from physical measurement of the concentration of the species of interest (Section 20-6). The denominator is determined from the **absorbed intensity** of radiation I_a (known as dose rate in radiochemistry).

$$I_a = \frac{\text{amount (or energy) of activator absorbed}}{\text{volume (or mass) of sample} \times \text{time of irradiation}} \qquad \textbf{(20-68)}$$

with common units being number or moles for amount; ergs, joules, or electron volts for energy, cubic centimeters or cubic decimeters for volume, grams or kilograms for mass, and seconds for time. Photochemists use amount or energy and volume in Eq. (20-68) (see Example 20-4), whereas radiochemists use energy and mass or volume. The absorbed amount per unit mass (dose) is often measured in rads $= 10^{-5} \text{ J g}^{-1}$.

EXAMPLE 20-4

In a certain experiment irradiation of biacetyl gas with UV radiation (253.7 nm) for 1.80×10^4 s produced $6.33 \times 10^{-8} \text{ mol cm}^{-3}$ of CO. The intensity of the radiation absorbed by the biacetyl was $1.71 \times 10^{-6} \text{ J cm}^{-3} \text{ s}^{-1}$ [11, p. 330]. What is the quantum yield of CO?

Solution: The energy of one photon

$$h\nu = hc/\lambda = \frac{(6.63 \times 10^{-34})(3.00 \times 10^8)}{253.7 \times 10^{-9}} = 7.84 \times 10^{-19} \text{ J}$$

The intensity I_a (Eq. 20-68) converted to number of photons

$$I_a = \frac{1.71 \times 10^{-6}}{7.84 \times 10^{-19}} \text{ cm}^{-3} \text{ s}^{-1} = 2.18 \times 10^{12} \text{ cm}^{-3} \text{ s}^{-1}$$

The number of photons absorbed is

$$I_a t = (2.18 \times 10^{12})(1.80 \times 10^4) \text{ cm}^{-3} = 3.93 \times 10^{16} \text{ cm}^{-3}$$

Number of CO molecules

$$6.33 \times 10^{-8} \times N = (6.33 \times 10^{-8})(6.02 \times 10^{23}) \text{ cm}^{-3} = 3.81 \times 10^{16} \text{ cm}^{-3}$$

The quantum yield, from Eq. (20-67)

$$\Phi = \frac{3.81 \times 10^{16}}{3.93 \times 10^{16}} = 0.97 \quad \bullet$$

What is the significance of the yield concept? Unlike rate laws, determination of yields does not involve time. Rather, yields are determined by analyzing the system *after* the reactions are over. A comparison of the yields of different products tells us which reaction paths are faster, which are slower (Table 20-4). The yield of a species characterizes the efficiency with which the activator causes reaction. Such values may be obtained for the total yield of a reactive species during the irradiation, even though at the time of analysis the species is no longer present. One way of getting this information is to add a scavenger—a substance known to react more rapidly with the unstable species of interest than any other substance in the reactive system. For example, H atom yields can be determined from the amount of HD produced by the reaction

$$H + DCOO^- \rightarrow HD + \cdot COO^-$$

The formate ion which may be added to H_2O in small amounts, is the scavenger. The radical yields in Table 20-4 have been determined in this or in other indirect ways.

20-5.2 Deposition of Energy

The measurement of absorbed energy is a problem particular to the study of activator-initiated reactions. In photochemistry the measurement is known as **actinometry** (from Greek *aktis*, ray of light), whereas in radiochemistry the term **dosimetry** is used. The methods for measuring absorbed radiation energy can be broadly classified as physical or chemical. The physical methods provide absolute primary standards. They are thermoelectric measurements in photochemistry and calorimetric measurements in radiochemistry. When relative measurements (secondary standards) suffice, physical methods such as photomultipliers are used for low energy radiation and various types of ionization measurements for high energy radiation.

The most widely used secondary standards are chemical actinometers and dosimeters. In these devices the amount of energy absorbed is determined from the amount of chemical reaction caused. The best reactions for this purpose are those which are not too sensitive to parameters other than the absorbed energy. Such other parameters include temperature, wavelength, and radiation intensity. The yields of actino- or dosimetric reactions must be well known. The usual procedure is to put the actinometric or dosimetric solution in the same (or equivalent) reaction vessel (Figure 20-11) as is used for the reacting system studied.

For radiochemical systems, the energy absorbed is proportional to the density of the absorbing medium. Thus it is simple to calculate the amount absorbed by the reacting system once the amount absorbed by the dosimetric system and the densities of both systems are known. The Fricke dosimeter is the best known. It contains a 10^{-3} M $FeSO_4 \cdot (NH_4)_2SO_4 + 0.4$ M H_2SO_4 solution saturated with air. It yields Fe^{3+} ions in amounts that depend not only upon the total energy absorbed but also upon the ionizing efficiency—LET (see page 672)—of the radiation. For ^{60}Co γ rays the yield is 15.6 Fe^{3+} ions per 100 eV.

The most commonly used solution actinometer for the UV region is an acidic solution of $K_3Fe(C_2O_4)_3$. The UV reaction reduces the ferric iron to Fe^{2+}. The latter is easily determined photometrically. Other necessary measurements are the fraction of light absorbed by the actinometric solution ($f = I_a/I_0$), and the exposure time t. When combined with quantum yields available in the literature[7] and used in Eqs. (20-67) and (20-68), these data yield the intensity of the light incident (I_0) upon the reaction vessel

$$I_0 = \frac{n_{Fe^{2+}}}{\Phi t f} \tag{20-69}$$

where $n_{Fe^{2+}}$ is the concentration of Fe^{2+}. A measurement of the fraction of light absorbed by the reacting system then easily yields I_a.

---

[7] $\Phi(253.7 \text{ nm}) = 1.25$; $\Phi(509.0 \text{ nm}) = 0.86$ [11, p. 784].

Before we can discuss what yields tell us about the mechanisms and elementary steps of activator-initiated reactions, we must describe the processes whereby activators deposit their energy in a system.

Figure 20-12 shows the initial processes and products. We see that among the products may be new activators—x and γ rays, and (secondary) electrons—as well as the original charged particles. If these have enough energy they may excite or ionize a new reactant molecule. With high energy radiation we then have a cascade of reactions that leads to an increasing number of the chemically interesting excited and ionized species R^*, R^+, or R^{+*}. For a 1 MeV electron, this cascade continues for about 10^{-11} s, until it and the secondary electrons as well no longer have sufficient energy to excite or ionize a reactant molecule.

How do we know that the deposition products shown in Figure 20-12 are the only products possible? Why, for instance, did we leave out the dissociation of a molecule? The answers are somewhat uncertain since the initial species, especially at the higher energies, are seldom directly observable. Their existence must be inferred

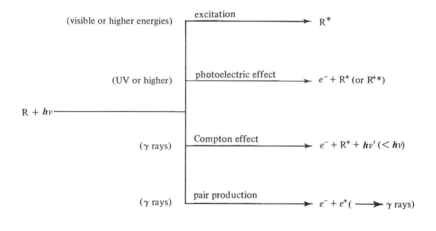

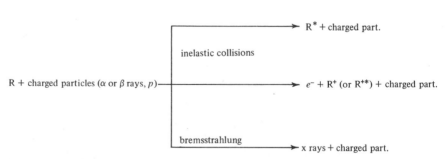

Figure 20-12 Activator-initiated primary processes. In each of the two "trees" the energy required for a process increases from top to bottom. Bremsstrahlung—radiation due to deceleration of charged particles when passing near nuclei—is important for electrons above 100 keV and for heavy particles above 1 GeV. High energy neutrons cause these reactions indirectly, by first producing protons and γ rays. R^*, R^{+*} indicate neutral, respectively ionic species in excited electronic states.

from indirect evidence. Besides indirect experimental evidence, this evidence exists primarily in the form of **selection rules**; that is, statements about the relative probability with which diverse processes may occur. Such statements are based on fundamental physical laws. While discussing selection rules, we shall also learn about some of the specific conditions that favor, for example, ionization over excitation.

The most important selection rule is the law of energy conservation. An activator must supply the energy required by a given process. In Table 20-5 we list typical

Table 20-5 The energies of typical atom and diatomic molecule processes compared with energies of various radiations

Typical processes	Process energy ($\mathscr{E}$) or related quantity	$\dfrac{\mathscr{E}}{kJ\,mol^{-1}}$	$\dfrac{T}{K} = \left(\dfrac{\mathscr{E}}{R}\right)K^{-1}$	Radiation energy ($\mathscr{E}$) or related quantity	Approximate limits of different radiation types
		0.0004	0.05	1 GHz	microwaves, lower limit
$Na^{35}Cl$, rotational transition[a]	10.56 GHz	0.0042			
		0.04	5	100 GHz	microwave–far IR boundary
OH, rotational transition[a]	1118 GHz	0.45			
I_2, vibrational transition[a]	214 cm^{-1}	2.6			
		3.6	430	300 cm^{-1}	far IR–IR boundary
		39	4600	4000 cm^{-1}	IR–near IR boundary
H_2, vibrational transition[a]	4405 cm^{-1}	53			
$Li_2 \rightarrow 2\,Li$	1.14 eV	110			
		160	1.9×10^4	4×10^{14} Hz	near IR–visible boundary
		320	3.8×10^4	8×10^{14} Hz	visible–UV boundary
$Cs \rightarrow Cs^+ + e^-$	3.9 eV	380			
$CO \rightarrow C + O$	9.14 eV	880			
$NO \rightarrow NO + e^-$	9.5 eV	920			
$F_2 \rightarrow F_2^+ + e^-$	17.8 eV	1700			
		2000	2.4×10^5	5×10^{15} Hz	UV–x ray boundary
$F^{4+} \rightarrow F^{5+} + e^-$	114 eV	1.1×10^4			
$F^{7+} \rightarrow F^{8+} + e^-$	954 eV	9.2×10^4			
		3.9×10^6	4.7×10^8	40 keV	x and γ ray boundary
		2.0×10^8	2.4×10^{10}	2 MeV[b]	β rays
		2.9×10^8	3.6×10^{10}	3 MeV[b]	γ rays
		4.9×10^8	6.0×10^{10}	5 MeV[b]	α rays

[a] From the ground to first excited state.
[b] Approximate upper limits for common natural sources.
SOURCES: G. Herzberg: *Molecular Spectra and Molecular Structure*, Vols. I and II. D. Van Nostrand, Princeton, N.J., 1950; R. C. Weast (ed.): *The Handbook of Chemistry and Physics*, 49th ed. The Chemical Rubber Co., Cleveland, 1968, p. E70; C. H. Bamford and C. F. H. Tipper (eds.): *Comprehensive Chemical Kinetics*. Elsevier, Amsterdam, 1969, Vol. I, pp. 65–67.

processes of atoms and diatomic molecules and compare the required energies with the energies carried by an activator particle or photon. Table 20-5 shows that there are three activator energy ranges, each with its typical processes:

1. Below about 100 kJ mol^{-1} a photon (or particle) will not have enough energy to break typical chemical bonds (however, it could break a hydrogen bond, $E \approx 20$ kJ mol^{-1}). Typical processes are rotational and vibrational excitation.

2. The chemical energy range typified in Table 20-5 by dissociations of diatomic molecules, extends from 100 kJ mol^{-1} to about 5×10^3 kJ mol^{-1}. For reasons discussed later, most activators have small probability for initiating direct chemical change, for example

$$AB \xrightarrow{\text{activator}} A + B$$

The processes with large probabilities are electronic excitation and ionization. They are important precursors to chemical reactions. The possibilities for electronic excitations are more numerous in the chemical range than for ionizations. Therefore the former are observed more often, especially at the lower end of the range.

3. Above $\approx 5 \times 10^3$ kJ mol^{-1} the deposition products vary only slightly with the activator energy. We see evidence for this in Table 20-4. More evidence comes from mass spectrometry: The final products and their relative yields are only slightly dependent on the activator energy. This can be so only if the initial products are the same in all cases. The deposition products are thought to be, in about equal amounts, singly ionized, excited or unexcited cations plus electrons and excited neutral molecules. When the activator particles slow down to energies in the chemical range, they begin to undergo reactions typical of that range.

Table 20-5 also lists temperatures T that are equivalent to the energies $\mathscr{E}$. These temperatures suggest why activator initiated reactions are important: if, for example, one attempted to dissociate Li_2 by heating a sample of lithium gas, one would have to approach an impractically high temperature, about 10,000 K.

A second selection rule that helps predict deposition products is the law of momentum conservation. It imposes limits on the fraction f of the energy of the activator particle that may be transferred upon a head-on (collinear) hard sphere collision.

$$f = E_t / E_i = 4 m_a m_r / (m_a + m_r)^2 \qquad \text{(20-70)}$$

where E_t is the energy transferred, E_i is the initial energy of the activator particle, and m_a, m_r are the masses of the activator and the reactant particles, respectively. [For a derivation of Eq. (20-70) see Problem 20.]

Let us apply Eq. (20-70) to a collision between an electron with energy E_i in the chemical range, and the bound atoms or electrons of a molecule. Since for an atom $m_r \gg m_a = m_e$, Eq. (20-70) reduces to $f \approx 4 m_a / m_r \ll 1$; that is, an electron can transfer but a small fraction of its energy to the atom. Since the atom is bound in the molecule with an energy about equal to E_i, the transferred energy is not enough to dissociate the molecule. If the reactant particle is a bound electron, Eq. (20-70) reduces to $f \approx 1$; all of the energy may be transferred in a head-on collision. These conclusions apply to all low-mass, chemical energy range activators, for example, photons.[8] We have thus explained why the low-mass activators can primarily cause the excitation and ionization of electrons rather than the dissociation of molecules. For similar reasons atoms with energies in the chemical range should be expected to enter into chemical reactions directly. Indeed, the reactions of such, so-called **hot atoms**, have been observed. The observations lead to some interesting conclusions about energy dependence of reaction rates (Chapter 22).

The third and final selection rule has limited validity. It is known in photochemistry as the Stark–Einstein law: a single molecule absorbs a single photon. The law may be generalized to include radiochemistry: an activator deposits *all* of its energy in a single molecule. In this form the law applies to all activators with energies in the chemical range and below. The law breaks down under two conditions. In the first condition the activator radiation has a very high intensity. Until recently it was not possible to reach these high intensities. With the advent of lasers (Chapter 13) high enough intensities have been reached to make the simultaneous absorption of two photons observable.

The generalized Stark–Einstein law also breaks down as the energy of the activator increases beyond the chemical range. Since this breakdown encompasses all of radiochemistry, we will discuss the deposition by a high energy activator in some detail. The breakdown implies that of two alternative energy deposition mechanisms—multiple ionization and excitation of the electrons of a single atom *or* the ionization and excitation of a single electron in many atoms—the second is the one that actually occurs. What are the physical reasons for this? Although momentum conservation would permit large amounts of energy to be transferred to the electrons of an atom and hence ionize the atom, the probability that a charged activator particle will actually come close enough (less than 1 Å) to an electron to impart the necessary energy via the Coulomb force is sufficiently small to make multiple ionization improbable. The faster the activator particle moves, the shorter the time interval during which the Coulomb force is experienced; therefore, the closer the activator particle must come to an electron in order to ionize it. Fast activator particles behave like bullets. Heavy particles are more efficient activators than light particles of the same energy (Table 20-4). Electromagnetic radiation can also exert a Coulomb force upon electrons via its oscillating electrical field with a spatial range of effectiveness about equal to its wavelength.

20-5.3 Mechanisms of Photochemical Reactions

With these selection rules in mind we can proceed to describe and illustrate the reactions that follow the deposition of energy. Let us begin with processes in the chemical energy range. Assuming that the absorbed radiation is monochromatic (has a narrow energy spread) the Stark–Einstein law predicts that reactant molecules would be excited to a singly excited electronic energy level [or, for somewhat higher energies, form a singly charged (excited) ion]. We designate the excited molecule by AB*. The possible

---

[8] Eq. (20-70) itself is not applicable to photons. It can be applied to nonrelativistic particles only ($v_{particle} < 0.9 \, c$).

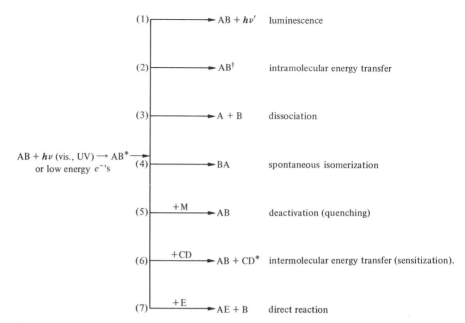

Figure 20-13 The primary elementary processes of electronically excited species produced by low energy activators. AB*, AB† indicate different electronic excited states. The released energy of exothermic processes is converted into heat immediately or eventually.

elementary processes it may undergo are shown in Figure 20-13. Other, less frequently encountered processes, may occur. Some of the primary reactions shown produce reactive species, for example, AB‡ or CD*, making secondary reactions possible. For example, CD* could undergo all of the reactions shown for AB* in Figure 20-13; if A and B are free radicals, they can undergo reactions typical of free radicals, and so on.

Next, we shall consider the quantitative relations that exist between I_a, the various rates, and the yields Φ (Eq. 20-67) of the reaction scheme in Figure 20-13. The Stark–Einstein law implies that the amount of activator absorbed equals the amount of AB reacted or of AB* produced. Therefore, we have from Eq. (20.68)

$$I_a = \frac{dAB^*}{dt} \tag{20-71}$$

If the only reaction of AB is with the activator, then $I_a = -dAB/dt$, $\Phi_{-AB} = 1$, where the minus sign in the subscript of the symbol Φ_{-AB} means that AB is *destroyed* by the activator. From Eq. (20-67) we see that the Stark–Einstein law can be expressed as $\Phi_{-AB} = \Phi_{AB^*} = 1$.

From atom conservation, it follows that for the yields Φ_i of the seven primary processes in Figure 20-13

$$\sum_{i=1}^{7} \Phi_i = \Phi_{AB^*} = 1 \tag{20-72}$$

This result which implies that $\Phi_i < 1$ helps to distinguish the products of primary and secondary processes. For example, if the latter involve chain reactions, yields much higher than 1 can be observed for the products of the chains.

To illustrate the application of these results let us analyze the mechanism of the photolysis of $O_3 + O_2$ mixtures by red light. The following mechanism has been proposed [44, p. 14].

$$O_3 + h\nu_{red} \rightarrow O_3^* \qquad r_1 = I_a \qquad\qquad\qquad\qquad\qquad \text{(20-73a)}$$

primary reactions

$$O_3^* \rightarrow O_2 + O \qquad r_2 = k_2 \cdot O_3^* \qquad\qquad\qquad\qquad\qquad \text{(20-73b)}$$

$$O + O_3 \rightarrow 2\,O_2 \qquad r_3 = k_3 \cdot O \cdot O_3 \qquad\qquad\qquad\qquad \text{(20-73c)}$$

secondary reactions

$$O + O_2 + M \rightarrow O_3 + M \qquad r_4 = k_4 \cdot O \cdot O_2 \cdot M \qquad\qquad \text{(20-73d)}$$

where M is a chemically inert (third) body, for example, a noble gas atom or O_3. Assuming steady states for O_3^* and O, we have for the net reaction rates of O_3^*, O, and O_3,

$$r_{O_3}^* = I_a - k_2 O_3^* \approx 0 \qquad\qquad\qquad\qquad \text{(20-74a)}$$

$$r_O = k_2 \cdot O_3^* - k_3 \cdot O \cdot O_3 - k_4 \cdot O \cdot O_2 \cdot M \approx 0 \qquad \text{(20-74b)}$$

$$r_{O_3} = -I_a - k_3 \cdot O \cdot O_3 + k_4 \cdot O \cdot O_2 \cdot M \qquad \text{(20-74c)}$$

Combination and rearrangement of Eqs. (20-74) gives

$$\frac{-r_{O_3}}{I_a} = \frac{2k_3 \cdot O_3}{k_3 \cdot O_3 + k_4 \cdot O_2 \cdot M} \qquad\qquad \text{(20-75)}$$

If r_{O_3} is constant,

$$\frac{-r_{O_3}}{I_a} = -\frac{\Delta[O_3]/\Delta t}{(\text{amount of photons absorbed})/\Delta t V} = \Phi_{-O_3} \qquad \text{(20-76)}$$

according to the definition in Eq. (20-67). Substitution of this result in Eq. (20-75) and subsequent inversion yields

$$\frac{1}{\Phi_{-O_3}} = \frac{1}{2}\left(1 + \frac{k_4 \cdot O_2 \cdot M}{k_3 \cdot O_3}\right) \qquad\qquad \text{(20-77)}$$

Figure 20-14 is a plot of $1/\Phi_{-O_3}$ versus the quotient of concentrations $O_2 \cdot M/O_3$ from experimental results. The agreement with Eq. (20-77) indicates that the proposed mechanism is a possible one. The intercept (where O_2 or M is zero) gives $\Phi_{-O_3} = 2$. Inspection of Eqs. (20-73a–c) shows that under these conditions every photon does indeed cause the disappearance of two ozone molecules, one of them by a secondary process. The value also implies that reaction (20-73b) is the only consumer of O_3^*. If there were another consumer, for example,

$$O_3^* + M' \rightarrow O_3 + M' \qquad r_5 = k_5 O_3^* \cdot M' \qquad \text{(20-73e)}$$

then, for constant concentrations of O_3^* and M', the primary yields, by the same argument as in Eq. (20-76), would be

$$\Phi_2 = r_2/I_a \qquad \text{and} \qquad \Phi_5 = r_5/I_a \qquad\qquad \text{(20-78)}$$

From Eq. (20-72) $\Phi_2 + \Phi_5 = 1$. Therefore a nonzero value for r_5 leads to a decrease in r_2 which, in turn, must lead to a decrease in r_3 and the overall yield Φ_{-O_3}.

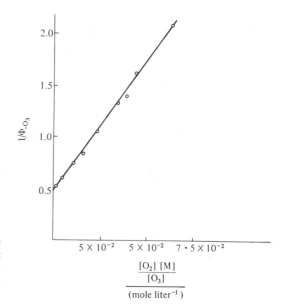

Figure 20-14 Plot of results for the photolysis of ozone by red light. See text for explanation of plotted quantities. [From R. P. Wayne: *Photochemistry*. Butterworth, London, 1970, p. 15.]

$$\frac{[O_2]\,[M]}{[O_3]}$$

$$(mole\ liter^{-1})$$

20-5.4 Mechanisms of Radiochemical Reactions

Let us now consider reactions caused by activators with energies above the chemical energy range. The energy-deposition pattern that follows from the bullet like nature of high energy radiation, discussed in connection with the high energy breakdown of the Stark–Einstein law, has an important effect on the nature of the follow-up reactions.

The deposition results in a very heterogeneous distribution of ionized and excited species. According to the spur model, they are initially localized in very small regions called **spurs** which in turn are grouped into narrow cylinders—**tracks** (Figure 20-15). The different energy deposition efficiencies of diverse radiation, implied by the two tracks in Figure 20-15, are measured quantitatively by the **linear energy transfer** (*LET*) quotient. LET is defined as the energy lost by an ionizing particle per unit length of track. (For typical values of LET see Table 20-4).

According to the **spur-diffusion** model, the first reactions after energy deposition occur within the tracks and spurs where high concentrations of reactive species exist. The results for the radiolysis of liquid water (Table 20-4) are consistent with the following mechanism for the deposition of energy and the subsequent spur-track reactions.

$$H_2O \overset{\text{\tiny\char`\~\char`\~}}{\longrightarrow} H_2O^+ + e^- \qquad \text{energy deposition}^9 \qquad \textbf{(20-79a)}$$

$$H_2O \overset{\text{\tiny\char`\~\char`\~}}{\longrightarrow} H_2O^* \qquad\qquad\qquad\qquad\qquad \textbf{(20-79b)}$$

$$H_2O^+ + H_2O \rightarrow H_3O^+ + OH \qquad \text{spur-track reactions} \qquad \textbf{(20-79c)}$$

$$H_2O^* \rightarrow H + OH \qquad\qquad\qquad\qquad\qquad \textbf{(20-79d)}$$

$$OH + OH \rightarrow H_2O_2 \qquad\qquad\qquad\qquad\qquad \textbf{(20-79e)}$$

$$H + H \rightarrow H_2 \qquad\qquad\qquad\qquad\qquad\qquad \textbf{(20-79f)}$$

$$H + OH \rightarrow H_2O \qquad\qquad\qquad\qquad\qquad \textbf{(20-79g)}$$

---

[9] $\overset{\text{\tiny\char`\~\char`\~}}{\longrightarrow}$ symbolizes absorption of high energy radiation.

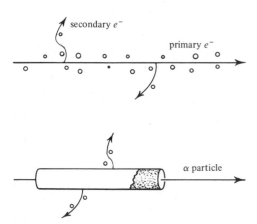

Figure 20-15 Drawings of simplified electron and α-particle tracks. The actual tracks of electrons throughout and of α particles near the end of the track are tiny zigzags about the smooth lines shown. Such tracks are seen in the Wilson cloud chamber. Track lengths (the *range*): about 10^{-3} cm in water, 1 cm in air for 1 MeV α particles, considerably longer for 1 MeV β, or γ rays. •, ○ are called spurs. Each spur originates from one primary ionization. Due to secondary ionization the spur may contain zero to several additional electron-cation pairs and excited species, as shown by the varying size of ○. Initial diameter of a spur is a few angstroms. The concentration of the reactive species in the spur rapidly decreases as the spur expands due to diffusion. Best defined and persistent spurs are found in high density media—liquids and solids.

Reactions (20-79e–g) are the only ones where rates depend upon the *squares* of the concentrations of the reactive species—the free radicals H and OH. These reactions therefore occur to an observable extent only under conditions where the average concentrations of H and OH are rather high, during the early life of a spur or when the average concentration of spurs within a track is so high that reactions between radicals in two different spurs become important. The latter is apparently the case with 5.3 MeV α rays for which the ratio of the radical to molecule yields is much lower than for the lower LET radiations (Table 20-4). In addition the net total yield—the sum of all yields—is lower for α rays since reaction (20-79g) replaces a part of the water consumed by reactions (20-79a–d). Similar considerations would lead us to predict that, at dose rates sufficiently high to make the tracks practically contiguous, the radical yields should decrease. This effect has been observed.

In order to complete the description of spur-track reactions we must discuss the fate of the secondary electron released by reaction (20-79a). It looses its energy rapidly via further ionization and excitation of H_2O molecules (Figure 20-15). When its energy is no longer sufficient to cause electronic excitation, further energy losses (to vibrations, rotations, translations) occur at a slower rate until the electron reaches thermal energies. Then it begins to polarize the H_2O molecules and becomes hydrated, just like any ion, in about 10^{-11} s. The identification of the hydrated electron [42], $e^-(aq)$, as a rather stable species, with a half-life in pure water of about 2×10^{-4} s, is a great triumph of radiochemistry. The hydrated electron, the simplest reducing agent possible, exhibits reactivity toward many organic, biologic, and inorganic substances. Thus it may be expected to play a role in the electron transfer processes of aqueous systems. In radiolyzed pure water its most important reactions, which can occur in tracks and spurs, are[10]

$$e^-(aq) + e^-(aq) \rightarrow H_2 + 2\ OH^- \qquad\qquad \textbf{(20-80a)}$$

$$e^-(aq) + H \rightarrow H_2 + OH^- \qquad\qquad \textbf{(20-80b)}$$

$$e^-(aq) + OH \rightarrow OH^- \qquad\qquad \textbf{(20-80c)}$$

$$e^-(aq) + H_3O^+ \rightarrow H \qquad\qquad \textbf{(20-80d)}$$

Other than $e^-(aq)$ these reactions do not involve any reactants or products not considered previously. Therefore, our earlier qualitative predictions of the products of spur-track reactions are still valid.

The high radical concentrations responsible for spur reactions actually last for a rather short time. Due to diffusion (Section 20-8.2) of the radicals throughout the water, their local concentrations decrease some million times in approximately 10^{-8} s and a homogeneous distribution is approached. Since gaseous diffusion is some 10,000 times faster, spur reactions are much less important in gases.

The radiolysis of water thus consists of two stages. The spur-track reactions are the first stage. The surviving reactive species react in the second stage, which lasts about 1 s. The radical concentrations are now so low that only radical-plus-stable-molecule-or-ion (including added solutes) reactions are important. These reactions may frequently be studied by the powerful direct methods described in Section 20–6.

The first stage mechanisms, however, are determined from indirect evidence. Apart from the selection rules discussed earlier, the important evidence is the primary yields of molecules and radicals (see Table 20-4) that we used in our qualitative discussion of the spur-track reactions. These primary yields of the first stage products (g_X) are obtained indirectly by combining the final molecular yields (G_X) determined at the end of stage two with atom conservation relations. The relations established earlier for photochemical yields are valid for the determination of the g_X with one exception: Eq. (20-72). This equation is not valid because the Stark–Einstein relation does not apply. A rough equivalent of Eq. (20-72) is the number of ions and or excited species formed per 100 eV. The number of ions can be measured only for gases. For excited species and for any species in liquids, only estimates are available. For gases $g_{M^+} = 2.5$–4 ions per 100 eV. This is about half the number expected on the basis of the first ionization potentials (Table 20-5). The remaining energy is assumed to go into electronic excitation. For liquids the same or slightly lower values of g_{M^+} are expected to apply.

Let us now discuss in detail the determination of the primary yields g_X. To obtain the necessary atom conservation relations, we first summarize the net decomposition of water (from Eqs. 20-79 and 20-80):

$$H_2O \rightarrow OH, H, H_2O_2, H_2, e^-(aq)(= H^+\ \text{excess}) \qquad \textbf{(20-81)}$$

The equality of the $e^-(aq)$ and excess H^+ (above its equilibrium concentration in

---

[10] H_2O on the left side of the equations is implicit in the parenthetical (aq).

unradiolyzed water) follows from charge balance. From the conservation of hydrogen atoms the net yield of H_2O decomposed at the end of stage one is

$$2g_{-H_2O} = g_{OH} + g_H + 2g_{H_2O_2} + 2g_{H_2} + g_{e^-(aq)} \qquad \textbf{(20-82)}$$

From conservation of O atoms

$$g_{-H_2O} = g_{OH} + 2g_{H_2O_2} \qquad \textbf{(20-83)}$$

By eliminating g_{-H_2O} between Eqs. (20-82) and (20-83) we have

$$2g_{H_2O_2} + g_{OH} = 2g_{H_2} + g_H + g_{e^-(aq)} \qquad \textbf{(20-84)}$$

Therefore, if any four yields are known, the fifth can be obtained from Eq. (20-84).

The four yields can be determined experimentally from the final molecular yields, G_X. Consider the determination of g_{H_2}. During stage two H_2 disappears by reaction with OH

$$H_2 + OH \rightarrow H_2O + H \qquad r_{85} = k_{85} \cdot H_2 \cdot OH \qquad \textbf{(20-85)}$$

This reaction causes g_{H_2} to be larger than G_{H_2}. If the reaction could be stopped, g_{H_2} would equal G_{H_2}. Reaction (20-85) is stopped by adding Br^- ($\approx 10^{-4}$ M) which reacts with OH

$$OH + Br^- \rightarrow Br + OH^- \qquad r_{86} = k_{86} \cdot OH \cdot Br^- \qquad \textbf{(20-86)}$$

Assuming that r_{85}/r_{86} is constant during stage two, that is, that the concentrations of Br^- and H_2 are constant, we have for the fraction of OH consumed by H_2

$$f_{H_2} = \frac{r_{85}}{r_{85} + r_{86}} = \frac{k_{85} \cdot H_2}{k_{85} \cdot H_2 + k_{86} \cdot Br^-} \qquad \textbf{(20-87)}$$

From pulse radiolysis (next section) $k_{85} = 4.5 \times 10^7$, $k_{86} = 10^9 \, M^{-1} s^{-1}$, $H_2 \approx 10^{-10}$ M. Hence $f_{H_2} \approx 10^{-8}$. Thus we see that reaction (20-86) does indeed protect H_2 from attack by reaction (20-85), and therefore $g_{H_2} = G_{H_2}$ in the presence of Br^-.

To determine $g_{H_2O_2}$ the radiolysis water is saturated with O_2 which scavenges the reducing agents H and $e^-(aq)$

$$O_2 + H \rightarrow H^+ + O_2^- \quad \text{and} \quad O_2 + e^-(aq) \rightarrow O_2^- \qquad \textbf{(20-88)}$$

The O_2^- produced reacts to give H_2O_2

$$2\,H_2O + 2\,O_2^- \rightarrow H_2O_2 + O_2 + 2\,OH^- \qquad \textbf{(20-89)}$$

In effect each H or $e^-(aq)$ produces half a molecule of H_2O_2. Each OH, on the other hand, prevents the formation of $\frac{1}{2} H_2O_2$

$$OH + O_2^- \rightarrow O_2 + OH^- \qquad \textbf{(20-90)}$$

or destroys $\frac{1}{2} H_2O_2$

$$OH + H_2O_2 \rightarrow O_2^- + H^+ + H_2O \qquad \textbf{(20-91)}$$

The net amount of H_2O_2 at the end of stage two, based on an H atom balance, is

$$G_{H_2O_2} = g_{H_2O_2} + \frac{1}{2}(g_H + g_{e^-(aq)}) - \frac{1}{2}g_{OH} \qquad \textbf{(20-92)}$$

This assumes that reaction (20-85) does not consume any OH. If Br^- is added, no OH will be consumed by H_2. The Br produced in reaction (20-86) reacts with O_2^- and

H_2O_2 in the same way as does OH itself. Thus Eq. (20-92) is valid in presence of Br^-. Substitution of the conservation equation (20-84) and $g_{H_2} = G_{H_2}$ in Eq. (20-92) gives

$$g_{H_2O_2} = \tfrac{1}{2}(G_{H_2O_2} + g_{H_2}) = \tfrac{1}{2}(G_{H_2O_2} + G_{H_2}) \qquad (20\text{-}93)$$

The primary yield of OH may be determined by combining data from O_2 and $O_2 + H_2$ saturated waters (Problem 24). A method for determining g_H was mentioned on page 664; $g_{e^-(aq)}$ can then be obtained from Eq. (20-84). The assumed mechanisms in the different solutions may be checked in a variety of ways. We mention two possibilities. The time dependence of the concentration $e^-(aq)$ may be determined in pulse radiolysis (Section 20-7) from the absorption spectrum of the hydrated electron, or the variation of scavenger concentrations may reveal unsuspected side reactions.

In recent years the spur-diffusion model just described has been shown to be inadequate. There is strong evidence [13, and references therein] that the electron and H_2O^+ produced by reaction (20-79a) can participate in electron transfer reactions prior to hydration, although at this time (early 1974) the precise nature of these reactions is not known.

We now compare the determination of mechanisms from yields and from rate laws. Yields are as ambiguous about mechanisms as rate laws. Yields have the advantage that they may be applied to both fast and slow reactions; on the other hand, yields determine ratios rather than absolute values of rate constants (for example, Eq. 20-77). Clearly, if the method of rate laws could be applied to the fast reactions of reactive intermediates, the postulated elementary reactions that make up a reaction mechanism could be observed directly. The development of these direct methods has proved to be rather difficult. Success has come only in the last 25 years. The direct methods are described in Sections 20-6 and 20-7.

20-6   THE MEASUREMENT OF FAST REACTION RATES

To observe directly all elementary reactions within a reacting system it is necessary to determine the concentrations of all species in the system as a function of time. For this purpose methods for measuring the rates of processes with lifetimes as short as those discussed in Section 20-1 are needed. In this section we shall describe the more important methods used to measure the rates of reactions with half-lives of less than 1 s.

To measure concentrations as a function of time, one can repeatedly quench the reacting system (for example, cool it suddenly or rapidly add a substance that will stop the reaction) or small aliquots of it can be quenched and the concentration can be obtained by chemical (for example, titration) or chromatographic means. These methods disturb the reacting system and are inherently slow, since the withdrawal of aliquots and the quenching take time. They are indispensable, however, if the stoichiometry of the reaction must be determined.

Methods that do not disturb the system and can be rather fast depend upon the measurement of a concentration-dependent physical property. Useful physical properties are absorbance, optical rotation or index refraction of electromagnetic radiation at a particular wavelength, ultrasonic absorption, electrical conductance (of solutions), thermal conductivity, heat released or absorbed, volume, pressure, dielectric constant, and so on. The general relation between concentrations of species i in a thermally and mechanically equilibrated system and a physical property Z can be derived from Eq. (8-16)

$$Z = \sum_i \bar{\mathcal{Z}}_i n_i \qquad \text{(8-16)}$$

Dividing by the known volume of the reacting system we get

$$Z/V = \sum_i \bar{\mathcal{Z}}_i C_i \qquad \text{(20-94)}$$

where C_i is the concentration of the ith species. If the partial molal quantities $\bar{\mathcal{Z}}_i$ are constants, the relation between Z and C_i is a simple linear one. (Use of nonlinear relations is seldom practical.)

Eq. (20-94) is most simply applied if all but one $\bar{\mathcal{Z}}_i$ are equal to zero. This is most often the case with absorbance at a particular wavelength. Using absorbance A_i of species i at wavelength λ as an example, we get

$$A_i = \bar{\mathcal{A}}_i V C_i = a_i d C_i \qquad \text{(20-95)}$$

This is Beer–Lambert's law, with $\bar{\mathcal{A}}_i$ being the partial molal absorbance; V, the volume; a_i, the absorptivity; and d, the thickness of the absorbing solution. If $\bar{\mathcal{A}}_i \approx 0$ (or $a_i \approx 0$) the ith species does not absorb at this particular wavelength. From measurements at N wavelengths, the concentrations of all N species may be determined. It is not even necessary that at every wavelength all but one $\bar{\mathcal{A}}_i$ be zero; one can determine the absorptivity of each species at all wavelengths at which measurements will be carried out and then, from measurements on the reacting system at N wavelengths, get N independent equations of the type of Eq. (20-94) from which one can solve for the N concentrations.

If several $\bar{\mathcal{Z}}_i$ have nonzero values and if only a single measurement of a given property is possible (for example, the volume and most of the other properties listed earlier), then there is a unique relation between the property and the concentrations only if the system obeys a single stoichiometric equation, $\sum_i \nu_i C_i = 0$, so that the change in the number of moles between times t and zero is $n_{it} - n_{i0} = \nu_i \zeta$. Expressing n_{it} in Eq. (8-16) in terms of the extent of reaction ζ (page 255), dividing the result by the volume V, and rearranging yields

$$\frac{Z}{V} - \frac{Z_0}{V} = \frac{\sum (\nu_i \zeta + n_{i0}) \bar{\mathcal{Z}}_i}{V} - \frac{\sum_i n_{i0} \bar{\mathcal{Z}}_i}{V} = (\sum_i \nu_i \bar{\mathcal{Z}}_i) \mu = c\mu \qquad \text{(20-96)}$$

where Z_0 is the value of Z at $t = 0$ and c is a constant. In order to determine the concentration μ (defined on page 634) at any time t we must know not only the values of Z at the selected times but also the initial value Z_0. Additionally, we must determine (by measuring Z of solutions of known μ) the constant c.

For first order reactions, a different formulation is possible. Assuming a constant volume, the concentration ratio A/A_0 is

$$\frac{A}{A_0} = 1 - \frac{a\mu}{A_0} = 1 - \left(\frac{a}{A_0 c}\right)\left(\frac{Z}{V} - \frac{Z_0}{V}\right) \qquad \text{(20-97)}$$

where the relation $A = A_0 - a\mu$ and Eq. (20-96) have been used ($a = \nu_A$, the stoichiometric coefficient). When $t = \infty$, $A = 0$; therefore $A_0/a = -\mu_\infty$. We can substitute for μ_∞ from Eq. (20-96) into Eq. (20-97) to yield, after cancellations

$$A/A_0 = 1 - (Z - Z_0)/(Z_\infty - Z_0) = (Z - Z_\infty)/(Z_0 - Z_\infty) \qquad \textbf{(20-98)}$$

Eq. (20-98) expresses the fact that the concentration ratio A/A_0 can be determined *without* a knowledge of the actual values A and A_0 [that is, without knowing c in Eq. (20-96)]. If the species A is a reactive intermediate, this fact has great practical importance for the determination of rate constants because it obviates the preparation of solutions with known concentrations of A.

As a somewhat atypical but simple example, consider the determination of the rate r for the reaction

$$CH_3CH(OC_2H_5)_2 + H_2O \rightarrow CH_3CHO + 2\,C_2H_5OH \qquad \textbf{(20-99)}$$

by a dilatometric method (see Figure 20-16), that is, by measuring the increase in volume that accompanies the reaction (that is, $Z = V$). From Eq. (20-96) we get

$$1 - V_0/V = c\mu \qquad \textbf{(20-100)}$$

By writing for μ an equation analogous to Eq. (20-6) we obtain for the rate

$$r = \frac{d\mu}{dt} + \mu \cdot \frac{d \ln V}{dt} \qquad \textbf{(20-101)}$$

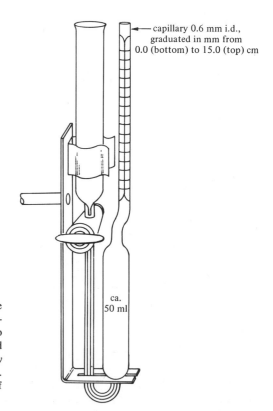

capillary 0.6 mm i.d., graduated in mm from 0.0 (bottom) to 15.0 (top) cm

ca. 50 ml

Figure 20-16 Dilatometer. During the run the stopcock is closed. The dilatometer is immersed in a well-regulated (to 0.001°C or better) thermostat. [Adapted from *Experiments in Physical Chemistry* by D. P. Shoemaker and C. W. Garland. Copyright 1962. Used with permission of McGraw-Hill Book Co.]

We may eliminate $d\mu/dt$ and μ with the help of Eq. (20-100) to get

$$r = c^{-1}\left(\frac{V_0}{V} + \frac{V - V_0}{V}\right)\frac{d \ln V}{dt} \qquad (20\text{-}102)$$

The second term in the parenthesis is the correction for the change in concentration due to the changing volume of the reacting system (page 633). For the diethyl acetal (initial concentration 0.4 M) reaction the correction is maximally 0.5%, that is, the primary reason for the concentration change is the varying number of moles of reactants and products.

In the preceding example the student may discern three problems that are generally encountered when one attempts to determine a rate law. The first is the **initiation** problem. By initiation of the reaction we mean bringing the reacting system to the conditions of homogeneity-heterogeneity, temperature, and so on, at which the reaction rate is to be measured. A classic initiation, used in the dilatometric method, is to mix the reactants together and insert them into a thermostat. The problem arises if the reactions of interest proceed appreciably during the time interval (t_i) required to initiate the reaction; that is, if the reaction time $t_r(\approx t_{1/2})$ is not such that $t_r \gg t_i$. It is then impossible to distinguish the effects of the initiation from those of the reaction.

A second problem, shared with any measurement on time dependent systems, involves the time **interval for measurement** of a property (t_m). Unless $t_r \gg t_m$ the reaction would be largely or completely finished before the first measurement of concentration or rate could be completed.

Finally, in common with all measurements, there is the **sensitivity** problem. Unless the measurement of concentrations is sensitive enough, it is not possible to determine the functions of concentration on the right hand side of Eqs. (20-8), (20-11), or (20-12). Even the rate itself cannot be measured unless the concentration of at least one species can be determined as a function of time.

For the dilatometric method the time interval $t_m \approx 1$ s if the liquid level in the capillary of the dilatometer (Figure 20-16) is determined by the human eye. A barely acceptable minimum sensitivity would be a volume change during the run of 10^{-3} cm^3 (equal to about 1 mm change of the liquid level). Assuming a precision for determining the liquid level of about 0.2 mm, this sensitivity would allow the measurement of five concentrations during the course of the reaction. However, if the rate of change of the liquid level were faster than about 0.2 mm s^{-1}, the sensitivity would decrease. The connection between the time interval for measurement and the sensitivity is quite general.

Except for first order reactions, the sensitivity problem is connected to both of the other two problems. This can be inferred from the formulas in Table 20-1 for the half-lives, which are approximate measures of t_r. Consider, for example, the very important second-order reactions for which $t_r \propto A_0^{-1}$. To eliminate the initiation and measurement problems, the reaction time has to be long. Long reaction times can be achieved by decreasing the concentrations of the reactants. If this is done, however, the concentrations of the reactants may become immeasurably low.

There are several different experimental methods designed to overcome the initiation problem to various degrees. According to the manner in which these methods deal with it, we differentiate between classical, fast mixing, field jump, pulse, and life-time methods. The methods of the last four types have largely overcome a long-standing bottleneck which had prevented the study of fast reactions. Additional information on the five experimental methods may be found in references [36], [37].

20-6.1 Classical Methods   A classical method is any method where the reaction is initiated in an easily accomplished way, usually simply by pouring the reactants from two or more separate containers into the reaction vessel and then stirring and equilibrating the resulting mixture in a thermostat. The conditions $t_r \gg t_i$, $t_r \gg t_m$ are achieved by choosing to study only slow reactions. Dilatometry (Figure 20-16), discussed earlier, is an example of a classical method. Classical methods can be used to study reaction kinetics in gases, liquids, solids, and heterogeneous systems. The majority of existing rate law data were probably obtained by such methods.

20-6.2 Fast Mixing Methods   A typical apparatus for a **fast mixing** (or **flow**) method is shown in Figure 20-17. The first use of this method was the study of the reaction of oxygen with hemoglobin (by Hartridge and Roughton in 1923). In the subsequent 50 years, a variety of specialized techniques have been developed.

In the **continuous flow** method, the reacting solution flows continuously along a tube beyond the mixing chamber. Observations.along the tube at different points show the reaction at different stages. Since this method needs large volumes of reacting solutions, two methods less wasteful of possibly difficult-to-obtain reactants have been developed. Both methods observe the reaction at a single point as close as possible to the outlet of the mixing chamber. The **accelerated flow** method measures the concentration of a reactant or product as a function of the flow speed, beginning with an initial speed of zero and ending when the speed has reached a maximum. The **stopped-flow** method (Figure 20-17), the most widely used of the three, follows the reaction as a function of the time after mixing once the flow has been suddenly stopped by the filling up of a receiving chamber. The fast mixing methods can be used for the study of reactions in both gas and liquid phases. Gas phase studies are harder to do because larger volumes are needed and because temperature equilibration and pressure control are more difficult to achieve than in the liquid phase.

Reactions of atoms and free radicals in the gas phase, as well as inorganic, organic, and enzyme reactions in water and nonaqueous solvents have been studied by the flow methods (see Figure 20-22 and Table 20-7).

20-6.3 Field-jump Methods   The field-jump methods, often called (chemical) relaxation methods, were conceived and largely developed by Manfred Eigen (1954) and his coworkers at the Max Planck Institute for Physical Chemistry, Göttingen. Eigen shared the 1967 Nobel prize in chemistry for this work. In this method the inherently slow initiation—by mixing—has been replaced by a faster process: the rapid (as short as 10^{-6} s or even less) changing of a "field" to which the reacting system is exposed. The "field" is most commonly temperature; less often it may be an electric field (for ionic or highly polar reactants) or pressure (for pressure-dependent reactions).

If temperature-jump is to be used, the reaction must be exo- or endothermic. Alternatively, the exo- or endothermic reaction should provide one of the reactants

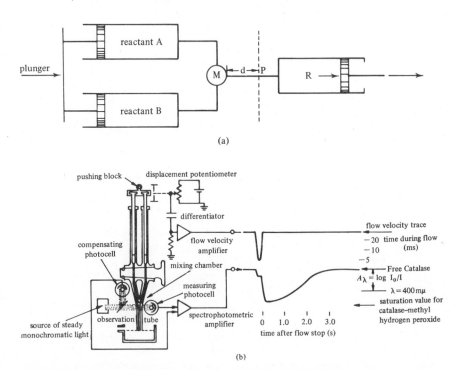

Figure 20-17 (a) Schematic diagram of apparatus for the stopped-flow method. Plunger simultaneously forces both reactants into the mixing chamber, M. Reacting mixture flows past the observation point, P, until stopped by the filling up of the receiving chamber, R. Absorbance at a particular wavelength λ, A, is then recorded as a function of time. Mixing achieved via multiple inlet jets in M arranged near-tangentially around the outlet orifice, plus velocities, v, high enough to give turbulent flow. Minimum $t_i = d/v_{max} = 0.005 \text{ m}/10 \text{ m s}^{-1} = 5 \times 10^{-4}$ s sets a limit to the fastest measurable rates (t_i defined in text); v_{max} determined by the onset of cavitation (bubble formation). Slowest measurable rates determined by diffusion into the region around P (page 633).

(b) Apparatus for the stopped or accelerated flow methods with a typical oscillograph record of the formation and decomposition of the catalase–methyl hydrogen peroxide complex in the presence of ethanol. [From B. Chance in S. L. Friess, E. S. Lewis, and A. Weissberger (eds.): *Techniques of Organic Chemistry*. John Wiley & Sons, 1963, Vol. VIII, p. 737.]

(for example, H^+) to the reaction under investigation. According to Eq. (9-11) the effect of a temperature change is to change the equilibrium constant, K. (At room temperature a temperature change of $10°$ would change K by 1.5% if $\Delta H \approx 1 \text{ kJ mol}^{-1}$). Before the temperature-jump, the reacting system is in equilibrium at the initial temperature. After the jump the system "relaxes" toward a *new* equilibrium state at the final temperature. By following the concentration of one species during the relaxation, we may investigate the rate law. The method is illustrated in Figure 20-18.

The magnitude of a typical temperature-jump (or any field-jump) causes a *small perturbation* upon the reacting system. "Small perturbation" in this context means that (1) the perturbation does not create measurable concentrations of new species, it merely changes the concentrations of already existing species; and (2) the concentration changes are so small that all reactions become effectively first order.

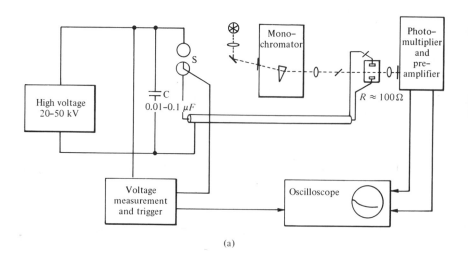

(a)

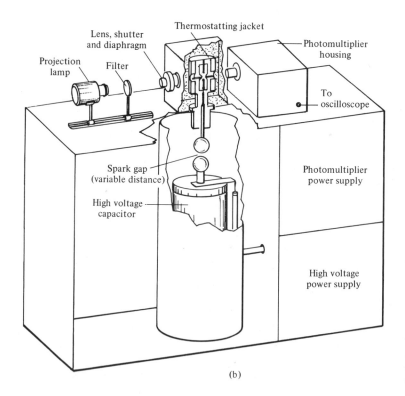

(b)

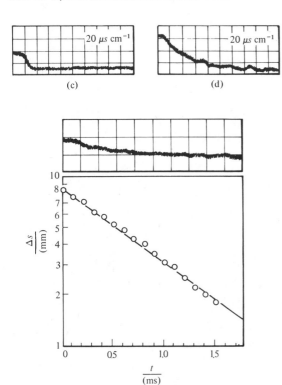

(c)

(d)

(e)

Figure 20-18 (a) Schematic diagram for temperature-jump method. High voltage source charges the capacitor C. After the breakdown voltage across spark gap S is reached. C discharges through sample cell and heats it. The discharge triggers the oscilloscope trace (at a preset rate). On the vertical axis of the oscilloscope the output of the photomultiplier is recorded. The output may be related to absorbance (as in the above arrangement), to fluorescence intensity, or the degree of polarization depending upon the system under study. The oscilloscope face may be photographed for a permanent record (see Figures 20-18c to 20-18e).

Minimum $t_i \approx 1 \mu s > RC$. The resistance R may be adjusted by adding an inert electrolyte, for example, 0.1 M KNO_3, to the cell. Slowest measurable reaction times (≈ 1 s) set by rise of convection currents and diffusion between heated and unheated parts of cell solution. δT determined by direct measurement or by calculation from the energy input. In a typical operation $\delta T \approx 6°C$, sample volume is ≈ 1 cm^3. Microwaves and even light have been used for heating the solutions. If the light is a pulsed laser beam then a minimum $t_i = 0.02 \mu s$ is possible [56].

(b) Simple temperature-jump arrangement.

(c) Oscillogram pattern: temperature jump (measured with highly concentrated glycine buffer and phenolphthalein as indicator; no relaxation).

(d) Oscillogram pattern: relaxational response in the Ca^{2+}/ATP system. [Parts (a)–(d) from M. Eigen and L. De Maeyer in S. L. Friess, E. S. Lewis, and A. Weissberger, (eds.): *Techniques of Organic Chemistry*. John Wiley & Sons, 1963, Vol. VIII, pp. 978–980.]

(e) Oscilloscope trace of the relaxation effect characteristic of ribonuclease isomerization and plot of logarithm of the signal amplitude versus time. 9.65×10^{-5} M ribonuclease, 2×10^{-5} M phenol red, 0.15 M KNO_3, pH = 7.0, 25°C. Time scale on trace is 500 μs per large horizontal division, $\tau = 1.03$ ms. [Part (e) from T. C. French and G. G. Hammes in *Methods in Enzymology*. Academic Press, New York, N.Y., 1969, Vol. XVI, p. 28.]

The second point requires demonstration. Consider the following equilibrium

$$2\,A \underset{k_-}{\overset{k_+}{\rightleftharpoons}} B \tag{20-103}$$

We define the concentration μ by the relations $A = A_e - 2\mu$ and $B = B_e + \mu$, where A_e, B_e are the equilibrium concentrations *after* the jump. This means that $\mu = 0$ at the end of the reaction and that it may be negative or positive during the reaction. Initially, $A_0 = A_e - 2\mu_0$ and $B_0 = B_e + \mu_0$. If the reaction is exothermic then, according to Le Chatelier's principle, $A_0 < A_e$ and $B_0 > B_e$ since the temperature increases during the jump (see Eq. 9-11). The rate r (Eq. 20-7) is negative. The reverse would be true for endothermic reactions. Assuming that the indicated forward and reverse reactions are elementary, the net rate r is

$$
\begin{aligned}
r = r_+ - r_- &= k_+ A^2 - k_- B \\
&= k_+ (A_e - 2\mu)^2 - k_- (B_e + \mu) \\
&= k_+ A_e^2 - k_- B_e - (k_+ \cdot 4A_e + k_-)\mu + 4k_+\mu^2 \\
&= -(k_+ \cdot 4A_e + k_-)\mu + 4k_+\mu^2 \\
&\cong -(k_+ \cdot 4A_e + k_-)\mu
\end{aligned}
\tag{20-104}
$$

The fourth line follows because at equilibrium the forward and reverse reaction rates are equal (Section 20-4). The approximate equality of the last line will be valid if μ is sufficiently small. Separation of variables in Eq. (20-104) and integration yields

$$
\begin{aligned}
\ln \mu/\mu_0 &= -(k_+ \cdot 4A_e + k_-)t \\
&= -k't = t/\tau
\end{aligned}
\tag{20-105}
$$

The last lines of Eq. (20-104) and Eq. (20-105) are, respectively, the differential and the integrated versions of a rate law that is first order with respect to the variable μ. The time τ is known, following traditional nomenclature in the physics of analogous processes, as the **relaxation time**. During the time interval τ the concentration μ decreases by the factor e^{-1} (e is the base of natural logarithms).

A similar treatment of other elementary reaction combinations shows (see Table 20-6) that all of them reduce to first order if μ is sufficiently small. It is possible to treat the exact equations (either analytically or numerically, with the help of a

Table 20-6 Reciprocal relaxation times for various combinations of elementary forward and reverse reactions

$A \underset{k_-}{\overset{k_+}{\rightleftharpoons}} B$	$1/\tau = k_+ + k_-$
$2\,A \rightleftharpoons A_2$	$1/\tau = 4k_+ A_e + k_-$
$A + B \rightleftharpoons C$	$1/\tau = k_+ B_e + k_-$
(B = buffered)	
$A + B \rightleftharpoons C + D$	$1/\tau = k_+(A_e + B_e) + k_-(C_e + D_e)$
$A + C \rightleftharpoons B + C$	$1/\tau = (k_+ + k_-)C_e$
(C = buffered)	

computer) of the type of Eq. (20-104) without making any approximations, but this is seldom necessary.

Figure 20-18e shows the exponential variation characteristic of a first-order process of a variable which is proportional to concentration:

$$\mu = \mu_0 \exp\left(-t/\tau\right) \tag{20-106}$$

[Eq. (20-106) is an alternative form of Eq. (20-105).] The same figure also shows a graphical determination of the relaxation time. Eq. (20-104) implies that the rate constants k_+, k_- can be determined by measuring the relaxation times at various concentrations of A_e. When both the forward and reverse reactions are first order, it is not possible to measure both rate constants. Only their sum may be measured. However, the sum may be combined with the equilibrium constant to determine both rate constants.

The field-jump techniques for initiating a reaction have been widely used to study reactions in liquids. Important results have been obtained in the study of enzyme catalysis, isomerizations, redox reactions, complex formation, ionic reactions, and acid-base reactions. Some results are shown in Table 20-7 and Figure 20-22.

The field-jump methods can be modified in a variety of ways. Three such modifications follow: (1) The field-jump methods can be extended to steady states (other than equilibria) by combining them with the stopped-flow method. If a steady state can be maintained for more than a few milliseconds then, after stopping the flow through a T-jump cell, a T-jump may be applied and the decay toward a new steady state can be observed in a manner entirely analogous to the equilibrium experiment. (2) Initiation times can be shortened to as low as 10 ns with laser-caused T-jumps. (3) Resonance effects due to reactions can be observed by using a periodic field-jump at a rate in the neighborhood of the reaction rate to be measured. An ultrasonic sound wave is probably the most common source of the periodic field (that is, pressure).

20-6.4 Pulse Methods

A pulse method initiates a reaction by creating new reactive species—excited electronic states, radicals, ions—in the system under study. The initiator is a pulse of electromagnetic radiation (light, UV, even x ray) or of charged particles (usually electrons) as short as a few picoseconds (10^{-12} s). If intense visible or ultraviolet radiation is used, the method is often called "flash photolysis." Flash photolysis was the first pulse method and was developed by Norrish and Porter in 1949 at the University of Cambridge. In 1967 Norrish and Porter shared the Nobel prize in chemistry with M. Eigen.

The pulse of particle or electromagnetic radiation can create reactive species in two ways—adiabatically or isothermally. In the adiabatic mode of action the energy of the pulse heats a sample with small heat capacity under nearly adiabatic conditions. (The heat losses are minimal because the duration of the pulse is so short.) The initial concentrations of the reactive species approach the equilibrium concentrations at the maximum temperature of the sample. For example, a high energy UV flash (1000–4000 J) can heat a gas at low pressure (1 torr) to above 3000°C. At this temperature chlorine gas will be completely dissociated.

If instead the heat capacity of the sample is increased by adding an inert (nonabsorbing) gas, the temperature rise will be minimal. The radiation pulse can act

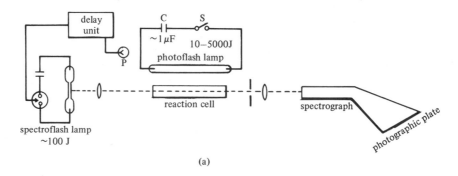

(a)

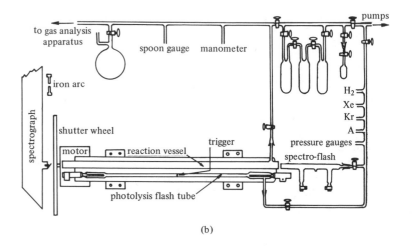

(b)

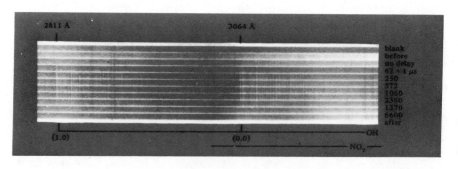

(c)

nearly isothermally, that is, as an activator, to create new reactive species by photolysis or radiolysis. A second method for achieving nearly isothermal conditions is to decrease the pulse energy. Under isothermal conditions only a few reactive species at low concentrations are likely to be created. The relaxation of the system toward an equilibrium may then be followed and the data analyzed by the methods used for field-jump techniques. However, particularly under adiabatic conditions, taking the entire visible (or UV) spectrum may be necessary to detect the various reactive species present and possibly to measure their concentrations. The technique for doing this as a function of time is known as **kinetic spectroscopy**. Figure 20-19 shows a typical flash photolysis-kinetic spectroscopy experiment. The same methods of detection are applicable to all pulse methods; the differences are in the generation of the pulses.

The **shock tube** is a pulse technique. In the shock tube, high concentrations of reactive species at temperatures as high as tens of thousands of degrees are created in a small volume by puncturing a diaphragm that separates a high pressure region from a low pressure region. As the high pressure driver gas expands, it compresses the low pressure gas and heats it up. The compression near the high-low pressure boundary is nearly adiabatic. This method is used almost exclusively to study gas reactions.

Most pulse methods can be applied equally well to gases and liquids. Many reactive intermediates have been detected and some reaction rates have been measured by this method. Important applications include gaseous combustion reactions, reactions of the hydrated electron, and primary processes in photosynthesis. Some results are given in Table 20-7 and Figure 20-22.

20-6.5 Lifetime Methods The lifetime (or spectral line-broadening) methods that utilize the effects of chemical reactions upon spectral line shapes and frequencies were first introduced in NMR spectroscopy by H. Gutowsky and coworkers in 1953 at the University of Illinois, Urbana. The method is most often applied to NMR

Figure 20-19 (a) Schematic diagram of the flash photolysis initiation plus kinetic spectroscopy detection apparatus. Closing of the switch S discharges capacitor C through the photoflash lamp. The resulting photoflash is absorbed by the reactant mixture. The photomultiplier P detects the photoflash and initiates the discharge of the spectroflash lamp via the delay unit, which fires a preset time interval after receiving the photoflash signal. The spectroflash absorption by the reactant mixture is recorded on the photographic plate after dispersion in the spectrograph. Spectra of the reacting system are obtained at various stages of the reaction by varying the preset time interval. To obtain background or pure reactant spectra, the spectroflash is discharged alone. Minimum $t_i = 1$ μs is set by flash duration. By using lasers to replace flash lamps in a modified apparatus t_i equal to 1 ns or even 1 ps may be achieved.

(b) Detailed diagram of the original flash photolysis apparatus, showing (1) a motor-driven shutter wheel which was used to synchronize, via electrical brush contacts, the shutter opening position (not visible) with the photo- and spectroflashes, and (2) the gas-handling system used for filling the reaction vessel and the gas discharge flash lamps. [Part (b) from G. Porter: *Proc. Roy. Soc.* (*London*), **A200**:284 (1950).

(c) Spectra of an exploding system that initially contained $NO_2(2 \text{ torr}) + H_2(20 \text{ torr}) + O_2(10 \text{ torr}) + N_2(15 \text{ torr})$. The explosion was initiated by flash photolysis of the NO_2 with a flash energy of 3300 J. Note the growth and decay of the OH concentration. [Part (c) from J. E. Nicholas and R. G. W. Norrish: *Proc. Roy. Soc.* (*London*), **A309**:174 (1969).

absorption lines, somewhat less frequently to ESR spectra, and is, in principle, applicable to any spectral line. The method avoids the initiation problem altogether. It does so by measuring the average lifetime of a chemically reacting species in an equilibrium system.

We illustrate the method by applying it to an equilibrium between two isomers

$$A \underset{k_-}{\overset{k_+}{\rightleftharpoons}} B \qquad (20\text{-}107)$$

The **residence time** $\tau_A = k_+^{-1}$ is a measure of the lifetime for the species A. Similarly, for B, $\tau_B = k_-^{-1}$. Assume that A and B contain a spin whose magnetic resonance absorption is measurable. The chemical reaction will affect the relaxation of the spin. This additional relaxation is taken into account by modifying the Bloch equations (page 495) for the magnetic moments M of each of the two species A, B. The modified equations for the M_X, M_Y components are

$$\frac{dM_{XA}}{dt} = -\frac{M_{XA}}{T_{2A}} - (\omega - \omega_A)M_{YA} - k_+M_{XA} + k_-M_{XB} \qquad (20\text{-}108a)$$

$$\frac{dM_{XB}}{dt} = -\frac{M_{XB}}{T_{2B}} - (\omega - \omega_B)M_{YB} - k_+M_{XB} + k_-M_{XA} \qquad (20\text{-}108b)$$

and

$$\frac{dM_{YA}}{dt} = -\frac{M_{YA}}{T_{2A}} + (\omega - \omega_A)M_{XA} + \gamma_n H'M'_{ZA} - k_+M_{YA} + k_-M_{XB}$$
$$(20\text{-}109a)$$

$$\frac{dM_{YB}}{dt} = -\frac{M_{YB}}{T_{2B}} + (\omega - \omega_B)M_{XB} + \gamma_n H'M'_{ZA} - k_+M_{YB} + k_-M_{YA}$$
$$(20\text{-}109b)$$

The subscripts A, B signify a quantity that belongs to the species A or B, respectively. ω_A, ω_B are the precessional frequencies ω_0 of the spin in the two species. All other symbols have been defined previously (pages 495–497). Note that we have used M'_{ZA} in Eqs. (20-109); that is, we have assumed an H' so low that $M_{ZA} \approx M'_{ZA}$ [see Eq. (h), page 495].

Eqs. (20-108) and (20-109) may be interpreted as rate expressions of the form of Eq. (20-11) in which the usual concentrations have been replaced by magnetic moments. Each term on the right-hand side of Eqs. (20-108) and (20-109) represents one of several parallel first-order processes that affects the relaxation rate of one of the components M_{XA}, M_{XB}, M_{YA}, M_{YB}. As an example consider the magnetization rate of the species A along the X axis, Eq. (20-108a). The first two right-hand side terms of this equation represent physical processes. The first, the spin–spin relaxation term, describes the decay of M_{XA} due to the randomization of the directions in which the individual spins point. This is usually caused by the mutual interactions of spins. The second term is nonzero whenever the rotation frequency ω of the XY axes (page 496) is different from the spin precession frequency ω_A. The motion of the xy plane component of the spins relative to the XY axes gradually transforms M_{YA} into M_{XA}.

The third term is the loss of M_{XA} due to the reaction $A \rightarrow B$, and the fourth term is the gain of M_{XA} due to the reaction $B \rightarrow A$. Both terms assume that the reaction probability is independent of the spin direction. Other than for free radical reactions (see Section 20-7) this is a generally valid assumption. The last term assumes, in addition, that during the transition from B to A the spin does not precess appreciably. (When this assumption is not valid, there is no necessary relation between dM_{XA}/dt and M_{XB}.) Modified Bloch equations, based on the preceding assumptions, for systems with more than two reacting species can be written by a straightforward extension of Eqs. (20-108) and (20-109). These equations embody the effects of chemical reactions (or "chemical exchange") upon magnetic resonance spectra.

In the most general case the equations can be solved numerically for $M_Y(\omega)$ (a quantity proportional to the absorption intensity) with the help of a computer. In order to get a tractable analytical expression for $M_Y(\omega)$ we have limited ourselves to a special case for which (1) spin–spin relaxation is negligible compared to the chemical relaxation, that is, $1/T_{2A} \approx 1/T_{2B} \approx 0$; (2) the concentrations of A and B are equal, that is, $[B]/[A] = k_+/k_- = 1$. For this case one can derive (see Example 20-5) the following line-shape expression.

$$M_Y = M_{YA} + M_{YB} = \tfrac{1}{2}\gamma_n H'M_Z' \cdot \frac{k(\omega_A - \omega_B)^2}{(\omega - \omega_A)^2(\omega - \omega_B)^2 + 4k^2(\omega - \bar{\omega})^2}$$

$$(20\text{-}110)$$

where $k = k_+ = k_-$ and $\bar{\omega} = \tfrac{1}{2}(\omega_A + \omega_B)$.

Eq. (20-110) is most readily applied to actual spectra if we rewrite it in terms of the parameter $R = 2^{3/2}k/(\omega_A - \omega_B)$ and the reduced difference coordinate S (shown in Figure 20-20a). The result is

$$I(\omega) \propto M_Y/[\tfrac{1}{2}\gamma_n H'M_Z'k(\omega_A - \omega_B)^{-2}] \propto 1/[(S^2 - 1)^2 + 2R^2 S^2] \quad (20\text{-}111)$$

Figures 20-20b through 20-20i show plots of Eq. (20-111) for various values of R. Figure 20-21 shows NMR spectra to which Eq. (20-111) should apply, namely, the spectra of the axial and equatorial protons in the cyclohexane $C_6D_{11}H$ at various temperatures. From Figure 20-20 one can see that the axial-equatorial exchange reaction appreciably affects the line shapes for a range of R values from about 0.1 to about 10. Thus, the range of rate constants that can be measured in a given system is rather limited. The absolute values of rate constants within the range depend on $\omega_A - \omega_B$ which, for NMR, is proportional to the difference in the chemical shifts of A and B (Eq. (16-48), page 498 and Table 16-3). In some cases the rate constant may be brought into the measurable range by varying the temperature (as in Figure 20-21) or, in case of apparent first-order rate constants [for example, the one in Eq. (20-104)], by changing the concentration of a reactant.

The actual methods for determining the rate constant depend upon the value of R and upon the accuracy desired. The most accurate method determines R by requiring that the theoretical expression, Eq. (20-111), be a least squares fit of the experimental spectrum.

Two other methods, rather than depending upon *all* of the information contained in the experimental spectrum, utilize either the peak-to-peak separation or the line-width at half intensity. From a determination (Problem 32) of the maxima of Eq.

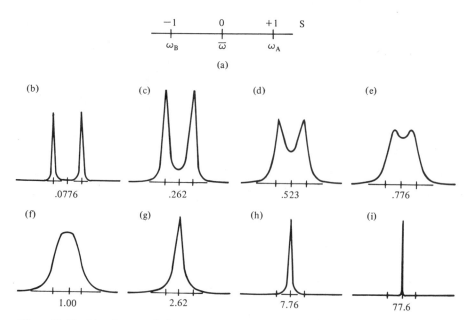

Figure 20-20 Line shapes, calculated from Eq. (20-111). (a) The ω and S coordinates compared.

$$S = (\omega - \bar{\omega})[\tfrac{1}{2}(\omega_A - \omega_B)]^{-1} = (\nu - \bar{\nu})[\tfrac{1}{2}(\nu_A - \nu_B)]^{-1}$$

(b) to (i) The line shapes, calculated for various values of R (given under each figure), that are applicable to the $C_6D_{11}H$ spectra in Figure 20-21. [Parts (b)–(i) from F. A. Bovey: *Nuclear Magnetic Resonance Spectroscopy*. Academic Press, New York, 1969, p. 188. Copyright 1969, Academic Press, Inc.]

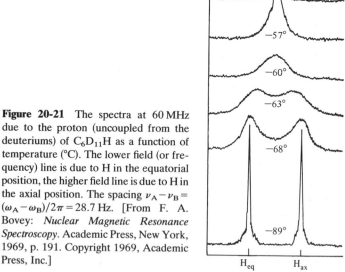

Figure 20-21 The spectra at 60 MHz due to the proton (uncoupled from the deuteriums) of $C_6D_{11}H$ as a function of temperature (°C). The lower field (or frequency) line is due to H in the equatorial position, the higher field line is due to H in the axial position. The spacing $\nu_A - \nu_B = (\omega_A - \omega_B)/2\pi = 28.7$ Hz. [From F. A. Bovey: *Nuclear Magnetic Resonance Spectroscopy*. Academic Press, New York, 1969, p. 191. Copyright 1969, Academic Press, Inc.]

(20-111) one finds the peak-to-peak separation ΔS to be

$$\Delta S = 2\sqrt{1-R^2} \qquad \text{(20-112)}$$

When the two peaks just coalesce (Figure 20-20f), $R = 1$ and $k = 2^{-3/2}(\omega_A - \omega_B)$. For separated peaks $R < 1$. Usually one cannot expect to determine values of $R < 0.1$ with any degree of accuracy.

A measurement of the line-width at an intensity equal to one half of the peak intensity may be applied to either the slow exchange (low R, Figure 20-20b) or the fast exchange (high R, Figures 20-20g and h) limits. For the slow exchange limit it is best to use Eq. (20-110). Consider the line due to A, that is, $\omega \approx \omega_A$. Then $\omega - \omega_B \approx \omega_A - \omega_B$ and $\omega - \bar{\omega} \approx \frac{1}{2}(\omega_A - \omega_B)$. Therefore Eq. (20-110) reduces to

$$I(\omega) \propto M_Y/(\tfrac{1}{2}\gamma_n H' M_Z') \approx \frac{k}{(\omega - \omega_A)^2 + k^2} \qquad \text{(20-113a)}$$

At peak intensity $\omega = \omega_A$ and the right-hand side of Eq. (20-113a) reduces to $1/k$. The two frequencies at the half-intensity points on either side of the line are the roots of the equation

$$\tfrac{1}{2}k^{-1} = k/[(\omega - \omega_A)^2 + k^2] \qquad \text{(20-113b)}$$

The difference $\Delta\omega$ between the two roots gives the line-width at half intensity

$$\Delta\omega = 2k \qquad \text{(20-114)}$$

For the fast exchange limit we need an expression valid near $S = 0$ where Eq. (20-111) reduces to

$$I(\omega) \propto 1/(1 + 2R^2 S^2) \qquad \text{(20-115)}$$

At half intensity this yields

$$\Delta S = 2^{1/2} R^{-1} \qquad \text{or} \qquad \Delta\omega = (\omega_A - \omega_B)^2 (4k)^{-1} \qquad \text{(20-116)}$$

Eqs. (20-114) and (20-116) represent the contribution of chemical exchange to the line-width. The line may have additional broadening due to physical processes.

In spite of the limitations noted, NMR spectra have been used to determine the rate constants of many reactions in solution (Figure 20-21). Among these reactions the most important have been ring inversions, rotations about single bonds, inter- and intramolecular proton transfers. Electron spin resonance (ESR) spectra have been used to determine rate constants of intra- and intermolecular electron or atom transfers in solution. Some examples are given in Table 20-7.

The method of slow passage through the spectrum that we described depends upon the establishment of a steady state and can be supplemented by various transient methods. These methods can extend the range of rate constants accessible by the slow passage method. The transient methods have not been used as extensively as the method discussed here.

EXAMPLE 20-5
Given the modified Bloch equations (20-108) and (20-109), derive the line shape expression, Eq. (20-110).

Table 20-7 Rate constants of some fast (probably) elementary reactions

Type	Reaction	Rate Constant[a]	Conditions	Experimental Method
second order gas	$H + C_6H_5CH_3 \rightarrow CH_3\dot{C}_6H_6$	1.0×10^8	in 11 or 54 atm Ar, 0.8 or 7 atm H_2, 0.03 atm toluene	pulse radiolysis
	$O + C_2H_4 \rightarrow$ products	$(4.5-5.5) \times 10^{11}$	room temp.	flash photolysis and fast flow
	$O + NO_2 \rightarrow NO + O_2$	$(2.7-4.2) \times 10^{12}$	room temp.	fast flow and flash photolysis
second order solution e^- transfer	$Fe(phen)_3^{3+} + Fe^{2+} \rightarrow Fe(phen)_3^{2+} + Fe^{3+}$	3.0×10^5	25°C	fast flow
	$Cu^+ + Cu^{2+} \rightarrow Cu^{2+} + Cu^+$	10^8	0.5 M H_2SO_4 12 M HCl	NMR of ^{63}Cu
	$C_{10}H_8 + C_{10}H_8^- \xrightarrow{Na^+} C_{10}H_8^- + C_{10}H_8$	$(0.3-4) \times 10^8$	30-5°C dimethoxyethane	ESR
	$e^-(aq) + O_2 \rightarrow O_2^-$	1.9×10^{10}	pH = 7	pulse radiolysis

H transfer	$H_3O^+ + OH^- \rightarrow 2H_2O$	1.4×10^{11}	25°C	el. field-jump
	$H_2O + OH^- \rightarrow OH^- + H_2O$	5×10^9	25°C	NMR
	$CH_3OH + OH^- \rightarrow CH_3O^- + H_2O$	2.6×10^6	22°C	NMR
ligand exchange	$Ca^{2+}(H_2O) + ATP^{4-} \rightarrow Ca(ATP)^{2-} + H_2O$[b]	$\geqslant 10^9$	room temp.	T-jump
	$Ni^{2+}(H_2O) + HP_3O_4^{4-} \rightarrow Ni(HP_3O_4)^{2-} + H_2O$	6.3×10^6	room temp.	T-jump
enzyme + substrate	hemoglobin·$(O_2)_3 + O_2 \rightarrow$ hemoglobin·$(O_2)_4$	2×10^7	room temp.	stopped flow
other	$I_2 + I^- \rightarrow I_3^-$	6.8×10^9	25°C, 0.02 M ionic strength	laser T-jump
	$I + I \rightarrow I_2$	6.9×10^9	room temp. in CCl$_4$	flash photolysis
	$OH + C_6H_6 \rightarrow (OH)\dot{C}_6H_6$	4.3×10^9	room temp. in H$_2$O	pulse radiolysis
conformational change	C_6H_{12}: chair → chair	88	−65°C	NMR
	C_6H_{12}: chair → chair	10^5	room temp.	ultrasonic
	$(C_2H_5)_3N$: rotn. about C—C bonds	10^8	room temp.	ultrasonic

[a] The units of the second-order rate constants are $dm^3 \, mol^{-1} \, s^{-1}$, those of the first order (conformational change) rate constants are s^{-1}.
[b] ATP is adenosine triphosphate.

DATA SOURCES: E. F. Caldin: *Fast Reactions in Solution*. Wiley, New York, 1964, for all reactions except for pulse radiolysis results which are from M. S. Matheson and L. M. Dorfman: *Pulse Radiolysis*. M.I.T. Press, Cambridge, Mass., 1969; the flash photolysis and fast flow gas reactions are from A. A. Westenberg: *Ann. Rev. Phys. Chem.*, **24**:77 (1973); the hemoglobin reaction is from F. J. W. Roughton: *Z. Electrochem.*, **64**:3 (1960); and the ligand reactions are from R. G. Wilkins and M. Eigen in E. S. Gould (ed.): *Mechanisms of Inorganic Reactions*. American Chemical Society, Washington, D.C., 1965.

Solution: Although the derivation can be done in the same manner as for the line-shape expression *without* chemical reaction (page 494) the algebraic manipulations are much simpler if complex magnetic moments and angular frequencies are introduced:

$$\hat{M}_A = M_{XA} + iM_{YA} \tag{a}$$

$$\hat{M}_B = M_{XB} + iM_{YB} \tag{b}$$

$$\hat{\omega}_A = \omega_A - i/T_{2A} \tag{c}$$

$$\hat{\omega}_B = \omega_B - i/T_{2B} \tag{d}$$

If Eq. (20-109a) is multiplied by i and the resulting equation is added to Eq. (20-108a), then, by introducing the definitions (a) and (c) and using the identity $i^2 \equiv -1$ we obtain

$$\frac{d\hat{M}_A}{dt} = i(\omega - \hat{\omega}_A)\hat{M}_A + i\gamma_n H' M'_{ZA} - k_+ \hat{M}_A + k_- \hat{M}_B \tag{e}$$

Eqs. (20-108b) and (20-109b), if combined in the same fashion, yield

$$\frac{d\hat{M}_B}{dt} = i(\omega - \hat{\omega}_B)\hat{M}_B + i\gamma_n H' \hat{M}'_{ZB} - k_- \hat{M}_B + k_+ \hat{M}_A \tag{f}$$

Since a magnetic resonance experiment measures the components of the total magnetic moment at any frequency, we must eliminate the separate contributions of species A and B from Eqs. (e) and (f). To eliminate M'_{ZA} and M'_{ZB}, define the fraction of spins in each species in terms of the concentrations A and B.

$$f_A = \frac{A}{A+B} = \frac{k_-}{k_+ + k_-} \quad \text{and} \quad f_B = \frac{B}{A+B} = \frac{k_+}{k_+ + k_-} \tag{g}$$

where the second equality in each equation follows from the existence of the equilibrium $B/A = k_+/k_-$. Therefore we have the relations

$$M'_{ZA} = f_a M'_Z \quad \text{and} \quad M'_{ZB} = f_B M'_Z \tag{h}$$

where M'_Z is the Z component of the total magnetic moment. For the usual magnetic resonance experiment, a steady state can be safely assumed. The following steady state equations result from Eqs. (e) and (f) when the substitutions indicated in Eq. (h) are made

$$[i(\omega - \hat{\omega}_A) - k_+]\hat{M}_A + k_- \hat{M}_B + if_A \gamma_n H' M'_Z = 0 \tag{i}$$

$$[i(\omega - \hat{\omega}_B) - k_-]\hat{M}_B + k_+ \hat{M}_A + if_B \gamma_n H' M'_Z = 0 \tag{j}$$

To eliminate $\hat{M}_A$ and $\hat{M}_B$, define a complex magnetic moment

$$\hat{M} = \hat{M}_A + \hat{M}_B \tag{k}$$

Eqs. (i) and (j) can be solved for $\hat{M}_A$ and $\hat{M}_B$. The resulting expressions when substituted in Eq. (k) yield

$$\hat{M} = -\gamma_n H'_Z M'_Z \frac{f_A(\omega - \hat{\omega}_B) + f_B(\omega - \hat{\omega}_A) + i(k_+ + k_-)}{(\omega - \hat{\omega}_A)(\omega - \hat{\omega}_B) + i[k_+(\omega - \hat{\omega}_B) + k_-(\omega - \hat{\omega}_A)]} \tag{l}$$

Into general expression (l) we now introduce the assumption that spin–spin relaxation is negligible ($1/T_{2A} \approx 1/T_{2B} \approx 0$). This means, according to Eqs. (c) and (d), that $\hat{\omega}_A = \omega_A$ and $\hat{\omega}_B = \omega_B$. Therefore Eq. (l) reduces to

$$\hat{M} = -\gamma_n H' M'_Z \frac{f_A(\omega - \omega_B) + f_B(\omega - \omega_A) + i(k_+ + k_-)}{(\omega - \omega_A)(\omega - \omega_B) + ik_+(\omega - \omega_B) + k_-(\omega - \omega_A)} \tag{m}$$

To get the line-shape expression we have to extract $M_Y(\omega)$ from Eq. (m). By definition (see Eqs. a and b)

$$\hat{M} = M_X + iM_Y \tag{n}$$

Eq. (m) does not have the form of Eq. (n). Instead it has the form

$$\hat{M} = \frac{a + ib}{c + id} \tag{o}$$

When Eq. (o) is multiplied by $(c - id)/(c - id)$, it reduces to the form of Eq. (n).

$$\hat{M} = \frac{a + ib}{c + id} \frac{c - id}{c - id} = \frac{(ac + bd) + i(bc - ad)}{c^2 + d^2} \tag{p}$$

Therefore, by equating the corresponding terms in Eqs. (n) and (p),

$$M_Y = \frac{bc - ad}{c^2 + d^2} = -\gamma_n H' M'_Z$$

$$\times \frac{(k_+ + k_-)(\omega - \omega_A)(\omega - \omega_B) - [f_A(\omega - \omega_B) + f_B(\omega - \omega_A)][k_+(\omega - \omega_B) + k_-(\omega - \omega_A)]}{(\omega - \omega_A)^2(\omega - \omega_B)^2 + [k_+(\omega - \omega_B) + k_-(\omega - \omega_A)]^2}$$

$$= \gamma_n H' M'_Z \frac{k_+ f_A(\omega_A - \omega_B)^2}{(\omega - \omega_A)^2(\omega - \omega_B)^2 + [k_+(\omega - \omega_B) + k_-(\omega - \omega_A)]^2} \tag{q}$$

The second equality in Eq. (q) is obtained when a, b, c, and d are substituted from Eq. (m). The third equality follows from the equilibrium relation $k_+/k_- = B/A = f_B/f_A$ and from $f_A + f_B = 1$.

Finally, if we assume equal concentrations of A and B (that is, $f_A = f_B = \frac{1}{2}$ and $k = k_+ = k_-$), Eq. (q) reduces to

$$M_Y = \frac{1}{2}\gamma_n H' M'_Z \frac{k(\omega_A - \omega_B)^2}{(\omega - \omega_A)^2(\omega - \omega_B)^2 + 4k^2(\omega - \bar{\omega})^2} \tag{r}$$

where $\bar{\omega} = \frac{1}{2}(\omega_A + \omega_B)$. Eq. (r) is the same as Eq. (20-110). ●

The methods for the measurement of reaction rates discussed so far do not exhaust all possibilities. To mention briefly only one possibility, the methods of Section 20-5 can be combined with one of the methods of this section. From the measurement of yields for two reactions of the same reactant (see Section 20-5) we can get the ratio of the two rate constants. If one constant can be measured by one of the methods of this section, the other may be determined from the ratio. Many of the rate constants of radiolysis reactions have been determined by combining the data of pulse radiolysis with measurements of radiolysis yields.

Figure 20-22 surveys some of the experimental methods used to study fast reactions and presents some general conclusions from such studies about the speeds

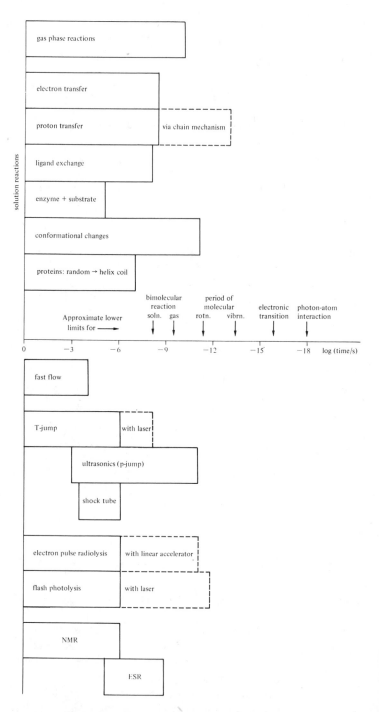

Figure 20-22 Approximate experimentally determined time ranges (indefinitely extensible to longer times than shown in the figure) and shortest possible times in which typical physicochemical processes occur. The time-measurement capabilities of various experimental methods.

The approximate lower limits of the times during which chemical reactions occur have been defined as half-lives. The shortest possible half-lives for bimolecular reactions have been calculated by assuming as

of important classes of fast reactions, for example, gas phase reactions of small molecules, electron (redox) and proton (acid-base) transfer reactions, protein reactions, ligand exchange reactions, and conformational changes. Important types of reactions, such as decompositions or isomerizations that involve bond breaking, are generally too slow to need study by fast reaction methods and therefore do not appear in Figure 20-22. Representative rate constants, measured by the various methods, are listed in Table 20-7.

The maximum rate of bimolecular gas phase reactions is determined by the rate of bimolecular collisions in the gas (see Section 21-2.3). In solutions the maximum bimolecular reaction rate is set by the rate at which the reactants diffuse toward each other (Section 21-2.5). Some proton transfer reactions in water exceed the diffusion limit by what is thought to be a chain mechanism, for example, for the $H_3O^+ + OH^-$ reaction the mechanism is

$$H_2OH^+ \cdots HOH \cdots HOH \cdots OH^- \rightarrow H_2O \cdots H_2O \cdots H_2O \cdots HOH.$$

Figure 20-22 shows that modern methods are sufficiently rapid to measure the rates of many important types of elementary reactions. However, adequate speed may not be sufficient by itself to allow the measurement of the rates of all elementary reactions in a system. The special problems of complex reacting systems are discussed in Section 20-7.

20-7 INTERMEDIATES

The methods described in the previous section apply with fewest complications to simple reacting systems in which all of the significant reacting species are stable molecules or ions. In many complex reacting systems significant intermediates will fall into one or more of four groups of unstable species: (1) unstable molecules; (2) free radicals and atoms; (3) ions; and (4) electronically, vibrationally, and/or rotationally excited species. Since pure samples of these intermediates cannot be prepared, their properties cannot be determined by the usual methods, namely, measurements on a pure sample or a sample of known concentration.

Unstable intermediates present three problems. They must be (1) identified, (2) detected within the reacting system under study; and (3) their concentrations must be measured as a function of time. In this section we shall discuss the three problems and describe some of the solutions.

20-7.1 Identification The first problem is to identify a species that has not been observed before and is part of a mixture. By far the most powerful tools for the solution of this problem are electromagnetic and mass spectroscopies. A successful

---

typical a pseudo first-order reaction with the concentration of one reactant (the stable one) equal to 0.01 M (≈ 0.2 atm for a gas at room temperature), much larger than the concentration of the second unstable reactant. (A half-life equal to 7×10^{-9} s corresponds to a bimolecular rate constant equal to $10^{11} M^{-1} s^{-1}$.)

identification proves that a particular chemical species is the source of an observed spectrum. To increase the probability of success, one attempts to maximize the concentration of the species to be observed by either of two methods.

The first method is to increase the production rate (r_p) of the reactive species relative to its disappearance rate (r_d). This is generally done in one of two ways: (1) increase the temperature, which will cause a rapid increase in the rate with the higher activation energy (that is, r_p); (2) in an activator-initiated reaction, increase the intensity of activator radiation. The second method is to decrease the disappearance rate of the reactive species to zero. This is done by trapping the species in an inert matrix (usually solid noble gases or N_2) at a temperature of 20 K or less. At these low temperatures diffusion becomes the rate-determining step for reactions among the reactive species. Example 20-6 describes the identification of methylene produced by the above methods. Table 20-8 lists a few of the many unstable gaseous species that have been identified so far.

Table 20-8 Some unstable species with known spectra[a]

Unstable Species	Where Observed
OH, NH, NH$_2$, CH, CN, C$_2$	high temperature flames
as above plus:	flash photolysis
SH, TeH, BH, PH, ClO, TeO, FeO, SnO, C$_3$, CF$_2$, HNO	
XH$_n$(X = C, Si; n = 1–3), CF, CN, C$_2$, HO$_2$, HCO, HCCl$_2^+$	low temperature matrix
OH, NH$_2$, CH$_2$, CH$_3$, CHO, CH$_3$CO, C$_4$H$_9$ (all isomers), C$_6$H$_5$	mass spectrometer

[a] ESR, IR, and UV are the usual known electromagnetic spectra.

EXAMPLE 20-6

How is an unstable species identified?

The identification of methylene (CH$_2$) has been the subject of much effort [14, Vol. I, pp. 294, 335]. The interest in CH$_2$ stems in part from the fact that it is an intermediate in the photochemical reaction between ketene (CH$_2$CO) or diazomethane and alkanes.

$$CH_2N_2 + CH_3(CH_2)_3CH_3 \xrightarrow{UV} CH_3(CH_2)_4CH_3 + CH_3CH_2CH_2\underset{|}{C}HCH_3 +$$

$$CH_3$$

$$(48\%) \qquad\qquad (35\%)$$

$$+ CH_3CH_2\underset{|}{C}HCH_2CH_3 + N_2$$

$$CH_3$$

$$(17\%)$$

These results are consistent with the production of methylene by photolysis of diazomethane

$$CH_2N_2 \xrightarrow{UV} CH_2 + N_2$$

followed by the insertion of CH_2 at random in all $C-H$ bonds.

$$CH_2 + -\overset{|}{\underset{|}{C}}-H \rightarrow -\overset{|}{\underset{|}{C}}-CH_2-H$$

The earliest attempts to identify CH_2 produced by electrical discharges through methane gas met with apparent success. However, later it was shown that the spectrum that had been assigned to CH_2 actually belongs to C_3. Finally, in 1961 Herzberg observed absorption due to singlet methylene (paired spins, $^1CH_2$) produced by flash photolysis of CH_2N_2. The triplet (two parallel spins) ground state ($^3CH_2$) was also observed. The conclusion that the observed spectra in the visible and the UV belong to CH_2 was based primarily on the rotational structure of the band spectra and on the agreement with theoretical predictions. The conclusion was confirmed by showing that CH_2N_2, $CDHN_2$, CD_2N_2 produce *three* different but closely related spectra. (Only two spectra would be expected, for example, for CH and CD.) Both techniques—flash photolysis and isotopic substitution—have helped to identify many other reactive species.

Identification of CH_2 has been obtained in a low temperature matrix. In the first (1960) attempts CH_2N_2 was photolysed in Ar and N_2 matrices at 20 K and certain IR and UV absorptions were assigned to CH_2. However, these assignments were later found to conflict with Herzberg's gas phase spectra of CH_2. In 1970 the ESR spectrum of the triplet spin state of CH_2 was observed in a Xe matrix in which diazirine

$$H_2C\begin{array}{c} \diagup N \\ \diagdown \parallel \\ N \end{array}$$

had been photolysed at 4 K. The identification was confirmed by the observation of the spectra of CHD and CD_2. The H–C–H angle was determined to be 136°, in agreement with reliable quantum mechanical calculations [15].

Methylene, produced by the pyrolysis of CH_2N_2, has also been observed in a mass spectrometer.

The identification of CH_2 has led to a knowledge of (1) its mass spectrum; (2) the vacuum UV absorption spectrum (near 140 nm) of $^3CH_2$ (the ground state); (3) absorption spectra of $^1CH_2$ in the red and in near-IR, as well as near-UV (very weak); and (4) the ESR spectrum of $^3CH_2$. In principle, these spectra can be used to detect CH_2 in any reacting system. ●

20-7.2 Detection in Reacting Systems The detectability problem arises because the sensitivity (see Section 20-6) of the available methods may be too low to meet the requirements of actual reacting systems. An idea of the requirements for the reaction scheme $A \xrightarrow{k_0} B \xrightarrow{k_1} C$ may be gained from the first two columns of Table 20-9

Table 20-9 A comparison of representative intermediate concentrations with detectability limits for some spectroscopic methods

k_1/k_0[a]	Maximum[a] Concentration of B (mol dm^{-3})	Type of Absorption Spectroscopy[b]
10	0.77×10^{-1}	NMR, IR
10^3	0.99×10^{-3}	visible, UV
10^6	1.00×10^{-6}	ESR
10^9	1.00×10^{-9}	IR laser magnetic
10^{15}	1.00×10^{-15}	resonance (LMR)
10^{18}	1.00×10^{-18}	visible and UV tunable laser limit

[a] $A \xrightarrow{k_0} B \xrightarrow{k_1} C$; $A_0 = 1$ mol dm^{-3}; concentration B is the maximum attained during the course of the reaction.

[b] This column lists spectroscopies that can detect concentrations equal to and higher than the one given on the same line in column two. The detectability limits are very rough but conservative for the established methods (above dashed line). Intensely absorbing substances, long absorption paths, or other special methods can lower the limits by as much as two orders of magnitude. The methods listed below the dashed line are under development. The limits are best achievements or expectations. For LMR see J. S. Wells and K. M. Evenson: *Rev. Sci. Inst.*, **41**:226 (1970). For estimates of detection limits by absorption spectroscopy with tunable lasers see K. P. G. Sulzman, J. E. L. Lowder, and S. S. Penner: *Combustion and Flame*, **20**:177 (1973).

where we list the ratio k_1/k_0 (which measures the relative reactivity of B and A) and the corresponding maximum concentration which the intermediate B reaches during the reaction. The range of the listed k_1/k_0 encompasses the values encountered in actual reacting systems. From column three of Table 20-9 we may conclude that routine spectroscopic methods are not sensitive enough to detect B in many A → B → C systems. The conclusion is also valid for other reaction schemes.

Although it is true that the detectability limits of routine spectroscopic measurements may be lowered by various modifications, the improvement is seldom more than a factor of 100. From Table 20-9 we see that this leaves much room for further improvement. Considerable current research activity is devoted to the development of more sensitive detection methods that utilize new light sources (lasers) and new detectors (solid state devices). The potential of two of the newer methods is indicated in Table 20-9.

In the past, the greatest successes in the detection of intermediates have come through the routine spectroscopic study of systems where the intermediate concentration is high, that is, high temperature or activator-initiated reacting systems. As an example we will discuss the studies of the low pressure oxidation of fuels (H_2, nonmetallic hydrides, hydrocarbons) by oxygen with the flash photolysis–kinetic spectroscopy method. In a typical experiment (see Norrish in [16]) the photolysis cell was filled with NO_2, H_2, O_2, and the inert gas N_2 at partial pressures of 2, 20, 10, and 15 torr, respectively. To initiate the oxidation the cell is flashed. The sensitizer gas, NO_2, absorbs the flash energy and dissociates to $NO + O$. The O atoms react with H_2 to yield $OH + H$. As shown in Figure 20-19c, the spectrum of the OH radical can be observed for about 5 ms after the flash. The spectrum appears faintly on the very first spectrogram (within a few microseconds after initiation), explosively increases in intensity to reach a maximum in about $\frac{1}{4}$ ms, and finally decays to zero on the spectrogram taken at 6.6 ms. During the disappearance of the OH spectrum one can observe the rapid appearance of H_2O in a different region of the spectrum. These observations are consistent with the chain mechanism of the $H_2 + O_2$ reaction given in Eq. (20-42).

The system exhibits first and second explosion limits similar to the ones shown in Figure 20-9. However, here the lower limit cannot depend upon wall termination reactions because it is not affected by the size of the reaction vessel. The reason the explosion occurs is the explosive and large increase in temperature above the initial flash-caused temperature of about 500°C. This large temperature rise makes the branching rate exceed the chain destruction rate (see Eq. 20-45) and thus causes an explosion. The lower explosion limit can then be seen as the condition when the heat losses to the surroundings balance the heat produced by the reaction.

During the product formation stage the branching rate drops rapidly because O_2 has been used up (Eq. 20-42a) and the propagation reaction (20-42d) dominates. The variation of the OH concentration is consistent with its role as a propagator of the oxidation. The time-dependent behavior of the OH concentration in the oxidation of fuels other than hydrogen is quite similar to the behavior just described. Therefore OH is implicated as the propagator in most if not all oxidations of fuels by oxygen.

20-7.3 Measurement of Concentrations This description of the $H_2 + O_2$ reaction is qualitative rather than quantitative because in Norrish's experiment the concentrations of the reacting species could not be measured as a function of time. The reason for this, the final of the three problems with the direct study of unstable intermediates, is obvious from Eq. (20-95); in order to get the concentration C from the absorbance A the absorptivity a must be known.

Usually a is determined from measurements of the absorbance of a system with a known concentration of the species of interest. If the determination is done under conditions similar to the determination of the unknown concentration, instrumental factors will cancel out to yield a reliable result for the unknown concentration. For reacting systems under other than steady state conditions, this similarity is rather difficult to arrange. More serious, however, is the fact that the concentrations of reactive intermediates are rather difficult to determine. A determination is possible from the thermodynamic functions of the intermediate, if they are known. For example, the concentration of OH is calculable from the known equilibrium constant of the equilibrium $H_2O(g) \rightleftharpoons H(g) + OH(g)$. Using the calculated concentration, the absorptivity a has been measured for OH at 1600°C. Alternatively, one can calculate a theoretically and correct for the instrumental factors. Accurate theoretical calculation is feasible for magnetic resonance spectra. For example, the OH concentration in some reacting flow systems has been determined by ESR from theoretical calculations combined with a calibration of the apparatus against a known concentration of the stable free radical NO.

If the absolute concentration of a reactive intermediate B cannot be determined, one can still study it in systems where B undergoes a first order or pseudo first-order reaction by determining (using Eq. 20-98) the concentration ratio $[B]/[B_0]$ as a function of time. Reactions in the near-equilibrium systems encountered with the field-jump method are invariably pseudo first order, as are reactions where the reactive intermediate reacts only with stable molecules whose concentrations are much in excess of the concentration of the intermediate. On the other hand, elementary reactions that involve more than one unstable species can be studied only if the absolute concentrations are known.

20-7.4   Inferences from Indirect Methods   Determination of reaction mechanisms by indirect methods other than those based on the rate law (Section 20-3) supplements in an important way the somewhat limited possibilities for the direct observation of reactive intermediates. Indirect *chemical* methods—determination of the reaction products and their relative yields, both from normal reactants and from isotopically tagged reactants—have been much used to draw conclusions about reaction mechanisms. The student will remember examples both from Section 20-5 and from his introductory courses in organic chemistry. We will describe only one such method—**metallic mirror** removal. Paneth and Hofeditz in 1929 used this method to give the first convincing proof for the existence of gaseous alkyl radicals. The radicals can be produced by heating an appropriate gas (for example, CH_3 from $Pb(CH_3)_4$ or CH_2 from CH_2N_2) mixed with a large amount of a carrier gas. After heating, the gases are allowed to flow past a thin layer (the "mirror") of a metal. It is found that mirrors made of certain metals are removed under these conditions and that the expected products are formed. Two examples of the reactions of radicals with mirrors are

$$CH_3 + M \rightarrow M(CH_3)_n$$

where M = Zn, Cd, Bi, Tl, Pb; n = valence number of M; and

$$CH_2 + Te \rightarrow (CH_2Te)_x$$

$(CH_2Te)_x$ is polymeric telluroformaldehyde.

Among the indirect *physical methods*, the most generally used are probably those which relate the spatial configurations of the products to those of the reactants. Since the student was taught about the stereochemical study of mechanisms in his organic chemistry courses, we will not discuss it further. Instead, to illustrate the power of indirect physical methods, we will discuss **chemically induced dynamic nuclear polarization (CIDNP)**.

The CIDNP effect causes an alteration of the normal absorption intensity pattern in the NMR spectra (Chapter 16) of the products of a reaction between two free radicals in solution (Figure 20-23). Therefore, the observation of CIDNP is unequivocal evidence that the reaction proceeds via radical intermediates. Furthermore, the identity of the products may suggest the identity of the intermediates. The effect may also be used, in some cases, to distinguish between alternate mechanisms involving free radicals. Finally, the theory of CIDNP explains in detail the mecha-

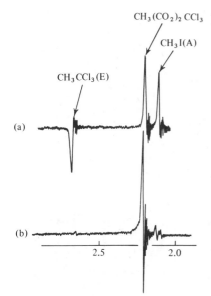

$CH_3(CO_2)_2 CCl_3$

$CH_3I(A)$

$CH_3CCl_3(E)$

(a)

(b)

2.5 2.0

Figure 20-23 Spectra of a CCl_4 solution in which the following reaction occurs

$$CH_3\overset{O}{\overset{\|}{C}}OO\overset{O}{\overset{\|}{C}}CCl_3 + I_2 \rightarrow$$

$$CH_3I + CH_3CCl_3 + CCl_3I + 2 CO_2$$

Identification with the generalized symbols of Figure 20-24: $R_1 = CH_3$, $R_2 = CCl_3$, $X = 2 CO_2$, $Sc = I$. (a) Spectrum at 50°C, the reaction is running. (b) Spectrum at 0°C immediately following scan (a), reaction has stopped.

$g_{CH_3} - g_{CCl_3} = -0.007 \qquad a_{CH_3} = -23 \text{ G}.$

The symbols (A) and (E) mark enhanced absorption and emission, respectively. [From H. R. Ward: *Accounts Chem. Res.*, 5:18 (1972). Copyright by The American Chemical Society.]

nism that controls the relative yields of products via the interaction of proton and electron spins in free radicals.

What is the explanation of the CIDNP effect? A simple general reaction scheme in which CIDNP may be observable is shown in Figure 20-24a. After the small molecule X has been expelled from the reactant molecule, the pair of radicals R_1 and R_2 occupies a cage formed by the solvent molecules. The two electron spins, one on each radical, are coupled to form either a singlet (S) or a triplet (T) state. Usually R_1XR_2 will be in a singlet state (all spins paired). Therefore, by conservation of total spin, the radical pair will also be in a singlet state. Only from this state can the singlet product R_1R_2 be formed. A transition of the radical pair to the triplet state would decrease the yield of R_1R_2 and increase the yields of R_1Sc and R_2Sc.

The transition $S \rightarrow T_0$ (where T_0 signifies the triplet state with the z component m of the total electron spin equal to zero) occurs when the two radicals drift a few angstroms apart. The consequences of this separation upon the energy levels of the radical pair are shown in Figure 20-24e. At small separations the S–T energy levels are far apart, whereas at large separations the S–T spacings become comparable to the nuclear spin level spacings. S–T transitions are now energetically feasible. All we need is a transition mechanism.

The required mechanism can be readily visualized: different frequencies of precession about the z axis of the two electron spins in a radical pair will cause the transitions $S \rightarrow T_0$ and $T_0 \rightarrow S$ (Figures 20-24c and 20-24d). The classical equations for the motion of a magnetic dipole such as the electron predict that the precession frequency is proportional to the local magnetic field H_{loc}. H_{loc} is the sum of the applied field H, the field due to the induced orbital motion of the electrons, and the field due to nuclear spins. For R_1, with one proton and one electron, the precession frequency ω_1 is given by (see Eq. 16-61)

$$\omega_1 = 2\pi\nu = \mu_0\hbar^{-1}(g_1H + a_1m_1) \tag{20-117}$$

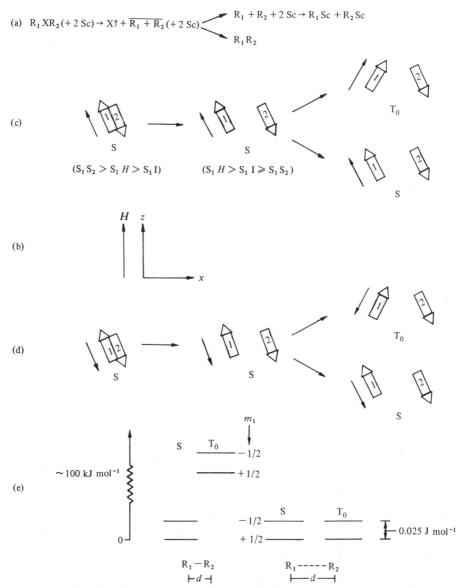

Figure 20-24 Radical reactions: enhancement of NMR signals by means of the CIDNP effect.

(a) A general scheme showing the production and the reactions of a radical pair (R_1, R_2) in solution. R_1 contains a single proton, R_2 has none. Sc is a scavenger of radicals. The bar over $R_1 + R_2$ indicates that the radicals are in a "cage" formed by the solvent molecules.

(b) Coordinate system and direction of applied magnetic field H for the spin diagrams in (c) and (d).

(c) and (d) The separation of the radicals in the cage and the consequent singlet S and triplet T_0 state mixing in the proton spin (symbolized by thin arrow) states $m_1 = +\frac{1}{2}$ and $m_1 = -\frac{1}{2}$, respectively. The electrons with spins (fat arrows) S_1 and S_2 are on radicals R_1, R_2, respectively. All three spins precess about H. The diagrams show them at a moment when their x components have maximum values. The x component of the total electron spin is zero for S state, nonzero for T_0 state. S_1S_2, S_1H, S_1I in (c) stand for the strength of the electron spin–spin, spin-H, and electron spin–proton spin couplings (that is, magnitude of interaction energies), respectively.

(e) Energy level diagram for the radical pair at two separations d (10–20 and 50–100 nm). Typical spacings (at 60 MHz) between levels are shown. The S–T_0 degeneracy at large d is approximate.

where all symbols are as defined in Chapter 16. For R_2, with one electron and no protons, we get

$$\omega_2 = \mu_0 \hbar^{-1} g_2 H \tag{20-118}$$

The difference between the two precessional frequencies is

$$\omega_1 - \omega_2 = 2\delta_{m_1} = \mu_0 \hbar^{-1}[(g_1 - g_2)H + a_1 m_1] \tag{20-119}$$

The probability $P_{m_1 T}(t)$ of finding the radical pair in the triplet state is given as a function of time t by

$$P_{m_1 T}(t) \approx \delta^2_{m_1} t^2 \tag{20-120}$$

where it has been assumed that $P_{m_1 T}(0) = 0$ and $P_{m_1 S}(0) = 1$. Eq. (20-120) is valid for $P_{m_1 T}(t) \ll 1$. The singlet probability is

$$P_{m_1 S}(t) \approx 1 - \delta^2_{m_1} t^2 \tag{20-121}$$

For typical values of $g_1 - g_2$ ($\approx 10^{-3}$), $H (\approx 10^4$ gauss) and $a_1 (\approx 10$ gauss), Eq. (20-119) yields $2\delta_{+1/2} = 1.3 \times 10^8 \, \text{s}^{-1}$ and $2\delta_{-1/2} = 4.4 \times 10^7 \, \text{s}^{-1}$. Since the typical radical pair has a diffusion-limited lifetime of 10^{-9} to 10^{-10} s, we obtain (Eq. 20-120) for the maximum probabilities

$$P_{+1/2T}(10^{-9} \, \text{s}) = 4 \times 10^{-3} \qquad P_{-1/2T}(10^{-9} \, \text{s}) = 5 \times 10^{-4}$$

These probabilities are related to the intensity of the NMR lines of the reaction products. The intensity of a line will be proportional to the rate dE/dt at which energy is gained (for absorption lines) from or lost (for emission lines) to the radiation field by the protons. From Eq. (16-40), we see that dE/dt is proportional to the population difference n between the lower and upper proton energy levels. This difference is simply related to the probabilities calculated previously

$$n_T = N_{+1/2T} - N_{-1/2T} = N_{tot}(P_{+1/2T} - P_{-1/2T}) \tag{20-122}$$

where $N_{+1/2T}$ and $N_{-1/2T}$ are the populations of the triplet state lower and upper proton energy levels, respectively. N_{tot} is the total number of protons in the system. A similar equation holds for the population difference n_S of the singlet state. Eq. (20-122) is valid for the radical pair at the time t for which the probabilities were calculated. If the radical pair separates after a time t, then the population difference n_T will relax toward the stationary state value $n(\infty)$. Whenever the product $R_1 Sc$ (Figure 20-24a) is formed before the relaxation is complete, there will be an enhanced absorption (A) by or emission (E) from the product $R_1 Sc$.

From n_T and $n(\infty)$ one can calculate maximum expected ratio of line intensities *with* and *without* reaction. The intensity without reaction is proportional to the stationary state value of the population difference, $n(\infty)$ (see Eq. 16-41) which, for weak radiation fields, is given by the zero-field equilibrium distribution. By combining Eqs. (16-31) and (16-33)

$$n(\infty) = (N_{+1/2T} + N_{-1/2T})[\exp(\hbar\omega/kT) - 1][\exp(\hbar\omega/kT) + 1]^{-1}$$
$$\approx (N_{+1/2T} + N_{-1/2T})(x/2)$$
$$= (P_{+1/2T} + P_{-1/2T})(N_{tot}/2)x \tag{20-123}$$

where the second approximate equality follows from the expansion (see Appendix III) of the exponential functions in terms of the variable $x = (\hbar\omega/kT) \ll 1$. The third equality follows from the relation $N_{m_1 T} = P_{m_1 T} \cdot N_{tot}$. Dividing Eq. (20-122) by Eq. (20-123) yields the desired ratio.

$$n_T/n(\infty) = [(P_{+1/2T} - P_{-1/2T})/(P_{+1/2T} + P_{-1/2T})](2/x) \tag{20-124}$$

A similar equation can be readily derived for $n_S/n(\infty)$.

At $\omega/2\pi = 60$ MHz and $T = 300$ K, $x = 10^{-5}$. This means that the maximum expected enhancement of intensity [when the magnitudes of the two probabilities in Eq. (20-124) are very different] is about 2×10^5. Due to relaxation, the observed enhancements are at least 10^2 times smaller. For the typical case for which we calculated that $P_{+1/2T} \approx 10 P_{-1/2T}$, the maximum possible enhancement of the population of the lower energy level ($m_1 = +1/2$, see Figure 20-24e) would be somewhat less than 2×10^5, there would be enhanced absorption by the product derived from the triplet state, that is, $R_1 Sc$, and enhanced emission from the singlet state product, $R_1 R_2$. Figure 20-23 shows a case in which there is enhanced singlet absorption and triplet emission because the signs of $g_1 - g_2$ and a_1 are the opposite of our typical case.

The general expression for the difference of the precession frequencies of the two electrons in a radical pair is

$$\omega_1 - \omega_2 = 2\delta = \mu_0 \hbar^{-1}[(g_1 - g_2)H + \sum a_{1i} m_{1i} - \sum a_{2i} m_{2i}] \tag{20-125}$$

where the sums are over all protons in radical 1 and 2, respectively. The symbols were defined in Chapter 16.

The most important implication of the above equation is that a CIDNP effect is possible even when $g_1 - g_2 = 0$. The simplest system for which this **multiplet** effect is possible is analyzed in Figure 20-25. The analysis shows that CIDNP changes the relative intensities of transitions that belong to the same multiplet.

Most radical pairs exhibit a combination of the Δg effect discussed earlier with the multiplet effect. Among the classes of reactions in which the CIDNP effect has been observed are:

1. $RCOOCOOR' + I_2 \xrightarrow{heat} RCO_2R' + RI + R'I + RR' + CO_2$
2. $RLi + R'I \rightarrow RR' + \text{scavenging products}$
3. photochemical reduction of aryl ketones and aldehydes, for example

$$\Phi_2CO + CH_3\Phi \xrightarrow{h\nu} \Phi_2\dot{C}OH + \Phi\dot{C}H_2 \rightarrow \Phi_2\underset{\underset{OH}{|}}{C}CH_2\Phi$$

In this section we have discussed the direct detection and concentration measurement of reactive intermediates. We have seen the problems, as well as some solutions which work in particular instances. Note that all examples of successful identification, detection, or concentration measurement of reactive intermediates were taken from gas phase reactions. There are fewer examples which could have been taken from liquid solution reactions (one possible example is the aqueous electron discussed in Section 20-5). The situation with respect to solutions has been well put in a recent article [17] on the mechanisms of homogeneous catalysis in solution: "If a chemist is able to [detect] a reactive intermediate . . . , the chances are good that he has found one of the less reactive intermediates." The overwhelming number of

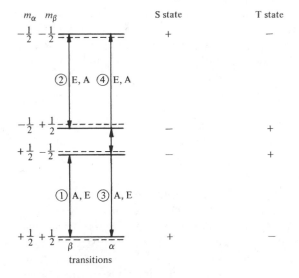

population increase ($+$) or decrease ($-$), with respect to thermal, for the radical pair in the

Figure 20-25 Multiplet effect for a caged radical pair $R_1 + R_2$ in which R_2 has no protons and R_1 (which is $-\overset{|}{\underset{H_\beta}{C}}-\overset{|}{\underset{H_\alpha}{C}}\cdot$) has two nonequivalent protons with $a_\alpha < 0$ and $a_\beta < 0$. $g_1 - g_2 = 0$. The quantities $2\delta_{m_\alpha m_\beta}$, which determine the indicated population changes of the different proton energy levels according to Eqs. (20-120) and (20-121), are obtained from Eq. (20-125) $|2\delta_{\pm\frac{1}{2}\mp\frac{1}{2}}| = \frac{1}{2}\mu_0\hbar^{-1}(|a_\alpha| + |a_\beta|)$ and $|2\delta_{\pm\frac{1}{2}\pm\frac{1}{2}}| = \frac{1}{2}\mu_0\hbar^{-1}(|a_\alpha| - |a_\beta|)$. The symbols next to each transition arrow, for example, ①, A, E, number the transitions in order of increasing frequency or resonant magnetic field H and indicate (first for the singlet, then for the triplet state) whether absorption (A) or emission (E) is enhanced.

The energy level spacings are drawn according to Eqs. (A-15), p. 512, with $\gamma_\alpha > \gamma_\beta$, $J_{\alpha\beta}^2 \ll H^2(\gamma_\alpha - \gamma_\beta)^2$, $J_{\alpha\beta} > 0$. The dashed lines represent the energy levels for $J_{\alpha\beta} = 0$.

reactive intermediates mentioned in this section are free radicals reflecting their "starring roles" as intermediates.

The methods for direct detection and concentration measurement of reactive intermediates, although under rapid development, will not soon displace the indirect methods for deducing reaction mechanisms such as CIDNP, or the use of rate laws (Section 20-3), yields (Section 20-5) or thermodynamics, (Section 20-4).

There is one additional and important method for checking a proposed reaction mechanism. This entails the comparison of the empirically determined Arrhenius parameters and activation energies of the proposed elementary reaction with the corresponding theoretically calculated quantities. In Chapter 21 we will present the theories that are most commonly used to calculate the necessary theoretical values of rate constants.

20-8   HETEROGENEOUS REACTIONS

Reactions are heterogeneous if more than one phase is essential for their occurrence. Special considerations are necessary for reactions whose rate-determining steps (1) occur at the interface between two phases or (2) involve material transport toward the interface. Microscopically the interface is a phase boundary region of finite thickness measured on a molecular scale of lengths. If the boundary is very sharp—as in gas–solid interfaces—we may simply say that a surface (of the solid) separates the gas from the bulk solid. Macroscopically the surface is characterized by its area; microscopically it is characterized by its atomic composition and structure.

In this section we shall examine the mechanisms of heterogeneous reactions and the basic nature of their elementary steps. We shall also survey the study and knowledge of surface intermediates.

20-8.1   Types of Reactions   The variety of interface reactions is very great. For many of them we have only a very general idea about their mechanisms. A generally accepted classification scheme does not exist. Without claiming to be comprehensive, we shall describe four classes of interface reactions. These will reveal the great variety of processes occurring at interfaces.

The first class of interface reactions is phase transitions: one phase is transformed into one or more new phases. The transitions observed in one-component systems—freezing, melting, vaporization, condensation, allotropic change—are examples of reactions in this class. In multicomponent systems, transformations into more than one phase may be observed, as in incongruent melting (page 241). The reaction mechanism of these reactions usually has two stages: the initial or induction period and the growth period.

During the induction period microscopic amounts of the new phase are formed. This process is known as **nucleation**. We showed in Section 7-6 that small droplets of a liquid are thermodynamically less stable than bulk liquid. This is true of nuclei of any phase. Therefore, nucleation is not a favored process if the old and new phases have equal thermodynamic stability. However, if the chemical potentials of the old phase and of the nuclei for the new phase become equal, nucleation will occur and thereafter the bulk new phase will grow with explosive speed. Thus, very pure water can be superheated or supercooled for many degrees above its normal boiling point or below its normal freezing point. In practice, phase transitions near the equilibrium transition temperatures are induced by seeding that is, by introducing the nuclei of the new phase into the old one, or by relying on minute amounts of the new phase or nucleation-inducing surfaces (such as those of dust particles) that may be naturally present.

During the growth period the rate-determining steps may be of two kinds. If the old and the new phases have rather different densities, **mass transfer** rates may be rate-determining. If the densities are similar, as in most liquid–liquid, liquid–solid, or solid–solid transformations, then the rearrangement of atoms or molecules and the associated bond breaking and formation will determine the rate at which the interface is propagated through the system.

The second class of interface reactions is created by and exists only during the reaction. For example, the flame of a Bunsen burner has two gas phases, an inner cone containing the reactants and an outer cone containing the products, separated by an interface—the reaction zone. The reaction zone does not move if flow and consumption rates of the reactants are equal, producing a stable flame.

There are two kinds of flames that have been used in kinetics studies [14, Vol. I, p. 167]: premixed flames and diffusion flames. The flame of a Bunsen burner is an example of a flame where the reactants are premixed. For kinetics studies one obtains a conveniently flat reaction zone by making the premixed gas flow through a fine wire mesh or a porous plate. In diffusion flames one of the reactants diffuses through a nozzle into the other reactant. A stable flame is achieved when the consumption and diffusion rates of the reactants are balanced. By either method it is possible to achieve high stationary concentrations of reactants and intermediates in the reaction zone and to measure them at leisure, allowing the study of fast reactions. The reactions of H_2 and various hydrocarbons with O_2 have been studied by the first method; those of halogen and alkyl halide reactions with sodium vapor by the second.

A third class of reactions is heterogeneous catalysis. Its importance is attested by the following examples: N_2 and H_2 form NH_3 in the presence of iron oxide (Haber process); petroleum is cracked on alumina (Al_2O_3); hydrocarbons hydrogenate, dehydrogenate, and isomerize on platinum; platinum or nickel catalyze electrode reactions involving hydrogen. The specificity of heterogeneous catalysis is suggested by the reactions of ethanol: over alumina above 300°C the products are C_2H_4 and H_2O; in the range 260–300°C, they are $(C_2H_5)_2O$ and H_2O; and over copper in the range 200–300°C, CH_3CHO, and H_2.

The elementary reactions that make up the mechanisms of heterogeneous catalysis belong to five broadly defined temporal stages: (1) transfer of reactants to surface; (2) adsorption of reactants on surface; (3) reactions on the surface; (4) desorption of products; (5) transfer of products away from the surface. The rate-determining step may be found in any stage. Present experimental methods do not always allow a clear resolution of the five stages of an actual reaction.

Reactions in the fourth and final class produce a solid which forms a layer at the interface. Examples are oxidations of solid metals in various fluids. The mechanism of such a reaction has four stages. The first two stages are the same as for heterogeneous catalysis. The third stage is the diffusion of one of the reactants through the product layer to the other reactant. The fourth stage is the formation of products.

Other classes of heterogeneous reactions, for example, dissolution reactions, could be described, but the four mentioned are sufficient to identify the processes specific to these reactions. These processes are mass transfer, adsorption-desorption, and surface reactions. Our plan is to discuss each of these processes in turn. We shall concentrate almost exclusively on fluid-solid surfaces.

20-8.2 Mass Transfer Mass transfer to and from a reaction zone—such as a fluid-solid surface—may occur by straight-line motions of molecules (in gases at very low pressures, see page 58), by diffusion, by convection, and by laminar or turbulent flow. In a particular system several of these processes, occurring simultaneously, may be responsible for the total mass transferred to or from the surface.

Since the situation is complex and the phenomena of mass transfer hold no fundamental interest to the reaction kineticist, it is best to arrange experiments so that mass transfer is not rate-determining. This may be accomplished by stirring the fluid, especially if the solid is suspended in the fluid as fine particles, or by using a flow system in which the fluid passes over a bed of solid particles. Porous particles should be avoided because diffusion through pores is slow.

Simple tests can determine whether mass transfer is rate-determining. One merely has to increase the mass transfer rate, that is, the flow rate or stirring rate, while keeping the exposure—(mass of solid exposed) × (exposure time)—of a given amount of fluid constant. When the reaction rate becomes constant, mass transfer effects may be assumed negligible.

It is not always possible to neglect mass transfer with respect to reactive species, for example, atoms and radicals. Because their surface reactions are so fast, mass transfer rates may become rate-determining with such species. We shall consider two possible rate-determining steps: (1) direct encounters with and (2) diffusion to the solid surface. Complications due to the macroscopic mass transfer processes mentioned previously will be ignored.

The physical situation in a gaseous system from which reactive species disappear by wall reactions is as follows. At pressures near atmospheric, the concentration of the reactive species in bulk is approximately constant throughout at a value $\bar{C}$. Near the walls, however, the concentration has a value C, much lower than $\bar{C}$, due to the disappearance of the reactive species via direct encounters with the wall. Since the reactive species reacts as soon as it gets to the wall, diffusion to the wall is rate-determining. However, as the pressure is lowered, diffusion speeds up and C approaches $\bar{C}$. Eventually the two concentrations become equal, and the rate of encounters with the wall then becomes rate-determining. A quantitative formulation of the two limiting cases follows.

For an ideal gas the encounter rate with a unit area of surface perpendicular to x is given by Eq. (2-60).

$$n(x) = n\left(\frac{RT}{2\pi M}\right)^{1/2} = \tfrac{1}{4}\bar{c}n \qquad\qquad (2\text{-}60)$$

Consider the gas to be contained between two parallel solid walls as shown in Figure 20-26. The encounter rate per unit volume of the gas, r_e (moles volume^{-1} time^{-1}) is then

$$r_e = n(x) \cdot S/V \cdot N^{-1} = \left(\frac{RT}{2\pi M}\right)^{1/2} Cd^{-1} \qquad\qquad (20\text{-}126)$$

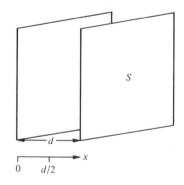

Figure 20-26 Dimensions and labels for gas molecule–wall encounter problems.

where $S/V = d^{-1}$ is the surface to volume ratio for the gas sample in Figure 20-26 and C has been defined previously.

An approximate rate of diffusion may be derived starting with Eq. (3-37).

$$\overline{x^2} = 2D\tau \tag{3-37}$$

(The variable y in Eq. (3-37) has been replaced with x in order to conform with Figure 20-26.) When diffusion is rate-determining there will be a concentration gradient of the reactant molecules. Most molecules will be located near $x = d/2$ (Figure 20-26). Therefore, it is reasonable to assume that molecules that diffuse to the walls have on the average traveled a distance of about $d/2$, that is, $\overline{y^2} \approx d^2/4$. When this value for $\overline{y^2}$ is substituted in Eq. (3-37) τ becomes the time for an average molecule to diffuse to the wall. Therefore, the average rate r_d ($= \bar{C}/\tau$) at which molecules in a unit volume diffuse to the wall is $(8D/d^2)\bar{C}$. A more rigorous derivation [14, Vol. II, pp. 107–112] changes only the numerical constant.

$$r_d = 9.9(D/d^2)\bar{C} \tag{20-127}$$

This equation is valid for any fluid in a rectangular container. It may be used to estimate the maximum possible rate of diffusion in any system, given the value for D.

Under what conditions is diffusion, rather than encounters with the wall, rate-determining? Let us start by calculating the ratio r_e/r_d for H_2 at 1 atm and 300 K. Assume that $d = 10$ cm; $D \approx 1$ cm^2 s^{-1}; $M = 2$ g mol^{-1}; $R = 8.3 \times 10^7$ erg K^{-1} mol^{-1}; and $C = \bar{C}$. Substitution of these values into Eqs. (20–126) and (20-127) gives

$$r_e/r_d \approx 10^5$$

Therefore, under these conditions diffusion would clearly be rate-determining. However, from Eqs. (3-26) and (2-66) we can conclude that the diffusion constant varies inversely as the pressure, that is, $D = D_0/P$ where D_0 is the diffusion constant at unit pressure P. On the other hand, the constants in Eq. (20-126) are independent of pressure. Thus we see that at about 10^{-5} atm the two rates would become equal. This is the pressure at which the free mean path $\lambda \approx d$; that is, where diffusion ceases to be a meaningful concept. We conclude that except at very low pressures Eq. (20-127) should be used to calculate the limiting rate for mass transfer in gases.

This conclusion may be checked experimentally. By changing the dimensions of the container, one can determine whether the rate is a function of d^{-1} or d^{-2}. Such a check reveals that our conclusion is flawed because a d^{-1} dependence, characteristic of surface encounter rate, is frequently found at pressures somewhat above 10^{-5} atm. The apparent reason is that not every collision is a sticky one, even for a reactive species. Rebound collisions can be included by replacing Eq. (20-126) with

$$r_e = s\left(\frac{RT}{2\pi M}\right)^{1/2} d^{-1}C \tag{20-128}$$

where s is the sticking coefficient. If $s = 10^{-3}$ (one in a thousand colliding molecules sticks), the pressure at which $r_e \approx r_d$, would increase to 10^{-2} atm.

How do the rates of surface reactions compare with the homogeneous rates of the same reactive species? Again, let us discuss a specific example, the $H_2 + O_2$ reaction analyzed earlier (pages 651-654). The mechanism of this reaction is a branched chain mechanism in which, at low pressures, the termination step is the reaction of H on the wall of the reaction vessel. The rate of the termination step in vessels with KCl-coated walls varies as d^{-1}. Therefore, we can identify r_4 from Eq. (20-42) and the concentration of H with r_e and C, respectively, of Eq. (20-128). Hence the rate constant

$$k_4 = s\left(\frac{RT}{2\pi M}\right)^{1/2} d^{-1} \tag{20-129}$$

from Eqs. (20-42) and (20-128). If diffusion were rate-determining, a similar identification would yield from Eqs. (20-42) and (20-127).

$$k_4 = 9.9 D/d^2 \tag{20-130}$$

Values of k_4 from Eqs. (20-129) and (20-130) may be compared to the rate for the homogeneous disappearance of H via reaction $H + O_2$ (Eq. 20-42) which is $k_1 \cdot O_2$. The three quantities are directly comparable because all three are equal to the relative rate of consumption of H, that is, $-H^{-1} dH/dt$. We assume the following values: $R = 8.3 \times 10^7$ erg K^{-1} mol^{-1}; $M = 1$ g mol^{-1}; $d = 10$ cm; $s = 1.5 \times 10^{-3}$; $T = 800$ K; $k_1 = 7 \times 10^9$ cm^3 mol^{-1} s^{-1}; total pressure $P = 1$ torr, $p_{H_2} = 2p_{O_2}$. We roughly estimate $D_H = D_0 T^{3/2} P^{-1}$ from $D_{H_2} = 1$ cm^2 s^{-1} (at 300 K and 760 torr), taking into account the smaller mass and diameter of H, as about 10^4 cm^2 s^{-1}. Calculated values of k_4 from (Eq. 20-129), (Eq. 20-130), and of $k_1 \cdot O_2$ are 15, 10^3, 50 s^{-1}, respectively. Thus we see that the encounter rate is lower (and therefore limiting) than the diffusion rate and that the encounter rate is barely competitive with the homogeneous rate of disappearance of H. At higher pressures the latter rate would be much higher, whereas the mass transfer rates would either stay constant or decrease and become negligible. In general, we can say that the fastest surface reactions are competitive with the fastest homogeneous reactions only at pressures below atmospheric. The converse is that surface reactions will be most important in competition with *slow* homogeneous reactions. For example, solid catalysts are used to speed up reactions that have high activation energies (that is, are slow) for the homogeneous path. The catalysts work by lowering the activation energy, not by increasing the Arrhenius parameter A.

These considerations can be readily extended to liquids. In liquids the diffusion coefficient is typically 10^{-5} cm^2 s^{-1} near room temperature. Diffusion, rather than encounters with the surface, is expected to be rate-determining. The fastest homogeneous reactions in liquid solutions and in gases have comparable rates, making fast surface reactions even less competitive with fast homogeneous reactions in liquids than was the case for gases.

20-8.3 Adsorption and Desorption It is well known that fluids and solutes are adsorbed on solid and liquid surfaces: active carbon in various forms such as activated charcoal or carbon black adsorbs many gases (as in gas masks) and substances from solution (as in purification and decolorization of sugar); platinum black adsorbs hydrogen and other substances; water adsorbs detergents—organic molecules with a hydrophilic group. Indeed, all surfaces we encounter in everyday life are "dirty" with substances adsorbed from the atmosphere or from other media with which the surface has been in contact. The substance that adsorbs is the

adsorbent (or **substrate**). The substance that has been adsorbed is the **adsorbate**. A good adsorbent can hold large amounts of adsorbate because (besides the attraction between the adsorbent and the adsorbate species) the adsorbent has a large ratio of surface area to mass.

Two extensive variables—amount of adsorbate and the surface area of the adsorbent—are necessary for a macroscopic description of adsorption–desorption processes and of surface reactions as well. This section describes the measurement of these two variables and their use, in conjunction with various models, in describing the equilibria and kinetics of adsorption–desorption. The emphasis will be on the models and aspects of the description commonly used for treating surface reactions (see Section 20-8.5).

The amount of adsorbate is most often determined indirectly. A measured volume of a gas (or solution) at a given pressure (or concentration) and temperature is allowed to equilibrate with a known mass of the adsorbent. The adsorbed amount is calculated from the decrease in the volume of the gas or concentration of solution. This method can routinely measure adsorbed amounts as low as 10^{-4} mole. However, since a typical small molecule occupies an area equal to about 10^{-1} nm^2 (10 Å^2) this number of moles can cover a total area of some 6 m^2 with a monolayer (a layer of molecules one molecule thick). If the adsorbent sample has a smaller area then a direct method—weighing with a microbalance (sensitive to 1 μg or less)—can lower the limit on the measurable amount of adsorbate some 10^4 times.

A common method for determining the surface area of the adsorbent is also an indirect one. One necessary experimental datum is an **adsorption isotherm**, a relation between the amount of adsorbate and the equilibrium pressure (or concentration) of the gas (or solution). Figure 20-27 shows six typical isotherm shapes that have been observed in various experiments.

When interpreted in terms of a model, the adsorption isotherm will yield the amount of adsorbate necessary to cover the adsorbent with a monolayer. For example, if V_m is the volume of gas necessary to form a monolayer upon adsorption then a common isotherm (suggested by Langmuir in 1918) states that

$$\theta = V/V_m = KP/(1 + KP) \qquad \textbf{(20-131)}$$

where V is the actual volume of gas adsorbed when in equilibrium with the gas at the pressure P. Therefore, θ is the fraction of surface covered with a monolayer. K is a constant. Figure 20-27a is a plot of Eq. (20-131).

An experimental isotherm that obeys the Langmuir equation (Eq. 20-131) may be used in a variety of ways for determining V_m. If $KP \gg 1$ then Eq. (20-131) reduces to

$$V \approx V_m \qquad \textbf{(20-132)}$$

that is, V_m is given directly by the constant volume measured above the pressure P which causes saturation of the adsorbent with respect to the adsorbate (see Figure 20-27a). If $KP \ll 1$ then Eq. (20-131) becomes analogous to Henry's law (page 207)

$$V \approx V_m KP \qquad \textbf{(20-133)}$$

In order to determine V_m from this equation, K must be obtained from an independent measurement for the same adsorbent-adsorbate pair.

When the approximations that lead to Eqs. (20-132) and (20-133) are not valid, it is best to rearrange Eq. (20-131) so that a straight line is obtained. Two possible

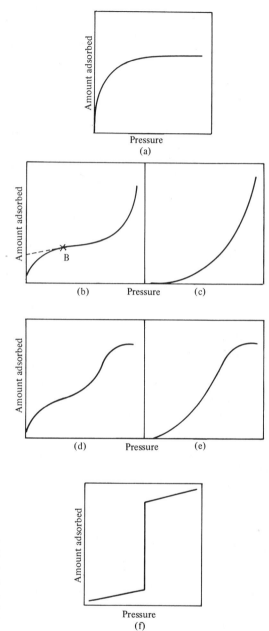

Figure 20-27 Types of adsorption isotherms. (a) monomolecular layer; (b) and (c) multimolecular layers; (d) and (e) multimolecular layers and condensation in pores; (f) surface phase transition of a monomolecular layer on a homogeneous surface.

rearrangements are

$$V^{-1} = (V_m K)^{-1} P^{-1} + V_m^{-1} \qquad (20\text{-}134)$$

or

$$P/V = V_m^{-1} P + (V_m K)^{-1} \qquad (20\text{-}135)$$

Eqs. (20-134) and (20-135) are equations of straight lines for the variable pairs (V^{-1}, P^{-1}) and (P/V, P), respectively. Their slopes are $(V_m K)^{-1}$ and V_m^{-1}; their intercepts

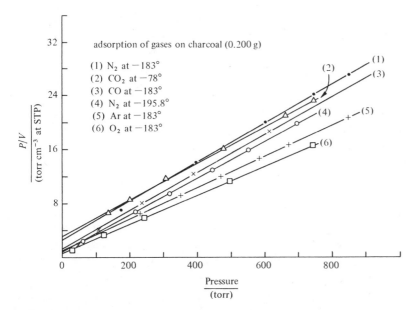

Figure 20-28 Adsorption isotherms of gases on 0.200 g of charcoal plotted according to Langmuir, Eq. (20-135). (1) N_2 at 89 K(173); (2) CO_2 at 194 K(185); (3) CO at 89 K(179); (4) N_2 at 74.4 K(181); (5) Ar at 89 K(215); (6) O_2 at 89 K(235). The quantities in parentheses are $V_m/(cm^3 \ g^{-1}$ at STP). Note that the data do not extend to very low or very high values of θ. [From S. Brunauer, P. H. Emmett, and E. J. Teller: *J. Am. Chem. Soc.*, **60**:309 (1938). Copyright by The American Chemical Society.]

are V_m^{-1} and $(V_m K)^{-1}$. Data for various gases adsorbed on charcoal are plotted according to Eq. (20-135) in Figure 20-28. The calculation of V_m and K from the slope and intercept in Figure 20-28 is shown in Example 20-7.

From V_m the number of moles of adsorbate N can be calculated. The specific surface area Σ of the sample is then given by

$$\Sigma = \sigma \cdot N \cdot N/m \tag{20-136}$$

where σ is the cross-sectional area of one adsorbate molecule and m is the mass of adsorbent. A sample calculation is shown in Example 20-7.

EXAMPLE 20-7

Given the N_2 on charcoal adsorption isotherm curve (1) of Figure 20-28 and the density of liquid N_2 (0.751 $g \ cm^{-3}$), both at 90 K, calculate the constants of the Langmuir equation V_m, K, the cross-sectional area of N_2 and the specific surface area of the charcoal Σ.

Solution: From curve (1) of Figure 20-28, the y-axis intercept is 2.6 torr $(cm^3$ at STP$)^{-1}$, and the slope is $(20.0-2.6)/(600-0) = 0.0290 \ (cm^3$ at STP$)^{-1}$. Therefore, from Eq. (20-135) $V_m = 1/0.0290 = 34.5 \ cm^3$ at STP. This is the value for 0.200 g of charcoal. $V_m/m = 34.5/0.200 = 173 \ (cm^3 \ g^{-1}$ at STP). Similarly from Eq. (20-135) $K = slope/intercept = 0.0290/2.6 = 0.011 \ torr^{-1}$. More precise values of V_m/m and K could be determined from a least squares analysis.

The cross-sectional area of an N_2 molecule is calculated by assuming that (1) the N_2 molecule may be represented by a sphere with diameter d, and (2) the spheres are packed as densely as possible in both the liquid and the adsorbed monolayer.

A three-dimensional closest-packing arrangement (12 nearest neighbors) of spheres with diameter d is the face-centered cubic lattice (Figure 17-4b). A diagonal of a cube face equals $2d$, therefore the edge of the cube a is related to d by $2a^2 = (2d)^2$ or $a = 2^{1/2}d$. The volume of the cube is $V = a^3 = 2^{3/2}d^3$. The number of spheres in this cube is four (each of eight spheres on cube corners is shared equally between eight cubes, each of six spheres on the faces is shared equally between two cubes). Therefore, the density of the liquid is $\rho = m/V = 4(M/N)/(2^{3/2}d^3)$ where M is the mass per mole of N_2.

A two-dimensional closest-packing arrangement of spheres is the hexagonal lattice (Figure 20-37). The edge of its elementary cell equals d. The area of the cell is $d^2 \sin 60° = 0.866d^2$. Each of the four spheres on the corners of the cell is shared with four other cells. Therefore the cell area is equal to the area occupied by one sphere; that is, the area is σ. Combining the equations for the volume and the area to eliminate d yields

$$\sigma = 0.866(2^{1/2}M/\rho N)^{2/3}$$
$$= 0.866[2^{1/2} \times 28.0/(0.751 \times 6.02 \times 10^{23})]^{2/3} \text{ cm}^2$$
$$= 1.71 \times 10^{-15} \text{ cm}^2$$
$$= 0.171 \text{ nm}^2$$

The specific surface area is calculated from Eq. (20-136) and the molar volume of an ideal gas 22,400 cm^3 at STP:

$$\Sigma = \sigma NN/m = \sigma(V_m/m)N/22,400$$
$$= 0.171 \times 173 \times 6.02 \times 10^{23}/22,400$$
$$= 7.95 \times 10^{20} \text{ nm}^2 \text{ g}^{-1}$$
$$= 795 \text{ m}^2\text{g}^{-1} \qquad \bullet$$

An isotherm that is used for the determination of surface areas much more often than Langmuir's was suggested by Brunauer, Emmett, and Teller in 1938. This, the so-called BET isotherm, is

$$\frac{x}{V(1-x)} = (V_m c)^{-1} + (c-1)(V_m c)^{-1}x \tag{20-137}$$

where $x = P/P_{vap}$ (with P_{vap} being the vapor pressure of the adsorbed liquid at the isotherm temperature) and c is a constant. The other symbols were introduced with Eq. (20-131). The most common shape ($c > 2$) of the BET isotherm is shown in Figure 20-27b. The less common shape (when $c \leq 2$) appears in Figure 20-27c.

The standard BET procedure for determining the surface area of a weighed adsorbent sample is as follows: (1) determine the adsorption isotherm for N_2 as adsorbate in the range $0.05 < x < 0.35$; (2) compute the least squares slope and intercept from the region of the isotherm where the plot of $x/[V(1-x)]$ versus x is linear (this region will usually extend over most of the experimental range of x); (3) calculate V_m from the slope and intercept; (4) use Eq. (20-136) with $\sigma_{N_2} = 0.162 \text{ nm}^2 = 16.2 \text{ Å}^2$ to calculate the surface area.

Isotherm methods do not exhaust the possibilities for measuring surface areas. The **gas permeability** method calculates the surface area from the rate of gas flow, caused by a known pressure gradient, through the pores and interstices of a compacted bed of adsorbent. The **electron microscope** can determine the size distribution of adsorbent particles from which, assuming a definite particle shape, one can calculate the surface area (excluding surface areas of pores within the particles).

Table 20-10 lists some of the surface areas measured by various methods. The accuracy of the measurements is rather low. One reason for this is the approximations made in deriving the isotherms.

Table 20-10 Surface areas of representative adsorbents

Adsorbent	Surface Area $\frac{}{(\text{m}^2 \text{ g}^{-1})}$	Method
carbon blacks		
P-33(2700°)	10–15	various
(nearly graphitic		
microcrystals)		
Statex A	90	electron microscope
	100	gas permeability
	140	N_2 adsorption, BET
Carbolac	1000	
alumina (Al_2O_3)	100–300	
porous nickel film	13	
platinum on alumina	240	

We shall now describe the Langmuir adsorption model, derive Eq. (20-131) from it, and discuss conditions and tests for its validity and the ways it may fail. A statistical derivation of the Langmuir isotherm is found in Section 19-11.

The Langmuir model assumes that (1) for a given adsorbate the adsorbent surface contains a certain number of adsorption sites; (2) each site can bind only one adsorbate molecule (that is, an immobile monolayer is implied); (3) the interaction energy is the same for all sites; and (4) the lateral interactions between molecules on different sites are equal to zero. There will be an equilibrium between the ideal gas A (at pressure P) in contact with the adsorbent surface, the unoccupied sites S, and the sites AS that contain an adsorbed molecule A:

$$A + S \underset{k_-}{\overset{k_+}{\rightleftharpoons}} AS \qquad \textbf{(20-138)}$$

Assumptions 3 and 4 imply that both surface components in equilibrium (20-138) will behave ideally. Therefore the Langmuir equilibrium constant K is given in terms of concentrations (rather than activities) by

$$K = \frac{k_+}{k_-} = \frac{AS}{P \cdot S} \qquad \textbf{(20-139)}$$

where *AS, S* are the equilibrium surface concentrations (amount/area). The fraction of surface θ covered with adsorbate molecules is

$$\theta = \frac{AS}{AS+S} = \frac{V}{V_m} \tag{20-140}$$

When the ratio AS/S is eliminated between Eqs. (20-139) and (20-140) and the result is solved for θ we obtain the Langmuir isotherm, Eq. (20-131).

The limitations of the Langmuir isotherm derive from the immobile monolayer assumption and from the assumption of ideality of the monolayer. The ideality assumption implies constancy of V_m for a given adsorbent regardless of temperature and near-constancy of V_m for various adsorbate molecules of similar size. A constancy test of V_m is shown in Figure 20-28. Ideality of the monolayer implies (if the gas A is assumed ideal), in addition, that K should be truly a constant. We can write for K (Eqs. 9-4 and 6-34)

$$K = \exp\left(-\Delta\mathscr{G}^\circ/\boldsymbol{R}T\right) = \exp\left(\Delta\mathscr{S}^\circ/\boldsymbol{R}\right) \cdot \exp\left(-\Delta\mathscr{H}^\circ/\boldsymbol{R}T\right) \tag{20-141}$$

where $\Delta\mathscr{G}^\circ$, $\Delta\mathscr{H}^\circ$, and $\Delta\mathscr{S}^\circ$ are the ideal free energy, enthalpy, and entropy of adsorption, respectively. Usually, all three quantities are negative.

If the monolayer is not ideal, both K and $\Delta\mathscr{G}$ will vary with θ. The variation is best understood by considering the changes in $\Delta\mathscr{H}$ and $\Delta\mathscr{S}$ separately. Considering the isosteric (θ = constant) enthalpy of adsorption first we can see that it could vary with θ for two reasons: (1) heterogeneity of the surface, which would violate assumption 3 and (2) lateral interactions between neighboring adsorbate molecules (violates assumption 4). The attractive adsorbate-adsorbent interaction will, on the average, decrease with increasing θ on a heterogeneous surface because the more strongly interacting sites will tend to be filled first. The magnitude of lateral adsorbate-adsorbate interaction will increase with increasing θ. If the lateral interaction is attractive, a cancellation will result that may lead to a nearly constant $\Delta\mathscr{H}$. Quite generally, the magnitudes of $\Delta\mathscr{H}$ and $\Delta\mathscr{S}$ vary in the same direction. Therefore, we may expect further cancellation between the $\Delta\mathscr{S}$ and $\Delta\mathscr{H}$ terms in $\Delta\mathscr{G}$. Thus we can see how the Langmuir isotherm may pass an experimental test in a limited range (see Figure 20-28) of θ, especially if the accuracy is not too high, even when two of its assumptions are not valid.

Since the adsorbed layer has a characteristic equilibrium "vapor" pressure, just like a pure liquid, measurements of $(d \ln P/dT)_\theta$ as a function of temperature can be used in conjunction with the Clausius–Clapeyron equation (7-19) to determine experimental values of the isosteric enthalpy of adsorption. The results of such determinations are shown in Table 20-11 and Figure 20-29. We see from Table 20-11 that $-\Delta\mathscr{H}(\theta)$ may decrease by a factor of 2 or more in going from $\theta = 0$ to $\theta = 1$. This implies that surface heterogeneity dominates the changes in the isosteric heat of adsorption. Figure 20-29, on the other hand, shows $-\Delta\mathscr{H}(\theta)$ for a highly homogeneous surface where the lateral interactions dominate.

Another technique that demonstrates surface heterogeneity is flash desorption. In this technique the adsorbate-covered adsorbent, usually a thin metallic filament, is very rapidly heated to a selected temperature. Complete or partial desorption results. The experiment may be repeated at several temperatures. If the adsorbate is held both weakly and strongly, the former variety will desorb at a low temperature

Table 20-11 Adsorption enthalpies for representative adsorbate–adsorbent pairs

Adsorbate–Adsorbent Pair	$\dfrac{T}{(K)}$	Isosteric Heat of Adsorption $[-\Delta\mathcal{H}(\theta)/(\text{kcal mol}^{-1})]$	
		$\theta = 0$	$\theta = 1$
N_2–Cr_2O_3	90	8.2	3.0
N_2–TiO_2 (anatase)	78	4.4	2.2
N_2–Graphon[a]	78	2.9	1.7
n-$C_{10}H_{22}$–Graphon[a]	373	21.0	17.1
H_2–W films	room	45	5
O_2–W films	room		194 ($\theta \approx 1.3$)

[a] A carbon black.

and the latter at a high temperature. Figure 20-30 shows the results of a flash desorption experiment.

The results just discussed, which show that the effects of both heterogeneity of surfaces and of lateral interactions are readily observable, are by no means atypical. Therefore, there is need for isotherms that do not assume ideal monolayers.

If we allow for lateral adsorbate interactions (a constant interaction energy ω for 1 mole of adsorbate molecules each interacting with a molecule on one of z nearest sites, no interaction with nonnearest neighbors) while keeping the other assumptions (page 717) of the Langmuir isotherm, we get the Fowler–Guggenheim isotherm (Eq.

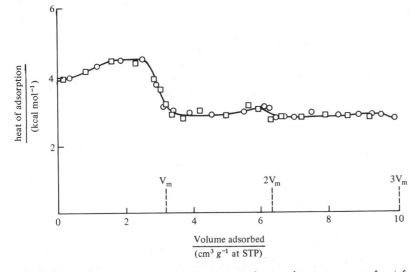

Figure 20-29 Heats of adsorption of krypton on the near-homogeneous surface of P-33(2700°) carbon at −183°C. [Reproduced by permission of the National Research Council of Canada from C. H. Amberg, W. B. Spencer, and R. A. Beebe, *Can. J. Chem.*, **33**:305 (1955).]

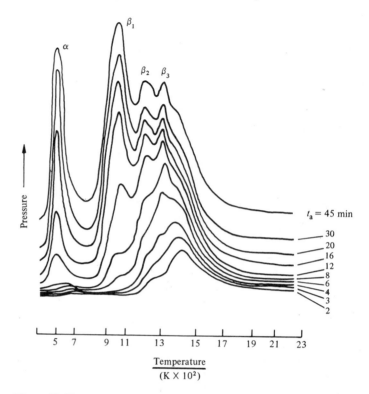

Figure 20-30   Flash desorption spectra of CO from a tungsten filament
after various times t_a of adsorption at 300 K. (The zero levels of the curves
have been shifted slightly to prevent overlap.) α, β_i are different adsorption
sites. The sites adsorb in order of decreasing heats of adsorption. (From P.
A. Redhead: *Trans. Faraday Soc.*, **57**:641 (1961).]

20-142, Table 20-12). If, in addition, we treat the adsorbate monolayer as a
two-dimensional fluid (rather than crystalline) phase (that is, relax assumption 2 on
page 717 to allow for surface mobility of the absorbate molecules), we get isotherms
such as Eq. (20-143) in Table 20-12. The adsorbate–adsorbent interaction potentials
which would lead to crystalline or fluid adsorbate phases are shown in Figures 20-31a
and 20-31b.

Isotherms derived from various models of heterogeneous surfaces have been
studied extensively. Perhaps the most important conclusion one can make from the
studies is that experimental isotherms do not discriminate readily between theoretical isotherms based on various models of heterogeneous or homogeneous surfaces.
For example, it has been shown that an isotherm derived by assuming a mobile film
(Eq. 20-143) and a highly heterogeneous surface can be fitted to the Langmuir
isotherm if coverage is in the range $\theta = 0.5$ to 0.8.

The final assumption of the Langmuir model to be relaxed is the monolayer
assumption (page 717). The best known treatment of multilayered adsorption leads
to the BET isotherm, Eq. (20-137). The model from which the BET isotherm is
derived has these features: (1) Each adsorbate layer individually obeys the Langmuir
isotherm. (2) The free energy of adsorption of the second and higher layers ($\Delta \mathcal{G}_{i>1}$) is

Table 20-12 Adsorption of gases and solutes[a]: isotherms for monomolecular films on homogeneous surfaces

Equation Number	Isotherm	Comments
(20-131)	$KP = \theta/(1-\theta)$	immobile film without lateral interactions (Langmuir isotherm)
(20-142)	$KP = \theta/(1-\theta) \cdot \exp\left[-(z\omega\theta)/RT\right]$	immobile film with lateral interactions (Fowler–Guggenheim, symbols defined in text)
(20-143)	$KP = \theta/(1-\theta) \cdot \exp\left[\theta/(1-\theta) -2\alpha\theta/kT\beta\right]$	mobile molecules that interact laterally according to a two-dimensional van der Waals equation of state: $(\pi + \alpha/\delta^2)(\delta-\beta) = kT$ where δ-surface area, π-surface pressure, α, β-van der Waals constants.

[a] Replace the pressure P by the solute concentration C to describe the adsorption of a solute out of a solution.

equal to the free energy of liquefaction of the adsorbed vapor $\Delta\mathscr{G}_L$ at the isotherm temperature. (3) The free energy of adsorption of the first layer is more negative than $\Delta\mathscr{G}_L$. The model thus implies that, starting with the second layer, the adsorbate is indistinguishable from bulk liquid. Experiment does not agree with this (see Figure 20-29). Nevertheless, although there are minor disagreements with experiment, the BET model recognizes the major fact about multilayered adsorption: as the pressure of the vapor that is adsorbed approaches the equilibrium vapor pressure, adsorption becomes indistinguishable from condensation.

Before leaving the subject of adsorption equilibria we wish to mention briefly the experimental distinction between **physical** and **chemical adsorption**, also known as **physisorption** and **chemisorption**. Theoretically this distinction is based upon the nature of the forces responsible for adsorption. Physisorption is due to van der Waals forces (pages 10 and following); chemisorption results from adsorption forces that are chemical (that is, chemical bonds are formed between adsorbate and adsorbent molecules). It is not always easy to distinguish the two kinds of adsorption on basis of experimental studies of adsorption equilibria. Multilayered adsorption is clearly physical. A heat of adsorption with a magnitude equal to that of chemical bond energies obviously belongs to chemisorption. However, chemisorption reactions, just like homogeneous reactions, may have reaction enthalpies that are near zero or even positive; physisorption is always exothermic. For example, hydrogen chemisorbs on some adsorbents as follows:

$$H_2 + 2\,S \rightarrow 2\,HS \tag{20-144}$$

This process can be slightly endothermic (if the HS bonds are weak) and still spontaneous if the entropy of the reaction is positive. For slightly exothermic

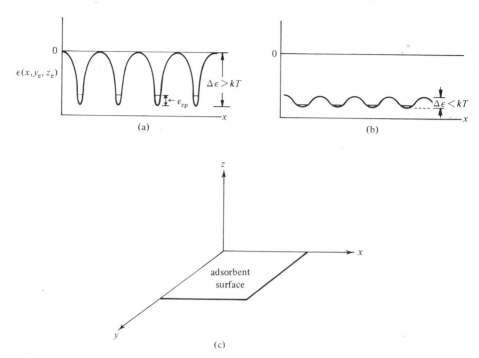

Figure 20-31   The adsorbate molecule-adsorbent surface (homogeneous, regular lattice) potential $\varepsilon(x, y, z)$ along the surface direction x (with y, z at their equilibrium distances y_e, z_e) for (a) crystalline ($|\varepsilon| > kT$), (b) gaseous adsorbate monolayer ($\Delta\varepsilon < kT$). $\varepsilon(x_e, y, z_e)$ would usually be very similar. ε_{zp} is the zero-point energy (p. 315). $\varepsilon(x, y, \infty) = 0$. (c) The x, y, z coordinate system.

reactions, the shape of the adsorption isotherm may indicate chemisorption. Reaction (20-144), for example, might obey an isotherm that derives from the Langmuir model plus the assumption of the dissociation of H_2. This isotherm is readily shown to be (Problem 42)

$$KP = [\theta/(1 - \theta)]^2 \tag{20-145}$$

Next, let us consider the kinetics of adsorption–desorption, in particular the rates of adsorption and desorption. Physical adsorption and desorption processes are elementary, whereas chemical adsorption and desorption reactions are complex. The rates of elementary adsorption and desorption processes have not been determined experimentally by direct means. However, the processes are simple enough to make theoretical considerations fruitful.

Consider the physisorption of an ideal gas. Its rate r_S (amount in moles per surface area per time) can be calculated from Eq. (2-60)

$$r_S = sn(x) = sn\left(\frac{RT}{2\pi M}\right)^{1/2} = s(2\pi MRT)^{-1/2}P \tag{20-146}$$

where s is the sticking or condensation coefficient (page 711). The third equality results when n is substituted from the ideal gas law. Let us apply Eq. (20-146) to the calculation of the time t necessary to cover a surface area S with a monolayer.

$$t = (r_S\sigma)^{-1}(S/S_u) = (2\pi MRT)^{1/2}S/(S_u s\sigma P) \tag{20-147}$$

where σ is the molecular area (page 715) and S_u is the unit area ($S_u = 1\ m^2 = 10^4\ cm^2$ in SI units). Nitrogen gas at $P = 10^{-7}\ Pa \approx 10^{-5}$ torr, $T = 300$ K, $M = 28\ g\ mol^{-1} = 0.028\ kg\ mol^{-1}$, with $\sigma = 16.2\ Å^2$ per molecule $= 9.75 \times 10^4\ m^2\ mol^{-1}$ will cover a surface area $S = 1\ cm^2$ of a metal (with $s \approx \frac{1}{4}$) in ≈ 1 s. This representative result shows that adsorption is a fast process. It is clear that surfaces will not stay clean ($<1\%$ coverage) for periods longer than 1 min except under conditions of ultrahigh vacuum ($P < 10^{-8}$ torr) or when the ambient gases are not adsorbed by the surface.

The usual expression for the desorption rate r_d of the species A is

$$r_d = kA = (\tau)^{-1} A = (\tau_0)^{-1} \exp(-\mathscr{E}_a/RT) \cdot A \approx (\tau_0)^{-1} \exp(\Delta\mathscr{H}_{ads}/RT) \cdot A \tag{20-148}$$

where A is the surface concentration, τ is the (average) **residence time** of a molecule on the surface, τ_0^{-1} is the Arrhenius parameter, and $\Delta\mathscr{H}_{ads}$ is the molar enthalpy of adsorption (a negative quantity). The last approximate equality follows from the two considerations that the activation energy of desorption has to be high enough to overcome the adsorbate–adsorbent attraction (equal to $-\Delta\mathscr{H}_{ads}$), and that for van der Waals forces there is no conceivable reason why it has to be any higher. Values of τ_0, calculated by statistical methods, range from about 10^{-12} to 10^{-14} s. For a typical physical adsorption enthalpy in Table 20-11, equal to about $-6\ kcal\ mol^{-1}$ or $-24\ kJ\ mol^{-1}$ we can estimate the magnitude of the residence time as 10^{-9} s at 300 K and 0.1 s at 100 K. The short residence times at room temperature and higher is a reason why physisorption is not important in heterogeneous catalysis (page 737).

A second use for Eq. (20-148) is to estimate the residence time at a particular site (that is, the mobility of the adsorbate). For this purpose the activation energy $\mathscr{E}_a = N\Delta\varepsilon < |-\Delta\mathscr{H}_{ads}|$ ($\Delta\varepsilon$ is defined in Figure 20-31b).

Chemical adsorption is best thought as at least two elementary processes in succession

$$A + S \underset{k_{-0}}{\overset{k_{+0}}{\rightleftharpoons}} A\text{---}S \underset{k_{-1}}{\overset{k_{+1}}{\rightleftharpoons}} AS \tag{20-149}$$

where the zeroth step represents physisorption and the first step is a chemical reaction of adsorbate A with the surface species. In general the chemical reaction can be expected to have the higher $\mathscr{E}_a$ and be rate-determining. Therefore the physisorption step may be in equilibrium. If so, application of the methods of Section 20-3 to mechanism (20-149) will show that the overall rate law of either adsorption or desorption will have the same form as Eq. (20-148). This fact is a partial justification, in the absence of experimentally measured rates of (fast) elementary heterogeneous reactions, for applying Eq. (20-148) to chemisorption. Chemisorption, in relation to heterogeneous catalysis, will be discussed in Section 20-8.5.

A shortcoming of the sorption studies discussed so far is the absence of a molecular characterization of the surface. Instead, we have encountered a vague concept, the adsorption site, and a quantity that is difficult to measure accurately, the surface area. We must now remedy the shortcoming by discussing the composition (kinds of atoms and the number of each kind) and structure (the arrangement of atoms) of surfaces.

20-8.4 Composition and Structure of Surfaces

The molecular definition of the term **surface** is, the plane in which the outermost atomic, molecular, or ionic layer of a solid lies. (Occasionally we will refer somewhat loosely to several near-surface layers as the "surface.") The **surface composition** is the percentage or amount of species (atoms, ions, or molecules) of each kind found on the surface. The **surface structure** is the orderly or disorderly (amorphous) spatial arrangement of the surface species. A disorderly arrangement means that the **surface phase** is a fluid. The surface fluid may consist of mobile species, in which case we have a surface gas if the fluid is dilute or a surface liquid if the fluid is dense. If the species have very low mobility, we have a viscous surface liquid or a glass. An orderly arrangement of surface species means that the surface is crystalline. (The student is urged to review the material on crystals, in Chapter 17, before going further.) The crystalline surface structure is formed by associating with every point of a two-dimensional lattice a **basis**—a unique assembly of atoms. All bases of a given crystalline surface are identical in atomic composition, arrangement, and orientation.

The experimental study of surface composition and structure has progressed far in the last two decades as a result of the work of many scientists. The progress is continuing. The experimental study of surfaces is difficult for the following reason: A single crystal of diamond in the form of a cube (1 cm^3) contains about 10^{23} carbon atoms. If we assume that each atom has a cross-sectional area equal to 10–20 Å^2 (0.1–0.2 nm^2) then the six faces of the cube contain about 10^{15} atoms. Therefore, a successful experimental surface study must detect what is essentially a minor impurity (its atomic concentration is $10^{15}/10^{23} \times 100 = 10^{-6}\%$). The essential principles of the more important methods of surface studies are discussed in the following paragraphs.

Present knowledge of the three types of surfaces most important for catalytic reactions is uneven. The most extensively studied are the crystalline surfaces of transition metals that primarily catalyze reactions involving hydrogen. Much less studied are the surfaces of semiconductors, mostly metallic oxides that catalyze oxidations and reductions, and of insulators, mostly acidic metal oxides that catalyze the cracking and isomerization of hydrocarbons (via carbonium ion formation). We shall present the most important results on the composition and structure of clean and adsorbate-covered metallic surfaces. The student should be cautioned that the details of some of these results and of their interpretations may, perhaps, have a higher probability of being discarded in the future than the results and interpretations in more established areas of chemistry.

Clean surfaces (whose composition is identical to the bulk composition) may be prepared by removal of impurities. This is done by heating, by bombarding the surface with ions or electrons, and by chemical reaction. An example of the third method is the removal of carbon from a platinum surface by oxidation with oxygen and subsequent heating to desorb the excess oxygen. Virginally clean surfaces are prepared by ultra high vacuum evaporation of a metal (usually) to form a film onto a substrate such as glass, mica, or metal or else by cleaving a crystal along a suitable crystal plane. In order to protect a clean surface from contamination, it must emerge in an ultra high vacuum ($<10^{-8}$ torr) environment. Presently attainable ultra high vacuum (10^{-10} torr is routinely produced and measured; the low pressure frontier is below 10^{-14} torr) will preserve clean surfaces for hours. This may not be true, however, at high temperatures where surface contamination by preferential diffusion of bulk impurities can occur.

To monitor surface impurities and, even more importantly, to study adsorbate species, a variety of methods are used. A destructive method for determining surface composition is to analyze desorbed species mass spectrometrically. If the desorption is done by bombardment with a narrow ($\approx 1 \, \mu m^2$ cross section) ion beam which can scan the surface, composition may be determined as a function of the surface position. Mass spectrometric methods are sensitive enough to detect small fractions of a monolayer.

Electromagnetic radiation or electron spectroscopy can determine surface composition nondestructively. IR and visible-UV spectroscopy depend upon a finely dispersed (high surface to volume ratio) adsorbent for sufficient sensitivity. Absorption spectra are commonly obtained by passing radiation through adsorbent thinly spread on a transparent plate which is inserted in the sample cell. To prevent scattering of radiation, the adsorbent particles have to be smaller than the wavelength of the radiation.

The absorption frequencies of physically adsorbed substances are quite similar to their condensed phase frequencies. Intensities, on the other hand, may be quite different due to the effects of van der Waals forces on the magnitudes of transition moments (page 401) and, even more importantly, due to changed selection rules (page 401) because symmetry has been lost upon adsorption. Thus, IR absorption due to the vibration of physically adsorbed H_2 has been observed. The chemical bonds formed during chemisorption may cause entirely new vibration or electronic transition bands to appear.

Electron spectroscopy determines the energy spectrum of electrons emitted by various activator-initiated processes (Figures 20-12 and 20-32). The electrons will be emitted only from surface layers if (1) the electrons emitted deep inside the crystal are not energetic enough to penetrate to the surface; (2) if the activation is done with electrons having a small momentum normal to the surface. For example, a 100-eV electron beam which moves normal to the surface will be scattered inelastically by valence electrons or, to a lesser extent, elastically by the atomic cores within the first 2–4 atomic layers. The same fate will befall a high energy beam of electrons

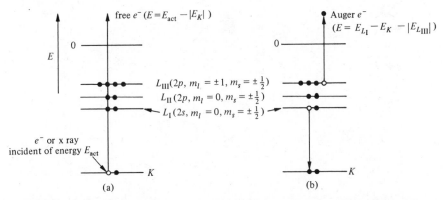

Figure 20-32 Two sample activator-initiated processes that can be used for characterization of surfaces. (a) Primary emission of an electron is the basis of **photoelectron spectroscopy** or **electron spectroscopy for chemical analysis (ESCA)**. (b) The **Auger process** (designated KL_IL_{III}) is a secondary process that involves two electrons. In these two examples E_K, respectively, $|E_{L_I} - E_K| - |E_{L_{III}}|$, both of which are characteristic of an element, would be measured. Energy levels other than the ones shown may be involved. The spectrum in both cases is taken by counting the number of emitted electrons as a function of energy.

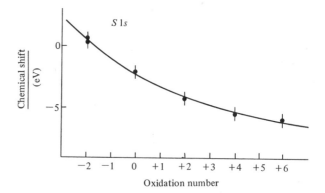

Figure 20-33 Chemical shift of the 1s electron shell in sulfur atoms as a function of oxidation state in inorganic sulfur compounds. [From G. A. Somorjai: *Principles of Surface Chemistry*, © 1972. Reprinted by permission of Prentice-Hall, Inc., Englewood Cliffs, N.J.]

$(10^3–10^5 \text{ eV})$ that approaches the surface at a glancing angle such that the normal component of the momentum of the electrons equals that of the 100-eV electrons. If the activators are deeply penetrating x rays, only condition 1 can be readily fulfilled. Either of the two processes shown in Figure 20-32 (namely, photoelectron and Auger electron emission) can detect as little as 1% of a monolayer.

    The identification of the surface atoms of a particular element is done by means of their characteristic energy levels, which are known from atomic spectroscopy. For

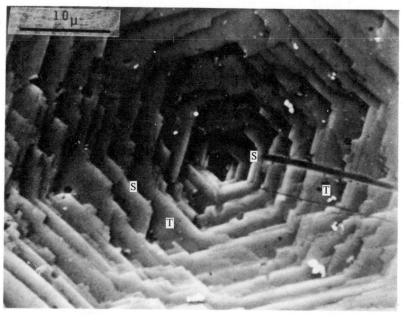

(a)

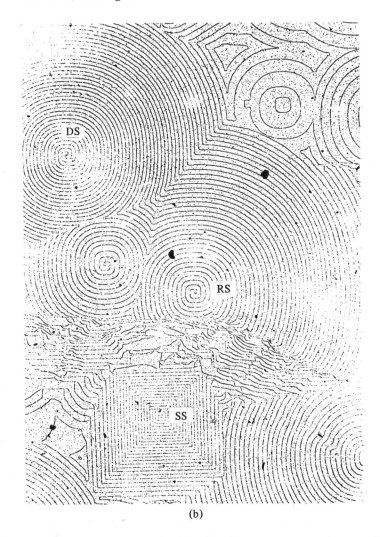

(b)

Figure 20-34   (a) Electron microscope picture of the basal plane of zinc magnified $10^5 \times$. T, identifies terraces; S, some of the steps which can be only one atom high. [From G. A. Somorjai: *Principles of Surface Chemistry.* © 1972. Reprinted by permission of Prentice-Hall, Inc., Englewood Cliffs, N.J.]

(b) The (100) surface of an NaCl crystal. The surface was created during evaporation in vacuum at 400°C for 90 min. The step heights of the double (DS) and simple round (RS) spirals are 0.281 nm (that is, one half of the elementary cell vector length), and the height of the square spiral (SS) is 0.562 nm. The steps are made visible to the electron microscope by "decorating"—deposition of gold atoms which preferentially adsorb at the bottom of the steps. [From H. Bethge in *Molecular Processes on Solid Surfaces* edited by E. Drauglis, et al. Copyright 1969. Used with permission of McGraw-Hill Book Company.]

example, the energy $(-E_K)$ necessary to remove an electron from the $K(1\,s)$ level of the fluorine atom is 693 eV. Thus activators with $E_{act} > 693$ eV will cause photoelectrons to be emitted with energy equal to $(E_{act} - 693)$ eV. Observation of photoelectrons with this energy is evidence for the presence of fluorine. When the atom is part of a molecule or is chemically bound to a surface, the core energy levels such as the K level will shift slightly due to the modified electric field of the valence electrons.

The corresponding change in the energies of the emitted electrons is known as the **chemical shift** (not to be confused with the same term used in NMR). The chemical shift can identify the molecule in which an atom is bound. An example of chemical shifts is shown in Figure 20-33.

The identification of surface species must be followed by studies of surface structure. We wish to know (1) whether the surface is rough or smooth to within an atomic diameter (that is, whether it is "atomically smooth"); and (2) the structure—amorphous or crystalline—of the one or more surface phases.

The surface of a single crystal which has been polished by mechanical or electrochemical means appears perfectly smooth to the eye. However, when the surface is viewed under an optical microscope with moderate magnification, surface features as high as 10^{-3} cm (10^4 nm) are revealed. When an electron microscope is pushed to the limits of resolution (5–10 nm or even less laterally; one atomic layer normal to the surface), ledge-dominated landscapes such as the ones shown in Figure 20-34 may appear. The steps leading from one terrace to the next are one or several atoms high. The terraces are probably important for surface reactions because they contain most of the surface atoms. Since the electron microscope does not resolve possible terrace features that may be tens or even hundreds of atoms wide nor does it show the ordering of surface atoms, another technique is needed.

The most important method for this purpose at present is **low energy** (<500 eV) **electron diffraction** (**LEED**). The diffracted electrons are the elastically scattered fraction of a beam that impinges normal to the surface. The basis of the method is the wavelike behavior of an electron (Chapter 18). According to the de Broglie relation (page 538) a 150-eV electron has a wavelength of about 0.1 nm. If on the surface there are domains that contain ordered rows of atoms (or other bases), these will act as diffraction gratings in two dimensions to produce, on a fluorescent screen, bright spots due to the diffracted beam. If the domains are less than about 5 nm in diameter, the spots may be too diffuse to observe. The minimum spot size, determined by the coherence of the incident electron beam, is reached when the domains have diameters of about 50 nm or larger. Many surfaces of metals and semiconductors with or without an adsorbate monolayer have been found to produce diffraction spots (Figure 20-35). Therefore, such surfaces, largely or totally, consist of ordered domains (see Table 20-13). They also could contain low concentrations of various imperfections. Although high probability of multiple scattering in LEED complicates the detailed interpretation of the diffraction patterns, nevertheless, the elementary cells (Chapter 17) of many crystalline surface phases have been identified and their spatial relation to the bulk phase cells established.

In order to understand these results we have to learn the language that crystallographers use to describe crystal surfaces. Consider an *ideal* surface created by the imaginary cleaving of an ideal (perfectly ordered) crystal between two atomic planes so that the newly exposed planes do not change their positions relative to planes in the bulk (Figure 20-36). There are only five different patterns in which the surface atoms (or other bases such as ions or molecules) may be arranged. These five patterns—the two-dimensional (Bravais) lattices and their elementary cells—are shown in Figure 20-37. The LEED pattern of platinum in Figure 20-35 demonstrates that in one surface of a platinum crystal the atoms are arranged according to the hexagonal lattice.

(a)

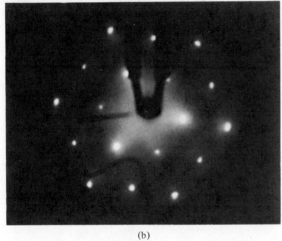

(b)

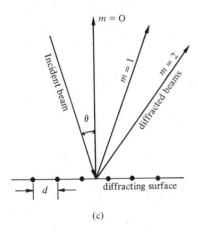

(c)

Figure 20-35 LEED patterns of the (111) face of platinum single crystal. The incident electron beam energies, E, are (a) 51 and (b) 181 eV. (c) The main features of the pattern can be understood qualitatively with the help of the Bragg equation (Eq. 18-47). [From G. A. Somorjai: *Principles of Surface Chemistry.* © 1972. Reprinted by permission of Prentice-Hall, Inc., Englewood Cliffs, N.J.]

Table 20-13 Structures of adsorbates on surfaces of platinum single crystals

Adsorbate	Temperature (°C)	Structure of Clean Surface		
		Pt(111)-(1×1)	Pt(100)-(5×1)	Pt(stepped) (see Figure 20-40)
H_2	25	disordered	disordered	disordered
	600–1000			2(1 dim.)-H
O_2	25	disordered	(2×2)-H	2(1 dim.)-O
	600–800			disordered + PtO_2
$H_2 + O_2$		($\sqrt{3} \times \sqrt{3}$)-R30°		
CO	25	c(4×2)-CO, (2×2)-CO	(1×1)-CO, c(4×2)-CO	disordered
$H_2 + CO$	25		c(2×2)-(H_2 + CO)	?
C	700–900	graphitic	graphitic	graphitic
C_2H_2		(2×1)-C_2H_2	c(2×2)-C_2H_2	
C_2H_4	25	(2×2)-C_2H_4	c(2×2)-C_2H_4	
C_3H_6; cis,trans-2-butene; butadiene; isobutylene	700	graphitic carbon disordered	graphitic carbon disordered	poorly ordered (2×2) pseudocubic (6×6)-R13°

SOURCES: G. A. Somorjai, R. W. Joyner and B. Lang: *Proc. Roy. Soc. Lond.*, **A331**:335 (1972); G. A. Somorjai, *Principles of Surface Chemistry*, Prentice-Hall, Englewood Cliffs, N.J., 1972, Table 5.4.

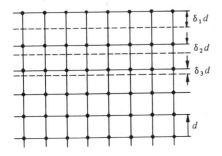

Figure 20-36 Positions of near-surface atomic planes of an ideal (– – – –) and a perfect (——) crystal.

On a *real* surface one might expect some imperfections. Some of the possible imperfections, as they exist in the square lattice, are described in Figure 20-38. Their role in surface reactions, although suspected to be important, is poorly known.

The spatial relation between surface and bulk structures of a single crystal has two aspects: (1) the orientation of the surface plane with respect to the three-dimensional elementary cell of the bulk crystal; and (2) the size, placement, and orientation of the surface elementary cell relative to the bulk cell.

The orientation (and the absolute location, which does not interest us) of a plane can be specified by the coordinates of three noncolinear points lying in the plane. If each point is chosen to be on a different coordinate axis then three numbers are sufficient to specify the orientation because two of the coordinates of a point are automatically zero. In crystallography three numbers known as **Miller indices**

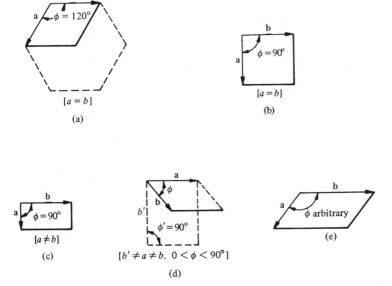

Figure 20-37 Two-dimensional space (Bravais) lattices in order of decreasing symmetry: (a) hexagonal; (b) square; (c) rectangular; (d) centered rectangular; (e) oblique. The solid lines and unprimed quantities refer to primitive (smallest possible) elementary cells. The dashed lines and primed quantities refer to other commonly used elementary cells.

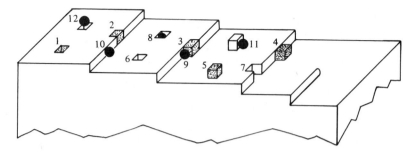

Figure 20-38 A schematic surface of an imperfect crystal. The shaded cubes are atoms of the crystal numbered in order of decreasing stability of their positions. Atom 1, in the surface; 2, in a step; 3, in a step at a kink; 4, self-adsorbed against a step; 5, self-adsorbed on a surface site; 6 and 7 are vacancies. Black spheres are adsorbed atoms numbered in order of decreasing stability of their positions. [From W. J. Dunning in E. A. Flood (ed.): *The Solid-Gas Interface.* Marcel Dekker, 1967, Vol. I, p. 283.)

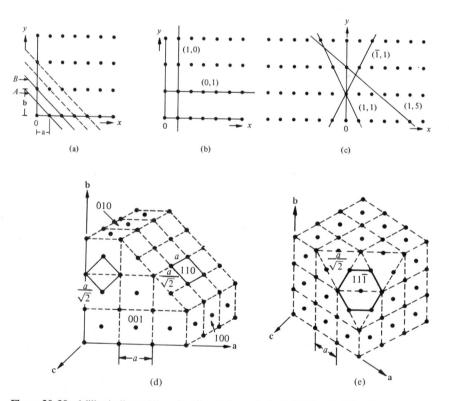

Figure 20-39 Miller indices of lines (h, k) and planes (h, k, l). (a) The $(1, 2)$ family of parallel lines. (b) Lines that belong to the vertical $(1, 0)$ and horizontal $(0, 1)$ families of lines. (c) Lines from three different families. A minus above the number [as in $(\bar{1}, 1)$] designates a negative intercept. (d) and (e) show the Miller indices of the common surface planes of a face-centered cubic crystal. The solid lines outline the elementary cells of the substrate surface lattices.

specify, with respect to the elementary cell, the orientation of a family of parallel planes such as crystal faces which contain the same arrangement of atoms or other bases.

For simplicity let us first consider the determination of the two Miller indices of a line passing through a two-dimensional lattice. The procedure is as follows (see Figure 20-39a): (1) select the origin 0 of the coordinate axes (which are parallel to the vectors **a, b**) at the corner of a cell; (2) draw a line (B) of the desired orientation as *near* as possible to the origin (but not through it) so that it intercepts each coordinate axis at a lattice point; (3) measure the nonzero coordinates x', y' of the intercepts using the vector lengths a, b as units; (4) calculate the Miller indices (h, k) from $(1/h) : (1/k) = x' : y' = (1/y') : (1/x')$ (for line B $h = y' = 1$, $k = x' = 2$). The indices (h, k) are the smallest possible integers that obey the condition that $1/h, 1/k \leqq 1$, where $1/h, 1/k$, are the intercept coordinates of a line (A in Figure 20-39a) which passes through the elementary cell next to the origin. If the line is parallel to one of the coordinate axes, say x, then the intercept $x' = \infty$. In this case the relation in step (4) does not give a unique result. We interpret the relation so that it gives the smallest possible integers for h, k that is, $(1/h) : (1/k) = (1/0) : (1/1) = x' : y' = \infty : 1$ (Figure 39b). Examples of the Miller indices for lines with various orientations are shown in Figure 20-39c.

The Miller indices of a plane or a set of parallel planes ($h, k, 1$) are determined in a manner entirely analogous to that of a line, the expression in step (4) being replaced by $(1/h) : (1/k) : (1/1) = x' : y' : z'$. Examples for a face-centered cubic (fcc, see Figure 17-4b) crystal are shown in Figures 20-39d and e.

It will be apparent from Figure 20-39 that the density of atoms is greatest on lines and planes with all Miller indices 0 or 1 and that the density progressively decreases as the average Miller index of a face increases. Since for the faces with the greatest density of atoms the number of nearest neighbors is most similar to that of the bulk crystal, these should be the most stable faces. Although crystals have a high activation energy for rearrangement and therefore do not, in general, have equilibrium faces and shapes, very large departures from equilibrium are not too likely. LEED studies confirm this. Most crystal surfaces are largely made up of low Miller index faces. Surfaces with high Miller indices often turn out to be **vicinal** surfaces (Figure 20-40), that is, surfaces made up of small terraces and steps composed of two or more low index faces.

Real crystalline surfaces fall in two classes with respect to the size and in-plane orientation of their elementary cells relative to the bulk cell. The surfaces of the first class are formed when the near-surface layers of atoms relax in a direction normal to the surface (Figure 20-36). Such surfaces are simply projections of the bulk structure onto the surface (see Figures 20-39d and e). Such surfaces are said to be "perfect" or to have the "substrate structure." The surfaces of the second class have structures that have resulted, apart from relaxation perpendicular to the surface, from a rearrangement of the surface atoms within the surface plane. However, frequently the elementary cells of the rearranged surface bear a simple relation to the (hypothetical) substrate structure cell. Therefore a single system of description can be applied to both classes of surface structures.

A clean surface with substrate structure is designated (1×1). It is specified by the chemical formula of the substrate for example, Pt, and the Miller indices of the surface plane, for example, (111). Thus we have Pt(111)-(1×1) (Figure 20-39e). If

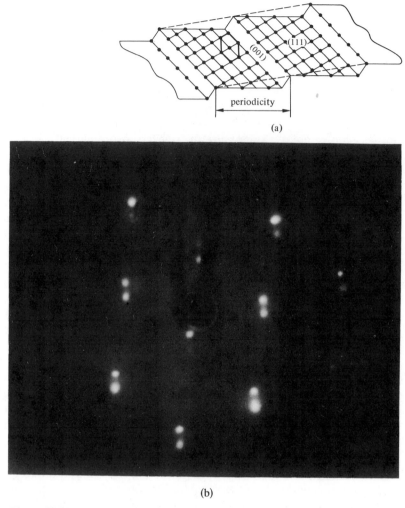

(a)

(b)

Figure 20-40 A vicinal surface of platinum. (a) Schematic representation showing a high Miller index plane (dashed lines) which consists of terraces—the (111) surfaces—separated by steps—the (001) surfaces. (b) Diffraction pattern of the surface in (a). [From G. A. Somorjai: *Principles of Surface Chemistry.* © 1972. Reprinted by permission of Prentice-Hall, Inc., Englewood Cliffs, N.J.]

the surface contains an adsorbate monolayer that has the substrate structure, we add the chemical formula (A) of the adsorbate to the substrate structure designation, that is, (1×1)-A. (For example, Pt(100)-(1×1)-CO; see Table 20-13.) From the description and from a knowledge of the bulk structure (which one can look up, given the chemical formula of the substrate), it is possible to determine the type of elementary cell and its dimensions. This is best done with the help of models or drawings such as are shown in Figures 20-39d and e.

The elementary cells of the structure on a rearranged surface frequently have dimensions a, b that are integral multiples n, m of the (hypothetical) substrate structure cell with dimensions a_s, b_s, that is, $a/a_s = n$, $b/b_s = m$. Such a rearranged structure is designated $(n \times m)$, for example, Pt(100)-(5×1). (Further examples are

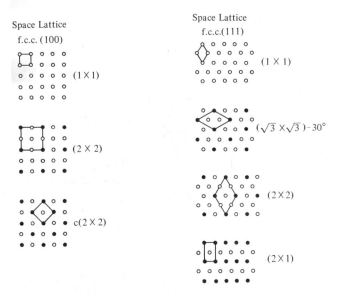

Figure 20-41 Schematic representations of some of the surface structures listed in Table 20-13. Solid dots represent the adsorbed atoms or molecules.

shown in Figure 20-41 and in Table 20-13.) The diffraction pattern from the Pt(100)-(5×1) surface is shown in Figure 20-42a. Since the diffraction pattern does not readily give information on the arrangement of atoms in a direction normal to the surface, an unambiguous assignment of the structure of layers immediately below the top one is not possible. Such layers are of interest because they may have been created by a "buckling" of the original surface layer. One possible rearrangement involving buckling that would lead to the observed periodicity of the diffraction pattern is shown in Figure 20-42b.

In addition to changes in cell dimensions rearrangement of the surface atoms may lead to a rotation of the elementary cell with respect to the (hypothetical) substrate structure cell. The rotation is specified by the symbol R followed by the number of degrees through which rotation has occurred, for example, Pt(100)-(2×2)-R45°-C_2H_2 also designated Pt(100)-c(2×2)-C_2H_2 (see Figure 20-41 and Table 20-13).

Next, we make some general observations about the structures of adsorbate-covered surfaces as determined in LEED experiments. These structures are the intermediates of heterogeneous reactions. Knowledge of them, therefore, suggests possible reaction mechanisms for surface reactions. Furthermore, it informs us about some important factors—adsorbate–adsorbent and lateral interactions—that govern sorption.

Even though most surface reactions have been studied at higher ambient pressures than those used in LEED experiments, there is no reason to expect that these higher pressures change the structures radically. Table 20-13 lists some of the structures that have been observed at various ambient pressures and temperatures on several kinds of surfaces of platinum. Similar, although not identical, results have been obtained for other face-centered cubic crystals of metals (Ni, Pd, Cu, Rh). The conclusions about these structures given below apply, in addition, to results from

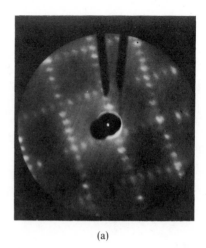

(a)

Top Views

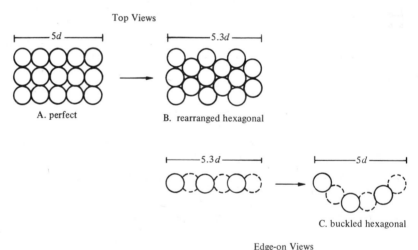

Edge-on Views

(b)

Figure 20-42   The Pt(100)-(5×1) surface. (a) Diffraction pattern at 124 eV. (b) The Pt(100)-(5×1) surface may consist of a buckled hexagonal surface layer, C, of Pt atoms. The schematic rearrangement of the (100)-layer, A, of the face-centered cubic Pt crystal to the layer C through an intermediate arrangement B is shown. [Part (a) from G. A. Somorjai: *Principles of Surface Chemistry*. © 1972. Reprinted by permission of Prentice-Hall, Inc., Englewood Cliffs, N.J.]

studies of body-centered metallic crystals (W, Ta, α-Fe, Mo) and crystals with the diamond structure (Si, C, Ge).

The ordered structures, closely related to the substrate structure, exhibited by the layer of most small chemisorbed molecules (see Table 20-13) testify to the existence of definite adsorption sites largely determined by the adsorbent. The forces responsible for the sites are probably best considered as modifications of the forces that form the chemical bonds in the corresponding gaseous metal hydrides and oxides or organometallic compounds. The modification is, of course, due to the surface structure that surrounds an atom or atoms at a site. These modifications are responsible for the variation of the adsorbate structure with the crystal face (Table 20-13) and hence for one kind of surface heterogeneity. A rather more important kind of surface heterogeneity for surface reactions (see Section 20-8.5) seems to be that of the stepped surface (Figure 20-40) where one can expect a gradual increase of the adsorbate–adsorbent bond strength as one proceeds from the bottom of one step to the top of the next one below it (see Figure 20-38).

Lateral interactions between adsorbate molecules can explain at least three features of adsorbate structures. The small size (Table 20-13) of the adsorbate elementary cells (wherein the adsorbate molecules are close-packed or nearly so) results from mutual attractions between adsorbate molecules. The disordered "structures" of the larger molecules probably have a similar cause: the interactions are large enough to prevent the reorientation that could lead to an ordered structure. The coadsorbed structures (for example, $H_2 + O_2$ and $H_2 + CO$ in Table 20-13) that would be stabilized by interactions due to induced polarization are obvious precursors to reaction. They suggest that reaction may be enhanced by conditions that favor the simultaneous adsorption of both species.

LEED has also been used to observe surfaces during oxide, carbide, and nitride formation from solid metals. For example, when a clean metal surface is dosed with oxygen, a new surface structure usually appears. As the temperature is increased, the clean metal structure reappears. The oxygen has not been desorbed. Rather, it has diffused into the bulk. Further dosing with oxygen at the higher temperature may repeat the cycle (as with nickel) or the diffusion may be so fast that the surface intermediate does not have time to form (as with aluminum). Finally, as the solubility limit of oxygen in the bulk metal is reached, one observes the formation of a permanent metal oxide structure. The new surface structures that form upon dosing with oxygen are of two kinds: (1) reconstructed surfaces (these have an ordered surface which seems to consist of both oxygen and metal atoms), for example, nickel plus oxygen; and (2) amorphous surfaces, for example, aluminum and oxygen. Further understanding of these processes awaits the determination of the atomic structure of near-surface layers.

The present knowledge of the structures of both clean and adsorbate-covered homogeneous and heterogeneous surfaces is clearly going to be enormously important to the understanding of surface reaction mechanisms (see Section 20-8.5). However, one should remember that the surface structures we have described are those with high surface concentrations. Reactive intermediates at very low concentrations are important for homogeneous reactions. Similarly, unstable surface structures (Figure 20-38), present at very low concentrations, may be important for heterogeneous reactions.

20-8.5 Mechanisms of Catalytic Surface Reactions   The mechanisms of catalytic surface reactions may be studied by determining the concentrations of surface intermediates or of reactants or products in the ambient gas or liquid solution as a function of time. These two approaches differ sufficiently from the corresponding methods for homogeneous reactions (Sections 20-3, 20-6, and 20-7) to merit separate discussion.

The measurement of surface concentrations as a function of time is a new but actively developing art. Generally, concentrations can be measured, with low accuracy, over a hundredfold range. Reliable measurement of reaction rates, especially those of fast reactions (see Section 20-6), is therefore not practical. However, one can gain a rough idea of the major surface species that participate in various stages of a reaction.

For example, the platinum-catalyzed dehydrocyclization reaction

$$n\text{-}C_7H_{16} \rightarrow C_7H_8 + 4\,H_2$$

has been studied on various platinum samples in a high vacuum chamber equipped for Auger electron spectroscopy, LEED, and mass spectrometry. Thus, surface composition and structure as well as composition of the ambient gas can be determined after each admission of n-heptane. The reaction is carried out at a temperature of 250–350°C and a total pressure of about 10^{-4} torr in the presence of added hydrogen (which, when coadsorbed on the surface, presumably suppresses total dehydrogenation of the n-heptane). The results in Table 20-14 indicate that

Table 20-14 Results on the reaction $n\text{-}C_7H_{16} \xrightarrow{Pt} C_7H_8 + 4\,H_2$

Surface Structure		Time (min) to Maximum Concentration of C_7H_8	Relative Concentrations[a] of C_7H_8 After Successive Doses of n-C_7H_{16}		
Initial, Clean	Formed During Reaction		First Dose	Second Dose	Third Dose
Pt, stepped,[b] (111)-terraces	graphitic C	4	2.6	2.4	1.8
Pt, stepped,[b] (100)-terraces	disordered C	8	2.6	1.4	0.8
Pt(111)	disordered C	12	0.4	0.2	0
Pt, polycrystal	disordered C	8	0.3	0	0

[a] Platinum surface area for all samples is about 1 cm^2.
[b] See Figure 20-40.
SOURCE: G. A. Somorjai and R. W. Joyner: *Proc. Roy. Soc. Lond.*, **A331**:335 (1972).

toluene forms only on hexagonally symmetric surfaces [that is, Pt(111)-(1×1), Pt(100)-(5×1), and graphitic carbon]. The surfaces that are being covered with an amorphous carbon layer gradually loose their reactivity. The high initial reactivity of the stepped surfaces indicates a role, in dehydrogenation perhaps, for the steps.

Undoubtedly, improved versions of experiments similar to the one just described will contribute much to the understanding of surface reactions.

From a knowledge of the time-dependent concentrations of reactants or products in the ambient gas or liquid solution, one can determine the rate law of a heterogeneous reaction in precisely the same way as for homogeneous reactions (Section 20-2). The derivation of the rate law from a postulated mechanism is also fundamentally the same. However, it is necessary to consider the definition of various reaction rates and the rate expression for the elementary surface reactions. The latter are assumed to be either unimolecular or bimolecular, the expressions being

$$r_s = -\frac{dN_{B,\,reacted}}{dt} = k\theta_B \text{(unimolecular)} \tag{20-150}$$

and

$$r_s = k\theta_B\theta_D \text{ (bimolecular)} \tag{20-151}$$

where θ_B and θ_D are the fractions of surface covered by two reactants B and D. The units of r_s are usually moles per second. The surface rate r_s is related to the heterogeneous reaction rate

$$r_h = r_s/VM \tag{20-152}$$

where V is the volume of the ambient gas or solution and M is the mass (or moles or surface area) of the adsorbent.

The mole conservation equation $-dN_{B,\,reacted} = dN_{B,\,ambient} + dN_{B,\,adsorbed}$ implies that changes in ambient concentrations measure the heterogeneous reaction rate only if the adsorbed B is in a steady state or if $N_{B,\,ambient} \gg N_{B,\,adsorbed}$. In either case $-dN_{B,\,reacted} \approx dN_{B,\,ambient}$, and therefore

$$r_h \approx r/M \tag{20-153}$$

where r is the ordinary homogeneous reaction rate defined in Eq. (20-4) or (20-5). Eqs. (20-152) and (20-153) relate the measured rate r (in the ambient gas or solution) to the rate r_s of the postulated surface reaction.

To treat the adsorption and desorption stages of a postulated mechanism the Langmuir model has been assumed in the overwhelming number of cases. This may seem surprising in view of the deficiencies of the Langmuir model. However, if the reaction occurs at a single class of sites, which may not be too close together (such as the steps on metallic surfaces), the Langmuir model may work rather well for the reaction even though it may work poorly for adsorption in general. Another reason for the popularity of the Langmuir model is the absence of compelling alternatives.

For most heterogeneous catalysis mechanisms (apart from mass transfer controlled radical recombination reactions (Section 20-8.2), the rate-determining steps are surface reactions.

A simple mechanism for which the rate law can be derived readily is the unimolecular surface reaction scheme

$$R+S \underset{k_-}{\overset{k_+}{\longrightarrow}} RS \overset{k_1}{\longrightarrow} P \tag{20-154}$$

where S, RS are adsorption sites and adsorbates, respectively. The steady state expression for the heterogeneous reaction rate of the adsorbate species, according to the Langmuir model (see Problem 44), is

$$r_{RS} = k_+ R(1-\theta) - k_-\theta - k_1\theta = 0 \tag{20-155}$$

For the product P the heterogeneous reaction rate r_h is obtained from Eq. (20-155) as

$$r_h = k_1\theta = k_1k_+R/(k_- + k_1 + k_+R) = k_1K'R/(1+K'R) \tag{20-156}$$

where the dimensions of k_1 are amount of P per volume of ambient gas (or solution) per mass (or amount or surface area) of adsorbent per time and $K' = k_+/(k_- + k_1)$. If

$k_- > k_1$, then $K' \approx K$ where K is the Langmuir equilibrium constant (Eq. 20-141). Since assumption of adsorption–desorption equilibrium leads to Eq. (20-156) with K' replaced by K, the rate law does not discriminate between the steady state and the equilibrium mechanisms unless K can be measured independently.

Eq. (20-156) and the Langmuir adsorption isotherm (Eq. 20-131) have identical forms. Therefore both equations reduce to the same limiting forms and can be subjected to the same testing methods. Among the reactions that obey mechanism (20-154), the most important are enzyme-catalyzed reactions. Indeed, an enzyme molecule with its one or at most a few identical adsorption sites conforms perfectly to the Langmuir model. [In enzymology mechanism (20-154) is named after Michaelis and Menten who proposed it before (1913) Langmuir! In the Michaelis–Menten equation $(K')^{-1}$ is known as the Michaelis constant. The catalyst mass becomes the total enzyme concentration.] Other reactions that obey Eq. (20-156) are some decompositions of hydrides on metal surfaces. Phosphine on tungsten displays both limiting behaviors (see Eqs. 20-132 and 20-133) around 700°C. Below 10^{-2} torr, the rate law is first order in PH_3; above 1 torr it is zero order. For intermediate pressures Eq. (20-156) must be used.

Other, related unimolecular mechanisms can be readily treated. For example, if the product P is also adsorbed, we get the rate law

$$r_h = k_1 K'_R R / (1 + K'_R R + K'_P P) \tag{20-157}$$

Thus, if $(1 + K'_P P) > K'_R R$ the rate law reduces to

$$r_h = k_1 K'_R R / (1 + K'_P P) \tag{20-158}$$

that is, the product inhibits the reaction. For example, the decomposition of nitric oxide on platinum is inhibited by the product oxygen. In enzyme reactions the inhibitors frequently are substances other than products.

The simplest observed bimolecular mechanism is the Langmuir–Hinshelwood mechanism

$$\left. \begin{array}{c} A + S \underset{k_-}{\overset{k_+}{\rightleftharpoons}} AS \\[2ex] B + S \underset{k'_-}{\overset{k'_+}{\rightleftharpoons}} BS \end{array} \right\} \xrightarrow{\ k_1\ } P + 2\,S \tag{20-159}$$

which leads to the rate law

$$r_h = k_1 K'_A K'_B AB / (1 + K'_A A + K'_B B)^2 \tag{20-160}$$

This rate reaches a maximum when $K'_A A = K'_B B + 1$ if the concentration of A is varied while that of B is held constant. The decrease at high concentrations of A or B occurs because the composition of the adsorbate layer approaches that of pure A or pure B. The reaction between cyclopropane and hydrogen on metals of group VIII behaves according to the Langmuir–Hinshelwood mechanism.

This concludes our examination of heterogeneous reactions, of their mechanisms, of the basic parameters of their elementary reactions, and of surface intermediates. Further understanding of heterogeneous reactions, as of homogeneous ones, will come from the theoretical and experimental studies of elementary reactions described in the next two chapters.

Problems

1: Obtain the expression in Table 20-1 for the rate constant of an nth order ($n \neq 1$) reaction from the rate law $r = kA^n$.

2. At 504°C dimethyl ether undergoes the decomposition

$$(CH_3)_2O \rightarrow CH_4 + H_2 + CO$$

The total pressure of the reacting system, measured as a function of time by Hinshelwood and Askey [18], is

time/(s)	0	390	777	1195	3155	∞
pressure/(torr)	312	408	488	562	779	931

(a) Calculate the partial pressures of $(CH_3)_2O$ and CH_4 at each time, t.
(b) Determine the order and the rate constant for the decomposition.

3. For the reaction

$$Co(NH_3)_6^{3+} + Cr(H_2O)_6^{2+} + Cl^- \rightarrow Co(NH_3)_6^{2+} + Cr(H_2O)_5Cl^{2+}$$

the rate constants at 25 and 37°C are 0.74 and 1.70 liter^2 mole^{-2} min^{-1}. Calculate (a) the activation energy and Arrhenius factor; (b) the approximate temperature at which $k = 10k_{25°C}$.

4. At what temperatures are the half-lives for the decomposition of the following compounds equal to 2 h?

	$\log A/(s^{-1})$	$\mathscr{E}_a/(kcal\ mol^{-1})$
CH_3CH_2Cl	13.45	55.0
$CH_3CHClCH_3$	13.4	50.5
t-butyl acetate	13.34	40.5

5. Gaseous ethylamine decomposes as follows:

$$C_2H_5NH_2 \rightarrow C_2H_4 + NH_3$$

At 500°C and an initial $C_2H_5NH_2$ pressure $P_0 = 55$ torr, in a constant volume vessel, the following data were obtained for the pressure increase, ΔP:

t/(min)	1	2	4	8	10	20	30	40
ΔP/(torr)	5	9	17	29	34	47	52	53.5

(a) Derive expressions, in terms of t, P_0, ΔP, for the one half, first-, and second-order rate constants.
(b) Determine which expression from (a) best fits the data and calculate the rate constant.

6. For the gas phase reaction $2\,NO + 2\,H_2 \rightarrow N_2 + 2\,H_2O$ the following initial rates were measured at 826°C:

$(P_{H_2})_0 = 400$ torr				$(P_{NO})_0 = 400$ torr			
$(P_{NO})_0/(\text{torr})$	359	300	152	$(P_{H_2})_0/\text{torr}$	289	205	147
$-(dP/dt)_0/(\text{torr s}^{-1})$	1.50	1.03	0.25		1.60	1.10	0.79

Determine the order of the reaction with respect to each reactant.

7. Data for the acid hydrolysis of methyl acetate at 25°C

$$CH_3CO_2CH_3 + H_2O \rightarrow CH_3CO_2H + CH_3OH$$
$$(A)$$

have been obtained.
(a) In an aqueous medium initial concentrations are 0.1 M HCl; 52.19 M H_2O; 0.7013 M A

$t/(\text{min})$	200	280	445	620	1515	1705
$(A_0 - A)/(\text{M})$	0.08455	0.1171	0.1727	0.2311	0.4299	0.4588

(b) In an acetone medium initial concentrations are 0.1 M H_2SO_4; 0.933 M H_2O; 2.511 M A

$t/(\text{min})$	60	120	180	240
$(A_0 - A)/(\text{M})$	0.1379	0.2611	0.3589	0.4177

For (a) and (b) determine the order of the reaction and calculate the rate constant, assuming the reverse reaction to be negligible.

8. For the gas phase reaction $CO + Cl_2 \rightarrow COCl_2$ the following data were obtained in a constant volume vessel at 25°C: with $(P_{Cl_2})_0 = 400$ torr and $(P_{CO}) = 4$ torr

$t/(\text{min})$	34.5	69.0	138	∞
$P_{COCl_2}/(\text{torr})$	2.0	3.0	3.75	4.0

and with $(P_{Cl_2})_0 = 1600$ torr and $(P_{CO})_0 = 4$ torr

$t/(\text{min})$	34.5	69.0	∞
$P_{COCl_2}/(\text{torr})$	3.0	3.75	4.0

Determine the order of the reaction with respect to each reactant and calculate the rate constant.

9. Which of the following reactions *must* have a complex mechanism?
(a) $H_2 + Br_2 \rightarrow 2\,HBr$

$$r = k \cdot H_2 \cdot Br_2^{1/2}/(k' + HBr/Br_2)$$

(b) $NH_4^+ + CNO^- \rightarrow (NH_2)_2CO$

$$r = k \cdot NH_4^+ \cdot CNO^-$$

(c) $C_6H_5NH_2 + I_3^- + OH^- \rightarrow IC_6H_4NH_2 + H_2O + 2 I^-$

$$r = k \cdot C_6H_5NH_2 \cdot OH^- \cdot I_3^- / (I^-)^2$$

(d) What are the formulas of the activated complexes in reactions (b) and (c)?

(e) What is the rate-determining step (see Problem 13) and the formula of the activated complex in reaction (a) when the Br_2 concentration is very low?

10. Derive the rate law for the second mechanism in Eq. (20-29).

11. Find the rate of product formation for the mechanism

$$R \xrightarrow{k_0} I_1 \xrightarrow{k_1} I_2 \xrightarrow{k_2} \cdots \xrightarrow{k_{n-1}} I_n \xrightarrow{k_n} P$$

Assume that all intermediates are in a steady state.

12. The rate laws for the isomerization of camphene hydrochloride in the presence and absence of camphene were given in the caption to Figure 20-2a.

(a) Propose a single mechanism for the isomerization to explain the two different observed rate laws. *Hint*: camphene hydrochloride (A) is known to equilibrate with camphene (B), that is, $A \rightleftharpoons B + HCl$.

(b) Express the rate constants in the two empirical rate laws in terms of the constants of the elementary reactions in the proposed mechanism.

13. For the reaction

$$H_2 + Br_2 \rightarrow 2 \, HBr$$

the following chain mechanism has been proposed:

$$Br_2 \xrightarrow{k_0} 2 \, Br$$

$$H_2 + Br \xrightarrow{k_1} HBr + H$$

$$H + Br_2 \xrightarrow{k_2} HBr + Br$$

$$H + HBr \xrightarrow{k_3} H_2 + Br$$

$$2 \, Br \xrightarrow{k_4} Br_2$$

Assuming the steady state approximation for Br and H, derive the experimental rate law given in Problem 9a.

14. What is the pressure of the lower explosion limit for a stoichiometric mixture of H_2 and O_2 gases at 773 K in a KCl-coated spherical vessel with a diameter of 5 cm? For the branching reaction (20-42a), the rate constant is $k_1 = 4.42 \times 10^9 \, cm^3 \, mol^{-1} \, s^{-1}$ and for the wall termination reaction (20-42d) it is $k_4 = 45 \, s^{-1}$.

15. The gas phase near-equilibrium reaction $H_2 + I_2 \overset{K}{\rightleftharpoons} 2 \, HI$ is initiated by

$$I_2 \overset{K_0}{\rightleftharpoons} 2 \, I$$

followed by two parallel mechanisms, A and B

Mechanism A:	Mechanism B:

$$2\,I + H_2 \underset{k_{-1}}{\overset{k_1}{\rightleftharpoons}} 2\,HI \qquad\qquad I + H_2 \underset{k_{-2}}{\overset{k_2}{\rightleftharpoons}} HI + H$$

$$H + I_2 \underset{k_{-3}}{\overset{k_3}{\rightleftharpoons}} HI + I$$

(a) Assuming that the zeroth forward and reverse steps are in equilibrium and that $dH/dt = 0$ show that the net rate for the overall reaction is

$$r = K_0 k_1 \left(H_2 \cdot I_2 - K^{-1}(HI)^2 \right)\left(1 + \frac{k_2}{k_1 K_0^{1/2}(I_2)^{1/2}} \cdot \frac{1}{1 + k_{-2}k_3^{-1} \cdot HI/I_2} \right)$$

Hint: Eq. (20-65) must be used.

(b) Mechanisms A and B are dominant at low and high temperatures, respectively. To what forms does the rate law reduce in the high and low temperature limits?

16. The reaction $HCO_2CH_3 + OH^- \rightarrow CH_3OH + HCO_2^-$ obeys a rate law first order in each of the reactants. The standard enthalpy and entropy of activation have been determined, by invoking the theoretical assumption that $f = kT/h$, as $\mathscr{H}_a^\ominus - \mathscr{H}_R^\ominus = 9.81$ kcal mol^{-1} and $\mathscr{S}_a^\ominus - \mathscr{S}_R^\ominus = -18.4$ eu. What is the rate constant at 25°C? *Note*: The standard states of the reactants and of the activated complex are 1 M concentrations for the enthalpy and entropy and hence for the rate constant.

17. Reactions

$$Cl + H_2 \underset{k_-}{\overset{k_+}{\rightleftharpoons}} HCl + H$$

are among the very few for which independent measurements of k_+, k_-, and the equilibrium constant K exist [19]: $k_+ = 1.2 \times 10^{13} \exp(-4300 \times 1.987^{-1}\,T^{-1})$ cm^3 mol^{-1} s^{-1} and $k_- = 2.3 \times 10^{13} \exp(-3500 \times 1.987^{-1}\,T^{-1})$ cm^3 mol^{-1} s^{-1}. The enthalpy of the reaction at 300 K is $\Delta\mathscr{H}^\ominus = 1.1$ kcal mol^{-1} the entropy is $\Delta\mathscr{S}^\ominus = 0.66$ cal mol^{-1} K^{-1}. Compare

(a) the ratio k_+/k_- with K at 300 K.

(b) $\Delta\mathscr{S}^\ominus$ and $\Delta\mathscr{H}^\ominus$ calculated from k_+/k_- with the measured values.

(c) What is the significance of the results?

18. The surface of Pioneer 10 spacecraft, during its 1974 passage through the radiation belt of Jupiter, received an electron plus proton radiation dose of 3.4×10^6 rad. If mercury were given an instantaneous dose of this magnitude, how much would its temperature rise? (Specific heat of mercury is 139 J kg^{-1}.)

19. An acid solution of $K_3Fe(C_2O_4)_3$ is used as an actinometer to measure the intensity of the light incident upon a vessel. After exposure to light of wavelength 253.7 nm for 10^3 s, the concentration of ferrous iron was found to be 5.0×10^{-5} M. The fraction of light absorbed by the actinometer solution is 0.1. What is the intensity of the incident light?

20. Assuming conservation of momentum and energy, as well as an elastic collision, derive the expression, Eq. (20-70), for the fraction of energy transferred from an activator to a reactant particle.

21. In aqueous solution monochloroacetic acid decomposes both thermally and photochemically as follows:

$$ClCH_2CO_2H + H_2O \rightarrow HOCH_2O_2H + HCl$$

Photolysis of 8.23 ml samples at 2537 Å, 25°C, and an initial monochloroacetic acid concentration of 0.5 M, yielded the following results for two different exposure times, t

t/(min)	837	2774
absorbed energy/(10^8 erg)	3.436	1.190
amount Cl^-/(10^{-5} mole)	2.325	0.863

Separate experiments in the dark showed that 3.5×10^{-10} mol Cl^- min^{-1} is formed thermally [20]. Calculate the quantum yield of the photochemical reaction.

22. A proposed mechanism for the gas phase photolysis of HBr is

$$HBr \xrightarrow{h\nu} H + Br \qquad\qquad r_1 = I_a(\phi \approx 1)$$

$$H + HBr \xrightarrow{k_2} H_2 + Br \qquad\qquad r_2 = k_2 \cdot H \cdot HBr$$

$$Br + Br + M \xrightarrow{k_3} Br_2 + M \qquad\qquad r_3 = k_3 \cdot Br^2 \cdot M$$

$$H + Br_2 \xrightarrow{k_4} HBr + Br \qquad\qquad r_4 = k_4 \cdot H \cdot Br_2$$

With help of the steady state approximation, derive the rate-law expression for the overall yield ($\Phi_{-HBr} = -r_{HBr}/I_a$) of the process.

23. Irradiation of cyclohexane with γ rays produces five radicals per 100 eV. The radicals are scavenged by iodine

$$\cdot C_6H_{11} + I_2 \rightarrow C_6H_{11}I + I$$

The unreacted atoms recombine $I + I \rightarrow I_2$. If a solution of iodine in cyclohexane (2×10^{-4} M) is irradiated with γ rays at a dose rate of 10^{16} eV g^{-1} s^{-1}, what is the concentration of I_2 after 1 min? (Density of cyclohexane is 0.78 g ml^{-1}.)

24. The primary yield of the hydroxyl radical, g_{OH}, can be determined from the final yields of H_2O_2 in ($O_2 + H_2$)-saturated and O_2-saturated waters, $G_{H_2O_2}$ (saturated with $O_2 + H_2$) and $G_{H_2O_2}$ (saturated with O_2).

(a) Taking into account the reactions that occur in ($O_2 + H_2$)-saturated water, that is, reactions (20-85), (20-88), and (20-89), express $G_{H_2O_2}$ (saturated with $O_2 + H_2$) in terms of the primary yields $g_{H_2O_2}$, g_H, $g_{e^-(aq)}$, and g_{OH}.

(b) Combine the result from (a) with Eq. (20-92) to derive an expression for g_{OH} in terms of $G_{H_2O_2}$ (saturated with $O_2 + H_2$) and $G_{H_2O_2}$ (saturated with O_2).

25. In the irradiation of water vapor containing a very small percentage of D_2, the reaction mechanism is believed to be

$$H_2O \xrightarrow{\quad\text{\Large$\leadsto$}\quad} H \rightarrow OH \qquad\qquad \textbf{(a)}$$

$$H_2O \xrightarrow{\quad\text{\Large$\leadsto$}\quad} H_2O^+ + e^- \qquad\qquad \textbf{(b)}$$

$$H_2O^+ + H_2O \rightarrow H_3O^+ + OH \qquad (c)$$

$$e^- + H^+ \rightarrow H \qquad (d)$$

$$OH + D_2 \rightarrow HDO + D \qquad (e)$$

$$H + D_2 \rightarrow HD + D \qquad (f)$$

$$2\,D \rightarrow D_2 \qquad (g)$$

If the yields for reactions (a) and (b) are 4.0 and 3.3 respectively, what is the yield of HD?

26. Lefort and Tarrago [21] irradiated acidic aqueous solutions with 5.3 meV α particles from polonium. In a solution containing 10^{-3} M Fe^{2+} and saturated with O_2 they obtained a final yield $G_{Fe^{3+}} = 5.5$ ions per 100 eV. In a second, degassed solution, which contained Ce^{4+}, the following final yields were determined: $G_{Ce^{3+}} = 3.10$, $G_{H_2} = 1.57$, $G_{O_2} = 1.57$ species per 100 eV. From an independent determination the primary yield of the HO_2 radical is 0.22 molecules per 100 eV.

Since in acidic solutions the hydrated electron is rapidly converted to the H atom via reaction (20-80d) (see text), only $(g_H + g_{e^-(aq)})$ can be determined from this experiment. During the second stage the reactions in the Fe^{2+} solution are

$$H + O_2 \rightarrow HO_2$$

$$HO_2 + Fe^{2+} \rightarrow Fe^{3+} + HO_2^-$$

$$HO_2^- + H^+ \rightarrow H_2O_2$$

$$H_2O_2 + Fe^{2+} + H^+ \rightarrow Fe^{3+} + H_2O + OH$$

$$OH + Fe^{2+} + H^+ \rightarrow Fe^{3+} + H_2O$$

In the Ce^{4+} solution the second stage reactions are

$$H + Ce^{4+} \rightarrow Ce^{3+} + H^+$$

$$H_2O_2 + Ce^{4+} \rightarrow Ce^{3+} + HO_2 + H^+$$

$$HO_2 + Ce^{4+} \rightarrow Ce^{3+} + H^+ + O_2$$

$$OH + Ce^{3+} + H^+ \rightarrow Ce^{4+} + H_2O$$

(a) Express the final yields $G_{Fe^{3+}}$, $G_{Ce^{3+}}$, G_{H_2}, and G_{O_2} in terms of the primary yields $(g_H + g_{e^-(aq)})$, g_{HO_2}, $g_{H_2O_2}$, g_{OH}, g_{H_2}.

(b) Calculate $(g_H + g_{e^-(aq)})$, $g_{H_2O_2}$, g_{OH}, g_{H_2}, and g_{-H_2O}.

27. The mutarotation of α-D-glucose is a reversible reaction, that is

$$\alpha\text{-D-glucose} \underset{k_-}{\overset{k_+}{\rightleftharpoons}} \beta\text{-D-glucose}$$

$$\text{(A)} \qquad\qquad\qquad \text{(B)}$$

which obeys the rate law

$$r = k_+ A - k_- B$$

The reaction may be followed dilatometrically.

(a) Integrate the rate law and show that, assuming a change in volume $(V - V_0)$ proportional to concentration B, a graph of $\ln (V' - V)$ versus t is a straight line with slope $m = -k = -(k_+ + k_-)$. (V is the volume at t and V' is the volume a constant interval t' later, that is, at $t + t'$.)

(b) Determine k from the following data [22] [obtained at 18°C in the presence of the catalysts sodium o-toluate (0.075 M) and o-toluic acid (0.0026 M)]:

$t/(\text{min})$	0	10	21	32	50
V (at t)/(arbitrary units)	4.395	4.76	5.09	5.37	5.72
V' (at $t + t'$)/(arbitrary units)	6.50	6.575	6.63	6.685	6.76

(c) The equilibrium constant for the mutarotation at 18°C is $K = 1.744$. Calculate the rate constants k_+ and k_-.

(d) Why is it advantageous to determine k from a graph of $\ln(V' - V)$ versus t, rather than from a graph of $\ln(V_\infty - V)$ versus t?

28. The oxidation of some aromatic amines by the enzyme ferroxidase (Fer) is mediated by iron. It is believed that the ferroxidase oxidizes Fe(II) to Fe(III) which, in turn, oxidizes the amine. The initial step of the oxidation is complex formation:

$$\text{Fer} + \text{Fe(II)} \xrightarrow{k} \text{Fer} \cdot \text{Fe(II)}$$

Osaki and Walaas [23] studied this reaction by a stopped flow method. The concentration of Fer was measured spectrophotometrically from the absorbance A at 610 nm

$t/(\text{ms})$	0	20	40
A	0.33	0.11	0.078

The initial concentrations of Fer and Fe(II) were 121 and 101 μM. Assuming the complex formation to be an elementary reaction calculate the rate constant k.

29. Consider the complexation reaction

$$M^{n+} + HL \rightleftharpoons ML^{(n-1)+} + H^+$$

The equilibrium constant for this reaction at 25°C is 1.0×10^{-3}. A solution is prepared at 25°C with initial concentrations of M^{n+} of 0.1 M and HL of 0.001 M, buffered at pH = 3.0. The solution is cooled to 15°C and a temperature-jump to 25°C is carried out. The reaction is monitored spectrophotometrically. The following data are obtained for the absorbance A:

$(A - A_{eq})/(\text{arb. units})$	Time/(ms)
0.80	0
0.66	0.4
0.54	0.8
0.43	1.2
0.37	1.6
0.29	2.0
0.25	2.4
0.20	2.8
0.17	3.2

(a) Calculate the forward and reverse rate constants for the reaction.
(b) Predict the relaxation time for this solution adjusted to pH $=6.0$. (The ligand exists in solution only as the fully protonated form indicated above.)

30. Spectrophotometric determination of the rate constant for a second-order reaction of an unstable species—the hydrated electron—from pulse radiolysis data.

(a) The absorptivity a_e^λ of the hydrated electron, $e^-(aq)$, at wavelength λ must be determined first. This has been done [24] from absorption measurements at 366 and 578 nm on pulse-radiolyzed pure water and on an aqueous solution of tetranitromethane, $C(NO_2)_4$, plus sucrose (to suppress some side reactions).

From absorbances, A_e^λ, measured in pulse-radiolyzed pure water one gets the ratio $E = a_e^{366}/a_e^{578} = A_e^{366}/A_e^{578} = 0.140$ (see Eq. 20-95).

In the $C(NO_2)_4$ solution the hydrated electron disappears, to a first approximation, by the reaction

$$e^-(aq) + C(NO_2)_4 \rightarrow C(NO_2)_3^- + NO_2$$

Therefore, the concentration of the nitroformate ion, $C(NO_2)_3^-$ or NF^-, at any time t after a pulse is $C_{NF} = C_{e,0} - C_e$ where $C_{e,0}$ and C_e are the $e^-(aq)$ concentrations at $t=0$ and $t=t$, respectively. It follows that there is a relation between the absorbances and absorptivities of NF^- ($a_{NF}^{366} = 10.2 \times 10^3$, $a_{NF}^{578} = 0 \, M^{-1} \, cm^{-1}$ at 24°C) and $e^-(aq)$. Show that, by assuming Eq. (20-95) for the two species at both wavelengths, the following equation may be derived.

$$\Delta A^{578} = \Delta A^{366}(a_r/(1 - a_r E))$$

where $\Delta A^\lambda = A_t^\lambda - A_{t'}^\lambda (t \neq t')$, and $a_r = a_e^{578}/a_{NF}^{366}$. From the following data determine graphically or by least squares $a_r/(1 - a_r E)$ and thence a_e^{578}:

ΔA^{366}	0.0500	0.100	0.150	0.200
ΔA^{573}	0.0682	0.140	0.211	0.277

(b) When aqueous sodium hydroxide, saturated with H_2 at 100 atm, is pulse-radiolyzed all primary hydrated electrons undergo a single reaction:

$$e^-(aq) + e^-(aq) \xrightarrow{k} H_2 + OH^-$$

Determine the second-order rate constant k from the following spectrophotometric data [25]:

A^{578}	1.0	0.40	0.36	0.25
$t/(\mu s)$	0	11.2	15.3	25.2

Assume that the absorptivity of $e^-(aq)$ is $10.6 \times 10^3 \, M^{-1}$ at 578 nm, and $d = 8$ cm. [This value, unlike the one determined in part (a), has been corrected for the effects of side reactions.]

31. From NMR spectra of $C_6D_{11}H$ (see Figure 20-21) at 60 MHz the following data were collected:

Temperature (°C)	Linewidths/Hz (at half-intensity)	Peak-to-peak Separation/Hz
−48.7	5.2	
−60.3	equatorial and axial peaks just merge	
−63.2		18
−67.3		26
−75.0	4.1	28.7

For each datum choose the appropriate formula from among the peak-to-peak separation (Eq. 20-112), the slow exchange (Eq. 20-114), and the fast exchange (Eq. 20-116) formulas and calculate the rate constants at different temperatures and the activation energy for the inversion $C_6D_{11}H_{ax} \rightarrow C_6D_{11}H_{eq}$.

32. From the line shape expression, Eq. (20-111), derive the expression, Eq. (20-112), for peak-to-peak separation of two NMR lines due to a chemically exchanging proton.

33. Gas phase reactions of atoms with stable molecules may be studied by the continuous and fast flow apparatus [26] shown in the diagram.

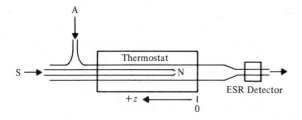

Atoms, for example, oxygen atoms, generated by a microwave discharge enter through A in a stream of He gas. The stable molecules are injected through S, which can be moved in the z direction. After mixing the A and S flows at nozzle N, reaction occurs. The unreacted atom concentration is measured by the ESR detector as a function of the position of the injector nozzle.

To derive an expression for the rate constant of a reaction in a well-mixed reactor consider a cylindrical volume element dV of cross sectional area a containing n_A atoms. The concentration of atoms entering and leaving the volume element is A and $A - dA$. The flow velocity through dV is u. Therefore, the rate at which atoms accumulate in dV due to flow is $ua\,dA$. The specific rate of destruction of A atoms through chemical reaction is r_A, yielding $r_A\,dV$ for the total destruction rate in dV. The net rate at which the amount of atoms in dV increases is

$$dn_A/dt = ua\,dA - r_A\,dV$$

(a) Show that for A independent of time (steady state) and $r_A = kA$

$$-u \frac{d \ln A}{dz} = k$$

Therefore, the rate constant k can be determined from a measurement of $d \ln A/dz$ by means of the apparatus shown in the diagram combined with a measurement of the flow velocity u.

(b) In a flow system that consists of the carrier gas He, mixed with oxygen atoms and SO_3, the following data were collected [27] at 385 K:

total pressure/(torr)	1.35	2.2	2.50	5.06
$-u(d \ln O/dz)/SO_3 / (10^{10} \text{ cm}^3 \text{ mol}^{-1} \text{s}^{-1})$	2.1	3.5	3.9	8.1

Assuming that the concentrations He, SO_3, and O are such that $He \gg SO_3 \gg O$ and that the only reaction in the system is $O + SO_3 + He \xrightarrow{k_0} SO_2 + O_2 + He$ determine k_0.

34. What is the half-life of the reaction between H^+ and OH^- in water at 25°C if the initial concentrations are $H^+ = OH^- = 10^{-3}$ M? (See Table 20-7 for the rate constant.)

35. Show that the hydrocarbon yields, given in Example 20-6, from the photolysis of diazomethane in the presence of n-pentane can be explained by assuming that CH_2 inserts at random in the CH bond of pentane.

36. CIDNP effect.

(a) Derive the expression, analogous to Eq. 20-124, for the ratio of the maximum difference (that is, n_S) in the two singlet-state populations of $R_1 R_2$ formed from $R_1 + R_2$ to the population difference found at equilibrium [that is, $n(\infty)$]. R_1 is a radical with one proton, R_2 has no protons.

(b) Calculate the maximum ratio $n_S/n(\infty)$ for a radical pair with a difference in g-factors $g_1 - g_2 = 10^{-3}$ and a proton-electron spin coupling constant of R_1 $a_1 = 10$ gauss in a magnetic field $H = 1.5 \times 10^4$ gauss (≈ 60 MHz) at 300 K. Assume 10^{-9} s for the lifetime of the radical pair.

37. For the proton NMR (60 MHz) spectra of CH_3CCl_3 and CH_3I in Figure 20-23 use the theoretical expressions given in the text to predict which product will absorb and which will emit, as well as to calculate the theoretical maximum emission and absorption intensities. Assume that the radical pair has a before-reaction lifetime of 10^{-9} s.

38. Consider a two-proton (AX) NMR spectrum of a diamagnetic molecule:

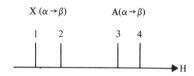

Suppose the molecule decomposes to form two radicals R_1 and R_2. The two protons are on R_1; $g_1 = g_2$; the nuclear-electron spin coupling constants are equal, $a_1 = a_2$. Predict which lines should be observed in absorption (A) and which in emission (E) in the spectra of (a) the recombination product R_1R_2; (b) the scavenging product R_1S.

39. One of several possible termination steps in the $H_2 + O_2$ chain reaction (see Section 20-3) is

$$H + wall \xrightarrow{\ k_4\ } destruction$$

In a spherical vessel the expressions for k_4 are

$$k_4 = \frac{S}{V} \cdot s(RT/2\pi M)^{1/2}$$

for a reaction which is limited by the frequency of reactive wall collisions, and $k_4' = 4\pi^2 D_H/d^2$ for a diffusion-limited reaction. (See Section 20-8.2 for the definition of symbols.) $D_H = 8.6 \times 10^3 \, cm^2 \, s^{-1}$. Which process is rate-determining in a KCl-coated ($s = 1.5 \times 10^{-3}$) spherical vessel with diameter $d = 5$ cm at 773 K and a total pressure of 1 torr?

40. Consider a vessel that contains two gaseous species A and B. The rate constant for the destruction of B at a wall is 10^5 times larger than the rate constant for the diffusion of B to a wall. In a steady state, what will be the ratio of the concentration of B in bulk to the concentration B at a wall?

41. Assuming that the adsorption process $X_2 + 2S \to 2XS$ obeys the Langmuir model, derive the adsorption isotherm, Eq. (20-145).

42. A gram of activated charcoal adsorbs the following volumes (at STP) V of a gas at various ambient pressures P:

$P/(torr)$	2.10	10	20	40	60
$V/(cm^3 \, g^{-1})$	0.392	1.82	3.33	5.71	7.5

Determine the Langmuir constants V_m and K.

43. Use the BET isotherm to determine V_m, c, and the specific surface area, Σ, of a substance from the following data:

x	0.1	0.15	0.25	0.30
$x[V(1-x)]^{-1}/(10^{-3} \, cm^{-3})$	3.77	4.02	4.51	4.76

where $x = P/P_{vap}$ and V is the adsorbed volume at STP of N_2 per gram of the substance. One N_2 molecule occupies a surface area $\sigma = 0.162 \, nm^2$.

44. The Michaelis–Menten mechanism for enzyme-catalyzed reactions is

$$E + S \underset{k_-'}{\overset{k_+'}{\rightleftharpoons}} ES \xrightarrow{\ k_1'\ } E + P$$

where E is the enzyme, S the substrate (or reactant), ES an enzyme-substrate complex, and P the product.

(a) Assuming steady state for ES derive the rate law (the Michaelis–Menten equation) for the formation of P in terms of S and the total enzyme concentration $E_0 = E + ES$.

(b) Express the Michaelis–Menten rate constants in terms of the corresponding rate constants in Eq. (20-156). (Note that $\theta = ES/E_0$.)

45. (a) What is the average linear dimension of Carbolac and or carbon black P-33 particles (see Table 20-10)? Assume solid cubic particles and a density $d = 2.3 \text{ g cm}^{-3}$.

(b) What is the ratio of the number of surface to total number of atoms for the average particle in P-33? Assume 140 pm for the diameter of a carbon atom.

46. The adsorption of H_2 on glass is spontaneous but endothermic ($\Delta \mathcal{H}_{ads} \approx 15 \text{ kcal mol}^{-1}$ near room temperature). Assuming that the H_2 dissociates upon adsorption and that the two H atoms have complete two-dimensional mobility (that is, a two-dimensional ideal gas) on the glass surface, calculate $\Delta \mathcal{S}$ for the reaction. Assume STP for H_2 gas and for the H atoms on the surface. A frequently used STP area covered by a mole of adsorbate species is $3.735 \times 10^7 \text{ m}^2 \text{ mol}^{-1}$ [35, p. 111]. Neglect the rotational and vibrational entropy of H_2. Does this model predict that the adsorption is spontaneous? If not, what additional assumptions can you invoke to explain the spontaneity of the process?

47. The two-dimensional van der Waals equation (Table 20-12) can describe mobile monomolecular layers such as those of fatty acids on water or *n*-heptane on ferric oxide. The two-dimensional pressure $\pi =$ force per unit length where the direction of the force is parallel to the adsorbent surface. For *n*-heptane the two-dimensional van der Waals constants, estimated from measured three-dimensional ones, are $\alpha = 5.5 \times 10^{-28} \text{ erg cm}^2 \text{ mol}^{-1}$ and $\beta = 47 \times 10^{-16} \text{ cm}^2 \text{ mol}^{-1}$.

(a) Plot π versus δ for *n*-heptane at 300, 251.2 (the two-dimensional critical temperature), and 200 K. Compare to the corresponding three graphs for the three-dimensional van der Waals equation.

(b) For *n*-heptane, compare the Langmuir isotherm, Eq. (20-131), to the van der Waals isotherm, Eq. (20-143), at room temperature (300 K) and at "high temperatures" [where the second term in the exponent of Eq. (20-143) is negligible] by plotting θ versus KP for each of the three cases.

48. The heat of adsorption for cesium atoms on a tungsten surface is $\Delta \mathcal{H}_{ads} = -55 \text{ kcal mol}^{-1}$. The activation energy for motion of the cesium atoms on the surface is $\mathcal{E}_a \approx 0.25 \Delta \mathcal{H}_{ads}$. Assuming the constant $\tau_0 = 10^{-13}$ s, calculate for a cesium atom

(a) the average residence time on the tungsten surface;

(b) the average residence time at one surface site when $T = 300$ and 3000 K.

49. (a) Suppose a line were drawn in Figure 20-39a such that it would intercept the *y* axis at $2.5b$ and the *x* axis at $2a$. What are the Miller indices of this line?

(b) Draw lines with the following Miller indices through each of the five two-dimensional space lattices in Figure 20-37: $(0, 1)$, $(2, 1)$, $(2, \bar{1})$, $(\bar{2}, 1)$, $(\bar{2}, \bar{1})$, and $(3, 2)$.

50. Consider the surface planes of a face-centered cubic crystal with Miller indices (310) and (210). Assuming that the surfaces retain the substrate structure, determine the two-dimensional space lattice and the elementary cell dimensions for each of the two planes.

51. An initial rates method (see Section 20-2) is used to determine the kinetic constants of an enzyme-catalyzed reaction under conditions where the concentration of the enzyme-substrate complex is in a steady state. For the decarboxylation of dihydroxyphenylalanine (DOPA) by DOPA decarboxylase the following initial rates were measured [28] as a function of the average concentration of the substrate S (DOPA) during the initial-rate-determination period:

$S/(10^{-4} M)$	20	10	7.5	5	4	2.5
$\dfrac{r}{(\mu \text{ liter } CO_2 \text{ per mg protein per hr})}$	30.4	25.5	21.9	19.2	17.5	13.3

Determine the maximum rate, r_{max}, and the Michaelis constant, K_m, for the reaction from a plot of r^{-1} versus $[S]^{-1}$. *Note*: Assume that the reaction obeys the rate law in Eq. (20-156). Hence, $r = r_h$, $r_{max} = k_1$, $K_m = (K')^{-1}$, and $S = R$. The reciprocal form of Eq. (20-156) (see Eq. 20-134), that is, the equation which is plotted, is known as the Lineweaver–Burk equation.

52. The enzyme (E) pepsinase (from swine kidneys) catalyzes the hydrolysis of the substrate (S) carbobenzoxy-L-glutamyl-L-tyrosine to yield L-tyrosine and a second product (P) carbobenzoxy-L-glutamic acid:

$$E + S \underset{k_-}{\overset{k_+}{\rightleftharpoons}} ES \overset{k_1}{\longrightarrow} E + P + \text{L-tyrosine}$$

The hydrolysis rate is measured manometrically by decarboxylating the L-tyrosine produced.

The product P competes for the enzyme, thus inhibiting the hydrolysis reaction:

$$E + P \overset{K_I^{-1}}{\rightleftharpoons} EP$$

(a) Assuming steady state for ES, derive an expression (see Eq. 20-157) for the rate of hydrolysis in terms of K_I, $K_m = (k_1 + k_-)/k_+$, and $E_0 = ES + EP + E$.

(b) Using plots of $1/r$ versus $1/S$ determine K_I and K_m from the data of Frantz and Stephen [29]:

$r/(\mu M\ CO_2\ \text{min}^{-1})$	43.4	52.6	71.3	111.1	13.3	14.7	26.6	58.1
$S/(\text{mM})$	4.7	6.5	10.8	30.3	4.7	6.5	10.8	30.3
$P/(\text{mM})$	←	— 0.0 —	→		←	—30.3—		→

53. (a) Assuming a steady state for the concentrations AS and BS, derive the rate law in Eq. (20-160).

(b) Show that, for concentration B constant, the rate is a maximum with respect to concentration A when $K'_A A + 1 = K'_B B$.

54. Hinshelwood and Burk [30] obtained the following results for the decomposition of ammonia on a quartz surface:

	$T = 1267$ K		$T = 1220$ K	
$P_0/(\text{torr})$	53.5	137.5	117	298
$t_{1/2}/(\text{s})$	43	44	196	191

(a) What is the order of the reaction? Assuming that the surface reaction is unimolecular, what would the order of this reaction be in the high pressure limit?

(b) What is the activation energy?

55. Yates, Taylor, and Sinfeldt [31] have determined that the heterogeneous catalytic reaction

$$C_2H_6 + H_2 \underset{\text{silica}}{\overset{\text{Ni on}}{\rightleftharpoons}} 2\ CH_4$$

obeys a rate law of the form $r = kp_{Et}^m p_{H_2}^n$ where m and n are integers. At 191°C the following data were obtained:

p_{H_2}/(atm)	0.10	0.20	0.40	0.20	0.20	0.20
p_{Et}/(atm)	0.030	0.030	0.030	0.010	0.030	0.10
r/r_0	3.10	1.00	0.20	0.29	1.00	2.84

where r_0 is the rate when $p_{H_2} = 0.20$ atm and $p_{Et} = 0.030$ atm.

(a) Determine m and n.

(b) Show that these results are consistent with the following mechanism:

$$C_2H_6 \rightleftharpoons (C_2)_{ads} + 3\ H_2$$
$$(C_2)_{ads} + H_2 \rightarrow 2\ CH \qquad \text{slowest step}$$
$$CH + \tfrac{3}{2}H_2 \rightarrow CH_4$$

56. Consider a reaction $A + B \rightarrow P$, which can proceed by a homogeneous or a heterogeneous mechanism. Assume that the heterogeneous reaction is slow enough to make an elementary surface reaction rate-determining, and that both mechanisms have rate-determining steps of the same molecularity. Under these circumstances the activated complex theory (see Section 21.5) predicts that the ratio of the Arrhenius parameters, A_{homo}/A_{hetero}, for the two rate-determining steps is of the order of 10^{12}. What must the differences in activation energies of the two steps be to make their rates equal at 300 and at 1000 K?

_____ *References and Recommended Reading*

[1] J. L. Bada, et al.: *Science*, **184**:791 (1974).

[2] S. G. Canagaratna: *J. Chem. Educ.*, **50**:200 (1973).

[3] S. W. Benson: *The Foundations of Chemical Kinetics*. McGraw-Hill, New York, 1960. A thorough survey of classical kinetics.

[4] M. H. Back and K. J. Laidler: *Selected Readings in Chemical Kinetics*. Pergamon Press, 1967. Twelve classic papers on chemical kinetics.

[5] W. J. Moore: *Physical Chemistry*, 4th ed. Prentice-Hall, Englewood Cliffs, N.J., 1972.

[6] J. H. Sullivan: *J. Chem. Phys.*, **46**:73 (1967).

[7] J. O. Edwards, E. F. Greene, and J. Ross: *J. Chem. Educ.*, **45**:381 (1968).

[8] K. K. Foo and C. H. Yang: *Combustion and Flame*, **17**:223 (1971).

[9] R. C. Tolman: *Principles of Statistical Mechanics.* Oxford University Press, Oxford, 1938.

[10] D. K. Bohme, et al.: *J. Chem. Phys.,* **58**:3504 (1974).

[11] J. G. Calvert and J. N. Pitts, Jr.: *Photochemistry.* Wiley, New York, 1964.

[12] E. J. Hart and M. Anbar: *The Hydrated Electron.* Wiley-Interscience, New York, 1970.

[13] R. K. Wolff, M. J. Bronskill, J. E. Aldrich, and J. W. Hunt: *J. Phys. Chem.,* **77**:1350 (1973).

[14] C. H. Bamford and C. F. H. Tipper (eds.): *Comprehensive Chemical Kinetics.* Elsevier, Amsterdam, 1969. A multivolume work.

[15] E. Wasserman, et al.: *J. Am. Chem. Soc.,* **92**:7491 (1970).

[16] S. M. Claesson (ed.): *Fast Reactions and Primary Processes in Chemical Kinetics.* Wiley-Interscience, New York, 1967. Fifth Nobel Symposium.

[17] C. A. Tolman and J. P. Jesson: *Science,* **181**:501 (1973).

[18] C. N. Hinshelwood and P. J. Askey: *Proc. Roy. Soc.,* **A115**:215 (1927).

[19] A. A. Westenberg and N. deHaas: *J. Chem. Phys.,* **48**:4405 (1968).

[20] R. N. Smith, et al.: *J. Am. Chem. Soc.,* **61**:2299 (1939).

[21] M. Lefort and X. Tarrago: *J. Phys. Chem.,* **63**:833 (1959).

[22] J. N. Brønsted and E. A. Guggenheim: *J. Am. Chem. Soc.,* **49**:2554 (1927).

[23] S. Osaki and O. Walaas: *J. Biol. Chem.,* **242**:2653 (1967).

[24] J. Rabani, W. A. Mulac, and M. S. Matheson: *J. Phys. Chem.,* **69**:53 (1965).

[25] M. S. Matheson and J. Rabani: *J. Phys. Chem.,* **69**:1324 (1965).

[26] A. A. Westenberg and N. deHaas: *J. Chem. Phys.,* **50**:707 (1969).

[27] A. A. Westenberg and N. deHaas: *J. Chem. Phys.,* **62**:725 (1975).

[28] J. H. Fellman: *Enzymologia,* **20**:366 (1959).

[29] I. D. Frantz, Jr. and M. L. Stephenson: *J. Biol. Chem.,* **169**:359 (1947).

[30] C. N. Hinshelwood and R. E. Burk: *J. Chem. Soc.,* **127**:1105 (1925).

[31] D. J. C. Yates, W. F. Taylor, and J. H. Sinfeldt: *J. Am. Chem. Soc.,* **86**:2996 (1964).

[32] A. W. Adamson: *Physical Chemistry of Surfaces,* 2nd ed. Wiley-Interscience, New York, 1967. Thorough survey of macroscopic surface studies.

[33] E. F. Caldin: *Fast Reactions in Solution.* Wiley, New York, 1964.

[34] F. S. Dainton: *Chain Reactions,* 2nd ed. Methuen, London, 1966.

[35] J. H. de Boer: *The Dynamical Character of Adsorption,* 2nd ed. Oxford University Press, 1968.

[36] A. R. Denaro and G. G. Jayson: *Fundamentals of Radiation Chemistry.* Butterworth, London, 1972.

[37] D. N. Hague: *Fast Reactions.* Wiley-Interscience, New York, 1971.

[38] K. J. Laidler: *Reaction Kinetics.* Pergamon, Oxford, 1963. Two volumes.

[39] L. D. Spicer and B. S. Rabinovitch: *Ann. Rev. Phys. Chem.,* **21**:349 (1970). Reviews the $H_2 + I_2$ reaction.

[40] G. A. Somorjai: *Principles of Surface Chemistry.* Prentice-Hall, Englewood Cliffs, N.J., 1972. Survey of recent results on reactions, composition, and structure of surfaces.

[41] J. M. Thomas and W. J. Thomas: *Introduction to the Principles of Heterogeneous Catalysis.* Academic Press, New York, 1967. Thorough survey of macroscopic studies of surfaces and surface reactions.

[42] J. C. Treacy and F. Daniels: *J. Am. Chem. Soc.,* **77**:2033 (1955).

[43] H. R. Ward: *Accounts Chem. Res.,* **5**:18 (1972). Chemically induced dynamic nuclear polarization (CIDNP).

[44] R. P. Wayne: *Photochemistry.* Butterworth, London, 1970.

[45] G. C. Webb and S. J. Thomson: *Heterogeneous Catalysis.* Wiley, New York, 1968.

[46] A. Weissberger (ed.): *Techniques of Chemistry,* 3rd ed. Wiley-Interscience, New York, 1974. Volume VI: *Investigation of Rates and Mechanisms of Reactions.* Part I, E. S.

Lewis (ed.): General considerations and reactions at conventional rates. Part II, G. G. Hammes (ed.): Investigation of elementary reaction steps in solution and very fast reactions.

[47] M. F. R. Mulcahy: *Gas Kinetics.* Wiley, New York, 1973.

[48] R. Weston and H. Schwartz: *Chemical Kinetics.* Prentice-Hall, Englewood Cliffs, N.J., 1972.

[49] A. Ault: *J. Chem. Educ.,* **51**:381 (1974). An introduction to enzyme kinetics.

[50] J. E. Finholt: *J. Chem. Educ.,* **45**:394 (1968), errata *ibid.,* p. 790. The temperature-jump method for the study of fast reactions.

[51] C. E. Hecht: *J. Chem. Educ.,* **39**:311 (1962). Chemical kinetics and thermodynamic consistency.

[52] W. S. Herndon: *J. Chem. Educ.,* **41**:425 (1964). Kinetics in gas-phase stirred flow reactors.

[53] L. S. Lobe and C. A. Bernardo: *J. Chem. Educ.,* **51**:723 (1974). Adsorption isotherms and surface reaction kinetics.

[54] B. Rabinovitch: *J. Chem. Educ.,* **46**:263 (1969). The Monte Carlo method—plotting the course of complex reactions.

[55] C. W. Ryun: *J. Chem. Educ.,* **48**:194 (1971). Steady-state and equilibrium approximations in chemical kinetics.

[56] P. D. Bartlett and I. Pöckel: *J. Am. Chem. Soc.,* **60**:1585 (1938).

[57] D. H. Turner, et al.: *J. Am. Chem. Soc.,* **94**:1554 (1972).

[58] R. R. Rye: *Accounts Chem. Res.,* **8**:347 (1975). Crystallographic dependence in the surface chemistry of tungsten.

chapter 21

Statistical Theories of Kinetics

21-1 INTRODUCTION

The theories of time-variant (nonequilibrium) systems, in particular of reacting systems, are not as advanced as the theories of time-invariant (equilibrium) systems. A comparison of the utility to chemists of the different theories will bear this out.

Whereas equilibrium thermodynamics is sufficiently developed to serve as one theoretical cornerstone of chemistry, irreversible thermodynamics is rarely used by chemists. (Recent important advances in the latter field suggest that this situation may soon change.) Whereas quantum chemistry of stationary states in its twofold role as the interpreter of spectroscopic and other experimental results and as the provider of more and more accurate ab initio molecular properties has been another cornerstone of chemistry, until recently theoretical reaction kinetics has made little use of molecular dynamics (which predict the relative motions of atoms within interacting-reacting molecules). Whereas statistical thermodynamics accurately calculates the macroscopic properties of gases from a knowledge of the energy levels of the constituent molecules, statistical theories of reaction rates are forced to assume very simple molecular models and to borrow heavily, without too much justification, from models of stable molecules.

The three theories most widely used to calculate Arrhenius parameters for elementary reactions are **collision theory** (Section 21-2), **activated-complex** or **absolute-rate** or **transition-state theory** (Section 21-5), and the **Rice–Ramsberger–Kassel–Marcus theory** of unimolecular reactions (Section 21-6). Other theories are discussed in Laidler [8]. Unlike equilibrium thermodynamics, there is no macroscopic theory that provides generally useful relations for the calculation of one macroscopic variable (for example, the reaction rate) given another. (See, however, Section 20-4 for some relations with limited utility.)

All three of the theories mentioned are statistical. This means that they must make assumptions about models for reacting molecules and about the form of the distribution (partition) functions. It is these assumptions that limit the areas of application and the accuracy of results.

Most activation energies of elementary reactions have been determined experimentally (Section 20-2.3) However, in some cases, semiempirical activation energies can be obtained from enthalpies (for endothermic reactions) or energy–bond order relations. Such methods [6] will not be discussed here. Purely theoretical, ab initio calculations of activation energies are making good progress, at least for the simpler reactions of light atoms (Section 21-3).

The theoretical interpretation of Arrhenius activation energy $\mathscr{E}_a = -\boldsymbol{R}(d \ln k/d(1/T))$, evaluated at temperature T, is based on an equation first derived by Tolman in 1920 from statistical mechanical considerations. At constant volume this equation is

$$E_a = E_T(R^\ddagger) - E_T(R)$$
$$= \Delta E_0 + \Delta E(T) \tag{21-1}$$

where $R^\ddagger$ represents those reactant molecules which can react. R symbolizes all reactant molecules. $E_T(R^\ddagger)$, $E_T(R)$ are the average energies of $R^\ddagger$ and R, respectively. At thermal equilibrium, these energies are the usual thermodynamic internal energies of $R^\ddagger$ and R. The second line of Eq. (21-1) shows that the activation energy may be decomposed into a temperature-independent part ΔE_0 (the difference in the internal energy at 0 K) and a temperature-dependent remainder, $\Delta E(T)$.

If the Arrhenius parameters and activation energies of all elementary reactions which have been assumed to make up a complex reaction are known, the assumptions can be tested by comparing the experimental rate law of the complex reaction to a calculated one. This is one way in which the results of this chapter may help to determine mechanisms.

An elementary reaction is the sum of all reactive encounters between reactant molecules. The elementary rate constant is an average over all encounters (Section 21-3). A reactive encounter can be a complicated physicochemical event involving transfer of energy to the reactant molecule(s) or from the product molecule(s) as well as rearrangement of chemical bonds (Section 21-4).

The complexity of elementary reactions has led to an unfortunate inconsistency in defining the molecularity of an elementary reaction. Consider an isomerization, for example, $CH_3NC \rightarrow CH_3CN$, and an atom combination, for example, $I + I \rightarrow I_2$. The isomerization is invariably called "unimolecular." The atom combination is frequently called "termolecular," because it requires a *third body* M to deexcite and stabilize the product, that is, $I + I + M \rightarrow I_2 + M$. However, the isomerization too requires a "second" body to energize the reactant so that it can break old and form new chemical bonds; that is, $CH_3NC + M \rightarrow CH_3CN + M$. In this chapter molecularity of an elementary reaction will refer to the number of reactant molecules necessary to form the products; that is the isomerization is unimolecular and the atom combination is bimolecular.

As the student proceeds further into this chapter he should be prepared to review his knowledge of statistical mechanics and molecular energy levels and (from physics) his knowledge of the mechanics of collisions between billiard balls.

21-2 COLLISION THEORY

Collision theory can predict the rate constants for elementary bimolecular reactions in the gas or liquid phase. It assumes that a collision between a pair of molecules can lead to reaction. In its simpler forms—those discussed in this section, which are based on classical mechanics—it further assumes that a reaction will always occur if the initial relative speed, v_0, of the two molecules equals or exceeds a limit, v_{0m}, and if, for the given speed v_0, their centers of mass approach within a distance, R, less than or equal to a limit, R_m. The relative motion of the molecules is, the theory further assumes, influenced by a spherically symmetric intermolecular potential $V(R)$.

21-2.1 Relative Motion of Two Molecules We wish to describe the relative motion of two molecules A and B, with masses m_A and m_B, in greater detail. The result (introduced for the two atoms of a diatomic molecule on page 415) that the relative motion of A and B is equivalent to the motion of a single particle with (reduced) mass $\mu = m_A m_B/(m_A + m_B)$ about a center of force at $R = 0$ applies to two molecules as well as to two atoms. Furthermore, for $R > R_m$ we assume that the collision is elastic; that is, that the energy E and the angular momentum, $\mathbf{L}$, of the relative motion of A and B are conserved. (Once $R \leqq R_m$, the reaction begins. The conserved quantities now are the total energy and angular momentum, which include the internal-rotational, vibrational, and electronic energy and angular momentum of the reactants. Simple collision theory cannot explicitly treat the reaction itself.) The conservation of $\mathbf{L}$ implies that the particle with mass μ moves, while $R > R_m$, in a plane perpendicular to the initial $\mathbf{L}$. This two-dimensional motion can be described by the two coordinates and velocities defined in Figure 21-1.

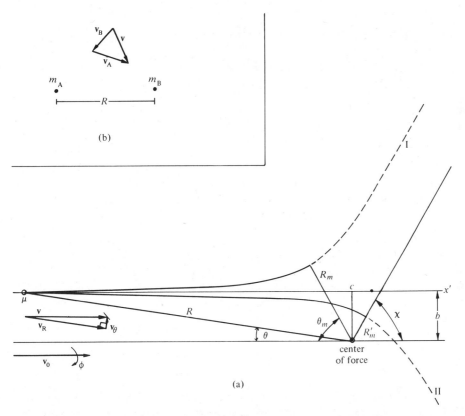

(b)

(a)

Figure 21-1 (a) Coordinates R, θ, and velocity components $\mathbf{v}_R$, $\mathbf{v}_\theta$ (parallel and perpendicular, respectively, to R) for μ particle motion; b, the impact parameter. Trajectories I and II of μ particles with identical impact parameters b and initial velocities $\mathbf{v}_0$ which are scattered elastically by repulsive and attractive, respectively, Coulomb forces. R_m, R'_m are the distances of closest approach (at the so-called classical turning points) for trajectory I and II, respectively. For trajectory I: θ_m is the angular coordinate of the classical turning point; $\chi = \pi - 2\theta_m$, ϕ, are the (elastic) scattering angles. If reaction occurs at R_m or R'_m the trajectory will not continue along the dashed lines.

(b) The coordinate R and velocities $\mathbf{v}_A$ and $\mathbf{v}_B$ of two particles A and B, the relative motion of which is represented by the μ-particle motion in (a) with velocity $\mathbf{v} = \mathbf{v}_A - \mathbf{v}_B$.

The energy E of the μ particle is equal to the kinetic energy associated with the two coordinates R and θ plus the potential energy $V(R)$. Remembering that the angular velocity $\dot{\theta} = d\theta/dt = v_\theta/R$ and designating initial quantities (when R is large and $V = 0$) by the subscript "0" the energy E is variously written as

$$
\begin{aligned}
E = E_0 = E_0(v_0) &= \tfrac{1}{2}\mu v_0^2 \\
&= \tfrac{1}{2}\mu v^2 + V(R) \\
&= \tfrac{1}{2}\mu(v_R^2 + v_\theta^2) + V(R) \\
&= \tfrac{1}{2}\mu(v_R^2 + R^2\dot{\theta}^2) + V(R)
\end{aligned}
\tag{21-2}
$$

These expressions are valid at any time before the reaction. The first line of Eq. (21-2) states that E is conserved; that is, that it remains constant, equal to the total initial energy E_0, and that E_0 is a function only of the initial relative speed v_0.

The magnitude of the angular momentum L is

$$
L = \mu R v_\theta = \mu R^2 \dot{\theta}
\tag{21-3}
$$

By conservation of angular momentum it is a constant. The following argument gives an expression for L in terms of the initial parameters v_0, b (Figure 21-1). If $V(R)$ were equal to zero for all R, the μ particle would follow a straight line path along xx' in Figure 21-1. At point c it would have $R = b$ and $v_\theta = v_0$. Therefore from Eq. (21-3)

$$
\begin{aligned}
L = L_0 &= \mu v_0 b \\
&= (2\mu E_0)^{1/2} b
\end{aligned}
\tag{21-4}
$$

where the second line results when v_0 is substituted from Eq. (21-2). Although Eq. (21-4) was derived for a particular potential $V(R)$, it must give L_0 for any potential because the angular momentum is conserved for any potential. Eq. (21-4) suggests the utility of the impact parameter b: for a given initial energy E_0 it is an easily visualized measure of the angular momentum L.

The constraints due to conservation of energy and of angular momentum can be combined in a single expression. From Eqs. (21-3) and (21-4) we get

$$
\dot{\theta} = d\theta/dt = L_0/(\mu R^2) = v_0 b/R^2
\tag{21-5}
$$

Eq. (21-5) when substituted into Eq. (21-2) yields

$$
\begin{aligned}
E_0 &= \tfrac{1}{2}\mu v_R^2 + L_0^2/(2\mu R^2) + V(R) \\
&= \tfrac{1}{2}\mu v_R^2 + (\tfrac{1}{2}\mu v_0^2)b^2/R^2 + V(R) \\
&= \tfrac{1}{2}\mu v_R^2 + E_0 b^2/R^2 + V(R)
\end{aligned}
\tag{21-6}
$$

The first term on the right-hand-side of Eq. (21-6) is the radial kinetic energy, the second term is the angular (rotational) kinetic energy, written in three ways by means of Eq. (21-4).

Eq. (21-6) can be solved for v_R to yield, after simplifying

$$
v_R = dR/dt = \pm v_0(1 - b^2/R^2 - V(R)/E_0)^{1/2}
\tag{21-7}
$$

Given the potential $V(R)$, the initial velocities or their equivalent (note that $\dot{\theta}_0 = L_0/\mu R_0^2 = v_0 b/R_0^2$), and the initial coordinates R_0 and θ_0 at $t = 0$, Eqs. (21-5) and (21-7) can be integrated to give the coordinates of the trajectory, R and θ, as a function of time. The differential equation that determines the trajectory directly is obtained if Eq. (21-7) is divided by Eq. (21-5).

$$dR/d\theta = v_R/\dot{\theta} = \pm R^2/b(1-b^2/R^2 - V(R)/E_0)^{1/2} \qquad \textbf{(21-8)}$$

The plus and minus signs in Eqs. (21-7) and (21-8) apply, respectively, to the outgoing and incoming trajectories.

Consider the qualitative features of μ-particle trajectories that are determined by the three factors appearing in Eq. (21-8): the initial energy E_0, the impact parameter b, and the potential energy $V(R)$. In the limit when $V(R)$ is small compared to the initial kinetic energy, that is, $V(R)/E_0 \approx 0$ everywhere along the line xx' in Figure 21-1, then the particle trajectory is the straight line xx'. The interesting trajectories are the ones for which at some point $V(R) \approx E_0$. Two such trajectories, for an attractive and a repulsive Coulomb potential $[V(R) = ZZ'e^2/R]$, are shown in Figure 21-1.

The trajectories in a potential that combines a weak long range attraction with a strong short range repulsion (for example, the frequently used Lennard-Jones potential, page 12) are shown for three different values of the impact parameter in Figure 21-2. Especially for trajectories with a large impact parameter (for example b_3 in Figure 21-2) an apparent repulsive force arises as a consequence of simultaneous angular momentum and energy conservation. Eq. (21-6) shows this most clearly. The angular kinetic energy, $E_0 b^2/R^2$, must increase as R decreases due to conservation of angular momentum. Energy conservation requires that this increase be compensated by a decrease in $V(R)$ or in the radial kinetic energy, $\frac{1}{2}\mu v_R^2$. Since kinetic energy of a particle is decreased if it encounters a repulsive force, the term $E_0 b^2/R^2$ acts, in effect, as a repulsive potential along R. Therefore, the last two terms of Eq. (21-6) may be combined to yield an effective potential

$$V_{\text{eff}}(R) = E_0 b^2/R^2 + V(R) = L_0^2/(\mu R^2) + V(R) \qquad \textbf{(21-9)}$$

The effective potential for which $V(R)$ is the Lennard-Jones potential (Eq. 1-21) is plotted in Figure 21-3 for various values of L_0. We see from the figure that at high values of L_0 the effective potential becomes entirely repulsive, whereas at lower values it has a maximum. The maximum is known as the **centrifugal barrier**.

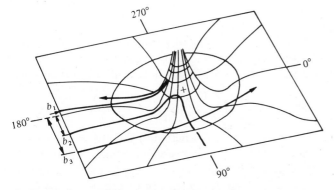

Figure 21-2 Schematic short-range-repulsive-long-range-attractive potential energy surface for μ-particle motion. Three representative trajectories are shown. The + indicates the center of force ($R = 0$).

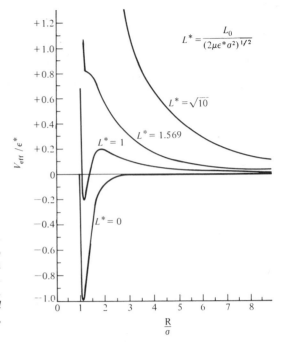

Figure 21-3 The effective Lennard-Jones potential (Eq. 1-21, substituted into Eq. 21-9) for various values of reduced angular momentum L^*. [From J. O. Hirschfelder, C. F. Curtiss, and R. B. Bird: *Molecular Theory of Gases and Liquids.* John Wiley & Sons, New York, 1954, p. 554.]

Introducing $V_{\text{eff}}(R)$ from Eq. (21-9) into Eq. (21-7), we have

$$v_R = \pm v_0 (1 - V_{\text{eff}}(R)/E_0)^{1/2} \qquad (21\text{-}10)$$

This is an equation for the motion of a particle in a single dimension. The minimum value of R, at the *classical turning point,* is R_m. It occurs when $v_R = 0$, or $V_{\text{eff}}(R_m)/E_0 = 1$. This result can be combined with Eq. (21-9) and solved for R_m to yield

$$R_m = b/(1 - V(R_m)/E_0)^{1/2} \qquad (21\text{-}11)$$

We see that R_m (but not θ_m, see Figure 21-1) is determined, for a given E_0, solely by the value of the potential at R_m. If this potential is repulsive, $R_m > b$; if it is attractive, $R_m < b$.

Depending upon the value of E_0, R_m will be located at different points of the $V_{\text{eff}}(R)$ curve. Different trajectories will result. The three major types of trajectories are shown in Figure 21-4. Collisions of types (b) and (c) are frequently called "close." Collision theory sometimes assumes that, of all possible collisions, only close collisions lead to reaction.

21-2.2   Reaction Cross section   We are now ready to derive collision theory rate constant expressions for a bimolecular reaction between molecules A and B. The rate constant is connected to the preceding results on collision trajectories via a very important quantity (see Section 18-5)—the **reaction cross section**, S_r, which is defined as

$$S_r(E_0) = \pi [b_m(E_0)]^2 \qquad (21\text{-}12)$$

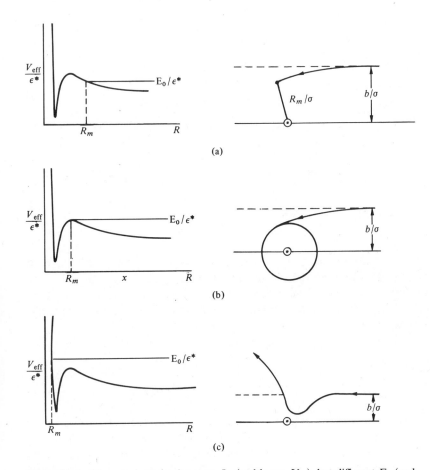

(a)

(b)

(c)

Figure 21-4   Three trajectories for same L_0 (and hence V_{eff}), but different E_0 (and hence b). [From R. Weston and H. Schwarz: *Chemical Kinetics.* © 1972. Reprinted with permission of Prentice-Hall Inc., Englewood Cliffs, N. J.]

Here b_m is the maximum impact parameter, at a given E_0, for which reaction can occur.

The meaning of S_r is best seen from Figure 21-1. If molecule A is the incoming "projectile" and B is the "target" which sits at the center of force, then S_r is the target area, in a plane perpendicular to v_0, that molecule A must hit in order to react.

The reaction cross section has the simplest form for a barrier potential similar to the hard sphere potential (Figure 1-5a):

$$V = E_{0m} \qquad \text{if } R \leqq R_m \qquad\qquad \textbf{(21-13)}$$

where E_{0m} is a constant.

We assume that reaction occurs whenever $R \leqq R_m$ and that $V(R > R_m)$ has no effect on the reaction. Our potential simulates fairly well the strong and steep repulsion that occurs between most stable molecules as they approach closely (Figure 1-5c) and primarily determines their reactivity.

Assuming that R_m in Eq. (21-13) is identical with the classical turning point radius, we substitute $V(R_m) = E_{0m}$ into Eq. (21-11) to obtain the maximum impact parameter as a function of E_0

$$b_m = R_m(1 - E_{0m}/E_0)^{1/2}$$
$$= \tfrac{1}{2}(\sigma_A + \sigma_B)(1 - E_{0m}/E_0)^{1/2} \qquad \textbf{(21-14)}$$

We have assumed that the distance of closest approach R_m is the average of the molecular diameters σ_A and σ_B. The error due to this rather arbitrary assumption should not be more than a factor of two. The increase in b_m with increasing E_0 reflects the decreasing importance of the centrifugal barrier. The reaction cross section is, from Eqs. (21-12) and (21-14)

$$S_r(E_0) = \tfrac{1}{4}\pi(\sigma_A + \sigma_B)^2(1 - E_{0m}/E_0) \qquad E_0 \geqq E_{0m}$$
$$S_r(E_0) = 0 \qquad\qquad\qquad\qquad\qquad\quad E_0 < E_{0m} \qquad \textbf{(21-15)}$$

The molecular diameters are usually taken to be those determined from hard sphere formulas at low temperatures (Table 3-2). Taking $\sigma_A = \sigma_B = 3$ Å $= 0.3$ nm, we calculate a maximum cross section (when $E_{0m} = 0$) $S_r(E_0) = 7$ Å$^2 = 0.07$ nm$^2 \approx$ 0.1 nm^2. We shall use the latter value for comparison with S_r calculated in other ways or determined experimentally.

21-2.3 The Rate Constant

The rate constant k is calculated in two steps beginning with Eq. (21-12). First, consider all collisions with the same $v_0 = (2E_0/\mu)^{1/2}$ (where $v_0 \geqq v_{0m}$). The rate at which the projectiles A hit their targets B is equal to the volume per time, $v_0 S_r$ (see Figure 2-10), multiplied by the projectile (with velocity v_0) concentration $A(v_0)$, that is, $v_0 S_r A(v_0)$ since all such projectiles within a cylindrical volume $v_0 S_r t$ next to their targets will hit the latter during the time t. The reaction rate is $v_0 S_r A(v_0)$ multiplied by the number of targets per unit volume $NB(v_0)$

$$r(v_0) = (Iv_0 S_r N)A(v_0)B(v_0)$$
$$= k(v_0)A(v_0)B(v_0) \qquad \textbf{(21-16)}$$

where N is Avogadro's number and B is the target concentration when the relative velocity is v_0. The constant $I = 1$ if the A and B molecules are different and $I = \tfrac{1}{2}$ if A and B are the same. In the latter case the constant I corrects for counting each collision twice: once when a molecule is a projectile and a second time when it is a target.

Although expression (21-16) is useful for the interpretation of specialized experiments where the relative translational velocity of reactants is selected (see Section 22-2), in the usual macroscopic reacting system a variety of relative speeds v_0 are present. Eq. (21-16) must be averaged over the distribution of speeds. If $N'(v_0)$ is the number of molecules with relative speeds v_0 and N' is the total number of molecules, the fraction of molecules with speeds in the range v_0 to $v_0 + dv_0$ is

$$\frac{dN'(v_0)}{N'} = f(v_0)\, dv_0 \qquad \textbf{(21-17)}$$

where $f(v_0)$ is the distribution function for relative molecular speeds (see Eqs. 2-29,

2-30, 2-31, and 2-38). Therefore, the rate constant averaged over all speeds is, from Eq. (21-16)

$$k = IN \int_0^\infty v_0 S_r(v_0) f(v_0) \, dv_0$$

$$= IN \int_{v_{0m}}^\infty S_r(v_0) v_0 f(v_0) \, dv_0 \tag{21-18}$$

where the second equality follows from $S_r(v_0 < v_{0m}) = 0$ (see Eq. 21-15). In terms of E_0, Eq. (21-18) is (see Section 4-1)

$$k = IN \int_{E_{0m}}^\infty S_r(E_0)(2E_0/\mu)^{1/2} f(E_0) \, dE_0 \tag{21-19}$$

Although a reacting system is not at equilibrium, collision theory frequently assumes that the distribution function $f(v_0)$ is the same as that for equilibrium systems, that is, Maxwellian (see Eq. 2-50).

$$f(v_0) = (2/\pi)^{1/2} (\mu/kT)^{3/2} v_0^2 e^{-\mu v_0^2 (2kT)} \tag{21-20}$$

Alternatively, the distribution function can be expressed in terms of E_0 (see Eq. 4-3)

$$f(E_0) = 2\pi^{-1/2} (1/kT)^{3/2} E_0^{1/2} e^{-E_0/kT} \tag{21-21}$$

Expressions for the barrier potential rate constant are obtained by substituting Eqs. (21-15) and (21-21) into Eq. (21-19) [or alternatively, Eqs. (21-15) and (21-20) into Eq. (21-18)]

$$k = \tfrac{1}{4}\pi(\sigma_A + \sigma_B)^2 IN \int_{E_{0m}}^\infty (1 - E_{0m}/E_0)(2E_0/\mu)^{1/2} f(E_0) \, dE_0$$

$$= \tfrac{1}{4}\pi(\sigma_A + \sigma_B)^2 IN[(8/\pi\mu)^{1/2}(1/kT)^{3/2} \int_{E_{0m}}^\infty (1 - E_{0m}/E_0)E_0 e^{-E_0/kT} \, dE_0] \tag{21-22}$$

If $E_{0m} = 0$, the integral on the first line of Eq. (21-22) becomes equal to the expression for the average relative speed $\bar{v}_0$ and the rate constant is

$$k = \tfrac{1}{4}\pi(\sigma_A + \sigma_B)^2 IN\bar{v}_0$$

$$= \tfrac{1}{4}\pi(\sigma_A + \sigma_B)^2 IN(8kT/\pi\mu)^{1/2} \tag{21-23}$$

where the formula for $\bar{v}_0$ is from Eq. 2-55. For a barrier potential cross section of 0.1 nm^2, $T = 300 \text{ K}$, and a relative molecular mass $M = N\mu = 2.5 \text{ g mol}^{-1} = 0.0025 \text{ kg mol}^{-1}$, Eq. (21-23) yields $k/I \approx 10^8 \text{ mol}^{-1} \text{ m}^{-3} \text{ s}^{-1} = 10^{11} \text{ mol}^{-1} \text{ dm}^{-3} \text{ s}^{-1}$. According to the barrier potential plus collision theory model, this is the maximum possible rate constant for the given values of S_r, T, and M.

If $E_{0m} \neq 0$, the integral in the brackets of Eq. (21-22) can be evaluated by a change of variables to $x = E_0/kT$. In terms of x the quantity in the brackets reduces to two standard integrals (see Appendix II) multiplied by constants:

$$(8kT/\pi\mu)^{1/2} \left(\int_{x_{0m}}^\infty x e^{-x} \, dx - x_{0m} \int_{x_{0m}}^\infty e^{-x} \, dx \right)$$

$$= (8kT/\pi\mu)^{1/2} (|-e^{-x}(x+1)|_{x_{0m}}^\infty - x_{0m}|-e^{-x}|_{x_{0m}}^\infty)$$

$$= (8kT/\pi\mu)^{1/2} e^{-x_{0m}} = \bar{v}_{0m} e^{-E_{0m}/kT} \tag{21-24}$$

Substitution of the result from Eq. (21-24) into Eq. (21-22) shows that the rate constant for $E_{0m} \neq 0$ is simply the rate constant for $E_{0m} = 0$, Eq. (21-23) multiplied by an exponential factor:

$$k = \tfrac{1}{4}\pi(\sigma_A + \sigma_B)^2 lN(8kT/\pi\mu)^{1/2} e^{-E_{0m}/kT} = BT^{1/2} e^{-\mathscr{E}_a'/RT} \qquad \textbf{(21-25)}$$

This expression has the near-Arrhenius form of the empirical Eq. (20-16) with $\mathscr{C}/R = \tfrac{1}{2}$ and the activation energy $\mathscr{E}_a' = NE_{0m}$. As shown by Eqs. (20-17) and (20-18) Eq. (21-25) can also be expressed, over a narrow temperature range, in the Arrhenius form.

Although our derivation of Eqs. (21-23) and (21-25) has been specific for reaction between A and B, it should not be too hard to see that these equations apply to any process that depends upon the rate at which A and B collide so that their centers come within a critical distance $[\leq (\sigma_A + \sigma_B)]$. For example, in Section 21-4 we will apply Eqs. (21-23) and (21-25) to the rate of vibrational energy transfer between molecules.

We have seen that collision theory predicts an empirically acceptable form for the rate constant. The theory fares less well in quantitative tests. It is not possible to compare theoretical and experimental activation energies because collision theory does not predict a value for the theoretical $\mathscr{E}_a$. Theoretical and experimental Arrhenius factors, however, may be compared. As an example consider two reactions ($\mathscr{E}_a$ given in calories per mole)

$$\text{H} + \text{H}_2 \rightarrow \text{H}_2 + \text{H} \qquad k = 10^{10.5} e^{-7500/1.99T} \text{ dm}^3 \text{ mol}^{-1} \text{s}^{-1} \quad \textbf{(21-26)}$$

and

$$\text{C}_6\text{H}_5 + \text{H}_2 \rightarrow \text{C}_6\text{H}_6 + \text{H} \qquad k = 10^{7.7} e^{-6500/1.99T} \text{ dm}^3 \text{ mol}^{-1} \text{s}^{-1} \quad \textbf{(21-27)}$$

According to Eq. (20-18), the Arrhenius factor $A = B(eT)^{1/2}$. We calculated previously a value of $BT^{1/2} = k$ (from Eq. 21-23) $= 10^{11} \text{ dm}^3 \text{ mol}^{-1} \text{s}^{-1}$. Since $e^{1/2} = (2.72)^{1/2} = 10^{0.22}$, we see that the A value for reaction (21-26) is rather low but, nevertheless, it can be readily explained by a distance of closest approach, R_m, somewhat smaller than that calculated from hard sphere diameters. The low Arrhenius factor of reaction (21-27) is impossible to understand in terms of Eq. (21-25). The reduced mass $\mu = 2 \times 77/(2 + 77) \approx 2$ of the reaction is smaller than that used to calculate the comparison value of $BT^{1/2}$ and therefore cannot be used to explain a low value of A. The only alternative explanation—a small R_m—would require the unreasonably low value of 0.1 Å.

The preceding example can be generalized: many bimolecular gas phase reactions between neutral species that require activation energy (and even some which do not) have Arrhenius factors that are too low to be explained by collision theory. The theory may be "saved" by multiplying the right-hand-side of Eq. (21-25) by a "fudge" factor P, known as the **steric factor.** The factor P is supposed to be equal to the fraction of molecular collisions in which the molecules A and B possess the relative orientations necessary for reaction. The problem with P is twofold: (1) there is no known method whereby it can be calculated reliably; and (2) it probably oversimplifies the actual situation (see Section 21-3).

21-2.4 Reactions without Activation Energy
Collision theory is most consistently successful for reactions without activation energy that occur between simple

species whose cross sections are governed by attractive potentials. Examples of such reactions are ion–ion reactions, $I_2^+ + I_2^-$ (see Problem 4), and ion–molecule reactions, such as $O_2^+ + N_2 \rightarrow NO^+ + NO$. These reactions are important in upper atmosphere chemistry, radiation chemistry, electrical discharge, flame reactions, and so on.

To derive cross sections for the reactions mentioned, consider potentials of the form

$$V(R) = a/R^n \qquad n > 2 \qquad \text{(21-28)}$$

For these potentials a centrifugal barrier can exist. To derive the cross section, we assume that reaction occurs for all distances of closest approach, R, which are inside the centrifugal (Figure 21-3) barrier. Therefore R_m is defined by differentiating the equation that results from the substitution of Eq. (21-28) into Eq. (21-9) and setting the derivative to zero. This yields

$$\left[\frac{dV_{\text{eff}}}{dR} \right]_{R=R_m} = 0 = -2E_0 b_m^2 / R_m^3 + na / R_m^{n+1}$$

or

$$2E_0 b_m^2 / R_m^2 = na / R_m^n \qquad \text{(21-29)}$$

Eq. (21-29) can be combined with Eq. (21-11) to eliminate R_m and to find b_m as an explicit function of E_0. First, from Eqs. (21-11) and (21-28), we get by substitution and squaring

$$b_m^2 / R_m^2 - 1 = (a/E_0)(1/R_m)^n \qquad \text{(21-30)}$$

Next, Eq. (21-29) may be substituted into Eq. (21-30) to get

$$b_m^2 = R_m^2 \cdot n/(n-2) \qquad \text{(21-31)}$$

or, finally from Eqs. (21-29) and (21-31)

$$R_m^n = (a/E_0)(n-2)/2 \qquad \text{(21-32)}$$

Eq. (21-32) is solved for R_m, and the result is substituted into Eq. (21-31) to yield

$$b_m^2 = \left(\frac{n}{n-2} \right) \left(\frac{a(n-2)}{2E_0} \right)^{2/n} \qquad \text{(21-33)}$$

Therefore the cross section is, from Eq. (21-12)

$$S_r(E_0) = \pi \left(\frac{n}{n-2} \right) \left(\frac{a(n-2)}{2E_0} \right)^{2/n} \qquad \text{(21-34)}$$

To illustrate the use of Eq. (21-34) we derive an expression for the rate constant between an ion and a molecule with a dipole moment induced by the ion. The potential (see the discussion on intermolecular forces on pages 9–13) for this case is

$$V(R) = -Z^2 e^2 \alpha / (2R^4) \qquad \text{(21-35)}$$

where Ze is the charge of the ion and α is the polarizability of the molecule. Therefore $a = -Z^2 e^2 \alpha / 2$, $n = 4$ and Eq. (21-34) becomes

$$S_r(E_0) = 2\pi Ze(\alpha/2E_0)^{1/2} \qquad \text{(21-36)}$$

Eq. (21-36) is substituted into Eq. (21-19). The result, with $E_{0m} = 0$, is easily integrated to yield the rate constant

$$k = 2\pi Ze(\alpha/\mu)^{1/2}N \tag{21-37}$$

Unlike all previous rate constant expressions, this one does not contain $\bar{v}_0$. That is, the rate constant is independent of the initial relative translational energy because its effect, on S_r, due to the centrifugal barrier, (Eq. 21-36) cancels exactly the effect on the rate constant (Eq. 21-19). This occurs only for a R^{-4} potential. Table 21-1 compares some theoretical and experimental rate constants for ion–molecule reactions. There is rather good agreement with the exception of one reaction that again needs a steric factor in the theoretical expression for k.

Table 21-1 Rate constants for some exothermic ion–molecule reactions

Reaction	$\alpha/(10^{-24}\ cm^3)$	Log $(k/(dm^3\ mol^{-1}\ s^{-1}))$	
		Observed	Calculated from Eq. (21-37)
$O_2^+ + NO \rightarrow NO^+ + O_2$	1.9	11.7	11.7
$O^+ + H_2 \rightarrow OH^+ + H$	0.8	12.1	12.0
$O^+ + N_2 \rightarrow NO^+ + N$	2.0	8.9	11.8
$O^- + H_2 \rightarrow H_2O + e^-$	0.8	12.0	12.0
Comparison k (Eq. 21-23) from barrier potential			11

SOURCE OF DATA: V. Cermak et al.: *Ion–Molecule Reactions.* John Wiley, New York, 1970.

21-2.5 Reactions in Liquids

Collision theory of chemical reactions in liquids differs considerably from that for gaseous reactions. Whereas in a gas near room temperature and atmospheric pressure an average molecule will travel about 10^{-5} cm in about 10^{-9} s between collisions (page 74), a molecule in a liquid interacts continuously with its nearest neighbors. These neighbors, of which there may be as many as 12,[1] form a cage around the molecule. The cage creates a potential well, within which the molecule can move back and forth. We may assume that this motion is equivalent to a rather soft vibration with a period of 10^{-13} to 10^{-12} s in each of three mutually perpendicular directions. The molecule changes direction twice within a period. The interval during which an average molecule in a liquid moves in a roughly straight line is therefore some 10^3 to 10^4 times shorter than the comparable time in gases. The maximum rate at which a liquid molecule can exchange momentum and energy and, therefore, react with its nearest neighbors, or decompose, is correspondingly 10^3 to 10^4 times greater in liquids. The amplitude of the vibration would be, typically, about $1\ \text{Å} = 10^{-8}$ cm or less, which is about 10^3 or more times shorter than the distance for straight-line motion in gases.

However, the fastest bimolecular reactions in solution are not appreciably faster than in gases because the rate at which the reactant molecules A and B diffuse

---

[1] This is the number of nearest neighbors in a closest-packing arrangement of equal-sized spheres.

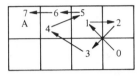

Figure 21-5 The formation of a caged AB pair via diffusion, schematically shown. The boxes schematically represent the cages formed by solvent molecules. The numbers indicate the sequence of possible steps by B that finally forms the AB pair.

together to form caged pairs is rate-determining. This process is shown schematically for a single pair of reactant molecules in Figure 21-5. In order to describe the average situation of reactant molecules more concretely consider the following reacting system.

Before reaction the concentrations are 55 M solvent (typical for water), 0.1 M A (randomly distributed), no B. At $t = 0$ 0.1 M B is introduced into the solution (for example, by flash photolysis from a precursor), distributed, we assume, at random (that is, homogeneously) throughout the solution. The molecules of B that happen to be next to an A (this would be about 2% of A's if each A has 12 nearest neighbors) immediately become caged pairs, and some pairs react or drift apart. As time goes on, pairs are formed from A and B molecules that were further and further apart initially. Eventually a steady state concentration gradient will be established such that the B concentration is highest for those B molecules that are least likely to encounter an A, that is, far from any A molecule and lowest at the boundaries of the cages with A inside.

The two processes—diffusion of B toward A and reaction of the caged AB pair—together constitute an elementary bimolecular reaction in solution (see definition in Section 20-3) that is

$$A + B \underset{k_{d-}}{\overset{k_{d+}}{\rightleftharpoons}} \overline{A + B_c} \overset{k_c}{\longrightarrow} \text{products} \qquad (21\text{-}38)$$

where k_{d+}, k_{d-} are the rate constants for steady state diffusion of B in and out, respectively, of the cages occupied by A. B_c is the concentration of molecules B in the same cage (indicated by the overline) with an A. Assuming that the bulk concentration $B (\approx \text{concentration of B far away from any A}) \gg B_c$, and that B_c is in a steady state, that is, $dB_c/dt \approx 0$; the rate of product formation is from Eq. (21-38)

$$r = kAB = k_c AB_c = k_{d+} AB/(1 + k_{d-}/k_c) \qquad (21\text{-}39)$$

The elementary rate constant is

$$k = k_{d+}/(1 + k_{d-}/k_c) \qquad (21\text{-}40)$$

If $k_{d-}/k_c \ll 1$ then $k \approx k_{d+}$. A reaction for which this occurs has a **diffusion-limited** rate.

It is clear from Eq. (21-40) that the diffusion-limited rate constant is an upper limit for the k of a bimolecular reaction in solution. We can derive a general formula for k_{d+} from Fick's first law of diffusion, Eq. 3-25. Assume that the concentration around an A is spherically symmetric. (Exact spherical symmetry could occur only for spherically symmetric molecules A at infinite dilution—when the concentration gradients of neighboring A's do not overlap.) Then the rate, r_s, at which molecules enter a sphere of radius R is equal to the surface area of the sphere times the flux across the surface

$$r_s = -4\pi R^2 \dot{n} = 4\pi R^2 D_{AB}(dB(R)/dR)_R \tag{21-41}$$

where the second equality follows from Fick's first law and D_{AB} is the coefficient of diffusion for B relative to A, given by

$$D_{AB} = D_A + D_B \tag{21-42}$$

where D_A and D_B are coefficients of diffusion for A and B relative to the solvent. Under steady state conditions the concentration, $B(R)$, of B at any R is constant in time. Therefore, Eq. (21-41) can be easily integrated to obtain $B(R)$

$$B(\infty) - B(R) = B - B(R) = r_s/(4\pi D_{AB}) \int_R^\infty dR/R^2$$

$$= r_s/(4\pi D_{AB}R) \tag{21-43}$$

If the radius of the cage for the AB pair is R_m and if $B \gg B(R_m)$ then we get, from Eqs. (21-42) and (21-43), the average rate at which B enters an A cage:

$$r_s = 4\pi(D_A + D_B)R_m B \tag{21-44}$$

The total rate of AB pair formation is then

$$r_{d+} = r_s I N A = 4\pi I N (D_A + D_B) R_m AB = k_{d+} AB \tag{21-45}$$

and

$$k \approx k_{d+} = 4\pi I N (D_A + D_B) R_m \tag{21-46}$$

The constant I has been defined previously, following Eq. (21-16).

The average cage radius R_m has the same significance as the R_m of gas reactions: it is the maximum distance of approach at which reaction can occur. When the two reactants are neutral or neutral plus ionic, the sum of the hard sphere radii gives R_m; for ions of opposite charge $R_m = Z_A Z_B e^2/(\varepsilon kT)$ where ε is the dielectric constant of the solvent ($R_m = 7$ Å for H_2O at 25°C); for ions of same charge R_m can become considerably smaller than the sum of the hard sphere radii. The diffusion coefficients of small molecules in nonviscous solvents near room temperature range from 10^{-9} to 10^{-8} m^2 s^{-1}. Thus, taking as typical of small molecules $D_A + D_B = 10^{-8}$ m^2 s^{-1}, $R_m = 3$ Å, we get for $k_{d+} = 2 \times 10^7$ m^3 mol^{-1} s$^{-1} \approx 10^{10}$ dm^3 mol^{-1} s^{-1}. This value compares to the $k = 10^{11}$ dm^3 mol^{-1} s^{-1} for a fast gas reaction. An experimental bimolecular rate constant near 10^{10} dm^3 mol^{-1} s^{-1} indicates a diffusion-controlled reaction (see Table 20-7).

Another indication of a diffusion-limited reaction is a small activation energy, usually 2–4 kcal mol^{-1}, about a third of the vaporization energy of the solvent. This contrasts with the fastest gaseous reactions, which have zero activation energies. A molecule diffuses by repeatedly "stepping" in random directions. In each step the molecule gets out of a potential well—its cage—either by exchanging places with a nearest neighbor or by distorting the cage "walls" enough to form a slightly different cage centered in a new position. The changes described should take less energy than removing a molecule (that is, vaporizing it) from liquid entirely.

The model just described suggests an equation, related to Eq. (20-148), for the average lifetime of a molecule in a cage

$$\tau = \tau_0 \, e^{\mathcal{E}_a/RT} \tag{21-47}$$

where τ_0 is the period of vibration in a cage discussed earlier. For $\tau_0 = 10^{-12}$ s and $\mathscr{E}_a/RT = 5$ we get $\tau = 10^{-10}$ s. A somewhat longer time (equivalent to several diffusion steps) will be necessary to establish the steady state described earlier within a few molecular diameters about all A. This indicates that Eq. (21-46) should not be applied to reacting systems with reaction times less than about 10^{-9} s. The increased reaction rates during the nonsteady state reaction regime have been observed.

The preceding treatment of liquid reactions could be made more rigorous, but this would not change the conclusions or make them more secure since all treatments ultimately rely on Fick's diffusion laws. These laws assume continuity of the solvent medium. Although continuity does not exist on a molecular scale, the relative motions of A and B molecules over only a few molecular diameters can lead to observable reaction (for example, in the 0.1 M A and B system initially about 40% of the A's have a B molecule within four molecular diameters).

Significantly improved versions of the simple collision theory models of bimolecular reactions discussed in this section are known for gaseous reactions. These refined models are discussed in Section 21-3.

21-3 MOLECULAR DYNAMICS: THE H + H₂ REACTION

In Section 21-2 we studied the collisions between reactant molecules, each regarded as a center of force. Although that study might be properly termed **molecular dynamics**, this term usually describes experimental and theoretical studies that include some aspect of the relative *atomic* motion within the colliding molecules. The appropriate molecular model is a set of mass points—atoms or groups of atoms—with a time-dependent spatial structure. Molecular dynamics will serve as the basis of an improved collision theory that can predict the effect of the initial internal (rotational, vibrational, electronic) energy of the reactants on the rate constant (discussed in this section) and on the internal energy of the products (Chapter 22).

The following theory of molecular dynamics leads to an improved expression for the rate constant. The theory will be tested against rate constants and other properties of reaction (21-26), measured either for its isotopic variants (for example, $D + H_2 \rightarrow HD + H$; $T + H_2 \rightarrow TH + H$) or for ortho- to para-$H_2$ conversion, which can proceed through reaction (21-26). o-H_2 and p-H_2 molecules differ in the relative orientation of the spins of their two protons: parallel and antiparallel, respectively.

The dynamics of the H$\cdots$H₂ system is best understood by a comparison with the dynamics of the μ particles described in Section 21-2. The relevant quantities are listed in Table 21-2. The number of coordinates necessary to describe the relative motion of three atoms can be calculated by the methods described on pages 79–82. We see from Table 21-2 that the number of coordinates is six for H$\cdots$H₂ compared to three for the μ particle. The three H atoms "move" in a six-dimensional mathematical space. The three dimensions directly involved in a chemical reaction are the interatomic distances R_{AB}, R_{AC}, R_{BC}. Figure 21-6 shows the time dependence of the interatomic distances for undefined orientational angles of two typical trajectories.

Apart from initial conditions (see below) a trajectory is determined by the potential V. From Table 21-2 we see that the potential for H$\cdots$H₂ has four terms.

Table 21-2 A comparison of μ particle and H + H₂ dynamics

What Is Compared	$\mu(A+B)$ particle[a]	$H+H_2$	Equation Number
coordinates	R, θ $\beta = \sphericalangle \mathbf{RL}_0 = 90°$	(R_{AB}, R_{AC}, R_{BC})[b], $(\theta, \phi\, \eta)$[c]	
potential energy	$V = V(R)$	$V = V(R_{AB}) + V(R_{AC})$ $+ V(R_{BC})$ $+ V(R_{AB}, R_{AC}, R_{BC})$	(21-48)
kinds of reactant particles	A, B	H and $H_2(J, v)$ with quantum numbers J, v variable	
conservation of energy	$E = E_0 = E_0(v_0)$ (Eq. 21-2)	$E = E_0 = E_0(v_0)$ $+ E_0(J, v)$	(21-49)
conservation of angular momentum	$\mathbf{L} = \mathbf{L}_0$ (Eq. 21-4)	$\mathbf{L} = \mathbf{L}_0 + \mathbf{J}_0$	(21-50)

[a] The symbols for the μ particle problem have been defined in Section 21-2.
[b] Defined in Figure 21-6. Transformed coordinates may be used, for example, R_{AB}, R_{BC}, γ (Figure 21-7).
[c] These angles define the orientation of A···BC in space, for example if the usual angles θ, ϕ (page 335) are used to define the orientation of BC, then a third angle, η, measured in a plane perpendicular to the BC axis, defines the orientation of the ABC triangle.

The term $V(R_{AB}, R_{AC}, R_{BC})$ is that part of V which cannot be separated into a sum of contributions from R_{AB}-, R_{AC}-, and R_{BC}-dependent terms. When A, for example, is far from the molecule BC then $V(R_{AB}) \approx V(R_{AC}) \approx V(R_{AB}, R_{AC}, R_{BC}) \approx 0$ and $V \approx V(R_{BC})$ where $V(R_{BC})$ now becomes the usual potential for the vibration of a diatomic molecule (Figure 14-3). When the three atoms are close together all four terms are nonzero.

We have shown (pages 412–413) that within the Born–Oppenheimer approximation, V at specified values of the interatomic coordinates is given by the total energy calculated from the electronic part of the Schrödinger equation for the species that is, H···H₂ or H₃ in our example. Therefore, a complete potential V may be obtained, in principle, from a quantum mechanical computation of V for various sets of carefully chosen interatomic distances R_{ij} plus interpolation for other values of the R_{ij}.

The complexity of such computations has prevented the computation of definitive (with errors ≈ 1 kJ mol^{-1}) potential energy functions in this way. (By the time the student reads this, such a computation will probably have been completed for H···H₂⁺ and for H---H₂.) Potential energy functions with errors of about 10–100 kJ mol^{-1} are available for H···H₂⁺, H···H₂, Li⁺···H₂, F···H₂, and a few other systems.

Most of the potential energy functions actually used in studies of molecular dynamics have been obtained semiempirically, that is, by determining parameters in quantum mechanical expressions for V from experimental data on the reactants, products, or activation energies. Such potential functions have errors which, in the best cases, are probably in the range of 10–100 kJ mol^{-1}. The potential function is

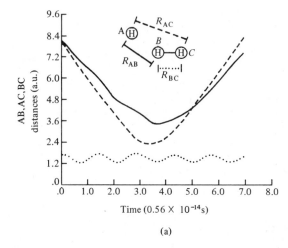

(a)

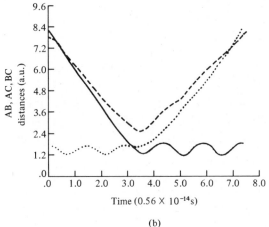

(b)

Figure 21-6 Typical H+H₂ collision trajectories with $J = 2$, $v = 0$, $v_0 = 1.96 \times 10^4 \, m\,s^{-1}$: (a) nonreactive; (b) reactive. Note that on both diagrams the vibration and rotation of H₂ can be seen. Atomic units: 1 au $= a_0$ (Bohr radius, page 336) $= 0.529$ Å $= 0.0529$ nm. [From M. Karplus, R. N. Porter, and R. D. Sharma: *J. Chem. Phys.*, **43**:3259 (1968).]

both a weak and an essential link in all important theories of molecular dynamics and reaction kinetics.

A potential function, no matter how calculated, is best "visualized" as an n-dimensional surface in $(n + 1)$-dimensional mathematical space (n interatomic distances plus V itself). For H···H₂, the space has four dimensions. To construct a physical model of the surface is clearly impossible. We can, however, reduce the problem to three dimensions if one relative atomic coordinate is held constant. A model of an H···H₂ surface for a constant bond angle γ is shown in Figure 21-7a. If the lines of constant V in Figure 21-7a are collapsed onto a plane perpendicular to the V axis, we obtain contour maps such as are shown in Figures 21-7b–d. In Figure 21-7e, V is plotted along the reaction coordinate (the products-to-reactions path of minimum energy on a potential energy surface) for various three-dimensional surfaces.

Once the potential energy function V is known, one can determine H···H₂ trajectories such as those shown in Figure 21-6. The time-dependence of each of the three interatomic coordinates is described by an equation analogous to Eq. (21-7) for μ-particle motion. Similarly, with each of the three orientational angles, there is

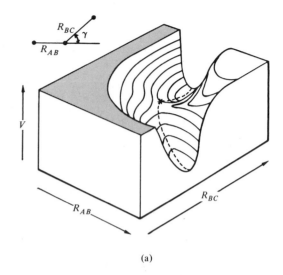

(a)

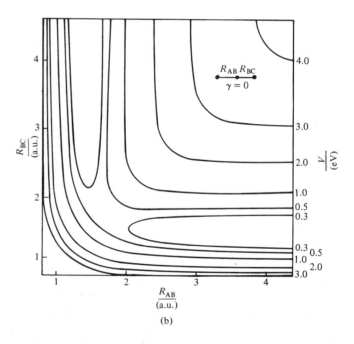

(b)

Figure 21-7 Potential energy (V) surfaces. (a) A three-dimensional model for an ABC system at constant angle γ. (*, saddle point.) Contour maps of a semiempirical (the PK) potential energy surface for H_3 (referenced to $V = 0$ for H and H_2): (b) linear configuration ($\gamma = 0$); (c) angular configuration ($\gamma = \pi/4$); (d) saddle-point detail for linear configuration ($\gamma = 0$); (e) $V(s)$: along reaction coordinate s (path of minimum V) for several angular configurations. The reaction coordinate is shown as a dashed line in Figures 21-7a and 21-7d. At the saddle point of Figures 21-7d and 21-7e the potential energy barrier $V(0)_{PK} = 38.2\ kJ\ mol^{-1} = 9.13\ kcal\ mol^{-1}$. A theoretical rigorous upper bound is $V(0)_{max} = 43.85\ kJ\ mol^{-1} = 10.48\ kcal\ mol^{-1}$, and an estimated lower bound is $V(0)_{min} = 39.7\ kJ\ mol^{-1} = 9.5\ kcal\ mol^{-1}$ [16]. The exact $V(0)$ is probably near the lower bound. [Parts (b)–(e) from R. N. Porter and M. Karplus: *J. Chem. Phys.*, **40**:1105 (1964).]

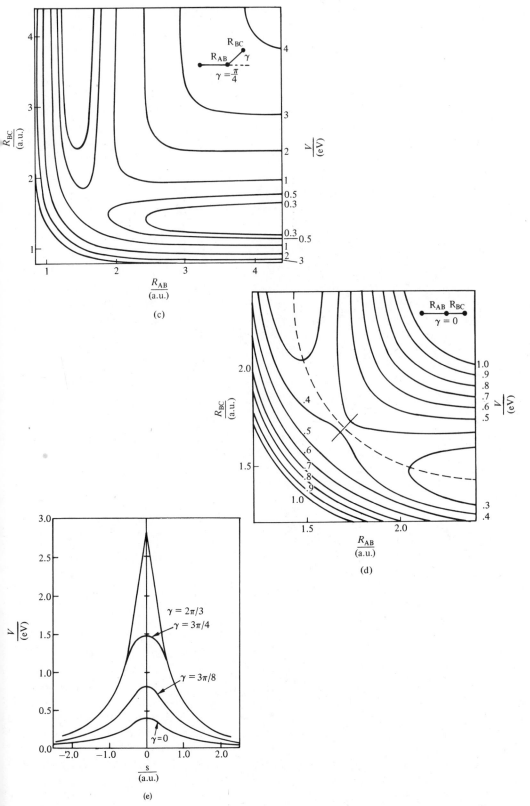

(c)

(d)

(e)

associated an equation analogous to Eq. (21-5). Before the six equations can be integrated, one has to specify an initial velocity or its equivalent along each coordinate and the initial value of each coordinate (see the μ-particle problem), 12 parameters in all.

Some initial parameters constrain a free H atom to collide with an H_2 molecule. Others are chosen for computational convenience. The most important parameters are those that specify an experimentally or theoretically interesting physical situation. For example, we will determine the reaction cross section as a function of three initial parameters: the velocity v_0 of the H atom relative to the center of mass of H_2, the initial rotational and vibrational quantum numbers J, v of H_2.

The remaining initial parameters, which specify the orientation (angles θ, ϕ) and the vibrational phase of the H_2 molecule, as well as the magnitude of the initial angular momentum L_0 (or impact parameter b) of the H atom relative to H_2, may be chosen at random in the range of possible values, for example, 0 to π for θ, or 0 to b_m for b. The random choice simulates the situation in most experiments. Given the initial parameters, the six differential equations may be integrated to determine the six coordinates as a function of time. The integration has to be done numerically on a computer. Two of the resulting trajectories can be seen in Figure 21-6.

Given a sufficient number of trajectories for specific values of v_0, J, and v the reaction probability $P_r(v_0, J, v)$ can be calculated to a good approximation. The reaction probability is defined as

$$P_r(v_0, J, v) = \lim_{N \to \infty} N_r(v_0, J, v)/N(v_0, J, v) \approx N_r(v_0, J, v)/N(v_0, J, v) \quad \textbf{(21-51)}$$

where N, N_r are the number of total and of reactive trajectories for the given values of parameters v_0, J, and v. The second, approximate equality applies when both N_r and N are reasonably large numbers, that is, a few hundred.

The reaction cross section is defined (see Eq. 21-12) as

$$S_r(v_0, J, v) = b_m^2 P_r(v_0, J, v) \quad \textbf{(21-52)}$$

where b_m is optimally chosen to be the impact parameter above which *all* trajectories are nonreactive. If b_m is chosen too large, more trajectories will have to be computed in order to get a reasonably accurate P_r.

The reaction cross sections for $H + H_2(J, v)$ with $J = 0, 5$; $v = 0$ (Figure 21-8) are shown as a function of v_0 in Figure 21-9. Reaction cross sections averaged over all J for $T + H_2$ and $T + D_2$ are shown in Figure 21-10.

What do the results in Figures 21-9 and 21-10 reveal about the $H + H_2$ reaction? First of all, let us consider the cross section near reaction threshold. Figure 21-7e shows that the potential energy barrier $V(0)_{PK}$ is least if the H atom approaches the H_2 molecule end-on. (This is a **collinear** collision, $b = 0$). Therefore the reactive trajectories near the threshold very probably belong to collinear collisions. The reaction cross sections of these collisions are very small (too small to show on the scale of Figures 21-9 and 21-10) since near-collinear collisions are a very small fraction of the total. Only when the total initial energy of reactants E_0 becomes large enough to make noncollinear reactive trajectories likely does the cross section grow to a value of about a Bohr diameter ($2a_0 = 1.06$ Å, page 336) squared. The reaction overwhelmingly proceeds via noncollinear collisions.

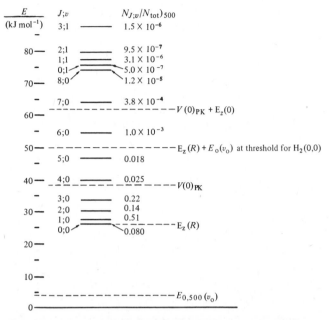

Figure 21-8 The rotation-vibration levels of the H_2 molecule [23]. Also shown are some other quantities relevant to the reaction $H + H_2 \rightarrow H_2 + H$. $\bar{E}_{0,500}(v_0)$ is the average relative kinetic energy of a pair of particles at 500 K. The semiempirical potential energy barrier is $V(0)_{PK}$ (defined in Figure 21-7). The quantum barrier is $V(0)_{PK} + E_z(0)$ where E_z is the zero point energy at $s = 0$; and the energy $E_z(R) + E_0(v_0)$ at reaction threshold for H_2 with $J = 0$, $v = 0$. [$E_z(R)$ is the zero point energy of the reactants.]

$(N_{J;v}/N_{tot})_{500}$ is the fraction of molecules in the Jvth level at 500 K, assuming a Boltzmann distribution. Calculated from Eq. (19-25) with $g = w_J(2J+1)$ and $w_J = 1$ if $J =$ even, $w_J = 3$ if $J =$ odd. [The w_J are degeneracies of proton spin levels. The Pauli exclusion principle allows p-H_2 (antiparallel spins) to exist only in even J levels while o-H_2 exists only in odd J levels.]

The energy requirements of the reaction can be interpreted in terms of energy and angular momentum conservation (Eqs. (21-49) and (21-50), Table 21-2). The total energy available for reaction, E, is the sum of the initial relative translational energy $E_0(v_0)$ plus the initial internal energy of the reactants, which for $H\cdots H_2$ is simply the rotational-vibrational energy of $H_2(J, v)$, that is, $E_0(J, v)$. For $H_2(0, 0)$ $E_0(0, 0) = E_z(R) = 25.9$ kJ mol^{-1} (from Figure 21-8). If we add to this $E_0(v_0)$ at threshold (equal to 24 kJ mol^{-1}, Figure 21-9), the result is $E = E_0 = 50$ kJ mol^{-1}. This value is shown in Figure 21-8. Because E_0 is some 10 kJ mol^{-1} higher than the potential energy barrier $V(0)_{PK}$, we conclude that even the collinear threshold reactions do not cross the barrier along the reaction coordinate s with zero kinetic energy. Part of the 10 kJ mol^{-1} is potential energy in excess of $V(0)_{PK}$; the rest is kinetic energy. The excess potential energy is due, in part, to a "bobsledding" effect. Because of the sharp curvature of the reaction coordinate near $s = 0$ (Figure 21-7d) the reacting system is forced to climb the outside bank of the potential energy valley. The rest of the 10 kJ mol^{-1} is part kinetic and part potential energy stored in vibrations in a

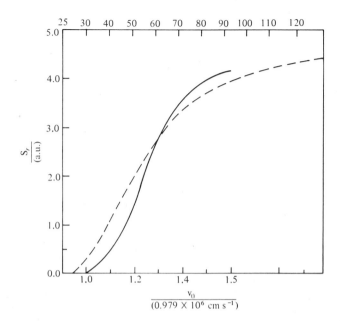

Figure 21-9 Reaction cross section $S_r(v_0, J, 0)$ for $H + H_2$. $J = 0$ (----); $J = 5$ (——). 1 au $= a_0^2 = (0.529 \text{ Å})^2 = 0.280 \text{ Å}^2 = 0.00280 \text{ nm}^2$. Threshold values (below which $S_r = 0$): $v_0 = 0.86(0.98 \times 10^4) \text{ m s}^{-1}$ or $E_0(v_0) = 24 \text{ kJ mol}^{-1}$ for $J = 0$; $v_0 = 0.965(0.98 \times 10^4) \text{ m s}^{-1}$ or $E_0(v_0) = 29.0 \text{ kJ mol}^{-1}$ for $J = 5$. [From M. Karplus, R. N. Porter, and R. D. Sharma: *J. Chem. Phys.*, **43**:3259 (1965).]

direction perpendicular to s and in bending vibrations. Noncollinear collisions have an increased $V(0)_{PK}$ (Figure 21-7e) and angular momentum. If the initial total angular momentum, L_0, is large, angular momentum conservation (Eq. 21-50) will require a larger fraction of the energy to be stored in rotation as the atoms pull closer together (see μ-particle motion).

The theoretical rate constant for reaction (21-26) can be calculated in two stages. In the first stage one calculates the rate constants for each internal energy level of H_2, $k(J, v)$, by Eqs. (21-18) or (21-19). In the second stage one sums over the rates for each internal energy level, $r(J, v)$, to obtain the total rate as a function of the concentrations H_2, and H

$$r = \sum_{Jv} r(J, v)$$

$$= \sum_{Jv} k(J, v) H_2(J, v) \cdot H = \sum_{Jv} k(J, v)(H_2(J, v)/H_2) H_2 \cdot H$$

$$= \sum_{Jv} k(J, v) f(J, v) \cdot H_2 \cdot H = k \cdot H_2 \cdot H \qquad \textbf{(21-53)}$$

where $f(J, v)$ is the fraction of H_2 molecules in the Jvth energy level. For H_2 molecules in thermal equilibrium, this fraction is given by the Boltzmann distribution (Eq. 19-25). Boltzmann fractions for a temperature of 500 K are given for the lowest

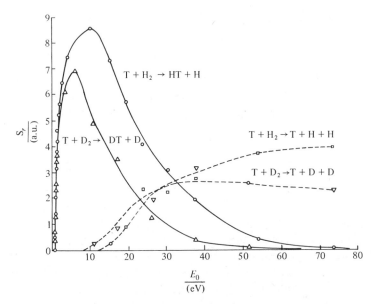

Figure 21-10 Cross sections as a function of relative translational energy $E_0(v_0)$ for exchange and dissociation reactions of tritium atoms with hydrogen and deuterium molecules. The cross sections are averaged over a Boltzmann rotational state distribution at 300 K. $1\,eV = 96.4\,kJ\,mol^{-1}$; $1\,au = 0.00280\,nm^2$. [From M. Karplus, R. N. Porter, and R. D. Sharma: *J. Chem. Phys.*, **45**:3871 (1966).]

13 levels in Figure 21-8. Over 99% of the H_2 molecules are in the lowest six levels at 500 K. To obtain a rate constant k with less than 1% error, no more than the first six terms $(J = 0–5, v = 0)$ in Eq. (21-53) are needed. The rate constants $k(J, 0)$ computed from Eq. (21-18) with an equilibrium distribution of v_0 are 0.321×10^8, 0.272×10^8, 0.122×10^8, 0.154×10^8, 0.099×10^8 and 0.091×10^8 cm^3 mol^{-1} s^{-1} for $J = 0–5$. These values multiplied by the Boltzmann fractions from Figure 21-8 yield $k = 2.26 \times 10^7$ dm^3 mol^{-1} s^{-1}, compared to an experimental value 1.7×10^7 dm^3 mol^{-1} s^{-1}. Values of k have been determined in the same way at other temperatures in the range 300–1000 K. If the logarithms of k are plotted against $1/T$, a straight line results, yielding an Arrhenius expression for $k = 4.3 \times 10^{10}\,e^{-7400/1.99T}$ cm^3 mol^{-1} s^{-1} with the activation energy in calories per mole. This result may be compared with the experimental one given with reaction (21-26).

The test of these theoretical results is not decisive because of errors in the experimental data, because of the approximate potential energy surface used, and because of hitherto unmentioned approximations in the calculation.

This type of calculation is frequently referred to as quasiclassical, "classical" because the trajectory is computed by classical mechanics, and "quasi" because the initial internal energy of H_2 is quantum mechanical. A quantum mechanical treatment would correct two shortcomings of the quasiclassical one: (1) It would take into account the zero point energy of H_3 everywhere (rather than for reactants only) along the reaction coordinate for the two "vibrational" degrees of freedom other than the reaction coordinate. (2) It would account for tunneling through the barrier. The zero point energy adds to $V(s)$ to yield an effective quantum potential energy surface.

The quantum barrier for motion without tunneling, $E_q = V(0) + E_z(0)$, has the value 62 kJ mol^{-1} (Figure 21-8) for the potential energy surface of Figure 21-7. The quantum barrier is therefore some 20 kJ mol^{-1} higher than the classical barrier $V(0)_{PK}$. Since tunneling would make the *effective* quantum barrier lower than 62 kJ mol^{-1}, the quasiclassical calculation could have about the "right" barrier height to give a good result. Definitive quantum calculations that would establish the importance of tunneling will probably be performed in the near future.

21-4 ENERGY TRANSFER

We have not yet explicitly considered the question of how molecules acquire the energy necessary to react. The use of Boltzmann (equilibrium) distribution functions to calculate average rate constants (as in Eqs. 21-18 and 21-53), assumes implicitly that processes which transfer the energy necessary to establish equilibrium are fast compared to chemical reaction. There are two important classes of elementary gas phase reactions—unimolecular isomerizations or decompositions and atom or small radical (re)combinations for which the assumption does not always hold.

21-4.1 Unimolecular Reactions Elementary unimolecular isomerizations and decompositions are slow (because of high activation energies) reactions that have been thoroughly studied by classical methods for measuring rates (Section 20-6). Four of many possible examples are

$$\underset{H_2C \triangle CH_2}{CH_2} \to CH_3CH=CH_2 \qquad k_\infty = 10^{15.5}\, e^{-65600/1.99T}\, s^{-1} \quad [9] \qquad \textbf{(21-54)}$$

$$CH_3NC \to CH_3CN \qquad k_\infty = 10^{13.6}\, e^{-38400/1.99T}\, s^{-1} \quad [9] \qquad \textbf{(21-55)}$$

$$C_2H_6 \to 2CH_3 \qquad k_\infty = 10^{16.0}\, e^{-86000/1.99T}\, s^{-1} \quad [9] \qquad \textbf{(21-56)}$$

$$C_2H_6^+ \to CH_2CH_2^+ + H_2 \quad or \quad CH_3CH^+ + H_2 \quad or \quad C_2H_5^+ + H$$

$$or \quad CH_3^+ + CH_3 \qquad \textbf{(21-57)}$$

The activation energies are in calories per mole. Reactions (21-57), initiated by activators, are usually observed in a mass spectrometer.

The early investigators of unimolecular reactions who determined the rate laws of unimolecular reactions to be first order, faced a very fundamental problem: How do the molecules acquire the energy to react? During the second decade of this century Perrin and others suggested that the molecules absorb ambient infrared radiation. The known low density of the radiation and the demonstrable photochemical ineffectiveness of infrared radiation soon made the suggestion untenable. Another early proposal, due to Lindemann in 1922 and Christiansen in 1921, was that the energy is acquired via collisions with other molecules. This proposal is the basis of modern theories.

The so-called **Lindemann theory** explains how energy acquisition via collisions, can lead to a first-order rate law. The theory assumes that a critical internal (primarily vibrational) energy, E_c, exists. Reactant molecules A, which have less than this energy, cannot react, whereas reactant molecules A^* with energy equal to or above E_c can react. Inelastic collisions with another molecule M, where M may be an added inert gas or A itself transfer energy to an A or from an A^*. The excited molecules A^* form the products P with a finite rate, comparable to the energy transfer rates.

Lindemann formulated the assumptions in the simplest possible way.

$$A+M \underset{k_d}{\overset{k_a}{\rightleftharpoons}} A^*+M \qquad (21\text{-}58)$$

$$A^* \xrightarrow{k_P} P \qquad (21\text{-}59)$$

(The notation ignores the energy loss or gain of M upon activation of A or deactivation of A^*.) The assumption that the pressure (or concentration) $A^* \ll A$ and that A^* is in a steady state $(dA^*/dt \approx 0)$ readily leads, by the methods of Section 20-3, to the rate law

$$\begin{aligned} r &= k_P A^* = k_P k_a MA/(k_d M + k_P) \\ &= (k_P k_a/k_d)A/[1+(k_P/k_d)M^{-1}] \\ &= [k_\infty/(1+(k_\infty/k_a)M^{-1})]A \\ &= kA \end{aligned} \qquad (21\text{-}60)$$

where we have defined the unimolecular rate constant (which is not a constant!)

$$\begin{aligned} k &= k_\infty/(1+(k_\infty/k_a)M^{-1}) \\ &= (k_P k_a/k_d)/(1+(k_P/k_d)M^{-1}) \end{aligned} \qquad (21\text{-}61)$$

and the high-pressure limit $(M \to \infty)$ of k

$$k_\infty = k_P k_a/k_d \qquad (21\text{-}62)$$

The significance of k_∞ is that in the high pressure limit $(k_\infty/k_a)M^{-1} \ll 1)$ the rate equation (21-60) reduces to

$$r = kA \approx k_\infty A = (k_a/k_d)k_P A \qquad (21\text{-}63)$$

Eq. (21-63) can also be derived by treating processes (21-58) as an equilibrium that precedes the rate-determining step, process (21-59). At high pressures when the collision rate is high, the molecules A and A^* *are* in equilibrium. This is the pressure range studied by the early workers who obtained first-order rate laws for unimolecular reactions.

The low pressure limit of rate law (21-60) results when $(k_\infty/k_a)M^{-1} \gg 1)$

$$r = kA \approx k_0 A = k_a MA \qquad (21\text{-}64)$$

In this limit the deactivation rate becomes immeasurably slow and the activation rate determines the overall rate. The region in which the unimolecular rate constant k decreases below its high pressure limit, k_∞, is known as the fall-off region.

The existence of the fall-off region experimentally distinguishes a unimolecular from a complex free radical reaction. Fall-off for reaction (21-55) is shown in Figure 21-11. The fall-off region also serves to test quantitative formulations of Lindemann theory. For example, k_a may be calculated from simple collision theory, Eq. (21-25), if we reasonably assume that all of the experimental activation energy of k_∞ (Eqs.

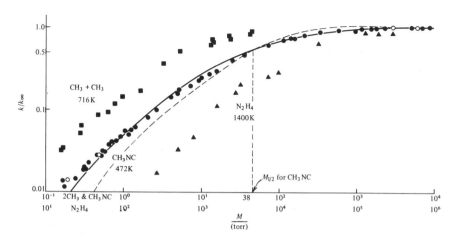

Figure 21-11 Pressure dependence of the experimental rate constants for two unimolecular and one radical combination reaction. For reaction $CH_3NC \rightarrow CH_3CN$: dashed curve is the plot of Eq. (26-61) with $M_{1/2\ theor} = M_{1/2\ exp}$ and $(k_\infty)_{theor} = (k_\infty)_{exp}$ (see text); solid curve is from RRKM theory with $(k_\infty)_{theor} = (k_\infty)_{exp}$ (see Example 21-2). [From M. F. R. Mulcahy: *Gas Kinetics*. Thomas Nelson & Sons, London, 1973, p. 15.]

21-54 to 21-56) belongs to k_a and that k_P, k_d have zero activation energies. Using the standard preexponential factor (page 765) $k_a = 10^{11}\ e^{-38400/1.99T}\ dm^3\ mol^{-1}\ s^{-1}$ for reaction (21-55).

This theoretical k_a may be compared with an experimental one obtained from Eq. (21-61). Note that when $k/k_\infty = \frac{1}{2}$, the pressure $M_{1/2}$ is

$$M_{1/2} = k_\infty / k_a \qquad (21\text{-}65)$$

$M_{1/2}$ from Figure 21-11 is about 40 torr. Combined with k_∞ from Eq. (21-55) this gives $k_a = k_\infty / M_{1/2} = 10^{12.0}\ e^{-38400/1.99T}\ s^{-1}\ torr^{-1} = 10^{16.5}\ e^{-38400/1.99T}\ dm^3\ mol^{-1}\ s^{-1}$ where we have converted from torr to moles per cubic decimeter by assuming the ideal gas law ($n/V = P/RT$).

The experimental k_a is nearly a million times higher than predicted by collision theory or, conversely, the experimental $M_{1/2}$ is 10^6 times less than the theoretical one. Similar calculations for other unimolecular reactions yield similar results. The larger the molecule the worse the disagreement.

The importance of internal (mostly vibrational) energy for the $H + H_2$ reaction suggests that internal energy might contribute to the energy stored in A^*. If so, then the previous estimate of k_a, which assumed that all of the internal energy of A^* comes from relative translational energy, would be low. The effect of internal energy on k_a may be formulated quantitatively if we assume that energy transfer occurs via "strong collisions." Strong collisions transfer amounts of energy that are so large that the activation and deactivation processes (21-58) are one-step, rather than correlated multistep processes. The net result is that the distribution of A^* and A molecules among their internal quantum states is random. The randomness of the A^* distribution can be maintained only if process (21-59) proceeds at random, independent of the internal quantum state of A^*. This is the random lifetime assumption. As a consequence of the strong collision and random lifetime assumptions, the distribution of A^* energies can be calculated statistically.

The statistical approach leads to a general steady state expression for the unimolecular rate constant k. If A_j^* (this generalizes the notation $H_2(J, v)$ used in Section 21-3) is the concentration of A^* molecules which are in the jth internal energy level, the fraction of A^* in level j is

$$f_j = A_j^*/A \qquad (21\text{-}66)$$

where A is the concentration of all A molecules.

Three rate constants describe the processes (21-58) and (21-59) involving A_j^*. The strong collision assumption means that A_j^* reverts to A in every collision. Therefore, k_d for all j is given to a good approximation by Eq. (21-23). The rate constant for the production of A_j^* by activation of A is k_{a_j},[2] and k_{P_j} is the rate constant for product formation from A_j^*. The net rate for A_j^* in the steady state is

$$r_{A_j}^* = k_{a_j}MA - k_d MA_j^* - k_{P_j}A_j^*$$
$$= k_{a_j}MA - k_d Mf_jA - k_{P_j}f_jA = 0 \qquad (21\text{-}67)$$

where we have substituted A_j^* from Eq. (21-66). According to the principle of detailed balancing (page 655), at equilibrium the first term in Eq. (21-67) would be

$$k_{a_j}MA = k_d MA_j^* = k_d Mf_{B_j}A \qquad (21\text{-}68)$$

where f_{B_j} is the Boltzmann fraction (Eq. 19-25) of A_j^*. Eq. (21-68) should also be valid under steady state (other than equilibrium) conditions since the equilibrium-perturbing process (21-59) does not involve either A or M. Combining Eqs. (21-67) and (21-68) yields, for the steady state fraction of A_j^*

$$f_j = f_{B_j}\frac{k_d M}{k_d M + k_{P_j}} \qquad (21\text{-}69)$$

Eq. (21-69) shows that, as expected, the steady state fraction f_j is less than f_{B_j} whenever the product formation rate is comparable to or greater than the deactivation rate of A_j^*. If $k_d M \gg k_{P_j}$, $f_j \approx f_{B_j}$. Eq. (21-69) obviates the difficult direct calculation of f_j under nonequilibrium conditions.

The total rate for the production of P is the sum of the rates for all j

$$r_P = \sum_j k_{P_j}A_j^* = \left(\sum_j k_{P_j}f_j\right)A = k_d M \sum_j \left(\frac{k_{P_j}f_{B_j}}{k_d M + k_{P_j}}\right)A \qquad (21\text{-}70)$$

where we have eliminated A_j^* and f_j by using Eqs. (21-66) and (21-69). Therefore the unimolecular rate constant becomes (see Eq. 21-61)

$$k = \sum_j \left[\frac{k_{P_j}f_{B_j}}{1 + (k_{P_j}/k_d)M^{-1}}\right] \qquad (21\text{-}71)$$

with its high-pressure form (see Eq. 21-62)

$$k_\infty = \sum_j k_{P_j}f_{B_j} = \sum_j k_{P_j}(k_{a_j}/k_d) \qquad (21\text{-}72)$$

---

[2] This constant is an average over all A states below the critical energy E_c.

and low pressure form (see Eq. 21-64)

$$k_0 = k_d\left(\sum_j f_{B_j}\right)M = \left(\sum_j k_{a_j}\right)M \tag{21-73}$$

where we have used the equilibrium constant relation $A_{e_j}^*/A_e = K = k_{a_j}/k_d$ and Eq. (21-66).

We now return to the question of the magnitude of k_a. Eq. (21-71) does not have the form of Eq. (21-61). It reduces to this form only if we assume all k_{P_j} equal, that is, $k_{P_j} = k_P$ for any j. Then we find, by comparing Eqs. (21-61) and (21-71) that

$$k_a = \sum_j k_{a_j} \tag{21-74}$$

As we shall see later (Figure 21-16), the number of vibrational quantum states just above E_c is very large. Thus the sum in Eq. (21-74) must be much larger than the k_a given by simple collision theory, and Eq. (21-74) should be in better agreement with experiment than the earlier collision theory calculation. The modification of Eq. (21-61) just described is essentially equivalent to one first proposed by Hinshelwood in 1927.

We know now how the experimental k_a and $M_{1/2}$ can be brought into agreement with the corresponding theoretical quantities. It remains to consider the shapes of the theoretical and experimental fall-off curves in Figure 21-11. The dashed curve is a plot of Eq. (21-61) made to agree with experiment at $M = M_{1/2}$ and at $M = \infty$. The curves from experiment and from Eq. (21-61) have different shapes. An improved theoretical curve will be obtained in Section 21-6 where we will apply the activated complex theory to the calculation of k_{P_j} and k.

21-4.2 Combination Reactions

Energy transfer via collisions is also important for exothermic combination reactions such as

$$I + I \rightarrow I_2 \tag{21-75}$$

or

$$CH_3 + CH_3 \rightarrow CH_3CH_3 \tag{21-76}$$

Figure 21-11 shows that the bimolecular reaction (21-76) behaves very much like a unimolecular reaction. What is the reason for this similarity? When two atoms combine, the released energy is initially in the reaction coordinate. Dissociation of the newly formed molecule will occur unless the energy is removed by collision. In radical combination the released energy, apart from removal by collision, may be transferred out of the reaction coordinate to other product molecule vibrations. However, if the product molecule is small, there are few vibrations and the probability of intramolecular transfer is low. Therefore, atom and small radical combinations can be explained by essentially the same elementary reaction.

The three processes that make up the elementary reaction are combination (k_c), dissociation (k_{di}), and deactivation (k_{de}).

$$A + B \underset{k_{di}}{\overset{k_c}{\rightleftharpoons}} AB^* \tag{21-77}$$

$$AB^* + M \xrightarrow{k_{de}} AB + M \tag{21-78}$$

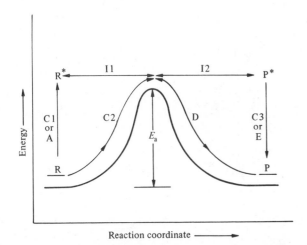

Energy ——→

Reaction coordinate ——→

Figure 21-12   The processes involved in an elementary reaction. Cn, bimolecular collisions; In, intramolecular processes; D, dissociation; A and E photon absorption and emission, respectively. The positions of the reactants (R, R*) and products (P, P*) on the ordinate correspond to their *internal* energies. Any combination of paths corresponds to a possible elementary reaction. Examples: a unimolecular isomerization is C1 (or A) + I1 + I2 + C3 (or E); H + H$_2$(0, 0) is C2 + D. Paths (not shown) between these two extremes are possible.

If the steady state is assumed for the excited molecule AB* we get for the rate

$$r = [k_c/(1 + (k_{di}/k_{de})M^{-1})]A \cdot B$$
$$= kA \cdot B \tag{21-79}$$

Obviously, the rate constant k varies with the pressure M in the same way as does the unimolecular rate constant.

In this section we have seen that an elementary reaction involves, in addition to rearrangement of atoms, energy transfer processes that may precede or follow or occur simultaneously with the atomic rearrangement. Some of the possibilities are summarized in Figure 21-12.

21-5   ACTIVATED COMPLEX THEORY

The **activated complex** theory, also called the **absolute rate** or **transition state** theory, was formulated in essentially its present form by H. Eyring in 1935. This theory is applicable not only to both uni- and bimolecular elementary reactions in gases, solutions, and on surfaces but to any rate process as long as the process has an appreciable activation energy ($\mathscr{E}_a \gtrsim 5\boldsymbol{R}T$). The theory may be derived rigorously (from a basis in molecular dynamics) under assumptions that restrict its applicability to certain gas reactions. Our derivation will follow a less rigorous path.

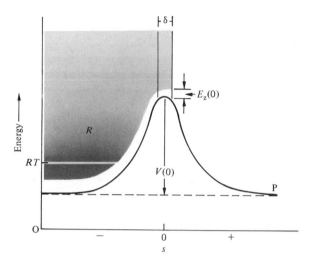

Figure 21-13 The distribution of reacting systems R at a temperature T as a function of total energy E and reaction coordinate s. The concentration of R within the ranges E to $E + dE$ and s to $s + ds$ is proportional to the depth of shading. It corresponds approximately to the Boltzmann distribution (Eq. 19-25, page 576) with the degeneracy of all energy levels equal to 1. (In actual systems with several atoms, the number of levels within the ranges E to $E + dE$ and s to $s + ds$ rapidly increases with E.) The curved edge of the shading is $V(s) + E_z(s)$—the sum of the potential energy and zero point energy (see Section 21-3). If tunneling were important, the shading would extend into the region below this edge.

The drawing might also be visualized as representing a less abstract but analogous system: a gas in equilibrium at T in a very strong gravitational field; originally confined to the R basin, shown moving out of it.

The spirit in which the activated complex theory approaches reaction rates may be seen from Figure 21-13. We derive the rate constant for the case where R in Figure 21-13 represents reactants A + B. The activated complex $X^{\ddagger}$ is then A and B at their maximum repulsion, that is, in the region of width δ in Figure 21-13. The reaction rate for

$$A + B \rightarrow P \qquad (21\text{-}80)$$

would then be, for an activated complex concentration $X^{\ddagger}$,

$$r = \nu X^{\ddagger} = kAB \qquad (21\text{-}81)$$

where ν is the average frequency with which the complexes move across the top of the barrier from left to right.

A crucial assumption, which makes the calculation of k from Eq. (21-81) possible is that the $X^{\ddagger}$ are distributed among their energy levels according to the Boltzmann distribution. For ideal species $X^{\ddagger}$, A, B, we have (Eq. 19-141 with $q_i = q_i^{\circ}$)

$$X^{\ddagger}/AB = (q_{\ddagger}/q_A q_B)\, e^{-\Delta \mathscr{E}_0 / RT} = K \qquad (21\text{-}82)$$

where the q are partition functions (see Eq. 19-35) based on the Boltzmann distribution. All specialized symbols in Eq. (21-82) were defined on pages 606–608. Substitution for $X^{\ddagger}$ in Eq. (21.81) shows that

$$k = \nu(q_{\ddagger}/q_A q_B) \, e^{-\Delta \mathscr{E}_0/RT} \qquad \qquad (21\text{-}83)$$

Eq. (21-83) is converted into a useful formula by treating the reaction coordinate s as a very low frequency normal mode of vibration whose partition function may be factored out of $q_{\ddagger}$ (see Eq. 19-96)

$$\begin{aligned}
q_{\ddagger} &= (1 - e^{-h\nu/kT})^{-1} q^{\ddagger} \\
&\approx 1/(1 - (1 - h\nu/kT)) q^{\ddagger} \\
&\approx (kT/h\nu) q^{\ddagger} \qquad \qquad (21\text{-}84)
\end{aligned}$$

where the approximate equality is valid for $h\nu/kT \ll 1$ and where $q^{\ddagger}$ contains one less vibrational frequency than the normal partition function $q_{\ddagger}$. Substituting Eq. (21-84) into Eq. (21-83)

$$k = kT/h(q^{\ddagger}/q_A q_B) \, e^{-\Delta \mathscr{E}_0/RT} \qquad \qquad (21\text{-}85)$$

The ratio kT/h, equal to 6.25×10^{12} Hz at 300 K, is a universal frequency associated with reactions. Eq. (21-85) is a fundamental equation of activated complex theory. It may be reformulated or modified for actual applications.

21-5.1 Thermodynamic Expression for Rate Constant

First consider the thermodynamic reformulation of Eq. (21-85). Comparison of Eqs. (21-82) and (21-85) shows that the latter contains a pseudoequilibrium constant.[3] Therefore Eq. (21-85) can be reformulated to give, for constant pressure conditions,

$$\begin{aligned}
k = kT/h \cdot K^{\ddagger} &= kT/h \cdot e^{-\Delta \mathscr{G}\ddagger/RT} \\
&= kT/h \cdot e^{\Delta \mathscr{S}\ddagger/R} \, e^{-\Delta \mathscr{H}\ddagger/RT} \qquad \qquad (21\text{-}86)
\end{aligned}$$

where $\Delta \mathscr{G}^{\ddagger}$, $\Delta \mathscr{S}^{\ddagger}$, $\Delta \mathscr{H}^{\ddagger}$ are the (standard) Gibbs free energy, the entropy, and the enthalpy of activation. If we assume that $\Delta \mathscr{H}^{\ddagger}$ is temperature-independent, simple approximate relationships of $\Delta \mathscr{H}^{\ddagger}$ and $\Delta \mathscr{S}^{\ddagger}$ to the experimental parameters of the Arrhenius equation, A and $\mathscr{E}_a$, follow from Eqs. (20-17) and (20-18):

$$\mathscr{E}_a = \Delta \mathscr{H}^{\ddagger} + RT \qquad \qquad (21\text{-}87)$$

and

$$A = ekT/h \cdot e^{\Delta \mathscr{S}\ddagger/R} \qquad \qquad (21\text{-}88)$$

These equations may be used to calculate $\Delta \mathscr{H}^{\ddagger}$ and $\Delta \mathscr{S}^{\ddagger}$ from experimental data. The units of A determine the units of $K^{\ddagger}$ and, hence, the standard states to which the enthalpy and entropy of activation refer. For example, if the units of A are cubic decimeters per mole per second, then the concentrations in the standard states of $X^{\ddagger}$, A, and B are 1 mol dm^{-3}. Some experimental values of $\Delta \mathscr{S}^{\ddagger}$, $\Delta \mathscr{H}^{\ddagger}$ are shown in Table 21-3.

---

[3] The constant is not a true equilibrium constant because it does not include contributions from the reaction coordinate.

Table 21-3 The dimerization of cyclopentadiene in various solvents[a]

Solvent	Temp. range, (°C)	$\dfrac{\Delta\mathscr{S}^{\ddagger}}{\text{(cal mol}^{-1}\text{ K}^{-1})^{\text{b}}}$	$\dfrac{\Delta\mathscr{H}^{\ddagger}}{\text{(kcal mol}^{-1})^{\text{b}}}$	$-\text{Log } k$, at 50°C
none, gas phase	79–150	-33 ± 2	15.9 ± 0.6	5.2
paraffin	-1–172	-28.2 ± 0.9	16.7 ± 0.3	4.7
C_6H_6	15–55	-33 ± 2	15.8 ± 0.6	5.0
CH_3CO_2H	25–70	-38 ± 3	14 ± 1	5.0

[a] From data in A. Wasserman: *Monatsh.*, **83**:543 (1952).
[b] Calculated according to Eqs. (21-87) and (21-88) for temperatures at the midpoint of the range given.

The data on cyclopentadiene dimerization in Table 21-3 show several aspects of gas and solution phase reactions. The large negative values of $\Delta\mathscr{S}^{\ddagger}$ reflect the replacement of two independent cyclopentadiene molecules with a single activated complex molecule. The somewhat similar values of $\Delta\mathscr{S}^{\ddagger}$ and $\Delta\mathscr{H}^{\ddagger}$ in the various media indicate similar activated complexes. The large magnitude of $\Delta\mathscr{S}^{\ddagger}$ in CH_3CO_2H indicates an activated complex that is more polar than the reactants and, therefore, is stabilized more than the reactants in a polar solvent. The $\Delta\mathscr{H}^{\ddagger}$ values reflect the same trends. Note that the variations of $\Delta\mathscr{S}^{\ddagger}$ and $\Delta\mathscr{H}^{\ddagger}$ from medium to medium cause the rate constant to vary in opposite directions so that the overall change of k is rather small.

21-5.2 Rate Constants from Molecular Properties   Activated complex theory can predict rate constants from molecular structure parameters and vibrational frequencies. The procedure is very similar to the calculation of equilibrium constants via statistical mechanics (pages 606–609). To discuss the procedure for calculating a rate constant it is best to rewrite Eq. (21-83) in a more explicit form.

$$k = l_{\ddagger}(kT/h)(q^{\ddagger}\,q_A^{-1}q_B^{-1})_{\text{tr}}(q^{\circ\ddagger}/q_A^{\circ}q_B^{\circ})_{\text{rot}}(q^{\ddagger}\,q_A^{-1}q_B^{-1})_{\text{vib}}$$
$$\times (q^{\ddagger}\,q_A^{-1}q_B^{-1})_{\text{el}}\, e^{-\Delta\mathscr{E}_0/RT} \tag{21-89}$$

The translational partition function for 1 mole of a chemical species (with mass m) in unit volume is $(2\pi mkT)^{3/2}/Nh^3$. The electronic and the vibrational partition functions (for a single degree of freedom) are defined in Eqs. (19-108) and (19-96). The total vibrational partition function of a species is the product of the partition functions for each degree of freedom. The activated complex has one less vibrational degree of freedom than that of a normal molecule with the same number of atoms, N, that is, $3N-7$ for a nonlinear complex and $3N-6$ for a linear one.

The rotational partition functions do not contain the symmetry number σ (see page 594)

$$q^{\circ}_{\text{rot}} = q_{\text{rot}} \times \sigma \tag{21-90}$$

where q_{rot} is defined in Eqs. (19-87) or (19-92). The symmetry of the reactants is taken into account by the statistical factor $l_{\ddagger}$, which is simply the number of equivalent ways in which the reactants may form the activated complex.

The reaction $H+H_2$ may form a linear activated complex $H-H-H$ by attaching the H atom on either end of the H_2 molecule and therefore $l_\ddagger = 2$. Other examples

$$H+D_2 \rightarrow H-D-D^\ddagger \rightarrow HD+D \qquad l_\ddagger = 2$$
$$O+CH_4 \rightarrow O-H-CH_3^\ddagger \rightarrow OH+CH_3 \qquad l_\ddagger = 4 \qquad \textbf{(21-91)}$$

In the computation of statistical factors, the following restriction on the symmetry of a hypothetical activated complex must be observed [8]: the complex *must* produce only one set of reactants and products. For example, the first reaction (20-91) cannot have a symmetrical complex $D'-H-D^\ddagger$, because it leads to two sets of products ($HD'+D$ and $HD+D'$).

The most difficult task in the calculation of rate constants from Eq. (21-89) is the determination of the moments of inertia and of the vibrational frequencies for the activated complex. If a potential energy surface in the vicinity of the activated complex is known, it is possible to calculate all necessary quantities unambiguously by procedures similar to the ones shown on pages 319, 323. The calculation of the rate constant from such data is shown in Example 21-1. In most cases when the potential energy surface is not known, one has to estimate the vibrational frequencies and moments of inertia by comparison with and extrapolation from related stable molecules. (Estimated properties of activated complex $CH_3NC^\ddagger$ are listed in Example 21-2.) Arrhenius factors that agree within an order of magnitude with experimental ones have been estimated in this way for a variety of reactions.

The estimation method is most readily applicable to the **isotope effect**, that is, to the calculation of rate constant ratios for isotopically substituted reactions such as

$$Cl + {}^iH^jH \rightarrow {}^iHCl + {}^jH \qquad (i \text{ or } j = 1, 2, 3) \qquad \textbf{(21-92)}$$

The major reason for the difference in the isotopic rate constants is that the reactants and activated complexes of the heavier isotopes have lower zero point vibrational energies than the lighter ones. This affects the activation energy $\mathscr{E}_a$. For example, $\mathscr{E}_a \approx \Delta\mathscr{E}_0$ for the dissociation of H_2 and D_2 is 103, and 105 kcal mol^{-1}, respectively. Less important contributions, except for reactants that consist entirely of hydrogen isotopes, come from the partition functions. The isotope effect is greatest if the substituted atom participates in a bond that is broken or formed during the reaction. Such effects are termed **primary**. The effects due to the isotopes of heavier atoms are smaller than those due to hydrogen isotopes. A calculation of the isotope effect is shown in Example 21-1. Some results for reaction (21-92) are shown in Figure 21-14.

EXAMPLE 21-1
From the given data calculate the activated complex theory rate constants, k and k', as well as k/k', for the reactions

$$H+H_2 \xrightarrow{k} H_2+H$$

and

$$H+D_2 \xrightarrow{k'} HD+D$$

at 500 K.

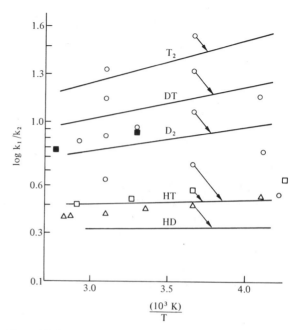

Figure 21-14 Isotope effect for the reaction of chlorine atoms with isotopic hydrogen molecules (reactions 21-92). k_1, rate constant for $Cl + H_2$ reaction; k_2, for $Cl +$ isotopic H_2 shown above each solid line. The solid lines are activated complex theory results (tunneling included) combined with semiempirical activation energies. $\bigcirc, \square, \triangle$ indicate experimental results. [From *Theories of Chemical Reaction Rates* by K. J. Laidler. Copyright 1969. Used with permission of McGraw-Hill Book Company.]

Given:

$$m_H = 1.0078 \text{ g mol}^{-1} = 1.673 \times 10^{-27} \text{ kg atom}^{-1}$$
$$m_D = 2.0141 \text{ g mol}^{-1} = 3.344 \times 10^{-27} \text{ kg atom}^{-1}$$

Moments of inertia:

$$I = 0.2771 \text{ g Å}^2 \text{ mol}^{-1} = 4.602 \times 10^{-48} \text{ kg m}^2 (H_2)$$
$$I' = 0.5538 \text{ g Å}^2 \text{ mol}^{-1} = 9.197 \times 10^{-48} \text{ kg m}^2 (D_2)$$
$$I_{\ddagger} = 1.742 \text{ g Å}^2 \text{ mol}^{-1} = 28.93 \times 10^{-48} \text{ kg m}^2 (H_3)$$
$$I'_{\ddagger} = 2.440 \text{ g Å}^2 \text{ mol}^{-1} = 40.50 \times 10^{-48} \text{ kg m}^2 (HD_2)$$

Equilibrium vibrational frequencies (s, stretch; b, bend):

$\omega = 4395 \text{ cm}^{-1} (H_2)$ $\qquad\qquad\qquad$ $\omega' = 3118 \text{ cm}^{-1} (D_2)$

$\omega_{s\ddagger} = 2046 \pm 25 \text{ cm}^{-1} (\leftarrow H - H - H \rightarrow)$ $\qquad$ $\omega'_{s\ddagger} = 1760 \text{ cm}^{-1} (\leftarrow H - D - D \rightarrow)$

$\omega_{b\ddagger} = 975 \pm 27 \text{ cm}^{-1} (H \overset{\uparrow}{-} H \overset{}{-} H)$ $\qquad\qquad$ $\omega'_{b\ddagger} = 745 \text{ cm}^{-1} (H \overset{\uparrow}{-} D - D)$
$\qquad\qquad\qquad\quad \downarrow \qquad \downarrow$ $\qquad\qquad\qquad\qquad\qquad\quad \downarrow \qquad \downarrow$

Potential energy barrier $V(0) = 39.7 \text{ kJ mol}^{-1}$.

The data for H, D, H_2, D_2 are experimental. The data involving the linear (corresponding to lowest possible potential energy barrier) activated complexes H_3 and HD_2 are calculated from G. W. Koeppl[24]. They are based on the best available theoretical potential energy surface (see Figure 21-7).

Notation: Properties of H, D, and the activated complexes are indicated by subscripts H, D, and ‡ respectively. Properties related to the deuterium reaction are primed.

Solution: The calculation of the rate constants is quite similar to the statistical calculation of equilibrium constants (see Example 19-6, page 608). First compute the vibrational factors $x = hv/kT = hc\omega/kT$ and the dissociation energies at 0 K, $\Delta\mathscr{E}_0$.

For the vibrational factors, using

$$h = 6.626 \times 10^{-27} \text{ erg} \qquad\qquad c = 2.998 \times 10^{10} \text{ cm s}^{-1}$$
$$k = 1.381 \times 10^{-16} \text{ erg K}^{-1} \text{ molecule}^{-1} \qquad T = 500 \text{ K}$$

we get for H_2

$$x = [(6.626 \times 10^{-27})(2.998 \times 10^{10})(4395)]/(1.381 \times 10^{-16})(500)$$
$$= 12.64$$

The six vibrational factors are

$$x = 12.64 \qquad x' = 8.970$$
$$x_{s\ddagger} = 5.886 \qquad x'_{s\ddagger} = 5.063$$
$$x_{b\ddagger} = 2.805 \qquad x'_{b\ddagger} = 2.143$$

The activation energy at 0 K for the H_2 reaction equals the potential energy barrier plus the difference in the zero point energies of the activated complex and H_2:

$$\Delta\mathscr{E}_0 = \mathscr{V}(0) + \mathscr{E}_{z\ddagger} - \mathscr{E}_z$$

The activated complex has three vibrational degrees of freedom, two of which (the bends) are degenerate. Since each degree contributes $\frac{1}{2}hv$ to the zero point energy (page 315) the result is

$$\Delta\mathscr{E}_0 = \mathscr{V}(0) + \tfrac{1}{2}Nhc(\omega_{s\ddagger} + 2\omega_{b\ddagger} - \omega)$$
$$= [39{,}700 + (\tfrac{1}{2}N)(6.626 \times 10^{-34})(2.998 \times 10^{10})$$
$$\times (2046 + 2 \times 975 - 4395)] \text{J mol}^{-1}$$
$$= 37{,}300 \text{ J mol}^{-1}$$

With the corresponding primed quantities, $\Delta\mathscr{E}'_0 = 40{,}500 \text{ J mol}^{-1}$.

The rate constant k is calculated from Eq. (21-89) with the ratios of the partition functions explicitly written for the first reaction (compare with Example 19-6, page

608)

$$k = l_{\ddagger}[(kT)/h]\left[Nh^3\left(\frac{3}{2\pi kT \cdot 2m_H}\right)^{3/2}\right](I_{\ddagger}/I)\left[\frac{(1-e^{-x})}{(1-e^{-x_{s\ddagger}})(1-e^{-x_{b\ddagger}})^2}\right]e^{-\Delta\mathscr{E}_0/RT}$$

$$= 2\left[\frac{(1.381\times10^{-23})(500)}{6.626\times10^{-34}}\right]\left\{(6.022\times10^{23})(6.626\times10^{-34})^3\right.$$

$$\times\left[\frac{3}{2\pi(1.381\times10^{-23})(500)(2)(1.673\times10^{-27})}\right]^{3/2}\right\}$$

$$\times\left[\frac{28.93\times10^{-48}}{4.602\times10^{-48}}\right]\left[\frac{(1-e^{-12.64})}{(1-e^{-5.886})(1-e^{-2.805})^2}\right]$$

$$\times(e^{-37,300/(8.314\times500)})m^3\,mol^{-1}\,s^{-1}$$

$$= (2.084\times10^{13})\times(5.200\times10^{-7})\times6.289\times1.136$$

$$\times(1.269\times10^{-4})m^3\,mol^{-1}\,s^{-1}$$

$$= 2k_{univ}\times k_{tr}\times k_{rot}\times k_{vib}\times k_{barrier}$$

$$= 9.820\times10^3\,m^3\,mol^{-1}\,s^{-1} = 9.820\times10^6\,dm^3\,mol^{-1}\,s^{-1} \qquad \textbf{(a)}$$

If Eq. (a) is divided by the analogous equation for the second reaction, the ratio

$$k/k' = \left[\frac{3m_Hm_D}{(m_H+2m_D)m_H}\right]^{3/2}\left[\frac{I_{\ddagger}I'}{I'_{\ddagger}I}\right]\left[\frac{(1-e^{-x})(1-e^{-x'_{s\ddagger}})(1-e^{-x'_{b\ddagger}})^2}{(1-e^{-x_{s\ddagger}})(1-e^{-x_{b\ddagger}})^2(1-e^{-x'})}\right]e^{-(\Delta\mathscr{E}_0-\Delta\mathscr{E}'_0)/RT}$$

$$= 1.314\times1.427\times0.8797\times2.159$$

$$= (k/k')_{tr}(k/k')_{rot}(k/k')_{vib}(k/k')_{barrier}$$

$$= 3.562 \qquad \textbf{(b)}$$

From Eqs. (a) and (b) we can calculate

$$k' = 9.820\times10^6/3.562\,dm^3\,mol^{-1}\,s^{-1}$$
$$= 2.8\times10^6\,dm^3\,mol^{-1}\,s^{-1}$$

These results agree fairly well with experimental data [10]:

$$k = 16.6\times10^6 \qquad k' = 3.402\times10^6\,dm^3\,mol^{-1}\,s^{-1} \qquad k/k' = 4.881$$

The two major sources of the remaining theoretical-experiment differences are thought to be

1. erroneously high experimental values of k, especially for $H+H_2$
2. neglect of tunneling in the theoretical k

We have factored the rate constant and the ratio k/k' so that the magnitude of the contributions from the universal frequency, from the translational, rotational, and vibrational degrees of freedom, and from the potential energy barrier can be compared. Note that k_{tr} is unit dependent; that is, it depends upon the volume of the "box" ($1\,m^3$ or $1\,dm^3$) in which the translational motion occurs. ●

21-5.3 Refinements: the Transmission Coefficient and the Activity

Two refinements are frequently introduced in the fundamental equation of activated complex theory. The first one is to multiply the right-hand side of Eq. (21-85) by the **transmission coefficient** κ. A general procedure for calculating κ does not exist. Qualitatively it accounts for any of the following four factors, important for some but not all reactions: (1) tunneling; (2) nonstandard shape of the potential energy surface; (3) products not deactivated; and (4) crossing to another potential energy surface.

Tunneling would allow an activated complex with an energy less than the potential energy barrier, $V(0)$, plus the zero point energy, $E_z(0)$. This additional contribution to product formation would give $\kappa > 1$. Tunneling should be relatively more important for reactions of light atoms and at low temperatures where the contributions from over-the-barrier reaction would be very small. If we assume that the discrepancy between the theoretical and experimental rate constants for the $H + D_2$ reaction (Example 21-1) is due to tunneling, $\kappa = k'_{exp}/k'_{theor} = 16.6 \times 10^6/2.8 \times 10^6 = 5.9$ at 500 K. For $D + H_2$ at 200 K, $\kappa = 12$ has been estimated.

The second factor (which would cause a transmission coefficient $\kappa < 1$) is a potential energy surface with a narrow, topographically complex valley leading from the reactants to the products or else a surface that stabilizes the activated complex (unlike the ones shown in Figure 21-7). In all such cases some trajectories would be reflected back to the reactant side after one or several limited excursions past the activated complex configuration toward the products. If the number of reflected trajectories is n_b then $\kappa = n_b/n_{tot}$. Products that retain the activation energy in the reaction coordinate (see Section 21-4) also have a chance to return to the reactants, and therefore the reaction would have $\kappa < 1$.

The final factor which causes $\kappa \ll 1$ is the existence of reactants and products in different electronic energy states, that is, on different potential energy surfaces. In this case a crossing from one to the other potential energy surface must occur with a characteristic transmission coefficient κ. The magnitude of κ can be predicted from selection rules analogous to (but not necessarily identical with) those for optical transitions (page 422). The rule easiest to apply is the electron spin conservation rule $|S_P| - |S_R| = 0$. Consider the two exothermic reactions

$$O_3(^1A) + NO(^2\Pi) \rightarrow NO_2(^2A) + O_2(^3\Sigma) \qquad \textbf{(21-93a)}$$

and

$$O_3(^1A) + CO(^1\Sigma) \rightarrow CO_2(^1\Sigma) + O_2(^3\Sigma) \qquad \textbf{(21-93b)}$$

where the electronic states of the molecules are designated in the parentheses. The superscripts specify the spin multiplet $(2S + 1)$. The total spin of the reactants or the products can be obtained by vector addition of the spins of the individual reactants or products. Thus for O_3 with $S = 0$ and NO with $S = \frac{1}{2}$ we get $|S_R| = \frac{1}{2}$. Similarly for the products of reaction (21-93a) $|S_P| = \frac{1}{2}$ and 3/2. Therefore, $|S_P - S_R| = \frac{1}{2} - \frac{1}{2} = 0$, and $\kappa \approx 1$. For reaction (21-93b) $S_R = 0$, $S_P = \frac{1}{2}$, therefore $\kappa \approx 0$.

Experimentally it is observed that reaction (21-93a) proceeds at a measurable rate whereas reaction (21-93b) is unmeasurably slow. However, reactions are known (for example, thermal decomposition of N_2O, some ion reactions, collisional deactivation of excited electronic states of molecules) that do not conserve spin and, nevertheless, are quite fast.

A second refinement of Eq. (21-85) is to assume that the reactants and the activated complex are nonideal. This means that activities, a, rather than concentrations appear in Eq. (21-82). Since, for example, $a_A = \gamma_A[A]$, and so on, Eq. (21-85) would change to

$$k = kT/h(q^{\ddagger}/q_A q_B)\, e^{-\Delta \mathscr{E}_0/RT} \cdot (\gamma_A \gamma_B / \gamma_{\ddagger})$$
$$= k^{\circ} \gamma_A \gamma_B / \gamma_{\ddagger} \tag{21-94}$$

where k° is the standard state rate constant (see Eq. 20-54). Eq. (21-94) is most frequently applied to ionic reactions in solution (Section 24-16).

The relation between gas phase and liquid phase rate constants of an elementary reaction may be elucidated with the help of Eq. (21-94). If we assume, for both phases, the gas phase to be the standard state for the reactants and the activated complex, k° is the same in gas and in liquid and the activity coefficients for the gas phase are equal to 1. Therefore,

$$k_{\text{liq}}/k_{\text{gas}} = \gamma_A \gamma_B / \gamma_{\ddagger} \tag{21-95}$$

where the activity coefficients are those for the liquid state. Since at equilibrium

$$\gamma_A A_{\text{liq}} = A_{\text{gas}} \tag{21-96}$$

the activity coefficients of reactants can be easily determined from the equilibrium concentrations. For diatomic gases in water they range from 30 (for O_2) to about 50 (for H_2). If we assume the same range of γ's for the activated complexes then, from Eq. (21-95) $k_{\text{liq}}/k_{\text{gas}}$ should range from about 20 ($= 30 \times 30/50$) to about 80 ($= 50 \times 50/30$). Few such ratios have been determined experimentally. A high ratio has been measured for $H + D_2 \rightarrow HD + D$ ($k_{\text{liq}}/k_{\text{gas}} = 40$ at 25°C), a low one for $OH + CO \rightarrow CO_2H$ (6 at 25°C) [5, p. 181]. With less polar (more gas like) liquids the rate constant differences in the two phases would be even less (see Table 21-3).

The validity of the activated complex theory rests on the assumption of Boltzmann ("equilibrium") distribution for the activated complex. Thermal equilibrium is established by energy exchanging collisions. In Section 21-2.5 we calculated that, on the average, such collisions occur every 10^{-12} to 10^{-13} s in liquids, and approximately every 10^{-9} s in gases. On the other hand, Figure 21-6 shows that the reactants pass through the activated complex region in about 10^{-14} s. How, then, can we assume, as we did in Example 21-1, that the activated complex theory applies to the $H + H_2$ reaction? The activated complexes, without being in thermal equilibrium, may simply retain the Boltzmann distribution of the reactants from which they were formed. It can be shown that retention does occur if the reactants stay in the same vibrational level relative to ground state throughout the reaction. For example, if the vibrational quantum numbers are v for H_2 and v_s, v_b, v_b' for the three modes of $H_3^{\ddagger}$, then $H + H_2$ ($v = 0$) $\rightarrow H_3^{\ddagger}$ ($v_s = v_b = v_b' = 0$) would maintain the Boltzmann distribution of the activated complex, where $H + H_2$ ($v = 0$) $\rightarrow H_3$ ($v_s = v_b = v_b' \neq 0$) would not. The Boltzmann distribution would be approximately maintained if the first process were considerably more frequent than the second. This is known to be true for the $H + H_2$ reaction. In other cases, especially in liquids, a sufficiently long-lived complex may exist to be in true thermal equilibrium.

21-6   UNIMOLECULAR REACTION THEORY

Simple collision theory provides a detailed expression (Eq. 21-71) for the unimolecular rate constant. This expression contains the unknown product formation rate constants k_{P_j}, which vary with the internal energy level j of the activated molecule A*. Since activated complex theory does take into account internal energy levels, it can be used to calculate k_{P_j} and, hence, k.

The earliest attempts to calculate k_{P_j} on the basis of a rather artificial molecular model for A* were made by Rice and Ramsperger in 1927 and by Kassel in 1928. Their model assumed that A* consists of s [$\leqq (3N-6)$] harmonic oscillators, all of same frequency, which randomly exchange energy until the oscillation along the reaction coordinate acquires energy in excess of the critical energy E_c and the products are formed. The number of oscillators s was treated as an empirically adjustable parameter. This model and theory is known by the initials RRK.

21-6.1 The RRKM Method

In 1952 Marcus refined the RRK theory by taking into account all vibrations and rotations of A* in terms of the actual vibration frequencies and rotation constants. This was done within the framework of the activated complex theory. The improved theory, known by the initials RRKM, is the most successful and refined statistical theory of reaction kinetics in existence.

The RRKM theory accepts the expression for the molecular rate constant, Eq. (21-71). The theory provides an expression for k_{P_j} which, combined with Eq. (21-71), yields an expression for calculating unimolecular rate constants.

In accordance with activated complex theory we assume

$$A_j^* \overset{K}{\rightleftharpoons} A_j^\ddagger \overset{\nu}{\longrightarrow} P \tag{21-97}$$

where $A_j^\ddagger$ is the activated complex with the same internal energy as A_j^*, K is a pseudoequilibrium constant, and ν is the frequency with which $A_j^\ddagger$ crosses to the product side.

The various internal energies of A_j^* and $A_j^\ddagger$ are shown and defined in Figure 21-15. In the simplest model the adiabatic energy of A_j^* is the rotational energy. The active internal energy E_j^* is the vibrational energy (above the zero point) of the $3N-6$ ($3N-5$ for a linear molecule) vibrations or internal rotations of the activated molecule in the jth energy level. In the activated complex $A_j^\ddagger$, which exists at and near the top of the potential energy barrier along the reaction coordinate, a vibration has been replaced by motion along the coordinate. The potential energy of the coordinate is not counted as active because it cannot be transferred to vibrations without destruction of the activated complex. The activated complex $A_j^\ddagger$ has considerably less active energy ($E^\ddagger$) than A_j^*, although both have the same total energy.

The exchange of energy between the active (vibrational) modes, including the reaction coordinate, occurs on a time scale set by the period of a molecular vibration, that is, typically 10^{-14} to 10^{-13} s. The exchange is assumed to be completely random so that after several periods the most probable (that is, Boltzmann) distribution of the active energy in A_j^* and $A_j^\ddagger$ is achieved. The resulting concentration ratio $A_j^\ddagger/A_j^*$ largely determines k_{P_j}.

The rate of product formation from A_j^* is

$$r_j = k_{P_j} A_j^* = \nu A_j^\ddagger = \nu K A_j^* \tag{21-98}$$

where the second and third equalities follow from Eq. (21-97) and the activated complex theory assumption (see Eq. 21-81). The rate constant k_{P_j} may be expressed

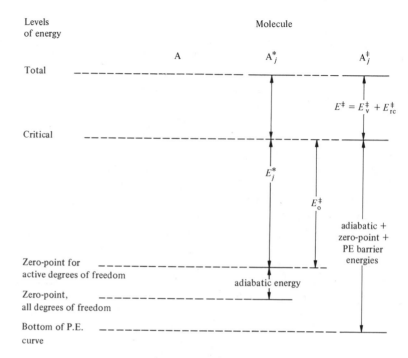

Figure 21-15 Relations among energies stored in the various forms—A, A_j^*, $A_j^\ddagger$—
of a unimolecular reactant. E_j^*, the active or nonfixed energy of A_j^* that can be
exchanged between different internal energy modes, including the reaction coordi-
nates $E^\ddagger$, same for activated complex; $E_0^\ddagger$, the part of E_j^* which becomes fixed in $A_j^\ddagger$
as potential energy of the reaction coordinate. $E_v^\ddagger$ and $E_{rc}^\ddagger$, variable amount of active
energy in the vibrations and in motion along the reaction coordinate of activated
complex. The **fixed** or **adiabatic energy** is internal energy that cannot flow into the
reaction coordinate. Rotational energy is fixed due to angular momentum conservation. In more detailed models part of rotational energy may be active.

in terms of partition functions (see Eq. 21-82)

$$k_{P_j} = \nu K = \nu q_{\ddagger j}/q_j^* = \nu l_\ddagger (q^{o\ddagger}/q_A^o)_{rot}(q_{\ddagger j}/q_j^*)_{vib} \qquad \textbf{(21-99)}$$

where $l_\ddagger$ is the statistical factor (page 788). The rotational partition functions were
defined in Eq. (21-90). Note that q_{rot}^o of A_j^* has been assumed to equal that of A. The
equal masses of A_j^* and $A_j^\ddagger$ make their translational partition functions cancel in Eq.
(21-99) (see Eq. 21-89).

21-6.2 Evaluation of Vibrational Partition Functions Since the vibrational
energy of either species is measured from the same level of energy (designated "zero
point for active degrees of freedom" in Figure 21-15), there is no $\Delta\mathscr{E}_0$ term in Eq.
(21-99) (see Eq. 21-89). Since the total energy of A_j^* and $A_j^\ddagger$ is the same, the
vibrational partition function ratio reduces to

$$(q_{\ddagger j}/q_j^*)_{vib} = g_{\ddagger j}/g_j^* \qquad \textbf{(21-100)}$$

where $g_{\ddagger j}$ and g_j^* are the degeneracies of the jth energy level of $A_j^{\ddagger}$ and A_j^*, respectively.

RRKM theory evaluates Eq. (21-100) as follows. For molecules with more than two or three atoms above room temperature, energy level spacings are nearly zero so that the $g_{\ddagger j}$ and g_j^* may be replaced with the average number of quantum states in an energy range dE^* centered on the jth level. If

$$G^*(E^*) = \left(\frac{dg^*}{dE^*}\right)_{E^* \text{ at } j} \qquad (21\text{-}101)$$

is the density of quantum states per unit energy at E_j^*, then the required average number of quantum states is $G^*(E^*)\,dE^*$. The replacement ($\rightarrow$) is

$$g_j^* \rightarrow G^*(E^*)\,dE^* = G^*(E^{\ddagger}+E_0^{\ddagger})\,dE^{\ddagger} \qquad (21\text{-}102)$$

where the equality follows from the relations $E^* = E^{\ddagger}+E_0^{\ddagger}$ (Figure 21-15) and $dE^* = dE^{\ddagger}$. Similarly,

$$g_{\ddagger j} \rightarrow G_{\ddagger}(E^{\ddagger})\,dE^{\ddagger} = G_{\ddagger}(E^*-E_0^{\ddagger})\,dE^* \qquad (21\text{-}103)$$

The contribution to degeneracy $g_{\ddagger j}$ from a particular combination of a reaction coordinate energy level n (with degeneracy $g_n^{\ddagger}$) and an energy level i (with degeneracy $g_i^{\ddagger}$) of the other active modes is $g_i^{\ddagger} g_n^{\ddagger}$ because any of the quantum states that belongs to n can combine with any quantum state of i.

The state density $G_{\ddagger}(E^{\ddagger})$ contains analogous contributions. Assuming that the n levels are continuous with density function $G_{\text{rc}}(E_{\text{rc}}^{\ddagger})$ (Figure 21-15) whereas the somewhat more widely spaced vibrational levels remain discrete, the contribution from the ith vibrational level is $g_i^{\ddagger} G_{\text{rc}}(E_{\text{rc}}^{\ddagger})$. For a reaction coordinate assumed to be a very low frequency harmonic oscillator the equation (derived from Eq. 11-16),

$$dE_{\text{rc}}^{\ddagger} = h\nu\,dn \qquad (21\text{-}104)$$

relates the number of energy levels dn to the range $dE_{\text{rc}}^{\ddagger}$. Hence,

$$G_{\text{rc}}(E_{\text{rc}}^{\ddagger}) = dn/dE_{\text{rc}}^{\ddagger} = (h\nu)^{-1} \qquad (21\text{-}105)$$

The allowed values of vibrational index i for a given j (or $E^{\ddagger}$, Figure 21-15) are determined by energy conservation. Since

$$E^{\ddagger} = E_v^{\ddagger} + E_{\text{rc}}^{\ddagger} \qquad (21\text{-}106)$$

the lower and upper limits of i must be

$$i = 0 \qquad \text{when } E_v^{\ddagger} = 0 \text{ and } E^{\ddagger} = E_{\text{rc}}^{\ddagger} \qquad (21\text{-}107)$$

and

$$i = i_{\max} \qquad \text{when } E^{\ddagger} = E_v^{\ddagger} \text{ and } E_{\text{rc}}^{\ddagger} = 0 \qquad (21\text{-}108)$$

Therefore, from Eqs. (21-105), (21-107), and (21-108)

$$G_{\ddagger}(E^{\ddagger}) = \sum_{i=0}^{i=i_{\max}} g_i^{\ddagger} G_{\text{rc}}(E^{\ddagger}) = \sum_{i=0}^{i=i_{\max}} g_i^{\ddagger}(h\nu)^{-1} = g^{\ddagger}(E^{\ddagger})/h\nu \qquad (21\text{-}109)$$

where the last equality defines the vibrational degeneracy $g^{\ddagger}(E^{\ddagger})$ of $A^{\ddagger}$ with energy $E^{\ddagger}$. Eq. (21-109) and replacements (21-102), (21-103) when introduced into Eq.

(21-100) yield

$$(q_{\ddagger j}/q_j^*)_{\text{vib}} = (h\nu)^{-1}g^{\ddagger}(E^* - E_0^{\ddagger})/G^*(E^*)$$
$$= (h\nu)^{-1}g^{\ddagger}(E^{\ddagger})/G^*(E^{\ddagger} + E_0^{\ddagger}) \qquad \textbf{(21-110)}$$

Substitution of Eq. (21-110) into Eq. (21-99) gives the rate constant

$$k_{P_j} = \frac{l_{\ddagger}}{h}\left(\frac{q^{\circ\ddagger}}{q_A^{\circ}}\right)_{\text{rot}} \frac{g^{\ddagger}(E^* - E_0^{\ddagger})}{G^*(E^*)} \qquad \textbf{(21-111)}$$

where the energies correspond to the jth level.

In the computation of the degeneracy, $g^{\ddagger}(E^{\ddagger})$, and the density of states, $G^*(E^*)$, all vibrational modes of A^* and $A^{\ddagger}$ are usually treated as harmonic oscillators. For this model $g^{\ddagger}(E^{\ddagger})$ will be simply the sum of all states (Eq. 21-109) whose energies, $E_v^{\ddagger}$, obey the condition

$$E^{\ddagger} \geq E_v^{\ddagger} = \sum_{i=0}^{3N-7}(h\nu_i)(n_i) \qquad \textbf{(21-112)}$$

The equality of Eq. (21-112) states that the total vibrational energy of the activated complex is a sum over the harmonic oscillator energies in each of the normal vibrations with frequencies ν_i and quantum numbers n_i. (For linear molecules the sum would have $3N - 6$ terms.) Example 21-2 shows an actual computation of $g^{\ddagger}(E^{\ddagger})$ for the activated complex of reaction (21-55), and discusses the closely related computation of $G^*(E^*)$ from $g^*(E^*)$. Figure 21-16 shows further results on $g^{\ddagger}(E^{\ddagger})$ and $g^*(E^*)$.

21-6.3 The RRKM Rate Constants

To introduce the quantum state density $G^*(E^*)$ in the expression (21-71) for the unimolecular rate constant, the Boltzmann population fraction of the jth level, f_{B_j}, has to be converted to a continuous form by replacing it with the average fraction in the energy range dE^* centered on the jth level. From Eqs. (19-25) and (21-102) we get the replacement

$$f_{B_j} = q_A^{-1}g_j^* e^{-E_j^*/kT} \rightarrow q_A^{-1}G^*(E^*)e^{-E^*/kT}dE^* \qquad \textbf{(21-113)}$$

where q_A is the vibrational partition function of A. Eqs. (21-111) and (21-113) substituted into Eq. (21-71) yield

$$k = l_{\ddagger}(hq_A)^{-1}\left(\frac{q^{\circ\ddagger}}{q_A^{\circ}}\right)_{\text{rot}}\int_{E_0^{\ddagger}}^{\infty} \frac{g^{\ddagger}(E^* - E_0^{\ddagger})e^{-E^*/kT}dE^*}{\left[1 + \dfrac{l_{\ddagger}q_{\text{rot}}^{\circ\ddagger}g^{\ddagger}(E^* - E_0^{\ddagger})}{hq_{A\text{rot}}^{\circ}G^*(E^*)k_dM}\right]} \qquad \textbf{(21-114)}$$

At high pressures $(M \rightarrow \infty)$ the integral can be readily shown to reduce to $kTq_{\text{vib}}^{\ddagger}e^{-E_0^{\ddagger}/kT}$. Therefore,

$$k_{\infty} = l_{\ddagger}(kT/h)(q^{\circ\ddagger}/q_A)_{\text{rot}}(q^{\ddagger}/q_A)_{\text{vib}}e^{-E_0^{\ddagger}/kT} \qquad \textbf{(21-115)}$$

The ratio k/k_{∞}, computed for reaction (21-55) from Eqs. (21-114) and (21-115) as described in Example 21-2, is shown as a function of the pressure of M in Figure 21-11. The agreement with experiment is seen to be good.

The low pressure form of Eq. (21-114) is easily proved to be

$$k_0 = k_d(q^*/q_A)_{\text{vib}}M \qquad \textbf{(21-116)}$$

Eqs. (21-115) and (21-116) are RRKM forms of Eqs. (21-72) and (21-73).

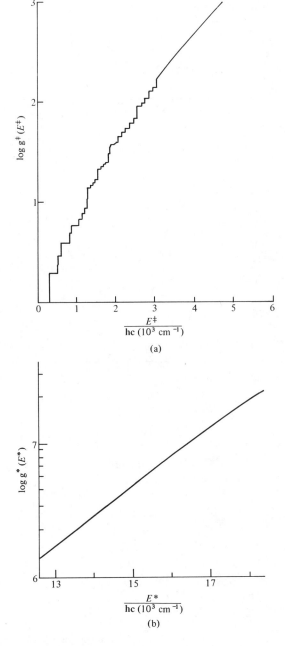

Figure 21-16 RRKM theory results for the unimolecular reaction $CH_3NC \rightarrow CH_3CN$ (see text and Example 21-2). (a) The degeneracy $g^{\ddagger}(E^{\ddagger})$ of the activated complex. (b) The degeneracy $g^*(E^*)$ for the activated molecule. [Reprinted with permission from F. W. Schneider and B. S. Rabinovitch: *J. Am. Chem. Soc.*, **84**:4215 (1962).]

EXAMPLE 21-2

In this example we illustrate some aspects of calculating the unimolecular rate constant for reaction (21-55) that is

$$CH_3NC \rightarrow CH_3CN$$

The data for CH_3NC and for the activated complex $CH_3NC^{\ddagger}$ are listed in Tables 21-4

Table 21-4 Vibration frequencies

Type of Vibration	CH₃NC $\tilde{\nu}/(cm^{-1})$	CH₃NC‡ $\tilde{\nu}/(cm^{-1})$	$\dfrac{Nh\tilde{\nu}c}{(kJ\ mol^{-1})}$	Mode Index i
C—H stretch	2998(3)[a]	2998(3)[a]	36	1, 2, 3
N≡C stretch	2166	1990	24	4
CH₃ deformation	1454(3)	1454(3)	17	5, 6, 7
CH₃ rocking	1129(2)	1129(2)	14	8, 9
C—N stretch	945	650	8	10
C—N—C deformation	263(2)	263	3	11, (12)
		reaction		
		coordinate		12

[a] The numbers in parentheses are the degeneracies.

Table 21-5 Bond lengths and angles

	C—H	C—N	C≡N	C—C	HCH
CH₃NC (r in nm)	0.1094	0.1427	0.1167		109° 46′
CH₃NC‡ (r in nm)	0.1094	161	0.120	0.194	109° 46′

Table 21-6 Moments of inertia

	I_A	I_B	I_C
CH₃NC (u nm²)[a]	0.032	0.503	0.503
CH₃NC‡ (u nm²)	0.116	0.316	0.400

The critical energy $\mathscr{E}_0^{‡} = 37.85$ kcal mol⁻¹ = 158 kJ mol⁻¹
The collision radius of CH₃NC = 0.450 nm

[a] u nm² = unified atomic mass unit × (nanometers)², see Table IX-5.

through 21-6. The data are taken from a definitive study of the reaction by F. W. Schneider and B. S. Rabinovitch [25] and B. S. Rabinovitch et al. [26].

The activated complex was assumed to be

The arrows show the atomic motions that correspond to the motion of the activated complex along the reaction coordinate, toward the product CH_3CN. The vibrational frequencies, bond lengths, and bond angles of the activated complex have been assumed the same as in CH_3NC for the methyl group (which is not directly involved in the reaction). The other frequencies and bond lengths are obtained from comparisons with related molecules and from empirical bond length–bond order–vibrational frequency relations. A final minor trial and error adjustment of the frequencies is made to make $(k/k_\infty)_{exptl} = (k/k_\infty)_{theor}$ at *one* pressure of M, or else to make $(k_\infty)_{exptl} = (k_\infty)_{theor}$.

Calculation of the factors in Eqs. (21-114) and (21-115):

1. The statistical factor $l_\ddagger = 3$ because three equivalent activated complexes may be formed by rotating the methyl group.

2. The ratio of the rotational partition functions. From Eq. 19-92, Eq. (21-90), and the data in Table 21-6

$$(q^{o\ddagger}/q_A^o)_{rot} = \left(\frac{I_A^\ddagger I_B^\ddagger I_C^\ddagger}{I_A I_B I_C}\right)^{1/2} = \left(\frac{0.116 \times 0.316 \times 0.400}{0.032 \times 0.503 \times 0.503}\right)^{1/2}$$

$$= 1.35$$

3. The vibrational partition functions. The frequencies in Table 21-4 are labelled with index i running from 1 to 12 for CH_3NC and from 1 to 11 for $CH_3NC^\ddagger$. Start the index at the highest frequency, label each degenerate mode separately, and end at the lowest frequency. Compute $x_i = (h\tilde{\nu}_i c/kT)$ for all i of both species. Sample computations at 472 K (as in Figure 21-11) for the C—H stretch frequencies of CH_3NC (see Appendix IX for the universal constants):

$$x_1 = x_2 = x_3 = (h\tilde{\nu}_1 c/kT) = \frac{(6.626 \times 10^{-34})(2998)(2.998 \times 10^{10})}{(1.381 \times 10^{-23})(472)}$$

$$= 9.134$$

The lowest possible x_i of CH_3NC is $x_{11} = x_{12} = 0.801$.

From Eq. 19-96 and the x_i computed as shown we get the partition function for each vibration of a molecule. The individual partition functions when multiplied together yield the total vibrational partition function for CH_3NC

$$q_{vib} = \prod_{i=1}^{12} (1 - e^{-x_i})^{-1} = (1 - e^{-9.134})^{-3} \cdots (1 - e^{-0.801})^{-2}$$

$$= (1.000)^3 \cdots (1.814)^2 = 3.86$$

and for $CH_3NC^\ddagger$

$$q_{vib}^\ddagger = \prod_{i=1}^{11} (1 - e^{-x_i})^{-1} = 2.33$$

4. The degeneracy $g^\ddagger(E^\ddagger)$ *and the state density* $G^*(E^*)$. Compute $g^\ddagger(E^\ddagger)$ of $CH_3NC^\ddagger$ for $\mathscr{E}^\ddagger = 10, 20, 30,$ and 40 kJ mol^{-1} by counting the number of vibrational quantum states that obey Eq. (21-112). Assume a harmonic oscillator model and use the approximate energies of the vibrational quanta listed in the next-to-last column of Table 21-4.

Table 21-7

Energy Range[a]	Vibrational Quanta						Number of States	$\mathscr{E}^{\ddagger}$ (highest state) $\overline{(\text{kJ mol}^{-1})}$
	n_i $(i=1\text{--}3)$	n_4	n_i $(i=5\text{--}7)$	n_i $(i=8,9)$	n_{10}	n_{11}		
I	0	0	0	0	0	0–3	4	9
II	0	0	0	0	0	4–6	3	18
III	0	0	0	0	0	7–10	4	30
IV	0	0	0	0	0	11–13	3	39
I	0	0	0	0	1	0	1	8
II	0	0	0	0	1	1–4	4	20
III	0	0	0	0	1	5–7	3	29
IV	0	0	0	0	1	8–10	3	38
II	0	0	0	0	2	0, 1	2	19
III	0	0	0	0	2	2–4	3	28
IV	0	0	0	0	2	5–8	4	40
III	0	0	0	0	3	0–2	3	30
IV	0	0	0	0	3	3–5	3	39
IV	0	0	0	0	4	0–2	3	38
IV	0	0	0	0	5	0	1	40
II	0	0	0	1[b]	0	0–2	6	20
III	0	0	0	1	0	3–5	6	29
IV	0	0	0	1	0	6–8	6	38
III	0	0	0	1	1	0–2	6	28
IV	0	0	0	1	1	3–6	8	40
III	0	0	0	1	2	0	2	30
IV	0	0	0	1	2	1–3	6	39
III	0	0	0	2[c]	0	0	3	28
IV	0	0	0	2	0	1–4	12	40
IV	0	0	0	2	1	0–1	6	39
II	0	0	1[d]	0	0	0, 1	6	20
III	0	0	1	0	0	2–4	9	29
IV	0	0	1	0	0	5–7	9	38
III	0	0	1	0	1	0, 1	6	28
IV	0	0	1	0	1	2–5	12	40
IV	0	0	1	0	2	0–2	9	39
IV	0	0	1	1	0	0–3	24	40
IV	0	0	1	1	1	0	6	39
III	0	1	0	0	0	0–2	3	30
IV	0	1	0	0	0	3–5	3	39
IV	0	1	0	0	1	0–2	3	38
IV	0	1	0	0	2	0	1	40
IV	1[d]	0	0	0	0	0–1	6	39

[a] See Table 21-8 for definitions.
[b] Two possible states $n_8 n_9$: 01 and 10 (the quantum may be in either of two degenerate states).
[c] Three possible states $n_8 n_9$: 20, 02, and 11.
[d] Three possible states $n_1 n_2 n_3$ or $n_5 n_6 n_7$: 100, 010, and 001.

Table 21-8

Range of $\mathscr{E}^{\ddagger}$/kJ mol^{-1}	Number of States (Entries from Table 21-7)	Total Number of States	$g^{\ddagger}(E^{\ddagger})$
$0 \leq \mathscr{E}^{\ddagger} \leq 10$ (I)	4, 1	5	5
$10 < \mathscr{E}^{\ddagger} \leq 20$ (II)	3, 4, 2, 6, 6	21	26
$20 < \mathscr{E}^{\ddagger} \leq 30$ (III)	4, 3, 3, 3, 6, 6, 2, 3, 9, 3, 6	48	74
$30 < \mathscr{E}^{\ddagger} \leq 40$ (IV)	3, 3, 4, 3, 3, 1, 6, 8, 6, 12, 6, 9, 24, 6, 3, 12, 9, 3, 1, 6	130	204

A systematic method for doing this is illustrated in Tables 21-7 and 21-8. As a first step count the number of quantum states found in each of the four energy ranges defined in the first column of Table 21-8. The results are shown in Table 21-7.

By systematically incrementing the quantum numbers, lowest frequency vibration first, the counting procedure obtains (from Eq. 21-112) the energies of all possible quantum states up to 40 kJ mol^{-1} and counts the number of states per range. For example, the first few entries in Table 21-7 are obtained by starting with the ground state, that is, $n_1 = n_2 = \cdots = n_{11} = 0$. Raise n_{11}, one by one, compute the energy (Eq. 21-112) and write down (in the next-to-last column of Table 21-7) the number of states in each energy range.

For $n_{11} = 0$ to 4 get from Eq. (21-112) the energies $E_v = h\nu_{11}n_{11}$ whose values are 0×3, 1×3, 2×3, 3×3, and 4×3 kJ mol^{-1}. The first four belong in energy range I (the energy of the highest state in this range, that is, $n_{11} = 3$ is listed in the last column of Table 21-7), therefore enter "4" in the number of states column. The fifth energy belongs to the lowest state of range II. Continue to increase n_{11} until $n_{11} = 14$ when the energy $\mathscr{E}_v = 14 \times 3 = 42$ which is above range IV. Next set $n_{10} = 1$ and repeat the procedure of varying n_{11}, and so on. Energies of $\mathscr{E}^{\ddagger}$ higher than about $(3N - 7)\boldsymbol{R}T \approx 40$ kJ mol^{-1} do not contribute appreciably to the rate constant.

Table 21-8 shows the computation of the degeneracy $g^{\ddagger}(E^{\ddagger})$ for $\mathscr{E}^{\ddagger} = 10, 20, 30, 40$ kJ mol^{-1}. The second column shows entries from Table 21-7: the number of quantum states in each energy range for the various sets of quantum numbers. The third column is the sum of the entries in the second column, that is, the total number of states in each range. The fourth column lists the four $g^{\ddagger}(E^{\ddagger})$. Figure 21-16a shows the results with the exact energies (Table 21-4) for the vibrational quanta of $CH_3NC^{\ddagger}$, up to about 60 kJ mol^{-1}.

The first step in the computation of $G^*(E^*)$ is to calculate the degeneracy $g^*(E^*)$ as a function of energy. The degeneracy $g^*(E^*)$ is the same function as $g^{\ddagger}(E^{\ddagger})$, but defined for the 12 vibrational modes of CH_3NC. Because E^* is much higher than $E^{\ddagger}$, the computation of $g^*(E^*)$ can be tackled only with the help of a computer or else by various approximation formulas. The result for $g^*(E^*)$ of CH_3NC is shown in Figure 21-16b. The density of states is obtained from Eq. (21-101)

$$G^*(E^*) = dg^*(E^*)/dE^* = 2.303g^*(E^*)(d \log g^*(E^*)/dE^*)$$

The range of E^* in which $G^*(E^*)$ makes significant contributions to the integral of Eq. (21-114) is from $\mathscr{E}_0^{\ddagger}$ to approximately $\mathscr{E}_0^{\ddagger} + (3N - 6)\boldsymbol{R}T$, that is, for CH_3NC at 472 K from 158 to 205 kJ mol^{-1}. From Figure 21-16b the average $d \log$

$g^*(\mathcal{E}^*)/d\mathcal{E}^* \approx 0.016 \text{ mol kJ}^{-1}$, $g^*(158) = 1.9 \times 10^6$; $g^*(205) = 1.4 \times 10^7$. Therefore, $G^*(158) \approx 7.0 \times 10^4$ and $G^*(205) \approx 5.3 \times 10^5 \text{ mol kJ}^{-1}$.

5. The deactivation constant. This constant is calculated from Eq. (21-23) and the collision diameter from Table 21-6 with

$$M = CH_3NC \text{ and } N\mu = \tfrac{1}{2} \times 41.05 \text{ g mol}^{-1} = 0.0205 \text{ kg mol}^{-1}$$
$$
\begin{aligned}
k_d &= \tfrac{1}{4}\pi(\sigma_M + \sigma_M)^2 IN(8kT/\pi\mu)^{1/2} \\
&= \tfrac{1}{4}\pi^{1/2}(0.9 \times 10^{-9})^2(\tfrac{1}{2})(6.022 \times 10^{23})[(8)(1.381 \times 10^{23})(472)/(3.40 \times 10^{-26}]^{1/2} \\
&= 1.34 \times 10^8 \text{ m}^3 \text{ mol}^{-1} \text{ s}^{-1} = 1.34 \times 10^{11} \text{ dm}^3 \text{ mol}^{-1} \text{ s}^{-1}
\end{aligned}
$$

6. The integral in Eq. (21-114) for each pressure of M is evaluated numerically using the quantities calculated in paragraphs 1 through 5. The result, divided by the calculated k_∞ is plotted in Figure 21-11.

21-6.4 Activation Energy, Fall-off Region, and RRKM Theory   RRKM theory can explain the lowering of the experimental Arrhenius activation energy with decreasing pressure, shown for CH_3NC in Figure 21-17. In general, the larger the molecule the greater the lowering of activation energy with decreasing pressure.

The phenomenon can be explained with help of Eq. (21-1) which can be rewritten in terms of the energy symbols used in this section.

$$E_a = E_0^\ddagger + \langle E^\ddagger \rangle - \langle E(A) \rangle \tag{21-117}$$

where $\langle E^\ddagger \rangle$, $\langle E(A) \rangle$ are the average vibrational energies of $A^\ddagger$ and A, respectively. Figure 21-17 shows that the RRKM theory predicts a decrease of $\langle E^\ddagger \rangle$ with decreasing pressure that parallels the decrease in the experimental E_a. Why does the average active energy of the reacting molecules decrease? Eq. (21-111) shows that

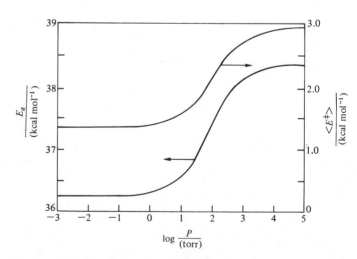

Figure 21-17   Plots showing the parallel variation of the experimental activation energy E_a and the theoretical (RRKM) average energy $E^\ddagger$ of the activated complex for reaction $CH_3NC \rightarrow CH_3CN$. [Reprinted with permission from F. Fletcher, B. S. Rabinovitch, K. Watkins, and D. J. Locker: *J. Phys. Chem.*, **70**:2823 (1966).]

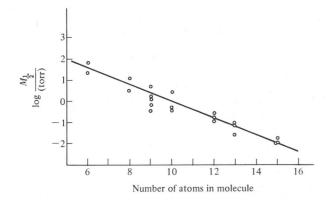

Figure 21-18 Dependence of fall-off pressure $M_{1/2}$ upon the number of atoms of the reacting molecule. ○, experimental points for 21 different molecules. [From P. J. Robinson and K. A. Holbrook: *Unimolecular Reactions*. Wiley, New York, 1972.]

k_{P_j} is proportional to the degeneracy $g^{\ddagger}(E^{\ddagger})$ which, according to Figure 21-16a, approaches zero as $E^{\ddagger} \rightarrow 0$. This means that k_{P_j} is very small at low active energy $E^{\ddagger}$. Low energy molecules are slow to react and have a better chance to be deactivated than high energy molecules. As pressure drops, the deactivation rate decreases. A larger and larger fraction of the low energy molecules react thus contributing to $\langle E^{\ddagger}\rangle$ which therefore decreases.

RRKM theory also explains the dependence of the fall-off region, characterized by the pressure $M_{1/2}$ (page 782), upon the number of atoms in the molecule. This approximate dependence is shown in Figure 21-18. We see that the fall-off pressure decreases with increasing number of atoms. From Eq. (21-71) it is clear that this is possible only if the average k_{P_j}/k_d decreases with the number of atoms so that $(k_{P_j}/k_d)M^{-1} > 1$ at ever lower pressures for increasingly large molecules. The average k_{P_j} decreases with the number of atoms because the larger molecule has more ways to store active energy without causing reaction.

The models of the molecular processes from which statistical theories of reaction rates are constructed cannot be thoroughly tested by comparison of the theoretical and experimental rate constants. Apart from experimental errors, the reason for this is the insensitivity of the theoretical rate constant (which is a complicated average over the models) to the finer details of the molecular models.[4] For example, the simple version of the RRKM theory assumed, with limited justification, a structure and vibrational frequencies for the activated complex and neglected the effect of anharmonicity on the rate constant. Tests show that the rate constant varies less than an order of magnitude if the vibrational frequencies of the activated complex are changed within reason or if anharmonicity is included in the model. Furthermore, a model is uncertain because the frequencies of the activated complex may be changed to compensate for neglect of anharmonicity. The energy randomization and strong

---

[4] The advantage of this is that models used to compute rate constants in agreement with experiment are easy to construct.

collision assumptions of the RRKM theory are similarly uncertain and difficult to test. The solutions to the problems just raised, such as they exist at present, are described in Chapter 22.

_____ *Problems*

1. In order to compare the strengths of various intermolecular interactions, compute, at separation distances of 0.5 nm and 1 nm, the interaction energy per mole for
   (a) the orientation-averaged dipole–dipole interaction in H_2O at 273 K
   (b) permanent dipole–induced dipole interaction in H_2O
   (c) dispersion interaction in HCl
   (d) ion–induced dipole interaction for $O_2^+ - NO$
   (e) positive–negative ion (singly charged) interaction
   (f) which interaction energies are higher than, equal to, and less than the average kinetic energy of a mole of particles at 273 K?
   For formulas and data, see Eq. (1-16), Table 1-2, and Section 1-3.

2. For two ions—with charges $+q$ and $-q$—that collide with an impact parameter b and initial relative translational energy E_0 derive
   (a) an expression for the minimum distance of approach R_m
   (b) an expression for the trajectory, that is, a relation between R and θ (defined in Figure 21-1). Assume that initially $R = \infty$ and $\theta = 0$
   (c) an expression for the deflection angle χ (Figure 21-1)
   (d) show that the expressions from (a) and (c) reduce to $R_m \approx b$ and $\chi \approx 0$ for large impact parameters ($2b \gg q^2/E_0$)
   (e) show that the expressions from (a) and (c) reduce to $R_m \approx b/(q^2/E_0)$, $\chi \approx \pi$ for small impact parameters ($2b \ll q^2/E_0$)
   (f) plot in arbitrary units the orbit for $q^2/E_0 = 2b$ for all $R \leq 10R_m$.

3. Calculate the relative velocity and kinetic energy, as well as the center-of-mass velocity and kinetic energy for
   (a) K and Cl_2 moving on paths that intersect at right angles with velocities 5.0×10^2 and 3.0×10^2 m s^{-1}, respectively.
   (b) H and Cl_2 moving as in (a).
   (c) K and Cl_2 moving on paths that intersect at 60° with same velocities as in (a).

4. The rate constant for the charge neutralization reaction $I_2^+ + I_2^-$ is, approximately, 7×10^{13} M^{-1} s^{-1} [17].
   (a) Assuming that all ion pairs that approach within a distance equal to or less than a critical distance R_m react, derive a theoretical expression for the rate constant.
   (b) Calculate R_m.
   (c) The electron transferred from I_2^- enters an excited state of the hydrogenlike atom $I_2^+ \cdots e^-$. Calculate the approximate quantum number n of a state for which the most probable radius of the electron is approximately $0.8 R_m$.

5. Calculate for the reaction $CH_3 + CH_3 \rightarrow CH_3CH_3$:
   (a) the reaction cross section
   (b) the rate constant at 100 and at 800 K.
   Assume that the hard sphere radius of CH_3 is 380 pm and that the energy barrier $E_{0m} = 0$.

6. The parameters of the Lennard-Jones potential between two methyl radicals are $\sigma = 390$ pm and $\varepsilon^*/k = 137$ K. For the reaction $CH_3 + CH_3 \rightarrow CH_3CH_3$ assume that the interaction of the two methyl radicals is governed by the attractive part of the Lennard-Jones potential.
   (a) Calculate the reaction cross section at 100 and at 800 K if the relative translational energy of the two radicals is $\frac{1}{2}kT$.
   (b) Derive an expression for the rate constant k as a function of temperature. Note that $\int_0^\infty x^{n-1} e^{-x}\, dx = \Gamma(n)$ and $\Gamma(5/3) = 0.9028$.
   (c) Calculate k at 100 and at 800 K.

7. Consider two ternary solutions: 55 M S (solvent), 0.1 M A, 0.1 M B and 55 M S, 1 M A, 1 M B.
   (a) For each solution calculate, assuming random distribution of A, B, and S, the percentage of A molecules in AB pairs if each A has 12 nearest neighbors. (Count ABA, BAB, and similar structures as a single pair.)
   (b) For the dilute solution calculate the percentage of A's that have the center of a B molecule within two molecular diameters of the center of the A. Assume that the solution is a closest-packed (page 716) structure of spherical molecules, all of the same diameter. (A more realistic assumption which, however, would complicate the calculation is that about 2–5% of the closest-packing sites are empty.) This structure has 54 molecules with centers within two diameters of the center of any given molecule.

8. The following data have been obtained at 24°C for two presumably diffusion-controlled aqueous reactions [18]:

$$OH + OH \rightarrow H_2O_2 \qquad k = 0.5 \times 10^{10}\, M^{-1}\, s^{-1}$$

$$OH + C_6H_6 \rightarrow C_6H_6OH \qquad k = 0.33 \times 10^{10}\, M^{-1}\, s^{-1}$$

   The diffusion constants are $D_{OH} = 2.6 \times 10^{-5}\, cm^2\, s^{-1}$ and $D_{C_6H_6} = 1.1 \times 10^{-5}\, cm^2\, s^{-1}$.
   (a) For each reaction calculate the cage radius R_m and compare it to the value calculated from the hard sphere radii 1.5 Å for OH and 2 Å for C_6H_6.
   (b) If the two values of R_m that were calculated in (a) for each reaction differ by a factor of two or more, estimate k_{d-}/k_c (Eq. 21-40) for that reaction.

9. The expression for the potential energy of the triatomic species ABC is given in Eq. (21-48). Assuming that $V(\infty, \infty, \infty) = 0$, calculate the value of $V(R_{AB}, R_{AC}, R_{BC})$ for the symmetric and linear activated complex of the $H + H_2$ reaction. *Data*: the H—H distance in the complex is equal to 93 pm; the height of the potential energy barrier at the complex is 0.424 eV; the potential energy of H_2 at the internuclear distances 74 (equilibrium value), 93, and 186 pm is respectively, -4.747, -4.309, and -0.820 eV.

10. The total cross section for scattering may be determined by measuring the attenuation of a particle beam which passes through a chamber containing a target gas. When a beam of K atoms was passed through a chamber (effective

length 5.43 cm) containing Ar gas at 1.2×10^{-4} torr and 300 K, the beam lost 80% of its initial intensity [19]. What is the total cross section? *Hint*: Note that this problem is very similar to that of the attenuation of a light beam by an absorbing medium.

11. Holbrook et al. [20] measured at 424.4°C the following rate constants for the unimolecular reaction

$P/$(torr)	100	74.0	51.1	28.5	9.16	3.84	2.33	1.47	1.00	0.443	0.232
$k/(10^{-4}\,\text{s}^{-1})$		10.4	10.7	10.3	9.95	8.57	6.79	6.23	5.25	4.60	3.62	2.84

(a) What is the transition pressure, $P_{1/2}$, and the high pressure rate constant, k_∞, for this reaction?

(b) The experimental high pressure activation energy is $\mathscr{E}_\infty = 57.81$ kcal mol^{-1}. What is the Arrhenius parameter, A_∞?

(c) Calculate the rate constant for activation, k_a, from collision theory and from experiment. (The collision diameter of 1,1-dichlorocyclopropane is $\sigma = 6.5 \times 10^{-8}$ cm.)

(d) On the same graph plot two curves of k/k_∞ versus $\log P/$(torr): (1) experimental, (2) with k/k_∞ calculated from Eq. (21-61) using k_a from experiment [see (c)]. Consider a third possible curve: k/k_∞ versus $\log P/$(torr) with k/k_∞ on your graph calculated from Eq. (21-61) using k_a from collision theory. Where on your graph will the third curve lie with respect to the first two curves?

12. The rate constants that appear in the Lindemann formulation of the mechanism for unimolecular reactions obey the Arrhenius expressions $k_a = A_a \exp(-\mathscr{E}_a/\boldsymbol{R}T)$, $k_d = A_d$, and $k_\infty = A_\infty \exp(-\mathscr{E}_a/\boldsymbol{R}T)$.

(a) Assuming that k_a and k_d are calculated from collision theory that is, that $A_a \approx A_d$) show that $k_P = A_\infty$.

(b) For the isomerization of CH_3NC compare the value of the theoretical k_P [calculated from the result in (a)] to the experimental k_P (calculated from the experimental $M_{1/2}$ and $k_d = 1.34 \times 10^{11}$ dm^3 mol^{-1} s^{-1}, see page 782).

Note: The result in (b) is representative. It shows that the excited molecules decompose to products much more slowly than predicted by the Lindemann mechanism plus collision theory. The RRKM theory (Section 21-6) accounts for both the slow decomposition of A* and the fast activation (page 782) of A.

13. (a) What is the expression, according to the activated complex theory, for the rate constant of the elementary reaction $OH + CO \rightarrow CO_2 + H$? Assume a linear activated complex and all electronic partition functions to be equal to one.

(b) What is the value of the temperature exponent $\mathscr{C}/\boldsymbol{R}$ in Eq. 20-16 for a tightly bound activated complex (that is, all of its vibrational frequencies $\nu_{\ddagger i}$ are such that $h\nu_{\ddagger i}/kT \gg 1$)?

(c) What is the value of $\mathscr{C}/\boldsymbol{R}$ in the limit of a very loosely bound activated complex?

14. Calculate the entropy and the enthalpy of activation for the reactions $N_2O \rightarrow N_2 + O$, $CO + OH \rightarrow CO_2 + H$, and $SO_2 + O + Ar \rightarrow SO_3 + Ar$ from the data in Table 20-2, assuming $T = 1000$ K.

15. Compute the statistical factors $l_{\ddagger}$ for four proton abstraction reactions

$$H + CH_4 \rightarrow CH_3 + H_2$$
$$H + CH_3D \rightarrow CH_2D + H_2$$
$$H + CH_2D_2 \rightarrow CHD_2 + H_2$$
$$H + CHD_3 \rightarrow CD_3 + H_2$$

16. Use activated complex theory to compute the rate constant at 500 K for the reaction $CH_3 + H_2 \rightarrow CH_4 + H$.

    Assume that the methyl radical is completely rigid, that is, equivalent to an atom. Use the data for H_2 in Example 21-1. Assume that the activated complex is linear with the following properties: vibrational frequencies are 2334 (symmetric stretch) and 1021 (doubly degenerate bend) cm^{-1}, moment of inertia 9.54×10^{-47} kg m^2, and barrier height 44.0 kJ mol^{-1}. Data from Johnston [21].

17. Use activated complex theory to compute the ratio of rate constants at 500 K for the isotopic reactions

$$Br + H_2 \xrightarrow{k} HBr + H$$

and

$$Br + D_2 \xrightarrow{k'} DBr + D$$

For H_2 and D_2 use the data in Example 21-1. The data for the linear activated complexes are shown in the following Table [22].

| | Vibrations | | Moments of |
	Stretch (cm^{-1})	Bend (doubly deg.) (cm^{-1})	Inertia (kg m^2)
H−H−Br	1313	540	13.0×10^{-47}
D−D−Br	933	383	25.5×10^{-47}

Assume that the potential energy barrier is equal to zero and that the ratio of transmission coefficients is $\kappa'/\kappa = 0.90$ (due to tunneling).

18. Compute the number of harmonic oscillator energy levels below 5000 cm^{-1} for N_2O (linear) from the approximate vibrational frequencies 1300, 590 (bending mode), and 2200 cm^{-1}. What effect would the correction for anharmonicity have on the number of energy levels below 5000 cm^{-1}?

19. Derive the low pressure RRKM expression for the unimolecular rate constant (Eq. 21-116) from the general expression for the unimolecular rate constant, Eq. (21-114).

    Note: The expression for the partition function of a system with closely spaced energy levels is $q = \int_0^\infty G e^{-E/kT} dE$.

20. Show that the high pressure RRKM expression for the unimolecular rate constant (Eq. 21-115) can be derived from the activated complex theory by the same procedure that leads to Eq. (21-83).

21. Two unimolecular reactions have the following Arrhenius parameters at high pressures and 600 K:

$$\log A/(s^{-1})$$

$(CH_3CO)_2 \rightarrow 2\ CH_3CO$	16.0
$(CH_3CO)_2O \rightarrow CH_2CO + CH_3CO_2H$	12.0

(a) Calculate the entropies of activation for the two reactions.

(b) What do the calculated entropies imply about differences in the structures and bond strengths of the activated complexes for the two reactions?

(c) What is the value of $\log A/(s^{-1})$ at 300, 600, 900 K of a reaction for which $\Delta S^{\ddagger} = 0$?

22. For the isomerization of CH_3NC at 472 K compute, from the results in Example 21-2 and Figure 21-16

(a) k_{∞}

(b) an approximate value of the contributions to the unimolecular rate constant at 40 torr from the following ranges of $\mathscr{E}^{\ddagger}$: $0-<3$, $9.5-10.5$, $24.5-25.5$, and $46.5-47.5$ kJ mol^{-1}.

_____ *References and Recommended Reading*

[1] C. H. Bamford and C. F. H. Tipper (eds.): *Comprehensive Chemical Kinetics.* Elsevier, Amsterdam, 1969. A multi-volume work.

[2] S. W. Benson: *The Foundations of Chemical Kinetics.* McGraw-Hill, New York, 1960. A thorough survey of classical kinetics.

[3] K. J. Laidler: *Reaction Kinetics.* Pergamon, Oxford, 1963. Two volumes.

[4] M. F. R. Mulcahy: *Gas Kinetics.* Wiley, New York, 1973.

[5] R. Weston and H. Schwartz: *Chemical Kinetics.* Prentice-Hall, Englewood Cliffs, N.J., 1972.

[6] S. W. Benson: *Thermochemical Kinetics: Methods for Estimation of Thermochemical Data and Rate Parameters.* Wiley, New York, 1968.

[7] R. D. Levine and R. B. Bernstein: *Molecular Reaction Dynamics.* Oxford University Press, New York, 1974.

[8] K. J. Laidler: *Theories of Chemical Reaction Rates.* McGraw-Hill, New York, 1969.

[9] P. J. Robinson and K. A. Holbrook: *Unimolecular Reactions.* Wiley-Interscience, New York, 1972.

[10] I. Shavitt: *J. Chem. Phys.,* **49**:4048 (1968).

[11] C. L. Arnot: *J. Chem. Educ.,* **49**:480 (1972). Activated complex theory of bimolecular gas reactions.

[12] E. F. Greene and A. Kupperman: *J. Chem. Educ.,* **45**:361 (1968). Chemical reaction cross sections and rate constants.

[13] J. O. Hirschfelder and C. A. Boyd: *J. Chem. Educ.*, **27**:127 (1950). A physical-chemical approach to reaction kinetics. Correlation of observed results with postulated models.

[14] B. H. Mahan: *J. Chem. Educ.*, **51**:308, 377 (1974). Collinear collision chemistry: I. A simple model for inelastic and reactive collision dynamics. II. Energy disposition in reactive collisions.

[15] L. M. Raff: *J. Chem. Educ.*, **51**:712 (1974). Illustration of reaction mechanism in polyatomic systems via computer movies.

[16] B. Liu: *J. Chem. Phys.*, **58**:1925 (1973).

[17] B. H. Mahan: *J. Chem. Phys.*, **40**:3683 (1964).

[18] L. M. Dorfman and M. S. Matheson: *Progr. Reaction Kinetics*, **3**:237 (1965).

[19] E. W. Rothe and R. B. Bernstein: *J. Chem. Phys.*, **31**:1619 (1959).

[20] K. A. Holbrook et al.: *Trans. Faraday Soc.*, **66**:869 (1970).

[21] H. S. Johnston: *Adv. Chem. Phys.*, **3**:131 (1960).

[22] R. B. Timmons and R. E. Weston, Jr.: *J. Chem. Phys.*, **41**:1654 (1964).

[23] G. Wolken, Jr., W. H. Miller, and M. Karplus: *J. Chem. Phys.*, **56**:4930 (1972).

[24] G. W. Koeppl: *J. Chem. Phys.*, **59**:3425 (1973).

[25] F. W. Schneider and B. S. Rabinovitch: *J. Am. Chem. Soc.*, **70**:4215 (1962).

[26] B. S. Rabinovitch et al.: *J. Phys. Chem.*, **70**:2823 (1966).

chapter 22

Resolution of Reactant and Product States

22-1 INTRODUCTION

The ideal experiment for the study of the collisions that constitute an elementary reaction between two molecules A and B is to "throw" an A at a B molecule, each in a definite translational-vibrational–rotational-electronic quantum state, and determine the resulting quantum states of the product molecules C.

The ideal experiment would yield information about an elementary reaction far beyond that contained in the statistical rate constants of Chapter 21. The nature of the information is seen if we undo, step by step, the averaging that leads to statistical rate constants. This unaveraging corresponds to resolving the reactant and product states experimentally so that experiments may be performed that involve only limited groups of states. In the limit of the ideal experiment, the "group" becomes a single state. In this chapter we shall describe experimental approaches to the ideal experiment.

We see from Eq. 21-53, that the statistical rate constant k is an average over the internal (rotational, vibrational) states of the reactants. Since the products of a reaction are in various internal states, k is also an average over the internal states of the products. Thus, the unaveraging of k leads to the detailed rate constants $k(\mathcal{R}', \mathcal{V}'|\mathcal{V}, \mathcal{R})$ where $\mathcal{R}$ represents the set of all rotational and $\mathcal{V}$ represents the set of all vibrational quantum numbers that specify the quantum state of each reactant. The primed quantities refer to the products. The detailed rate constant may belong to either of the two kinds of processes that make up an elementary reaction, that is, chemical process

$$A(\mathcal{R}_A, \mathcal{V}_A) + B(\mathcal{R}_B, \mathcal{V}_B) \rightarrow C(\mathcal{R}'_C, \mathcal{V}'_C) + D(\mathcal{R}'_D, \mathcal{V}'_D) \qquad (22\text{-}1)$$

or energy transfer

$$A(\mathcal{R}_A, \mathcal{V}_A) + M(\mathcal{R}_M, \mathcal{V}_M) \rightarrow A(\mathcal{R}'_A, \mathcal{V}'_A) + M(\mathcal{R}'_M, \mathcal{V}'_M) \qquad (22\text{-}2)$$

where $\mathcal{R} \equiv (\mathcal{R}_i, \mathcal{R}_j)$ and $\mathcal{V} \equiv (\mathcal{V}_i, \mathcal{V}_j)$. All collisions between particles, such as in process 1 (or between a particle and a photon), which lead to a major rearrangement of atoms, that is, bond breaking, electronic excitation, or at least geometric isomerization, are called **reactive**. Collisions, such as in process 2, which only transfer energy, are called **inelastic**.

The detailed rate constant is an average over the distribution of the relative velocities v_0 of reactants. When this average is undone, we get the detailed cross section (see Eq. 21-18) for reaction, $S_r(\mathcal{R}', \mathcal{V}'|v_0, \mathcal{R}, \mathcal{V})$ or for inelastic scattering, $S_i(\mathcal{R}', \mathcal{V}'|v_0, \mathcal{R}, \mathcal{V})$. For a given reaction the detailed cross section is not an independent function of the final relative product velocity v'. From energy conservation, the initial reactant energy E_0 must equal the final product energy $E(v', \mathcal{V}', \mathcal{R}')$ plus the difference between the product and reactant potential energies, equal to the standard enthalpy at 0 K ΔH_0°

$$E_0 = E_0(v_0) + E(\mathcal{R}, \mathcal{V})$$
$$= E(v', \mathcal{R}', \mathcal{V}') + \Delta H_0^{\circ}$$
$$= E(v') + E(\mathcal{R}', \mathcal{V}') + \Delta H_0^{\circ} \qquad \text{(22-3)}$$

Eq. (22-3) shows that $E(v')$ (and hence v') is a function of v_0, $\mathcal{R}$, $\mathcal{V}$, $\mathcal{R}'$, $\mathcal{V}'$.

The final average that may be undone is that over the scattering angles of the products. If we use the scattering angles χ and ϕ defined in Figure 21-1 and the solid scattering angle ω (page 552), the detailed cross section is (see Section 18-5).

$$S(\mathcal{R}', \mathcal{V}'|v_0, \mathcal{R}, \mathcal{V}) = \int_0^{2\pi} \int_0^{\pi} S(\mathcal{R}', \mathcal{V}', \omega|v_0, \mathcal{R}, \mathcal{V}) \sin \chi \, d\chi \, d\phi$$
$$= \int_0^{4\pi} S(\mathcal{R}', \mathcal{V}', \omega|v_0, \mathcal{R}, \mathcal{V}) \, d\omega \qquad \text{(22-4)}$$

The detailed differential cross section $S(\mathcal{R}', \mathcal{V}', \omega|v_0, \mathcal{R}, \mathcal{V})$ has the dimensions of area per solid angle with the solid angle being measured in steradians.

The experimental methods for measuring the detailed rate constants and total or differential cross sections are largely determined by the fact that we live in a world where a reactant has a Boltzmann, or nearly Boltzmann, distribution over its quantum states. In order to resolve the reactant states in this distribution we can either *enhance* the population of some states at the expense of others or else *select* certain states out of the Boltzmann population. Selection methods can be used to analyze product states as well.

The simplest enhancement method is to increase the populations of higher energy levels at the expense of the ground state level by increasing the temperature of the reactant sample. The resulting resolution of reactant states is generally poor and incomplete. An example of a simple selection method is to allow a gas to effuse into a vacuum through a series of coaxial holes. The series of holes would select translational states with near-zero momenta perpendicular to the axis of the holes.

Reaction studies in which molecular states are partially resolved are done either with **bulk** samples of mixed reactants or else with **crossed molecular beams** (Figure 22-1). In experiments with bulk samples the number of collisions between molecules is generally poorly controlled. Therefore, one must infer indirectly the nature and number of collisions which produce the observed distribution of product states from the distribution of reactant states.

The interpretation of crossed molecular beam experiments is simpler because they are run under conditions where a reactant molecule A in the beam undergoes no more than a single collision with a reactant molecule B (Example 22-1). Differential cross sections can be determined only by molecular beam experiments. Both types of experiments generally yield incomplete information, quantitative or qualitative, on detailed rate constants and total cross sections.

In the following sections we shall describe methods for resolving translational (Section 22-2) and rotational–vibrational (Sections 22-3 and 22-4) states and the resulting information on molecular models of reactive and inelastic collisions. Both the methods and the models are under rapid development. The inquiring student will find the latest developments in the current literature.

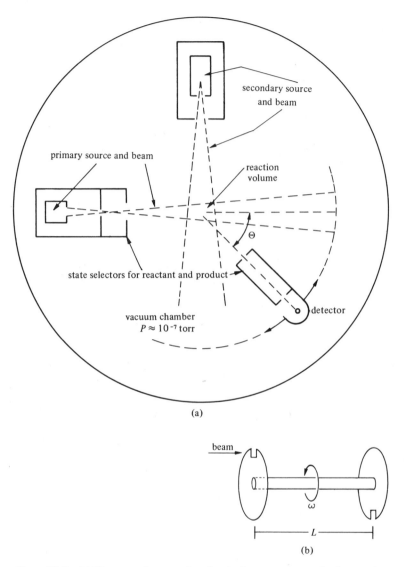

(a)

(b)

Figure 22-1 (a) Diagram of a crossed molecular beam apparatus. In the reaction volume reactants from the primary and secondary sources form products. The products, scattered at an angle Θ, are detected by the detector. Sources—see text and Figure 22-3. Variable temperature sources allow the populations of reactant states to be varied. Selectors—see text. Detectors—either an ionizing hot platinum or tungsten wire (for molecules that contain alkali atoms) or a mass spectrometer. Detector sensitivities limit the extent to which reactant and product states can be selected. Maximum sensitivities are about 10^{-17} A or 10^2 alkali atoms s^{-1} for the hot wire detector and about 10^3 molecules s^{-1} for mass spectrometer. To measure differential cross sections the detector-product selector assembly can be moved in and out (not shown in figure) of the plane defined by the reactant beams. Fast pumping and liquid nitrogen cooled walls keep interference by ambient gases to a minimum.

(b) Schematic of simple velocity selector. All molecules with velocities $v = L/t = L(2\omega)/n$ will pass through selector. Angular velocity $\omega =$ revolutions/t; n, the number of half-revolutions. State-of-art selectors have multiple slots (to increase intensity) on each disc and several discs (to select $n = 1$) on same rotating shaft.

EXAMPLE 22-1

From the data for a typical molecular beam experiment (Figure 22-1) (1) show that the molecules in the primary and secondary source beams and in the product beam are unlikely to experience more than one collision while in transit through the vacuum chamber; (2) calculate the percentage of primary beam molecules that collide with (a) the secondary beam and with (b) ambient gas; (3) calculate the number of product molecules formed in 1 s.

Data: ambient $P = 10^{-7}$ torr; reaction $V = 5 \times 10^{-2}$ cm^3; flux (primary beam) $n(x) = 10^{15}$ molecule cm^{-2} s^{-1}; concentration in reaction volume (secondary beam) $n = 10^{10}$ molecule cm^{-3}; reaction cross section $S_r = 0.1$ nm^2 molecule^{-1}; hard sphere collision cross section of N$_2$ $S_{hs} = 0.1$ nm^2 molecule^{-1}.

1. Does a beam molecule collide with the ambient residual gas? Assuming the ambient gas to be N$_2$ at 25°C, and the collision diameters of all molecules equal to that of N$_2$ we can use the value of 6.55×10^{-8} m for the free mean path $\bar{\lambda}'$, of N$_2$ at 25°C and $P' = 1$ atm (calculated in Example 3-2). From Eq. (2-66) we see that $\bar{\lambda} \propto P^{-1}$, therefore

$$\bar{\lambda}/\bar{\lambda}' = P'/P \quad \text{or} \quad \bar{\lambda} = \bar{\lambda}'(P'/P) = 6.55 \times 10^{-8}(760/10^{-7}) = 500$$

This free mean path is about 100 times longer than the characteristic dimensions of a molecular beam apparatus. Therefore, a given beam molecule is unlikely to collide with ambient gas molecules.

2(a) What percentage of primary beam molecules collides with the secondary beam? Assuming the secondary beam to be an essentially stationary target the maximum ratio of the reactive target area to the 1 cm^2 area presented by a 1 cm^2 face of a volume equal to 5×10^{-2} cm^3 would be

$$5 \times 10^{-2} \text{ cm} \times n \times S_r = 5 \times 10^{-2} \times 10^{10}(0.1 \times 10^{-18} \times 10^4) = 5 \times 10^{-7}$$

Therefore only 0.00005% of the primary beam molecules are likely to experience a single reactive collision.

(b) What percentage of primary beam molecules collides with the ambient gas? Assuming that the beam passes through a 50-cm length of ambient gas, the ratio of the hard sphere target area to the 1 cm^2 area presented by a 1 cm^2 end face of a 1 cm$^2 \times$ 50 cm cylindrical volume filled with ambient gas (N$_2$) at a concentration n_{amb} would be

$$50 \text{ cm} \times n_{amb} \times S_{hs} = 50 \text{ cm } P(RT)^{-1} N S_{hs}$$
$$= 50(10^{-7}/760)(82 \times 300)^{-1}(6 \times 10^{23})(0.1 \times 10^{-14})$$
$$= 1.5 \times 10^{-4}$$

About 0.02% of primary beam molecules experience a collision with the ambient gas. This scattering can be an important source of detector noise when the detection of primary beam molecules that are being scattered by the secondary beam is desired.

3. The number of product molecules formed in 1 s is

$$n(x)S_r n V = 10^{15}(0.1 \times 10^{-14})(10^{10})(5 \times 10^{-2}) = 5 \times 10^8 \text{ molecules}^{-1}$$

Since this is the number scattered to all angles, the number arriving at the detector positioned at a particular angle may be hundreds of times smaller.

22-2 TRANSLATIONAL STATES

In this section we shall discuss the production of reactants in specified translational states and some results obtained therefrom. Experimentally it is possible to resolve both (together or separately) the directions of the reactant momenta and their magnitudes (that is, translational energies). Similarly, one may determine both the direction of the product momenta and/or their translational energy.

22-2.1 Directions of Reactant and Product Momenta

The momentum directions of well-defined molecular beams fall within a narrow range. Reactive collisions between two such beams produce a nonisotropic distribution of product beams, corresponding to the variation of the differential cross section with the solid angle.

Such distributions distinguish two kinds of reactive collisions: (1) direct, and (2) indirect (a long-lived complex).

Direct collisions can be further subdivided into "stripping" and rebound collisions. In stripping collisions one reactant, while passing the second reactant, strips the latter of an atom (or a group of atoms) and continues in the original direction, as does the remnant of the second reactant. The $K + ICl$ reaction, as seen from Figure 22-2, proceeds predominantly by stripping collisions (top inset, Figure 22-2). Such reactions invariably have large total reaction cross sections, due to long range forces (ionic for the $K + ICl$ reaction) that permit large impact parameters. As can be seen by comparison with Figure 21-1, stripping collisions are similar to elastic collisions at large impact parameters; in either case small deflections are observed.

Rebound collisions dominate the $K + CH_3I$ reaction (Figure 22-2). The velocity of CH_3 relative to CH_3I reverses (hence "rebound") after the reactive collision. The K and I atoms depart together. Again, in analogy with elastic collisions, rebound collisions occur at small impact parameters and are associated with small reaction cross sections.

Both kinds of direct collisions are of short duration; less than half a typical molecular rotation period ($\approx 10^{-12}$ s). Complexes that survive (at least) several rotations and decompose randomly will produce more complex product distributions than the ones shown in Figure 22-2. Such distributions have served to identify long-lived ($>10^{-12}$ s) complexes in several alkali atom–alkali halide exchange reactions, such as $Rb + CsCl \rightarrow RbCl + Cs$.

22-2.2 Translational Energy and Reaction Cross Section

Reactants with well-defined translational energies are used to determine reaction thresholds and the cross sections S_r as a function of relative translational energy E_0. If our goal is the accurate determination of S_r for the $T + H_2$ reactions, we see from Figure 21-10 that we would need monoenergetic (energy spread <1 eV) reactants covering a range of translational energies from 0 to about 200 eV. This range of accessible energies, as well as the energy spread, is characteristic of the needs in an ideal experiment.

For bulk experiments in the multiple collision regime, a variety of **hot atom** (but not molecule) sources have been used. Hot atoms have kinetic energies considerably in excess of the average thermal kinetic energy of the surrounding medium. The two major sources of such atoms are nuclear reactions (for example, $_0^1n + _2^3He \rightarrow _1^1H + _1^3H$

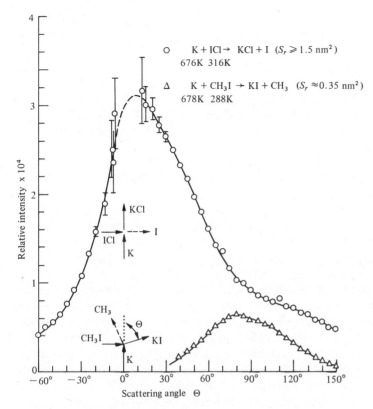

Figure 22-2 Product distributions for direct reactions. Reaction K+ICl is dominated by stripping collisions. Most collisions in reaction K+CH$_3$I are rebounds (as seen in the center of mass or μ particle coordinate system). The insets show the directions of the reactant beams and of the most intense scattering of products. The product directions shown by dashed arrows are determined indirectly from conservation equations. The total (summed over all solid angles) relative intensity of a scattered product is proportional to the total reactive cross section S_r. [After D. Herschbach: *Molecular Beams*. In J. Ross (ed.): *Adv. Chem. Phys.*, **10** (1966). Interscience Publishers.]

produces tritium with $E_{0T} = 1.92 \times 10^5$ eV) and photolysis (for example, for DI + $h\nu \rightarrow D + I$ we have $E_0 = \Delta E_0^\circ(DI) - h\nu$; the energy of the light D atom $E_{0D} \approx E_0$).

To study the reactions of hot atoms with various substrates, one mixes a small amount of the hot atom source material with an excess of the substrate, irradiates and, finally, analyzes for products. The primary reason for using radioactive hot atoms is the great sensitivity of radioactive methods for detecting small amounts of products.

The experimental results obtainable with the two kinds of hot atom sources differ considerably. Hot atoms, such as tritium, have kinetic energies much in excess of chemical reaction energies. These atoms lose some of their energy in inelastic collisions. Reaction begins when the energy of the hot atom decreases to the point where some reactions have appreciable cross sections. Some atoms never react while hot. Instead, they become thermalized. Thus hot atoms, by covering the whole range of chemical reaction energies, undergo all possible reactions. The products detected will be all those produced by reactions with measurable cross sections. It has been determined that hot tritium atoms have measurable S_r for three types of reactions with organic substrates (excited species marked by *):

1.   H atom abstraction $(T^* + RH \rightarrow HT + R)$
2.   substitution for another radical $R'(R'R + T \rightarrow RT^* + R')$
3.   addition to a π-bond system—olefins, acetylenes, aromatics $[T + RC{=}CR' \rightarrow (RCT{-}CR')^*]$.

More quantitative results, in particular on reaction thresholds, can be obtained with photolytically produced hot atoms because their energy can be controlled by varying the frequency of the radiation source. Available sources limit the photolytic hot atom energies to less than 4 eV. A threshold experiment determines the lowest energy of the hot atom at which the product can be detected. The threshold of the reaction $D + H_2$ has been determined to be 25 kJ mol^{-1}, which compares to the theoretical value (24 kJ mol^{-1}) for $H + H_2$ (Figure 21–9).

Methods for producing molecular beams with well-defined kinetic energies and for measuring the kinetic energies of the products differ considerably from hot-atom methods. The only important selection method utilizes the rotating coaxial slotted discs shown in Figure 22-1b. Enhancement is most easily done with ions that may be accelerated above their thermal velocities and formed into a beam (focused) by means of electric fields. (Magnetic fields are used for guiding the beam.) The major difficulty is achieving sufficient intensities at low energies ($\leqslant$5 eV) where ion–ion repulsion ("space charge" effect) defocuses the beam. The most important enhancement method for neutral reactants is adiabatic expansion (page 121) through a nozzle, which converts the thermal energy into directed kinetic energy and simultaneously lowers the reactant temperature as much as 20-fold. The lowered temperature means that the **nozzle beam** has a low spread of velocities, and a kinetic energy considerably higher than the average thermal energy in a single degree of freedom of the original reactant. The properties of beams from various sources are compared in Figure 22-3.

Crossed beams with well-defined velocities are useful for determining the translational energy dependence of reaction cross sections. For example, the cross section S_r for $K + HBr \rightarrow KBr + H$ has been determined in the range 0–5 kcal mol^{-1} [6]. The threshold for this reaction is about 0.4 kcal mol^{-1}. The shape of the S_r versus $E_0(v_0)$ curve is very similar to that shown in Figure 21-9a, Chapter 21. An S_r versus $E_0(v_0)$ curve of an exothermic ion-molecule reaction, $Ar^+ + D_2 \rightarrow ArD^+ + D$, is shown in Figure 22-4. This reaction has no threshold. The cross section decreases monotonically with increasing relative translational energy.

From studies of ion- and atom-molecule reactions, the qualitative features of the reaction cross section dependence upon the translational energy of the reactants can be summarized as follows. For reactions with a threshold the maximum, S_r is 10–100% higher than the average S_r in the region, just above the threshold, where the thermal reaction occurs. The reasons for the increase of S_r above threshold were discussed in Section 21-3. As the relative translational energy increases above the energy typical of a chemical bond (2–5 eV), the cross section of a reaction invariably decreases. One of the reasons is competition from the endoergic reactions, which

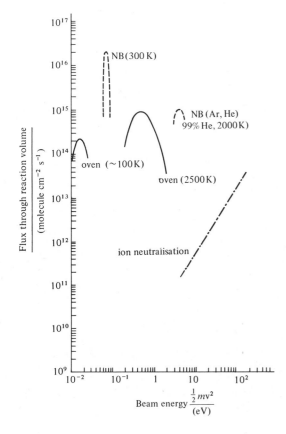

Figure 22-3 Very approximate maximum beam fluxes through a reaction volume 10 cm from various sources: ovens (sources with variable temperature), nozzle beams (NB), and ion beams or ion neutralization beams (neutral molecule beams prepared by neutralizing accelerated ion beams). The curve for the last two sources shows the maximum intensity of beams accelerated to a given energy. The other curves show the energy profiles of typical beams. A velocity-selected beam from an oven source will have a flux at the selected velocity equal to about 5% of that shown for the unselected source. [After J. P. Toennies in H. Hartmann (ed.): *Chemische Elementarprozesse.* Springer-Verlag, Berlin, 1968.]

become possible at higher energies (for example, for the reaction in Figure 22-4, two endoergic competitors are $Ar^+ + D_2 \rightarrow Ar^+ + 2\,D$ and $\rightarrow Ar + D^+ + D$). However, a decrease with increasing $E_0(v_0)$ of the sum over the S_r of *all* competing reactions is also observed. The reason is that the duration of the interaction between reactants decreases as the relative velocity increases. The weak, long-range forces become too slow-acting to transfer energy and to rearrange atoms to a degree sufficient to cause chemical reaction. Thus, as $E_0(v_0)$ increases, increasingly smaller impact parameters, that is, stronger forces, are necessary to cause chemical reaction.

22-3 INTERNAL STATES OF REACTANTS

Molecules in specified vibrational energy levels or states are produced primarily by enhancement. Selection has been used to produce molecules in specified rotational states. Selection methods of limited applicability exist for rotational–vibrational product analysis.

Electric fields can select polar molecules in rotational states with certain properties. Thus, in a molecular beam experiment, electric fields in a hexapolar configuration can select symmetric top molecules with either end (on the average) oriented

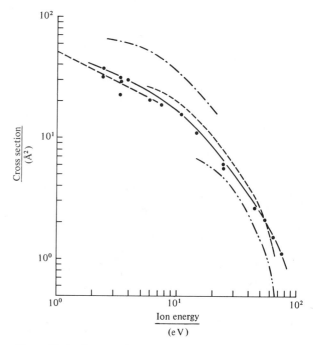

Figure 22-4   Cross section as a function of the kinetic energy of the incident ion for $Ar^+ + D_2 \rightarrow ArD^+ + D$ from several experimental determinations. Note the typically low accuracy of measurements. [From A. Henglein in Ch. Schlier (ed.): *Molecular Beams and Reaction Kinetics*. Academic Press, New York, 1970.]

toward a second reactant. Thus the dependence of reaction cross sections on the relative orientation of reactants may be determined. For example, the reaction

$$Rb + CH_3I \rightarrow RbI + CH_3$$

has been studied for the two average orientations $Rb \rightarrow \leftarrow ICH_3$ and $Rb \rightarrow \leftarrow H_3CI$. The first orientation has a cross section more than $1\frac{1}{2}$ times larger than the second [7].

   Chemical activation and **photoactivation** are two methods that can produce molecules with a definite amount of vibrational excitation. Both methods can be used for either molecular beam or bulk experiments. Chemical activation, in general, produces a higher degree of vibrational excitation than photoactivation.

   Chemical activation utilizes exothermic reactions to produce excited molecules in a narrow energy range (most molecules within $\pm 10\%$ of average energy) just above the activation energy threshold. Production of the same molecule via various reactions allows the average excitation energy $\mathscr{E}_{av}$ to be varied. Two examples are

$$H + \underset{\underset{CH_3}{|}}{CH}=\underset{\underset{CH_3}{|}}{CH} \xrightarrow[179\ K]{low\ P} C_2H_5\dot{C}HCH_3^* \qquad \mathscr{E}_{av} = 8.4\ kcal\ mol^{-1}$$

$$\text{(22-5)}$$

$$D + CH_3CH_2CH{=}CH_2 \xrightarrow[298\ K]{high\ P} CH_3CH_2\dot{C}HCH_2D^* \qquad \mathscr{E}_{av} = 14.2\ kcal\ mol^{-1}$$

and

$$C_2H_5 + F_2 \rightarrow C_2H_5F^* + F \qquad \mathscr{E}_{av} = 69 \text{ kcal mol}^{-1}$$
$$CH_2 + CH_3F \rightarrow C_2H_5F^* \qquad \mathscr{E}_{av} = 109 \text{ kcal mol}^{-1} \qquad \textbf{(22-6)}$$

Photoactivation uses electromagnetic radiation to produce vibrationally excited molecules. Ordinary black body or gas discharge sources may be combined with monochromators for this purpose. However, lasers (Section 13-4) are far superior sources because they have narrower band widths and are much more intense so that they can produce higher concentrations of excited molecules in a much shorter time. The latter allows the study of fast processes involving the excited molecules. For example, an infrared globar source plus a monochromator plus a mechanical chopper for producing pulses of radiation typically has a band width at half-intensity of 10^{-2} cm, a pulse length of 10^{-3} s, and an intensity of 10^{12} photons per pulse at a wave number of 10^4 cm^{-1}; the same quantities for a CO_2 laser are 10^{-3} cm, 10^{-6} s, and 10^{20} photons per pulse at 10^4 cm^{-1}.

The three major types of processes that a vibrationally excited molecule R^* may undergo are uni-(or bi-)molecular reaction, collisional deactivation, and spontaneous emission, that is,

$$R^*(+R') \rightarrow P \qquad \textbf{(22-7)}$$

$$R^* + M \rightarrow R + M \qquad \textbf{(22-8)}$$

$$R^* \rightarrow R + h\nu \qquad \textbf{(22-9)}$$

Process (22-9) is known as (thermal **luminescence** if R^* is produced by high temperatures, as **chemiluminescence** for R^* produced directly by a chemical reaction, and as (IR) **fluorescence** for R^* from photoactivation. It is generally slower than reaction or collisional deactivation. It is used in pulse experiments (Section 20-6.4) to monitor the R^* population changes caused by processes (22-7) and (22-8).

22-3.1 Reactions of Vibrationally Excited Species Vibrationally excited reactants have been used in molecular beam experiments to determine the dependence of the reaction cross section $S_r(v)$ upon the vibrational quantum number v. The reaction

$$K + HCl(v) \rightarrow KCl + H$$

with $v = 0, 1$ has been studied. The excited HCl was produced by crossing HCl and IR laser beams. It was found that $S_r(1)/S_r(0) \approx 100$.

Both chemical and photoactivation of reactants in bulk have been used to test the strong collision (page 782) and random lifetime (page 782) assumptions of the RRKM unimolecular reaction theory. Such tests are carried out with activated reactants R^* whose energies E_j^* exceed the critical energy $E_0^{\ddagger}$ (the symbols are defined in Figure 21-15).

If we neglect process (22-9) and the energy spread of R^*, the yield ratio of the pressures of the decomposition product P and the deactivation product R is, from RRKM theory (see Eqs. 21-58, 21–59, and 21–111),

$$P/R = k_{P_j}/(k_d M) \qquad \textbf{(22-10)}$$

Eq. (22-10) predicts that the ratio P/R, for an energy level j, should be inversely proportional to the pressure M if the strong collision assumption holds. Figure 22-5

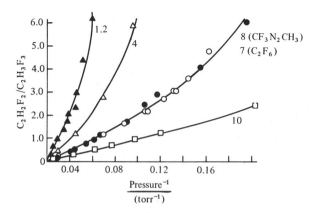

Figure 22-5 Test of the strong collision hypothesis for the reaction $CH_3CF_3 \rightarrow CH_2CF_2 + HF$. Experimental points for $M = H_2(\blacktriangle)$, $CH_3Cl(\triangle)$, $CF_3N_2CH_3(\bullet)$, $C_2F_6(\bigcirc)$, and $n\text{-}C_6F_{14}(\square)$. The solid lines are results of a theoretical calculation which assumes that energy is lost in several "weak" collisions rather than in a single strong collision. The average energy lost per weak collision (in kcal mol^{-1}) is indicated next to each curve. [From D. W. Setser in J. C. Polanyi (ed.): *Chemical Kinetics*, Butterworth, London, 1972, p. 12.]

shows typical experimental and theoretical results when R is $C_2H_3F_3$ and P is $C_2H_2F_2$ (the other product being HF). We see that the best straight line is obtained when M is $n\text{-}C_6F_{14}$ and the worst is when M is H_2. Thus $n\text{-}C_6F_{14}$ obeys the strong collision assumption rather well (that is, on the average it loses energy equal to or greater than $|E_j^* - E_0^{\ddagger}|$ per collision), whereas H_2 does not obey it [that is, energy lost $< |E_j^* - E_0^{\ddagger}|$]. The indicated trend is quite general: for large molecules (more than ten atoms) at thermal energies, the strong collision assumption is valid. One exception can occur when $M = R$. Experiment shows that, in general, collisions between two identical molecules transfer energy more efficiently than collisions between two different molecules of same size.

The assumption of random lifetimes may be tested by varying the energy E_j^* in the manner illustrated by reactions (22-5) and (22-6). According to Eq. (22-10) this should change the ratio P/R, because k_{P_j} depends on E_j^*. The experimental and RRKM [predicted from Eq. (21-111)] changes can be compared. This comparison has been carried out for $C_2H_5F^*$ and shows agreement between theory and experiment.

However, there are reactions where energy randomization does not occur. The best known is

$$
\begin{array}{c}
F_2 \diagdown \\
C \underset{\underset{H_2}{|}}{\overset{}{\longrightarrow}} C\text{-}C \underset{\underset{H_2}{|}}{\overset{F\ \ F}{\longrightarrow}} C \diagup^{F_2} \quad \rightarrow \quad C\text{-}C\text{-}C=CH_2 + CF_2
\end{array}
\qquad (22\text{-}11)
$$

An activated isotopic form of the reactant with the excess vibrational energy initially located in one of the rings is prepared by the reaction

$$
\begin{array}{c}
\text{F}_2 \diagdown \quad\quad \text{F} \;\; \text{F} \\
\text{C}\!-\!\!-\!\text{C}\!-\!\text{C}\!=\!\text{CF}_2 + \text{CD}_2 \;\to \\
\diagup \quad \diagdown \\
\text{C} \\
\text{H}_2
\end{array}
\qquad
\begin{array}{c}
\text{F}_2 \diagdown \quad\quad \text{F} \;\; \text{F} \\
\text{C}\!-\!\!-\!\text{C}\!-\!\text{C}\!-\!\!-\!\text{CF}_2^* \\
\diagup \quad \diagdown \quad\quad \diagdown \\
\text{C} \quad\quad \text{C} \\
\text{H}_2 \quad\quad \text{D}_2
\end{array}
\qquad \textbf{(22-12)}
$$

When the product of reaction (22-12) decomposes by reaction (22-11) a slight excess of $C_3F_3H_2CF\!=\!CD_2$ over $C_3F_3H_2CF\!=\!CH_2$ is formed. This is interpreted to mean that part of $C_3F_3H_2\!-\!C_3F_3D_2$ decomposes before the vibrational energy has been distributed randomly and equally throughout both rings. The relaxation time for energy randomization is found to be $\approx 10^{-12}$ s [8, 9].

Even though energy randomization does not occur in all unimolecular reactions, we do not know yet how to distinguish the reactions to which it does apply.

22-3.2. Intermolecular Energy Transfer Photoactivation, especially with lasers, is used extensively to study intermolecular energy transfer (process 22-8) in the absence of chemical reaction. Such studies build upon earlier work that used ultrasonic absorption and shock tubes to study energy transfer.

In general, a molecule may transfer its electronic, vibrational, rotational, and translational energy to one or more degrees of freedom, internal or translational, of a second molecule. Such transfers determine the distribution of reactant molecules among their energy levels and therefore are of obvious importance for reaction kinetics and dynamics. The processes that we can study with vibrationally excited molecules are intermolecular vibrational energy transfer (designated $V \to V$) and vibrational to rotational or translational energy transfer ($V \to R$ or T). These are also the best understood transfer processes.

Most studies have observed the transfer of a single vibrational quantum between diatomic or small polyatomic molecules. The experimental results are expressed as energy transfer relaxation times, cross sections, or probabilities P. The latter are defined as the ratio of the measured cross section to the hard sphere cross section.

Typical results are shown in Figure 22-6 for the exoergic process

$$CO_2(001) + N_2(0) \to CO_2(000) + N_2(1) \qquad \Delta E = -19 \text{ cm}^{-1} \qquad \textbf{(22-13)}$$

where the numbers in parentheses are the vibrational quantum numbers. For CO_2 they are in the order symmetric stretch, bend, antisymmetric stretch (at 2349 cm^{-1}).

The low temperature part of the curve in Figure 22-6 decreases with temperature. It is caused by energy exchange mediated via long range polar forces (quadrupole–quadrupole interactions for the CO_2-N_2 pair, dipole–dipole interactions for molecules with permanent dipoles, and so on). The decrease with temperature is due to shorter average interaction times as the relative velocity increases.

Figure 22-6 also shows that the process

$$CO(1) + NO(0) \to CO(0) + NO(1) \qquad \Delta E = -267 \text{ cm}^{-1} \qquad \textbf{(22-14)}$$

behaves differently at low temperature than process (22-13). The latter is typical of a

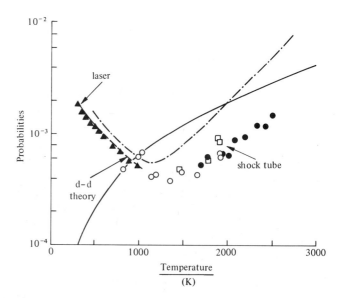

Figure 22-6 V → V energy transfer probabilities for process (22-13) (–·– are early shock tube results; experimental points are recent laser and shock tube results) and process (22-14) (solid line). [After C. B. Moore: *Adv. Chem. Phys.*, **23**:41 (1973).]

near-resonant ($\Delta E \approx 0$) V → V process, whereas the former process is nonresonant ($|\Delta E| > 0$). V → V transfer via attractive interactions is important only for near-resonant processes. The high probability of near-resonant V → V transfer between two identical molecules correlates well with the high relative probabilities for energy transfer between such molecules observed in tests of the strong collision assumption.

In Table 22-1 we summarize the characteristic (relaxation) times for various elementary processes obtained by diverse methods. These times determine the maximum possible reaction rates and the possible ways in which a distribution approaches the Boltzmann distribution.

The relaxation times for intermolecular energy transfer (excluding electronic energy) in Table 22-1 rank as follows in increasing order: T → T and R → R, R → T, V → V, V → R, T. Therefore, if vibrational energy, in the form of a very short pulse, is added to a gas in thermal equilibrium at a temperature T_0, in 10^{-9} to 10^{-5} s the extra vibrational energy will distribute itself among the vibrational degrees of freedom according to a Boltzmann distribution at an effective temperature $T_{vib} > T_0$. In 10^{-6} to 10^{-3} s the excess vibrational energy will flow into the rotational and translational degrees of freedom. At the end of 10^{-3} s the gas will be at a nearly uniform temperature T_{gas} such that $T_{vib} > T_{gas} > T_0$. Finally (in 1 s or less) the gas will return to T_0 as energy is lost to the walls. Effective temperatures, such as T_{vib}, can be measured by determining at least one population ratio for a pair of energy levels. These temperatures are very useful for comparing the local populations, which exist in specific energy levels, with the equilibrium population for the same levels.

Table 22-1 Very approximate characteristic times for particle and molecular motions; relaxation times for energy transfer (all times in seconds) of a gas

	Electronic (E)	Vibrational (V)	Rotational (R)	Translational (T)
Motion, through 1 nm[a]; translational energy 2.5 kJ mol^{-1} (300 K) ~10^2 kJ mol^{-1} (10^4 K)	2×10^{-16}–10^{-14}			8×10^{-15}–4×10^{-13} (H atom)
Period	10^{-16} (in H atom)	10^{-14}–10^{-13} ($\approx$5000–500 cm^{-1})	10^{-11}–10^{-10} (100–10 GHz)	
Energy transfer to electromagnetic radiation via spontaneous emission				
E^d ⇌	$10^{-7\,b}$–$10^{-2\,c}$		very long	
V^d ⇌	$>10^{-11}$	10^{-3}–1		
V ⇌ (intramolecular)		10^{-9}–10^{-5} (1–10^4 collisions) 10^{-12}		
R^d ⇌		$\left.\begin{array}{c} 10^{-6}\text{–}10^{-3} \end{array}\right\}$	$<10^{-9}$	$<10^{-9}$
T^d ⇌			10^{-9}–10^{-8}	10^{-3}–1
container wall ⇌				

[a] The free mean path at STP is approximately 100 nm.
[b] Electric dipole allowed transitions.
[c] Electric quadrupole allowed triplet $\rightarrow$ singlet.
[d] Intermolecular energy transfer.

22-4 INTERNAL STATES OF PRODUCTS

The distribution of products among internal states immediately after the reaction and prior to collisional deactivation can be studied experimentally and theoretically. Successful and systematic experimental studies have been based on observations of chemiluminescence from bulk samples. One of the more prominent chemiluminescence methods allows two reactant streams to mix and react in a low pressure chamber with liquid nitrogen cooled walls. Under these conditions, most product molecules are deactivated at the walls rather than by intermolecular collisions. The observed chemiluminescence comes from newly formed excited product molecules in transit to the walls. The frequencies of the chemiluminescence identify the internal states of the product molecules and the relative intensities determine the relative

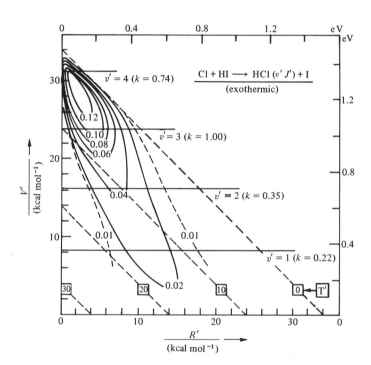

Figure 22-7   The distribution of the vibrational, rotational, and translational energies of excited HCl produced by the reaction $Cl + HI \rightarrow HCl + I$. All points on a contour have the same value of the detailed rate constant $k(R', V') = k(J', v')$ measured on a relative scale on which $k(v' = 3) = \sum_{J'} k(J', v' = 3) = 1$. The solid horizontal lines show the energies of the vibrational levels v' with $k(v')$ in parentheses. The contours have physical significance only on the solid lines because the molecule cannot have an amount of vibrational energy that does not correspond to that of a vibrational level. Translational energy T' (calculated from the rotational and vibrational energies using energy conservation—see Eq. 22-3) is constant along the dashed diagonal lines.

The diagram shows that most of the reaction energy appears as vibration and that, in the lower vibrational levels, rotational and translational excitation is higher than in level $v' = 4$. [From J. C. Polanyi: *Chemical Kinetics*. Butterworth, London, 1972, p. 148].

populations and rate constants of the states. The exoergic reactions that have been studied in this way are $H + X_2$, $X + H_2$, $X + HX'$ (X, X' are halogen atoms) and a few others. In Figure 22-7 the results for $Cl + HI \rightarrow HCl(J', v') + I$ are shown.

Theoretical trajectory studies (Section 21-3) of the triatomic reaction $A + BC \rightarrow AB + C$ have succeeded in identifying the feature of a potential energy surface that largely determines whether the product of an exoergic reaction will be vibration-ally excited. This feature—the position of the saddle point along the reaction coordinate—and its consequences are shown in Figure 22-8. We see that the product will be vibrationally excited if the reactants reach the saddle point early, that is, if the AB attraction begins while the BC distance is still near its equilibrium value. This kind of potential energy surface is called **attractive**. AB attraction that begins only when the BC distance is considerably larger than the equilibrium value causes the reaction energy to appear entirely as translational energy of the products. A potential energy surface with a late saddle point is called **repulsive**. (The repulsion is between A and C.)

The preceding description applies most rigorously when $m_A \ll m_C$. In this case the fast moving light atom A hits B before the heavier atom C can begin to move away from B. The reaction path therefore curves sharply near the saddle point. A more gradual curvature around the saddle point, that is, the simultaneous motion of atoms A and C with $m_A \gtrsim m_C$, would lead to vibrational excitation even on a repulsive surface (path 2 in Figure 22-8b).

Rotational excitation of products occurs when the impact parameter is large. The rotational energy comes from the relative translational energy of the reactants. A

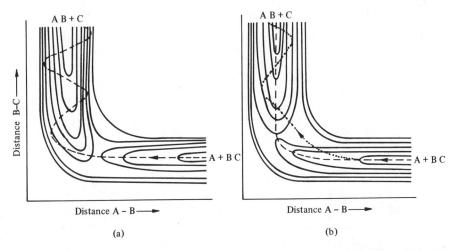

Figure 22-8 Schematic (a) attractive and (b) repulsive potential energy surfaces. Paths for exoergic reaction are shown by dashed or dotted lines. Reaction on the attractive surface and reaction on the repulsive surface for atomic masses $m_A \gtrsim m_c$ (path 2 · · · ·) lead to vibrational excitation of product AB, whereas $m_A \ll m_C$ (path 1– – – – –) on the repulsive surface leads to translational energy release.

By the principle of microscopic reversibility (Section 20-4) vibrational excitation of AB should be more effective than translational energy for the endothermic reaction AB + C on the attractive surface and along path 2. A large translational energy of AB relative to C should be more effective for path 1 on the repulsive surface.

second type of rotational excitation occurs at the expense of vibrational excitation on a repulsive surface for bent configurations of the activated complex when $m_A \gtrsim m_B$.

The most important application of the knowledge on vibrationally excited products is to the operation of chemical lasers, such as the CO_2 laser. These devices utilize inverted vibrational populations produced by a chemical reaction.

_____ *Problems*

1. A crossed beam study [10] of the reaction $K + HBr \rightarrow KBr + H$ was carried out at an initial relative kinetic energy of the reactants equal to 1.9 kcal mol^{-1}. It was determined that the most probable relative kinetic energy of the products is 2.4 kcal mol^{-1}. The standard enthalpy for the reaction is $\Delta \mathcal{H}_0^\circ = 4.7$ kcal mol^{-1}. The most probable rotational energy of HBr (at 260 K) is 0.3 kcal mol^{-1}. What is the most probable internal energy of the product KBr?

2. What are the relative translational energies of the products of the reaction $Cl + HI \rightarrow HCl$ $(J' = 0, v') + I$ at 300 K for $v' = 3$ and 4 if the reactants have no vibrational, $\mathcal{E}_a + RT$ translational, and $\frac{3}{2}RT$ rotational energy? The dissociation energies of the diatomic molecules are $D_0^\circ(HI) = 3.056$ and $D_0^\circ(HCl) = 4.430$ eV, the activation energy for the reaction is $\mathcal{E}_a = 3.0$ kcal mol^{-1} and the vibrational frequency of HCl is 2889 cm^{-1}. Assume that the HCl vibration is harmonic. What is the maximum v' for the given reactant energies?

3. In a crossed beam experiment on reactive scattering a mass spectrometer detector of effective area 1 mm^2 must detect a flux of product molecules as low as 10^4 molecule^{-1} s^{-1} mm^{-2}. If the ambient product molecules are not to contribute more than 10% of the minimum detector signal, what is the maximum permissible partial pressure of the product molecules (molecule mass 30 g mol^{-1}) within an ambient gas at 300 K?

4. (a) Calculate the number of product molecules formed per second by two crossed reactant beams with the following properties: flux of primary beam is 10^{14} molecule^{-1} s^{-1}; concentration of reactant molecules in the secondary beam is 10^{10} molecule cm^{-3} within the reaction volume of 10^{-4} cm^3; reaction cross section is 1 nm^2.

   (b) Assuming that the products scatter evenly into one hemisphere, how many product molecules per second strike a detector subtending an angle of $4\pi \times 10^{-5}$ steradians?

5. Consider the reaction $A + BC \rightarrow AB + C$. Assume that the initial velocity of A is v_A and of BC is $v_{BC} = 0$ and that BC is in the ground state.

   (a) In an ideal stripping collision, an A collides and reacts with a B without disturbing the C group of atoms, that is, $v_C = 0$. Use momentum and energy conservation to show that, in this case, the amount of energy converted from translational energy to internal energy of AB is given by

$$(\tfrac{1}{2} m_A v_A^2) \times m_B / (m_A + m_B)$$

(b) Calculate the amount of translational energy that an ideal stripping collision would convert into internal energy for the reaction $N_2^+ + CH_4 \rightarrow N_2H^+ + CH_3$. Assume that CH_4 has negligible internal energy and is at rest and that $\frac{1}{2}m_{N_2^+}v_{N_2^+}^2 = 25$ eV. *Note*: Gislason et al. [11] have shown that a large fraction of NH_2^+ is produced by nearly ideal stripping collisions.

6. Hot tritium atoms can undergo the following substitution reactions with deuterated methanes

$$CH_n D_{4-n} + T \begin{cases} \rightarrow CH_n D_{3-n} T + D \\ \rightarrow CH_{n-1} D_{4-n} T + H \end{cases}$$

The relative total yields from the substitution of 2.8 eV T atoms for H or D in methanes are $CH_4 : CH_3D : CH_2D_2 : CD_4 = 7.2 : 5.6 : 3.1 : 1.0$. Explain qualitatively the decrease in the substitution yield.

7. The symmetric stretching vibration of CO_2 has a first excited state (labeled 100) at 1388 cm^{-1}. The first excited state (001) of the antisymmetric stretch is at 2349 cm^{-1}. A CO_2 laser emits radiation due to the transition $001 \rightarrow 100$.

(a) Calculate the populations, relative to the ground state, of the two excited states in a sample of CO_2 gas in equilibrium at 300 K.

(b) Supposing that the gas sample absorbs enough radiation from a CO_2 laser to equalize the populations of the two excited states, calculate the effective temperature relative to the ground state for each of the two excited states. Discuss qualitatively the variation with time of the effective temperatures in the two states.

8. At 300 K, 1 torr total pressure, in pure CO_2, the first excited state of the symmetric stretch is deactivated via collisions at a rate of 330 quanta molecule^{-1} s^{-1} [12]. If the hard sphere diameter of CO_2 is 3.30 Å, what is the mean number of hard sphere collisions required to remove one quantum from the first excited state of the symmetric stretch?

References and Recommended Reading

See references [4], [5], [7], [8], and [9] at the end of Chapter 21.

[1] M. A. D. Fluendy and K. P. Lawley: *Chemical Applications of Molecular Beam Scattering*. Chapman and Hall, London, 1973.

[2] *Disc. Faraday Soc.*, **55** (1973). Molecular beam scattering. A symposium volume on molecular beams and related subjects.

[3] J. C. Polanyi: *Chemical Kinetics*. Butterworth, London, 1972. (Vol. IX, Series One, *Physical Chemistry*, in the MTP International Review of Science.)

[4] R. J. Conley, E. F. Greene, and S. M. McInnis: *J. Chem. Educ.*, **51**:620 (1974). A molecular beam apparatus for student experiments on reactive scattering.

[5] C. J. G. Raw: *J. Chem. Educ.*, **43**:480 (1966). Elementary theory of molecular beam scattering.

[6] D. Beck, E. F. Greene, and J. Ross: *J. Chem. Phys.*, **37**:2895 (1962).

[7] R. J. Beuhler, Jr., R. B. Bernstein, and K. H. Kramer: *J. Am. Chem. Soc.*, **88**:5331 (1966).

[8] J. D. Rynbrandt and B. S. Rabinovitch: *J. Chem. Phys.*, **54**:2275 (1971).

[9] J. D. Rynbrandt and B. S. Rabinovitch: *J. Phys. Chem.*, **75**:2164 (1971).

[10] A. E. Grosser et al.: *J. Chem. Phys.*, **42**:1268 (1965).

[11] E. A. Gislason et al.: *J. Chem. Phys.*, **50**:142 (1969).

[12] J. T. Yardley and C. B. Moore: *J. Chem. Phys.*, **46**:4491 (1967).

Part VI

Electrochemistry

chapter 23

Introduction to Solutions of Electrolytes

23-1 THE COLLIGATIVE BEHAVIOR OF ELECTROLYTIC SOLUTION

As is seen in Section 8-11.2 the laws of colligative properties apply to dilute solutions of nonvolatile solutes reasonably well. However, in 1885 van't Hoff, while following the colligative behavior of solutions, discovered that a large group of solutions does not obey the simple laws of colligative behavior, even when dilute. The change of the colligative properties in these solutions was always to be found greater than the theoretically expected change. To account for this anomalous behavior, van't Hoff introduced a factor i defined by the relation

$$i = \frac{\Delta z_{exp}}{\Delta z_{th}} = (\nu - 1)\alpha + 1 > 1 \qquad (23\text{-}1)$$

where Δz_{exp} and Δz_{th} are the experimentally found and theoretically expected changes in the colligative properties, α is the degree of dissociation of the solute in the solution, and ν is the number of particles formed by the dissociation of one molecule of solute. For NaCl, $\nu = 2$, (Na^+ and Cl^-), and for $Al_2(SO_4)_3$, $\nu = 5$, ($2\ Al^{3+}$ and $3\ SO_4^{2-}$).

Solutions belonging to this group resemble metals in one respect, in that they also conduct electricity. The difference between the two kinds of conductors is merely in the mode of transporting electricity. In a metallic conductor, current is transported by electrons and the conducting material does not undergo any chemical change. In a solution conductor, current is carried by positively charged ions, cations, and negatively charged ions, anions, formed in the solution by interaction of the solute with solvent molecules. Melts of such solutes also conduct electricity. When the molecules of a substance are dissociated in solutions or in melts into charged particles, that substance is called an **electrolyte**. Commonly, the ions are already present in the solid solute. Interaction with solvent molecules or an increased thermal energy helps to keep the ions separated from each other and thus able to conduct electricity.

The branch of chemistry dealing with the formation and behavior of ions in solutions and in melts, as well as with the behavior of ions on the boundaries with other phases, solid or liquid, is known as electrochemistry[1, 2]. Strictly speaking, on a microscopic level chemistry always deals with charged particles (electrons, protons); electrochemistry however, is concerned only with systems able to conduct electrical current in solutions or in melts.[1]

In order to explain the strange behavior of this group of solutions, called electrolytic solutions, Arrhenius in 1887 proposed the dissociation theory.

---

[1] The plasma state, a highly ionized gaseous state of matter, is generally not the subject of electrochemistry.

According to this theory molecules of electrolytes dissociate to a greater or lesser extent into ions and equilibrium is established between the nondissociated (neutral) molecules and their ions. However, on the basis of observations made by many experimentalists, Debye and Hückel concluded, in 1923, that some electrolytes may undergo a complete dissociation into ions. In such solutions there are no neutral molecules present and consequently no equilibrium can be established between the neutral molecules and their ions. Consequently, solutions of this kind, called strong electrolytes, cannot follow Arrhenius' theory. In contrast to these solutions, partially dissociated electrolytes, called weak electrolytes, obey the Arrhenius theory quite well.

Arrhenius ignored the influence of the solvent on the process of dissociation. It was recognized later, that the solvent significantly affects the dissociation process and the behavior of the ions in the solution. Ions in solutions do interact with each other and also with the neutral molecules of the solvent. For this reason, an electrolyte dissociated strongly in one solvent may undergo a very slight dissociation in another solvent, and vice-versa.

23-2 ION–SOLVENT INTERACTIONS

The nature and extent of interaction of ions with solvent molecules, known as solvation, is not fully understood theoretically. The question of how many molecules of the solvent can arrange themselves around an ion in the strongly bound first layer and in the less tightly bound second and further layers is difficult to answer quantitatively. At constant temperature, however, this number will depend on the nature of the solvent, acidity, basicity, dielectric constant, polarizability, and so on, as well as on the charge, size, and shape of the ion itself. Since the electron charge density is larger around a small ion of a given charge, it is reasonable to expect the extent of interaction—that is, solvation—to be greater with small ions. This has been proved to be true. For example, the primary hydration (solvation in water as solvent) numbers of Li^+, Na^+, K^+, and Rb^+ are 5, 4, 3, and 3, respectively. Those solvent molecules that are strongly bound, that is, that have lost their translational degrees of freedom completely, form the inner or first solvation layer, their number being the primary solvation number.

The phenomenon of solvation of ions is clearly shown in the case of transport of electricity by electrolytic solutions. The measurement of transference numbers (see page 844) indicates that during the passage of electrical current through solutions not only the ions but also solvent molecules forming the solvation layer are transported.

In the simplest theory of ion–solvent interaction, Born (1920) considers the ions to be spheres of radius r_i and charge $Z_i e$ [3, 4, 5]. According to this theory, the Gibbs free energy change associated with the transfer of 1 mole of ions from a vacuum into a solvent of a dielectric constant D, called the Gibbs free energy of solvation, is expressed as[2]

$$\Delta \mathscr{G}_{\text{solv}} = -\frac{N(Z_i e)^2}{8\pi\varepsilon_0 r_i}\left(1 - \frac{1}{D}\right) \tag{23-2}$$

[2] For units see pages 836, 951.

where N is Avogadro's number, ε_0 is the permittivity in vacuum and is equal to $8.8542 \times 10^{-12} \; C^2 \, J^{-1} \, m^{-1}$.

The entropy of solvation follows directly from Eq. (23-2)

$$\Delta \mathscr{S}_{solv} = -\frac{\partial \Delta \mathscr{G}_{solv}}{\partial T} = \frac{N(Z_i e)^2}{8 \pi \varepsilon_0 r_i D} \frac{\partial \ln D}{\partial T} \tag{23-3}$$

and the enthalpy of solvation is obtained from combination of Eqs. (23-2) and (23-3) as

$$\Delta \mathscr{H}_{solv} = \Delta \mathscr{G}_{solv} + T \, \Delta \mathscr{S}_{solv} = -\frac{N(Z_i e)^2}{8 \pi \varepsilon_0 r_i} \left[1 - \frac{1}{D} \left(1 + \frac{\partial \ln D}{\partial \ln T} \right) \right] \tag{23-4}$$

There are two oversimplifications in the Born theory. One is the assumption that the dielectric constant of the solvent is the same in the immediate surroundings of the ion as in the bulk of the solution (the same approximation is used by Debye, page 857); the second ignores the work connected with the compression of the solvent molecules around the ion. In spite of these simplifications, the Born equation is very useful and has often been used to compare the solvating effects of different solvents.

When 1 mole of ions is transferred from a solvent of dielectric constant D_1 into a solvent of dielectric constant D_2, Eq. (23-2) assumes the form

$$\Delta \mathscr{G} = \frac{N(Z_i e)^2}{8 \pi \varepsilon_0 r_i} \left(\frac{1}{D_2} - \frac{1}{D_1} \right) \tag{23-5}$$

where $\Delta \mathscr{G}$ is the difference of the Gibbs free energy of solvation of 1 mole of ion i in solvent 2 and 1, respectively. The Gibbs free energies of solvation of selected ions in water, liquid ammonia, and ethanol are listed in Table 23-1.

Table 23-1 Solvation Gibbs free energies of ions in three different solvents [in kcal mol^{-1}]

Ion	Solvent		
	$H_2O(l)$	$NH_3(l)$	Ethanol(l)
H^+	258.0	281.0	252.0
Li^+	117.0	124.0	115.0
Na^+	96.0	99.0	90.0
K^+	78.0	79.4	73.2
Rb^+	74.4	73.3	—
Cs^+	64.0	65.6	—
Cl^-	74.0	65.5	71.3
Br^-	68.0	62.8	66.2
I^-	59.4	57.0	58.5

The data listed in Table 23-1 refer to infinitely dilute solutions at 25°C. However, it has been proved experimentally that the Gibbs free energy of solvation is only slightly concentration dependent.

23-3 ION–ION INTERACTIONS

As opposed to ion–solvent interactions, ion–ion interactions are strongly concentration dependent.

The attractive part of the interaction energy between two oppositely charged ions, ε_a, can be derived from Coulomb's law[3] as

$$\varepsilon_a = -\frac{Z_i Z_j e^2}{4\pi\varepsilon_0 D r_{ij}} \tag{23-6}$$

where Z_i, Z_j, refer to the number of charges on the respective ions, e is the charge of an electron, r_{ij} is the distance between the centers of the two ions, and D and ε_0 have the same meaning as before.

Experimental facts indicate that, at close distance, repulsion may exist even between two oppositely charged particles. The energy corresponding to the repulsion is assumed to follow the same relationship as for neutral molecules (see page 11).

$$\varepsilon_r = \frac{B}{r_{ij}^n}$$

Thus, the total interaction energy between two ions equals

$$\varepsilon = \varepsilon_a + \varepsilon_r = -\frac{Z_i Z_j e^2}{4\pi\varepsilon_0 D r_{ij}} \cdot \frac{B}{r_{ij}^n} \tag{23-7}$$

The constant B may be evaluated from the equilibrium condition: at $r_{ij} = r_{ij}^*$, $\partial\varepsilon/r_{ij} = 0$. The result is

$$B = \frac{Z_i Z_j e^2 r_{ij}^{*(n-1)}}{4\pi\varepsilon_0 D n} \tag{23-8}$$

---

[3] For the interaction of two charges q_i and q_j in a vacuum, Coulomb's law has the form

$$f = \frac{q_i q_j}{k r_{ij}} = \frac{Z_i e Z_j e}{k r_{ij}}$$

where f is the force acting between the two charges separated by a distance r_{ij}.

There are two choices of units recommended for the constant k. If k is set equal to unity with the proper units in the cgs system ($f = 1$ dyne, $r_{ij} = 1$ cm), Coulomb's law defines the unit of charge in the cgs system, 1 esu or 1 statcoulomb. However, if we express the force in Newtons (N), the distance in meters (m), and the charge in Coulombs C, the value of k is found to be 1.1126×10^{-10} $C^2 J^{-1} m^{-1} (C^2 N^{-1} m^{-2})$. For practical purposes k is related to another constant, called the permittivity of vacuum, ε_0:

$$k = 4\pi\varepsilon_0$$

The value of ε_0 is 8.8542×10^{-12} $C^2 J^{-1} m^{-1}$. In a medium different from vacuum, Coulomb's law must be modified to

$$f = \frac{q_i q_j}{4\pi\varepsilon_0 D r_{ij}^2}$$

where D is the dielectric constant of the medium (in vacuum equal to 1).

Following from this equation, the energy is

$$.\varepsilon_a = -\frac{\partial f}{\partial r_{ij}} = -\frac{q_i q_j}{4\pi\varepsilon_0 D r_{ij}}$$

and Eq. (23-7) transforms into

$$\varepsilon = -\frac{Z_i Z_j e^2}{4\pi\varepsilon_0 D r_{ij}^*}\left[\left(\frac{r_{ij}^*}{r_{ij}}\right)-\frac{1}{n}\left(\frac{r_{ij}^*}{r_{ij}}\right)^n\right] \tag{23-9}$$

The value of the exponent n may be found experimentally, usually from compressibility data. It varies from about 9 to 12, with 12 being common for monatomic ions.

According to the approximate formula (23-9), the interaction energy between two ions in a certain solvent depends on the distance between the centers of the two ions only: $\varepsilon = f(r_{ij})$. It is reasonable to expect that this will be the case for symmetrical ions only. In solutions containing unsymmetrical ions, the shape of the ion affects the shape of the electrical field around it. Consequently, the magnitude of the interaction will vary not only with the distance but also with the angle of approach of ions. The interaction energy function is therefore more complex than indicated by Eq. (23-9) and, at present, no theory can predict its exact form.

In 1923 Debye and Hückel (see page 857) developed a theory, based on the assumption of complete dissociation of electrolytes, that is able to predict the behavior of ions in very dilute solution. The theory accounts for the ion–solvent as well as ion–ion interactions. This theory was later extended to more concentrated solutions by using Bjerrum's assumption about the association of ions (into pairs, triplets, and so on) in solutions [5, 6].

As is obvious from Eq. (23-9), the dielectric constant of the solvent strongly affects the interionic energy. In solvents of low dielectric constant, the ionic interaction extends over much greater distances than in solvents of high dielectric constant. Since nonideality of ionic solutions is usually attributed to ion–ion interactions, it follows that solutions of high dielectric constant solvents exhibit greater conformity to ideal behavior than those at the same ionic concentration in solvents of low dielectric constant. The degree of dissociation of a given electrolyte will be greater in solvents of high dielectric constant.

The reliability of Eq. (23-9) can be tested on a simple system, such as an ionic crystal. An ionic crystal may be considered as a solution of positively and negatively charged particles distributed in orderly manner in a medium of low dielectric constant ($D \approx 1$). The distance between adjacent ions is a constant (neglecting the change due to vibrations about the equilibrium position), characterized by the crystal lattice structure only, $r_{ij} = r_{ij}^*$. Under such conditions, Eq. (23-9) reduces to the form

$$-\varepsilon = \varepsilon_c^* = \frac{Z_i Z_j e^2}{4\pi\varepsilon_0 r_{ij}^*}\left(1-\frac{1}{n}\right) \tag{23-10}$$

ε_c^* is the cohesion energy per molecule.

It must be recognized, that the interaction in crystals extends not only to the nearest neighbors but spreads out to the more distant particles. These interactions, known as higher order interactions, depend on the lattice structure only. They have been evaluated for numerous crystal lattices and are expressed in terms of the **Madelung constant**, A (see also page 520). Introducing this constant into Eq. (23-10), we get for 1 mole of a crystal

$$\mathscr{E}_c^* = \frac{A N Z_i Z_j e^2}{4\pi\varepsilon_0 r_{ij}^*}\left(1-\frac{1}{n}\right) \tag{23-11}$$

where N is Avogadro's number and $\mathscr{E}_c^*$ is the molar cohesion energy.

This approximate equation (the electron–electron interactions, the zero-point energies of the ions, and the dispersion and polarization energies are ignored) supplies reasonable results, as is seen from Table 23-2, where lattice energies calculated from this equation and from the Born–Haber cycle are compared. Direct measurements of the lattice energy are rather difficult to make.

Table 23-2 Comparison of the lattice energies obtained from Eq. (23-11) and from the Born–Haber cycle (kcal mol^{-1})

Salt	n	$\mathscr{E}_c^*$	$\mathscr{E}_{BH}$	$\mathscr{E}_c^* - \mathscr{E}_{BH}$
LiCl	7.0	193.3	201.0	−7.7
NaCl	8.0	180.4	190.0	−9.6
KCl	9.0	164.4	166.0	−1.6
CsCl	10.5	148.9	153.0	−4.1

The Born–Haber cycle includes the following steps (see also Chapter 17):

Steps	Process	Energy
1. $M(s) + \frac{1}{2}X_2(g) \rightarrow M(s) + X(g)$	dissociation of $X_2(g)$	$\frac{1}{2}\mathscr{D}$
2. $M(s) + X(g) \rightarrow M(g) + X(g)$	sublimation of $M(s)$	$\Delta\mathscr{H}_{subl}$
3. $M(g) + X(g) - e^- \rightarrow M^+(g) + X(g)$	ionization of $M(g)$	IE
4. $M^+(g) + e^- + X(g) \rightarrow M^+(g) + X^-(g)$	electron affinity of $X(g)$	EA
5. $M^+(g) + X^-(g) \rightarrow MX(s)$	condensation of gaseous ions to solid MX	$-\mathscr{E}_c^*$

Summation of the five reactions results in the reaction of formation of MX(s):

$$M(s) + \tfrac{1}{2}X_2(g) = MX(s); \ \Delta\mathscr{H}_f$$

From the first law of thermodynamics it follows, that

$$\mathscr{E}_c^* = -\Delta\mathscr{H}_f + \tfrac{1}{2}\mathscr{D} + \Delta\mathscr{H}_{subl} + IE - EA \qquad (23\text{-}12)$$

in which the enthalpy of sublimation, $\Delta\mathscr{H}_{subl}$, dissociation energy, $\mathscr{D}$, ionization energy, IE, and electron affinity, EA, can be obtained experimentally.

The agreement is fairly good. Better results are obtained when corrections for dispersion and polarization forces as well as for electron–electron forces are added.

EXAMPLE 23-1

The dielectric constants of vacuum and water at 25°C are 1 and 78.5, respectively. Making the assumption that repulsion, dispersion and polarization do not contribute to the interaction energy, calculate the energy required to separate a sodium and a chloride ion from a distance of 2 Å to infinity in vacuum and in water. (1 Å = 10^{-8} cm = 10^{-10} m).

Solution: In the cgs system of units, Coulomb's law may be written as

(a) In vacuum

$$\Delta\varepsilon = \int_{2\text{Å}}^{\infty} f\,dr_{ij} = \int_{2\text{Å}}^{\infty} \frac{Z_i Z_j e^2}{r_{ij}^2}\,dr_{ij}$$

$$= -\frac{Z_i Z_j e^2}{r_{ij}}\Bigg]_{2\times10^{-8}}^{\infty} = \frac{(4.8\times10^{-10})^2}{2\times10^{-8}}$$

$$= 1.15\times10^{-11}\,\text{erg} = 1.15\times10^{-18}\,\text{J} = 7.20\,\text{eV}$$

where $1\,\text{eV} = 1.6\times10^{-12}\,\text{erg} = 1.6\times10^{-19}\,\text{J}$.

(b) In water

$$\Delta\varepsilon = \int_{2\text{Å}}^{\infty} f\,dr_{ij} = \int_{2\text{Å}}^{\infty} \frac{Z_i Z_j e^2}{Dr_{ij}^2}\,dr_{ij}$$

$$= -\frac{Z_i Z_j e^2}{Dr_{ij}}\Bigg]_{2\times10^{-8}}^{\infty} = \frac{(4.8\times10^{-10})^2}{78.5\times2\times10^{-8}}$$

$$= 1.467\times10^{-13}\,\text{erg} = 1.467\times10^{-20}\,\text{J} = 0.0917\,\text{eV} \quad\bullet$$

As is seen from Example 23-1, much less energy is required to separate the ions in water; thus, water favors the dissociation of NaCl more than vacuum. This explains why gaseous ions formed from neutral gaseous molecules by x-ray radiation recombine to form uncharged molecules whereas in aqueous solutions the ions are stable.

It seems to be useful to distinguish between the process of ionization and dissociation. Ionization is the process in which ions are formed

$$Na(s) \rightarrow Na^+ + e^-$$
$$Cl(g) + e^- \rightarrow Cl^-$$
$$NH_3(g) + CH_3COOH(l) \rightarrow CH_3COO^-NH_4^+(l)$$

The process of dissociation refers to the separation of the oppositely charged ions in the solution (or melt) from each other

$$Na^+Cl^-(s) + x\,H_2O(l) \rightarrow Na^+(H_2O)_y + Cl^-(H_2O)_z$$

The indices y and z denote the hydration numbers of the cation and the anion, respectively.

Ionization is a preliminary step to dissociation.[4] Sodium chloride is already completely ionized in its solid form; its dissociation into ions depends, however, on the nature of the solvent and also on the temperature and concentration of the solution.

In low dielectric constant solvents, association of ions into larger aggregates may be observed, especially at high concentrations [5, 6]. If ions come sufficiently close together, a distance d can be reached at which electrostatic attraction is greater than the thermal energies tending to disorder the ions ($d \approx Z_i Z_j e^2/2DkT$). In such a case a new entity could be formed in which the two ions are recombined into a neutral

---

[4] The process $CaCO_3(s) \rightarrow CaO(s) + CO_2(g)$ is sometimes also called dissociation. We, however, refer to similar processes as decomposition.

species, known as an ion pair:

$$X^+ + Y^- \rightarrow (X^+Y^-)^0$$

$(X^+Y^-)^0$ is not the same as an undissociated molecule because all the operative forces are purely electrostatic in character. An ion pair is only a temporary arrangement due to inelastic collisions, whereas an undissociated molecule is a permanent arrangement.

Making the assumptions that ions are rigid unpolarizable spheres and that all the interactions between ions are of Coulombic type only, Bjerrum (in 1926) derived a theory dealing with the association of ions. Because of its complexity the theory will not be presented in this book [1, 3, 4, 7, 8].

--------------------------------- *References and Recommended Reading*

[1] H. S. Harned and B. B. Owen: *Physical Chemistry of Electrolytic Solutions.* Reinhold, New York, 1958.

[2] B. E. Conway and M. J. Salomon: *J. Chem. Educ.,* **44**:554 (1967). Electrochemistry: its role in teaching physical chemistry.

[3] J. O'M. Bockris and A. K. K. Reddy: *Modern Electrochemistry,* Vols. I and II. Plenum Press, New York, 1970.

[4] J. Koryta, J. Dvořák, and V. Boháčkova: *Electrochemistry.* Methuen, London, 1970.

[5] J. E. Prue: *J. Chem. Educ.,* **46**:12 (1969). Ion pairs and complexes: free energies, enthalpies, and entropies.

[6] G. Corsaro: *J. Chem. Educ.,* **39**:622 (1962). Ion strength, ion association and solubility.

[7] C. A. Kraus: *J. Chem. Educ.,* **35**:324 (1958). The present state of the electrolyte problem.

[8] R. J. Parsons: *J. Chem. Educ.,* **45**:390 (1968). Electrochemical dynamics.

chapter 24

Transport and Thermodynamic Properties in Solutions of Electrolytes

24-1 ELECTROLYSIS AND FARADAY'S LAW

In solutions containing ions, electrical current is conducted by the migration of positive and negative ions. The ions are set in motion by an applied electric potential difference between two electrodes. (If there is a concentration gradient in the solution, ions can migrate even in the absence of an electrical field, by diffusion.) The positively charged cations move towards the negative electrode, the cathode, and the negatively charged anions migrate toward the positive electrode, the anode (see Figure 24-1).

Since the mean free path of molecules in liquids is very short, ions collide frequently with molecules of the solvent. Consequently, the velocity of net migration under the influence of an applied field[1] is very small; in an applied field of 1 V cm^{-1}, the velocity is only of the order $10^{-4} \text{ cm s}^{-1}$, which is very small when compared to the velocity of thermal motion, about 10^4 cm s^{-1} at room temperature (see page 76). After the ions have reached their respective electrodes, they may undergo a reaction with the electrodes. Some examples of basic types of electrode reactions follow.

> 1. **Processes taking place at the cathode—reduction.** Reduction of ions, complexes, or molecules to a lower oxidation state. Deposition of metals on the electrode after the discharge of ions.
>
> In a hydrochloric acid solution
>
> $$H^+ + e^- \rightarrow \tfrac{1}{2} H_2(g)$$
>
> In a copper sulfate solution
>
> $$\tfrac{1}{2} Cu^{2+} + e^- \rightarrow \tfrac{1}{2} Cu(s)$$
>
> In a sodium chloride solution
>
> $$H^+ + e^- \rightarrow \tfrac{1}{2} H_2(g)$$
>
> The H^+ ions present in low concentration from the dissociation of water are discharged more easily than the Na^+ ions. The sodium ions are, however, the carriers of electricity. Consequently, the solution around the cathode is basic.

---

[1] In the SI system of units a unit electric field exerts a force of 1 N (Newton) on a charge of 1 C (Coulomb). The unit is therefore

$$N \, C^{-1} = \frac{N \, m \, C^{-1}}{m} = \frac{J \, C^{-1}}{m} = \frac{V \, C \, C^{-1}}{m} = V \, m^{-1}$$

volts per meter.

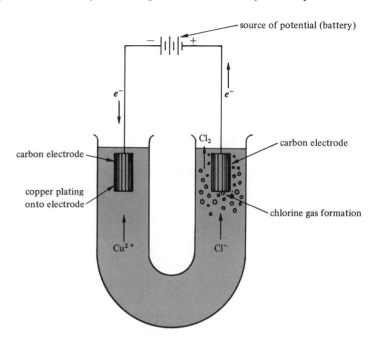

Figure 24-1 Electrolytic cell: electrolysis of $CuCl_2$.

In a solution of the cadmium cyanide complex containing excess cyanide ions, the following reaction takes place

$$Cd(CN)_4^{2-} + 2\,e^- \rightarrow Cd(s) + 4\,CN^-$$

2. Processes taking place at the anode—oxidations. Oxidation of ions, complexes, or molecules to species in a higher oxidation state. Ionization of the electrode material to form soluble ions or complexes.

In a hydrochloric acid solution

$$Cl^- - e^- \rightarrow \tfrac{1}{2}Cl_2(g)$$

In a copper sulfate solution

$$OH^- - e^- \rightarrow \tfrac{1}{2}H_2O(l) + \tfrac{1}{4}O_2(g)$$

The OH^- ions formed from water are discharged more easily than are the SO_4^{2-} ions. The SO_4^{2-} ions are, however, the carriers of electricity. As a result, the solution around the anode is acidic.

In a sodium chloride solution

$$Cl^- - e^- \rightarrow \tfrac{1}{2}Cl_2$$

Faraday studied the quantitative relation between the amount of chemical change observed at the electrodes and the quantity of electricity passed through the solution.[2] The results of Faraday's investigation can be summarized in the following two laws (published in 1834) [5]:

1. The amount of a substance formed or consumed at the electrodes during the process of electrolysis is proportional to the quantity of electricity passed.
2. The amounts of different substances resulting from the electrode reactions by the same amount of electricity (charge) are in the ratio of their electrochemical equivalents, $M/|Z|$.

More concisely the two laws may be written as

$$w = \frac{M}{|Z|\mathfrak{F}} I\tau = \frac{M}{|Z|\mathfrak{F}} Q \tag{24-1}$$

Here w is the weight of the substance deposited or removed on or from the electrode (the total amount discharged), M is the molecular weight, $|Z|$ is the magnitude of the charge of the ions in terms of the electronic charge as a unit, and $\mathfrak{F}$, called Faraday's constant, is the quantity of charge that will generate exactly 1 gram-equivalent (g-eq) weight of any substance. It is the charge of an Avogadro's number of electrons; $\mathfrak{F} = Ne = (1.60210 \times 10^{-19}\,C)(6.02252 \times 10^{23}) = 96{,}487\,C\,mol^{-1}$.

Faraday's laws give excellent results for the yields of electrode reactions. They cannot, however, give any information about the actual course, that is, the mechanism, and the rate of the reaction [7]. This depends on factors like the current density, the shape and nature of the electrode, the polarization of the electrode, the magnitude of the field, the concentration gradient, and on many other factors [3, page 932]. For example, the electrolytic reduction of aldehydes and ketones at a mercury electrode under specific conditions gives hydroxy compounds, called the pinacones

$$2\,C_6H_5 \cdot CHO + 2\,e^- + 2\,H^+ \rightarrow C_6H_5 \cdot CHOH \cdot CHOH \cdot C_6H_5$$

The mechanism by which this reaction proceeds is the subject of much discussion but is beyond the scope of this book; we shall also avoid here detailed discussion of rates of electrode reactions.

The concentration gradients formed during the process of electrolysis can be easily eliminated by adequately stirring the solution in the electrolytic cell. The polarization of the electrodes, however, is much more difficult to avoid. Polarization may have a decisive influence on the reaction mechanism and on the rate-determining step (see page 924).

A coulometer is a device whose function is fully based on Faraday's law. It is used to evaluate the charge passed through a solution from the experimentally found amount of chemical change at the electrodes or from the concentration changes in the solution near the electrodes. The most precise equipment of this kind is the silver

---

[2] The amount of electrical charge, Q, and the current, I, are related by

$$Q = \int_{\tau_1}^{\tau_2} I\,d\tau$$

where τ denotes the time.

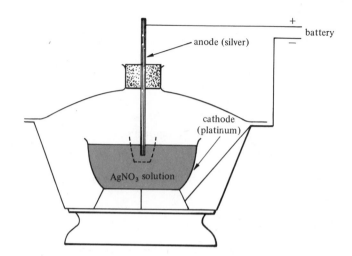

Figure 24-2   Silver coulometer.

coulometer, Figure 24-2. It is a small electrolytic cell containing two platinum or silver electrodes in a solution of aqueous silver nitrate. From the gain in mass of the cathode, the quantity of electricity passed through the electrolytic cell may be found from Eq. (24-1). Since 1 $\mathcal{F}$ (96,487 C) deposits 107.870 g of silver, 1 C is equivalent to $107.870/96,487 = 1.118 \times 10^{-3}$ g of silver.

24-2   TRANSFERENCE NUMBERS

According to Faraday's law, the number of equivalents of cations and anions chemically changed at the respective electrodes must be the same. However, the concentration changes in the solutions near the two electrodes are not necessarily the same (see Table 24-1). The explanation of this experimentally proved fact lies in the different velocities,[3] v, with which ions migrate in solutions. At a fixed concentration and a constant applied field, the velocity will depend on the size and charge of the ion. The more solvated the ions, the larger they are; consequently, their velocity will be smaller. Because of their different velocities, the quantity of electricity transferred by cations and anions is different.

The fraction of the charge carried by an ion is called its transference number $t_\pm$ [8]

$$t_\pm = \frac{q_\pm}{Q} \qquad (24\text{-}2)$$

---

[3] Velocity v, and mobility u, are related through the equation

$$u = \frac{v}{\Delta\Phi/\Delta x}$$

where $\Delta\Phi/\Delta x$ is the potential gradient or the applied electrical field. Mobility is thus the velocity in an applied field of unit strength.

where $q_\pm$ is the charge transported by the cation or anion and Q is the total amount of electricity carried through the solution.

In order to find a relation between the velocity and transference number of an ion let us assume a hypothetical experiment in which an imaginary plane is placed perpendicular to the surface of the solution and that the total amount of electricity crossing that plane is $1\mathfrak{F}$. It is logical to expect that the charge transferred by the cation, q_+, will be proportional to the velocity of the cation, v_+

$$q_+ = kZ_+m_+v_+ \tag{24-3a}$$

and the charge carried by the anion, q_-, is related directly to the velocity of the anion, v_-

$$q_- = k|Z_-|m_-v_- \tag{24-3b}$$

m_+, m_-, denote the molality of the cation and anion, respectively, and Z_+, $|Z_-|$ are the number of positive and negative charges per ion. The constant k depends on the applied electrical field only and must therefore be the same in both equations.

Since $Q = q_+ + q_-$, combination of Eqs. (24-2), (24-3a) and (24-3b) yields

$$t_\pm = \frac{q_\pm}{q_+ + q_-} = \frac{Z_\pm m_\pm v_\pm}{Z_+m_+v_+ + |Z_-|m_-v_-}$$

In addition, the solution is electrically neutral, $Z_+m_+ = |Z_-|m_-$, and therefore

$$t_\pm = \frac{v_\pm}{v_+ + v_-} = \frac{u_\pm}{u_+ + u_-} \tag{24-4}$$

The second equality in Eq. (24-4) follows from the fact that the electric field at the plane is the same for the cation and the anion; u_+, u_- denote the mobility of the cation and anion respectively.

As easily seen from the preceding equations

$$t_+ + t_- = 1 \tag{24-5}$$

and

$$\frac{t_+}{t_-} = \frac{v_+}{v_-} = \frac{u_+}{u_-} \tag{24-6}$$

The migration of ions and their interaction with the solvent as well as with their respective electrodes result in concentration changes in the solution near the electrodes. In order to find a relationship between these concentration changes and the transference, let us consider the following sample cell:

1. The electrolytic cell is divided into three parts: the cathode, the anode, and the middle compartment.
2. The solution contains 16 equivalents of a uni-univalent electrolyte.
3. The total charge passed through the solutions equals $4\mathfrak{F}$.
4. The velocity of the anion is three times that of the cation.

The results before and after electrolysis are summarized in Table 24-1.

Table 24-1 Concentration changes in the solution around the electrodes caused by the migration of ions and their interaction with the electrodes

	Cathode Compartment (g-eq)		Middle Compartment (g-eq)		Anode Compartment (g-eq)	
	Cation	Anion	Cation	Anion	Cation	Anion
original state	5	5	6	6	5	5
increased by transfer	+1	0	+1	+3	0	+3
decreased by transfer	0	−3	−1	−3	−1	0
decreased by the chemical changes at the electrodes	−4	0	0	0	0	−4
final state	$5+1-4=2$	$5-3=2$	$6+1-1=6$	$6+3-3=6$	$5-1=4$	$5+3-4=4$
concentration change	$5-2=3$	$5-2=3$	$6-6=0$	$6-6=0$	$5-4=1$	$5-4=1$

The results indicate that after the electrolysis the solution is electrically neutral in all three compartments. The concentration changes at the two electrodes are, however, different and are inversely proportional to the velocities of the ions.

$$\frac{\Delta n_{ca}}{\Delta n_{an}} = \frac{v_-}{v_+} = \frac{u_-}{u_+} = \frac{t_-}{t_+} \tag{24-7}$$

where Δn_{ca} is the change at the cathode and Δn_{an} is the change at the anode.

Thus, from the concentration changes in gram-equivalents around the cathode, Δn_{ca}, and anode, Δn_{an}, respectively the transference number of the ions can be evaluated. In the case shown in Table 24-1, $t_-/t_+ = \Delta n_{ca}/\Delta n_{an} = \frac{3}{1}$; since $t_+ + t_- = 1$, $t_+ = 0.25$ and $t_- = 0.75$.

The method proposed by Hittorf for the determination of the transference numbers is based on the principle just described [6, 14]. The more accurate the analytical procedure used to determine the concentration changes, the better the results obtained for the transference numbers. Since ions in solution are more or less solvated, the concentration changes will also depend to a certain extent on the solvation number of the ions.

The more frequently used moving boundary method, which is presented next, is based on direct measurements of the velocity (mobility) of ions in solutions. The method involves the use of two ionic solutions, having one ion in common, which are placed carefully into a vertical tube so that a sharp boundary develops between the two solutions. The position of the boundary is then observed. Current passing through the tube will make the boundary move up or down, depending on the ions present in the solution and the polarity of the electrodes. The transference number

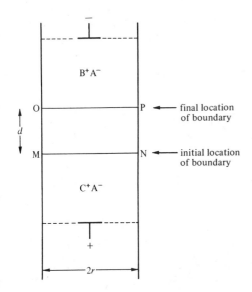

B^+A^-

O — P ← final location of boundary

d

M — N ← initial location of boundary

C^+A^-

+

$2r$

Figure 24-3 Transference number determination by the moving boundary method.

can be calculated from the distance the boundary has moved in unit time, from the cross-sectional area of the tube, and from the amount of electricity that has passed through the cell.

An electrolytic cell of this kind is seen in Figure 24-3. The two ionic solutions, B^+A^- and C^+A^- respectively, are separated by the boundary MN. After n_e equivalents of electricity have passed through the cell, the boundary has moved a distance d. Consequently, the volume of solution B^+A^- between the cross sections MN and OP, V, has been replaced by an equal volume of solution C^+A^-.

If the normality of solution C^+A^- is N, then NV equivalents of C^+ passed through the cross-sectional area of the tube, carrying a charge equal to $t_+ n_e$. Thus[4]

$$n_e t_+ = NV \qquad \text{(24-8a)}$$

Hence

$$t_+ = \frac{NV}{n_e} \qquad \text{(24-8b)}$$

All the quantities on the right-hand side of Eq. (24-8b) can be obtained experimentally; V is the volume (in liters) of a cylinder of radius r and height d, N is the normality of the solution C^+A^- and n_e $(\equiv Q/\mathfrak{F})$ is the amount of electricity in equivalents that have passed through the cell.

The moving boundary method supplies good results for the transference numbers (precision ± 0.0002). Some values are listed in Tables 24-2 and 24-3.

The transference numbers are temperature as well as concentration dependent. They also vary with the nature of the counter-ion in the solution and with the nature of the solvent. The values usually vary between 0.4 and 0.6, except for H^+ and OH^-

[4] It is assumed that convection at the boundary is negligibly small and that the shift of the boundary does not depend on the concentration of solution C^+A^-. The convection is minimized by choosing the lower solution to be the one with the higher density.

Table 24-2 Temperature and concentration dependence of the transference number of the potassium ion in aqueous potassium chloride solutions

t (°C)	Concentration/(g-eq liter^{-1})					
	0 (extrapolated)	0.005	0.01	0.02	0.05	0.1
15	0.4928	0.4926	0.4925	0.4924	0.4923	0.4921
25	0.4906	0.4903	0.4902	0.4901	0.4900	0.4900
35	0.4889	0.4887	0.4886	0.4885	0.4885	0.4888
45	0.4872	0.4869	0.4868	0.4868	0.4869	0.4873

Table 24-3 The transference numbers of some cations in aqueous solutions of different concentrations

Salt	Concentration/(g-eq liter^{-1})					
	0 (extrapolated)	0.01	0.02	0.05	0.1	0.2
HCl	0.8209	0.8251	0.8266	0.8292	0.8314	0.8337
LiCl	0.3364	0.3289	0.3261	0.3211*	0.3168	0.3112
NaCl	0.3963	0.3918	0.3902	0.3876	0.3854	0.3821
KCl	0.4906	0.4902	0.4901	0.4899	0.4898	0.4894
KBr	0.4849	0.4833	0.4832	0.4831	0.4833	0.4841
KI	0.4892	0.4884	0.4883	0.4882	0.4883	0.4887
KNO$_3$	0.5072	0.5084	0.5087	0.5093	0.5103	0.5120
Na$_2$SO$_4$	0.3860	0.3848	0.3836	0.3829	0.3828	0.3828
K$_2$SO$_4$	0.4790	0.4829	0.4848	0.4870	0.4890	0.4910
CaCl$_2$	0.4380	0.4264	0.4220	0.4140	0.4060	0.3953

Table 24-4 The transference number of the cation in aqueous solutions of CdI$_2$ at 18°C

Concentration (g-eq liter^{-1})	t_+	Concentration (g-eq liter^{-1})	t_+
0.0200	0.443	0.0819	0.343
0.0250	0.442	0.1245	0.281
0.0314	0.427	0.1610	0.223
0.0415	0.407	0.2497	0.075
0.0621	0.381	0.4975	−0.003

ions. The values in Table 24-3 also prove that the smaller an ion is, the lower its transference number will be; this is an experimental proof that the smaller an ion, the more solvated it is.

Solutions containing ions able to form complexes exhibit very strange behavior. As seen from Table 24-4, the transference number of the cation in such solution is much more concentration dependent than in the case of simple ionic solutions.

The anomalous behavior is explained by the formation of complex ions, such as CdI_3^- and CdI_4^{2-}, that migrate toward the positive anode under the influence of an applied field. Consequently, the concentration of CdI_2 in the cathode area is reduced. At low concentrations the complexes disintegrate into simple ions, Cd^{2+} and I^{-1}, and the transference numbers assume their normal values.

24-3 ELECTROLYTIC CONDUCTANCE

The resistance R, offered by any homogeneous portion of matter to the passage of electricity, is directly proportional to the length of the resistor l and inversely proportional to the area A across which the current flows:

$$R = \rho \frac{l}{A} \tag{24-9}$$

The proportionality constant ρ is the resistance offered by a cube of unit length; it is called the resistivity, or specific resistance (units, ohm cm).[5] The reciprocal of ρ is the specific conductance (units, $ohm^{-1} cm^{-1}$); denoting it by χ, we write

$$\chi = \frac{1}{\rho} = \frac{1}{R} \frac{l}{A} \tag{24-10}$$

With the aid of Ohm's law

$$R = \frac{\Phi}{I} \tag{24-11}$$

Eq. (24-10) assumes the form

$$\chi = \frac{I/A}{\Phi/l} = \frac{Q/\tau A}{\Phi/l} = \frac{i}{E} \tag{24-12}$$

where I is the current in amperes,[6] Φ is the potential difference between the two ends of the resistor, and τ is time. The specific conductance is thus the amount of electricity transported across a unit area in unit time ($\equiv i$) under a unit electric field ($\equiv E$).

---

[5] In the SI system of units, ohm m.

[6] In the SI system of units, an ampere (1 A) is defined as the current that produces a force between the conductors equal to $2 \times 10^{-7} N m^{-1}$ when maintained in two parallel rectilinear conductors of infinite length and negligible circular cross section at a distance apart of 1 m in a vacuum. All the other electrical units used in electrochemistry are derived: potential, Φ, in volts, V, where $V = watt A^{-1}$ and resistance, R, in ohms, Ω, where $\Omega = V A^{-1}$.

The first measurement of conductance of electrolytic solutions indicated a strong dependence of the resistance on the applied potential difference. This rather strange behavior, not observed in metallic conductors, was largely due to the polarization resulting from the use of direct current. As soon as reliable conductance data measured with an alternating current on ionic solutions were available, it became apparent that the resistance of these solutions follow Ohm's law exactly, at least at low and medium electric field strengths.

In principle, the specific conductance may be obtained directly from Eq. (24-10) if the distance l between the electrodes and the areas A of the electrodes are known. However, it seems to be more convenient to standardize the conductance cell—to find the ratio l/A—with a solution of known specific conductance, for example, aqueous KCl solution. Denoting l/A by C, we have

$$\chi = \frac{C}{R} \qquad (24\text{-}13)$$

where the cell constant C remains fixed as long as the distance between the two electrodes and their effective area is unchanged.

The conductance of ionic solutions is always determined by measuring the resistance of the solution in a cell, Figure 24-4a, connected into one arm of an alternating current (ac) bridge circuit, shown schematically in Figure 24-4b. With ac frequencies in the audio range, 1000–4000 Hz, the direction of the current changes so rapidly, that polarization effects are practically eliminated. In Figure 24-4b, G is the source of the alternating current, R_x is the unknown resistance of the cell filled with the solution, R_1, R_2 are standard resistors of known resistances, R_3 is an adjustable resistor and B is a detector, that is, a cathode-ray oscilloscope. R_3 is varied until the detector indicates no potential difference between points a and b. The bridge is then balanced. The circuit must be balanced for both resistance and reactance. Hence

$$\frac{R_x}{R_3} = \frac{R_2}{R_1} \qquad (24\text{-}14)$$

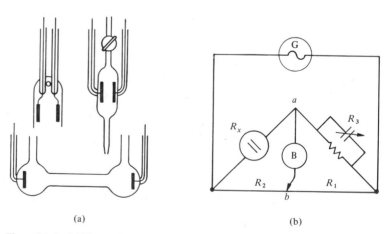

(a)   (b)

Figure 24-4   (a) Types of conductometric cells. (b) Measurement of conductance of electrolytes.

and the unknown resistance, R_x, is evaluated. To reduce the polarization to a minimum, the electrodes are made from platinized platinum.

The conductance of a particular ionic solution depends on the concentration, on the temperature, and also on nature of the solvent.

For convenience and for better comparison of conductance data of different solutions, Kohlrausch[7] introduced the equivalent conductance, Λ_c, defined as

$$\Lambda_c = \frac{1000\chi}{c} \qquad (24\text{-}15)$$

in which c denotes the normality of the solution and χ is the observed specific conductance of the solution minus that of the pure solvent. The specific conductance of pure water is negligibly small $(4.3 \times 10^{-8} \text{ ohm}^{-1} \text{ cm}^{-1})$ and may be ignored in the majority of cases. It must, however, be taken into consideration for solutions containing sparingly soluble solutes.

The variation of the equivalent conductance with concentration is seen from the data listed in Table 24-5. As is obvious, the change is more remarkable for solutions containing less dissociated solutes [9]. The equivalent conductance of aqueous hydrochloric acid solutions—strongly dissociated—varies in a certain concentration range from 426.16 to 391.32 $\text{ohm}^{-1} \text{ cm}^2 \text{ eq}^{-1}$, whereas for an aqueous acetic acid solutions—less dissociated—the change is between 390.71 and 5.20 $\text{ohm}^{-1} \text{ cm}^2 \text{ eq}^{-1}$.[8]

Figure 24-5 illustrates the influence of the solvent on the conductance. The large conductance of the HCl-water system (curve 1) proves that HCl is strongly dissociated in this solvent; the relatively much lower conductance of the HCl-anhydrous acetic acid solution (curve 2) indicates that HCl is much less dissociated in this system. Of the five acids, $HClO_4$, HBr, H_2SO_4, HCl, and HNO_3 in anhydrous acetic acid solutions, $HClO_4$ has the highest conductance. Even in this case, however, the conductance is just a small fraction of that which $HClO_4$ would exhibit in aqueous solutions.

From a large number of conductance experiments Kohlrausch observed that the equivalent conductance of aqueous as well as nonaqueous solutions at concentrations not exceeding 0.01 molar may be well expressed by the relation

$$\Lambda_c = \Lambda_\infty - Ac^{1/2} \qquad (24\text{-}16)$$

Λ_∞ is a very important property of the solution; it represents the equivalent conductance of an electrolyte at infinite dilution. Its value may be easily obtained graphically, extrapolating the Λ_c versus $c^{1/2}$ curve to zero concentration, or numerically from Kohlrausch's square root law

$$\lim_{c \to 0} \Lambda_c = \Lambda_\infty = \lim_{c \to 0} (\Lambda_\infty - Ac^{1/2})$$

The value of the constant A is determined from experimental data. As is shown later it can also be evaluated theoretically.

---

[7] From 1869 to 1880, F. Kohlrausch and his coworkers published a series of excellent papers dealing with conductance studies.

[8] Since many conductance data in the literature are expressed in the cgs system of units, the theory of conductance as well as the data used in this chapter are expressed in the same system of units.

Table 24-5   Equivalent conductances of various electrolytes in water at 25°C/(ohm^{-1} cm^2 eq^{-1})

Normality	NaCl	KCl	NaI	KI	HCl	AgNO$_3$	CaCl$_2$	NaC$_2$H$_3$O$_2$	HC$_2$H$_3$O$_2$
0.0000	126.5	149.9	126.9	150.3	426.1	133.4	135.8	91.0	(390.6)
0.0005	124.5	147.8	125.4	—	422.7	141.4	131.9	89.2	67.7
0.001	123.7	146.9	124.3	—	421.4	130.5	130.4	88.5	49.2
0.005	120.6	143.5	121.3	144.4	415.8	127.2	124.2	85.7	22.9
0.01	118.5	141.3	119.2	142.2	412.0	124.8	120.4	83.8	16.3
0.02	115.8	138.3	116.7	139.5	407.2	121.4	115.6	81.2	11.6
0.05	111.1	133.4	112.8	135.0	399.1	115.7	108.5	76.9	7.4
0.10	106.7	129.0	108.8	131.1	391.3	109.1	102.5	72.8	5.2

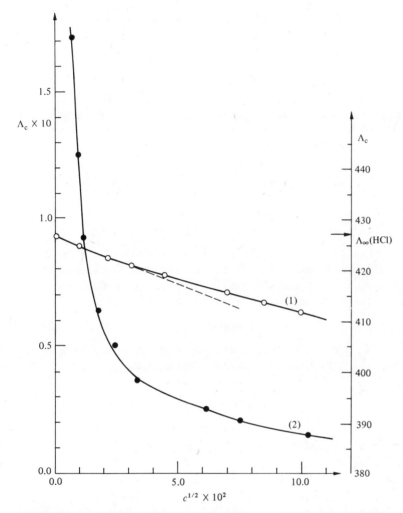

Figure 24-5 Equivalent conductance of HCl in water (curve 1, scale on the right) and in anhydrous acetic acid (curve 2, scale on the left).

The extrapolation procedure cannot be applied for weakly dissociated electrolytes. Such solutions have lower conductances at higher concentrations, but the values increase greatly with increasing dilution. Because of this steep increase in Λ_c at high dilution, the extrapolation to zero concentration is uncertain and may result in large errors.

For this reason, Kohlrausch recommended a different procedure for weakly dissociated solutes. He discovered that the difference in Λ_∞ for pairs of electrolytes having a common ion is always approximately a constant. For example (in units of ohm^{-1} cm^2 eq^{-1})

$$(\Lambda_\infty)_{KCl} = 149.86 \qquad (\Lambda_\infty)_{KI} = 150.38 \qquad (\Lambda_\infty)_{KClO_4} = 140.04$$
$$(\Lambda_\infty)_{NaCl} = 126.45 \qquad (\Lambda_\infty)_{NaI} = 126.94 \qquad (\Lambda_\infty)_{NaClO_4} = 116.48$$
$$\Delta = 23.41 \qquad\qquad \Delta = 23.44 \qquad\qquad \Delta = 23.56$$

From this observation Kohlrausch came to the following conclusion: the equivalent conductance at infinite dilution is composed of two independent quantities that are characteristic of the cation and the anion under consideration. Consequently

$$\Lambda_\infty = \lambda_\infty^+ + \lambda_\infty^- \tag{24-17}$$

where λ_∞^+ and λ_∞^- are the ionic equivalent conductances at infinite dilution. Values of Λ_∞ computed from Eq. (24-17) agree to within 0.4% with the observed values.

The application of Eq. (24-17) to the weakly dissociated aqueous acetic acid solution yields

$$\begin{aligned}
(\Lambda_\infty)_{\text{HAc}} &= (\Lambda_\infty)_{\text{NaAc}} + (\Lambda_\infty)_{\text{HCl}} - (\Lambda_\infty)_{\text{NaCl}} \\
&= \lambda_\infty^{\text{Na}^+} + \lambda_\infty^{\text{Ac}^-} + \lambda_\infty^{\text{Cl}^-} - \lambda_\infty^{\text{Na}^+} - \lambda_\infty^{\text{Cl}^-} + \lambda_\infty^{\text{H}^+} \\
&= \lambda_\infty^{\text{H}^+} + \lambda_\infty^{\text{Ac}^-} \\
&= 91.00 + 426.16 - 126.45 = 390.71 \text{ ohm}^{-1} \text{ cm}^2 \text{ eq}^{-1}
\end{aligned}$$

Thus, from the conductances of completely dissociated sodium acetate, hydrochloric acid, and sodium chloride in water solutions, the equivalent conductance of weakly dissociated acetic acid in water solution at infinite dilution can be obtained.

Walden found a simple relation between Λ_∞ and the viscosity of the solvent, η:

$$\Lambda_\infty \eta = \text{constant} \tag{24-18}$$

This rule is best obeyed by electrolytes consisting of large symmetrical ions, such as tetramethylammonium picrate, in water, methyl cyanide, acetone, and some other solvents.

24-4 ARRHENIUS' THEORY AND THE EQUIVALENT CONDUCTANCE

According to Arrhenius, electrolytes dissociate into ions ($B^+C^- \rightarrow B^+ + C^-$) and these ions are responsible for the transfer of electricity through the solution. Since the solution must always be neutral, the number of units of electricity (not t_+ and t_-) carried by the cation and the anion should be equal. If we denote by α the degree of dissociation of the electrolyte, then the concentrations of the cation and anion are $n_+ = n\alpha$ and $n_- = n\alpha$, respectively, where n is the number of molecules of solute per cubic centimeter of solution. Consequently, the total amount of electricity transported per second across a plane area A, perpendicular to the direction of flow is

$$\begin{aligned}
Q/\tau = I &= An_+ v_+ Z_+ e + An_- v_- |Z_-| e \\
&= AnZe(v_+ + v_-)\alpha
\end{aligned} \tag{24-19}$$

v_+ and v_- are the velocities of the cation and anion, Z_+ and $|Z_-|$ are the charges of the cation and anion, respectively; $Z = Z_+ = |Z_-|$ for a symmetrical electrolyte.

Using the current density, i (current per unit area), Eq. (24-19) may be rewritten into

$$\begin{aligned}
i = I/A &= nZe(v_+ + v_-)\alpha \\
&= nZeE(u_+ + u_-)\alpha
\end{aligned} \tag{24-20}$$

since the mobilities u_+, u_- and velocities v_+, v_- are related by the equations

$$v_+ = Eu_+ \tag{24-21a}$$

$$v_- = Eu_- \tag{24-21b}$$

where E is the strength of the applied field. Combination of Eqs. (24-12) and (24-20) yields

$$\chi = i/E = nZe(u_+ + u_-)\alpha$$

$$= \frac{NZe(u_+ + u_-)\alpha c}{1000}$$

$$= \frac{\mathfrak{F}(u_+ + u_-)\alpha c}{1000} \tag{24-22}$$

where N is Avogadro's number, $c(\equiv 1000n/N)$ is the concentration in gram-equivalents per liter, and the product NZe of a uni-univalent electrolyte is equal to Faraday's constant $\mathfrak{F}$.

Taking into consideration Eq. (24-15), we conclude

$$\Lambda_c = \mathfrak{F}(u_+ + u_-)\alpha \tag{24-23}$$

Similarly at infinite dilution ($\alpha \approx 1$)

$$\Lambda_\infty = \mathfrak{F}(u_\infty^+ + u_\infty^-) \tag{24-24}$$

From the last two equations and from Kohlrausch's law of independent mobilities of ions at infinite dilution it follows that

$$u_\infty^+ = \frac{\lambda_\infty^+}{\mathfrak{F}} \tag{24-25a}$$

$$u_\infty^- = \frac{\lambda_\infty^-}{\mathfrak{F}} \tag{24-25b}$$

and

$$\alpha = \frac{\Lambda_c}{\Lambda_\infty} \tag{24-26}$$

Arrhenius' theory makes possible the evaluation of the degree of dissociation from conductance measurements. Eq. (24-26) is, however, an approximate relationship only; the lower the concentration of the solution, the better the results.

Arrhenius' theory has been further applied by Ostwald to calculate the dissociation constant of an electrolyte from conductance data. Ostwald treated the dissociation like any chemical reaction and expressed the equilibrium constant in the following way

$$K = \frac{c_{B^+} c_{A^-}}{c_{BC}} \tag{24-27}$$

c_{B^+} and c_{C^-} are the equilibrium concentrations of the cation and anion respectively, and c_{BC} is the equilibrium concentration of the undissociated electrolyte. The use of concentration instead of activity restricts Eq. (24-27) to extremely dilute solutions.

Clearly, $c_{BC} = (1-\alpha)c$; $c_{B^+} = c_{C^-} = \alpha c$ and therefore

$$K = \frac{\alpha^2}{1-\alpha}c \tag{24-28}$$

where c is the original concentration of the solute (before dissociation took place). On replacing α by Arrhenius' ratio (Eq. 24-26), we have

$$K = \frac{c\Lambda_c^2}{\Lambda_\infty(\Lambda_\infty - \Lambda_c)} \tag{24-29}$$

which is the Ostwald dilution law. Eq. (24-29) can be linearized into

$$c\Lambda_c = -K\Lambda_\infty + \frac{K\Lambda_\infty^2}{\Lambda_c} \tag{24-30}$$

If $c\Lambda_c$ is now plotted as a function of $1/\Lambda_c$, the intercept is $-K\Lambda_\infty$, the slope is $K\Lambda_\infty^2$, and K is readily obtained. Figure 24-6 and Table 24-6 present data for aqueous acetic acid solutions. The linear law is seen to be satisfactorily obeyed. We find that $K\Lambda_\infty = 7.00 \times 10^{-3}$ and $K\Lambda_\infty^2 = 2.74$. Consequently, $\Lambda_\infty = 391.6$ ohm^{-1} cm^2 eq^{-1}) and $K = 1.787 \times 10^{-5}$ g-eq liter^{-1}, in good agreement with Ostwald's original results. This procedure represents another simple way of getting Λ_∞ of weakly dissociated electrolytes.

As is seen, while the concentration is increased by a factor over 200, α decreases by a factor of 10, and K remains approximately constant at an average value of 1.772×10^{-5}. The small variation of K with concentration is explained by the

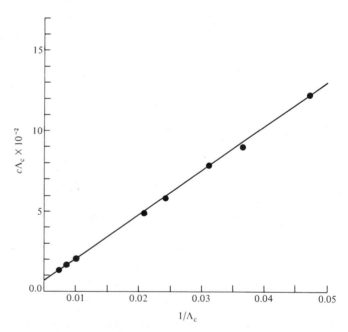

Figure 24-6 Evaluation of the dissociation constant of acetic acid (in water) from conductance measurements.

Table 24-6 Determination of the dissociation constant of acetic acid in aqueous solution at 25°C from conductance measurements

$c \times 10^5$ (mol liter^{-1})	Λ_c (ohm^{-1} cm^2 eq^{-1})	Percentage Dissociation $100\alpha = 100\dfrac{\Lambda_c}{\Lambda_\infty}$	$K \times 10^5$ (mol liter^{-1})
2.80	210.32	53.69	1.744
11.14	127.71	32.60	1.756
15.32	112.02	28.60	1.755
21.84	96.47	24.63	1.799
102.83	48.13	12.28	1.770
136.34	42.22	10.77	1.774
241.40	32.21	8.22	1.778
334.07	27.19	6.94	1.782
591.15	20.96	5.35	1.788

inaccuracy of Eq. (24-26) and also by the fact that Eq. (24-27) is expressed in terms of concentrations rather than activities.

For strongly dissociated solutes Ostwald's theory cannot be applied because $\alpha \approx 1$ at all concentrations.

Because the degree of dissociation and consequently the conductance of an electrolyte at constant temperature depends on the nature of the solvent, particularly on its dielectric constant, [the viscosity effect of the solvent is obvious from Walden's rule, Eq. (24-18)], the equilibrium constant of an electrolyte in different solvents will be different. A very rough approximation, based on Coulomb's law of interaction, leads to the formula

$$\left[\frac{\partial \ln K}{\partial (1/D)}\right]_{T,P} = \frac{Z_i Z_j}{(r_i + r_j)kT} \tag{24-31}$$

where r_i and r_j are the radii of the respective ions and Z_i and Z_j their charges. In mixed solvents, such as water–ethyl alcohol, hundredfold changes in K have been observed, resulting from the change of D with the concentration of the solvent. Eq. (24-31) enables a rough estimation of $(r_i + r_j)$ from conductance measurements in solvents of different dielectric constants.

24-5 THE THEORY OF IONIC INTERACTIONS

Arrhenius' theory proved to be invalid when applied to strongly dissociated electrolytes. It is also unable to explain the variation of the conductance and transference number with concentration.

The first successful theory of strong electrolytes was suggested by Debye–Hückel (in 1923) and later improved by Onsager (in 1926). The Debye–Hückel theory is based on a few assumptions, of which the following seem to be of great importance [1, 2, 4, 16]:

1. The dissociation of the solute molecules is complete. Consequently, it applies to dilute solutions only.
2. Due to the fact that the dominant forces acting between ions are of Coulombic nature, ions are not equivalent to molecules in their influence on transport and thermodynamic properties.
3. Since the solution as a whole is always electrically neutral—the total positive charge and the total negative charge have equal magnitude—it might be incorrectly assumed that, on the average, the mutual action between ions results in no net accumulation of positive and negative charges in the solution. This is not so; as the following paragraph will show.

    There are two opposing (competing) factors operating in ionic solutions: the Coulombic force factor, which arranges the ions into a certain organized structure, and the factor arising from the thermal motion of the solvent molecules and ions, which tends to prevent any kind of organized structure in the solution. As a result of thermal collisions, no perfect organization can ever be reached in the solution. The higher the temperature the less organized the structure will be. Because of these two opposing factors, ions tend toward an arrangement in which negative ions will predominate as the nearest neighbors of any chosen central (positive) ion and vice-versa. Around a certain cation there will be more anions than cations, and around a certain anion there will be more cations than anions (see Figure 24-7). A central ion is surrounded by a group of ions called the **ionic atmosphere**.
4. Since ions and solvent molecules are in a steady motion, the central ion–ionic atmosphere concept refers to the time average of the ionic configurations. Solutions of strongly dissociated electrolytes can therefore be treated as systems composed of central ions and ionic atmospheres in solvents.

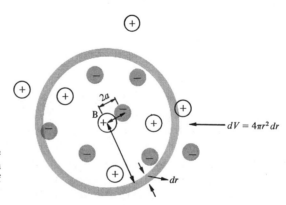

Figure 24-7 The distribution of positive and negative ions of radius a around a positive central ion B. The concept of ionic atmosphere.

In our presentation of the Debye–Hückel theory, we first derive an expression for the electrostatic potential in the vicinity of any arbitrarily chosen central ion. To achieve our goal, we consider a very dilute solution of a strongly dissociated electrolyte of a volume V, containing N_+^0 cations of valence Z_+ and N_-^0 anions of valence $|Z_-|$. If the value of the electrostatic potential (yet to be determined) at a distance r from the central (positive) ion is Φ, the potential energy at this distance is

$$\varepsilon_p = Z_+ e \Phi \qquad (24\text{-}32)$$

The evaluation of Φ requires the knowledge of the distribution of ions in the solution. To find this, we apply the Maxwell–Boltzmann distribution law from page 50.

$$dN_+ = C e^{-\varepsilon_p/kT} \, dV$$
$$= C e^{-Z_+ e \Phi/kT} \, dV \qquad (24\text{-}33)$$

where dN_+ is the number of cations in a volume element dV, located at a distance r from the central ion. The significance of the proportionality constant C is obvious from the boundary conditions: at a large distance from the central ion, the electrostatic potential is zero and, consequently, the distribution of the ions becomes uniform.[9] Thus, C equals the average number of cations in a unit volume of the solution.

$$C = \frac{dN_+}{dV} = n_+^0 = \frac{N_+^0}{V} \qquad (24\text{-}34)$$

In a similar way, we write for the anions

$$dN_- = n_-^0 \, e^{-(-Z_-)e\Phi/kT} \, dV \qquad (24\text{-}35)$$

Next, we apply Eqs. (24-33) and (24-35) to evaluate the charge density ρ, which is defined as the charge per unit volume:

$$\rho = \frac{Z_+ e \, dN_+ + (-Z_-)e \, dN_-}{dV}$$
$$= Z_+ e n_+^0 \, e^{-Z_+ e\Phi/kT} - Z_- e n_-^0 \, e^{-(-Z_-)e\Phi/kT} \qquad (24\text{-}36)$$

In a dilute solution, however, $Z_+ e\Phi \ll kT$ and $|Z_-|e\Phi \ll kT$, consequently[10]

$$\rho = -\frac{e^2 \Phi}{kT} (n_+^0 Z_+^2 + n_-^0 Z_-^2) \qquad (24\text{-}37)$$

When more than two different kinds of ions are present in the solution

$$\rho = -\frac{e^2 \Phi}{kT} \sum_{i=1}^{k} n_i^0 Z_i^2 \qquad (24\text{-}38)$$

---

[9] $e^{-Z_+ e\Phi/kT} = e^0 = 1$

[10] When $x \ll 1$, $e^{-x} = 1 - x$ and $e^x = 1 + x$. This assumption represents a very serious approximation in the Debye–Hückel model, since $Ze\Phi \ll kT$ only becomes true at a distance from the central ion about equal to the diameter of one H_2O molecule.

The potential Φ and the charge density ρ are related by the well-known Poisson differential equation from electrostatics

$$\nabla^2\Phi = \frac{-\rho}{\varepsilon_0 D} \tag{24-39}$$

where ε_0 is the permittivity of vacuum ($\equiv 8.8542 \times 10^{-12}\,C^2\,g^{-1}\,m^{-1}$), D is the dielectric constant of the bulk solvent,[11] and ∇^2 (read "del squared") is the Laplacian operator.

According to Eq. (24-39), the net electrical field coming out of a small region of space is proportional to the charge within this region. In the Cartesian coordinate system ∇^2 is given by [15]

$$\nabla^2 \equiv \frac{\partial^2}{\partial x^2} + \frac{\partial^2}{\partial y^2} + \frac{\partial^2}{\partial z^2} \tag{24-40}$$

and in the polar coordinate system

$$\nabla^2 \equiv \frac{\partial^2}{\partial r^2} + \frac{2}{r}\frac{\partial}{\partial r} + \frac{1}{r^2}\frac{\partial^2}{\partial \theta^2} + \frac{\cos\theta}{r^2\sin\theta}\frac{\partial}{\partial\theta} + \frac{1}{r^2\sin^2\theta}\frac{\partial^2}{\partial\phi^2} \tag{24-41}$$

In a spherically symmetrical field, $\partial\Phi/\partial\theta$ and $\partial\Phi/\partial\phi$ vanish and combination of Eqs. (24-38), (24-39), and (24-41) results in

$$\frac{1}{r^2}\frac{\partial}{\partial r}\left(r^2\frac{\partial\Phi}{\partial r}\right) = \left(\frac{e^2}{\varepsilon_0 DkT}\sum_{i=1}^{k} n_i^0 Z_i^2\right)\Phi = K^2\Phi \tag{24-42}$$

where r denotes the distance from the central ion and K^{-1} has the dimension of length; it is called the Debye length. It is the distance over which the electrostatic field of an ion extends with appreciable strength. K^{-1} is a measure of the thickness of the ionic atmosphere.

The easiest way to solve Eq. (24-42) is to rewrite it in the form

$$\frac{\partial}{\partial r}\left(r^2\frac{\partial\Phi}{\partial r}\right) = K^2 r^2\Phi \tag{24-43}$$

or

$$\frac{\partial^2(r\Phi)}{\partial r^2} = K^2(r\Phi) \tag{24-44}$$

and to substitute $v = r\Phi$. Thus

$$\frac{\partial^2 v}{\partial r^2} = K^2 v \tag{24-45}$$

The solution of this simple differential equation is

$$v = k_1 e^{-Kr} + k_2 e^{Kr} \tag{24-46}$$

Since $\Phi r = v$, we have

$$\Phi = k_1\frac{e^{-Kr}}{r} + k_2\frac{e^{Kr}}{r} \tag{24-47}$$

---

[11] The effective dielectric constant of the solvent near an ion is lower (due to its charge) than the dielectric constant of the bulk solvent. The neglecting of this important fact, represents another very serious problem in the Debye–Hückel treatment of ionic solutions.

The boundary conditions are again applied to evaluate the two integration constants, k_1 and k_2, respectively. If $r \to \infty$, then $\Phi \to 0$, whence $k_2 = 0$.[12] Following from this, we have

$$\Phi = k_1 \frac{e^{-Kr}}{r} \tag{24-48}$$

The constant k_1 is evaluated from the combined Eqs. (24-38) and (24-48) and from the condition that the ion–ionic atmosphere system is electroneutral as a whole. Hence, by integration of the space charge around the central ion over the entire volume, except for the effective volume of the ion, a charge of equal magnitude but of opposite sign to that of the central ion is obtained.

The effective volume is determined from the closest distance between the central ion and an ion in the ionic atmosphere, a (see Figure 24-7). Upon expressing the volume element as $dV = 4\pi r^2\, dr$, the space charge is obtained as

$$
\begin{aligned}
\int_a^\infty 4\pi r^2 \rho\, dr &= -\int_a^\infty 4\pi r^2 \frac{e^2 \Phi}{kT} \sum_{i=1}^k n_i^0 Z_i^2\, dr \\
&= -\int_a^\infty 4\pi r^2 \frac{e^2}{kT} \sum_{i=1}^k (n_i^0 Z_i^2) k_1 \frac{e^{-Kr}}{r}\, dr \\
&= -4\pi\varepsilon_0 D k_1 \int_a^\infty K^2 r\, e^{-Kr}\, dr \\
&= -Z_i e
\end{aligned}
\tag{24-49}
$$

The constant k_1 is equal to

$$k_1 = \frac{Z_i e}{4\pi\varepsilon_0 D} \frac{e^{Ka}}{1 + Ka} \tag{24-50}$$

and the final expression for the potential Φ is

$$\Phi = \frac{Z_i e}{4\pi\varepsilon_0 D} \frac{e^{Ka}}{1 + Ka} \frac{e^{-Kr}}{r} \tag{24-51}$$

This is the potential of a central ion surrounded by its ionic atmosphere. Two factors contribute to its value: the contribution arising from the isolated ion itself, Φ', ($\equiv Z_i e / 4\pi\varepsilon_0 Dr$), and the contribution due to the ionic atmosphere, Φ''. Because the linear superposition of fields applies to electrostatics, the two contributions can be added algebraically. Thus

$$\Phi'' = \Phi - \Phi' = \frac{Z_i e}{4\pi\varepsilon_0 Dr} \left(\frac{e^{K(a-r)}}{1 + Ka} - 1 \right) \tag{24-52}$$

Eq. (24-52) holds for any $r \geq a$. At distance $r < a$, no ion of the ionic atmosphere can exist, and the potential remains constant and equal to the value for $r = a$.

$$\Phi''_{r=a} = -\frac{Z_i e}{4\pi\varepsilon_0 D} \frac{\cdot K}{1 + Ka} \tag{24-53}$$

In extremely dilute solutions, as the data in Table 24-7 indicate, $K \approx 10^8\ \mathrm{m}^{-1}$, so that

---

[12] Since, in the limit e^{Kr}/r goes to infinity as r goes to infinity.

with $a \approx 10^{-10}$ m, $Ka \ll 1$ and Eq. (24-53) reduces to

$$\Phi''_{r=a} = -\frac{Z_i e K}{4 \pi \varepsilon_0 D} \qquad (24\text{-}54)$$

The ionic atmosphere contribution to the potential is a very important factor in electrochemistry; some phenomena observed in ionic solutions can be clearly explained on the basis of this extra potential.

As is seen from Eq. (24-42) K is a function of temperature, dielectric constant of the solvent, and the valence number as well as of the concentration of the ions in the solution. Data listed in Table 24-7 indicate its variation with the valence type and concentration of the ions.

Table 24-7 Effective radii of ionic atmospheres, K^{-1} (in meters) at various concentrations at 298 K

Concentration	Valence Numbers of the Ions in the Electrolyte			
(mol liter^{-1})	uni-univalent 1:1	uni-bivalent 1:2	bi-bivalent 2:2	uni-trivalent 1:3
0.0001	304×10^{-10}	176×10^{-10}	152×10^{-10}	124×10^{-10}
0.0010	96×10^{-10}	55×10^{-10}	48.1×10^{-10}	39.3×10^{-10}
0.0100	30.4×10^{-10}	17.6×10^{-10}	15.2×10^{-10}	12.4×10^{-10}
0.1000	9.6×10^{-10}	5.5×10^{-10}	4.8×10^{-10}	3.9×10^{-10}

Attention must also be given to the concentration scale in which K is expressed. On the molarity concentration scale, K is given as

$$K = \left[\frac{e^2 \mathbf{N}^2 2}{\varepsilon_0 D \, \mathbf{R} T 1000} \frac{1}{2} \sum_{i=1}^{k} (c_i Z_i^2) \right]^{1/2} \qquad (24\text{-}55)$$

and in terms of molality[13]

$$K = \left[\frac{e^2 \mathbf{N}^2 d_1 2}{\varepsilon_0 D \mathbf{R} T 1000} \frac{1}{2} \sum_{i=1}^{k} (m_i Z_i^2) \right]^{1/2}$$

$$= A I^{1/2} \qquad (24\text{-}56)$$

m_i denotes the molality of ion i, d_1 is the density of the pure solvent, and I is the ionic strength of the solution:

$$I = \frac{1}{2} \sum_{i=1}^{k} (m_i Z_i^2) \qquad (24\text{-}57)$$

The proportionality constant in Eq. (24-56), A, depends on the temperature, on the nature of the solvent and of the applied concentration scale.

[13] In a dilute solution, $c_i = m_i d_1$. Consequently, the constant A expressed on the molality and molarity scale will have different values. Since the density of pure water is approximately 1 g cm^{-3}, in dilute aqueous solutions A will be independent on the applied concentration scale.

24-6 THE IONIC ATMOSPHERE AND THE THEORY OF CONDUCTANCE

The ionic atmosphere concept is used next to explain, at least qualitatively, some observations made on the conductance of electrolytic solutions.

Eq. (24-24) indicates that the relationship between the conductance and ionic mobility (velocity) in dilute solutions is linear. Thus, a decrease in the mobility results in a lowering of the conductance. The following three factors are to be taken into consideration when dealing with the theory of conductance: (1) the viscous effect, (2) the electrophoretic effect, and (3) the relaxation effect. The mathematical treatment of conductance is the most difficult part of the theory of solutions of strong electrolytes. In view of this, only qualitative results will be given in this book. [For more details see references 1, 2, 4, 16.]

1. Viscous effect. The effect that the solvent itself exhibits on the velocity of ions in solutions. In an applied electrical field, ions are moving in the direction of the field. Opposing the electrical force is the frictional viscous drag of the solvent. For an ion of a given shape, size, and valence, the frictional drag will be larger in more viscous solvents; consequently, the velocity and conductance of the ion in such solvents will be lower. The frictional drag is usually estimated from Stokes' law: $f = 6\pi a\eta v$, where f is the frictional force, a is the radius of the ion, η is the viscosity coefficient of the solvent, and v denotes the velocity of the ion. The viscous effect would exist even if only a single ion were present in the solution.

Due to the fact that the number of ions is very large even in very dilute solution, the ion–ionic atmosphere interactions will result in a further lowering of the velocities of the ions.

2. The electrophoretic effect. Caused by simultaneous movement of the central ion in the direction of the applied electrical field while the ionic atmosphere around the central ion moves in the opposite direction. The central ion as well as the ions of the ionic atmosphere are solvated and carry with them their associated solvent molecules. Since the ionic atmosphere is comprised of more than one ion, more solvent molecules will be carried in the direction opposite to the applied field. Thus, the central ions are forced to move against a stream of solvent and, consequently, their velocities will be reduced. It can be proved that this reduction depends on the thickness of the ionic atmosphere, K^{-1}.

3. The relaxation-time effect. Originates in the finite time delay—the relaxation time—required for the renewal of the spherical symmetry of the ionic atmosphere around the central ion at any location along its route toward the respective electrodes. The relaxation time is of the order 10^{-7} to 10^{-9} s, depending on the concentration of the solution. For better understanding of this phenomenon let us consider only one central ion and its atmosphere in a very dilute solution. On its way to the electrode, the ion constantly leaves its ionic atmosphere behind. Since the atmosphere has a charge opposite to that of the central ion, there is a net accumulation of opposite charge behind the ion. As a result of the interaction between the ion and its atmosphere, the originally symmetric atmosphere becomes asymmetric (see Figure 24-8). The asymmetric atmosphere exerts an electrostatic drag on the ion, thus reducing its velocity in the direction of the field.

In 1927 Onsager took into consideration the three aforementioned effects and derived an equation for the concentration dependence of the equivalent conductance

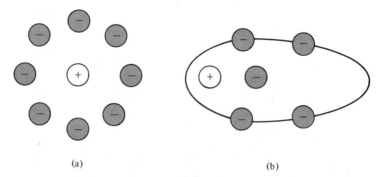

(a)   (b)

Figure 24-8   The relaxation-time effect. (a) Ionic atmosphere when field is off. (b) Asymmetric cloud around moving ion when field is on.

in dilute solutions:

$$\Lambda_c = \Lambda_\infty - \left[\frac{29.142\,(Z_+ + |Z_-|)}{(DT)^{1/2}\eta} + \frac{9.903 \times 10^5}{(DT)^{3/2}}\Lambda_\infty\omega\right](c_+ Z_+^2 + c_- Z_-^2)^{1/2} \quad \text{(24-58)}$$

which for a uni-univalent type of electrolyte reduces to the form

$$\Lambda_c = \Lambda_\infty - \left[\frac{82.42}{(DT)^{1/2}\eta} + \frac{8.204 \times 10^5}{(DT)^{3/2}}\Lambda_\infty\right]c^{1/2} \quad \text{(24-59)}$$

c denotes the molarity of the solution, Z_+ and Z_- the valence numbers of the cation and anion respectively, D is the dielectric constant of the bulk solvent, Λ_∞ is the equivalent conductance at infinite dilution, T is the absolute temperature, and ω is a parameter defined as

$$\omega = (Z_+|Z_-|)\frac{2q}{1 + q^{1/2}} \quad \text{(24-60)}$$

in which

$$q = \frac{[Z_+|Z_-|]}{Z_+ + |Z_-|} \times \frac{\lambda_\infty^+ + \lambda_\infty^-}{Z_+\lambda_\infty^- + |Z_-|\lambda_\infty^+} \quad \text{(24-61)}$$

In a uni-univalent electrolyte $q = 0.5$.

Onsager's equation, which in this simple form is identical with the empirical formula proposed by Kohlrausch (Eq. 24-16), provides excellent results for dilute solutions of uni-univalent electrolytes, see Figure 24-9. It fails, however, when applied to solutions of concentrations of $c > 0.01$ and to solutions of polyvalent electrolytes. The poor results in these cases can be explained by the fact that the Debye–Hückel and Onsager theories apply only to very dilute ionic solutions. At higher concentrations, individual charged ions able to conduct electricity combine (associate) into electroneutral ion pairs $(B^+A^-)^0$ which are unable to conduct electricity.[14] As a result of the combination of the charged particles, the conductance is reduced.

Nevertheless, after reaching a certain concentration, charged triple-ions $(B^+A^-B^+)^+$ are formed from some of the ion pairs in the solution, resulting in an

[14] The theory of ion-pair formation, called the association theory, has been developed independently by Bjerrum and by Fuoss and Kraus [1, 2, 9, 11].

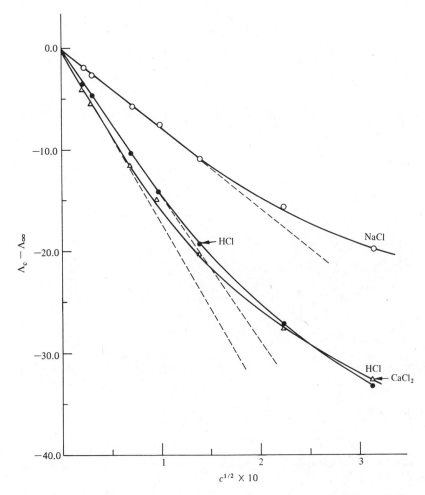

Figure 24-9 Test of the Onsager (Kohlrausch) equation. $\bigcirc$, NaCl; $\bullet$, HCl; $\triangle$, CaCl$_2$.

increase of the conductance. A typical conductance-concentration curve, exhibiting a minimum, is plotted in Figure 24-10. The formation of ion pairs, ion triplets, and so on is favored in solvents of low dielectric constant. Association, however, also depends to a large extent on the ionic radii and on the charge of the ions. Bjerrum's calculations indicate that in a $1\ M$ aqueous solution, 13.8% of those uni-univalent ions having a diameter of 2.82×10^{-10} m are associated; and of those with a diameter of 1.76×10^{-10} m, 28.6% are associated. Such a large association may drastically affect the thermodynamic behavior of ionic solutions.

24-7 WIEN AND DEBYE—FALKENHAGEN EFFECTS

The conductance of completely dissociated electrolytes is independent of the electrical field strength in the range of moderate field strengths. At higher field

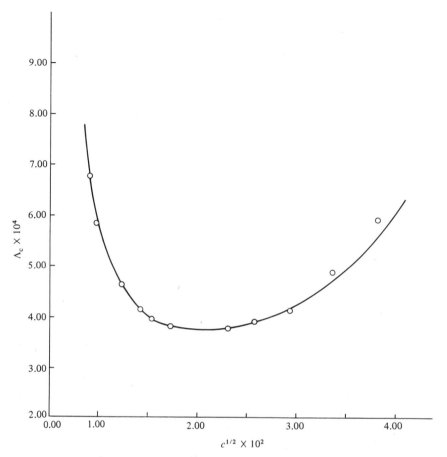

Figure 24-10 Equivalent conductance of tetraisoamylammonium nitrate in dioxane–water solvent (1.26% water).

strengths, of the order 10^7 V m^{-1}, however, Wien (in 1929) observed an increase in the conductance with increasing field strength. The increase is more intense for more concentrated solutions and for polyvalent ions. The conductance always approaches a limiting value with increasing field strength.

The effect of field strength on the conductance, called the Wien effect, can easily be interpreted by making use of the ion–ionic atmosphere interaction model. Under conditions of high field strength, the ionic mobility is so large that the ionic atmosphere is unable to rearrange itself fully around the moving central ion. Consequently, both the electrophoretic and relaxation-time effects exert no influence upon the moving ion, and the observed conductance rises above the value found for low field strength.

A second phenomenon, also observed by Wien under conditions of high field strength, deals with the influence of the electrical field on the dissociation constant of weak electrolytes. A high field strength affects the dissociation positively and, consequently, affects the dissociation equilibrium constant of such an electrolyte. The theoretical interpretation of this very important observation, as given by

Onsager (in 1934) and Bass (in 1968) [1, 2, 4, 16], is based on the fact that at high electrical field strengths the probability of ion-pair and triple-ion formation is much lower than at low field strengths.

As mentioned before, the time-relaxation effect originates in the time delay required for renewal of the spherical symmetry of the ionic atmosphere around the moving central ion. Debye and Falkenhagen (in 1928) developed a theory predicting that, in an applied alternating electrical field of high frequency, the ionic atmosphere will not be able to adopt an asymmetric structure corresponding to a moving central ion and thus will have no influence on the conductivity of the ion. In 1928 Sach proved experimentally that this is really so at frequencies exceeding 5 MHz, when the equivalent conductance approaches a value somewhat lower than Λ_∞. The value must be lower than Λ_∞ because the high-frequency field eliminates the time-relaxation effect only; it has no influence on the electrophoretic effect.

24-8 TEMPERATURE DEPENDENCE OF THE EQUIVALENT CONDUCTANCE

Experimental data indicate that both temperature and pressure affect the conductance positively; the temperature effect is large, whereas the pressure affects the conductance only slightly. Temperature and pressure influence the viscosity of the solvent in opposite ways. Thus, one may speculate that lowering the viscosity of the solvent is responsible for the increase of the conductance with temperature (2% per 1°C).

In a narrow temperature interval, the temperature dependence of the ionic equivalent conductance may be expressed as

$$\lambda_{t_2} = \lambda_{t_1}\left[1 + \frac{1}{\lambda}\frac{d\lambda}{dt}(t_2 - t_1)\right] \tag{24-62}$$

where λ_{t_2} and λ_{t_1} are the conductances at temperature t_2 and t_1 respectively. $(1/\lambda)(d\lambda/dt)$ is the conductance temperature coefficient.

The conductance and its temperature coefficient for some ions at different temperatures are listed in Table 24-8.

24-9 SOME APPLICATIONS OF CONDUCTANCE MEASUREMENTS

Conductance is very widely used in different areas of chemistry. The solubility of sparingly soluble salts can be determined from conductance measurements. It may serve as a criterion of purity of some chemical compounds. Conductance measurements represent extremely useful means in the study of chemical reactions (acid-base titration, precipitation reactions, and so on). They are often applied in the field of reaction kinetics to study the mechanism and to evaluate the rate constants of chemical reaction (that is, condensation of polymers, saponification of esters, and so on). Conductance measurements may also provide very important information about the properties and structure of chemical complexes, sizes of ions, and solvation of

Table 24-8 Conductances and conductance temperature coefficients of some ions at different temperatures

Ion	15°C		25°C		35°C		45°C	
	λ_∞ (ohm^{-1} cm^2 eq^{-1})	$\frac{1}{\lambda_\infty}\frac{d\lambda_\infty}{dt}$ (K^{-1})	λ_∞ (ohm^{-1} cm^2 eq^{-1})	$\frac{1}{\lambda_\infty}\frac{d\lambda_\infty}{dt}$ (K^{-1})	λ_∞ (ohm^{-1} cm^2 eq^{-1})	$\frac{1}{\lambda_\infty}\frac{d\lambda_\infty}{dt}$ (K^{-1})	λ_∞ (ohm^{-1} cm^2 eq^{-1})	$\frac{1}{\lambda_\infty}\frac{d\lambda_\infty}{dt}$ (K^{-1})
H_3O^+	300.6	0.01655	349.82	0.01385	397.0	0.0155	441.4	0.00970
Na^+	39.75	0.02445	50.10	0.02180	61.53	0.01930	73.77	0.01693
K^+	59.66	0.02235	73.50	0.01950	88.21	0.01705	103.49	0.01500
Cl^-	61.42	0.02340	76.35	0.02020	92.22	0.01765	108.90	0.01570
Br^-	63.15	0.02285	78.14	0.01930	94.03	0.01730	110.68	0.01535

ions. Diffusion coefficients as well as velocities of ions are also obtainable from conductance data. Some industrial plants are fully automatized on the basis of conductance measurements.

Recently, conductance measurements have also been applied in medical research. Damaged and healthy human skin can be distinguished on the basis of their conductance. Some theories have been developed recently relating conductance to some physiological phenomena.

For a sparingly soluble salt, like silver chloride in water for example, $\Lambda_c \approx \Lambda_\infty$ and therefore the combined formulas (24-15) and (24-17) may be written as

$$\Lambda_c = \Lambda_\infty = \lambda_\infty^{Ag^+} + \lambda_\infty^{Cl^-} = \frac{1000(X - X_{H_2O})}{c} \qquad (24\text{-}63a)$$

Consequently

$$c = \frac{1000(X - X_{H_2O})}{\lambda_\infty^{Ag^+} + \lambda_\infty^{Cl^-}} \qquad (24\text{-}63b)$$

$$= \frac{1000(2.61 - 0.60) \times 10^{-6}}{61.92 + 73.34}$$

$$= 1.486 \times 10^{-5} \text{ eq. liter}^{-1}$$

$X = 2.61 \times 10^{-6}$ and $X_{H_2O} = 0.60 \times 10^{-6}$ ohm^{-1} cm^{-1} are the specific conductances of the saturated solution and pure water, respectively. The conductance value for water indicates that water was not purified sufficiently; the purest water so far prepared has $X = 5.5 \times 10^{-8}$ ohm^{-1} cm^{-1}.

Due to the slight dissociation of water into ions, pure water may be considered as a very dilute solution of both H_3O^+ (hydrated hydrogen ions) and OH^- ions. The concentration of both ions can be obtained from Eq. (24-63a)

$$c_{H_3O^+} = c_{OH^-} = \frac{1000 X_{H_2O}}{\lambda_\infty^{H_3O^+} + \lambda_\infty^{OH^-}}$$

$$= \frac{1000 \times 5.5 \times 10^{-8}}{349.82 + 197.6} = 1.0 \times 10^{-7} \text{ eq. liter}^{-1}$$

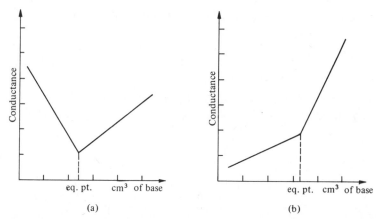

Figure 24-11 Conductometric titration curves. (a) Titration of strong acid with strong base. (b) Titration of weak acid with strong base.

Titrations followed conductometrically are often used in chemistry. During such titrations, ions of certain velocities are replaced by ions of higher velocities. This replacement results in a continuous change of the conductance during the titration. For example, during an acid-base titration of HCl with NaOH the following reaction takes place

$$H_3O^+ + Cl^- + Na^+ + OH^- + x\ H_2O \rightarrow Na^+ + Cl^- + (x+2)\ H_2O$$

As seen, the fast moving hydrogen ions, H_3O^+, are replaced by slower moving sodium ions. After reaching the equivalence point, fast moving OH^- will be present in excess in the solution; consequently, the conductance will again increase. The equivalence point is determined from the break in the curve (see Figure 24-11).

24-10   DIFFUSION IN SOLUTIONS OF ELECTROLYTES

As discussed in Section 8-6, neutral species migrate from one region in the system to another region if there is a chemical potential difference between the two regions. Such migration is generally referred to as diffusion. Ions also migrate in solutions even in the absence of an external electrical field if there is a chemical potential difference in the solution.

The driving force of diffusion, f_i acting upon an ion i in the y direction is given by

$$f_i = -\frac{1}{N}\left(\frac{\partial \mu_i}{\partial y}\right)_T \tag{24-64}$$

where N is Avogadro's number, and $(\partial \mu_i/\partial y)_T$ is the molar chemical potential gradient in the y direction at constant temperature. The rate of diffusion depends on the velocity of the ion, v_i, which under the action of unit force is given as

$$v_i = -\frac{u_i}{N|Z_i|e}\left(\frac{\partial \mu_i}{\partial y}\right)_T \tag{24-65}$$

here u_i and $|Z_i|$ are the mobility and valence number of the ion, respectively.

For the net flow of ions through a unit area in the y direction per unit time, S_{iy}, we write

$$S_{iy} = c_i v_i = -\frac{u_i c_i}{N|Z_i|e}\left(\frac{\partial \mu_i}{\partial y}\right)_T \qquad (24\text{-}66)$$

The chemical potential gradient in a very dilute solution—ideal behavior—may be evaluated from the equation

$$\mu_i = \mu_i^\bullet + RT \ln c_i$$

Thus

$$\left(\frac{\partial \mu_i}{\partial y}\right)_T = RT\left(\frac{\partial \ln c_i}{\partial y}\right)_T = RT\frac{1}{c_i}\left(\frac{\partial c_i}{\partial y}\right)_T \qquad (24\text{-}67)$$

Substitution of Eq. (24-67) into Eq. (24-66) results in

$$S_{iy} = -\frac{u_i c_i}{N|Z_i|e} RT\frac{1}{c_i}\left(\frac{\partial c_i}{\partial y}\right)_T$$

$$= -\frac{u_i kT}{|Z_i|e}\left(\frac{\partial c_i}{\partial y}\right)_T \qquad (24\text{-}68)$$

Upon comparison of Eq. (24-68) with Fick's law of diffusion Eq. (3-1)

$$S_{iy} = -D_i\left(\frac{\partial c_i}{\partial y}\right)_T \qquad (24\text{-}69)$$

we find that the diffusion coefficient D_i, is related to the mobility of the ion by

$$D_i = \frac{kT}{|Z_i|e}u_i \qquad (24\text{-}70)$$

According to Eq. (24-25), at infinite dilution $u_i = \lambda_i/\mathfrak{F}$; consequently

$$D_i = \frac{kT\lambda_i}{|Z_i|e\mathfrak{F}} = \frac{NkT\lambda_i}{|Z_i|Ne\mathfrak{F}} = \frac{RT}{|Z_i|\mathfrak{F}^2}\lambda_i \qquad (24\text{-}71)$$

where $\mathfrak{F}$ is Faraday's constant ($\equiv Ne$).

This equation was derived by Nernst in 1888. It enables one to evaluate the mobility (conductivity) of ions from diffusion experiments. A disadvantage of this equation is that it deals with single-ion behavior only (see next page).

24-11   CHEMICAL POTENTIAL AND THE STANDARD STATE IN SOLUTIONS OF ELECTROLYTES

An electrolyte of the type $B_{\nu_+}^{Z_+}A_{\nu_-}^{Z_-}$ dissociates in a solvent according to the equation

$$B_{\nu_+}^{Z_+}A_{\nu_-}^{Z_-} = \nu_+ B^{Z_+} + \nu_- A^{Z_-}$$

Consequently, a dilute solution of such an electrolyte can be considered as a solution in which the ionic species B^{Z+} and A^{Z-} are dissolved in a solvent. Making use of Eq. (8-48) the chemical potentials of the individual ions can be expressed as

$$\mu_{B^{Z+}} = \left(\frac{\partial G}{\partial n_{B^{Z+}}}\right)_{T,P,n_{A^{Z-}},n_s} \tag{24-72}$$

and

$$\mu_{A^{Z-}} = \left(\frac{\partial G}{\partial n_{A^{Z-}}}\right)_{T,P,n_{B^{Z+}},n_s} \tag{24-73}$$

where $n_{A^{Z-}}$, $n_{B^{Z+}}$ are the number of moles of the anion and cation respectively, and n_s denotes the number of moles of solvent.

Realize that the last two equations have no physical meaning; $n_{B^{Z+}}$ cannot vary while $n_{A^{Z-}}$ is kept constant and vice versa. Briefly, single ions can neither be added to nor removed from a solution. An electrolyte as a whole, however, can be added to or removed from a solution. The Gibbs free energy change due to adding of dn moles of an electrolyte to a fixed amount of solvent, at constant temperature and pressure, is given by

$$
\begin{aligned}
dG &= \left(\frac{\partial G}{\partial n_{B^{Z+}}}\right)_{T,P,n_{A^{Z-}},n_s} dn_{B^{Z+}} + \left(\frac{\partial G}{\partial n_{A^{Z-}}}\right)_{T,P,n_{B^{Z+}},n_s} dn_{A^{Z-}} \\
&= \left(\frac{\partial G}{\partial n_{B^{Z+}}}\right)_{T,P,n_{A^{Z-}},n_s} \nu_+ \, dn + \left(\frac{\partial G}{\partial n_{A^{Z-}}}\right)_{T,P,n_{B^{Z+}},n_s} \nu_- \, dn \\
&= [\nu_+ \mu_{B^{Z+}} + \nu_- \mu_{A^{Z-}}] \, dn \\
&= \mu_{el} \, dn
\end{aligned}
\tag{24-74}
$$

since stoichiometry requires that

$$dn_{B^{Z+}} = \nu_+ \, dn \tag{24-75a}$$

and

$$dn_{A^{Z-}} = \nu_- \, dn \tag{24-75b}$$

The term in the brackets of Eq. (24-74) represents the chemical potential of the electrolyte as a whole. As is evident, it can be expressed in terms of the chemical potentials of the individual ions. Similar conclusions can also be drawn about the other thermodynamic properties of solutions of electrolytes.

In analogy with nonelectrolytes, we write for the concentration dependence of the chemical potentials of the individual ions

$$\mu_{B^{Z+}} = \mu_{B^{Z+}}^{\bullet} + RT \ln a_{B^{Z+}} \tag{24-76a}$$

and

$$\mu_{A^{Z-}} = \mu_{A^{Z-}}^{\bullet} + RT \ln a_{A^{Z-}} \tag{24-76b}$$

Eqs. (24-76a) and (24-76b) are of little use at present because the exact activities of the individual ions are experimentally unattainable.[15]

---

[15] Recent research done in the field of single-ion behavior indicates that, by means of some extrathermodynamical methods, information on single-ion properties can be obtained.

The chemical potential of the electrolyte as a whole μ_{el} is given by

$$
\begin{aligned}
\mu_{el} &= \nu_+\mu_B{}^{z_+} + \nu_-\mu_A{}^{z_-} \\
&= \nu_+\mu_B^\bullet{}^{z_+} + \nu_+RT \ln a_B{}^{z_+} \\
&\quad + \nu_-\mu_A^\bullet{}^{z_-} + \nu_-RT \ln a_A{}^{z_-} \\
&= \mu_{el}^\bullet + RT \ln a_{B^{z_+}}^{\nu_+} a_{A^{z_-}}^{\nu_-}
\end{aligned}
\tag{24-77}
$$

Upon introducing the mean activity, $a_\pm$, defined as

$$
a_\pm^\nu = a_{B^{z_+}}^{\nu_+} a_{A^{z_-}}^{\nu_-} = a_{el} = ({}^m\gamma_\pm m_\pm)^\nu
\tag{24-78}
$$

we have

$$
\begin{aligned}
\mu_{el} &= \mu_{el}^\bullet + \nu RT \ln a_\pm \\
&= \mu_{el}^\bullet + \nu RT \ln {}^m\gamma_\pm + \nu RT \ln m_\pm
\end{aligned}
\tag{24-79}
$$

in which a_{el} represents the activity of the electrolyte as a whole, ν denotes the number of ions formed by the dissociation of one molecule of the electrolyte, and ${}^m\gamma_\pm$ is the mean molal activity coefficient expressed as ${}^m\gamma_\pm = ({}^m\gamma_{B^{z_+}}^{\nu_+} {}^m\gamma_{A^{z_-}}^{\nu_-})^{1/\nu}$; the mean molality, $m_\pm$, is related to the molality of the electrolyte as a whole, m, by the equation

$$
m_\pm = m(\nu_{B^{z_+}}^{\nu_+} \nu_{A^{z_-}}^{\nu_-})^{1/\nu}
\tag{24-80}
$$

If the standard state is chosen at infinite dilution, then

$$
\lim_{m \to 0} \frac{a_\pm}{m_\pm} = \lim_{m \to 0} {}^m\gamma_\pm = 1
\tag{24-81}
$$

and Eq. (24-79) reduces to the form

$$
\mu_{el}^\bullet = \lim_{m \to 0} (\mu_{el} - \nu RT \ln m_\pm)
\tag{24-82}
$$

This limit really exists; however, the $\lim_{m \to 0}(\mu_{el} - RT \ln m)$, which would apply if the electrolyte were treated as a nonelectrolyte, does not exist.

A few examples relating the mean molality $m_\pm$ to molality m are given in Table 24-9.

Table 24-9 Mean molality and molality relation for some electrolytes

Electrolyte	ν_+	ν_-	$\nu = \nu_+ + \nu_-$	$(\nu_+^{\nu_+}\nu_-^{\nu_-})^{1/\nu}$	$m_\pm = m(\nu_+^{\nu_+}\nu_-^{\nu_-})^{1/\nu}$
NaCl	1	1	2	1	m
K_2SO_4	2	1	3	$(2^2 \times 1)^{1/3}$	$4^{1/3}m$
$MgSO_4$	1	1	2	1	m
Na_3PO_4	3	1	4	$(3^3 \times 1)^{1/4}$	$27^{1/4}m$

Molality is the most frequently used concentration scale in electrochemistry. However, equations similar to those expressed in terms of molality may also be derived for the other concentration scales. Remember that the activities as well as the activity coefficients based on different standard states differ in their values. The chemical potential itself is independent of the standard state used; however, the difference in the chemical potentials $(\mu_i - \mu_i^\bullet)$ depends on the concentration scale used; see Example 24-1.

As ions are electrically charged particles, their activities differ from the corresponding concentrations at much lower concentrations than in the case of nonelectrolytes. The electrostatic forces of interaction act at longer distances than the van der Waals forces which are mainly responsible for the nonideal behavior of nonelectrolytes.

EXAMPLE 24-1

Relate the molal activity coefficient, $^m\gamma_i$, to the activity coefficient based on the mole fraction concentration scale γ_i. Assume that the standard state is taken at infinite dilution. Also derive the other relationships among the activity coefficients of the mole fraction, molality and molarity concentration scales.

Solution: Classical thermodynamics gives the following relation for the chemical potential

$$\mu_i = \mu_i^\bullet + RT \ln (x_i\gamma_i) = {^m}\mu_i^\bullet + RT \ln (m_i{^m}\gamma_i) \tag{a}$$

For solutions in the limit of infinite dilution, we may write

$$x_i \approx \frac{m_i M_1}{1000} \tag{b}$$

where M_1 is the molecular weight of the pure solvent and

$$\lim_{m_i \to 0} {^m}\gamma_i = \lim_{x_i \to 0} \gamma_i = 1 \tag{c}$$

Consequently

$$^m\mu_i^\bullet = \mu_i^\bullet + RT \ln \frac{M_1}{1000} \tag{d}$$

Substitution of Eq. (d) into Eq. (a) results in

$$\mu_i = \mu_i^\bullet + RT \ln (x_i\gamma_i) = \mu_i^\bullet + RT \ln \frac{M_1}{1000} + RT \ln (m_i{^m}\gamma_i) \tag{e}$$

From this

$$\gamma_i \approx {^m}\gamma_i \frac{m_i M_1}{1000 x_i} \tag{f}$$

It can be shown that Eq. (f) is the same as Eq. (l) derived in Example 8-3 on page 198.

For dilute ionic solutions

$$x_\pm = \frac{m_\pm}{\nu m + 1000/M_1} \tag{g}$$

and

$$\gamma_\pm = {^m}\gamma_\pm \left[1 + \frac{\nu m M_1}{1000}\right] \tag{h}$$

In a similar way it may also be proved (see Problem 24-19) that

$$\gamma_{\pm} = {}^{c}\gamma_{\pm} \left[\frac{d}{d_1} + \frac{c(\nu M_1 - M)}{1000 d_1} \right] \tag{i}$$

and

$$^{m}\gamma_{\pm} = {}^{c}\gamma_{\pm} \left[\frac{d}{d_1} - \frac{cM}{1000 d_1} \right] = {}^{c}\gamma_{\pm} \frac{c}{m d_1} \tag{j}$$

where c is the molarity, d and d_1 the densities of the solution and pure solvent, respectively, M and M_1 the molecular weight of the solute and pure solvent, respectively. At low concentrations, the second terms in the brackets approach zero, $d_1 = d$, and thus, all three different activity coefficients are equal.

24-12   PARTIAL MOLAL ENTHALPIES OF IONS IN SOLUTIONS

Differentiation of the equation

$$\mu_{el} = \nu_{+}\mu_{+} + \nu_{-}\mu_{-} = \nu_{+}\mu_{+}^{\bullet} + \nu_{-}\mu_{-}^{\bullet} + \nu RT \ln m_{\pm} + \nu RT \ln {}^{m}\gamma_{\pm}$$

with respect to T at constant pressure and composition results in

$$\left[\frac{\partial(\mu_{el}/T)}{\partial T} \right]_{P,m} = \nu_{+} \left[\frac{\partial(\mu_{+}/T)}{\partial T} \right]_{P,m} + \nu_{-} \left[\frac{\partial(\mu_{-}/T)}{\partial T} \right]_{P,m}$$

$$= \nu_{+} \left[\frac{\partial(\mu_{+}^{\bullet}/T)}{\partial T} \right]_{P} + \nu_{-} \left[\frac{\partial(\mu_{-}^{\bullet}/T)}{\partial T} \right]_{P} + \nu R \left(\frac{\partial \ln {}^{m}\gamma_{\pm}}{\partial T} \right)_{P,m} \tag{24-83}$$

Since

$$\left[\frac{\partial(\mu/T)}{\partial T} \right]_{P,m} = -\frac{\mathcal{H}}{T^2}$$

Eq. (24-83) assumes the form

$$-\frac{\mathcal{H}_{el}}{T^2} = -\nu_{+} \left(\frac{\bar{\mathcal{H}}_{+}}{T^2} \right) - \nu_{-} \left(\frac{\bar{\mathcal{H}}_{-}}{T^2} \right)$$

$$= -\nu_{+} \left(\frac{\bar{\mathcal{H}}_{+}^{\bullet}}{T^2} \right) - \nu_{-} \left(\frac{\bar{\mathcal{H}}_{-}^{\bullet}}{T^2} \right) + \nu R \left(\frac{\partial \ln {}^{m}\gamma_{\pm}}{\partial T} \right)_{P,m} \tag{24-84}$$

After rearrangement, we have

$$\left(\frac{\partial \ln {}^{m}\gamma_{\pm}}{\partial T} \right)_{P,m} = \frac{-\nu_{+}\bar{\mathcal{H}}_{+} - \nu_{-}\bar{\mathcal{H}}_{-} + \nu_{+}\bar{\mathcal{H}}_{+}^{\bullet} + \nu_{-}\bar{\mathcal{H}}_{-}^{\bullet}}{\nu RT^2}$$

$$= \frac{-\nu_{+}(\bar{\mathcal{H}}_{+} - \bar{\mathcal{H}}_{+}^{\bullet}) - \nu_{-}(\bar{\mathcal{H}}_{-} - \bar{\mathcal{H}}_{-}^{\bullet})}{\nu RT^2}$$

$$= \frac{-\nu_{+}\bar{\mathcal{L}}_{+} - \nu_{-}\bar{\mathcal{L}}_{-}}{\nu RT^2} \tag{24-85}$$

In these equations $\bar{\mathcal{H}}_{+}$, $\bar{\mathcal{H}}_{-}$ denote the partial molal enthalpies of the ions in the solution, and $\bar{\mathcal{H}}_+^\bullet$, $\bar{\mathcal{H}}_-^\bullet$ represent the same quantities in the standard state. The differences $(\bar{\mathcal{H}}_+ - \bar{\mathcal{H}}_+^\bullet)$ and $(\bar{\mathcal{H}}_- - \bar{\mathcal{H}}_-^\bullet)$ are the relative partial molal enthalpies of the ions in solution, $\bar{\mathcal{L}}_+$ and $\bar{\mathcal{L}}_-$, respectively.

As is obvious, the rate of change of the activity coefficient with temperature can provide some information on the behavior of the ions in solution.

The standard enthalpy of formation of ions is a very important ionic property. It enables one to calculate the standard enthalpy of a reaction in which the ion participates. In order to evaluate the standard enthalpy of formation of an ion, we will agree to the convention that the standard enthalpy of formation of the hydrated hydrogen ion, $H^+(aq)$, is zero at any temperature.

The evaluation of the standard enthalpies of formation of ions is shown in Example 24-2.

EXAMPLE 24-2
The following chemical reactions and their standard enthalpy changes are given.

(a) $H^+(aq) + OH^-(aq) = H_2O(l)$ $\qquad\qquad$ $\Delta\mathcal{H}^\circ = -13{,}360$ cal mol^{-1}

(b) $H_2(g) + \frac{1}{2} O_2(g) = H_2O(l)$ $\qquad\qquad$ $\Delta\mathcal{H}^\circ = -68{,}317$ cal mol^{-1}

(c) $NaOH(s) + n\ H_2O(l) = Na^+(aq) + OH^-(aq)$ $\quad$ $\Delta H^\circ = -10{,}250$ cal

In addition the $\Delta\mathcal{H}_f^\circ$ of $NaOH(s)$ is given as $-101{,}990$ cal mol^{-1}.

Find the enthalpy of formation of the $OH^-(aq)$ and $Na^+(aq)$ ions.

Solution: Subtract Eq. (a) from Eq. (b) to obtain

$$H_2(g) + \tfrac{1}{2} O_2(g) = H^+(aq) + OH^-(aq) \qquad \Delta H^\circ = -54{,}957 \text{ cal}$$

Consequently

$$\Delta H^\circ = -54{,}957 = (\Delta\mathcal{H}_f^\circ)_{H^+(aq)} + (\Delta\mathcal{H}_f^\circ)_{OH^-(aq)} - (\Delta\mathcal{H}_f^\circ)_{H_2(g)} - \tfrac{1}{2}(\Delta\mathcal{H}_f^\circ)_{O_2(g)}$$

By our previous definition, however

$$(\Delta\mathcal{H}_f^\circ)_{H_2(g)} = (\Delta\mathcal{H}_f^\circ)_{O_2(g)} = (\Delta\mathcal{H}_f^\circ)_{H^+(aq)} = 0$$

and thus, the standard enthalpy of formation of the $OH^-(aq)$ ion is $-54{,}957$ cal mol^{-1}.

Now if we substitute this value into the expression for the enthalpy of reaction (c) and take into consideration the fact that the number of moles of water on both sides of this reaction is extremely large, we may write

$$-10{,}250 = (\Delta\mathcal{H}_f^\circ)_{OH^-(aq)} + (\Delta\mathcal{H}_f^\circ)_{Na^+(aq)} - (\Delta\mathcal{H}_f^\circ)_{NaOH(s)}$$
$$= -54{,}957 + (\Delta\mathcal{H}_f^\circ)_{Na^+(aq)} - (-101{,}990)$$

From this

$$(\Delta\mathcal{H}_f^\circ)_{Na^+(aq)} = -57{,}283 \text{ cal mol}^{-1}$$

The standard enthalpies of formation of all the other ions may be found in an analogous way. A similar method may be also applied to calculate the standard Gibbs free energy of formation of ions. In this case however, $(\Delta\mathcal{G}_f^\circ)_{H^+(aq)} = 0$.

Values of $\Delta\mathcal{H}_f^\circ$ and $\Delta\mathcal{G}_f^\circ$ of some ions are listed in Table 24-10.

Table 24-10 Standard enthalpies and standard Gibbs free energies of formation of some ions at 25°C [in cal mol^{-1}]

Substance	$\Delta \mathscr{H}_f^\circ$	$\Delta \mathscr{G}_f^\circ$	Substance	$\Delta \mathscr{H}_f^\circ$	$\Delta \mathscr{G}_f^\circ$
$H^+(aq)$	0.0000	0.0000	$OH^-(aq)$	$-54,957$	$-37,595$
$Na^+(aq)$	$-57,283$	$-62,589$	$F^-(aq)$	$-78,660$	$-66,080$
$K^+(aq)$	$-60,040$	$-67,466$	$Cl^-(aq)$	$-40,020$	$-31,350$
$Ag^+(aq)$	25,310	18,430	$Br^-(aq)$	$-28,900$	$-24,574$
$Ba^{2+}(aq)$	$-128,670$	$-134,000$	$I^-(aq)$	$-13,370$	—
$Zn^{2+}(aq)$	$-36,430$	$-35,184$	$CN^-(aq)$	36,100	39,600
$Pb^{2+}(aq)$	390	$-5,810$	$SO_4^{2-}(aq)$	$-216,900$	$-177,340$
$Fe^{2+}(aq)$	$-21,000$	$-20,300$	$NO_3^-(aq)$	$-49,372$	$-26,410$
$Fe^{3+}(aq)$	$-11,400$	$-2,520$	$CO_3^{2-}(aq)$	$-161,631$	$-126,220$
$NH_4^+(aq)$	$-31,740$	$-19,000$	$Cr_2O_7^{2-}(aq)$	$-349,100$	$-300,500$

SOURCE: F. D. Rossini et. al.: *Selected Values of Chemical Thermodynamic Properties, N.B.S. Circular* 500. U.S.G.P.O., Washington, D.C., 1952.

24-13 EXPERIMENTAL DETERMINATION OF ACTIVITY COEFFICIENTS

Solutions of electrolytes behave nonideally even when extremely dilute. The knowledge of the activity coefficients of the constituents of an electrolytic solution is therefore even more desirable than in the case of nonelectrolytic solutions.

A number of different experimental methods for the determination of activity coefficients are available in the literature. Some of these methods are derived from the colligative behavior of electrolytic solutions. As a representative of these methods we shall present the freezing point depression method.

For a solvent in a nonideal solution, the basic colligative equation (see page 213) can be modified into

$$d \ln a_1 = \frac{\Delta \mathscr{H}_{f,1}}{RT^2} dT \qquad (24\text{-}86)$$

where a_1 and $\Delta \mathscr{H}_{f,1}$ are the activity and enthalpy of fusion of the solvent, respectively. Since we are interested in the behavior of the solute rather than in the behavior of the solvent, we replace a_1 by the activity of the solute, a_2. The two are related to each other through the Gibbs–Duhem relationship (page 182)

$$n_1 \, d\mu_1 = -n_2 \, d\mu_2$$

Since $\mu_i = \mu_i^\bullet + RT \ln a_i$, and at constant temperature $d\mu_i = RT \, d \ln a_i$, the preceding relationship may also be written as

$$n_1 \, d \ln a_1 = -n_2 \, d \ln a_2 \qquad (24\text{-}87)$$

Eq. (24-87) is independent of the choice of standard state or the particular activity scale being used. It is possible therefore to express it in terms of molality, m:

$n_1 = 1000/M_1$ and $n_2 = m$. Consequently

$$\frac{1000}{M_1} d \ln a_1 = -m \, d \ln a_2 \qquad (24\text{-}88)$$

The activity of the electrolyte, a_2, however, is related to the mean activity of its ionic constituents $a_\pm$ by $a_2 = a_\pm^\nu$ and therefore

$$d \ln a_\pm = -\frac{1000}{M_1 m \nu} d \ln a_1 \qquad (24\text{-}89)$$

Substitution of Eq. (24-86) into Eq. (24-89) yields

$$d \ln a_\pm = -\frac{1000}{M_1 m \nu} \frac{\Delta \mathscr{H}_{f,1}}{RT^2} dT \qquad (24\text{-}90)$$

We now limit our derivation to dilute solutions. In such solutions, we can set $T = T^*$, the freezing point of pure solvent. Rearrangement of Eq. (24-90) results in

$$d \ln a_\pm = -\frac{1}{K_f m \nu} dT$$

$$= \frac{1}{K_f m \nu} d(\Delta T) \qquad (24\text{-}91)$$

where K_f is the molal freezing point constant, $K_f = M_1 RT^{*2}/1000 \Delta \mathscr{H}_{f,1}$, $\Delta T = (T^* - T)$ is the freezing point depression, and $dT = -d(\Delta T)$. Replacing $a_\pm$ by $^m\gamma_\pm \, m_\pm$ and realizing that $m_\pm$ differs from m by a constant factor only ($d \ln m_\pm = d \ln m$), we obtain

$$d \ln {}^m\gamma_\pm = \frac{1}{K_f m \nu} d(\Delta T) - d \ln m \qquad (24\text{-}92)$$

For mathematical reasons it is more convenient to introduce a new variable

$$j \equiv 1 - \frac{\Delta T}{K_f m \nu} \qquad (24\text{-}93)$$

prior to the integration of Eq. (24-92). In an ideal solution $\Delta T/K_f m \nu = 1$ and consequently $j = 0$. Solving for ΔT, followed by differentiation, considering both m and j as variables, yields

$$d(\Delta T) = K_f \nu[(1-j) \, dm - m \, dj] \qquad (24\text{-}94)$$

Eliminating $d(\Delta T)$ between Eqs. (24-92) and (24-94), we obtain

$$d \ln {}^m\gamma_\pm = (1-j)\frac{dm}{m} - dj - d \ln m$$

$$= -\frac{j}{m} dm - dj \qquad (24\text{-}95)$$

Eq. (24-95) may now be integrated. Realizing that at infinite dilution, $m \to 0$, $j \to 0$, and $^m\gamma_\pm \to 1$, we have

$$\ln {}^m\gamma_\pm = -\int_0^m \frac{j}{m} dm - j \qquad (24\text{-}96)$$

The remaining integral must be evaluated graphically from experimental ΔT versus molality data.

Eq. (24-96) is often expressed in a different form

$$\ln {}^{m}\gamma_{\pm} = -(1-g_{m}) - \int_{0}^{m}(1-g_{m})\frac{dm}{m} \qquad (24\text{-}97)$$

where g_{m} is the molal osmotic coefficient defined by

$$g_{m} = \frac{\pi}{\pi^{*}} \qquad (24\text{-}98)$$

π is the osmotic pressure of a solution of a given composition and π^{*} is the hypothetical osmotic pressure of an ideal solution of the same composition. As is obvious

$$j = 1 - g_{m} \qquad (24\text{-}99)$$

and $g_{m} = 1$ for an ideal solution. For a 2 M KCl solution, $g_{m} = 0.912$.

Mean molal activity coefficients of some electrolytes at 25°C are plotted in Figure 24-12. The characteristic behavior of ${}^{m}\gamma_{\pm}$ is a relatively sharp drop from a value of 1 at very low molalities to a minimum, followed by increasing values at higher molalities. At present, there is no theory available able to predict quantitatively the

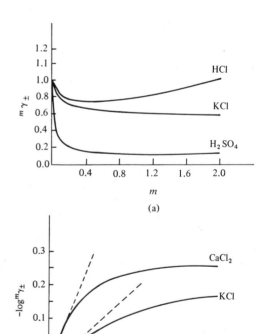

Figure 24-12 Mean activity coefficients of strong electrolytes. (a) ${}^{m}\gamma_{\pm}$ as a function of molality. (b) $-\log{}^{m}\gamma_{\pm}$ as a function of the square root of molality.

shape of the experimental curves in a wide range of concentration. However, a qualitative explanation can be given in terms of the solvation of ions: As the ions tie up more and more solvent molecules, the effective concentration, that is, the moles of solute per mole of free solvent, becomes greater than the concentration calculated as moles of solute per mole of solvent. The solute in the apparently more concentrated solution has a higher chemical potential than expected and, consequently, also a higher activity coefficient. The formation of ion pairs and triplets at higher concentrations also has a significant influence on the shape of the curves.

Another method for determining mean activity coefficients is presented in Example 24-3.

24-14 THEORETICAL CALCULATIONS OF ACTIVITY COEFFICIENTS: THE DEBYE–HÜCKEL LAW

The chemical potential of an ion in a solution is given as

$$\mu_i = \mu_i^{\bullet} + RT \ln m_i + RT \ln {}^m\gamma_i$$
$$= \mu_i(\text{ideal}) + \mu_i^E(\text{nonideal}) \tag{24-100}$$

Strong electrolytes in dilute solutions are assumed to be completely dissociated into ions, and the presence of undissociated (neutral) molecules in such solutions need not be considered. Thus, the deviation of the solution from ideality is caused entirely by the electrical interactions:

$$\mu_i^E(\text{nonideal}) = \mu_i(\text{electrical}) = RT \ln {}^m\gamma_i \tag{24-101}$$

and for one ion in the solution

$$\frac{\mu_i(\text{electrical})}{N} = kT \ln {}^m\gamma_i \tag{24-102}$$

This excess chemical potential—its value would be zero in sufficiently dilute ionic solutions—must be equal to the electrical work done, at constant T and P, against the field of the charges of the ions in the ionic atmosphere as the charge of the central ion i is gradually increased from 0 to q ($\equiv Z_i e$). This work is equal to

$$w = \int_0^q \Phi'' \, dq \tag{24-103}$$

where $q = Z_i e$ and $dq = e \, dZ_i$. Making use of Eq. (24-54), we may write

$$w = \int_0^{Z_i e} \Phi'' e \, dZ_i = -\int_0^{Z_i e} \frac{Z_i e^2 K}{4\pi\varepsilon_0 D} dZ_i$$
$$= -\frac{Ke^2 Z_i^2}{8\pi\varepsilon_0 D} \tag{24-104}$$

Setting Eqs. (24-102) and (24-104) equal to each other, we obtain

$$\ln {}^m\gamma_i = -\frac{Ke^2 Z_i^2}{8\pi\varepsilon_0 DkT} \tag{24-105}$$

$^m\gamma_i$ is the molal activity coefficient of a single ion. Such a quantity, however, cannot be determined exactly at present and it seems convenient to replace it by the mean activity coefficient, defined as $^m\gamma_\pm = {}^m\gamma_+^{\nu+}\,{}^m\gamma_-^{\nu+}$. Following from this, we have

$$\nu \ln {}^m\gamma_\pm = (\nu_+ + \nu_-) \ln {}^m\gamma_\pm = \nu_+ \ln {}^m\gamma_+ + \nu_- \ln {}^m\gamma_- \tag{24-106}$$

Introducing Eq. (24-106) into Eq. (24-105) and assuming electroneutrality ($\nu_+ Z_+ = \nu_-|Z_-|$), we have

$$\ln {}^m\gamma_\pm = -|Z_+ Z_-| \frac{e^2 K}{8\pi\varepsilon_0 DkT} \tag{24-107}$$

Converting to common logarithms and introducing the value for K from Eq. (24-56), we finally obtain

$$\log {}^m\gamma_\pm = -1.825 \times 10^6 \, (Z_+|Z_-|) \frac{d_1}{(DT)^{3/2}} I^{1/2}$$

$$= -A(Z_+|Z_-|)I^{1/2} \tag{24-108}$$

In this equation I is the ionic strength. The values of $e = 1.602 \times 10^{-19}$ C and $R = 8.314$ J K^{-1} mol^{-1} were used for the evaluation of the numerical constant. For water at 298 K, $D = 78.54$, $d_1 = 0.998$ kg liter^{-1} and the equation becomes

$$\log {}^m\gamma_\pm = -0.509|Z_+ Z_-|I^{1/2} \tag{24-109}$$

This is the limiting Debye–Hückel law. As is clear from the assumptions made during its derivation, it can give good results only in extremely dilute solutions, up to 0.01 M for univalent electrolytes.

Better results are obtained by substituting the value for Φ'' from Eq. (24-53). In such a case the result is

$$\log {}^m\gamma_\pm = -\frac{A|Z_+ Z_-|I^{1/2}}{1 + BaI^{1/2}} \tag{24-110}$$

where a is the apparent ionic radius of the solute ions and A and B are constants dependent on the temperature and nature of the solvent.

The Debye–Hückel theory can further be improved by considering the results of the theory of ionic association. An equation of the form

$$\log \gamma_\pm = -\frac{A|Z_+ Z_-|I^{1/2}}{1 + BaI^{1/2}} + CI \tag{24-111}$$

can be derived. The term CI accounts for the short range interactions. It was originally introduced as an empirical term and ascribed to the salting-out effect. The ions become solvated; they orientate the solvent molecules and, hence, the electrostatic forces become weaker. Consequently, the activity coefficients of ions increase with the increasing ionic strength.

It is interesting to compare the equations relating the logarithm of the activity coefficient to the concentration for solutions of nonelectrolytes and electrolytes, respectively.

For nonelectrolytes, we have (see page 227).

$$\log \gamma_i = Ax_j^2 + Bx_j^3 + \cdots \tag{24-112}$$

and

$$\lim_{x_j \to 0} \frac{\partial \log \gamma_i}{\partial x_j} = 0 \tag{24-113}$$

The limiting value of the slope is zero.

For electrolytes at infinite dilution, we have derived

$$\log \gamma_\pm = -A' I^{1/2} \tag{24-109}$$

and

$$\lim_{I \to 0} \frac{\partial \log \gamma_\pm}{\partial I} = -\infty \tag{24-114}$$

The limiting value of the slope is $-\infty$.

EXAMPLE 24-3

The solubility of silver chloride in water in the presence of different electrolytes was determined by Neumann [17]. Some of his measurements, dealing with the solubility of this salt in the presence of potassium nitrate at 25°C, are given in Table 24-11.

Table 24-11 The solubility of silver chloride in aqueous potassium nitrate solutions at 25°C

c_{KNO_3} (mol liter^{-1})	$c_{AgCl} \times 10^5$ (mol liter^{-1})	$^c\gamma_\pm$ (calculated)
0.000000	1.273	0.9958
0.013695	1.453	0.8725
0.016431	1.469	0.8630
0.020064	1.488	0.8519
0.027376	1.516	0.8361
0.033760	1.537	0.8247
0.040144	1.552	0.8168

Calculate the mean activity coefficient of silver chloride in solutions having the given concentrations and verify the accuracy of the numerical value of the constant A in the Debye–Hückel equation; its value at 25°C is 0.509.

Solution: The activity of a solute in a saturated solution is constant at a given pressure and temperature:

$$a_0 = a = \text{constant}$$
$$c_0 \, ^c\gamma_{\pm 0} = c \, ^c\gamma_\pm = \text{constant} \tag{a}$$

The subscript 0 refers to silver chloride in pure water. By taking the logarithm, Eq. (a) can be rearranged to give

$$-\log \, ^c\gamma_\pm = \log \frac{c}{c_0} - \log \, ^c\gamma_{\pm 0} \tag{b}$$

The activity coefficient of the dissolved substance in the presence of an electrolyte can be calculated from this equation if the activity coefficient of the solute in the pure solvent and the corresponding solubility are known.

Since the solution is very dilute (sparingly soluble salt), $^c\gamma_{\pm 0}$ can be calculated from the Debye–Hückel limiting law

$$-\log {}^c\gamma_{\pm 0} = A|Z_+ Z_-|I_0^{1/2} \tag{c}$$

For a uni-univalent electrolyte $I = c$; consequently

$$\log {}^c\gamma_{\pm 0} = -0.509 \times (1.273 \times 10^{-5})^{1/2} = -0.00181$$

and

$${}^c\gamma_{\pm 0} = 0.9958$$

By substituting the calculated value of $^c\gamma_{\pm 0}$ and the solubility data from Table 24-11 into Eq. (b), the activity coefficients at all concentrations can be calculated. For example

$$\log {}^c\gamma_\pm = -\log \frac{1.453 \times 10^{-5}}{1.273 \times 10^{-5}} - 0.00181 = -0.0592$$

and

$${}^c\gamma_\pm = 0.8725$$

The activity coefficients, calculated for the other concentrations in an analogous way, are given in Table 24-11.

Now, by combining Eq. (b) with the Debye–Hückel equation, we get

$$\log \left(\frac{c}{c_0} \right) = \log {}^c\gamma_{\pm 0} - \log {}^c\gamma_\pm$$
$$= A|Z_+ Z_-|(I^{1/2} - I_0^{1/2}) \tag{d}$$

Here I is the ionic strength of the solution containing potassium nitrate and silver chloride, and I_0 is the ionic strength of the silver chloride solution in water. By plotting $\log c/c_0$ against $I^{1/2} - I_0^{1/2}$ we get a line that passes through the origin. The slope of this gives the value of $A|Z_+ Z_-|$. The dependence is shown in Figure 24-13 and the slope of the line is 0.500, which is the value of the constant A; $(|Z_+ Z_-| = 1)$.

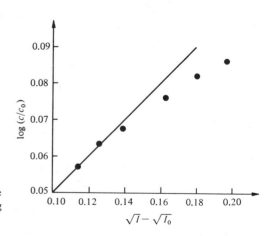

Figure 24-13 Determination of the value of A in the Debye–Hückel limiting law from solubility measurements.

The agreement is satisfactory, particularly if we consider that the concentration of the indifferent electrolyte is relatively large so that the use of the limiting Debye–Hückel law is not really justified. The value of the constant A expressed on the molarity and molality scale for aqueous solutions at room temperature ($d_1 \approx$ 1 kg liter^{-1}) is the same.

24-15 HYDROGEN ION CONCENTRATION: THE pH, pa_H, AND ptH SCALES

Ionic solutions deviate from ideal behavior even when very dilute. Thus, in all the relationships involving concentration, the activity rather than concentration must be used. The equilibrium constants for such processes as dissociation, hydrolysis, solubility, and so on, must be expressed in terms of activity. As we shall see later, this is also true for the rate of ionic reactions.

It is a well known experimental fact that many chemical reactions, including those in biological systems, are very sensitive to the hydrogen ion concentration in the medium. To permit a convenient means of expressing the concentration of hydrogen ions without involving negative exponents, such as 10^{-2}, 10^{-7}, 10^{-14}, and so on, Sørensen suggested the use of the pH scale (*puissance d'hydrogen*). On this scale, the pH of any solution is defined as

$$\mathrm{pH} \equiv -\log c_{H^+} = -\log c_{H_3O^+} \qquad \text{(24-115)}$$

where $c_{H_3O^+}$ is the molar concentration of the hydrated hydrogen ions ($H^+ + H_2O \rightarrow H_3O^+$). For reasons given previously, it is more accurate to define pH in terms of activity:

$$\mathrm{p}a_H = -\log a_{H_3O^+} \qquad \text{(24-116)}$$

In order to give this definition an exact meaning, the concentration scale to which the activity is referred must be exactly specified. Using the molarity scale, we write

$$\mathrm{p}a_H = -\log a_{H_3O^+} = -\log (c_{H_3O^+}{}^c\gamma_{H_3O^+}) \qquad \text{(24-117)}$$

Here ${}^c\gamma_{H_3O^+}$ is the single ion activity coefficient. The problem with this thermodynamically correct equation is that a neutral solution must always contain both positively and negatively charged ions. Consequently, single ion activity coefficients are not exactly measurable quantities. As a result, the definition of a pa_H scale that has an exact thermodynamic meaning cannot be used from a practical standpoint, and it is necessary to resort to nonthermodynamic concepts in order to derive a useful pH scale (see page 917).

Various attempts have been made to circumvent the difficulties by defining pH in terms of quantities that do have thermodynamic significance. An example of such a definition is the thermodynamic scale, ptH, defined as

$$\mathrm{ptH} \equiv -\log (c_{H_3O^+}{}^c\gamma_\pm) \qquad \text{(24-118)}$$

where ${}^c\gamma_\pm$ is the mean molar activity coefficient. Since both ${}^c\gamma_\pm$ and the molarity of the hydrogen ion are determinable quantities, the ptH of the solution will now have thermodynamic meaning. However, relation (24-118) is limited in its applicability; it gives reasonable results for single uni-univalent electrolytes only.[16]

24-16   THE EFFECT OF INTERIONIC FORCES ON REACTION RATES

In this section we shall discuss the interionic electrostatic forces in solutions of electrolytes and see that they strongly influence the rate constants of ionic reactions.

The magnitude of the electrostatic force between two ions, with charges $Z_i e$ and $Z_j e$, separated by a distance r, is given by Coulomb's law

$$f = \frac{Z_i Z_j e^2}{4\pi\varepsilon_0 D r^2} \tag{24-119}$$

Following from this, the work necessary to decrease the distance between the two ions by dr is

$$dw = -f\,dr = -\frac{Z_i Z_j e^2}{4\pi\varepsilon_0 D r^2}\,dr \tag{24-120}$$

and the total work required to bring the two ions from infinite separation to their closest possible distance, σ, (collision diameter), is

$$w = -\int_\infty^\sigma \frac{Z_i Z_j e^2}{4\pi\varepsilon_0 D r^2}\,dr$$

$$= \frac{Z_i Z_j e^2}{4\pi\varepsilon_0 D \sigma} \tag{24-121}$$

The influence of this work on the rate constant can be estimated from the activated complex theory, which leads to the equation (Eq. 21-86)

$$k' = \frac{kT}{h} e^{-\Delta\mathscr{G}^\ddagger/RT}$$

or

$$\ln k' = \ln\left(\frac{kT}{h}\right) - \frac{\Delta\mathscr{G}^\ddagger}{RT} \tag{24-122}$$

Here k' is the rate constant, k is Boltzmann's constant, h is Planck's constant, and $\Delta\mathscr{G}^\ddagger$ is the activation Gibbs free energy per mole. $\Delta\mathscr{G}^\ddagger$ can be assumed to be made up of two parts: the nonelectrostatic and electrostatic parts, respectively. Thus, for 1 mole, we write

$$\Delta\mathscr{G}^\ddagger = \Delta\mathscr{G}_0^\ddagger + \Delta\mathscr{G}_{el}^\ddagger$$

$$= \Delta\mathscr{G}_0^\ddagger + \frac{Z_i Z_j e^2 N}{4\pi\varepsilon_0 D \sigma} \tag{24-123}$$

[16] For more on pH see page 917.

where $\Delta \mathscr{G}_{el}^{\ddagger}$ has been substituted from Eq. (24-121) and N is Avogadro's number $(\Delta \mathscr{G}_{el}^{\ddagger} = wN)$.

Combination of Eqs. (24-122) and (24-123) yields

$$\ln k' = \ln \left(\frac{kT}{h}\right) - \frac{\Delta \mathscr{G}_0^{\ddagger}}{RT} - \frac{Z_i Z_j e^2}{4\pi\varepsilon_0 D\sigma kT} \tag{24-124}$$

This, however, is equivalent to

$$\ln k' = \ln k_0 - \frac{Z_i Z_j e^2}{4\pi\varepsilon_0 D\sigma kT} \tag{24-125}$$

The subscript zero denotes the properties at zero electrostatic force. The second term on the right hand side of Eq. (24-125) is responsible for the effect of the nonideality of the solution (electrostatic forces) on the rate constant.

As is obvious from relation (24-125), the rate constants and consequently the reaction rate between ions of the same sign of charge is lower and the reaction rate between ions of opposite sign is higher when compared with similar reactions in which one or both of the reactants are uncharged. The dielectric constant of the solvent D, also has a strong influence on the rate constants of ionic reactions.

Next, we will discuss the effect of various kinds of salts on ionic reactions. According to the transition-state theory, the reaction between two ions, B^{Z_B} and C^{Z_C}, can be written as

$$B^{Z_B} + C^{Z_C} \rightleftharpoons BC^{\ddagger(Z_B + Z_C)} \rightarrow \text{products} \tag{24-126}$$

Since the complex $BC^{\ddagger}$ is in equilibrium with the reactants, the equilibrium constant has the form

$$K^{\ddagger} = \frac{a_{BC^{\ddagger}}}{a_B a_C} = \frac{c_{BC^{\ddagger}}}{c_B c_C} \frac{{}^c\gamma_{BC^{\ddagger}}}{{}^c\gamma_B \, {}^c\gamma_C} \tag{24-127}$$

${}^c\gamma$ denotes the molar activity coefficients.

The rate of the reaction is assumed to be proportional to the concentration of the activated complex; thus

$$-\frac{dc_B}{d\tau} = -\frac{dc_C}{d\tau} = k'' c_{BC^{\ddagger}} \tag{24-128}$$

The second-order rate constant for the overall reaction is defined by

$$-\frac{dc_B}{d\tau} = -\frac{dc_C}{d\tau} = k' c_B c_C \tag{24-129}$$

From Eqs. (24-128) and (24-129), it follows that

$$k' = k'' c_{BC^{\ddagger}} / c_B c_C \tag{24-130}$$

Comparison of this equation with relation (24-127) shows that

$$k' = k'' K^{\ddagger} \frac{{}^c\gamma_B \, {}^c\gamma_C}{{}^c\gamma_{BC^{\ddagger}}} = k_0 \frac{{}^c\gamma_B \, {}^c\gamma_C}{{}^c\gamma_{BC^{\ddagger}}} \tag{24-131}$$

$k_0 (\equiv k'' K^{\ddagger})$ is the value the second-order rate constant would have at zero ionic strength, where ${}^c\gamma_B = {}^c\gamma_C = {}^c\gamma_{BC^{\ddagger}} = 1$.[17]

---

[17] It is assumed that the nonideal behavior of the solution is caused only by electrostatic forces.

In dilute solutions the activity coefficients as function of ionic strength are given by the limiting Debye–Hückel law

$$\log {}^c\gamma_i = -0.509 Z_i^2 I^{1/2} \qquad \textbf{(24-109)}$$

where I is the ionic strength. Taking the logarithm of (24-131) and substituting in it from the Debye–Hückel expression, we get

$$\log k' = \log k_0 - 0.509[Z_B^2 + Z_C^2 - (Z_B + Z_C)^2] I^{1/2}$$
$$= \log k_0 + 1.018 Z_B Z_C I^{1/2} \qquad \textbf{(24-132)}$$

According to this equation (suggested by Brønsted) a plot of $\log k'$ versus $I^{1/2}$ should be linear. The slope of the straight line depends upon the product $Z_B Z_C$. Experimental data shown in Figure 24-14 prove the validity of Eq. (24-132) for solutions of low ionic strength.

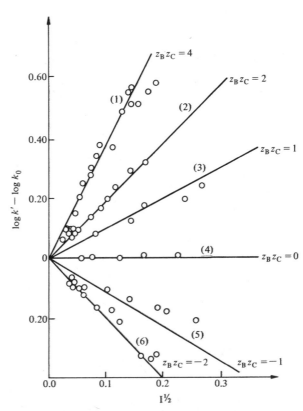

Figure 24-14 Variation of rates of ionic reactions with ionic strengths.

(1) $2Co(NH_3)_5Br^{2+} + Hg^{2+} \rightarrow$
(2) $S_2O_8^{2-} + 2I^- \rightarrow$
(3) $[NO_2NCO_2C_2H_5]^- + OH^- \rightarrow$
(4) $C_{12}H_{22}O_{11} + OH^- \rightarrow$
(5) $H_2O_2 + H^+ + Br^- \rightarrow$
(6) $[Co(NH_3)_5Br]^{2+} + OH^- \rightarrow$

Unreactive salts present in the solution affect the ionic strength of the solution and, consequently, also affect the rate constant. The variation of k' with I is called the primary kinetic salt effect. As mentioned before the rate constant k' may increase, decrease, or remain unchanged with increasing ionic strength; this depends on the signs of the charges Z_+ and Z_-.

We shall next discuss another important experimentally proved fact, namely, the effect of the ionic strength on the rate of some homogeneously catalyzed reactions. Of particular interest here are the reactions accelerated by hydrogen ions, specific hydrogen ion catalysis, or by hydroxyl ions, specific hydroxyl ion catalysis. There are also reactions that can be catalyzed by any substance that may act in the sense of the Brønsted theory as an acid or a base—generalized acid or generalized base catalysis. The essential feature of catalysis by an acid is the transfer of a proton from the acid to the substrate (the substance subject to catalysis); catalysis by a base involves the acceptance of a proton by the base.

In the case of a generalized acid catalysis, the rate of the reaction will depend on all the Brønsted acids present in the solution. For a reaction taking place in an aqueous acetic acid and sodium acetate solution, the following Brønsted acids are present: H_2O, H_3O^+, and CH_3COOH (HAc). Thus, the rate of the reaction is

$$r = k_{H_2O}c_{H_2O}c_s + k_{H_3O^+}c_{H_3O^+}c_s + k_{HAc}c_{HAc}c_s$$

$$= c_s(k_{H_2O}c_{H_2O} + k_{H_3O^+}c_{H_3O^+} + k_{HAc}c_{HAc}) \qquad \textbf{(24-133)}$$

where the subscript s denotes the substrate. From data measured at properly chosen conditions, the catalytic coefficients, k_{H_2O}, $k_{H_3O^+}$, and k_{HAc}, and the total rate of the reaction can be found.

Similar conclusions can be made about reactions in which the substrate is subject to generalized base catalysis.

The rate of a general acid-base catalyzed reaction varies with the activity of both hydrogen and hydroxyl ions. These depend, through their activity coefficients, on the ionic strength of the solution. A salt added to the solution affects through the ionic strength the activity coefficients of the reactants and the activated complex and, consequently, the rate of the reaction. This effect is called the secondary kinetic salt effect. Unlike the primary effect, this does not alter the rate constant provided the constant is calculated from the true H_3O^+ or OH^- concentration.

A few homogeneously catalyzed reactions are listed in Table 24-12.

Table 24-12 Some homogeneously catalyzed reactions

Reaction	Catalyst	Solvent
decomposition of H_2O_2	HCl, HBr	H_2O
inversion of sucrose	H_3O^+	H_2O
hydrolysis of esters	H_3O^+	H_2O
decomposition of triacetone alcohol	OH^-	H_2O
oxidation of phosphorous acid by iodine	generalized acid	H_2O
isomerization of nitromethane	generalized base	H_2O
mutarotation of glucose	generalized acid-base	H_2O

Problems

1. To calibrate an ammeter, a hydrogen coulometer and a silver coulometer were connected in series into the electrical circuit. When current passed through, (a) 95.0 cm^3 of hydrogen were collected in a gas burette in 1 hr ($t = 19°C$, $P = 744$ mm Hg). (b) In the same time 0.845 g of silver were deposited in the silver coulometer. What is the value of the current?

2. An electric current passes through two coulometers connected in series. The electrodes in both coulometers are platinum; the first coulometer contains a solution of metallic nitrate, the second dilute sulphuric acid. After a certain time 0.675 g of metal is deposited on the cathode of the first coulometer and 73.1 cm^3 of hydrogen gas at 16°C and 770 mm Hg is evolved on the cathode of the second coulometer. Calculate the equivalent weight of the metal.

3. In a moving boundary experiment, with a 0.1 N KCl solution, the boundary moved 4.64 cm during 67 min when a current of 5.21 mA was used. The cross-sectional area of the tube was 0.230 cm^2, and the specific conductance of the solution, $\chi = 0.0129\Omega^{-1}$ cm^{-1} at 25°C. Calculate (a) the electrical field strength and (b) the mobility of the potassium ion.

4. A solution of silver nitrate, containing 0.0074 g of AgNO$_3$ per gram of water and silver electrodes were used to determine transference numbers. After the experiment, 25 g of the anodic solution contained 0.2553 g of AgNO$_3$. The series-connected silver coulometer precipitated 0.0785 g of silver during the experiment. Calculate the transference numbers of the silver and nitrate ions.

5. A 0.01 M solution of KCl has a specific conductance of $1.4088 \times 10^{-3}\,\Omega^{-1}$ cm^{-1}. A conductance cell filled with this solution has a resistance of 4.2156 Ω.
   (a) What is the cell constant?
   (b) The same cell filled with a solution of HCl has a resistance of 1.0326 Ω. What is the conductance of the HCl solution?

6. Solutions of silver nitrate in water have the following equivalent conductances as a function of molarity at 25°C:

$\frac{\Lambda_c}{(\Omega^{-1}\,\text{cm}^2\,\text{eq}^{-1})}$	109.14	115.24	121.41	124.76	127.20	130.51	131.36
molarity	0.1	0.05	0.02	0.01	0.005	0.001	0.00

From these data determine (a) the equivalent conductance of silver nitrate at infinite dilution. (b) Find out whether the data follow the Onsager conductance equation.

7. When determining the dissociation constant of propionic acid, D. Belcher [18] measured the equivalent conductance of aqueous solutions of propionic acid as a function of concentration. The results obtained at a temperature of 25°C are as follows:

$\dfrac{c \times 10^3}{(\text{mol liter}^{-1})}$	$\dfrac{\Lambda_c}{(\Omega^{-1}\,\text{cm}^2\,\text{eq}^{-1})}$
0.56685	55.320
0.87116	45.348
1.8650	31.657
4.8026	20.099
8.8839	14.903
15.401	11.373

The equivalent conductances of sodium propionate, hydrochloric acid, and sodium chloride at infinite dilution are $85.92\ \Omega^{-1}\,\text{cm}^2\,\text{eq}^{-1}$, $426.04\ \Omega^{-1}\,\text{cm}^2\,\text{eq}^{-1}$, and $126.42\ \Omega^{-1}\,\text{cm}^2\,\text{eq}^{-1}$, respectively. Calculate the thermodynamic dissociation constant of propionic acid (*Hint*: use Ostwald's dilution law, and evaluate the activity coefficient from the limiting Debye–Hückel law).

8. From Eq. (23-2) derive expressions for the enthalpy of transfer and entropy of transfer. Assume that only the dielectric constant is a function of temperature.

9. Estimate the Gibbs free energy of transfer of Na^+ and K^+ ions from an environment having a dielectric constant of 80 into a medium having a dielectric constant of 5. The ionic radii are 0.95 Å for sodium and 1.33 Å for potassium.

10. Calculate the specific conductance of 0.100 N NaCl solution at 25°C from the mobilities of the sodium and chloride ions at this concentration, which are 42.6×10^{-5} and $68.0 \times 10^{-5}\ \text{cm}^2\,\text{V}^{-1}\,\text{s}^{-1}$, respectively.

11. A saturated solution of AgCl has the specific conductance of $1.80 \times 10^{-6}\ \Omega^{-1}\,\text{cm}^{-1}$ after subtracting out the conductance of the water. The ionic conductances of the silver and chloride ions at infinite dilution are $\lambda^0_{Ag^+} = 61.92\ \Omega^{-1}\,\text{cm}^2\,\text{eq}^{-1}$ and $\lambda^0_{Cl^-} = 73.64\ \Omega^{-1}\,\text{cm}^2\,\text{eq}^{-1}$, respectively. Calculate the solubility of silver chloride in water in grams per liter.

12. The following solubility data for the sparingly soluble tritrivalent salt $[Co(NH_3)_6][Fe(CN)_6]$ in the presence of KNO_3 were obtained at 25°C:

m_{KNO_3}	Solubility $\times 10^5$ (moles in 1000 g of solvent)
0.0000	2.900
0.0005	3.308
0.0010	3.586
0.0020	4.080

Show that the data are in agreement with the Debye–Hückel theory and calculate the mean activity coefficients of the $[Co(NH_3)_6]^{3+}$ and $[Fe(CN)_6]^{3-}$ ions at each concentration.

13. An improved form of the Debye–Hückel limiting law can be written as

$$\log {}^m\gamma_M = -\frac{AZ_M^2 I^{1/2}}{1+I^{1/2}} + \sum_i B_{MX_i} m_{X_i}$$

where A is the usual Debye–Hückel constant, I is the ionic strength, the summation is over all the ions present with charge opposite to M, B_{MX_i} is a parameter characteristic of the MX_i pair, and m_{X_i} is the molality of X_i. Show that for the electrolyte $M_{\nu_+}X_{\nu_-}$ the mean ionic activity coefficient is given by

$$\log \gamma_\pm = -\frac{A|Z_M Z_X| I^{1/2}}{1+I^{1/2}} + \frac{\nu_+}{\nu_+ + \nu_-}\sum_i B_{MX_i} m_{X_i} + \frac{\nu_-}{\nu_+ + \nu_-}\sum_i B_{MX_i} m_{M_i}$$

14. The solubility product constant of AgCl is 1.71×10^{-10} at 25°C.
  (a) What is the mean ionic activity on the molality scale of AgCl in a saturated aqueous solution at 25°C?
  (b) What is the mean ionic activity on the mole fraction scale, referred to the infinitely dilute solution of AgCl in a saturated aqueous solution at 25°C?

15. From the cryoscopic data given in the following table, calculate the mean activity coefficient of a 0.005 M solution of lithium chloride.

Concentration $(m \times 10^3)$	Freezing Point Depression $(\Delta T \times 10^3)$
0.815	2.99
1.000	3.58
1.388	5.03
1.889	6.87
3.350	12.12
3.706	13.45
5.982	21.64
10.810	38.82

16. The solubility of silver iodine in pure water at 25°C is 1.771×10^{-4} mol liter^{-1}. Calculate the solubility of silver iodide in a solution of potassium nitrate that contains 0.3252×10^{-2} moles of KNO_3 per liter. The constant of the Debye–Hückel law at 25°C is 0.509.

17. The standard Gibbs free energy of formation of the following species is given: $\Delta\mathscr{G}^\circ_{f,Br^-} = -24.6$ kcal mol^{-1}, $\Delta\mathscr{G}^\circ_{f,Tl^+} = -7.8$ kcal mol^{-1}, and $\Delta\mathscr{G}^\circ_{f,TlBr} = -39.7$ kcal mol^{-1}. Calculate the solubility product of TlBr. Compare the result with the experimentally found value 3.71×10^{-6}.

18. The solubility of $AgBrO_3$ at 25°C in solutions containing $LiCO_3$ is

molarity of $LiCO_3$	0	0.025	0.050	0.075
molarity of $AgBrO_3 \times 10^3$	8.09	8.83	9.09	9.39

  (a) Using a graphical method obtain the thermodynamic solubility product of $AgBrO_3$ at 25°C.

(b) Calculate the mean ionic activity coefficient of $AgBrO_3$ in saturated solution of this salt.

19. Derive the equations

$$^m\gamma_\pm = {}^c\gamma_\pm \left(\frac{d}{d_1} - \frac{cM_2}{1000d_1} \right)$$

$$\gamma_\pm = {}^c\gamma_\pm \left(\frac{d}{d_1} + \frac{c(\nu M_1 - M_2)}{1000d_1} \right)$$

relating the mean ionic activities on the molality (m), molarity (c), and mole fraction concentration scales. Here d and d_1 are the densities of the solution and solvent and M_1 and M_2 are the molecular weights of the solvent and solute, respectively.

20. Calculate the energy per mole required to separate a positive and negative charge from a distance of 4×10^{-10} m to infinity.
(a) in vacuum;
(b) in a solvent of dielectric constant 10;
(c) in water (dielectric constant $= 80$) at $t = 25°C$.

_____ *References and Recommended Reading*

[1] J. O'M. Bockris and A. K. K. Reddy: *Modern Electrochemistry*, Vol. I, II. Plenum Press, New York, 1970.

[2] J. Koryta, J. Dvořák, and V. Boháčkova: *Electrochemistry*. Methuen, London, 1970.

[3] R. M. Fuoss and F. Accascina: *Electrolytic Conductance*. Interscience, New York, 1959.

[4] H. S. Harned and B. B. Owen: *Physical Chemistry of Electrolytic Solutions*. Reinhold, New York, 1958.

[5] R. G. Ehl and A. J. Ihde: *J. Chem. Educ.*, **31**:226 (1954). Faraday's electrochemical laws and the determination of equivalent weights.

[6] A. L. Levy: *J. Chem. Educ.*, **29**:384 (1952). Difficulties in the Hittorf method of determining transfer numbers.

[7] E. S. Amis: *J. Chem. Educ.*, **30**:351 (1953); *ibid.*, **29**:337 (1952); *ibid.*, **28**:635 (1951). Coulomb's law and the quantitative interpretation of reaction rates.

[8] M. Spiro: *J. Chem. Educ.*, **33**:464 (1956). The definition of transference numbers in solutions.

[9] J. T. Stock: *J. Chem. Educ.*, **31**:410 (1954). The variation of equivalent conductance with concentration.

[10] B. E. Conway and M. J. Salomon: *J. Chem. Educ.*, **44**:554 (1967). Electrochemistry: its role in teaching physical chemistry.

[11] C. A. Kraus: *J. Chem. Educ.*, **35**:324 (1958). The present state of the electrolyte problem.

[12] G. Corsaro: *J. Chem. Educ.*, **39**:622 (1962). Ion Strength, Ion Association and Solubility.

[13] J. E. Prue: *J. Chem. Educ.*, **46**:12 (1969). Ion pairs and complexes; free energies, enthalpies and entropies.

[14] G. A. Lonergan and D. C. Pepper: *J. Chem. Educ.*, **42**:82 (1965). Transport numbers and ionic mobilities by the moving boundary method.

[15] H. Hameka: *Introduction to Quantum Theory*. Harper and Row, New York, 1967.

[16] R. J. Parsons: *J. Chem. Educ.*, **45**:390 (1968). Electrochemical dynamics.

[17] E. W. Neumann: *J. Am. Chem. Soc.*, **54**:2195 (1932).

[18] D. Belcher: *J. Am. Chem. Soc.*, **60**:2744 (1938).

chapter 25

Galvanic Cells

25-1 INTRODUCTION

Electrochemical cells can be grouped into two categories:

1. Electrolytic cells
2. Galvanic cells

The two kinds of cells differ in their function. An electrolytic cell is a device in which an external supply of electrical energy is necessary in order to bring about a physical or chemical change in the cell (see electrolysis, page 842). In a galvanic cell, the free energy released by a physical or chemical change is transformed into electrical energy. Thus, a galvanic cell can serve as a source of electricity, whereas an electrolytic cell cannot. Electrolytic cells have been discussed in Section 24-1. We will now concentrate on galvanic cells.

Galvanic cells have proved to be of great value, both as practical sources of electrical energy and as sources of large amounts of information of interest to chemists. A galvanic cell consists of two electrodes—the half-cells. The potential difference (see page 896) between two half-cells, when measured while no current is drawn from the external leads, is the **electromotive force (emf)** of the cell.

25-2 HALF-CELL POTENTIALS

Consider an inert metal, like platinum, immersed in a solution containing the oxidized and reduced forms of a chemical species, designated as ox and red. These are called electroactive species because they participate in the charge transfer reactions with the electrons of the metal:[1]

$$ox + Ze^- \rightarrow red$$
$$red \rightarrow ox + Ze^-$$

The first reaction is known as reduction or cathodic charge transfer, and the second reaction represents oxidation or anodic charge transfer.

---

[1] The electroactive species could both be ionic, as ferric and ferrous ions, or both uncharged, as quinone and hydroquinone with charge balance maintained by hydrogen ions, or ox could be uncharged and red could be charged negatively, as iodine and iodide, and so on. When the electrode metal, for example, copper, participates in the charge transfer reactions, ox is the metal ion and red the metal. The discussion is somewhat simpler for inert metals like platinum, where the metal serves only as a source or sink for electrons.

Before the piece of metal and the solution are brought into contact they are electrically uncharged—electroneutral—but on contact, when charge transfer begins, some electrons will be removed from the metal by the cathodic reaction and some supplied to the metal by the anodic reaction. If the rates of the two processes are not equal at the moment of contact, the electrode (metal) becomes charged. Thus, when the cathodic rate is dominant, electrons will be removed from the metal faster than they can be supplied, producing an electron deficiency and therefore a positive electrode potential. The existence of an electrode potential causes changes in the charge transfer rate; a positive potential acts to decrease the rate of the cathodic reaction and increase the rate of the anodic process, compared to the initial rates when the electrode was uncharged. As the charge transfer proceeds, the potential changes until the anodic and cathodic rates become equal, thus establishing an equilibrium.

The kinetics of electrode reactions, like other chemical reactions, can be described by equations relating reaction rates to concentration (or preferably to activity) and to activation energy (Sections 8-8.2 and 20-2.3). In electrode reactions, however, the activation energy is potential dependent and the rate of the charge transfer reaction, in moles per second, is proportional to the electrical current, in coulombs per seconds [3]. A reaction rate of M moles per second at an electrode produces a current of I coulombs per second (amperes)

$$I = Z\mathfrak{F}\frac{dM}{d\tau} \tag{25-1}$$

$\mathfrak{F}$ is the Faraday, Z is the number of equivalents per mol of M, and τ is time.

In the absence of an electrode potential, a cathodic reaction has some characteristic activation energy, $\mathscr{G}_c^\ddagger$. For the cathodic reaction

$$\mathrm{ox} + Ze^- \rightarrow \mathrm{red}$$

the first-order reaction rate, in moles per second, when the potential is zero, is given by

$$-\left(\frac{dM_{\mathrm{ox}}}{d\tau}\right)_c = \frac{I_c}{Z\mathfrak{F}} = k_c c_{\mathrm{ox}}\, e^{-\mathscr{G}_c^\ddagger/RT} \tag{25-2a}$$

and for the reverse, anodic reaction, when the potential is zero, we write

$$+\left(\frac{dM_{\mathrm{ox}}}{d\tau}\right)_a = \frac{I_a}{Z\mathfrak{F}} = k_a c_{\mathrm{red}}\, e^{-\mathscr{G}_a^\ddagger/RT} \tag{25-2b}$$

where I_c and I_a are the cathodic and anodic currents, k_a and k_c are the specific rate constants for the anodic and cathodic reactions and c_{ox}, c_{red} are the concentrations of the oxidized and reduced form. $\mathscr{G}_c^\ddagger$ can be viewed as a barrier between ox in the solution and ox combined with electrons ($\equiv$ red) at their zero-point energy level at the electrode surface, namely, at potential equal to zero. An increase in the electron energy level subtracts from the height of the activation barrier $\mathscr{G}_c^\ddagger$, thereby increasing the rate of the cathodic charge transfer process. A negative potential has just such an effect, giving cathodic currents greater than the zero potential value; whereas a positive potential gives smaller currents. For anodic reactions the effect is opposite; the activation barrier is greater than $\mathscr{G}_a^\ddagger$ for negative potentials and smaller for positive potentials.

A change in the potential produces a proportional change in the Gibbs free energy at the electrode,[2] $\Delta \mathcal{G}_{el}$:

$$\Delta \mathcal{G}_{el} = -Z \mathcal{F} \Phi \tag{25-3}$$

where Φ is the potential. Of the total $\Delta \mathcal{G}_{el}$, the fraction $\alpha \Delta \mathcal{G}_{el}$ or $-\alpha Z \mathcal{F} \Phi$ subtracts from $\mathcal{G}_c^{\ddagger}$, giving a cathodic activation energy of $\mathcal{G}_c^{\ddagger} + \alpha Z \mathcal{F} \Phi$. The remainder of $\Delta \mathcal{G}_{el}$ or $-(1-\alpha) Z \mathcal{F} \Phi$ adds to $\mathcal{G}_a^{\ddagger}$, giving an anodic activation energy of $\mathcal{G}_a^{\ddagger} - (1-\alpha) Z \mathcal{F} \Phi$. Here α is the so-called charge transfer coefficient.

The cathodic current as function of potential is obtained from Eq. (25-2a) after substituting the correct value for the activation energy $(\mathcal{G}_c^{\ddagger} + \alpha Z \mathcal{F} \Phi)$:

$$I_c = Z \mathcal{F} k_c c_{ox} \, e^{-(\mathcal{G}_c^{\ddagger} + \alpha Z \mathcal{F} \Phi)/RT} \tag{25-4}$$

$\mathcal{G}_c^{\ddagger}$, the cathodic activation energy at zero potential, can be considered as a constant and can be eliminated from Eq. (25-4) by substituting $k_c' = k_c \, e^{-\mathcal{G}_c^{\ddagger}/RT}$. Since current is proportional to the electrode area, A, we introduce a new constant defined, as

$$k_c'' = \frac{k_c'}{A} \tag{25-5}$$

consequently

$$I_c = Z \mathcal{F} A k_c'' c_{ox} \, e^{-(\alpha Z \mathcal{F} \Phi)/RT} \tag{25-6a}$$

Similarly for the anodic current, we obtain

$$I_a = Z \mathcal{F} A k_a'' c_{red} \, e^{(1-\alpha)(Z \mathcal{F} \Phi)/RT} \tag{25-6b}$$

When the concentration c, is in moles per cubic centimeter, the rate constant has units of velocity, centimeters per second $(cm \, s^{-1})$.

Since at equilibrium, $I_c = I_a$, it follows from Eqs. (25-6a) and (25-6b) that

$$(Z \mathcal{F}/RT)\Phi = \ln \frac{k_c''}{k_a''} + \ln \frac{c_{ox}}{c_{red}} \tag{25-7}$$

Replacing the concentration by activity, $c = a/{}^c\gamma$, and rearranging Eq. (25-7), we have

$$\Phi = \frac{RT}{Z \mathcal{F}} \ln \frac{k_c''{}^c \gamma_{red}}{k_a''{}^c \gamma_{ox}} + \frac{RT}{Z \mathcal{F}} \ln \frac{a_{ox}}{a_{red}} \tag{25-8}$$

Let $\Phi = \Phi^\circ$ when $a_{ox} = a_{red} = 1$; then

$$\Phi^\circ = \frac{RT}{Z \mathcal{F}} \ln \frac{k_c''{}^c \gamma_{red}}{k_a''{}^c \gamma_{ox}} \tag{25-9}$$

and consequently

$$\Phi = \Phi^\circ + \frac{RT}{Z \mathcal{F}} \ln \frac{a_{ox}}{a_{red}} \tag{25-10}$$

This is the **Nernst equation**, relating the potential of an electrode to the activities of the reactants and products of the reaction. Φ° is the standard electrode potential. For

---

[2] At constant T and P, $\Delta G = w$; when only electrical work is considered, $w = -Z \mathcal{F} \Phi$, and therefore $\Delta G_{el} = -Z \mathcal{F} \Phi$.

a general reaction of the form

$$b \text{ ox} + Z e^- \to d \text{ red}$$

the Nernst equation becomes

$$\Phi = \Phi^\circ - \frac{RT}{Z\mathfrak{F}} \ln \frac{a_{red}^d}{a_{ox}^b} \qquad (25\text{-}11)$$

Substituting the values $R = 8.314 \text{ J K}^{-1} \text{ mol}^{-1}$, $T = 298$ K, $\mathfrak{F} = 96{,}487$ C, and $\ln = 2.303 \log$ results in[3]

$$\Phi = \Phi^\circ - \frac{0.05916}{Z} \log \frac{a_{red}^d}{a_{ox}^b} \qquad (25\text{-}12)$$

The potential of a single electrode can be discussed but cannot be measured,[4] except by introducing a second—that is, a reference—electrode as a point of zero or known absolute potential. The measurable potential difference between the two electrodes, which is the electromotive force of the cell when no current is drawn from it, is

$$E = \Phi - \Phi_{ref} \qquad (25\text{-}13)$$

The absolute potential of a reference electrode Φ_{ref} is, however, also inaccessible, so an arbitrary scale of electrode potentials has been established by assigning a zero potential to the standard hydrogen electrode, SHE (see page 908 and [6]). The reaction taking place on this electrode is

$$H_2(P = 1 \text{ atm}) \rightleftharpoons 2 H^+(a = 1) + 2e^- \qquad (25\text{-}14)$$

and $\Phi_{H_2/H^+}^\circ = 0$ at all temperatures.

Thus, the potential of a single electrode on the hydrogen scale is the electromotive force of a cell, E, consisting of the electrode in question and the standard hydrogen electrode [6]

$$E = \Phi - \Phi^\circ(\text{SHE}) = \Phi \qquad (25\text{-}15)$$

The reaction on the hydrogen electrode [Eq. (25-14)] represents oxidation and consequently the reaction on the other electrode must be reduction; the potential of this electrode is the reduction potential. Thus, the electromotive force of any cell is equal to

$$E = [\Phi - \Phi^\circ(\text{SHE})]_{right} - [\Phi - \Phi^\circ(\text{SHE})]_{left}$$
$$= E_{right} - E_{left} \qquad (25\text{-}16)$$

where the subscripts "right" and "left" refer to the location of the electrodes in the cell scheme (see next page). In the difference the potential of the SHE cancels out. For this reason we could assign any value to the SHE potential.

Eq. (25-15) defines the magnitude of a single electrode potential when compared to the SHE potential; its sign, however, must be decided by convention (see

---

[3] The Nernst equation can be derived in a simpler but less illustrative thermodynamic way by substituting $\Delta G^\circ = -Z\mathfrak{F}\Phi^\circ$ and $\Delta G = -Z\mathfrak{F}\Phi$ into Eq. (9-3).

[4] Gibbs (in 1899) pointed out that it is impossible to design an experimental device that would be able to measure a difference in electric potentials between two points in media of different chemical composition, for example, a metal electrode and its surrounding ionic solution (see [1]) on pages 891, 932.

references [1] and [4] at the end of this chapter). The presently accepted convention (IUPAC, Stockholm, 1953), is as follows: The E of a cell is given a positive sign if the electrode on the left-hand side of the cell scheme is negative and that of the right-hand side is positive. Thus, oxidation will occur spontaneously on the left-hand side electrode and reduction on the right-hand side electrode; electrons are moving through the external circuit from the left to the right.

The Stockholm convention rules may be illustrated by applying them to the well-known Daniell cell, shown in Figure 25-1. The scheme of the cell may be written as

$$Zn|ZnSO_4(a_1)\|CuSO_4(a_2)|Cu$$

The metal-electrolyte boundaries are marked by one vertical line $|$. The location of the liquid–liquid boundary is also marked by a vertical line. If, however, the formation of a potential across this boundary is avoided or neglected (see page 906), the junction between the two electrolytes is denoted by double vertical lines $\|$.

The metal wires, for example, the copper wires attached to the electrodes during the measurement of E are omitted from the scheme.

Oxidation must take place on the left electrode:

$$Zn - 2e^- \rightleftharpoons Zn^{2+}$$

Substituting this into the Nernst equation, we have

$$E_{Zn^{2+}/Zn} = E^\circ_{Zn^{2+}/Zn} + \frac{0.05916}{2} \log \frac{a_{Zn^{2+}}}{a_{Zn}}$$

$$= E^\circ_{Zn^{2+}/Zn} + \frac{0.05916}{2} \log a_{Zn^{2+}} \qquad \textbf{(25-17)}$$

because the activity of the metallic zinc is by definition equal to 1.

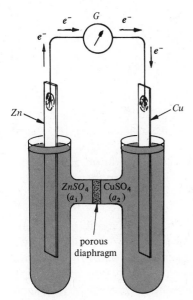

Figure 25-1 The Daniell electrochemical cell.

Reduction must occur on the right electrode:

$$Cu^{2+} + 2e^- \rightleftharpoons Cu$$

and

$$E_{Cu^{2+}/Cu} = E^\circ_{Cu^{2+}/Cu} + \frac{0.05916}{2} \log \frac{a_{Cu^{2+}}}{a_{Cu}}$$

$$= E^\circ_{Cu^{2+}/Cu} + \frac{0.05916}{2} \log a_{Cu^{2+}} \qquad \textbf{(25-18)}$$

again $a_{Cu} = 1$.

The overall cell reaction is

$$Zn + Cu^{2+} \rightleftharpoons Cu + Zn^{2+}$$

and the E of the cell can be expressed as

$$E = E_{right} - E_{left}$$

$$= E^\circ_{Cu^{2+}/Cu} - E^\circ_{Zn^{2+}/Zn} - \frac{0.05916}{2} \log \frac{a_{Zn^{2+}}}{a_{Cu^{2+}}}$$

$$= E^\circ - \frac{0.05916}{2} \log \frac{a_{Zn^{2+}}}{a_{Cu^{2+}}} \qquad \textbf{(25-19)}$$

E° is the standard electromotive force of the cell ($\equiv E^\circ_{Cu^{2+}/Cu} - E^\circ_{Zn^{2+}/Zn}$). The E of the cell must be positive according to the convention; therefore the Gibbs free energy of the cell reaction must be negative, $\Delta G = -Z\mathscr{F}E$. Consequently, the overall cell reaction will proceed spontaneously.

We may now turn our attention to the definition of the sign of a single electrode potential. The Stockholm convention provides that the SHE be written on the left-hand side of the cell scheme. The half-cell reaction corresponding to the potential of the electrode in question is thus always formulated as a reduction process. Consequently

$$E = E_{right} - E^\circ(SHE) = E_{right}$$

Since E of the cell must be positive, E_{right}, the potential of the electrode in question, must also be positive.

Standard reduction potentials of some half-cells, at 25°C, are presented in Table 25-1. A positive electrode potential signifies that the electrode system is a better oxidizing agent than hydrogen ions ($a = 1$) in contact with hydrogen gas ($p_{H_2} = 1$ atm) on platinum. A negative electrode potential means that the system is a better reducing system than hydrogen on platinum at the specified condition. Because the half-cell potentials depend on the activities of the ions involved in the reaction, the values of the standard potentials cannot always be used as indicators of the direction of half-cell reactions.

The experimental determination of the electromotive force of a cell, E, is usually carried out by balancing E of the cell against a known potential drop supplied by a potentiometer, see Figure 25-2. When the two opposing electromotive forces are exactly balanced, no reaction takes place in the cell and no current flows through the galvanometer. When E of the cell exceeds the opposing E very slightly, a small

Table 25-1 Standard reduction potentials at 25°C ($E° = E_{cell}$ when reaction is combined with $H_2(1\ atm) \rightarrow 2\ H^+(1\ M) + 2e^-$)

Reaction	$E°$(V)	Reaction	$E°$(V)
$F_2 + 2e^- \rightarrow 2F^-$	2.87	$Cu^{2+} + e^- \rightarrow Cu^+$	0.15
$Co^{3+} + e^- \rightarrow Co^{2+}$	1.82	$Sn^{4+} + 2e^- \rightarrow Sn^{2+}$	0.15
$H_2O_2 + 2H^+ + 2e^- \rightarrow 2H_2O$	1.77	$2H^+ + 2e^- \rightarrow H_2$	0.000
$PbO_2 + 4H^+ + SO_4^{2-} + 2e^- \rightarrow PbSO_4 + 2H_2O$	1.70	$Pb^{2+} + 2e^- \rightarrow Pb$	−0.13
$Ce^{4+} + e^- \rightarrow Ce^{3+}$	1.61	$Sn^{2+} + 2e^- \rightarrow Sn$	−0.14
$MnO_4^- + 8H^+ + 5e^- \rightarrow Mn^{2+} + 4H_2O$	1.51	$2\,CuO + H_2O + 2e^- \rightarrow Cu_2O + 2\,OH^-$	−0.15
$Au^{3+} + 3e^- \rightarrow Au$	1.50	$CuI + e^- \rightarrow Cu + I^-$	−0.17
$Cl_2(g) + 2e^- \rightarrow 2Cl^-$	1.36	$Ni^{2+} + 2e^- \rightarrow Ni$	−0.25
$Cr_2O_7^{2-} + 14H^+ + 6e^- \rightarrow 2Cr^{3+} + 7H_2O$	1.33	$Co^{2+} + 2e^- \rightarrow Co$	−0.28
$MnO_2 + 4H^+ + 2e^- \rightarrow Mn^{2+} + 2H_2O$	1.23	$PbSO_4 + 2e^- \rightarrow Pb + SO_4^{2-}$	−0.31
$O_2(g) + 4H^+ + 4e^- \rightarrow 2H_2O$	1.23	$Cu_2O + H_2O + 2e^- \rightarrow 2\,Cu + 2\,OH^-$	−0.34
$2\,IO_3^- + 12H^+ + 10e^- \rightarrow I_2 + 6H_2O$	1.20	$Cd^{2+} + 2e^- \rightarrow Cd$	−0.40
$Br_2 + 2e^- \rightarrow 2Br^-$	1.09	$Fe^{2+} + 2e^- \rightarrow Fe$	−0.44
$OCl^- + H_2O + 2e^- \rightarrow Cl^- + 2\,OH^-$	0.94	$Cr^{3+} + 3e^- \rightarrow Cr$	−0.74
$2\,Hg^{2+} + 2e^- \rightarrow Hg_2^{2+}$	0.92	$Zn^{2+} + 2e^- \rightarrow Zn$	−0.763
$Cu^{2+} + I^- + e^- \rightarrow CuI$	0.85	$2\,H_2O + 2e^- \rightarrow H_2 + 2\,OH^-$	−0.828
$Ag^+ + e^- \rightarrow Ag$	0.80	$Mn^{2+} + 2e^- \rightarrow Mn$	−1.18
$Hg_2^{2+} + 2e^- \rightarrow 2Hg$	0.79	$Al^{3+} + 3e^- \rightarrow Al$	−1.66
$Fe^{3+} + e^- \rightarrow Fe^{2+}$	0.771	$H_2 + 2e^- \rightarrow 2H^-$	−2.25
$O_2 + 2H^+ + 2e^- \rightarrow H_2O_2$	0.68	$Mg^{2+} + 2e^- \rightarrow Mg$	−2.37
$Cu^{2+} + Cl^- + e^- \rightarrow CuCl$	0.566	$Ce^{3+} + 3e^- \rightarrow Ce$	−2.48
$I_2 + 2e^- \rightarrow 2I^-$	0.54	$Na^+ + e^- \rightarrow Na$	−2.71
$Cu^{2+} + 2e^- \rightarrow Cu$	0.34	$Ca^{2+} + 2e^- \rightarrow Ca$	−2.87
$Hg_2Cl_2 + 2e_- \rightarrow 2Hg + 2Cl^-$	0.270	$Ba^{2+} + 2e^- \rightarrow Ba$	−2.90
$Hg_2Cl_2 + 2e^- \rightarrow 2Hg + 2Cl^-$ (sat. KCl)	0.244	$Cs^+ + e^- \rightarrow Cs$	−2.92
$AgCl + e^- \rightarrow Ag + Cl^-$	0.223	$K^+ + e^- \rightarrow K$	−2.93
		$Li^+ + e^- \rightarrow Li$	−3.05

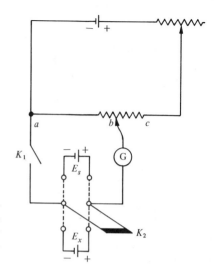

Figure 25-2 A direct reading potentiometer. K_1, main switch; K_2, double throw switch; E_s and E_x, the electromotive forces of the standard cell and the unknown cell, respectively.

current flows and the chemical reaction accompanied by a decrease in the Gibbs free energy takes place in the cell. When the opposing electromotive force slightly exceeds the E of the cell, the reverse chemical reaction occurs in the cell, one which is accompanied by a positive Gibbs free energy change (electrolytic cell).

Good results are also obtained by using a vacuum-tube voltmeter, since the input resistance of these instruments can be made so high that the current they draw is scarcely detectable.

The Weston cell shown in Figure 25-3, is the most frequently used source of a known, constant, and reproducible value of E. Schematically it can be written as

$$Cd(Hg)|CdSO_4 \cdot \tfrac{8}{3} H_2O(s)|(s)Hg_2SO_4|Hg$$

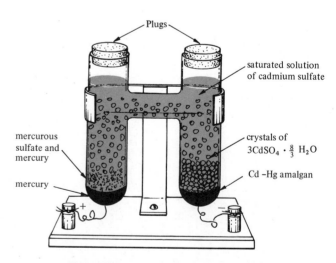

Figure 25-3 The Weston standard electrochemical cell.

The cell reaction is

$$Cd(s) + Hg_2SO_4(s) + \tfrac{8}{3} H_2O(l) \rightarrow CdSO_4 \cdot \tfrac{8}{3} H_2O(s) + 2\ Hg(l)$$

The E of the normal Weston cell (the cell in which the $CdSO_4 \cdot \tfrac{8}{3} H_2O$ solution is saturated) as function of temperature is

$$E = 1.018646 - 4.06 \times 10^{-5}(t - 20) - 0.95 \times 10^{-6}(t - 20)$$

t is the temperature expressed in centigrade and E is expressed in absolute volts.

The Weston cell is reversible; it is not subject to permanent damage due to the passage of current through it; and it has a very low temperature coefficient of E, dE/dt.

25-3 DETERMINATION OF THE STANDARD ELECTROMOTIVE FORCE OF CELLS

It is neither practical nor necessary to measure the $E°$ of a cell directly. Such a method would require cells in which all the products and reactants are present at unit activity. Although this may be possible in some cases, it is inconvenient, and is therefore not recommended.

The commonly used method for determining the standard electromotive force of cells takes advantage of the fact that, by choosing the standard state at infinite dilution, $a \rightarrow m$ as $m \rightarrow 0$. Electromotive force measurement is made at a series of concentrations and the value of $E°$ is obtained graphically by extrapolation to infinitely dilute solution.

In 1932 Harned and Ehlers investigated the following cell at various concentrations of HCl and at 25°C

$$Pt\text{-}H_2(g)(1\ atm)|HCl(m)|AgCl(s)|Ag$$

The reaction on the left electrode is

$$H_2(g) \rightarrow 2\ H^+ + 2e^-$$

and consequently its potential is $(E^\circ_{H_2/H^+} = 0)$

$$E_{(H_2/H^+)} = -\frac{0.05916}{2} \log \frac{a^2_{H^+}}{a_{H_2(g)}} \tag{25-20}$$

The reaction on the right electrode proceeds according to

$$2\ AgCl(s) + 2e^- \rightarrow 2\ Ag(s) + 2\ Cl^-$$

and the potential has the form

$$E_{(AgCl/Ag)} = E^\circ_{(AgCl/Ag)} + \frac{0.05916}{2} \log \frac{a^2_{AgCl(s)}}{a^2_{Ag(s)} a^2_{Cl^-}} \tag{25-21}$$

Then, for the overall cell reaction, we write

$$H_2(g) + 2\ AgCl(s) \rightarrow 2\ H^+ + 2\ Ag(s) + 2\ Cl^-$$

and

$$E = E^\circ_{(AgCl/Ag)} - \frac{0.05916}{2} \log (a^2_{H^+} a^2_{Cl^-}) \qquad (25\text{-}22)$$

because $a_{AgCl(s)} = a_{Ag(s)} = a_{H_2(g)} = 1$.

As it turns out, there is no thermodynamically valid way to measure the activity of a single ion. We always measure the behavior of the two ions together; therefore, we substitute from Eq. (24-78) into Eq. (25-22) in order to get

$$E = E^\circ_{(AgCl/Ag)} - 0.05916 \log a^2_{\pm} \qquad (25\text{-}23)$$

and this can be further changed to yield

$$\begin{aligned} E &= E^\circ_{(AgCl/Ag)} - 0.05916 \times 2 \log a_{\pm} \\ &= E^\circ_{(AgCl/Ag)} - 0.11832 \log m - 0.11832 \log {}^m\gamma_{\pm} \end{aligned} \qquad (25\text{-}24)$$

At this point, several different approaches can be used to determine the $E^\circ_{(AgCl/Ag)}$ value. The most common of these utilize some form of the Debye-Hückel equation for the activity coefficients. In dilute aqueous solutions the dependence of ${}^m\gamma_{\pm}$ on the ionic strength, I, can be expressed by the limiting Debye–Hückel law[5]

$$\log {}^m\gamma_{\pm} = -AI^{1/2} = -Am^{1/2} = -0.509\, m^{1/2} \qquad (25\text{-}25)$$

and consequently, we obtain ,

$$E^\circ_{(AgCl/Ag)} = E + 0.11832 \log m - 0.06021\, m^{1/2} \qquad (25\text{-}26)$$

If the Debye–Hückel limiting law were a precise description of the dependence of ${}^m\gamma_{\pm}$ on m at finite concentration, the left-hand side of Eq. (25-26) would be independent of m. This is not the case, as shown in Figure 25-4, but fortunately this quantity varies linearly with m permitting a precise extrapolation to $m = 0$ for a value of $E^\circ_{(AgCl/Ag)}$. The intercept is 0.2224 V. Since the other electrode in this cell is the standard hydrogen electrode, $E^\circ_{(AgCl/Ag)} = E^\circ_{cell}$.

25-4 EVALUATION OF SOME THERMODYNAMIC PROPERTIES FROM ELECTROMOTIVE FORCE MEASUREMENTS

As was mentioned before, much important information can be obtained from the study of galvanic cells. From the relationship

$$\Delta G^\circ = -Z\mathfrak{F}E^\circ \qquad (25\text{-}27)$$

the standard Gibbs free energy change of the cell reaction can be obtained.

By differentiating Eq. (25-27) with respect to T, we obtain the standard entropy change of the reaction (see page 141)

$$\left(\frac{\partial \Delta G^\circ}{\partial T}\right)_P = -Z\mathfrak{F}\left(\frac{\partial E^\circ}{\partial T}\right)_P = -\Delta S^\circ \qquad (25\text{-}28)$$

---

[5] In a uni-univalent electrolyte $\dot{m} = I$.

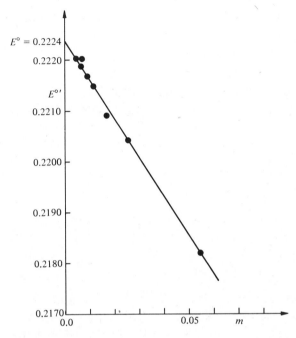

Figure 25-4 Determination of the standard potential of the silver–silver chloride half-cell.

At constant T

$$\Delta H^\circ = \Delta G^\circ + T\Delta S^\circ \qquad (25\text{-}29)$$

and therefore

$$\Delta H^\circ = -Z\mathfrak{F}\left[E^\circ - T\left(\frac{\partial E^\circ}{\partial T}\right)_P\right] \qquad (25\text{-}30)$$

As is obvious from the Eqs. (25-28) through (25-30) the temperature coefficient of E°, $(\partial E^\circ/\partial T)_P$, is required for the evaluation of the standard entropy change as well as for the standard enthalpy change of the cell reaction. Since the cell system does electrical work in addition to the $P\text{-}V$ work, $\Delta H^\circ \neq q_P$ (see page 101).

In calculating the equilibrium constant, we use Eq. (9-4)

$$\Delta G^\circ = -\mathbf{R}T \ln K$$

Combining this with Eq. (25-27), we obtain

$$Z\mathfrak{F}E^\circ = \mathbf{R}T \ln K \qquad (25\text{-}31)$$

and consequently at 25°C

$$\log K = \frac{Z\mathfrak{F}E^\circ}{2.303\,\mathbf{R}T} = \frac{ZE^\circ}{0.05916} \qquad (25\text{-}32)$$

This equation can be employed to evaluate equilibrium constants for any kind of cell reaction. Its application will be shown on pages 904, 910.

The standard Gibbs free energy and the standard entropy of ions in solution can also be found from cell measurements. For illustration let us consider a cell reaction of the type

$$X + 2 H^+ \rightarrow X^{2+} + H_2$$

If all the reactants and products are in their standard states, then

$$\Delta G^\circ = -Z\mathfrak{F}E^\circ = \mu^\circ_{X^{2+}} + \mu^\circ_{H_2} - 2\mu^\circ_{H^+} - \mu^\circ_X \qquad \textbf{(25-33)}$$

According to our previously introduced convention the Gibbs free energies of elements are taken as zero in their standard states (see page 260). Since $\mu^\circ_{H^+}$ is also zero by convention (see page 875), it follows that

$$\mu^\circ_{X^{2+}} = -Z\mathfrak{F}E^\circ = -2\mathfrak{F}E^\circ \qquad \textbf{(25-34)}$$

If the value of the standard electromotive force of the cell is known, the standard Gibbs free energy of the ion, X^{2+}, can be found.

The entropy change for the same reaction is

$$\Delta S^\circ = \mathscr{S}^\circ_{X^{2+}} + \mathscr{S}^\circ_{H_2} - 2\mathscr{S}^\circ_{H^+} - \mathscr{S}^\circ_X \qquad \textbf{(25-35)}$$

$\mathscr{S}^\circ_X$ and $\mathscr{S}^\circ_{H_2}$ can be evaluated from calorimetric or spectroscopic measurements. According to our convention $\mathscr{S}^\circ_{H^+}$ is zero. Now if ΔH° is measured calorimetrically, the equation, $\Delta S^\circ = (\Delta H^\circ - \Delta G^\circ)/T$, enables to evaluate ΔS° for the reaction and consequently $\mathscr{S}^\circ_{X^{2+}}$.

The measurement of E of cells is the most powerful method we have for obtaining values of activity coefficients of electrolytes. When compared to the previously described colligative properties method (Section 24-13) it has the advantages of being easier to handle and useful over a wide range of temperatures. The method is illustrated for the cell

$$\text{Pt-H}_2(g)(1 \text{ atm})|\text{HCl}(m)|\text{AgCl(s)}|\text{Ag}$$

The electromotive force of this cell as function of concentration has already been derived (Eq. 25-24):

$$E = E^\circ - 0.11832 \log m - 0.11832 \log {}^m\gamma_\pm$$

Having determined E° by extrapolation as described in Section 25-3, we see that the experimental values of E determine the value of ${}^m\gamma_\pm$ for each value of m.

Another application of E of cells is shown in Example 25-1.

EXAMPLE 25-1
The formation of anodic sludge during the electrolytic refining of copper is explained by the reaction

$$2 \text{ Cu}^+ = \text{Cu} + \text{Cu}^{2+}$$

in the main part of the electrolyte during electrolysis.

Calculate the relative representation of both, cuprous and cupric, ions in the electrolyte during the electrolysis of copper and explain why anodic sludge is produced according to this equation. The following data are given:

$$\text{Cu}^{2+} + 2e^- = \text{Cu} \qquad E^\circ = 0.340 \text{ V}$$
$$\text{Cu}^{2+} + e^- = \text{Cu}^+ \qquad E^\circ = 0.167 \text{ V}$$

Solution: It follows from thermodynamics that the Gibbs free energy change connected with the transition of the metal into an ionic state with a higher valency, ΔG_2°, is equal to the sum of the Gibbs free energy change that accompanies the transition of the metal into a state with lower valency, ΔG_1°, and the Gibbs free energy change connected with the oxidation of the ion with lower valency into an ion with higher valency, ΔG_{12}°:

$$\Delta G_2^\circ = \Delta G_1^\circ + \Delta G_{12}^\circ$$

If we consider reduction instead of oxidation and substitute for ΔG°

$$\Delta G^\circ = -Z\mathfrak{F}E^\circ$$

we get

$$Z_2\mathfrak{F}E_2^\circ = Z_1\mathfrak{F}E_1^\circ + (Z_2-Z_1)\mathfrak{F}E_{12}^\circ$$

and after rearranging

$$Z_2 E_2^\circ - Z_1 E_1^\circ = (Z_2-Z_1)E_{12}^\circ$$

where Z_2 denotes the higher ionic valence, Z_1 the lower ionic valence, E_1°, E_2°, and E_{12}° are the standard reduction potentials of the different electrode processes, and $\mathfrak{F}$ is Faraday's constant.

The last equation shows that the values of the potentials are not mutually independent. If, for example, the standard reduction potentials of the systems $Cu|Cu^{2+}$ and $Cu^+|Cu^{2+}$ at a temperature of 25°C are 0.340 and 0.167 V, we have for the standard potential of the system $Cu|Cu^+$ ($Z_1 = 1$, $Z_2 = 2$)

$$2 \times 0.340 - E_1^\circ = 0.167$$

from which

$$E_1^\circ = 0.513 \text{ V}$$

If the electrode is submerged in a solution containing M^{Z_1} and M^{Z_2} ions, the following reaction takes place until equilibrium is attained:

$$(Z_2-Z_1)M + Z_1 M^{Z_2} = Z_2 M^{Z_1}$$

The value of the equilibrium constant, $K = a_1^{Z_2}/a_2^{Z_1}$, is calculated on the basis of the fact that at equilibrium the potential of the metal M must be equal both with respect to the M^{Z_1} ion and the M^{Z_2} ion. We can thus write

$$E_1^\circ + \frac{RT}{Z_1\mathfrak{F}} \ln a_1 = E_2^\circ + \frac{RT}{Z_2\mathfrak{F}} \ln a_2$$

From this we get

$$\ln K = \ln \frac{a_1^{Z_2}}{a_2^{Z_1}} = (E_2^\circ - E_1^\circ)\frac{Z_1 Z_2 \mathfrak{F}}{RT}$$

At a temperature of 25°C we obtain

$$\log K = (E_2^\circ - E_1^\circ)\frac{Z_1 Z_2}{0.05916}$$

In our case $E_1^\circ = 0.513$ and $E_2^\circ = 0.340$ V so that the equilibrium constant of the reaction

$$Cu + Cu^{2+} = 2\ Cu^+$$

at 25°C has the value

$$\log K = \log \frac{a_{Cu^+}^2}{a_{Cu^{2+}}} = \frac{(0.340 - 0.513)1 \times 2}{0.05916} = -5.85$$

and

$$K = 1.41 \times 10^{-6}$$

From this result it follows that, in the presence of metallic copper, only an insignificant amount of Cu^+ ions can exist in equilibrium with Cu^{2+} ions. If the concentration of Cu^+ ions exceeds the value derived from the equilibrium constant (concentration gradients are formed in the solution as a result of imperfect mixing), the univalent ions are freed from their supersaturated state so that the following reaction takes place

$$2\ Cu^+ = Cu + Cu^{2+}$$

and metallic powder copper is produced, which sinks to the bottom of the elec-trolyzer as anodic sludge.   ●

25-5 LIQUID JUNCTION POTENTIAL

All the applications of electromotive force measurements discussed in the preced-ing paragraphs yield exact results only if, during the measurement of E, no current is drawn from the cell and, thus, the reaction in the cell is performed reversibly. This is, however, not always the case.

The boundary between the two electrolyte solutions—the two half-cells with different chemical composition or concentration of electrolyte—is the most common source of irreversibility in measuring the electromotive force of cells. The potential difference developed at such a boundary, called the **liquid junction** or **diffusion potential**, results from the Gibbs free energy change of transferring the migrating ions from one electrolyte solution to the other, a typical irreversible process. If, for example, a concentrated solution of hydrochloric acid forms a junction with a dilute hydrochloric acid solution, both hydrogen ions and chloride ions diffuse from the concentrated solution to the dilute solution. The hydrogen ion moves faster; thus, the dilute solution becomes positively charged because of an excess of positive hydrogen ions. The more concentrated solution is left with an excess of negative chloride ions and thus acquires a negative charge. The potential difference across the double layer—positively charged layer in the dilute solution and negatively charged layer in the concentrated solution—may in some cases affect the electromotive force of the cell significantly.

In general, it may be stated that the difference of potential resulting from the junction of two solutions is caused by the difference in the rate of diffusion (mobilities, transference numbers) of the oppositely charged ions, the more dilute solution acquiring a charge corresponding to that of the faster moving ion. The diffusion potential may affect the measured electromotive force of the cell positively or negatively, depending on the relative mobilities of the cation and anion.

Although the liquid junction potential can seldom be eliminated completely, it can be decreased sufficiently in many cases so that the data evaluated from electromotive force measurements are meaningful. The most convenient way of minimizing the liquid junction potential is by means of a salt bridge of potassium chloride connecting the two solutions of different electrolytes. Gels or porous barriers (in the case of the Daniell cell) are also used to minimize liquid junction potentials.

The mechanism of the salt bridge is difficult to analyze quantitatively. It permits flow of current by migration of ions and reduces the net junction potential to a very small value. Most of the current is carried by the K^+ and Cl^- ions, which migrate with approximately the same mobility, and this makes potassium chloride the best electrolyte to use for a salt bridge.

A good way to avoid the formation of liquid junction potentials is to use cells with one liquid electrolyte (common to both electrodes) such as the Weston cell, which has been discussed.

The liquid junction potential, E_L, is a complicated function of the activities, a_i, and transference numbers, t_i, of the several ionic species present in the junction (see also page 914).

$$E_L = -\frac{RT}{\mathcal{F}} \int_1^2 \sum_i \frac{t_i}{Z_i} \, d \ln a_i \tag{25-36}$$

An exact evaluation of E_L from this equation is not possible, since it requires the knowledge of single ion activities. The liquid junction potential can, however, be estimated from the Henderson equation (see references [5], [7], and [8] at the end of this chapter). For the boundary

$$HCl(c = 0.042)|KCl(c = 4.2)$$

the value of E_L estimated from the Henderson equation is approximately 0.0035 V. For a further discussion on liquid junction potentials see concentration cell with transfer, page 913.

The slight concentration changes in the electrolyte near the electrodes resulting from the cell reaction are other small sources of irreversibility.[6] They can be eliminated by proper stirring of the electrolyte solutions.

25-6 CLASSIFICATION OF HALF-CELLS

Although an electrochemical cell requires two half-cells for an electrochemical reaction to occur and its electrical consequences to be measured, the nature of the cell that can be constructed is best understood in terms of the individual half-cells. The more important half-cells are discussed next.

---

[6] The electrode surface itself may also be a source of irreversibility.

25-6.1 Metal–metal ion half-cells The metal–metal ion half-cell consists of a piece of metal immersed in a solution containing the metal ion of activity a (see Daniell's cell on page 897). All these electrodes operate on the basis of the reaction[7]

$$M^{Z+}(a) + Ze^- \rightarrow M(s)$$

for which the electrode potential is given by

$$E_{M^{z+}/M} = E^\circ_{M^{z+}/M} + \frac{RT}{Z\mathfrak{F}} \ln a_{M^{z+}} \tag{25-37}$$

since $a_{M(s)}$ is by definition chosen as unity.

The potential of each of these electrodes is sensitive to, and is determined by, its own ions in the solution in which the metal is immersed.

25-6.2 Amalgam half-cells In an amalgam electrode, the metal partaking directly in the electrode reaction is dissolved in mercury so that the variables include not only the concentration of the metal ions in the solution but also the concentration of the metal in the amalgam (see Weston cell on page 900).

The half-cell reaction can be written as

$$M^{Z+}(a) + Ze^- \rightarrow M(Hg, a)$$

and the electrode potential equation has the form

$$E_{M^{z+}/M(Hg)} = E^\circ_{M^{z+}/M} - \frac{RT}{Z\mathfrak{F}} \ln \frac{a_{M(Hg)}}{a_{M^{z+}}}$$

$$= E^\circ_{M^{z+}/M(Hg)} + \frac{RT}{Z\mathfrak{F}} \ln a_{M^{z+}} \tag{25-38}$$

where $a_{M(Hg)}$, the activity of the metal in the amalgam (different from unity), is included in $E^\circ_{M^{z+}/M(Hg)}$ and $a_{M^{z+}}$ is the activity of the ion in the solution.

25-6.3 Gas–ion half-cells Gas electrodes consist of a gas bubbling about an inert metal (wire or foil) immersed in a solution containing ions to which the gas is reversible. The function of the inert metal, which usually is platinized platinum or graphite, is to facilitate the establishment of equilibrium between the gas and its ions in the solution (acts as a catalyst) and also to serve as the electron carrier for the electrode.

We intend to discuss only two of the many gas–ion electrodes: the hydrogen electrode, which is reversible to hydrogen ions (see Figure 25-5), and the chlorine electrode, reversible to chloride ions.

For the hydrogen electrode, the half-cell reaction is

$$2 H^+(a) + 2e^- \rightarrow H_2(g)$$

---

[7] The potentials used here correspond to a reduction process. Thus, the standard hydrogen electrode represents the left-hand side half-cell whereas the electrode in question is on the right-hand side of the cell.

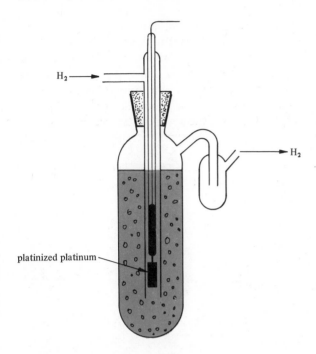

Figure 25-5   The hydrogen electrode.

and the potential is expressed by

$$E_{\mathrm{H^+/H_2}} = E^{\circ}_{\mathrm{H^+/H_2}} - \frac{RT}{2\mathfrak{F}} \ln \frac{a_{\mathrm{H_2(g)}}}{a^2_{\mathrm{H^+}}} \qquad (25\text{-}39)$$

As already mentioned, the standard potential of the hydrogen electrode, that is, the value of the potential at $P_{\mathrm{H_2(g)}} = 1$ atm and $a_{\mathrm{H^+}} = 1$, was conventionally chosen to be zero. As a result of this, Eq. (25-39) can be rewritten into

$$E_{\mathrm{H^+/H_2}} = - \frac{RT}{2\mathfrak{F}} \ln \frac{P_{\mathrm{H_2(g)}}}{a^2_{\mathrm{H^+}}} \qquad (25\text{-}40)$$

since at moderate and low pressures $a_{\mathrm{H_2(g)}} = f_{\mathrm{H_2(g)}} = P_{\mathrm{H_2(g)}}$.

The hydrogen electrode is one of those half-cells suitable for the measurement of pa_{H} (see page 917).

For the chlorine electrode we may write analogously

$$\mathrm{Cl_2(g)} + 2e^- \rightarrow 2\,\mathrm{Cl^-}(a)$$

and

$$E_{\mathrm{Cl_2(g)/Cl^-}} = E^{\circ}_{\mathrm{Cl_2/Cl^-}} - \frac{RT}{2\mathfrak{F}} \ln \frac{a^2_{\mathrm{Cl^-}}}{P_{\mathrm{Cl_2(g)}}} \qquad (25\text{-}41)$$

In order to avoid corrosion by chlorine the platinum metal is often replaced by graphite.

25-6.4   Metal–insoluble salt–anion half-cells   The common characteristic of electrodes of this group is that they all consist of a metal in contact with one of its sparingly soluble salts and a solution containing the ion present in the salt other than the metal. For instance, a silver–silver chloride–chloride half-cell is composed of a silver wire coated with silver chloride and immersed in a solution of chloride ions. Electrodes of this type are extremely important in electrochemistry, and they are used very frequently. We cite only two examples: the silver–silver chloride–chloride electrode, Figure 25-6, and the mercury–mercurous chloride (calomel)–chloride electrode, Figure 25-7.

Let us consider first the silver–silver chloride–chloride half-cell. In this half-cell silver is in contact with silver ions (supplied by AgCl) and therefore it acts like a metal–metal ion electrode:

$$Ag^+(a) + e^- \rightarrow Ag(s)$$

The potential of the half-cell is

$$E_{AgCl/Ag} = E^\circ_{Ag^+/Ag} - \frac{RT}{\mathscr{F}} \ln \frac{a_{Ag(s)}}{a_{Ag^+}} \tag{25-42}$$

However, the activity of the Ag^+ ions is determined by the solubility product of silver chloride.

$$K_{sp} = a_{Ag^+} a_{Cl^-} \tag{25-43}$$

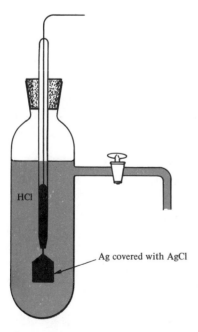

Figure 25-6   The silver–silver chloride half-cell.

HCl

Ag covered with AgCl

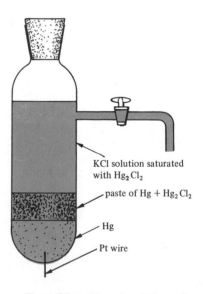

Figure 25-7   The calomel electrode.

KCl solution saturated with Hg_2Cl_2

paste of $Hg + Hg_2Cl_2$

Hg

Pt wire

Substitution for a_{Ag^+} from Eq. (25-43) into Eq. (25-42) yields

$$E_{AgCl/Ag} = E^\circ_{Ag^+/Ag} + \frac{RT}{\mathscr{F}} \ln K_{sp} - \frac{RT}{\mathscr{F}} \ln a_{Cl^-}$$

$$= E^\circ_{AgCl/Ag} - \frac{RT}{\mathscr{F}} \ln a_{Cl^-} \tag{25-44}$$

where

$$E^\circ_{Ag^+/Ag} + \frac{RT}{\mathscr{F}} \ln K_{sp} = E^\circ_{AgCl/Ag} \tag{25-45}$$

Obviously, from the known standard potentials of the silver and silver–silver chloride–chloride electrodes it is possible to calculate the solubility product of silver chloride.

In an analogous way, for the calomel electrode we derive

$$Hg_2Cl_2(s) + 2e^- \rightarrow 2\,Hg(l) + 2\,Cl^-(l)$$

and

$$E_{Hg_2Cl_2/Hg} = E^\circ_{Hg_2Cl_2/Hg} - \frac{RT}{\mathscr{F}} \ln a_{Cl^-} \tag{25-46}$$

again $a_{Hg(l)} = a_{Hg_2Cl_2(s)} = 1$.

The calomel electrode, which has a constant and reproducible potential, is used most frequently. However, the silver–silver chloride–chloride half-cell is the most suitable for accurate measurements.

25-6.5   Oxidation-reduction half-cells   Although each half-cell reaction involves oxidation or reduction, the term oxidation-reduction half-cells is used to denote electrodes in which the potential results from the tendency of ions to pass from one state of oxidation into another more stable state. It is found that when an inert metal collector, usually platinum, is immersed in a solution containing ions in two different states of oxidation, the metal acquires a potential.

For the ferrous-ferric electrode, $Pt/Fe^{3+}, Fe^{2+}$, we write

$$Fe^{3+} + e^- \rightarrow Fe^{2+}$$

and the potential is given by

$$E_{Fe^{3+}, Fe^{2+}} = E^\circ_{Fe^{3+}, Fe^{2+}} - \frac{RT}{\mathscr{F}} \ln \frac{a_{Fe^{2+}}}{a_{Fe^{3+}}} \tag{25-47}$$

The potential of the electrode depends on the ratio of the activities of both ferrous and ferric ions.

EXAMPLE 25-2
The following overall cell reaction is given

$$2\,Fe^{3+} + Sn^{2+} \rightleftharpoons 2\,Fe^{2+} + Sn^{4+}$$

(a)   Write schematically a cell in which the reaction could take place.

(b) Write the two half-cell reactions.
(c) Calculate the standard electromotive force of the cell at 25°C. Given: $E^\circ_{Sn^{4+},Sn^{2+}} = 0.15$ V; $E^\circ_{Fe^{3+},Fe^{2+}} = 0.771$ V.
(d) Evaluate the equilibrium constant of the given reaction and interpret the result.

Solution

(a) From the values of the standard reduction potentials of the two electrodes it follows that the reduction of the Fe^3 ions is more spontaneous than the reduction of the Sn^{4+} ions. Consequently, the iron system will be located on the right-hand side of the cell

$$Pt|Sn^{2+},Sn^{4+}\|Fe^{3+},Fe^{2+}|Pt$$

(b) The reaction on the left-hand side electrode is

$$Sn^{2+} - 2e^- \rightarrow Sn^{4+} \qquad E^\circ_{Sn^{4+},Sn^{2+}} = 0.15 \text{ V}$$

and the reaction on the right-hand side electrode is

$$2\,Fe^{3+} + 2e^- \rightarrow 2\,Fe^{2+} \qquad E^\circ_{Fe^{3+},Fe^{2+}} = 0.771 \text{ V}$$

It is easy to recognize that the sum of the two half-cell reactions results in the overall cell reaction given.

(c)
$$E^\circ = E^\circ_{Fe^{3+},Fe^{2+}} - E^\circ_{Sn^{4+},Sn^{2+}}$$
$$= 0.771 - 0.15 = 0.62 \text{ V}$$

(d)
$$\Delta G^\circ = -2\mathfrak{F}E^\circ = -\boldsymbol{R}T\,2.303 \log K$$
$$\log K = \frac{2\mathfrak{F}E^\circ}{2.303\,\boldsymbol{R}T} = \frac{2 \times 0.62}{0.05916} = 20.96$$

and

$$K = 9.1 \times 10^{20}$$

The large value of K indicates that stannous ions will reduce ferric ions quantitatively. ●

25-7 TYPES OF GALVANIC CELLS

Any combination of two suitable half-cells result in a galvanic cell. In general there are two different types of galvanic cells:

1. Chemical cell
2. Concentration cell

In a chemical cell, such as the Daniell cell or Weston cell already discussed, the electromotive force arises from a chemical reaction within the cell. Chemical cells can be with or without transference. In a cell without transference the boundary between the two electrolytic solutions does not contribute to the E of the cell.

In a concentration cell the electromotive force is a consequence of transport of the electrolyte from the more concentrated solution to the more dilute solution (physical process). Again, we have concentration cells with and without transference.

Let us first discuss concentration cells with transference. Consider the cell

$$Pt\text{-}H_2(g)(1 \text{ atm})|HCl(a_2)|HCl(a_1)|(1 \text{ atm})H_2(g)\text{-}Pt$$

where $a_1 > a_2$.

After 1 equivalent of electricity has passed through the cell the following changes have occurred on the electrodes:

Left-hand side electrode (2)

$$\tfrac{1}{2}H_2(g) \rightarrow H^+(a_2) + e^-$$

As a result of this the hydrogen ion concentration is increasing at the left electrode.

Right-hand side electrode (1)

$$H^+(a_1) + e^- \rightarrow \tfrac{1}{2}H_2(g)$$

The concentration of the hydrogen ion is lowered at the right electrode.

In addition, at the boundary of the two electrolytes, a fraction t_+ of current is carried by H^+ and a fraction t_- is carried by Cl^-. The fractions t_+ and t_- are the transference numbers of the cation and anion, respectively. One equivalent of positive current passing through the boundary requires t_+ equivalents of H^+ ions to be moved from the solution (labeled solution 1) with activity a_1 to the solution (labeled solution 2) with activity a_2 and t_- equivalents of Cl^- ions to be moved in the opposite direction, from solution 2 to solution 1. Thus, at the boundary

$$t_+H^+(1) \rightarrow t_+H^+(2)$$
$$t_-Cl^-(2) \rightarrow t_-Cl^-(1)$$

The total change within the cell is the sum of the changes at the individual electrodes and at the boundary:

$$t_+H^+(2) + t_-Cl^-(1) + H^+(1) \rightarrow t_+H^+(1) + t_-Cl^-(2) + H^+(2)$$

Since $t_+ = 1 - t_-$, this can be rearranged into

$$t_-H^+(1) + t_-Cl^-(1) \rightarrow t_-H^+(2) + t_-Cl^-(2)$$

The cell reaction represents the transfer of t_- equivalents of HCl from solution (1) to solution (2). The Gibbs free energy change associated with this physical process is

$$\Delta G = t_-[(\mu^{\circ}_{H^+} + \mu^{\circ}_{Cl^-}) + RT \ln (a_{H^+}a_{Cl^-})]_2 - t_-[(\mu^{\circ}_{H^+} + \mu^{\circ}_{Cl^-}) + RT \ln (a_{H^+}a_{Cl^-})]_1$$

$$= t_-RT \ln \frac{(a_{H^+}a_{Cl^-})_2}{(a_{H^+}a_{Cl^-})_1} \tag{25-48}$$

Substituting, $\Delta G = -\mathfrak{F}E$, results in

$$E_{wt} = -t_-\frac{RT}{\mathfrak{F}} \ln \frac{(a_{H^+}a_{Cl^-})_2}{(a_{H^+}a_{Cl^-})_1} \tag{25-49}$$

where E_{wt} is the electromotive force of a concentration cell with transference.[8] This equation is exactly valid only when the transference number does not vary within the concentration range of the two electrolytic solutions.

---

[8] The two electrodes differ only in the concentration of the electrolyte. Consequently, the standard potentials of the two electrodes is the same and $E^{\circ} = 0$.

Eq. (25-49) is inconvenient to use because it relates E_{wt} to single ion properties. In order to make the equation applicable, two changes are recommended: (1) to express the single ion activities from the Debye–Hückel law (only in dilute solutions) or (2) to replace the ionic activities by mean activities, $(a_{H^+} a_{Cl^-}) = a_{\pm}^2$. In this case, we have

$$E_{wt} = -t_- \frac{2\,RT}{\mathfrak{F}} \ln \frac{(a_\pm)_2}{(a_\pm)_1} \tag{25-50}$$

If the boundary between the two solutions did not contribute to the E of the cell (liquid junction potential eliminated completely, see [7] and [8]), then, the only change in the cell would be that contributed by the two electrodes:

$$H^+(1) \rightarrow H^+(2)$$

The corresponding value of ΔG would then be

$$\Delta G = (\mu^\circ + RT \ln a_{H^+})_2 - (\mu^\circ + RT \ln a_{H^+})_1$$
$$= RT \ln \frac{(a_{H^+})_2}{(a_{H^+})_1}$$

and replacing ΔG by $-\mathfrak{F}E$ yields the electromotive force of a concentration cell without transference, E_{wot}:

$$E_{wot} = -\frac{RT}{\mathfrak{F}} \ln \frac{(a_{H^+})_2}{(a_{H^+})_1} \tag{25-51}$$

In order to make this equation applicable, the single ion activities can be either evaluated from the Debye theory or replaced by mean activities. In the latter case, we get the approximate relation

$$E_{wot} = -\frac{RT}{\mathfrak{F}} \ln \frac{(a_\pm)_2}{(a_\pm)_1} \tag{25-52}$$

The difference in the electromotive force of cells with (Eq. 25-49) and without (Eq. 25-51) transference must be the contribution of the boundary between the two solutions to the E of the cell, that is, the liquid junction potential:

$$E_L = E_{wt} - E_{wot} = (1 - t_-)\frac{RT}{\mathfrak{F}} \ln \frac{(a_{H^+})_2}{(a_{H^+})_1} - t_- \frac{RT}{\mathfrak{F}} \ln \frac{(a_{Cl^-})_2}{(a_{Cl^-})_1} \tag{25-53}$$

Due to the fact that the single-ion activities are unknown, the liquid junction potential can never be determined accurately.[9] By replacing the single ion activities by mean ionic activities, an approximate relationship for the liquid junction potential can be obtained

$$E_L = E_{wt} - E_{wot} = (1 - 2t_-)\frac{RT}{\mathfrak{F}} \ln \frac{(a_\pm)_2}{(a_\pm)_1} \tag{25-54}$$

It is obvious that, in the case of a uni-univalent electrolyte, the liquid junction potential is small if t_- is near 0.5. This is the reason why KCl solutions are used in salt bridges. Eq. (25-54) defines the liquid junction potential with a certain inaccuracy, depending on the error introduced by replacing the single activities by mean ionic activities.

[9] A more accurate form of relation (25-53) is Eq. (25-36).

In our discussion we have dealt with concentration cells in which the liquid junction potential has been completely eliminated. Are such cells available?

A cell without a liquid junction potential can be schematically represented by

$$\text{Pt-H}_2(g)(1\ \text{atm})|\text{HCl}(a_2)|\text{AgCl}|\text{Ag-Ag}|\text{AgCl}|(a_1)\text{HCl}|(1\ \text{atm})\text{H}_2(g)\text{-Pt}$$

where $a_1 > a_2$. This is a double cell connected in series rather than a single cell (see Figure 25-8).

The reaction in the left-hand side cell is

$$\text{H}_2(1\ \text{atm}) + \text{AgCl} \rightarrow \text{H}^+(a_2) + \text{Cl}^-(a_2) + \text{Ag}(s)$$

and the reaction in the right-hand side cell is

$$\text{H}^+(a_1) + \text{Cl}^-(a_1) + \text{Ag}(s) \rightarrow \text{H}_2(1\ \text{atm}) + \text{AgCl}(s)$$

The overall cell process is then

$$\text{H}^+(a_1) + \text{Cl}^-(a_1) \rightarrow \text{H}^+(a_2) + \text{Cl}^-(a_2)$$

Hydrochloric acid is transferred from the more concentrated solution 1 to the less concentrated solution 2.

The electromotive force of this double cell is

$$E = -\frac{RT}{\mathfrak{F}} \ln \frac{(a_{\text{H}^+}a_{\text{Cl}^-})_2}{(a_{\text{H}^+}a_{\text{Cl}^-})_1} = -\frac{2RT}{\mathfrak{F}} \ln \frac{(a_\pm)_2}{(a_\pm)_1} \qquad (25\text{-}55)$$

Since $(a_\pm)_1 > (a_\pm)_2$, the E of the cell is positive and, consequently, ΔG of the cell reaction is negative.

Another type of concentration cell that does not have a liquid junction potential is the electrode-concentration cell. In this cell two electrodes, differing in their concentration, are immersed in one electrolytic solution.

The simplest cell of this type is the one containing two gas electrodes saturated with the same gas at different pressures:

$$\text{Pt-H}_2(g)(P_1)|\text{HCl}(a)|(P_2)\text{H}_2(g)\text{-Pt}$$

where $P_1 > P_2$.

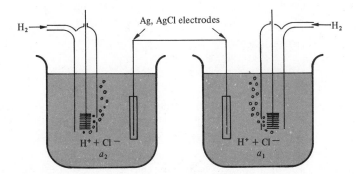

Figure 25-8 Electrochemical cell without liquid junction potential.

The reaction at the left-hand half-cell is

$$\tfrac{1}{2} H_2(g)(P_1) \rightarrow H^+(a) + e^-$$

and the corresponding reaction at the right-hand half-cell is

$$H^+(a) + e^- \rightarrow \tfrac{1}{2} H_2(g)(P_2)$$

The overall cell process, which is independent of the concentration (activity) of the electrolyte, is

$$\tfrac{1}{2} H_2(g)(P_1) \rightarrow \tfrac{1}{2} H_2(g)(P_2)$$

Thus, during the cell process, hydrogen gas expands from the higher to the lower pressure. The Gibbs free energy corresponding to the gas expansion is

$$\Delta G = (\mu^\circ + RT \ln a_{H_2}^{1/2})_2 - (\mu^\circ + RT \ln a_{H_2}^{1/2})_1$$

$$= RT \ln \frac{(a_{H_2})_2^{1/2}}{(a_{H_2})_1^{1/2}} \tag{25-56}$$

because $(a_{H_2})_1 > (a_{H_2})_2$, ΔG is negative. Consequently, we may write for E of the cell

$$E = -\frac{RT}{2\mathfrak{F}} \ln \frac{(a_{H_2})_2}{(a_{H_2})_1} = -\frac{RT}{2\mathfrak{F}} \ln \frac{P_2}{P_1} \tag{25-57}$$

At low and moderate pressures $a = f = P$. Since $P_2 < P_1$ the electromotive force of the cell is positive.

Another cell of this kind consists of two amalgam electrodes. For example

$$Pb\cdot Hg(a_1) | PbSO_4(a) | (a_2) Hg\cdot Pb$$

where $a_1 > a_2$. The mercury acts only as an inert carrier for the lead metal whose concentration can be varied over a considerable range.

The left-hand side half-cell reaction is

$$Pb(a_1) \rightarrow Pb^{2+}(a) + 2e^-$$

The right-hand side half-cell reaction is

$$Pb^{2+}(a) + 2e^- \rightarrow Pb(a_2)$$

and for the overall cell process, we write

$$Pb(a_1) \rightarrow Pb(a_2)$$

The Gibbs free energy corresponding to the transfer of lead from the high activity amalgam to that of low activity is

$$\Delta G = (\mu^\circ + RT \ln a_{Pb})_2 - (\mu^\circ + RT \ln a_{Pb})_1$$

$$= RT \ln \frac{(a_{Pb})_2}{(a_{Pb})_1} \tag{25-58}$$

Thus, for the electromotive force of this cell it follows that

$$E = -\frac{RT}{2\mathfrak{F}} \ln \frac{(a_{Pb})_2}{(a_{Pb})_1} \tag{25-59}$$

As expected the E of such a cell depends only on the activities—that is, concentrations—of the lead in the two amalgam electrodes.

25-8 FUEL CELLS

A class of cells of great potential importance that has received much attention recently is the so-called fuel cell [10]. It is an electrochemical cell in which the substance oxidized is capable of being used as a fuel in a heat engine, that is, capable of being burned, for example, hydrogen, methane, kerosene, garbage, and so forth. A fuel cell then is an electrochemical cell whose cell reaction is the reaction of a fuel with an oxidizer. The construction of the cell provides that the fuel and the oxidizer can be continuously supplied to the cell and the products can be continuously removed from the cell at appropriate rates.

Let us now turn our attention to the hydrogen–oxygen fuel cell. Schematically it can be written

$$\text{Pt-H}_2(g)|\text{NaOH(aq)}|\text{O}_2(g)\text{-Pt}$$

If the cell performs under reversible conditions, the half-cell reactions are

$$4\,\text{OH}^-(aq) + 2\,\text{H}_2(g) \rightarrow 4\,\text{H}_2\text{O(l)} + 4e^-$$
$$\text{O}_2(g) + 2\,\text{H}_2\text{O(l)} + 4e^- \rightarrow 4\,\text{OH}^-(aq)$$

and the overall cell reaction

$$2\,\text{H}_2(g) + \text{O}_2(g) \rightarrow 2\,\text{H}_2\text{O(l)}$$

represents the combustion of hydrogen to water.

For the electromotive force of the cell, we write

$$E = E^\circ - \frac{RT}{4\mathfrak{F}} \ln \left[\frac{a^2_{\text{H}_2\text{O(l)}}}{a^2_{\text{H}_2(g)} a_{\text{O}_2(g)}} \right] \tag{25-60}$$

After making some simplifying assumptions, Eq. (25-60) can be transformed into

$$E = E^\circ + \frac{RT}{4\mathfrak{F}} \left[\ln P^2_{\text{H}_2} P_{\text{O}_2} + \frac{4mg_m}{55.516} \right] \tag{25-61}$$

where m is the molality of NaOH(aq) and g_m is the molal osmotic coefficient of water in NaOH(aq) at T. E° of this cell at 25°C is 1.229 V.

The advantage of fuel cells is that they convert chemical energy directly into electrical energy. This method of conversion is potentially much more efficient than the alternate route of burning fuel under a steam boiler and converting the thermal energy to the mechanical energy of a dynamo, which then converts the mechanical energy into electrical energy. Since appreciable amounts of power are drawn from such cells, the cells do not perform reversibly. Consequently, the electrical energy available from the cells is less than ΔG.

25-9 DETERMINATION OF pa_H

The electrometric determination of pa_H is essentially the evaluation of a quasi thermodynamic property[10]—which has been designed to be a measure of the acidity

[10] This is due to the fact that single-ion activities cannot be measured accurately.

of a solution—from a measurement of the electromotive force of a suitable galvanic cell, such as [5].

Cell type A

$$\text{Pt-H}_2(g)(\text{soln. X containing H}^+)|\text{KCl (sat. soln.)}|\text{Hg}_2\text{Cl}_2(s)|\text{Hg}$$

$$\uparrow$$

$$\text{liquid junction}$$

The half-cell reactions are

$$H_2(g) \rightarrow 2\,H^+(\text{soln. X}) + 2e^-$$

$$Hg_2Cl_2(s) + 2e^- \rightarrow 2\,Hg(l) + 2\,Cl^-(\text{sat. KCl soln.})$$

and the overall cell reaction is

$$H_2(g) + Hg_2Cl_2(s) \rightarrow 2\,H^+\,(\text{soln. X}) + 2Hg(l) + 2\,Cl^-(\text{sat. KCl soln.}) \pm \text{ion transfer}$$

The liquid junction potential is not eliminated; therefore, ions are crossing the boundary between the two electrolytes.

At a hydrogen pressure of 1 atm, the electromotive force of the cell is given by

$$E_{(X)} = E° + E_{L(X)} - \frac{RT}{\mathfrak{F}} \ln a_{H^+(X)} - \frac{RT}{\mathfrak{F}} \ln a_{Cl^-(\text{KCl soln.})} \tag{25-62}$$

The product $(a_{H^+(X)} a_{Cl^-(\text{KCl soln.})})$ is not measurable because it involves single ion activities in different solutions.[11] Eq. (25-62) can be rearranged into

$$-\log a_{H^+(X)} = \frac{\mathfrak{F}(E_{(X)} - E° - E_{L(X)})}{2.303\,RT} + \log a_{Cl^-(\text{KCl soln.})} \tag{25-63}$$

A similar equation can also be written for a standard solution (S) of an exactly assigned $pa_{H(S)}$ value.

$$-\log a_{H^+(S)} = \frac{\mathfrak{F}(E_{(S)} - E° - E_{L(S)})}{2.303\,RT} + \log a_{Cl^-(\text{KCl soln.})} \tag{25-64}$$

Subtraction of Eq. (25-64) from Eq. (25-63) results in

$$pa_{H(X)} = pa_{H(S)} + \frac{(E_{(X)} - E_{(S)})\mathfrak{F}}{2.303\,RT} - \frac{(E_{L(X)} - E_{L(S)})\mathfrak{F}}{2.303\,RT} \tag{25-65}$$

The activity of the chloride ion is a constant at constant temperature; consequently, it cancels out by subtraction.

It is obvious that meaningful pa_H measurements are possible when $(E_{L(X)} - E_{L(S)})$ deviates slightly from zero[12] and pa_H values are assigned to the standards. This, however, in no way implies that it is possible to calculate unambiguously the concentration of H^+ ions from pa_H measurements. Suppose we define $pa_H = -\log a_{H^+}$, where $a_{H^+} = m_{H^+}\,{}^m\gamma_?$. The interpretation given to ${}^m\gamma_?$ is necessarily conventional and depends on how the $pa_{H(S)}$ values for the standards are assigned.

---

[11] If the ions were in the same solution then $a_{H^+} a_{Cl^-} = a_\pm^2$; $a_\pm$ is measurable.

[12] E_L values are not known nor can they be reliably estimated in general.

To illustrate how the $pa_{H(S)}$ values are assigned, we shall next outline the National Bureau of Standards Buffer Standardization Procedure [5]. The cell used for this purpose is

Cell type B

$$\text{Pt-H}_2(g)|(\text{buffer soln. contg. Cl}^-\text{ions})|\text{AgCl(s)}|\text{Ag}$$

This cell contains one electrolyte only; therefore there is no liquid junction.

The cell reaction is

$$\tfrac{1}{2}\,\text{H}_2(g) + \text{AgCl(s)} \rightarrow \text{Ag(s)} + \text{H}^+(aq) + \text{Cl}^-(aq)$$

For a cell with a fixed hydrogen pressure, we derive

$$-\log a_{H^+} = \frac{(E - E'^\circ)\mathfrak{F}}{2.303\,RT} + \log a_{Cl^-} \tag{25-66}$$

where

$$E'^\circ = E^\circ + \frac{RT}{2\mathfrak{F}} \ln a_{H_2} \tag{25-67}$$

After substituting $a_{Cl^-} = m_{Cl^-}\,{}^m\gamma_{Cl^-}$, Eq. (25-66) can be transformed into

$$-\log a_{H^+} = \frac{(E - E'^\circ)\mathfrak{F}}{2.303\,RT} + \log m_{Cl^-} + \log {}^m\gamma_{Cl^-} \tag{25-68}$$

Following from this, we have

$$\text{p}(a_H \cdot {}^m\gamma_{Cl^-}) = \frac{(E - E'^\circ)\mathfrak{F}}{2.303\,RT} + \log m_{Cl^-} \tag{25-69}$$

where $\text{p}\,{}^m\gamma_{Cl^-}$ has been substituted for $-\log {}^m\gamma_{Cl^-}$. This is the basic equation for pa_H measurements.

The following procedure is adopted:

1. Determination of $\text{p}(a_H \cdot {}^m\gamma_{Cl^-})$ for several m_{Cl^-} at fixed buffer concentration ($m_{buf} \gg m_{Cl^-}$).
2. Extrapolation of a graph of $\text{p}(a_H \cdot {}^m\gamma_{Cl^-})$ versus m_{Cl^-} to $m_{Cl^-} = 0$ yields $\text{p}(a_H \cdot {}^m\gamma_{Cl^-})^0$.
3. Computation of pa_H from $\text{p}(a_H \cdot {}^m\gamma_{Cl^-})^0$ by introduction of a conventional individual ionic activity coefficient, ${}^m\gamma_{Cl^-}$ (the Bates–Guggenheim convention).[13]

    $$\log {}^m\gamma_{Cl^-} = -\frac{AI^{1/2}}{1 + 1.5I^{1/2}}$$

    into

    $$pa_H = \text{p}(a_H \cdot {}^m\gamma_{Cl^-})^0 + \log {}^m\gamma_{Cl^-} \tag{25-70}$$

4. For certain selected reference solutions, chosen for their reproducibility, stability, buffer capacity, and ease of preparation, set $pa_H = pa_{H(S)}$.

---

[13] It is obvious that different conventions yield pa_H values differing by more than $0.01\ pa_H$ unit if $I > 0.1$.

This procedure enables one to measure pa_H to within ± 0.01 pa_H units,[14] provided that the composition and ionic strength of the electrolyte solution whose pa_H is to be determined are not much different from that of the solution used for the standardization of the instrument. The uncertainties arising from the liquid junction potential of cell type A may affect the accuracy of the pa_H reading largely, especially for concentrated solutions ($I > 1$) and for solution containing ions with $|Z| > 2$.

Questions still remain to be answered: Is it possible to apply the concept of pa_H to nonaqueous solutions? Can the aqueous pa_H scale already discussed be used for nonaqueous solutions?

The definition, $pa_H = -\log a_{H^+}$, contains nothing that would restrict it to aqueous solutions only. Thus, it can be used in both, aqueous and nonaqueous solutions. The aqueous pa_H scale is, however, of little use for nonaqueous solutions. Although very little has yet been accomplished toward establishing pa_H scales that are valid in nonaqueous media, it would appear that such scales are possible, at least, for waterlike solvents. Each solvent must have its own pa_H scale. Analogously to the aqueous scale, meaningful nonaqueous pa_H scales also require, that the term $E_{L(X)} - E_{L(S)}$ be as close to zero as possible. If this requirement is fulfilled, the problem then reduces to one developing standard solutions. For more details see reference [5].

25-10 GLASS ELECTRODE

The hydrogen electrode, even when very precise, is not always suitable for pa_H measurements because some substances poison the platinized platinum surface so that the electrode cannot function well. In addition to this, reactions beside the desired one can occur in the solution; some substances present in the solution may undergo reduction by the adsorbed hydrogen gas. Therefore, for routine work the more universal but less precise glass electrode is employed.

The capability of the glass electrode to indicate H^+ ions is explained by the fact that the very thin glass membrane from which the electrode is made is permeable to hydrogen ions but not to the other ions in the solution.

In one of its most common forms, the glass electrode consists of a thin glass bulb (see Figure 25-9) filled partially with a solution of known pa_H (1) and immersed in a solution whose pa_H (2) is to be measured. Two reference electrodes—for instance, saturated calomel electrodes—are immersed in solution (1) and (2) respectively.

The cell, or rather the double cell, can be schematically written as

$$\text{membrane}$$

$$\text{Hg}|\text{Hg}_2\text{Cl}_2(\text{s}),\text{KCl(sat. soln.)}|\text{soln.}(1)(a_{H^+} = a_1) \quad \vdots$$

$$\text{soln.}(2)(a_{H^+} = a_2)|\text{KCl(sat. soln.)}, \text{Hg}_2\text{Cl}_2|\text{Hg}$$

---

[14] The fact that the electromotive force of a cell can be measured to better than 0.1 mV in no way implies that the pa_H has a reliability of 0.1 mV/59 mV ≈ 0.002 pa_H units.

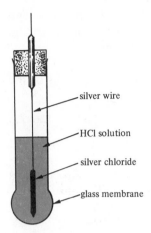

Figure 25-9　The glass electrode.

The only ion that can pass through the membrane is the H^+ ion. Consequently, $t_{H^+} = 1$ $(t_- = 0)$ and the electromotive force of the cell becomes

$$E = \frac{RT}{\mathscr{F}} \ln a_1 - \frac{RT}{\mathscr{F}} \ln a_2 \qquad (25\text{-}71)$$

Since the activity of the solution inside the glass bulb is fixed, the first term is a constant for a certain electrode and Eq. (25-71) assumes the form

$$E = \text{const.} + 0.05916 \, pa_H \qquad (25\text{-}72)$$

The pa_H of the solution in question can be found from measuring E of the cell if the value of the constant is known. To find this a solution of definite pa_H (buffer) is used and the E of the cell is measured.[15]

Because of the extremely high internal resistance of the glass electrode, which may amount to as much as 100 million ohms, ordinary potentiometers cannot be used to measure the E of cells containing glass electrodes. Vacuum tube voltmeters, which require practically no current for operation, must be employed. For this purpose vacuum tube circuits have been developed—pH meters—that are not only sensitive to 0.01 units of pH, or better but also are portable and very rugged.

25-11　MEMBRANE EQUILIBRIUM

Consider two electrolytic solutions of different concentrations separated from each other by a rigid and heat-conducting membrane. The membrane is permeable to all species, including the solvent, except at least one species. As a result of diffusion through the membrane, the diffusible components will undergo a change in concentration whereas the concentration of the indiffusible species will remain constant on

---

[15] The value of the constant depends primarily on the pa_H of the solution inside the glass bulb. In addition to this, it also depends on the so-called asymmetry potential arising from the fact that the two surfaces (inside and outside) of the glass membrane do not behave exactly alike.

both sides of the membrane. Finally, equilibrium will be established in the system; due to the presence of the indiffusible species there will be a definite concentration difference and consequently a potential difference across the boundary between the two solutions, the so-called **membrane potential**.

Since the membrane is heat-conducting, the temperature on both sides of the membrane will be the same at equilibrium:

$$T^{(1)} = T^{(2)} = T \tag{25-73}$$

The parenthetical superscripts refer to the solutions on the two sides of the membrane.

For all components to which the membrane is permeable the electrochemical potential must be the same on both sides of the membrane at equilibrium

$$\mu_{i,\text{el}}^{(1)}(T, P^{(1)}) = \mu_{i,\text{el}}^{(2)}(T, P^{(2)}) \tag{25-74}$$

Because the membrane is also permeable to the solvent, the equilibrium condition will, in general, require a pressure difference between the two sides of the membrane: $P^{(1)} - P^{(2)} = \pi$, where π is the osmotic pressure (see page 217).

The electrochemical potential is defined by

$$\mu_{i,\text{el}} = \mu_i + Z_i \mathfrak{F} \Phi$$
$$= \mu_i^\circ(T, P) + \mathbf{R}T \ln a_i + Z_i \mathfrak{F} \Phi \tag{25-75}$$

The extra term, $Z_i \mathfrak{F} \Phi$, is the reversible work required to transport a charge Z_i from zero potential to Φ potential.[16]

The standard chemical potential, $\mu_i^\circ(T, P)$ may·be split up into two terms

$$\mu_i^\circ(T, P) = \mu_i^*(T, P \to 0) + \int_{P \to 0}^{P} \mathcal{V}_i^\circ \, dP \tag{25-76}$$

where $\mathcal{V}_i^\circ$ is the molar volume in the standard state. If the compressibility of the solution can be ignored, integration of Eq. (25-76) yields

$$\mu_i^\circ(T, P) = \mu_i^*(T, P \to 0) + \mathcal{V}_i^\circ P \tag{25-77}$$

Upon combining Eqs. (25-74), (25-75), and (25-77), we have

$$\mathcal{V}_i^\circ P^{(1)} + Z_i \mathfrak{F} \Phi^{(1)} + \mathbf{R}T \ln a_i^{(1)} = \mathcal{V}_i^\circ P^{(2)} + Z_i \mathfrak{F} \Phi^{(2)} + \mathbf{R}T \ln a_i^{(2)} \tag{25-78}$$

Since the solvent is the same on both sides of the membrane, $\mu_i^{*(1)}(T, P \to 0) = \mu_i^{*(2)}(T, P \to 0)$, at equilibrium.

Rearrangement of Eq. (25-78) yields

$$\Delta \Phi = \Phi^{(2)} - \Phi^{(1)} = \frac{\mathbf{R}T}{Z_i \mathfrak{F}} \left(\ln \frac{a_i^{(1)}}{a_i^{(2)}} + \frac{\pi \mathcal{V}_i^\circ}{\mathbf{R}T} \right) \tag{25-79}$$

This is the osmotic membrane equilibrium equation where $\Delta \Phi$ is the membrane potential and $\pi = P^{(1)} - P^{(2)}$ is the osmotic pressure. An exponential form of Eq. (25-79) is

$$\frac{a_i^{(2)}}{a_i^{(1)}} = e^{(\pi \mathcal{V}_i^\circ - Z_i \mathfrak{F} \Delta \Phi)/\mathbf{R}T} \tag{25-80}$$

---

[16] It is analogous, for example, to the chemical potential of a substance in a gravitational field; the extra term, in that case is mgh, where h is the height.

Eq. (25-80) holds for all components to which the membrane is permeable. For neutral components, however, $Z_i = 0$, and the only term remaining in the exponential is that containing the osmotic pressure:

$$\frac{a_i^{(2)}}{a_i^{(1)}} = e^{\pi \mathcal{V}_i^\circ / RT} \tag{25-81}$$

It can be shown that at certain conditions this equation is identical with Eq. (8-159) on page 217.

At high dilution, $P^{(1)} \approx P^{(2)}$, the osmotic pressure π is very small, and Eq. (25-80) assumes the form

$$\frac{a_i^{(2)}}{a_i^{(1)}} = e^{-Z_i \mathcal{F} \Delta \Phi / RT} \tag{25-82}$$

This is the nonosmotic membrane equilibrium condition.

Writing equations for the cation and anion similar to Eq. (25-79), we have

$$\Delta \Phi = \frac{RT}{Z_+ \mathcal{F}} \left[\ln \frac{a_+^{(1)}}{a_+^{(2)}} + \frac{\pi \mathcal{V}_+^\circ}{RT} \right] \tag{25-83a}$$

and

$$\Delta \Phi = \frac{RT}{|Z_-| \mathcal{F}} \left[\ln \frac{a_-^{(1)}}{a_-^{(2)}} + \frac{\pi \mathcal{V}_-^\circ}{RT} \right] \tag{25-83b}$$

In dilute solutions the osmotic terms of the two different kinds of ions contribute very little to $\Delta \Phi$. Thus, elimination of $\Delta \Phi$ from the last two equations yields

$$[a_+^{(2)}]^{|Z_-|} [a_-^{(2)}]^{Z_+} = [a_+^{(1)}]^{|Z_-|} [a_-^{(1)}]^{Z_+} \tag{25-84}$$

or in particular cases for univalent, divalent, and so on, cations and anions:

$$\frac{a_+^{(2)}}{a_+^{(1)}} = \left(\frac{a_{2+}^{(2)}}{a_{2+}^{(1)}} \right)^{1/2} = \left(\frac{a_{3+}^{(2)}}{a_{3+}^{(1)}} \right)^{1/3} = \cdots = \frac{a_-^{(1)}}{a_-^{(2)}}$$

$$= \left(\frac{a_{2-}^{(1)}}{a_{2-}^{(2)}} \right)^{1/2} = \left(\frac{a_{3-}^{(1)}}{a_{3-}^{(2)}} \right)^{1/3} = \cdots = \lambda \tag{25-85}$$

The constant λ is called the **Donnan distribution coefficient**.

Substituting λ into Eq. (25-82), we obtain the final expression for the Donnan membrane potential:

$$\Delta \Phi = -\frac{RT}{\mathcal{F} Z_i} \ln \lambda \tag{25-86}$$

Membrane potentials contain individual ionic activities that cannot be measured exactly, however, within the concentration range in which the Debye–Hückel limiting law is valid, they can be replaced by mean activities or evaluated directly from the Debye–Hückel theory.

For the measurement of membrane potentials a cell consisting of a permeable membrane, the two solutions under investigation, and two identical reference electrodes is used.

$$\text{Ag}|\text{AgCl,KCl}|\text{solution (1)}|\text{solution (2)}|\text{KCl,AgCl}|\text{Ag}$$

$$\begin{array}{ccc} \text{liquid} & \text{membrane} & \text{liquid} \\ \text{junction} & & \text{junction} \end{array}$$

Since the two reference electrodes are identical, the electromotive force of this cell is given by the sum of the membrane potential and the two liquid junction potentials. The latter are usually neglected.

Numerous biological membranes, in particular cell membranes, act like Donnan membranes. The permeability of the cell membranes is often changed by a chemical action in the living cell, so that the "resting value" of the membrane potential may differ in magnitude as well as in sign from the value during the active period of the cell.

Membrane equilibria are often used for the separation of inorganic ions from solutions of proteins, nucleic acids, and other biologically important polyelectrolytes by means of dialysis. Very often membrane hydrolysis occurs; this permits a transformation of a protein into an acidic form without any direct chemical interference with the system.

25-12 ADDITIONAL COMMENTS ON ELECTRODE PROCESSES

Earlier in this chapter we discussed the process of establishing of a potential between a metal and its ions in a solution. This process, like any other heterogeneous reaction (Section 20-8), can be viewed as a succession of partial steps, the slowest of which, called the rate-determining step, controls the reaction rate of the overall process and the electrical current.

When a current is passed through an electrochemical cell, the potentials of both electrodes are shifted by varying amounts from their equilibrium values, that is, their values when no current is flowing through the cell. This phenomenon is called the polarization of electrodes. When only one electrode reaction takes place at an electrode at a given current density, the overpotential is said to be a measure of the polarization. The overpotential is defined by the equation

$$\eta = \Phi(i) - \Phi(0) \tag{25-87}$$

where $\Phi(0)$ is the electrode potential at zero current density—the equilibrium potential—and $\Phi(i)$ is the electrode potential at current density i.

At equilibrium the cathodic and anodic current densities are equal $(i = 0)$; following from Eqs. (25-6a) and (25-6b) we derive

$$i_0 = i_c = i_a = Z\mathfrak{F}k_c'' c_{ox} e^{-\alpha Z\mathfrak{F}\Phi/RT}$$

$$= Z\mathfrak{F}k_a'' c_{red} e^{(1-\alpha)Z\mathfrak{F}\Phi/RT} \tag{25-88}$$

where i_0 is the exchange current density at the equilibrium electrode potential.

We may now express the current density in terms of i_0 and the overpotential, $\eta = \Phi(i) - \Phi(0)$, as

$$i = i_a - i_c = i_0(-e^{-\alpha Z\mathfrak{F}\eta/RT} + e^{(1-\alpha)Z\mathfrak{F}\eta/RT}) \tag{25-89}$$

This is a rather fundamental equation. It is termed the **Butler–Volmer equation** and it shows how the current across a metal-ionic solution interface depends on the overpotential.

The magnitude of i determines the deviation of the electrode potential from its equilibrium value when an external current is passed. The larger the deviation, the slower the electrode reaction.

The dependence of current density on overpotential is shown in Figure 25-10. As is obvious, the shape of the curves depends on the transfer coefficient α.

If the overpotential is low, $\eta \ll RT/Z\mathfrak{F}$, we can expand the exponentials and neglect all terms except the first two, giving

$$i = \frac{i_0 Z\mathfrak{F}\eta}{RT} \tag{25-90}$$

If the overpotential has large positive values, $\eta \gg RT/Z\mathfrak{F}$, $i_c \approx 0$, and the Butler–Volmer equation reduces to

$$\ln i = \ln i_a = \ln i_0 + \frac{(1-\alpha)Z\mathfrak{F}}{RT}\eta \tag{25-91a}$$

If $\eta \ll 0$, $i_a \approx 0$ and

$$\ln i = \ln i_c = \ln i_0 - \frac{\alpha Z\mathfrak{F}}{RT}\eta \tag{25-91b}$$

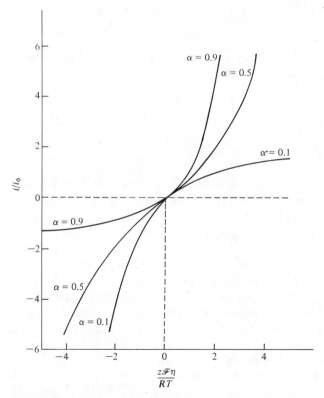

Figure 25-10   The dependence of current density on overpotential for various values of charge transfer coefficients.

These are the famous **Tafel equations,**[17] relating the current density to the overpotential. The transfer coefficient α, may be obtained from a graph in which $\ln i$ is plotted against η.

It may also be shown from the Butler–Volmer equation that

$$\lim_{\eta \to 0} \left(\frac{\partial i}{\partial \eta} \right) = \frac{i_0 Z \mathfrak{F}}{RT} \tag{25-92}$$

The reciprocal value of this differential quotient

$$R_{\mathrm{p}} = \frac{RT}{Z \mathfrak{F} i_0} \tag{25-93}$$

is called the polarization resistance at the equilibrium electrode potential.

Figure 25-11 illustrates the typical interdependence of i and η for metals at which the following reaction occurs:

$$2\,H_3O^+ + 2e^- \rightarrow H_2 + 2\,H_2O$$

The linear part of the curve is in accord with the Tafel equation (25-91). When H_2 is generated on metals different from platinum, slight deviations from linearity are observed especially at high current densities.

The reactions, one of which may be rate limiting, at a hydrogen cathode are

1. The formation of H atoms on vacant metal surface sites

$$H_3O^+ + e^- \rightarrow H + H_2O$$

2. The reaction between a hydrogen atom already adsorbed on a metal surface site and a hydrogen ion in the solution

$$H + H_3O^+ + e^- \rightarrow H_2(g) + H_2O$$

3. The reaction between two hydrogen atoms already adsorbed on the metal surface

$$H + H \rightarrow H_2(g)$$

Which one of these steps is rate limiting will depend on the electrode material, and, to a large extent, on the condition of its surface. Some sites on the surface may provide active centers having a high catalytic activity for a particular reaction step. On smooth, shiny, and polished surfaces, the overvoltage is invariably greater than on rough, pitted, or etched surfaces.

The cause of overvoltage is not yet clear. No theory has been proposed that can account satisfactorily for all the observed phenomena.

The existence of overvoltage makes it possible to plate some metals electrolytically even though the evolution of hydrogen gas would be expected from a consideration of reversible electrode potentials. For example, if a solution of a zinc salt is electrolyzed in a cell with inert electrodes, metallic zinc is deposited on the cathode even though zinc is above hydrogen in the electromotive series; that is, the standard potential of a zinc electrode is negative with respect to the SHE.

---

[17] The empirical form of the Tafel equation is $\eta = a + b \log i$

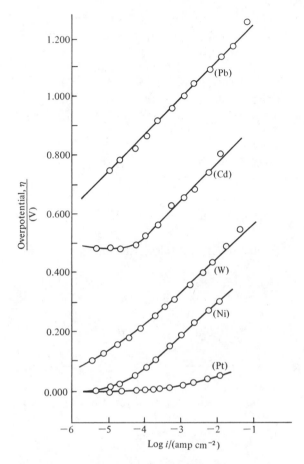

Figure 25-11 The change of the overpotential with current density for various metals.

It is worthwhile mentioning that charging the lead storage battery would not be possible if it were not for a huge hydrogen overvoltage on the negative plate, which permits the reaction

$$PbSO_4 + 2e^- \rightarrow Pb + SO_4^{2-}$$

to occur with high efficiency. A high oxygen overvoltage on the positive plate, PbO_2, is required so that the reaction

$$PbSO_4 + 2\,H_2O \rightarrow PbO_2 + SO_4^{-2} + 4\,H^+ + 2e^-$$

can also occur with high efficiency. If these overvoltages were not large, it would not be possible to polarize the plates to the potential required for the charging reactions to occur, and passage of the charging current would only decompose the water into hydrogen and oxygen.

Corrosion reactions are another group of reactions that depend critically on the presence or absence of a significant hydrogen overvoltage on the surface of metals.

The presence of overvoltage is sometimes undesirable. It increases the voltage and consequently the electrical energy required for many industrial electrolytic processes.

In some cases the diffusion process rather than the overpotential may become the rate-limiting step for the overall reaction. When this happens, the electroactive species are discharged at the electrode at a faster rate than they are supplied to the electrode by diffusion. Following from this, a concentration gradient is developed across the thin layer adjacent to the electrode. This results in a new kind of polarization, called the **diffusion overpotential.**[18]

The thin layer adjacent to the electrode acts like a concentration cell and, consequently, the diffusion overpotential, η, can be expressed as

$$\eta = \frac{RT}{|Z|\mathfrak{F}} \ln \frac{a_1}{a_0} = \frac{RT}{|Z|\mathfrak{F}} \ln \frac{c_1{}^c\gamma_1}{c_0{}^c\gamma_0} \tag{25-94}$$

where a_1 and a_0 are the activities of the electroactive species in the bulk solution and at the electrode surface, respectively.

The rate of diffusion as a function of the concentration difference across the layer can be expressed by making use of Fick's first law. Assuming steady state diffusion— the concentration gradient is linear—and that the diffusion coefficient, D, is independent of the concentration, the rate of diffusion may be expressed as

$$-\text{rate} = -\frac{dn}{d\tau} = DA\frac{c_0 - c_1}{\delta} \tag{25-95}$$

where A is the area of the electrode, δ is the thickness of the layer, and $dn/d\tau$ is the number of moles diffused per unit time.

Following from this the current to the electrode (cathode) is

$$I = -|Z|\mathfrak{F}\frac{dn}{d\tau} = |Z|\mathfrak{F}DA\frac{(c_0 - c_1)}{\delta} \tag{25-96}$$

Combination of Eq. (25-96) and Eq. (25-94) results in

$$\eta = \frac{RT}{|Z|\mathfrak{F}} \ln \left[\frac{{}^c\gamma_1}{{}^c\gamma_0} \left(1 - \frac{\delta I}{Ac_0|Z|\mathfrak{F}D} \right) \right] \tag{25-97}$$

If every charged species colliding with the electrode surface discharges, then $c_1 = 0$, and the maximum limiting current becomes

$$I_{\max} = \frac{|Z|\mathfrak{F}DAc_0}{\delta} \tag{25-98}$$

Relationships (25-96) and (25-97) are only approximate. A more quantitative theory must take into consideration the shape of the electrode, the dependence of D on concentration, and the nonsteady state of diffusion. Different results are also obtained for moving and stationary electrodes.

---

[18] The student must be careful to distinguish between the term diffusion overpotential and the term diffusion or liquid junction potential.

_____ *Problems*

1. Calculate the E of the cell

$$\text{Pt-H}_2(\text{g})(0.30 \text{ atm})|\text{HCl}(m = 0.010)|\text{Cl}_2(\text{g})(0.60 \text{ atm})\text{-Pt}$$

at 25°C. The mean activity coefficient of a 0.010 m HCl solution is 0.904.

2. Calculate the E of the following cell at 25°C:

$$\text{Ag}|\text{AgCl}|\text{NaCl}(a_1 = 0.0100)|\text{NaCl}(a_2 = 0.0250)|\text{AgCl}|\text{Ag}$$

The transference number of Na^+ is 0.392. The liquid junction potential in such a cell can be estimated from the simplified Henderson equation

$$\Phi_L = \frac{RT}{\mathcal{F}} \left(\frac{\lambda_+^\circ - \lambda_-^\circ}{\lambda_+^\circ + \lambda_-^\circ}\right) \ln \frac{(a_\pm)_2}{(a_\pm)_1}$$

where λ_i° is the ionic conductance at infinite dilution. $\lambda_{\text{Na}^+}^\circ = 50.11$ $\Omega^{-1} \text{cm}^2 \text{eq}^{-1}$, $\lambda_{\text{Cl}^-}^\circ = 76.34 \ \Omega^{-1} \text{cm}^2 \text{eq}^{-1}$.

3. The E of the cell

$$\text{Pt-H}_2(\text{g})(1 \text{ atm})|\text{HCl}(m)|\text{AgCl}|\text{Ag}$$

was measured at 20°C as a function of the concentration of hydrochloric acid.

m	$\dfrac{E}{(\text{V})}$
0.005314	0.49395
0.008715	0.46987
0.013407	0.44899
0.021028	0.42726
4.0875	0.12307

From these data (a) determine the standard potential of the silver chloride electrode at 20°C; and (b) calculate the mean activity coefficient of hydrochloric acid in a 4.0875 m solution.

4. The E of the cell

$$\text{Ag}|\text{AgNO}_3(a_1 = 0.001)|\text{AgNO}_3(a_2 = 0.01)|\text{Ag}$$

at 25°C is 63.1 mV. The two solutions are separated by a porous partition. Calculate the mean transference number of the silver ion.

5. The E of the cell

$$\text{Pt-H}_2(\text{g})(1 \text{ atm})|\text{HClO}_4(m_1)\|\text{KCl,KNO}_3\|\text{HReO}_4(m_2)|\text{ReO}_3(\text{s})\text{-Pt}$$

was measured as a function of concentration of both perchloric and perrhenic acids. Some of the measured values of E at 25°C are

$m_1 = m_2$	$E/(V)$	$m_1 = m_2$	$E/(V)$
0.0540	0.500	0.0135	0.487
0.0478	0.494	0.00680	0.468
0.0290	0.501	0.00106	0.400
0.0270	0.492	0.00054	0.373

Calculate the standard reduction-oxidation potential of the electrode system ReO_4^-/ReO_3 at 25°C.

6. For the E of the cell

$$Mo|MoS_2(s)|H_2S(g)(1\ atm)|KCl(m = 0.01)|HCl(m = 0.01)|H_2(g)\text{-Pt}$$

the following data were obtained

$t/(°C)$	15	25	35
$E/(V)$	0.41483	0.41187	0.40874

(a) Determine what reaction takes place in the cell.
(b) Derive the relation giving the temperature dependence of the equilibrium constant of this reaction and calculate its enthalpy of reaction.
(c) Calculate the entropy change of the reaction and compare it with the value obtained from the absolute entropies of the different substances:

substance:	$H_2(g)$	$H_2S(g)$	$Mo(s)$	$MoS_2(s)$
$\mathscr{S}_{298.15}^{\circ}/(cal\ K^{-1}\ mol^{-1})$:	31.21	49.15	6.83	14.96

7. The temperature dependence of the E of the cell

$$Hg|Hg_2Br_2|KBr(c = 0.1)\|KCl(c = 0.1)|Hg_2Cl_2|Hg$$

is given by the equation

$$E = 0.1318 - 1.880 \times 10^{-4}t \qquad \langle 9.5\text{–}40°C \rangle$$

(a) Calculate ΔH for the cell reaction at 298.15 K. If E is measured to ± 0.0001 V, and t to $\pm 0.010°C$, how precise is the value of ΔH?
(b) Calculate the solubility product of mercurous bromide at a temperature of 25°C if the reduction potential of 0.1 N calomel electrode at 25°C is 0.3335 V and the mean activity coefficient of 0.1 N KBr at 25°C is 0.772. The standard reduction potential of the Hg/Hg_2^{2+} system is 0.799 V.

8. The E of a concentration cell with transference is related to the mean ionic activities, $a_\pm$, by the equation

$$dE = \pm 2t_\pm \frac{RT}{\mathscr{F}} d \ln a_\pm$$

where t_i is the transference number. The following empirical formula was obtained for the E of the cell

$$Ag|AgCl|LiCl(c_1)|LiCl(c_2)|AgCl|Ag$$
$$E = -43.865 + 45.363(\log A) - 1.4902(\log A)^2$$

where A is 10^4 times the mean ionic activity of LiCl at concentration c_1; $a_{\pm}$ at concentration c_2 is 10^{-3}—this choice merely fixes the constant term in E—and E is given in millivolts. The activity coefficient of a $0.010\ m$ solution of LiCl is 0.905. Evaluate t_{Li^+} for a $0.01\ m$ solution of LiCl.

9. The standard electrode potentials of the systems Sn^{4+}/Sn^{2+} and Sn^{2+}/Sn are given: $E^\circ_{Sn^{4+}/Sn^{2+}} = 0.15\ V$, $E^\circ_{Sn^{2+}/Sn} = 0.14\ V$. Evaluate the standard electrode potential of the system Sn^{4+}/Sn.

10. The concentration of Ni^{2+} ions in a solution of a nickel salt is to be reduced to a value of 10^{-3} mole liter^{-1} without the evolution of hydrogen. What is the lowest value of the pH of this solution for which this process is possible? The hydrogen overvoltage on nickel is $0.21\ V$.

11. The experimentally measured exchange current density for the electrode system Pt/Fe^{2+}, Fe^{3+} is $i_0 = 0.50\ A\ cm^2$ at $298\ K$. $\Delta\mathcal{H}^{\ddagger}$ is $8719\ cal\ mol^{-1}$ and the transfer coefficient, α, is 0.58. Calculate at $298\ K$ the relative current density, i/i_0, as a function of overpotential, η, from -1.0 to $+1.0\ V$.

12. A pH meter gives a reading of $200\ mV$ when measuring a solution whose hydrogen ion activity is considered to be 10^{-4}. What is the hydrogen ion activity of a solution for which the meter reads $100\ mV$? (assume $t = 25°C$)

13. The E of the cell

$$Ag|AgCl|NaCl(m)|NaHg(amalgam)\|NaHg(amalgam)|NaCl(0.100\ m)|AgCl|Ag$$

at $25°C$ is as follows:

m	0.200	0.500	1.000	2.000	3.000	4.000
$E/(V)$	0.03252	0.07584	0.10955	0.14627	0.17070	0.19036

The mean activity coefficient of a $0.100\ m$ NaCl solution is 0.773. Calculate the mean activity coefficients of NaCl in the given solutions and plot the values against $\sqrt{m}$.

14. At $25°C$, we are given the following values for the E of the cell

$$Pb|PbSO_4(s)|H_2SO_4(m)|H_2(1\ atm)\text{-}Pt$$

m	0.00100	0.00200	0.00500	0.0100	0.0200
$E/(V)$	0.1017	0.1248	0.1533	0.1732	0.1922

Obtain $E°$ for this cell by extrapolation. From this value calculate the thermodynamic solubility product of $PbSO_4$.

15. At various pressures and at $25°C$ we are given the following values for the E of the cell

$$Pt\text{-}H_2(P)|HCl(m = 0.1)|AgCl|Ag$$

$P/(atm)$	10	37.9	51.6	110.2	286.6	731.8	1035.2
$E/(mV)$	399.0	445.6	449.6	459.6	473.4	489.3	497.5

Calculate the fugacity coefficients ($\nu = f/P$) and plot them as a function of P over a range of 1 to 1000 atm.

16. Consider a membrane of a plant cell permeable to Na^+, Cl^-, and H_2O but not to proteins. Suppose that initially there is a NaCl solution of concentration 0.050 moles liter^{-1} on each side of the membrane, and a protein P of concentration 1×10^{-3} mol liter^{-1}, which ionizes to give P^{Z+} ($Z = 10$) plus 10 Cl^- ions. Calculate the Donnan potential across the cell wall at equilibrium (approximate activities by concentrations).

17. Calculate the equilibrium constant for the reaction at 25°C

$$2H^+ + D_2(g) \rightleftharpoons H_2(g) + 2D^+$$

from the electrode potential of Pt-D_2 (1 atm)/D^+, which is -3.4 mV at 25°C.

_____*References and Recommended Reading*

[1] F. C. Anson: *J. Chem. Educ.*, **36**:394 (1959). Electrode sign convention.

[2] O. Robbins, Jr.: *J. Chem. Educ.*, **48**:737 (1971). The proper definition of standard electromotive force.

[3] K. J. Laidler: *J. Chem. Educ.*, **47**:600 (1970). The kinetics of electrode processes.

[4] T. S. Licht and A. J. de Bethume: *J. Chem. Educ.*, **34**:433 (1957). Recent developments concerning the sign of electrode potentials.

[5] R. G. Bates: *Determination of pH, Theory and Practice.* Wiley, New York, 1964.

[6] T. Biegler and R. Woods: *J. Chem. Educ.*, **50**:604 (1973). The standard hydrogen electrode: a misrepresented concept.

[7] P. A. Rock: *J. Chem. Educ.*, **47**:683 (1970). The design of electrochemical cells without liquid junction.

[8] J. J. Campion: *J. Chem. Educ.*, **49**:827 (1972). An alternative approach to liquid junction potential.

[9] J. O'M. Bockris: *J. Chem. Educ.*, **48**:352 (1971). Overpotential: a lacuna in scientific knowledge.

[10] A. K. Vijh: *J. Chem. Educ.*, **47**:680 (1970). Electrochemical principles involved in fuel cell.

Partial Differentiation and Exact Differentials

Let us consider the following continuous function

$$u = f(x, y) \tag{I-1}$$

According to this equation to each pair of x, y values (independent variables) there is only one corresponding value of u (dependent variable). The function can be thought of as a surface in a three-dimensional space (Figure I-1).

When u changes infinitesimally from a point characterized by x and y values $[u(x, y)]$ to a point characterized by $(x + dx)$ and $(y + dy)$ values $[u(x + dx, y + dy)]$, the change in u can be mathematically expressed by

$$du = \left(\frac{\partial u}{\partial x}\right)_y dx + \left(\frac{\partial u}{\partial y}\right)_x dy \tag{I-2}$$

The first partial derivative term on the left-hand side of Eq. (I-2) takes account of the change in u with x at constant y; the second partial derivative term is responsible for the change in u with y at constant x. Thus, the total change in u can be expressed as a sum of the two partial changes. Applying this to Figure I-1, it is obvious that the change in u in going from A to C is expressed as the sum of the changes in going from

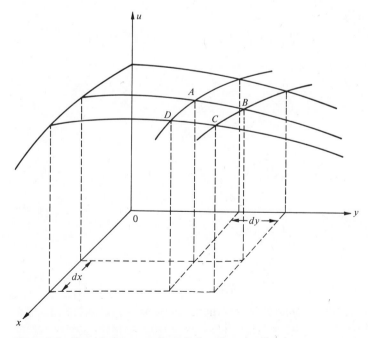

Figure I-1 Graphical presentation of a function, $u = f(x, y)$.

933

A to *D* and from *D* to *C*, or alternatively from *A* to *B* and from *B* to *C*; *D* has the same value of *y* as *A* and the same value of *x* as *C*; *B* has the same value of *x* as *A* and the same value of *y* as *C*.

Eq. (I-2) is the basic equation of partial differentiation. It can be applied only when the function *u*, and its derivatives, $(\partial u/\partial x)_y$, $(\partial u/\partial y)_x$ are continuous.

Partial derivatives may be given the same physical interpretation as the ordinary derivatives except that all but the one independent variable are held fixed, so that the change in the dependent variable is attributed to a change in only one of the independent variables.

When *u* is held constant, $du = 0$, and Eq. (I-2) can be arranged into

$$\left(\frac{\partial x}{\partial y}\right)_u = -\frac{(\partial u/\partial y)_x}{(\partial u/\partial x)_y} \tag{I-3}$$

If another function, such as

$$v = f'(x, y) \tag{I-4}$$

exists, then it can be shown that $(\partial u/\partial x)_v$ and $(\partial u/\partial x)_y$ are related in a simple way to each other. By dividing Eq. (I-2) by dx, we obtain

$$\frac{du}{dx} = \left(\frac{\partial u}{\partial x}\right)_y + \left(\frac{\partial u}{\partial y}\right)_x \frac{dy}{dx} \tag{I-5}$$

and applying the restriction of constant *v* Eq. (I-5) assumes the form

$$\left(\frac{\partial u}{\partial x}\right)_v = \left(\frac{\partial u}{\partial x}\right)_y + \left(\frac{\partial u}{\partial y}\right)_x \left(\frac{\partial y}{\partial x}\right)_v \tag{I-6}$$

A necessary and sufficient condition that

$$du = M\,dx + N\,dy \tag{I-7}$$

be an exact differential is given by Euler's theorem:

$$\left(\frac{\partial N}{\partial x}\right)_y = \left(\frac{\partial M}{\partial y}\right)_x \tag{I-8}$$

If this condition is met then *du* is an exact differential (*du*) and

$$M = \left(\frac{\partial u}{\partial x}\right)_y \tag{I-9a}$$

$$N = \left(\frac{\partial u}{\partial y}\right)_x \tag{I-9b}$$

Consequently

$$\frac{\partial^2 u}{\partial x\,\partial y} = \frac{\partial^2 u}{\partial y\,\partial x} \tag{I-10}$$

This equation states in a mathematical form that if *du* is an exact differential, the order of taking the derivatives is immaterial. The change in *u* depends on the initial and final state only and in no way on the path along which the system is passing from the initial to the final state. The integral of du is therefore

$$\int_1^2 du = \Delta u = u(x_2, y_2) - u(x_1, y_1) \tag{I-11}$$

and u is a function of state. If, however, du is an inexact differential, the integral

$$\int_1^2 du = \Delta u = u(x_2, y_2) - u(x_1, y_1)$$

exists only when the path of integration is specified and u is not a function of state (heat and work in thermodynamics).

If a factor t is able to transform an inexact differential du into an exact differential du, the factor t is called an integrating factor:

$$du = t\,du = tM\,dx + tN\,dy \tag{I-12}$$

Temperature, for instance, transforms the inexact differential of heat dq into an exact differential of entropy dS ($dS = dq_{rev}/T$), so that temperature is an integrating factor.

If there are more than two independent variables, for instance, $u = g(x, y, z)$, Eq. (I-2) can be easily extended into

$$du = \left(\frac{\partial u}{\partial x}\right)_{y,z} dx + \left(\frac{\partial u}{\partial y}\right)_{x,z} dy + \left(\frac{\partial u}{\partial z}\right)_{x,y} dz$$
$$= M\,dx + N\,dy + L\,dz \tag{I-13}$$

If

$$u = f(x, y) \tag{I-14}$$

but

$$x = g(z, v) \tag{I-15}$$

and

$$y = h(z, v) \tag{I-16}$$

it can be shown that

$$\left(\frac{\partial u}{\partial z}\right)_v = \left(\frac{\partial u}{\partial x}\right)_y \left(\frac{\partial x}{\partial z}\right)_v + \left(\frac{\partial u}{\partial y}\right)_x \left(\frac{\partial y}{\partial z}\right)_v \tag{I-17}$$

and

$$\left(\frac{\partial u}{\partial v}\right)_z = \left(\frac{\partial u}{\partial x}\right)_y \left(\frac{\partial x}{\partial v}\right)_z + \left(\frac{\partial u}{\partial y}\right)_x \left(\frac{\partial y}{\partial v}\right)_z \tag{I-18}$$

Derivatives and Integrals

Note that u and v are functions of x only and a is a constant.

(*Derivatives*)

$$\frac{d(a)}{dx} = 0 \qquad\qquad \frac{d(au)}{dx} = a\frac{du}{dx}$$

$$\frac{d(x^n)}{dx} = nx^{n-1} \qquad\qquad \frac{d(u^n)}{dx} = nu^{n-1}\frac{du}{dx}$$

$$\frac{d(e^x)}{dx} = e^x \qquad\qquad \frac{d(e^u)}{dx} = e^u\frac{du}{dx}$$

$$\frac{d(a^x)}{dx} = a^x \ln a \qquad\qquad \frac{d(\ln x)}{dx} = \frac{1}{x}$$

$$\frac{d(a^u)}{dx} = a^u \ln a\frac{du}{dx} \qquad\qquad \frac{d(\log_{10} x)}{dx} = \frac{1}{2.3026\,x}$$

$$\frac{d(\ln u)}{dx} = \frac{1}{u}\frac{du}{dx} \qquad\qquad \frac{d(\log u)}{dx} = \frac{1}{2.3026\,u}\frac{du}{dx}$$

$$\frac{d(u+v)}{dx} = \frac{du}{dx} + \frac{dv}{dx}$$

$$\frac{d(uv)}{dx} = u\frac{dv}{dx} + v\frac{du}{dx}$$

$$\frac{d(u/v)}{dx} = \frac{1}{v}\frac{du}{dx} - \frac{u}{v^2}\frac{dv}{dx}$$

$$\frac{d(\sin x)}{dx} = \cos x \qquad\qquad \frac{d(\sin u)}{dx} = \cos u\frac{du}{dx}$$

$$\frac{d(\cos x)}{dx} = -\sin x \qquad\qquad \frac{d(\cos u)}{dx} = -\sin u\frac{du}{dx}$$

(*Integrals*)

$$\int dx = x + C \qquad\qquad \int x^n\,dx = \frac{x^{n+1}}{n+1} + C \qquad (n \neq 1)$$

$$\int \frac{dx}{x} = \ln x + C \qquad\qquad \int e^x\,dx = e^x + C$$

$$\int a^x\,dx = \frac{a^x}{\ln a} + C \qquad\qquad \int \ln x\,dx = x \ln x - x + C$$

936

$$\int \sin x \, dx = -\cos x + C$$

$$\int \cos x \, dx = \sin x + C$$

$$\int (ax+b)^n \, dx = \frac{(ax+b)^{n+1}}{a(n+1)} + C \qquad (n \neq 1)$$

$$\int \frac{dx}{ax+b} = \frac{\ln(ax+b)}{a} + C$$

$$\int \frac{x \, dx}{ax+b} = \frac{x}{a} - \frac{b}{a^2}\ln(ax+b) + C$$

$$\int \frac{x^2 \, dx}{ax+b} = \frac{1}{a^3}\left[\frac{(ax+b)^2}{2} - 2b(ax+b) + b^2\ln(ax+b)\right] + C$$

$$\int au \, dx = a\int u \, dx$$

$$\int (u+v) \, dx = \int u \, dx + \int v \, dx$$

$$\int u \, dv = uv - \int v \, du$$

$$\int e^{ax}x^n \, dx = \frac{n!e^{ax}}{a^{n+1}}\left[\frac{(ax)^n}{n!} - \frac{(ax)^{n-1}}{(n-1)!} + \frac{(ax)^{n-2}}{(n-2)!} + (-1)^r\frac{(ax)^{n-r}}{(n-r)!} + \cdots + (-1)^n\right] + C$$

The value of the definite integral $\int_0^\infty e^{-ax^2}x^n \, dx$ is different for odd and even values of n

Even Values of n	Odd Values of n
$\int_0^\infty e^{-ax^2} \, dx = \frac{1}{2}\sqrt{\pi/a}$	$\int_0^\infty e^{-ax^2}x \, dx = 1/2a$
$\int_0^\infty e^{-ax^2}x^2 \, dx = \frac{1}{4}\sqrt{\pi/a^3}$	$\int_0^\infty e^{-ax^2}x^3 \, dx = 1/2a^2$
$\int_0^\infty e^{-ax^2}x^4 \, dx = \frac{3}{8}\sqrt{\pi/a^5}$	$\int_0^\infty e^{-ax^2}x^5 \, dx = 1/a^3$

The general formula of the integral for even values of n is

$$\int_0^\infty e^{-ax^2}x^n \, dx = 1\cdot 3\cdot 5\cdots(n-1)\frac{(\pi a)^{1/2}}{(2a)^{(1/2)(n+1)}}$$

and for odd values of n

$$\int_0^\infty e^{-ax^2}x^n \, dx = \frac{[1/2(n-1)]!}{2a^{(1/2)(n+1)}}$$

Frequently Used Expansions

(*Binomial*)

$$(1+x)^n = 1 + nx + \frac{n(n-1)}{2!} \cdot x^2 + \frac{n(n-1)(n-2)}{3!} \cdot x^3 + \cdots$$

$$(1-x)^n = 1 - nx + \frac{n(n-1)}{2!} \cdot x^2 - \frac{n(n-1)(n-2)}{3!} \cdot x^3 + \cdots$$

$$(1+x)^{-n} = 1 - nx + \frac{n(n+1)}{2!} \cdot x^2 - \frac{n(n+1)(n+2)}{3!} \cdot x^3 + \cdots$$

$$(1-x)^{-n} = 1 + nx + \frac{n(n+1)}{2!} \cdot x^2 + \frac{n(n+1)(n+2)}{3!} \cdot x^3 + \cdots$$

$$(1+x)^{-1} = 1 - x + x^2 - x^3 + \cdots$$

$$(1-x)^{-1} = 1 + x + x^2 + x^3 + \cdots$$

(*Logarithmic*)

$$\ln(1+x) = x - \tfrac{1}{2}x^2 + \tfrac{1}{3}x^3 - \tfrac{1}{4}x^4 + \cdots$$

$$\ln(1-x) = -(x + \tfrac{1}{2}x^2 + \tfrac{1}{3}x^3 + \tfrac{1}{4}x^4 + \cdots$$

$$\ln\frac{1+x}{1-x} = 2\left(x + \frac{x^3}{3} + \frac{x^5}{5} + \cdots\right)$$

(*Exponential*)

$$e^x = 1 + x + \frac{x^2}{2!} + \frac{x^3}{3!} + \cdots$$

$$e^{-x} = 1 - x + \frac{x^2}{2!} - \frac{x^3}{3!} + \cdots$$

$$e^{ix} = 1 + ix - \frac{x^2}{2!} - \frac{ix^3}{3!} + \frac{x^4}{4!} + \frac{ix^5}{5!} + \cdots$$

$$e^{-ix} = 1 - ix - \frac{x^2}{2!} + \frac{ix^3}{3!} + \frac{x^4}{4!} - \frac{ix^5}{5!} + \cdots$$

(*Trigonometric*)

$$\sin x = x - \frac{x^3}{3!} + \frac{x^5}{5!} - \cdots$$

$$\cos x = 1 - \frac{x^2}{2!} + \frac{x^4}{4!} - \cdots$$

$$\tan x = x + \tfrac{1}{3}x^3 + \tfrac{2}{15}x^5 + \cdots$$

$$\operatorname{cosec} x = \frac{1}{x}(1 + \tfrac{1}{6}x^2 + \tfrac{7}{360}x^4 + \cdots)$$

$$\sec x = 1 + \tfrac{1}{2}x^2 + \tfrac{5}{24}x^4 + \cdots$$

$$\cot x = \frac{1}{x}(1 - \tfrac{1}{3}x^2 - \tfrac{1}{45}x^4 - \cdots)$$

(*Hyperbolic*)

$$\sinh x = \tfrac{1}{2}(e^x - e^{-x}) = x + \frac{x^3}{3!} + \frac{x^5}{5!} + \cdots$$

$$\cosh x = \tfrac{1}{2}(e^x + e^{-x}) = 1 + \frac{x^2}{2!} + \frac{x^4}{4!} + \cdots$$

$$\tanh x = x - \tfrac{1}{3}x^3 + \tfrac{2}{15}x^5 - \cdots$$

$$\operatorname{cosech} x = \frac{1}{x}(1 - \tfrac{1}{6}x^2 + \tfrac{7}{360}x^4 - \cdots)$$

$$\operatorname{sech} x = 1 - \tfrac{1}{2}x^2 + \tfrac{5}{24}x^4 - \cdots$$

$$\coth x = \frac{1}{x}(1 + \tfrac{1}{3}x^2 - \tfrac{1}{45}x^4 + \cdots)$$

Taylor's Expansion of $f(x)$

$$f(x) = f(a) + f'(a)(x-a) + f''(a)\frac{(x-a)^2}{2!} + f^{(3)}(a)\frac{(x-a)^3}{3!} + \cdots + f^{(k)}(a)\frac{(x-a)^k}{k!} + \cdots$$

Maclaurin's Expansion of $f(x)$

$$f(x) = f(0) + f'(0)x + f''(0)\frac{x^2}{2!} + f^{(3)}(0)\frac{x^3}{3!} + \cdots + f^{(k)}(0)\frac{x^k}{k!} + \cdots$$

The first, second, kth derivatives of $f(x)$ are denoted by $f'(x)$, $f''(x)$, and $f^{(k)}(x)$. As is obvious the Taylor expansion and the Maclaurin expansion are equal when $a = 0$.

It must be understood that series are valid only for a certain range of values called the range of convergence.

Complex Numbers

A complex number c is composed of two real numbers, a and b, and a pure imaginary unit i:

$$c = a + ib \tag{IV-1}$$

where $i^2 = -1$.

The algebra of complex numbers is very much the same as the algebra of real numbers. To prove this let us consider the following two complex numbers: $c_1 = a_1 + ib$, and $c_2 = a_2 + ib_2$.

The sum of the two numbers is

$$c_1 + c_2 = (a_1 + a_2) + i(b_1 + b_2) \tag{IV-2}$$

Their product is

$$c_1 c_2 = (a_1 a_2 - b_1 b_2) + i(a_1 b_2 + a_2 b_1) \tag{IV-3}$$

Their ratio is [multiply the numerator and denominator by $(a_2 - ib_2)$].

$$\frac{c_1}{c_2} = \frac{a_1 + ib_1}{a_2 + ib_2} = \frac{a_1 a_2 + b_1 b_2}{a_2^2 + b_2^2} + i\left[\frac{a_2 b_1 - a_1 b_2}{a_2^2 + b_2^2}\right] \tag{IV-4}$$

A frequently applied formula of complex algebra is the Euler equation:

$$e^{\pm i\phi} = \cos\phi \pm i \sin\phi \tag{IV-5}$$

Obviously, any complex number $a + ib$ can be expressed alternatively in the form $A e^{i\phi}$ where A and ϕ are real, since a solution always exists for the relation

$$A e^{i\phi} = A \cos\phi + i(A \sin\phi) = a + ib \tag{IV-6}$$

Complex numbers may be plotted into a graph $(x - y$ plane), Figure IV-1, x coordinate being the real and the y coordinate the imaginary part. A and ϕ are the

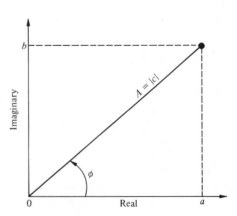

Figure IV-1 Graph of a complex number, $(a + bi)$, in a plane.

polar coordinates of the point representing a complex number, whereas a and b are the rectangular coordinates of the same point. Multiplication by a real number moves a point directly away from the origin: Multiplication by $e^{i\phi}$ rotates a point through an angle ϕ counterclockwise.

The complex conjugate c^* of a complex number c is obtained by replacing i by $-i$. Thus (c^*c) is a real number; its square root is called the magnitude, or absolute value of c, and is often represented by the symbol $|c|$. Obviously $|c| = A$ for the complex number in Eq. (IV-6).

The Error Function

The error function is defined as

$$\text{erf}(x) = \frac{2}{\sqrt{\pi}} \int_0^x e^{-u^2} \, du \tag{V-1}$$

If the upper limit is extended to $x \to \infty$, the integral is $\frac{1}{2}\sqrt{\pi}$ so that $\text{erf}(\infty) = 1$. It follows from this that, as x varies from zero to infinity, the erf (x) varies from zero to unity. By adding to both sides of Eq. (V-1) the integral from x to ∞ multiplied by $2/\sqrt{\pi}$, we obtain

$$\text{erf}(x) + \frac{2}{\sqrt{\pi}} \int_x^\infty e^{-u^2} \, du = \frac{2}{\sqrt{\pi}} \left[\int_0^x e^{-u^2} \, du + \int_x^\infty e^{-u^2} \, du \right] = \frac{2}{\sqrt{\pi}} \int_0^\infty e^{-u^2} \, du = 1 \tag{V-3}$$

Consequently

$$\frac{2}{\sqrt{\pi}} \int_x^\infty e^{-u^2} \, du = 1 - \text{erf}(x) \tag{V-4}$$

We define the co-error function, erfc (x), by

$$\text{erfc}(x) = 1 - \text{erf}(x)$$

Thus

$$\int_x^\infty e^{-u^2} \, du = \frac{\sqrt{\pi}}{2} \text{erfc}(x) \tag{V-5}$$

The Method of Lagrange Undetermined Multipliers

Let us consider a continuous function $f(x_1, x_2, x_2, \ldots, x_n)$ of a set of n variables, $x_1, x_2, x_3, \ldots, x_n$. For an infinitesimal change in this function we write

$$df = \left(\frac{\partial f}{\partial x_1}\right)_{x_{i\neq 1}} dx_1 + \left(\frac{\partial f}{\partial x_2}\right)_{x_{i\neq 2}} dx_2 + \cdots + \left(\frac{\partial f}{\partial x_n}\right)_{x_{i\neq n}} dx_n$$

$$= \sum_{i=1}^{i=n} \left(\frac{\partial f}{\partial x_i}\right)_{x_{j\neq i}} dx_i \qquad \text{(VI-1)}$$

If the function has an extremum (maximum, minimum) df is zero for a change of any of the independent variables, dx_i. Consequently, the location of an extremum can be found by solving the n equations:

$$\left(\frac{\partial f}{\partial x_1}\right)_{x_2, x_3, \ldots, x_n} = 0 \qquad \text{(VI-2a)}$$

$$\left(\frac{\partial f}{\partial x_2}\right)_{x_1, x_3, \ldots, x_n} = 0 \qquad \text{(VI-2b)}$$

$$\left(\frac{\partial f}{\partial x_n}\right)_{x_1, x_2, x_3, \ldots} = 0 \qquad \text{(VI-2c)}$$

The Lagrange method of undetermined multipliers is a technique for maximizing and minimizing a function $f(x_1, x_2, \ldots, x_n)$ when relations (constraints) among the variables $x_1, x_2, \ldots, x_n$ exist. If constraining conditions, such as

$$g(x_1, x_2, \ldots, x_n) = 0 \qquad \text{(VI-3)}$$

and

$$h(x_1, x_2, \ldots, x_n) = 0 \qquad \text{(VI-4)}$$

exist, $x_1, x_2, \ldots, x_n$ cannot all be independently varied; the dx_i's cannot take on any value; and consequently all $(\partial f/\partial x_i)_{x_{j\neq i}}$ cannot be equal to zero. If there are n variables and n' relations among them, $(n > n')$, then only $(n - n')$ variables are independent.

The Lagrange method asserts that solution of the set of $n + 2$ equations of the type

$$\left(\frac{\partial f}{\partial x_i}\right) + \lambda_1 \frac{\partial g}{\partial x_i} + \lambda_2 \frac{\partial h}{\partial x_i} = 0 \qquad \text{(for } i = 1, 2, \ldots, n) \qquad \text{(VI-5)}$$

$$g(x_1, x_2, \ldots, x_n) = 0 \qquad \text{(VI-3)}$$

$$h(x_1, x_2, \ldots, x_n) = 0 \qquad \text{(VI-4)}$$

will yield values of $x_1, x_2, \ldots, x_n$, that locate an extremum subject to the constraints.

λ_1 and λ_2 are the Lagrange undetermined multipliers that are constants, not functions of x_i's.

For better understanding of the method read Example VI-1 carefully.

EXAMPLE VI-1

The elevation of a mountain as a function of the Cartesian coordinates of the ground is given by the equation

$$z = e^{-(x^2+y^2)} \qquad \text{(a)}$$

(a) Find the x and y coordinates at the highest point on the mountain (no constraints).

(b) If a path characterized by the equation

$$y + x - 2 = 0 \qquad \text{(b)}$$

goes across the mountain (not across the summit), what is the highest point reached?

Solution:

(a) To find the highest point simply take the derivative of the function z (or of the natural logarithm of z since the maxima of $f = \ln z$ and z coincide) with respect to x at constant y and with respect to y at constant x and solve the resulting equations for x and y.

$$\left(\frac{\partial f}{\partial x}\right)_y = \left(\frac{\partial \ln z}{\partial x}\right)_y = -2x = 0$$

$$\left(\frac{\partial f}{\partial y}\right)_x = \left(\frac{\partial \ln z}{\partial y}\right)_x = -2y = 0$$

From this $x = 0$ and $y = 0$.

(b) Differentiate the equation characterizing the mountain shape to obtain

$$-d \ln z = df = 2x\, dx + 2y\, dy = 0 \qquad \text{(c)}$$

The mountain path restriction leads to

$$dx + dy = 0 \qquad \text{(d)}$$

Multiplying the last equation by λ and adding it to Eq. (c), we obtain

$$(2x + \lambda)\, dx + (2y + \lambda)\, dy = 0 \qquad \text{(e)}$$

Since dx and dy are now independent Eq. (e) can be true for any x and y if and only if

$$(2x + \lambda) = (2y + \lambda) = 0$$

Consequently

$$x = -\frac{\lambda}{2} \qquad y = -\frac{\lambda}{2}$$

Substitution of these values of x and y into Eq. (b) yields $\lambda = -2$, $x = 1$, and $y = 1$.
The highest point, subject to the path restriction, is at $x = 1$ and $y = 1$.   ●

Stirling's Approximation

To evaluate the factorial of a large number, Stirling's formula is employed. Its derivation follows:

$$\ln N! = \ln N + \ln (N-1) + \ln (N-2) + \cdots + \ln 2 + \ln 1 = \sum_{i=1}^{N} \ln i$$

For large value of N the sum can be replaced by an integral to obtain

$$\ln N! \approx \int_{1}^{N} \ln x \, dx$$

where x is a continuous function.

Integration by parts, with $u = \ln x$, $du = d \ln x$ and $v = x$, $dv = dx$, results in

$$\ln N! \approx x \ln x]_{1}^{N} - \int_{1}^{N} x \frac{1}{x} \, dx$$

$$\approx N \ln N - N$$

or

$$N! \approx N^{N} e^{-N}$$

A more accurate procedure leads to

$$N! \approx N^{N} e^{-N} \sqrt{2N\pi} \left[1 + \frac{1}{12N} + \frac{1}{288N^{2}} + \cdots \right]$$

Method of Least Squares

Suppose the values y and x are determined experimentally in n trials giving n pairs of values of y and x. If the function relating x to y is linear

$$y = ax + b$$

the method of least squares enables one to find the best values for a and b consistent with the experimental data. If we introduce a deviation Δ, defined by the equation

$$\Delta = y - (ax + b)$$

then, for the n trials we can write, after squaring

$$\sum \Delta^2 = \sum y^2 + a^2 \sum x^2 + nb^2 - 2a \sum xy - 2b \sum y + 2ab \sum x$$

The best values for a and b are obtained by minimizing $\sum \Delta^2$ with respect to a and b; that is, by setting the first derivatives of Δ^2 with respect to a and b, respectively, equal to zero. We find

$$\frac{\partial \sum \Delta^2}{\partial a} = 2a \sum x^2 - 2 \sum xy + 2b \sum x = 0$$

and

$$\frac{\partial \sum \Delta^2}{\partial b} = 2nb - 2 \sum y + 2a \sum x = 0$$

Solution of these equations for a and b results in

$$a = \frac{\sum y \sum x - n \sum yx}{\left(\sum x\right)^2 - n \sum x^2}$$

and

$$b = \frac{\sum yx \sum x - \sum y \sum x^2}{\left(\sum x\right)^2 - n \sum x^2}$$

where n is the number of trials,

$$\sum y = y_1 + y_2 + \cdots + y_n$$
$$\sum x = x_1 + x_2 + \cdots + x_n$$
$$\sum yx = y_1 x_1 + y_2 x_2 + \cdots + y_n x_n$$
$$\sum x^2 = x_1^2 + x_2^2 + \cdots + x_n^2$$

946

SI Units and Fundamental Constants

A Physicochemical Quantities and Units In order to improve communication across interdisciplinary and international boundaries, international scientific organizations, among them the International Union of Pure and Applied Chemistry (IUPAC), are engaged in working out standardized names, symbols, and units for physicochemical quantities.

All measurements of any unknown physicochemical quantity q_u are comparisons of q_u to a known quantity q_k. The result of any measurement is a pure number—the ratio q_u/q_k. We may regard a particular known quantity q_k as a unit ($q^{\ominus}$) of the quantity. Hence, for all quantities an algebraic relation exists between the symbol of it, the number which is a measure of it, and the symbol for its unit:

$$\text{symbol for quantity} = \text{number} \times \text{symbol for unit}$$

As an example, consider the length $l = 5 \times 10^2 \times l^{\ominus} = 5 \times 10^2$ cm. The expression can be written alternatively as $l/\text{cm} = 5 \times 10^2$ or as $l/(10^2\,\text{cm}) = 5$. Therefore, if 5×10^2 or 5 is an entry in a column of a table, the proper heading of the column is l/cm, respectively $l/(10^2\,\text{cm})$. The same considerations apply to the identification of the numbers that label the coordinates of graphs and to equations that contain numerical values, for example

$$\ln(p/\text{MPa}) = 8.7582 - 2099.4/(6.41T/\text{K})$$

The symbol for a physicochemical quantity should be a single letter of the Greek or Roman alphabet, printed in italic (sloping) type. Symbols for units, are printed in Roman (upright) type. (In this text all of these recommendations are followed only in part.)

B International System of Units (SI)[1] The SI units were adopted by the General Conference of Weights and Measures in 1960 for use in science and technology.

The SI system has seven fundamental ("base") units (Table IX-1) from which one can construct various derived units. The most common derived units have special names and symbols (Table IX-2). Standard multiples and submultiples of base and derived units are formed with the help of the prefixes in Table IX-3. The special names of some nonstandard multiples and submultiples are given in Table IX-4. The SI equivalents of some common non-SI units are listed in Table IX-5.

---

[1] SI is the international abbreviation derived from the official name in French "le Système International d'Unités."

Table IX-1 Names and symbols of the SI base units[a]

Physical Quantity	Name of SI Unit	Symbol for SI Unit
length	meter	m
mass	kilogram	kg
time	second	s
electric current	ampere	A
thermodynamic temperature	kelvin	K
amount of substance	mole	mol

[a] The base unit of luminous intensity—the candela (cd)—is not considered in this appendix.

Table IX-2 SI derived units having special names and symbols

Physical Quantity[a]	Name of Unit	Symbol for Unit	Definition of Unit
frequency	hertz	Hz	s^{-1}
energy	joule	J	$m^2 \cdot kg \cdot s^{-2}$
force	newton	N	$m \cdot kg \cdot s^{-2} = J \cdot m^{-1}$
power	watt	W	$m^2 \cdot kg \cdot s^{-3} = J \cdot s^{-1}$
pressure	pascal	Pa	$m^{-1} \cdot kg \cdot s^{-2} = N \cdot m^{-2} = J \cdot m^{-3}$
electric charge	coulomb	C	$s \cdot A$
electric potential difference	volt	V	$m^2 \cdot kg \cdot s^{-3} \cdot A^{-1} = J \cdot s^{-1} A^{-1}$
electric resistance	ohm	Ω	$m^2 \cdot kg \cdot s^{-3} \cdot A^{-2} = V \cdot A^{-1}$
electric conductance	siemens	S	$m^{-2} \cdot kg^{-1} \cdot s^3 \cdot A^2 = \Omega^{-1}$
electric capacitance	farad	F	$m^{-2} \cdot kg^{-1} \cdot s^4 \cdot A^2 = C \cdot V^{-1}$
magnetic flux	weber	Wb	$m^2 \cdot kg \cdot s^{-2} \cdot A^{-1} = V \cdot s$
inductance	henry	H	$m^2 \cdot kg \cdot s^{-2} \cdot A^{-2} = V \cdot s \cdot A^{-1}$
magnetic flux density	tesla	T	$kg \cdot s^{-2} \cdot A^{-1} = V \cdot m^{-2} \cdot s$

[a] The named physical quantity is merely representative; a unit is equally appropriate for any other physical quantity having the same dimension.

Table IX-3 SI prefixes

Submultiple	Prefix	Symbol	Multiple	Prefix	Symbol
10^{-1}	deci	d	10	deca	da
10^{-2}	centi	c	10^2	hecto	h
10^{-3}	milli	m	10^3	kilo	k
10^{-6}	micro	μ	10^6	mega	M
10^{-9}	nano	n	10^9	giga	G
10^{-12}	pico	p	10^{12}	tera	T
10^{-15}	femto	f			
10^{-18}	atto	a			

Table IX-4 Decimal fractions and multiples of SI units having special names

Physical Quantity	Name of Unit	Symbol for Unit	Definition of Unit
length	Ångstrom	Å	10^{-10} m
length	micron[a]	μ	10^{-6} m $= \mu$m
area	barn	b	10^{-28} m^2
force	dyne	dyn	10^{-5} N
pressure	bar	bar	10^5 N·m^{-2}
energy	erg	erg	10^{-7} J
kinematic viscosity	stokes	St	10^{-4} m^2·s^{-1}
viscosity	poise	P	10^{-1} kg·m^{-1}·s^{-1}
magnetic flux[b]	maxwell	Mx	10^{-8} Wb
magnetic flux density[b] (magnetic induction [B])	gauss	G	10^{-4} T
absorbed dose (ionizing radiation)	rad	rd	10^{-2} J·kg^{-1}

[a] The use of the word micron and the symbol μ for μm should be abandoned immediately.

[b] The units for these quantities are not dimensionally identical in the cgs system and in SI. [See C. H. Page: *Amer. J. Phys.*, **38**:421 (1970).]

Table IX-5 Non-SI units

Physical Quantity	Name of Unit	Symbol for Unit	Definition of Unit
length	inch	in.	2.54×10^{-2} m
mass	pound (avoir dupois)	lb	$0.453\ 592\ 37$ kg
force	kilogram-force	kgf	$9.806\ 65$ N
pressure	atmosphere	atm	$101\ 325$ N·m^{-2}
pressure	torr	torr	$(101\ 325/760)$ N·m^{-2}
pressure	conventional millimeter of mercury	mm Hg	$13.5951 \times 980.665 \times 10^{-2}$ N·m^{-2}
energy	British thermal unit	BTU	1055.056 J
energy	kilowatt-hour	kWh	3.6×10^{6} J
energy	thermochemical calorie	cal$_{th}$	4.184 J
activity (radioactive nuclides)	Curie	Ci	3.7×10^{10} s^{-1}
exposure (x or γ rays)	Roentgen	R	2.58×10^{-4} C·kg^{-1}
energy	electron volt	eV	$\approx 1.6022 \times 10^{-19}$ J
mass	unified atomic mass unit	u	$\approx 1.660\ 41 \times 10^{-27}$ kg
electric dipole moment	Debye	D	$\approx 3.335\ 64 \times 10^{-30}$ A·m·s
distance (astronomical)	parsec	pc	$\approx 3.085\ 68 \times 10^{16}$ m
charge per mol	Faraday	F	$\approx 9.648 \times 10^{4}$ C mol^{-1}

C Fundamental Constants and Energy Conversion Factors   *Note*: The following tables contain only SI symbols for physicochemical quantities and units. In the text different symbols have been used for the fundamental constants which appear frequently. These different symbols (with the corresponding SI symbol in parentheses) are: $c(c)$, $e(e)$, $h(h)$, $\hbar(\hbar)$, $N(N_A)$, $\mathscr{F}(F)$, $\mathscr{V}(V_m)$, $R(R)$, and $k(k)$.

Table IX-6 Fundamental constants

Compiled by E. R. Cohen and B. N. Taylor under the auspices of the CODATA Task Group on Fundamental Constants. This set has been officially adopted by CODATA and is reprinted from J. Phys. Chem. Ref. Data, Vol. 2, no. 4, p. 741 (1973) and CODATA Bulletin no. 11 (December 1973).

Quantity	Symbol	Value[a]	Uncert. (ppm)	SI ←Units→	cgs[b]
speed of light in vacuum	c	299792458(1.2)	0.004	$m \cdot s^{-1}$	$10^2 \, cm \cdot s^{-1}$
permeability of vacuum	μ_0	4π		$10^{-7} \, H \cdot m^{-1}$	
		$=12.5663706144$		$10^{-7} \, H \cdot m^{-1}$	
permittivity of vacuum, $1/\mu_0 c^2$	ε_0	8.854187818(71)	0.008	$10^{-12} \, F \cdot m^{-1}$	
fine-structure constant, $[\mu_0 c^2/4\pi](e^2/\hbar c)$	a	7.2973506(60)	0.82	10^{-3}	10^{-3}
	a^{-1}	137.03604(11)	0.82		
elementary charge	e	1.6021892(46)	2.9	$10^{-19} \, C$	$10^{-20} \, emu$
		4.803242(14)	2.9		$10^{-10} \, esu$
Planck constant	h	6.626176(36)	5.4	$10^{-34} \, J \cdot s$	$10^{-27} \, erg \cdot s$
	$\hbar = h/2\pi$	1.0545887(57)	5.4	$10^{-34} \, J \cdot s$	$10^{-27} \, erg \cdot s$
Avogadro constant	N_A	6.022045(31)	5.1	$10^{23} \, mol^{-1}$	$10^{23} \, mol^{-1}$
atomic mass unit, $10^{-3} \, kg \cdot mol^{-1} \, N_A^{-1}$	u	1.6605655(86)	5.1	$10^{-27} \, kg$	$10^{-24} \, g$
electron rest mass	m_e	9.109534(47)	5.1	$10^{-31} \, kg$	$10^{-28} \, g$
		5.4858026(21)	0.38	$10^{-4} \, u$	$10^{-4} \, u$
proton rest mass	m_p	1.6726485(86)	5.1	$10^{-27} \, kg$	$10^{-24} \, g$
		1.007276470(11)	0.011	u	u
ratio of proton mass to electron mass	m_p/m_e	1836.15152(70)	0.38		
neutron rest mass	m_n	1.6749543(86)	5.1	$10^{-27} \, kg$	$10^{-24} \, g$
		1.008665012(37)	0.037	u	u
electron charge to mass ratio	e/m_e	1.7588047(49)	2.8	$10^{11} \, C \cdot kg^{-1}$	$10^7 \, emu \cdot g^{-1}$
		5.272764(15)	2.8		$10^{17} \, esu \cdot g^{-1}$
magnetic flux quantum, $[c]^{-1}(hc/2e)$	Φ_0	2.0678506(54)	2.6	$10^{-15} \, Wb$	$10^{-7} \, G \cdot cm^2$
	h/e	4.135701(11)	2.6	$10^{-15} \, J \cdot s \cdot C^{-1}$	$10^{-7} \, erg \cdot s \cdot emu^{-1}$
		1.3795215(36)	2.6		$10^{-17} \, erg \cdot s \cdot esu^{-1}$
Josephson frequency-voltage ratio	$2e/h$	4.835939(13)	2.6	$10^{14} \, Hz \cdot V^{-1}$	
quantum of circulation	$h/2m_e$	3.6369455(60)	1.6	$10^{-4} \, J \cdot s \cdot kg^{-1}$	$erg \cdot s \cdot g^{-1}$
	h/m_e	7.273891(12)	1.6	$10^{-4} \, J \cdot s \cdot kg^{-1}$	$erg \cdot s \cdot g^{-1}$

Table IX-6 Fundamental constants—(*Continued*)

Quantity	Symbol	Value[a]	Uncert. (ppm)	SI ← Units → cgs[b]		
Faraday constant, $N_A e$	F	9.648456(27)	2.8	10^4 C·mol^{-1}	10^3 emu·mol^{-1}	
		2.8925342(82)	2.8		10^{14} esu·mol^{-1}	
Rydberg constant, $[\mu_0 c^2/4\pi]^2 (m_e e^4/4\pi\hbar^3 c)$	R_∞	1.097373177(83)	0.075	10^7 mol^{-1}	10^5 cm^{-1}	
Bohr radius, $[\mu_0 c^2/4\pi]^{-1}(\hbar^2/m_e e^2)$ $=\alpha/4\pi R$	a_0	5.2917706(44)	0.82	10^{-11} m	10^{-9} cm	
classical electron radius, $[\mu_0 c^2/4\pi](e^2/m_e c^2)$ $=\alpha^3/4\pi R_\infty$	$r_e=\alpha\lambdabar_C$	2.8179380(70)	2.5	10^{-15} m	10^{-13} cm	
Thomson cross section, $(8/3)\pi r_e^2$	σ_e	0.6652448(33)	4.9	10^{-28} m^2	10^{-24} cm^2	
free electron g-factor, or electron magnetic moment in Bohr magnetons	$g_e/2 =$ μ_e/μ_B	1.0011596567(35)	0.0035			
free muon g-factor, or muon magnetic moment in units of $[c](e\hbar/2m_\mu c)$	$g_\mu/2$	1.00116616(31)	0.31			
Bohr magneton, $[c](e\hbar/2m_e c)$	μ_B	9.274078(36)	3.9	10^{-24} J·T^{-1}	10^{-21} erg·G^{-1}	
electron magnetic moment	μ_e	9.284832(36)	3.9	10^{-24} J·T^{-1}	10^{-21} erg·G^{-1}	
gyromagnetic ratio of	γ_p'	2.6751301(75)	2.8	10^8 s^{-1}·T^{-1}	10^4 s^{-1}·G^{-1}	
protons in H$_2$O	$\gamma_p'/2\pi$	4.257602(12)	2.8	10^7 Hz·T^{-1}	10^3 Hz·G^{-1}	
γ_p' corrected for	γ_p	2.6751987(75)	2.8	10^8 s^{-1}·T^{-1}	10^4 s^{-1}·G^{-1}	
diamagnetism of H$_2$O	$\gamma_p/2\pi$	4.257711(12)	2.8	10^7 Hz·T^{-1}	10^3 Hz·G^{-1}	
magnetic moment of protons in H$_2$O in Bohr magnetons	μ_p'/μ_B	1.52099322(10)	0.066	10^{-3}	10^{-3}	
proton magnetic moment in Bohr magnetons	μ_p/μ_B	1.521032209(16)	0.011	10^{-3}	10^{-3}	
ratio of electron and proton magnetic moments	μ_e/μ_p	658.2106880(66)	0.010			
proton magnetic moment	μ_p	1.4106171(55)	3.9	10^{-26} J·T^{-1}	10^{-23} erg·G^{-1}	
magnetic moment of protons in H$_2$O in nuclear magnetons	μ_p'/μ_N	2.7927740(11)	0.38			
μ_p'/μ_N corrected for diamagnetism of H$_2$O	μ_p/μ_N	2.7928456(11)	0.38			
nuclear magneton, $[c](e\hbar/2m_p c)$	μ_N	5.050824(20)	3.9	10^{-27} J·T^{-1}	10^{-24} erg·G^{-1}	
ratio of muon and proton magnetic moments	μ_μ/μ_p	3.1833402(72)	2.3			
muon magnetic moment	μ_μ	4.490474(18)	3.9	10^{-26} J·T^{-1}	10^{-23} erg·G^{-1}	
ratio of muon mass to electron mass	m_μ/m_e	206.76865(47)	2.3			
muon rest mass	m_μ	1.883566(11)	5.6	10^{-28} kg	10^{-25} g	
		0.11342920(26)	2.3	u	u	
Compton wavelength of the electron, $h/m_e c=\alpha^2/2R_\infty$	λ_C	2.4263089(40)	1.6	10^{-12} m	10^{-10} cm	
$\lambdabar_C=\lambda_C/2\pi$ $=\alpha a_0$		3.8615905(64)	1.6	10^{-13} m	10^{-11} cm	

Table IX-6 Fundamental constants—(*Continued*)

Quantity	Symbol	Value[a]	Uncert. (ppm)	SI ← Units →	cgs[b]
Compton wavelength of the proton, $h/m_p c$	$\lambda_{C,p}$ $\chi_{C,p} =$ $\lambda_{C,p}/2\pi$	1.3214099(22) 2.1030892(36)	1.7 1.7	10^{-15} m 10^{-16} m	10^{-13} cm 10^{-14} cm
Compton wavelength of the neutron, $h/m_n c$	$\lambda_{C,n}$ $\chi_{C,n} =$ $\lambda_{C,n}/2\pi$	1.3195909(22) 2.1001941(35)	1.7 1.7	10^{-15} m 10^{-16} m	10^{-13} cm 10^{-14} cm
molar volume of ideal gas at STP	V_m	22.41383(70)	31	10^{-3} m$^3 \cdot$ mol^{-1}	10^3 cm$^3 \cdot$ mol^{-1}
molar gas constant, $p_0 V_m/T_0$ ($T_0 \equiv 273.15$ K; $p_0 \equiv 101325$ Pa $\equiv 1$ atm)	R	8.31441(26) 8.20568(26)	31 31	J·mol^{-1}·K^{-1} 10^{-5} m^3·atm ·mol^{-1}·K^{-1}	10^7 erg·mol^{-1}·K^{-1} 10 cm^3·atm ·mol^{-1}·K^{-1}
Boltzmann constant, R/N_A	k	1.380662(44)	32	10^{-23} J·K^{-1}	10^{-16} erg·K^{-1}
Stefan-Boltzmann constant, $\pi^2 k^4/60\hbar^3 c^2$	σ	5.67032(71)	125	10^{-8} W·m^{-2}· K^{-4}	10^{-5} erg·s^{-1} ·cm^{-2}·K^{-4}
first radiation constant, $2\pi hc^2$	c_1	3.741832(20)	5.4	10^{-16} W·m^2	10^{-5} erg·cm^2 ·s^{-1}
second radiation constant, hc/k	c_2	1.438786(45)	31	10^{-2} m·K	cm·K
gravitational constant	G	6.6720(41)	615	10^{-11} m^3·s^{-2} ·kg^{-1}	10^{-8} cm^3 ·s^{-2}·g^{-1}
ratio, kx-unit to ångström, $\Lambda = \lambda(\text{Å})/\lambda(\text{kxu})$; $\lambda(CuK\alpha_1) \equiv 1.537400$ kxu	Λ	1.0020772(54)	5.3		
ratio, Å* to ångström, $\Lambda^* = \lambda(\text{Å})/\lambda(\text{Å}^*)$; $\lambda(WK\alpha_1) \equiv 0.2090100$ Å*	Λ^*	1.0000205(56)	5.6		

[a] Note that the numbers in parentheses are the one-standard-deviation uncertainties in the last digits of the quoted value computed on the basis of internal consistency and that the unified atomic mass scale $^{12}C \equiv 12$ has been used throughout. In cases where formulas for constants are given (for example, R_∞), the relations are written as the product of two factors. The second factor, in parentheses, is the expression to be used when all quantities are expressed in cgs units, with the electron charge in electrostatic units. The first factor, in brackets, is to be included *only if* all quantities are expressed in SI units.

[b] In order to avoid separate columns for "electromagnetic" and "electrostatic" units, both are given under the single heading "cgs Units." When using these units, the elementary charge e in the second column should be understood to be replaced by e_m or e_e, respectively.

Table IX-7 Energy conversion factors

Quantity	Value[a]	Units	Uncert. (ppm)
1 kg	5.609545(16)	10^{29} MeV	2.9
1 u	931.5016(26)	MeV	2.8
electron mass	0.5110034(14)	MeV	2.8
muon mass	105.65948(35)	MeV	3.3
proton mass	938.2796(27)	MeV	2.8
neutron mass	939.5731(27)	MeV	2.8
1 electron volt	1.6021892(46)	10^{-19} J	2.9
		10^{-12} erg	2.9
	2.4179696(63)	10^{14} Hz	2.6
	8.065479(21)	10^5 m^{-1}	2.6
		10^3 cm^{-1}	2.6
	1.160450(36)	10^4 K	31
voltage-wavelength conversion	1.2398520(32)	10^{-6} eV·m	2.6
		10^{-4} eV·cm	2.6
Rydberg constant, R_∞	2.179907(12)	10^{-18} J	5.4
		10^{-11} erg	5.4
	13.605804(36)	eV	2.6
	3.28984200(25)	10^{15} Hz	0.075
	1.578885(49)	10^5 K	31
Bohr magneton, μ_B	5.7883785(95)	10^{-5} eV·T^{-1}	1.6
	1.3996123(39)	10^{10} Hz·T^{-1}	2.8
	46.68604(13)	m^{-1}·T^{-1}	2.8
		10^{-2} cm^{-1}·T^{-1}	2.8
	0.671712(21)	K·T^{-1}	31
nuclear magneton, μ_N	3.1524515(53)	10^{-8} eV·T^{-1}	1.7
	7.622532(22)	10^6 Hz·T^{-1}	2.8
	2.5426030(72)	10^{-2} m^{-1}·T^{-1}	2.8
		10^{-4} cm^{-1}·T^{-1}	2.8
	3.65826(12)	10^{-4} K·T^{-1}	31

[a] Note that the numbers in parentheses are the one-standard-deviation uncertainties in the last digits of the quoted value computed on the basis of internal consistency.

References

[1] National Bureau of Standards: *J. Chem. Educ.*, **48**:469 (1973). Policy for NBS usage of SI units.

[2] M. L. McGlashan: *Ann. Rev. Phys. Chem.*, **24**:51 (1973). Internationally recommended names and symbols for physicochemical quantities and units.

(Chapter 1)

1. Yes, it does   **2.**(a) At $P \to 0$ both gases have the same value of α: $3.6609 \times 10^{-3}, T^{-1}$; (b) $-273.15°C$
3. 74.19   **4.** 752.1 mm Hg   **5.** 1779°C   **6.** 54.0 kg   **7.** 136; $C_8H_8O_2$ **8.** 0.999   **9.** -21.1
$cm^3 mol^{-1}$. The negative sign indicates that oxygen cannot be treated as a hard sphere gas without
attraction.   **10.** $B = \frac{2}{3}\pi N\sigma^3$ **11.** 2.13×10^{-8} cm   **12.** 712 K   **13.** 33.0 atm   **14.**(a) 17.62 atm;

(b) 17.3 atm   **15.** 30.005   **16.** $\Delta \mathscr{V}_{res} = -\left(B' - \dfrac{C' - B'}{RT}P + \cdots\right)$; $\lim_{P \to 0} \Delta \mathscr{V}_{res} = -B'$ **17.** 159 liter

18. $\alpha \approx \dfrac{1}{T} + \left(\dfrac{2a}{RT} - b\right)/T\mathscr{V}$; $\beta \approx \dfrac{1}{P} + \dfrac{2a - 2bRT}{R^2T^2}$ **19.**(a) $T_B = (A/bR)^{2/3}$; (b) $T_B = \sqrt{6T_c}$ **20.** No
critical point   **24.** 0.278 liter mol^{-1}

25.

Constituent	Weight Per cent	Mole Per cent	Volume Per cent	Mole Fraction
H_2	6.43	50.0	50.0	0.500
CO	67.82	38.0	38.0	0.380
N_2	10.71	6.0	6.0	0.060
CO_2	14.02	5.0	5.0	0.050
CH_4	1.02	1.0	1.0	0.010
$\rho = 0.4260$ g liter^{-1}				

26. $P_{N_2} = 38.67$ atm; $P_{CH_4} = 150.54$ atm; $P = 189.21$ atm

(Chapter 2)

3. (a) 0.465 atm; (b) 5×10^{-44} atm   **4.** $\ln \dfrac{P}{P_0} = \dfrac{Mg}{Rb}\ln\left(\dfrac{T_0 - bz}{T_0}\right)$ **5.** 3.46×10^{-4} **6.** (a) 1839 m s^{-1};
1694 m s^{-1}; 1502 m s^{-1} (b) 952 m s^{-1}; 874 m s^{-1}; 775 m s^{-1} (c) 3970 m s^{-1}; 3646 m s^{-1}; 3231 m s^{-1}
7. (a) 6.84×10^2 m s^{-1}; (b) 290°C   **8.** (a) 7.72×10^{-21} J; 1.85×10^{-21} cal   (b) 5.65×10^{-21} J;
1.35×10^{-21} cal   (c) 1.51×10^{-21} J; 3.61×10^{-22} cal   **9.** 2.04×10^2 m s^{-1} **10.** 7.5×10^{-3} Torr
11. 1.336×10^{-5} atm   **13.** (a) 2.23×10^{27} s^{-1} cm^{-3}; (b) 6.93×10^5 cm   **15.** 2.33×10^{-4} atm

(Chapter 3)

1. $\eta = \dfrac{2r^2(\rho_s - \rho_l)g}{9\upsilon}$ (where ρ_s and ρ_l are the densities of sphere and liquid, respectively, g is gravitational
acceleration and υ is the velocity of fall of the sphere)   **2.** (a) 9×10^{-3} m; (b) 1.6×10^{-4} g cm^{-1} s^{-1}
3. (a) 2.07 Å; (b) 11.13 cm^3 mol^{-1} **4.** 57.1 g mol^{-1} **5.** 3.52 and 4.25 cal cm^{-1} s^{-1} k^{-1} **7.** 7.94 m
8. 5.1×10^{-9} m   **10.** 18.95% oxygen   **11.** 7.47×10^{-10} m

(Chapter 4)

1.(a) 36.8%; (b) 13.5%   **3.** by 44.4 degrees   **5.**(a) $\frac{31}{30}$; (b) $\frac{4}{3}$ **6.** 9 (the vibrational degrees of freedom
are not excited at room temperature)   **9.**(a) 0.083; (b) 0.0046

(Chapter 5)

5.(a) 64.33 J; (b) 64.33×10^7 ergs; (c) 15.37 cal   **6.**(a) 2.43 J K^{-1} mol^{-1}; (b) 111.6 J K^{-1} mol^{-1}
7. 15.92 kJ   **9.**(a) $\mathscr{C}_P = 35.98 + 4.00 \times 10^{-2}t$ (J °C^{-1} mol^{-1}); (b) 41.97 J K^{-1} mol^{-1}; (c) -7995.6 J;
(d) -8394 J (error 4.9%)   **10.**(a) 3146.4 J; (b) 2225.9 J   **13.** -423.00 kJ mol^{-1} **14.** -2058.1 kJ,
19.98 kJ   **15.** 16.07 kJ $\pm$ 13%   **16.** $-35,794$ kJ   **17.**(a) 894.1 J; (b) -83.68 J   **18.** -49.08 kJ
mol^{-1} **19.** 262.76 kJ mol^{-1} **21.** -1624.7 kJ   **22.** $\Delta \mathscr{H}_T^\circ = 9992.0 + 0.18T - 2.90 \times 10^{-3}T^2 +$
$9.93 \times 10^{-7}T^3$ (cal)   **23.** $\Delta \mathscr{C}_P = -0.603 - 1.350 \times 10^{-4}T + 1.001 \times 10^5T^{-2}$ **24.** 587 K   **25.**
1800 K   **26.** air/water = 8.89/1   **27.** nitrogen   **28.** 476°C   **30.**(a) 315,000 kg; (b) 168°C
31. 798.7 K; 34.3 atm   **32.** -1251.4 J   **35.** -47.40 kJ   **36.**(a) 162.63 kJ; (b) 10 atm; (c) -163.63 kJ
37. $-q = w' = 72.8$ J

(Chapter 6)

1. 0.375   **2.** $w'_1 = -2145.8$ J; $w'_2 = -1159.0$ J; $w'_3 = 1824.7$ J; $w'_4 = 1159.0$ J; $w' = 321.1$ J; 15%, $\Delta U = 0$ **3.** -1.209 J K^{-1} g^{-1} **4.**(a) 22.00 J K^{-1} mol^{-1}; (b) 25.02 J K^{-1} mol^{-1} **5.** 124.06 J K^{-1} mol^{-1} **6.** 95.17 J K^{-1} **7.** 282.38 J K^{-1} mol^{-1} **8.** 1.12 J K^{-1} mol^{-1} **9.**(a) -1728.0 J; (b) 576 J K^{-1} mol^{-1}; (c) Entropy is the criterion of the reversibility of processes only for an isolated system. Here, we would also have to take the entropy change of the surroundings into consideration. For the total entropy change, we have $\Delta S = 0$.   **10.** 0; 0; $\gamma P/V$ **11.** -0.00334 K atm^{-1} **12.** 1285 K; 866 K   **13.** 48.32 J K^{-1} mol^{-1} **14.** 10.33 J K^{-1} mol^{-1} **15.** $\Delta \mathscr{G} = RT[\ln(200/100) + B(200 - 100) + (C/2)(200^2 - 100^2)$ **16.** -326.3 J   **17.** $\Delta \mathscr{G} = \int \mathscr{V}\, dP = RT \ln(P_2/P_1)$; $\Delta \mathscr{A} = -\int P\, d\mathscr{V} = -RT \ln(\mathscr{V}_2/\mathscr{V}_1) = RT \ln(P_1/P_2)$ **18.** -3100.3 J   **19.** $w' = 3100.3$ J mol^{-1}; $q_P = \Delta \mathscr{H} = -40.67$ kJ mol; $\Delta \mathscr{U} = -3.76$ kJ mol^{-1}; $\Delta \mathscr{G} = 0$; $\Delta \mathscr{A} = 3100.3$ J mol^{-1}; $\Delta \mathscr{S} = -108.8$ J K^{-1} mol^{-1}

(Chapter 7)

1. 280.08 kJ kg^{-1} **2.** 4226.3 J mol^{-1} **5.** 818°C, 36.0 atm   **6.**(a) 47.93 kJ mol^{-1}; (b) water: 0.3168 mm Hg K^{-1}; ice: 0.3615 mm Hg K^{-1} **7.** $A = 19.305$; $B = -2003.4$; $C = -3.984$ **8.**(a) $\Delta \mathscr{H}_{\text{vap}} = 22{,}176 - 29.197T$ (cal mol^{-1}); (b) 13,473 cal mol^{-1} **9.** 721.8 mm Hg   **10.**(a) 34.85 kJ mol^{-1}; (b) 34.63 kJ mol^{-1} **11.**(a) 31.52 kJ mol^{-1}; (b) $\langle 31.40 - 32.89 \rangle$ kJ mol^{-1}; (c) 30.54 kJ mol^{-1} **12.** 40.28 mm Hg

13.

Temperature Interval	k
85–105	1.75
105–125	1.81
125–145	2.02

(Chapter 8)

1.(a) 35.00; (b) 4.23; (c) 4.85; (d) 0.0803; (e) 8.03   **2.**(a) -8.368 kJ mol^{-1}, -63.84 kJ mol^{-1}; (b) 55.23 kJ mol^{-1}, 11.51 kJ mol^{-1} **4.** $\mathscr{V}_1 = 58.36 + 42.98N_2^2 - 117.54N_2^3 + 70.35N_2^4$; $\mathscr{V}_2 = 18.06 + 7.37N_1^2 - 70.06N_1^3 + 70.35N_1^4$ **5.** 63.6 atm   **6.** 199 atm   **7.** 82.3 atm   **10.** 112.4°C   **11.** 4.16 J K^{-1} **12.** 4.31×10^{-5} g   **13.** 177.1; $C_{14}H_{10}$ **14.** 164.6   **15.** $(C_6H_5COOH)_2$ **16.** 6.565×10^{-4} atm   **17.** $\gamma_1 = 1.086$; $\gamma_2 = 1.780$ **18.** $\ln \gamma_1 = 0.331x_2^2 + 1.171x_2^3$; $\ln \gamma_2 = 2.088x_1^2 - 1.171x_1^3$ **21.** 59.15 g; 90.85 g   **22.** 45.20 g   **23.**(a) 95.4°C; (b) 61.4%   **24.** 6.55 kg   **25.** one   **26.** 0.192 g

(Chapter 9)

1. 5.64×10^{35} **2.**(a) From left to right; (b) From right to left   **3.** 54,635 cal mol^{-1} **4.** 0.431   **5.** 43.4%   **6.** CO_2 and H_2 35.1 mol%, CO 29.8%   **7.** 2.47×10^{-2}%   **8.**(a) log $K_p = 2100/T - 6.195$; (b) $\Delta \mathscr{H}° = -9600$ cal mol^{-1}; (c) 23 mol %   **9.** 0.307   **10.** $K_1 = 6.32 \times 10^{-11}$; $K_2 = 4 \times 10^{-21}$ **12.** 3.463×10^3; **13.**(a) log $K = 23.305 - 6680/T$; (b) 100.5°C; (c) 30,568 cal   **14.** $K_3 = 6.125$; 0.49   **15.** 9.56 cal K^{-1} mol^{-1} **17.** 11,061 kcal   **18.** CH_4, 1.12%; H_2O, 44.15%; CO, 3.57%; H_2, 43.07%; CO_2, 8.09%   **19.** -57, 354 cal mol^{-1} **20.**(a) -21.03 kcal mol^{-1}; (b) -16.8 cal mol^{-1} **21.** 6.93 cal K^{-1} mol^{-1} **22.** 46.4 cal K^{-1} mol^{-1}

(Chapter 10)

1. Experimental values for λ_N are $N = 2$, 1,3-butadiene $\lambda = 2170$ Å (theor. 2008); $N = 3$, 1,3,5-hexatriene $\lambda = 2570$ Å (theor. 3270); $N = 4$, 1,3,5,7-octatetraene $\lambda = 2900$ Å (theor. 4462); $N = 5$, 1,3,5,7,9-decapentadiene $\lambda = 3340$ Å (theor. 5704); Agreement between theory and experiment is poor.   **2.** $E_2 - E_1 = 3h^2/8ma^2$ is 1.128×10^2 eV (electron) and 6.141×10^{-2} eV (proton)   **3.** $H_2^+: \mathfrak{H} = -(\hbar^2/2m)\Delta - (e^2/r_A) - (e^2/r_B)$;   He : $\mathfrak{H} = -(\hbar^2/2m)\Delta_1 - (\hbar^2/2m)\Delta_2 - (2e^2/r_1) - (2e^2/r_2) + (e^2/r_{12})$; $H_2: \mathfrak{H} = -(\hbar^2/2m)\Delta_1 - (\hbar^2/2m)\Delta_2 - (e^2/r_{A1}) - (e^2/r_{B1}) - (e^2/r_{A2}) - (e^2/r_{B2}) + (e^2/r_{12})$; the pairs of subscripts 1 and 2, A and B designate electrons and nuclei, respectively.   **4.** 4 eV corresponds to 3099 Å; 80 kcal mole^{-1} corresponds to 3572 Å.   **5.** $\Delta x \cdot \Delta p > h$; $\Delta x \approx a$, $\Delta p \approx h/a$; $E \approx p^2/2m \approx h^2/2ma^2$.   **6.** car: $\Delta x \cdot \Delta v = 6.624 \times 10^{-32}$, $\Delta v = 2.682 \times 10^3$ cm s^{-1}, $\Delta x = 2.47 \times 10^{-35}$ cm; bullet: $\Delta x \cdot \Delta v = 6.624 \times 10^{-28}$, $\Delta v = 6.096 \times 10^4$ cm s^{-1}, $\Delta x = 1.09 \times 10^{-32}$; $H_2: \Delta x \cdot \Delta v = 1.979 \times 10^{-3}$; $\Delta v = 2 \times 10^6$ cm s^{-1}; $\Delta x = 9.90 \times 10^{-10}$ cm   **7.** According to Eq. (10-105) $N = 7$, hence $C_{14}H_{16}$ **8.** Two throws, relative weights are 2(1), 3(2), 4(3), 5(4), 6(5), 7(6), 8(5), 9(4), 10(3), 11(2), 12(1). Three throws 3(1), 4(3), 5(6), 6(10), 7(15), 8(21), 9(25),

10(27), 11(27), 12(25), 13(21), 14(15), 15(10), 16(6), 17(3), 18(1) **9.** $\int_0^a \sin(\pi x/a) \sin(3\pi x/a)\, dx = (a/\pi) \times$ $\int_0^\pi \sin y \sin 3y\, dy = (a/2\pi)[\int_0^\pi \cos 2y\, dy - \int_0^\pi \cos 4y\, dy] = 0$ **10.** (d^2/dx^2) is Hermitian; (d/dx) is not Hermitian **11.** Only the first four lines of the Balmer series, 15,233 cm^{-1}, 20,565 cm^{-1}, 23,032 cm^{-1}, 24,373 cm^{-1} **12.** 10,999 Å **13.** $(x^2 - 4x + 5)^{-1} = [(x - 2)^2 + 1]^{-1}$; $\int (t^2 + 1)^{-1} dt = \arctan t$; Prob. between $X_0 + 1$ and $X_0 - 1$ is $90°/180° = 0.5$; Prob. between $X_0 + \frac{1}{2}$ and $X_0 - \frac{1}{2}$ is $(26° 34'/90°) \approx$ 0.3 $(\tan 26° 34' = \frac{1}{2})$ **14.** $x_m = 6$; $x_0 = 7$. At $x_0 = 7$.

(Chapter 11)

1. H_2: $\omega = 8.278 \times 10^{14}$, $k = 5.73 \times 10^5$ dyne cm^{-1}; HBr: $\omega = 4.991 \times 10^{14}$, $k = 4.12 \times 10^5$ dyne cm^{-1} **2.** $\mathfrak{H}\psi_0 = (-\hbar^2/2m)(4\alpha^2 x^2 - 2\alpha)\psi_0 + \frac{1}{2}kx^2\psi_0 = \varepsilon\psi_0$; $\mathfrak{H}\psi_1 = (-\hbar^2/2m)(4\alpha^2 x^2 - 6\alpha)\psi_1 + \frac{1}{2}kx^2\psi_1 = \varepsilon\psi_1$; $\alpha = (km)^{1/2}/2\hbar$; $\varepsilon_0 = 2\alpha\hbar^2/2m$; $\varepsilon_1 = 6\alpha\hbar^2/2m$ **3.** $\omega = (k/m)^{1/2}$; $\omega_D = \omega_H/\sqrt{2} = 3108$ cm^{-1} **4.** $\omega_{DF} = 2999$ cm^{-1} **5.** $\langle x^2 \rangle = \hbar/2(km)^{1/2}$; $\langle p^2 \rangle = 2(km)^{1/2}\hbar$; $\Delta x \cdot \Delta p = \hbar$ **6.** $(\Delta q)^2 = \hbar/2(km)^{1/2} = \hbar\lambda/4\pi mc$; $\Delta q = 6.521 \times 10^{-10}$ cm **7.** On the HCl axis, 0.0354 Å away from the Cl nucleus. **8.** According to Eq. (11-70), $r(H_2) = 0.7414$ Å; $r(HD) = 0.7410$ Å; $r(D_2) = 0.7411$ Å; differences are minor **9.** $^1H^{11}B$, rotational constant $B = 12.02$ cm^{-1} **10.** Ground state $R_0 = 1.262$ Å, excited state $R_1 = 1.304$ Å **11.** $F_3(x, y, z) = c_1(x^3 - 3xz^2) + c_2(x^3 - 3xy^2) + c_3(y^3 - 3xy^2) + c_4(y^3 - 3yz^2) + c_5(z^3 - 3zx^2) + c_6(z^3 - 3zy^2) + c_7xyz$ **12.** $v_0 = 2.187 \times 10^8$ cm s^{-1}; $p_0 = \hbar/a_0 = 1.992 \times 10^{-9}$ g cm s^{-1}; $T_0 = p_0^2/2m = e^2/2a_0 = -$ground state energy of H atom **13.** $(h/i)(x\, \partial/\partial y - y\, \partial/\partial x)(x \pm iy) = \pm\hbar(x \pm iy)$; $(M_z)_{op}z = 0$ **14.** $E_J = (\hbar^2/2ma_0^2)J(J + 1) = (e^2/2a_0)J(J + 1)$; $E_1 - E_0 = 2 \cdot (e^2/2a_0) =$ two times minus the ground state energy of the hydrogen atom **15.** Take m as electron mass, M_1 as proton mass and M_2 as mass of He nucleus. Reduced masses μ_H and μ_{He} are $\mu_H = m(1 + m/M_1)^{-1}$ and $\mu_{He} = m(1 + m/M_2)^{-1}$. Define $a_0 = \hbar^2/me^2$ and $\varepsilon_0 = e^2/a_0$. Units of length and energy are $a_H = (1 + m/M_1)a_0$; $a_{He} = (1 + m/M_2)(a_0/2)$; $\varepsilon_H = (1 + m/M_1)^{-1}\varepsilon_0$; $\varepsilon_{He} = (1 + m/M_2)^{-1}(4\varepsilon_0)$ **16.** $e^2/a_0 = 27.2$ eV $= 4.358 \times 10^{-11}$ erg mol^{-1} **17.** $R_H = 2\pi^2 e^4\mu/ch^3$ cm$^{-1} = 109,100$ cm^{-1}; exp $R_H = 109,677.581$ cm^{-1} **18.** $\int_0^\infty e^{-r}(r - 2)e^{-r/2}r^2\, dr = \int_0^\infty (r^3 - 2r^2)e^{-3r/2}\, dr = (2^4 \cdot 3!/3^4) - (2 \cdot 2^3 \cdot 2!/3^3) = 0$ **19.** $\phi(4s) = F_0(x, y, z)g_{4,0}(r) = c(r^3 - 24r^2 + 144r - 192)\exp(-r/4)$; $\phi(4p) = F_1(x, y, z)g_{4,1}(r) = (ax + by + cz)(r^2 - 20r + 80)\exp(-r/4)$; $\phi(4d) = F_2(x, y, z)g_{4,2}(r) = [c_1(x^2 - z^2) + c_2(y^2 - z^2) + c_3xy + c_4xz + c_5yz](r - 12)\exp(-r/4)$; $\phi(4f) = F_3(x, y, z)g_{4,3}(r) = [c_1(x^3 - 3xz^2) + c_2(x^3 - 3xy^2) + c_3(y^3 - 3xy^2) + c_4(y^3 - 3yz^2) + c_5(z^3 - 3zx^2) + c_6(z^3 - 3zy^2) + c_7xyz]\exp(-r/4)$ **20.** Zero; inconsistent with classical model **21.** $\langle \mathcal{H} \rangle = -(1/2) + (1/2)\delta^2$; minimum $- 1/2$ for $\delta = 0$ **22.** $\langle \mathcal{H} \rangle = (3/2) \times [\sqrt{\alpha} - (2\sqrt{2}/3\sqrt{\pi})]^2 - (4/3\pi)$; $\mathcal{H}_{min} = -(4/3\pi)$; $\alpha_{min} = (8/9\pi)$

(Chapter 12)

1. Electron spin S is 0 or 1. Orbital angular momentum $L = 2$. There is a singlet state $J = 2$ (fivefold degenerate) and three triplet states $J = 3$ (sevenfold degenerate), $J = 2$ (fivefold degenerate) and $J = 1$ (threefold degenerate) **2.** Orbital angular momentum can be 2, 1, or 0. Spin angular momentum can be 0 or 1. We have triplet states with $J = 0, 1, 2$ and singlet states with $J = 0, 1, 2$ ($L = 2$ must have $S = 0$). **3.** He: $\exp(-1.70r)$; O: $\exp(-7.70r)$; I: $\exp(-52.70r)$ **4.** $mc^2 = 81.8 \times 10^{-8}$ ergs mol$^{-1} = 1.88 \times 10^4(e^2/a_0)$; $E(1s) = -1.445(e^2/a_0)$ for He; $-29.645(e^2/a_0)$ for O; $-0.139 \times 10^4(e^2/a_0)$ for I. Relativistic effects become noticeable in iodine. **5.** $S_x(\alpha + \beta) = (\hbar/2)(\alpha + \beta)$; $S_x(\alpha - \beta) = (-\hbar/2)(\alpha - \beta)$ **6.** $S^2 = S_1^2 + S_2^2 + 2(S_{1x}S_{2x} + S_{1y}S_{2y} + S_{1z}S_{2z})$; $S_1^2\alpha_1\alpha_2 = S_2^2\alpha_1\alpha_2 = \frac{3}{4}\hbar^2\alpha_1\alpha_2$; $S_{1z}S_{2z}\alpha_1\alpha_2 = \frac{1}{4}\hbar^2\alpha_1\alpha_2$; $(S_{1x}S_{2x} + S_{1y}S_{2y})\alpha_1\alpha_2 = 0$; $S^2\alpha_1\alpha_2 = 2\hbar^2\alpha_1\alpha_2$ **7.** $S_1^2\alpha_1\beta_2 = S_2^2\alpha_1\beta_2 = \frac{3}{4}\hbar^2\alpha_1\beta_2$; $S_{1z}S_{2z}\alpha_1\beta_2 = -\frac{1}{4}\hbar^2\alpha_1\beta_2$; $(S_{1x}S_{2x} + S_{1y}S_{2y})\alpha_1\beta_2 = \frac{1}{2}\hbar^2\beta_1\alpha_2$; $S^2\alpha_1\beta_2 = \hbar^2(\alpha_1\beta_2 + \beta_1\alpha_2)$ **8.** Even: (1, 2, 3, 4), (1, 3, 4, 2), (1, 4, 2, 3), (2, 1, 4, 3), (2, 3, 1, 4), (2, 4, 3, 1), (3, 1, 2, 4), (3, 2, 4, 1), (3, 4, 1, 2), (4, 1, 3, 2), (4, 2, 1, 3), (4, 3, 2, 1). Odd: (1, 2, 4, 3), (1, 3, 2, 4), (1, 4, 3, 2), (2, 1, 3, 4), (2, 3, 4, 1), (2, 4, 1, 3), (3, 1, 4, 2), (3, 2, 1, 4), (3, 4, 2, 1), (4, 1, 2, 3), (4, 2, 3, 1), (4, 3, 1, 2) **9.**(a) $e^{-pr_1} \cdot r_2 e^{-qr_2} - r_1 e^{-qr_1} \cdot e^{-pr_2}$; (b) $e^{-pr_1} \cdot (r_2 e^{-qr_2} - \lambda e^{-pr_2}) - (r_1 e^{-qr_1} - \lambda e^{-pr_1}) \cdot e^{-pr_2}$ **10.** 1s: $(Z^3/\pi)^{1/2} e^{-Zr}(Z = 2)$; 2s: $(q^5/3\pi)^{1/2} r e^{-qr}$, $q = 0.575$; $\langle 1s|2s \rangle = 24(Z + q)^{-4} \times (Z^3q^5/3)^{1/2} = 0.223489$ **11.** (1s): $\exp(-Z_1 r)$, $Z_1 = 3.70$; (2s): $r\exp(-\frac{1}{2}Z_2 r)$, $Z_2 = 1.95$ **12.** (1s): $\exp(-Z_1 r)$, 2s: $r\exp(-Z_2 r/2)$, $2p_z$: $z\exp(-Z_2' r/2)$; (3s): $r^2\exp(-Z_3 r/3)$, $3p_z$: $2r\exp(-Z_3' r/3)$, 4s: $r^3\exp(-Z_4 r/4)$; Na: $Z_1 = 10.70$, $Z_2 = Z_2' = 6.85$, $Z_3 = 2.20$; K: $Z_1 = 18.70$, $Z_2 = Z_2' = 14.85$, $Z_3' = 7.75$, $Z_4 = 2.20$ **13.** $\Psi_{1/2} = (1/\sqrt{6})\sum_P P\delta_P[(1s)_1\alpha_1(1s)_2\beta_2(2s)_3\alpha_3]$; $\Psi_{-1/2} = (1/\sqrt{6})\sum_P P\delta_P \times [(1s)_1\alpha_1(1s)_2\beta_2(2s)_3\beta_3]$ **14.** $L = 3\sqrt{6}\hbar$, $L = \sqrt{2}\hbar$, and $L = 0$ **15.** $Z_1 = 10.70(1s)$ and $Z_2 = 6.85(2s, 2p)$; $\langle 2p|r|2p \rangle = 5/6.85 = 0.73$ **16.** Consider the point P and the small solid angle $d\Omega$ in Figure 12-2. The corresponding cone slices out two small areas on the surface, namely $dS_1 = R_1^2\, d\Omega$ at the bottom and $dS_2 = R_2^2\, d\Omega$ at the top. The field strengths F_1 and F_2 due to these areas are $F_1 = dS_1/R_1^2 = d\Omega$ and $F_2 = dS_2/R_2^2 = d\Omega$ and they cancel. Consequently all areas on the surface pair up into cancelling field strengths and the field strength at the arbitrary point P is zero.

(Chapter 13)

1. $E = (0, E, 0)$, $E = A \cos [2\pi(\sigma x - \mu t)]$; $H = (0, 0, H)$, $H = A \cos [2\pi(\sigma x - yt)]$ **2.** $1 : E = (0, 0, E)$, Pol. direction $(0, 0, 1)$; II: $E = (-\frac{1}{2}E\sqrt{2}, \frac{1}{2}E\sqrt{2}, 0)$, Pol. direction $(-\frac{1}{2}\sqrt{2}, \frac{1}{2}\sqrt{2}, 0)$ **3.** $v_0 = 4.837 \times 10^{14} \, s^{-1}$, $\lambda_0 = 6197.7$ Å   **4.** $\Delta E = 11,932.5 \, cm^{-1}$, $v = 7.2137 \times 10^7 \, cm \, s^{-1}$ **5.** 6000 Å: $(A/B) = 7.7076 \times 10^{-13}$ (cgs units); 3 cm: $(A/B) = 6.1661 \times 10^{-27}$ (cgs units)   **6.** $\mu_z = 0.24831 ea_0$; $B = 1.2167 \times 10^{19}$ **7.** $(A/B) = 9.26638 \times 10^{-11}$; $A = 1.1226 \times 10^9$ **8.** It is forbidden because the transition moment is zero   **9.** Population inversion and resonance condition   **10.** See section 13-4

(Chapter 14)

1. In terms of M, $^{12}C = 12 M$, $M = 1.66043 \times 10^{-24}$ g; H_2: $\mu = 0.5039125$, $H^{35}Cl$: $\mu = 0.979592$; HI: $\mu = 0.9998845$, $^{14}N^{14}N$: $\mu = 7.001537$ **2.** $v = 0, R_0 = 1.2839$ Å; $v = 1, R_0 = 1.3030$ Å   **3.** transition moment parallel: $^1\Sigma_g \to {}^1\Sigma_u$; transition moment perpendicular: $^1\Sigma_g \to {}^1\Pi_u$ **4.** $\mu = 11.38401 \times 10^{-24}$ g; $R_0 = 1.1282$ Å; $R_1 = 1.2351$ Å   **5.** H_2: 4 K, $1 : 10^{-57} : 10^{-4597}$, 300 K, $1 : 0.1739 : 10^{-15}$; N_2: 4 K, $1 : 0.01308 : 10^{-38}$, 300 K, $1 : 0.9438 : 0.3116$ **6.** HCl, $k = 5.158 \times 10^5$ dyne cm^{-1}; DCl, $\omega = 2144.25 \, cm^{-1}$ **7.** See Section 14-2   **8.** Because the nuclear separation R depends on the electronic state   **9.** Microwave spectrum   **10.** (a) $P_{0,0} = \iint \cos \theta \sin \theta \, d\theta \, d\phi = 0$; (b) $P_{0,0} = \iint \sin \theta \exp (\pm i\phi) \times \sin \theta \, d\theta \, d\phi = 0$ **11.** $B_{0,v} = (\hbar^2/2\mu) \int f_v(q)(R_0 + q)^{-2} f_v(q) \, dq = (\hbar^2/2\mu R_0^2) \int f_v(q)[(1 + 3(q/R_0)^2] f_v(q) \, dq = (\hbar^2/2\mu R_0^2) + (3h^2/2\mu R_0^4) \int f_v(q) q^2 f_v(q) \, dq$ **12.** $k = (d^2 U/dR^2)_0 = 2b^2 D$; $b = (k/2D)^{1/2}$ **13.** $R_0(H_2) = 0.74165 \times 10^{-8}$ cm; $R_0(D_2) = 0.74164 \times 10^{-8}$ cm   **14.** Usually the force constant is smaller.   **15.** Electronic dipole moment and all its derivatives with respect to R are zero.   **16.** Transition moment is parallel to molecular axis, consequently there is no Q branch.   **17.** It is due both to anharmonicity effects and to higher order terms in the power series expansion of the electronic dipole moment in terms of the displacement coordinate q.

(Chapter 15)

1. Because the wave function must be symmetric in z, that is symmetric in v.   **2.** $R = 2$, $S = 0.586453$, $P_A = P_B = 0.315168$, $P_{AB} = 0.369663$; $R = 2.5$, $S = 0.458308$, $P_A = P_B = 0.369663$, $P_{AB} = 0.314273$ **3.** The VB method because it includes some correlation and because the MO method overrepresents ionic structures.   **4.** In the limit both approaches should give the same result.   **5.** $R = 1, S = 0.858385$; $R = 2$, $S = 0.586453$; $R = 4$, $S = 0.189262$; $R = 10$, $S = 0.002013$ **6.** On a proton, $\psi = 0.359601$, at midpoint $\psi = 0.233041$.   **7.** Because it includes correlation.   **8.** If we place the electrons in molecular orbitals, we end up with two electrons to be accommodated in the doubly degenerated $1\pi_u$ orbital; the configuration with the lowest energy has the electrons with parallel spins, each in a different $1\pi_u$ orbital.   **9.** The CH_2 radical has 8 electrons, one pair each occupies the $1s$ carbon orbital and the two C—H bond orbitals; this leaves two electrons for the doubly degenerate carbon π orbital; the configuration with the lowest energy has then two electrons in different π orbitals with parallel spins.   **10.** If CH_2 is bent the two π orbitals of the linear configuration have slightly different energies, but we can still place one electron each in these orbitals with parallel spins and have a triplet state.   **11.** The NH_3 dipole moment is due to the lone pair of electrons on the nitrogen, the NH bond moments have the opposite directions but they are small. The NF bond moments are larger than the NH bond moments and they cancel the lone pair moments more or less.   **12.** The Coulomb repulsion between the NH bonds is smaller than the Coulomb repulsion between a lone pair and a bond. Starting with a tetrahedral configuration the Coulomb repulsion pushes the bonds away from the lone pair and squeezes them together so that the H—N—H bond angle is smaller than the tetrahedral angle.   **13.** Effective nuclear charge $Z = 3.25$; $2s$ orbital: $(Z^5/96\pi)^{1/2} r \exp (-Zr/2)$; $2p_x$ orbital: $(Z^5/32\pi)^{1/2} x \exp (-Zr/2)$ **14.** Effective nuclear charge $Z = 3.90$; $\langle z \rangle = \langle 2s|z|2p_z \rangle = 5/Z\sqrt{3} = 0.7402$ (atomic units)   **15.** The relative magnitudes are $\mu(O—H) > \mu(N—H) > \mu(C—H)$. Bond moments depend on the electronegativity difference, the bond length, and the hybridization of the atomic orbitals. Here the electronegativity is the major effect.   **16.** The carbon hybridized orbitals are sp^3, sp^2 and sp respectively.   **17.** There are 10 valence electrons and we have an electron pair each in the C—H and C—N σ bond orbitals and the nitrogen sp lone pair. There are two mutually perpendicular π bonding C—N orbitals with a pair of electrons in each of them.   **18.** The radical can have π molecular orbitals extending through the whole molecule, whereas in the molecule they are confined to the benzene rings. The increase in resonance energy stabilizes the radical with respect to the molecule.   **19.** The Coulomb repulsion between the nitrogen lone pair and the σ bonds is larger than between the two σ bonds, hence the σ bonds are squeezed together and the C—N—C bond angle is slightly less than 120°.   **20.** $E_n = \alpha + 2\beta x_n$; $x_n = \cos (n\pi/7)$; $x_1 = 0.9010$,

$x_2 = 0.6234$, $x_3 = 0.2226$, $x_4 = -0.2226$, $x_5 = -0.6234$, $x_6 = -0.9010$; resonance energy $= 0.988\beta$
21. $-xa_1 + a_8 + a_9 = 0$; $-xa_6 + a_7 + a_{10} = 0$; $-xa_2 + a_3 + a_9 = 0$; $a_6 - xa_7 + a_8 = 0$; $a_2 - xa_3 + a_4 = 0$; $a_1 + a_7 - xa_8 = 0$; $a_3 - xa_4 + a_5 = 0$; $a_1 + a_2 - xa_9 + a_{10} = 0$; $a_4 - xa_5 + a_{10} = 0$; $a_5 + a_6 + a_9 - xa_{10} = 0$ **22.** 1.428 Å   **23.** $-xa_1 + a_2 = 0$; $a_1 - xa_2 + a_3 = 0$; $a_2 - xa_3 + a_4 = 0$; $a_3 - xa_4 = 0$. The fifth carbon atom is not part of the conjugated system. Equations are the same for butadiene, solutions are too.

(Chapter 16)

1. The samples had unsuitable relaxation times.   **2.** 3398.9 gauss   **3.** 28.00 m   **4.** 1496.8 g
5. 2.043×10^{17} at 300 K and 3.064×10^{18} at 20 K   **6.** See Section 16-2   **7.** $\alpha n(t) - (n'/T_1) = [\alpha n(0) - (n'/T_1)] \exp(-\alpha t)$; $\alpha = 2W - (1/T_1)$ **8.** See Section 16-2   **9.** See Section 16-3   **10.** See Section 16-3   **11.** 2.238×10^2 Hz $= 2.238 \times 10^{-4}$ MHz   **12.** The electron density is largest on HI and smallest on HF because of the electronegativity differences between the halogens.   **13.** No, see Section 16-4.   **14.** No. Either from calculations or from HD.   **15.** The shifts due to proton shielding are proportional to H and the spin–spin couplings are independent of H. Chemical shift effects are predominant at high fields and spin–spin couplings at low fields.   **16.** In Tucson, Ar., H $= 0.528$ gauss, $\lambda = 1.33 \times 10^7$ cm $= 133$ km   **17.** There are 7 resonance lines with intensity ratios $1:6:15:20:15:6:1$.
18. The values of the unpaired electron's wave function at the positions of the various nuclei.   **19.** Because of configuration interaction, see Section 16-5.   **20.** $\mathfrak{H} = \gamma H(I_{1z} + I_{2z} + I_{3z}) + J[(\mathbf{I}_1 \cdot \mathbf{I}_2 + \mathbf{I}_1 \cdot \mathbf{I}_3 + \mathbf{I}_2 \cdot \mathbf{I}_3)] = \gamma H I_z + (1/2)J(I^2 - I_1^2 - I_2^2 - I_3^2)$.

(Chapter 17)

1. Two   **2.** Four Na^+ and four Cl^- **3.** $R(Na–Cl) = 2.813 \times 10^{-8}$ cm; $a = 5.626 \times 10^{-8}$ cm
4. $(\sqrt{3}/6)(4a^2 + 3c^2)^{1/2}$ **5.** Argon, molecular crystal; SiC, covalent crystal; LiH, ionic crystal; ICE is basically covalent; SnS is covalent or ionic   **6.** Germanium because it has the smallest energy gap
7. $R(Na–Cl) = 2.813 \times 10^{-8}$ cm $= 5.3156$ au; $U_0 = 8.946$ eV unit^{-1} **8.** $R(K–Br) = 6.226$ au; $U_0 = 7.64$ eV unit^{-1} **9.** $R(Cs–Cl) = 3.559$ Å $= 6.726$ au; $U_0 = 7.131$ eV unit^{-1} **10.** $\varepsilon_0 = \alpha + 2\beta$; $\varepsilon_1 = \varepsilon_{-1} = \alpha + \beta\sqrt{2}$; $\varepsilon_2 = \varepsilon_{-2} = \alpha$; $\varepsilon_3 = \varepsilon_{-3} = \alpha - \beta\sqrt{2}$; $\varepsilon_4 = \alpha - 2\beta$ **11.** 4.6336×10^{22} d cm^{-3} **12.** $E_F(0) = 3.117$ eV   **13.** See Section 17-3   **14.** See Section 17-4   **15.** See Section 17-4

(Chapter 18)

1.(a) $(1.3–0.27) \times 10^{-23}$ (photon), $(1.4–0.77) \times 10^{-22}$ (rel. electron), $(1.3–0.17) \times 10^{-23}$ (neutron) kg m s^{-1}; (b) 3×10^8 (photon), $(1.3–0.81) \times 10^8$ (electron), 7700–990 (neutron) m s^{-1}; (c) see Table 18-1   **2.**(a) 6.3×10^8 (laser), 63 (x-ray tube) W m^{-2}; (b) 6.9×10^5 (laser), 220 (x-ray tube) N C^{-1} or V m^{-1} **5.** $I_e/I_0 = 8.0 \times 10^{-28}$ (at 0°), 4.0×10^{-28} (at 90°)   **9.** $\lambda = 7.42$ pm at 90°   **10.** $\sigma = e^4/(6\pi\varepsilon_0^2 m^2 c^4) = 6.7 \times 10^{-29}$ m^2 **11.**(a) $y = 5134, 3461, 945$ when $x = 0, \pi/2, \pi$; (b) 228.9 pm (average)

13.

For HCl:	$\langle \xi \rangle_v$	$\langle \xi^2 \rangle_r^{1/2}$	$\langle R \rangle_v$	R_v	$\langle R^{-3} \rangle_v^{-1/3}$
$v = 0$	0.0124	0.0595	129.04	128.36	128.14 pm
$v = 1$	0.0373	0.1031	132.21	130.19	129.50 pm

(Chapter 19)

1. 12,600   **2.** 13.7, 10.8%   **4.**(a) 0.0337; (b) 3×10^{-74} **6.** 9.75×10^{23} **7.** $\ln Q = 6.95 \times 10^{24}$
10. 126.1 J K^{-1} mol^{-1}

11.

Substance	$\dfrac{\mathscr{C}_V}{(\text{J K}^{-1}\,\text{mol}^{-1})}$	$\dfrac{-\left(\dfrac{\mathscr{G} - \mathscr{H}_0}{T}\right)}{(\text{J K}^{-1}\,\text{mol}^{-1})}$	$\dfrac{\mathscr{S}}{(\text{J K}^{-1}\,\text{mol}^{-1})}$
Ne	12.47	125.43	146.22
Ar	12.47	133.95	154.73
Kr	12.47	143.19	163.98

12. 0.111   **13.** $\ln Q = 2.38 \times 10^{24}$; $\ln Q = 2.80 \times 10^{24}$ **15.** 1.23×10^5 K   **17.** $(\mathcal{G} - \mathcal{H}_0)/T = -72.13$ J K^{-1} mol^{-1}; $(\mathcal{H} - \mathcal{H}_0)/T = 12.47$ J K^{-1} mol^{-1}; $\mathcal{C}_P = 12.47$ J K^{-1} mol^{-1}; $\mathcal{S} = 84.6$ J K^{-1} mol^{-1} **18.** $(\mathcal{G} - \mathcal{H}_0)/T = 212.72$ J K^{-1} mol^{-1}; $(\mathcal{H} - \mathcal{H}_0)/T = 35.38$ J K^{-1} mol^{-1}; $\mathcal{S} = 248.1$ J K^{-1} mol^{-1}; $\mathcal{C}_P = 39.82$ J K^{-1} mol^{-1} **19.** $(\mathcal{G} - \mathcal{H}_0)/T = 38.62$ J K^{-1} mol^{-1}; $(\mathcal{H} - \mathcal{H}_0)/T = 8.31$ J K^{-1} mol^{-1}; $\mathcal{C}_P = 8.31$ J K^{-1} mol^{-1}; $\mathcal{S} = 47.22$ J K^{-1} mol^{-1} **20.** $[(\mathcal{G} - \mathcal{H}_0)/T]_{vib} = 1.17$ J K^{-1} mol^{-1}; $[(\mathcal{H} - \mathcal{H}_0)/T]_{vib} = 2.55$ J K^{-1} mol^{-1}; $(\mathcal{C}_P)_{vib} = 5.96$ J K^{-1} mol^{-1}; $\mathcal{S}_{vib} = 3.72$ J K^{-1} mol^{-1} **21.** $\mathcal{U} - \mathcal{U}_0 = 13.53$ kJ mol^{-1}; $\mathcal{C}_V = 12.95$ J K^{-1} mol^{-1}; $\mathcal{S} = 184.9$ J K^{-1} mol^{-1} **22.** 7.43   **23.** 0.607   **24.** 3.27   **28.** 12.7 J K^{-1} mol^{-1} **29.** Debye's theory fits better, θ_D varies between 313 and 320 K; θ_E varies much more

(*Chapter 20*)

2.(a)

$t/$(s)	0	390	777	1195	3155	∞
$P_{(CH_3)_2O}/$(torr)	312	264	224	187	78.5	2.5
$P_{CH_4}/$(torr)	0	48	88	125	233.5	309.5

(b) $n = 1$, $k = 4.4 \times 10^{-4}$ s^{-1} **3.(a)** $\mathcal{E}_a = 12.4$ kcal mol^{-1}, $A = 1.52 \times 10^9$ liter^2 mol^{-2} min^{-1}; **(b)** 60°C (from Figure 20-5)   **4.** 415, 360, and 237°C   **5.(a)** $k_{1/2} = 2(P_0^{1/2} - (P_0 - \Delta P)^{1/2})/t$, $k_1 = t^{-1} \cdot \ln(P_0/(P_0 - \Delta P))$, $k_2 = \Delta P/[tP_0(P_0 - \Delta P)]$; **(b)** $k_1 = 0.0924$ min^{-1} **6.** $n_{NO} = 2$, $n_{H_2} = 1$ **7.(a)** Aqueous: $n = 1$ (MeAc), $k = 6.20 \times 10^{-4}$ min^{-1}; (b) Acetone: $n = 1$ (MeAc and H_2O), $k = 10.0$ M min^{-1} **8.** $n_{CO} = 1$, $n_{Cl_2} = \frac{1}{2}$, $k = 1.0 \times 10^{-3}$ torr$^{-1/2}$ min^{-1} **9.** (a) and (c) are complex; (d) NH_4CNO for (b), $C_6H_5NH_2OHI$ for (c); (e) HBr_2, step 3   **10.** Same rate law as first mechanism   **11.** $r_n = k_0R$ **14.** 0.74 torr   **15.(b)** low T: $r = K_0k_1[H_2 \cdot I_2 - K^{-1}(HI)^2]$; high T: $r = K_0^{1/2}k_2 \cdot \dfrac{H_2 \cdot I_2^{1/2} - K^{-1}(HI)^2/I_2^{1/2}}{1 + k_{-2}k_3^{-1} \cdot HI/I_2}$

16. 38.0 M^{-1} s^{-1} **17.(a)** $k_+/k_- = 0.14$, $K = 0.22$; **(b)** calculated $\Delta\mathcal{H}^{\ominus} = 0.80$ kcal mol^{-1}, $\Delta\mathcal{S}^{\ominus} = -1.3$ eu   **18.** 245°C   **19.** 1.89×10^{-5} J cm^{-3} s^{-1} **21.** 0.31   **22.** $\Phi_{-HBr} = \phi + \phi \times \dfrac{1 - k_4 \cdot Br_2/(k_2 \cdot HBr)}{1 + k_4 \cdot Br_2/(k_2 \cdot HBr)}$ **23.** 1.8×10^{-4} M   **24.(a)** $G_{H_2O_2}(\text{sat. w. } O_2 + H_2) = g_{H_2O_2} + \frac{1}{2}(g_H + g_{e^-(aq)} + g_{OH})$; **(b)** $g_{OH} = G_{H_2O_2}(\text{sat. w. } O_2 + H_2) - G_{H_2O_2}(\text{sat. w. } O_2)$ **25.** 7.3   **26.(a)** $G_{Fe^{3+}} = 3(g_H + g_{e^-(aq)}) + 2g_{H_2O_2} + 3g_{HO_2} + g_{OH}$, $G_{Ce^{3+}} = (g_H + g_{e^-(aq)}) + 2g_{H_2O_2} - g_{OH} + g_{HO_2}$, $G_{H_2} = g_{H_2}$, $G_{O_2} = g_{H_2O_2} + g_{HO_2}$; **(b)** $(g_H + g_{e^-(aq)}) = 0.58$, $g_{H_2O_2} = 1.35$, $g_{OH} = 0.40$, $g_{H_2} = 1.57$, and $g_{-H_2O} = 3.54$ **27.(b)** $k = 0.01415$ min^{-1}; **(c)** $k_+ = 0.00899$ min^{-1}, $k_- = 0.00516$ min^{-1} **28.** 1.27×10^6 M^{-1} s^{-1} **29.(a)** $k_+ = 4.4 \times 10^2$, $k_- = 4.4 \times 10^5$ M^{-1} s^{-1}; **(b)** 23 ms   **30.(a)** $a_e^{578} = 1.2 \times 10^3$ M^{-1} cm^{-1}; **(b)** 1.0×10^{10} M^{-1} s^{-1} **31.** 250(-48.7°C), 64(-60.3°C), 50(-63.2°C), 27(-67.3°C) s^{-1}; $\mathcal{E}_a = 12$ kcal mol^{-1} **33.(b)** 3.8×10^{17} cm^6 mol^{-2} s^{-1} **34.** 7 ns   **36.(a)** $\dfrac{2}{x} \times \dfrac{P_{+(1/2)S} - P_{-(1/2)S}}{P_{+(1/2)S} + P_{-(1/2)S}}$; **(b)** -560 **37.** CH_3CCl_3: -2.9×10^4 (E); CH_3I: 1.4×10^5 (A)

38.

	Spectrum Line			
	1	2	3	4
R_1R_2	A	E	A	E
R_1Sc	E	A	E	A

39. Reactive wall collisions   **40.** $B_{bulk}/B_{wall} = 10^5$ **42.** $V_m = 20$ cm^3 (at STP) g^{-1}; $K' = 0.010$ torr^{-1} **43.** $V_m = 122$ cm^3; $c = 2.5$; $\Sigma = 531$ m^2 g^{-1} **44.(a)** $r = k_1'E_0S/(K_m + S)$ where $K_m = (k_-' + k_+')/k_+'$; **(b)** $k_1 = k_1'$ and $K' = K_m^{-1}$ **45.(a)** Carbolac, 430 pm; P-33, 4300 pm; **(b)** 0.18   **46.** $\Delta\mathcal{S} = 14.7$ cal K^{-1} mol^{-1} **48.(a)** 10^{27} (300 K) and 10^{-9} (3000 K); **(b)** 10^{-3} (300 K) and 10^{-12} (3000 K) s   **49.(a)** (5, 4)   **50.** (310)-plane: rectangular $a \times \sqrt{10}a/2$; (210)-plane: rectangular $a \times \sqrt{5}a$ **51.** r_{max} K 36 μliter CO_2 per mg protein per hr; $K_m = 4.25$ M   **52.(a)** $r = (k_1/K_m)S/(1 + S/K_m + P/K_1)$; **(b)** $K_1 = 9.9 \times 10^{-3}$ M; $K_m = 1.2 \times 10^{-2}$ M   **54.(a)** 1st order, 0 order at $P = \infty$; **(b)** 99 kcal mol^{-1} **55.(a)** $m = 1.0$, $n = -2.0$; **56.** 16.5 (at 300 K), 55 (at 1000 K) kcal mol^{-1}

(Chapter 21)

1.

	(a)	(b)	(c)	(d)	(e)	(f)
0.5 nm	-1.9	-1.0×10^{-3}	-1.0	-0.50	-66	average KE $=$
1 nm	-0.016	-1.6×10^{-4}	-0.016	-0.031	-33	0.81 kcal mol^{-1}

2.(a) $R_m = (-q^2/E_0 + c)^{1/2}/2$; (b) $\theta = -\sin^{-1}[(aR - 2b^2)/(cR)] + \sin^{-1}(a/c)$; (c) $x = \pi + 2\sin^{-1}[(aR - 2b^2)/cR] - 2\sin^{-1}(a/c)$ where $a = q^2/E_0$ and $c = q^4/E_0^2 + 4b^2$ **3.**(a) relative $v = 5.8 \times 10^2\,m\,s^{-1}$, KE $= 7.1 \times 10^{-21}$ J; center-of-mass $v = 2.6 \times 10^2\,m\,s^{-1}$, KE $= 6.3 \times 10^{-21}$ J; (b) relative $v = 5.8 \times 10^2\,m\,s^{-1}$ and KE $= 2.8 \times 10^{-22}$ J, center-of-mass $v = 3.0 \times 10^2\,m\,s^{-1}$ and KE $= 1.8 \times 10^{-21}$ J; (c) relative $v = 4.4 \times 10^2\,m\,s^{-1}$, KE $= 8.1 \times 10^{-21}$ J, center-of-mass $v = 3.2 \times 10^2\,m\,s^{-1}$, KE $= 9.3 \times 10^{-21}$ J **4.**(a) $k = \left(\dfrac{8\pi kT}{\mu}\right)^{1/2} R_m^2\left(1 + \dfrac{e^2}{R_m kT}\right)$: (b) $R_m = 2.4$ nm; (c) $n \approx 6$ **5.**(a) 0.43 nm^2; (b) 7.0×10^9 (at 100 K) and 19×10^9 (at 800 K) M^{-1}s^{-1} **6.**(a) 2.0 (at 100 K) and 0.97 (at 800 K) nm^2; (b) $k = 1.354 N\sigma^2(2\pi k/\mu)^{1/2}(12\varepsilon/k)^{1/3}T^{1/6}$; (c) 2.63 (at 100 K) and 3.7 (at 800 K) 10^{11} M^{-1} s^{-1} **7.**(a) 2.1%(dilute), 19%(concentrated); (b) 9.2% **8.**(a) OH + OH : cage $R_m = 2.5$ Å, hard sphere $R_m = 3.0$ Å, OH + C$_6$H$_6$: cage $R_m = 1.2$ Å, hard sphere $R_m = 3.5$ Å; (b) $k_{d-}/k_c = 2.0$ for second reaction **9.** 5.115 eV **11.**(a) $P_{1/2} = 1.47$ torr, $k_\infty = 10.4 \times 10^{-4}$ s^{-1}; (b) $A_\infty = 1.26 \times 10^{15}$ s^{-1}; (c) $k_a = 2.82 \times 10^{-7}$ M^{-1}s^{-1} (collision theory), $k_a = 30.7$ M^{-1} s^{-1} (experiment) **12.** $k_P = A_\infty = 10^{13.6}$ s^{-1} (theoretical); $k_P = 2.0 \times 10^8$ s^{-1} (expt) **13.**(b) $\mathscr{C}/R = -3/2$; (c) $\mathscr{C}/R = -15/2$

14.

Reaction	$\Delta H^{\ddagger}$/kcal	$\Delta S^{\ddagger}$/(eu)
N$_2$O $\rightarrow$ N$_2$ + O	57.7	-11.9
CO + OH $\rightarrow$ CO$_2$ + H	-1.4	-10.4
SO$_2$ + O + Ar $\rightarrow$ SO$_3$ + Ar	-1.0	-0.67

15. 4, 3, 2, 1 **16.** 1.54×10^6 M^{-1}s^{-1} **17.** 3.95 **18.** 114 **21.**(a) 11 and -7.0 eu; (c) $A/(10^{13}$ s$^{-1}) = 1.7, 3.40, 5.1$

(Chapter 22)

1. 4.5 kcal mol^{-1} **2.** 38.1 (for $v' = 3$) and 3.5 (for $v' = 4$) kJ mol^{-1} **3.** 10^{-15} torr **4.**(a) 10^7; (b) 200 molecules s^{-1} **7.**(a) $n_{100}/n_{000} = 1.3 \times 10^{-3}$, $n_{001}/n_{000} = 1.2 \times 10^{-5}$; (b) T/K = 270 (100 level), 490 (001 level) · **8.** 1.8×10^4 collisions quantum^{-1}

(Chapter 24)

1.(a) 0.208 A; (b) 0.210 A **2.** 108.1 g **3.**(a) 1.76 V cm^{-1}; (b) 65.5×10^{-5} cm^2 V^{-1}s^{-1} **4.** $t_{Ag^+} = 0.421$; $t_{NO_3^-} = 0.579$ **5.**(a) 5.9390×10^{-3} cm^{-1}; (b) $5.7518 \times 10^{-3}\,\Omega^{-1}$ cm^{-1} **6.** $133.36\,\Omega^{-1}$ cm^2 eq^{-1} **7.** 1.334×10^{-5} mol liter^{-1} **10.** $0.01067\,\Omega^{-1}$ cm^{-1} **11.** 1.88 g liter^{-1} **14.**(a) $m_{a\pm} = 13.1 \times 10^{-5}$; (b) $x_{a\pm} = 2.36 \times 10^{-7}$ **15.** 0.920 **17.** 4.5×10^{-6}

(Chapter 25)

1. 1.580 V **3.**(a) 0.2256 V; (b) 1.86 **4.** 0.466 **5.** 0.771 **6.**(a) H$_2$S(g) + $\frac{1}{2}$Mo(s) $\rightleftharpoons$ H$_2$(g) + $\frac{1}{2}$MoS$_2$(s); (b) log $K = 5049.6/T - 3.049$; (c) $\Delta\mathscr{S}^\circ = -13.88$ cal K^{-1} mol^{-1} (excellent agreement with the value calculated from the absolute entropies, -13.88 cal K^{-1} mol^{-1}). **7.**(b) 5.52×10^{-23} **8.** 0.333 **9.** 0.145 V **12.** 2.31 **16.** 2.50 mV

Absolute rate theory. *See* Activated complex theory
Absolute zero, 211–212
  principle of unattainability of, 271
Absorption
  band, 410
  coefficient, 538
  spectrum, 399
Acentric factor, 27
Acetone, 277
Acetylene, 465, 468, 476
Actinometry, 665
Activated complex, 644
  determination of formula, 648
  electrode reactions and, 894–895
  theory, 785–794
Activation energy, 641, 757–759
Activators, 662, 666, 668, 670
Activity, 194–200
  coefficient, 195–200, 220–229, 872–883
  concentration dependence of, 198
  concept and definition of, 194
  mean ionic, 872
  pressure dependence of, 198
  standard state conventions, 106, 195–197, 870
  temperature dependence of, 198
Adiabatic
  calorimeter, 100
  demagnetization, 271
  energy, 796
  process, 99
  temperature of reaction, 114
  work, 99
Additivity rule, 460
Adsorbate
  immobile film of, 717
  mobile film of, 720, 723
  monolayer, 717–718
  multilayer, 720
  rate of adsorption of, 722
Adsorbent, 713
  surface area, 715–717
Adsorption, 614, 713–722
  chemical, 721
  isotherms
    BET, 716
    Fowler-Guggenheim, 721
    Gibbs, 249
    Langmuir, 615, 713, 722, 740
  physical, 721
Advancement of reaction, 255, 634
  equilibrium constant and, 255
  reaction rate and, 634

Alkali metals, 518–519
  electronegativity, 518–519
  ionic radii of, 518–519
Allowed transitions, 401
Aluminum, 103, 611
Amagat's law, 31–32
  compressibility factor and, 32
Amberg, C. H., 719
Ammonia, 144, 153, 157, 262, 264–265, 458, 461, 462, 835
Ammonia maser, 402
Ammonium chloride, 157, 265
Ammonium fluoride, 562
Ampere (unit), 849
Amplitude, 541–544
Andrew, T., 21
Ångstrom (unit), 295
Angular
  frequency, 313
  momentum, 326, 334, 357
    eigenfunctions, 330–332
    magnitude, 326, 329
    projection, 330
    vector, 327, 357
  velocity, 321
Anharmonic oscillator, 596
Aniline, 231, 245
Anion, 833
Anisotropy, 531
Anode. *See* Electrodes
Anodic
  current, 895
  density, 924
  reaction(s), 842
Anthracene, 530
Antibonding orbital, 442, 452
Antisymmetric wave function, 379–380
Aristotle, 389
Aromatic molecules, 474
Arrhenius, S., 641, 787, 837, 854, 857
Arrhenius
  dissociation theory, 833–835, 855, 857
  parameter, 641; *see also* Activated complex theory
Arthur, C. J., 8
Association of ions, 839
  Bjerrum's theory, 865
  Fuoss and Kraus' theory, 865
Asymmetric top, 437
Atom(s)
  energies of typical processes in, 667
  enthalpies of formation of, 108
    table of values of, 108
  vector model of, 374

Atomic
   configurations, 369, 371
     of ground state, 371
   form factor, 547–548, 551
   orbital, 382, 441
   spectra, 369
Auger process, 725
Austenite, 242
Average
   kinetic energy per molecule, 43
   square speed, 42, 43; *see also* Speed
   value of property, 51, 304
Avogadro, A., 4, 10, 43, 44, 61
Avogadro
   number, 4, 10, 43, 61
   principle, 4, 44
Azeotrope, 221, 222, 223

Bamford, C. H., 640, 667
Band
   head, 431
   spectrum, 410
Barometric formula, 48
Basis set
   double-zeta (DZ), 565–566
   extended (EBS), 565–566
Bass, L., 867
Bauer, S. H., 562–564
Baxter, G. P., 34
Beebe, R. A., 719
Beer-Lambert's law, 677
Benzene, 206, 439, 469–470, 477, 499, 530
Berkeley-Hartley osmotic pressure apparatus,
   218
Bernoulli, D., 40
Berthelot, M., 129
Bethge, H., 727
Bird, R. B., 74, 762
Bjerrum, N., 864
Black body radiation, 393–394
Bloch, F., 484, 493, 523, 530, 688, 691, 694
Bloch
   equations, 493–494, 688, 691, 694–695
   functions, 525–526, 532–533
Blue, R. U., 282
Bodenstein, M., 660
Body-centered cubic lattice, 520–521
Bohr, N., 398
Bohr
   atom, 295–297
   megneton, 485
   radius, 336
   theory, 295, 297, 398
Boiling point
   elevation. *See* Colligative properties
   normal, 160
Boltzmann, L., 10, 40, 43, 49, 138, 573, 576, 588

Boltzmann
   constant, 10, 43
   distribution law, 398, 528, 575; *see also*
     Maxwell-Boltzmann statistics
   factor, 49
   equation, 138, 588
Boltzmannons, 573
Bond
   angles, 462, 464, 562, 564–565, 800
   energy, 107–111, 455, 458, 465, 467, 469
     table of values, 109
   length, 465, 474, 476, 460–463, 800
   moment, 463, 465
   order, 474–476
Bonding orbital, 441, 452
Born, M., 412–413, 437–438, 521, 565, 834, 838
Born-Haber cycle, 521, 838
Born-Oppenheimer approximation, 412–413,
   437–438, 565
Bosons, 573
Boundary, 88
   adiabatic, 99, 130
   conditions, 303, 574
   diathermal, 99
Bovey, F. A., 690
Boyle, R., 3, 4, 5, 6, 10, 37, 40, 93
Boyle's
   temperature, 37
   law, 3, 5, 40, 93
Bragg equation, 560
Branched chain, 652
Brewer, L., 58
Brockway, L. O., 558
Broglie, L. de, 538
Brown, R. A., 581
Brunauer, S., 715, 716
Buffer solution(s), 919, 921
   standardization procedure, 919
Burns, M. L., 581
Butadiene, 308, 473, 475, 476
t-Butyl chloride, 554

Cadmium dimethyl, 600
   free internal rotation, 600
Caged pairs, 769
Calcium carbonate, 157
Calcium oxide, 157
Caldin, E. F., 693
Calomel electrode, 910
Calorie (unit), 101
Calorimeter, 99, 100
Calorimetry, 99
Carbon dioxide
   liquefaction isotherms, 22–23
   normal modes of vibration, 598–599
Carnot's cycle, 131
Cast iron, 292

Catalysis, 57, 135–138, 709, 887
  heterogeneous, 709, 737–740
  homogeneous, 887
    generalized acid, 887
    generalized base, 887
Cathode. *See* Electrodes
Cathodic
  current, 895
    density, 924
  reactions, 841
Cation, 833
Cementite, 241
Center of gravity, 320, 415
Center of inversion, 421
Centrifugal barrier, 761
Cermak, V., 768
Cesium chloride, 519–521
CH_3NC, molecular properties of, 800
Chain center, 650
Chain length, 651
Chance, B., 681
Chapman, S., 72
Charge density, 859, 860
Chemical activation, 820
Chemical potential
  activity and, 194–200
  of charged particles, 870–874
  equilibrium and, 188
  fugacity and, 189, 190
  an intensive property, 186–188
  partition function and, 607
Chemical shift (ESCA), 728
Chemical shift (NMR), 497–500
Chemically induced dynamic nuclear
  polarization, (CIDNP), 702–707
  multiplet effect in, 706
Chemiluminescence, 821
Chlorine
  gas electrode, 909
  molecule, 15, 323, 557, 594
  rotational constant of, 925
  thermodynamic properties of, 603–604
Chloroform, 226
Christiansen, J. A., 780
Clapeyron equation, 164–166
Classical mechanics, 287, 289, 293
  one-dimensional motion, 287
  three-dimensional motion, 293
Classical turning point, 762
Clausius, R., 40, 79, 96, 130, 137, 163, 164
Clausius
  Clapeyron equation, 164–166
  equipartition principle, 79
  first law of thermodynamics, 96
  second law of thermodynamics, 130, 137
Closed-shell state, 381
Clusius, K., 602

Cluster integral, 617
Coefficient(s)
  absorption, 538
  activity, 195–200, 219–220, 872–874, 879–883
  compressibility, 34
  diffusion, 72–74
  Donnan distribution. *See* Membrane
    equilibrium
  fugacity, 190–192
  induced emission, 398
  Joule-Thomson, 150–152
    table of values, 151
  Nernst distribution, 246
  spontaneous emission, 398
  thermal conductance, 71
  thermal expansion, 5, 6, 34
  virial, 15–18, 32–33, 618
  viscosity, 65–70
    table of values, 66
Coherent scattering, 538–539
Cohesive energy
  in crystals, 837–838; *see also* Born-Haber
    cycle
  in gases, 18, 146
Cohesive pressure. *See* Internal pressure
Colligative properties
  boiling point elevation, 214
  of electrolytic solutions, 833
  freezing point depression, 215
  osmosis, 217–218; *see also* Osmotic pressure
  vapor pressure lowering, 212
Collision(s)
  close, 762
  collinear, 776–777
  diameter, 12–14, 69, 71, 74
  direct, 816
  elastic, 40
  frequency, 59, 60
  indirect, 816
  inelastic, 812
  mean free path and, 61
  reactive, 812
  rebound, 816
  stripping, 816
  theory of reaction rate, 757
  transport properties and, 68–75
  with solid surface, 57; *see also* Effusion
Combination rule, 296
Combined first and second law, 138
Combustion
  calorimeter, 100
  energy, 104
  enthalpy, 105
Complete set, 348
Complex
  conjugate, 293
  function, 292

Complex (*Cont.*):
    number, 940
    set, 348
Component(s) of system, 157
Compressibility factor, 13, 14
    critical, 25, 26, 29
    diagram, 28
    fugacity coefficient and, 191
    of mixtures, 31, 32
    principle of corresponding state and, 26, 27
Compton, A. H., 548
Concentration cell(s); *see also* Electrochemical
cell(s)
    with transference, 913
    without transference, 914
Concentration units
    mean molality, 872
    molality, 180
    molarity, 180
    mole fraction, 179
Condensed systems, 229–233, 235–247
Conductance
    application of, 868
    equivalent, 850
    ionic, 854–855
        at high fields. *See* Wien's effect
        at high frequencies. *See* Debye-
        Falkenhagen effect
    measurement of, 850
    in nonaqueous solvents, 851, 866; *see also*
        Walden's rule
    specific, 849–850
    temperature dependence of, 867
Conductometric titration, 869
Configuration interaction, 449, 510
Congruent melting point, 239
Conjugated hydrocarbons, 309
Conjugated molecules, 474
Constant(s)
    boiling point elevation, 214
    Boltzmann, 10, 43
    coupling, 506, 509–510
    critical, 22–28
    dielectric, 835–836
    Dirac, 294
    electrolytic cell, 850
    Faraday, 843
    freezing point depression, 215
    force, 81; *see also* Laws, Hooke
    fundamental, 947
    gravitational, 46
    Henry's law, 207–211
    ideal gas, 7
    Madelung, 520, 837
    of motion, 327, 358
    Planck, 11, 290
    Poisson, 118
    proton shielding, 499

Rydberg, 296
    shielding, 497–499
    specific rate, 634; *see also* Rate reaction law
    spin-spin coupling, 502–505
    Sutherland, 69
    van der Waals, 18–20
Constant phase, 391
Coolidge, A. S., 448
Cooling curve, 236, 237
Coordinate(s)
    cartesian, 290
    elliptical, 441, 477
    normal, 438
    polar, 331, 335, 416
    position, 616
Copper sulfate pentahydrate, 265
Copper sulfate trihydrate, 265
Cork screw rule, 327
Correlation energy, 448
Corresponding state, 26–28, 167
    Pitzer's acentric factor and, 27
    Lyndersen, Greenhorn, Hougen and, 27
Coulomb's law, 836
Coulometer, 843–844
Coulson, C. A., 474
Covalent crystals, 518, 522
Crist, R. H., 660
Critical
    compressibility factor, 25, 29
    pressure, 22
    temperature, 21
        of solution, 231–233
    volume, 23
        Hakala's method of evaluation, 24
Cross product, 357
Cross section, 61, 550–551, 762–764, 812–813
    for attractive potentials, 767
    of reactions, 762–764, 812–820
    orientational dependence of, 820
    translational dependence of, 818
Crystal structures, 516
Cubic expansion coefficient, 5, 6, 34; *see also*
    Coefficient, thermal expansion
Cubic lattice, 517
Cuprous chloride, 240
Current density, 849, 924
    exchange, 924
Curtiss, C. F., 54, 762
Cyclopentadiene dimerization, 788
Cylindrical symmetry, 422

Dalton, J., 30, 31, 32, 44, 204, 605
Dalton's law of partial pressures, 30, 31, 44, 204
    compressibility factor of mixtures and, 31
    kinetic molecular theory and, 44
Daniell cell. *See* Electrochemical cell(s)
Davy, H., 96
Davydov splitting, 531

Debye, P., 266, 270–271, 396, 611, 613, 834, 835, 837, 857–860, 864, 879–883, 923
Debye
  cubic law. *See* Law(s)
  Falkenhagen effect, 864
  Hückel theory, 834, 837, 857, 879–883, 923
  theory of heat capacity, 611
DeDonder, T., 256
Degenerate eigenfunction(s), 303, 571
Degenerate energy level(s), 571, 572, 599
Degree(s) of freedom
  rotational, 79, 81
  translational, 71, 79
  vibrational, 79, 81
Delta function, 503
Demagnetization. *See* Adiabatic demagnetization
Derivative(s), 936
Desorption, 712
Detergent(s), 250
Determinant, 472
Deuterium molecule, 594
Dewar structures of benzene, 439
Dialysis, 924
Diamond, 528
Diatomic molecule(s)
  eigenfunctions, 420
  eigenstate symbols, 422
  electronic energy levels, 408
  electronic Hamiltonian, 411
  energies of typical processes in, 667
  energy levels, 419, 426, 431
  equilibrium bond distance, 408
  rotational constants, 408
  rotational energy level, 407
  separation of translational and internal motion, 434
  spectra, 410
  vibrational energy levels, 408
  vibrational frequencies, 409
Dielectric constant, 834, 860
  effect on dissociation, 857
Differential
  exact, 91
  operator, 302
  partial, 92, 93, 94
Diffusion
  coefficient, 72, 928
  current, 928
  distance, 73; *see also* Fick's laws
  in gases, 72
  in ionic solutions, 869–870
  overpotential, 928
  potential. *See* Liquid(s), junction potential
  rate, 928
Dilatometry, 678
Diphenylpicryl hydrazine, 510

Dipole
  dipole interaction, 10
  moment,
    electric, 457, 463, 465
    induced, 10
    operator, 401, 457
    permanent, 10
    values for hydrogen halides, 457
Dirac constant, 294
Dispersion forces, 11
Dissociation, 429, 441, 455, 833, 839
  energy, 429, 441, 455
  solvent effect on, 839–840
  of strong electrolytes. *See* Debye-Hückel theory
  of weak electrolytes. *See* Arrhenius dissociation theory
Distillation
  column, 235
  diagrams, 204, 206, 222, 232, 233
  fractional, 234
  steam, 233
Distribution
  Bose-Einstein. *See* Laws
  Fermi-Dirac. *see* Laws
  Maxwell-Boltzmann. *See* Laws
  of molecular energies, 77
  of molecular velocities, 48–54
    experimental verification of, 55
Donnan equilibrium. *See* Membrane equilibrium
Doppler effect, 410
Dorfman, L. M., 693
Dosimetry, 665
Double bond, 465
Drauglis, E., 727
Du Noüy tensiometer, 170
Dulong and Petit law, 83
Dunning, W. J., 732
Dymock, J. B., 278

Easterman, Simpson, and Stern's method of verifying the Maxwell distribution law, 55
Ebulliometer, 173
Ebulliometry. *See* Colligative properties, boiling point elevation
Effective nuclear charge, 367, 383, 442
Effective potential, 366
Efficiency, heat engines. *See* Carnot's cycle
Effusion, 58, 59
  Knudsen method of, 58
Eigen, M., 680, 683, 685, 693
Eigenfunction(s), 303
Eigenvalue(s), 303
Eigenvalue spectrum, 522
Einstein, A., 393, 395–398, 401, 610, 611
Einstein coefficient(s), 397–398, 401
Elastic scattering, 539
Electric dipole operator, 422

Electrical
  charge, 843
  conductance, 528, 849
  current, 849
  field, 389, 841, 849
  resistance, 849
Electrochemical cell(s)
  chemical, 912
  concentration, 912
  Daniell, 897
  electrolytic, 842, 893
  fuel, 917
  galvanic, 893
  standard EMF, determination, 901
    thermodynamic properties from, 902
  Weston, 900
Electrode(s)
  amalgam, 908
  anode, 842
  cathode, 842
  gas-ion, 908–909
  glass, 920
  kinetics of, 893–895
  metal-insoluble salt-anion, 910
  metal-metal ion, 908
  polarization of, 843, 923
  potential, 893–900
    convention, 897–898
    standard, 899
    reactions, 841–843, 893–895
  reference, 896
Electrolysis, 841; *see also* Faraday, laws of
  electrolysis
Electrolytes, 833; *see also* Conductance
  Arrhenius' dissociation theory of, 833–834,
    855, 857
  chemical potential of, 870
  Debye-Huckel theory of, 834, 837, 857
  standard state of, 870
Electromagnetic radiation, 389–396
Electron
  affinity, 455, 838
  correlation, 446, 448
  diffraction, 545, 550–551, 554, 556–559
  gas, 619
  nuclear spin-spin coupling, 507
  spectroscopy for chemical analysis (ESCA).
    See Photoelectron spectroscopy
  spin, 364, 372, 484
    eigenfunction(s), 375
    resonance, 484–485, 506–511
Electronegativity, 454–456, 518
  of alkali metals, 518
  of halogens, 518
Electronic
  partition function, 600
  quantum number, 422

transition, 424
  moment, 428
Electrophoretic effect, 863
Elementary cell, 516–517, 731
Emission
  spectrum, 398
  spontaneous, 397–398
  stimulated, 397–398
Emmett, P. H., 715, 716
Endothermic reactions, 104
Energy
  of activation, 68, 641
  average per molecule, 43
    equipartition principle, 79
  band(s), 404, 523, 525–527
  bonding, 107
    table of values, 109
  conversion factors, 950
  degradation of, 130, 137
  density, 393, 398
  distribution law in terms of, 77
  first law of thermodynamics and, 95–98
  gap, 525, 528
  internal, 95
  ionization, 13
  kinetic, 95
  lattice, 838; *see also* Born-Haber cycle
  leak, 490
  mass equivalent of, 96
  rotational, 79, 80, 594
  translational, 79, 590
  vibrational, 79, 80, 597
Enthalpy
  of adsorption, 718–720
  of atomization, 108
    table of values, 108
  of chemical reactions. *See* Thermochemistry
  of combustion, 106
  of formation, 106
    table of values, 107
  function, 113, 273
    table of values, 275
  of fusion, 164
  of mixing, 201, 606
  partial molal, 184, 874
  pressure dependence of, 117
  of solvation, 835
  of sublimation, 164
  temperature dependence. *See* Kirchhoff's
    law
  of vaporization, 164
Entropy
  of adsorption, 718
  Caratheodory principle and, 136
  as criterion of equilibrium, 146
  as criterion of spontaneity, 136
  evaluation of changes of, 136–137, 142–144

irreversible processes, 137
reversible processes, 136
and heat death of Universe, 130, 137; *see also*
   Clausius
of mixing, 202, 228, 606
probability and. *See* Boltzmann, equation
second law of thermodynamics and, 129–139,
   585–588
of solvation, 835
statistical evaluation of, 585
third law of thermodynamics and, 266–268,
   585, 588
   table of values, 268
Equation(s)
Antoine, 165
Arrhenius, 641
barometric, 47–48
Beattie-Bridgman, 21
Berthelot, 19–21
   reduced, 27
Bloch, 493–494, 688, 691, 694–695
Boltzmann, 138, 580, 588
Bragg, 560
Butler-Volmer, 924
Clapeyron, 164
Clausius-Clapeyron, 164
Debye, solid heat capacity, 613
Dieterici, 19
Donnan, 923
Duhem-Margules, 226
Einstein, 96
   solid heat capacity, 611
Gibbs, 169
Gibbs-Duhem, 182, 184, 198, 223–224
Henderson, 907
Hückel, 471, 478
ideal gas, 7–8, 93
Kammerlingh-Onnes, 15; *see also* Virial
   equation
Kelvin, 171
Maxwell, 389
of motion, 287, 294
Nernst diffusion, 870
Nernst, EMF of cell, 895
Onsager, 864
Poiseuille, 67
Poisson, 119
Poisson-Boltzmann, 860
Ramsay-Shield, 171
Rayleigh-Jeans, 612
Reddlich-Kwong, 19
Sackur-Tetrode, 591
Schrödinger
   time dependent, 399
   time independent, 293, 297
Stirling, 574, 945
Sutherland, 69

Taffel, 925–926
Tait, 34
Van der Waals, 18
   reduced, 26
van't Hoff, 258
virial, 15, 618
Equilibrium
   constant(s), 255–265
      partition function and, 606–609
      pressure dependence of, 257
      standard E of cells and, 903
      standard Gibbs free energy and, 256
      temperature dependence of, 257, 274
   thermodynamic criterion of, 146, 188, 191
Equipartition principle, 78–79
Equivalent conductance, 851
   applications of, 867
   Arrhenius' theory and, 854
   and mobility of ions, 855
   and specific conductance, 851
   of strong electrolytes, 853
   Walden's rule and, 854
   of weak electrolytes, 853
Ethane, 105, 109, 465, 476, 600
   rotation of methyl groups in, 600
Ethyl alcohol, 24, 199, 227, 488, 497, 835
Ethylbenzene, 276
Ethylene, 105, 465, 466, 476, 505, 598, 600
   torsional vibration, 599
Ethylene glycol monoethyl ether, 206
Ethylene glycol monomethyl ether, 206
Euler
   polynomial, 324–325, 355, 360
   theorem, 92, 934
Eutectic
   halt, 237
   mixture, 236
   temperature, 236
Eutectoid, 242
Even permutations, 380
Evenson, K. M., 670
Exact differential(s), 91, 933
Exciton, 532–534
Exclusion principle, 378
Exothermic reaction(s), 104
Expansions in mathematics, 938
Expectation value, 304–305
Explosion limit, 652
Extensive variables, 88, 89
Extent of reaction, 255; *see also* Advancement
   of reaction
Extinction, 537
   coefficient, 537
Eyring, H., 562, 564, 785

Face-centered cubic structure, 519–521
Fall-off region, 781, 805

Faraday, M., 841, 843
Faraday
  constant, 843
  laws of electrolysis, 841, 843
Felsing, U. A., 8
Fermi, E., 503, 528, 573, 620
Fermi
  contact potential, 503–505
  Dirac statistics, 573–575
  energy level, 528–529, 620
  temperature, 528
Fermions, 573
Ferric chloride, 240
Ferrite-α, 242
Fick's laws
  first, 64, 72
  second, 73, 870
Film(s), 250, 713, 718
First law of thermodynamics, 87–128
  statistical interpretation of, 586
Flames, 708–709
Flanagan, G. N., 103
Flash
  desorption, 720
  photolysis, 685–687, 696
Fletcher, F., 804
Flood, E. A., 732
Flooding method. *See* Isolation method
Flow
  heat, 64
  laminar, 65, 67
  mass, 64
  molecular, 58
  momentum, 64
Fluid(s), 3, 64, 65
Fluorescence, 821
Fluorine molecule, 453
  population of electronic levels, 601
Fluorine monoatomic, 601
Fock, V. A., 383, 499, 566
Forbidden transitions, 401
Force(s)
  constant, 318
  definition, 40–41, 293
  intermolecular, 9–13
    attractive, 10–11
    coulombic, 836
    dispersion, 11
    induction, 10
    long range, 10
    orientation, 10
    repulsive, 10
    short range, 10
    van der Waals, 11
Fowler, R. H., 721
Franck-London, 430
  approximation, 430
  factors, 430

Freezing point depression. *See* Colligative
  properties
Frequency
  parameter. *See* Arrhenius parameter
  and period, 390
  and wavelength, 390
Friess, S. L., 681, 683
Frost, A. U., 278
Fuel(s)
  cell(s). *See* Electrochemical cell(s)
  oxidation of, 700–701
Fugacity, 189–192
  coefficient, 189
    diagram, 192
  concentration dependence of, 195
  and equilibrium, 191
  temperature dependence of, 195
Fuoss, R. M., 864
Fractional distillation, 234

g factor, 486–487
Galvanic cell(s). *See* Electrochemical cell(s)
Gas(es)
  electron, 619–621
  ideal, 3–9
    application of the first law to, 116–122
    equation of state of, 7
    kinetic molecular theory of, 40
    lattice, 614–615
    universal constant, 7
    statistical treatment of, 589
  laser, 404
  mixtures of, 30–33, 604–606
  nonideal, 9–13
    equations of state of, 15–35, 615
    fugacity of, 189
    intermolecular forces in. *See* Forces
    liquefaction of, 21–26
  transport properties of, 64–75
    diffusion, 72
    thermal conductance, 71
    viscosity, 64
Gaussian function, 428
Gay-Lussac, J., 5, 6, 7, 116
Gay-Lussac's law, 5
Gerade, 422
Germanium, 528
Giauque, W. F., 271, 278
Gibbs, J. W., 139, 140, 142, 158, 169, 170, 173,
  174, 182, 187, 189, 219, 224, 226, 229, 230, 248,
  249, 255, 256, 258, 260, 269, 273, 274, 604, 718,
  787, 835, 871, 884, 895, 896, 898, 902, 903, 905
Gibbs
  activation energy, 787, 894–895
  adsorption isotherm, 249
  Duhem equation. *See* Equation(s)
  equation, 169
  free energy

of adsorption, 718
criterion of equilibrium and, 146
EMF of cells and, 902–905
equilibrium constant and, 256
excess, 219–220, 224, 228
of formation, 259–260
function, 273
of mixing, 202, 219, 229–230
partial molal. *See* Chemical potential
partition function and, 585, 591
pressure dependence of, 141
of solvation, 834–835
of surface, 171
temperature dependence of, 141
phase rule, 157–158, 188–189
Glass electrode. *See* Electrode(s)
Goudsmit, S., 372–373
Gradient
concentration, 64, 72; *see also* Fick's laws
temperature, 64
velocity, 64, 65
Graham's law, 43, 58
Graphite, 58, 516, 522, 611
Gravitational acceleration, 46–47
Gray tin, 528
Group theory, 420
Guggenheim, E. A., 721
Guseva, A. N., 280
Gyromagnetic ratio, 487

Haber, F., 264, 521, 834
Hakala's evaluation of critical volume, 24
Half-life, 634, 637
Halogens, 518–519
electronegativity, 518–519
ionic radii, 518–519
Hamiltonian, 399, 411, 616
operator, 302
Hansen, 484
Hard sphere model, 12–13, 46, 59, 68, 72, 74
potential, 12–13
Hare, W. A., 76
Harmonic
approximation, 419, 433, 438
oscillator. *See* Laws, Hooke
classical treatment of, 79–81
extent of motion of, 318–319
probability density of, 317
quantum mechanics of, 313–319
statistical thermodynamics of, 313–319,
610–613
Harned, H. S., 901
Hartman, H., 819
Hartree, D., 384–385
Hartree-Fock
distances, 566
energies, 566
limit, 447

method, 383–384, 447–448
orbitals, 383
Hartridge, H., 680
HD molecule, 665
Heat
of adsorption, 718–720
capacity, 101
Debye's cubic law of, 266, 270–271, 613
Debye's theory of, 611–613
Dulong-Petit's law of, 83
Einstein's theory of, 610–611
of electron gas, 529, 620–621
equipartition principle and, 82
molar, 102–103, 147
partition function and, 584
specific, 102
engine. *See* Carnot's cycle
of fusion, 164
of reaction. *See* Thermochemistry
of sublimation, 164
of vaporization, 164, 167
Hedberg, K., 562
Heisenberg, W., 290
Heisenberg's uncertainty principle, 290
Heitler-London method, 445
Helium
atom, 364, 374
energy level diagram, 370
cubic expansion coefficient, 6
heat capacity, 175–176
neon laser, 404
phase diagram, 175–176
quantum mechanics of, 364–369
Helmholtz, H., 139, 140, 141, 146, 187, 586, 590,
598, 601, 605
Helmholtz free energy
criterion of equilibrium and, 146
definition of, 139
partition function and, 586, 590, 598, 601, 605
temperature dependence of, 140–141
volume dependence of, 141
Hemolysis, 219
Henglein, A., 820
Henning, F., 35
Henry's law, 207–211
Hermite polynomials, 316, 352
Hermitian operator, 306–307
Herschbach, R., 562, 564, 817
Herzberg, G., 562, 667
Herzfeld, K. F., 650
Hess, G. H., 105
Heuse, W., 35
Hexagonal close-packed structure, 521
Hexagonal structure, 517
Hildebrand, J. H., 167, 228
Hildebrandt, R. L., 555
Hill, T. I., 614
Hinshelwood, G. N., 651–652

Hirschfelder, J. O., 74, 762
Hittorf's method, 846
Holbrook, A., 805
Holden, A. N., 562
Homogenous polynomial(s), 324, 355
Hood, J. J., 641
Hot atoms, 669, 816–818
Hückel, E., 470, 471, 478, 510
Hückel
  equation, 471, 478
  MO theory, 474–476, 478–480, 510
Huygens, C., 389
Hybridization, 458
Hybridization parameters, 468
Hybridized orbitals, 451, 459
Hydrate, 265
Hydration number, 834
Hydrogen
  atom, 332–346
    d functions of, 338, 340, 344–346
    eigenfunctions of, 341, 359–361
    p functions of, 338–339, 343
    probability density of, 342–344
    quantum numbers, 334, 337, 369
    radial eigenfunction, 336
    s, p, d states, 337
    solution of Schrödinger equation for, 332–346
  bromide, 323
  chloride, 157, 265, 594
  fluoride, 454
  gas electrode, 909
    standard, 896–899
  iodine reaction, 645, 660
  molecular ion, 346, 438–448
    bond length, 443
    dissociation energy, 443
  molecule, 6, 8, 259, 262, 264, 323, 347, 444–449
    bond length, 444–446
    characteristic rotational temperature, 594
    dissociation energy, 444–446
    possible electronic structure, 347
    rotational constant, 323–324
  peroxide–hydrogen iodide reaction, 631
Hydrostatic pressure, 47
Hypertonic solution(s), 218
Hypotonic solution(s), 218

Ideal crystal, 610–613
  Debye's theory of heat capacity of, 611–613
  Einstein theory of heat capacity of, 610
Ideal gas, 1–9; *see also* Gas(es)
  equation of state, 7–10, 93
  first law of thermodynamics and, 116
  molecular interpretation of, 40–44
  statistical thermodynamics of, 589
  temperature scale, 90, 91
Ideal lattice gas, 614

Ideal solution, 181, 200–219
  general characteristics of, 181, 200–202
  vapor-liquid equilibrium in, 202; *see also*
    Raoult's law
Immiscible liquids, 233
  distillation diagram of, 233
  steam distillation of, 233, 234
  vapor pressure of, 233
Impact parameter, 759–760
Incoherent scattering, 538–539
Incongruent melting point, 240
Induced emission, 489
Inelastic scattering, 539
Infrared spectrum, 407, 432, 598
Initial rate method, 636, 638
Integrals, 936–937
Intensity(ies), 397
  addition of, 540
  electrical current, 849
  of infrared and microwave transitions, 432–433
  relative, of electronic transitions, 431
  scattered by atoms, 547–549
  scattered by electrons, 546
  of scattered electrons, 550
Intensive variable(s), 64
  concentration, 64
  density, 64
  pressure, 64
  temperature, 64
Interfaces, 708
Intermolecular forces. *See* Forces
Intermolecular potential energy
  hard sphere molecules, 12
    with attraction, 12
    without attraction, 12
  Lennard-Jones, 12, 13
Internal energy. *See also* First law of
  thermodynamics
  partition function and, 584
  variation with volume, 116, 146
Internal pressure, 18, 33, 45, 146
Internuclear distance, calculation of, 323
Ion(s)
  activity coefficients of, 872–875
  association of, 839, 865–866
  conductance of, 854–855
  diffusion of, 869–870
  enthalpy of formation of, 875–876
  entropy of formation of, 877
  Gibbs free energy of formation of, 874–876
  ion interaction, 834; *see also* Debye, Hückel
    theory
  mobility of, 844–846, 854–855, 869–870
  single, 871
  solvation of, 833
  solvent interaction with, 834–836
  velocity of, 844–846, 854–855

Ionic
  atmosphere. *See* Debye, Hückel theory
    table of radii, 862
  character, 447
  crystal, 516, 837
  equilibrium, 833
  interaction
    attractive. *See* Coulomb's law
    Bjerrum's theory of, 837
    Debye-Hückel theory of, 834, 837, 857
  potential, 859–862
    repulsive, 836
  radii, 519
  strength, 862
    effect on nonideality of solutions, 880–882
    effect on reaction rate, 886
  structure, 348, 446
Ionization potential, 455, 838
Iron-carbon phase diagram, 242
Irreversible process, 96
  criterion of, 139, 140
  evaluation of entropy changes, 144
Isobaric process, 100
Isochoric process, 99
Isolated system, 88
Isolation method, 636
Isoteniscope, 172
Isotherm(s), 4–5, 121
  liquefaction of CO_2, 23
Isotonic solution, 219
Isotope effect, 789

James, H. M., 446–448
Joule, J. P., 40, 96, 116, 149, 271
Joule
  law, 116–117
  Thomson experiment, 149, 271
  Thomson coefficient, 151
    table of values of, 151
  unit of energy, 7
Joyner, R. W., 730
Justice, J. L., 103

Karplus, M., 773, 774, 778, 779
Kassel, L. S., 278, 795
Keil, R. G., 581
Kékulé structures of benzene, 439
Kelvin, Lord. *See* Thomson, W.
Kelvin, 6, 171
  equation, 171
  temperature scale, 6
Kinetic spectroscopy, 687
Kirchoff's law, 110, 259
Kistiakowsky, G. B., 167
Kistiakowsky's rule, 167
Knudsen's effusion method, 58, 59, 173
Koeppl, G. W., 791
Kolos, W., 448

Kraus, C. A., 864
Kuhn, W., 308

Ladder operator, 353
Lagrange's method of undetermined multipliers,
  575, 580, 943
Laidler, K. J., 790
Lang, B., 730
Langmuir, I., 613, 713, 720, 739–740
Langmuir adsorption isotherm. *See* Adsorption
Laplace operator, 302, 860
Larson, A. T., 264
Laser, 402–404, 578, 821
Lattice energy, 521, 838
Laue equations, 560
Laurie, V. W., 562, 564
Law(s)
  Amagat, 31, 32
  of atmosphere, 46
  average values, 51
  Avogadro, 4, 44
  Bose-Einstein distribution, 575
  Boyle, 4, 5, 6, 41, 42, 93
  Charles, 6
  of conservation of energy, 333
  Coulomb, 836
  Currie, 283
  Dalton, 30, 31, 44, 204, 220
  Debye cubic, 266, 270–271, 613
  Debye-Hückel limiting, 879–880
  Faraday, 843
  Fermi-Dirac distribution, 573, 620
  Fick's first and second, 64, 72–73, 870
  First, of thermodynamics, 95–122
  Gay-Lussac, 5, 6
  Graham, 44, 58
  Henry, 207–211
  Hess, 105
  Hooke, 81, 313
  Joule, 116–117
  Kirchhoff, 110, 259
  Kohlraush, 851, 854
  Lenz, 497
  Maxwell-Boltzmann, 51, 575
  Nernst distribution, 246
  Newton, 40
  Ohm, 849
  Oswald dilution, 855–856
  Raoult, 204
  Second, of thermodynamics, 129–156
  Stark-Einstein, 672
  Third, of thermodynamics, 266–273
  Zeroth, of thermodynamics, 89
LCAO (linear combination of atomic orbitals),
  441
LCAO MO method, 445
Le Chatelier's principle, 264
Lead storage battery, 927

Least squares method, 946
LEED (low energy electron diffraction), 729
Lennard-Jones potential function, 11, 12, 761– 762
LET (linear energy transfer quotient), 672
Lever rule, 204
Lewis, E. S., 681, 683
Lewis, G. N., 190, 395
Lide, D. R., 562, 565
Lindemann, F. A., 780–781
Line spectrum, 295
Linear molecule(s), 82
Linewidth, 410
Liquefaction of gases, 21
Liquid(s)
    crystal(s), 175
    diffraction by, 561
    equation of state of, 35
    junction potential, 897, 906–907, 914, 918
    immiscible, 233
    partially miscible, 229, 243
    statistical theory of, 618–619
    surface tension of, 170
    vapor pressure of, 164–165
    viscosity, 67–68
Lithium, 522, 529
Localized exciton, 532
Locker, D. J., 804
London forces. *See* Forces
Lone electron pair, 462
Lone pair orbitals, 452
Lorentzian function, 494
Lowder, J. E. L., 700
Luminescence, 821
Lydersen, Greenhorn, Hougen's principle of
  corresponding state, 27

Mack, E., 62, 76
Madelung constant, 518, 837
Magnetic
    field strength, 389
    moment of electron, 373, 484–485
    resonance, 484
    susceptibility, 529–530
Magnetization, 530
Marcus, R., 795
Margules expansion(s), 224
Maser, 402–403
Mass-energy equivalence, 96
Mass transfer, 72, 708–712
Matheson, M. S., 693
Maximum work, 139, 140
Maxwell, J. C., 40
Maxwell
    Boltzmann statistics, 573–577
    distribution of molecular energies, 77
    distribution of molecular velocities, 48–53
        Estermann, Simpson, and Stern method of
          verification of, 55

equations, 389
    mean free path, 59, 61
    viscosity dependence on pressure, 68–69
Maxwellons, 573
Mayer, J. R., 96
Mayer, De, L., 683
McGlashan, M. L., 954
Mean, 42, 56–57, 59, 61
    free path, 59, 61
    speed, 56, 57
    square speed, 42
Mechanical equivalent of heat, 101
Meibom, S., 562
Membrane, 216–218, 921–923
    equilibrium, 921–923
    of glass electrode, 920–922
    potential, 922
    semipermeable, 216–218
Menten, M. L., 740
Merritt, F. R., 562
Metal(s), 521–530
    electronegativity of, 518–519
    ionic radii of, 518–519
Metallic mirror, 702
Metastable, 160–162, 404
    phase equilibria, 160–162
Methane, 109, 115, 458, 462, 465
Methanol, 227
Methylene radical, 698–699
Michaelis, L., 740
Microscopic reversibility, 655
Microwave spectrum, 407, 410, 433
Miller indices, 731–733
Molality, 179, 180
Molarity, 179, 180
Mole fraction, 179, 180
Molecular
    basis of equation of state, 40
    crystals, 175, 518, 530–534
    diameter. *See* Collision, diameter
    dynamics, 771
    energy levels, 419
    flow, 58
    orbital, 441, 475, 470
    properties, measurement of, 536
    structure
        ab initio, 565–566
        average, 564
        effect of vibration on, 561–564
        effective, 563–564
        equilibrium, 564–565
        from NMR, 561–562
        from quantum mechanics, 565–566
    wave function, 411
Moment, 10, 79, 437, 592, 604
    dipole, 10–11
    of inertia, 79, 80, 437, 592–595, 604
    magnetic, 10

Momentum, 40–41, 64–68
  angular, 326, 334, 357
    eigenfunctions, 330–332
    magnitude, 326–329
    projection, 330
    vector, 327, 357
  coordinates, 616
  transfer of, 64–68
Monochromatic light, 392
Monoclinic structure, 161, 517, 525
Monolayer(s), 250, 713, 718
Moore, C. B., 824
Morgan, J., 555
Morgan, J. H., 280
Most probable, 56, 300
  speed, 56–57
  value, 304
Moving boundary method, 846–847
Mulcahy, M. F. R., 782
Mulliken, R. S., 449, 455
Multiple bonds, 465
Multiplicative operator, 302

Napthalene, 104, 530–531, 533
Nash, L. K., 581
Negative absolute temperature, 578
Nernst, W., 246, 267, 870, 895
Nernst
  distribution law, 246
  equation, EMF of cell, 895
  heat theorem, 267; *see also* Third law of
    thermodynamics
Newton, I., 389
Newton (unit). *See* Appendix IX
Newtonian mechanics, 287
Nicholas, J. E., 687
Nicotine, 232
Nitrogen fluoride, 269–270
Nitrogen molecule, 6, 8, 210, 262, 364, 453, 594,
  608–609
NO molecule, 594
NO_2 molecule, 263
N_2O_4 molecule, 263
Nodal plane, 342
Nonaqueous solvents, 180, 851, 853–854, 866
Nondegenerate, 303
Nonideal solution(s), 219
  Gibbs free energy of mixing, 219
  excess Gibbs free energy of, 219, 220
  distillation diagrams of, 221–223
  vapor-liquid equilibrium in, 220
    experimental determination of, 221
Nonlinear optics, 402
Normal boiling point, 160, 161, 167
Normal mode, 82, 438, 598–600
Normalizable, 298
Norrish, G. N., 685, 687
Nozzle beam, 818

Nuclear
  charge, effective, 367–368
  magnetic resonance, 484, 487–494
  magneton, 486–487
  spin, 484, 486–487
Nucleation, 708
Number of degree(s) of freedom, 158, 160, 161,
  189

n-Octane, 69–70
Odd permutations, 380
Ohm (unit), 849, 948
Onsager, L., 857, 863–864, 867
Onsager's equation for conductance, 863–864
Open system, 88
Operator, 302, 380, 422
Optical pumping, 404
Orbit, 288
Orbital(s)
  antibonding, 422
  bonding, 441
  of diatomic molecules, 453–454
  general discussion on, 369, 381
  Hartree-Fock, 383
  hybridized, 451–452, 458–469
  lone pair, 462
  molecular, 441
  self-consistent field (SCF), 383
  Slater, 385–386
  π, 467
  σ, 467
Orthogonal, 306
Orthonormal set, 349
Orthonormality, 459
Orthorhombic structure, 517
Oscillator
  anharmonic, 596
  approximations, 419, 433, 438, 610–613
  classical, 79–81
  extent of motion of, 318–319
  harmonic. *See* Law(s), Hooke
  partition function of, 595–597
  probability density of, 317
  quantum mechanics of, 313–319
Osmosis. *See* Colligative properties
Osmotic coefficient, 878
Osmotic pressure, 217, 878, 922–923; *see also*
  Van't Hoff, law
  apparatus of Berkeley and Hartley, 218
Ostwald, W., 67, 855
Ostwald
  dilution law, 855
  viscosimeter, 67
Othmer, D. F., 167
Overlap, 459
  charge, 443–444
  integral, 426–427, 452, 467–468, 471
Overpotential, 924–928

Overstreet, R., 278
Oxidation-reduction electrode(s), 911
Oxidation-reduction reaction. *See*
  Electrochemical cell(s), galvanic
Oxygen molecule, 211, 259, 265, 422, 453

P branch, 431
Pace, F. L., 269
Packard, M. E., 484
Page, S. H., 949
Pair interaction, 17, 617
Partial
  differential, 92, 933
  molal properties, 120
    determination of, 183; *see also* Equation(s),
      Gibbs-Duhem pressure, 30–31, 44, 204,
      217, 220, 223; *see also* Dalton's law of
      partial pressures, and Raoult's law
      volume 31; *see also* Amagat's law
Particle in a box, 298, 528
Partition function
  classical, 616
  configurational, 616
  definition of, 576
  discussion of, 579
  electronic, 600
  of ideal crystal
    Debye, 612
    Einstein, 610
  of ideal lattice gas, 614
  molar, 583
  molecular, 576
  nuclear, 602
  rotational, 591
  thermodynamic functions and, 583–586
  translational, 589
  vibrational, 595
Pascal (unit), 91, 948
Pascal triangle, 509
Paul, M. A., 562, 565
Pauli, W., 369, 378, 381, 777
Pauli exclusion principle, 308, 369, 378, 381,
  777
Pauling, L., 455
Pearlite, 242
Penner, S. S., 700
Period, 390
Peritectic phase reaction, 240
Permittivity of vacuum, 836, 860
Permutation, 378–381
  even, 380
  odd, 380
  operator, 380
Pertubation theory, 312, 349–351
pH, paH, ptH, 883–884
  determination of, 917–920; *see also* Hydrogen
    gas electrode

Phase
  definition, 157
  diagrams, 158–160, 162, 204, 206, 222, 231–
    233, 237–238, 240–245, 247
  rule. *See* Gibbs phase rule
  shift, 540–541
Phenol, 245
Phosphine, 273
Phosphorus pentachloride, 15
Phosphorus system phase diagram, 162
Phosphorus trichloride, 15
Photoactivation, 820
Photochemistry, 662
Photoconductivity, 527
Photodissociation, 430
Photoelectric effect, 393, 395
Photoelectron spectroscopy, 725
Photon, 395, 399
Physical adsorption, 721
Pierce, L., 269
Pitzer, K. S., 27
Pitzer's accentric factor, 27
Planck, M., 11, 290, 297, 392–395, 398
Planck's constant, 11, 290
Plane wave, 389
Plasma, 3, 833
Poise (unit), 65, 949
Poiseville, J. L., 67
Polanyi, J. C., 822, 826
Polarizability, 10, 11
Polarization, 391, 442, 521, 994
  direction, 391
  of electrodes, 924
Polymorphism, 161
  enantiotropy, 161
  monotropy, 162, 163
Population inversion, 403, 518; *see also* Laser
  and Negative absolute temperature
Porter, G., 685, 687
Porter, R. N., 773, 774, 778, 779
Post, B., 555
Potassium, 241
Potential
  chemical, 187, 870
    and equilibrium, 188
  diffusion. *See* Liquid(s), junction potential
  electrochemical, 922
  electrode, 893–895
  energy
    curves, 12, 413, 419, 427–429, 437, 445
    functions, 12, 413, 417
    surface, 565–566, 772–775, 827
    table of standard values, 899
  ionization, 455, 838
  membrane, 922
  over-. *See* Equation(s), Butler-Volmer and
    Taffel

Potentiometer, 900
Pound, R. V., 484
Preexponential parameter. *See* Arrhenius, parameter
Pressure, 41, 91
  critical, 22
  effect
    on boiling point, 164
    on equilibrium constant, 254
    on vapor pressure, 168
  fugacity and, 190
  hydrostatic, 47
  internal, 35, 45, 146
  kinetic molecular theory and, 41, 42
  over curved surfaces, 169
  reduced, 26
  statistical interpretation of, 585, 591
  surface, 250
Principle
  Avogadro's, 4, 44
  of Caratheodory, 136
  of corresponding states, 26, 167
  of degradation of energy, 130
  equipartition, 78
  of Le Chatelier, 264
  superposition, 398
  of unattainability of absolute zero. *See* Third law of thermodynamics
Probability, 292
  density function, 49–50, 292, 297
  distribution function, 48–52, 291
  entropy and. *See* Boltzmann, equation
  mathematical, 138
  thermodynamic, 138
Progress variable, 255
Promotion energy, 461
Propagation direction, 391
Properties of system, 88, 90, 180
  extensive, 88
  intensive, 88
  partial molal, 180
  thermometric, 90
Propyne, 110
Proton
  magnetic moment, 498
  magnetic resonance, 487
  magnetic shielding, 497
  proton coupling, 501
  transfer, 693, 696–697
Purcell, E. M., 484
Pyridine, 225
π electrons, 308, 470
π orbitals, 451, 467
π states, 421

Q branch, 431
Quantization rule, 394

Quantum
  effect, 594
  mechanics, 287–568
    accuracy of measurement, 287
    calculations on H_2^+, 440–444
    calculations on H_2, 444–449
    calculation on helium atom, 364–369
    calculations on molecules, 438
    of one-dimensional motion, 287
    probability in, 291–293, 303–305
    three-dimensional motion, 293–294
  numbers, 422
  state, 813
    enhancement of population in, 813
    selection of, 813

R branch, 431
Rabinovitch, B. S., 799, 800, 804
Radar waves, 484
Radiation, 389, 392, 662
  chemistry, 662
  field, 389, 392
Radiolysis of water, 672–676
Raman, C. V., 552
Raman
  effect, 552
  spectra, 552, 598
Ramsay-Shield equation, 171
Ramsperger, H. C., 795
Randall, M., 277
Random life-time assumption, 782, 822–823
Raoult's law, 204
Rate
  constant(s)
    activated complex theory and, 787–788
    for barrier potential, 765
    collision theory of, 764–766
    detailed, 813–814
    of electrode reactions, 894–895
    equilibrium and, 656–660
    expressions for, 641–642, 657–658, 691
    gas and liquid phase relation for, 794
    standard state and, 657
    temperature dependence of, 640
    unimolecular, 781, 783–784, 798
  determining step, 646, 926
  reaction law, 739–740
Rayleigh-Jeans theory, 612
Reaction(s)
  bimolecular, 644
  bulk sample studies of, 813
  $n\text{-}C_7H_{16} \xrightarrow{Pt} C_7H_8 + 4\ H_2$, 738
  $CH_3NC \rightarrow CH_3CN$, 780, 799–804
  $Cl + {}^iH^jH \rightarrow {}^iHC + {}^jH$ (i or $j = 1, 2, 3$), 789–791
  combination, 784–785
  complex, 644

Reaction(s) (*Cont.*):
  consecutive, 645
  with conservation of spin, 793
  coordinates, 644
  diffusion limited, 711–712, 769–771
  between electronic states, 799–804
  elementary, 644, 758
  fast, 629, 692–693, 696
  $H + H_2 \rightarrow H_2 + H$, 766, 771–780, 789–792, 818
  $H_2 + I_2 \rightarrow 2$ HI, 645, 660
  $2 H_2 + O_2 \rightarrow 2 H_2O$, 651–654, 711–712
  heterogeneous
    Langmuir model for, 739–740
    Langmuir-Hinshelwood mechanism of, 740
    rate law for, 739–740
    types of, 708–709
  initiation of, 679–680, 685
  ion-molecule, 767–768
  ionic, 884–887
  mechanism, 644
    consecutive steps, 645, 658–659
    cyclic, 659
    parallel steps, 645
  order, 634
  oscillating, 632
  of radicals, 702–706
  rate laws, 631, 634, 639
    by electronic pulse radiolysis, 696
    by ESR, 696
    by flow methods, 680–681, 696
    by NMR, 687–691, 696
    from physical properties, 676–677
    by temperature jump method, 680–683, 696
    by ultrasound, 696
  recombination. *See* Reaction, combination
  slow, 630
  in solution, 768–771
  termolecular, 644
  unimolecular, 644, 780–784, 794–806
    Lindemann theory of, 780–782
  of vibrationally excited species, 821–823
Redhead, A., 720
Reduced equations of state, 26–27
Reduced mass, 61, 80–81, 318, 321–322, 333, 368, 407, 415, 603
Reduced variable(s), 26–28
Reference electrode, 896
Regular solution(s), 229
  activity coefficients of, 229
  entropy of, 228
  partial molal heat capacity of, 229
  partial molal volume of, 229
Relaxation, 491
  mechanism, 496
  time, 493

  effect on conductance, 863
  spin-lattice, 494
  transverse or spin-spin, 494
Reservoir, 131, 490
Resonance, 439
  condition, 403
  energy, 472–473
  structures, 454
    for butadiene, 476
    for C_6H_6, 439
    for H_2^+, 440
    for H_2, 447
Reversible process, 96, 98
  entropy change of, 139
  Gibbs free energy change of, 140
  Helmholtz free energy change of, 139
Rhombic structure of sulfur, 161–162
Riccoboni, R., 602
Rice, O. K., 650, 795
Rice-Ramsperger-Kassel-Marcus theory, 795
Richards, T. W., 266
Rigid rotator
  classical energy, 79
  degree of freedom, 79
  partition function of, 593
  quantum mechanics, 320–326
    eigenfunctions, 323, 355–357
    eigenvalue, 324–326, 416
    energy, 321–322, 328, 416
Ritz, W., 296
Ritz combination rule, 296
Robinson, P. J., 805
Root mean square speed. *See* Speed
Roothaan, C. C., 446, 448, 449
Roothaan SCF method, 449
Rossini, F. D., 260, 268
Rotational
  characteristic temperature, 593
  constant, 323, 408, 410, 420
  degree of freedom, 79, 80
  energies, 409
  energy levels, 407, 593
  fine structure, 433
  Hamiltonian, 416
  motion, 79–80, 320, 407
  partition function, 593
  quantum numbers, 323, 593
  spectroscopy. *See* Microwave spectrum
  structure, 430
Roughton, F. J. W., 680, 693
Ruby laser, 403
Ruedenberg, K., 449
Rule(s)
  Kistiakowsky, 167
  lever, 205
  phase, 158, 188

Ritz combination, 296
  selection, 401–402, 420, 422–432, 507, 532
  Slater, 386
  Trouton, 167
  Walden, 854
Rutherford atom model, 296
Rydberg constant, 296
Rymer, T. B., 557

Salt
  bridge, 907; *see also* Liquid(s), junction
    potential
  effect, 887
    primary, 887
    secondary, 887
    on solubility of gases, 209
Scaling parameter, 350
Scattering
  angles, 557–558
  coefficient, 538
  Compton, 551
  length, 546, 548, 550–551
  of neutrons, 551–552
SCF, 383–384
  method, 384
  orbitals, 383
Schaefer, H. F., 565
Schlier, Ch., 820
Schneider, F. W., 799, 800
Scholfield, K., 642
Schrödinger, E., 293, 294, 297, 301, 312, 399
Schrödinger equation, 293–294, 297, 301, 312, 399
  time-dependent, 399
  time-independent, 293–294, 297, 301, 312
Schwarz, H., 763
Screening constant, 447
Second law of thermodynamics, 129–155; *see also* Thermodynamics
  Carnot's statement, 133
  Clausius' statement, 137
  statistical interpretation of, 587–588
  Thomson's statement, 133–134
Second order phase transition(s), 173–176
Selecton rules, 401–402, 420, 422–432, 507, 532
Semenov, N., 651–652, 660
Semiconductor(s), 521–530
Setser, W., 822
Sharma, 778–779
Shielding, 367
Shock tube, 687, 696
SI units, 947
$SiCl_4$, electron diffraction of, 558–559
Silver, 611
Silver-silver chloride half cell, 910, 911
Single-valued function, 331

Singlet state, 376
Slater, J. C., 386
Slater, 386
  orbitals, 386
  rules, 386
Snyder, L. C., 562
Sodium, 241
Sodium bromide, 184
Sodium chloride, 519–521, 555
Solid-melt phase diagrams, 237–242
Solubility
  of gases in liquids, 207
    Henry's law, 207
    pressure dependence of, 209
    temperature dependence of, 209
Solubility product, 911
Solvation
  enthalpy, 834
  entropy, 835
  Gibbs free energy, 834–835
  number, 832
Somorjai, G. A., 726, 727, 729, 730, 734, 736
sp hybridization, 469
sp^2 hybridization, 467
sp^3 hybridization, 467
Specific
  conductance, 849, 850, 851
  heat, 102, 529
  resistance, 849
Spectra
  atomic, 364
  band, 410
  line, 295
    of helium, 295
    of hydrogen, 296
  molecular, 407–437
    infrared, 407, 562
    microwave, 407
    Raman, 552, 562, 598
    ultraviolet, 407, 562
    visible, 407, 562
Spectral linewidth, 410
Speed
  distribution of molecular, 48–50
    experimental verification of, 55
  mean, 42, 56–57
  mean square, 42–43
  most probable, 56–57
  relative, 60
  root mean square, 42–43
Spencer, H. M., 103
Spencer, U. B., 719
Spherical top molecule, 437
Spin
  angular momentum, 373
  functions, 485

Spin (*Cont.*):
  lattice relaxation time, 494
  magnetic moment, 374
  multiplicity, 422
  operator, 375
  selection rule, 401
  spin coupling, 500, 502–506
  spin interactions, 490, 500
  spin relaxation time, 496
  states, 375
Spontaneity, thermodynamic criteria of, 136, 139, 140, 146
Spontaneous emission, 397, 489
Spur-diffusion model, 672, 676
Spur-track reactions, 672
Standard
  cell, 901
  electrode potential, 895, 899
    table of values, 899
  electromotive force of cell, 901
  enthalpy of formation, 107
    table of values, 107
  entropy, 268
    table of values, 268
  Gibbs free energy of formation, 258
    table of values, 260
  hydrogen electrode, 896, 898, 909
  states, 106, 195–200, 870–874
Stanley, H. M., 278
Stannic chloride, 105
Stannous chloride, 105
Stark-Einstein law, 669–670
Starkweather, H. W., 34
State density, 797, 801–804
Stationary states, 294, 297, 399
Statistical factor, 788–789
Statistical thermodynamics, 571–625
Steady state approximation, 648
Steam distillation, 233, 234
Steel, 242
Steric factor, 766
Stevenson, D. P., 280
Stimulated emission, 397
Stirling's approximation, 874, 880, 945
Strong collisions, 782, 821–822
Stull, D. R., 274
Styrene, 276
Sublimation, 158, 162, 164
  curve, 158, 162
  enthalpy, 164
Sulfur, 161–162, 273
  phase diagram, 161–162
Sulfur dioxide, 112–113
Sulfur trioxide, 112–113
Sullivan, J. H., 660
Sulzman, K. P. G., 700

Superposition, 348, 392
  principle, 348
Surface
  area per molecule, 250
  composition, 724–728
  Gibbs free energy, 171
  ideal, 728
  platinum crystal, 729, 734–736
  pressure, 250
  reaction on a metallic, 737
  real, 731–732
  structure, 724, 726–727, 730–731, 734–736
  tension, 170, 249
    effect on vapor pressure. *See* Kelvin, equation
    of mixtures, 249
    table of values, 170
    temperature dependence of, 170
    unit of, 170
  vicinal, 733
Surroundings, 88
Sutherland equation, 69
Swan, H. T., 62
Symmetric, 378–380, 421, 437
  eigenfunctions, 378–380, 421
  Hamiltonian, 421
  top molecules, 437
Symmetry, 420
System, 88
  closed, 88
  isolated, 88
  open, 88
σ orbitals, 450, 467
Σ states, 421
Σ-π approximation, 469

Taylor, H. H., 660
Teller, E. J., 715, 716
Temperature
  adiabatic reaction, 114
  average kinetic energy and, 43
  broadening of the Maxwell distribution curve and, 52
  characteristic, 593, 596, 610, 612–613
  critical, 21
  dependence of enthalpy of reaction. *See* Kirchhoff's law
  dependence of entropy, 143–144
  dependence of equilibrium constant, 257, 274
  dependence of fugacity, 194
  dependence of gas solubility, 209–211
  dependence of Gibbs free energy, 140–141
  dependence of Helmholtz free energy, 140–141
  dependence of reaction rate constant, 640–643
  dependence of surface tension. *See* Ramsay-Shield equation

dependence of vapor pressure. *See* Clapeyron equation and Clausius, Clapeyron equation
dependence of viscosity, 68–69
eutectic, 236
inversion, 151
jump method, 680
lower critical solution, 230–231
negative absolute, 580
reduced, 26
scales
  absolute. *See* Kelvin, temperature scale
  centigrade, 91
  ideal gas, 91
  international practical, 91
  thermodynamic, 6, 134
thermal equilibrium and, 188
upper critical solution, 230–231
zeroth law of thermodynamics and, 89
Terenter, A. P., 280
Term(s). *See* Ritz combination rule
Ternary system(s), 242–247
Tetrachloroethylene, 204–205, 225
Tetragonal structure, 517
Thermal
  conductivity, 71
    Chapman-Enskog approach, 72
  equilibrium, 89
  expansion coefficient. *See* Cubic expansion coefficient
  relaxation, 492
Thermochemistry, 104–116
  endothermic reactions, 104
  exothermic reactions, 104
  Hess' law, 106
  Kirchhoff's law, 113
Thermodynamic
  consistency test, 224–226
  equation of state, 146
  functions and partition function, 583
  probability, 138, 572–573
    and entropy, 138, 588
Thermodynamics
  electrolytic solutions, 870
    chemical potential of, 871–872
    mean activity of, 872–874
    mean activity coefficient of, 872–874
    partial molal enthalpies in, 874–875
    partial molal Gibbs free energies in, 875
  first law of, 87–128
    application to gases, 116–122
    calorimetry, 99–103
    thermochemistry, 104–116
  galvanic cells, 902–906
    enthalpy change of reactions in, 903
    entropy change of reaction in, 903
    equilibrium constant of reaction in, 903

Gibbs free energy change of reactions in, 902–903
  phase equilibria, 157–247
    in multicomponent systems, 179–247
    in single-component systems, 157–178
  statistical, 571–625
    and distribution laws, 574–578
    of electron gas, 619–621
    of ideal crystal, 610–613
    of lattice gas, 614–615
    partition function, 576, 579–580, 583–619
  of surfaces, 169–171, 247–250
    effect on vapor pressure, 169–171
    Gibbs adsorption isotherm. *See* Adsorption
  third law of, 266–283
    Nernst heat theorem, 267
  zeroth law of, 89
Thermometric
  equation, 90
  properties, 90
Thermometry, 90
Third law of thermodynamics, 266–283
  absolute entropy and, 267
  consequences of, 269
  Nernst heat theorem and, 267
  principle of unattainability of absolute zero and, 271
  statistical interpretation of, 587–588
  verification of, 272
Thompson, B., 96
Thomsen, J., 129
Thomson, J. J., 149
Thomson, W., 133, 148, 171
Tie line, 230, 244
Time
  average, between collisions, 768
  dependent Schrödinger equation, 399
  independent Schrödinger equation, 293–294, 297, 301, 312
  interval between two collisions, 768
  particle and molecular motion, 768, 825
  relaxation, 684, 823–825, 863
  residence, 688, 723
  a variable in reaction kinetics, 629–631
Tin, 273
Tipper, C. F. H., 640, 667
Titration, conductometric, 869
Toennies, J. S., 819
Toluene, 198, 206, 499
Torrey, H. C., 484
Total differential, 92, 933–936
Townes, C. H., 562
Transference number
  boundary moving method of determination, 847
  definition, 845

Transference number (*Cont.*):
  EMF of cells and, 913
  Hittorf's method of determination of, 846
  mobility of ions and, 845
  tables of values of, 848
  velocity of ions and, 845
Transition
  moment, 401, 422–432
  probability, 398, 422
  state theory. *See* Activated complex theory
Translational
  energy, 77, 95, 300, 411, 591
    degree of freedom of, 79
    distribution of. *See* Distribution, molecular
      energies
  entropy, 591
  Gibbs free energy, 591
  heat capacity, 590
  partition function, 589
  symmetry, 516, 533
Transport properties, 64, 71–72
  diffusion, 72
  thermal conductivity, 71
  viscosity, 64
Triclinic structure, 517
Triethylamine, 231
Trimethylamine, 8
Triple bond, 467
Triple point, 90, 161
Triplet state, 377, 422, 453
Trouton's rule, 167
Tunneling, 719, 793
Turoff, R. D., 581

Uhlenbeck, G. E., 372–373
Ungerade, 422
Unstable species, 698, 700
  detectability limits, 700

Valence bond method, 445
Van der Waals, J. C., 10, 18, 23, 25, 44, 152
Van der Waals
  constants, 18–20
    and critical state, 25
  equation, 18, 21
    molecular interpretation of, 44–46
    reduced, 26
  forces. *See* Forces
  isotherms of CO_2, 23
Van't Hoff, J. H., 218, 258, 833
Van't Hoff
  equation, 258
  i factor, 833
  law, 218; *see also* Colligative properties,
    osmosis
Vapor pressure
  of pure liquids, 160–173

experimental determination of, 172
  pressure dependence of, 168
  surface and size dependence of, 169
  table of values, 165
  temperature dependence. *See* Equation(s)
    Clapeyron and Clausius-Clapeyron
  of solutions, 202, 220
    composition phase diagrams, 204, 206, 222
    lowering. *See* Colligative properties
Vaporization
  enthalpy, 164, 165
  entropy, 166, 167
Variation principle, 347–349, 369
Vector model of atom, 328, 372, 375, 377
Velocity, molecular distribution, 48–52
Vibrational; *see also* Harmonic oscillator
  characteristic temperature, 593, 610
  degree of freedom, 79–81
  energy, 407, 409, 597
  energy levels, 408, 595
  entropy, 598
  force constant. *See* Laws, Hooke
  frequency, 409, 800
  Gibbs free energy, 597
  molar heat capacity, 597
  normal mode, 81, 598–599
  partition function, 595
  spectroscopy. *See* Infrared spectrum
  transition moment, 433
Vibronic, 426, 530–531
  energy levels, 426
  states, 426
  transitions, 530–531
Virial
  coefficients, 15–18
    of mixtures, 32–33
    temperature dependence of, 17
  equation, 15, 148, 203
    statistical derivation of, 17, 18, 615–618
Viscosimeter, 67
Viscosity, 64–70
  coefficient, 65
  table of values, 66
  temperature dependence of, 68–69
Viscous effect, 863
Volt (unit), 849, 948
Volume
  critical, 22; *see also* Hakala's evaluation of
  excluded, 45, 61
  of mixing, 201, 606
  partial. *See* Amagat's law
  partial molal, 184, 201, 229
  reduced, 26–27

Walden's rule, 854
Wall, F. T., 558
Ward, H. R., 703

Warren, B. E., 555
Wasserman, A., 788
Water, 102, 157, 160–161, 184, 210–211, 231,
    233, 245, 247, 259, 458, 463–464, 661, 672–676,
    697, 835, 839
    conductance of, 851, 868
    dissociation of, 841–842
    irradiation of, 661
    phase diagram of, 160–161
    proton transfer reactions in, 697
    radiolysis of, 672–676
Watkins, K., 804
Watt (unit), 849, 948
Wave
    functions, 290–295, 298
        antisymmetric, 378–380
        complete set of, 348
        normalized, 301
        orthogonal, 306
    length, 390
    mechanics, 287
    number, 296, 390, 599, 601
Weak electrolyte. *See* Electrolyte(s)
Weissberger, A., 681, 683
Wells, J. S., 700
Westenberg, A. A., 693
Weston, R., 763
Weston cell. *See* Electrochemical cell(s)

Wheatstone bridge, 850
Wien's effect, 865
Wieser, J. D., 555
Wilkins, R. G., 693
Wilson, W. E., 642
Work
    adiabatic, 99
    compression, 97
    convertibility of heat into. *See* Carnot's cycle
    electrical, 97, 879, 884, 895, 903
    expansion, 97
    first law of thermodynamics and, 95
    reversible. *See* Maximum work
    surface, 97, 170–171

X-ray diffraction, 544–545, 554–557
Xenon, entropy of, 602–603

Yield
    of chemical reaction, 262
        effect of inert gas on, 262
        effect of pressure on, 262
        effect of temperature on, 262
    of photochemical reactions, 662
Youell, J. E., 278

Zavoiski, E., 484